Indira's

OBJECTIVE AGRICULTURAL BIOTECHNOLOGY

2nd Edition

Other Books

Indira's

OBJECTIVE AGRICULTURAL BIOTECHNOLOGY

2nd Edition

Dr. R.L. Arya
Senior Scientist
Central Tobacco Research Institute Research Station,
Dinhata,
Coochbehar (W.B.)

Sonam Arya
Ph.D. Research Scholar (Agril. Biotechnology)
Sardar Ballabh Bhai Patel University of
Agriculture and Technology,
Meerut (UP.).

Dr. Alok Das
Scientist (Plant Biotechnology)
Indian Institute of Pulses Research,
Kanpur (U.P.)

Dr. Vaishali
Assistant Professor
Department of Biotechnology
College of Agriculture
Sardar Ballabh Bhai Patel University of
Agriculture and Technology,
Meerut (UP.).

Competition Tutor

An imprint of Scientific Publishers (India)

Published by:
Competition Tutor
5-B, New Pali Road, P.O. Box 91
Jodhpur 342 001 (India)

E-mail: competitiontutor@gmail.com
Website: www.competitiontutor.com

Print: 2017

An imprint of Scientific Publishers (India)

ISBN: 978-81-7233-938-8

Printed in India

PREFACE

India is a developing country and making significant strides to promote research and development activities to promote Biotechnology. As a result of emergence of biotechnology, novel products and services that encourage the human activities and welfare are realized. Though modern tools and techniques are of recent origin however, biotechnology existed since time immemorial, much before the other realms of technology came into practice. The fermentation technology of alcoholic beverage and setting of curd from milk are to cite a few. Biotechnology is becoming as the one of the most favorite subject for the students of biology and gradually emerged as a subject in schools, subsequently universities offering undergraduate, post graduate, M.Phil and Doctoral programmes with the aim to develop trained human resources. The competitive examinations have become a routine procedure admissions and recruitment for higher position in learning and education. These examinations mostly comprise multiple choice questions (MCQs) to test the knowledge within a specified period of time.

A quick reference manual covering all aspects of biotechnology is a long felt need for students and aspirants appearing in different competitive examinations. Towards this end, efforts have been made by us to prepare and presents all facets of biotechnology in easy, short and understandable manner. **"Indira's Objective Agricultural Biotechnology 2nd Edition"** for competitive exams in agricultural biotechnology discipline contains 23 chapters covering all related disciplines. These chapters include : Introduction, Cell Biology, Genetics and Cytogenetics, Plant Physiology and Molecular Biology, Plant Biochemistry, Agricultural Microbiology, Ecology, Diversity of life and ecological principles, Environmental Biotechnology, Natural and Forest Conservation, Genomics and Recombinant DNA technology, Transgenics, Molecular markers, Plant Tissue Culture, Bioinformatics, Seed Biotechnology, Plant Protection Biotechnology, Industrial Biotechnology, Food and Beverage Biotechnology, Intellectual Property Rights, Patenting and Biosafety, Animal Biotechnology. Model test papers and previous years solved papers have been given due importance at the end of the book present a general guidance of examination pattern. Each chapters contains multiple choice questions covering every aspects and total about 12000 objective questions with multiple choices have been framed and arranged sequentially. Recent technological development in the field of agriculture also been incorporated in the text. This book is primarily intended to serve as a ready reference for those appearing in competitive examinations of undergraduate, post graduate, M. Phil and doctorate programmes in Biotechnology of various Universities. We believe that this book will also be useful for teachers, researchers and development officers for references and easy answering of many complicated questions. Every possible effort has been made to accommodate previous year's questions of various competitive examinations, on memory basis. The chapters are chosen in view to cover the course contents of competitive examinations like IAS, IFS, ARS, PCS, Banking, SLETs, UGC-NET and others.

In this edition, there could be some shortcomings for which valuable and informative suggestions will be gratefully appreciated. These shortcomings, if found in the text may be incorporated in forthcoming edition for the benefit of readers with your critical suggestions, so that book may be more useful to the students.

We convey our gratitude to the entire Management of Scientific Publishers, Jodhpur for taking all pains in the publication of the book. The authors specially thank Dr. Kalish Chaudary, Lachoo College, Jodhpur for critically suggesting the corrections in the first edition of the book.

Authors are be grateful to all those persons as well as various books, periodicals, magazines, newsletters, journals etc., that helped in the preparation of this book. The authors/publishers look forward to suggestions from all readers, especially for experienced researchers and scholars for further improving the subject content as well as the presentation of this book.

2015, Kanpur

R. L. ARYA
SONAM ARYA
ALOK DAS
VAISHALI

CONTENT

Chapter 1

INTRODUCTION

1. Biotechnology is:
(A) Technology used in biology
(B) Use of genetic engineering
(C) Development of technology for medical application
(D) Use of microorganisms in industrial processes

2. Who introduced the term Biotechnology?
(A) Marshall Nirenberg
(B) Karl Ereky in 1919
(C) John H. Northrop & Kunitz
(D) Van Helmont

3. Biotechnology is a multidisciplinary field that encompasses inventions that are related to:
(A) Basic science (B) Applied science
(C) Both (A) and (B) (D) None of these

4. Transgenic plants containing genes for drought and salt tolerance, improved quality for value addition and increase shelf life are produced through:
(A) Genetic transformation
(B) Genetic drift
(C) Genetic transduction
(D) Genetic transfection

5. Agrobacterium – mediated transformation an indirect method of transformation that involves:
(A) *Agrobacterium tumefaciens*
(B) *E. coli*
(C) *S. cerevisiae*
(D) None of these

6. Lamda phage infects bacteria *E. coli* and replicates by a:
(A) Lysogenic pathway (B) Lytic pathway
(C) Both (A) and (B) (D) None of these

7. The vectors that are hybrids of lambda phages and plasmids, and their DNA can replicates in the cell like that of a plasmid or be packaged like that of a phage is called:
(A) Phagemids (B) Fosmids
(C) Plasmids (D) Cosmids

8. Cosmid DNA can be packaged into phage particles by using the:
(A) In vitro system (B) In vivo system
(C) Both of these (D) None of these

9. Plasmids capable of propagating and transferring (shuling) genes between two different organism, one of which is typically a prokaryote (*E. coli*) and other a eukaryote (yeast) is called:
(A) Cloning vector insertion vector
(B) Shuttle vector
(C) Viral vector (D) None of these

10. Recombinants can therefore, be identifies because the characteristics coded by the inactivated gene is no longer displayed by the host cells is called:
(A) Insertional inactivation
(B) Transduction
(C) Transfection (D) None of these

11. Virus mediated gene transfer, in which the foreign DNA is packaged inside a virus particle is called:
(A) Insertional inactivation (B) Transduction
(C) Transfection (D) None of these

12. The use of chemical or physical methods to persuade cells to take up DNA from the culture medium or direct physical introduction of DNA into the cell is called:
(A) Insertional inactivation (B) Transduction
(C) Transfection (D) None of these

13. A fundamental tool in molecular genetics which takes advantage of the ability of individual single-stranded nucleic acid molecules to form double stranded molecules is called:
(A) Nucleic acid hybridization
(B) In situ hybridization
(C) Both (A) and (B)
(D) None of these

14. Nucleic acid hybridization can be used to detect homologous DNA or RNA sequences not only in cell extracts, but also in chromosomes or intact cells – a procedure is called:
(A) Nucleic acid hybridization
(B) *In situ* hybridization
(C) Both (A) and (B)
(D) None of these

15. The technique is used to detect the presence of a known piece of DNA in any given DNA preparation from an organism is known as:
(A) Southern blot (B) Northern blot
(C) Both (A) and (B) (D) None of these

16. A method of estimating the size and abundance of a particular transcript is known as:
(A) Southern blot (B) Northern blot
(C) Both (A) and (B) (D) None of these

17. The determine the 5'-end of transcript, the most reliable method is:
(A) Primer extension (B) Reporter assays
(C) Expression vector (D) None of these

18. To find out the expression patterns of a gene within the cell, its promoter can be fused to one of the several reporter genes and transfected inside living cells by:
(A) Primer extension (B) Reporter assays
(C) Expression vector (D) None of these

19. To maximize the expression of heterologous genes – these genes that are taken from different organisms and being expressed in bacterial, yeast, animal or plant cells - it is necessary to design a cloning vector, which allows optional transcription of the genes. Such vectors are also called:
(A) Primer extension (B) Reporter assays
(C) Expression vector (D) None of these

20. Recombinants are hybrid DNA molecules consisting of autonomously replicating DNA segment plus inserted elements. Such hybrid molecules are also called:
(A) Chimera (B) Ligation
(C) Transformation (D) Cloning

21. For plasmid purification the bacterial cells extract is deproteinized, the RNA removed, and the DNA concentrated by:
(A) Methane precipitation
(B) Ethanol precipitation
(C) CO_2 precipitation
(D) Ammonia precipitation

22. Autonomously replicating genetic element used to carry a fragment of genomic DNA into a host cell for the purpose of gene cloning:
(A) Cloning vector insertion vector
(B) Shuttle vector (C) Viral vector
(D) None of these

23. The chromosome of the transgenic animal includes an endogenous coding sequence, preferably the c-myc gene, which is substantially the same as the:
(A) Oncogene sequence
(B) Coding sequence (C) Viral sequence
(D) Homologous endogenous sequence

24. The transcription of the oncogene sequence is under the control of a promoter sequence different from the promoter sequence (or a synthetic promoter sequence) controlling transcription of the endogenous:
(A) Oncogene sequence (B) Coding sequence
(C) Viral sequence
(D) Homologous endogenous sequence

25. The term oncogene encompasses the:
(A) Viral sequence
(B) Homologous endogenous sequence
(C) Both (A) and (B)
(D) Coding sequence

26. The promoter sequence controlling transcription of the oncogene sequence is:
(A) Oncogene sequence (B) Coding sequence
(C) Viral sequence (D) Inducible

27. Introduction of the oncogene sequence at the fertilized-oocytes stage ensures that the oncogene sequence will be present in transgenic animals is:
(A) Germ cell (B) Somatic cells
(C) Both (A) and (B) (D) None of these

28. Introduction of the oncogene sequence at a later embryonic stage might result in the oncogene's absence from some somatic cells of the founder animal, but the descendants of such an animal that inherit the gene will carry the activated oncogene in all of their:
(A) Germ cell (B) Somatic cells
(C) Both (A) and (B) (D) None of these

29. The process that promotes gene transfer between bacteria by direct cell-to-cell contact, comes closest to sex in higher organisms is called:
(A) Conjugation (B) Transduction
(C) Transformation (D) Transposons

30. The transfer of bacterial DNA from one cell (the donor) to another (the recipient) via the agency of bacteriophage is called:
(A) Conjugation (B) Transduction
(C) Transformation (D) Transposons

31. The mechanism of gene transfer involves uptake of free DNA from killed donor cells by a recipient cell is known as:
(A) Conjugation (B) Transduction
(C) Transformation (D) Transposons

32. The discrete genetic elements with the ability to move from one genomic position to other non homologous positions is known as:
(A) Conjugation (B) Transduction
(C) Transformation (D) Transposons

33. These single-stranded sequences (GGGCGGCGA-CCT/AGGTCGCCGCCC) are complementary to each other and are called:
(A) COS (Cohesive sites) (B) Cis
(C) CS trans (D) Chi

34. Lambda is representative of a large group of bacteriophages that can propagate either through the standard lytic cycle or through a process called:
(A) Lysogeny (B) Caspid
(C) Both (A) and (B) (D) None of these

35. In mature virus particles, these genomes are packaged inside a protein coat called:
(A) Lysogeny (B) Caspid
(C) Both (A) and (B) (D) None of these

36. The represented phage genome in lysogenic cells is called:
(A) Prophage or provirus (B) Metaphage
(C) Transphage (D) Post phage

37. Who describes relative position of various genes on the chromosome(s) of an organism is?
(A) Genetic map (B) Synaptic map
(C) Physical map (D) None of these

38. Who is the first person to use the term molecular biology is?
(A) William Astbury in 1945
(B) John C Kendrew in 1950
(C) Hans Krebs in 1969
(D) K.C. Mehta in 1975

39. The *E. coli* chromosome is a circular duplex DNA of about:
(A) 4.6 x 10^5 base pairs
(B) 4.6 x 10^6 base pairs
(C) 4.6 x 10^7 base pairs
(D) 4.6 x 10^8 base pairs

40. The process of DNA replication in eukaryotes is ----- complex than that in prokaryotes:
(A) For more (B) Less
(C) Same (D) Unknown

41. Pure plasmid DNA was isolated from a bacterium. Restriction enzyme digestion of this plasmid with either Bam HI or Eo RI resulted in two DNA fragments. A double digestion of the same plasmid with both these enzymes resulted in three DNA fragments. From this we an conclude that the isolated plasmid DNA is:
(A) Double stranded and linear
(B) Double stranded and circular
(C) Single stranded and linear
(D) Single stranded and circular

42. The first ever vaccination was against which disease:
(A) Chicken pox (B) Rabies
(C) Small pox (D) Pneumonia

43. For cloning eukaryotic gene in prokaryotes, genes should be isolated from:
(A) cDNA library (B) Genomic library
(C) Eukaryotic host (D) None of these

44. In order to move from a molecular marker to the linked gene of interest, which of the following approaches can be used?
(A) Chromosome jumping
(B) Chromosome walking
(C) Colony hybridization
(D) Chromosome sorting

45. Essential constituents of a cloning vector are:
(A) Replicon, promoter and selectable marker
(B) Replicon, unique restriction site and selectable marker
(C) Promotor, operatop and restriction site
(D) Replicon, unique restriction site, selectable marker and promoter

46. Which is correct about heavy metal?
(A) Heavy metals refer to the metals which are present in the soil with a density greater than 5 or 6 g m^{-3}
(B) As, Cd, Cr, Hg, Cu, Ni, Pb are examples of heavy metals
(C) They are harmful for humans as well as plants
(D) All of these

47. Essential constituents of an expression vector are:
(A) Replicon, promoter and selectable marker
(B) Replicon, unique restriction site, selectable marker and promoter
(C) Promtor, operator and restriction site
(D) Replicon, unique restriction site and promoter

48. Algae have emerged as one of the most promising sources, especially for biodiesel production is:
(A) Algae an grow in places away from the farmlands and forests
(B) The yields of oil from algae are orders of magnitude higher than those from traditional oilseeds
(C) Algae can be grown in sewage(s) and next to power-plant smokestacks where they acid in pollution management
(D) All of these

49. Plant viral diseases can be controlled by
(A) Antisense RNA technology
(B) By using ribozymes (C) RNAi technology
(D) Settellite RNA (E) All of these

50. For patenting a plant variety it should be:
(A) NDUS (B) DUD
(C) DDS (D) SUS

51. Micro propagation can be done by:
(A) Through shoot bud proliferation
(B) Through adventitious buds
(C) Meristem tip culture
(D) All of these

52. Products will be applicable for patenting only if they are:
(A) Novel (B) Inventive
(C) Useful (D) All of these

53. A CsCl gradient will separate DNA molecules by
(A) Adhesion (B) Adsorption
(C) Density (D) Absorption

54. IPR stands for:
(A) Intellectual Property Rights
(B) Indigenous Product Right
(C) Indian Property Rights
(D) International Product Rights

55. Which is correct about nutraceuticals:
(A) A combination of nutrition and pharmaceutical
(B) It refers to extracts of foods claimed to have a medicinal effect on human health
(C) It is usually contained in a medicinal format such as powder, tablet or capsule in a prescribed dose
(D) All of these

56. Virus free plants can be obtained by:
(A) Meristem culture (B) Pollen culture
(C) Ovary culture (D) Shoot tip culture

57. Selection of protoplast fusion hybrids can be done by:
(A) Visual selection (B) Fluorescent labels
(C) FACS
(D) Any of the above methods

58. Structural DNA nanotechnology is for:
(A) Nanoelectronics (B) Nanorobotics
(C) Smart material (D) All of these

59. In vitro techniques for germplasm storage are useful for:
(A) Clonally propagated crops
(B) Crops producing recalcitrant seeds
(C) Crops producing orthodox seeds
(D) Both (A) and (B)

60. Embryo therapy involves:
I. In vitro fertilization of egg
II. Production of several copies of the normal allele of defective gene
III. Introduction of the normal allele into zygote, or in the cells of developing embryo
IV. Integration of the normal allele (DNA), preferably in the place of the defective allele so that it may function normally

Options
(A) I, II, III, IV (B) I, II, III
(C) II, III, IV (D) I, III, IV

61. The copyright of a book may be held by:
(A) Author(s) (B) Editor(s)
(C) Publisher (D) Any of these

62. Which advantages is offered by in vitro technique of germplasm storage?
(A) The material can be stored over long periods
(B) They are ideal for germplasm exchange
(C) The risk of errors in labeling are minimized
(D) All of these

63. Which statement is correct?
(A) When available ESTs assembled in contiguous overlapping clusters are described as contigs
(B) When ESTs appear as singles and which can not be assembled in contigs are described as singletons
(C) Unigene set is a combined set of contigs and singletons
(D) All of these

64. Which of the following enzyme is used in brewing?
(A) β amylase (B) Cellukase
(C) Dextranase (D) Pectinase

65. The combination of biology and IT is:
(A) Nanotechnology (B) Biotechnology
(C) Bio-informatics (D) Information technology

66. Bioinformatics can be used for :
I. Genome mapping
II. Identification of genes from gene sequences
III. Determination of protein structure and simulation of protein interactions
IV. Discovery of new therapeutic targets and designing of medicines aimed at specific target sites

Options
(A) I, II, III (B) II, III. IV
(C) I, II, III, IV (D) I, II

67. Technique for sequencing DNA and proteins is:
(A) NMR spectroscopy (B) X-ray crystallography
(C) Electron microscopy (D) All of these

68. Which is correct about biochemical oxygen demand (BOD)?
(A) The amount of O_2 required by aerobic microorganisms to decompose the organic matter present in a sample of water
(B) It is used to measure the degree of water pollution
(C) It is also called biological oxygen demand
(D) All of these

69. Which molecular techniques can be used to characterized germplasm?
I. Techniques that simply describe DNA polymorphism
II. Techniques based on DNA linked to a gene of interest
III. Techniques that characterized the gene of interest itself

Options
(A) I, II (B) I, III
(C) I, II, III (D) II, III

70. Immobilization of enzymes is done by:
(A) Physical adsorption (B) Cross linking
(C) Entrapment (D) All of these

71. Photorespiration occurs in
(A) Mitochondria (B) Peroxisomes
(C) Chloroplast (D) All of these

72. Important constituents of biogas:
(A) Methane (B) Ethane
(C) Carbon monoxide (D) SO_2

73. In USA, which systems are available for protection of IPRs is considered to be the most powerful and the most expansive in scope of coverage?
(A) The Plant Patents Act of 1930
(B) The Plant Variety Protection Act of 1970
(C) The Utility Patents Act of 1985
(D) All of these

74. Cosmid DNA can be packaged into phage particles by using the:
(A) *In vitro* system (B) Transduction
(C) Both (A) and (B) (D) None of these

75. DNA chip carries a large number of:
(A) Oligonucleotides (B) cDNAs
(C) DNA probes (D) All of these

76. Which is correct about bulbosum technique of haploid production?
I. The technique was developed with the cross *Hordeum bulbosum* x *H. vulgare*
II. It is based on the chromosome elimination during early embryo development
III. Generally embryo rescue is necessary
IV. This technique has been used with bread wheat
Options
(A) I, II, III, IV (B) I, II, III
(C) II, III, IV (D) I, III, IV

77. Biofertilizers can be used to achieve:
(A) To fix atmospheric nitrogen
(B) To solubilize soil nutrients like phosphate
(C) To stimulate plant growth
(D) All of these

78. Which plant is used for bio-diesel production?
(A) Jatropha (B) Soybean
(C) Hevea (D) Wolfia

79. DNA chip is a wafer of
(A) Silicon (B) Al
(C) Carbon (D) Phosphorus

80. Biosensers are used as
(A) Purified enzymes (B) Antibodies
(C) Whole microbial cells (D) All of these

81. In India, which of the following overseas the large-scale commercial use of GMOs and recombinant DNA products:
(A) Department of Atomic Energy
(B) Council of Scientific and Industrial Research
(C) Ministry of Environment and Forest
(D) Department of Biotechnology

82. Burning of plastic bags, emits toxic gasses leading to:
(A) Bronchitis (B) Lung infection
(C) Respiratory problems (D) All of these

83. Antibiotic resistance is used as a selectable marker for:
(A) Selection of vector
(B) Selection of recombinant vectors
(C) Selection of transformed cells
(D) Selection of resistant bacteria

84. Which vector was used for development of hepatitis vaccine?
(A) pBluescript (B) PUC 18
(C) Cosmid (D) Yeast expression vector

85. The plants, which carry additional, stably integrated and expressed, foreign gene(s) from transpecies is known as:
(A) Genetic counseling
(B) Genetic engineering
(C) Gene therapy
(D) Transgenic plants

86. Genetic engineering of industrially useful micro-organisms aims at developing organisms that have which of the following features?
I. Capacity to produce such compounds, which they are normally incapable of producing
II. Ability to produce substances in quantities beyond their natural capabilities
III. Elimination of their undesirable characteristics
IV. None of above
Options
(A) I, II, III, IV (B) I, II, III
(C) II, III, IV (D) I, III, IV

87. A C-terminal KDEL motif will most often ensure:
(A) The protein to be folded by KSC70
(B) The protein to be degraded by the polyubiquitine pathway
(C) ER-retention of the protein
(D) Secretion of the protein

88. Which of the following vector can be used for cloning bigger DNA fragments?
(A) YACs (B) Cosmids
(C) Plamids (D) Phagemids

89. Type 2 diabetes mellitus occur due to
(A) Deficiency of insulin
(B) Inactivation of insulin recepters
(C) Both of the above (D) None of the above

90. Which is correct about HIV?
I. HIV uses receptors, viz., CD4 and CCR5, to infect T cells
II. CD4 is essential for T cells function, but CCR5 appears to be dispensable
III. Due to lack of CCR5 receptor in about 1% Caucasian population, such individuals naturally show immunity to AIDS
IV. Inhibition of CCR5 expression through appropriate siRNA molecules reduces the infection of T cells by HIV
Options
(A) II, III, IV (B) I, II
(C) I, IV (D) I, II, III, IV

91. Which plant is used to accumulate copper from factory waste?
(A) *Baccillus subtilis* (B) *Pseudomonas putida*
(C) *Zoogloea ramigera* (D) *E. coli*

92. Commonly used host in gene cloning is:
(A) *E. coli* (B) *Pseudomonas*
(C) Fungi (D) Bacillus

93. Which among the following microorganism is used as bio fertilizer?

(A) *Rhizobium* (B) *Bacillus*
(C) *Pseudomonas* (D) *E. coli*

94. Marker free selection can be done by:

(A) Co-transformation (B) Deletion
(C) Inversion (D) Insertion

95. Bacteria possessing restriction enzymes are not affected because:

(A) These restriction enzymes are in inactive form
(B) Their genome is methylated
(C) These enzymes are present in less concentration
(D) Their genome lack unique restriction sites for enzymes

96. DNA Plant Technology faced difficulties during release of Endless Summer due to the use of:

(A) Selectable marker *npt*II
(B) CaMV 35S peromoter and the ACC synthase gene
(C) Antisense RNA technology and *Agrobacterium tumerfaciens* mediated genetic transformation
(D) All of these

97. Indian has created a National Biodiversity Authority with headquarters at:

(A) New Delhi (B) Kolkata
(C) Chennai (D) Hyderabad

98. The specialized structures which cap the ends of eukaryotic chromosomes is known as:

(A) Centromere
(B) Telomeres
(C) DNA replication
(D) Transgenic plants

99. The copyright suffers from:

(A) It provides protection for a specific period and only from reproduction as such of the copyright material
(B) It does not prevent another person from using either the idea or he information contained in a copyright material
(C) In case of DNA sequence, one may get around this protection by designing an alternative sequence to encode the same protein
(D) All of these

100. Why have transgenic proteins present in crops often been tested in the presence of stomach acid before the plants were released?

(A) To see whether the protein will be mutated into a more harmful protein (e.g. prions)
(B) To see whether the protein can be digested to provide substantially equivalent nutrients.
(C) To see whether the protein will be digestible by an average consumer.
(D) To see whether the protein will persist long enough to cause an allergic reaction.

101. Molecular probes can be used:

I. Diagnosis of diseases and DNA fingerprinting
II. Screening of genomic and cDNA libraries
III. Location of genes or DNA sequences on chromosomes
IV. Identification of varieties of crop species

Options

(A) II, III, IV (B) I, II
(C) I, IV (D) I, II, III, IV

102. Transmission of chromosomal DNA from generation to generation is crucial to cell propagation is called:

(A) DNA conservation (B) Genetic engineering
(C) DNA replication (D) Transgenic plants

103. Carotenoids (carotene and xanthophylls) obtained from cyanobacteria are commercially used as :

(A) Food colouring
(B) Food additives to enhances flesh colour of salmon fish
(C) To enhance colour of egg yolk
(D) All of these

104. Which statement is correct?

(A) Convention on biological diversity (CBD) is concerned with conservation of whole biodiversity while PGR is concerned with conservation of plant diversity
(B) SBBs stands for state biodiversity Boards
(C) BMCs stands for biodiversity management committees
(D) All of these

105. There is a group of additional histones, all loosely related to one another and collectively called:

(A) Linker histone (B) Linker nonhistone
(C) DNA replication (D) Chromatosomes

106. The PPVFR Authority (PPVFRA) began registration of plant varieties in which years?

(A) 2005 (B) 2006
(C) 2007 (D) 2008

107. Fusion protein are also called:

(A) Hybrid protein (B) Chimeric protein
(C) Both of these (D) None of these

108. Which is an example of geographical indications?

(A) Champagne wine (B) Scotch whiskey
(C) California wine (D) All of these

109. The most useful methods for studying protein-protein interactions are:

(A) Phage display (B) Yeast two hybrid system
(C) Both of these (D) None of these

110. Three restriction endonucleases P, Q and R recognize 4bp, 6bp and 8 bp sequences respectively. The relative frequency of occurrence of these sequence on a bacterial genome is:

(A) $P > Q > R$ (B) $P > R < Q$
(C) $R < Q > P$ (D) $Q < R > P$

111. Consider the following statements about embryo culture?

I. It is routinely used for the recovery of hybrid plants from distant crosses

II. It is particularly useful where embryos fail to develop due to endosperm degeneration

III. It has been used extensively in the extraction of haploid barley plants from *H. bulbosum x H. vulgare* crosses

IV. Embryo culture is regularly used in orchid propagation and in breeding of those species that show dormancy, e.g., Iris

Options

(A) II, IV (B) I, II
(C) I, III (D) I, II, III, IV

112. Choose the right combination of components requires to set up a polymerase chain reaction from the following:

(A) Template DNA, two primers, dNTPs and DNA ligase
(B) Template DNA, two primers, NTPs and DNA ligase
(C) Template RNA, two primers, NTPs and DNA polymerase
(D) Template DNA, two primers, dNTPs and DNA polymerase

113. People become tolerant to malaria if they are carriers for:

(A) Sickle cell anemia (B) Thalisemia
(C) Leishmenia (D) Kala azar

114. Which is an example of fortified food?

(A) Addition of iodine to table salt (iodized salt)
(B) Addition of calcium to fruit juice
(C) Addition of vitamin D to such foods as margarine
(D) All of these

115. A nucleic acid sample is resistant to digestion with exonuclease. When heated it does not show typical melting curve of a linear double stranded DNA. On CsCl-ethidium bromide equilibrium density centrifugation it settles at the bottom of the centrifuge tube. The nuclei aid is:

(A) ccc pBR322
(B) Bacteriophage P22 DNA
(C) rRNA
(D) RFIIM13DNA

116. Genetic map is one that:

(A) Establishes site of the genes on a chromosomes
(B) Establishes the various stages in gene evolution
(C) Shows the stages during the cell division
(D) Shows the distribution of various species in a region

117. Geographical indications includes:

(A) Natural products
(B) Agricultural goods
(C) Goods of handicraft (D) All of these

118. Which one of the following elements need not to present in an expression vector?

(A) Selection marker to select for host cells containing the vector
(B) Two different origins of replication
(C) Promoter sequence upstream of the
(D) Unique restriction enzyme sites for insertional cloning

119. The in-vitro strategies of germplasm storage are useful under:

I. When crop species are asexually propagated and/or at least some of their genotypes do not produce seed

II. When crop species produce recalcitrant seeds

III. When asexually propagated species are highly heterozygous

IV. When materials developed by genetic transformation show instability

Options

(A) II, III, IV (B) I, II
(C) I, IV (D) I, II, III, IV

120. Microbes have a relatively -------that allows the chemical conversions possible:

(A) High metabolic rate
(B) Large surface area
(C) High multiplication rate
(D) All of these

121. Chromosome walking involves:

I. Selection of a clone of interest from genomic library identified by a probe and subcloning of a small fragment from one end of the DNA insert from this clone

II. The subclones fragment of the selected clone is then hybridized with other clones in the library to identify a clone having DNA insert overlapping with that of the first one

III. The restriction map of each selected clone is prepared and compared to know the region of overlapping

IV. Identification of few overlapping restriction sites will amount to walking along the chromosome or along a long chromosome segment

Options

(A) II, III, IV (B) I, II
(C) I, IV (D) I, II, III, IV

122. Which can be produced from dead bodies of animals?

(A) Calmodulin and aprotinin
(B) Myelin and ubiquitin
(C) Gelatin and protein kinase
(D) All of these

123. Disorder, which occur due to single nucleotide change:

(A) Sickle cell anemia
(B) Thalisemia
(C) Leishmenia
(D) Anemia

124. Cybrids are produced:
(A) By *in vitro* fusion of gemetes
(B) By fusion of two cells and contain full nuclear genomes of the two parents
(C) By fusion of two cells and contain full nuclear genome of one and partial nuclear genome of the other parent
(D) By fusion of two cells and contain nuclear genome of one parent and cytoplasmic genome of the other parent

125. An otherwise excellent transgenic variety of which crops developed during 1990a could not be released due to IPR problems?
(A) Potato (B) Tomato
(C) Tobacco (D) All of these

126. PPVFR Act (2001) is comparable to UPOV Act (1991) with respect to:
(A) Requirement of novelty, distinctiveness uniformity and stability for registration of a variety
(B) Protection extended to commercial use of all the material of the protected variety
(C) Essentially-derived varieties being subjected to PBR protection granted to the concerned initial varieties
(D) Both (A) and (C)

127. Which one of the following statements is true?
(A) All microarrays are DNA microarrays
(B) Complete genome sequence should be known to make a microarray
(C) All the microarrays use radioisotopes
(D) Microarrays can be used to measure mRNA levels

128. Which of the following can be used for transferring the DNA into the host cells?
P. Transformation
Q. Sonication
R. Transfection
S. Electroporation
(A) Only P can be used
(B) Only Q and R can be used
(C) Only Q, R and S can be used
(D) Only P, R and S can be used

129. Cryopreservation is done to store cells/cultures etc. for a longer period at:
(A) High humidity (B) Very low temperature
(C) Low humidity (D) High temperature

130. Which of the following is most likely in the present in a BAC?
(A) cDNA (B) EST
(C) LB,RB (D) UTR

131. DNA is present in:
(A) Nucleus (B) Mitochondria
(C) Chloroplast (D) All of these

132. Microsatellites are also known as:
(A) VNTRs (B) RFLPs
(C) STRs (D) RAPDs

133. Match the column A and B carefully and select the correct option from those given below:

Column A	Column B
A. Proteomics	I. Single nucleotide polymorphism
B. DNA marker	II. mRNA content
C. Genomics	III. Proteins
D. Transcriptome	IV. DNA sequencing

Options
(A) A - II, B – I, C – IV, D – III
(B) A - IV, B – III, C – II, D – I
(C) A - III, B – I, C – IV, D – II
(D) A - II, B – III, C – IV, D - I

134. Which of the following enzyme is used to overcome oxidative stress?
(A) SOD (B) PDH
(C) LDH (D) G6PDH

135. Which of the following pairs of species best demonstrates the c-value paradox?
(A) Arabidopsis and rice (B) Rice and maize
(C) Rice and poplar (D) Rice and wheat

136. Chromosme number of Arabidopsis is:
(A) 4 (B) 5
(C) 6 (D) 7

137. Complete set of cloned DNA fragments of an organism is known as:
(A) Genomic library (B) cDNA library
(C) mRNA library (D) Any of the above

138. During the process of reverse transcription:
(A) mRNA is synthesized from DNA
(B) DNA is synthesized from DNA
(C) Proteins are formed from mRNA
(D) DNA is synthesized from mRNA

139. Ribosomes are synthesized in
(A) Nucleus (B) Nucleolus
(C) Cytoplasm (D) All the above

140. Match the column A and B carefully and select the correct option from those given below:

Column A	Column B
A. Sexual hybridization	I. mRNA content
B. Protoplast fusion	II. Transgene organism
C. Recombinant DNA technology	III. Limited to closely related species
D. Transcriptome	IV. Difficult to predict outcome

Options
(A) A - II, B – I, C – IV, D – III
(B) A - IV, B – III, C – II, D – I
(C) A - III, B – IV, C – II, D – I
(D) A - II, B – III, C – IV, D - I

141. Which one is the first to be used as a source of energy?

(A) Carbohydrates (B) Proteins
(C) Lipids (D) Fats

142. Which is not correct about geographical indications (GIs):

(A) Once enforced by legislation, GIs exclude others from using a GI as trademark
(B) The chief requirement for GI protection is that a given quality, reputation or some other characteristics of the product be essentially attributable to the locality of its origin
(C) The GIs do not cover manufactured products and even food products
(D) Indian has enacted the Geographical Indication Act (1999) to claim GI for a variety of goods, including 'Basmati rice'

143. **Assertion (A) :** Consumption of oils containing high amount of oleic acid is beneficial for human beings

Reason (R) : It has cholesterol lowering propeties

Options

(A) Both A and R are correct and R adequate explanation of A
(B) Both A and R are correct, but R does not explain A
(C) Only A is correct
(D) Only R is correct

144. Vaccine against bird flu have been developed by using complete sequence of – gene from an H5N1:

(A) Haemagglutinin (HA) (B) Protease gene
(C) Protease inhibitor gene (D) RNAase

145. Golden rice is a transgenic crop of the future with the following improved trait:

(A) High insect resistance
(B) High lysine content
(C) High protein content
(D) High vitamin – A content

146. Which is correct about reverse transcriptase?

I. This enzyme was discovered by Temin and Baltimore
II. It is generally obtained from avian myeloblastosis virus
III. It is RNA dependent DNA polymerase
IV. This enzyme performs similar functions as DNA polymerase and has an absolute requirement for a primer with a free 3'-OH

Options

(A) II, III, IV (B) I, II
(C) I, IV (D) I, II, III, IV

147. Cloning is a means to:

(A) Replace original genotype
(B) Preserve genotype
(C) Production of HGH gene in *Escherichia coli*
(D) None of the above

148. Match the column A and B carefully and select the correct option from those given below:

Column A	Column B
A. Probe	I. Recombinant DNA technology
B. Transgene	II. Gene isolation
C. Embryo rescue	III. Cell fusion
D. Somatic hybridization	IV. Distant hybridization

Options

(A) A - II, B – I, C – IV, D – III
(B) A - IV, B – III, C – II, D – I
(C) A - III, B – I, C – IV, D – II
(D) A - II, B – III, C – IV, D – I

149. The medium is added continuously to the fermentor tank to replace that, which has been fermented. This is true in case of:

(A) Batch technique
(B) Continuous flow technique
(C) Both are true
(D) None is true

150. Eukaryotic DNA replication is carried out by

(A) DNA Pol α (B) DNA Pol β
(C) DNA Pol μ (D) DNA Pol δ

151. **Assertion (A) :** The gene encoding thaumatin has been transferred and expressed in *E. coli* and yeast with a view to obtain thaumatin at commercially accepted prices.

Reason (R) : The cost of production remains unacceptably high due to rather low levels of expression

Options

(A) Both A and R are correct and R adequately explains A
(B) Both A and R are correct, but R does not explain A
(C) Only A is correct
(D) Only R is correct

152. ABO blood groups is a good example for both------- and multiple alletism

(A) Co dominance (B) Dominance
(C) Recessive ness (D) Over dominance

153. Match the column A and B carefully and select the correct option from those given below:

Column A	Column B
A. Micropropagation	I. Haploid
B. Regeneration	II. Axillary bud break
C. Androgenesis	III. Interspecific hybrid
D. Embryo rescue	IV. Organogenesis and somatic embryogenesis

Options

(A) A - II, B – IV, C – I, D – III
(B) A - IV, B – III, C – II, D – I
(C) A - III, B – I, C – IV, D – II
(D) A - II, B – III, C – IV, D - I

154. The ability to create transgenic organisms allows genetic characteristics to be transferred beyond:
(A) Genus
(B) Species lines
(C) Across kingdom boundaries
(D) All of these

155. Which advantage is offered by *in vitro* techniques of germplasm storage?
(A) Require relatively very small space
(B) Help in storage of germplasm free from diseases, insects etc.
(C) Materials are learn source of nucleus seed
(D) All of these

156. An enzyme, aminoglycoside-3-phosphotransferase II. [APH(3)II] is commonly called :
(A) ß-glucuronidase
(B) Penicillinase
(C) Neomycin phosphotransferase II
(D) Streptomycin phosphotransferase

157. Why is the gene that was used by Sheller *et al.* to produce spider silk in transgenic plants called a "synthetic" gene?
(A) It produced silk fibre, which is a synthetic product in plants.
(B) It is assembled from synthesized DNA oligomers.
(C) Some of the amino acids encoded do not occur in nature
(D) It is encodes synthetic components such as an epitope tag.

158. Nucleic acid hybridization can be used to detect homologous DNA or RNA sequences not only in cell extracts, but also is chromosomes or intact cells- a procedure is called:
(A) *In situ* hybridization
(B) *Ex situ* hybridization
(C) Both (A) and (B)
(D) None of these

159. DNA synthesis can be specially measured by estimating the incorporation of radio-labelled:
(A) Thymidine (B) Deoxyribose sugar
(C) Uracil (D) Adnine

160. In 1970, Halmilton Smith and his co-worker first isolated a restriction enzyme from the bacterium of:
(A) Haemophilus influenzae strain Rd
(B) 200 base-pairs
(C) 256 base-pairs (D) 4096 base-pairs

161. Undesirable pleiotropic effect of a transgene on the transgenic plants may be corrected by :
(A) Changing the genetic background of transgene through hybridization
(B) Targetting of the transgene product into an appropriate cellular compartment
(C) Mutagenesis in the transgene
(D) Both (A) and (B)

162. Which is a soluble dietary fibre products, and is used as a nutraceutical to prevent cancer?
(A) Psyllium seed husk
(B) Broccoli (Sulgoraphane)
(C) Soy or clover products
(D) Reseveratrol

163. Which life forms can not be patented in India under the provisions of Indian Patents Act (1970)?
(A) Microorganisms (B) Animals
(C) Plants (D) All of these

164. Monitoring and tackling the IPR aspects of inventions has:
(A) Demand time
(B) Enhances cost
(C) May act as a distinctive for R and D efforts
(D) All of these

165. Which of the following attempts to increase the utility of fungal herbicides succeeded?
(A) Expression of a phytoxic protein at abnormally high levels in Fusarium.
(B) Expression of a phytotoxic protein from Fusarium in another fungal species.
(C) Expression of a phytoxin resistance protein from Fusarium in a crop plant.
(D) Expression of a phytotoxic protein from Fusarium in a weed.
(E) Expression of a phytotoxic protein from Fusarium in a crop plant.

166. Which is a soluble dietary fibre products, and is used reduce hypercholesteremia?
(A) Psyllium seed husk
(B) Broccoli (Sulgoraphane)
(C) Soy or clover products
(D) Reseveratrol

167. **Assertion (A) :** Deficiencies of micronutrients are common in the diet of developing countries.
Reason (R) : Deficiencies of micronutrients can be overcome by fortification/biofortification
Options
(A) Both A and R are correct and R adequately overcome the A
(B) Both A and R are correct, but R does not explain A
(C) Only A is correct
(D) Only R is correct

168. Which is a soluble dietary fibre products, and is used as:
(A) Psyllium seed husk
(B) Broccoli (Sulgoraphane)
(C) Soy or clover products (D) Reseveratrol

169. Which scientists showed that *Saccharomyces cerevisiae* causes fermentation forming products such as beer and butter milk?
(A) Louis Pasteur (B) Alexander Fleming
(C) Selman Waksman (D) Schatz

170. The hereditary material present in the bacterium *E. coli* is:
(A) Single standard DNA
(B) Double standard DNA
(C) DNA (D) RNA

171. *Saccharomyces cerevisiae* is employed in production of
(A) Idli (B) Beer
(C) Bread (D) All of these

172. A compound produced by an organism that inhibits the growth of another micro-organisms is called:
(A) Antiseptic (B) Antibiotic
(C) Anticoagulent (D) Antibodies

173. The codons causing chain termination are:
(A) AGT, TAG, UGA (B) UAG, UGA, UAA
(C) TAG, TAA, TGA (D) GAT, AAT, AGT

174. A restriction endonuclease is a bacterial enzyme that cuts double-stranded DNA by leaving the phosphodiester bonds at specific nucleotide sequences known as:
(A) Restriction site (B) 200 base-pairs
(C) 256 base-pairs (D) 4096 base-pairs

175. Bacteria are present everywhere except in:
(A) Ice (B) Soil
(C) Sea water (D) Distilled water

176. Microbiology is a discipline of science dealing with the study of:
(A) Forms and structure of organism
(B) Malignant tissues
(C) Microbes (D) Parasites

177. The plant virus was first isolated in crystalline form
(A) TMV (B) Bactriophages
(C) Canophages (D) None of these

178. Genetically engineered human insulin is manufactured by the use of:
(A) *E. coli* (B) *Rhizopus*
(C) *Pseudomonas* (D) *Ashbya gossypii*

179. Introduction of a tobacco mosaic virus transgene into plants in 1986 was used to demonstrate:
(A) Cross-expression (B) Co-inoculation
(C) Co-transformation (D) Cross-protection
(E) Cross–inoculation

180. Which of the following are not possible uses for fructans as presented in class?
(A) Natural insecticides (B) Fat substitute
(C) Paper production (D) Sweetners

181. Which of the following plants produce endosperm following fertilization?
(A) Wheat (B) Canola
(C) Pine (D) Both (A) and (B)
(E) None of these

182. Which of the following statements is true regarding the production of transgenic cassava?
(A) The release of cyanide was inhibited by immunomodulation.
(B) The release of cyanide was inhibited by antisense gene regulation
(C) The release of cyanide was inhibited by immunolocalization
(D) The production of cyanide was enhanced by over expression of an enzyme
(E) A non-toxic cyanide derivative was produced by metabolic engineering

183. Which method enhances vitamin C production?
(A) Polymerization of a series of simple sugars
(B) Partitioning of ascorbate into seed storage proteins
(C) Dorivutization of tocopherols from the shikimate pathway
(D) Increased enzymatic activity in the beta-carotenoid pathway
(E) Enzymatic reduction of a component of pectin

184. Alanine is found in:
(A) Skin (B) Muscles
(C) Brain (D) All of these

185. Which of the following statements concerning the RNA/DNA hybrid oligonucleotides is incorrect?
(A) They form a specific hairpin structure that is recognized by the host cell.
(B) They may be synthesized to contain a mutation that will be incorporated into the host genome.
(C) They provide a mechanism for targeted alteration of the plant genome.
(D) They are incorporated using a mechanism that involves host DNA repair machinery.
(E) They hybrid molecule must be introduced into the cell using a binary vector.

186. The molecular scissors used to cut DNA into specific genes of interest are called:
(A) Exonucleases
(B) Restriction endonucleases
(C) Ligases
(D) Polymerases

187. Enzyme TPA or tissue plasminogen activator is used for :
(A) Dissolving blood clots
(B) Maintaining plasma content
(C) Clearing turbidity of juices
(D) Stimulating thromboplastin production

188. In most of the cancer cases which gene is mutated?
(A) p53 (B) p21
(C) p16 (D) p15

189. Which of the following is involved in seedless watermelon production?
(A) Apomixes (B) Interspecific crosses
(C) Tetraploidy (D) Both (A) and (B)

190. What is a disadvantage of the current use of transgenic antibody fragments to induce resistance to the herbicide picloram?
(A) Each antibody molecul can neutralize only one herbicide molecule.
(B) Picloram is an auxinic herbicide, thus the antibodies will neutralize endogenous auxin.
(C) The antibodies induce morphological defects (e.g. epinasty)
(D) The antibodies are encoded by two separate polypeptide chains.
(E) The antibodies are quickly degraded by host enzymes, so they have little effect on picloram.

191. For cloning of the gene, which is not important?
(A) Vector (B) Host
(C) Restriction enzymes (D) DNase

192. Which of the following was an unexpected result connected to the transgenic production of the terpenoid linalool?
(A) Increased production of linalool resulted in lower levels of GA hormones.
(B) Increased transgene expression was not strongly correlated with increased levels of linalool.
(C) Transgene expression resulted in production of unexpected enantiomers of linalool.
(D) Linalool accumulation was localized to the nectary.
(E) Increased transgene expression resulted in higher levels of herbivory.

193. Interferon is:
(A) Antibacterial (B) Antiviral
(C) Antianimal cell (D) Bacterium

194. The bond between sugar and nitrogenous base in case of DNA is known as:
(A) Glycosidic bond (B) Phosphodiester bond
(C) H- bond (D) Peptide bond

195. Transplastomics:
(A) Targets genes in the chloroplast
(B) Provides exceptionally low yields of protein products
(C) Produces genes that are released in pollen
(D) Offers little opportunity for practical use

196. Somatic embryogenesis is:
(A) Germ line cells developing into embryos
(B) Non-germ line cells developing into embryos
(C) Embryos developing from zygotes
(D) Embryonic tissue becoming somatic

197. Manipulation of the genetic material towards a desired end in a directed and predetermined way is called:
(A) Genetic engineering
(B) Gene cloning
(C) Recombinant DNA technology
(D) All of these

198. Which of the following methods would be the most precise and accurate information about where and when a given gene is expressed?
(A) *In situ* hybridization (B) DNA microarray
(C) Protein microarray
(D) Reporter gene fusion including introns

199. Star Link is closely associated with which of the following terms?
(A) Glufosinate (B) Event 176
(C) Cry9 (D) Cry1 (E) Glyphosate

200. Third generation vaccines are:
(A) Attenuated pathogen (B) Pathogen relative
(C) Monoclonal antigen (D) Synthetic antigen

201. Which of the following is present on a Ti plasmid, but not on any component of a binary vector system?
(A) Vir genes (B) GUS genes
(C) LB, RB (D) MCS
(E) Opine genes

202. Genetic modification of plants can be achieved by :
(A) Somatic hybridization (B) Genetic engineering
(C) Somaclonal variation (D) All of these

203. Why might you find two kanamycin resistance genes on different parts of a T-DNA plasmid?
(A) One of the genes could be used for positive selection, and one could be used for negative selection.
(B) One of the genes could be used for selection in plants, and one could be used for selection in bacteria.
(C) This arrangement facilities site-specific recombination for marker-free selection.
(D) Each gene might have been introduced by a independent co-transformation event.

204. RNA molecules involve in splicing are:
(A) Sn rRNA (B) 5s rRNA
(C) 16s RNA (D) tRNA

205. AIDS lead to destruction of immune system due to depletion of
(A) B-cells (B) T helper cells
(C) Macrophages (D) Plasma cells

206. Which of the following is true of glyphosate?
(A) It as an animal LD 50 of 0.4 g/hg
(B) It is more selective than 2,4-D
(C) It can be manufactures without a license from the patent holder
(D) It inhibits the synthesis of branch chain amino acids

207. Which peoples are considered under farmer's rights for sharing the profit earned from the devel-opment of high yielding varieties evolved from the use of raw materials/plant genetic resources?
(A) Tribal people (B) Rural communities
(C) Traditional farming families
(D) All of these

208. Which countries has 30 years protection for inbred lines of maize and for clovers and a few grasses?
(A) France (B) Germany
(C) USA (D) UK

209. According to the assigned reading, in terms of total area of transgenic crops cultivated in 2011, the four major producers of transgenic crops (ranked highest to lowest) are:
(A) USA, Argentina, Canada, China
(B) USA, China, Canada, India
(C) USA, Canada, China, Argentina
(D) USA, China, India, Canada

210. Restriction site for cloning should be:
(A) Tendemly repeated
(B) Palindromic in nature
(C) Repeated sequence (D) Hexanucleotide

211. The quickest way to produce homozygous breeding lines from heterozygous parents is through:
(A) Half-seed technique (B) Doubled haploids
(C) Introgression (D) Aneuploidy

212. In a -----protocol, bacteria with engineered abilities to detoxify pollutants are intentionally released in an area.
(A) Microcosm establishment (B) Hybridization
(C) Bioremediation (D) Rhizosecretion

213. What does the CT-value (cutoff threshold) in real – time PCR experiments mean?
(A) The lower the value, the higher the transcript amount.
(B) The higher the value, the higher the transcript amount
(C) No transcript, if the value is larger than 30
(D) No transcript, if the value is lower than 30

214. Which of the following organelle lack membrane?
(A) Mitochondria (B) Ribosomes
(C) Microtubules (D) Peroxisomes

215. Enzyme immobilization is
(A) Conversion of an active enzyme into inactive form
(B) Providing enzyme with protective covering
(C) Changing a soluble enzyme into insoluble state
(D) Changing pH so that enzyme is not able to carry out its function

216. Which organelles is/are semi autonomous in nature?
(A) Mitochondria (B) Chloroplast
(C) Nucleus (D) Both (A) and (B)

217. Prions are microorganism, which consist of---------:
(A) DNA and protein (B) DNA and RNA
(C) DNA, RNA and Proteins (D) Proteins only

218. Virions are ----:
(A) DNA viruses (B) RNA viruses
(C) Naked DNA viruses (D) Naked RNA viruses

219. Prokaryotic RNA Polymerase comprises five subunits, which are:
(A) αβμδε (B) αεδδσ
(C) σεδωω (D) αεδδθ

220. AFLP molecular markers are ------ in nature:
(A) Dominant (B) Co-dominant
(C) Recessive (D) Neutral

221. Which is an example of a prokaryote does not contain nucleus?
(A) Blue green algae (B) Sieve cell of phloem
(C) Fungi (D) Yeast

222. During splicing of proteins which part is removed?
(A) Exons (B) Introne
(C) Exteins (D) Inteins

223. Which of the following technique is used for amplification of DNA?
(A) HPLC (B) PCR
(C) TLC (D) ELISA

224. How many microorganism are generally present per gram of biofertilizers?
(A) $10^7 - 10^9$ (B) $10^2 - 10^3$
(C) 10^5- 10^6 (D) $10^{12} - 10^{15}$

225. Taq polymerase used in PCR is:
(A) Thermotolerant (B) Highly processive enzyme
(C) Osmotolerant (D) All of these

226. In Bt cotton, expression of crystal protein occurs in:
(A) Only in stem (B) In leaves
(C) Systemic expression (D) Both (A) and (B)

227. Most commonly used method for transformation of animals is:
(A) Agrobacterium mediated transformation
(B) Protoplast method
(C) Electrotransformation
(D) Micro injection

228. Which is an example of a prokaryote?
(A) Blue green algae (B) Sieve cell of phloem
(C) Fungi (D) Yeast

229. Franken food is the 'nick name' of:
(A) Fast food (B) Frozen food
(C) GM food (D) Fortified food

230. Colchicine arrest cells in Mitosis, because:
(A) It binds to monomeric tubulin and thus depolymerizes microtubules
(B) It prevents chromosome condensation by inhibiting histone-histone interactions
(C) It dephosphorylates tubulin monomers and thus depolymerises microtubules
(D) It binds to the + end of F-actin and prevents further addition of G-actin to actin filaments

231. Transgenic expression study can be done by:
(A) Transcription profiling (B) RT-PCR
(C) Western hybridisation (D) All of these

232. Which is an example of a eukaryote may contain more than one nucleus per cell?
(A) Blue green algae (B) Sieve cell of phloem
(C) Fungi (D) Yeast

233. Hybridomas are employed for:
(A) Production of somatic hybrids
(B) Killing cancer cells
(C) Synthesis of antibiotics
(D) Synthesis of monoclonal antibodies

234. Which statements is correct?
(A) Earthworm (*Eisenia foetida*) has the ability to detoxify the toxic thermal power waste fly ash
(B) Uptake of high concentration of cadmium cause itai-itai disease of humans
(C) Excess of fluorine in drinking water causes dental and skeletal problems
(D) All of these

235. Which of the following are biological mutagens?
(A) UV radiation (B) Transposons
(C) Ionic radiation (D) Ethidium bromide

236. Which organisms is not used for preparation of biofertilizers?
(A) Azotobacter (B) Pseudomonas
(C) Azolla (D) Azospirillum

237. The first field tests of what genetically altered organism?
(A) Bt Corn
(B) Vaccinia virus containing a gene from the rabies virus
(C) The "flavsavr" tomato
(D) Strawberry seedlings sprayed with "ice-minus" bacteria

238. Which is associated with transgenic cotton varieties producing coloured cottons?
(A) They would reduced pollution
(B) They are attacked by large number of pests due to their colour
(C) They require higher doses of inorganic fertilizers
(D) They would have enhanced fibre strength

239. When pyrimidine is substituted by purines and vice- versa, this type of mutation is known as:
(A) Transition (B) Transversion
(C) Non sense mutation (D) Miss sense mutation

240. Usually which of the following RNA is used for production of cDNA?
(A) rRNA (B) mRNA
(C) tRNA (D) siRNA

241. Most commonly used method for transformation of plants is:
(A) Agrobacterium mediated transformation
(B) Protoplast method
(C) Electrotransformation
(D) Micro injection

242. Which organisms is used for production of single cell protein:
(A) Pseudomonas (B) Azospirillum
(C) Spirulina (D) BGA

243. Bacterial genes lack:
(A) Exons (B) Introns
(C) Promoters (D) Operator

244. In Bt cotton expression of crystal protein occurs:
(A) On surface of leaves (B) Systemic expression
(C) Only in stem part (D) Stem and leaves

245. The use of chemical or physical methods to persuade cells to take up DNA from the culture medium or direct physical introduction of DNA into the cell is called:
(A) Transfection (B) Transduction
(C) Both (A) and (B) (D) None of these

246. Which is not correct about breeder's exemption?
(A) The use of a protected variety for the development of new varieties is exempted from protection
(B) The PBR for a new variety evolved from the initial variety will be of the breeder who developed the new variety and of the holder of PBR title of the initial variety as well
(C) Under the UPOV Act, 1978 all new varieties evolved using a protected variety were exempted from protection under this provision
(D) The UPOV Act, 1991 has limited breeders exemption in that 'essentially-derived varieties' are covered by the PBR granted to the initially variety

247. Which is the main considerations for the development of PBR systems?
(A) It allows breeders to benefit from the varieties developed by them
(B) It encourages plant breeding activities
(C) Private sector is encouraged to invest in plant breeding and seed industry
(D) All of these

248. In phytoremediation plants are used to:
(A) Control microbial attack on plants
(B) Control microbial contamination of soil
(C) Accumulates pollutants from soil thus clears soil from pollutants
(D) All of these

249. Which of the following statements regarding manipulation of the flavonol bio-synthetic pathway in tomato as to true?
(A) Increased flavanol production led to decreased carotenoid production.
(B) Increased flavanol production led to increased carotenoid production.
(C) Increased flavanol production did not affect carotenoid production.
(D) Decreased flavanol production did not increase carotenoid production.

250. The roots of a plant are converted into drug-producing structures in a process called:
(A) Microcosm establishment
(B) Mibridization
(C) Bioremedition
(D) Rhizosecretion

251. For clearing oil spills which microorganism is used?
(A) *Pseudomonas putida* (B) *Bacillus* sp.
(C) *Rhizobium* (D) *E. coli*

252. Which statement is not correct?
(A) Fatty acid biosynthesis in plants is very similar to that in bacteria
(B) In case of plants, fatty acids are synthe-sized in oxysomes
(C) High oleic acid oils are nutritionally more beneficial
(D) Transgenic variety Laurical has high lauric acid content

253. Bt cotton is resistant to bollworm, it is due to:
(A) Bt is repellent to bollworm
(B) Bt is contact poison for bollworm
(C) Bt is stomach poison; it changes the perm-eability of stomach due to generation of pores in alkaline environment of stomach
(D) Bt is causing death due to asphagsiation of insect

254. The PPVFR Authority (PPVFRA) began regist-ration of plant varieties in which years?
(A) 2005 (B) 2006
(C) 2007 (D) 2008

255. Which is not correct about gamma-linolenic acid (GLA)?
(A) It is a non-essential fatty acid
(B) It is the first intermediate in the conversion of linoleic acid to arachidonic acid
(C) It is produced by desaturation of linoleic acid by the enzyme fatty acid desaturase
(D) It is beneficial in alleviation of many physio-logical and pathological conditions such as hypercholestereremia, cancer and diabetes

256. Which among the following is not used as vector for gene cloning?
(A) Cosmid (B) Phagemid
(C) PAC (D) MAC

257. Which is not correct about protoplasts?
(A) They may be cultured on a suitable medium in a variety of ways
(B) They regenerate cell wall within 2-4 days of culture
(C) The first division ion protoplasts occurs generally after 2-7 days and multicellular colonies appears after 2-3 weeks
(D) Protoplasts are cultured in light

258. Which is correct about map based cloning?
(A) A critical prerequisite is that the target gene must be localized definitely with respect to a molecular markers
(B) It starts with an interesting phenotype observed at whole organism level and proceeds to clone the gene via genetic mapping and chromosome walking
(C) Non-availability of tightly linked flanking markers to the gene of interest in a problem
(D) All of these

259. GEM stands for:
(A) Genetically Engineered Metazoa
(B) Genetically Engineered Microbes
(C) Genetically Engineered Metabolism
(D) Genetically Engineered Mustard

260. During RNA isolation which chemical is used for breaking of plasma membrane?
(A) Phenol (B) Isopropanol
(C) CTAB/SDS (D) Ethanol

261. Bioreactor are used for :
(A) Production of enzymes
(B) Production of secondary metabolites
(C) Production of cell cultures
(D) All of these

262. Genetic improvement of crop plants through biotechnology needs conventional breeding due to:
(A) The elite cultivars developed by plant bree-ding will be parents of the next generation of trasgenics
(B) Field testing across locations or cropping systems and over years will be needed to select the best transgenic event
(C) Breeding may be required to minimize/ eliminate pleitropic effects
(D) All of these

263. Which statements is correct about thaumatin?
(A) Gene encoding thaumatin is isolated from African shrubs, *Thaumatococcus danielli*
(B) It is a protein that is 100000 times sweeter than sugar on a molecular basis and has a flavour enhancing effect as subsweetness levels
(C) The gene producing thaumatin has been transferred and expressed in *E. coli* and yeast
(D) All of these

264. Who said "Environment today is a question of survival. Our rivers are drying, cities are choking, underground water tables are depleting, the air is not fit for breathing. Only a strong people's move-ment can bring a change in our present situation. Unless we have a green vote bank, the attitude of political parties towards the environment will not change":
(A) Rajendra Prasad (B) M.C. Mehta
(C) R.T. Doshi (D) K. C. Mehta

265. Which acids is commonly used in foods?
(A) Fumaric and lactic acids
(B) Citric acid and vinegar
(C) Maleic acid and tartaric acid
(D) All of these

266. Synthetic seeds are:
(A) Protoplasts (B) Somaclones
(C) Cybris (D) Somatic embryos

267. DNA Plant Technology faced difficulties during release of Endless Summer due to the use of:
(A) Selectable marker *npt*II
(B) CaMV 35S peromoter and the ACC synthase gene
(C) Antisense RNA technology and *Agrobacterium tumer faciens* mediated genetic transformation
(D) All of these

268. Geographical indications includes:
(A) Natural products (B) Agricultural goods
(C) Goods of handicraft (D) All of these

269. Cytoplasmic hybrids are known as:
(A) Hybrids (B) Cybrids
(C) Both (A) and (B) (D) None of the above

270. Protein purification can be done by
(A) Affinity chromatography
(B) Ion exchange chromatography
(C) Sephadex column
(D) All of these

271. Which phyto hormone is used for development of synthetic seeds?
(A) Auxin (B) GA
(C) ABA (D) Ethylene

272. During stress which of the following amino acid is accumulated?
(A) Proline (B) Tryptophan
(C) Tryosine (D) Lysine

273. Transfer of genes (gene flow) in between different species is known as:
(A) Vertical gene flow
(B) Horizontal gene flow
(C) Lateral gene flow
(D) None of these

274. Which is not correct about Scarlet?
(A) It is a somaclonal variety of potato
(B) It was selected from shoot-tip cultire derived plants
(C) This variety is comparable to the parent cultivar in yield and disease resistance
(D) It shows darker and more stable skin colour, which is a desirable quality trait

275. Horizontal gene flow can be reduced by expressing genes in:
(A) Nucleus (B) Chroloplast
(C) Both (A) and (B) (D) None of these

276. 'Kwashiorkar' is concerned with :
(A) Severe protein deficiency
(B) Food poising
(C) Flourine toxicity (D) Nitrate toxicity

277. Most commonly used organism in SCP is:
(A) *Bacillus* (B) *E. coli*
(C) *Spirulina* (D) *Pseudomonas*

278. Double stranded DNA can be isolated from sDNA or RNA by using:
(A) Ion exchange column
(B) Sephadex column
(C) Hydroxyapatite column
(D) Affinity chromatography

279. Which sigma factor of RNA polymerase comes into play during stress?
(A) σ^{32} (B) σ^{70}
(C) σ^{14} (D) σ^{60}

280. RNA interference is used to control different viral diseases in case of plants and animals. The active fragment involving in interference is known as:
(A) Si RNA (B) Micro RNA
(C) Both (A) and (B) (D) None of these

281. Which is not correct about geographical indications (GIs):
(A) Once enforced by legislation, GIs exclude others from using a GI as trademark
(B) The chief requirement for GI protection is that a given quality, reputation or some other characteristics of the product be essentially attributable to the locality of its origin
(C) The GIs do not cover manufactured prod-ucts and even food products
(D) Indian has enacted the Geographical Indication Act (1999) to claim GI for a variety of goods, including 'Basmati rice'

282. Development of high density molecular linkage map of a particular mapping population can be used for:
(A) MAS (B) Map based cloning
(C) Species characterization (D) All of these

283. Which one of the following is used in genetic engineering?
(A) RNA polymerase (B) DNA polymerase
(C) Restriction endonuclease (D) Nuclease

284. Which is not correct about virus cross protection?
(A) The phenomenon of crosss protection was first observed by McKinney in 1929 in case of tobacco mosaic virus
(B) Plants infected with a mild strain of virus resist the development of symptoms upon infec-tion by a more severe strain of the same virus
(C) A mild protecting strain effective for one region may be severe for another region
(D) Selection of mild protecting strain is rather easy

285. Genetically engineered bacteria are become employed for production of:
(A) Human insulin (B) Thyroxine
(C) Cortisol (D) Epinephrine

286. During RNA isolation, which chemical is used to prevent RNAase contamination?
(A) DMSO (B) DEPC
(C) Glycerol (D) All of these

287. Which of the histone protein is known as linker protein?
(A) HI (B) H2A
(C) H2B (D) H3

288. During stress which of the following phyto-hormone is accumulated?
(A) Auxin (B) GA
(C) ABA (D) Ethylene

289. Star activity of restriction enzyme means
(A) It cuts the restriction site more efficiently
(B) It cuts at only one site
(C) It does not cut at any site
(D) Due to change in ionic conc. or other components it cuts at more sites generating more number of fragments

290. Plasmids are used in genetic engineering because they are:
(A) Easily available (B) Able to replicate
(C) Able to integrate with host chromosomes
(D) Rarely available

291. Trade secret can not be applied for:
(A) Plant varieties (B) Equipment design
(C) Books (D) All of these

292. Government of India has now established the PPVFR Authority (PPVFRA) with head-quarters at:
(A) Chennai (B) Kolkata
(C) New Delhi (D) Mumbai

293. Giant mouse has been produced through:
(A) Tissue culture
(B) Gene differentiation
(C) Gene manipulation (D) Gene interaction

294. Yeast two hybrid system is used to:
(A) Determine interaction between DNA and RNA
(B) Determine interaction between two domains of proteins (i.e., DNA binding and protein binding domains)
(C) Determine interaction between RNA and proteins
(D) Determine interaction between DNA and proteins

295. In transgenics, expression of transgene in target tissue is determined by:
(A) Enhancer (B) Transgene
(C) Promoter (D) Reporter

296. In tissue culture regeneration of shoot and root occurs by manipulating the balance of:
(A) Auxin and ABA (B) Auxin & Cytokinin
(C) Cytokinin and ABA (D) ABA and Ethylene

297. Which category of microbes are used in retting process of jute?
(A) Microbes tolerant to high temperature
(B) Which process high amount of hydrolytic enzyme activity
(C) Anaerobic microbes only
(D) Aerobic microbes only

298. Bacterium commonly used in plant genetic engineering is:
(A) *Agrobacterium* (B) *Cryonebacterium*
(C) *Bacillus subtilis* (D) *Salmonella*

299. Which factor is responsible for detoxification of drugs in case of humans?
(A) p53 (B) P450
(C) p21 (D) All of these

300. Minisatellite are also known as:
(A) VNTRs (B) RFLPs
(C) STRs (D) RAPDs

301. During the World war II, technology for citric acid production was based on:
(A) *Aspergillus niger*
(B) *Aspergillus oryzae*
(C) *Rhizobium* sp.
(D) *Clostridium acetobutylium*

302. Among the given macromolecules, which one is the richest source of energy?
(A) Carbohydrates (B) Proteins
(C) Lipids (D) Nucleic acids

303. Which amino acid provide twists in tertiary structure of proteins?
(A) Serine (B) Glycine
(C) Proline (D) Tryptophan

304. The techniques to clone the first gene were developed in:
(A) 1972 (B) 1974
(C) 1976 (D) 1977

305. Urokinase enzyme is produced by:
(A) *Baccillus fastidious*
(B) *Baccillus subtilis*
(C) *Baccillus thuriengenesis*
(D) *Pseudomonas putida*

306. Which among the following is released in photorespiration?
(A) Serine (B) Cysteine
(C) Proline (D) Tryptophan

307. Which life forms can not be patented in India under the provisions of Indian Patents Act (1970):
(A) Microorganisms (B) Animals
(C) Plants (D) All of these

308. Monitoring and tackling the IPR aspects of inventions has:
(A) Demand time
(B) Enhances cost
(C) May act as a distinctive for R & D efforts
(D) All of these

309. Which programmes converts the submitted amino acid sequence into nucleotide sequence and compares it with nucleotide sequence database?
(A) $BLAST_x$ (B) $BLAST_n$
(C) $tBLAST_n$ (D) $tBLAST_x$

310. Which enzyme plays important role in reverse transcription?
(A) DNA polymerase (B) RNA polymerase
(C) Reverse transcriptase (D) Ribonuclease

311. GAATIC is restriction sequence of CTTAAG:
(A) Eco RI (B) Bam HI
(C) Sau 3 (D) Nco I

312. In tissue culture disease resistance can be obtained by:
(A) Somaclonal variation (B) Meristem culture
(C) Somatic hybridization (D) Anther culture

313. In case of phytoremediation, plants are used for which of the following?
(A) To accumulate pollutants from soil
(B) To control microbial attack on plants
(C) To control microbial contamination of the water
(D) All of these

314. Which amino acid plays an important role in splicing of proteins?
(A) Serine (B) Glycine
(C) Proline (D) Tryptophan

315. Which is not correct about somaclonal varieties?
(A) A low neurotoxin somaclonal variety, Ratan, of khesari has been released for commercial cultivation in M.P.
(B) Pusa Jaikisan is a somaclonal variety of *B. juncea*
(C) A somaclonal variety Bio-13 of citronella java developed by NBRI, Lucknow
(D) Pusa Jaikisan was derived from Varuna

316. Bioreactors are used for:
(A) Production of ethanol
(B) Production of enzymes on large scale
(C) Production of cell culture on large scale
(D) All of these

317. Which of the following techniques is used to inactivate a gene by altering the DNA encoding it:
(A) Homologous recombination
(B) Antisense nucleic aid blocks
(C) Antibody microinjection
(D) Introduction of dominant inhibitory mutants

318. Mathanogenic bacteria are:
(A) Anaerobic in nature
(B) Aerobic in nature
(C) Micro aerophillic in nature
(D) Saprophitic in nature

319. First of all coloured cotton developed in India in the state of:
(A) Madhya Pradesh (B) Maharastra
(C) Gujarat (D) Andhra Pradesh

320. First hybrid variety of sorghum was evolved in the year:
(A) 1946 (B) 1956
(C) 1964 (D) 1974

321. Which yeast strain is naturally used for ethanol production?
(A) *Saccharomyces cerevicae*
(B) *S. diastaticus*
(C) Both of these (D) None of these

322. Example of tandem gene cluster is
(A) rRNA genes (B) Histone gene cluster
(C) Both (A) and (B) (D) None of these

323. What are the components of PCR reaction?
(A) Buffer, Taq polymerase, dNTPs, primer and template
(B) Buffer, dNTPs, primer and template
(C) Buffer, Taq polymerase, primer an d template
(D) Buffer, Taq polymerase, dNTPs and template

324. The lineal use for the recombinant DNA technology is to have rapidly replicating bacteria produce large amounts of specific proteins (e.g., hormones). Expression of a eukaryotic gene in prokaryotes involves:
(A) A shine-Delgarno (SD) sequence in mRNA
(B) Absence of introns
(C) Regulatory elements upstream of the gene
(D) All of these

325. Stable transformation of genes in case of plants can be done by:
(A) Biolistic gun method
(B) Agrobacterium mediated transformation
(C) Protoplast fusion
(D) Microinjection

326. Artificial seed refers to:
(A) Seeds of a plant with desired characteristics synthesized with recombinant DNA technology
(B) Somatic embryos encapsulated in a gel
(C) Seeds of synthetic plants
(D) All of these

327. Which of the following is a dominant marker?
(A) RAPD (B) RFLP
(C) SSR (D) AFLP

328. Chaperon play an important role in:
(A) Protein folding (B) Protein denaturation
(C) Protein trafficking (D) Protein synthesis

329. Methods for which can not be patented?
(A) Treatment of disease (B) Diagnosis of disease
(C) Prevention of disease (D) All of these

330. PPVFR Act 2001 allows registration of:
(A) Farmer's varieties (B) Extant varieties
(C) New Varieties (D) All of these

331. First hybrid variety of bajra was evolved in the year:
(A) 1946 (B) 1956
(C) 1965 (D) 1974

332. What is the role of SDS in case of SDS PAGE?
(A) It increases speed of proteins in gel
(B) It protects proteins from denaturation
(C) It provides negative charge to all the proteins so that separation can be done only on the basis of molecular weight
(D) SDS is detergent so is used to emulsify contaminating fats if any in samples

333. In lac operon, allolactose act as:
(A) Inducer (B) Gratituous inducer
(C) Repressor (D) Substrate

334. High density molecular map can be developed by:
(A) RFLP (B) RAPD
(C) AFLP (D) SNPs

335. Translation of histone protein occur during:
(A) S phase of cell division only
(B) G1 and S phase
(C) G2 phase (D) M phase only

336. Which is concerned with computer based analysis of large amounts of biological data sets?
(A) Molecular biology (B) Computer science
(C) Information technology (D) Bioinformatics

337. Which among the following contains only one type of nucleic acid?
(A) Bacteria (B) Virus
(C) Dead virus (D) Fungi

338. Histone octomer comprise of:
(A) HI, H2A, H2B and H3
(B) HI, H2A, H2B, H3 and H4
(C) HI, H2B, H3 and H4
(D) H2A, H2B, H3 and H4

339. Operon comprise of
(A) Operator (B) Structural gene
(C) Promoter (D) Repressor
(E) All of these

340. International Centre for Genetic Engineering and Biotechnology is located at :
(A) Chennai (B) New Delhi
(C) Bangalore (D) Both (A) and (B)

341. Histone proteins are ----- in nature.
(A) Acidic (B) Basic
(C) Neutral (D) Any of these

342. You want to know whether the enzyme you have introduced as a transgene in poplar is being expressed in the intended cells at the intended development stage. Which technique would be most informative?
(A) Microarray
(B) Immunolocalization
(C) RNA in situ hybridization
(D) Promoter fusion
(E) QRT-PCR

343. Chaperone present in ER is:
(A) Hsp 60 (B) Hsp 43
(C) Hsp 80 (D) Bip protein

344. Which is correct about B cells?
(A) B cells develop and mature in bone marrow
(B) Mature B cells leave the bone marrow and migrate via blood stream to the secondary lymphoid organs or tissues, where they ultimately become plasma cells
(C) B cells play an important role in immune system of body
(D) All of these

345. What strategy can be adopted to overcome the chances of resistance development in case of insects against transgenic crops?
(A) Growing of refuge crops
(B) Tissue specific expression of genes
(C) Temporal gene expression
(D) Any of these

346. After transformation of plant cells through Agrobacterium, which antibiotic is used to carry out Agrobacterium free selection of cells/tissues?
(A) Ampicillin (B) Rifamycin
(C) Tetracyclin (D) Carbencillin/Kenamycin

347. Which of the following statements about basal promoters is true?
(A) Basal promoters can be located in the 3 UTR
(B) Basal promoters direct assembly of the core translation machinery
(C) Basal promoters are sufficient for expression in the shoot apical meristem
(D) Basal promoters are necessary for enhancer trapping
(E) Basal promoters are necessary for heterologous expression.

348. Removal of the topological strain by inducing the negative supercoilling is carried out by:
(A) Topoisomerases (B) DNA gyrases
(C) Primases (D) DNA polymerases

349. Which gene is used for development of salt tolerant crops?
(A) Heat shock proteins
(B) Genes involved in proline biosynthesis
(C) LEA genes
(D) All of these

350. In terms of base sequence of genomic DNA, we are more than 98% similar to:
(A) Mice (B) Chimpanzee
(C) Rabbit (D) Monkey

351. Chromosomal aberration in which two breaks occur in a chromosome and the intercalary segment reunites in a reverse order is categorized as:
(A) Duplication (B) Deletion
(C) Inversion (D) Translocation

352. In case of PPVFR Act, 2001, the gene fund shall be used for:
(A) Paying compensation to communities for their contributions to the development of a variety
(B) Conservation of genetic resources
(C) Sustainable use of genetic resources
(D) All of these

353. In the first instance, PPVFR began registration of varieties in how many of the crops?
(A) 6 (B) 12
(C) 18 (D) 20

354. Which form of RNA has a structure resembling clover leaf:
(A) rRNA (B) hnRNA
(C) mRNA (D) tRNA

355. **Assertion** (A) : During preparation of embryo for culture, care should be taken to excise the young embryos with their suspensors intact
Reason (R) : Suspensor supplies GA_3 to young embryos
Options:
(A) Both A and R are correct and R adequately overcome the A
(B) Both A and R are correct, but R does not explain A
(C) Only A is correct
(D) Only R is correct

356. The term transformation is used to?
(A) Conversion of a normal vertebrate cell into a tumorigenic cell
(B) Uptake of DNA by a cell
(C) Delivery of DNA into a bacterial cell by a virus
(D) Both (A) and (C)

357. Which of the following statements about astaxanthin is false?
(A) It is a carotenoid
(B) It is normally produced by salmon
(C) It is a red-orange pigment
(D) It is produced from terpenoids
(E) It has anti-oxidant properties

358. Which is called natural genetic engineer?
(A) *Agrobacterium tumefaciens*
(B) *Escherichia coli*
(C) *Neurospora crassa*
(D) *Pseudomonas putida*

359. Which non-infectious diseases is associated with SCP industries?
(A) Asthama (B) Allergy
(C) Dermatitis (D) All of these

360. Which is called stress hormone?
(A) GA3 (B) Ethylene
(C) ABA (D) Cytokinin

361. The unique properties of each amino acid are determined by its particular:
(A) R-groups (B) H-bonds
(C) Peptide (D) Amino groups

362. During catabolism, glucose is converted to G.6.P using A.T.P., while in anabolism:
(A) Glucose is formed from phosphate ester by hydrolysis.
(B) Glucose is formed from pyruvate.
(C) Pyruvate is produced from its enol phosphate by transphosphorylation to ADP
(D) None of these

363. Which ministries established Department of Biotechnology in 1986?
(A) Ministry of Human Resource Development
(B) Ministry of Environment
(C) Ministry of Science and Technology
(D) Ministry of Agriculture

364. Which organisms was discovered by Fleming to produce the antibiotic penicillin?
(A) *Aspergillus niger*
(B) *Penicillium notatum*
(C) *Penicillium griseofulvum*
(D) *Oenicillium chrysogenum*

365. When producing human serum albumin (HAS) in transgenic plants, what would be an advantage of incorporating the transgenic into the chloroplast genome?
(A) HAS forms inclusion bodies if expressed in cytoplasm
(B) HAS becomes allergenic if expressed in cytoplasm
(C) HAS is toxic to plants if expressed in cytoplasm
(D) HAS is degraded by plant enzymes if expressed in cytoplasm
(E) HAS diverts too many amino acids if expressed in cytoplasm

366. Which enzymes is used for the production of glucose from starch?
(A) Lipase (B) Glucoamylase
(C) α- amylase (D) Both (B) and (C)

367. Totipotency is the property of :
(A) Plant cells (B) Animal cells
(C) Cancer cells (D) Both (A) and (B)

368. Which of the following transgenic methods were used (in separate experiments) in an attempt to increase the prevalence of certain amino acids in seeds?

(A) Altered tRNA gene; overexpression of sunflower seed albumin (SSA).

(B) Altereed tRNA gene; overexpression of human serum albumin (HAS)

(C) Altered rRNA gene; overexpression of sunflower seed albumin (SSA)

(D) Altereed rRNA gene; overexpression of human serum albumin (HAS)

(E) Altereed tRNA gene; repressionof sunflower serum albumin (SAS) expression

369. Common food yeast is:

(A) Saccharomyces (B) Cryptococcus
(C) Torulopsis (D) Endomyces

370. Tissue plasminogen activator is:

(A) A vitamin (B) An electric device
(C) An enzyme
(D) A chemical stimulates tissue differentiation

371. How was luciferase (LUC) with an artificial stop codon used as a control in the experiments involving altered tRNA?

(A) Diminished LUC expression demonstrated that the altered tRNA could properly recognize stop codons.

(B) LUC expression demonstrated that the gene was being induced by Lys.

(C) LUC expression demonstrated that Lys was being substituted for the stop codon by the altered tRNA.

(D) Loss of LUC expression demonstrated that a transversion had occurred, in order to create a stop codon.

(E) The artificial LUC is an example of a synthetic gene.

372. Monoclonal antibodies are produced by:

(A) Sarcoma cells (B) Germ cells
(C) Somatic cells
(D) Hybrid of carcinogen cell and bone cell

373. Transgenic animal has:

(A) Foreign DNA in all its cells
(B) Foreign RNA in all its cells
(C) Foreign DNA in some of the cells
(D) Both (B) and (C)

374. Which is not correct about abscisic acid?

(A) It is a naturally occurring growth inhibitor
(B) Precursor of ABA is maleic acid
(C) It is synthesized in leaves
(D) ABA counteracts the effects of GA and promotes dormancy in buds and seeds

375. In human beings, interferons are:

(A) Leucocytic (B) Immune
(C) Fibroblastic (D) All of these

376. Which methods is used for building of phylogenetic trees?

(A) Distance matrix methods
(B) Maximum parsimony methods
(C) Maximum likelihood methods
(D) All of these

377. The microbes play an important role in the catalytic complicated chemical reactions of:

(A) Wine production (B) Fermentation
(C) Dairy product (D) All of these

378. In the breakdown of glucose to pyruvic acid, how many enzymes catalyze these successive steps?

(A) 8 (B) 9
(C) 10 (D) 11

379. Virus mediated gene transfer, in which the foreign DNA is packaged inside a virus particle is called:

(A) Transfection (B) Transduction
(C) Both (A) and (B) (D) None of these

380. Which drugs is produced through biotechnology?

(A) Erythropocitin (B) Fibroblast interferon
(C) Insulin (D) All of these

381. Which one of the following makes use of RNA template to synthesize DNA?

(A) DNA polymerase
(B) RNA polymerase
(C) Reverse transcriptase
(D) DNA dependent RNA polymerase

382. In the area of biotechnology, which materials can be kept as trade secret?

(A) Cell lines
(B) Strains of microorganism
(C) Production processes
(D) All of these

383. Which biotechnological inventions can be patented?

(A) Microorganisms, cell lines, plant/animal lies obtained through biotechnological approaches
(B) DNA sequences/genes and the proteins encoded, if any, by them
(C) Biotechnological process/technologies for modification of the properties of various organisms
(D) All of these

384. *Bacillus thuringiensis* (Bt) strains have been used for designing novel:

(A) Biofertilizers
(B) Bio-metallurgical
(C) Bio –mineralization process
(D) Bio-insecticidal plants

385. Optical isomerism arises from the presence of:

(A) An systematic carbon atom
(B) A centre of symmetry
(C) A line of symmetry
(D) None of these

386. What is the best method to identify the cellular location of a protein?

(A) Place a reporter gene next to the promoter of the gene encoding the portion, and identify the cellular location of the reporter protein
(B) Use of labeled antibody to identify the cellular location of the protein
(C) Separate the cellular compartments by centrifugation and screen the different compartments with n antibody
(D) Tag the protein with fluorescent amino acids and identify the cellular location of fluorescent microscopy

387. Border sequences need to be incorporated into the design of plasmid vectors for agrobacterium mediated transformation to ensure:

(A) Greater promoter efficiency
(B) Oncogene deactivation
(C) Efficient replication of the plasmid
(D) Integration of the genes of interest into the host gene

388. What is a genomic library?

(A) A collection of recombinant molecules with inserts that contain all of the genes of an organism
(B) A collection of recombinant molecules with inserts that contain all of an organism's genome
(C) A collection of recombinant molecules that express all of the genes of an organism
(D) A collection of recombinant molecules that have been sequenced

389. Which of the following portions of a longer duplex DNA segment are likely to be recognition sequences of a restriction enzyme?

5'-AGTC -3'	5'-ATCG – 3'
(A) 3'-TCAG-5'	(B) 3'-TAGC -5'
5'-ACCT-3"	5'-ACGT-3'
(C) 3'-TGGA-5'	(D) 3'-TGCA-5'

390. Which non-infectious diseases is associated with SCP industries?

(A) Asthama (B) Allergy
(C) Dermatitis (D) All of these

391. Success in germplasm storage is determined by:

(A) Inherent longevity and physiological storage behaviour of the species
(B) Initial quality (e.g., moisture content) of the material stored
(C) Storage methods and conditions used
(D) All of these

392. Lipids are source of :

(A) Vitamins
(B) Structural components of cell organization
(C) Energy
(D) All of these

393. Which is not correct about verminator technology?

(A) Zeneca, U.K. has developed what it calls improved plant germplasm invention
(B) Improved plant germplasm invention has been dubbed as verminator technology by RAFI, Canada
(C) This technology uses a gene from the fat tissues of a goat, the product of which interferes with normal plant growth
(D) If the transgenic crop is sprayed with a specific chemical, plants show normal growth

394. A restriction endonuclease has the recognition sequence G/AATTC, where "/" indicates the cut site. This sequence is found, on average, one every 'x' residues in a chromosome. 'x'=

(A) 146 base-pairs (B) 200 base-pairs
(C) 256 base-pairs (D) 4096 base-pairs

395. Which is not correct about patent?

(A) It is the right of an individual or a group of individuals to gain profit from a particular invention or unique manufacturing process
(B) It is granted by the Government of a country to the applicant for an unlimited period of time
(C) When a patent is granted, the inventor becomes the owner of the patent like any other form of property
(D) Patents are granted for an invention (including a product), innovation/improvement in an invention, process/product of an invention and a concept

396. Which features is not concerned with the PPVFR Act, 2001?

(A) The Central Government is to constitute a National Gene Fund
(B) Citizens of convention countries will have the same rights as citizens of India under the Act
(C) Application for registration of a variety may be made in India within 2 years from the date of application for registration of the same plant variety made in the convention country
(D) The rights of PBR holder shall not be deemed infringed by a farmer who was not aware of the existence of such right

397. In case of medicine, biotechnology can be used for :

(A) Production of synthetic vaccines
(B) Diagnosis of disease
(C) Production of rare drugs
(D) All of these

398. Which is correct about food additives?

(A) These are substances added to food to preserve flavour or improve taste and appearance
(B) They may be natural or artificial
(C) To regulate these additives, and perform consumers, each additive is assigned a unique number, initially these were E numbers used in Europe for all approved additives
(D) All of these

399. Bacterial artificial chromosomes (BACs), cosmids, phages, plasmids, and yeast artificial chromosomes (YACs) are all commonly used cloning vectors that differ in their cloning capacities, with a range from approximately 100 bp to 3000 kb. Which of the following is the proper order for these vectors in terms of increasing cloning capacity?

(A) BAC, cosmid, phage, plasmid, YAC
(B) YAC, BAC, cosmid, phage, plasmid
(C) Plasmid, phage, cosmid, BAC, YAC
(D) Plasmid, cosmid, phage, BAC, YAC

400. Which statement is not correct about cryopreservation?

(A) It has proved to be the most reliable method for long term preservation of cell cultures
(B) It consists of choice of explants preculture and pretreatment, cryoprotection, freezing, storage, thawing and reculture
(C) DMSO is rarely used
(D) A general protocol applicable to all species and explants is not available

401. Using the strain of *E. coli* named A, a recombinant plasmid P was amplified. When plasmid DNA was isolated from strain A and was used to transform another *E. coli* strain B, no transformants could be obtained. Transformations using control DNA were fine. When the plasmid DNA from strain A was used to transform another strain, colonies were obtained. This is because:

(A) Strain B has an intact restriction-modification system whereas A do not
(B) Strain B is not capable or recombination
(C) A and C are exonuclease free whereas B is not
(D) Strain B is antibiotic sensitive

402. Many developing countries pay huge subsidies (e.g., USA pays – 50 billion dollars per year) to their farmers is not growing which of the following?

(A) Medicinal plants (B) Food crops
(C) Horticultural crops (D) Fodder crops

403. Which of the following is considered as primary physical containment?

(A) Containment equipment
(B) Special laboratory design
(C) Laboratory practices
(D) Both (A) and (C)

404. Which of the following organisms has the minimal genome?

(A) *E. coli*
(B) *Heliobacter pylori*
(C) *Methanococcus jannaschii*
(D) *Mycoplasma genitalium*

405. DNA replication occur in which stage of cell division?

(A) G1 phase (B) G2 phase
(C) S phase (D) M phase

406. Which of the following technique is used for DNA fingerprinting?

(A) HPLC (B) RFLD
(C) TLC (D) ELISA

407. Immobilisation protests an enzyme against:

(A) High temperature (B) Low temperature
(C) Amylases (D) Protease

408. cDNA synthetic DNA made from mRNA with the use of a special enzyme called:

(A) Reverse transcriptase
(B) Expression vector
(C) Promoter gene (D) Reporter gene

409. Proceeded cheese is:

(A) Soft cheese (B) Blended cheese
(C) Ripe cheese (D) Hard cheese

410. Gene producing a protein that can be detected and quantified using a simple assay is called:

(A) Structural gene (B) Reporter gene
(C) Promoter gene (D) Eucaryotic gene

411. Physical containment relates to:

(A) Laboratory design
(B) Containment equipment
(C) Special laboratory procedures
(D) All of these

412. A gene or DNA sequence having a known location on a chromosome and associated with a particular trait or gene is used as a:

(A) Genetic marker (B) Promoter gene
(C) Genetic map (D) Genetic engineering

413. Recombinants can therefore be identified because the characteristics coded by the inactivated gene is no longer displayed by the host cells is called:

(A) Transfection (B) Transduction
(C) Insertional inactivation
(D) None of these

414. Immobilised yeast is:

(A) Preserved for future use
(B) Little active in fermentation
(C) 20 times more efficient
(D) Food yeast

415. Recombinants are hybrid DNA molecules consisting of autonomously replicating DNA segment plus inserted elements. Such hybrid molecules are also called:

(A) Chimera (B) Ligation
(C) Transformation (D) Cloning

416. A fundamental in molecular genetic which takes advantage of the ability of individual single – stranded nuleic acid molecules to form double-stranded molecules is called:

(A) Nuleic acid hybridization
(B) ex situ hybridization
(C) in situ hybridization (D) None of these

417. A DNA microarray also known as:
(A) Gene chip (B) DNA chip
(C) Gene array (D) All of these

418. Which among the following is not unicellular?
(A) Bacteria (B) Yeast
(C) Fungi (D) None of these

419. Phage display method was first described by:
(A) George Smith (B) Ian Wilmut
(C) Flemming (D) Watson and Crick

420. Trade secret relates to:
(A) Certain materials, e.g., bacterial strains
(B) Processes
(C) Formulae
(D) Any one of these

421. Which biotechnology related matters can be patented?
(A) Method/process of production
(B) Products, etc
(C) Applications of the various processes/products
(D) Any of these

422. Hybridoma cells are developed by fusion of:
(A) T cells with myeloma cells
(B) B cells with myeloma cells
(C) T cells with B cells (D) B cells with APCs

423. The display of a heterologous target protein under study on phage particle is known as:
(A) Phage display (B) Chimeric protein
(C) Both of these (D) None of these

424. Which bond in the DNA provides in the backbone?
(A) Glycosidic bond (B) Phosphodiester bond
(C) H- bond (D) Peptide bond

425. During meiosis actual reduction of chromosomes occur in which sub stage?
(A) Anaphase (B) Metaphase
(C) Telophase (D) Prophase

426. Which of the following amino acid participate in urea cycle?
(A) Arginine (B) Lysine
(C) Tryptophan (D) Histidine

427. Sequence tagged sites that are derived from DNA are called:
(A) Sequence tagged sites mapping
(B) Expressed sequence tags
(C) Both of these (D) None of these

428. Which factor is necessary for maintaining lysogeny in lemda phase?
(A) Cro (B) C1 repressor
(C) Both (D) None

429. Which sequence in case of eukaryotes is important for mRNA tailing?
(A) GAGAGA (B) GAATTC
(C) UACGAC (D) UACUAAC

430. Shine Dalgarno sequence present on mRNA in case of prokaryotes plays as important role in protein synthesis. This sequence binds with ----- during translation initiation:
(A) 16s RNA (B) tRNA
(C) 23S RNA (D) 5s RNA

431. Synthetic vaccines have been developed for:
(A) Typhoid and rabies
(B) Cancer (Feline leukaemia virus)
(C) Cholera and foot and mouth disease of cattle
(D) All of these

432. Which is used as a nutraceutical?
(A) Soluble dietary fibre products such as Psyllium seed husk, broccoli (sulforaphane), soy or clover (isoflavonoids)
(B) Reseveratrol from red grape products
(C) Alpha-linolenic acid from flax seeds
(D) All of these

433. The technique for correcting defective genes responsible for disease development is called:
(A) Gene therapy (B) Gene inhibition
(C) Gene cloning (D) All of these

434. The technique used to distinguish between individual of the same species using only samples of their DNA is called:
(A) DNA fingerprinting (B) DNA testing
(C) DNA typing/DNA profiling
(D) All of these

435. Department of biotechnology promotes human resource development through:
(A) Master's programmes
(B) Post-doctoral programmes
(C) Certificate/Diploma programmes
(D) All of these

436. The ability to transfer or reprogram cultured cells with new genetic material (DNA and genes) has provided a major tool to molecular biologists wishing to study the cellular effects of the expression of these genes (new proteins) is called:
(A) Genetic counseling (B) Genetic engineering
(C) Gene therapy (D) All of these

437. Which transgenic crops is being grown on commercial scale in India?
(A) Bt-cotton (B) Bt-brinjal
(C) Bt-rice (D) Bt-tobacco

438. Which is correct about dietary supplement?
(A) It is also called nutritional or food supplement
(B) It is a preparation intended to provide nutrients such as fatty acids, vitamins, minerals or amino acids that are missing or are not consumed in sufficient quantity in diet
(C) The hormones DHEA, pregnenolone and pineal hormone melatonin are marketed as dietary supplements in the USA
(D) All of these

439. Amniocentesis, a diagnosis technique that enables doctors to remove and culture fetal cells from pregnant women, has given doctors an important tool for the early diagnosis of fetal disorders is known as:

(A) Genetic counseling (B) Genetic engineering
(C) Gene therapy (D) Gene mapping

440. Plants cells tissue has the ability to develop into a complete plant. This property of plants is known as:

(A) Pluropotency (B) Totipotency
(C) Multipotency (D) None of these

441. The common reporter genes obtained from *E coli* is:

(A) Chloramphenicol acetyl transferase (CAT)
(B) ß -galactosidase
(C) ß-glucuronidase (GUS) (D) All of these

442. In cell culture medium commonly used carbon source is:

(A) Fructose (B) Sucrose
(C) Glucose (D) Mannose

443. Sieve tubes of phloem tissues lack:

(A) Nucleus (B) Cell wall
(C) Mitochondria (D) Ribosomes

444. Which of the following DNA structure forms left hand helix?

(A) DNA A (B) DNA B
(C) DNA C (D) DNA Z

445. For molecular characterization of plant species which of the following molecular markers are used?

(A) RAPD (B) AFLP
(C) SSR/STMS (D) All of these

446. The whole process of transgenic plants development involving introduction, integration and expression of foreign gene(s) in the host is called:

(A) Genetic counseling
(B) Genetic engineering
(C) Genetic transformation
(D) Transgenic plants

447. Barr and Bertram observed a darkly staining body in interphase nerve cells of female acts that was absent in similar cells of males. This body described as:

(A) Sex chromatic body (B) Barr body
(C) Both of these (D) None of these

448. Sterilisation of medium through autoclaving is done at:

(A) 10 pounds (B) 15 pounds
(C) 20 pounds (D) 25 pounds

449. For plasmid purification the bacterial cells extract is deproteinized, the RNA removed, and the DNA concentrated by:

(A) Ethanol precipitation
(B) Methane precipitation
(C) Carbon precipitation
(D) Ammonia precipitation

450. **Problem (P):** An inherent problem associated with the map-based cloning is the non-availability of tightly linked flanking markers to the gene interest

Solution (S): Transposon tagging provides solution for the isolation of useful genes

Options

(A) Both P and S are correct and S adequately explains P
(B) Both P and S are correct, but S does not explain P
(C) Only P is correct
(D) Only S is correct

451. The vectors that are hybrids of lambda phages and plasmids, and their an replicate in the cell like that of a plasmid or be packaged like that of a phage is called:

(A) Cosmids (B) Fosmids
(C) Plasmids (D) Phagemids

452. Sterlization of antibiotics is done by:

(A) Autoclaving (B) Dry heat
(C) Use of filter (D) Any of the above methods

453. tRNA synthesis is done by:

(A) RNA Polymerase I (B) RNA Polymerase II
(C) RNA Polymerase III (D) All of these

454. Autonomously replicating genetic element used to carry a fragment of genomic DNA into a host cell for the purpose of gene cloning:

(A) Cloning vector (B) Insertion vector
(C) Shuttle vector (D) Viral vector

455. DNA is fit for making tools in case of nanotechnology. This is due to:

(A) Branching nature of DNA
(B) Flexibility in DNA conformation
(C) Small size of DNA
(D) All of these

456. The short tagged tracts of DNA sequence (200 to 500 base pairs) that is operationally unique and has a single occurrence in the genome is called:

(A) Sequence tagged sites mapping
(B) Expressed sequence tags
(C) Both of these (D) None of these

457. Lambda phage infest bacteria *E. coli* and replicates by a:

(A) Lysogenic pathway (B) Lytic pathway
(C) Both (A) and (B) (D) None of these

458. A genetically engineered organism that carries one or more genes in its chromosomes that has been made inoperative is known as:

(A) DNA fingerprinting (B) DNA testing
(C) DNA typing/DNA profiling
(D) Gene Knockout

459. DNA replication is:

(A) Conservation (B) Semi-conservative
(C) None of these (D) Both of the above

460. mRNA usually is being extracted using by:
(A) Poly T resin (B) RNase P
(C) a polyA resin (D) an rRNA trap

461. A set of disease resistance mechanisms that are not specific to a particular pathogen comes under
(A) Adaptive immunity (B) Innate immunity
(C) Passive immunity (D) Active immunity

462. In USA, which systems are available for protection of IPRs related to plants?
(A) The Plant Patents Act (1930)
(B) The Plant Variety Protection Act (1970)
(C) The Utility Patents Act (1985)
(D) All of these

463. Which of these is/are the characteristic/s of the specific immunity?
(A) Antigenic specificity
(B) Diversity and self/nonself recognition
(C) Immuninologic memory
(D) All of these

464. Who coined the term Green revolution:
(A) Williamn Gadd (B) M.S. Swaminathan
(C) Norman E. Borlaoug (D) C. Subramanian

465. Diploid species of cotton are:
(A) Desi (B) American
(C) Both (A) and (B) (D) None of these

466. Tetraploid species of cotton are:
(A) Desi (B) American
(C) Both (A) and (B) (D) None of these

467. Silent valley has been preserved as it has:
(A) Recreational value
(B) Valuable timber plants
(C) Natural forest
(D) Rare plant and animals

468. Silent valley having rare plants and animals is located in:
(A) Kerala (B) Karnataka
(C) Jammu and Kashimir (D) Andhra Pradesh

469. Which one is connected with conservation of forests?
(A) Gir (B) Silent valley
(C) Kaziranga (D) Bharatpur

470. New approach to conservation is establishment of:
(A) Reserve forests (B) Santuries
(C) Biosphere reserves (D) National parks

471. Social forestry is:
(A) Management of forests by cooperative societies
(B) Growing and management of useful plants on common lands
(C) Growing one type of trees
(D) Growing different types of plants together

472. Decrease in species diversity in tropical countries is mainly due to:
(A) Urbanisation (B) Pollution
(C) Deforestation (D) Soil erosion

473. Adulteration of mustard oil with *Argemone mexicana* seed oil caused in the recent past:
(A) Tuberclosis (B) Diarrhoea
(C) Dropsy (D) None of these

474. Wheat is a:
(A) Monoecious plant (B) Diocious plant
(C) Hermaphrodite plant (D) None of these

475. A genetic process called 'gene conversion' is observed in:
(A) Yeast (B) Fungi
(C) Other eukaryotes (D) All of these

476. Which means that one of the allele is being converted from one type to other, possibly by DNA recombination during meiosis after fusion of the gametes is called:
(A) Gene conversion (B) Gene transcription
(C) Gene promoter (D) Gene disruption

477. The first step of gene expression when one strand of the double-stranded DNA encompassing the gene is copied to the functional RNA (synthesized from 5' to 3' direction) is called:
(A) Gene conversion (B) Gene transcription
(C) Gene promoter (D) Gene disruption

478. The part of the gene which controls its transcription is called ------ that is located upstream of the coding sequence is called:
(A) Gene conversion (B) Gene transcription
(C) Gene promoter (D) Gene disruption

479. The technique to introduce gene-specific mutations to chromosomes in intact organism is called:
(A) Gene conversion (B) Gene transcription
(C) Gene promoter (D) Gene disruption

480. Intervening the noncoding sequences are called:
(A) Exons (B) Introns
(C) Both (A) and (B) (D) None of these

481. Intervening the coding sequences are called
(A) Exons (B) Introns
(C) Both (A) and (B) (D) None of these

482. Introns are more common for -------- coding genes:
(A) mRNA (B) sRNA
(C) tRNA (D) DNA

483. The set of Indian guidelines was also issued in 1990 by the Department of Biotechnology, Ministry of Science and Technology covering:
(A) Genetically engineered organisms
(B) Genetic transformation of plants and animals
(C) Containment measures at the laboratory level under different risk group
(D) Modalities of implementation

484. The cloned piece of disrupted genomic DNA is called a:
(A) Gene conversion (B) Gene targeting DNA
(C) Gene promoter (D) Gene disruption

485. DNA measures are available for comparative determination of sizes of:
(A) Single stranded RNA molecules
(B) Gene targeting DNA
(C) Single stranded DNA molecules
(D) None of these

486. Mixtures of *in-vitro* transcribed RNA molecules are also available in the molecular weight ranges of:
(A) 0-500 bases (B) 500-9000 bases
(C) 9000-10000 bases
(D) More than 10000 bases

487. Plasmids are extra chromosomal double stranded circular DNA molecules present in:
(A) Bacteria (B) Lower eukaryotes
(C) Both (A) and (B) (D) None of these

488. Ribosomal RNA constitutes 80-85% of total cellular RNA, transfer RNA and small nuclear RNA constitute about 10-15%, while messenger RNA constitutes only:
(A) 1-5% (B) 5-10%
(C) 10-15% (D) 15-20%

489. Which of the following organisms is used in the steroid synthesis?
(A) *E. coli*
(B) *Rhizopus stolonifer*
(C) *Ashbya gossypii* (D) *Pseudomonas*

490. In the conversion of pyruvic acid to acetyl co-enzyme A, NAD^+ is
(A) Reduced (B) Oxidized
(C) Hydroxylated (D) None

491. The technique of insertion of a desired gene into DNA of plasmid vecter is:
(A) Gene splicing (B) Gene dressing
(C) Gene cloning (D) Gene drafting

492. One of the main technique in recombinant DNA technology is:
(A) DNA cloning (B) DNA molecules
(C) Both of these (D) None of these

493. Which one is used in genetic engineering ?
(A) RNA polymerase (B) DNA polymerase
(C) Restriction endonuclease (D) Nuclease

494. Pyruvate is used to produce:
(A) Glucose (B) Fructose
(C) Sucrose (D) Maltose

495. DNA fingerprinting is useful for study of:
(A) Forensics (B) Polymorphism
(C) Identify and relationships
(D) All of these

496. Gene mouse has been produced through:
(A) Tissue culture
(B) Gene differentiation
(C) Gene manipulation (D) All of these

497. Which is preferred for making artificial seeds?
(A) Potassium arginate (B) Sodium alginate
(C) Agar (D) Gel-Rite

498. A map of the locations of identifiable landmarks on chromosomes is known as:
(A) Structural gene (B) Genetic marker
(C) Genetic map (D) Eucaryotic gene

499. Plasmids capable of propagating and transferring (shuttling) genes between two different organism, one of which is typically a prokaryote (*E. coli*) and other a eukaryote (yeast) is called:
(A) Shuttle vector (B) Transduction
(C) Both (A) and (B) (D) None of these

500. The technique of obtaining large number of plantlets by tissue culture method is called:
(A) Plantlet culture (B) Organ culture
(C) Micropropagation (D) Macropropagation

501. Who is the father of plant tissue culture?
(A) P. Maheshwari (B) G. Haberlandt
(C) P.R. White (D) R.J. Gautheret

502. Transfection is a:
(A) Transfer of host genes from one cell to another by a virus
(B) Genetic transformation of animal cells
(C) Conversion of a normal animal cell into a cancerous cell
(D) A type of virus infection

503. In India, which of the following overseas that implements the research and development experiments utilizing GMOs and recombinant DNA:
(A) Department of Atomic Energy
(B) Council of Scientific and Industrial Research
(C) Ministry of Environment of Forest
(D) Department of Biotechnology

504. Strains of which of the following bacteria is used for the degradation of petroleum and management of oil spills?
(A) *Pseudomonas putida*
(B) *Clostridium acetobutylicum*
(C) *Methylophilus methylotrophus*
(D) *Pseudomonas fluorescence*

505. The name magic bullets refers to:
(A) Antibiotics (B) Monoclonal antibodies
(C) Vaccines (D) Interferons

506. The biological materials which stimulate immunity against a pathogen when introduced into the body is known as:
(A) Vaccines (B) Single cell protein (SCP)
(C) Antibodies (D) Antibiotics

507. Which factor is necessary for maintaining lysogeny in lemeda phage?
(A) Cro (B) CI repressor
(C) Both (A) and (B) (D) None of these

508. Shine Dalgamo sequence present on mRNA in case of prokaryotes plays an important role in protein synthesis. This sequence binds with .... during translation initiation:
(A) 16s RNA (B) tRNA
(C) 23s RNA (D) 5s RNA

509. Which sequence in case of eukaryotes is important for mRNA tailing?
(A) GAGAGA (B) GAATTC
(C) UACGAC (D) UACUAAC

510. Cryopreservation refers to germplasm protection by:
(A) Reducing in moisture
(B) Tissue culture
(C) Low temperature treatment
(D) Breeding with wild varieties

511. In phytoremediation plants are used to:
(A) Control microbial attack on plants
(B) Control microbial contamination of soil
(C) Accumulate pollutants from soil thus clears soil from pollutants
(D) All of these

512. Plants cells/tissue has the ability to develop into a complete plant. This property of plants is known as:
(A) Pluropotency (B) Totipotency
(C) Multipotency (D) Both (A) and (B)

513. During meiosis actual reduction of chromosomes occur in which sub stage:
(A) Anaphase (B) Metaphase
(C) Telophase (D) None of these

514. Which of the following are biological mutagens?
(A) UV radiation (B) Ethylene
(C) Transposons (D) Ethidium bromide

515. When pyrimidine is stimulated by purines and vice versa, this type of mutation is known as:
(A) Transition (B) Transversion
(C) None sense mutation (D) Mis sense mutation

516. Operon comprise of:
(A) Operator (B) Structural gene
(C) Promotor (D) All of these

517. In lac operon, allolactose act as:
(A) Substrate (B) Gratituous inducer
(C) Repressor (D) None of these

518. Star activity of restriction enzyme means:
(A) It cuts the restriction site more efficiently
(B) It cuts at only one site
(C) It does not cut at any site
(D) Due to change in ionic concentration or other components it cuts at more sites generating more number of fragments

519. Prions are microorganisms, which consist of ....:
(A) DNA and protein
(B) RNA and protein
(C) DNA, RNA and proteins
(D) Proteins only

520. Which sigma factor of RNA polymerase comes into play during stress:
(A) α^{32} (B) α^{14}
(C) α^{60} (D) α^{70}

521. During DNA isolation which chemical is used for breaking of plasma membrane:
(A) Phenol (B) CTAB/SDS
(C) Isopropanol (D) Ethylene

522. What is the role of SDS in case of SDS PAGE?
(A) It increases speed of proteins in gel
(B) It protects proteins from denaturation
(C) It provides negative charge to all the proteins so that separation can be done only on the basis of molecular weight
(D) SDS is detergent so is used to emulsify contaminating fats if any in samples

523. RNA interference is used to control different viral diseases in case of plants and animals. The active fragment involving in interference is known as:
(A) Si RNA
(B) Micro RNA
(C) Both (A) and (B)
(D) None of these

524. In most of the cancer cases which gene is mutated:
(A) p^{53} (B) p^{21}
(C) p^{16} (D) p^{450}

525. Which factor is responsible for detoxication of drugs in case of humans?
(A) p^{53} (B) p^{450}
(C) p^{21} (D) None of these

526. mRNA usually is being extracted using:
(A) Poly T resin (B) RNase P
(C) a poly A resin (D) an rRNA trap

527. Which of the following statements is true regarding the production of transgenic cassava:
(A) a non-toxic cyanide derivative was production by metabolic engineering
(B) The release of cyanide was inhibited by antisense gene regulation
(C) The release of cynide was inhibited by immunolocatization
(D) The production of cynide was enhanced by overexpression of an enzyme

528. Technique for sequencing DNA and protein is:
(A) NMR spectroscopy
(B) X-ray crystallography
(C) Electron microscopy
(D) All of these

529. Example of tandem gene cluster is:
(A) rRNA genes
(B) Histone gene cluster
(C) Both (A) and (B)
(D) None of these

530. Which vector was used for development hepatitis vaccine?
(A) Phagemids
(B) PUC 18
(C) Cosmid
(D) Yeast expression vector

531. The medium is assess continuously to the fermentor tank to replace that, which has been fermented. This is true in case of:
(A) Batch technique
(B) Continuous flow technique
(C) Both (A) and (B)
(D) None of these

532. Removal of the topological strain by inducing to the negative supercoiling is carried out by:
(A) Topoisomerases (B) DNA gyrase
(C) Ligases (D) DNA polymerases

533. One of the following would be called VNTR:
(A) TTTT CCCC (B) GT GT GT GT
(C) GG TT GG TT (D) GGGG TTTT

534. Translation of histone proteins occur during:
(A) S phase of cell division only
(B) G1 phase only
(C) G1 and G2 phase
(D) M phase only

535. Chaperone present in ER is:
(A) Hsp 60 (B) Hsp 43
(C) Hsp 80 (D) Bip protein

536. Chaperon play an important role in:
(A) Protein folding
(B) Protein denaturation
(C) Protein trafficking
(D) Protein synthesis

537. Which of the histone protein is known as linker protein:
(A) H1 (B) H2A
(C) H2B (D) H3

538. Histone octomer comprise of:
(A) H1, H2A, H2B and H3
(B) H1, H2A, H2B, H3 and H4
(C) H1, H2B, H3 and H4
(D) H2A, H2B, H3 and H4

539. Chromosomal aberration in which two breaks occur in a chromosome and the intercalary segment reunites in a reverse order is categorized as:
(A) Translocation (B) Deletion
(C) Inversion (D) None of these

540. Which of the following statements regarding manipulation of the flavonol bio-synthetic pathway in tomato is true?
(A) Decreased flavanol production increased carotenoid production
(B) Increased flavonol production led to increased carotenoid production
(C) Increased flavonol production did not affect carotenoid production
(D) Decreased flavonol production did not increase carotenoid production

541. Which method enhances vitamin C production?
(A) Polymerization of a series of simple sugars
(B) Partitioning of ascorbate into seed storage proteins
(C) Derivatization of tocopherols from the shikimate pathway
(D) Enzymatic reduction of a component of pectin

542. Which of the following statements concerning the RNA/DNA hybrid oligonucleotides is incorrect?
(A) They hybrid molecule must be introduced into the cell using a binary vector
(B) They may be synthesized to contain a mutation that will be incorporated into the host genome
(C) They provide a mechanism for targeted alteration of the plant genome
(D) They are incorporated using a mechanism that involves host DNA repair machinery

543. Which of the following statements about basal promoters is true?
(A) Basal promoters are necessary for heterologous expression
(B) Basal promoters direct assembly of the core translation machinery
(C) Basal promoters are sufficient for expression in the shoot apical meristem
(D) Basal promoters are necessary for enhancer trapping

544. Immobilization of enzymes is done by:
(A) Physical adsorption
(B) Cross linking
(C) Entrapment
(D) All of these

545. Why might you find two kanamycin restance genes on fidderent parts of a T-DNA plasmid?
(A) One of the genes could be used for positive selection, and one could be used for negative selection
(B) One of the genes could be used for selection in plants, and one could be used for selection in bacteria
(C) This arrangement facilities for marker-free selection
(D) Each gene might have been introduced by a independent co-transformation event

546. Why have transgenic proteins present in crops often been tested in the presence of stomach acid before the plants were released?

(A) To see whether the protein will be mutated into a more harmful protein (e.g. prions)
(B) To see whether the protein can be digested to provide substantially equivalent nutrients
(C) To see whether the protein will be digested by an average consumer
(D) To see whether the protein will persist long enough to cause an allergic reaction

547. Haploid production cannot be done through:

(A) Pollen culture (B) Callus culture
(C) Anther culture (D) Hybridization

548. Which of the following was an unexpected results connected to the transgenic production of the terpenoid linalool:

(A) Inccreased production of linalool resulted in lower levels of GA hormones
(B) Increased transgene expression was not strongly correlated with increased levels of linalool
(C) Increased transgene expression resulted in higher levels of herbivory
(D) Linalool accumulation was localized to the nectar

549. Which of the following methods would give you the most precise and accurate information about where and when a given gene is expressed?

(A) in situ hybridization
(B) ex situ hybridization
(C) Protein microarray
(D) Reporter gene fusion including introns

550. Somatic embryogenesis is:

(A) Gern line cells developing into embryos
(B) Non-germ line cells developing into embryos
(C) Embryos developing from zygotes
(D) Embrypnic tissue becoming somatic

551. tRNA synthesis is done by:

(A) RNA Polymerase I (B) RNA Polymerase II
(C) RNA Polymerase III (D) Both (A) and (B)

552. The antibody protein is called:

(A) Single cell proteins (B) Gelation
(C) Immunoglobuline (D) Gamma globulins

553. During stress which of the following amino acid is accumulated?

(A) Proline (B) Tryptophane
(C) Tyrosine (D) Alanine

554. During stress which of the following phytohormone is accumulated?

(A) Auxin (B) GA
(C) ABA (D) IAA

ANSWERSHEET

1. INTRODUCTION

1. (D)	2. (B)	3. (C)	4. (A)	5. (A)	6. (C)	7. (D)	8. (A)	9. (B)	10. (A)
11. (B)	12. (C)	13. (A)	14. (B)	15. (A)	16. (B)	17. (A)	18. (B)	19. (C)	20. (A)
21. (B)	22. (A)	23. (A)	24. (B)	25. (C)	26. (D)	27. (C)	28. (C)	29. (A)	30. (B)
31. (C)	32. (D)	33. (A)	34. (A)	35. (B)	36. (A)	37. (A)	38. (A)	39. (B)	40. (A)
41. (A)	42. (B)	43. (A)	44. (A)	45. (B)	46. (D)	47. (B)	48. (D)	49. (E)	50. (A)
51. (D)	52. (D)	53. (C)	54. (A)	55. (D)	56. (D)	57. (D)	58. (D)	59. (D)	60. (A)
61. (D)	62. (D)	63. (D)	64. (A)	65. (C)	66. (C)	67. (D)	68. (D)	69. (C)	70. (D)
71. (D)	72. (A)	73. (C)	74. (A)	75. (D)	76. (A)	77. (D)	78. (A)	79. (A)	80. (D)
81. (C)	82. (D)	83. (C)	84. (D)	85. (D)	86. (B)	87. (C)	88. (A)	89. (B)	90. (D)
91. (C)	92. (A)	93. (A)	94. (A)	95. (B)	96. (D)	97. (C)	98. (B)	99. (D)	100. (D)
101. (D)	102. (C)	103. (D)	104. (D)	105. (A)	106. (C)	107. (C)	108. (D)	109. (C)	110. (A)
111. (D)	112. (D)	113. (A)	114. (D)	115. (C)	116. (A)	117. (D)	118. (B)	119. (D)	120. (D)
121. (D)	122. (D)	123. (A)	124. (D)	125. (B)	126. (D)	127. (D)	128. (C)	129. (B)	130. (D)
131. (D)	132. (C)	133. (C)	134. (A)	135. (D)	136. (B)	137. (A)	138. (D)	139. (B)	140. (C)
141. (A)	142. (C)	143. (A)	144. (A)	145. (D)	146. (D)	147. (B)	148. (A)	149. (B)	150. (D)

151. (A) 152. (A) 153. (A) 154. (D) 155. (D) 156. (C) 157. (B) 158. (A) 159. (A) 160. (A)
161. (D) 162. (B) 163. (D) 164. (D) 165. (B) 166. (C) 167. (A) 168. (A) 169. (A) 170. (B)
171. (D) 172. (B) 173. (B) 174. (A) 175. (D) 176. (C) 177. (A) 178. (A) 179. (D) 180. (A)
181. (C) 182. (E) 183. (E) 184. (B) 185. (E) 186. (B) 187. (A) 188. (A) 189. (A) 190. (A)
191. (D) 192. (B) 193. (B) 194. (A) 195. (A) 196. (B) 197. (D) 198. (D) 199. (C) 200. (D)
201. (E) 202. (D) 203. (B) 204. (A) 205. (B) 206. (C) 207. (D) 208. (A) 209. (A) 210. (B)
211. (B) 212. (C) 213. (A) 214. (B) 215. (B) 216. (D) 217. (D) 218. (D) 219. (B) 220. (A)
221. (B) 222. (D) 223. (B) 224. (A) 225. (A) 226. (C) 227. (D) 228. (A) 229. (C) 230. (A)
231. (D) 232. (C) 233. (D) 234. (D) 235. (B) 236. (B) 237. (C) 238. (A) 239. (B) 240. (B)
241. (A) 242. (C) 243. (B) 244. (B) 245. (A) 246. (B) 247. (D) 248. (C) 249. (C) 250. (D)
251. (A) 252. (B) 253. (C) 254. (C) 255. (A) 256. (D) 257. (D) 258. (D) 259. (B) 260. (C)
261. (D) 262. (D) 263. (D) 264. (B) 265. (D) 266. (D) 267. (D) 268. (D) 269. (B) 270. (D)
271. (C) 272. (A) 273. (B) 274. (A) 275. (B) 276. (A) 277. (C) 278. (C) 279. (A) 280. (C)
281. (C) 282. (D) 283. (C) 284. (D) 285. (A) 286. (B) 287. (A) 288. (C) 289. (D) 290. (B)
291. (D) 292. (C) 293. (C) 294. (B) 295. (D) 296. (B) 297. (B) 298. (A) 299. (B) 300. (A)
301. (A) 302. (C) 303. (C) 304. (A) 305. (A) 306. (A) 307. (D) 308. (D) 309. (C) 310. (C)
311. (A) 312. (A) 313. (A) 314. (A) 315. (C) 316. (D) 317. (A) 318. (A) 319. (B) 320. (C)
321. (A) 322. (C) 323. (A) 324. (D) 325. (B) 326. (B) 327. (A) 328. (A) 329. (D) 330. (D)
331. (C) 332. (C) 333. (B) 334. (D) 335. (A) 336. (D) 337. (B) 338. (D) 339. (B) 340. (B)
341. (B) 342. (B) 343. (D) 344. (D) 345. (D) 346. (D) 347. (D) 348. (B) 349. (D) 350. (B)
351. (C) 352. (D) 353. (B) 354. (D) 355. (A) 356. (D) 357. (B) 358. (A) 359. (D) 360. (C)
361. (A) 362. (A) 363. (C) 364. (B) 365. (D) 366. (D) 367. (A) 368. (A) 369. (C) 370. (C)
371. (C) 372. (D) 373. (A) 374. (B) 375. (D) 376. (D) 377. (D) 378. (D) 379. (B) 380. (D)
381. (C) 382. (D) 383. (D) 384. (D) 385. (A) 386. (A) 387. (D) 388. (B) 389. (D) 390. (D)
391. (D) 392. (D) 393. (C) 394. (D) 395. (B) 396. (C) 397. (D) 398. (D) 399. (C) 400. (C)
401. (A) 402. (B) 403. (D) 404. (D) 405. (C) 406. (B) 407. (D) 408. (A) 409. (B) 410. (B)
411. (D) 412. (A) 413. (C) 414. (C) 415. (A) 416. (A) 417. (D) 418. (C) 419. (A) 420. (D)
421. (D) 422. (B) 423. (A) 424. (B) 425. (A) 426. (A) 427. (B) 428. (B) 429. (D) 430. (A)
431. (D) 432. (D) 433. (A) 434. (D) 435. (D) 436. (B) 437. (A) 438. (D) 439. (A) 440. (B)
441. (D) 442. (B) 443. (A) 444. (D) 445. (D) 446. (C) 447. (C) 448. (B) 449. (A) 450. (A)
451. (A) 452. (C) 453. (C) 454. (A) 455. (D) 456. (A) 457. (C) 458. (D) 459. (B) 460. (A)
461. (B) 462. (D) 463. (D) 464. (A) 465. (A) 466. (B) 467. (D) 468. (A) 469. (B) 470. (C)
471. (B) 472. (C) 473. (C) 474. (A) 475. (D) 476. (A) 477. (B) 478. (C) 479. (D) 480. (B)
481. (A) 482. (A) 483. (C) 484. (B) 485. (A) 486. (B) 487. (C) 488. (A) 489. (B) 490. (A)
491. (B) 492. (A) 493. (C) 494. (A) 495. (D) 496. (C) 497. (B) 498. (C) 499. (A) 500. (C)
501. (B) 502. (B) 503. (D) 504. (A) 505. (B) 506. (A) 507. (B) 508. (A) 509. (D) 510. (C)
511. (C) 512. (B) 513. (A) 514. (C) 515. (B) 516. (D) 517. (B) 518. (D) 519. (D) 520. (A)
521. (B) 522. (C) 523. (C) 524. (A) 525. (B) 526. (A) 527. (D) 528. (D) 529. (C) 530. (D)
531. (B) 532. (B) 533. (B) 534. (A) 535. (D) 536. (A) 537. (A) 538. (D) 539. (C) 540. (C)
541. (D) 542. (A) 543. (D) 544. (D) 545. (B) 546. (D) 547. (B) 548. (B) 549. (D) 550. (B)
551. (C) 552. (C) 553. (A) 554. (C)

CHAPTER - 2

CELL BIOLOGY

1. Malignant cancer cells have all of the following properties except:
(A) Unregulated cell division
(B) Inhibition of angiogenesis
(C) Resistance to apoptosis
(D) Cellular immortality

2. The theoretical value of the resting potential E calculated from the Nernst equation is more negative than observed values of E because:
(A) There is a slow leak of K^+ ions
(B) There is a slow leak of Na^+ ions
(C) There is a slow leak of Mg^+ ions
(D) There are fewer K^+ channels in the membrane than the Nernst equation predicts

3. Dosage compensation in Drosophila is achieved by:
(A) Hypoactivation of both x chromosomes
(B) Hyperactivation of Y chromosomes
(C) Hyperactivation of maternal chromosome
(D) Hyperactivation of paternal chromosome

4. True statements about retrotransposons include which of the following?
P. They replicate through an RNA intermediate
Q. They utilize reverse transcriptase for transposition
R. They may contain introns
(A) P (B) R
(C) P and R (D) P, Q and R

5. What changes are made in B cells when they switch from producing IgM or IgG immuno-globulins?
(A) This change is accomplished by alternative splicing of the RNA transcripts
(B) This change occurs in the proteins as the IgM/IgD constant regions are proteolytically removed from the IgG protein
(C) This change occurs in the renome as the genes encoding the constant regions for IgM and IgD are deleted by the RAG1 and RAG 2 proteins
(D) This changes occurs in the genome as the genes encoding the constant regions for IgM and IgD are deleted independently of the RAG proteins

6. Which of the following process is not powered directly by a transmembrane proton gradient?
(A) Beating of eukaryotic cilia
(B) Generation of ATP in mitochondria
(C) Rotation of bacterial flagella
(D) Uptake of lactose by *E. coli*

7. When two carbohydrates are epimers:
(A) They rotate plane-polarized light in opposite directions
(B) They differ in length by one carbon
(C) One is an aldose, the other a ketose
(D) They differ only in the configuration about one carbon atom

8. Which of the following is/are correct about homeotic selector genes?
(A) Present in chromosome 3 in Drosophila
(B) Expression is regulated by the gap and pair-rule genes
(C) Defines anterior and posterior polarity
(D) Both (A) and (B)

9. Which one of the following pairs is not correctly matched?
(A) Inbreeding depression : homozygosity
(B) Hybrid vigour : heterzygosity
(C) Apogamy : fertilization
(D) Male sterility : cross pollination

10. Self incompatibility:
(A) Prevents cross-pollination
(B) Involves the S-locus that contains as many as 50 genes
(C) Occurs when the pollen secretes RNase that hydrolyze RNA in the carpel
(D) Occurs when pollen and stigma have different S-locus alleles

11. Which of the following amino acids is considered both ketogenic and glucogenic?
(A) Aspartate (B) Alanine
(C) Proline (D) Tyrosine

12. During the reduction of FAD^+:
(A) A flavin group is transferred
(B) An equivalent of a hydride ion is transferred
(C) The isoalloxazine becomes charged
(D) Two hydrogen atoms are added to the isoalloxazine ring

13. Telomerase activity was monitored in the following cell types. The highest amount of telomerase activity was found in the combination of:
(A) Embryonic and hematopoitic stem cells
(B) Nerve cells and muscle cells
(C) Erythrocytes ad macrophages
(D) Hepatocytes and easinophiles

14. A neuron's resting membrane potential is established and maintained by:

1. A high concentration of K^+ in the extracellular fluid and a high concentration of Na^+ in the cytosol
2. The plasma membranes higher permeability to Na^+ because of the presence of numerous Na^+ leakage channels
3. Differences in both ion concentrations and electrical gradients
4. The fact that there are numerous large, non-diffusible anions in the cytosol
5. Sodium-potassium pumps that help to maintain the proper distribution of sodium and potassium

(A) 1, 2 and 5 (B) 1, 2 and 3
(C) 2, 3 and 4 (D) 3, 4 and 5

15. The carbon and nitrogen atoms in the pyrimidine ring system are derived from:
(A) Glycine and carbamoyl phosphate
(B) Aspartate and glycine
(C) Glutamine and glycine
(D) Aspartate and carbamoyl phosphate

16. Which of the following properties is not exhibited by T_H cells?
(A) Stimulate division of B cells
(B) Stimulate division of cytotoxic T cells (CTLs)
(C) Are cytotoxic for other cells
(D) Stimulate migration of microphages

17. What characteristics domain is found in eukaryotic proteins that enable them to enter the endoplasmic reticulum?
(A) Stop transfer domain
(B) Signal sequence
(C) Signal sequence receptor
(D) Signal recognition protein

18. Glycogen is a branched polymer of glucose. It has:
(A) No reducing ends
(B) No non-reducing ends
(C) One reducing end and several non-reducing ends
(D) One non-reducing end and several reducing ends

19. Choose the false statement(s) in mammals
P. Fertilization occurs in the ampulla of oviduct
Q. Oocytes complete meiosis before sperm entry
R. Cleavage is rotational type
S. Cleavage show asynchrony (not divide at the same time)
(A) Q (B) P and S
(C) Q, R and S (D) P, Q, R and S

20. In serine protease, serine transfers a proton to a histidine residue. Serine is acting as a:
(A) Specific acid (B) General acid
(C) Specific base (D) General base

21. In the placenta of some animals, like cats and dogs the villi are developed in the form of a belt or girdle like band around the middle of their blastocyst or horionic sac, which is more or less elliptical in shape. Such placentae are called
(A) Zonary (B) Diffuse
(C) Disocidal (D) Cotyledonary

22. An example of competitive inhibition of an enzyme is the inhibition of:
(A) Succinic dehydrogenase by malonic acid
(B) Cytochrome oxidase by cyanide
(C) Hexokinase by glucose-6-phosphate
(D) Carbonic anhydrase by carbon dioxide

23. Epiboly involves:
(A) Inward movement of macromeres
(B) Overgrowth of micromeres over macromeres
(C) Rapid proliferation of cells at the rim of the blastopore
(D) Invagination of cells at the blastopore

24. Loss - of - function mutations in the suspensor gene in Arabidopsis led to the development of two embryos in a seed. After analyzing the expression of this gene in early embryos, you find high levels of suspensor mRNA in the developing suspensor cells. What is the likely function of the suspensor protein?
(A) Suspensor protein likely stimulates development of the embryonic tissue
(B) Suspensor protein likely stimulates development of the suspensor tissue
(C) Suspensor protein likely inhibits embryonic development in the suspensor
(D) Suspensor protein likely inhibits suspensor development in the embryo

25. During fertilization, polyspermy is prevented by:
(A) Zona pelluida in the presence of Na^+ ions
(B) Vitelline membrane in the presence of Ca^{++} ions
(C) Cortial granules in the presene of Ca^{++} and Na^{++} ions
(D) Influx of Na^+, Ca^{++} and Mg^{++} ions

26. Which of the following is the important reactive group of glutathione in its role as an antioxidant?
(A) Serine (B) Sulfhydryl
(C) Tyrosine (D) Acetyl-coA

27. Which of the following statements about the structure of ATP is correct?
(A) It contains three phosphoanhydride bonds
(B) It contains two phosphate ester bonds
(C) The nitrogenous base is called adenosine
(D) The active form is usually in a complex with Mg^{++} or Mn^{++}

28. Mobile IS elements, in addition to the transposase gene, must carry:
(A) A drug resistance gene
(B) Inverted terminal repeat sequence
(C) Replication genes (D) Host genes

29. Match the following:

P. *Plasmodium vivax* 1. Malignant tertian malaria

Q. *Plasmodium malariae* 2. Benign tertian malaria

R. *Plasmodium faliparum* 3. Quartan malaria

S. *Plasmodium ovale* 4. Mild tertian malaria

(A) P – 2, Q – 3, R – 1, S – 4
(B) P – 2, Q – 1, R – 3, S – 4
(C) P – 1, Q – 4, R – 2, S – 3
(D) P – 3, Q – 2, R – 4, S – 1

30. In which of the following is the enzyme correctly paired with its allosteric effector?

(A) Hexokinase : ATP
(B) Phosphofructokinase : Glucose – 6- Phosphate
(C) Pyruvate kinase : Alanine
(D) Glucokinase : Fructose 2, 6-bisphosphate

31. The p21 and p15 protons are example s of:

(A) Cdk inhibitors (B) Cyclins
(C) Oncogenes (D) Growth factors

32. Which of the following assumptions are used in the Briggs-Haldane derivation of the Michaelis-Menten equation?

(A) The concentration of the ES complex rapidly reaches a steady state
(B) Substrate concentration is greater than enzyme concentration
(C) At saturating concentrations of substrate, nearly all of the enzyme is found in the ES complex
(D) All of these were used in deriving the Michaelis-Menten equation

33. Cell –mediated immune responses are:

(A) Enhanced by depletion of complement
(B) Suppressed by corticosteroids
(C) Enhanced by depletion of T cells
(D) Enhanced by depletion of macrophage

34. Which of the following answer is not correct? Glutathione

(A) Cycles between oxidized and reduced forms in the cell
(B) Is involved in the detoxification of H_2O_2 and organic peroxides
(C) Is a tripeptide
(D) Contains a Se atom

35. Allosteric inhibition of an enzyme involves which of the following?

(A) Binding of a negative modulator to a site other than the substrate binding site
(B) Binding of a negative modulator competitively to the substrate binding site
(C) Binding of a negative modulator noncompetitively to the substrate binding site
(D) Cooperative binding of substrate to an enzyme

36. Which of the following statement is false about short interspersed elements (SINEs)?

(A) The major family of SINEs contain the Alu sequences
(B) They arose by reverse transcription of small RNAs
(C) They are transposable elements
(D) They encode reverse transcriptase

37. Upon studying the folding of RNase A, Anfinsen concluded that:

(A) Protein denaturation and renaturation does not violate basic laws of thermodynamics
(B) A steine residue can form a disulfide bond only with the nearest cysteine residue
(C) Attainment of native conformation by a protein is a spontaneous process
(D) One a protein is denatured, it an only be partially renatured

38. Integrins are characterized by which of the following properties?

P. Are heterodimeric protein
Q. Bind to the tripeptide sequence Arg-Gly-Asp
R. Are peripheral membrane proteins
S. Are expressed in a cell specific manner

(A) P and Q (B) Q and S
(C) P, Q and S (D) P, Q, R and S

39. Which part of coenzyme A is found attached to a serine residue in the acyl carrier protein of the fatty acid synthase?

(A) Phosphopantetheine (B) Pantothenate
(C) Phosphopantothenate (D) Betaalanine

40. Apomixis is a type of reproduction in plants in which:

(A) Fertilization does not take place
(B) Male nucleus takes part in fertilization
(C) Pollen fusion takes place
(D) Generative nucleus takes part in fertilization

41. Which of the following statement is not true of fatty acids found in most mammalian cells?

(A) They usually contain 12 to 20 carbons
(B) Polyunsaturated fatty acids contain conjugated double bonds
(C) All double bonds are in the cis configuration
(D) Free fatty acids are ionized at physiological pH

42. Two nitrogen atoms are eliminated per urea cycle. One of these comes from ammonium ion via carbamoyl phosphate. Which of the following compounds brings in the other nitrogen atom?

(A) Arginine (B) Citrulline
(C) Aspartate (D) Ornithine

43. The difference between the molecular weight of sucrose and that of the sum of the molecular weights of its components (glucose and fructose) is:

(A) 0 (B) 1
(C) 16 (D) 18

44. Which of the following is a true statement regarding p 53 tumor suppressor gene?
(A) Inherited mutations are the commonest cause of loss of function
(B) p53 does not have a role in the regulation of the expression of genes involved in the control of apoptosis
(C) p53 is activated when DNA damage occurs
(D) p53 phosphorylation results in loss of its ability in blocking entry of the cell into S-phase

45. In the cell cycle:
(A) M phase is both the most complex and the longest phase
(B) There is a G_0 phase in equilibrium with the G_1 phase
(C) Quiescent cells cannot be induced to re-enter the cell cycle
(D) Microtubule spindles form during the S phase

46. Differential distribution of substances in the egg most typically results in:
(A) Differences in gene expression which may establish a pattern in the embryo as the cells divide
(B) Amplification of specific genes during development
(C) Development of polyploidy tissues
(D) Loss of specific genes during development

47. Which of the following is not a dietary antioxidant?
(A) Vitamin C (B) Vitamin K
(C) Beta-carotene (D) Vitamin E

48. All of the following are characteristics of both MHC class I and class II molecules except:
(A) They are expressed codominantly
(B) They are expressed constitutively on all nucleated cells
(C) They are glycosylated polypeptides with tertiary domain structure
(D) They are involved in presentation of antigen fragments to T cells

49. If an average molecular weight of one amino acid is 110, the molecular weight of a peptide made up of 10 amino acids is expected to be:
(A) 1100 (B) 938
(C) 876 (D) 744

50. Which of the following covalent bond types are found in the structure of ATP?
(A) N-glyosidic, thioester, phosphodiester
(B) Phosphoanhydride, phosphomonoester, N-glyosidic
(C) Ester, ether, phosphoanhydride
(D) Ether, thioester, phosphomonoester

51. The termination of gastrulation is indicated by:
(A) Closure of natural tube
(B) Closusre of blastopore
(C) Obliteration of archenteron
(D) Obliteration of blastocoel

52. The nucleolus is the site where:
(A) Ribosomal RNA is transcribes and ribosomes are assembled
(B) DNA replication occurs
(C) Proteins recently imported from the cytoplasm are deposited
(D) Translation occurs

53. During the gluconeogenic conversion of pyruvate into glucose in the liver, all of the following are involved except:
(A) Pyruvate carboxylase
(B) Phosphoenolpyruvate carboxylase
(C) Phosphoenolpyruvate carboxykinase
(D) Glucose 6- phosphatase

54. The post office is involved in processing, packaging & distribution. Because of these functions, it could be used as an analogy for which cellular organelle?
(A) Plasma membrane (B) Nucleolus
(C) Golgi apparatus (D) Mitochondria

55. A polar molecule:
(A) Is slightly negative at one end and slightly positive at the other end
(B) Has an extra electron, giving it a negative charge
(C) Has an extra neutron, making it weight more
(D) Has covalent bonds

56. N-linked oligosaccharides on secreted glycoproteins are attached to:
(A) Nitrogen atoms in the polypeptide backbone
(B) The serine or threonine in the sequence Asn-X-Ser/Thr
(C) The amino terminus of the protein
(D) The asparagine in the sequence Asn-X-Ser/Thr

57. Which of the following is/are example of A-B exotoxins?
(A) Cholera toxin (B) Anthrax toxin
(C) Botulium toxins (D) All of these

58. Primary and secondary antibody responses differ in:
(A) The predominant isotype generated
(B) The number of lymphoytes responding to antigen
(C) The speed at which antibodies appear in the serum
(D) All of these

59. Aspirin, used as a common analgesic, antipyretic and anti-inflammatory agent, inhibits the synthesis of which one of the following?
(A) Arachidonic acid (B) Prostaglandins
(C) Glucocortioids (D) Histamine

60. The product of a ras gene:
(A) Is located in the nucleus
(B) Has tyrosine kinase activity
(C) Binds guanine nucleotides
(D) Acts as tumor-suppressor protein

61. Active transposable elements have which of the following feature(s)?
P. Repeated sequences at the ends of the transposable element
Q. Different numbers and chromosomal positions in different species of a single genus
R. The ability to alter the phenotype of an organism
(A) P only (B) Q only
(C) R and Q (D) P, Q and R

62. Fruits are complex organs that are specialized for disposal of seeds. Which of the following plant tissues does not contribute to mature fruit?
(A) Sporophytic tissue from the previous generation
(B) Gametophytic tissue from the previous generation
(C) Sporophytic tissue from the next generation
(D) Gametophytic tissue from the next generation

63. Number of nucleotide present in FMN is:
(A) Two (B) One
(C) Four (D) Zero

64. Before part of the neural tube is differentiated into the spinal cord there are two important regions of the tube that influence what types of nerves will originate from the spinal cord. The roof (top) plate and the floor (bottom) plate. These two regions secrete bone morphogenic protein 4 (BMP4) and Sonic Hedgehog (Shh) respectively. What types of neurons originate from the floor plate of the neural tube (future spinal cord)?
(A) Sensory neurons
(B) Neurons of the autonomic nervous system
(C) Interneurons
(D) Motor neurons

65. In angiosperm, triple fusion produces:
(A) Polar nucleus
(B) Secondary nucleus
(C) Primary endospermic nucleus
(D) Zygote nucleus

66. If a reaction is at equilibrium, the free energy, ΔG change is:
(A) 1 (B) 0
(C) 10 (D) 0.1

67. Which of the following is/are true about bird?
P. Merobastic cleavage
Q. Mesolecithal egg
R. Discoidal cleavage
S. Centrolecithal egg
(A) P and S (B) Q and S
(C) P and R (D) R and S

68. The gene for which of the following would be considered a protooncogene?
(A) p53
(B) Rb protein
(C) p21 (a Cdk inhibitor protein)
(D) Ras

69. The correct sequence of events during fertilization in mammals is:
(A) Capacitation, activation of egg, acrosomal reaction and entry of sperm in egg
(B) Capacitation, acrosomal reaction, activa-tion of egg and entry of sperm in egg
(C) Acrosomal reaction, activation of egg, capacitation and entry of sperm in egg
(D) Activation of egg, acrosomal reaction and entry of sperm in egg and capacitation

70. Match each inhibitor in the left column with its primary effect from the right column:

Column I	Column II
A. Azide	P. Inhibition of electron transport
B. Atractyloside	Q. Uncoupling of electron transport and oxidative phosphorylation
C. Rotenone	R. Inhibition of ADP-ATP translocation
D. Dinitrophenol	S. Inhibition of ATP synthase
E. Carbon monoxide	
F. Oligomycin	
G. Antimycin A	

(A) A – P, B – R, C - P, D – Q, E – P, F – S, G – P
(B) A – S, B – R, C - P, D – Q, E – P, F – R, G – P
(C) A – P, B – S, C - P, D – Q, E – Q, F – S, G – Q
(D) A - Q, B - P, C - R, D - S, E - Q, F - R, G - P

71. Bacterial cell division is:
P. Asymmetric
Q. Termed as binary fission
R. Similar to budding of budding yeast
S. Faster than yeast cell division
(A) Q (B) P and Q
(C) Q and S (D) P, Q and R

72. Which of the following property is not related to water?
(A) High dielectrical constant
(B) Dipolar nature
(C) Has bond angle of 104.5^0
(D) Low heat of vaporization

73. Which of the following is not true about creatine phosphate?
(A) It is guanidine derivatives
(B) Found in vertebrates
(C) Energy storage molecules in muscle tissue
(D) Standard free energy change is -3 kcal/mol

74. Which of the following statement about retrotransposons is correct?
(A) They transpose via an RNA intermediate
(B) They contain genes for ribosomal proteins
(C) They possess a gene for RNA-dependent RNA polymerase
(D) They are found only in bacteria

75. The peptide bond in proteins is:
(A) Nonpolar, but rotates to three preferred dihedral angles
(B) Planar, but rotates to three preferred dihedral angles
(C) Nonpolar, and fixed in a trans configuration
(D) Planar and usually found in the trans-configuration

76. Specific recognition between cell types such as leukocytes and endothelial cell of blood vessels is mediated by cell surface glycoproteins called:
(A) Selectins (B) Fibronectins
(C) Spectrins (D) Porins

77. Heptens:
(A) Require carrier molecules to be immunogenic
(B) React with specific antibodies when homologous carriers are not employed
(C) Interact with specific antibody, even if the haptens are monovalent
(D) All of these

78. The ability of a single B cell to express both IgM and IgD molecules on its surface at the same time is made possible by:
(A) Allelic exclusion
(B) Isotype switching
(C) Simultaneous recognition of two distinct antigens
(D) Selective RNA splicing

79. Which one of the following is the major force of attraction that stabilizes the three dimentional structure of globular proteins?
(A) Peptide bond (B) Van der Waals interaction
(C) Hydrogen bond (D) Hydrophobic interaction

80. The most appropriate definition of seed in terms of homologous organs found in spore producing land plant is:
(A) A seed of an integumented sporangium with a single functional spore
(B) A seed of an integumented indehisent sporangium containing an egg
(C) A seed of an integumented gymnosporangium
(D) A seed of an integumented bud containing an egg

81. Which of the following statement is not true?
(A) The pI is the pH value at which a protein has net charge
(B) At a pH value equal to its pI, a protein will not move in the electric field of an electrophoresis
(C) An acidic protein will have a pI less than 7
(D) A basic protein will have a pI greater than 7

82. Sex chromosome-based compensation in human in brought about by:
(A) Inactive of one X-chromosome in females
(B) Hyperactivity of single x chromosome in males
(C) Hypoactivity of both x – chromosome in females
(D) Hyperactivity of autosomes in females

83. Which of the following amino acid if added to a acidic solution will have buffer action on solution at the physiological pH?
(A) Alanine (B) Glycine
(C) Histidine (D) Arginin

84. Which of the following is not correct statement?
P. Ametabolons insects undergo direct development
Q. In ametabolons insects, there is larval stages
R. Hemimetabolons insects pass through nymph stages
S. Holometabolons insects pass from larva to pupa to sexually mature adult
(A) Q (B) P and Q
(C) S (D) R and S

85. When a plant of chromosomal type aa pollinates a plant AA, what chromosome constitution of embryo and endosperm is expected in the resulting seeds?
(A) Diploid zygote of type Aa and triploid endosperm of type AAa
(B) Diploid zygote of type aa and triploid endosperm of type Aaa
(C) Diploid zygote of type AA and triploid endosperm of type AAa
(D) Diploid zygote of type Aa and triploid endosperm of type aaa

86. The correct equation for the reduction of $NADP^+$ is:
(A) $NADP^+ + 2H^+ \longrightarrow NADPH^+ + H^+$
(B) $NADP^+ + H^+ \longrightarrow NADPH$
(C) $NADP^+ + H^+ \longrightarrow NADPH$
(D) $NADP^+ + 2H^+ \longrightarrow NADPH_2$

87. Which of the following amino acids have side chains that are negatively charged under physiological conditions (ie., near pH 7)?
(A) Asp (B) His
(C) Trp (D) Cys

88. Which of the following drives directional transport through nuclear pore complexes?
(A) ATPase (B) GTPase
(C) Ran GTPase (D) AB transporter

89. The name for the type of cleavage in the mammalian embryo is called?
(A) Planar cleavage (B) Unequal cleavage
(C) Rotational cleavage (D) Radial cleavage

90. Tracing of a cell lineage during development means that:
(A) The cells giving rise to and derived from a specific cell are known
(B) The sequence of the enhancers for developmental genes is known
(C) The regulatory genes for the organism have been genetically mapped
(D) Cell components in the membrane involved in signaling have been isolated

91. A blastula composed of densely-packed by larger sized and relatively small number of cells without or externally small blastocoelic space in the centre is called:
(A) Stereoblastula
(B) Coeloblastula
(C) Discoblastula
(D) All of these

92. The trans Golgi network is:
(A) The intermediate compartment between the ER and the Golgi
(B) The part of the Golgi where fusion of vesciles from the ER occurs
(C) Where sorting of proteins to the lysosomes, plasma membrane and cell exterior occurs
(D) The network of vesicles that transport proteins between Golgi cisternae

93. Which of the following is not true about RNA?
(A) RNA is chemically more reactive than DNA
(B) 2'-OH prevents RNA to adopt B-conformation
(C) RNA was discovered by Hoppe-Seyler
(D) Feulgen reaction is performed by both DNA and RNA

94. Most protection against viral disease in the body takes place through the activities of:
(A) Penicillin molecules
(B) Antibody molecules
(C) Antigen molecules
(D) Inteferon molecules

95. The secretory IgA was electrophoresed on SDS-PAGE under reduced and denaturing conditions. The number of polypepetide bands detected on the gel will be:
(A) 2 (B) 3
(C) 4 (D) 5

96. Ca^{++} is important is skeletal muscle contraction because it:
(A) Activates the myosin ATPase by binding to it
(B) Binds to troponin to remove a constant inhibition of cross – bridge attachment
(C) Prevents the formation of bonds between the myosin cross bridges and the actin filament
(D) Is required to detach the myosin head from the actin filament

97. Which of the following statements is true about intermediate filaments?
(A) Rather than consisting of a single type of protein, they can be made up of a number of different proteins
(B) They are involved in cell movement
(C) The basic structure of an intermediate filament protein is a globular head and a long alpha-helical tail
(D) Like microfilaments, they exhibit treadmilling

98. Which statement is true about enzyme inhibition?
(A) In competitive inhibition, the inhibitor binds to the active site of the enzyme
(B) In non-competitive inhibition, the inhibitor binds to the other than active site of the substrate
(C) In irreversible inhibition, a poison binds to the enzyme so that it can never work again
(D) All of these

99. Terminator seeds produced by seed companies using techniques of genetic engineering are in fact:
(A) Hybrid seeds which germinate in the farmer's field but do not produce seeds in the next generation
(B) Hybrid seeds which give a very high yield
(C) Seeds which germinate and produce incompatible gametes
(D) Seeds that develop into fertile plants which produce non germinable seeds

100. What is the role of the bicold gene in Drosophila development?
(A) To determine the anterior end of a fly embryo
(B) To determine the thoratic region
(C) To determine even numbered segments
(D) To determine odd numbered segments

101. Which of the following pair is incorrect?

Inhibitor	Function
(A) FCCP	Make membran permeble proton
(B) Malonate	Prevent oxidation of succinate
(C) Cyanide	Inhibit cytochrome oxidase
(D) Aetoacetate	Make enolase inactive

102. All of the following statements about actin and myosin are true except:
(A) The globular head section of myosin has domain for binding ATP except
(B) Actin is the major protein of the thick filament
(C) Binding of ATP to the actin myosin complex promotes dissociation of actin and myosin
(D) F-actin is stabilized when tropomyosin is bound to it

103. Which of the following is not true about beta oxidation of fatty acid containing even number of carbons?
(A) Endproduct – acetyl CoA
(B) Cofactor required – NAD^+ and FAD
(C) Decreases during starvation
(D) Site – mitochondria

104. Which of the following amino acid residue is likely to be bound on the inside of a water-soluble protein?
(A) Arg (B) His
(C) Asp (D) Ile

105. The Z-DNA helix:
(A) Has fewer base pairs per turn than the B-DNA
(B) Is favored by an alternating GC sequence
(C) Tends to be found at the 3'-end of genes
(D) Is inhibited by methylation of the bases

106. C-Value paradox suggests us about:
(A) Colinearity between genome size and complexity of organism
(B) Non-colinearity between genome size and complexity of organism
(C) Dosage compensation
(D) Number of chromosome

107. Choose the mismatch:
(A) Vitamin A - Synthesis of rhodopsin
(B) Vitamin E - Antioxidant
(C) Vitamin K - Activates blood clotting factors
(D) Vitamin B1 - Electron transfer reaction

108. Which of the following is not a class of segmentation gene in Drosophila?
(A) Gap genes
(B) Pair-rule genes
(C) Homeotic selector genes
(D) Segment polarity genes

109. Immediately after fertilization in animals, the first structural and biochemical changes in the egg are initiated by:
(A) New gene transcription
(B) The release of Ca^{++} from internal reservoirs
(C) The initiation of DNA synthesis
(D) A lowering of cytosolic pH

110. The Oral Polio Vaccine (OPV), administered to children in India, is:
(A) Inactivated polio vaccine
(B) Live attenuated (weakened) poliovirus
(C) Subunit vaccine specific for oral route
(D) Recombinant subunit vaccine

111. The aspartic acid residue (ASP^{52}) at the active site of lysozyme serves which of the following purposes as part of the catalytic mechanism?
(A) General acid catalysis
(B) General base catalysis
(C) Electrostatic stabilization of a carbonium ion
(D) Covalent catalysis

112. Intracellular receptors:
(A) Usually bind hydrophobic ligands
(B) May be located either in the cytosol or nucleus in unbound state
(C) When bound to ligand regulat gene transcripton
(D) All of these

113. Caspases are involved in the process of:
(A) DNA replication
(B) Mutation and recombination
(C) Antibody synthesis
(D) Apoptosis

114. Which of the following statement is false?
(A) Proteins tend to be least soluble in water at their isoelectric point
(B) Sickle cell Hb and normal Hb have same value of negative charge
(C) Amino acids are made visible on the chromatogram by treatment with ninhydrin
(D) The net charge on an amino acid is a function of the pH of the solution

115. In humans:
(A) Gene for the testes-determining factor resides on the short arm of the Y chromosome
(B) SRY is found in normal XX females
(C) SOX9, a gene present on x chromosome is involved in sex determination
(D) None of these

116. Which of the following statement is correct?
(A) The absence of introns in prokaryotic genes indicates that introns were added to the eukaryotic line sometime after the evolutionary separation of prokaryotes and eukaryotes
(B) Satellite DNA sequences are generally not transcribed and are most often located in the heterochromatin associated with the entromeric regions of chromosome
(C) Since *Alu* sequence are transcribed by RNA polymerase II, their continued movement in the genome depends on their insertion in the vicinity of a polymerase II promoter
(D) The two X- chromosomes in a female mammalian cell, only one of which is active, are replicated at the same time during the S phase

117. Which one of the following bacterial toxins does not have ADP ribosylation activity?
(A) Diphtheria toxin
(B) Cholera toxin
(C) *S. aureus* alpha-toxin
(D) *Pseudomonas* exotoxin

118. Choose the incorrect pair:

Element	Enzyme
(A) Mo	Nitrogenase
(B) Cu	Cytochrome oxidase
(C) Se	Nitrate reductase
(D) Zn	Alcohol dehydrogenase

119. Which of the following statement(s) is/are correct?
P. Migration of single cells into the blastocoel at the time of gastrulation is called invagination
Q. During embryogenesis discoidal cleavage occurs in chicken
R. During fertilization in frog the sperm always penetrates the egg in the animal hemisphere
(A) P (B) Q
(C) P and Q (D) P, Q, and R

120. In humans, the SRY gene
P. Codes testes-determining factor
Q. Present on the Y chromosome
R. Activates (either directly or indirectly) the SF1 gene
S. Inhibits the WNT4 gene
(A) P and Q (B) P and S
(C) P, Q and R (D) P, Q, R and S

121. Choose the incorrect statement with regard to collagen molecule?
(A) It is formed from three - chains
(B) Glycine accounts for one-third of amino acids found within collagens
(C) It contains less hydroxyproline than elastin
(D) Ascorbate is essential for hydroxylation of proline

122. Apoptosis or programmed cell death occurs in all of the following cases except:
(A) In virus – infected cells
(B) In cells damaged by injury
(C) In cells with potentially cancer-causing mutations
(D) During the elimination of tissue between the digits in the formation of human fingers

123. If a polypeptide has 400 amino acid residues, what in its approximate mass?
(A) 11000 daltons (B) 44000 daltons
(C) 22000 daltons (D) 88000 daltons

124. Which of the following statements are correct?
P. All immunoglobulins possess two paratopes per molecule; i.e., the are bivalent
Q. All classes of immunoglobulins can be synthesized in a membrane-bound form as well as in a secreted form
R. Most MHC loci coding for class I and class II molecules are highly polymorphic within each individual
S. Immunoglobulin class switching does not involve light chains
(A) P and Q (B) Q and S
(C) P, Q and R (D) P, Q, R and S

125. Eukaryotic genome differs from prokaryotic genome in:
(A) DNA is circular and single-stranded in prokaryotes
(B) Intervening sequences are present in eukaryotic DNA
(C) DNA is complexed with histones in prokaryotes
(D) DNA is organized into operons in eukaryotes

126. All of the following statements about the enzyme complex that carries out the synthesis of ATP during oxidative phosphorylation are correct except:
(A) It is inhibited by oligomycin
(B) It contains a proton channel
(C) It can activate ATPase activity
(D) It can bind molecular oxygen

127. Choose the mismatch:
(A) SRP - Bring proteins and the ribosomes that are translating them from the cytosol to the rough ER
(B) KDEL - Targets proteins emerging from the golgi apparatus to rough ER
(C) Chaperones - Assist in protein folding
(D) eIFs - Help in translocation of ribosome on mRNA

128. What is the [S] for the enzyme catalyzed reaction which has a initial velocity of 1.62 mole/litre/min. and the maximum velocity of 21.85 mole/litre/min. and a Michaeli-Menten constant of 3.88 mole/litre?
(A) 2.51 mole/litre (B) 5.30 mole/litre
(C) 0.25 mole/litre (D) 4.78 mole/litre

129. Receptor tyrosine kinase:
P. Are seven-transmembrane-helix receptors
Q. Are integral membrane enzyme
R. Activate their targets via G protein casade
S. Are often activated by ligand-induced dimerization
(A) P and Q (B) R and S
(C) P and R (D) Q and S

130. Which of the following element is least abundant in our body?
(A) Carbon (B) Oxygen
(C) Phosphorus (D) Hydrogen

131. Which of the following statements about the MHC is incorrect?
(A) It codes for complement components
(B) It codes for both chains of the MHC class I molecule
(C) It codes for both chains of the MHC class II molecule
(D) It is associated with susceptibility and resistance to different diseases

132. Which of the following is not a commonly observed consequence of binding of a signaling molecule to its cell surface receptor?
(A) Receptor dimerization
(B) Receptor phosphorylation
(C) Conformational changes in the receptor
(D) Increased synthesis of the receptor

133. Indicate which is not a correct statement:
(A) The innate immune response is not antigen specific
(B) The innate immune response does not generate memory cells
(C) The innate immune response is mounted only by invertebrates
(D) The genes for the Toll receptors involved in mounting innate immune response are present in the germ line DNA

134. What is the charge on the oligonucleotide 5' pGpGpApCpT 3' at neutral pH?

(A) -5 (B) - 6
(C) -7 (D) - 8

135. Which of the following statements are correct?

P. Polarity of water makes it an excellent solvent
Q. Water has high tensile strength
R. Cohesive property of water is due to H-bonding
S. Water has high dielectric constant

(A) P and S (B) Q and R
(C) Q, R and S (D) P, Q, R and S

136. Which of the following is incorrect statement?

(A) Actin filaments, microtubules and intermediate filaments each have a different arrangement in a eukaryotic cell and in combination with a variety of accessory proteins, provide different function
(B) Microtubules are highly dynamic structures that can shorten as well as lengthen; in general, they grow by addition of subunits to the plus end and shrink by removal of subunits from their minus end
(C) In an isolated cell fragment, microtubules will reorganize themselves to form a star like array emanating from a new microtubule organizing centre at the enter of the cell fragment
(D) The normal locations of the ER and Golgi are thought to be determined by receptor proteins on their cytosolic surface that bind specific microtubule-dependent motors a plus – end-directed kinesin for the ER and a minus-end-directed dynein for the Golgi apparatus

137. The neural crest cells migrate:

(A) Dorsolaterally into the ectoderm, where they become melanocytes
(B) Ventrally to become sympathetic and parasympathetic neurons
(C) Ventrally to give adrenal medulla cells
(D) All of these

138. Reassociation experiments revealed that eukaryotic DNA does not renature uniformly. These experiments are performed by fragmentating total DNA from an organism into pieces about 1000 base pairs long. The DNA is then melted into single strands and allowed to reform complementary base-paired double helices. Which class of DNA would be expected to have the most rapid rate if renaturation (reassociation)?

(A) Single-copy DNA (B) Pseudogenes
(C) Simple sequence DNA
(D) Small multigene families

139. Which of the following is/are correct about caspases?

(A) A cyteine protease
(B) Cleave their target proteins at specific aspartic acids
(C) A serine protease
(D) Both (A) and (B)

140. Which of the following statement about the cations or targets of the second messengers of the phosphoinositide aside is incorrect?

(A) DAG activates protein kinase (PKC)
(B) Most of the effects of IP3 and DAG are antagonistic
(C) DAG increases the affinity of PKC for Ca^{++}
(D) PKC requires Ca^{++} for its activity

141. The MEK kinase is unusual in that:

(A) It is activated by a kinase
(B) It lies downstream of G-protein coupled recepters
(C) It is a dual-specificity kinase, having the ability to phosphorylate both threonines and tyrosines
(D) It activates a kinase

142. Myoglobin shows a hyperbolic response, while hemoglobin shows a sigmoidal response for oxygen binding. Which of the following statements are true with respect to this observation?

P. Hemoglobin binds 2,3-BPG while myoglobin does not
Q. Hemeglobin exists in two different conformational states while myoglobin does not
R. Hemeglobin is a tetramer while myoglobin is a monomer
S. Hemoglobin is present in RBCs while myoglobin is present in the muscle

(A) R and S (B) S and P
(C) P and Q (D) Q and R

143. The following is a list of subcellular structures and their functions choose the option that correct matches the subcellular structures to their functions:

P. Tanoplast A. Lipid biosynthesis
Q. Perosisomes B. Protein degradation
R. Lysosome C. Storage of starch
S. Spherosome D. Removal of free radicals

(A) P – A, Q – B, R - C, S – D
(B) P – C, Q – B, R - C, S – B
(C) P – C, Q – B, R - D, S – A
(D) P – A, Q – B, R - C, S – C

144. Which of the following signal molecule is responsible for quorum sensing?

(A) Acyl homoserine lactone
(B) Maltose (C) Acyl choline
(D) Brassinosteroid

145. The clonal selection theory best accounts for:

(A) The production of memory cells
(B) The mechanis whereby antigen & antibody unit
(C) The maturation of B cells into plasma cells
(D) The specificity of the immune response

146. Which of the following pairs is inter-converted in the process of mutarotation?

(A) α-D-glucose and β–D-glucose
(B) D-glucose and L-glucose
(C) D-Glucose and D-fructose
(D) α -D-Glucopyranose and β –D-glucofuranose

147. A homeotic mutation is one which:
(A) Is present in only one form in an individual
(B) Substitutes one body part for another in development
(C) Results in development of a tumor
(D) Is wild type at one temperature and abnormal at another

148. Cell cycle is controlled by:
P. Certain cyclins
Q. Certain cyclin-dependent kinase
R. Certain inhibitory proteins
S. Certain phosphatases
(A) P and Q (B) R and S
(C) P and R (D) P, Q, R and S

149. In order for viruses to replicate in their host cells, all the following must occur except:
(A) The genome must be released in the host cell cytoplasm
(B) ATP must be synthesized within the virus
(C) The virus must unite with the correct host cell
(D) The host cell must contain ribosomes for the synthesis of protein

150. Match the embryonic cleavage patterns with the corresponding organisms

P. Isolecithal	1. Chick
Q. Mesolecithal	2. Drosophila
R. Telolecithal	3. Human
S. Centrolecithal	4. Frog

(A) P -1, Q – 2, R – 3, S – 4
(B) P -2, Q – 1, R – 3, S - 4
(C) P -3, Q – 4, R – 1, S – 2
(D) P -2, Q – 2, R – 1, S – 3

151. Cholera toxin causes massive and often fatal diarrhea by:
(A) Inactivating G_1 protein
(B) Irreversibly activating adenylate cyclase
(C) Locking Gs protein into an inactive form
(D) Only correctly folded proteins are translated in the endoplasmic reticulum

152. When haemoglobin is converted from the deoxy form to oxyhemoglobin:
(A) It becomes more acid and release protons
(B) Carbamino formation is promoted
(C) Binding of BPG is favoured
(D) Bound No is transferred to glutathione

153. SH2 domains are:
(A) Protein domains that binds phosphotyrosine-containing peptides
(B) The domains on receptor protein-tyrosine kinases that contain the phosphorylated tyrosine
(C) Domains that mediate the dimerization of receptor protein-tyrosine kinase
(D) The domains on receptor protein-tyrosine kinases that possess the kinase activity

154. Which of the following is not correct with regard to ketone bodies?
(A) They consist of water soluble ketones
(B) The ketones can be used as energy sources
(C) The brain will use ketone bodies during starvation conditions
(D) The ketones in ketone bodies cannot be used as fuels

155. A typical animal cell (nucleated) membrane contains glycolipids and glycoproteins in the plasma membrane. To determine its topological distribution, lectin is used as a probe. The following interactions may be the basis of the probing method:
P. Protein – protein interaction
Q. Protein – sugar interaction
R. Protein – lipid interaction
S. Protein – sterol interaction
The appropriate answer is
(A) Only P (B) Only Q
(C) P, Q and R (D) Only S

156. Evidence that mitochondria originated from an endosymbiotic relationship between aerobic bacteria and ancestral eukaryotic cells includes all of the following except:
(A) DNA in mitochondria is not packaged by histones
(B) Mitochondria DNA is circular like prokaryotic DNA
(C) Protein synthesis in mitochondria is inhibited by antibiotics as in bacterial protein synthesis
(D) Ribosomal RNAs in mitochondria code for prokaryotic like ribosomal proteins

157. Identify the correct statement regarding gastrulation
P. Formation of archenteron surrounded by the endoderm
Q. Little growth so that overall size of the embryo does not increase significantly
R. Greater activity of the nucleus and the formation of new proteins
S. Complex but orderly and irreversible morphogenetic movements
(A) S (B) P and Q
(C) Q and R (D) P, Q, R and S

158. Which of the following is common in both *E coli* and aukaryotic chromosomes?
(A) DNA is circular
(B) DNA is negatively supercoiled
(C) DNA is contained in the nucleus
(D) DNA is packaged into nucleosomes

159. Which of these events occur first in organogenesis?
(A) Notohord development
(B) Gastrulation
(C) Neural fold development
(D) Neural plate development

160. Henderson's equation can be represented as:

(A) $pH = pK_a + \log = \frac{[Acid]}{[Salt]}$

(B) $pH = pK_a + \log = \frac{[Salt]}{[Acid]}$

(C) $pK_a = pH + \log \frac{[Salt]}{[Acid]}$

(D) $pK_a = pH - \log \frac{[Salt]}{[Acid]}$

161. Integrins are transmembrane proteins that connect:

(A) The nuclear lamina to cytoplasmic kinases
(B) The extracellular matric to the cytoskeleton
(C) Foal adhesions to hemidesmosomes
(D) Microtubules to actin filaments

162. Which of the following is not true about NO?

(A) Acts as intracellular signaling molecule as well as neurotransmitter
(B) Regulates the vasodilation
(C) Synthesized from L-arginine
(D) Induces adenylate cylase, which catalyzes AMP formation

163. Delayed type hypersensitivity is classified as an immune response mediated by:

(A) Antibodies (B) T cells
(C) B cells (D) Mast cells

164. The equilibrium constant, k and change in free energy, ΔG^o are indicated by the reaction:

(A) $\Delta G^o = RT \text{ Ink}$ (B) $\Delta G^o = RT \log_{10} k$
(C) $\Delta G^o = -RT \text{ Ink}$ (D) $\Delta G^o = R \text{ Ink } T$

165. Homeobox sequences:

(A) Are conserved sequences present in the genome of many animal species that regulate development
(B) Are found in prokaryotes but not in eukaryotes
(C) Were identified as the integration sites for bacterial viruses
(D) Represent integration sites for transposable elements

166. All of the following statements about telomerase are correct except:

(A) The RNA component acts as a template for the synthesis of a segment of DNA
(B) It adds telomeres to the 5'-ends of the DNA strands
(C) It provides a mechanism for replicating the ends of linear chromosomes
(D) It reoganizes a G-rich single strand of DNA

167. p53 protein is associated with all of the following except:

(A) Tumor suppression
(B) Programmed cell death
(C) Transcription
(D) Post-trascriptional modifications

168. Membrane carrier proteins differ from membrane channel proteins by which of the following characteristics?

(A) Carrier proteins are glyoproteins, while channel proteins are lipoproteins
(B) Carrier proteins transport molecules down their electrochemical gradient, while channel proteins transport molecules against their electrochemical gradient
(C) Carrier proteins can mediate active transport, while channel proteins cannot
(D) Carrier proteins are synthesized on free cytoplasmic ribosomes, while channel proteins are synthesized on ribosomes bound to the endoplasmic reticulum

169. If the ΔG of the reaction $A \longrightarrow B$ is -3.0 kal/mol, which of the following statement is correct?

(A) The reaction will proceed spontaneously from left to right at the given conditions
(B) The reaction will proceed spontaneously from right to left at standard conditions
(C) The equilibrium constant favours the formation of B over the formation of A
(D) The value of $\Delta G^{O'}$ is also negative

170. Consider the following statements regarding cleavage:

P. All division are mitotic
Q. There is no growth in the resulting blastomers and the total size and volume of the embryo remains same
R. Oxygen consumption is greatly increased

Identify the true statements

(A) P and Q (B) P and R
(C) Q and R (D) P, Q and R

171. In an experiment involving planer bilayer, a solution of table salt (sodium and chloride ions in water) is added on the left side of the membrane while pure water is added on the right side. After 30 minutes the researchers test for the presence of ions on each side of the membrane by noting whether the solutions are capable of conducting electricity. The right side tests negative for ions. What can you conclude:

(A) The experiment failed
(B) The water somehow blocked the movement of ions across the membrane
(C) The left side would probably also test negative for ions
(D) Ions cannot cross planar bilayers

172. Effective chemotherapeutic agents are difficult to develop for the treatment of fungal infections because:

(A) Fungi have cell wall
(B) Fungi have better mechanism to inactivat drugs
(C) Fungi are eukaryotic cells and their cellu-lar machinery is similar to that of the host
(D) Fungal pathogens typically infect organs inaccessible for antibiotic treatment

173. Which are the non-covalent bonds responsible for the high melting and boiling points of water?
(A) H-bond (B) Van der Waals force
(C) Hydrophobic force (D) None of these

174. Which of the following would most likely explain a 46, xx individual with development of testes?
(A) Mutation of SOX9
(B) Presence of excessive androgen during in utero development
(C) Translocation of SRY from the Y chromosome to an autosome
(D) Chimerism with a 46, XY cell line

175. Which of the following is part of the first law of thermodynamics?
(A) Energy cannot be created or destroyed
(B) The entropy of the universe is decreasing
(C) The entropy of the universe is constant
(D) Energy cannot be transferred or transformed

176. Match the correct combination of toxin and the mode of action:

Toxin	Mode of action
P. Pertussis toxin	1. Prevents release of glycine by nerve end
Q. Diptheria toxin	2. Blocks G-protein signal transduction
R. Botulinum toxin	3. Induces fluid loss from intestinal ells
S. Tetanus toxin	4. Inhibits protein synthesis in eukaryotes
	5. Causes haemolysis
	6. Blocks release of acetylcholine by nerve end

(A) P–3, Q–2, R–5, S–4 (B) P–2, Q–4, R–6, S–1
(C) P–5, Q–6, R–3, S–2 (D) P–1, Q–3, R – 2, S–5

177. Which of the following is false? If the pH of a solution of HA, which dissociates to $H^+ + A^-$, equals the pK_a, then:
(A) The pH of the solution is 7
(B) The concentration of the acid form of the compound (HA) equals the concentrations of the dissociated form of the compound (A^-)
(C) The solution has a greater capacity for buffering than at any other pH
(D) The hydrogen ion concentration is equal to describing the dissociation of the acid, $HA \leftrightarrow H^+ + A^-$

178. Protein phosphorylation is part of most pathways of intracellular signal transduction. All of the following statements about phosphorylation are correct except:
(A) Phosphorylation can be on a serine or threonine
(B) Phosphorylation changes the structure and function of the protein
(C) Protein kinase asides result in amplification of the signal
(D) Tyrosine kinases are always part of the receptor protein

179. Which of the following statement is/are correct?
P. During the stimulation of B cells, a type of T cell called the suppressorT cell is involved
Q. The antibody identified as IgE functions in many allergic reaction
R. Antibody-mediated immunity is dependent upon the activity of T lymphoytes
S. The largest antibody molecule, consisting of five monomer units, is IgD
(A) P and S (B) Only Q
(C) P and R (D) P, Q, R and S

180. Which of the following is/are correct about *Botulinum* toxins?
P. Phase-encoded neurotoxins
Q. Binds to neuroreceptor ganglicosides on holinergic neurons
R. Exotoxin
S. Inhibits release of the stimulatory neurotransmitter, acetylcholine
(A) P and Q (B) R and S
(C) P, Q and R (D) P Q, R and S

181. Identify the only correct statement. The solubility of fatty acids in water increases as:
(A) The number of carbon atoms increases
(B) The number of carbon atoms decreases, and the number of double bonds increases
(C) The number of carbon atoms increases, and the number of double bonds decreases
(D) The number of carbon atoms increases, and the number of double bonds increases

182. Which of the following is false about myosin?
(A) It is involved in muscle contraction
(B) It has a long alpha-helical tail through which it forms homodimers
(C) It does not act as a molecular motor
(D) It links the actin bundles to the plasma membrane in the microvilli of intestinal cells

183. Which of the following statement about ribozymes is false?
(A) Ribozymes are capable of self-replication, but cannot catalyze other types of reactions
(B) Manfred Eigen found that RNA's could replicate themselves in solution without acid of proteins
(C) In tetrahymena thermophila, an intron was found that carried out its own excison splicing
(D) A tRNA-processing enzyme containing RNA was found in which the RNA portion provided the catalysis

184. The KDEL sequence, found on luminal proteins of the ER, is responsible for:
(A) Translocation of proteins into the ER lumen
(B) Insertion of protein into the membran of the ER
(C) Recognition by signal peptidase of the signal sequence
(D) Retrieval of ER luminal proteins from the Golgi

185. Which one of the following proteins is a major cell cycle regulator?

(A) E2F (B) E47
(C) EcoR1 (D) Pit-1

186. Choose the correct pairs from the following

P. Isocitrate lyase	1. Conversion of amino acid into glucose
Q. PEP carboxykinase	2. Biotin
R. Pyruvate dehydrogenase complex	3. Synthesis of glucose from acetate
S. Phosphofrutokinase	4. Lipoic acid
T. Pyruvate carboxylase	5. An allosteric enzyme

(A) P–1, Q–2, R–4, S–5, T-3
(B) P–3, Q–1, R–4, S–5, T-2
(C) P–3, Q–1, R–4, S–2, T-5
(D) P–2, Q–5, R–1, S–4, T-3

187. In scurvy, defective collagen is due to sufficient vitamin, which

(A) In ordinarily incorporated into crosslinks between tropocollagen molecules
(B) Is usually involved in the hydroxylation of proline residues
(C) Inhibits the oxidative degradation of collagen
(D) Is required for the conversion of lysyl residues into aldehydes

188. CheA, a protein that participates in bacterial hemotaxis:

P Is a histidine kinase
Q Shows autophosphorylation
R Transfers phosphoryl groups to conserved aspartate residues in the cheY
S Is a tyrosine kinase

(A) P and Q (B) R and S
(C) P, Q and R (D) P, Q, R and S

189. Pepsin hydrolysis of IgG molecule will result in the production of:

(A) One FC fragment and one F(ab') 2 fragment
(B) One FC fragment and one Fab fragment
(C) One FC fragment and one Fab fragment
(D) One FC (ab') 2 fragment and one Fab fragment

190. Cytokinesis is animal cells is caused by:

(A) The sliding movements of a band of micro-tubules around the circumference of the cell
(B) The contraction of a band of actin filam-ents around the circumference of the cell
(C) The movement of the mitotic spindle fibres
(D) Endocytosis of the plasma membrane around the equator of the cell

191. The primary function of cortical reaction or egg activation during fertilization is to:

(A) Secrete hyaluroidase enzyme
(B) Block the polyspermy
(C) Form the zygote nucleus
(D) Changes the pH of egg cytoplasm

192. Which of the following chemotherapeutic agents works by impairing de novo purine synthesis?

(A) Acyclovir (acyloguanosine)
(B) 5-fluorouracil (antimetabolite)
(C) Methotrexate (antifolate)
(D) Hydroxyurea

193. How would a loss – of – function mutation in the α–amylase gene affect seed germination?

(A) The seed could not imbibe water
(B) The embryo would starve
(C) The seed coat would not rupture
(D) The seed would germinate prematurely

194. Treatment of root tip, meristem cells with the microtubule inhibitor colchicine results in all of the following except:

(A) Induction of polyploidy
(B) Prevention of cytokinesis
(C) Inhibition of mitotic spindle assembly
(D) Cessation of DNA replication

195. A characteristics of homologous chromosomes is that:

(A) They carry alleles for the same genes in the same relative positions
(B) They regularly exchange parts by crossing over at meiosis
(C) They physically pair at meiosis
(D) All of these

196. Which of the false about ketogenesis?

(A) Acetyl CoA acts as precursor
(B) HMG CoA synthase participate during reaction
(C) Occurs in the liver
(D) Acetone is the only product

197. If a mouse is repeatedly immunized with a complex protein such as Bovine Serum Albumin (BSA), the resulting serum antibodies will be of:

(A) Identical isotype and identical antigen binding sites
(B) Various isotype but identical antigen binding sites
(C) Identical isotype various antigen binding sites
(D) Various isotype and various antigen binding sites

198. Which of the following cofactor(s) is/are electron carrier(s)?

(A) Only NAD^+ (B) Only $NADP^+$
(C) NAD^+, $NADP^+$ & FAD (D) $NADP^+$ not FAD

199. Targeting of enzymes to lysosomes involved which of the following steps?

1. Dissociation of ligand from receptor
2. Binding to a mannose-6-phosphate receptor
3. Decrease in pH
4. Fusion with late endosome
5. Addition of phosphate to mannose

(A) 2 and 5 (B) 2, 3 and 4
(C) 2, 4 and 5 (D) 1, 2, 3, 4 and 5

200. Which of the following statements about the different types of enzyme inhibition is incorrect?

(A) Competitive inhibition occurs when a substrate competes with an enzyme for binding to an inhibitor protein

(B) Competitive inhibition occurs when the substrate and the inhibitor compete for the same active site on the enzyme

(C) Noncompetitive inhibition of an enzyme cannot be overcome by adding large amounts of substrate

(D) Competitive inhibitors are often similar in chemical structure to the substrate of the inhibited enzyme

201. Immediately after fertilization in animals, the first structural and biochemical changes in the egg are initiated by:

(A) New gene transcription

(B) The release of Ca^{++} from internal reservoir

(C) The initiation of DNA synthesis

(D) A lowering of cytosolic pH

202. The production of ATP by oxidative phosphorylation is driven by energy from:

(A) Coenzyme A

(B) Isomerization of the cytochromes

(C) The formation of NADH

(D) The movement of protons from the intermembrane space to the matrix of the mitochondrion

203. Which of the following statements correctly describes p53?

(A) It is a membrane receptor that binds to a cell growth hormone

(B) Cell division stops until p53 binds to DNA and repairs the damage

(C) p53 is an example of an oncogene, and its expression causes uncontrolled cell division

(D) Activation of p 53 leads to inhibition of the S-phase cyclin-CDK complex

204. Why does over expression of BC1-Z gene as a conseq-uence of a translocation result in increased cell proliferation in low grade B-cell lymphomas?

(A) It blocks the pro-apoptotic gene bax

(B) It blocks p53 expression

(C) It blocks expression of the mismatch repair gene HMLH-1

(D) It stimulates lymphocyte growth factor receptors

205. Which of the following combinations of features is true for sprophytic self-incompatibility?

(A) 3-called pollen, dry stigma and rejection reaction in the style

(B) 2-called pollen, short pollen viability, rejection reaction on the stigma

(C) 2-called pollen, wet stigma, rejection reaction in the style

(D) Short pollen viability, dry stigma, rejection reaction on the stigma

206. Identify the correct pairs for the primary functions of the different enzyme classes:

P. Kinases	1. Cleave bonds by elimination
Q. Lyases	2. Make large molecules from small molecule
R. Synthases	3. Transfer phosphate group to biomolecules
S. Phosphatases	4. Remove phosphate group from biomolecules

(A) P–4, Q–2, R–3, S–1

(B) P–3, Q–1, R–2, S–4

(C) P–4, Q–1, R–2, S–3

(D) P–3, Q–2, R–1, S–4

207. All of the following receptors may belong to the steroid receptor gene superfamily except:

(A) Aryl hydrocarbon receptor

(B) Tocopherol receptor

(C) Retinoic acid receptor

(D) Thyroid hormone receptor

208. Which of the following statement is not true about the amino acids?

(A) Only around 20 amino acids are used in protein synthesis

(B) All amino acids exist in two stereoisomeric forms

(C) Amino acids have an N-terminus, C-terminus and R groups

(D) There are more than 300 different kinds of amino acids present in the cell

209. An uncoupler of oxidative phosphorylation such as dinitrophenol:

(A) Inhibits electron transport and ATP synthesis

(B) Allows electron transport to proceed without ATP synthesis

(C) Inhibits electron transport without impairment of ATP synthesis

(D) Specifically inhibits cytochrome b

210. Which of the following best supports the endosymbiotic theory of the evolutionary origin of mitochondria?

(A) Mitochondria, chloroplasts and prokaryotes contain electron carriers

(B) Genes for mitochondrial pyruvate dehydrogenase subunits are found in the nuclear DNA

(C) Mitochondrial and bacterial ribosomal functions are inhibited by the same antibiotics

(D) The outer mitochondrial membrane contains the protein porin

211. Retinoblastoma is a tumor suppressor protein and is directly involved in:

(A) Cell cycle progression

(B) Transmission of extra-cellular signals to the nucleus

(C) Cytoskeletal rearrangement

(D) General inhibition of various oncoproteins

212. Which of the following is a prime example of a ribonucleoprotein enzyme, the catalytic activity of which (under non-physiological conditions), residues solely in its RNA component?

(A) Reverse transcriptase (B) Cytochrome C
(C) Bacterial RNA polymerase (D) RNase P

213. Angiogenesis contributes to cancer development by:

(A) Providing nutrients and oxygen to tumors, and by facilitating metastasis
(B) Inhibiting a mutation in a gene that causes uncontrolled cell growth
(C) Inhibiting apoptosis (programmed cell death)
(D) Preventing the normal inhibition of growth that occurs between cells upon contact

214. When the self compatibility S allele of the pollen grain matches either of the S alleles of the stigma then it is termed as:

(A) Gametophytic self-incompatibility
(B) Sporophytic self-incompatibility
(C) Interspecific incompatibility
(D) None of these

215. Which one of the following statements incorrectly describes lysosomes:

(A) Hydrolytic enzymes move to the lysosomes via vesicular transport
(B) Materials phagocytosed into the cell go to the lysosomes
(C) The pH of lysosomes is more acidic than that of the cytoplasm
(D) Hydrolytic enzymes are transported to lysosomes directly from the cytosol

216. Which of the following is not true? When RNA undergoes for alkali-catalyzed cleavage:

(A) Transesterifiation reaction occurs
(B) Cyclic 2'-3'-nucleoside monophosphate forms
(C) Cyclic 5'-3'-nucleoside monophoshate forms
(D) 2'- OH is converted into 2'-alkaxide ion

217. Which of the following is used to treat ovarian cancer?

(A) Neomysin (B) Taxol
(C) Gentamycin (D) Cylosporin

218. The following are the DNA molecules with base composition. Which one of them requires high melting temperature for denaturation?

(A) %A – 28, %T – 28, %G – 22, % C - 22
(B) %A – 15, %T – 15, %G – 35, % C - 35
(C) %A – 35, %T – 35, %G – 15, % C - 15
(D) %A – 30, %T – 30, %G – 20, % C – 20

219. In Xenopus, a type of frog and most vertebrates there is a certain factor that arrests the secondary oocyte in metaphase II of meiosis. What is the name of this factor?

(A) Mitosis promoting factor (MPF)
(B) Vitelloenesis factor (VF)
(C) Cytostatic factor (SF)
(D) Sonic Hedgehog (SH)

220. Homeobox sequences:

(A) Are found from fruit files to human beings
(B) Control segmentation and basic patterns of development of body parts
(C) Are very different but present among nearly all eukaryotes
(D) A and B are correct

221. What mechanism is responsible for movements of eukaryotic cilia and flagella?

(A) Kinesin moving on microfilaments
(B) Dynein moving on microfilaments
(C) Dynein moving on microtubules
(D) Myosin moving on microfilaments

222. Which of the following statement is false about collagen?

(A) It is the most abundant protein in animal tissues
(B) It forms a double helix, where two collagen molecules are wrapped around one another in a ropelike structure
(C) It contains an amino acid called hydroxyproline
(D) It is a secreted protein

223. All of the following processes occur in the pathway leading to regulated protein secretion in animal cells except:

(A) Formation of transport vesciles from the rough endoplasmic reticulum
(B) An increase in the concentration of cytosolic calcium ions prior to secretion
(C) Synthesis of an amino-terminal signal sequence
(D) Phosphorylation of a mannose residue in a glycoprotein

224. Match the biological functions mentioned to Column A with the enzymes given in Column B:

Column A	Column B
P. Diayglycerol synthesis	1. Protein kinase A
Q. REB phosphorylation	2. ras
R. GTP hydrolysis	3. Phospholipase
	4. Phospholipase D
	5. Protein kinase G

(A) P – 3, Q – 1, R – 5 (B) P – 4, Q 1, R - 2
(C) P – 3, Q – 1, R – 2 (D) P – 3, Q – 5, R - 2

225. Which of the following is not found within the lipid bilayer?

(A) Fatty acid esters (B) Cholesterol
(C) Oligosaccharides (D) None of these

226. The topological features of circular DNA may not affect one of the following:

(A) The electrophoretic mobility of the DNA
(B) The sedimentation properties of the DNA
(C) The affinities toward proteins that bind to the DNA
(D) The susceptibilities of the DNA to the action of DNA ligase

227. Identify the pathways where the following reactions occur:

Group I

P. Fatty acid + GTP= CoA → Acyl-CoA + GDP + Pi

Q. NH_3 + glutamate + ATP → Glutamate+ADP+Pi

R. Succinate + E –FAD → Fumarate + E – $FADH_2$

S. Malonyl – S – CoA + AP-SH → Malonyl-S-AP + CoA-SH

Group II

1. Fatty acid synthesis
2. Fatty acid oxidation
3. Oxidative phosphorylation
4. Citric acid cycle
5. Gluconeogenesis
6. Amino acid biosynthesis

(A) P – 5, Q – 3, R – 4, S – 6
(B) P – 1, Q – 2, R – 5, S – 1
(C) P – 3, Q – 2, R – 6, S – 6
(D) P – 2, Q – 6, R – 4, S – 1

228. The sodium-potassium pump is called an electrogenic pump because it:
(A) Ionizes sodium and potassium atoms
(B) Pump hydrogen ions out of the cell
(C) Is used to drive the transport of other molecules against a concentration gradient
(D) Contributes to the membrane potential

229. Which of the following statement about glycogen, amylase and amylopectin is false?
(A) All are homopolymers of glucose
(B) Amylose is unbranched, amylopectin and glycogen contain many (1-6) branches
(C) All are stored intracellularly as semi-soluble granules
(D) Amylose serves primarily as structural elements in plants cell walls

230. Which of the following is unique to internal fertilization?
(A) Many zygote are formed at one time
(B) An egg changes one it is fertilized so that no more than one sperm an enter it
(C) Every zygote into an embryo
(D) Sperm are deposited inside the female body

231. Somatic mutation of VDJ coding regions in DNA allows the progeny of a single B lymphocyte to produce immunoglobulins of different:
(A) Classes (B) Sub-classes
(C) Idiotypes (D) Allotypes

232. Acetyl-CoA, the cytoplasmic substrate for fatty acid synthesis, is formed in mitochondria. The inner mitochondrial membrane is impermeable to acetyl CoA. Which of the following compounds is the form in which the carbon of acetyl-CoA is transported to the cytoplasm?
(A) Malate (B) Acetate
(C) Citrate (D) Puruvate

233. The earliest stages of B-cell differentiation:
(A) Occur in the embryonic thymus
(B) Require the presence of antigen
(C) Involve rearrangement of k–chain gene segments
(D) Involve rearrangement of heavy – chain gene segments

234. Rhodopsin, - aderenergic receptors and musarinic acetylcholine receptors share which of the following features?
(A) Each causes an inhibitory intracellular response
(B) Each activates a tyrosine kinase cascade
(C) Each is composed of an αß-dimer
(D) Each functions through a heterotrimeric G-protein

235. Pararine hormone are those which:
(A) At on the same cells which produce them
(B) At on the cells in the immediate vicinity of the cells which produce them
(C) At on the cells which are distantly located form the cells where they are produced
(D) Do not require any second messenger to transmit their respective signals into the target cell

236. IP3 initially causes Ca^{++} to be released into the cytoplasm from:
(A) Mitochondria
(B) Lysosome
(C) The endoplasmic reticulum
(D) The plasma membrane (from extracellular to intracellular)

237. Both class I and class II MHC molecules are:
(A) Composed of α and ß chains with variable and constant regions
(B) Expressed consititutively on all nucleated cells
(C) Expressed on the B cell membrane
(D) Part of the T cell receptor for antigen

238. Drugs that either stabilize or depolymerize microtubules can be used in cancer hemotherapy. Which of the following is correct concerning such drugs?
(A) They stimulate the immune system
(B) They prevent chromatin condensation
(C) They prevent movement of tumor cells into other tissues
(D) They interfere with mitosis

239. The α–helix structure:
(A) Is maintained by hydrogen bonding between amino acid side chains
(B) Makes up about the same percentage of all proteins
(C) Is stabilized by hydrogen bonding between the amino & carboxyl termini of polypeptide chain
(D) Is stabilized by hydrogen bonds between backbone N-H and C = O groups of polypeptide chains

240. The difference between a benign and a malignant tumor is:
(A) Malignant tumor are painful and benign tumors are not
(B) A malignant tumor has the ability to spread to other tissues and to initiate tumor at secondary sites, whereas benign tumor do not spread
(C) Benign tumors cause a less severe form of cancer than malignant tumors
(D) Malignant tumors are caused by viruses, but benign tumors arise spontaneously

241. Which of the following change in cellular property is not associated with transformation?
(A) Loss of contact inhibition
(B) Decrease angiogenesis
(C) Increase in mobility of surface proteins
(D) Density independent cell prolification

242. If for the biochemical reaction $A \longrightarrow B$, $\Delta H < O$ and $\Delta S > O$, then:
(A) The reaction is spontaneous
(B) The Reaction is endothermic
(C) $\Delta G = O$
(D) The disorder in the system will decrease if the reaction proceeds

243. During fertilization, the movement of the pollen tube towards the ovule is guided by a protein released from:
(A) Egg (B) Synergid
(C) Antipodal cell (D) Polar nuclei

244. Failure of products Mullerian-inhibitory substance in a genetic male would result in:
(A) Failure of formation of a testes
(B) Presence of a uterus
(C) Breast development at puberty
(D) Failure of virilization of the phallus

245. Which of the following statement is true about ion channels?
(A) They are opened either by binding of ligands or by changes in electric potential across the membrane
(B) They require ATP
(C) They are open most of the time
(D) The rate of transport is slow compared to the rate of transport via carrier proteins

246. Somatic embryogenesis is a procedure in plant tissue culture methodology described best as:
(A) Formation of both shoot and root meristem
(B) Formation of embryos
(C) Formation of axillary buds
(D) None of these

247. Lysosomal protein targeting takes place through:
(A) COP-coated vesicles
(B) Clathrin coated vesicles
(C) Liposome
(D) Receptor mediated endocytosis

248. The isoelectric point of a protein is defined as:
(A) The pH at which the net charge on the molecule is zero
(B) The pH at which all groups are protonated
(C) The pH at which all groups are unprotonated
(D) The pH at which each acidic group is protonated and each basic group is unprotonated

249. Actin filaments are found in all of the following except the:
(A) Flagella of bacteria
(B) Sacromeres of skeletal muscle cells
(C) Microvilli of the intestinal brush border
(D) Contractile rings of dividing animal cells

250. Which of the following is not true about fatty acid synthesis?
(A) Cofactors – NADPH
(B) Precursor – Two carbon unit
(C) Site – Cytosol
(D) Carrier - Carnitine

251. The common pathway of entry into the endoplasmic reticulum (ER) of secretary, lysosomal and plasma membrane proteins is best explained by which of the following?
(A) Binding of their mRNAs to a special class of ribosomes attached to the ER
(B) Addition of a common sorting signal to each type of protein after completion of synthesis
(C) Addition of oligosaccharides to all three types of proteins
(D) Presence of a signal sequence that targets each type of protein to the ER during synthesis

252. These samples of antibodies were electrophoresed under denaturing and reducing conditions on a 15% arylamide gel, followed by staining with coomassie blue dye. Samples 1, 2 and showed two, three and four stainable bands respectively. Which one of the following conclusions can be made from these observations?
(A) Sample 1 is IgG, 2 is IgA and 3 is IgM
(B) Sample 1 is IgA, 2 is IgM and 3 is IgG
(C) Sample 1 is IgG, 2 is IgM and 3 is IgA
(D) Sample 1 is IgA, 2 is IgG and 3 is IgM

253. A mammalian protein, src:
P. Has a viral counterpart, v-src, that is oncogenic
Q. Is a product of proto-oncogene
R. Is a component of a signaling pathway for cell growth and differentiation
S. Is a protein tyrosine kinase
Which of the above statements are correct?
(A) P and Q (B) P and R
(C) P, Q and S (D) P, Q, R and S

254. A mutation depleting an upstream activating sequence for a single gene would be expected to be:
(A) Polar (B) Trans-dominant
(C) Cis-dominant (D) Silent

255. Which of the following combination of statements regarding maturation promoting factor (MPF) is true?

P. MPF in amphibians regulates cell cycle

Q. MPF in amphibians is compose of cyclin & DKs

R. MPF in amphibians undergoes periodic degradation

S. MPF in amphibians undergoes periodic phosphorylation

(A) Q and R (B) P and S

(C) P, Q and R (D) P, Q, R and S

256. Proteins that are to be used outside the cell in eukaryotes are synthesized:

(A) In the mitochondria

(B) On the rough endoplasmic reticulum

(C) On the smooth endoplasmic reticulum

(D) On free ribosomes

257. Which of the following statement are true for steroid hormones?

P. Increase the enzymatic activity of pre-existing target enzyme

Q. At act cell nucleus

R. Interact with the plasma membrane receptors of target cells

S. Form a complex with receptor and acts as transcriptional enhancers

(A) P, R (B) Q, S

(C) P, Q (D) R, S

258. Which of the following is not true about NO:

(A) Acts as intracellular signaling molecule as well as neurotransmitter

(B) Regulates the vasodilation

(C) Synthesized form L-arginine

(D) Induces adenylate cyclase, which catalyzes AMP formation

259. Which of the following statement is incorrect? A monoclonal antibody

(A) Could theoretically be isolated from serum

(B) Is active against multiple epitopes on one protein

(C) Is produced by cells that must be grown in HAT protein

(D) Is produced by immortal cells

260. Choose the correct statement(s). In Mammals:

(A) Primary sex determination is chromosomal

(B) A person with five X chromosomes and one Y chromosome would be male

(C) Presence of Y chromosome results in synthesis of anti-Mullerian duct hormone

(D) A, B and C

261. Nuclear localization signals are composed primarily of:

(A) Hydrophobic amino acids

(B) Acidic amino acids

(C) Basic amino acids

(D) Sulphur containing amino acids

262. Which of the following is a false statement?

(A) Like microtubules and actin filaments, intermediate filaments form polarized structures and their function depends on this polarity

(B) A single epithelial cell can make a variety of keratins, all of which copolymerize into a single keratin filament system

(C) Keratines and vimentin-related proteins do not co-polymerize with each other: when they are expressed in the same cell, they form separate filament systems

(D) The most dramatic example of the importance of phosphorylation in the disassembly of intermediate filaments is provided by the nuclear lamins, which are phosphorylated and disassembled each time the cell enters mitosis

263. Which of the following is the means to provide nutrition to the developing embryo?

(A) Supply the embryo with yolk

(B) Form a larval feeding stage

(C) Formation of placenta

(D) All of these

264. Secretion of Mullerian-inhibiting substance by Sertolic cells:

(A) Causes the Mullerian ducts to develop into the oviducts

(B) Causes the Mullerian ducts to develop into the Wolffian ducts

(C) Causes the Wolffian ducts to develop into the vas deferens

(D) Causes the Mullerian ducts to regress by apoptosis in males

265. If the egg cell of a megagametophyte of a flowering plant has a marker gene designate as X and the sperm cell of a compatible pollen grain has a marker gene designated as Y, which of the following represents the endosperm that will result from fertilization in most flowering plants?

(A) XX (B) YY

(C) XY (D) XXY

266. Which of the following peptide sequences is likely to be present in the plasma membrane?

(A) Asp – Glu – Asp – Glu – Asp - Glu

(B) Val – Leu – Ile – Val – Leu - Ile

(C) Arg – His – Lys – Arg – His - Lys

(D) Ser – Thr – Ser – Thr – Ser – Thr

267. Estrogen and testosterone are steroid hormones, and are most likely to bind to:

(A) Membrane ion channel

(B) Enzyme linked membrane receptor

(C) G protein linked membrane receptor

(D) Cytoplasmic receptors

268. An antimicrobial agent can be used as an antifungal drug because of its ability to bind to sterols in the membrane and change membrane fluidity is:

(A) Amphotericin B (B) Chloranphenicol

(C) Streptomycin (D) Rifompin

269. The correct sequence of expression of genes during the embryonic development of Drosophila is:

(A) Maternal effect genes → Gap genes → Pair rule genes → Segment polarity genes

(B) Maternal effect genes → Pair rule genes → Gap genes → Segment polarity genes

(C) Maternal effect genes → Segment polarity genes → Pair rule genes → Gap genes

(D) Gap genes → Segment polarity genes → Pair rule genes → Maternal effect genes

270. Kinesin is correctly described as:

(A) An enzyme that phosphorylates intermediates filament proteins

(B) A co-factor with cyclin-B in the control of cell cycle events

(C) A motor protein that conducts vesicles along microtubules

(D) An actin-associated protein involved in the control of movement

271. Cholera toxin exerts its harmful effects in gut cells by:

(A) Inhibiting the receptor part of the CFTR protein

(B) Inducing the onset of apoptosis

(C) Stimulating the production of cylic AMP

(D) Blocking the chloride channel in the cell membrane

272. Whether a microtubule shrinks or grows is determined by:

(A) The rate of GTP-bound tubulin addition relative to the rate of tubulin GTP hydrolysis

(B) The phosphorylation state of beta-tubulin

(C) The rate of ATP hydrolysis relative to the rate of ATP-bound tubulin addition

(D) The presence or absence of gamma-tubulin

273. The distribution of transmembrane protein in the plane of a cell membrane can best be visualized by which of the following:

(A) Thin-section electron microscopy

(B) Freeze-fracture electron microscopy

(C) Scanning electron microscopy

(D) SDS gel electrophoresis

274. Mutations in hormeotic genes can lead to what type of developmental defect?

(A) The anterior portion of the embryo does not develop

(B) Several adjacent segments will be missing from an otherwise intact embryo

(C) The embryo will develop with every other segment failing to form

(D) The development of segments will be changed completely from their normal identity to that of a different segment

275. Proteoglycans are a group of cell-surface an extra-cellular matrix substances that:

(A) Are variable in molecular weight

(B) Are highly positively charged

(C) Have a very low molecular weight

(D) Have a very long life span

276. Tumor suppressor genes on their products an

P. Act as dominant negative genes

Q. Act as recessive genes

R. Prevent cell-cycle progression

S. Regulate apoptosis

(A) P and Q (B) Q and S

(C) P, Q and R (D) P, Q, R and S

277. Each individual antigenic determinant of the variable region of the antibody is referred to as:

(A) Paratope (B) Epitope

(C) Agretope (D) Idiotope

278. Which of the following are the mechanism of action of antibiotics?

P. Interfere with bacterial cell wall synthesis

Q. Inhibit bacterial transcription of DNA

R. Inhibit translation of bacterial mRNA

S. Change the permeability of plasma membrane

(A) P and Q (B) R and S

(C) P, Q and R (D) P, Q, R and S

279. When the experimentally isolated parts of an embryo develop according to their fate map, it is referred to as:

(A) Mosaic development

(B) Regulative development

(C) Clonal development

(D) Syncytial development

280. The antigen specificity of a particular B cell:

(A) Is induced by interaction with antigen

(B) Is determined only by the L-chain sequence

(C) Is determined by H+L-chain variable region sequences

(D) Changes after isotype switching

281. Tumor suppressor genes:

(A) Have the same effect as protonognes

(B) Contribute to cancer if only one allele is inactivated by mutation

(C) Produce protein products that promote apoptosis

(D) Produce products that promote cancer

282. Signal hypothesis refers to

(A) The hormonal control of eukaryotic proteins synthesis

(B) The synthesis of secreted proteins by membrane bound ribosomes

(C) Characteristics of 70S ribosomes involved in the synthesis of proteins for export by the cell

(D) Sequence of amino acids at the carbonyl end of a secretary proteins which is leaved off after the secretion

283. Longer fatty acids make membranes less fluid because:
(A) The long chains tend to get interwined tangled
(B) This allows greater hydrogen-bonding between the methyl groups
(C) Longer fatty aids bind rather tenaciously to membrane proteins
(D) There are more van der Walls interactions between the chain

284. A mutation in a representative dicot plant resulted into having altered floral organs as sepal, sepal, carpel, carpel. According to AB model of flowering which class of mutation is there in the plant?
(A) A class gene mutant
(B) B class gene mutant
(C) A & B class double mutant
(D) Class gene mutant

285. After interacting with a cell – surface receptor that signals activation to a trimeric G-protein, the first event that happens is:
(A) GTP is hydrolyzed to produce GDP
(B) Tyrosine on the G-protein become phosporylate
(C) The G-protein dissociates into a β– and a γ-subunits
(D) GTP replaces GDP in the α –subunit

286. The role of calcium in muscle contraction is to:
(A) Break the cross-bridges as a cofactor in the hydrolysis of ATP
(B) Bind with troponin, changing its shape so that the actin filament is exposed
(C) Transmit the action potential across the neuromuscular junction
(D) Spread the action potential through the T tubules

287. Which of the following would most likely explain a 46, XX individual with development of testes?
(A) Mutation of SOX9
(B) Presence of excessive androgen during in utero development
(C) Translocation of SRY from the Y-chromo-some to an autosome
(D) Chimerion with a 46, XY cell line

288. Which of the following are not correct pairs of characteristics that differentiate benign and malignant tumors?

Benign	Malignant
(A) Polyclonal	Monoclonal
(B) May result in death, but rarely	Frequently results in death
(C) Cannot invade	Can invade
(D) Cannot metastasize	May metastasize

289. Satellite DNA is made up of:
(A) Unique sequences of DNA
(B) Interspersed repeated sequences
(C) Randomly repeated sequences
(D) Minichromosomes

290. If a loss-of-function mutation occurs in the B-type genes of Arabidopsis, what will be the composition of the flower whorls (from whorl 1 to whorl 4)?
(A) Sepals–petals–stamens-carpels
(B) Sepals-sepals-stamens-carpels
(C) Sepals-sepals-carpels-carpels
(D) Petals-petals-stamens-stamens

291. In human, an individual with an XXY chromosomal complement will develop as:
(A) A male (B) A female
(C) An hermaphrodite (D) A super-female

292. Which of the following processes leads to formation of polytene chromosome?
(A) Non-disjunction of chromatids during meiosis
(B) Recombination between adjacent chromosome segments
(C) Sister chromatid exchange
(D) Repeated replication without separation of chromatids

293. With regard to p53 tumor suppressor gene, which of the following statement is false?
(A) It is located on chromosomes 17
(B) It encodes 53 kDa protein
(C) It holds the cell cycle in the G1 phase
(D) None of these

294. Amphibian metamorphosis is controlled by:
(A) Thyroid hormone (B) Parathyroid hormone
(C) Oxytocin (D) Ecdysone

295. The number of distinct disaccharides that can be formed from two molecules of glucose is:
(A) 11 (B) 6
(C) 5 (D) 1

296. Response of a tissue almost stopped when exposed for prolonged time to a chemical. After washing and leaving it for sometime, the tissue started responding to the same chemical at the same concentration The reason for the reduced response is likely to be:
(A) Increased apoptosis of the cells
(B) Increased necrosis of the cells
(C) Desensitization/down-regulation of the receptors
(D) Fixation of the cells

297. Mobile genetic elements present in human genome are:
P. Long interspersed elements (LINEs)
Q. Short interspersed elements ((SINEs)
R. P elements
S. IS elements
(A) Q and R (B) P and Q
(C) P and R (D) Q and S

298. Ingression is:
(A) Migration of individual cells from the sur- face layer into the interior of the embryo
(B) Movement of epithelial sheets
(C) Inward movement of outer layer
(D) Splitting of one cellular sheet into two or more

299. X-chromosome inactivation is:

P. Yielding a completely inactive x-chromosome
Q. Yielding a mostly inactive x-chromosome
R. Controlled via two special non-coding RNAs
S. Controlled via two special proteins

(A) P only (B) R only
(C) Q and R (D) P, R and S

300. Which of the following is not correct about an embedded segment of an integral membrane protein:

(A) A typically an α – helical segment long enough about 20-25 amino acids, to span the lipid bilayer
(B) Typically has polar C = O and NH groups in contact with the fatty acyl chains of phospholipids
(C) Typically has any charged residues facing the interior of the molecule
(D) Often contains hydrophobic amino acids, flanked by positively charged residues that interact with phospholipids head groups

301. Which one of the following statements does not correctly describes G-proteins?

(A) Energy for their activity comes from GTP molecules
(B) They become activated by interaction with '7-pass' receptor
(C) G-proteins are membrane associated proteins
(D) G-protein activation always increase the concentration of cAMP

302. The establishment of the anterior-posterior or dorso -ventral body axes is called:

(A) Division (B) Pattern formation
(C) Morphogenesis (D) Differentiation

303. The main difference between the genomes of the bacteria *H. influenzae* and *M. genitalium* is:

(A) That the genome of *H. influenzae* is a circular molecule and that of *M. genitalium* is linear
(B) That the *H. influenzae* genome contains fewer genes
(C) That the *H. influenzae* genome encodes more metabolic enzymes
(D) That the genome of *H. influemzae* encodes more proteins involved in DNA replication

304. Which of the following is incorrect about amphibian?

(A) Cleavage is holoblastic
(B) Cleavage is unequal
(C) Cleavage is non-radial and asymetrical
(D) Yolk is present

305. What is pseudogene?

(A) A gene that is only expressed at certain developmental stages
(B) A nonfunctional gene
(C) A gene that contains a mutation but is still functional
(D) A sequence of DNA that is slowly evolving to become an active gene

306. Psychrotroph bacteria:

(A) Can grow at 0 -7ºC has optimum growth between 20-30ºC
(B) Can grow at 0ºC and has an optimum growth temperature of 10ºC
(C) Have optimum growth between 20-45ºC
(D) Can grow at 55ºC or higher

307. Digitoxigenin, a cardiotonic steroid that inhibits ATPase when applied on extra cellular face of membrane, helps in accumulation of Ca^{++} inside the cardiac muscle cells by:

(A) Activating Na^+ - K^+ pump and blocking Na^+ - Ca^{++} exchanger
(B) Inhibiting Na^+ - K^+ pump and blocking Na^+ - Ca^{++} exchanger
(C) Having no effect on Na^+ - K^+ pump
(D) Increasing passive diffusion

308. Which of the following is/are true about competitive enzyme inhibitors?

P. They are usually small molecules
Q. They are usually water-soluble
R. They are usually resembling transition state intermediates
S. Their inhibition can be reversed if concentration of substrate is increased

(A) P, Q and R (B) P and S
(C) P and R (D) P, Q, R and S

309. Choose the correct statement(s):

P. In many unicellular organisms mitosis occurs without breakdown of the nuclear envelope
Q. The muscle like sliding of actin and myosin filaments in the contractile ring generates the fore required for cleavage

(A) P (B) Q
(C) P and Q (D) None of these

310. Match each compounds in the left column with its characteristic from the right column

P. G proteins	1. Exchange with GDP on G_a subunits
Q. Adenylate cyclase	2. Binds epinephrine
R. IP3	3. Is a second messenger arising from PIP_2
S. DAG	4. Is a small G protein GTPase
T. Ras	5. Transduces hormone stimulus from an activated 7TM membrane receptor to adenylate cyclase
U. β–Adrenergic receptor	6. Is activated by G_a-GTP
	7. Activates protein kinase
	8. Has inducible tyrosine kinase activity

(A) P -5, Q – 6, R – 3, S – 7, T – 4, U - 2
(B) P -5, Q – 7, R – 3, S – 1, T – 4, U - 8
(C) P -2, Q – 5, R – 3, S – 1, T – 4, U - 8
(D) P -8, Q – 4, R – 3, S – 7, T – 6, U - 1

311. Which of the following processes are not occurring during mitosis (M-phase) of the cell cycle?
P. DNA replication
Q. Crossing-over
R. Sister chromatid segregation
S. DNA damage control
(A) P and Q (B) Q and R
(C) R and S (D) P, Q and S

312. Protein transport into mitochondria takes place:
(A) Co-translationally
(B) Post-translationally
(C) Through ER-Golgi pathway
(D) Via peroxisomes

313. Which of the following is found in mammals during male gamete formation?
(A) Two successive centrosome duplications during miosis
(B) Accumulation of yolk during gamete formation
(C) Generation of a polar during meiosis I
(D) Formation of four functional gametes from a primary germ cell

314. Which of the following is not useful in identifying the amino-terminal residue of a protein?
(A) Cyanogen bromide (B) Dabsyl chloride
(C) Fluorodinitrobenzene (D) Phenyl isothiocyanate

315. A peptide has the sequence Glu-His-Trp-Ser-Gly-Leu-Arg-Pro-Gly. What is the net charge of the molecule at pH 11.
(A) + 2 (B) -1
(C) 0 (D) - 2

316. Which of the following sequence of events are correct in mammalian fertilization?
P. Sperm activation in female reproductive tract
Q. Acrosome reaction
R. Binding of sperm of zona pellucida
S. Fusion of sperm and egg membrane
(A) P, Q, R, S (B) P, R, Q, S
(C) P, S, R, Q (D) P, S, Q, R

317. What would be a likely explanation for the existence of pseudogenes?
(A) Gene duplication
(B) Gene duplication and mutation events
(C) Unequal crossing over
(D) Evolutionary pressure

318. What does not happen during gastrulation?
(A) Three embryonic germ layers are established
(B) Large-scale migration of cells to organize the three germ layers
(C) The embryo increases in size
(D) Developmental fates of the germ layers are decided

319. Cell cycle is controlled by:
(A) Change in the concentration of cdks
(B) Change in Ca^{++} levels
(C) Cyclins and cdks
(D) P53 and p21

320. Which of the following statements is false about the glucose transport?
(A) It transports glucose across the plasma membrane via mechanism called facilitated diffusion
(B) It has 12 alpha-helical transmembrane segments
(C) A conformational change in the transporter is involved in the transport process
(D) Flow through it is unidirectional

321. Enzyme Q can catalyze the conversion of molecule A to A' and the conversion of molecule B to B'. Both of these conversion reactions follow Michaelis–Menten kinetics and both have the same V_{max} However, the K m for the A → A' conversion is three times s large as the K_M for the B → B' conversion. What is the relationship between the substrate concentrations, [A] and [B], when both of these reactions are proceeding at a reaction velocity equal to ½ V_{max}?
(A) [A] > [B] (B) [B] > [A]
(C) [A] = [B] (D) Not enough information

322. The major advantage of sexual reproduction is that it:
(A) Allows rapid generation of new genotypes due to recombination
(B) Allows rapid transmission of hereditary traits compared to asexual reproduction
(C) Prevents transmission of defective traits
(D) Prevents loss of homozygosity

323. The Hox gene complexes of mice and other mammals have evolved by:
(A) The splitting of one cluster into two
(B) The duplication of a single ancestral Hox gene such as the one found in insects
(C) The duplication of the ancestral cluster
(D) The loss of three clusters from the ancestral condition of four clusters

324. In human, the Barr body is an:
(A) Active x chromosome in females
(B) Active x chromosome in males
(C) Inactive Y chromosome in males
(D) Inactive x chromosome in females

325. At what stage in development are human oocytes arrested between fetal life and puberty? When does the second meiotic division occur in the oocyte?
(A) Zygotene, at ovulation
(B) Diplotene, after fertilization
(C) Diplotene, at ovulation
(D) Zygotene, after fertilization

326. The proteins encoded by photo-onogenes participate in various metabolic processes including:
(A) Regulation of transcription
(B) Cell-to cell signaling
(C) Intracellular signaling transduction
(D) All of these

327. Match list –I with list – II and select the correct answer using the codes given below the lists:

List I (Structures)	List II (Originated from)
P. Retina	1. Endodermal cells
Q. Nephric tubules	2. Ectodermal cells
R. Pigment pattern of skin	3. Neural rest cells
S. Allantosis	4. Mesodermal cells

(A) P-1, Q–2, R–3, S–4 (B) P-4, Q–3, R–2, S–1
(C) P-2, Q–4, R–3, S–1 (D) P-3, Q–4, R–1, S–2

328. Which of the following statements about histones is incorrect?
(A) They are highly basic because they contain many positively charged amino acid side chains
(B) They are extensively modified after their translation
(C) In combination with DNA, they are the primary constituents of chromatin
(D) They account for approximately one-fifth of the mass of a chromosome

329. Broker, how, Roberts and P. Sharp hybridized mRNA from adenovirus to denatured adenovirus DNA and examined the resultant duplex molecules by electron microscopy. Which of the following did they observed?
(A) mRNA and DNA formed a continuous duplex throughout the length of the RNA
(B) mRNA and DNA formed a continuous duplex throughout the length of the DNA
(C) On one strand there were looped out regions of single-stranded RNA
(D) On one strand there were looped out regions of single-stranded DNA

330. During embryonic development, which germ layer gives rise to the nervus system?
(A) Octoderm (B) Mesoderm
(C) Endoderm (D) Blastoderm

331. Match list–I (extra embryonic layers in rabbit) with list–II (characteristic function of the extra embryonic layers) and select the correct answer using the codes given below the lists:

List I	List II
P. Amnion	1. Somatopleure protects the embryo from mechanical shocks
Q. Chorion	2. Splanchnopleure supplies O_2 and nutrients to the foetus
R. Allantois	3. Splanchnopleure develops villi
	4. Somatopleure develops villi

(A) P-2, Q–4, R–1 (B) P-2, Q–1, R–3
(C) P-1, Q–3, R–2 (D) P-1, Q–4, R–2

332. Most human cells are diploid with total DNA content of 2C. The DNA content increases to 4 C before the onset of mitosis. At anaphase, the DNA content of each cluster will be:
(A) 4 C (B) 2 C
(C) 1 C (D) 3 C

333. Chemoautotroph:
(A) Is a nutritional category that describes organisms that obtain energy from chemicals and carbon from CO_2
(B) Is a nutritional category found only among bacteria
(C) Is a nutritional category that describes organisms which pass electrons from reduced inorganic molecules through the electron transport chain
(D) All of these

334. All the following may be methods for the inhibition of microbial growth by antibiotics except:
(A) Antibiotics distrupt cell wall synthesis
(B) Antibiotics interfere with cell membrane function
(C) Antibiotics prevent the release of energy from ATP
(D) Antibiotics inhibit the synthesis of protein

335. FOS, JUN and MYC are:
(A) Proteins expressed on the surface of cancerous cells
(B) Protein kinase that phosphorylate transcription factors in cancerous cells
(C) Proteins involved in regulation of express- ion of genes involved in growth promotion
(D) Proteins involved in ion transport in cancerous cells

336. All the following cellular events involve actin filaments except:
(A) Amoeboid movement
(B) Cytokinesis in animal cells
(C) Flagellar movement in bacteria
(D) Contraction of smooth muscle

337. SNARE proteins are found in the membranes of all of the following compartments except:
(A) Mitochondria (B) Golgi complex
(C) Early endosome
(D) Endoplasmic reticulum

338. Asexual reproduction process of budding occurs in:
(A) All fungi
(B) Yeasts
(C) Fungi undergoing sexual reproduction
(D) *Bacillus subtilis*

339. Match list –I (for membranes) with list – II (functional of foetal membranes) and select the correct answer using the codes given below the lists:

List I	List II
P. Yolk sac	1. Elimination of metabolic wastes
Q. Chorion	2. Shock absorption
R. Allantocis	3. Food storage and digestion
S. Amnion	4. Exchange of substance between mother and embryo

(A) P-4, Q–3, R–1, S–2 (B) P-4, Q–3, R–2, S–1
(C) P-3, Q–4, R–2, S–1 (D) P-3, Q –4, R–1, S–2

340. Which amongst the following is not a characteristics of gap junctions?

(A) They allow small ions and molecules to migrate from a cell to an adjoining cell

(B) They are constituted of a class of membrane associated proteins named connexin

(C) They are found only in plant cells

(D) They are channel-like structures connecting the cytoplasm of adjoining cells

341. In bacterial cell division:

P. MreB, similar to actin helps in chromosome movement

Q. FtsZ, a tubulin homologue, helps in cytokinesid

R. ATP provides the energy

S. Microtubules make

(A) P and Q (B) Q and R

(C) R, Q and R (D) P, Q, R and S

342. Diploid chromosomes number in human is 46. However, if you count chromatids which are visible during mitotic or meiotic cell divisions, you would be able to count different sets of chromatid numbers depending on the stages and the type of the cell division in question. Match the chromatid numbers given on the left with cell division stage shown on the right:

A. 46 1. Mitotic metaphase

B. 23 2. Aneuploid meiotic telophase II

C. 92 3. Meiotic telophase I

D. 24 4. Meiotic anaphase II

(A) A–3; B–4; C-1; D–2 (B) A–2; B–4; C-1; D–3

(C) A–1; B–4; C-2; D–3 (D) A–1; B–2; C-3; D–4

343. If phosphoglyceride A has a higher T_m than phosphoglyceride B, which of the following differences between A and B may exist? (In each case only one parameter-either chain length of double bonds-is compared)

P. A has shorter fatty acids chains than B

Q. A has longer fatty acids chains than B

R. A has more unsaturated fatty acid chains than B

S. A has more saturated fatty acid chains than B

(A) P and Q (B) Q

(C) Q and S (D) R and S

344. Which of the following process(s) occur(s) during mitosis?

P. During anaphase kinetochore microtubules become shorter and polar microtubules elongate

Q. Kinetochore microtubules, much like polar microtubules, are in a state of dynamic flux

(A) P (B) Q

(C) P and Q (D) None of these

345. All of the following occur during the cleavage stage of animal development except:

(A) An increase in the nuclear–to-cytoplasmic ratio of the cells

(B) An increase in the number of blastomeres

(C) An increase in the mass of the embryo

(D) An increase in the surface – to volume ratio of the cells

346. All of the following statements about membrane structure and function are true except:

(A) Both sides of a membrane are identical in structure and function

(B) Special membrane proteins can o-transport two solutes by coupling diffusion with active transport

(C) Diffusion, osmosis and facilitated diffusion do not require any energy input from the cell

(D) Voltage across the membrane depends on an unequal distribution of ions across the PM

347. Which of the following statement about myosin is not true?

(A) Myosin binds to polymerized actin

(B) In vitro, myosin assembles spontaneously into the thin filaments

(C) Myosin is an ATPase

(D) Myosin as domains that interact with one another to effect its physiological functions

348. Which of the following statements are true in case of fluid-mosaic model for cell membranes?

P. Between 5-8 nm thick and appear trilaminar when viewed in cross-section under electron microsope

Q. Less than 1 nm thick and consist of a layer of protein sandwitched between two layers of phospholipids

R. In the lipid bilayer, proteins are embedded at irregular intervals and held by hydrophilic interactions between lipids and hydrophilic domains of the proteins

S. The protein domains exposed on one side of the lipid bilayer are different from those exposed on the other side

(A) P and Q (B) P and S

(C) Q and S (D) P and R

349. Match list –I with list – II and select the correct answer using the codes given below the lists:

List I Type of placenta	List II Mammalian examples
P. Diffuse	1. Cat
Q. Cotyledonary	2. Bat
R. Zonary	3. Deer
S. Discoidal	4. Horse

(A) P-4, Q–3, R–1, S–2 (B) P-2, Q–1, R–3, S–4

(C) P-4, Q–1, R–3, S–2 (D) P-2, Q–3, R–1, S–4

350. A feature common to almost all transmembrane proteins is:

(A) A phosphorylated exterior domain

(B) A structure consisting almost exclusively of beta-sheets

(C) An amino acid sequence rich in acidic residues

(D) An alpha – helical region of about 20 to 25 hydrophobic amino acids

351. Binding of mannose 6-phosphate to protein lead to its localization to:

(A) Lysosome (B) Golgibody

(C) Plasma membrane (D) Nucleus

352. Viral encoded Ras oncogene transforms normal mammalian cells into cancer cells. Viral Ras protein differs from its normal counterpart by:
(A) Diminished GTPase activity
(B) Increased GTPase activity
(C) Diminished ATPase activity
(D) Increased ATPase activity

353. During receptor-mediated endocytosis of LDL bond to its receptor:
(A) Both receptor an ligand are degraded
(B) The receptor is degraded and the ligand is recycled
(C) Both are recycled
(D) The ligand is degraded and the receptor is recycled

354. Evidence indicating that chloroplasts were originally free-living prokaryotes which subsequently evolved:
(A) Similarities of rRNA sequences between chloroplasts and free-living prokaryotes
(B) Similarities of structures between chloroplasts and some contemporary free-living prokaryotes
(C) Presence of circular DNA in chloroplasts and in free-living prokaryotes
(D) Ability of choloroplasts to synthesize all their own protein

355. What is one of the ways that the membranes of winter wheat are able to remain fluid when it is extremely cold?
(A) By increasing the percentage of unsatur-ated phospholipids in the membrane
(B) By decreasing the number of hydrophobic proteins in the memLrane
(C) By increasing the percentage of cholesterol molecules in the membrane
(D) Both (A) and (B)

356. Holocentric chromosomes are which of the following?
(A) Chromosomes with multiple centromeres
(B) Additional chromosomes possessed by some individuals in a population
(C) Short chromosomes with man genes, as found in chicken
(D) Circular chromosomes found in some lower eukaryotes

357. The spemann organizer of the amphibians are:
(A) Mesodermal in origin
(B) Endodermal in origin
(C) Ectodermal in origin
(D) Epidermal in origin

358. In many tumors, there is a growth of blood vassels in the direction of tumors. The phenomenon is known as:
(A) Necrosis (B) Carinogenesis
(C) Angiogenesis (D) Morphogenesis

359. Match list –I with list – II and select the correct answer using the codes given below the lists:

List I	List II
P. Triple fusion	1. Transfer of pollen grains to the stigma
Q. Pollination	2. Fusion of male and female gametes
R. Parthenocarpy	3. Seedless development of fruit
S. Syngamy	4. Fusion of polars and male gametes

(A) P-1, Q–2, R–3, S–4 (B) P-4, Q–1, R–3, S–2
(C) P-2, Q–1, R–3, S–4 (D) P-4, Q–1, R–2, S–3

360. Which of the following is false about mitochondrial genome?
(A) Mitochondrial genome with few exceptions is circular
(B) Mitochondrial genome codes for some nuclear proteins
(C) Mutation rate is high in mitochondrial genome
(D) Genes present in mitochondrial genome are mostly interrupted

361. The rotatory motion of bacterial flagellum is driven by:
(A) Energy obtained through hydrolysis of ATP
(B) Trans-membrane electrochemical potential liked to protein pumping
(C) Direct uptake of extra-cellular nutrients
(D) Concentration gradient of nutrients in the environment

362. The common pathway of entry into the endoplasmic reticulum (ER) of secretary, lysosomal and plasma membrane proteins is best explained by which of the following?
(A) Binding of their mRNAs to a special class of ribosomes attached to the ER
(B) Addition of a common sorting signal to each type of protein after completion of synthesis
(C) Addition of oligosaccharides to all three types of proteins
(D) Presence of a signal sequence that targets each type of protein to the ER synthesis

363. Which of the following are true about transposition except:
(A) Transposons move from one location to a different one within a chromosome
(B) Both the donor and target sites must be homologous
(C) Transposons have insertion sequences that are recognized by transposasses
(D) The transposon may either be excised & moved, or be replicated with the replicated piece moving

364. Which one of the following modifications leads to protein degradation?
(A) Methylation (B) Acetylation
(C) Phosphoryation (D) Ubiquitination

365. What is the primary reason that red blood cells are used in the study of the plasma membrane?

(A) They can easily obtained

(B) They contain more plasma membrane than any other cell type

(C) Being made up of a lipid monolayer, they are of particular interest

(D) They lack nuclei and membrane–bounded organelles

366. Malignant tumour is characterized by:

(A) Slow simple expansion of cells

(B) A typical tissue structure & uncontrolled growth

(C) Protoconogenes expression

(D) No chromosomal abnomalities

367. The main difference between active transport and facilitated diffusion is that:

(A) In active transport the molecules move from areas of high to areas of low concentration

(B) Carrier protein is involved only in case of active transport

(C) In active transport energy is consumed to move molecules against a concentration gradient

(D) In active transport only water molecules are transported

368. Inosphores are:

(A) The gating mechanisms associated with the transport of ions

(B) Intrinsic proteins that passively transport ions

(C) Chemicals that form pores in the plasma membrane and allow ions to cross

(D) Intrinsic proteins that actively transport ions

369. Drugs that either stabilize or depolymerize micro-tubules can be used in cancer chemotherapy. Which of the following is correct concerning such drugs?

(A) They stimulate the immune system

(B) They prevent chromatin condensation

(C) They prevent movement of tumor cells into other tissues

(D) They interfere with mitosis

370. Cytokinesis in animal cells is caused by:

(A) The sliding movements of a and of microtubules around the circumference of the cell

(B) The contraction of a band of actin filam-ents around the circumference of the cell

(C) The movement of the mitotic spindle fibres

(D) Endocytosis of the plasma membrane around the equator of the cell

371. Protein phosphateses:

(A) Catalyze the addition of phosphate residues to proteins

(B) Catalyze the removal of phosphate resid-ues from proteins

(C) Catalyze the addition of glycosylphosphatidyli-nositol to proteins

(D) Are proteins that specifically bind phosphorylated proteins

372. Which of the following is incorrect concerning MHC class II molecules?

(A) B cells may express different allelic forms of MHC class molecules on their surface

(B) MHC class II molecules are synthesized in the endoplasmic reticulum of many cell types

(C) Genetically different individuals express different MHC class II alleles

(D) MHC class II molecules are associated with β_2-microglobulin on the cell surface

373. Which of the following events can induce a tran-sient arrest in the translation of a secretary protein?

(A) Binding of a polysome to an ER receptor

(B) Binding of SRP to an N-terminal signal sequence

(C) Binding of snRNPs to the large ribosomal subunit

(D) Presence of a stop-transfer sequence in the polypeptide

374. Which is not correct about chloroplasts?

(A) Grana are membrane components of the thylakoids

(B) The membranes that are stacked in the thylakoid are called stroma

(C) Photo-harvesting apparatus are embedded in the thylakoid membrane

(D) Like mitochondrial membrane, chloroplast membranes are also capable of generating ATP using proton motif force

375. What properties is/are expected to be significantly different between hot-spring bacteria (that live at 120ºC) and regular intestinal *E coli*?

P. Number of cysteinesin the proteins

Q. Number of methionine in the protein

R. Molecular weight of the protein

S. GC richness of the genomic DNA

T. Richness of saturated fatty acids in plasma membrane

(A) P and T (B) Q and R

(C) R and S (D) P and S

376. Cholera toxin exerts its harmful effects in gut cells by:

(A) Inhibiting the receptor part of the CFTR protein

(B) Inducing the onset of apoptosis

(C) Stimulating the production of cyclic AMP

(D) Blocking the chloride channel in the cell membrane

(D) A significantly shorter life span

377. In sporophytic self-incompatibility, rejection of the male gametophyte occurs at the level of:

(A) Ovary

(B) Stigma surface

(C) Trasmitting tract of the stigma

(D) Stylar canal

378. Match list –I (type of egg) with list – II (group of animals) and select the correct answer using the codes given below the lists:

List I	List II
P. Isolecithal egg	1. Amphibia
Q. Moderately telolecithal egg	2. Cephalochordata
R. Telolecithal egg	3. Insects
S. Centrolecithal egg	4. Bony fishes

(A) P-2, Q–1, R–3, S–4 (B) P-2, Q–1, R–4, S–3
(C) P-, Q–4, R–3, S–2 (D) P-3, Q–2, R–4, S–1

379. The ability of an individual B cell to express only one H chain allotype is called:
(A) Allelic exclusion
(B) Co-dominant expression
(C) Isotypic exclusion
(D) Nonproductive rearrangement

380. Which of the following is true about growth hormone receptor?
(A) Activates Gs
(B) Requires dimerization to exerts its effects
(C) Must be internalized to exerts its effects
(D) Resembles epinephrine and ACTH receptors

381. In a xenogenic cell-based therapy, the donor and recipient belong to different species. In which of the following, do not donor and recipient belong to the same species?
P. Autologous
Q. Allogenic
R. Syngeneic
(A) Only P (B) Only P and Q
(C) Only P and R (D) P, Q and R

382. Which of the following is not a similarity between the cellular and humoral immune responses?
(A) Antigen-specific lymphoytes undergo clonal selection and expansion
(B) Cytokine signals promote efector cell differentiation
(C) Memory cells are generated
(D) Receptor isotype switching occurs

383. Which of the following statement is correct?
(A) The plasma membrane is highly impermeable to all charged molecules
(B) All membrane transport proteins so far known traverse the lipid bilayer, and their polypeptide backbones generally extend back and forth across the membrane a number of times
(C) Ribosomal RNAs are produced in the nucleolus, a specialized region of the nucleus, and are then transported into the cytoplasm, where they are packaged with ribosomal proteins to form ribosomes
(D) Sine the Na^+ - K^+ ATPase exchanges equal numbers of Na^+ and K^+ ions in each pumping cycle, it is electrically neutral

384. Antiviral drugs are used to:
(A) Inhibit the replication of viruses within cells
(B) Prevent the formation of enzymes within the viral genome
(C) Inhibit the formation of the viral envelope
(D) Stimulate the production of antibodies the body

385. In Drosophila, a homeotic mutation would be the most likely explanation for which of the followings:
(A) A decrease in the number of parasegments
(B) A leg developing where an antenna would normally be located
(C) Forked and dense body bristles instead of straight and sparse ones
(D) Rapidly hydrolyzing G protein GTP to HDP

386. Drugs and xenobiotics for clearance from our body primarily by:
(A) Proteasomal degradation
(B) Lysosomal degradation
(C) Cytochrome P_{450}
(D) Nuclear receptors

387. Which of the following are the second messengers that are produced by the phosphoinositide aside?
P. Phosphatidyl inositol 4, 5-bisphosphate (PIP_2)
Q. Inositol 1, 4, 5-trisphosphate (IP3)
R. Inositol 4-phosphate
S. Diacylglycerol (DAG)
(A) P and Q (B) R and S
(C) Q and S (D) P and R

388. Which of the following combination of statements regarding gastrulation is true?
P. It is marked by coordinated mass cell movement
Q. It results in the formation of organizer
R. It results in the formation of three germ layers
S. It makes the end of neurulation
(A) P and R (B) P and Q
(C) Q and R (D) P, Q, R and S

389. Match the following

List I	List II
P. Microlecithal	1. Insects
Q. Mesolecithal	2. Tunicates
R. Macrolecithal	3. Lung fishes
S. Centrolecithal	4. Protherian mammals

(A) P-2, Q–4, R–3, S–1 (B) P-2, Q–3, R–4, S–1
(C) P-1, Q–2, R–4, S–1 (D) P-1, Q–4, R–2, S–3

390. Microsatellite are used more commonly than minisatelliite as DNA markers because:
(A) Minisatellites are present in too many locations within the genome
(B) Restriction enzymes can be used to type microsatellites but not minisatellite
(C) There are very few microsatellite in eukaryotic genome they are easily identified and analyzed
(D) Microsatellite are present throughout eukaryotic genomes and are easily ampli-fied using PCR

391. The mechanism of signal transduction by steroid hormones differs from amine and peptide hormones because:
(A) Steroids use small, water soluble second messengers
(B) They bind with specific receptor proteins on target – cell plasma membranes
(C) They bind to cytoplasmic or nuclear receptors and effect gene expression
(D) They are secreted from exocrine glands

392. Siderophores:
P. Make complex with Fe^{+++}
Q. Help in iron transport
R. Secrete by microbes
S. Contain heme prosthetic group
(A) P, Q and R (B) R and S
(C) P and R (D) P, Q, R and S

393. Skin cancer is induced by which type of DNA damage caused by exposure to harmful UV rays in sunlight:
(A) Alkylation (B) Deamination
(C) Depurination
(D) Pyrimidine diminer formation

394. What is the acrosomal reaction?
(A) The acrosomal reaction is a repulsive interaction between the sperm and the egg
(B) The acrosomal reaction is digestion of the acrosome by the sperm when it encounters an egg
(C) The acrosomal reaction is the digestion of the zona pellucida, permitting the sperm to contact the egg
(D) The acrosomal reaction is the fusion of the sperm and egg plasma membranes

395. Match list –I (cell type) with list – II (characteristics) and select the correct answer using the codes given below the lists:

List I	List II
P. Spermatogonium	1. Does not divide
Q. Spermatocyte	2. Divides mitotically
R. Spermatid	3. Divides meiotically
S. Sertoli cell	4. Divides both mitotically and meiotically
	5. Nourishes other cell types

(A) P-4, Q–3, R–1, S–2 (B) P-4, Q–3, R–2, S–1
(C) P-3, Q–4, R–2, S–1 (D) P-3, Q–4, R–1, S–2

396. The mode of action of a steroid hormone involves:
(A) Binding to a cell membrane receptor
(B) Activation of protein kinases
(C) Covalent modification of enzyme activity
(D) Modifying gene transcription

397. The folding of sheets of cells, the migration of cell death are all mechanisms of:
(A) Cleavage division (B) Pattern division
(C) Morphogenesis (D) Differentiation

398. Antigen binding to the B cell receptor
(A) Transduces a signal through the antigen biding chains
(B) Invariably leads to B-cell activation
(C) Transduces a signal through the Igα and Igβ molecules
(D) Results in macrophage activation

399. Which of the following is/are correct about plants?
P. Plants do not gastrulate
Q. Plants have specific meiosis rather than gamete meiosis
R. The life cycle of land plants includes both diploid and haploid multicellular stages
(A) P and Q (B) Q and S
(C) P and R (D) P, Q and R

400. HIV can kill CD4 T cells by:
(A) Expressing gp120 on T cell membranes to make them targets for ADCC
(B) Inducing apoptosis by binding to CD-4 without being presented on class II MHC
(C) Inducing their lysis by HIV-specific CTL
(D) All of these

401. The ion product of water is:
(A) Is independent of temperature
(B) Has a numerical value of 1×10^{-14} at 25°C
(C) Is the equilibrium constant for the reaction

$$HOH \longleftrightarrow H^+ + OH^-$$

(D) Requires that $[H^+]$ & $[OH^-]$ always be identical

402. cAMP activates cAMP-dependent protein kinase by:
(A) Stimulating its phosphorylation
(B) Stimulating the dimerization of kinase subunits
(C) Increasing the expression by stimulating the release of a translational inhibitory protein bunds to its mRNA
(D) Binding regulatory subunits and inducing their release from the catalytic subunits

403. If bicoid mRNA is injected into the posterior pole of a wild type Drosophila embryo, the expected phenotype is:
(A) Head structures at both the posterior and anterior poles
(B) Head structures at the posterior pole only
(C) Head structure at the anterior pole only
(D) No head structure at all

404. Which of the following is/are correct about apoptosis?
P. It is mediated by an intracellular proteolytic cascade
Q. It is mediated by family of proteases called caspases
R. Cytochrome plays role in apoptosis
S. Gene Bcl-2 inhibits apoptosis
(A) P and Q (B) Q and S
(C) P, Q and S (D) P, Q, R and S

405. Ontogeny is:
(A) Development of the ear
(B) The evolutionary history of a species or group
(C) The study of cancer
(D) The development of an individual organism

406. Tyrosine kinase receptors
P. Should be trimerized to be active
Q. Play important role in activating signal transduction asides
R. Have phosphate-removing active sites in their cytosolic tails
S. Can be oncogenes
(A) Q (B) P and Q
(C) Q and S (D) P, Q and R

407. The majority of oncogene proteins are:
(A) Metabolic enzymes
(B) Structural proteins, such as nuclear lamins
(C) Proteins involved in cell sorting
(D) Components of signaling pathways that regulate cell proliferation

408. Although multiple disulfide bonds are possible during the formation of the tertiary structure of some secretary proteins, only the correct ones are found in the secreted product. This is primarily due to the fact that:
(A) Incorrectly folded proteins are degraded by lysosomes
(B) Processing and folding is continued in the endosomes
(C) A protein facilitates the formation of correct disulfide bonds in the endoplasmic reticulum
(D) Only correctly folded proteins are translated in the endoplasmic reticulum

409. Which of the following statement(s) about class switching is/are correct?
P. Plasma cells that initially synthesize IgM switch to form IgG with the same antigen specificity
Q. Class switching does not affect the variable region of the H chains
(A) P (B) Q
(C) P and Q (D) None of these

410. Tumor-suppressor genes on their products an:
P. Act as dominant negative genes
Q. Act as recessive genes
R. Prevent cell-cycle progression
S. Regulate apoptosis
(A) P and Q (B) Q and S
(C) P, Q and R (D) P, Q, R and S

411. In mammals, dosage compensation is carried out by:
(A) Decreasing the transcription of the x in males
(B) Increasing the transcription of the x in males
(C) Inactivating one x chromosomes in fema-les by condensation as heterochromatin
(D) Inactivating one x chromosomes in males by condensation as heterochromatin

412. Which of the following maternal mRNAs encode proteins that directly influence early development in Drosophila?
P. Torso Q. Bioid
R. Knirps S. Nanos
(A) P and Q (B) Q and R
(C) P, Q and S (D) P, Q, R and S

413. Which of the following is a correct statement about NK cells?
(A) They proliferate in responses to antigen
(B) They kill target cells by phagoytosis and intracellular digestion
(C) They are a subset of polymorphonuclear cells
(D) They kill target cells in an axtracellular fashion

414. Choose the mismatch:
(A) Tropoblast - become chorion
(B) Trophoblast – forms embryonic structure
(C) Inner cell mass – become the amnion and embryo
(D) Chorion – forms part of placenta

415. Which of the following statement is incorrect conce-rning antigen-specific receptors on both B and T cells?
(A) They are locally distributed transmembrane molecules
(B) They have extensive cytoplasmic domains that interact with intracellular molecules
(C) They consist of polypeptides with variable and constant regions
(D) They are associated with signal transduction molecules at the cell surface

416. Choose the correct statement(s) about homeod-omain:
P. Proteins are a family of transcription factors
Q. Characteristics by a 60 amino acid domain
R. Encoded by a 180 bp DNA sequence called homeobox
S. Highly conserved
(A) P and Q
(B) Q, R and S
(C) P, Q, R and S
(D) None of these

417. Which of the following is least likely to lead to autoimmunity?
(A) Loss of suppressor T cells
(B) Release of sequestered self antigen
(C) Genetic predisposition
(D) Increased clearance of immune complexes

418. Choose the incorrect statement. transcriptionally inactive genes:
(A) May be located within heterochromatin
(B) May be located within nucleosomes
(C) Often are methylated
(D) Are sensitive to DNase 1

119. Which of the following maternal is/are correct pair(s)?

P. Radical cleavage - echinoderms

Q. Spiral cleavage - annelids

R. Bilateral cleavage - tunicates

S. Rotational cleavage - mammal

(A) P and S (B) Q, R and S

(C) Q and R (D) P, Q, R and S

420. Which statement about antigen epitopes is false?

(A) An epitope may be shared by two different antigens

(B) A protein molecule usually contains multiple epitopes

(C) B cells bind only processed antigen epitopes

(D) Epitopes may be linear or assembled

421. How are the eight human globin genes organized?

(A) They are randomly distributed on the human chromosomes

(B) They are present in two clusters, one on chromosome 11 and the other on chromosome 16

(C) They are in a single cluster on the x chromosome

(D) They are in four clusters, alpha, beta, gamma and delta

422. Which of the following statement is correct?

P. The plasma membrane is highly impermeable to all charged molecules

Q. All membrane transport proteins so far known traverse the lipid bilayer, and their polypeptide backbones generally extend back and forth across the membrane a number of times

R. Ionophores operated by shielding the charge of the transported ion so that it can penetrate the hydrophobic interior of the lipid bilayer

S. Since the Na^+ - K^+ ATPase exchanges equal numbers of Na^+ and K^+ ions in each pumping cycle, it is electrically neutral

(A) P and Q (B) Q and R

(C) R and S (D) P and S

423. Cell division in animal cells is made possible by a beltlike bundle of actin filaments and myosin-II filaments known as the:

(A) α-actin

(B) Tropomyosin

(C) Contractile ring

(D) Actin binding protein

424. The role of the antigen-presenting cell in the immune response is all of the following except:

(A) The initial catabolism of polypeptide antigens

(B) To allow selective associaton of MH gene products and peptides

(C) To supply second signals required to fully activate T cells

(D) To present non-self peptides associated with MHC class II molecules to B cells

425. Which of the following is not correct about secreted proteins:

(A) Are all present in one type of vescle

(B) Are often synthesized as a precursor and leaved before release

(C) May be continuously exocytosed

(D) May require an acidic compartment for maturation

426. Cytochalasins are drugs that interfere with actin polymerization into microfilaments. If you add cytochalasin to cultured mammalian cells that have just begun mitosis, what is most likely to happen?

(A) The cells will arrest at mitotic metaphase

(B) The cells will ease metabolism and die

(C) The cells will complete mitosis and arrest at cytokinesis

(D) The cell will arrest at mitotic anaphase

427. Initial density of a culture of bacteria with a generation time of 30 minutes was 1×10^5 cells/ml. After 5 hours of incubation, what serial dilution will you have to plate out to get ~100 colonies per ml?

(A) 10^3 (B) 10^4

(C) 10^5 (D) 10^6

428. Identify the correct set of three statement for cytoskeletal protein filaments form the following list:

A. Actin filament is about 8 nm wide

B. Actin filament is 25 nm wide

C. Intermediate filaments have size intermediate between atin filaments and microtubules

D. Protofilaments of microtubules are composed of alpha/beta tubulin heterodimer

E. Colchicine binds to the tubulin subunits in spindle microtubule causing disassembly to free units

(A) C, D, E (B) B, C, D

(C) A, C, D (D) A, B, C

429. DNA is present in:

(A) Nucleus (B) Mitochondria

(C) Chloroplast (D) All of these

430. Ribosomes are synthesized in:

(A) Nucleus (B) Nucleolus

(C) Cytoplasm (D) Chloroplast

431. Which of the following sequence is correct in meiotic prophase:

(A) Leptonema – Diplonema – Zygonema – Pachynema – Diakinesis

(B) Leptonema – Zygonema - Pahcynema - Diplonema – Diakinesis

(C) Pachynema – Zygonema - Leptonema – Diplonema – Diakinesis

(D) Diplonema – Pachynema – Zygonema – Diakinesis – Leptonema

432. Meiosis makes possible of:
(A) The interchange of material between homologous chromosomes
(B) The distribution of the member of each chromosome pair to each gamete
(C) Both (A) and (B)
(D) None of these

433. Which of the following sequence is correct in mitotic cell division:
(A) Prophase – Interphase – Metaphase – Anaphase – Telophase
(B) Interphase – Prophase – Anaphase – Metaphase –Telophase
(C) Metaphase – Anaphase – Prophase –Telophase – Interphase
(D) Interphase – Prophase –Metaphase – Anaphase – Telophase

434. DNA contains ribose sugar and RNA contains:
(A) Deoxyribose sugar (B) Ribose sugar
(C) Both (A) and (B) (D) None of these

435. The classical unit of inheritance is:
(A) Tetrad (B) Gene
(C) Polar body (D) Antibody

436. Antibiotics are product of:
(A) Bacteria and fungi
(B) Fungi and parasites
(C) Parasites and bacteria
(D) Fungi and bacteria

437. A totipotent cell means:
(A) An undifferentiated cell capable of developing into a system or entire plant
(B) An undifferentiated cell capable of developing into complete embryo
(C) An undifferentiated cell capable of developing into a organ
(D) Cell which lacks the capability differentiate into an organ or system

438. In terms of basic cell structure, what do an elephant and the oak tree have in common ?
(A) They both are eukaryotes
(B) They both have a cell nucleus
(C) They both have mitochondria
(D) All of these

439. The simplest way to differentiate a prokaryotic cell from a eukaryotic one is to:
(A) Look for a plasma membrane
(B) See if a nucleus is present
(C) Check for the presence of DNA
(D) Determine if the cell is an entire organism or not

440. In which of the following cell types more cellular detail can be observed using an electron microscope compared to a light microscope?
(A) Animal (B) Bacterial
(C) Protiest (D) All of these

441. Which of the following structures is expected in a bacterium?
(A) Nucleus
(B) Plasma membrane
(C) Golgi apparatus
(D) Endoplasmic reticulum

442. A certain cell organelle which is made of a double phospholipid bilayer that has many large pores in it, is most likely:
(A) The nuclear envelope
(B) The plasma membrane
(C) The mitochondrion
(D) The cytoskeleton

443. Which of the following is present in both prokaryotes and eukaryotes?
(A) Lysosome (B) Chloroplast
(C) Plasma membrane (D) Vacuole

444. In which of the following cell organelles, it is expected to find the biochemical reactions that harness energy from the breakdown of sugar molecules to synthesize large amounts of ATP?
(A) Lysosome (B) Vesicles
(C) Mitochondria (D) Plasma membrane

445. Microtubules, motor proteins and actin filaments are all part of:
(A) The mechanism of photosynthesis that occurs in chloroplasts
(B) The rough ER (endoplasmic reticulum) in prokaryotic cells
(C) The cytoskeleton of eukaryotic cells
(D) The process that moves small molecules across cell membrane

446. Clearly defined nucleus, mitochondria in a large central vacuole and chloroplasts, indicate the group of organisms from:
(A) Bacteria (B) Fungi
(C) Plants (D) Animals

447. Which of the following macromolecules are found in the plasma membrane?
(A) Lippids and proteins only
(B) Lipids, proteins and carbohydrates
(C) Proteins and carbohydrates onle
(D) Proteins only

448. Which of the following are responsible causing the diseases?
(A) T cells (B) Pathogens
(C) Macrophages (D) Lymphocytes

449. An organism's first line of defense against attack by an invader such as a virus or bacterium is usually:
(A) To flee or hide
(B) A specific immune response
(C) Its body wall
(D) A nonspecific immune response

450. Which of the following cell organelles are expected to be associated with motor proteins?

(A) Smooth ER (B) Vesicles
(C) Plasma membrane (D) Chloroplasts

451. What do dystrophin, utrophin, actin and tubulin have to do with eukaryotic cell structure and function?

(A) They are components in the reaction of photosynthesis
(B) They all participate in the degradation of large amounts of ATP
(C) They all participate in the production of large amounts of ATP
(D) They are all embedded proteins in plasma membrane

452. Where in a eukaryotic cell, DNA can be found?

(A) Nucleus (B) Mitochondria
(C) Vacuole (D) Both (A) and (B)

453. According to the fluid mosaic model of cell membranes:

(A) The most common type of molecules in the membrane are proteins
(B) Basic membrane structure results from how the proteins interact with water
(C) The membrane is a highly mobile mixture of phospholipids and proteins
(D) The unique properties of cell types are determined by their phospholipids

454. Which of the following is the best criterian for deciding whether a cell is prokaryotic or eukaryotic?

(A) The cell came from a single –celled or multicelled organism
(B) The cell has a nucleus or not
(C) The cell has cytosol or not
(D) DNA is present in the cell or not

455. Eukaryotic cells are more efficient than prokaryotes because their internal compartmentalization:

(A) Makes each compartment nutritionally independent of all others
(B) Allows for specialization through the subdivision of particular tasks
(C) Allows for specialization through merging of different tasks
(D) Reduces overall cell size

456. Which of the following organelles is directly connected to the outer membrane of the nucleus in a eukaryotic cell?

(A) Mitochondrion
(B) Lysosome
(C) Golgi apparatus
(D) Endoplasmic reticulum

457. Which of the following two organelles look most alike structurally?

(A) Nucleus and vesicle
(B) Golgi apparatus and smooth endoplasmic reticulum (ER)
(C) Vacuole and cytoskeleton
(D) Lysosome and chloroplast

458. The highly folded membrane found in such eukaryotic organelles as mitochondria and chloroplasts:

(A) Increase the surface area where key chemical processes can occur
(B) Help the cell against physical damage
(C) Make it possible to package large amounts of DNA within the cell
(D) Assist with cell movement

ANSWERSHEET

2. CELL BIOLOGY

1. (B)	2. (B)	3. (D)	4. (D)	5. (D)	6. (A)	7. (D)	8. (D)	9. (C)	10. (D)
11. (D)	12. (D)	13. (A)	14. (D)	15. (D)	16. (C)	17. (B)	18. (C)	19. (A)	20. (B)
21. (A)	22. (A)	23. (B)	24. (C)	25. (C)	26. (B)	27. (D)	28. (B)	29. (A)	30. (C)
31. (A)	32. (D)	33. (B)	34. (D)	35. (A)	36. (D)	37. (C)	38. (C)	39. (A)	40. (A)
41. (B)	42. (C)	43. (D)	44. (C)	45. (B)	46. (A)	47. (B)	48. (B)	49. (B)	50. (B)
51. (D)	52. (A)	53. (D)	54. (C)	55. (A)	56. (D)	57. (D)	58. (D)	59. (B)	60. (C)
61. (D)	62. (D)	63. (D)	64. (D)	65. (C)	66. (B)	67. (C)	68. (D)	69. (B)	70. (A)
71. (C)	72. (D)	73. (D)	74. (A)	75. (D)	76. (A)	77. (D)	78. (D)	79. (D)	80. (A)
81. (A)	82. (A)	83. (C)	84. (A)	85. (A)	86. (C)	87. (A)	88. (C)	89. (C)	90. (A)
91. (A)	92. (C)	93. (D)	94. (D)	95. (B)	96. (B)	97. (C)	98. (D)	99. (D)	100. (A)

101. (D)	102. (B)	103. (C)	104. (D)	105. (B)	106. (B)	107. (D)	108. (C)	109. (B)	110. (B)
111. (C)	112. (D)	113. (D)	114. (B)	115. (A)	116. (B)	117. (C)	118. (C)	119. (D)	120. (D)
121. (C)	122. (B)	123. (B)	124. (B)	125. (B)	126. (D)	127. (D)	128. (B)	129. (D)	130. (C)
131. (B)	132. (D)	133. (C)	134. (B)	135. (D)	136. (D)	137. (D)	138. (C)	139. (D)	140. (B)
141. (C)	142. (A)	143. (C)	144. (A)	145. (D)	146. (A)	147. (B)	148. (D)	149. (B)	150. (C)
151. (B)	152. (A)	153. (A)	154. (D)	155. (B)	156. (D)	157. (D)	158. (B)	159. (B)	160. (B)
161. (B)	162. (D)	163. (B)	164. (C)	165. (D)	166. (B)	167. (D)	168. (C)	169. (A)	170. (D)
171. (D)	172. (C)	173. (A)	174. (C)	175. (A)	176. (B)	177. (A)	178. (D)	179. (B)	180. (D)
181. (B)	182. (C)	183. (A)	184. (D)	185. (A)	186. (B)	187. (B)	188. (C)	189. (A)	190. (B)
191. (B)	192. (C)	193. (B)	194. (D)	195. (D)	196. (D)	197. (D)	198. (C)	199. (A)	200. (A)
201. (B)	202. (D)	203. (D)	204. (B)	205. (D)	206. (B)	207. (B)	208. (B)	209. (B)	210. (C)
211. (A)	212. (D)	213. (A)	214. (A)	215. (D)	216. (C)	217. (B)	218. (B)	219. (C)	220. (D)
221. (C)	222. (B)	223. (D)	224. (C)	225. (C)	226. (D)	227. (D)	228. (D)	229. (D)	230. (D)
231. (C)	232. (C)	233. (D)	234. (D)	235. (B)	236. (C)	237. (C)	238. (D)	239. (D)	240. (B)
241. (B)	242. (A)	243. (B)	244. (B)	245. (A)	246. (B)	247. (B)	248. (A)	249. (A)	250. (D)
251. (D)	252. (C)	253. (D)	254. (C)	255. (D)	256. (D)	257. (B)	258. (D)	259. (B)	260. (D)
261. (C)	262. (A)	263. (D)	264. (D)	265. (D)	266. (B)	267. (D)	268. (A)	269. (A)	270. (C)
271. (C)	272. (A)	273. (B)	274. (D)	275. (A)	276. (D)	277. (A)	278. (D)	279. (A)	280. (C)
281. (C)	282. (B)	283. (D)	284. (B)	285. (D)	286. (B)	287. (C)	288. (A)	289. (C)	290. (C)
291. (A)	292. (D)	293. (D)	294. (A)	295. (A)	296. (C)	297. (B)	298. (A)	299. (C)	300. (B)
301. (D)	302. (B)	303. (C)	304. (C)	305. (B)	306. (A)	307. (B)	308. (D)	309. (C)	310. (A)
311. (D)	312. (B)	313. (D)	314. (A)	315. (D)	316. (B)	317. (B)	318. (C)	319. (C)	320. (D)
321. (A)	322. (A)	323. (C)	324. (D)	325. (B)	326. (D)	327. (C)	328. (D)	329. (D)	330. (A)
331. (C)	332 (B)	333. (A)	334. (C)	335. (C)	336. (C)	337. (A)	338. (B)	339. (D)	340. (C)
341. (C)	342. (A)	343. (C)	344. (C)	345. (C)	346. (A)	347. (B)	348. (B)	349. (A)	350. (D)
351. (A)	352. (B)	353. (D)	354. (D)	355. (A)	356. (A)	357. (C)	358. (C)	359. (B)	360. (B)
361. (B)	362. (D)	363. (B)	364. (D)	365. (D)	366. (B)	367. (C)	368. (C)	369. (D)	370. (B)
371. (B)	372. (D)	373. (B)	374. (B)	375. (B)	376. (C)	377. (B)	378. (B)	379. (A)	380. (B)
381. (D)	382. (D)	383. (B)	384. (A)	385. (B)	386. (C)	387. (C)	388. (A)	389. (A)	390. (D)
391. (C)	392. (A)	393. (D)	394. (C)	395. (D)	396. (D)	397. (C)	398. (C)	399. (D)	400. (D)
401. (B)	402. (D)	403. (A)	404. (D)	405. (D)	406. (C)	407. (D)	408. (C)	409. (C)	410. (D)
411. (C)	412. (C)	413. (D)	414. (B)	415. (B)	416. (C)	417. (B)	418. (D)	419. (D)	420. (C)
421. (B)	422. (B)	423. (C)	424. (D)	425. (A)	426. (C)	427. (D)	428. (C)	429. (D)	430. (B)
431. (B)	432. (C)	433. (D)	434. (A)	435. (B)	436. (A)	437. (A)	438. (D)	439. (B)	440. (D)
441. (B)	442. (A)	443. (C)	444. (C)	445. (C)	446. (C)	447. (B)	448. (B)	449. (C)	450. (B)
451. (C)	452. (D)	453. (C)	454. (B)	455. (B)	456. (D)	457. (B)	458. (A)		

Chapter - 3

GENETICS AND CYTOGENETICS

1. The branch genetics started in
(A) 1857 (B) 1866
(C) 1900 (D) 1905

2. Sum total of characters which are transmitted from generation to generation constitute:
(A) Genetics (B) Phenocopy
(C) Heredity (D) Variation

3. Rediscovery of Mendel's work was made in:
(A) 1866 (B) 1900
(C) 1898 (D) None of these

4. The term Genetics was first used by:
(A) Johannsen (B) Punnett
(C) Bateson (D) Darlington

5. Genetics deals with:
(A) Genes (B) Heredity
(C) Variations (D) Heredity and Variations

6. The heredity variations are due to:
(A) Asexual reproduction
(B) Sexual reproduction
(C) Vegetative reproduction
(D) All of these

7. Concept of 'Pangenes' was given by:
(A) Wolff (B) Swammerdam
(C) Johannsen (D) Darwin

8. 'Genotype-phenotype concept' was given by:
(A) Johennsen (B) Charles Darwing
(C) Mendel (D) Bateson

9. When two genotype produce the same phenotype due to different environments one is called ----- of other :
(A) Phenotype (B) Allelomorph
(C) Phenocopy (D) None of these

10. Mendel worked on:
(A) Pigeonpea (B) Garden pea
(C) Sweet pea (D) Chickpea

11. Mendel originally belonged to:
(A) Sweden (B) Austria
(C) Australia (D) Czechoslovakia

12. Mendel published the results of his experiments in:
(A) 1857 (B) 1866
(C) 1868 (D) 1900

13. Mendel studies how many pairs of contrasting characters?
(A) 1 (B) 2
(C) 7 (D) 10

14. Which of the following characters was not studied by Mendel?
(A) Plant height (B) Position of flower
(C) Colour of pod (D) Length of pod

15. The term 'Allele' was proposed by:
(A) Johennsen (B) Bateson
(C) Mendel (D) Punnett

16. Emasculation is achieved by:
(A) Removal of stigma
(B) Removal of anthers
(C) Removal of sepals and petals
(D) All of these

17. Alleles are:
(A) Homologous chromosomes
(B) Linked genes
(C) Dominant genes
(D) Alternate forms of gene

18. If a plant heterozygous for tallness is selfed, the F_2 generation has both tall and dwarf plants. This proves the principle of:
(A) Dominance (B) Independent assortment
(C) Segregation (D) Incomplete dominance

19. In an experiment on pea, plants with yellow round seeds were crossed with plants with green wrinkled seeds; the phenotypic ratio of F_1 progeny is:
(A) 9 yellow round : 3 round green : 3 wrinkled yellow : 1 green wrinkled
(B) All wrinkled green
(C) All yellow round
(D) 1 Yellow round :1 Green round : 1 Yellow wrinkled : 1 Green wrinkled

20. Red (RR) flower plant is crossed with white (WW), all the offsprings are pink, this indicates that R gene is:
(A) Hybrid (B) Recessive
(C) Incomplete dominant (D) Mutant

21. Cross between a heterozygous individual and homozygous recessive parent is called:
(A) Monohybrid cross (B) Dihybrid cross
(C) Back cross (D) Test cross

22. An organism with two identical alleles of gene in a cell is:
(A) Dominant (B) Homozygous
(C) Hterozygous (D) Recessive

23. In F_2 progeny of dihybrid cross, the expected genotypic proportions of individuals homozygous for both dominant characters is:
(A) 3/16 (B) 9/16
(C) 12/16 (D) 1/16

24. Mendel performed his work on pea plants during:
(A) 17th century (B) 18th century
(C) 19th century (D) 20th century

25. Who is known as "father of Genetics"?
(A) C. Darwing (B) G. Mendel
(C) Hugo de Vries (D) Weismann

26. When dominant epistasis is operative between two gene loci, the classical 9 : 3 : 3 : 1 ratio becomes modifies into:
(A) 9 : 3 : 4 ratio (B) 9 : 6 : 4 ratio
(C) 12 : 3 : 1 ratio (D) 15 : 1 ratio

27. One complete set of chromosomes inherited as a unit from one parent is known as:
(A) Genotype (B) Genome
(C) Gene pool (D) Karyotype

28. The geometrical device that helps in visualizing all the possible combinations of male and female gametes is known as:
(A) Bateson square (B) Morgan square
(C) Punnett square (D) Mendel square

29. The condition in which only one allele of a pair is present is called:
(A) Homozygous (B) Hemizygous
(C) Heterozygous (D) Incomplete dominance

30. A pure tall pea plant was reared in soil poor in the nutrition and reached the size of a pure dwarf pea plant, if this plant is selfed, then the genotype in the F_1 gen. is most likely to be:
(A) All tall (B) 50% tall : 50% dwarf
(C) All dwarf (D) Not possible to predict

31. A cross between a plant heterozygous for two factors and a plant recessive for both the factors, gives a phenotypic ratio of:
(A) 9 : 1 : 1 : 7 (B) 9 : 3 : 3 : 1
(C) 1 : 1 : 1 : 1 (D) 1 : 7 : 7 : 1

32. Which is back cross in a monohybrid cross?
(A) 9 : 3 : 3 : 1 (B) 1 : 1
(C) : 2 : 1 (D) 3 : 1

33. Which term represents a pair of contrasting characters?
(A) Homozygous (B) Heterozygous
(C) Complementary genes (D) Allelomorphs

34. In monohybrid cross, the F_2 phenotypic ratio is:
(A) 1 : 2 (B) 2 : 1
(C) 3 : 1 (D) 1 : 3 : 1

35. The term heterosis represents:
(A) Hybrid incompatibility (B) Hybrid vigour
(C) Hybrid sterility (D) Structural hybridity

36. Genes which in homozygous condition result in non-viable progeny are:
(A) Duplicate genes (B) Linked genes
(C) Lethal genes (D) Epistatic genes

37. Which is called a 'pleiotropic gene'?
(A) Control characters of human beings only
(B) Control character of plants
(C) Control one character in association with other
(D) Control more than one character

38. The genotype of a dominant parent is determined by crossing it with a hybrid; this cross is called:
(A) Back cross (B) Test cross
(C) Long cross (D) Red cross

39. Allosomes are:
(A) Bead like structures in chromosomes
(B) Node like structures on chromosomes
(C) Sex –chromosomes
(D) None of the above

40. 'F_1' generation represents:
(A) First fertile generation
(B) First final generation
(C) First filial generation
(D) First federation generation

41. Independent assortment of genes occurs due to the orientation of chromosomes at:
(A) Metaphase of mitosis
(B) Metaphase I of meiosis
(C) Metaphase II of meiosis
(D) Any phase of the cell division

42. Incomplete dominance was first reported in 4 O' Clock plant by:
(A) Mendel (B) Morgan
(C) Batesen (D) Correns

43. Same genotypic and phenotypic ratios in F_2 generations are obtained in:
(A) Complete dominance
(B) Blending inheritance
(C) Codominance (D) None of these

44. In which, incomplete dominance is seen?
(A) Wheat (B) Maize
(C) Mouse (D) Snapdragons

45. Lethal factors were reported in plants by:
(A) Cuenot (B) Baur
(C) Batesen (D) Morgan

46. If 2n = 20 in a plant, then how many linkage groups are present in this?
(A) 20 (B) 10
(C) 5 (D) 5 or 10

47. The most common cause of the pleiotropic effect of a gene is due to:
(A) The same product of the given gene being involved in different metabolic pathways
(B) The gene making very different products in different cell types
(C) The DNA sequence of the gene getting changed in cell specific manner
(D) The gene not functioning in some cells

48. Gene interaction in which either of dominant gene is enough to give rise to same phenotypic effect is:
(A) Complementary gene (B) Duplicate gene
(C) Epistatic gene (D) Supplementary gene

49. When two pairs of genes interact to give 9 : 7 F_2 ratio, they are:
(A) Supplementary factors
(B) Complementary factors
(C) Epistatic factors (D) Duplicate factors

50. 12 : 3 : 1, F_2 ratio is obtained in interaction of ...... Factors:
(A) Complementary (B) Supplementary
(C) Epistatic (D) Inhibitory

51. Recombination percentage in a diploid cannot exceed:
(A) 100 (B) 75
(C) 50 (D) 25

52. Exception to Mendel's law of independent assortment was first discovered by:
(A) Morgan (B) Sturtevant
(C) Bateson and Punnett (D) None of these

53. Phenomenon of linkage was first discovered in:
(A) Sweet pea (B) Garden pea
(C) Mirabilis (D) Snapdragon

54. Two linked genes a and b show 20% recombination. The individuals of a dihybrid cross between ++/++ x ab/ab shall show gametes:
(A) ++ 80 : ab : 20 (B) ++ 50 : ab : 50
(C) ++ 40 : ab 40 : + a10 : + b : 10
(D) ++ 30 : ab 30 : + a 20 : + b : 20

55. Coupling and repulsion phenomenon are concerned with:
(A) Crossing over (B) Linkage
(C) Mutation (D) None of these

56. Theory of linear arrangement of genes on chromosomes was formulated by:
(A) Sturtevant (B) Morgan
(C) Bateson (D) Mendel

57. When blue flower long pollen plant (BBLL) was crossed with red flowered round pollen plant (bbll) in Lathyrus and their F_1 was crossed with red round parent, then what is the ratio of blue long and red round obtained in this cross?
(A) 1/16 : 1/16 (B) 7/16 : 7/16
(C) 7/16 : 1/6 (D) 9/16 : 3/16

58. In a diploid organism, the genes A, B and C are present on the same chromosome in that order. The AB interval; is 10 map units and B interval is 20 map units. In an ABC/abc heterozygous individual, what will be the proportion of gametes that carry the genotype ABC?
(A) 1% (B) 10%
(C) 20% (D) 30%

59. Somatic crossing over was discovered in Drosophila by:
(A) Pontecorvo (B) Morgan
(C) Stern (D) Serebrousky

60. G. Pontecorvo observed non-germinal cover in:
(A) Maize (B) Drosophila
(C) Aspergillus (D) None of these

61. Most accepted theory of crossing over is:
(A) Copy choice theory (B) Precosity theory
(C) Contact first theory
(D) Breakage first theory

62. Phenotypic expression of the heterozygote being greater than that of either homozygote is called:
(A) Dominance (B) Midparent value
(C) Overdominance (D) Pseudodominance

63. Pseudoautosomal gene is present on:
(A) Both X-and Y- chromosomes
(B) Both X- and autosomes
(C) Both Y- and autosomes
(D) Only autosomes

64. Exchange of chromosomal segments between non-homologous chromosomes is:
(A) Crossing over (B) Inversion
(C) Translocation (D) None of these

65. Which of the following most accurately explains the cause for the abnormal numbers of chromosomes during human reproduction that an result in Down syndrome, Turner's syndrome, or Klinefelter syndrome?
(A) The occurrence of non-disjunction of homologous chromosomes during meiosis
(B) The duplicative production of extra chromosomes during DNA replication
(C) The abnormal pairing of non-homologous chromosomes during prophase of miosis I
(D) The selective loss of particular chromosomes from the sex cells after formation of the mature gamete

66. The process which involves reverse order of genes in chromosome is:
(A) Deficiency (B) Aneuploidy
(C) Inversion (D) Translocation

67. 'x' represents:
(A) Gametic number
(B) Basic chromosome number
(C) Haploid number (D) All of these

68. Presence of chromosome number, which is different from the multiple of basic chromosome number is called?
(A) Hyperploidy (B) Heteroploidy
(C) Aneuploidy (D) Euploidy

69. 2n – 2 is a:
(A) Monosomic (B) Nullisomic
(C) Tetrasomic (D) Trisomic

70. If haploid chromosome no. in an organism is 12, what is the chromosome no. present in its trisomic?
(A) 11 (B) 13
(C) 25 (D) 23

71. In a three-factor cross among genetic markers A, B and C, the relative frequencies of recombinants were AB (1), AC (0.2) and BC (0.3). Which of the following is the most probable order of these markers on the genome?
(A) ABC (B) BCA (C) CAB
(D) The data do not reflect the order of genes

72. Reduction in size of vegetative and floral parts is found in:
(A) Autopolyploids (B) Allopolyploids
(C) Haploids (D) Diploids

73. 'Gigantism' is mostly found in:
(A) Autopolyploids (B) Allopolyploids
(C) Haploids (D) None of these

74. Polyploidy can be artificially induced by:
(A) 0.1 M NaCl solution (B) Zeatin
(C) 0.25 M sucrose solution (D) Colchicine

75. In Drosophila Y-chromosome is:
(A) Essential for male sex determination
(B) Required for development of secondary sex-characters
(C) Required for dosage compensation
(D) Not required for male sex determination

76. Colchicine was first obtained from Colchicines autumnale by:
(A) Dustin (B) Houde
(C) E.R. Sears (D) Blakeslee

77. Colchicine is a:
(A) Steroid (B) Alkaloid
(C) Cholesterol (D) None of these

78. Natural autopolyploidy is found in:
(A) Cynodon (B) Grapes
(C) Wtermelon (D) None of these

79. One of the following statements is most appropriate about the Y chromosome in mammals:
(A) Essential for the development of embryo
(B) Non-essential for the development of the embryo but important for male sex determination
(C) Non-essential for sex determination by important for male fertility
(D) Essential for male sex determination and fertility

80. Complete linkage is observed in:
(A) ♂ Drosophila (B) ♀ Drosophila
(C) Birds (D) Maize

81. Percentage of crossing over will be more if:
(A) Linked genes are located far away from each other
(B) Linked genes are located close to each other
(C) Both (a) and (b)
(D) None of the above

82. What are the sexual phenotypes of the following genotypes in Drosophilla : XXY, XO?
(A) Male, female (B) Male, male
(C) Intersex, female (D) Female, male

83. Who is associated with Raphanobrassica?
(A) Muntzing (B) E.R. Sears
(C) Karpechenko (D) None of these

84. 'Triticale' is derived from cross between?
(A) Wheat and aegilops (B) Wheat and rice
(C) Wheat and maize (D) Wheat and rye

85. Why is the gene for pattern baldness in man exhibits dominance in man but acts recessively in women? Because:
(A) Baldness gene expression in regulated by hormones
(B) Baldness gene is autosomal
(C) Baldness gene in on X-chromosome
(D) Baldness gene in on Y-chromosome

86. In common hexaploid bread wheat, the three diploid genomes are:
(A) A, B and C (B) A, C and D
(C) B, C and D (D) A, B and D

87. The three genomes of common hexaploid wheat show:
(A) Homologous relationship
(B) Non-homologous relationship
(C) Homoeologous relationship
(D) None of the above

88. Each of the following chromosome abnormalities involves a 20 megabase region of the long arm of chromosome 5 (5q). Which abnormality is most likely to cause severe disease?
(A) Deletion of the region
(B) Duplication of the region
(C) A balanced translocation involving the region (i.e., in the translocation carrier)
(D) Pericentric inversion

89. Mutation was first discovered in:
(A) *Drosophila* (B) *Neurospora*
(C) *E. coli* (D) *Oenothera*

90. The term 'Sport' was given by:
(A) Hugo de Vries (B) Darwin
(C) Lamarck (D) None of these

91. The change due to structural changes in chromosomes:
(A) Can be reversed (B) Cannot be reversed
(C) Is not known (D) None of these

92. Which of the following will not cause triploidy:
(A) Fusion of an egg and polar body with subsequent fertilization by sperm cell
(B) Meiotic failure, producing diploid sperm or egg
(C) Dispermy
(D) Mitosic failure in the early embryo

93. Which of the following is called Drosophila of plant kingdom?
(A) *E. coli* (B) *Acetabularia*
(C) *Neurospora* (D) *Saccharomyces*

94. In mice, a mutation results in the production of yellow oat colour, as opposed to the normal gray oat. Crossing yellow mice with normal mice results in equal proportions of normal and yellow mice, whereas crossing yellow mice amongst themselves results in a ratio of 2 yellow : 1 normal progeny. The results suggest that:
(A) More than one gene is involved in conferring the yellow phenotype
(B) The yellow allele is lethal under homozygous condition
(C) The yellow allele exhibits cytoplasmic heredity
(D) The results cannot be explained by any genetic principles

95. A mutation in sheep called 'Ancon sheep' was recorded by:
(A) Hugo de Vries (B) Darwin
(C) Seth Wright (D) None of these

96. Mutagen is:
(A) Organism in which mutation has occurred
(B) Mutation inducing agent
(C) Both (a) and (B)
(D) None of the above

97. Frequency of mutation:
(A) Varies with character and organism
(B) Can be increased by X-rays
(C) Is greatly affected by environmental factors
(D) All of these

98. A cross between unlike organisms is called:
(A) Test cross (B) Heterosis
(C) Back cross (D) Hybrid

99. If a man who is colour-blind marries a woman who is homozygous for normal colour vision, the probability of their son being colour-blind is:
(A) 0 (B) 1
(C) 0.5 (D) 0.75

100. In nature, the rate of mutation is?
(A) 1×10^5 (B) 1×10^8
(C) 1×10^{-5} (D) 1×10^{12}

101. Mutations are
(A) Mostly useful (B) Always useful
(C) Rarely useful (D) Always harmful

102. Who used X-rays to cause mutation?
(A) Morgan (B) Muller
(C) Alterberg (D) Auerbach

103. Consider two linked loci, labeled A and B. Each locus has two alleles, labeled 1 and 2. The frequencies of the alleles in a population are: A1 = 0.6, A2 = 0.4, B1 = 0.7, B2 = 0.3. If there is linkage equilibrium between these two loci in the population, what is the expected frequency of chromosomes carrying a combination of the A1 and B2 alleles?
(A) 0.24 (B) 0.12
(C) 0.18 (D) 1.00

104. Changes due to mutation are:
(A) Small (B) Negligible
(C) Large (D) All of these

105. Pseudodominance is:
(A) There is incomplete dominance
(B) Dominance of recessive alleles in a single generation
(C) Deficiency in which a recessive allele will behave like a dominant allele
(D) There is no such phenomenon

106. Gene sanctuary refers to:
(A) An area of diversity protected against interference from man
(B) Gene collection
(C) Gene evoalution (D) Gene mutation

107. In a test cross of an individual heterozygous for each of three linked genes, the most frequent lasses of progeny were ABc/abc and abC/abc, and the least frequent lasses were ABC/abc and abc/ abc. What is the order of the genes:
(A) A - C - B (B) B – A - C
(C) A – B - C (D) None of these

108. If an organism is having genotype AaBb, then what are the possible combinations of gametes after meiosis?
(A) AB, ab (B) Aa, Bb
(C) AB, aB, Ab, ab (D) Not possible to tell

109. Suppose DNA has base sequence as UAC, GAC, AGC, CGC, ACA, AAA and due to mutation only first base is deleted, what is the effect on coding?
(A) First amino acid will be different
(B) There will be one amino acid less
(C) Complete change in type and sequence of amino acids
(D) No change

110. Which of the following is non-ionizing radiation?
(A) X- rays (B) β- rays
(C) γ- rays (D) UV-rays

111. Why are some autosomal dominant disorders (e.g., Marfan syndrome) seen more commonly in the offspring of older fathers?
(A) Replication errors accumulate as sperm-producing stem cells continue to divide
(B) Rate of nondisjunction increases in older males
(C) Recombination rates increase in older males
(D) None of these

112. Certain types of cancers can be correlated with specific changes in chromosome structure. In patients suffering from myelogenous leukemia, the abnormal chromosome detected was termed Philadelphia chromosome. Which of the following chromosomes is altered in this disease?
(A) Chromosome 10 (B) Chromosome 11
(C) Chromosome 20 (D) Chromosome 22

113. UV-rays as mutagenic agents were shown by:
(A) Auerbach (B) Stadler
(C) Altenberg (D) Muller

114. Chemicals as mutagenic agents were first shown by:
(A) Altenberg (B) Auerbach
(C) Stadler (D) None of these

115. People who carry an allele for normal hemoglobin and an allele for sickle cell hemoglobin are resistant to malaria. This phenomenon is called:
(A) Heterozygote advantage
(B) Polyploidy
(C) Mutation selection balance
(D) Recessive superiority

116. Hypophosphatemia is caused by an X-linked dominant gene in humans. If a hypophosphatemia man marries a normal woman, which of their children will have hypophosphatemia?
(A) All daughters
(B) Half of their daughters
(C) All of the sons
(D) Half of their sons

117. Which of the following statements are correct:
P. Down's syndrome/mongolism is due to extra 21st chromosome
Q. Down's syndrome is due to non-disjunction of chromosomes
R. Trisomy has chromosome complement of 2n+1
S. Monosomics are 2n - 1
(A) P, Q (B) R, S
(C) P, Q, R (D) P, Q, R, S

118. Base substitution caused by 5-bromo uracil and BudR are called:
(A) Inversions (B) Transitions
(C) Transversions (D) Translocations

119. Proflavin and Acridine orange induce:
(A) Transitions (B) Transversions
(C) Frameshift mutations (C) Inversions

120. Which of the following statements are correct:
P. Cri-du-chat syndrome is due to chromosomal change involving deletion
Q. Formation of multivalents in meiosis is due to reciprocal translocation
R. Klinefelter's syndrome has 44 + XXY
S. In Down's syndrome of a male child, the sex complement is XY
(A) P, Q (B) R, S
(C) P, Q, R (D) P, Q, R, S

121. Which is not associate with mutations?
(A) Better progenies
(B) Death of organisms
(C) Change in genetic constitution
(D) Hybrid vigour

122. 'Mutations which are recessive are not lost by a gene pool' is called:
(A) Survival of dominance
(B) Survival of recessive
(C) Hardy Weinberg law
(D) None of these

123. Cytoplasmic inheritance is responsible for:
(A) Inheritance of kappa particles
(B) Iojap inheritance in corn
(C) Petite character in yeast
(D) All of these

124. A decentric chromosome is unstable because:
(A) It cannot resynthesize its telomeres during replication
(B) It pairs with non-homologous chromosomes in meiosis
(C) It pairs with non-homologous chromosomes in mitosis
(D) It is often simultaneously drawn to opposing spindle poles in mitosis

125. Who is credited to have differentiated between heritable variation and environmental variation?
(A) Darwin (B) Morgan
(C) Hugo de Vries (D) Bateson

126. The resistance developed by some insects against insecticides is due to:
(A) Hybrid vigour
(B) Genetic mutations
(C) Variations due to sexual reproduction
(D) Genetic recombinations

127. Who coined the term Heterosis?
(A) Mendel (B) G.H. Shull
(C) Schmidt (D) Schleiden

128. 'Gene Battery Model' was given by:
(A) Jacob and Monod
(B) McClinktock and Creigheten
(C) Bawden and Pierre
(D) Bridden and Davidson

129. The force that can change the frequency of an allele in a population are:
(A) Forward mutation, gene conversion, neutral evolution and recombination
(B) Selection, mutation, migration, inbreeding and random genetic drift
(C) Dominance, family selection, fitness and diversification
(D) Gene interaction, gene transfer, gene mutation and out breeding

130. Which of the following crosses would result in 1 : 2 : 1 F_2 ratio?
(A) Yg x gg (B) Yg x yg
(C) Yg x yy (D) YY x gg

131. The first 'hybrid' was produced by:
(A) Camerarius (B) Fairchild
(C) Lamarck (D) N. Grew

132. Sex determination was first of all studied in which plant?
(A) Coccinia (B) Melandrium
(C) Rumex (D) Humulus

133. Which of the following shows slowest cell division?
(A) Haploid (B) Diploid
(C) Polyploid (D) Same in all

134. 2n = x = 7, represents a:
(A) Haploid (B) Diploid
(C) Monoploid (D) Polyploid

135. In wheat 2n=6x=42. No. of possible monosomics is:
(A) 42 (B) 21
(C) 7 (D) 6

136. Translocation was first reported in:
(A) Datura (B) Maize
(C) Oenothera (D) Doob grass

137. In snapdragon, when yellow leaved var. (Aurea) x Aurea, what is the ratio of yellow: green in progeny?
(A) 3 : 1 (B) 9 : 3
(C) 1 : 1 (D) 2 : 1

138. Mutations are mostly:
(A) Dominant (B) Incompletely dominant
(C) Codominant (D) Recesive

139. Suppose DNA has base sequence AAG GAG GAC CAA CCA, which one represents frameshift mutations?
(A) AAG GAG GAC CAA CCA
(B) AAG AAG GAC CAA CCA
(C) AAG GAG ACC AAC CAA
(D) AAG GAG GAC CAA CCA

140. No. of linkage groups is equal to:
(A) Diploid no. of chromosomes
(B) Haploid no. of chromosomes
(C) Tetraploid no. of chromosomes
(D) None of the above

141. Who discovered Mendel's principle in 1900 A.D.?
(A) Avery, McLeod, McCarty
(B) Sutton, Morgan, Bridges
(C) Bateson, Punnett and Bridges
(D) de Vries, Correns and Tschermak

142. Heterosis is also called:
(A) Hybrid vigour (B) Hybrid sterility
(C) Mutation (D) Variation

143. Why are haploids superior to diploids in study of mutations?
(A) They have shorter life time
(B) Smaller no. of chromosomes
(C) They allow expression of recessive mutation immediately
(D) Obtained in large number

144. The individuals having same maternal and paternal parents are called?
(A) Epistasis (B) Hypostasis
(C) Siblings (D) Gynander

145. When a pyrimidine is replaced by purine, it is called:
(A) Frameshift mutation (B) Transition
(C) Transversion (D) Reversible mutation

146. Picking up plants with superior phenotype for further propagation is:
(A) Mass selection (B) Pure line selection
(C) Hybrid vigour (D) Introduction

147. Picking up only homozygous plants for further cultivation is known as:
(A) Mass selection (B) Pure line selection
(C) Clone (D) Aclimitization

148. Homozygosity is obtained after inbreeding for:
(A) Two generations (B) Four generations
(C) Eight generations (D) Ten generations

149. PV-18, Kalyan sona and S-227 wheat varieties have developed through:
(A) Gamma ray irradiation (B) Multiple crossing
(C) Selection (D) Genetic engineering

150. Importing better varieties and plants from outside and acclimitising them to local environment is:
(A) Selection (B) Cloning
(C) Introduction (D) Heterosis

151. Heterosis is:
(A) Appearance of spontaneous mutations
(B) Induction of mutations
(C) Mixture of two or more traits
(D) Superiority of hybrids over their parents

152. Cutting of anthers from the intersexual flowers of female parents is:
(A) Emasculation
(B) Male sterile line
(C) Artificial pollination
(D) Selectivity or differential reproduction

153. Male sterile is due to:
(A) Cytoplasmic inheritance
(B) A recessive nuclear gene
(C) Interaction of nuclear gene & cytoplasmic factor
(D) All the above

154. Heterosis cannot be maintained in sexually reproducing plants as it disappears on:
(A) Outbreeding (B) Inbreeding
(C) Cross breeding (D) None of the above

155. Polyploidy gives:
(A) Hybrid vigour (B) Imrpoved varieties
(C) Gigas effect (D) All the above

156. Polyploidy involves:
(A) Increases in number of chromosomes
(B) Increases in number of genomes
(C) Increase in number of plasmons
(D) Deletion of part or whole of a chromosomes

157. Polypoidy is induced through:
(A) Colchicine (B) Irradiation
(C) Mutagenic chemicals (D) Ethylene

158. Best varieties of Sugar Beet are:
(A) Tetraploids (B) Diploids
(C) Triploids (D) Allopolyploids

159. Most varieties of sugarcane are:
(A) Tetraploids (B) Triploids
(C) Hexaploids (D) Aneuploids

160. Main varieties of potato are:
(A) Tetraploids (B) Triploids
(C) Aneuploids (D) Diploids

161. Bread Wheat is:
(A) Autopolyploids (B) Allopolyploids
(C) Aneuploids (D) Trisomic

162. Seedless Water Melon is:
(A) Hexaploid (B) Tetraploid
(C) Triploid (D) Pentaploid

163. The first domesticated Wheat was:
(A) *Triticum boeoticum* (B) *T. dicocum*
(C) *T. dicoccoies* (D) *T. monococcum*

164. Induced mutation have produced a rice variety:
(A) Pelita-1 (B) Reimei
(C) Pelita-2 (D) All the above

165. Explant is:
(A) Plant collected after harvesting
(B) Exploited part of a plant
(C) Small part of the plant meant for tissue culture
(D) Uprooted for transplantation

166. The growth kinetics that results from metabolizing one sugar before another is referred to as:
(A) Exponential growth
(B) Diphasic growth
(C) Diauxic growth (D) Chemotaxis

167. A technique of micropropagation is:
(A) Multiple shoot production
(B) Multiple shoot production and somatic embryogenesis
(C) Growth of micro-organisms on culture medium
(D) Somatic embryogenesis

168. Pathogen free plants can be obtained through:
(A) Clean cultivation
(B) Embryoid cultivation
(C) Bud or shoot up propagation on culture medium
(D) Grafting

169. Embryoid is:
(A) Nonzygotic embryo
(B) Nonfunctional embryo
(C) Parthenogenetic embryo
(D) An early stage in callus differentiation

170. Androgenic haploid culture was first performed by:
(A) Guha and Maheswari
(B) Steward
(C) Skoog and Miller
(D) Halperrin and Wetherell

171. Name the tissue culture technique used to grow nonviable inter-specific hybrids:
(A) Androgenic haploids
(B) Shoot tip culture
(C) Somatic embruogenesis
(D) Embryo rescue

172. Variations appearing in tissue culture are:
(A) Culture variations
(B) Auxotrophs
(C) Somaclonal variations
(D) Pangenetic variations

173. Pollution tolerant variants can be produced through:
(A) Large scale growth in polluted areas
(B) Tissue culture
(C) Protoplast technique
(D) Genetic engineering

174. Somatic hybridization is achieved through?
(A) Grafting
(B) Protoplast fusion
(C) Conjugation
(D) Recombinant DNA technology

175. Which one is required for protoplast fusion?
(A) Treatment with cellulose and pectinase
(B) Electrofusion or PEG treatment
(C) Both (A) and (B)
(D) Recombinant DNA technology

176. Restriction endonucleases are useful in:
(A) Breaking DNA at particular places
(B) Producing sticky DNA ends
(C) Both (A) and (B)
(D) Crossing over

177. The number of known species in genus Gossypium have been estimated at:
(A) 10 (B) 15
(C) 20 (D) 25

178. Colchicine brings about:
(A) Gene mutations
(B) Chromosome aberrations
(C) Quick replication
(D) Duplication of chromosomes

179. The principle male sterile line introduced for hybrid pearlmillet production in India was:
(A) Tift-23 A (B) BJ-104
(C) MHB -118 (D) CO-2

180. Mutations are generally induced by means of:
(A) Bet rays (B) Alpha rays
(C) Gamma rays (D) UV rays

181. In crop improvement programme, haploids are important because they:
(A) Require one half of nutrients
(B) Are helpful in study of meiosis
(C) Grow better under adverse conditions
(D) Form perfect homozygous individuals on diploidisation

182. What would be the effect of decreasing the number of tryptophane codons from 2 to 0 in the sequence of the tryptophan operon that codes for the leader peptide?
(A) The production of full – length m RNA would no longer be dependent upon the trp concentration
(B) When trp levels were low, higher amounts of full-length mRNA would be produced
(C) When trp levels were low, lower amounts of full-length mRNA would be produced
(D) Both (A) and (B)

183. Restriction endonucleases are used as:
(A) Molecular scalpels for cutting DNA at specific sites
(B) Molecular cemet for combining DNA segments
(C) Molecular primer for building nucleotides in tendem
(D) Molecular degradation to DNA break up

184. Polyploidy is induced by:
(A) Kinetin (B) GA_3
(C) Colchicine (D) IAA

185. A plant with multiple uses (Kalpabraksha) is:
(A) Cotton (B) Coconut
(C) Jute (D) Sunflower

186. Which one is hexaploid wheat?
(A) *Triticum durum* (B) *T. monococcum*
(C) *T. tugidum* (D) *T. aestivum*

187. Heredity variations can be got with the help of:
(A) X- rays (B) DDT
(C) Auxin (D) Gibberellin

188. What is the role of the I gene of phage lambda?
(A) It encodes a protein that enables the phage to enter lysis
(B) Its product helps the prophage to excise from the host DNA
(C) It encodes the lambda repressor which represses the lytic cycle
(D) Its gene product helps the prophage to integrate into the host DNA

189. A clone is:
(A) Homozygote produced by sexual reproduction
(B) Heterozygote produced by sexual method
(C) Homozygote obtained asexually
(D) Heterozygote obtained asexually

190. Inbreeding is possible between two members of:
(A) Order (B) Family
(C) Genus (D) Species

191. Improved wheat variety having genes for dwarfness and higher percentage of protein/lysine is:
(A) Kalayan (B) Sharbati Sonora
(C) Sonalika (D) Lerma Rojo

192. Which of the following statements are true with respect to a special complex called 'dicot'?
P. It consists of deoxyribonuclease and DNA fragments
Q. It consists of ribonuclease and RNA fragments
R. It involved in gene silencing
S. It triggers apoptosis
(A) P, R (B) Q, R
(C) P, S (D) Q, S

193. Majority of improved Indian rice varieties have been arisen from a cross between:
(A) *Oryza sativa indica X Oryza nivara*
(B) *Oryza sativa indica X Oryza sativa japonica*
(C) *Oryza sativa japonica X Oryza nivara*
(D) *Oryza rufipogon X Oryza nivara*

194. Hybrid vigour is due to:
(A) Crossing over
(B) Linkage
(C) Heterozygosity
(D) Chiasma

195. Crop improvement is possible through:
(A) Judicious combination of selection, introduction and hybridization
(B) Selection
(C) Scientific improvement of cultivated plants
(D) Introduction

196. Which of the following is not directly related to gene regulation?
(A) Acetylation of histones
(B) Spicing
(C) Glycosylation in the endoplasmic reticulum
(D) Silencer

197. A breeder involving disease resistant variety will start with:
(A) Working out yield of different varieties
(B) Go through the subject in library
(C) Selection of parents
(D) Hybridisation

198. Emasculation of flowers is carried out through removal of:
(A) Stigma (B) Sepals and petals
(C) Anthers (D) Entire organs

199. Regulations governing movement of diseased plant material are called:
(A) Crop protection (B) Quarantine
(C) Plant regulation (D) Rotation

200. A tool in crop improvement involving identification of genes, transfer and integration is:
(A) Protoplast fusion and tissue culture
(B) Somaclonal hybridization
(C) Gene bank technology
(D) Genetic engineering

201. Aneuploidy is:
(A) Zygote having multiple of two gametic number
(B) Zygote having four times the number of gametic chromosomes
(C) Zygote having abnormal number of chromosomes
(D) Zygote having three times the number of gametic chromosomes

202. Intact duplex DNA is a substrate for:
(A) DNA pol I (B) DNA pol III
(C) RNA polymerase (D) DNA pol IV

203. Haploid plants are preferred over diploids for mutation study because in haploids
(A) Recessive mutations express immediately
(B) Induction of mutations is easier
(C) Culturing is easier
(D) Dominant mutations express immediately

204. Axenic culture is
(A) Culture without contamination
(B) Culture without a nutrient
(C) Culture of a gene
(D) Culture of a tissue

205. Plants can be made disease resistant through:
(A) Colchicine treatment
(B) X-ray treatment
(C) Breeding with wild relatives
(D) Hormone treatment

206. Which of the following is not directly associated with regulation of eukaryotic gene expression?
(A) Acetylation of histones
(B) Methylation of DNA
(C) Alternative splicing
(D) Activation of caspases

207. Most of our crop plants are:
(A) Autopolyploid in origin
(B) Allopolyploid in origin
(C) Mixed genotypic in origin
(D) Heterozygous in origin

208. In Rho-independent transcription termination, stem and loop structure is formed in:
(A) The transcript (B) The template DNA
(C) Coding strand (D) Both (A) and (B)

209. *Triticum aestivum* is:
(A) Hexaploid (B) Alloautopolyploid
(C) Heteropolyploid (D) Autopolyploid

210. Bombay green banana has developed from:
(A) Mass selection (B) Pure line selection
(C) Clonal selection (D) Hybridisation

211. Removal of anthers of some flowers during plant breeding is:
(A) Anthesis (B) Emasculation
(C) Pollination (D) For collection of pollen

212. Hybrid potato and tomato is:
(A) Tomapa (B) Pomato
(C) Potamato (D) None of the above

213. Androgenic haploids are used for
(A) Raising stock for micropropagation
(B) Induction of mutations
(C) Knowing effect of all genes
(D) Preparation of homozygous diploids

214. The 'A' fragment of diphtheria toxin catalyzes the addition of an ADP-ribose group to the eukaryotic elongation factor EF-2. The source for the ADP-ribose group is:
(A) FAD (B) FMN
(C) NAD^+ (D) $NADP^{++}$

215. Crosses involving plants of the same variety are:
(A) Intravarietal (B) Interspecific
(C) Intervarietal (D) Inta-generic

216. An old breeding technique is
(A) Introduction (B) Selection
(C) Mutation breeding (D) Hybridisation

217. There are important classes of pattern-regulating genes specify the basic features and functions during embryonic development are:
(A) Maternal effect genes (B) Segmentation genes
(C) Homeotic genes (D) All of these

218. Aim of plant breeding:
(A) Control pollution (B) Keep soil fertile
(C) Produce improved varieties
(D) To maintain wild plants

219. Technique mainly used in agriculture biotechnology is:
(A) Plant breeding (B) Transformation
(C) Dual replication (D) Tissue culture

220. Important use of agriculture biotechnology is to:
(A) Produce pest resistance
(B) Increase nitrogen content
(C) Increase plant weight
(D) Decrease seed number

221. Pure line breeds refer to:
(A) Homozygosity & independent assortment
(B) Homozygosity only
(C) Heteozygosity only
(D) Heterozygosity and linkage

222. The nucleotide sequence of DNA involved in binding to a transcription factor can be determined by:
(A) DNA footprinting
(B) S1 nuclease treatment
(C) DNA fingerprinting
(D) Northern hybridization

223. miRNA based silencing of genes is a type of:
(A) Transcription gene silencing
(B) Post transcription gene silencing
(C) Translation gene silencing
(D) Post translation gene silencing

224. The *Triticum aestivum* is a wheat:
(A) Diploid (B) Tetraploid
(C) Hexaploid (D) Monoploid

225. Gossypium hirsutum is:
(A) New World tetraploid
(B) Old World tetraploid
(C) New World diploid
(D) Old World diploid

226. How many maternal effect genes are critical to the formation of the anterior-posterior axis?
(A) 3 (B) 4
(C) 5 (D) 6

227. New varieties of plants are produced by:
(A) Selection and hybridization
(B) Mutation and Selection
(C) Introduction and mutation
(D) Selection and introduction

228. Wheat revolution in green revolution has been due to introduction of disease resistant, fertilizer responsive, strong stemmed and dwarf varieties imparted from:
(A) California (B) Beijing
(C) Mexico (D) Uzbekistan

229. Haploid plants are obtained by culturing:
(A) Root tip (B) Endosperm
(C) Young leaves (D) Pollen grain

230. Farmers connot use seeds of plants showing hybrid vigour in successive years because:
(A) Hybrid vigour is lost on inbreeding
(B) Dunkel's proposals
(C) They are not allowed to sow their own seeds
(D) It is cheaper to purchase fresh seeds

231. A major use of embryo culture is in:
(A) Induction of somaclonal variations
(B) Overcoming hybridization barriers
(C) Production of alkaloids
(D) Clonal propagation

232. The primer of the lagging strand during DNA replication is removed by:
(A) 3' to 5' exonuclease activity of DNA pol III
(B) DNA primase
(C) 5' to 3' exonuclease activity of DNA pol I
(D) 3' to 5' exonuclease activity of DNA pol I

233. 'Norin gene' of dwarfness in Wheat originated through spontaneous mutation in:
(A) India (B) Japan
(C) Mexica (D) USSR

234. In postulating the theory of evolution by natural selection Darwin was greatly influenced by:
(A) Mutation theory of Hugo de Vries
(B) Lamarck's theory of acquired characters
(C) Malthus idea of population control
(D) Environmental factors

235. How many maternal genes involved in pattern formation have been identified?
(A) 15 (B) 20
(C) 25 (D) 30

236. The present giraffe has a long neck as compared to its ancestors. Lamarck believed it could be due to:
(A) Natural selection
(B) Isolation
(C) Inheritance of acquired characters
(D) Speciation

237. A species inhabiting different geographical areas is known as:
(A) Sympatric (B) Allopatric
(C) Sibling (D) Biosphere

238. Charles Darwin toured in a ship for 5 years:
(A) Beagle (B) Vikrant
(C) Phillips (D) Alexander

239. Who gave the principle that population tends to multiply more rapidly than food supply?
(A) Darwin (B) Malthus
(C) Lamarck (D) Haldane

240. Role of isolation in evolution is:
(A) Differentiation of species
(B) Maintenance of species
(C) Evolution of species
(D) Extermination of species

241. The ultimate source of continuous variations is:
(A) Natural selection
(B) Sexual reproduction
(C) Hormonal action
(D) Mutations

242. Four mutant strains of *E. coli* having the following characteristics were isolated:

Mutant strain 1: lac-repressor is not able to bind to operator because of a mutation in the operator region of the lac operon

Mutant strain 2: Allolactose is not able to bind to repressor because of a mutation in the repressor – encoding gene

Mutant strain 3: Catabolite activator protein (AP) is not able to bind to the promoter because of a mutation in the promoter region

Mutant strain 4: AMP is not able to bind to AP because of a mutation in the AP-encoding gene.

Lac genes are expressed in strain(s)

(A) 1 only
(B) 1, 3 and 4 only
(C) 1 and 4 only
(D) 1 and 2 only, but in the presence of IPTG

243. The evolution of a species is based upon sum total of adaptive changes preserved by:

(A) Natural selection (B) Isolation
(C) Speciation (D) Human conservation

244. Which of the following concept is attributed to Charles Darwin?

(A) Use and disuse of organs is of great importance in evolution
(B) Every cell must come from a preexisting cell
(C) In the struggle for existence, the fittest would survive
(D) The gametes will carry only one of a pair of contrasting characters

245. Key point of Lamarck's view about organic evolution is that every offspring:

(A) Is similar to its parents
(B) Inherits characters acquired by the parental generation
(C) Shows struggle for existence
(D) Repeats phylogeny in its ontogeny

246. The pair-rule gene proteins activate the transcription of the :

(A) Gap genes
(B) Pair rule genes
(C) Segment-polarity genes
(D) All of these

247. The fountain head of evolution is:

(A) Mutations
(B) Acquired characters
(C) Natural selection
(D) Sexual reproduction

248. The idea of natural selection as the fundamental process for evolutionary change was reached:

(A) Independently by Charles Darwin and Alfred Russel Wallace in 1900
(B) By Charles Darwin in 1866
(C) By Alfred Russel Wallace in 1901
(D) Independently by Charles Darwin and Alfred Russel Wallace in 1858

249. Match the terms in group I with their possible explanation in group 2:

Group 1	**Group 2**
P. Orthologs	1. A cell or an organism having foreign gene
Q. Paralogs	2. The complement of a protein expressed by a genome
R. Proteome	3. Genes from different species related to each other
S. Transgenic	4. Genes from same species related to each other

(A) P-2, Q–4, R–1, S-3 (B) P-4, Q–3, R–2, S-1
(C) P-3, Q–4, R–2, S-1 (D) P-1, Q–2, R–3, S–4

250. To be evolutionary successful, mutation must occur in:

(A) Somatic RNA (B) Plasma protein
(C) Germplasm DNA (D) Centrioles

251. Some organism resemble other organisms and thus escape from enemies. This phenomenon is known as:

(A) Homology (B) Variation
(C) Analogy (D) Mimicry

252. For food, light and space, the greatest competition is between two:

(A) Closely related species of the same niche
(B) Closely related species of the different niche
(C) Unrelated species of the same niche
(D) Unrelated species of the different niche

253. The protein products of the gap genes interact with neighboring gap gene proteins to activate the transcription of the:

(A) Gap genes (B) Pair rule genes
(C) Segment-polarity genes (D) All of these

254. The first attempt to solve the problem of organic evolution was done by:

(A) Mendel (B) Haeckel
(C) Lamarck (D) Darwin

255. Lamarck's theory of evolution is known as:

(A) Natural selection
(B) Inheritance of acquired characters
(C) Survival of fittest
(D) Mutation theory

256. Lamarck's theory of evolution fails to explain:

(A) Indian women have been getting their ears pierced for thousands of years, but not a single baby with pierced ears is born
(B) Giraffes have long necks because they have to eat leaves from the tall trees on their habitat
(C) Deer have the capacity to escape from agile enemies because these can run very fast
(D) High altitude pines have branches which slope down so that snow fails to collect upon them in winter.

257. The greatest weakness in Darwin theory was his failure to explain:
(A) Over production
(B) Struggle for existence
(C) Survival of the fittest
(D) Variations

258. The ultimate source of all variations is:
(A) Sexual reproduction (B) Hormonal action
(C) Natural selection (D) Mutations

259. A culture of bacteria (all of one species) is grown on solid culture medium containing the antibiotic streptomycin. At first no growth is apparent, although bacteria on a similar medium without streptomycin grow rapidly. However, in a short time, two colonies of bacteria begin to grow on the streptomycin-containing medium. What is the most likely explanation?
(A) At least two bacteria adjusted to the streptomycin due to exposure to it. This streptomycin resistance was passed on to their offspring
(B) At least two bacteria were resistant to streptomycin before being exposed to it. These cells generated streptomycin resistant colonies
(C) Streptomycin induced mutations in at least two bacterial cells that made them resistant. These mutated cells produced resistant colonies
(D) All bacteria will eventually grow in the presence of streptomycin if given long enough

260. Sudden heritable changes are called:
(A) Mutations (B) Variations
(C) Recombinations (D) Evolution

261. "Philosophic Zoologique" was written by:
(A) Linnaeus (B) Lamarck
(C) Wallace (D) Darwin

262. Continuity of germplasm was given by:
(A) Lamarck (B) Weismann
(C) Haeckel (D) Darwin

263. Genetic drift is on account of:
(A) Variations (B) Mutation
(C) Increase in population
(D) Decrease in population

264. A species inhabiting different geographical areas is known as:
(A) Allopatric (B) Sympatric
(C) Sibling (D) Biospecies

265. Segmentation genes are expressed after:
(A) Ovulation (B) Fertilization
(C) Maturity (D) All of these

266. Lamarckism connot explain:
(A) Webbed toes in aquatic birds
(B) Weak muscles in the son of a wrestler
(C) Long narrow and limbless body of snakes
(D) Heterophylly

267. Darwinism connot explain:
(A) Mimicry (B) Adaptation
(C) Struggle for existence (D) Tusks of elephants

268. Zinc tolerant plant of Zwar mines is:
(A) Agrostis (B) Astragalus
(C) Haplopappus (D) Impatiens

269. According to Neo-Darwinism, natural selection operates through:
(A) Fighting between organisms
(B) Killing the weaker organism
(C) Differential reproduction
(D) Variations

270. The genes divide the embryo into broad regions and are so named because their mutations result in gap in the embryo's segmentation pattern is called:
(A) Gap genes (B) Pair rule genes
(C) Segment-polarity genes (D) All of these

271. Inspite of great disadvantageous nature of sickle cell anaemia it has a high frequency in some parts of Africa. It is because of the reason that
(A) Sickle cell anaemics are resistant to malaria
(B) In South Africa oxygen concentration is high
(C) It is a case of genetic drift
(D) All the above

272. Survival of fittest used by:
(A) Charles Darwin (B) Herbert Spencer
(C) Jean Baptiste Lamarck (D) Hugo de Vries

273. For its entry into gene pool a mutation must occur in
(A) Centriole (B) Plasma proteins
(C) Somatic RNA (D) Germplasm DNA

274. Two genetic loci produce identical phenotypes in *cis* as well as *trans* position. They are:
(A) Pseudoalleles (B) Multiple alleles
(C) Different genes (D) Parts of same gene

275. Who believed that an organ in disuse will become vestigial?
(A) Lamarck (B) De Vries
(C) Mendel (D) Darwin

276. The relation between the fraction of repetitive DNA and the time required for reassociation of its denatured strands in a specific genome is:
(A) Direct (B) Inverse
(C) Hyperbolic (D) None of these

277. A very important factor in evolution of a new species is:
(A) Extensive inbreeding
(B) Extensive out breeding
(C) Reproductive isolation
(D) Immigration

278. "Continuity of germplasm" theory was given by:
(A) De Vries (B) Weismann
(C) Darwin (D) Lamarck

279. The σ factor found in many bacteria is best described as a:
(A) Subunit of DNA pol that allows for synthesis in both 5' to 3' and 3' to 5' directions
(B) Subunit of RNA pol responsible for the specificity of the initiation of transcription of RNA from DNA
(C) Subunit of the 3S ribosome to which mRNA binds
(D) Subunit of the 50S ribosome that catalyzes peptide bond synthesis

280. A cross involving parents differing only in one trait is:
(A) Haploid (B) Diploid
(C) Monohybrid (D) Dihybrid

281. In mendelian crosses the genes responsible for different characters separate through:
(A) Fusion (B) Aggregation
(C) Segregation (D) Distribution

282. Galpagos Islands were visited by:
(A) Wallace (B) Darwin
(C) Lamarck (D) De Vries

283. The major objection against the theory of Natural selection is that it:
(A) Stresses on interspecific competition
(B) Explains natural calamities to take heavy toll of lives
(C) Explains adaptation
(D) Stresses on slow small variations

284. Density gradient centrifugation of genomic DNA of a specific organism revealed a lower density peak in addition to the main peak. This is due to:
(A) AT rich satellite DNA
(B) GC rich satellite DNA
(C) Equal proportion of AT & GC rich satellite DNA
(D) None of these

285. Most modern breeds of domestic Dog have evolved through:
(A) Sexual selection (B) Artificial selection
(C) Natural selection (D) Isolation

286. Most important theory of general biology was proposed by:
(A) Mendel and Morgan (B) Beadle and Tatum
(C) Watson and Crick (D) Darwin & Wallace

287. Hugo de Vries worked on the plant:
(A) Garden pea/*Pisum sativum*
(B) Sweet pea/*Lathyrus odoratus*
(C) *Primula sinensis*
(D) Evening Primrose/ *Oenothera lamarckiana*

288. T.R. Malthus is famous for his book on:
(A) Population
(B) Mathematics
(C) Geography
(D) Genetics

289. Which one of the following is false for prokaryotic genes?
(A) Promoter of a gene in a genome cannot be present within the upstream open reading frame
(B) The ribosome binding site can be either upstream or downstream of the TATA box (Pribnow box)
(C) All prokaryotic promoters must have a – 10 and a–35 region in order to allow transcription
(D) Sigma factors bind to the – 10 and -35 region of promoters

290. Weismann cut off tails of mice generation after generation but tails neither disappeared nor shortened showing that:
(A) Darwin as correct
(B) Tail is an essential organ
(C) Mutation theory is wrong
(D) Lamarckism was wrong in inheritance of acquired characters

291. C_0 t analysis provides an estimate of the:
(A) G + C content of the DNA
(B) T_m of the DNA
(C) Complexity of the genome
(D) Hyperchromic shift of the genome

292. Genetic drift is change of:
(A) Gene frequency in same generation
(B) Appearance of recessive genes
(C) Gene frequency from one generation to next
(D) None of the above

293. Sum total of genes in a population is:
(A) Genotype
(B) Karyotype
(C) Genetic constitution
(D) Gene pool

294. Which of the following best describes the function of the sigma subunit in the RNA polymerase of *E. coli*?
(A) It is essential for elongation of the RNA transcript
(B) It is essential for the recognition of and binding to the promoter sequences
(C) It increases RNA polymerase binding to any DNA template
(D) It is required for transcription termination

295. Improved race of Pigeon developed due to:
(A) Environmental selection
(B) Natural selection
(C) Artificial selection
(D) Protective selection

296. Evolution will not occur if:
(A) There is no environmental influence
(B) Genes are without effect
(C) Somatic variations are not heritable
(D) There is no genetic variations in individ-uals of a populations

297. Major defect of Darwinism was:
(A) Nondescription of survival of fittest
(B) Nondescription of inheritance of fittest
(C) Nondescription of overproduction of young ones
(D) Nondescription of reason for variations

298. Which book written by Charles Darwin?
(A) Origin of species (B) Species Plantarum
(C) Genera Plantarum
(D) De Naturien Familien

299. Which one does not favour Lamarckian concept of inheritance of acquired characters?
(A) Lack of pigment in cave dwellers
(B) Absence of limbs in snakes
(C) Presence of webbed toes in aquatic birds
(D) Melanisation of Peppered Moth in industrial areas

300. Which one provides correct sequence of events in origin of new species according to Darwinism?
1. Natural selection
2. Variations and their inheritance
3. Survival of fittest
4. Struggle for existence
(A) 1, 2, 3, 4 (B) 2, 3, 1, 4
(C) 3, 4, 1, 2 (D) 4, 2, 3, 1

301. Which of the following species of denatured DNA will denature most rapidly in solution, under appropriate conditions of ionic strength, pH and temperature?
(A) Human live nuclear DNA
(B) Vaccinia virus DNA
(C) *E. coli* DNA
(D) Yeast nuclear DNA

302. Darwin is associated with:
(A) Natural selection (B) Mutation
(C) Inorganic evolution (D) All the above

303. In which respect, Darwin's theory is wrong:
(A) High efficiency of reproduction
(B) Origin of species
(C) Survival of fittest
(D) Arrival of fittest

304. Specificity of transcription by RNA polymerases holoenzyme in *E. coli* is altered by changing the:
(A) Alpha subunit (B) Sigma subunit
(C) Beta subunit (D) Beta prime subunit

305. Closely related, morphologically similar, reproductively isolated sympatric population are designated as:
(A) Clones (B) Clines
(C) Demes (D) Sibling species

306. Frequency of gene in a population will increase if the gene is:
(A) Lethal (B) Dominant
(C) Recessive (D) Favourable selected

307. Objection to natural selection theory of Darwin is:
(A) Vestigial and over-specialised organs
(B) Inheritance of acquired characters
(C) Struggle for existence
(D) Continuity of germplasm

308. Presence of vestigial organs supports:
(A) Natural selection
(B) Germplasm theory
(C) Evolution but not Lamarck's theory
(D) Synthetic theory

309. Which of the following is true of the relationship between genomes and physical complexity of organisms?
(A) Among mammals, genome sizes are roughly proportional to the mass of the adult animals
(B) The number of chromosomes is usually proportional to the complexity of an animal's adult form
(C) Species with very similar body plans often have genomes of very different sizes
(D) Larger genomes typically have fewer repeated sequences

310. Hexaploid Wheat developed through:
(A) Hybridomas (B) Chromosom doubling
(C) Hybridisation
(D) Hybridisation and chromosome doubling

311. Lederberg's replica experiment is connected with:
(A) Adaptations (B) Evolution
(C) Genetic (D) Artificial selection

312. Which are immortal?
(A) Germ cells (B) Somatic cells
(C) Pituitary cells (D) Glomerular cells

313. Xeroderma pigmentosum is an autosomal recessive genetic disease that causes a variety of phenotypic changes in skin cells exposed to sunlight. This disease involves a defect in which of the following DNA repair mechanisms?
(A) Base excision repair
(B) Double-strand break repair
(C) Mismatch repair
(D) Nucleotide excision repair

314. Hybrid breakdown is failure of:
(A) Hybrid zygote to develop into offspring
(B) Hybrid adult to produce fertile offspring
(C) Fusion failure of ova and sperms of two species
(D) All the above

315. Chief agent of evolution is:
(A) Natural selection (B) Mutations
(C) Acquired characters (D) Sexual reproduction

316. Genetic drift is found in:
(A) Small population with or without mutated genes
(B) Large population with random mating
(C) Animal population
(D) Plant population

317. The Alu element found in the human genome is:
(A) A 70S RNA homologue
(B) A good RNA polymerase III template
(C) About 300 bp long repeat
(D) All of these

318. Who is not related to organic evolution?
(A) T. R. Malthus (B) Erasmus Darwin
(C) Darlington (D) Charles Darwin

319. Modern theory of organic evolution stresses on:
(A) Mutations (B) Isolation
(C) Population (D) All the above

320. Presence of different types of beaks in finches of Galapagos islands adapted to different feeding habits provides evidence for:
(A) Intraspecific variations
(B) Natural selection
(C) Intraspecific competition
(D) Interspecific competition

321. The SOS repair mechanism is activated by:
(A) 5 - bromourail (B) 2 - aminopurine
(C) Hydroxylamine (D) Thymine dimmers

322. Theory of pangenesis was given by:
(A) Darwin (B) Lamarck
(C) Hugo de Vries (D) Oparin

323. Phenomenon of industrial melanism demonstrates:
(A) Geographical isolation (B) Natural selection
(C) Reproductive isolation (D) Induced mutation

324. Genetic drift:
(A) Is random change in gene frequency
(B) Has nothing in common with inbreeding
(C) Is an orderly change in gene frequency
(D) Produces greatest fluctuations in large population

325. Darwin wrote "Origin of Species" in:
(A) 1530 (B) 1975
(C) 1859 (D) 1895

326. Which is related to reproductive isolation?
(A) Genetic isolation
(B) Behavioural isolation
(C) Temporal isolation
(D) All the above

327. Gene pool of a population tends to remain stable if the population is large, without large scale mutations, without migration and with:
(A) Random mating
(B) Moderate environmental changes
(C) Natural selection
(D) Reduction in predators

328. Bacterial cells that can be lysogenised by lambda phase, possess in their genome:
(A) 80S ribosomal genes (B) attB sites
(C) attP sites (D) attL sites

329. The theory that environment can bring about changes in individuals was proposed by:
(A) Mendel (B) Darwin
(C) Lamarck (D) Lyell

330. Darwin theory states that:
(A) Characters are acquired through inheritance
(B) Species change morphologically with time
(C) Nature selects organisms which can adapt
(D) Evolution is due to effect of environment

331. The animal that explains industrial melanism is:
(A) Lizard (B) Pepper Moth
(C) Monkey (D) Frog

332. During DNA replication, which of the following is not true at the replication fork?
(A) One core DNA polymerase III replicates the leading strand and another molecule replicates the lagging strand
(B) One DNA polymerase III molecule replica-tes the leading strand and one DNA poly-merase I molecule replicates the lagging strand
(C) The sliding lamp protein is behind the DNA polymerase at the primer binding site
(D) A ring of DNA B helicase is placed between two polymerases

333. Chromosomes found in the salivary gland of Drosophila are:
(A) Lampbrush (B) Polytene
(C) Supernumerary (D) B-chromosomes

334. A giant chromosome with a number of chromonemata is:
(A) Lampbrush chromosome
(B) Heterochromosome
(C) Supernumerary chromosome
(D) Polytene chromosome

335. Lampbrush chromosomes occur in:
(A) Salivary gland (B) Lymph gland
(C) Cancer cells (D) Oocytes

336. Chromosomes ends are called:
(A) Satellites (B) Telomeres
(C) Centromeres (D) Kinetochores

337. The genome of a typical bacterium contains about 5 x 10^6 base pairs and an be replicated in about 41 minutes. The human genome is 600 x larger (3 x 10^9 base pairs) and at the rate of a bacterium would require 300 hours (~12 days) to be replicated; yet the entire human genome can be replicated with several hours. How is this possible?
(A) Eukaryotic DNA is simpler to replicate than prokaryotic DNA
(B) Human DNA polymerases work much faster than those of prokaryotes
(C) The nucleosomes of eukaryotic DNA allow for faster DNA replication
(D) Human DNA contains more origins of replication than prokaryotic DNA

338. A Chromosome with sub-terminal centromere is:
(A) Acentric (B) Acrocentric
(C) Metacentric (D) Telocentric

339. A chromosome with centromere near the middle is called:
(A) Metacentric (B) Submetacentric
(C) Acrocentric (D) Telocentric

340. Puffs or balbiani ring in salivary gland chromosomes are sites of:
(A) DNA replication (B) DNA duplication
(C) RNA synthesis (D) Protein synthesis

341. Chromatin fibres condense to form chromosomes during:
(A) G1-Phase (B) S-Phase
(C) Metaphase (D) Prophase

342. The stage of counting of chromosomes is:
(A) Early prophase (B) Mid prophase
(C) Late prophase (D) Metaphase

343. The genes which are present on the same chromosomes are called:
(A) Identical genes (B) Allelomorphs
(C) Linked genes (D) Dominant genes

344. Drosophila has four homologous pairs of chromosomes. What is the number of linkage groups in this animal?
(A) 4 (B) 2
(C) 8 (D) Uncertain

345. Genes for colour blindness in man is located on:
(A) X-chromosome only
(B) Y-chromosome only
(C) Either X or Y chromosome
(D) Both X and Y chromosomes

346. A colour blind daughter may be born if the:
(A) Father is normal and mother is colour blind
(B) Father is colour blind and mother is normal
(C) Father is normal and mother is carrier
(D) Father is colour blind and mother is a carrier

347. A somatic cell in human male contains:
(A) No gene on sex chromosomes
(B) Only one sex linked gene for each character
(C) Two genes for every sex linked character
(D) Genes only on sex chromosomes

348. Which of the following disease is sex linked?
(A) Colour blindness (B) Malingnancy
(C) Hepatitis (D) Leukemia

349. The genes for the eye colour and size of the wing in Drosophila are located on the same chromosome. They can be separated by:
(A) Non-disjunction
(B) Crossing over
(C) Hybridization
(D) Not be separated at any stage

350. The sex linked characters are those:
(A) Which are related to sexual physiology
(B) The genes of which are present on the sex chromosomes
(C) Which appear in male or in female
(D) Which are controlled by sex hormone

351. The chromosomes other than sex chromosomes are called:
(A) Autosomes (B) Heterosomes
(C) Karyosomes (D) None of the above

352. The points where the crossing over occurs between the homologous chromosomes of a pair are called:
(A) Centromeres (B) Chromomeres
(C) Chromocenters (D) Chiasmata

353. If the crossing over had occurred at two-strand stage in Neurospora, the ascospores would be arranged in:
(A) 1-1 position (B) 2-2 position
(C) 4-4 position (D) None of these

354. Complete linkage is found in:
(A) Birds (B) Snakes
(C) Female Drosophila (D) Male Drosophila

355. A linkage group is defined as:
(A) All the linked genes of a chromosomal pair
(B) Different groups of genes present on different chromosomes
(C) All the genes located on the same chromosome
(D) None of these

356. A phenomenon which works opposite to the linkage is:
(A) Independent assortment (B) Crossing over
(C) Segregation (D) Mutation

357. When two genes are situated very close to one another at a chromosome:
(A) The percentage of crossing over between them is very high
(B) Hardly any cross-overs are produced
(C) No crossing over can take place
(D) Only double cross-overs can occur between them

358. Greater is the distance between the two genes on a chromosome:
(A) Greater is the linkage strength
(B) Lesser is the linkage strength
(C) Linkage strength remains unchanged
(D) There is no relationship between the two

359. Crossing over occurs at
(A) Single-strand stage of chromosomes during prophase
(B) Two-strand stage during zygotene
(C) Four-strand stage of pachytene
(D) Metaphase II of meiosis

360. The term coupling and repulsion signify the same phenomenon which is termed:
(A) Synapsis (B) Disjunction
(C) Linkage (D) Non-disjunction

361. The linked characters would always inherit together till they are:
(A) Delinked due to segregation
(B) Masked by dominance
(C) Mutated
(D) Separated due to crossing over

362. During meiosis, the crossing over which occurs between the homologous chromosomes is of:
(A) No genetic significance
(B) Importance to bring about recombination of characters
(C) Importance because it produces nutrition
(D) Importance because it causes reorganization of cytoplasm

363. A white eyed (recessive) and long winged male Drosophila was crossed to a vestigial winged (recessive) red eyed female. The two characters are linked in this animal. When a female F_1 was test crossed, the F_2 generation produced:
(A) Mostly white eyed with long wings
(B) Mostly red eyed with vestigial wings
(C) Mostly white eyed with long wings and red eyed with vestigial wings
(D) All white eyed and vestigial winged

364. *E. coli* DNA ligase uses NAD^+ as a co factor. During the actual ligation reaction (joining of the 5' phosphate and 3' hydroxyl to form a phosphodiester bond), the leaving group is:
(A) NAD^+ (B) NMN
(C) ATP (D) AMP

365. The genes of different traits located on different loci on the same chromosome are:
(A) Alleles (B) Linked
(C) Pleiomorphic (D) Mutated

366. In Four O'Clock plant, normal leaves (A) and variegated leaves (B) occur in different plants. If (B) male is crossed with (A) female, the hybrid has normal leaves, but when (B) female is crossed with (A) male, the hybrid has variegated leaves. It is a case of:
(A) Mutation
(B) Cytoplasmic inheritance
(C) Complementary genes
(D) Supplementary genes

367. Crossing over occurs between:
(A) Two non-sister chromatids of a homologous pair of chromosomes
(B) Two chromatids of any chromosome
(C) Two chromatids of same chromosome
(D) All the foregoing

368. Choose the correct match from a, b, c, and d

Group I	Group II
P. 5' aping of mRNA	1. 5-methyl guanosine
Q. Ribozyme	2. Polyadenylate transferase
R. Promoter	3. Spliceosome
S. Poly-A-tail	4. 7-methyl guanosine
	5. RNA polymerase
	6. Catalytic RNA
	7. Polyadenylate polymerase

(A) P-1, Q-6, R-5, S-2 (B) P-4, Q-6, R-5, S-7
(C) P-1, Q-5, R-3, S-7 (D) P-4, Q-3, R-6, S-2

369. The frequency of crossing over would be higher if:
(A) Two genes are located closely
(B) Two genes are far apart on a chromosome
(C) Two genes are not located on the same chromosome
(D) None of the above

370. Mendelian recombinations are due to:
(A) Mutation (B) Linkage
(C) Crossing over
(D) Independent assortment of characters

371. The blue green algae and bacteria contain:
(A) One linked group (B) Two linked groups
(C) Three linked groups (D) None of these

372. Which one of the following character in man is controlled by a recessive gene?
(A) Colour blindness (B) Woolly hair
(C) Brachydactyly (D) Curly hairs

373. What is one conclusion that can be drawn from the observation that the genetic code is nearly identical in all cells on earth?
(A) Life on earth evolved only one
(B) Only four types of nitrogenous bases occurs in DNA
(C) Higher forms of life evolved from lower forms
(D) Natural selection

374. Crossing over may result into:
(A) Gene mutation (B) Genomatic mutation
(C) Genetic recombination
(D) None of these

375. Maize has 10 pairs of chromosomes. How many linkage group does it have?
(A) 20 (B) 10
(C) 5 (D) 40

376. In flowering plants meiosis occurs at the time of:
(A) Germination of seed
(B) Formation of endosperm
(C) Formation of embryo
(D) Formation of pollens

377. Most of the mutations remain unexpressed because they are:
(A) Too small (B) Very significant
(C) Very few (D) Mostly recessive

378. Which of the following statement is correct concerning a homeobox?

(A) It is part of the promoter in eukaryotic genes that code for proteins involved in segmentation

(B) It is a conserved proteins structure found in glycotytic pathway enzymes

(C) It is a conserved protein structure found in tRNA-binding proteins

(D) It is a conserved DNA sequence found in genes that code for proteins that regulate development

379. In transversion:

(A) Purine is replaced by another purine

(B) Purine is replaced by pyrimidine or vice versa

(C) Some nitrogen bases are eliminated from a gene

(D) Some nitrogen bases are added in a gene

380. Colour blindness is more likely to occur in males than females because:

(A) Genes for this character are located on the X-chromosomes

(B) The trait is dominant in males and recessive in females

(C) Some males suffer from deficiency of vitamin A

(D) The Y-chromosome of males have the genes for distinguishing colours

381. Suppose that out of 1000 offspring in the F+ generation of a dihybrid cross. 748 are tall and hairy, 6 are tall and smooth, 4 are short and hairy, 242 are short and smooth. The reason is that the genes controlling these characters are:

(A) Lethal genes

(B) Complementary genes

(C) Linked genes

(D) Epistatic genes

382. Yellow colour is dominant over green colour. In a generation of 50% yellow and 50% green individuals, the parents are:

(A) Both homozygous

(B) Both heterozygous

(C) One of them is heterozygous

(D) None of these

383. In vitro translation of a polycistronic RNA generally results in formation of unequal amounts of protein products form the individual cistrons. The most likely reason is:

(A) Different strength of each promoter

(B) Rapid RNA degradation

(C) Rapid degradation of proteins

(D) Different strength of each of the ribosome binding sites

384. Ame's test is performed to determine:

(A) Carcinogenicity of chemicals

(B) Pathogenicity of a bacterium

(C) Viability of a bacterial culture

(D) Sporulation of bacterium

385. An exchange of segments between the two non-homologous chromosomes is called:

(A) Polyploidy (B) Chromosomal aberration

(C) Translocation (D) Inversion

386. If a part of a chromosome gets separated and reattached in reverse position to the same chromosome, the mutation is called:

(A) Inversion (B) Transversion

(C) Translocation (D) Gene mutation

387. The ribosome is involved in all of the following, except:

(A) Peptide bond formation

(B) Aminoacylation of tRNA

(C) Binding of protein factors during elongation

(D) Binding of aminoacyl tRNA to mRNA

388. A sudden or spontaneous change in the structure and action of a particular gene is called:

(A) Linkage (B) Variation

(C) Mutation (D) Allelomorph

389. In case of an abnormal crossing over, the two homologous chromosomes happen to exchange unequal length of segments. The phenomenon is called:

(A) Transition (B) Duplication

(C) Inversion (D) Gene mutation

390. The first step of 'proof-reading' during protein synthesis is carried out by:

(A) Ribosomes

(B) m-RNA

(C) t-RNA

(D) Aminoacyl t-RNA synthetase

391. Mutation did not result in:

(A) Death of organism

(B) Better progenies

(C) Change in genetic constitution of the cell

(D) Hybrid vigour

392. Gene mutations are those which involve:

(A) The change in nature and sequence of base triplets of DNA

(B) The change in genome

(C) The change in all the genes

(D) The disappearance of certain part chromosome

393. Ame's test is based on a hypothesis that if a substance is a mutagen, it will increase the rate at which Salmonella-auxotrophs revert to:

(A) Lysine synthesizers

(B) Arginine synthesizers

(C) Histidine synthesizers

(D) Tryptophan synthesizers

394. Aneuploidy is the term applied for the:

(A) Gene mutation

(B) Chromosomal mutation

(C) Chromosomal mutations involving the addition or loss of one or more chromosome

(D) Chromosomal mutation involving the addition of one or more complete set of chromosomes

395. **What term is applied to the gene mutation where a base is replaced by another base?**
(A) Duplication (B) Aneuploidy
(C) Euploidy (D) Substitution

396. **A monosomic organism can best be represented as:**
(A) 2n +1 (B) 2n +2
(C) n + 1 (D) 2n -1

397. **Insulin is secreted as a dimmer consisting of two covalently linked polypeptides. These polypeptides are derived:**
(A) By alternative splicing of the mRNA from a single gene
(B) By proteolytic processing of a single protein
(C) From two genes
(D) None of these

398. **A flower was fertilised by pollen grains having the constitution (n-1). If the ovules are normal, which of the following is applicable to the seeds so produced?**
(A) Allopolyploidy (B) Monosomy
(C) Nullisomy (D) Both (A) and (B)

399. **Nullisomy is the term used for the condition when an organism has:**
(A) An additional chromosome
(B) One chromosome less than normal
(C) A complete set of chromosomes except one homologous pair
(D) None of these

400. **In genetics, suppression of a mutation refers to:**
(A) Restoration of the original phenotype due to a second mutation
(B) Restoration of the original DNA sequence by mutation
(C) Prevention of expression of the mutant gene by metabolic regulation
(D) Appearance of the recessive phenotype in a heterozygous diploid

401. **A classical example of allopolyploidy is:**
(A) Brassica (B) *Raphano-brassica*
(C) Raphanus (D) All the above

402. **The mutation which returns to the original state is called?**
(A) Reversible mutation (B) Lethal mutation
(C) Backward mutation (D) Abnormal mutation

403. **Mutations are:**
(A) Always useful
(B) Seldom useful
(C) Always harmful
(D) May be useful or harmful

404. **The drug chloramphenicol blocks:**
(A) Cell-wall synthesis
(B) Translation-termination release factors
(C) Polypeptide chain elongation
(D) Polypeptide chain initiation

405. **DNA polymerase contains a lysine residue that important for binding to DNA. Mutations were found that converted this lysine to either glutamate, glycine, valine or arginine. Which mutations would be practices to be the most and least harmful to the ability of the enzyme to bind DNA?**

Most	Least
A. Glycine	Arginine
B. Arginine	Glycine
C. Glutamate	Valine
D. Glutamate	Arginine

406. **Euploidy is the term applied for those mutations in which:**
(A) There is multiplication of chromosomal number
(B) Involves addition of complete set of chromosomes
(C) There is loss of a few chromosomes
(D) None of these

407. **Which is not applicable to mutations?**
(A) They are inheritable genetic changes
(B) Many of them remain undetected
(C) They can be induced artificially
(D) They are possible only in diploid cells

408. **Autopolyploids are those which have:**
(A) More than one set of chromosomes
(B) Only one additional set of chromosomes
(C) The same basic set of chromosomes
(D) All of these

409. **How many types of gametes are required for F_2 generation of a monohybrid cross?**
(A) 4 (B) 8
(C) 16 (D) 32

410. **When chromosome sets are present in multiple of 'n' the condition is termed:**
(A) Diploidy (B) Haploidy
(C) Euploidy (D) Aneuploidy

411. **Which of the following is not an aneuploidy?**
(A) Monoploid (B) 2n -1
(C) Trisomic (D) 2n +2

412. **The reason of genetic variability in sexually reproducing organisms is:**
(A) Temperature (B) Mutations
(C) Certain chemicals (D) Evolution

413. **Most of the high yielding varieties of wheat are:**
(A) Imported (B) Monosomic
(C) Trisomic (D) Mutants

414. **The smallest segment of a gene which can undergo mutation is called:**
(A) Muton (B) Recon
(C) Cistron (D) Interferon

415. **Most of the mutations are:**
(A) Forward (B) Backward
(C) Reversible (D) Lethal

416. Which one of the following statements describes the ubiquitin-mediated degradation of proteins in the cytosol?
(A) One molecule of ubiquitin binds to the protein to be degraded
(B) The process is catalyzed by a single enzyme
(C) The process depends on edenosine triphosphate (ATP)
(D) The N-terminal residue of ubiquitin becomes covalently attached to the protein to be degraded

417. Mutation theory of evolution by Hugo de Vries was based on his studies on:
(A) *Pisum sativum*
(B) *Oenothera lamarkiana*
(C) *Acetabularia*
(D) *Lathyrus odoratus*

418. Mutagenes are:
(A) The genes which can mutate
(B) The organism which show mutation
(C) Genes that regulate mutations
(D) The agents which cause mutations

419. The mutations which occur in germ cells can be detected phenotypically in:
(A) Parents (B) Next generation
(C) Third generation (D) None of these

420. The frequency of mutation in a species can be increased by the use of:
(A) X-rays (B) UV - rays
(C) Nitrous acid (D) All of these

421. A pleiotropic mutation is the one which:
(A) Affects one character at a time
(B) Affects many characters at a time
(C) Is always lethal
(D) Easily gets preserved in natural conditions

422. Down's syndrome is a typical case of:
(A) Trisomy (B) Nullisomy
(C) Gene mutation (D) Monosomy

423. Which of the following statements can be used to support the endosymbiotic theory for the origin of mitochondria?
P. Mitochondrial DNA has a circular genome
Q. Mitochondrial are about the same size as bacteria
R. Mitochondrial ribosomes are 70S
S. Mitochondrial ribisomes are inactivated by the antibiotic chloramphenicol
(A) P and Q (B) Q and R
(C) R and S (D) P, Q, R and S

424. Datura is a classical example of study of:
(A) Monosomics (B) Trisomics
(C) Triploids (D) Nullisomics

425. Which one is a hexaploid species of wheat?
(A) *Triticum monococcum* (B) *T. aestivum*
(C) *T. durum* (D) *T. turgidum*

426. When chromosome sets are present in multiple of 'n', the condition is called as:
(A) Diploidy (B) Aneuploidy
(C) Haploidy (D) Euploidy

427. Persons carrying the dominant gene(s) an taste a particular chemical and are called tastes. The others are nontasters. In a large random-mating human population at the Hardy-Weinberg equilibrium, there are 36% tasters and 64% non-tasters. The frequency of hetrozygotes in the population is:
(A) 6% (B) 8%
(C) 32% (D) 48%

428. X-rays cause mutations by:
(A) Breaking spindle
(B) Rupturing of nuclear membrane
(C) Changing the chromosome morphology
(D) Inducing karyokinesis

429. Colchicine interferes in:
(A) Spindle organization (B) DNA replication
(C) Chromosome condensation (D) Polyploidy

430. Gene mutation occur in:
(A) Chromosome (B) DNA
(C) Nucleus (D) RNA

431. It is thought that the Galapagos Islands were originally colonized by one species of finch. Today these Islands contain many species of finches that occupy diverse niches. This has most likely resulted from which of the following?
P. Biological magnification
Q. Convergent evolution
R. Adaptive radiation
(A) P and R (B) Q and P
(C) R only (D) Q and R

432. Haploids are used for researches because:
(A) They contain only one chromosome
(B) They contain two sets of chromosomes
(C) They contain three sets of chromosomes
(D) They contain only one set of chromosomes

433. Cytoplasmic inheritance differs from nuclear inheritance in the absence of:
(A) Similarity of reciprocal crosses
(B) Biparental contribution
(C) Effect on backcrossing
(D) All of these

434. Which of the following is characteristics of allelopatric speciation?
(A) Large populations
(B) Isolation through adaptation of alleles
(C) Geographic isolation
(D) Asexually reproducing populations

435. Cytoplasmic inheritance is also called maternal inheritance because it is:
(A) Without sexual reproduction
(B) Only female parent takes part in multiplication
(C) Most of the cytoplasm of the zygote is provided by ovum
(D) All of these

436. Number of chromatids present in a metaphasic chromosome are:
(A) 4 (B) 2
(C) 3 (D) 1

437. Number of chromatids per chromosome complex at metaphase is:
(A) One in mitosis and two in meiosis
(B) Two in mitosis and one in meiosis
(C) Two both in mitosis and meiosis
(D) Two in mitosis and four in meiosis

438. Choose the correct statement(s):
P. Similar traits resulting from similar selection pressures acting on similar gene pools is parallel evolution
Q. The evolution in one population in response to the evolution of another population, and vice versa is called co-evaluation
(A) P (B) Q
(C) P and Q (D) None of these

439. Chromosomes are formed by:
(A) DNA, RNA and proteins (B) DNA only
(C) RNA only (D) Proteins

440. Which of the following does not explain the term survival of the fittest?
(A) Fittest animals leave higher number of progeny than those which perish
(B) Fittest group of animals out populates its competitors
(C) Fittest animals are best predators
(D) Fittest group is rich in genetic variations

441. Centromere is connected with:
(A) Spindle formation
(B) Poleward movement of chromosomes
(C) Chromosome segregation
(D) DNA replication

442. Centromere is also called:
(A) Metamere
(B) Satellite
(C) Chromocentre
(D) Primary constriction

443. A chromosome with terminal centromere is:
(A) Telocentric (B) Metacentric
(C) Polycentric (D) Acrocentric

444. The terminal end of chromosome is:
(A) Chromomere (B) Centromere
(C) Telomere (D) Metamere

445. Which of the following statements is not consistent with the principle of totipotency?
(A) Cell specilization is based on position
(B) Plant cell regenerate entire plants by mitosis
(C) Plant cell can differetiate into any cell type
(D) Cell specialization is based on gene content

446. An octamer of 4 histones complexed with DNA forms:
(A) Centromere (B) Nucleosome
(C) Mesosome (D) Endosome

447. A nucleosome is made of:
(A) Histone wrapped octameric core of nucleic acid
(B) Histone
(C) DNA wrapped over octameric core of histone
(D) DNA

448. Choose the correct sequence of evolutionary events in one form of allopatric speciation, using the codes given below:
P. Geographical isolation
Q. Ecological isolation
R. Increased pre-mating reproductive isolation
S. Increased genetic divergence
T. Selection completed
(A) R – S - Q - T (B) P – S – R – T
(C) R – Q – S – T (D) P – Q – R - T

449. Polytene chromosomes were seen by
(A) Ruckert (B) Balbiani
(C) Wilson (D) Heitz

450. Haploids are able to express both recessive and dominant alleles/mutations because there are:
(A) Many alleles for each gene
(B) Two alleles for each gene
(C) Only one allele for each gene in the individual
(D) Only one allele in a gene

451. Which of the following best describes pseudogenes?
(A) Two functional genes within an organism that arose from the duplication of one gene
(B) Genes that share the same ancestral sequence, but are found in different organisms
(C) Sequences of DNA that are very similar to functional genes, but do not produce a functional product
(D) Sequences of DNA that are very similar to a native genes, but do produce a functional product

452. The phenomenon of exchange of segments between paternal and maternal chromosome is called:
(A) Segregation (B) Recombination
(C) Crossing over (D) Linkage

453. Which one can have sex chromosomes?
(A) Unisexual flower
(B) Bisexual flower
(C) Unisexual plant
(D) Hermaphrodite plant

454. Crossing over produces:
(A) Recombination of linked genes
(B) Synapsis of linked genes
(C) Expression of recessive genes
(D) Linkage of dominant genes

455. The hypothesis that evolution occurs in spurts, with great amounts of evolutionary change followed by periods of stasis, is
(A) Puntuated equilibrium
(B) Allopatric speciation
(C) Gradualism
(D) Hardy – Weinberg equilibrium

456. Out of 8 ascospores formed in Neurospora, the arrangement is 2a : 4A : 2a showing:
(A) No crossing over
(B) Some meiosis
(C) Second generation division
(D) First generation division

457. Chromosomes connected with sex determination are:
(A) Autosomes (B) Heterosomes
(C) Oxysomes (D) B-chromosomes

458. An autosome is:
(A) Chromosome half (B) Sex chromosome
(C) Chromosome other than sex
(D) None of these

459. Linkage in plants was first shown in:
(A) *Lathyrus odoratus* (B) *Zea mays*
(C) *Oenothera lamarkiana* (D) *Pisum sativum*

460. A DNA vaccine is:
(A) A DNA molecule that is recognized by an antibody
(B) A vaccine that works by stimulating the immune system to recognize pathogen DNA sequences
(C) A vaccine that is administered as DNA; the DNA is then expressed to produce a protein, which stimulates an immune response
(D) All of these

461. Association of parental combination of characters in offspring in excess of dihybrid cross is due to:
(A) Polygeny (B) Linkage
(C) Pseudoalleles (D) Codominance

462. In a dihybrid cross, two recessive genes showed 10% recombinants. The distance between two genes are:
(A) 10 map units (B) 20 map units
(C) 30 map units (D) 40 map units

463. An individual with cd genes was crossed with wild type + +. On test crossing F_1, the progeny was + c 105, + d 115, cd 880 and ++ 900. Distance between cd genes is:
(A) 11 map units (B) 22 map units
(C) 44 map units (D) 88 map units

464. Which of the following may be part of the reason why thymine replaced uracil in DNA?
(A) Cytosine spontaneously deaminates to uracil
(B) Thymine is unstable
(C) Thymine synthetic pathway is easier
(D) Thymine has a stronger binding affinity for adenine than does uracil

465. Plant in which chromosomal basis of sex determination was discovered first is:
(A) Rumex (B) Melandium
(C) Coccinia (D) Sphaerocarpus

466. Sex linked characters are generally:
(A) Lethal (B) Recessive
(C) Dominant (D) Not inherited

467. Trisomy has chromosome complement of:
(A) 2n - 1 (B) 2n-1-1
(C) 2n+1+1 (D) 2n+1

468. The first successful gene therapy was done to which of the following enzyme:
(A) Homogentisic acid oxidase
(B) Adenosine decarboxylase
(C) Adenosine deaminase
(D) Adenyl cyclase

469. Mutations are commonly:
(A) Incomplete (B) Codominant
(C) Recessive (D) Dominant

470. Haploid chromosome number is 12. What is the number in a monosomic?
(A) 23 (B) 24
(C) 25 (D) 26

471. Gene mutation is due to:
(A) Change in base sequence
(B) Change in sequence of cistrons in DNA
(C) Linkage and crossing over
(D) Random segregation

472. Mitochondrial DNA is advantageous for evolutionary studies because:
(A) It is inherited only through the female parent and thus evolves in a way that allows trees of relationship to be easily constructed
(B) It is inserted into the X chromosome
(C) It first appeared in humans and is not found in other animals
(D) It evolves more slowly than the genes in the nucleus

473. Name the mutagen:
(A) SO_2 (B) CO_2
(C) NO (D) HNO_2

474. Transversion is:
(A) Insertion of nitrogen base
(B) Replacement of a purine with a purine
(C) Replacement of a purine with pyrimidine
(D) Replacement of a pyrimidine with pyrimidine

475. Which of the following is not useful for phylogenetic analysis of human populations
(A) Ribosomal RNA genes
(B) Microsatellites
(C) Mitochondrial DNA
(D) Multiallelic genes

476. Recessive mutations are expressed in:
(A) Homozygous condition
(B) heterozygous condition
(C) Next generation (D) Same generation

477. Monosomic are:
(A) n- 1 - 1 (B) 2n+1
(C) 2n -2 - 1 (D) 2n -1

478. The plant on which de Vries worked in preparation of mutation theory was:
(A) Pansy (B) Evening Primerose
(C) Garden pea (D) Sweet pea

479. Sudden inheritable genetic change is:
(A) Mutation
(B) Inheritance of acquired trait
(C) Independent assortment
(D) Natural selection

480. Mutation is a change that is:
(A) Never inherited
(B) Inherited only in F2 generation
(C) Inherited
(D) Responsible for plant growth

481. An autosomal recessive disorder has a population frequency of 1:40000. The carrier frequency is therefore:
(A) 1/50 (B) 1/75
(C) 1/100 (D) 1/150

482. Colchicine interferes with:
(A) Chromosome replication
(B) Organisation of spindle
(C) Chromosome condensation
(D) Incorporation of nitrogen bases

483. Exact multiple of haploid number is:
(A) Euploid (B) Hyperploid
(C) Heteroploid (D) Aneuploid

484. Peptide nucleic acid (PNA):
(A) A polynucleotide analog in which the sugar phosphate backbone is replaced by amide bonds
(B) The use of molecular data to understand the genetic basis to variable phenotypes
(C) The use of molecular data to infer the evolutionary relationship between genomes
(D) The application of rigorous mathematical methods to analysis of variable characters

485. Haploid number in 10. What shall be tetrasomic number?
(A) 22 (B) 25
(C) 30 (D) 40

486. In pea, seed coat colour changes from grey to white. This is an example of:
(A) Induced mutation
(B) Spontaneous mutation
(C) Pleiotropic mutation
(D) Transformation

487. Polyploidy means occurrence of:
(A) Haploid sets of chromosomes
(B) Diploid sets of chromosomes
(C) More than diploid sets of chromosomes
(D) All the above

488. If the frequency of males affected with an x-linked recessive condition in a human population is 0.10 (one in ten), what will be the expected frequency of affected females?
(A) 0.00001 (B) 0.0001
(C) 0.001 (D) 0.01

489. A polyploidy with genomes derived from same original species is:
(A) Amphidiploid (B) Autopolyploid
(C) Autoallopolyploid (D) Allopolyploid

490. A polyploidy derived from F_1 hybrid between two species is:
(A) Autopolyploid (B) Autodiploid
(C) Amphidiploid (D) Hexaploid

491. The likelihood of an individual in a population carrying two specific alleles of a human DNA marker, each of which has a frequency of 0.2, will be:
(A) 0.32 (B) 0.16
(C) 0.10 (D) 0.08

492. An auxotroph is:
(A) Plant capable of synthesizing own carbohydrates
(B) Plant showing quick bending response to sunlight
(C) A mutant having lost the ability to synthesize one or more nutrients
(D) An organism dependent on another for nutritional requirements

493. Ultimate source of organic variations is:
(A) Mutations (B) Natural selection
(C) Hormonal activity (D) Isolation

494. A trisomic individual possesses extra chromosomes:
(A) 1 (B) 2
(C) 3 (D) 4

495. If haploid chromosome number is 10. The monosomic number shall be:
(A) 9 (B) 12
(C) 18 (D) 19

496. Which of the following mutations are not hereditary:
(A) Gametic (B) Somatic
(C) Genetic (D) Germinal

497. Aneuploidy is zygotic chromosome number:
(A) Thrice of gametic number
(B) Twice of gametic number
(C) Quadruple of gametic number
(D) Abnormal

498. Both extra-nuclear as well as nuclear materials are involved in transmission of hereditary information as:
(A) Mitochondria and plastids are having DNA
(B) Both cytoplasm and nuclear material are equally involved in hereditary
(C) Each type of organisms have particular mitochondria or plastids
(D) Some cells are procaryotic

499. When present in tissue culture medium, gibberellins:
(A) Helps to break dormancy of buds and bulbs
(B) Promotes dormancy development in buds and bulbs
(C) Is regarded as plant growth inhibitor
(D) Prevent normal recognition of auxin molecule

500. An example of covergent evolution is:
(A) Australian marsupials and placental mammals
(B) The flippers in fish, penguins and dolphins
(C) The wings in birds, bats and insects
(D) All of these

501. Cytoplasmic male sterility is passed down:
(A) Biparentally (B) Paternally
(C) Maternally (D) None of these

502. Cytosterility in the pollen of maize is due to:
(A) Extra chromosomal inheritance
(B) Complementary genes
(C) Nuclear inheritance
(D) Hypostasis

503. Cytoplasmic male sterility in maize is manifestation of interaction between:
(A) Chloroplast and nuclear genes
(B) Mitochondrial and nuclear genes
(C) Chloroplast and mitochondrial genes
(D) Cytoplasmic factors & male sterile genes

504. Which is not an example of cytoplasmic inheritance?
(A) Female sterility in maize
(B) Plastic inheritance
(C) Kappa particle inheritance
(D) Stigma particle inheritance

505. When a certain character is inherited only through female parent, it probably represents:
(A) Multiple plastid inheritance
(B) Cytoplasmic inheritance
(C) Incomplete dominance
(D) Mendelian nuclear inheritance

506. The two eukaryotic organelles responsible for cytoplasmic inheritance are:
(A) Lysosomes and mitochondria
(B) Chloroplasts and lysosomes
(C) Mitochondria and chloroplasts
(D) Mitochondria and Golgi complex

507. Which crop variety is not due to induced mutations?
(A) Reimei of Rice (B) Prabhat of arhar
(C) Sharbati Sonora of wheat (D) Aruna of castor

508. Haploids are preferred over diploids for mutation studies because in haploids:
(A) Recessive mutations express immediately
(B) Dominant mutations express immediately
(C) Mutations are readily induced
(D) Tissue culture is easy

509. Vir genes are important for gene transfer to plants because of which of the following reason?
(A) They act on the origin of replication of the Ti plasmid
(B) They prevent oncogenes from working
(C) They are responsible for the mobilization of T-DNA
(D) They enable Agrobacterium to penetrate through intact plant surfaces

510. The cranial capacity of which one of the following prehistoric humans was almost the same as that of the modern man?
(A) Neanderthal man (B) Peking man
(C) Java ape man (D) Australopithecus man

511. Which of the following situations might create a population bottleneck?
(A) A small number of animals migrating from one geographic location to another, leaving behind the rest of the population
(B) Overhunting or overfishing of a particular species over a short period of time
(C) Natural disasters that leave behind a fraction of a population
(D) All of these statements describe situations that may lead to a population bottleneck

512. Orthologos:
P. Are always located on the same chromosome
Q. Have similar amino acid sequence
R. Have identical promoters
S. Have similar functions
Which of the above statements are correct?
(A) P (B) R
(C) Q and R (D) Q and S

513. Alleles of different genes found on same chromosome may separate due to:
(A) Epistasis (B) Crossing over
(C) Continuous variations (D) Pleiotropy

514. Which of the following is suitable for experiment on linkage?
(A) aaBB x aaBB (B) AABB x aabb
(C) AaBb x AaBb (D) AAbb x AaBB

515. Consider the following:
1. Tyrannosaurus (dinosaur)
2. Lobe-finned fishes
3. Great apes
4. Eohippus (Horse)
5. Trilobites
Which of the following represents the proper sequence, from the earliest in the fossil record to the latest, of the animals listed above?
(A) 1, 2, 5, 4, 3 (B) 2, 5, 1, 3, 4
(C) 2, 5, 1, 4, 3 (D) 5, 2, 1, 4, 3

516. Eucaryotic chromosome is made of:
(A) DNA (B) DNA + protein
(C) DNA + lipids (D) RNA + lipids

517. A strong mutagene is:
(A) Cold (B) Heat
(C) Ice (D) X-ray

518. The number of base substitutions possible in amino acid codons is:
(A) 461 (B) 494
(C) 525 (D) 549

519. Out of A-T, G-C pairing, bases of DNA may exist in alternate valency state owing to arrangement called:
(A) Anologue substitution
(B) Tautomerisational mutation
(C) Frameshift mutation
(D) Point mutation

520. Choose the false statements:
P. No circular RNA viral genome are known to exist (exception is Hepatitis-delta virus)
Q. No DNA viruses are known to have segmented genome
R. Although oxygenic photosynthesis evolved long aerobic respiration, the endosymbiotic theory has the evolution of mitochondria receding the evolution of chloroplasts
S. According to Carl Woese, eubacteria are likely to have evolved from an archaebacterial ancestor
(A) Q (B) S
(C) P and Q (D) Q and S

521. Linked genes separate due to:
(A) Mutation (B) Recombination
(C) Crossing over (D) None of these

522. Two genes situated very close on the chromosome show:
(A) High crossing over
(B) Hardly any crossing over is detected
(C) No crossing over
(D) Only double cross-over can occur

523. Genes located on Y-chromosome are:
(A) Autosomal genes (B) Mutant genes
(C) Sex linked genes (D) Holandric genes

524. Point at which polytene chromosomes appear attached is:
(A) Centromere (B) Chromocentre
(C) Chromomere (D) Centriole

525. Trisomy of 21st chromosomes results in:
(A) Down's syndrome (B) Sickle cell anaemia
(C) Turner's syndrome (D) Klinefelter's syndrome

526. Match list – I with list – II and select the correct answer using the codes given below the lists:

List – I	List – II
P. Annual rhythm	1. The ability to time and repeat the functions at interval of 15 days
Q. Lunar rhythm	2. The ability to time and repeat the functions at interval of 24 hours
R. Ciradian rhythm	3. The ability to time and repeat the functions every year
S. Tidal rhythm	4. The ability to time and repeat the functions based on tidal phase

(A) P-1, Q–2, R–3, S–4 (B) P-2, Q–1, R–3, S-4
(C) P-3, Q–2, R–4, S–1 (D) P-3, Q–1, R–2, S-4

527. Which histone is not component of *nu* particle?
(A) H2 (B) H1
(C) H3 (D) H4

528. Complete haploid set of chromosomes of a species is:
(A) Genome (B) Allele
(C) Genetic code (D) Genotype

529. Circannual rhythm in an organism is:
P. Regulated by temperature
Q. Regulated by photoperiod
R. An innate character imprinted upon the organism by the environment
S. An acquired character developed by the animals in response to the frequent changes in the environment
Which of the above statements are correct?
(A) P and Q (B) Q and R
(C) Q and S (D) P and S

530. Which one is not related to chromosomal abnormality?
(A) Klinefelter's syndrome (B) Euploidy
(C) Aneuploidy (D) AIDS

531. Point mutation is:
(A) Loss of gene
(B) Change in a base of gene
(C) Addition of a gene
(D) Deletion of a segment of gene

532. Colchicine brings about:
(A) Polyploidy (B) Cell division
(C) Cell elongation (D) Cell differentiation

533. A man made allopolyploid is:
(A) Water melon (B) Gossypium
(C) Triticale (D) Triticum

534. Shoot organogenesis by culture results into:
(A) A bipolar structure that has no vascular connection with the explant
(B) A monopolar structure that has a strong connection with the pre-existing vascular tissue of the explant
(C) A monopolar structure that has no vascular connection with the explant
(D) A bipolar structure that has a strong connection with the pre-existing vascular tissue of the explant

535. Genome comparisons have suggested that mouse DNA has mutated about twice as fast as human DNA. What is a possible explanation for his discrepancy?
(A) Mice are much smaller than humans
(B) Mice live in much less sanitary conditions than humans and are therefore exposed to wider range of mutation-causing substances
(C) Mice have a smaller genome size
(D) Mice have a much shorter generation time

536. Cis-trans expression of genes is an example of:
(A) Mutation
(B) Intergenic crossing over
(C) Intragenic crossing over
(D) Cytoplasmic inheritance

537. Match list – I with list – II and select the correct answer using the codes given below the lists:

List – I	List – II
P. Temperature affects size of animals	1. Gloger's rule
Q. Temperature affects colouration of animals	2. Jordan's rule
R. Temperature affects size of extremities of certain animals	3. Bergman's rule
S. Temperature affects number of vertebrae in certain fish	4. Allen's rule

(A) P-3, Q–1, R–4, S-2 (B) P-3, Q–4, R–1, S-2
(C) P-2, Q–4, R–1, S-3 (D) P-2, Q–1, R–4, S-3

538. The substance which causes a definite change in genes is called:
(A) Mutagene (B) Toxin
(C) Cytotoxin (D) Alkaloid

539. Recessive mutagen is not expressed in:
(A) Homozygous male (B) Heterozygous male
(C) Heterozygous female (D) Homozygous female

540. Which is an example of cytoplasmic inheritance?
(A) Eye color in Drosophila (B) Flower colour in pea
(C) Height in pea (D) Sterile pollen

541. Inclusive fitness:
P. Is the sum of the direct and indirect fitness effects of an individual's behaviour
Q. Provides an evolutionary explanation for altruistic behaviour among animal communication
(A) P (B) Q
(C) P and Q (D) None of these

542. In a chromosome the protein content is:
(A) Nil (B) Trace
(C) Half of DNA (D) Same as DNA

543. Number of linkage in a polynucleotide would be:
(A) Same as number of nucleotide
(B) Twice the number of nucleotides
(C) One less than the number of nucleotides
(D) One

544. What causes mutations?
(A) Colchicine (B) Cosmic rays
(C) Gamma rays (D) Crossing over

545. Match Group – I with Group – II and select the correct answer using the odes given below the lists:

Group – I	Group – II
P. Insect resistant cotton	1. Bt
Q. Golden rice	2. Round up
R. 'Flavr – Savr' tomato	3. 2, 4-D
S. Herbicide tolerant soyabean	4. Carotenoids
	5. Ferritin
	6. A -deaminase

(A) P-2, Q–5, R–1, S–3 (B) P-1, Q–4, R–6, S-2
(C) P-1, Q–4, R–6, S–3 (D) P-2, Q–4, R–6, S–1

546. Number of chromosomes can increase or decrease due to:
(A) Genetic repetition (B) Mutation
(C) Nondisjunction (D) All of these

547. Gene is formed of:
(A) Polynucleotide (B) Histone
(C) Hydrocarbon (D) Lipoprotein

548. Which one is a triploid?
(A) Guava (B) Apple
(C) Orange (D) Banana

549. RNA is supposed to be the first genetic material in the evolutionary process. It performs both catalytic and genetic functions. Which of the following is/are correct about RNA as catalyst (ribozyme)?
P. Display self splicing
Q. Catalyses peptide bond formation
R. RNase P is an example of ribozyme
S. Catalyzes formation of DNA
(A) P and Q (B) Q, R and S
(C) P, Q and R (D) P, Q, R and S

550. Mutation are responsible for:
(A) Extinction of organisms
(B) Variations in population
(C) Increase in population
(D) Decrease in population

551. Which of the following statements are true on DNA delivery methods during plant transformations?

P. Single stranded nicks are made in T-DNA border repeat by the VirD1, VirD2 and VirD3 protein complex

Q. VirA Gene products from the export apperaratus on the membrane for the transfer of T-DNA

R. Gold/Tungsten particles are used as microprojectiles in biolistic method

S. Acceleration of DNA-coated microprojectiles is carried out with compressed CO_2

(A) P and S (B) R and S
(C) P and R (D) Q and S

552. Hexaploid or modern wheat developed through:
(A) Chromosome doubling
(B) Hybridomas
(C) Hybridisation
(D) Hybridisation and chromosome doubling

553. A fruitfly heterozygous for sex linked genes is mated with normal female fruitfly. Male specific chromosome will enter the egg cells in proportion of:
(A) 1 : 1 (B) 2 : 1
(C) 3 : 1 (D) 4 : 1

554. Genetic identity of human male is known by:
(A) Reverse mutation (B) Frameshift mutation
(C) Autosomes (D) Sex chromosomes

555. A mutation at one base of first codon of a gene forms a nonfunctional protein. It is:
(A) Nonsense mutation
(B) Nucleolus
(C) Cell organelles
(D) Mis-sense mutation

556. Which of the following comparison between RFLPs, RAPDs and AFLPa is/are correct?

	RFLPa	RAPDs	AFLPs
P. PCR based	No	Yes	Yes
Q. Polymorphism	Moderate	High	High
R. Codominant	Yes	No	No
S. Reproducibility	High	Medium	High

(A) P (B) Q
(C) P and S (D) P, Q, R and S

557. Process of altering code in DNA where message is reversed is:
(A) Deletion (B) Insertion
(C) Inversion (D) Movement

558. A change in chromosomal number is called:
(A) Polyploidy
(B) Aneuploidy
(C) Chromosomal mutation
(D) Somatic mutation

559. *Triticum aestivum* (Bread wheat) is:
(A) Tetraploid (B) Hexaploid
(C) Diploid (D) Triploid

560. Which of the following statements are true about glyphosate tolerant transgenic plants?

P. Transgenic plants detoxify glyphosate

Q. Transgenic plants produce an altered enzyme that is not affected by glyphosate

R. Transgenic plants sequester glyphosate in vacuoles

S. Transgenic plants overcome the inhibition amino acid biosynthesis

(A) P and Q (B) P and R
(C) Q and S (D) R and S

561. Numerical change in chromosome number which is not the exact multiple of haploid genome is:
(A) Haploid (B) Autopolyploid
(C) Allopolyploid (D) Aneuploid

562. Lampbrush chromosomes are also called diplotene chromosomes because they:
(A) Show chiasmata
(B) Resemble diplotene chromosomes
(C) Are the permanent diplotene stage
(D) All the above

563. Phytomediation processes includes:

P. **Phytoextration:** Uptake and concentration of substances from the environment into the plant biomass

Q. **Phytostabilization:** Reducing the mobility of substances in the environment

R. **Phytotransformation:** chemical modification of environment substances as a direct result of plant metabolism

S. **Phytostimulation:** Enhancement of soil microbial activity for the degradation of contaminants

(A) P and Q (B) P, Q and R
(C) P, Q and S (D) P, Q, R and S

564. Mutation which are not inherited are:
(A) Somatic (B) Gametic
(C) Zygotic (D) Genetic

565. Which one is de novo mutation?
(A) TT x TT $\rightarrow$ TT (B) tt x tt $\rightarrow$ Tt
(C) Tt x tt $\rightarrow$ Tt (D) Tt x Tt $\rightarrow$ TT, Tt, tt

566. Lampbrush chromosomes were described by:
(A) Balbiani (B) Benzer
(C) Dupraw (D) Ruckert

567. Basic structural unit of eucaryotic chromosome is:
(A) Chromatin fibre (B) DNA
(C) Chromonema (D) Nucleosome

568. Which of the following are commonly used as reporter genes?

P. NPT gene
Q. Luciferase gene
R. CFTR gene
S. GFP gene

(A) Q and S (B) P and R
(C) R and S (D) P and Q

569. A mutation results in change in:
(A) Sequence of amino acids in a protein
(B) tRNA sequence of nucleotides
(C) rRNA of ribosomes
(D) None of these

570. The two chromatids of a chromosome are attached in the region of:
(A) Telomere (B) Centromere
(C) Chromomere (D) None of these

571. Free mating is due to:
(A) Sex reversal by gene
(B) Environmental control of sex
(C) Hormonal control of sex
(D) Sex determination by chromosomes

572. Down's syndrome is due to:
(A) Autosome (B) Sex chromosome
(C) Sex linked disease (D) Duplication

573. A nucleosome is formed of:
(A) Histone and linker DNA
(B) Histone with wrapper DNA
(C) DNA (D) RNA

574. Rearrangement of genes occurs due to:
(A) Translocation and duplication
(B) Translocation and deficiency
(C) Deletion and deficiency
(D) Translocation and inversion

575. Colchicine arrests cell division at:
(A) Prophase (B) Metaphase
(C) Anaphase (D) Telophase

576. DNA is associated with basic protein:
(A) Albumin (B) Nonhistone
(C) Histone (D) Both (A) and (B)

577. Hyperchromism is presence of:
(A) Some chromosome more than once
(B) Some type of chromosomes less than once
(C) Variable chromosomes in nucleus
(D) None of these

578. What are the assumptions of Hardy-Weinberg equilibrium?
(A) Small population size, random mating, no selection, no migration, no mutation
(B) Large population size, random mating, no selection, no migration, no mutation
(C) Large population size, random mating, heterozygote survive the best, no migration, mutation
(D) Large population size, like individuals mate, no selection, no migration, no mutation

579. Euploidy is:
(A) One chromosome more than haploid set
(B) One chromosome more than diploid set
(C) One chromosome less than haploid set
(D) Exact multiple of haploid set of chromosomes

580. Chromosomes are made of:
(A) DNA + Pectin (B) RNA + DNA
(C) DNA + Histone (D) DNA only

581. Mutation refers to sudden change in:
(A) Genotype (B) Phenotype
(C) Metabolic rate (D) Genetic make up

582. In the endosymbiont hypothesis for the origin of eukaryotic cells, which of the following is proposed as the role of cyanobacteria?
(A) They become the chlorophyll of plant cells
(B) They are the ancestor of mitochondria
(C) They are the ancestor of the chloroplasts
(D) They close their DNA and become the vacuoles of plant cells

583. How many genomes are present in a typical green plant cell?
(A) 1 (B) 2
(C) 3 (D) 4

584. Formation of multivalents in meiosis is due to:
(A) Deletion (B) Inversion
(C) Monosomy (D) Reciprocal translocation

585. DNA parts which can switch their position are:
(A) Exons (B) Introns
(C) Cistron (D) Transposons

586. The concept of random fluctuations in gene frequency as an evolutionary force was originally proposed by:
(A) Muller (B) Hardy & Weinberg
(C) Sewall Wright (D) Wallace

587. Chromosome aberration occurs due to:
(A) Aneuploidy (B) Polyploidy
(C) Physical effects (D) All the above

588. Cytoplasmic inheritance is also called:
(A) Maternal inheritance
(B) Cytoplasmic association
(C) Clonal inheritance (D) None of these

589. Linkage decreases frequency of:
(A) Recessive allele (B) Dominant allele
(C) Hybridisation (D) Both (B) and (C)

590. Which of the following statements defines anagenesis?
(A) Anagenesis describes the divergence between members of the same species into races
(B) Anagenesis is the process that occurs when one species gives rise to a new distinct species
(C) Anagenesis is the process whereby one species gives rise to two new species
(D) Anagenesis is the calculations used to develop a phylogenetic tree

591. Gene located on differential region of Y-chromosome are called:
(A) Mutant genes (B) Holandric genes
(C) Autosomal genes (D) XY linked genes

592. Exchange of one part of a chromosome with a part of the same or another chromosome is:
(A) Linkage (B) Crossing over
(C) Translocation (D) Inversion

593. A material which arrests cell division is obtained from:
(A) Colchicine (B) Chrysanthemum
(C) Cactus (D) Dalbergia

594. A reason for maternal inheritance is due to genes present in:
(A) Cytoplasm (B) Mitochondria
(C) Golgibody (D) Nucleolus

595. The genus *Rhagoletis* is a group of small, brightly coloured flies. Each species in the genus feeds, during its larval stage, on the fruit of just one plant family, and when the larvae hatch into adults, they court and mate on the same fruit. The species that feed on fruits from different plant families are reproductively isolated by:
(A) Ecological isolation (B) Temporal isolation
(C) Behavioral isolation
(D) Mechanical isolation

596. Which is the main category of mutations?
(A) Genetic mutation (B) Zygotic mutation
(C) Somatic mutation (D) All of these

597. Diagram representing relative features and number of chromosomes is:
(A) Idiotype (B) Idiogram
(C) Karyogram (D) Karyotype

598. Number of linkage groups in Maize with ten pairs of chromosomes is:
(A) 5 (B) 10
(C) 15 (D) 20

599. Homology in antomical parts helps in determining evolutionary kinship because:
(A) Homologous body parts invariably perform similar functions
(B) Undergo similar genetic changes
(C) Display evolutionary adaptations
(D) Have common embryological origin

600. Cross over frequency is proportional to:
(A) Phenotypic recombinant frequency
(B) Diploid number of chromosomes
(C) Haploid number of chromosomes
(D) Genotypic recombinant frequency

601. Chromosome complement 2n -1 is:
(A) Monosomy (B) Nullisomy
(C) Trisomy (D) Tetrasomy

602. The phenomenon of movement of DNA segment from one chromosome to another is:
(A) DNA replication
(B) DNA transposition
(C) DNA recombination
(D) DNA hybridization

603. Determination of percentage of crossing over between two linked genes is important in:
(A) Maintaining heterozygosity in population.
(B) Indicating relative position of genes in chromosomes
(C) Fixation of heterosis in organism
(D) Explaining the phenomenon coupling repulsion

604. When two related populations that have been geographically isolated come back into contact, retain their reproductive isolation, and compete for resources, there may be a divergence in traits known as:
(A) Character convergence
(B) Sibling speciation
(C) Phyletic evolution
(D) Character displacement

605. Transition mutation is due to replacement of:
(A) GC by TA (B) AT by CG
(C) CG by GC (D) AT by GC

606. In Drosophila, XXY is female. In Humans it represents an abnormal male because:
(A) Y-chromosome induces male traits in humans
(B) Y-chromosome is essential for female sex in Drosophila
(C) Y-chromosome is not essential for male sex in human
(D) None of these

607. In Drosophila, during organ differentiation, one organ can be replaced by another like wings by legs. Genes responsible for it are:
(A) Plastid genes
(B) Homeotic genes
(C) Complementary genes
(D) Supplementary genes

608. Green fluorescent protein (GFP) cloned from jellyfish has now wide application in biological research. The fluorescence emitted by GPF is due to:
(A) Presence of two zinc ions in GPF molecule
(B) Heme, that serves as a prosthetic group in GPF molecule
(C) Three amino acids residues within GPF molecule
(D) Whole GPF molecule

609. Sex linked disorder is:
(A) Diphtheria (B) Colour blindness
(C) Tuberculosis (D) Leprosy

610. Extra chromosomal genetic material is:
(A) Prion (B) Cosmid
(C) Phasmid (D) Plasmid

611. A colour blind person cannot distinguish:
(A) Red and green (B) Green and blue
(C) Yellow and white (D) Black and yellow

612. Down's syndrome is due to extra chromosome:
(A) 20 (B) 21
(C) 24 (D) 26

613. The allele associated with sickle cell anemia apparently reached a high frequency in some human populations due to:
(A) Random mating
(B) Superior fitness of heterozygotes in areas where malaria was present
(C) Migration of individuals with the allele into other populations
(D) A high mutation rate at that specific gene

614. Place of gene on a chromosome is called:
(A) Locus (B) Chromomere
(C) Focus (D) Centrosome

615. One of the following is a random process:
(A) Variations (B) Adaptations
(C) Evolution (D) Mutations

616. Is it possible to say when a gene will mutate:
(A) Always (B) Sometimes
(C) Never (D) Rarely

617. What is the difference between a PAC and a BAC?
(A) One is a plasmid –based vector and the other is derived from a yeast chromosome
(B) One is derived from bacteriophage P1 can another is derived from bateriophage B1
(C) One is derived from the *E. coli* F-plasmid and other is derived from bacteriophage P1
(D) None of these

618. Turner's syndrome is due to:
(A) Trisomy of chromosome 21
(B) Trisomy of chromosome 18
(C) Autosomal recessive gene
(D) Absence of one sex chromosome

619. Genes located on mitochondrial DNA bring about:
(A) Paternal inheritance
(B) Maternal inheritance
(C) Biparental inheritance
(D) There is no inheritance

620. T-DNA is:
(A) DNA of plasmid origin which is transferred to the Agrobacterium chromosome
(B) DNA from the chromosome of Agrobacterium species which is transferred to the plant genome
(C) DNA of plasmid origin which is transferred to the plant genome
(D) Telomeric DNA of linear chromosome

621. Lyon's hypothesis is connected with:
(A) Number of barr bodies
(B) Genetic compatibility
(C) Genetic incompatibility
(D) Centromere position

622. Gene which increases frequency of mutations of other genes is:
(A) Mutagen (B) Complementary gene
(C) Mutator gene (D) Hypostatic gene

623. Frequency of recessive allele is 0.4. What is the frequency of heterozygous carriers in the population of 200?
(A) 96 (B) 100
(C) 116 (D) 126

624. The ratio of phenotypes in F_2 of a monohybrid cross is:
(A) 3 : 1 (B) 1 : 2 : 1
(C) 9 : 3 : 3 : 1 (D) 2 : 1 : 1

625. A pure tall pea was crossed with a pure dwarf pea. All the plants of F_1 were found to be tall. This is due to:
(A) Dominance
(B) Disappearance of factor for dwarfness in F_1 generation
(C) Coordination
(D) Segregation of factors

626. The monohybrid genotypic ratio 1:2:1 in F_2 generation indicates
(A) Segregation
(B) Dominance
(C) Independent assortment
(D) Incomplete dominance

627. Which of the following gene-therapy vectors preferentially infest nerve cells?
(A) Adeno-associated virus (B) Retrovirus
(C) Herpes virus (D) Adenovirus

628. The F_2 generation of a cross produced identical phenotypic and genotypic ratio. It is not an expected Mendelian result, and can be attributed to:
(A) Independent assortment
(B) Linkage
(C) Incomplete dominance
(D) Complete dominance

629. The law of independent assortment can be related to one of the following:
(A) Both homologous chromosomes of each pair are received by a single gamete
(B) The genes present on the same chromosome get randomly redistributed
(C) The non-homologous chromosomes show random distribution during anaphase-I of meiosis
(D) The exchange of segments between the non-homologous chromosomes

630. If organism is triploid, then Hardy-Weinberg theorem applicable will be:
(A) $(p + q)^2$ (B) $(p + q)$
(C) $(p + q + r)^2$ (D) $(p + q - r)^2$

631. The factors which represent the contrasting pairs of characters are called:
(A) Dominants (B) Alleles
(C) Homologous pairs (D) Dominant & recessive

632. Which of the following virus is not used as gene therapy vectors?
(A) Retroviruses
(B) Adenoviruses
(C) Adeno-associated viruses
(D) Polyomavirus

633. Chromosome walking is best described as:
(A) Aligning DNA sequences by computer to generate contigs
(B) Generating a map along a chromosome in a step-by-step manner
(C) Identifying clones whose inserts overlap to generate a library of clones that over a given segment of DNA
(D) Sequencing a genome at a time to ensure that no gaps are present at the end of the project

634. One of the following did not constitute the seven contrasting pairs of characters noticed by G.J. Mendel in pea:
(A) Colour of pod (B) Shape of the leaves
(C) Shape of pod (D) Height of the plants

635. In genetics the term test cross means:
(A) The crossing of F_1 individual with homozygous recessive
(B) Crossing an F_1 individual with either of the two parents
(C) Crossing F_1 individual with another F_1 individual
(D) Crossing F_1 individual with that of F_2

636. Which of the following is most likely to prevent extinction of a rare allele in which homozygosity represents a genetic lethal?
(A) New mutation (B) Genetic drift
(C) Founder effect
(D) Balanced polymorphism

637. The hexokinase (Hk) locus in Indian sparrow has allelic frequencies of 0.8 and 0.2, respectively, for the alleles Hk-A and Hk-B. What is the probability that two heterozygous individuals will male, given that Hk is in Hardy-Weinberg equilibrium?
(A) 0.01 (B) 0.10
(C) 0.16 (D) 0.32

638. How many types of gametes are expected from the organism with genotype AABBCC?
(A) 1 (B) 2
(C) 4 (D) 6

639. Which of the following is not a feature of molecular polygenetics?
(A) The use of molecular data to reconstruct a phylogenetic tree
(B) The use of molecular data to understand the genetic basis to variable phenotypes
(C) The use of molecular data to infer the evolutionary relationship between genomes
(D) The application of rigorous mathematical methods to analysis of variable characters

640. The basis for development of leukemia in some children treated with a retrovirus–derived genetherapy vector is:
(A) Insertion of the viral genome adjacent to an oncogene
(B) Use of a virus that contains an oncogene
(C) Insertional mutation of tumor suppressor gene
(D) Stimulation of immune response to cell surface antigens introduced by an inserted gene

641. A plant with a genotype Aa Bb is crossed with a plant having the genotype aa bb. The genotype of F_1 would be:
(A) AaBb, AABB
(B) aabb, aaBb
(C) aaBB, AABb
(D) AaBb, Aabb, aaBb, aabb

642. How many types of gametes are expected from a plant with genotype PpQq, provided there occurs independent assortment?
(A) 4 (B) 6
(C) 8 (D) 10

643. *Agrobacterium rhizogenes* can be used to genetically engineer plants because of which of the following statement?
(A) It contains a Ti plasmid
(B) Infected plants produce a good root system
(C) It infest all type of plants
(D) It causes gene transfer to plants

644. The lengths of the branches of a phylogenetic tree constructed from DNA sequence data indicate:
(A) The length of time since the organisms diverged
(B) The number of synonymous changes between the genes
(C) The degree of different between the genes represented by the nodes
(D) None of these

645. Which of the following crosses would produce a genotypic ratio of 1 : 2 : 1 in F_2?
(A) AB x AB (B) ab x ab
(C) Ab x Ab (D) ab x ab

646. In *Mirabilis jalapa* when two F_1 pink flowered plants were crossed with each other, the F_2 generation produced 40 red, 80 pink and 40 white flowering plants. This is a case of:
(A) Duplicate genes (B) Lethal genes
(C) Incomplete dominance (D) Epistasis

647. Genotype-phenotype concept was first proposed by:
(A) Bateson (B) Johannsen
(C) Sutton and Boveri (D) Punnet

648. The Biological clock in higher vertebrates is regulated by:
(A) The pituitary gland (B) Cerebral cortex
(C) Supra-chiasmatic nucleus in hypothalamus
(D) Thymus

649. Which of the following methods for studying loss of gene function does not involve any modification of the genome?
(A) RNA interference by injection of double stranded RNA
(B) Expression of an integrated antisense transgene
(C) Gene knockout by homologous recombination
(D) None of these

650. The phenomenon which defines the independent assortment is:
(A) Segregation (B) Crossing over
(C) Dominance (D) Linkage

651. A pea with white flowers was crossed to another pea which is also white flower plant. When selfed the F_2 generation produced purple and white in the ration 9 : 7. The reason for the results is that:
(A) It is typical monohybrid Mendelian ratio
(B) Purple flower colour is dominant over the white
(C) It is a complementary factor
(D) None of these

652. Possession by two or more species of a similar or identical character state that has not been derived by both species from their common ancestor is called:
(A) Homoplasy (B) Homology
(C) Homozygote (D) Homoerotic mutation

653. The law of segregation of characters is also called the law of purity of gametes because:
(A) Gametes have only one of the two alleles for each character
(B) Gametes cannot be contaminated
(C) Gametes are different type of cells
(D) It was just another name adopted accidentally

654. Identify the incorrect statement:
(A) 2, 4-dichlorophenoxyacetic acid is the most commonly used chemical analog of indole-3-acetic acid
(B) In somatic embryogenesis, embryo initiation needs a high concentration of 2, 4-D
(C) Crown-gall disease depends on the presence of Ti-plasmid in *Agrobacterium tumefaciens*
(D) *Agrobacterium tumefaciens* is responsible for hairy root formation in plants

655. The genes for same trait present on monohomologous chromosomes are:
(A) Alleles (B) Linked genes
(C) Multiple alleles (D) None of these

656. Recombinant live attenuated vain against hepatitis B was prepared from:
(A) Plasma of infected individual
(B) Recombinant yeast expressing hepatitis B surface antigen
(C) Recombinant vaccinia virus expressing hepatitis B surface antigen
(D) Transgenic plants expressing hepatitis B surface antigen

657. In a coral island in Atlantic Ocean, a natural calamity killed most of its population. The island is repropulated with the surviving individuals and their progeny. The present day population in the island shows high incidence of a rare recessive genetic disorder. What could be the most probably evolutionary force is responsible for this phenomenon?
(A) Geographic isolation
(B) Genetic drift
(C) Selective advantage of the recessive disorder
(D) Natural selection

658. Mendel formulated the law of heredity considering seven pairs of contrasting characters in the pea plant. If he had studied an eight pair, the law which would have been altered is:
(A) Law of segregation
(B) Law of dominance
(C) Law of independent assortment
(D) Law of purity of gametes

659. Medel was successful in formulating the law of inheritance whereas his predecessors were not because:
(A) He studied one clear–cut character at a time
(B) The characters studied by him were present on separate chromosomes
(C) Of the right choice of material
(D) He kept accurate records of his experiments

660. Appearance of hidden character in some progeny of F_2 population indicates:
(A) Law of purity of gametes
(B) Law of independent assortment
(C) Law of dominance
(D) Law of unit characters

661. Medel's law of segregation is based upon the F_2 ratio of:
(A) 2 : 1 : 2 (B) 9 : 3 : 3 : 1
(C) 1 : 2 : 1 (D) 3 : 1

662. The difference between transfection and transduction is:
(A) In transfection, the transgene is inserted in a plasmid, while in transduction, the transgene is inserted in a viral genome
(B) Transfection involves the transfer of naked DNA into the cell while transduction involves packaging the DNA into a virus particle, which then infest the cell
(C) There is no difference the terms are synonymous
(D) None of these

663. If dwarf pea plant was treated with GA_3 it grew as tall as the pure tall pea plant. If this treated plant is crossed with a pure tall plant the phenotypic ratio F_1 is likely to be:
(A) All tall
(B) All dwarf
(C) 50% tall and 50% dwarf
(D) 75% tall and 25% dwarf

664. From an evolutionary point of view, plant tendrils derived from leaves and those derived from stems are example of:
(A) Analogous structures resulting from convergence only
(B) Homologous structures resulting from convergence only
(C) Analogous structures resulting from divergence only
(D) Homologous structures resulting from divergence only

665. In a dihybrid cross AABB x aabb, F_2 progeny of AABB, AABb, AaBB and AaBb occurs in the ratio of:
(A) 1 : 1 : 1 : 1 (B) 9 : 3 : 3 : 1
(C) 1 : 2 : 2 :1 (D) 1 : 2 : 2 : 4

666. A microarray is a large collection of specific DNA oligonucleotides spotted in a defined pattern on a microscope slide. What is the most useful experiment that can be done with such a tool?
(A) Predicting the presence of specific metabolites in a cell
(B) Comparing newly synthesized nuclear RNA with cytoplasmic RNA to locate introns
(C) Comparing RNA produced under two different physiological conditions to understand patterns of gene expression
(D) Comparing proteins produced under two different physiological conditions to understand their function

667. Genes do not occur in pairs in:
(A) Zygote (B) Endosperm cells
(C) Somatic cell (D) Gametes

668. Sewall Wright's metaphor of evolution on an adaptive landscape describes a population as occupying a field of genotypes graded by contours delimiting gene combinations of equal fitness. The evolutionary mechanism most likely to move a population consistently "uphill" toward adaptive peaks on this landscape is:
(A) Chance (B) Directed mutation
(C) Natural selection
(D) Gene flow between populations

669. Segregation of genes take place during:
(A) Metaphase (B) Anaphase
(C) Prophase (D) Embryo formation

670. A cross between hybrid and a parent is known as:
(A) Test cross (B) Back cross
(C) Monohybrid cross (D) Reciprocal cross

671. A pleiotropic gene is one which:
(A) Affects one character
(B) Affects more than one character
(C) Requires another gene for expression
(D) Supplements the effect of another gene

672. In an experiment on pea plant, pure plants with yellow round seeds were crossed with plants producing green wrinkled seeds. What will be phenotypic ratio of F_1 progeny?
(A) 9 yellow round : 3 round green : 3 wrinkled yellow : 1 green wrinkled
(B) All yellow round
(C) 1 yellow round : 1 round green : 1 wrinkled yellow : 1 green wrinkled
(D) All wrinkled green

673. Which one of the following are not the criteria for using ribosomal RNAs as evolutionary chronometers?
(A) The large size of ribosomal RNAs gives large number of possible sequence combinations
(B) Ribosomal RNAs are functionally invariant
(C) Ribosomal RNAs are universally distributed
(D) Ribosomal RNAs have enzyme activity

674. When a wheat variety of red kernels (homozygous for two nonallelic and independent dominant genes) is crossed with white kernelled wheat (homozygous for two recessive nonallelic independent genes), the phenotypic ratio in F_2 generation would be:
(A) 9 : 7 (B) 1 : 10 : 4 : 1
(C) 1 : 4 : 6 : 4 : 1 (D) 1 : 2 : 4 : 2 : 4 : 2 : 1

675. In multiple allele system a gamete possesses:
(A) One alleles (B) Two alleles
(C) Three allele (D) Several alleles

676. Sum total of all the genetic information in the breeding members of a population at a given time is known as:
(A) Gene pool (B) Gene frequency
(C) Genome (D) Genetic Drift

677. The spread of genes from one breeding population to another by migration which may result in changes in gene frequency is called:
(A) Genetic Drift (B) Gene flow
(C) Genetic clone (D) Gene frequency

678. A genetic clone is:
(A) Plants produced by asexual means
(B) Heterozygous plant produced by sexual means
(C) Homozygous plant produced by sexual means
(D) Hybrid produced by sexual means

679. Gaseous mixture used by Miller for synthesis of amino acids through heat and electric discharge included:
(A) Methane, ammonia, hydrogen and water vapours
(B) Methane, ammonia, nitrogen and water vapours
(C) Ammonia, carbon dioxide, hydrogen and water vapours
(D) Methane, carbon dioxide, nitrogen and water vapours

680. A pleitropic gene is:
(A) I^A (B) Hb^S
(C) Hb^A (D) I^B

681. Dominant gene for tallness is T and for yellow colour is Y. A plant heterozygous for both the traits is selfed, then the ratio of pure homozygous dwarf and green offspring would be:
(A) 4/16 (B) 3/16
(C) 2/16 (D) 1/16

682. A gene that shows its effect on more than one character is:
(A) Multiple gene (B) Pleiotropic gene
(C) Multi gene (D) Polygene

683. A dihybrid ratio of 1 : 4 : 6 : 4 : 1 is obtained instead of 9 : 3 : 3 : 1. This is an example of:
(A) Complementary genes
(B) Pleiotropic genes
(C) Polygenic inheritance
(D) Supplementary genes

684. A device for the detection of an analyte that combines a biological component with a physiochemical detector component is known as:
(A) Biosensor (B) Piezoelectric
(C) Biomimic (D) Bioremediation

685. Epistatic gene differs from dominant gene in:
(A) Epistatic gene is nonallelic
(B) Epistatic gene never express itself independently
(C) Epistatic and hypostatic genes are present at different loci
(D) All of these

686. Genotype is:
(A) Genetic composition of an organism
(B) Genetic composition of plastids
(C) Genetic composition of germ cells
(D) Genetic composition of an individual

687. The slope of the beaks of Darwin's flinches, industrial melanism, and the changes in horse teeth are all examples of:
(A) Artificial selection
(B) Natural selection
(C) Covergent evolution
(D) Homologous structure

688. Tall red flowered pea plant crossed to dwarf white flowered plant yields only tall red flowered plants. A test cross shall give a ratio of:
(A) 1 : 1 : 1 (B) 3 : 1 : 1
(C) 1 : 1 : 1 : 1 (D) 1 : 2 : 4 : 6 : 4 : 2 : 1

689. RR (red) *Antirhinum* is crossed with white (WW) one. Offspring RW are pink. This is an example of:
(A) Dominant – recessive
(B) Incomplete dominance
(C) Hybrid
(D) Supplementary genes

690. Dominant lethal gene is one which:
(A) Determines sex of offspring
(B) Allows the organism to survive but not reproduce
(C) Allows the organism to survive and reproduce
(D) Kills the organism

691. Which of the following is not true regarding adeno-associated virus?
(A) It contains double-stranded DNA
(B) It may integrate into the host chromosome
(C) It may exist in the ell as an episome
(D) It may be associated with insertional mutagenesis when used as a gene therapy vector

692. Which of the following is true of Mendelism but not of Darwism?
(A) It influenced human thought tremendously
(B) It was modified immediately
(C) It gave well defined principles even in early stage
(D) It was based on insufficient data

693. The three principles of Mendelism are:
(A) Dominance-recessiveness, segregation and independent assortment
(B) Linkage, segregation and independent assortment
(C) Linkage, dominance-recessiveness, and segregation
(D) Linkage, dominance-recesiveness and independent assortment

694. In a plant, the alleles of neither black flower colour nor white flower color are dominant. When both are present, the flower colour is grey. If two grey flowered plants are crossed the progeny will appear in the ratio of:
(A) All grey
(B) Either black or white
(C) 2 grey : 1 black : 1 white
(D) 1 black : 1 white

695. A pure tall pea plant is crossed with pure dwarf pea plant. The progeny is self pollinated. The ratio of true breeding tall pea plants to true breeding dwarf pea plants shall be:
(A) 2 : 1 (B) 1 : 1
(C) 3 : 1 (D) 1 : 2 : 1

696. Evalutionary changes in one species prompt corresponding changes in other species with which the former interacts ecologically. This process is known as:
(A) Coevaluation (B) Complementation
(C) Parallel evolution (D) Microevolution

697. Heterozygous tall (Tt) and dwarf (tt) pea plants are crossed. The ratio in the offspring shall be:
(A) All tall (B) All dwarf
(C) 3 tall : 1 dwarf (D) 1 tall : 1 dwarf

698. Alleles of a gene are found on:
(A) Same chromosome
(B) Homologous chromosomes
(C) Nonhomologous chromosomes
(D) Any chromosomes

699. Parents are heterozygous albino. What will be the first three children?
(A) Some normal, heterozygous and albino
(B) All normal
(C) No normal (D) All heterozygous albino

700. Arabidopsis is advantageous for plant genetic research because:
(A) It is commercially important as a food crop
(B) It is an endangered species
(C) It is the closest to humans of any existing plant
(D) It is a small plant with a small genome size which can be raised inexpensively

701. Which is a dominant trait?
(A) Colour blindness (B) Albinism
(C) Haemophilia (D) Rh factor

702. TT mates with Tt. What will be characteristic of offspring?
(A) 25% recessive (B) 50% recessive
(C) 75% recessive (D) All dominant

703. The allele which is unable to express its effect in the presence of another is called:
(A) Codominant (B) Supplementary
(C) Complementary (D) Recessive

704. Lethal genes are:
(A) Causative for appearance of ancestral traits
(B) Always recessive
(C) Gene present on different chromosomes but influencing single trait.
(D) Killer in homozygous state

705. The first vertebrates appeared in which one of the following periods of the Palaeozoic era?
(A) Ordovician (B) Silurian
(C) Devonian (D) Mississippian

706. The procedure adopted by Mendel which was different from his predecessors is:
(A) Keeping breeding records
(B) Employing many traits at one time
(C) Differentiating inherited traits
(D) Quantitative analysis of data

707. Mendel did not propose:
(A) Dominance (B) Incomplete dominance
(C) Segregation (D) Independent assortment

708. Genotype of F_1 individuals can be tested by:
(A) Backcross with homozygous recessive parent
(B) Reciprocal crossing
(C) Backcrossing with heterozygous parent
(D) Backcrossing with homozygous dominant parent

709. A dihybrid condition is:
(A) tt Rr (B) Tt rr
(C) tt rr (D) Tt Rr

710. Which one of the following statements about circadian rhythm is correct?
(A) Circadian rhythm, dependent on endogen- ous clock, is restricted to birds & mammals
(B) An animal's circadian rhythm has a period of exact 24 hours
(C) The period of circadian rhythm is not exactly 24 hours
(D) Circadian rhythm is necessary because earth is titled on its axis

711. Which contributed to the success of Mendel?
(A) Qualitative analysis of data
(B) Observation of distinct inherited traits
(C) His knowledge of biology
(D) Consideration of one character at one time

712. Pleiotropic gene has:
(A) Single genotype (B) Multiple genotype
(C) Single phenotype (D) Multiple phenotype

713. An organism with two identical alleles is:
(A) Dominant (B) Hybrid
(C) Heterozygous (D) Homozygous

714. Hybrid of pure red flowered and white flowered pea is crossed to pure red flowered plant. The progeny shall be:
(A) White flowered
(B) Red flowered
(C) Both red and white flowered plants
(D) Mostly red flowered plants

715. A heterozygous round seeded plant is crossed to recessive wrinkled seeded plant. The progeny would be:
(A) 303 rounded : 301 wrinkled
(B) 301 rounded : 100 wrinkled
(C) 200 rounded : 100 wrinkled
(D) 99 rounded : 301 wrinkled

716. In a genetic cross having recessive epistasis, F_2 phenotypic ratio would be:
(A) 9 : 6 : 1 (B) 15 : 1 : 1
(C) 9 : 3 : 4 (D) 12 : 3 : 1

717. What is fate of most duplicated genes?
(A) Gene activation
(B) Gain of a novel function through subsequent mutation
(C) They are transferred to a new organism using lateral gene transfer
(D) They become orthologs

718. The contrasting pairs of factors in Mendelian crosses are called:
(A) Multiple alleles (B) Allelomorphs
(C) Alloloci (D) Paramorphs

719. In *Mirabilis/Antirrhinum,* hybrid between red and white flowered plants is pink flowered. This is:
(A) Heterosis (B) Segregation
(C) Incomplete dominance (D) Dominance

720. Mendel's laws apply only when:
(A) Characters are linked
(B) Parents are pure breeding
(C) F1 monohybrid ratio shows two types of individuals
(D) First pair of contrasting characters is dependent upon other pairs

721. An altruistic behavior is one in which an individual (the donor) performs an action that helps another animal (the recipient) with no apparent advantage to itself. This is known as:
(A) Kin selection (B) Allelopathy
(C) Mutualism
(D) Competitive exclusion principle

722. Cross between hybrid and recessive parent is:
(A) Back cross (B) Test cross
(C) Monohybrid cross (D) Dihybrid cross

723. An exception to Mendel's principles/Mendel did not include in his discoveries:
(A) Dominance
(B) Purity of gametes/segregation
(C) Linkage
(D) Independent assortment

724. In a cross between AABB x aabb, the ratio of F_2 genotypes between AABB, AaBB, Aabb and aabb would be:
(A) 9 : 3 : 3 : 1 (B) 2 : 1 : 1 : 2
(C) 1 : 2 : 2 :1 (D) 7 : 5 : 3 : 1

725. Mendel formulated the law of purity of gametes on the basis of :
(A) Test cross (B) Back cross
(C) Monohybrid cross (D) Dihybrid cross

726. A cross between hybrid and its parent is:
(A) Back cross (B) Reciprocal cross
(C) Monohybrid cross (D) Dihybrid cross

727. Stabilizing selection favours:
(A) Both extreme forms of a trait
(B) Intermediate forms of a trait
(C) Environmental differences
(D) One extreme form over the other extreme form and over intermediate forms of a trait

728. Test cross is a cross between:
(A) Hybrid x Dominant parent
(B) Hybrid x Recessive parent
(C) Hybrid x Hybrid
(D) All of these

729. Multiple alleles control inheritance of:
(A) Phenylketonuria (B) Colour blindness
(C) Sickle cell anaemeia (D) Blood groups

730. A gene pair hides the effect of another. The phenomenon is:
(A) Epistasis (B) Dominance
(C) Mutation (D) None of these

731. Orthologous genes are which of the following?
(A) Genes that do not share a common ancestral origin
(B) Homologous genes that are present in the genomes of different organisms
(C) Homologous genes that are present in the same genomes
(D) Nonhomologous genes that arose from convergent evolution

732. In a dihybrid cross, F_2 ratio of 15 : 1 is due to:
(A) Supplementary genes (B) Dominant epistasis
(C) Duplicate genes (D) Recessive epistasis

733. Polymorphism is mainly due to:
(A) Monogenic inheritance
(B) Polygenic inheritance
(C) Both of the above
(D) None of these

734. An allele is dominant if it is expressed in:
(A) Both homozygous and heterozygous states
(B) Second generation
(C) Heterozygous combination
(D) Homozygous combination

735. Percentage of heterozygous individuals obtained from selfing of Rr individuals is:
(A) 100 (B) 75
(C) 50 (D) 25

736. Hardy-Weinberg genetic equilibrium holds good for:
(A) Panmictic population
(B) Mendelian population
(C) Assortative mating population
(D) Dissortative mating population

737. Law of dominance-recessiveness is proved by:
(A) Back cross
(B) Incomplete dominance
(C) Monohybrid cross
(D) Dihybrid cross

738. In a test cross, the individual is crossed with:
(A) Homozygous dominant individual
(B) Heterozygous individual
(C) Homozygous recessive
(D) None of these

739. Which of the following is required to see a founder effect?
(A) High rate of new mutation
(B) Population bottleneck in which the popul- ation is reduced to a relatively small number of breeding individuals
(C) Lethality of homozygotes with a mutant alleles
(D) Selection if fevor of heterozygotes

740. A yellow seeded true breeding pea plant is crossed with green seeded true breeding plant. F_1 plants were yellow seeded showing that allele for yellow colour is:
(A) Heterozygus (B) Dominant
(C) Lethal (D) Epistatic

741. RrYy yields gamete types:
(A) One (B) Two
(C) Three (D) Four

742. Heterozygous organism for two genes shall be:
(A) RRYY (B) RrYY
(C) RrYy (D) RRYy

743. Mendel's principle of segregation is based on separation of alleles during:
(A) Gamete formation
(B) Seed formation
(C) Pollination
(D) Embryonic development

744. Mendel's studied inheritance of seven pairs of traits in pea which can have 21 possible combinations. If you are told that in one of these combinations, independent assortment is not observed in later studies, your reaction will be:
(A) Independent assortment principle may be wrong
(B) Mendel might not have studied all the combinations
(C) It is impossible
(D) Later studies may be wrong

745. The Hardy-Weinberg law describes:
(A) Genotype frequencies of a population when evolutionary forces are not acting
(B) How sexual reproduction would change the relative gene frequencies in a population
(C) How mutations occur and balance each other
(D) Genotype frequencies of a population when evolutionary forces are acting

746. Hybrids are generally superior to parents due to:
(A) Heterosis
(B) Homozygosity
(C) Heterozygosity
(D) Parents are generally weak

747. Ratio 9 : 3 : 3 : 1 is due to:
(A) Duplicate genes (B) Lethal genes
(C) Monohybrid cross (D) Dihybrid cross

748. Ratio 9 : 7 is due to:
(A) Supplementary genes (B) Lethal genes
(C) Complementary genes
(D) Epistatic genes

749. Monohybrid cross involves:
(A) Individuals of two different parents
(B) Individuals different in one traits
(C) Individuals different in two traits
(D) Individual with parent

750. Assuming Hardy-Weinberg equilibrium, the genotype frequency of heterozygotes, if the frequency of the two alleles at the gene being studied are 0.6 and 0.4, will be:
(A) 0.80 (B) 0.64
(C) 0.48 (D) 0.32

751. Mendel chose contrasting traits in pea:
(A) 4 (B) 5
(C) 6 (D) 7

752. How many contrasting traits of pea pod were chosen by Mendel?
(A) 1 (B) 2
(C) 3 (D) 3

753. Monohybrid ratio is:
(A) 3 : 1 (B) 9 : 7
(C) 1 : 2 (D) 9 : 3 : 3 : 1

754. Which of the following is not an assumption of the Hardy-Weinberg equilibrium?
(A) No mutation
(B) No more than two alleles at a locus
(C) Random mating with respect to genotype
(D) No selection on the basis of genotype

755. A cross between F_1 hybrid and a recessive parent (Tt x tt) gives a ratio of:
(A) 1 : 1 (B) 2 : 1
(C) 3 : 1 (D) 4 : 1

756. Different forms of a gene are called:
(A) Heterozygotes (B) Alleles
(C) Complementary genes
(D) Supplementary genes

757. A homozygous red flowered pea plant is crossed with homozygous white flowered pea plant. The offspring would be:
(A) 50% red flowered (B) 50% pink flowered
(C) All red flowered (D) All pink flowered

758. Genetically identical progeny is produced when the individual:
(A) Performs self fertilization
(B) Produces identical gametes
(C) Practices inbreeding without meiosis
(D) Practices reproduction

759. If two populations are merged, each with different frequencies of an allele at a locus, and randomely mating occurs immediately, how long will it take to achieve a Hardy-Weinberg equilibrium in the new population?
(A) One generation
(B) Two generations
(C) Variable-depends on allele frequencies
(D) The population will never achieve equilibrium

760. Cross AABb x aaBb yields AaBB : AaBb:Aabb : aabb offspring in the ratio o:
(A) 0 : 3 : 1 : 1 (B) 1 : 2 : 1 : 0
(C) 1 : 1 : 1 : 1 (D) 1 : 2 : 1 : 1

761. The process of mating between closely related individuals is:
(A) Self breeding (B) Inbreeding
(C) Hybridisation (D) Cross breeding

762. A cross between pure tall pea plant with green pods an dwarf pea plant with yellow pods will produce short F_2 plants out of 16:
(A) 2 (B) 3
(C) 4 (D) 5

763. The physical similarity of body shape in dolphins, sharks and penguns results from:
(A) Parallel evolution
(B) Geographic isolation
(C) Convergent evolution
(D) A property of a common ancestor

764. A cross between homozygous recessive and heterozygous plant is:
(A) Monohybrid cross (B) Dihybrid cross
(C) Test cross (D) Back cross

765. In case of incomplete dominance, F_2 generation has:
(A) Genotypic ratio equal to phenotypic ratio
(B) Genotypic ratio is 3 : 1
(C) Phenotypic ratio is 3 : 1
(D) None of these

766. Mendel's principles are related to:
(A) Variation (B) Reproduction
(C) Evolution (D) Heredity

767. Which of the following statements regarding Darwin and his theory of evolution is false?
(A) Darwin noticed that South American species of birds more closely resembled species found in other temperate regions that European species of birds
(B) Darwin proposed that species adapt over time
(C) Darwin proposed that natural selection produces changes in species over time
(D) Darwin was the original and sole developer of the theory of natural selection

768. Dihybrid cross is related to:
(A) Principle of dominance
(B) Principle of independent assortment
(C) Principle of segregation
(D) Principle of purity of gametes

769. Phenotypic dihybrid ratio is:
(A) 9 : 3 : 3 : 1 (B) 15 : 1 : 1
(C) 9 : 6 : 1 (D) 1 : 2 : 1

770. Mendel enunciated principles of inheritance :
(A) 2 (B) 3
(C) 4 (D) 5

771. In red-white flowered cross of *Mirabilis jalapa*, F_2 generation has red, pink and white flowered plants in the ratio of:
(A) 1 : 2 : 1 (B) 1 : 0 : 1
(C) 2 : 1 : 1 (D) 1 : 1 : 2

772. An increase in the inbreeding coefficient, F, is likely to result in:
(A) Reduced likelihood of heterozygotes being present in a population
(B) Higher proportion of genes that show linkage
(C) Higher proportion of genes with introns
(D) Lower level of difference between proteins in two daughter cells

773. Phenomenon of an allele of one gene suppressing the activity of allele of another gene is called:
(A) Dominance (B) Epistasis
(C) Suppression (D) Inactivation

774. Recombinations and variations in eucaryotic organisms are due to:
(A) Mitosis and meiosis
(B) Fertilization and mitosis
(C) Meiosis and fertilization
(D) Meiosis and amitosis

775. Mendel's law of independent assortment is based on F_2 ratio of:
(A) 1 : 2 : 1 (B) 9 : 3 : 3 : 1
(C) 2 : 1 : 1 (D) 3 : 1 : 1

776. Pure red flowered and white flowered plants were crossed. It produced 120 offspring with :
(A) 90 white flowered and 30 red flowered
(B) 90 red flowered and 30 white flowered
(C) 60 red flowered and 60 white flowered
(D) All red flowered

777. A balanced polymorphonism occurs when:
(A) There is selection against the homozygous recessive individuals
(B) There is selection against heterozygotes
(C) There is selection against all genotypes
(D) There is selection against all homozygotes

778. Red seed colour shows an F_2 ratio of 1 : 4 : 6 : 4 : 1. It is due to:
(A) Two polygenes
(B) Multiple poly gene
(C) Different number of recessive genes
(D) Supplementary genes

779. Two dominant independently assorting genes react with each other. They are:
(A) Complementary (B) Supplementary
(C) Duplicate (D) Collaborative

780. Polygenes show:
(A) Different phenotypes (B) Different genotypes
(C) Both of the above (D) None of these

781. Epistasis in which dihybrid ratio of 9 : 3 : 3 : 1 between Aa Bb x Aa Bb is modified is:
(A) Dominance of one allele on another allele at he same locus
(B) Dominance of one allele on another allele at both its loci
(C) Interaction of two alleles at different loci
(D) Interaction between two alleles at same

782. Genes located on same locus but having different expressions are:
(A) Multiple alleles (B) Oncogenes
(C) Polygenes (D) Codominants

783. Which is true:
(A) Etiolation is genetic while albinism is physiological
(B) Etiolation is not reversible
(C) Etiolation is physiological but albinism is genetic
(D) Etiolation and albinism are synonyms

784. Allelomorphic pair implies:
(A) A pair of contrasting characters
(B) A pair of noncontrasting characters
(C) Any two characters
(D) Sex-linked characters

785. In a cross of red and white flowered Snapdragon, F_1 plants will have flowers:
(A) Red (B) White
(C) Both (A) and (B) (D) Pink

786. Cross between AaBB and aaBB will form:
(A) 1 AaBB : 1 aaBB (B) All AaBB
(C) 3 AaBB : 1 aaBB (D) 1 AaBB : 3 aaBB

787. Four daughter cells obtained from a single meiosis differ from one another because of:
(A) Crossing over
(B) Independent assortment
(C) Change in chromosome number
(D) Both (A) and (B)

788. Incomplete dominance produces in F_2 generation:
(A) Phenotypic ratio of 3 : 1
(B) Genotypic ratio of 3 : 1
(C) Phenotypic ratio similar to genotypic ratio
(D) No segregation occurs

789. Hybrid pea plant with yellow round seeds (YyRr) is self pollinated. Phenotypic ratio of next generation would be:
(A) 13 : 3 (B) 9 : 7
(C) 1 : 4 : 6 : 4 : 1 (D) 9 : 3 : 3 : 1

790. A cross between AaBb and aabb yields:
(A) AaBb
(B) AaBb, Aabb, aaBb and aabb
(C) AaBb
(D) Aabb

791. Heterosis is:
(A) Hybrid weakness
(B) Hybrid weakness and vigour
(C) Hybrid vigour
(D) Neither weakness nor vigour of hybrid

792. Mendel found certain traits not to assort independently. It is due to:
(A) Dominance (B) Linkage
(C) Crossing over (D) Amitosis

793. In *Lathyrus odoratus*, cross between two purple flowered plants gives a pink/white progeny. It is due to:
(A) Codominance (B) Incomplete dominance
(C) Epistatsis (D) Segregation

794. A pure red flowered pea plant is crossed with pure white flowered pea plant. The offspring will be:
(A) All red flowered (B) Half red flowered
(C) Half white flowered (D) Half pink flowered

795. How many type of gametes will be produced by individuals of AABbcc genotype?
(A) 2 (B) 4
(C) 6 (D) 9

796. If Mendel had studied 7 traits using a plant with 12 chromosomes instead of 14, he would have?
(A) Discovered incomplete dominance
(B) Not discovered the law of independent assortment
(C) Discovered sex linkage
(D) Mapped the chromosome

797. Crossing over in diploid organism is responsible for:
(A) Dominance of genes
(B) Linkage between genes
(C) Recombination of linked genes
(D) Segregation of alleles

798. Mendel's laws of heredity can be explained with the help of:
(A) Mitosis (B) Meiosis
(C) Cloning (D) Both (A) and (B)

799. Number of phenotypes possible from AaBbCc x AaBbCc is:
(A) 16 (B) 18
(C) 32 (D) 36

800. Phenotypic ratio of a monohybrid F_2 progeny of incomplete dominance is:
(A) 1 : 1 : 1 : 1 (B) 2 : 1 : 1
(C) 1 : 2 :1 (D) 3 : 1 : 1

801. Genotypic ratio of a dihybrid cross is:
(A) 12 : 3 : 1 (B) 15 :1 : 1
(C) 1 : 2 : 1 (D) 1 : 2 : 2 : 4 : 1 : 2 : 1 : 2 :1

802. Number of gamete types formed by AaBbCc individual is:
(A) 4 (B) 6
(C) 8 (D) 16

803. Which is connected with multiple phenotypes?
(A) Epistasis (B) Pleiotropy
(C) Polygenic inheritance (D) Mutations

804. In *Mirabilis jalapa*, crossing of red flowered and white flowered plants, yields pink flowered plants. This is due to:
(A) Crossing over
(B) Complete dominance
(C) Incomplete dominance (D) Epistasis

805. Mendel did not observe linkage due to:
(A) Mutation (B) Synapsis
(C) Crossing over
(D) Independent assortment

806. Organisms phenotypically similar but genetically different are:
(A) Heterozygus (B) Homozygous
(C) Monozygous (D) Multizygous

807. Alleles are:
(A) Genes found on allosomes
(B) Pair of genes governing a specific character like height
(C) Genes governing eye characters
(D) None of these

808. Incomplete dominance is found in:
(A) *Neurospora crassa* (B) *Lathyrus odoratus*
(C) *Pisum sativum* (D) Antirrhinum

809. Hybrid vigour is due to:
(A) Mixing of traits of male and female
(B) Homozygosity in gamete state
(C) Superiority of genes
(D) Heterozygosity

810. If both parents are albino, all the offspring shall be:
(A) Albino
(B) Some albino, some normal
(C) Homozygous normal
(D) Heterozygous normal

811. Two individuals with similar external appearance but different genetic make up have the similar:
(A) Genotype (B) Phenotype
(C) Heterozygote (D) Homozygote

812. Out of two alleles of the same gene, one finds morphological expression. The phenomenon is:
(A) Epistasis (B) Codominance
(C) Incomplete dominance (D) Dominance

813. In *Cucurbita*, W is epistatic over Y and y gene normally responsible for yellow and green colour fruits. It produces white flowers. What is the ratio of fruits in the progeny of cross WwYy x wwYy?
(A) 9 white : 7 yellow : 0 green
(B) 3 white : 4 yellow : 1 green
(C) 4 white : 3 yellow : 1 green
(D) 2 white : 1 yellow : 1 green

814. A recessive trait in Garden pea is:
(A) Wrinkled seeds (B) Tall stem
(C) Round seeds (D) Coloured seed coat

815. A cross involving two pairs of alleles is:
(A) Crossing over (B) Genetics
(C) Dihybrid cross (D) Monohybrid cross

816. Which is correct about traits chosen by Mendel?
(A) Terminal pod is dominant
(B) Constricted pod is dominant
(C) Green coloured pod is dominant
(D) Tall plants are recessive

817. Tallness (T-) is dominant over dwarfness (tt) while red flower colour (R-) is dominant over white colour (rr). A plant with genotype TtRr is crossed with plant of genotype ttrr. Percentage of progeny having tall plants with red flowers is:
(A) 25% (B) 50%
(C) 75% (D) 100%

818. A cross between pure tall pea plant with green pods and dwarf pea plant with yellow pods will produce tall F_2 plants, out of 16:
(A) 15 (B) 13
(C) 12 (D) 7

819. "Gametes are never hybrid". It is a statement of law of:
(A) Dominance
(B) Segregation
(C) Independent assortment
(D) Random fertilization

820. Mendel could not find recombination and crossing over as:
(A) Traits he chose were either present on different chromosomes or were far apart
(B) Traits chosen by him were not influenced by genes
(C) He did not have a high power microscope
(D) He selected only pure types

821. Most favourite and ideal material for researches in genetics is:
(A) Housefly (B) Mosquito
(C) Frog (D) Fruitfly

822. The average life of a crop variety is:
(A) 25-30 years (B) 5-15 years
(C) 15-25 years (D) 1-5 years

823. Gene resistance to Brown Plant Hopper (*Nilaparvata lugens*) of modern rice has been obtained from:
(A) Wild rice
(B) African rice
(C) Old rice varieties from Kerala
(D) IR-8

824. A species restricted to a given area is:
(A) Endemic species (B) Allopatric species
(C) Sympatric species (D) Sibling species

825. Endangered plant species are conserved through:
(A) Herbarium (B) Gene library
(C) Gene bank (D) Reducing pollution

826. Which of the following are homologous organs?
(A) Wings of birds, flippers of whales and forearms of man
(B) Wings of butterfly, pterygium of flying squirrel and fins of flying fishes
(C) Tails of snake, scorpion and wall lizard
(D) Lungs of rabbit, trachea of cockroach and book of scorpion

827. Resistance to herbicide chlorsulfuron in plants is due to a change in:
(A) Glutamine synthetase
(B) Threonine deaminase
(C) Acetolactate synthase
(D) DNA polymerase

828. Which of the following may be part of the reason why replaced uracil in DNA?
(A) Cytosine spontaneously deaminates to uracil
(B) Thymine synthetic pathway is easier
(C) Thymine is unstable
(D) Thymine has a stronger binding affinity for adnine than does uracil

829. Consider the following statements:
P. Wings of insects and birds are homologous organs
Q. Wings of insects and bats are analogous organs
R. Wings of insects and birds are analogous organs
S. Wings of birds and bats are homologous organs
Which of the above statements are correct?
(A) P and Q (B) P and S
(C) Q, R and S (D) P, Q and S

830. The late Mesozoic era (the cretaceous period) was characterized by:
(A) Rise of dinosaurs, rise of toothed birds and rise of flowering plants
(B) Extreme specialization of reptiles, extinction of great reptiles and extinction of toothed birds
(C) Rise of archai mammals, rise of dinosaurs and appearance of first toothed birds
(D) Extinction of archai mammals, rise of birds based appearance of first dinosaurs

831. Gene with multiple effects is:
(A) Epistatic (B) Pleiotropic
(C) Codominant (D) Supplementary

832. Monohybrid test cross ratio is:
(A) 1 : 2 : 1 (B) 3 : 1
(C) 1 : 1 (D) 2 : 1

833. Intraallelic gene suppression is called:
(A) Epistatis (B) Hypostatis
(C) Dominance (D) None of these

834. Mendel may not be able to establish the law of independent assortment, if both characters were located on:
(A) Separate non-homologous chromosomes
(B) Same homologous chromosomes
(C) Separate chromosome but at different loci
(D) Sex chromosomes

835. Who coined the term linkage?
(A) Bateson and Punnett
(B) T.H. Morgan
(C) Correns
(D) de Vries

836. How many linkage groups are present in cyanophyceae?
(A) 0 (B) 1
(C) 2 (D) 4

837. Crossing over is more frequent in:
(A) Males (B) Females
(C) Equal in both (D) None of these

838. Bar eye character in Drosophila is due to:
(A) Defficiency (B) Duplication
(C) Translocation (D) Polyploidy

839. Which is called 'illegitimate crossing over'?
(A) Inversion (B) Translocation
(C) Transposons (D) None of these

840. Trisomic were first obtained in:
(A) Chinese spring wheat (B) Pigweed
(C) Jimson weed (D) Maize

841. Colchicine is used in what concentration for arresting and breaking the spindle?
(A) 0.001-0.05% (B) 0.1-0.5%
(C) 0.01-0.5% (D) 1-5%

842. More the distance between the two genes on a chromosome:
(A) More is the linkage strength
(B) Lesser is the linkage strength
(C) Linkage strength remains unchanged
(D) None of the above

843. Rephanobrassica is a classical example of:
(A) Autopolyploidy (B) Allopolyploidy
(C) Segmental polyploidy (D) None of these

844. Which is hexaploid spieces of wheat?
(A) *Triticum durum* (B) *T. monococcum*
(C) *T. aestivum* (D) *T. turgidum*

845. Mutation is:
(A) Change that is inherited
(B) Change in parents not inherited
(C) Plant growth controlling factor
(D) Change which affects the offsprings of F_2 gen.

846. Earliest record of point mutation goes back to:
(A) 1691 (B) 1791
(C) 1910 (D) 1940

847. Suppression of action of genes by genes not allelomorphic to those suppressed is called:
(A) Dominance (B) Episome
(C) Codominance (D) Epistasis

848. Which is point of mutation in a cell?
(A) Nucleus (B) Chromosome
(C) Centromere (D) Gene

849. Most of mutations are:
(A) Forward (B) Reverse directional
(C) Directionless (D) None of these

850. Cytoplasmic male sterility in maize in inherited:
(A) Paternally (B) Maternally
(C) Both (a) and (B) (D) None of these

851. Induced mutations are most common in:
(A) Coarse cereals (B) Pulses
(C) Higher plants (D) Cereals

852. Auxotrophs in *Neurospora* were produced by which radiation Beadle and Ttum?
(A) γ- rays (B) α-rays
(C) UV-rays (D) IR-rays

853. Which of the following mutagens can be best used in inducing mutation in micro-organisms?
(A) X- rays (B) ß- rays
(C) UV-rays (D) γ- rays

854. Which is the most widely used chemical mutagen in plant mutation breeding now-a-days?
(A) Mustard gas (B) EMS
(C) MMS (D) Nitrous acid

855. Who is associated with 'Green Revolution' in India?
(A) E.J. Butler (B) M.S. Swaminathan
(C) R.S. Paroda (D) Panjab Singh

856. Which of the following is an achievement of mutations?
(A) Todd's Mitcham var. (B) Aruna var
(C) Sharbati sonora var (D) All of these

857. Which type of mutations is more frequent?
(A) Transitions (B) Transversions
(C) Both equal (D) Not known

858. F_1 generation is produced by:
(A) Crossing F_1 progeny with one of the parents
(B) Selfing the progeny of two individual parents
(C) Selfing the parents
(D) Recessive cross between individual parents

859. Four inbred/pure lines of maize are crossed. The cross is:
(A) Double cross
(B) Dihybrid cross
(C) Tetrahybrid cross
(D) Tetraploid cross

860. In high yielding hybrid crop varieties, hybrid vigour can be exploited only when hybrid seeds are purchased afresh every year because:
(A) Govt. of India has accepted Dunkel's proposals
(B) Hybrid vigour is lost due to inbreeding depression
(C) Farmers are not allowed to grow their own seeds
(D) It is cheaper to purchase fresh seed

861. Which of the following is not used for crop improvement?
(A) Inbreeding (B) Introduction
(C) Hybridization (D) Mutations

862. Heterosis is:
(A) Hybrid sterility (B) Hybrid vigour
(C) Hybrid incompatability (D) Hybrid inviability

863. Best method to increase crop yield is (e.g., Wheat):
(A) Using tractors
(B) Sowing seeds of improved varieties
(C) Eradication of weeds
(D) Reduce ration holders

864. Improved rice variety IR-8 has been introduced in India from:
(A) Taiwan (B) Japan
(C) Phillipines (D) Bangladesh

865. Method of selection in vegetatively propagated plants is:
(A) Pure line selection (B) Pedigree selection
(C) Mass selection (D) Clonal selection

866. Which variety is not due to induced mutations?
(A) Aruna of Castor (B) Prabhat of Arhar
(C) Reimi of Rice
(D) Sharbati Sonora of wheat

867. The desired varieties of economically useful crops are raised by:
(A) Vernalisation (B) Mutations
(C) Natural selection (D) Hybridisation

868. The indica varieties of rice are crossed with japonica varieties as they are:
(A) High yielding (B) Cheaper
(C) Short life cycles (D) Resistant to diseases

869. A scientist wants to study the effect of virus on a plant. Which part he should exclude for inoculation:
(A) Shoot apex (B) Stem cortex
(C) Phloem (D) Pith

870. Mass selection is preferred in case of:
(A) Vegetatively propagated plants
(B) Cross pollinated plants
(C) Self pollinated plants
(D) Both (B) and (C)

871. Plant with similar genotype are:
(A) Haploids (B) Autopolyploids
(C) Clone (D) All the above

872. Quarantine regulation is meant for:
(A) Preventing entry of diseases plants in the country
(B) Spraying diseases plants with insecticides
(C) Promoting dry farming
(D) Growing fruits trees in all the states

873. Hybrid vigour is best maintained in vegetatively reproducing plants because they are:
(A) Resistant to diseases
(B) Easily propagated
(C) With long life span
(D) Little liable to lose vigor due to absence of sexual reproduction

874. The first transgenic crop was:
(A) Pea (B) Tobacco
(C) Flax (D) Cotton

875. The use and disuse principles of evolution or theory of "inheritance of acquired characters" was proposed by:
(A) Hugo de Vries (B) Lamarck
(C) Weismann (D) Darwin

876. For evolutionary success, mutation must occur in:
(A) Somatic RNA (B) Somatic DNA
(C) Germplasm DNA (D) Germplasm RNA

877. A starfish with six arms may be a case of:
(A) Variation (B) Evolution
(C) Autotomy (D) Mutation

878. There is more competition for survival between:
(A) Different animals of same niche
(B) Same animals in same niche
(C) Different animals of different niche
(D) Same animals of different niche

879. Who wrote the "Origin of species"?
(A) Lamarck (B) Darwin
(C) De Vries (D) Mendel

880. 'Theory of natural selection' was proposed by:
(A) Lamarck (B) Darwin
(C) De Vries (D) Haeckel

881. The evolution of species is based upon sum total of adaptive changes preserved by:
(A) Hybridization (B) Isolation
(C) Human conservation (D) Speciation

882. According to Neo-Darwinism variations which are the raw material of evolution occur:
(A) In an individual and remain restricted to him
(B) In gene pool of a population
(C) In all the membranes of one species
(D) In all the members of all species

883. Neo-Darwinism believes that new species develop through:
(A) Mutations with natural selection
(B) Continuous variations with natural selection
(C) Hybridisation
(D) Mutations

884. Theory of inheritance of acquired characters was given by:
(A) Wallace (B) Lamarck
(C) Darwin (D) Dè Vries

885. 'Hugo de Vries' contribution is:
(A) Theory of natural selection
(B) Theory of mutations
(C) Law of Dominance
(D) Law of segregation

886. Darwin was most influence by:
(A) Lamarck's theory of acquired characters
(B) Weismann's theory of germplasm
(C) Wallace's theory of origin of species
(D) Essay on population by Malthus

887. Theory of natural selection dwells on:
(A) Role of environment in evolution
(B) Natural selection acting on favourable variations
(C) Changes in gene complex resulting in heritable variations
(D) Inheritance of acquired characters

888. Frequency of a character increases when it is:
(A) Recessive (B) Dominant
(C) Inheritable (D) Adaptable

889. Natural selection theory of Darwin states that:
(A) Environment does not play any role in evolution
(B) Heritable variations arise through changes in gene complex of a species
(C) Natural selection acts on favourable variations which appear among individuals of a species
(D) Characters acquired during the life of an individual are inherited by its offspring

890. Light coloured Peppered Moth/*Biston betularia* gets changed to its darker carbonaria variety due to:
(A) Translocation of block of genes in response to heavy carbons
(B) Deletion of gene segment due to industrial pollution
(C) Mutation of single mendelian gene for survival in smoke laden industrial environment
(D) Industrial carbon deposited on wings

891. De Vries theory of mutation is:
(A) Opposed to natural selection theory
(B) Not opposed to natural selection theory
(C) Opposed to germplasm theory
(D) Opposed to Lamarck's theory

892. Gene frequency in a population remains stable unless and until there is:
(A) Random drift (B) Random mating
(C) Selection (D) Mutation

893. Darwin's Natural Selection is based on:
(A) Use and disuse
(B) Inheritance of acquired characters
(C) Variations
(D) Prodigality, struggle for existence and survival of fittest

894. Genetic drift operates in:
(A) Small population
(B) Large population
(C) Island population
(D) Mendelian population

895. Naturalist who sailed round the world in ship Beagle was:
(A) Charles Lyell (B) Charles Darwin
(C) Alfred Wallace (D) Lamarck

896. Lederberg's replica experiment explains:
(A) Lamarck's theory (B) Mutation theory
(C) Darwin's theory (D) None of the above

897. Chromatid is:
(A) One half of chromosome
(B) Haploid chromosome
(C) Complete chromosome
(D) Duplicate chromosome

898. Mendel did not get any linkage in his experiments on pea. One of the reasons was that:
(A) He did not keep an exact record
(B) There is no linkage in pea
(C) He did not have means to detect linkage
(D) All the seven characters selected by him were present on different chromosomes or showed 50% cross-overs

899. *Neurospora crassa* is widely used in genetical studies because of all the following except one:
(A) It is a haploid plant and mutations can be easily detected
(B) It can be easily cultured
(C) Life cycle short
(D) Spores are not affected by mutagens

900. Plotting of specific genes on the chromosome is known as:
(A) Chromosome map (B) Linkage map
(C) Genetic map (D) All the above

901. In gene mutation, adenine is replaced by guanine. It is known as:
(A) Substitution (B) Point mutation
(C) Transition (D) Transversion

902. In Neurospora, both recessive and dominant alleles express their effect because:
(A) Two genes control each character
(B) There are two alleles for each character
(C) The organism contains only one allele for a gene
(D) Each gene only one allele

903. Genes located on X-chromosomes are called:
(A) Sex influenced (B) Sex limited
(C) Sex linked (D) Maternal genes

904. Mutation is:
(A) A change in phenotype of an organism caused by some environmental factor
(B) An inheritable change in genetic material
(C) A temporary change in structure of the nucleus
(D) Any one of the above

905. 'Which of the following is a case of autoallopolyploidy?
(A) AA, AA, A (B) AA, AA, AB
(C) AA-1 (D) 2n -2

906. Illegitimate crossing over' is another term for:
(A) Transition
(B) Transversion
(C) Reciprocal translocation
(D) None of the above

907. A mutation changes the original base sequence of DNA, GATAGCCG to a new sequence GGTAGCCG. What is the type of mutation?
(A) Transition (B) Transversion
(C) Translocation (D) Inversion

908. A genomatic mutation is the mutation involving:
(A) Change in gene
(B) Change in chromosomal structure
(C) Change in the number of chromosomes
(D) All of these

909. When a purine is replaced by a pyrimidine in a part of DNA, the mutation so produced is called:
(A) Transition (B) Transversion
(C) Deletion (D) Reversal

910. Mutations were first induced in Drosophila by the use of:
(A) X-ray (B) DDT
(C) High temperature (D) UV –rays

911. The mutations which prove fatal for the organism are called:
(A) Spontaneous (B) Induced
(C) Deleterious (D) Lethal

912. Cytoplasmic inheritance is due to:
(A) Plastids (B) Mitochondria
(C) Cytoplasmic particles (D) All of these

913. A trisomic individual has:
(A) One extra chromosome
(B) One less chromosome
(C) Two extra chromosomes
(D) One pair of extra chromosomes

914. It is a common observation that some insects have developed resistance to DDT. The reason may be attributed to:
(A) Poor quality of DDT
(B) Genetic recombination
(C) Genetic mutation
(D) Hybrid vigour

915. The frequency of mutations:
(A) Can be increased by recombination
(B) Varies in the character and the organism
(C) Is affected by environmental factors
(D) Can be increased by x-rays

916. Genome represents:
(A) Haploid chromosome set
(B) Complete chromosome set
(C) Diploid chromosome set
(D) All the genes present in the population

917. Centromere is part of:
(A) Cytoplasmic organelles (B) Ribosome
(C) Chromosome (D) Nucleolus

918. Allosomes are the name of:
(A) Sex chromosomes
(B) Swellings on the chromosomes
(C) Chromosomes other than the ones which determine sex
(D) Nucleolus organising regions of chromosomes

919. Down's syndrome is due to:
(A) Crossing over
(B) Linkage
(C) Sex-linked inheritance
(D) Nondisjunction of chromosomes

920. Crossing over takes place in:
(A) One strand stage (B) Two strand stage
(C) Three strand stage (D) Four strand stage

921. Two linked genes a and b show 20% recombination. The individuals of a dihybrid cross between ++/++ x ab/ab shall show gametes:
(A) ++ 80 : ab : 20
(B) ++ 50 : ab : 50
(C) ++ 40 : ab : 40 : + a10 : + b : 10
(D) ++ 30 : ab : 30 : + a 20 : + b : 20

922. Chromosome with median centromere and equal arms is:
(A) Acrocentric (B) Submetacentric
(C) Telocentric (D) Metacentric

923. Balbiani rings occur in:
(A) Polytene chromosomes
(B) Lampbrush chromosomes
(C) Polysomes
(D) Heterosomes

924. Which is a base analogue?
(A) 5-Bromouracil (B) Caffeine
(C) Colchicine (D) Nitrous acid

925. Mutations used in agriculture are commonly:
(A) Induced (B) Spontaneous
(C) Lethal (D) Recessive and lethal

926. The major source of variations are:
(A) Polyploidy (B) Mutations
(C) Chromosome aberrations (D) Segregation

927. The gene that controls the rate of mutation of another gene is:
(A) Regular gene (B) Inducer gene
(C) Mutable gene (D) Mutator gene

928. In Neurospora, 8 ascospores are formed instead of 4. This indicates:
(A) One meiosis
(B) Two meiosis
(C) Two mitosis
(D) Meiosis followed by mitosis

929. Which one can induced polyploidy?
(A) Colchicine (B) Acridines
(C) Ethylene (D) Maleic hydrazide

930. Harmful mutations do not get eliminated from gene pool because:
(A) They have survival value
(B) Genetic drift
(C) They are formed repeatedly
(D) They are recessive and carried by heterozygous individuals

931. Mutations can be brought about by:
(A) Aniline dye (B) X-rays
(C) Auxins (D) DDT

932. A normal green male maize is crossed with albino female. The progeny is albino because:
(A) Trait for albinism is dominant
(B) The albinos have biochemical to destroy plastids derived from green male
(C) Plastids are inherited from female parent
(D) Green plastids of male must have mutated

933. Which one of the following carries extra nuclear genetic material?
(A) Golgi apparatus (B) Ribosomes
(C) Chromosome (D) Plastids

934. Smallest segment of genetic material affected by mutation is:
(A) Recon (B) Cistron
(C) Muton (D) Exon

935. Chromosome pairing in salivary gland chromosomes resembles zygotene pairing in:
(A) It is followed by crossing over
(B) It can occur in nonhomologous segments
(C) Pairing occurs in homologous regions only
(D) It is followed by reductional separation

936. Two dominant nonallelic genes are 50 map units apart. The linkage is:
(A) Cis type (B) Trans type
(C) Absent/Incomplete (C) Complete

937. Which arrangement of ascospores in Neurospora does not represent second generation segregation?
(A) aaaaAAAA (B) AAaaAAaa
(C) AAaaaaAA (D) aaAAAAaa

938. Wheat plant is 6n=42. What will be the number of chromosomes in its monosomic, haploid number and monoploid?
(A) 43, 21 and 7 (B) 41, 21 and 7
(C) 15, 7 and 7 (D) 13, 7 and 7

939. In a cross between genotype AB and ++. 650 out of 1000 individuals were of parental type. The distance between A and B is:
(A) 35 map units (B) 45 map units
(C) 15 map units (D) 30 map units

940. Mutation can be induced in bacteria by:
(A) Growing different strains in same culture
(B) Starvation
(C) Providing growth substances
(D) High energy radiation

941. Mendel found that some traits do not assort independently. Later workers found it due to:
(A) Linkage (B) Crossing over
(C) Amitosis
(D) Dominance of one trait over another

942. Number of genes on a chromosome is:
(A) 1 (B) 10
(C) 35 (D) Infinite

943. Different types of chromosomes can be recognised by the position of:
(A) Genes (B) Centromere
(C) Centrosome (D) Spindle

944. Exchange of segments between non-homologous chromosomes is:
(A) Translocation (B) Inversion
(C) Crossing over (D) Tetrasomy

945. In order to remain linked, the distance between two genes should not increase beyond:
(A) 10 map units (B) 20 map units
(C) 40 map units (D) 50 map units

946. Which pteridophyte has the maximum chromosome number?
(A) *Ophioglossum reticulatum*
(B) *Azolla pinnata*
(C) *Lycopodium cernuum* (D) *Selaginella apus*

947. Lampbrush chromosomes are seen during :
(A) Meiotic metaphase (B) Mitotic metaphase
(C) Mitotic prophase (D) Meiotic prophase

948. Sudden change which breeds true is:
(A) Mutation
(B) Law of inheritance
(C) Inheritance of acquired characters
(D) Natural selection

949. Taylor demonstrated chromosome replication to be semiconservative in:
(A) *Vicia faba* (B) HeLa cell
(C) Pea (D) Mouse liver cell

950. Drosophila melanogaster possesses:
(A) 3 pairs autosomes + 1 pair sex chromosomes
(B) 2 pairs autosomes + 2 pairs sex chromosomes
(C) 1 pair autosomes + 3 pairs sex chromosomes
(D) 2 pairs autosomes + 1 pair sex chromosomes

951. In Down's syndrome of a male child, the sex complement is
(A) XO (B) XY
(C) XX (D) XXY

952. Mutation is:
(A) Change in gene frequency
(B) Genetic drift
(C) Change in base pairs in DNA molecule
(D) Environmental mechanism of evolution

953. A gene is:
(A) Three letter code in DNA
(B) Equivalent to chromosome
(C) Part of chromosome producing a set of enzymes (protein)
(D) Part of chromosome producing only one enzyme (protein)

954. Albinism and phenyl ketonuria are disorders due to:
(A) Recessive autosomal genes
(B) Dominant autosomal genes
(C) Dominant sex genes
(D) Recessive sex genes

955. If there is complete linkage in F_2 generation:
(A) Parental types and recombinations appear in equal ratio
(B) Recombinants are less than parental types
(C) Recombinants are more than parental types
(D) There will be only parental types

956. Lampbrush chromosomes occur inside:
(A) Nucleus of human cells
(B) Oocytes of frog
(C) Salivary glands of Drosophila
(D) Salivary gland of silkworm

957. Down's syndrome is due to trisomy of 21st chromosome caused by:
(A) Nondisjunction during egg formation
(B) Nondisjunction during sperm formation
(C) Addition of extra chromosome during mitosis of zygote
(D) Either A or B

958. Distance between two linked genes is measured in map units that depict:
(A) Ratio of crossing over between them
(B) Cross – over value
(C) Number of genes between them
(D) Both (B) and (C)

959. Extra 18th autosomal chromosome results in:
(A) Edward's syndrome (B) Patau's syndrome
(C) Down's syndrome (D) None of these

960. Queen Victoria of England was:
(A) Haemophiliac carrier (B) Colour blind
(C) AIDS patient (D) Deaf

961. Extra nuclear genes are present in:
(A) Cytoplasm
(B) E.R. and Cytoplasm
(C) Ribosome and Cytoplasm
(D) Mitochondria and cytoplasmic particles

962. The back cross is:
(A) A cross between F_1 individual and F_2 individual
(B) A cross between an F_1 individual with another F_1 individual
(C) Cross between F_1 & one of the two parents
(D) Cross between F_2 with one of the parents

963. A monohybrid cross is the one in which:
(A) Only a single plant is involved for the experiment
(B) A single pair of contrasting characters is considered for the genetic results
(C) A hybrid is crossed to a homozygous
(D) None of these

964. Even harmful mutations do not get eliminated from gene pool due to:
(A) Genetic drift
(B) Higher frequency due to dominant nature
(C) Being recessive and persisting in hetero-zygous condition
(D) Survival value

965. Turner's syndrome is represented by:
(A) XYY (B) XO
(C) XXY (D) XXX

966. The law of segregation of characters postulated by Mendel can be related to:
(A) The presence of two genes for each character in a somatic cell
(B) A gamete receiving only one of the two homologous chromosomes during meiosis
(C) Presence of both genes on the same chromosome
(D) None of these

967. Mendel selected pea as material for his experiments because:
(A) It is an annual plant with comparatively short life cycle
(B) The flower are self pollinated
(C) The number of seeds produced is quite large
(D) All the above

968. At which stage is the fate of genetic constitution of gametes finally decided?
(A) Metaphase – I (B) Anaphase – I
(C) Anaphase – II (D) Interkinesis

969. How would you test a pea plant whether it is a pure or hybrid for tallness?
(A) Cross it with another tall pea plant of unknown genotype
(B) Cross it with a pure tall pea plant
(C) Cross with a homolozygous dwarf pea
(D) Cross it with any pea plant

970. According to the law of Independent assortment in a dihybrid cross:
(A) There are four genotypes in F_2
(B) F_2 contains 16 phenotypes
(C) There is a single individual which is homozygous recessive for both the characters
(D) It is not possible to forecast the different phenotypes

971. The number of different type of gametes produced from a plant with genotype Aa Bb Cc is:
(A) 2 (B) 4 (C) 8 (D) 16

972. How many types of gametes are expected from the plants with genotypes AA, BB, aa, bb?
(A) 3 (B) 4 (C) 8 (D) 16

973. A haploid set of all the genes present in a gamete is called:
(A) Genotype (B) Phenotype
(C) Genome (D) Linkage group

974. The mendelian principle, which has always stood true is:
(A) The law of independent assortment
(B) The law of segregation
(C) The law of dominance (D) All the above

975. A tobacco plant heterozygous for albinism (a recessive character) is self pollinated and 1200 seeds are subsequently germinated. How many seedlings would have the parental phenotype?
(A) 400 (B) 600
(C) 900 (D) 1200

976. Mendel is popular for postulating:
(A) Origin of species (B) Cell theory
(C) Linkage theory (D) Law of inheritance

977. 'Like begets like' an important and universal phenomenon of life, is due to:
(A) Eugenics (B) Inheritance
(C) Dominance (D) Crossing over

978. Why pea plants more suitable than dogs for mendelian experiments?
(A) Pea plants can be self pollinated
(B) All pea plants are diploid
(C) There were no pedigree records of dogs
(D) Dogs have many genetic traits

979. Multiple alleles are present:
(A) At different loci in the same chromosome
(B) In different chromosomes
(C) At the same locus in one type of chromosomes
(D) None of these

980. A dwarf pea plant was treated with GA. The plant became tall. The treated plant was then crossed with a homozygous tall pea. The results in F_2 are expected to be:
(A) All tall (B) Tall & dwarf in 3 : 1 ratio
(C) 50% tall (D) All dwarf

981. Mendel observed red flowers in F_1 when he crossed red and white because of:
(A) Dominance
(B) Recessive gene
(C) Law of independent Assortment
(D) Law of segregation

982. The genotypic ratio of a monohybrid cross will be:
(A) 3 : 1 : 1 (B) 1 : 1 : 1
(C) 1 : 2 : 1 (D) 2 : 1 : 1

983. A cross between unlike organisms is called:
(A) Test cross (B) Back cross
(C) Composite (D) Hybrid

984. Had Mendel decided to study those traits together which are determined by linked genes, he would not have found out:
(A) Principle of purity of gamete
(B) Principles of Codominance
(C) Principle of segregation
(D) Principle of independent assortment

985. A pure tall plant is reared in a soil poor in nutrition and reached the size of dwarf plant. If this plant is selfed, the phenotype in the F1 generation is most likely to be:
(A) All tall plants
(B) 50% tall & 50% dwarf
(C) All dwarf
(D) Data insufficient

986. A modified dihybrid mendelian ratio of 9 : 3 : 4 indicates:
(A) Supplementary genes
(B) Epistatic genes
(C) Complementary genes
(C) Lethal genes

987. Genetics deals with:
(A) Heredity & Variation
(B) Heredity
(C) Mutations
(D) Variation

988. In a dihybrid cross, F_2 phenotypic ratio is 13 : 3. It is case of:
(A) Complementary genes
(B) Epistatic genes
(C) Multigenic inheritance
(D) Incomplete dominance

989. Mendel conducted hybridization experiments on:
(A) Pigeonpea (B) Garden pea
(C) Wild pea (D) Sweet pea

990. F_2 generation of a cross between two white flowered strains of sweet pea yields 9 purple flowered plants: 7 white flowered plants. This is an example of:
(A) Supplementary genes
(B) Epistasis
(C) Complemetary genes
(D) Gene inhibition

991. The body feature are:
(A) Phenotype
(B) Genotype
(C) Sex of the individual
(D) Controlled by environment

992. Nonallelic genes having similar phenotypic effect, interact to produce a different trait and a ratio of 9 : 7 in F_2 generation are:
(A) Epistatic genes
(B) Hypostatic genes
(C) Supplementary genes
(D) Complementary genes

993. Mendel's second law is:
(A) Segregation
(B) Dominance
(C) Independent assortment
(D) Polygenic inheritance

994. When two hybrids rrTt and Rrtt are crossed, the phenotypic ratio of offspring shall be:
(A) 3 : 1 : 1 (B) 1 : 1 : 1 : 1
(C) 1 : 1 : 1 (D) 9 : 3 : 3 : 1

995. Gametes of AaBb individual can be:
(A) Aa, Bb (B) AB, ab
(C) AB, ab, aB (D) AB, Ab, aB, ab

996. Segregation of Mendelian factors (no linkage, no crossing over) occurs during:
(A) Anaphase I
(B) Anaphase II
(C) Metaphase I
(D) Metaphase II

997. Back cross is:
(A) F_1 x F_1
(B) F_1 x Recessive parent
(C) F_1 x Dominant parent
(D) F_1 x Any parent

998. Mendel's second law of independent assortment can be demonstrated by:
(A) Test cross (B) Back cross
(C) Monohybrid cross (D) Dihybrid cross

999. In a cross 45 tall and 14 dwarf plants were obtained. Genotype of parents was:
(A) TT x TT (B) TT x Tt
(C) Tt x Tt (D) TT x tt

1000. Which one is exception to Mendel's principle of dominance?
(A) Wild pea (B) Mirabilis
(C) Garden pea (D) Pigeonpea

1001. Genes controlling seven traits in pea studied by Mendel were actually located on:
(A) Two chromosomes
(B) Three chromosomes
(C) Four chromosomes
(D) Five chromosomes

1002. Breeding method for transferring monogenic disease resistance from coild species to improved variety of wheat is:
(A) Pedigree method
(B) Bulk method
(C) Back cross method
(D) Pure line selection

1003. Opaque-2 maize composites are richer as compared to normal maize is:
(A) Tryptophan
(B) Lysin
(C) Tryptophan and Lysin both
(D) Protein

1004. Cabbage is cross pollinated owing to:
(A) Floral morphology
(B) Protandry
(C) Protogyny
(D) Self incompatibility

1005. Haploid are having:
(A) Somatic chromosome number
(B) Small chromosomas
(C) Very long chromosomes
(D) Genetic chromosome number

1006. Bulk population breeding is suitable for:
(A) Fruit crops
(B) Vegetable crops
(C) Small grain crops
(D) Flower crops

1007. In sugarcane breeding the initial selection after hybridization is done in the generation
(A) F0 (B) F1
(C) F2 (D) F6

1008. Self compatibility is helpful in the development of:
(A) Syntehtic variety (B) Hybrid variety
(C) Pureline variety (D) Composite variety

1009. Seedless variety of Guava is:
(A) Haploid (B) Diploid
(C) Anuploid (D) Tetraploid

1010. Multiplines are produced by mixing equal amounts of seeds of :
(A) Genetically dissimilar lines
(B) Isogenic lines
(C) Inbred lines
(D) Synthetic lines

1011. Ratio obtained in Inhibitory Gene Action is:
(A) 9 : 7 (B) 9 : 3 : 4
(C) 15 : 1 (D) 13 : 3

1012. Pedigree selection is used in:
(A) Cross pollinated plants
(B) For crosses at intorgeneric level
(C) Self pollinated crops
(D) Plants difficult to be hybridization

1013. Gener are present:
(A) Near the chromosomes
(B) On the chromosomes
(C) In the chromosomes
(D) In between the chromosomes

1014. Chromosome number is reduced to half is:
(A) Meiosis I (B) Meiasis II
(C) Mitosis (D) Cytokinins

1015. Self incompatibility is helpful in the development of:
(A) Synthetic variety
(B) Hybrid variety
(C) Pureline variety
(D) Composite variety

1016. Appropriate breeding method for transferring disease-resistance from wild donor source to an established in:
(A) Pureline method
(B) Pedigree method
(C) Back cross method
(D) Bulk method

1017. Specific combining ability of the parents should be high in order to develop:
(A) Pureline (B) Synthetics
(C) Composites (D) Hybrids

1018. Name a cereal variety developed by mutation breeding in India:
(A) Sonalika (B) Jagannath
(C) Kiran (D) Savita

1019. The cornerstone of all plant breeding practices is:
(A) Introduction (B) Selection
(C) Hybridization (D) Selfing

1020. Which is the Rereditory material in chromosomes:
(A) RNA (B) ATP
(C) DNA (D) ADP

1021. Production of a new variety or a hybrid by crossing together two genotypically different individuals is referred to as:
(A) Mass selection (B) Recombination
(C) Hybridization (D) None of these

1022. Clonal selection is a common practices in which of the following:
(A) Mango (B) Rice
(C) Sugarcane (D) Banana

1023. Condition when a gene governs more than one trust is:
(A) Multiple attack (B) Polygene
(C) Pleiotropy (D) None of these

1024. *Brassica juncea* is a :
(A) Anupoly ploid (B) Haploid
(C) Autotetra ploid (D) Allotetraploid

1025. Horizontal resistance is:
(A) Resistance against one race of pathogen
(B) Resistance against all the races of pathogen
(C) Resistance against lodging
(D) None of these

1026. Clonal selection can be used to:
(A) Wheat (B) Groundnut
(C) Potato (D) Mustard

1027. Single cross is a cross between:
(A) Two open pollinated varieties
(B) Two inbreds
(C) Inbred and variety
(D) Hybrid and variety

1028. The increase in vigor or growth of a hybrid over the parents is known as:
(A) Hybrid (B) Hereditary
(C) Heterosis (D) None of these

1029. The basic mechanism of hereditary transmission is:
(A) Asexual reproduction (B) Polyploidy
(C) Mitotic division
(D) Splitting of chromosome

1030. Adenine of RNA will pair with:
(A) Thymine (B) Uracil
(C) Cytosine (D) Adenine

1031. The first artificial hybrid was made by:
(A) Mendel (B) Thomas Fairchild
(C) Koelreuter (D) Morgan

1032. Systemic hybridization on a scientific basis began with the work:
(A) Shull – maize (B) Patel – cotton
(C) Koelreuter – Nicotiana (D) Knight – fruits

1033. An individual with gametic chromosome number is:
(A) Euploid (B) Haploid
(C) Dihaploid (D) Polyhaploid

1034. The common form of DNA present in the living organisms is the:
(A) A form (B) Z form
(C) C form (D) B form

1035. DNA polymerase I was discovered by:
(A) Harshley and Chase (B) Yanofsky
(C) Kornberg (D) Crick

1036. The DNA replicating enzyme in bacteria:
(A) DNA Pol I (B) DNA Pol II
(C) DNA Pol III (D) None of these

1037. Protein involved in keeping the DNA single stranded during replication is:
(A) Ligase (B) Helicase II
(C) Topoisomerase (D) SSB protein

1038. Triticale is:
(A) Interspecific cross (B) Intraspecific cross
(C) Intergeneric cross (D) Intrageneric cross

1039. In a synthetic variety of maize produced by combining 10 lines, the loss of excess vigour in F2 generation will be:
(A) 1/10 (B) 1/5
(C) ¼ (D) ½

1040. Okazaki fragments are formed due to synthesis of DNA in direction:
(A) 5'-3' (B) 3'-5'
(C) Both (A) and (B) (D) None of these

1041. Maximum absorbance of DNA molecule is at ------ OA wave length:
(A) 260 (B) 380
(C) 400 (D) 530

1042. The resulting two daughter cells produced by bridge-breakage fusion cycle will have:
(A) Duplication (B) Deficiency
(C) Both (A) and (B) (D) None of these

1043. Nucleolus plays an important role in production of:
(A) rRNA (B) tRNA
(C) mRNA (D) All of these

1044. An amino acid which has related to drought tolerance in plants:
(A) Glutamic acid (B) Proline
(C) Aspartic acid (D) Hydroproline

1045. Inhibitors to cell membrane synthesis are:
(A) Polymixin
(B) Gramicidin and Ristocatin
(C) Vanomycin (D) All of these

1046. Inhibitors to mRNA synthesis (Transcription) are:
(A) Actinomycin – D (B) Rifampicin
(C) 5-bromouracil (D) All of these

1047. Seedless water melon is a:
(A) Diploid (B) Monoploid
(C) Triploid (D) Pentaploid

1048. Inbreed lines in maize are maintained by:
(A) Self pollination (B) Open pollination
(C) Cross pollination (D) None of these

1049. Mutation due to change in gene because of:
(A) Loss (B) Degeneration
(C) Addition (D) All of these

1050. A cybrid is a:
(A) Sexual hybrid
(B) Hybrid formed by cell division
(C) Plant formed by tissue culture
(D) Synonymous to hybrid

1051. Reversion occur due to in:
(A) Gene interaction (B) Mutation
(C) Both (A) and (B) (D) None of these

1052. The capacity of a cell to develop into a complete plant is called:
(A) Tissue culture (B) Polyploidy
(C) Totipotency (D) Haploidy

1053. The term cybrid refers to the:
(A) Somatic hybrids (B) Gametic hybrids
(C) Both (A) and (B) (D) None of these

1054. During mitosis, chromosomal DNA duplication takes place in:
(A) Interphase (B) Prophase
(C) Metaphase (D) Anaphase

1055. rDNA analysis provides information on:
(A) Phylogenetic relationship
(B) Strain differences
(C) Biotype differences
(D) None of these

1056. Cryopreservation is not suitable for preserving organisms which are:
(A) Frost sensitive (B) Frost resistant
(C) Aerobic (D) None of these

1057. Wheat is naturally:
(A) Auto polyploids (B) Auto-allopolyploids
(C) Segmental allopolyploids
(D) Segmental autopolyploids

1058. Reverse transcription is also known as:
(A) Teminism (B) Temin
(C) Translation (D) All of these

1059. Translation involves:
(A) Chromosomes from other species
(B) Non-homologous chromosomes
(C) Homologous chromosomes
(D) None of these

1060. The tissues which nourish microspores is:
(A) Tapatum (B) Nucleus
(C) Endodermis (D) Mesophyll

1061. The cDNA libraries can be prepared by isolating:
(A) tRNA (B) mRNA
(C) DNA (D) rDNA

1062. Aflatoxin are produced by:
(A) Infected plants (B) Fungi
(C) Infected bacteria (D) Nematodes

1063. DNA can be denatured by heating at a temperature of:
(A) 94^0C (B) 86^0C
(C) 90^0C (D) 80^0C

1064. Self splicing introns are found in:
(A) tRNA genes (B) rRNA genes
(C) Structural genes (D) All of these

1065. Superiority of F1 hybrid over a standard commercial variety is called:
(A) Heterobeltiosis (B) Luxuraince
(C) Economic heterosis (D) Relative heterosis

1066. The smallest unit of genetic material that can not be divided but is interchangeable among units refers to:
(A) Cistron (B) Muton
(C) Recon (D) Plasmon

1067. The shortest of mitotic phases is the:
(A) Telophase (B) Interphase
(C) Metaphase (D) Anaphase

1068. The homologous chromosomes follow the process of synopsis in the:
(A) Laptotene (B) Zygotene
(C) Pachytene (D) Diplotene

1069. A plasmid used to introduce BT toxin into a plant cell:
(A) Carries an antibiotic resistance gene that makes the plant resistant to antibiotics
(B) Causes the plant to produce opines
(C) Is toxic for the plant cell
(D) Carries a bacterial gene with a plant promoter

1070. During meiosis crossing over takes place at:
(A) Zygotene stage (B) Diplotene stage
(C) Diakinesis stage (D) Pachytene stage

ANSWERSHEET

3. GENETICS AND CYTOGENETICS

1. **(C)**	2. **(C)**	3. **(B)**	4. **(C)**	5. **(D)**	6. **(B)**	7. **(D)**	8. **(A)**	9. **(C)**	10. **(B)**
11. **(B)**	12. **(B)**	13. **(C)**	14. **(D)**	15. **(B)**	16. **(B)**	17. **(D)**	18. **(C)**	19. **(C)**	20. **(C)**
21. **(D)**	22. **(D)**	23. **(D)**	24. **(C)**	25. **(B)**	26. **(C)**	27. **(B)**	28. **(C)**	29. **(B)**	30. **(A)**
31. **(C)**	32. **(B)**	33. **(D)**	34. **(C)**	35. **(B)**	36. **(C)**	37. **(C)**	38. **(B)**	39. **(C)**	40. **(C)**
41. **(B)**	42. **(D)**	43. **(B)**	44. **(D)**	45. **(A)**	46. **(B)**	47. **(A)**	48. **(B)**	49. **(B)**	50. **(C)**
51. **(C)**	52. **(C)**	53. **(A)**	54. **(C)**	55. **(B)**	56. **(A)**	57. **(B)**	58. **(A)**	59. **(C)**	60. **(C)**
61. **(B)**	62. **(C)**	63. **(A)**	64. **(A)**	65. **(A)**	66. **(C)**	67. **(B)**	68. **(C)**	69. **(B)**	70. **(C)**

71. (D) 72. (C) 73. (A) 74. (D) 75. (D) 76. (B) 77. (B) 78. (B) 79. (D) 80. (A)
81. (A) 82. (D) 83. (C) 84. (D) 85. (A) 86. (D) 87. (C) 88. (A) 89. (A) 90. (B)
91. (B) 92. (D) 93. (C) 94. (B) 95. (C) 96. (B) 97. (D) 98. (D) 99. (A) 100. (C)
101. (C) 102. (B) 103. (C) 104. (D) 105. (C) 106. (A) 107. (A) 108. (C) 109. (C) 110. (D)
111. (A) 112. (D) 113. (C) 114. (B) 115. (A) 116. (A) 117. (D) 118. (B) 119. (C) 120. (D)
121. (D) 122. (C) 123. (D) 124. (D) 125. (C) 126. (B) 127. (B) 128. (D) 129. (B) 130. (D)
131. (B) 132. (D) 133. (C) 134. (C) 135. (B) 136. (C) 137. (D) 138. (D) 139. (C) 140. (B)
141. (D) 142. (A) 143. (C) 144. (C) 145. (C) 146. (A) 147. (B) 148. (D) 149. (C) 150. (C)
151. (D) 152. (A) 153. (D) 154. (B) 155. (C) 156. (B) 157. (A) 158. (C) 159. (D) 160. (A)
161. (B) 162. (C) 163. (D) 164. (D) 165. (C) 166. (C) 167. (B) 168. (C) 169. (A) 170. (A)
171. (D) 172. (C) 173. (B) 174. (B) 175. (C) 176. (C) 177. (C) 178. (D) 179. (A) 180. (C)
181. (D) 182. (A) 183. (A) 184. (C) 185. (B) 186. (D) 187. (A) 188. (C) 189. (C) 190. (D)
191. (B) 192. (B) 193. (B) 194. (C) 195. (A) 196. (C) 197. (C) 198. (C) 199. (B) 200. (D)
201. (C) 202. (C) 203. (A) 204. (A) 205. (B) 206. (D) 207. (A) 208. (A) 209. (A) 210. (C)
211. (B) 212. (B) 213. (D) 214. (C) 215. (A) 216. (D) 217. (D) 218. (C) 219. (D) 220. (A)
221. (A) 222. (A) 223. (B) 224. (C) 225. (A) 226. (B) 227. (A) 228. (C) 229. (D) 230. (A)
231. (B) 232. (C) 233. (B) 234. (C) 235. (D) 236. (C) 237. (B) 238. (A) 239. (B) 240. (C)
241. (B) 242. (A) 243. (B) 244. (C) 245. (B) 246. (C) 247. (A) 248. (D) 249. (C) 250. (C)
251. (D) 252. (A) 253. (B) 254. (C) 255. (B) 256. (A) 257. (D) 258. (D) 259. (B) 260. (A)
261. (B) 262. (B) 263. (D) 264. (A) 265. (B) 266. (B) 267. (D) 268. (D) 269. (C) 270. (A)
271. (A) 272. (B) 273. (D) 274. (A) 275. (A) 276. (B) 277. (C) 278. (B) 279. (B) 280. (C)
281. (C) 282. (B) 283. (C) 284. (A) 285. (B) 286. (D) 287. (D) 288. (A) 289. (B) 290. (D)
291. (C) 292. (C) 293. (D) 294. (B) 295. (B) 296. (D) 297. (D) 298. (A) 299. (A) 300. (D)
301. (A) 302. (A) 303. (D) 304. (B) 305. (D) 306. (D) 307. (A) 308. (C) 309. (C) 310. (D)
311. (A) 312. (A) 313. (D) 314. (B) 315. (B) 316. (A) 317. (D) 318. (C) 319. (D) 320. (B)
321. (D) 322. (A) 323. (B) 324. (A) 325. (C) 326. (D) 327. (A) 328. (B) 329. (C) 330. (C)
331. (B) 332. (B) 333. (B) 334. (D) 335. (D) 336. (B) 337. (D) 338. (B) 339. (B) 340. (C)
341. (D) 342. (D) 343. (C) 344. (A) 345. (A) 346. (D) 347. (B) 348. (A) 349. (B) 350. (B)
351. (A) 352. (D) 353. (C) 354. (D) 355. (A) 356. (B) 357. (B) 358. (B) 359. (C) 360. (C)
361. (D) 362. (B) 363. (C) 364. (D) 365. (B) 366. (B) 367. (A) 368. (B) 369. (B) 370. (D)
371. (A) 372. (A) 373. (A) 374. (C) 375. (B) 376. (D) 377. (D) 378. (D) 379. (B) 380. (A)
381. (C) 382. (C) 383. (D) 384. (A) 385. (C) 386. (A) 387. (B) 388. (C) 389. (B) 390. (D)
391. (D) 392. (A) 393. (C) 394. (C) 395. (D) 396. (D) 397. (B) 398. (B) 399. (C) 400. (A)
401. (B) 402. (A) 403. (D) 404. (C) 405. (D) 406. (B) 407. (D) 408. (C) 409. (C) 410. (C)
411. (A) 412. (B) 413. (D) 414. (A) 415. (A) 416. (C) 417. (B) 418. (D) 419. (B) 420. (D)
421. (B) 422. (A) 423. (D) 424. (B) 425. (B) 426. (D) 427. (D) 428. (C) 429. (A) 430. (B)
431. (C) 432. (D) 433. (D) 434. (C) 435. (C) 436. (B) 437. (D) 438. (C) 439. (A) 440. (C)
441. (B) 442. (D) 443. (A) 444. (C) 445. (D) 446. (B) 447. (C) 448. (D) 449. (B) 450. (C)
451. (C) 452. (C) 453. (C) 454. (A) 455. (A) 456. (C) 457. (B) 458. (C) 459. (A) 460. (C)
461. (B) 462. (A) 463. (A) 464. (A) 465. (B) 466. (B) 467. (D) 468. (C) 469. (C) 470. (A)
471. (A) 472. (A) 473. (D) 474. (C) 475. (A) 476. (A) 477. (D) 478. (B) 479. (A) 480. (C)
481. (C) 482. (B) 483. (A) 484. (A) 485. (A) 486. (B) 487. (C) 488. (D) 489. (B) 490. (C)

491. (D) 492. (C) 493. (A) 494. (A) 495. (D) 496. (B) 497. (D) 498. (A) 499. (A) 500. (D)
501. (C) 502. (A) 503. (D) 504. (A) 505. (B) 506. (C) 507. (D) 508. (A) 509. (C) 510. (A)
511. (D) 512. (D) 513. (B) 514. (B) 515. (D) 516. (B) 517. (D) 518. (D) 519. (B) 520. (B)
521. (C) 522. (B) 523. (D) 524. (B) 525. (A) 526. (D) 527. (B) 528. (A) 529. (C) 530. (D)
531. (B) 532. (A) 533. (C) 534. (B) 535. (D) 536. (B) 537. (A) 538. (A) 539. (C) 540. (D)
541. (C) 542. (D) 543. (D) 544. (C) 545. (C) 546. (C) 547. (A) 548. (D) 549. (C) 550. (B)
551. (C) 552. (D) 553. (A) 554. (D) 555. (D) 556. (D) 557. (C) 558. (C) 559. (B) 560. (C)
561. (D) 562. (D) 563. (D) 564. (A) 565. (B) 566. (D) 567. (D) 568. (A) 569. (A) 570. (B)
571. (C) 572. (A) 573. (B) 574. (D) 575. (B) 576. (C) 577. (A) 578. (B) 579. (D) 580. (C)
581. (D) 582. (C) 583. (B) 584. (D) 585. (D) 586. (C) 587. (D) 588. (A) 589. (C) 590. (B)
591. (B) 592. (C) 593. (A) 594. (B) 595. (A) 596. (A) 597. (B) 598. (B) 599. (D) 600. (A)
601. (A) 602. (B) 603. (B) 604. (D) 605. (D) 606. (A) 607. (B) 608. (C) 609. (B) 610. (D)
611. (A) 612. (B) 613. (B) 614. (A) 615. (D) 616. (C) 617. (C) 618. (D) 619. (B) 620. (C)
621. (A) 622. (C) 623. (A) 624. (A) 625. (A) 626. (A) 627. (C) 628. (C) 629. (C) 630. (C)
631. (B) 632. (D) 633. (C) 634. (B) 635. (A) 636. (A) 637. (B) 638. (A) 639. (B) 640. (A)
641. (D) 642. (A) 643. (D) 644. (A) 645. (C) 646. (C) 647. (B) 648. (C) 649. (A) 650. (D)
651. (C) 652. (A) 653. (A) 654. (D) 655. (D) 656. (B) 657. (B) 658. (C) 659. (A) 660. (A)
661. (D) 662. (B) 663. (A) 664. (A) 665. (D) 666. (C) 667. (D) 668. (C) 669. (B) 670. (B)
671. (B) 672. (B) 673. (A) 674. (C) 675. (A) 676. (A) 677. (B) 678. (A) 679. (A) 680. (B)
681. (D) 682. (B) 683. (C) 684. (A) 685. (D) 686. (D) 687. (B) 688. (C) 689. (B) 690. (D)
691. (A) 692. (C) 693. (A) 694. (C) 695. (B) 696. (A) 697. (D) 698. (B) 699. (A) 700. (D)
701. (D) 702. (B) 703. (D) 704. (D) 705. (A) 706. (D) 707. (B) 708. (A) 709. (D) 710. (A)
711. (D) 712. (D) 713. (D) 714. (B) 715. (A) 716. (C) 717. (A) 718. (B) 719. (C) 720. (B)
721. (A) 722. (B) 723. (C) 724. (C) 725. (C) 726. (A) 727. (B) 728. (B) 729. (D) 730. (A)
731. (B) 732. (C) 733. (B) 734. (A) 735. (C) 736. (A) 737. (C) 738. (C) 739. (B) 740. (B)
741. (D) 742. (C) 743. (A) 744. (B) 745. (A) 746. (A) 747. (D) 748. (C) 749. (B) 750. (C)
751. (D) 752. (D) 753. (A) 754. (B) 755. (A) 756. (B) 757. (C) 758. (B) 759. (A) 760. (B)
761. (B) 762. (B) 763. (C) 764. (C) 765. (A) 766. (D) 767. (D) 768. (B) 769. (A) 770. (B)
771. (A) 772. (A) 773. (B) 774. (C) 775. (B) 776. (D) 777. (D) 778. (A) 779. (A) 780. (A)
781. (B) 782. (A) 783. (C) 784. (A) 785. (D) 786. (A) 787. (D) 788. (D) 789. (D) 790. (B)
791. (C) 792. (B) 793. (D) 794. (A) 795. (A) 796. (B) 797. (C) 798. (B) 799. (A) 800. (C)
801. (D) 802. (C) 803. (B) 804. (C) 805. (D) 806. (A) 807. (B) 808. (D) 809. (C) 810. (A)
811. (B) 812. (D) 813. (C) 814. (A) 815. (C) 816. (C) 817. (B) 818. (C) 819. (B) 820. (A)
821. (D) 822. (B) 823. (C) 824. (A) 825. (C) 826. (A) 827. (C) 828. (A) 829. (C) 830. (C)
831. (B) 832. (C) 833. (C) 834. (B) 835. (B) 836. (B) 837. (B) 838. (B) 839. (B) 840. (C)
841. (C) 842. (B) 843. (B) 844. (C) 845. (A) 846. (B) 847. (D) 848. (D) 849. (A) 850. (B)
851. (D) 852. (C) 853. (C) 854. (B) 855. (B) 856. (D) 857. (A) 858. (B) 859. (C) 860. (B)
861. (A) 862. (B) 863. (B) 864. (C) 865. (D) 866. (B) 867. (D) 868. (A) 869. (A) 870. (C)
871. (C) 872. (A) 873. (C) 874. (B) 875. (B) 876. (C) 877. (A) 878. (B) 879. (B) 880. (B)
881. (B) 882. (B) 883. (A) 884. (B) 885. (B) 886. (D) 887. (B) 888. (D) 889. (C) 890. (C)
891. (B) 892. (C) 893. (D) 894. (A) 895. (B) 896. (C) 897. (A) 898. (D) 899. (D) 900. (D)
901. (C) 902. (C) 903. (C) 904. (B) 905. (B) 906. (C) 907. (B) 908. (C) 909. (B) 910. (A)

911. (D) 912. (D) 913. (A) 914. (C) 915. (D) 916. (A) 917. (C) 918. (A) 919. (C) 920. (D)
921. (C) 922. (D) 923. (A) 924. (A) 925. (A) 926. (B) 927. (D) 928. (D) 929. (A) 930. (D)
931. (B) 932. (C) 933. (D) 934. (C) 935. (C) 936. (C) 937. (A) 938. (B) 939. (A) 940. (D)
941. (A) 942. (D) 943. (B) 944. (A) 945. (C) 946. (A) 947. (D) 948. (A) 949. (A) 950. (A)
951. (B) 952. (C) 953. (D) 954. (A) 955. (D) 956. (B) 957. (D) 958. (B) 959. (A) 960. (A)
961. (D) 962. (C) 963. (B) 964. (C) 965. (B) 966. (B) 967. (D) 968. (C) 969. (C) 970. (C)
971. (D) 972. (B) 973. (C) 974. (B) 975. (B) 976. (D) 977. (B) 978. (A) 979. (C) 980. (B)
981. (A) 982. (C) 983. (D) 984. (D) 985. (A) 986. (A) 987. (A) 988. (B) 989. (B) 990. (C)
991. (A) 992. (D) 993. (C) 994. (B) 995. (D) 996. (B) 997. (D) 998. (D) 999. (C) 1000. (B)
1001. (C) 1002 (C) 1003 (B), 1004 (C) 1005 (D) 1006 (D) 1007 (C) 1008 (C) 1009 (C) 1010 (B)
1011 (D) 1012 (C) 1013 (B) 1014 (A) 1015 (C) 1016 (C) 1017 (C) 1018 (B) 1019 (B) 1020 (C)
1021 (C) 1022 (D) 1023 (C) 1024 (D) 1025 (B) 1026 (C) 1027 (B) 1028. (C) 1029. (D) 1030. (B)
1031. (B) 1032. (C) 1033. (B) 1034. (D) 1035. (C) 1036. (C) 1037. (D) 1038. (C) 1039. (A) 1040. (A)
1041. (A) 1042. (C) 1043. (A) 1044. (B) 1045. (D) 1046. (D) 1047. (C) 1048. (A) 1049. (D) 1050. (B)
1051. (B) 1052. (C) 1053. (A) 1054. (A) 1055. (A) 1056. (C) 1057. (C) 1058. (A) 1059. (C) 1060. (A)
1061. (B) 1062. (B) 1063. (A) 1064. (B) 1065. (C) 1066. (C) 1067. (D) 1068. (B) 1069. (D) 1070. (D)

Chapter - 4

PLANT PHYSIOLOGY AND MOLECULAR BIOLOGY

1. Photochemical reaction occurs in:
   (A) The plasma membrane of green plants
   (B) The membranes of lysosomes
   (C) The outer membrane of chloroplasts
   (D) The thylakoid membranes

2. Which of the following events involved in stomatal opening:
   (A) Have CAM pathway
   (B) True drought tolerant
   (C) High water use efficiency
   (D) True drought avoider

3. Vernalization is a:
   (A) Reversible process
   (B) Low temperature treatment for seed germination
   (C) Auxin dependent phenomenon
   (D) Process in which genetic makeup of plants change

4. What amount of water absorbed by plants is lost in transpiration?
   (A) 85% (B) 90%
   (C) 95% (D) 100%

5. Which statement is not correct regarding transpiration?
   (A) Living cells are involved
   (B) V.P., D.P., O.P. involved
   (C) Causes dryness of surface
   (D) Physiological process

6. Stomatal transpiration accounts for what % of water loss?
   (A) 75% (B) 80-90%
   (C) 90% (D) 95%

7. Which of the following is not correct ? CAM and C4 plants as compared to C3 plants are:
   (A) Characteristics of hot arid areas
   (B) More rapid photosynthesizers
   (C) More economical in their use of water
   (D) Characterized by reaching maximum photosynthetic rates at high radiation intensities

8. Which of the following lacks stomata?
   (A) Angiosperms (B) Bryophytes
   (C) Algae (D) Pteridophytes

9. Number of stomata/square cm of leaf is:
   (A) 500-1000 (B) 1000-2000
   (C) 1000-60000 (D) None of these

10. Fully open stomata of leaf occupies what percent of leaf area?
    (A) 0.05% (B) 0.5%
    (C) 0.95% (D) 1-2%

11. Barbell shaped guard cells are found in:
    (A) Oilseeds (B) Cucurbits
    (C) Grasses (D) Pulses

12. Which of the following is false about glyoxylate cycle?
    (A) It is a variant of citric acid cycle
    (B) It results in the formation of succinate from acetyl-CoA
    (C) It occurs in glyoxisome
    (D) It occurs in all eukaryotic cells

13. When stomata are equally distributed on both surfaces of leaf tissue, are:
    (A) Potato type (B) Oat type
    (C) Apple type (D) Lily type

14. 'Photosynthetic production in guard cells' theory was put forwarded by:
    (A) Scarth (B) Lloyd
    (C) Von Mohl (D) Sayre

15. Which ions are responsible for stomatal movement according to Levitt?
    (A) Fe^{2+} (B) Zn^{2+}
    (C) K^{+} (D) Na^{+}

16. Scotoactive stomata are found in:
    (A) Aloe (B) Salsola
    (C) Sedium (D) All of these

17. Maximum opening of stomata occurs at:
    (A) 8.00 AM (B) 12.00 Noon
    (C) 10.00 AM (D) 4.00 PM

18. Cooperation of the two photosystems of the chloroplast is required for:
    (A) ATP synthesis
    (B) Reduction of $NADP^{+}$
    (C) Oxidation of the reaction centre of photosystem - I
    (D) Generation of a proton–motive force

19. On per unit area basis, transpiration rate is more in:
    (A) Smaller plants (B) Larger plants
    (C) Equal in both (D) Not known

20. Which of the following statement about ubiquinol is correct?
(A) It is the mobile electron carrier between cytochrome oxigoreductase and cytochrome oxidase
(B) It is an integral membrane protein
(C) Its oxidation involves the simultaneous transfer of two electrons to the Fe-S centre of cytochrome reductase
(D) It is a lipid –soluble molecule

21. Transpiration is maximum in:
(A) Blue light (B) Red light
(C) Far red light (D) Green light

22. Transpiration is a 'necessary evil' called by:
(A) Kozlowski (B) Kramer
(C) Curtis (D) None of these

23. The photosynthetic process is an oxidation-reduction reaction. Which of the following is an oxidation reaction in photosynthesis?
(A) Conversion of phosphoglyceric acid to glyceraldehyde phosphate
(B) Carboxylation of RuBP using Rubisco
(C) Splitting of H_2O to form oxygen
(D) Phosphorylation

24. Which of the following is not an antitranspirants?
(A) ABA (B) Asprin
(C) PMA (D) Auxin

25. Transpiration is measured by:
(A) Porometer (B) Potometer
(C) Photometer (D) Manometer

26. Guttation occurs through:
(A) Lenticels (B) Hydathodes
(C) Stomata (D) Epidermis

27. Guttation is due to:
(A) O.P (B) S.P.
(C) Root pressure (D) T.P.

28. Which is wrong regarding guttation?
(A) Loss in liquid form (B) Through hydathodes
(C) Due to root pressure (D) Pure water

29. Which one of the following statements about sieve tube elements in plants is not correct?
(A) They are supported by companion cells
(B) They must die to become functional
(C) They translocate many organic nutrients
(D) They link end to end forming sieve tubes

30. Guttation occurs when:
(A) Warm days followed by cool nights
(B) Cool days followed by warm nights
(C) Plants growing in dry, warm soil
(D) Plants grown in humids

31. The highest loss of water through guttation takes place in:
(A) *Tropoeolum najus* (B) Oat
(C) *Colocasia antiquorum* (D) Barley

32. The stomatal movement is due to:
(A) Pressure of water vapour inside
(B) Turgor pressure of guard cells
(C) Genetic constitution (D) None of the above

33. Transpiration efficiency refers to:
(A) Water absorption, transpiration ratio
(B) Transpiration in relation to ascent of sap
(C) Weight of dry matter produced per kg of water transpired
(D) None of the above

34. When relative humidity of atmosphere is cent per cent, then stomata will:
(A) Open (B) Closed
(C) Partially open (D) Partially closed

35. Which statement about the Mn centre of photosystem II is incorrect?
(A) The Mn centre has four possible oxidation states
(B) Electrons are transferred from the Mn centre to P 680
(C) A tyrosine residue on the D1 protein is an intermediate in electron transfer
(D) The O_2 released by the Mn centre comes from the oxidation of water

36. Blue green algae (cyanobacteria) form heterocysts when deprived of ammonia and nitrate. In this form, the algae lack nuleic and are attached to adjacent vegetative cells. Heterocysts have photosystem I activity but are entirely devoid of photosystem II activity. What is their role?
(A) N_2 fixation (B) Denitrification
(C) Proline synthesis (D) Spore formation

37. Guard cells differ from adjacent of epidermal cells is having:
(A) Mitochondria (B) Golgibodies
(C) Cell wall (D) Chloroplasts

38. Which of the following statement about the thylakoid membrane is incorrect?
(A) It contains photosystems I and ATP synthase in the unstaked regions
(B) It contains the cytochrome b6f complex in the unstaked regions only
(C) It contains photosystem II mostly in the stakd regions
(D) It facilitates communication between photosystem I and II by the circulation of plastoquinones and plasocyanines in the thylakoid space

39. The most important factors affecting transpiration is:
(A) Atmospheric humidity (B) Temperature
(C) Wind (D) Light

40. A nonliving continuum of cell walls and intercellular spaces is:
(A) Ground complex (B) Alburnum
(C) Apoplast (D) Desmotubule

41. Match list - I with list – II and select the correct answer using the codes given below the lists:

List I	List II
P. Richmond Lang effect	1. Reversal of red-drop
Q. Emerson effect	2. Inhibition of glycolysis by oxygen
R. Pasteur effect	3. Inhibition of photosynthesis by oxygen
S. Warburg effect	4. Senshene retardation in detached leaves

(A) P-1, Q–4, R–2, S–3 (B) P-4, Q–1, R–3, S–2
(C) P-1, Q–4, R–3, S–2 (D) P-4, Q-1, R–2, S-3

42. The phenomenon of absorption of water that depends on water loss from aerial parts of plants (Transpiration) is:
(A) Active absorption
(B) Passive absorption
(C) Both passive and active absorption
(D) None of the above

43. Root hairs occur in the zone of:
(A) Cell division (B) Cell elongation
(C) Cell maturation (D) Mature cells

44. Water lightly held to soil particles is:
(A) Gravitational water (B) Capillary water
(C) Hygroscopic water (D) Free water

45. Which of the following is not correct about tarpene?
(A) It is the largest class of secondary metabolite
(B) Phenylalanine acts as a precursor
(C) It is synthesized by pathway mevalonic acid
(D) It plays defensive role

46. Which of the following statement is not true with respect to the photosystem II (PSII)?
(A) PSII is located in the grana
(B) Chrolophyll molecule in the reaction centre for the PSII is bound to a 110 Kd integral membrane protein
(C) A complex of cytochromes b and f transport electrons from PSII to PSI
(D) PSII reaction centre is bound to two Mn^{++} ions

47. Root pressure may be caused by:
(A) Osmotic flow of water into xylem of absorbing part of root
(B) Loss of water from xylem of the plant due to transpiration
(C) Low water potential of leaves
(D) High water potential of leaves

48. The osmotic theory of active water absorption was first given by:
(A) Kramer (1941) (B) Eaton (1943)
(C) Arkins (1916) (D) Priestley (1923)

49. Force for passive water absorption develops in:
(A) Xylem (B) Aerial parts
(C) Root (D) Root hairs

50. The lowest water potential is found in the xylem channels of:
(A) Stem (B) Root
(C) Root in the root hairs zone (D) Leaves

51. Which of the following is not involved in maintaining oxygen homeostasis in nitrogen fixing nodules?
(A) Leghaemoglobin
(B) Cytochrome oxidase
(C) ATP dependent potassium pump
(D) Dinitrogenase reductase

52. If radioactive carbondioxide ($^{14}CO_2$) is used in calvin cycle then which carbon atom of 3-phosphoglyerate will appear radioactive:
(A) Carbon – 3 (B) Carbon - 1
(C) Carbon – 2 (D) Carbon – 4

53. Photophosphorylation is synthesis of:
(A) ADP from ATP
(B) Glucose 6-phosphate from glucose
(C) ATP from ADP
(D) $NADP^+$ from NAD^+

54. The two pathways required for the net synthesis of glucose from triglycerides in germinating groundnut seeds are:
(A) Hexose monophosphate shunt & gluoneogenesis
(B) Calvin cycle and glyoxalate cycle
(C) Glycolysis and cori cycle
(D) Glyoxalate cycle and gluoneogenesis

55. The initial stage of water absorption by root cell is by:
(A) Imbibition (B) Adsorption
(C) Osmosis (D) Desorption

56. The Pasteur effect refers to:
(A) An increase in hexokinase activity owing to increased oxidative phosphorylation
(B) An increase in ethanol formation from pyruvate on changing from aerobic to anaerobic metabolism
(C) An increase in glucose utilization on changing from anaerobic to aerobic metabolism
(D) A decrease in glucose utilization on changing from anaerobic to aerobic metabolism

57. Root cap has no role in water absorption because:
(A) It has no direct connection with the vascular system
(B) It has loosely arranged cells
(C) It has no cells containing chloroplasts
(D) It has no root hairs

58. The maximum absorption of water by roots occurs in the (region) zone of:
(A) Root cap (B) Cell division
(C) Cell elongation (D) Root hairs

59. Stomtal opening is measured by
(A) Planimeter (B) Potometer
(C) Porometer (D) Photometer

60. Transpiration cohesion tension theory operates in:
(A) Active absorption
(B) Passive absorption
(C) Both active and passive absorption
(D) None of the above

61. The phenomenon of uptake of water at the expense of energy by the cell and usually against the osmotic gradient is known as:
(A) Osmosis (B) Active absorption
(C) Passive absorption (D) Imbibition

62. Rate limiting enzyme in ethylene biosynthesis is:
(A) α-aminolevulinic acid synthase
(B) Aminocyclopropane carboxylic acid synthase
(C) Polygaturanases
(D) Myo-inositol tetrakine phosphate

63. Which of the following statement is not correct:
(A) Plants absorb excess quantity of water
(B) Plants take in small quantity of mineral salts through the soil water
(C) Water and inorganic salts are taken in simultaneously by root hairs
(D) Plants absorb only one thing at a time, water or inorganic salts

64. Root pressure can be measured by means of:
(A) Porometer (B) Potometer
(C) Auxanometer (D) Manometer

65. How many ATP molecules can be derived from each molecule of acetyl-CoA that enters the Kreb's cycle (consider 1 NADH and $FADH_2$ give 3 and 2 ATP, Respectively)?
(A) 6 (B) 12
(C) 18 (D) 38

66. Rapid water absorption takes place by:
(A) Passive absorption (B) Active absorption
(C) Expenditure of energy (D) Osmosis

67. Microorganisms responsible for nitrification are:
(A) *Nitrosomonas* and *Nitrobacter*
(B) *Nostoc* and *Anabaena*
(C) *Rhizobium* and *Azatobacter*
(D) *Clostridium* and *Pseudomonas*

68. Which is correct about RNA polymerase III?
(A) It occurs in nucleoplasm
(B) It provides ~ 10% of total polymerase activity
(C) It transcribes tRNA and 5SRNA genes
(D) All of these

69. Soil water is able to reach xylem elements of root because of:
(A) Imbibition
(B) Gradient of suction pressure
(C) Gradient of turgor pressure
(D) Ion concentration of water

70. Assertion (A): In bulk of eukaryotic mRNAs, a poly – (A) tail is added at the 3'-end of primary transcripts or pre-mRNAs
Reason (R): Addition of a poly – (A) tail at the 3'-end of primary transcript is essential for the processing of pre-mRNAs
Options
(A) Both A & R are correct and R adequately explains A
(B) Both A & R are correct but R does not explain A
(C) Only A is correct (D) Only R is correct

71. Which of the following statement about the mitochondrial protein gradient and membrane potential is correct?
(A) Either of them is sufficient to make ATP from ADP + Pi
(B) Both are required to make ATP
(C) Usually cancel one another since the system is at equilibrium
(D) Neither of them is required for ATP synthesis

72. Water potential found in root hair cells is generally:
(A) 1-2 atm (B) 1 -20 atm
(C) -1 to 4.0 atm (D) -1 to -4 atm

73. Path of water movement from soil to xylem is:
(A) Metaxylem → Protoxylem → Cortex → Soil → Root hair
(B) Cortex → Root hair → Endodermis → Pericycle → Protoxylem → Metaxylem
(C) Soil → Root hair → Cortex → Endodermis → Pericycle → Protoxylem → Metaxylem
(D) Pericycle → Soil → Root hair → Cortex → Endodermis → Protoxylem → Metaxylem

74. The reactions of the krebs cycle:
(A) Take place in the cytosol of eukaryotic cells
(B) Generate ATP also by substrate phosphorylation
(C) Are important for the metabolism of carbohydrates but not other molecules
(D) Both (A) and (B)

75. In soil the water available for root absorption is:
(A) Gravitational water (B) Capillary water
(C) Hygroscopic water (D) Combined water

76. Rate of water absorption can be increased through:
(A) Decreased transpiration
(B) Decreased ion absorption
(C) Increased photosynthesis
(D) Increased transpiration

77. Which of the following statements completes the sentence more correctly? The rate of flow of electrons through the electron transport chain is most directly regulated by:
(A) The ATP : ADP ratio
(B) The concentration of acetyl-CoA
(C) The rate of oxidative phosphorylation
(D) Feedback inhibition by H_2O

78. The movement of water from one cell of cortex to the next in the root is due to:
(A) Water potential gradient
(B) Chemical potential gradient
(C) Accumulation of inorganic salts in the cells
(D) Accumulation of organic salts in the cells

79. Exudation of xylem sap on cutting of a shoot is due to:
(A) Guttation (B) Root pressure
(C) Transpiration (D) Osmosis

80. Water entering root due to diffusion is part of:
(A) Endocytosis (B) Guttation
(C) Passive absorption (D) Active absorption

81. Which of the following is not a substrate of ribulose 1, 5-bisphosphate carboxylase?
(A) CO_2
(B) Glyceraldehyde – 3- phosphate
(C) Ribulose 1, 5-bisphosphate
(D) O_2

82. Stump of well watered potted plant shows exudation of sap due to:
(A) Root pressure (B) Capillarity
(C) Cohesion force (D) Transpiration pull

83. A gene is alternatively spliced, and its exon 1 joined to any one or more of exons produces a functional protein. How many different functional proteins will be produced if this gene has 3 exons?
(A) 1 (B) 2
(C) 3 (D) 4

84. **Assertion** (A): Plant species with C_4 pathway generally have higher photosynthetic rate than C_3 plant species, specially under high light intensity
Reason (R): The main factor, which increase photosynthetic efficiency of C_4 plant species is the lack of measurable photorespiration in them
Options
(A) Both A & R are correct & R adequately explains A
(B) Both A & R are correct but R does not explain A
(C) Only A is correct (D) Only R is correct

85. Water absorbed by root in order to meet the requirement of transpiration is due to:
(A) Transpiration pull (B) Osmosis
(C) Imbibition (D) Plasmolysis

86. Root hair absorbs water from soil through:
(A) Turgor pressure (B) Ion exchange
(C) Osmosis (D) D.P.D.

87. Which of the following inclusion bodies contain the enzymes responsible for carbon dioxide fixation in bacteria, blue green algae?
(A) Lysosomes
(B) Peroxisomes
(C) Metachromatic granules (D) Carboxysome

88. Prolonged water-logging kills plants due to:
(A) Stoppage of root respiration
(B) Dilution of soil nutrients
(C) Dilution of plant cell sap
(D) Leaching of nutrients

89. Rate of water absorption is slow near freezing point because:
(A) Water absorption is a metabolic process
(B) Cell growth stops
(C) Transpiration is reduced
(D) Cell membranes become more viscous

90. Arrange root hair cell, inner cortical cell and mesophyll cell in ascending order of D.P.D.:
(A) Mesophyll cell, Root hair cell and Cortical cell
(B) Cortical cell, Mesophyll cell and Root hair cell
(C) Root hair cell, Cortical cell and Mesophyll cell
(D) Root hair cell, Mesophyll cell and Cortical cell

91. The upward movement of water, against the downward pull of gravity in plants depends on the process of:
(A) Absorption (B) Photosynthesis
(C) Respiration (D) Transpiration

92. Strasburger rejected the vital force theory on the ground that:
(A) Living cells are incapable of translocation
(B) Water rises in dead cells
(C) Respiration occurs in living cells
(D) Living cells are capable of growth

93. The enzyme nitrogenase that fixes atmospheric nitrogen in the root nodules of leguminous plants comprises two sub-units the composition of each respectively being:
(A) Fe-protein & Fe-Mo-protein
(B) Mo-protein & Fe-protein
(C) Mg-protein & Fe-Mo-protein
(D) Fe-protein & Fe-protein

94. Which of the following statement about the enzyme complexes of the electron transport system is correct?
(A) They are located in the mitochondrial matrix
(B) They cannot be isolated from one another in functional form
(C) They have very similar visible spectra
(D) They are integral membrane proteins loc- ated in the inner mitochondrial membrane

95. The pressure that develops in roots due to the metabolic activities of living cells is:
(A) Turgor pressure (B) Osmotic pressure
(C) Root pressure (D) Diffusion pressure

96. A gene is alternatively spliced, and its exon 1 joined to any one or more of exons produces a functional protein. How many different functional proteins will be produced if this gene has 4 exons?
(A) 4 (B) 6
(C) 7 (D) 8

97. **Assertion (A):** Enhancer sequences increase the activity of a promoter lying in their vicinity.

Reason (R): Enhancers most likely function by increasing the concentration of transcription factors in the vicinity of promoters

Options

(A) Both A and R are correct and R adequately explains A
(B) Both A & R are correct but R does not explain A
(C) Only A is correct (D) Only R is correct

98. Which of the following statement is wrong in root-pressure concept of ascent of sap:
(A) Water can be raised to a height of 6' – 7'
(B) Water moves upward in the absence of root
(C) Water movement by this force is slow
(D) Root pressure operates in all the plants

99. Imbibitional theory was given by:
(A) Sachs (B) Boehm
(C) Scholander (D) Curtis

100. Which of the following statements about cyclic photophosphorylation is incorrect?
(A) It doesn't involve NADPH formation
(B) It is activated when ATP is limiting
(C) It does not generate O_2
(D) It involves a substrate-level phosphorylation

101. Cohesive force of water molecules is of the magnitude of (Dixon and Joly):
(A) 1-10 atm (B) 10-15 atm
(C) 45-200 atm (D) 15-45 atm

102. The following evidences rejected the atmospheric pressure theory. Which of them is incorrect:
(A) It can raise water to height of 33 feet only if complete vacuum is created
(B) Free surface of water is required for proper operation of atmospheric pressure which is readily available in plants
(C) Pressure falls below that of atmospheric pressure because of transpiration
(D) Water rises rapidly to compensate the water loss because of atmospheric pressure

103. Which of the following is not an example of N_2 fixation organism?
(A) *Frankia* (B) *Pisum sativum*
(C) *Rhizobium* (D) *Azospirillum*

104. Which of the following is not the least convincing in the imbibitional theory of ascent of sap:
(A) The imbibitional force is very high from 100-1000 atm
(B) The imbibitional force can raise the water to a height of 200' – 400' i.e. the tallest tree
(C) The movement of water is along the walls of the xylem vessels and not through the lumen (cavity) of the vessels
(D) The movement of water, no doubt is slow, but to some extent can keep pace with transpiration

105. The girding or ringing experiment is that:
(A) The metallic ring is tightly tied to a stem
(B) The ring tissue external to xylem is removed in a stem
(C) The area is simply marked in form ring in stem
(D) None of the above

106. The girdling or ringing experiment to prove that the water moves through the xylem vessels was performed by:
(A) Stocking (B) Malpighi
(C) Molisch (D) Askenasy

107. Which of the following proteins in the photosynthetic electron-transport chain is not a transmembranes protein?
(A) ATP synthase (B) LHC
(C) PSII (D) Ferodoxin

108. Capillary theory was proposed by:
(A) Unger (B) Sachs
(C) Bohm (D) Mac Dougal

109. The main function of xylem vessel is:
(A) To provide mechanical support to the plant
(B) To conduct organic food from one part of the plant to another
(C) To conduct water and minerals from one part of the plant to another
(D) None of the above

110. Oxidative deamination is:
(A) The conversion of an amino acid to a keto acid plus ammonia
(B) The conversion of an amino acid to a carboxylic acid plus ammonia
(C) The transfer of an amino group from an amino acid to a keto acid
(D) The transfer of an amino group from an amino acid to a carboxylic acid

111. A girdled plant will eventually die because of:
(A) Absence of downward movement of water
(B) Absence of upward movement of water
(C) Absence of upward supply of organic nutrients
(D) Absence of downward movement of organic nutrients

112. Plants exhibit subepidermal evaporation of water during:
(A) Photosynthesis (B) Guttation
(C) Transpiration (D) Respiration

113. Which is not a constituents of protoplasm?
(A) Cell wall (B) Cell membrane
(C) Vacuole (D) Both (A) and (C)

114. The cut end of a shoot is dipped in eosine solution. What will happen?
(A) Ascent of sap does not occur
(B) Ascent of sap does not occur but the leaves remain fresh
(C) Leaves wilt but ascent of sap continues
(D) Ascent of sap occurs and the tracheary elements get stained

115. Which of the following conditions does not stimulate the activity of allosteric enzyme Rubisco?
(A) Increases in pH
(B) Increases in Mg^{++}
(C) Increases in NADPH concentration
(D) Increases in the concentration of K^+

116. Root pressure theory of ascent of sap is unacceptable because:
(A) Water can ascend without root or root pressure
(B) Root pressure cannot explain ascent of sap beyond 10 meters
(C) Root pressure is more during early morning than afternoon
(D) Root pressure does not occur in spring

117. Transpiration cohesion theory explains that the upwards pull of water is transmitted from top to bottom by cohesion of molecules caused by:
(A) Hydrophilic cell walls (B) Hydrogen bonds
(C) Oxygen bonds (D) Surface tension

118. Rising column of water does not break its connection with xylem walls despite negative pressure or tension due to:
(A) Cohesion amongst water molecules
(B) Strong transpiration pull
(C) Adhesion
(D) Surface tension

119. Ascent of sap is:
(A) Upward movement of water in the pant
(B) Downward movement of organic nutrients
(C) Upward and downward movement of water in the plant
(D) Redistribution of inorganic substances in the plant

120. The synthesis of ATP by hemiosmosis:
(A) Is endergonic and is oupled to exergonic electron transport
(B) Is exergonic and is coupled to endergonic electron transport
(C) Is due to electrons shuttled down the ET to the final electron acceptor, which NAD^+
(D) Produces less ATP than the Kreb's cycle and glycolysis combined

121. Instrument that can be used to demonstrate pull due to vaporisation of water is:
(A) Potometer (B) Atmometer
(C) Auxanometer (D) Anemometer

122. Further desaturation of 18:0 molecule occurs in association with?
(A) Cell wall (B) Mitochondria
(C) Endoplasmic reticulum
(D) Peroxysomes

123. The transport of sap from root to top of the plant is:
(A) Ascent of sap (B) Conduction
(C) Transport (D) Translocation

124. If plants are grown for several days in an atmosphere containing $^{14}CO_2$ in place off $^{12}CO_2$, one would expect to find:
(A) Large amounts of radioactive oxygen released from the stomatas
(B) A large increase in ^{14}C in the starch
(C) A large decrease in the rate of carbon fixation in the guard cells
(D) An increase in the activity of RuBP carboxylase in the photosynthetic cells

125. In xylem, the ascent of sap takes place in:
(A) Tracheids with associated xylem parenchyma
(B) Walls of tracheary elements
(C) Xylem parenchyma
(D) Lumen of tracheary elements

126. The force which water molecules are held together in xylem is called:
(A) Imbibition pressure (B) Osmotic pressure
(C) Adhesion force (D) Cohesion force

127. Root pressure is unable to explain the ascent of sap because it is not found in:
(A) Bryophytes (B) All plants in all reasons
(C) Trees (D) Spring

128. The relay pump theory, in vital theories of ascent of sap, was given by:
(A) Strasburger (B) Dixon
(C) Godlewski (D) Unger

129. The apparatus that can measure root pressure is:
(A) Manometer (B) Potometer
(C) Auxanometer (D) Osmometer

130. Gibbrellic acid induced starch degradation in endosperm essentially requires gene expression, protein synthesis and protein transport. In an experiment to understand this, GFP tagged protein is seen to be moving from cytoplasm into the nucleus before alpha amylase could be detected. What do you think this GFP is tagged with?
(A) It is in fusion with a gene regulator
(B) It is in fusion with a signaling kinase
(C) It is in fusion with alpha-amylase
(D) It is tagged with a cell wall protein

131. In a gridled plant which dies first:
(A) Shoot (B) Root
(C) Both die simultaneously (D) Leaf

132. Guide RNA functions in:
(A) RNA processing
(B) RNA editing
(C) RNA degradation
(D) Transcription regulation

133. Most of the 18:0 molecules (stearic acid) are desaturated to 18:0 molecules (oleic acid) in which organelles?
(A) Plastids (B) Mitochondria
(C) Endoplasmic reticulum (D) Peroxysomes

134. A tree girdled upto xylem may survive for some time but ultimately dies because:
(A) Water does not move upwards
(B) Sugar does not move upwards
(C) Sugar does not move downwards
(D) Water does not move downwards

135. Water in plants is transported by or ascent of sap takes place through:
(A) Cambium (B) Phloem
(C) Xylem (D) Epidermis

136. The most widely accepted theory for ascent of sap in trees is:
(A) Capillarity
(B) Role of atmospheric pressure
(C) Pulsating action of living cells
(D) Transpiration pull and cohesion theory of Dixon and Joly

137. Cell A has osmotic potential of -16 bars and pressure potential of 8 bars, whereas cell B has osmotic potential of -12 bars and pressure potential 2 bars. The direction of flow of water will be:
(A) From cell B to A (B) From cell A to B
(C) No flow (D) In both directions

138. Water rises in the stem due to:
(A) Cohesion and transpiration pull
(B) Turgor pressure
(C) Osmotic pressure
(D) None of the above

139. Dixon and Joly are associated with:
(A) Light reaction and photosynthesis
(B) Anaerobic respiration
(C) Cohesion theory of ascent of sap
(D) Apical dominance

140. According to the vital force theory, ascent of sap is due to active pulsation of innermost layer of cortex. This theory was given by:
(A) J.C. Bose (B) Dixon
(C) Strasburger (D) Sachs

141. Root pressure is maximum when:
(A) Transpiration is high and absorption low
(B) Transpiration is very low and absorption high
(C) Transpiration is very high and absorption high
(D) Transpiration and absorption both are slow

142. Phototropism refers to how plants respond to sunlight by bending towards the light. This is the result of:
(A) Cell division on the side of the shoot apex closer to the light source
(B) Cell division on the side of the shoot apex away from the light source
(C) Cell differentiation on the side of the shoot apex away from the light source
(D) Cell elongation on the side of the shoot apex away from the light source

143. Xylem conducts sap from:
(A) Leaves to roots (B) Roots to leaves
(C) Roots to stems (D) Stems to roots

144. Which DNA polymerases functions in mitochondria?
(A) Pol β (B) Pol ε
(C) Pol γ (D) Pol δ

145. A plant will wilt when:
(A) A few roots are removed (B) Phloem is blocked
(C) Xylem is blocked (D) Both (B) and (C)

146. Cohesion-transpiration pull theory operates only in:
(A) Passive water absorption
(B) Active water absorption
(C) Conditions favouring transpiration
(D) Conditions restricting transpiration

147. Upward movement of water through xylem is best explained by:
(A) Cohesion theory (B) Pulsation theory
(C) Capillarity theory (D) Root pressure theory

148. Which contributes most to the transport of water from soil to the leaves of a tree:
(A) Root pressure
(B) Cohesion of water and transpiration pull
(C) Capillary rise of water inside xylem
(D) Hydrolysis of ATP

149. The principal pathway of water translocation in angiosperms is:
(A) Sieve cells
(B) Sieve tube elements
(C) Xylem vessel system
(D) Xylem and phloem

150. Cohesive force of water is due to:
(A) O-bonds (B) OH-bonds
(C) S-bonds (D) H-bonds

151. Water ascends in plants through:
(A) Xylem (B) Phloem
(C) Stele (D) Pith

152. Identify the correct statement:
(A) Cytokinins does not regulate cell division in plants
(B) Kinetin was discovered as a breakdown product of DNA
(C) Osmotic adjustment of cells does not help water balance in plants
(D) Cytokinins enhances leaf senescence

153. The force responsible for raising water in 100 feet tall plant is:
(A) Transpiration (B) Root pressure
(C) Air pressure (D) Capillary action

154. Water supply in the plant is due to:
(A) Osmosis (B) Guttation
(C) Cohesive force (D) Imbibition

155. The principle by which blotting paper absorbs water is:
(A) Capillary action (B) Transpiration pull
(C) Root pressure (D) Absorptive capacity

156. Which one of the following statements is correct with respect to the synthesis of starch and sucrose?
(A) Starch is synthesized in the mitochondria and sucrose in the cytosol
(B) Starch is synthesized in the cytosol and sucrose in the chloroplast
(C) Starch is synthesized in the chloroplast and sucrose in the mitochondria
(D) Starch is synthesized in the chloroplast and sucrose in the cytosol

157. In vascular plants, transport of sugar solutions is through tubes constructed of cells called:
(A) Vessel elements (B) Companion cells
(C) Tracheilds (D) Sieve-tube elements

158. If cohesion-tension transpiration pull theory is correct, a break in water column should:
(A) Increase water content of leaves
(B) Increase rate of photosynthesis
(C) Cause wilting of leaves
(D) No effect

159. Which is relevant for eukaryotes?
(A) Methylation of A in origin
(B) Methylation of C in origin
(C) Licensing factor
(D) DNA polymerase synthesis

160. The direction of water in Cycas leaflets is:
(A) Upward (B) Downward
(C) Lateral (D) Both (A) and (B)

161. Stomata open when guard cells are:
(A) Turgid (B) Flaccid
(C) Large (D) Small

162. Of the followıng four dorsiventral leaves which will show the maximum loss of weight:
(A) Smeared with Vaseline on both surfaces
(B) Smeared with Vaseline on the upper surface
(C) Smeared with Vaseline on the lower surface
(D) Unsmeared

163. Sir J.C. Bose proposed a theory of ascent of sap known as:
(A) Pulsation theory
(B) Transpiration pull theory
(C) Relay pump theory
(D) Capillary force theory

164. In alerone layer of graminaceous plants gibberellin promotes the synthesis of:
(A) Ribonuclease
(B) α- amylase
(C) β-1, 3-glucanase
(D) Tryptophan sunthase

165. Choose the correct set of words denoted by P, Q and R for the following statement. Pheophytin is a P molecule in which the central atom Q has been replaced by two atoms of R:
(A) P – Plastocyanin, Q – Copper, R - Hydrogen
(B) P - Plastocyanin, Q – Zinc, R - Oxygen
(C) P – Chlorophyll, Q – Magnesium, R - Oxygen
(D) P–Chlorophyll, Q–Magnesium R-Hydrogen

166. Potometer are made on the principle that:
(A) The amount of water transpired is approximately equal to amount of water absorbed
(B) The amount of water transpired is more than the amount of water absorbed
(C) The amount od water transpired is less than the amount of water absorbed
(D) Humidity causes reduction in transpiration

167. Which mRNAs lacks poly-(A) tail?
(A) Histone mRNA (B) Globin mRNA
(C) Antibody mRNA (D) Vicilin mRNA

168. In *E. coli* , DNA replication regulations is mostly likely based on:
(A) Methylation of A in origin
(B) Methylation of C in origin
(C) Licensing factor
(D) DNA polymerase synthesis

169. The process in which loss of water occurs in the form of water vapour is:
(A) Respiration (B) Guttation
(C) Transpiration (D) Exosmosis

170. Stomatal aperture is surrounded by guard cells and widens (opens) when guard cells are:
(A) Flaccid (B) Turgid
(C) Bean shaped (D) Dumb bell shaped

171. A mutant plant lacked plastocyanin. Upon illumination of this plant:
(A) Plastoquinol will remain in oxidized state
(B) Plastoquinol will be in reduced state
(C) Ferrodoxin will be in th reduced state
(D) NAD will be in reduced state

172. The maximum loss of water in transpiration is from:
(A) Lenticels (B) Cuticle
(C) Stomata (D) Hydathades

173. The stomata are called sunken when:
(A) Guard cells are in line with epidermal cells
(B) Guard cells are situated below epidermal cells
(C) Guard cells are situated above epidermal cells
(D) Guard cells are situated in lower epidermis

174. Transpiration is high under:
(A) Dry environment
(B) Low atmospheric pressure
(C) High temperature
(D) All the above

175. Sunken stomata:
(A) Increase transpiration
(B) Decrease transpiration
(C) Hinder transpiration
(D) Stop transpiration

176. Stomatal frequency indicates:
(A) Number of stomata per unit area
(B) Rate of water loss
(C) Rate of gaseous exchange
(D) Width of stomatal aperture

177. DCMU (Dichlorophenyl dimethyl urea) is a herbicide which kills the plant by:
(A) Inhibition of respiration
(B) Destroying the chloroplasts
(C) Inhibiting flow of electrons from water to $NADP^+$
(D) Inhibiting PSI and photolysis of water

178. In dorsiventral leaf, the number of stomata per unit area are generally:
(A) Same on both the surface
(B) More on lower surface (Epidermis)
(C) More on upper surface (Epidermis)
(D) Absent of upper surface (Epidermis)

179. Rate of transpiration is dependent upon:
(A) Negative turgor pressure
(B) Temperature
(C) O.P.
(D) Vapour pressure deficit

180. In isobilateral leaf, the number of stomata per unit area are:
(A) More on upper surface
(B) More on lower surface
(C) Approximately same on both the surfaces
(D) Absent on both the surfaces

181. In xerophytic leaf the stomata are situated:
(A) On both surfaces
(B) On upper surface
(C) On lower surface
(D) Absent from both surfaces

182. Assume a thylakoid is somehow punctured so that the interior of the thylakoid is no longer separated from the stroma. This damage will have the most direct effect on which of the following processes?
(A) The flow of electrons from photosystem II to photosystem I
(B) The reduction of $NADP^+$
(C) The splitting of water
(D) The synthesis of ATP

183. The loss of water in the form of water drops is called:
(A) Transpiration
(B) Respiration
(C) Guttation
(D) Exosmosis

184. Guttation is from:
(A) Uninjured edges of leaves near vein ending
(B) Epidermal layers of leaf surfaces
(C) Injured edges of leaves
(D) None of the above

185. Transpiration is unavoidable evil because of:
(A) Structure of leaf and harmful effect
(B) Beneficial and harmful effect
(C) Maintenance of turgidity for growth
(D) Gaseous exchange for photosynthesis and respiration

186. Drooping of leaves due to loss of turgor at noon but recovery in the evening is referred to as:
(A) Temporary wilting (B) Incipient wilting
(C) Permanent wilting (D) Midday desiccation

187. F_1F_0-ATPase in chloroplasts is located on the:
(A) Inner chloroplast membrane with F1 facing the stroma
(B) Inner chloroplast membrane with F1 facing the inter-membrane space
(C) Thylakoid membrane with F1 facing the stroma
(D) Thylakoid membrane with F1 facing the thylakoid lumen

188. Loss of water by cells without external sign of leaf drooping is called:
(A) Temporary wilting
(B) Nascent wilting
(C) Incipient wilting
(D) Permanent wilting

189. Which does not cause stomatal opening:
(A) Circadian rhythm
(B) Influx of K^+ ions
(C) Light
(D) High CO_2 concentration

190. Transpiration does not help in:
(A) Translocation of minerals
(B) Excretion of sap
(C) Ascent of sap
(D) Cooling

191. Plants with scotoactive stomata perform:
(A) C_4 photosynthesis
(B) CAM photosynthesis
(C) C_3 photosynthesis
(D) Anoxygenic photosynthesis

192. Which DNA replication does not require a primer?
(A) *E. coli*
(B) *Vicia faba*
(C) Production of replicative form of X174 genome
(D) Replication of RF form of X174 genome

193. Day time loss of water in the vapour form from stomata is a trade off for intake of:
(A) Minerals (B) Oxygen
(C) Carbon di oxide (D) Energy

194. For keeping stomata open, besides K^+ ions the guard cells require a constant supply of:
(A) ABA (B) ATP
(C) Organic acids (D) Protons

195. Presence of stomata on the under surface of dorsiventral leaf is a mechanism of:
(A) Reduction in transpiration
(B) Protection from dust
(C) Proper regulation of transpiration
(D) Increase in transpiration

196. An antitranspirant is:
(A) Phenyl Mercuric Acetate
(B) Abscisic acid (C) Salicylic acid
(D) All of these

197. Transpiration is a process related to:
(A) Osmosis (B) Diffusion
(C) Activated transport (D) Facilitated diffusion

198. Presence of hair on the leaf surface:
(A) Reduces transpiration
(B) Increases transpiration
(C) Helps in rapid exchange of gases
(D) Prevents guttation

199. Some plants possess modifications like phyllocl-ades, cladodes, phyllodes, scale leaves etc for:
(A) Differentiation and evolution
(B) Decreasing transpiration
(C) Increasing transpiration
(D) Storage of absorbed water

200. Splicing commonly occurs in processing of:
(A) rRNA (B) tRNA
(C) mRNA (D) All of these

201. Starch degradation activity of enzyme phosphor-ylase increases under:
(A) High pH (B) Low pH
(C) Neutral medium (D) Not connected with pH

202. According to Lloyd, the opening and closing of stomata is governed by:
(A) K^+ ion influx and effux
(B) Change in pH
(C) Starch and sugar interconversion
(D) Photosynthetic activity of guard cell chloroplasts

203. Who proposed that opening and closing of stomata is connected with the change in pH of guard cells:
(A) Lyoyd (B) Von Mohl
(C) Sayre/Scarth (D) Levitt

204. Stomata remain open when relative humidity is:
(A) Above 70% (B) 50-70%
(C) 30-50% (D) Below 30%

205. Stomata close at relative humidity of:
(A) 60-70% (B) 50-60%
(C) 50-70% (D) Less than 50%

206. Succinate dehydragenase is inhibited by malonate, oxalate and oxaloacetate. This is because they:
(A) Contain two anionic group and bind to a site other than the active site of the enzyme
(B) Are competitive inhibitors and bind to a site other than the active site of the enzyme
(C) Are non-competitive inhibitors and bind to the active site of the enzyme
(D) Contain two anionic groups and bind to the active site of the enzyme

207. The class of seeds which must maintain a relatively high moisture content in order to remain viable are known as:
(A) Orthodox seeds (B) Recalcitrant seeds
(C) Labyrinth seeds (D) Ex-albuminous seeds

208. Which of the following hypostomatous leaves would dry up last:
(A) Both surface greased
(B) Both surface ungreased
(C) Lower surface greased
(D) Upper surface greased

209. In majority of plants, the guard cells are:
(A) Dumb bell shaped (B) Reniform
(C) Rounded (D) Polygonal

210. Spraying plant surfaces with phenyl mercuric acetate and silicon emulsion will result in:
(A) Increased photosynthesis
(B) Increased transpiration
(C) Decreased transpiration
(D) Exosmosis

211. Stomatal opening in a leaf:
(A) Involves losing of K^+ channels
(B) Occurs when there is an increase in turgor pressure in the guard cells
(C) Occurs in the dark
(D) Occurs in response to abscissic acid
(E) SD plant will flower and LD plant will remain vegetative

212. Sunken stomata are found in:
(A) Mesophytes (B) Hydrophytes
(C) Xerophytes (D) Epiphytes

213. Transpiration occurs from:
(A) Leaves (B) Stem
(C) All aerial parts (D) Roots

214. Transpiration efficiency/Transpiration ratio is:
(A) Water absorbed to water transpired
(B) Unit weight of drymatter synthesized in relation to units of water transpired by the plant
(C) Unit weight of drymatter in relation to water consumed
(D) Unit weight of water transpired to unit weight of drymatter synthesized

215. A mesophytic plant growing in well watered soil shows decreased transpiration in the afternoon due to:
(A) Closure of many stomata
(B) Contraction of cuticle
(C) Less water availability from soil
(D) High rate of photosynthesis

216. Guard cells differ from epidermal cells having:
(A) Mitochondria (B) Vacuoles
(C) Cell wall (D) Chloroplasts

217. When a short day plant (SD) and a long day plant (LD), both kept under favourable conditions for flowering, are exposed to a short flash of red light in the middle of the dark period:
(A) Both SD and LD plants will flower
(B) Both SD and LD plants will remain vegetative
(C) SD plant will remain vegetative and LD plant will flower
(D) None of these

218. Which of the following is most likely cause for wider opening of stomata?
(A) Atmosphere outside the stomata is becoming less saturated
(B) Secretion of salt molecules by adjacent guard cells is taking place
(C) Water molecules enter the guard cells
(D) The night temperature is going to fall

219. During high wind velocity, the stomata:
(A) Open more widely
(B) Close down
(C) Remain unaffected
(D) Remain unaffected but lose more water due to mass action

220. Guttation is the process of elimination of water from plants through:
(A) Stomata (B) Hydathodes
(C) Lenticels (D) Wounds

221. Water drops present on leaf margins of Tropaeolum. Balsam and grassy in early morning are due to:
(A) Guttation (B) Dew
(C) Osmosis (D) Transpiration

222. What is the action spectrum of transpiration?
(A) Green and ultraviolet (B) Orange and Red
(C) Blue and far red (D) Blue and red

223. Jasmonic acid:
(A) Is synthesized from linolenic acid
(B) Is growth promoter
(C) Delay leaf senescence
(D) Only present in mosses and ferns

224. Dumb bell shaped guard cells are found in:
(A) Gymnosperms (B) Most dicot
(C) Cereals (D) Xerophytes

225. Transpiration occurs when the outer atmosphere:
(A) Is wet
(B) Has more moisture content than the substomatal cavities
(C) Has less moisture than substomatal cavities
(D) Possesses the same moisture content as in substomatal cavities

226. Stomatal opening is under the control of:
(A) Epidermal cells
(B) Palisade cells
(C) Spongy parenchyma cells
(D) Guard cells

227. Transpiration rate of a dorsiventral leaf is:
(A) Higher on the lower surface
(B) Higher of the upper surface
(C) Equal on both the surface
(D) Dependent upon species

228. Growing plants in zero gravity on the space shuttle prevents:
(A) Phototropism because it is dark in space
(B) Photomorphogenesis because only near-red light filters into space
(C) Gravitropism because there is little gravity in space
(D) Aerotropism because there is no oxygen in space

229. Maximum transpiration takes place from:
(A) Stem (B) Leaves
(C) Roots (D) Flowers and Fruits

230. Stomata open during day time because the guard cells:
(A) Produce osmotically active sugars or organic acids
(B) Are thin walled
(C) Are lean shaped
(D) Have to help in gaseous exchange

231. In which type, the stomata are present eclusively on the upper surface of the leaves:
(A) Potato type (B) Potamogeton type
(C) Barley type (D) Water lily type

232. Cobalt chloride is blue in dry state. In contact with moisture, it turns:
(A) Yellow (B) Pink
(C) Red (D) Green

233. Stomata open at night and close during day time in:
(A) Xerophytes (B) Mesophytes
(C) Succulents (D) Hydrophytes

234. Percentage of water left in the soil when plant wilts is known as:
(A) Turgidity
(B) Field capacity
(B) Wilting coefficient
(D) Water retaining power of soil

235. Which of the following physiological effects is caused in plants by gibberellic acid?
(A) Shortening of genetically tall plants
(B) Elongation of genetically dwarf plants
(C) Rooting in stem cuttings
(D) Yellowing of young leaves

236. The conditions under which transpiration would be much rapid:
(A) High humidity
(B) Excess of water in soil
(C) Low humidity, high temperature, guard cells are turgid (open), moist soil
(D) Low velocity of wind

237. In which of the following plants would metabolism be hindered if the leaves are coated with wax on their upper surface:
(A) *Hydrilla* (B) Lotus
(C) *Pistia* (D) *Vallisneria*

238. Wilting in plants occurs when:
(A) Phloem is blocked
(B) Xylem is removed
(C) Pith is removed
(D) A few leaves are removed

239. Number of stomata present per cm^2 of a common leaf is about:
(A) Less than 100 (B) More than 100000
(C) 1 Million (D) 1000

240. The following percentage of water absorbed by herbaceous plants is lost in transpiration:
(A) 80 (B) 90
(C) 99 (D) 100

241. Of the processes which occur in leaves the one which may lower their temperature (cooling effect) is:
(A) Respiration
(B) Photosynthesis
(C) Transpiration
(D) Hydrolysis

242. Which of the following statement provides a true example of both photomorphogenesis and phototropism?
(A) Phototropism is a growth towards blue light and photomorphogenesis is a growth towards red light
(B) Phototropism is a growth towards blue light and photomorphogenesis is a developmental process triggered by near–red light
(C) Phototropism is a growth towards blue light and photomorphogenesis is a germination triggered by blue light
(D) Phototropism is a movement towards blue light, that does not involve growth and photomorphogenesis is a movement towards red light that does involve growth

243. Transpiration from the two leaf surfaces are compared by dry paper strips preciously soaked in solution of:
(A) Sodium chloride (B) Calcium chloride
(C) Cobalt nitrate (D) Cobalt chloride

244. Transpiration is regulated by movements of:
(A) Guard cells (B) Subsidiary cells
(C) Epidermal cells (D) Mesophyll cells

245. Which of the following is the correct sequence of events in cytokinin signaling pathway?
(A) Receptor binding → Activtaion of His kinase → Phosphorylation of an aspartate residues → Phosphorylation of AHP protein
(B) Receptor binding → Phosphorylation of an aspartate residues → Activation of His kinase → Phosphorylation of AHP protein
(C) Receptor binding → Phosphorylation of AHP protein → Activation of His kinase → Phosphorylation of an aspartate residues
(D) Receptor binding → Phosphorylation of AHP protein → Phosphorylation of an aspartate residues → Activation of His kinase

246. Transpiration rate is dependent upon:
(A) Stomatal frequency (B) Position of stomata
(C) State of stomata (D) All the above

247. Cytokinins receptor is:
(A) Similar to bacterial two component histidine kinase
(B) Acidic cytosolic protein
(C) GPR
(D) Tyrosine kinase

248. Transpiration differs from evaporation in:
(A) Rate of water loss
(B) Transpiration is a physiological process while evaporation is a physical process.
(C) Transpiration is a physical process while evaporation is a physiological process.
(D) Frequency of water loss

249. If a cell with a solute potential of -0.2 MPa and a pressure potential of 0.4 MPa is placed in a chamber filled with pure water that is pressurized with 0.5 MPa, what will happen?
(A) Water will flow out of the cell
(B) Water will flow into the cell
(C) The cell will be rushed
(D) The cell will explode

250. Transpiration increases in:
(A) Hot, damp and windy conditions
(B) Cool, damp and windy conditions
(C) Cool, dry and still conditions
(D) Hot, dry and windy conditions

251. Common type of transpiration is:
(A) Stomatal (B) Cuticular
(C) Lenticular (D) Foliar

252. Match the terms in Group I with their definitions in Group II:

Group I	Group II
P. Ammonification	1. Conversion of atmospheric nitrogen to ammonia
Q. Denitrification	2. Conversion of organic nitrogen into ammonia
R. Nitrification	3. Conversion of nitrite or nitrate into atmospheric nitrogen
S. Nitrogen fixation	4. Conversion of ammonia into nitrite and nitrate

(A) P-, Q–3, R–1, S–4
(B) P-3, Q–2, R–4, S-1
(C) P-3, Q–2, R–1, S–4
(D) P-2, Q–3, R–4, S-1

253. Hot weather wilting is caused by:
(A) Reduced water absorption
(B) Excessive transpiration
(C) Excessive transpiration and reduced absorption
(D) Excessive water absorption

254. The movement of water and minerals through xylem is explained by the:
(A) Pressure – flow theory
(B) Translocation theory
(C) Bulk-flow theory
(D) Cohesion – tension theory

255. Actual water vapour content of atmosphere is:
(A) Relative humidity
(B) Differential humidity
(C) Absolute humidity
(D) Positive humidity

256. Which one keeps its stomata open during night and closed during day:
(A) Cactus (B) Water lily
(C) Ivy (D) Hibiscus

257. How many ATP molecules are required for the conversion of one N_2 to $2NH_4$ during biological N_2 fixation?
(A) 8 ATP (B) 10 ATP
(C) 12 ATP (D) 16 ATP

258. Which wall of the guard cells is thick:
(A) Lateral (B) Inner
(C) Outer (D) Middle

259. What is the strongest reducing agent in photosynthetic electron-transfer reactions?
(A) Plastoquinone (B) P 680*
(C) P 700* (D) P 700

260. Higher CO_2 concentration around the leaves results in:
(A) Reduced photosynthesis
(B) Complete stomatal closure
(C) Wider opening of stomata
(D) Partial closure of stomata

261. Rate of transpiration is reduced with:
(A) Rise in temperature
(B) Decrease in light intensity
(C) Increase in wind velocity
(D) Increase in water uptake

262. Which one gives the most valid explanation for guard cells movements
(A) Guard cells photosynthesis
(B) Starch hydrolysis theory
(C) Potassium influx and efflux
(D) Transpiration

263. Water movement against gravity is caused by:
(A) Imbibition (B) Transpiration
(C) Osmosis (D) Diffusion

264. When chrolophyll a is in its oxidized state:
(A) It fluorescence in the thylakoids
(B) O_2 is generated from sucrose
(C) It directly reduces $NADP^+$
(D) It results in protons being generated from H_2O

265. In thin leaved mesophytes, stomata open during the day and close during the night. They belong to:
(A) Barley type (B) Potato type
(C) Alfalfa type (D) Bean type

266. The large subunit of Rubisco is coded by:
(A) Chloroplast genome
(B) Nuclear genome
(C) Mitochondrial genome
(D) Both mitochondrial and nuclear genome

267. In terrestrial habitates, temperature and rainfall conditions are influenced by:
(A) Water transformations (B) Transpiration
(C) Thermoperiodism (D) Translocation

268. Low atmospheric pressure changes the rate of transpiration to:
(A) Decrease slowly (B) Increase
(C) Remain unchanged (D) Decrease rapidly

269. During wilting the sequence of events shall be:
(A) Endosmosis → Plasmolysis → Temporary wilting → Permanent wilting
(B) Exosmosis → Plasmolysis → Temporary wilting → Permanent wilting
(C) Exosmosis → deplasmolysis → Plasmolysis → Temporary wilting → Permanent wilting
(D) Exosmosis → Plasmolysis → Deplasmolysis → Temporary wilting → Permanent wilting

270. Escape of liquid from hydathodes (Uninjured leaf tips) is called:
(A) Evaporation (B) Transpiration
(C) Guttation (D) Osmosis

271. Conversion of starch to organic acids is required for:
(A) Stomatal opening (B) Stomatal closing
(C) Stomatal formation (D) Stomatal activity

272. Dibromothymoquinone, an inhibitor of cytochrome bf complex of photosynthetic electron transport chain was added to the green alga chlamydomonas. After illumination it was found that in the alga:
(A) Plastiquinone was in reduced state
(B) P 700 was in reduced state
(C) Plastocyanin was in reduced state
(D) Both plastoquinone and plastocyanin were in oxidized state

273. In which of the following organelles glycolate is oxidized to glyoxylate during photorespiration:
(A) Chloroplast (B) Peroxisome
(C) Mitochondria (D) Endoplasmic reticulum

274. Element involved in stomatal regulation is:
(A) Zinc (B) Magnesium
(C) Potassium (D) Iron

275. In the pentose phosphate pathway:
(A) Only the C-1 carbon of glucose is oxidized to CO_2
(B) All the carbons of glucose is oxidized to CO_2
(C) No decarboxylation occurs
(D) C-4 and C-5 of glucose is oxidized to CO_2

276. In hot summer day, plant cooling occurs due to:
(A) Transport of water from root to all parts of the plant
(B) Loss of liquid water
(C) Water loss from entire plant
(D) Loss of water vapours from foliar surface

277. Stomatal opening in a leaf:
(A) Involves closing of K^+ channels
(B) Occurs when there is an increase in turgor pressure in the guard cells
(C) Occurs in the dark
(D) Occurs in response to abscissic acid

278. Stomata generally open during the day because the guard cells have:
(A) Outer thin wall (B) Chlorophyll
(C) Kidney –shaped (D) Larger nuclei

279. Tendrillar movement are examples of::
(A) Phototropism (B) Thigmotropism
(C) Geotropism (D) Hydrotropism

280. Transpiration is helpful to plants in:
(A) Cooling
(B) Loss of excess nutrients
(C) Upward conduction/ascent of sap
(D) Loss of excess water

281. A carotenoid – less mutant plant was grown under normal sunlight, it will experience:
(A) Increased photosynthetic rate
(B) Increased chlorophyll biosynthesis
(C) Reduced photorespiration
(D) Increased chlorophyll oxidation and necrosis

282. Temporary wilting is due to:
(A) Photosynthesis (B) Transpiration
(C) Respiration (D) Absorption of water

283. Water potential in leaf tissue is positive (near zero) during:
(A) Low transpiration
(B) Excessive absorption
(C) Excessive transpiration
(D) Guttation

284. Plastocyanin is a mobile electron carrier between:
(A) PS II and cytochrome bf complex
(B) Cytochrome bf complex and PS I
(C) Oxygen evolving complex and PS II
(D) PS I and ferrodoxin

285. A leafy twig of mesophytic plant dipped in water would demonstrate:
(A) Photosynthesis (B) Transpiration
(C) Respiration (D) Guttation

286. An adaptation for better gaseous exchange in plant leaves is:
(A) Hair on lower surface
(B) Multiple epidermis
(C) Waxy cuticle
(D) Stomata on lower surface away from direct sun rays

287. Which is not related to transpiration?
(A) Absorption and distribution of minerals
(B) Circulation of water
(C) Temperature
(D) Bleeding

288. At constant temperature, the rate of transpiration will be higher at:
(A) At sea level (B) 1 km below sea level
(C) 1 km above sea level
(D) 1.5 km above sea level

289. Hydathodes occur on:
(A) Stem (B) Leaves
(C) Roots (D) All the above

290. The loss of water through cuticle reaches upto:
(A) 10% (B) 20%
(C) 30% (D) 40%

291. Treatment of intact mitochondria with an uncoupler, such as 2, 4-dinitrophenol, in the presence of ADP, P succinate, and oxygen would have which of the following effects on the rates of electron transport and ATP synthesis? (+= Stimulation) = No effect; – = inhibition)

	Rate of electron transport	Rate of ATP synthesis
(A)	+	+
(B)	–	–
(C)	+	–
(D)	–	+

292. A twig kept in water having some salt remains fresh for longer period due to:
(A) Decrease in bacterial degradation
(B) Exosmosis
(C) Decrease in transpiration rate
(D) Absorption of more water

293. When Darwin illuminated the coleoptile of a grass from one side, it bent towards the light source. This was due to:
(A) Diffusion of auxin from tip to the subterminal region
(B) Inreased cell division on dark side
(C) Diffusion of some chemicals from basal to the tip region
(D) Increased cell division on the illiminated side

294. Maximum transpiration occurs in:
(A) Algal cells (B) Xerophytic plants
(C) Hydrophytic plants
(D) Mesophytic plants

295. Which of the following statement is not true with respect to the Photosystem II found in the chloroplasts?
(A) PS II is located in the grana
(B) Chl molecule in the reaction centre for the PS II is bound to a 110 Kd integral membrane protein
(C) A complex of cytochromes a and f transports electrons form PS II and PS I
(D) PS II reaction enter is bound to two Mn^{++} ions

296. Nonfunctional stomata can be seen in:
(A) Mango leaf (B) Pea leaves
(C) Hydrilla (D) Lotus

297. Phytohormone connected with closing of stomata is:
(A) ABA (B) Kinetin
(C) GA (D) IBA

298. Latest explanation for closure of stomata is:
(A) Starch glucose theory
(B) Active K+ ions theory
(C) ABA theory
(D) None of these

299. Choose the mismatch:

	Inhibitor	Function
(A)	Malnate	Prevents oxidation of succinate
(B)	Cyanide	Inhibits cytochrome oxidase
(C)	Oligomycin	Inhibits ATP synthase
(D)	Rotenone	Blocks CoQH-cyt C oxidoreductase

300. In photorespiration, glycolate and glyoxylate are produced sequentially in the following organelles. Choose the correct sequence:
(A) Chloroplast and mitochondria
(B) Chloroplast and peroxisome
(C) Perosisome and mitochondria
(D) Peroxisome and chloroplast

301. When half the leaves are removed randomly transpiration will show:
(A) Higher magnitude but lower flux or rate per unit
(B) Lower magnitude but higher flux
(C) Both magnitude and flux increase
(D) Both magnitude and flux decrease

302. Which statement about seed germination in a diotuledon is false?
(A) The radicle elongates first and absorbs water needed for germination
(B) The hypocotyl elongates and carries the plumule into the light
(C) The plumule expands and functions in photosynthasis
(D) The shoot apical meristem is formed after germination

303. Who said that "Transpiration is a necessary evil"?
(A) Bose (B) Steward
(C) Anderson (D) Curtis

304. In C4 plants, phosphoenol pyruvate carboxylase is located in:
(A) Cytosol (B) Chloroplast
(C) Peroxisome (D) Mitochondrion

305. Rate of transpiration is related to:
(A) Light and temperature
(B) Light, temperature, atmospheric humidity and wind
(C) Light, temperature and wind
(D) Soil and temperature

306. Enzyme connected with opening and closing of stomata is:
(A) α amylase (B) Pyruvate kinase
(C) PEP carboxylase (D) RuDP carboxylase

307. In an attempt to create a nitrogen-fixing bioreactor, you set out to culture Rhizobium from the root nodules of pea plants. After months of work, you are still having trouble getting the rhizobium to grow alone and fix nitrogen. What might explain this failure?
(A) Rhizobium only grows in association with plant root hairs
(B) Rhizobium from root nodules lacks a cell wall, so it will not grow in culture
(C) You aedentally put a fungicide in your growth medium that kills all of the Rhizobium
(D) Rhizobium from root nodules lack chromosomal DNA, so it will not grow in culture

308. In guard cells when sugar is converted into starch, the stomatal pore:
(A) Closes completely (B) Opens partially
(C) Opens fully (D) Remains unchanged

309. High CO_2 concentration in leaf interior will cause:
(A) Stomatal opening (B) Stomatal closure
(C) Photorespiration (D) Stomata are destroyed

310. Which one will reduce rate of transpiration?
(A) Rise in temperature
(B) Increase in water uptake
(C) Increase in wind velocity
(D) Decrease in light intensity

311. Stomata can open at nigh in:
(A) Xerophyte (B) Hydrophytes
(C) Gametophytes (D) Both (B) and (C)

312. Rate of transpiration is highest when:
(A) Soil is wet and air is dry
(B) Soil is wet and air is humid
(C) Soil is dry and air is humid
(D) Both soil and air is dry

313. In the following biosynthetic pathway X and Y represent:
X $\rightarrow$ S-adenosyl methionine $\rightarrow$ 1-aminocyclo propne-1-carboxylic acid $\rightarrow$ Y
(A) $\rightarrow$X Tryptophan Y $\rightarrow$IAA
(B) $\rightarrow$X Methionine Y$\rightarrow$ Ethylene
(C) $\rightarrow$X Geranylgeranyl phospate Y$\rightarrow$ Gibberrellin
(D) $\rightarrow$X Tryptophan Y$\rightarrow$ Ethylene

314. Red drop is:
(A) Drop in oxygen yield
(B) Drop in quantum yield
(C) Drop in organic yield
(D) Drop in photosynthetic yield

315. If plants are grown for several days in an atmosphere containing $^{14}CO_2$ in place of $^{12}CO_2$, one would except to find:
(A) Large amounts of radioactive oxygen released from the stomatas
(B) A large increase in 14 in the starch
(C) A large decrease in the rate of carbon fixation in the guard cells
(D) An increase in the activity of RuBP carboxylase in the photosynthetic cells

316. Light energy is converted into chemical energy through the formation of:
(A) $NADPH_2$ (B) ATP
(C) ATP and $NADPH_2$ (D) None of the above

317. Which one of the following pairs is correctly matched?
(A) Nutation : Movement due to injury
(B) Traumatotropism : Movement due to change in turgor pressure
(C) Thigmotropism : Movement of tendrils due to contact
(D) Seismonasty : Movement of stem tips of twining plants

318. The first stable product in CAM/C_4 plants is:
(A) Starch (B) Oxalo-acetic acid
(C) Sugar (D) Malic acid

319. Photophosphorylation consists of:
(A) Cyclic and non-cyclic phosphorylation
(B) Oxidative phosphorylation
(C) Substrate phosphorylation
(D) None of the above

320. The by-product of photosynthesis is:
(A) Organic compounds (B) Oxygen
(C) Water (D) Energy

321. Plant shoot bend towards a light source s a result of:
(A) The increased mount of food synthesized by the leaves facing the light
(B) An unequal auxin distribution in their shoot apex
(C) The necessity of light for transcription
(D) Increased elasticity on the light-facing side

322. The co-operative photosynthesis is found in:
(A) C_4 plants (B) C_3 plants
(C) C_2 plants (D) Succulents

323. The sugar formed at photosynthetic site is:
(A) Utilised (B) Stored
(C) Transported (D) Assimilated

324. Stroma is the ground matrix of:
(A) Lysosomes (B) Oxysomes
(C) Ribosomes (D) Chloroplast

325. When chlorophyll absorbs light, it gets excited and emits (releases):
(A) Oxygen (B) Water
(C) Electrons (D) Energy rich compounds

326. During senescence of a plant:
(A) The plant displays sequential senescence of the leaves
(B) Metabolites like carbohydrates, protein etc. are transported from young leaves to the ageing leaves
(C) Metabolites like carbohydrates, protein etc. are transported out of the ageing leaves to the young leaves
(D) Secondary plant products accumulate in the tissues and organs during ageing

327. The excess energy of electron is used in the synthesis of:
(A) Organic compounds
(B) ATP from ADP and iP
(C) $NADPH_2$ from NADP
(D) None of the above

328. The excess energy of electron is used in the synthesis of ATP and the process is called:
(A) Oxidative phosphorylation
(B) Substrate phosphorylation
(C) Photophosphorylation
(D) Phosphorylation

329. Match list - I with list – II and select the correct answer using the codes given below the lists:

List I	List II
P. Boron	1. Die-back
Q. Copper	2. Interveinal chlorosis
R.Iron	3. Little leaf disease
S. Zinc	4. Terminal leaf necrosis

A) P-1, Q–4, R–3, S–2 (B) P-1, Q–4, R–2, S-3
(C) P-4, Q–1, R–3, S–2 (D) P-4, Q–1, R–2, S-3

330. Kranz anatomy is found in:
(A) Stems of C_4 plants (B) Stems of C_3 plants
(C) Leaves of C_4 plants (D) Leaves of C_3 plants

331. $C_{40}H_{56}$ is the empirical formula of:
(A) Chlorophyll (B) Carotene
(C) Xanthophyll (D) Anthocyanin

332. Identify the correct statements:
P. Vernalization is the process where flowering is promoted by heat shock
Q. The four different types of floral organs are initiated as separate whorls
R. The flowering stimulus is transported to the meristem via the xylem
S. Abscisic acid synthesis occurs via the acrotenoid biosynthetic pathway
(A) P and Q (B) Q and S
(C) R and S (D) P and S

333. Calvin cycle operates in chloroplasts. Where does Hatch-Slack pathway occur
(A) Mitochondria (B) Golgi bodies
(C) Chloroplasts (D) Cytoplasm

334. How many kreb cycle is needed for complete oxidation of one glucose molecule?
(A) 1 (B) 2
(C) 3 (D) 4

335. In case of eukaryoted, RNA processing is completed in:
(A) Cytopasm (B) Nucleolus
(C) Nucleus (D) Vacuole

336. Emerson effect is related to:
(A) Decrease in photosynthesis in presence of high light intensity
(B) Decrease in photosynthesis when lights of two different wavelengths are provided together
(C) Increase in photosynthesis in presence of monochromatic light
(D) Increase in photosynthesis when lights of two different wavelengths are provided together

337. $C_{40}H_{56}O_2$ is an empirical formula of a :
(A) Xanthophyll (B) Carotene
(C) Anthocyanin (D) Chlorophyll

338. Autumnal leaf colour is due to:
(A) Chlorophyll a (B) Carotenoids
(C) Anthrocyanin (D) Betacyanin

339. Which of the following equation sums up photosynthetic reaction most accurately:
(A) $CO + H_2O \longrightarrow CH_2O + O_2$
(B) $6CO_2 + 6H_2O \longrightarrow C_6H_{12}O_6 + 6O_2$
(C) $6CO_2 + 12H_2O \longrightarrow C_6H_{12}O_6 + 6O_2 + H_2O$
(D) $6CO_2+12H_2O \xrightarrow[\text{Chlorophyll}]{\text{Light}} C_6H_{12}O_6+6O_2+6H_2O$

340. Which one is directly involved in light reaction of photosynthesis:
(A) Chlorophyll a (B) Chlorophyll b
(C) Carotenoids (D) All of these

341. What is the order of the following metabolic reactions in catabolism of a glucose molecule?
P. Pyruvate → Acetyl-CoA
Q. Glucose → Gluose-6-phosphate
R. Citrate → Isocitrate
S. ADP → ATP
(A) Q, S, P and R (B) Q, S, R and P
(C) P, Q, R and S (D) Q, R, P and S

342. Grana are absent in the chloroplasts of:
(A) Shrubs (B) Herbs
(C) Algae (D) Trees

343. Photosynthetic pigments are located in the chloroplast in:
(A) Intrathylakoid
(B) Thylakoid membrane
(C) Intermembrane space
(D) Inner membrane of envelope

344. Phytochrome:
P. Is a blue protein pigment
Q. Show photoreversibility
R. Has a linear tetrapyrrole termed phytochromobilin
S. Regulates nytinastic movement of leaves
(A) P, Q and R (B) Q and R
(C) Q, R and S (D) P, Q, R and S

345. Chemosynthetic bacteria obtain energy from:
(A) Sun (B) Infra-red rays
(C) Organic substances
(D) Inorganic chemicals

346. In cyclic photophosphorylation the electron released by reaction centre is ultimately accepted by:
(A) Ferredoxin (B) $NADP^+$
(C) Reaction centre (D) Plastocyanin

347. Which organelles mainly synthesize amino acids in the cells?
(A) Mitochondria (B) Ribosomes
(C) Chloroplasts (D) Both (A) and (B)

348. In noncyclic photophoshorylation the electron released by P 680 is replaced by:
(A) Plastoquinone (B) Water
(C) CO_2 (D) $NADP^+$

349. Which is correct about RNA polymerase I?
(A) It is present in nucleolus
(B) It constitutes 50-70% of total activity
(C) It transcribes genes for rRNA
(D) All of these

350. Plants do not store carbohydrate as glucose but do so as starch because glucose:
(A) Is unstable
(B) Attracts herbivores
(C) Will change nucleic acids
(D) Alters osmotic balance

351. Electron carriers involved in photophosphorylation are located in:
(A) Outer chloroplast membrane
(B) Inner chloroplasts membrane
(C) Stroma
(D) Thylakoid membranes

352. Chlorophyll b differs from chlorophyll a in that it does not:
(A) Become oxidized
(B) Become reduced
(C) Become excited
(D) Produce excited electrons

353. NADH is utilized during:
P. Glycolysis
Q. Kreb's cycle
R. Electron transport chain
S. Fermentation
(A) P and Q (B) Q and R
(C) R and S (D) P and S

354. NADPH is produced in photosynthesis during:
(A) Dark reaction
(B) Noncyclic photophosphorylation
(C) Pseudocyclic photophosphorylation
(D) Cyclic photophosphorylation

355. Which is correct about plastoquinones?
(A) They are special quinine present in plastids
(B) They carry two electrons from PS II toward PS I
(C) They transport H+ from stroma into the thylakoid lumen
(D) All of these

356. The GC-AG rule does not apply to introns of:
(A) Chloroplast introns
(B) Nuclear introns of yeast
(C) Mitochondrial introns
(D) All of these

357. The process of photophosphorylation is similar to mitochondrial reaction called:
(A) Substrate-level phosphorylation
(B) Oxidative phosphorylation
(C) Carbohydrate phosphorylation
(D) Hydrolysis of phosphorylated compounds

358. Consider the following statements about alkaloids:
P. They are heterocyclic ring compounds containing nitrogen and have basic properties
Q. They occur rarely (barring a few exceptions) in cryptogams, gymnosperms or monocotyledons but they occur abundantly in dicotyledons
R. They are synthesized mostly in flowers
(A) P and Q (B) Q and R
(C) P and R (D) P, Q and R

359. Excited pigment molecule shifts and electron in an outer orbital having:
(A) Less energy (B) More energy
(C) Lesser spin (D) More spin

360. Source of protons within the chloroplast is:
(A) Water (B) Excited chlorophyll
(C) Carbon dioxide (D) Rubisco

361. Decline in quantumn yield was noticed by Emerson at a wavelength of:
(A) 400-450 nm (B) 600-650 nm
(C) 650-680 nm (D) Morethan 680 nm

362. Photosynthetic autotrophs get their energy requirement from:
(A) Heat (B) Light
(C) Inorganic chemicals (D) Organic chemicals

363. Energy required for ATP synthesis in PS II comes from:
(A) Proton gradient (B) Electron gradient
(C) Reduction of glucose (D) Oxidation of glucose

364. A heterotroph obtains its carbon supply from:
(A) Organic chemicals (B) CO_2
(C) Methane (D) Cyanide

365. Photosynthesis (nourishment in light) was discovered by:
(A) Stephen Hales (B) Priestley
(C) Ingenhousz (D) Van Mayer

366. Which statement is correct?
(A) PS I is located exclusively in stroma thylakoids and nonappressed region of grana that face the stroma
(B) The PS I core complex from barley contains 11 different polypeptides varying in size
(C) Out of the 11 polypeptide of PS I complex, six are coded by nuclear genes
(D) All of these

367. Which is correct about RNA polymerase II?
(A) It is present in nucleoplasm
(B) It constitutes 20-40% of total activity
(C) It transcribes all the genes that produce mRNA
(D) All of these

368. Chrolophyll was first extracted by:
(A) Van Mayer
(B) Pelletier and Caventou
(C) Willstatter and Stoll
(D) Senebier

369. In 1772, Joseph Priestley demonstrated that:
(A) Plants foul the air
(B) Animals purify the air during day time
(C) Plants and animals restore air for each other
(D) Plants get killed in the air produced by animals

370. Chlorophyll consists of:
(A) A head of phytol and tail of four pyrerole rings
(B) A head of linked carbons and tail of linked pyrrole rings
(C) A head of four pyrrole rings and tail of linked nitrogens
(D) A head of four pyrrole rings and an alcoholic phytol tail

371. The source of hydrogen for carbon assimilation is:
(A) NADPH (B) $FADH_2$
(C) H_2O (D) RuBP

372. Which of the following pairs are correctly matched:

P. 2, 4-dinitrophenol	Uncoupling agent in oxidative phosphorylation
Q. Oligomycin	Inhibitor of ATP formation in oxidative phosphorylation
R.Valinomycin	Ionophore carriers potassium through the mitochondrial membrane
S. Iodoacetate	Separates the phosphorylating F1-ATPas from the inner mitochondrial membrane

(A) P, Q and R (B) P, R and S
(C) Q, R and S (D) P, Q and R

373. The first hypothesis that oxygen released during photosynthesis is derived from water, was proposed by:
(A) Engelmann (B) Priestley
(C) Van Niel (D) Blackman

374. Consider the following statements regarding the salt resistants plants:
P. Some salt resistant plants have salt glands at the surface of the leaves
Q. Accumulation of solute like sorbitol, glycine, betacine, proline
Which statement(s) is/are correct
(A) P only (B) Q only
(C) P and Q (D) None of these

375. The number of carbon atoms present in ribulose diphosphate is:
(A) 4 (B) 5
(C) 6 (D) 8

376. Photosynthesis is a process:
(A) Reductive and exergonic
(B) Reductive and catabolic
(C) Reductive, endergonic and catabolic
(D) Reductive, endergonic and anabolic

377. Dark reactions of carbon assimilation occur in:
(A) Cytoplasmic matrix (B) Mitochondria
(C) Leucoplasts (D) Chloroplasts

378. ATP molecules required for synthesis of a glucose molecule in Calvin-Benson cycle are:
(A) 38 (B) 36
(C) 32 (D) 18

379. Which is not correct about starch?
(A) It is stored in photosynthetically inactive organs like roots, modified shoots and seeds
(B) The constituents of starch grains are amylase and amylopectin
(C) Both amylase and amylopectin are made up of D-glucose units
(D) Starch molecules have a definite molecular weight and are assembled into starch grains

380. Mitochondria and plastids present in eukaryotic cells contain RNA polymerases that are analogous to:
(A) Eukaryotic polymerase (B) Phage polymerases
(C) Bacteria; polymerases (D) Both (B) and (C)

381. Who discovered that green plant parts and light are essential for photosynthesis?
(A) Englemann (B) Ingenhousz
(C) Sachs (D) Priestley

382. Lignin
P. Is a polysaccharides
Q. Is a second most abundant organic compound
R. Is a polymer of aromatic alcohols like coumaryl, sinapyl alcohol
S. Synthetic pathway is peroxidase dependent
(A) P and S (B) Q and R
(C) Q, R and S (D) P, Q, R and S

383. Electron donor to PS I is:
(A) Ferredoxin (B) FeS centre
(C) Plastoquinone (D) Plastocyanin

384. Wavelength of light least effective in photosynthesis is:
(A) Violet (B) Blue
(C) Green (D) Red

385. In Photosynthesis light forms:
(A) ATP and Sugar
(B) Hydrogen, Oxygen and Sugar
(C) ATP, Hydrogen donor and Oxygen
(D) ATP, Hydrogen and Oxygen donor

386. Which one of the following is limiting rate of photosynthesis in nature:
(A) Light (B) CO_2
(C) O_2 (D) Chlorophyll

387. Consider the following events involved in stomatal opening:
P. Turgor pressure of guard cells increases
Q. K^+ ions move into guard cells
R. pH of guard cells decreases
S. Water moves into guard cells
What is the correct sequence of these events leading to stomatal opening?
(A) Q, S, R, P (B) R, Q, S, P
(C) Q, R, S, P (D) R,P, Q, S

388. In PS I, the carrier that pick up electrons from P_{700} is:
(A) Fe protein (B) Fe-S protein
(C) Fe-Cu protein (D) Fe-Mg protein

389. Rate of photosynthesis is:
(A) Equal to that of respiration
(B) Less than that of respiration
(C) Depends upon chlorophyll content
(D) Faster than that of respiration

390. Chloroplasts are agranal in:
(A) Algae
(B) Bundle sheath cells of C_4 plants
(C) Mesophyll cells of C_4 plants
(D) Both (A) and (B)

391. Brassinosteroids
P. Are steroid growth promoter
Q. Synthesized by short day plants only
R. Function is very similar to ABA
S. Are chemically similar to ecdysone hormone
(A) P and S (B) R and S
(C) P and Q (D) P and R

392. Illuminated isolated chloroplasts release O_2 in the absence of CO_2. Which will increase oxygen output?
(A) H_2O (B) $NADP^+$
(C) NAD^+ (D) Green light

393. Which helped in confirming that oxygen evolved in photosynthesis comes from water:
(A) $^{18}H_2O$ (B) $^{14}CO_2$
(C) $^{15}NO_3$ (D) ^3H_2O

394. Consider the following compounds
P. Phosphoenol pyruvate
Q. Malic acid
R. Pyruvic acid
S. Oxaloacetic acid
What is the correct sequence of formation of these intermediate compounds in C_4 pathways?
(A) P, Q, R, S (B) Q, S, R, P
(C) S, P, Q, R (D) S, Q, R, P

395. In an experiment on O_2 evolution by photosynthesizing *Hydrilla* plant, a pinch of sodium bicarbonate was added to water. The rate of photosynthesis or O_2 evolution will:
(A) Increase (B) Decrease
(C) Stop (D) Not be affected

396. Properties of starch molecules are mainly determined by:
(A) Degree of branching (including branch length)
(B) Types of bonds linking the glucose units
(C) Chain length
(D) All of these

397. Which consensus sequences is recognized by more than one transcription factors?
(A) -35 sequence (B) CAAT box
(C) Octamer sequence (D) GC box

398. An illuminated plant is provided with 0.03% CO_2. The plant will:
(A) Die
(B) Just survive
(C) Show normal photosynthesis
(D) Have reduced respiration

399. Photosynthesis will be highest when the plant is exposed to:
(A) Continuous strong light
(B) Continuous weak light
(C) Alternate strong and weak light
(D) Intermittent light

400. Which is correct for photosynthetic organs:
(A) Light energy $\rightarrow$ Heat $\rightarrow$ Chemical energy
(B) Chemical energy $\rightarrow$ Light energy $\rightarrow$ Heat
(C) Heat $\rightarrow$ Light energy $\rightarrow$ Chemical energy
(D) Light energy $\rightarrow$ Chemical energy $\rightarrow$ Heat

401. In photosynthesis:
(A) No ATP is formed
(B) ATP is formed in initial stages
(C) Water is not involved
(D) Oxygen comes from carbondioxide

402. Photolysis of water involves:
(A) Excitement of water
(B) Evolution of water
(C) Breakdown of water by light
(D) Splitting of water into ions, H^+ and OH^-

403. Which of the following statements are correct?
P. To remove seed dormancy by mechanical removal of seed-coat is called Scarification
Q. Vernallization is done at low temperature
R. The presence of auxins in a solution could be tested by *Avena sativa* stem tip test
S. The photoperiodic stimulus is perceived by leaves
(A) P and Q (B) P, Q and R
(C) Q and S (D) P, Q, R and S

404. The radiant energy absorbed by carotenoids is:
(A) Lost as heat
(B) Transferred to chlorophylls
(C) Be emitted as fluorescence or phosphorescence
(D) Used to perform photochemical act

405. High CO_2 – compensation point is found in:
(A) C_3 plants (B) C_4 plants
(C) CAM plants (D) Lichens

406. Consider the following statements regarding the plant growth substances:
P. Cytokinins make dormant buds active
Q. The richest concentration of cytokinins generally occur in endosperm
R. Ethylene is involved in the formation of aerenchyma in submerged roots and stems
S. Gibberellins are involved in the breaking of seed dormancy
Which of these statements are correct?
(A) P and S (B) Q, R and S
(C) P, Q and S (D) P, Q, R and S

407. The substrate for photorespiration is:
(A) Malic acid (B) Oxaloacetic acid
(C) Gycolic acid (D) PGA

408. Which is correct?
(A) Flower senescence is promoted by ethylene
(B) Ethylene biosynthesis involves two enzymes, viz., ACC synthase and ACC oxidase
(C) ACC synthase enzyme is encoded by acs gene and ACC oxidase is encoded by aco gene
(D) All of these

409. Which is usually surrounded by a GC rich sequence?
(A) TATA box (B) Pribnow box
(C) GC box (D) CAAT box

410. During photosynthesis O_2 is liberated by oxidation of:
(A) H_2O (B) CO_2
(C) Phosphoglyceraldihyde
(D) None of the above

411. The first food substance plants make in photosynthesis is:
(A) Sugar (B) Starch
(C) Proteins (D) Fats

412. With reference to sucrose, consider the following statements::
P. Sucrose is a non-reducing sugar
Q. The linkage in it is an α_1- β_1 glycosidic bond
R. It is a disaccharide having glucose and β - fructose
Which of the statement(s) given below is/are correct
(A) P only (B) Q and R
(C) P and R (D) P, Q and R

413. Hypothesis for oxygen coming from water was put forward by:
(A) Hill (B) Warburg
(C) Blackman (D) Mendel

414. The most effective wavelength of light participating in photosynthesis is
(A) 450-520 nm (B) 700-760 nm
(C) 660-700 nm (D) 520-560 nm

415. The first product of CO_2 fixation in C_4 plants is:
(A) Phosphoglyceric acid (PGA)
(B) Oxalo-acetate
(C) Malic acid
(D) Phosphoenol pyruvate

416. More free energy is released during the citric acid cycle than during glycolysis, but only 1 mole of ATP is produced for each mole of acetyl CoA that enters the cycle. Hat happens to most of the remaining free energy that is produced during the citric acid cycle?
(A) It is used to synthesize GTP
(B) It is used to reduce electron carriers
(C) It is lost as heat
(D) It is used to reduce pyruvate

417. The imperical formula for chlorophyll a is:
(A) $C_{28}H_{72}O_5N_4Mg$ (B) $C_{45}H_{72}O_6N_4Mg$
(C) $C_{55}H_{72}O_5N_4Mg$ (D) $C_{45}H_{72}O_6N_4Mg$

418. Gibberellin receptor
P. Is membrane bound
Q. Interacts with G-protein
R. Increases the concentration of GMP when binds with gibberellin
S. Is cytosolic in nature
(A) P and Q (B) R and S
(C) P, Q and R (D) Q, R and S

419. The process of taking in CO_2 by plants and releasing O_2 is component of:
(A) Transpiration (B) Respiration
(C) End-osmosis (D) Photosynthesis

420. The bye product of photosynthesis is:
(A) CO_2 (B) H_2O
(C) Glucose (D) O_2

421. Which is not correct about transcription unit?
(A) It is that sequence of a DNA molecule that is transcribed into a single RNA molecule
(B) In case of all eukayotes, each gene is a distinct transcription unit
(C) In case of prokaryotes, each operon functions as a single transcription unit
(D) Enhancer sequences are present within promoters

422. Which occupies a fixed position given in the promoter at -25?
(A) TATA box (B) Pribnow box
(C) GC box (D) CAAT box

423. Our present day view regarding photosynthesis is that it:
(A) Converts light energy into chemical energy
(B) Creates useful energy
(C) Fixes CO_2 into carbohydrates
(D) Reverses the action of respiration

424. ABA:
P. Is a sesquiterpenoid
Q. Synthesized within chroloplast
R. Synthesized from carotenoid violaxanthin
S. Transport occurs through phloem only
(A) P and Q (B) P, Q and R
(C) P, R and S (D) P, Q, R and S

425. In which of the following process the light energy is converted into chemical energy:
(A) Digestive action (B) Respiration
(C) Fermentation (D) Photosynthesis

426. The oxygen in photosynthesis is released from:
(A) CO_2 (B) H_2O
(C) Carbohydrate (D) Chlorophyll

427. Photosynthesis is the fastest in:
(A) Blue light (B) Sun light
(C) Red light (D) Green light

428. Photophosphorylation is the process in which:
(A) CO_2 and O_2 unite
(B) Phosphoglyceric acid is produced
(C) Aspartic acid is formed
(D) Light energy is converted into chemical energy through production of ATP

429. The following 4 different solutions are prepared by mixing the components of electron transport chain. Which among them is expected to cause a net transfer of electron to cytochrome C?
(A) It is used to synthesize GTP
(B) It is used to reduce electron carriers
(C) It is lost as heat
(D) It is used to reduce pyruvate

430. The first step in photosynthesis is the:
(A) Joining of 3-carbon atom to form glucose
(B) Formation of ATP
(C) Ionization of water
(D) Excitement of an electron of chrolophyll by photon of light

431. Which of the following statements are true for auxin?
P. IAA is the principal auxin in plants
Q. Auxin shows polar transport
R. Distribution of auxin in cell is pH dependent
S. TIBA and NPA block polar transport of auxin
(A) P and Q
(B) R and S
(C) P, Q and R
(D) P, Q, R and S

432. Which is correct about transcription factors?
(A) These factors are required for transcription initiation at all the promoters of eukaryotes
(B) They join RNA polymerase II to form a complex around the startpoint
(C) They determine the site of transcription initiation
(D) All of these

433. ATP formation during photosynthesis is:
(A) Phosphorylation
(B) Photophosphorylation
(C) Oxidative phosphorylation
(D) None of these

434. Which of these statements about thioredoxin is correct?
(A) It contains a heme that cycles between two oxidation states
(B) Its oxidised form predominated while light absorption is taking place
(C) It activates some biosynthetic enzymes by reducing disulfide bridges
(D) It activates some degradative enzymes by reducing disulphide bridges

435. Chlorophyll is present:
(A) On the surface of chloroplast
(B) Dispersed throughout chloroplast
(C) In the stroma of chloroplasts
(D) In the grana of chloroplasts

436. Ethylene receptor is:
P. Similar to bacterial two component system
Q. Consists of histidine kinase
R. Amino terminal end contains ethylene binding site
S. Contains Cu^{++} ion
(A) P and Q (B) Q and S
(C) P Q and R (D) P, Q, R and S

437. The process in which water is split during photosynthesis is:
(A) Photolysis (B) Hydrolysis
(C) Plasmolysis (D) Hemolysis

438. Which functional sequence/module is not found in eukaryotic promoters?
(A) TATA box (B) Pribnow box
(C) GC box (D) CAAT box

439. Which of the following wave length of light is absorbed maximum for photosynthesis:
(A) 660 nm (B) 440 nm
(C) 680 nm (D) 700 nm

440. The specific function of light energy in the process of photosynthesis is to:
(A) Activate chlorophyll
(B) Split of water
(C) Reduced carbon di oxide
(D) Synthesize glucose

441. The number of protons translocated by the various vectorial localized on the inner mitochondrial membrane of an organism was cytochrome aa_3 complex: 4 and F_1F_0-ATPas: 3. One proton is also required for the transport of inorganic phosphate into the mitochondrial matrix:

I. The number of ATP molecules that can be synthesized by the oxidation of one NADH molecule is:

(A) 2 (B) 2.5
(C) 3 (D) 3.3

II. If the cytosolic NADH is transported to the matrix by the glycerol 3- phosphate shuttle, then the number of ATPs synthesized is:

(A) 1.5 (B) 2
(C) 2.5 (D) 3.3

442. The basic structure of chlorophyll comprises of:

(A) Porphyrin system (B) Cytochrome system
(C) Plastocyanin system (D) Flavoprotein only

443. In pigment system I, active chlorophyll is:

(A) P_{700} (B) P_{600}
(C) P_{608} (D) P_{720}

444. Discovery of Emerson effect has clearly shown the existence of:

(A) Two distinct photochemical reactions or processes
(B) Light and dark reactions in photosynthesis
(C) Photophosphorylation
(D) Photo-respiration

445. Suppose the interior of the thylakoids of isolated chloroplasts were made acidic and then transferred in the dark to a pH -8 solution. What would be likely to happen?

(A) The Calvin cycle will be activated
(B) The isolated chroloplasts will make ATP
(C) Cyclic photophosphorylation will occur
(D) Only a and b will occur

446. The site for light reaction is:

(A) Grana (B) Stroma
(C) ER (D) Cytoplasm

447. When cell converts light energy into chemical energy, which of the following reaction would take place:

(A) ADP + iP = ATP (B) ATP + iP = ADP
(C) AMP + iP = ADP (D) GDP + iP = GTP

448. Which is often present in several copies in a single promoter?

(A) TATA box (B) Pribnow box
(C) GC box (D) CAAT box

449. Assimilatory power produced in Hill reaction and used in Blackman's reaction refers to:

(A) Generation of ATP and $NADPH_2$
(B) Reduction of CO_2
(C) Splitting of water
(D) Disintegration of plastids

450. Which of the following statement about photosystem II is incorrect?

(A) It is a multimolecular transmembrane assembly containing several polypeptides, several chlorophyll molecules, a special pair of chlorophyll (P680), pheophytin and plastoquinones
(B) It transfers electrons to photosystem I via the cytochrome bf complex
(C) It uses light energy to create a separation of charge whose potential energy can be used to oxidize H_2O and to produce a reductant, plastoquinol
(D) It uses an Fe^{++} - Cu^{+} centre as a charge accumulator to form O_2 without generating potentially harmful hydroxyl radicals, superoxide anions or H_2O_2

451. Water soluble pigment is:

(A) Chlorophyll (B) Carotene
(C) Anthocyanin (D) Xanthophyll

452. Photophosphorylation means synthesis of:

(A) ATP from ADP (B) NADP
(C) ADP from ATP (D) PGA

453. Which of the following statement regarding reduced glutathione is not true?

(A) It contains one γ–carboxyglutamate, one cysteine and one glycine residue
(B) It keeps the cysteine residues of proteins in their reduced states
(C) It is regenerated from oxidized glutathione by glutathione reductase
(D) It reacts with hydrogen peroxide and organic peroxides

454. In which wavelength of light, bacteria carryout photosynthesis:

(A) Utra violet (B) Blue
(C) Red (D) Far red

455. Pigment system is connected with:

(A) Flowering (B) Reduction of CO_2
(C) Photolysis of water (D) None of the above

456. Besides water and light which is more essential as a raw material for food formation:

(A) CO_2 (B) NAD
(C) O_2 (D) Mineral salts

457. In pigment system II, active chlorophyll is:

(A) P_{680} (B) P_{683}
(C) P_{700} (D) P_{673}

458. Chloroplast contains maximum quantity of:

(A) RuDP-carboxylase (B) Hexokinase
(C) Pyruvic carboxylase (D) None of the above

459. Which one is the precursor of chlorophyll

(A) Bacteriochlorophyll
(B) Bacterio-viridin
(C) Tryptophan
(D) Chlorophyllide (proto chlorophyll)

460. Which is not correct about ribulose-1, 5-bisphosphate carboxylase?
(A) Makes up about 50 per cent of the chloroplast protein
(B) Is located in the same compartment as the oxygen-generating complex
(C) Is encoded by a combination of nuclear and chloroplast DNA
(D) Is located in the compartment where starch is stored

461. Dark reaction in photosynthesis is called so because:
(A) It can occur in dark alone
(B) It does not require direct light energy
(C) It cannot occur during day time
(D) It occur more rapidly at night

462. During ATP synthesis, electrons pass through:
(A) CO_2 (B) O_2
(C) H_2O (D) Cytochromes

463. While the number of ATP molecules generated by HMP pathway is less, microbes have this pathway to generate:
(A) Amino acid
(B) Nuleic acid precursors
(C) Secondary metabolites
(D) Fatty acid precursors

464. In case of C_4 pathway:
(A) CO_2 combines with PGA
(B) CO_2 combines with PEP
(C) CO_2 first combines with RuBP
(D) CO_2 combines with RMP

465. Where does the primary photo-chemical ıeaction occur in chloroplast:
(A) Stroma (B) Periplast cavity
(C) Quantasomes
(D) Inner membrane of chloroplast

466. Ribulose biphosphate carboxylase enzyme catalyzes the reaction between:
(A) Oxaloacetic acid and acetyl CoA
(B) CO_2 and ribulose1,5,biphosphate
(C) Ribulose biphosphate and phosphoglyceraldehyde
(D) PGA and dihydroxy acetone phosphate

467. Which of the following statement is false? In citric acid cycle
(A) Four pairs of hydrogen atoms flow out of the cycle
(B) Two water molecules are formed
(C) One high energy compound (GTP) is produced
(D) Isocitrate is reduced to α–ketoglutarate and O_2

468. Carbon dioxide is formed in all the following except:
(A) Respiration in plants
(B) Burning of sugar
(C) Heating of lime stone
(D) Photosynthesis of plants

469. Which of the following enzyme is used for carboxylation of RuBP?
(A) Carboxydismutase
(B) Hexosekinase
(C) Peroxidase
(D) Phosphopentokinase

470. Grana refer to:
(A) Stacks of thylakoids in plastids of higher plants
(B) A constant inquantum equation
(C) Glycolysis of glucose
(D) By product of phorosynthesis

471. The reaction catalyed by phosphofructokinase – 1 (PFK -1)
(A) Is activated by high concentrations of ATP and citrate
(B) Uses fructose – 1- phosphate as substrate
(C) Is the rate-limiting reaction of the glycolytic pathway
(D) Is inhibited by fructose -2, 6-bisphosphate

472. During dark reaction of photosynthesis:
(A) Water is split
(B) CO_2 is reduced to organic compounds
(C) Chlorophyll is activated
(D) 6C sugar is broken down into 3C sugar

473. In C_3 plants first stable product of photosynthesis during dark reaction is:
(A) PGA (B) PGAI
(C) RuBP (D) RUDP

474. NADH → Q → X → Cyt1 → Y → $Cyt_{(a+a3)}$ → O_2
Sequence of electron transfer in oxidative phosphorylation is given above. complete the missing sequence:
(A) X – Cyta and Y – Cytb
(B) X – Cyta and Y - Cytc
(C) X – Cytb and Y – Cytc
(D) X – Cytb and Y – Cytb1

475. CO_2 is necessary of photosynthesis. The chemical used to remove this gas most effectively from entering a control apparatus is :
(A) Potassium hydroxide solution
(B) Calcium oxide (C) Sodium carbonate
(D) Distilled water

476. The isotope of carbon used extensively for studies in photosynthesis is:
(A) ^{12}C (B) ^{14}C
(C) ^{13}C (D) ^{16}C

477. Glycolate accumulates in chloroplasts when there is:
(A) High CO_2 (B) Bright light
(C) Low temperature (D) Low CO_2

478. Which of the following have high CO_2 compensation point:
(A) C_2 plant (B) C_3 plant
(C) C_4 plant (D) Alpine herbs

479. Better adapted individuals tend to produce more offspring than individuals within the same species less well adapted to that environment. This observation forms the principle of:
(A) Differential reproductive success
(B) Variation
(C) Heritability
(D) Use and disuse

480. Which of the following statements regarding Darwin and his theory of evolution is false?
(A) Darwin noticed that South American species of birds more loosely resembled species found in other temperate regions than European species of birds
(B) Darwin proposed that species adapt over time
(C) Darwin proposed that natural selection produces changes in species over time
(D) Darwin was the original and sole developer of the theory of natural selection

481. The C_4 plants are different from the C_3 plants with reference to the:
(A) The substance that accepts CO_2 in carbon assimilation
(B) Types of end product of photosynthesis
(C) The number of NADPH that are consumed in preparing sugar
(D) Types of pigments involved in photosynthesis

482. Red drop discovered by Emerson due to distruption of photochemical activity of:
(A) Carotenoids (B) PS I
(C) PS II (D) Both (B) and (C)

483. Which of the following is not an example of co-evolution?
(A) Adaptations of flowers and their exclusive pollinators
(B) Passion-flower vines and the butterfly *Heliconius*
(C) A parasite that is specific for one host
(D) Aposematic coloration of monarch butterflies and predators that learn not to eat them

484. The percentage of light energy utilized for photosynthesis by higher plants is:
(A) 100 (B) 50
(C) 10 (D) 1-2

485. $NADP^+$ is hydrogen acceptor of:
(A) Krebs cycle
(B) Photolysis
(C) Hexose monophosphate shunt
(D) Both (B) and (C)

486. The first carbon dioxide acceptor in C_4 plants is:
(A) Phosphoenol –pyruvate
(B) Ribilose-1-5 diphosphate
(C) Oxalo-acetic acid
(D) Phosphoglyceric acid

487. A plant is kept in 300 ppm CO_2 concentration. What will happen to it:
(A) Plant will die soon
(B) Plant will not grow but will not die
(C) Plant will show normal photosynthesis
(D) Respiration will be greatly decreased

488. Sometime, speciation can be quite abrupt after long periods of stability. This is called:
(A) Founder effect
(B) Crossing - over
(C) Macroevolution
(D) Punctuated equilibrium

489. The carbon dioxide acceptor in calvin cycle/C_3 plants is:
(A) Phosphoenol –pyruvate (PEP)
(B) Ribulose-1-5 diphosphate (RuBP)
(C) Phosphoglyceric acid (PGA)
(D) Ribulose-monophosphate (RMP)

490. In angiosperms, chlorophyll formation requires:
(A) Cytochrome (B) Carotenes
(C) Chloroplasts (D) Light

491. A prezygotic isolating mechanism prevents successful:
(A) Gamete production
(B) Fertilization
(C) Zygote development
(D) Reproduction of hybrids

492. Ninth-tenth of all photosynthesis of world is (85-90%) carried out by:
(A) Large trees with millions of branches and leaves
(B) Algae of the ocean
(C) Chlorophyll containing ferns of the forest
(D) Scientists in the laboratories

493. Which one of the following would not be limiting factors in photosynthesis?
(A) O_2 (B) CO_2
(C) Chlorophyll (D) Light

494. Which one directly transfer electrons in $NADP^+$ during light reaction?
(A) PS I (B) PS II
(C) Cytochromes (D) Plastocyanin

495. Of the following species concepts, which one emphasizes reproductive isolation as the major cause for speciation:
(A) Biological species concept
(B) Evolutionary species concept
(C) Morphological species concept
(D) Phylogenetic species concept

496. PEP is CO_2 acceptor in:
(A) C_4 plants (B) C_3 plants
(C) C_2 plants (D) Both (A) and (B)

497. Maize, Sugarcane and some other tropical plants have high efficiency of CO_2 fixation because they operate:
(A) Calvin cycle (B) Hatch-Slack cycle
(C) TCA cycle (D) PP pathway

498. Genetic drift occurs when a few individuals of a species colonize an island. This particular phenomenon is known as:
(A) The bottleneck effect
(B) The founder effect
(C) Assortative mating
(D) Random mating

499. Compensation point is the value of a factors where there is:
(A) Beginning of photosynthesis
(B) Little photosynthesis
(C) Photosynthesis equal to rate of respiration
(D) Neither photosynthesis nor respiration

500. Which one of the following is the common storage product of photosynthesis:
(A) Protein (B) Fat
(C) Starch (D) Glucose/sucrose

501. When populations of a plant species are located adjacent to one another because of abrupt changes in soil conditions, they are called:
(A) Allopolyploid (B) Autopolyploid
(C) Sympatric (D) Parapatric

502. Chlorophyll takes part in:
(A) Absorption of light
(B) Absorption of CO_2
(C) Absorption of electrons
(D) Absorption of light and photochemical splitting of water

503. Starch is detected in food by its:
(A) White appearance
(B) Blue reaction with iodine solution
(C) Presence as an energy store
(D) Granular form even if cooked

504. An essential process connected with photosynthesis is:
(A) Photolysis of water
(B) Synthesis of glucose
(C) Breakdown of glucose
(D) Photophosphorylation

505. Which of the following is least effective in photosynthesis:
(A) Sunlight (B) Red light
(C) Blue light (D) Green light

506. Light energy is changed to chemical energy in the process of:
(A) Bioluminescence
(B) Photosynthesis
(C) Both (A) and (B)
(D) Photorespiration 446. Chloroplasts fix:

507. An isolated population of humans, with approximately equal numbers of blue-eyed and brown-eyed individuals, was decimated by an earthquake. Only a few brown-eyed people remained to form the next generation. This kind of change in the gene pool is called a:
(A) Hardy-Weinberg equilibrium
(B) Blocked gene flow
(C) Bottleneck effect (D) Founder effect

508. NAD is:
(A) Nicotinamide adenine dinucleotide
(B) Nicotinamide adenosine dinucleotide
(C) Nicotine adenosine dinucleotide
(D) Nicotinamide adenine diphosphate

509. The experimental material that has largely been responsible for making rapid advances in research on photosynthesis is:
(A) *Chlamydomonas* (B) *Chlorella*
(C) Spinach leaf (D) *Hydrilla*

510. Rate of photosynthesis is independent of:
(A) Intensity of light (B) Duration of light
(C) Quality of light (D) Temperature

511. The ingredient required for ATP synthesis is:
(A) Ca (B) S
(C) P (D) Fe

512. Absorption of light brings about:
(A) Absorption of CO_2
(B) Liberation of O_2
(C) Oxidation of chlorophyll
(D) Reduction of chlorophyll

513. Energy changes in photosynthesis is:
(A) Electrical $\rightarrow$ Chemical
(B) Light $\rightarrow$ Chemical
(C) Chemical $\rightarrow$ Light $\rightarrow$ Electrical
(D) Light $\rightarrow$ Electrical $\rightarrow$ Chemical

514. Bacterial photosynthesis contains:
(A) PS I (B) PS II
(C) Both PS I and PS II (D) None of these

515. A taxonomic system based only on the traits that reflect the order in time in which branches arose in a phylogenetic tree is called:
(A) Phylogeny (B) Cladistics
(C) Classical evolutionary taxonomy
(D) Phenetics

516. What is true for photosynthesis:
(A) Both carbon dioxide and water are oxidized
(B) Both carbon dioxide and water are reduced
(C) Carbon dioxide is oxidized and water reduced
(D) Carbon dioxide is reduced and water oxidized

517. Ina chlorophyll molecule, magnesium is present:
(A) At one end of porphyrin
(B) Centre of porphyrin
(C) Phytol unit
(D) Between porphyrin head and phytol tail

518. The best test of the relatedness of two species is in the similarly of their:
(A) Anatomy (B) DNA and proteins
(C) Development (D) Courtship behaviors

519. In aquaria green plant are grown for:
(A) CO_2 (B) Fresh food
(C) Oxygen (D) Both (A) and (B)

520. Leaves are green because they:
(A) Absorb green light
(B) Do not absorb but reflect green light
(C) Utilise green light
(D) Absorb and reflect green light

521. Algae often float on surface of water during day but sink down during night due to:
(A) Evolution and trapping of oxygen bubbles during the day in their photosynthesis
(B) Becoming light as they consume most of their food in the night
(C) Warming action of sun during the day
(D) Release of absorbed air by warming of water

522. North American organisms are different from south American organisms, because the two continents:
(A) Have very different climates
(B) Have never been connected
(C) Were physically separated for about 35 million years, until about 2 million years ago
(D) Have very different geographies

523. Carbon dioxide joins the photosynthetic pathways:
(A) PS I
(B) PS II
(C) Light reaction
(D) Dark reaction

524. According to Lamarck, a giraffe has a long neck because:
(A) A reactor designed it that way
(B) Catastrophes eliminated short-necked forms
(C) Its ancestors stretched their necks to get food
(D) Ancestral giraffes with slightly longer necks than others got more food and left more surviving offspring

525. The Darwinian fitness of an organism is a measure of:
(A) Its ability, relative to others in the population, to pass its genes to the next generation
(B) The number of offspring it produces
(C) Its lifespan
(D) Its physical vigor

526. Hill reaction occurs in:
(A) Dark (B) Light
(C) Both light and dark (D) All the time

527. Nobel prize was awarded to the Scientist for discovering the pathway of carbon assimilation:
(A) Watson (B) Krebs
(C) Calvin (D) Parnas

528. Many bacteria are now resistant to penicillin, because:
(A) Penicillin causes gene mutations, some of which are beneficial
(B) Previously resistant forms survived and reproduced better than nonresistant forms
(C) The hospital environment inhibits competition among bacteria
(D) Penicillin triggers the synthesis of resistant proteins

529. The products of photochemical reaction are:
(A) O_2, ATP and NADPH
(B) O_2
(C) ATP and NADPH
(D) Organic compounds especially carbohydrates

530. When a short day plant (SD) and a long day plant (LD), both kept under favourable conditions for flowering, are exposed to a short flash of red light in the middle of the dark period:
(A) Both SD and LD plants will flower
(B) Both SD and LD plants will remain vegetative
(C) SD plant will remain vegetative and LD plant will flower
(D) SD plant will flower and LD plant will remain vegetative

531. Both respiration and photosynthesis require:
(A) Sunlight (B) Green cells
(C) Cytochrome (D) Organic substrate

532. A photosynthetic organism which does not release oxygen is:
(A) Blue green algae
(B) Green sulphur bacterium
(C) Green algae
(D) Algal component of lichen

533. The sap of a plant cell has an osmotic potential of -10 bars and there is a wall pressure of 2 bars. When this cell is placed in a solution with an osmotic potential of -3 bars the force causing water to enter cell is:
(A) - 8 bar (B) - 7 bar
(C) - 5 bar (D) - 3 bar

534. PSII contains a nonchlorophyllous pigment in blue green algae:
(A) β-carotene
(B) Zeaxanthin
(C) Phycocyanin
(D) Cytochrome c

535. Which is not required for synthesis of carbohydrates:
(A) Chlorophyll (B) Water
(C) Nitrogen (D) Carbon dioxide

536. Isotopes employed to study photosynthesis are:
(A) ^{11}C and ^{32}P (B) ^{15}C and ^{32}P
(C) ^{16}C and ^{15}N (D) ^{14}C and ^{18}O

537. The evidence that during photosynthesis oxygen comes from water:
(A) Photosynthetic bacteria employ H_2S and CO_2 to form carbohydrates, water and sulphur
(B) Isolated illuminated chloroplasts release oxygen if proided with potassium ferrocyanide
(C) Isotopic ^{18}C provided as $H_2{}^{18}O$ appears as $^{18}O_2$ liberated in photosynthesis
(D) All the above

538. Which one of the following pairs is correctly matched?
(A) Nutation: Movement due to injury
(B) Traumatotropism: Movement due to change in turgor pressure
(C) Thigmotropism: Movement of tendrils due to contact
(D) Seismonasty: Movement of stem tips of twining plants

539. Which one results in energy wastage:
(A) Photosynthesis (B) Photorespiration
(C) Dark respiration (D) Glycolysis

540. Which of the following constitutes Hill reaction:
(A) Photolysis by chloroplasts
(B) Photolysis of water by isolated chloroplasts forming ATP and NADPH
(C) Photolysis of water by isolated illuminated chloroplast causing reduction of some chemicals and liberation of oxygen
(D) Photolysis of water liberating oxygen and hydrogen by illuminated isolated chloroplasts

541. Thylakoids removed from chloroplasts were kept in illuminated culture having CO_2 and water. It did not produce sugar because of:
(A) Absence of enzymes
(B) Nonlinking of PS I and PS II
(C) Absence of light trapping
(D) Nonformation of assimilatory power

542. The substrate for photorespiration is:
(A) Phosphoglyceric acid (B) Glycolate
(C) Serine (D) Glycine

543. In C_4 plants, synthesis of sugar/final CO_2 fixation occurs in:
(A) Epidermal cells (B) Spongy cells
(C) Undifferentiated mesophyll cells
(D) Bundle sheath cells

544. Which is produced during water stress that brings about stomatal closure?
(A) Abscisic acid (B) Ethylene
(C) Ferulic acid (D) coumarin

545. As compared to chrolophyll a, bacteriochlorophyll has:
(A) One pyrrole ring with four hydrogen
(B) One pyrrole ring with three hydrogen
(C) One pyrrole ring with two hydrogen
(D) One pyrrole ring with one hydrogen

546. $NADP^+$ is reduced to NADPH in:
(A) PS I (B) PS II
(C) Calvin cycle
(D) Noncyclic photophosphorylation

547. When the rate of translocation is slow, the rate of photosynthesis shall:
(A) Increase (B) Decrease
(C) Remain unchanged (D) Become zero

548. Active transport of ions across the membrane of a root hair cell can be assumed to be taking place if:
P. The cell produces more glutathione
Q. The cell has mitochondria
R. The uptake of ions stop when cyanide is added
S. The uptake of ions is against the concentration gradient
(A) P, R (B) R, S
(C) Q, R (D) Q, S

549. Photosynthesis has two reaction complexes, one followed by the other. The second reaction complex:
(A) Traps light energy
(B) Synthesises starch
(C) Functions in the presence of light
(D) Fixes carbon dioxide

550. The chief source of nitrogen for green plants are:
(A) Atmospheric nitrogen
(B) Nitrates
(C) Ammonium salts
(D) Low molecular weight-organic nitrogenous compound

551. Quanta required for assimilation of cone molecules of CO_2 or liberation of the molecule of oxygen are:
(A) 16 (B) 8
(C) 4 (D) 2

552. Ribulose biphosphate carboxylase oxygenase is located inside:
(A) Chloroplasts (B) Mitochondria
(C) Golgibodies (D) Peroxisomes

553. Chemosynthesis and photosynthesis are similar in being:
(A) Found in embryophytes
(B) Associated with heterotrophs
(C) Dependent on solar energy
(D) Autotrophic in nutrition

554. Phytochrome is:
(A) Monomer (B) Multimer
(C) Homodimer (D) Heterodimer

555. In case of prokaryotes, the net gain of ATP from complete oxidation of one molecule of glucose in respiration?
(A) 24 (B) 30
(C) 36 (D) 38

556. Cytochrome b6-f complex consists of how many polypeptides?
(A) 1 (B) 2
(C) 3 (D) 4

557. Gardeners pinch off the tips of plants to stimulate their side growth. Which one of the following is implicated in this phenomenon?
(A) Abscisic acid (B) Auxin
(C) Cytokinin (D) Gibberellin

558. Greatest producers of organic matter are:
(A) Crop plants (B) Forests
(C) Plants of the land area
(D) Phytoplankton of oceans

559. Aerobic bacteria collect near illuminated phytoplankton due to:
(A) Manufactured food
(B) Light
(C) Oxygen
(D) Reduced CO_2 concentration

560. Which photosynthetic bacteria possess both PS I and PS II:
(A) Purple sulphur bacteria
(B) Cyanobacteria
(C) Purple nonsulphur bacteria
(D) Green sulphur bacteria

561. Which products of Hill reaction are used in Blackman's reaction:
(A) ATP, NADPH (B) ATP, NADH
(C) ADP, NAD (D) ATP, NAD

562. The light which stimulates flowering in an inducible plant:
(A) Is perceived by the apical meristem
(B) Is perceived throughout the plant
(C) Is perceived by the leaves
(D) None of the above are correct

563. Chlorophyll b is
(A) $C_{54}H_{70}O_6N_4Mg$ (B) $C_{55}H_{70}O_6N_4Mg$
(C) $C_{55}H_{72}O_5N_4Mg$ (D) $C_{45}H_{72}O_5N_4Mg$

564. The size of chlorophyll molecule is:
(A) Head 15 x 15 Å, tail 25 Å
(B) Head 20 x20 Å, tail 25 Å
(C) Head 15 x 15 Å, tail 20 Å
(D) Head 10 x 12 Å, tail 15 Å

565. Which statement about phloem transport is correct?
(A) Gravity influence phloem transport
(B) Phloem transport occurs unidirectionally
(C) Sugar transported in phloem as non-reducing sugar
(D) Ca^{2+} is most abundant action in phloem sap

566. Green light has a wavelength of:
(A) 400-450 nm (B) 500-580 nm
(C) 660-720 nm (D) 720-800 nm

567. During monsoon, the rice crop of eastern states of India shows lesser yield due to limiting factor of:
(A) CO_2 (B) Light
(C) Temperature (D) Water

568. Photosynthetic pigments found in the chloroplasts occur in:
(A) Thylakoid membranes
(B) Pllastoglobules
(C) Matrix
(D) Chloroplast envelop

569. Dark reaction of photosynthesis occur in:
(A) Granal thylakoid
(B) Stromal lamella membranes
(C) Stroma outside photosynthetic lamellae
(D) Periplastidial space

570. Which blue green algae does not form heterocytes?
(A) *Anabaena* (B) *Nostoc*
(C) *Gleotrichia* (D) *Scytonema*

571. A Very efficient converter of solar energy with net productivity of 2-4 kg/m^2 or more in the crop of:
(A) Wheat (B) Sugarcane
(C) Rice (D) Bajra

572. Formation of ATP in photosynthesis and respiration is an oxidation process which utilizes the energy from:
(A) Cytochromes (B) Ferredoxin
(C) Electrons (D) Carbon dioxide

573. Cryptochrome is the pigment that absorbs:
(A) Red light (B) Green light
(C) Blue light (D) Yellow light

574. Algae employed by Calvin et al in experiments on photosynthesis belong to:
(A) Euglena (B) Chara
(C) Chlamydomonas (D) Chlorella

575. C_4 plants are also known as:
(A) Calvin type
(B) Calvin-Bassltam type
(C) Hatch and Slack type (D) Emerson type

576. Path of dark reaction of photosynthesis was traced through the use of:
(A) ^{32}P (B) $^{14}CO_2$
(C) $^{18}O_2$ (D) X-rays

577. Conversion of PGA to PGAL involves
(A) Reduction (B) Oxidation
(C) Hydrolysis (D) Photolysis

578. The plant growth regulator that retards senescence is:
(A) Cytokinin (B) Gibberellic acid
(C) Indoleacetic acid (D) Ethylene

579. When a plant is not reproducing, most of its cytokinins are produced in its:
(A) Leaves (B) Lateral roots
(C) Shoot apex (D) Roots

580. Cyclic photophosphorylation is carried out by:
(A) PS I (B) PS II
(C) Both (A) and (B) (D) Photolysis and PS II

581. In C_4 plants, Calvin cycle operates in:
(A) Stroma of bundle sheath chloroplasts
(B) Grana of bundle sheath chloroplasts
(C) Grana of mesophyll chloroplasts
(D) Stroma of mesophyll chloroplasts

582. Photosynthetic enhancement with flashing light was first observed by:
(A) Benson and Calvin
(B) Hill and Calvin
(C) Hatch and Slack
(D) Emerson and Arnold

583. Which is sensitive to longer wavelengths of light:
(A) PS II (B) PS I
(C) Phosphorylation (D) Photolysis

584. Maximum O_2 evolution occurs from:
(A) Forests (B) Marine phytoplankton
(C) Crops (D) Land mass

585. During cyclic photophosphorylation, which is electron donor of PS II?
(A) CO_2 (B) Water
(C) PS I (D) Cytochrome

586. Photosynthetic rate is the highest in which plants?
(A) C_3 plants (B) C_4 plants
(C) CAM plants (D) Both (B) and (C)

587. Which technique has helped in investigation of Calvin cycle?
(A) X-ray crystallography
(B) X-ray technique
(C) Radioactive isotope technique
(D) Intermittent light

588.

Group I (Hormone)	Group II (Functions)
P. Gibberellin	1. Phototropism an gravitrophism
Q. IAA	2. Stomatal movement
R. Cytokinin	3. Delay of senescence
S. Ethylene	4. Combating water deficit
	5. Seed germination
	6. Ripening of fruits

(A) P -6, Q – 4, R – 1, S - 2
(B) P -5, Q – 1, R – 3, S - 6
(C) P -3, Q – 4, R – 5, S - 1
(D) P -4, Q – 3, R – 1, S – 5

589. Downward movement of organic and inorganic solutes from leaves occurs through:
(A) Phloem by diffusion
(B) Phloem by protoplasmic streaming
(C) Parenchymatous cells by diffusion
(D) Phloem by mass flow

590. C_4 plants belongs to:
(A) Gramineae
(B) Monocots
(C) Dicots
(D) Both monocots and dicots

591. From the given statements identify the incorrect one:
(A) GA involves in flowering
(B) Ethylene is produced during ripening of the seeds
(C) Auxin helps in cell elongation and formation of root
(D) Cytokinin helps in embryo development and prevent leaf senescence

592. Kranz anatomy is typical of:
(A) C_4 plants (B) C_3 plants
(C) C_2 plants (D) CAM plants

593. Site of O_2 evolution and photophosphorylation is:
(A) Granal stacks (B) Chloroplast surface
(C) Inner membrane of chloroplasts
(D) Stroma

594. Quantosomes occur in:
(A) Chloroplasts (B) Mitochondria
(C) Nucleus (D) Lysosomes

595. Sugarcane shows high efficiency of CO_2 fixation because it performs:
(A) Calvin cycle (B) EMP pathway
(C) Hatch and Slack pathway
(D) TCA cycle

596. Which one of the following does not perform C_4 photosynthesis?
(A) *Saccharum* (B) *Zea mays*
(C) *Triticum aestivum (=T. vulgare/Crotalaria*
(D) *Euphorbia milli (=E. splenedens)*

597. Photosystem II occurs in:
(A) Stroma (B) Cytochrome
(C) Grana (D) Mitochondrial surface

598. Which of the following acts as a precursor of IAA biosynthesis?
(A) Tryptophan (B) Methionine
(C) Putrescine
(D) Geranyl geranyl pyrophosphate

599. The mitochondrial genome is the smallest in:
(A) Mammals (B) Yeasr
(C) Arabidopsis (D) Tomato

600. Kreb's cycle starts with entry of:
(A) Acetyl-CoA (B) Oxaloacetate
(C) Both (A) and (B) (D) Pyruvic acid

601. Through which of the following does an electron pass to $NADP^+$ in light reaction:
(A) Cytochrome (B) Ferredoxin
(C) Plastoquinone (D) Plastocyanin

602. Pigment system I performs independently:
(A) Noncyclic photophosphorylation
(B) Cyclic photophosphorylation
(C) Oxidative phosphorylation
(D) Photolysis

603. The mitochondrial genome is the smallest in:
(A) Mammals (B) Yeast
(C) Arabidopsis (D) Tomato

604. In case of eukaryotes the net gain of ATP from the complete oxidation of one molecule of glucose in respiration?
(A) 24 (B) 30
(C) 36 (D) 38

605. Ferredoxin is a constituent of:
(A) PS I (B) PS II
(C) Hill reaction (D) P_{680}

606. ATP is:
(A) Adenine triphosphate
(B) Adenosine tetraphosphate
(C) Adenosine triphosphate
(D) Adenine tetraphosphate

607. How many ATPs (gross ATP) are produced from complete oxidation of one molecule of glucose in respiration?
(A) 36 (B) 38
(C) 30 (D) 24

608. Which is due to cpDNA alone?
(A) Photosystem I (B) Photosystem II
(C) Chlroplast RNA polymerase
(D) All of these

609. In case of eukaryotes, the net gain of ATP from complete oxidation of one molecule of glucose in respiration?
(A) 24 (B) 30
(C) 36 (D) 38

610. All type of plastids possess essentially the same structure because they:
(A) Perform the same function
(B) Store food materials like starch, fat and protein
(C) Occurs in aerial parts
(D) Can transform from one form to another

611. Chlorophyll is found in:
(A) Stroma of chloroplasts
(B) Grana of chloroplasts
(C) Covering of chloroplasts
(D) Dispersed from in the chloroplasts

612. For synthesis of a molecule of glucose, the requirement of ATP and NADPH is respectively:
(A) 15 and 10 (B) 33 and 22
(C) 12 and 8 (D) 18 and 12

613. Chlorophyll is occurs in:
(A) All photosynthetic autotrophs
(B) In all higher plants
(C) All oxygen liberating autotrophs
(D) All parts except fungi

614. Which is due to cpDNA alone?
(A) Photosystem I (B) Photosystem II
(C) Chloroplast RNA polymerase
(D) All of these

615. The enzyme that catalyses carbon dioxide fixation in C_4 plants is:
(A) RuBP carboxylase (B) PEP carboxylase
(C) Carbonic anhydrase (D) Carboxy dismutase

616. Photo-oxidation or protolysis of water (in photosynthesis) occurs in assimilation of:
(A) Cytochrome B (B) Plastocyanin
(C) PS II (D) PS I

617. The source of oxygen evolved during photosynthesis is:
(A) CO_2 (B) H_20
(C) Carbohydrate (D) Photosynthetic curve

618. Bacterial photosynthetic differs from that of higher plants in:
(A) Not liberating oxygen
(B) Nonrequirement of light
(C) Nonfixation of energy
(D) Requirement of host organism

619. Krantz type of leaf anatomy is found in:
(A) Maize (B) Amaranthus
(C) Sorghum (D) All of these

620. Chief functions of leaves are:
(A) Transpiration and photosynthesis
(B) Respiration and photosynthesis
(C) Respiration and digestion
(D) Respiration and transpiration

621. Food is transported to different parts of plant through:
(A) Xylem (B) Phloem
(C) Pith (D) Cortex

622. Krantz type of leaf anatomy is found in:
(A) Maize (B) Amaranthus
(C) Sorghum (D) All of these

623. How many ATPs (gross ATP) are produced from complete oxidation of one molecule of glucose in respiration?
(A) 24 (B) 30
(C) 36 (D) 38

624. Presence of organic substances in the phloem can be demonstrated by means of:
(A) Ringing of stem (B) Grafting
(C) Root pressure (D) Defoliation

625. Emerson found photosynthesis to increase with light flashes. Blackman observed increase in photosynthesis with the increase in light and CO_2 till temperature became limiting. What of the following is true in light of above observation:
(A) Light intensity is temperature dependent
(B) Light harvesting occurs only in the presence of CO_2
(C) Photosynthesis has two phases, light and dark
(D) One factor is limiting photosynthesis at one time

626. Translocation of carbohydrate nutrients usually occurs in the form of:
(A) Glucose (B) Maltose
(C) Starch (D) Sucrose

627. Crassulacean acid metabolism (CAM) occurs in:
(A) Agave (B) Pineapple
(C) Opuntia (D) All of these

628. Which one die first when a ring of bark is removed:
(A) Shoot (B) Root
(C) None
(D) Both shoot and root die simultaneously

629. At high oxygen concentration, the rate of photosynthesis decreases due to:
(A) Warburg effect (B) Pasteur effect
(C) Emerson effect
(D) Richmond Lang effect

630. Kreb's cycle is also known as:
(A) Tricarboxylic acid cycle (TCA)
(B) Citric acid cycle
(C) Organic aid cycle (D) All of these

631. Oxygen containing carotenoids are:
(A) Carotenes (B) Xanthophylls
(C) Phycobilins (D) Anthocyanins

632. In the two light reactions of photosynthesis:
(A) PS I produces strong oxidant while PS II a strong reductant
(B) PS I produces strong reductant NADPH while PS II a strong oxidant
(C) PS I produces ATP which is not formed by PS II
(D) PS I produces ATP which is not formed by PS II

633. Apical dominance is the result of:
(A) Auxin (B) Cytokinin
(C) Auxin and ABA (D) ABA

634. Crassulacean Acid metabolism (CAM) occurs in:
(A) Agave (B) Opuntia
(C) Pineapple (D) All of these

635. When day light hours are increased, the rate of photosynthesis:
(A) Increases (B) Decreases
(C) Remains unchanged (D) None of the above

636. Photosynthesis is:
(A) Photochemical process (B) Chemical process
(C) Oxidation process
(D) Reduction process

637. Apical dominance is the result of:
(A) Auxin (B) Cytokinin
(C) Auxin and ABA (D) ABA

638. Complete girdling ultimately kills the tree due to:
(A) Starvation of root
(B) Stoppage of photosynthesis
(C) Nonmovement of minerals
(D) Nonmovement of water

639. In CAM plants, formation of malic acid takes place during:
(A) Sunset (B) Morning
(C) Night (D) Afternoon

640. Photorespiration occurs in:
(A) Ribosomes (B) Mitochondria
(C) Peroxisomes (D) Lysosomes

641. How many ATPs are produced (if we start with Acetyl CoA) in Kreb cycle?
(A) 18 (B) 24
(C) 30 (D) 38

642. Kreb cycle is also known as:
(A) Tricarboxylic acid cycle (TCA)
(B) Citric acid cycle
(C) Organic acid cycle
(D) All of these

643. CAM occurs in:
(A) Thin green leaves with reticulate venation
(B) Thin green leaves with parallel venation
(C) Thin coloured leaves
(D) Fleshy green leaves

644. Minimum solar energy is trapped by:
(A) Planting trees
(B) Cultivating crops
(C) Growing algae in tanks
(D) Growing grasses

645. How many ATPs are produced (if we start with Acetyl CoA) in Kreb's cycle?
(A) 18 (B) 30
(C) 24 (D) 38

646. Which of the following trees will die first:
(A) Deciduous (B) Pruned
(C) Hollowed trunk (D) Girdled

647. Two types of photosynthetic pigments are:
(A) Chlorophyll a and carotenoids
(B) Chlorophyll b and carotenoids
(C) Chlorophyll and carotenoids
(D) Chlorophyll a and chlorophyll b

648. PEP carboxylation takes place in:
(A) Mesophyll cells of C4 plants
(B) Epidermal cells of C4 plants
(C) Bundle sheath cells of C4 plants
(D) Xylem cells of C4 plants

649. Main difference between chlorophyll a and chlorophyll b is:
(A) $-CH_3$ of chlorophyll a is replaced by $-CHO$ in chlorophyll b
(B) Chlorophyll a is linear while chlorophyll b is branched
(C) Chlorophyll a has no Mg
(D) All the above

650. Light energy is used in photosynthesis for:
(A) Breakdown of H_2O
(B) Breakdown of CO_2
(C) Activation of chlorophyll
(D) Breakdown of $C_6H_{12}O_6$

651. In CAM plants, formation of malic acid takes place during:
(A) Sunset (B) Morning
(C) Night (D) Afternoon

652. If plants stop photosynthesis which gas will disappear:
(A) CO_2 (B) N_2
(C) O_2 (D) NH_3

653. Glycolysis occurs in:
(A) Cytoplasm (B) Mitochondria
(C) Nucleus (D) Golgi bodies

654. Which plant shows chloroplast dimorphism:
(A) Wheat (B) Rice
(C) Sugar Beet (D) Sugar cane

655. Cytochromes are:
(A) O_2 acceptors (B) H_2 acceptors
(C) Electron acceptor (D) H_2O acceptors

656. .Electron transport system take place in which part of mitochondria?
(A) Oxysome (B) Outer membrane
(C) Matrix (D) Both (A) and (C)

657. During dark reaction the three carbon atoms of 3-PGA are derived from:
(A) RuBP only (B) CO_2 only
(C) RuBP + CO_2 (D) RuBP + CO_2 + PEP

658. PEP carboxylation takes place in:
(A) Mesophyll cells of C4 plants
(B) Epidermal cells of C4 plants
(C) Bundle sheath cells of C4 plants
(D) Xylem cells of C4 plants

659. First product of photorespiration is:
(A) Phosphoglycolate (B) Glycolate
(C) Glycine (D) None of the above

660. Assimilatory power formed in light reaction of photosynthesis is:
(A) ATP + $NADPH_2$ (B) ADP + $NADH_2$
(C) ATP (D) $NADPH_2$

661. Photoperiod is the most relevant for:
(A) Abscission (B) Flowering
(C) Shoot bud formation (D) Root growth

662. Which one is a C_4 plant:
(A) Maize (B) Sugarcane
(C) Sorghum (D) All the above

663. Balance between CO_2 and O_2 is maintained by:
(A) Transpiration (B) Photosynthesis
(C) Photorespiration (D) C_4 pathway

664. Which is correct about chlorophylls a and b leaves of higher plants:
(A) Both are present in equal proportion
(B) Chlorophyll a is more than chlorophyll b
(C) Chlorophyll a is less than chlorophyll b
(D) Chlorophyll b is ten times more than chlorophyll a

665. In C_4 plants, chloroplasts are also found in:
(A) Spongy parenchyma (B) Epidermal cells
(C) Guard cells
(D) Bundle sheath cells

666. In C_4 plants, fixation of carbon dioxide occurs in:
(A) Transfusion tissue (B) Palisade tissue
(C) Cortex of stem
(D) Spongy mesophyll and bundle sheath cells

667. Liberation of oxygen when green cells in water are exposed to sunlight in presence of suitable acceptor is:
(A) Emerson effect (B) Blackman's reaction
(C) Hill's reaction (D) Arnon reaction

668. During cyclic photophosphorylation, which is electron donor of PS II?
(A) CO_2 (B) Water
(C) PS I (D) Cytochrome

669. Electron transport system take place in which part of mitochondria?
(A) Oxysome (B) Ouetter membrane
(C) Matrix (D) Both (A) and (C)

670. ATP synthesis during light reaction is:
(A) Oxidative phosphorylation
(B) Photolysis
(C) Photophosphorylation
(D) Phosphorylation

671. Pigment acting as a reaction centre during photosynthesis is:
(A) Carotene (B) Phytochrome
(C) P_{700} (D) Cytochrome

672. In response to stress, which amino acid is accumulated?
(A) Proline (B) Glycine
(C) Lysine (D) Phenylalanine

673. Photosystem has main light harvesting pigments:
(A) Chlorophyll a -660, chl a-670, chl a-680, chl a 690, chl a-700
(B) Chlorophyll a-660, chl b-, chl-700, chl-800 and carotenoid
(C) Chlorophyll a-680, chl-685, chl-695 and xanthophylls
(D) Chlorophyll a-700, chl-880, chl b, chl-685 and phytocobilins

674. Glycolysis occurs in:
(A) Cytoplasm (B) Mitochondria
(C) Nucleus (D) Golgi bodies

675. Cytochrome b6-f complex consists of how many polypeptides?
(A) 1 (B) 3
(C) 4 (D) 6

676. C_4 cycle was discovered in:
(A) Groundnut (B) Sugarcane
(C) Chrysanthemum (D) Apple

677. Which one is the common product of both respiration and photosynthesis:
(A) ATP (B) Quinone
(C) Cytochrome (D) Chlorophyll

678. Chlorophylls absorb visible light of wavelengths:
(A) 400-500 nm only (B) 300-400 nm only
(C) 600-800 nm only
(D) 400-500 nm and 600-700 nm

679. Chlorophyll a has:
(A) -CHO group (B) -COOH group
(C) -CH3 groups (D) No functional group

680. RuBISCO has which activities?
(A) Carboxylase (B) Reductase
(C) Oxygenase (D) Both (A) and (C)

681. Calvin cycle involves:
(A) Oxidative carboxylation
(B) Reductive carboxylation
(C) Phosphorylation
(D) Oxidative phosphorylation

682. Chlorophyll is soluble in:
(A) Water (B) Organic solvents
(C) Both (A) and (B) (D) None of these

683. Photorespiration is favoured by:
(A) High oxygen and low carbon dioxide
(B) High carbon dioxide and low oxygen
(C) High temperature and low oxygen
(D) High humidity and temperature

684. In which, electron transport leading to phosphorylation is a common reaction?
(A) Chloroplasts (B) Mitochondria
(C) Both (A) and (B) (D) Golgi bodies

685. Mitochondrial DNA does not code for:
(A) tRNA
(B) Enzyme for DNA replication
(C) mRNA (D) rRNA

686. Photosynthesis is a process in which :
(A) ATP is generated
(B) NADH is reduced to NAD
(C) Oxidative phosphorylation occurs
(D) CO_2 is released

687. Carbon dioxide assimilation occurs through:
(A) Krebs cycle
(B) Calvin cycle
(C) Glycolysis
(D) Anaerobic respiration

688. Number of calvin cycles required to generate a molecule of hexose is:
(A) 2 (B) 4
(C) 6 (D) 8

689. Which one occurs both during cyclic and noncyclic modes of photophosphorylation:
(A) Involvement of both PS I and PS II
(B) Formation of ATP
(C) Release of O_2
(D) Formation of NADPH

690. Calvin cycle is:
(A) Dependent upon light (B) Independent of light
(C) Supported by light (D) Hindered by light

691. In C_4 plants, initial carbon dioxide fixation occurs in chloroplasts of:
(A) Palisade tissue
(B) Spongy parenchyma
(C) Guard cells
(D) Hypodermis and chlorenchyma

692. Phytochrome is a:
(A) Monomer (B) Multimer
(C) Homodimer (D) Heterodimer

693. Pyruvate kinase transfers a phosphate group from phosphoenolpyruvate to ADP, forming pyruvate and ATP. The reaction catalyzed by this enzyme is essentially irreversible. Which of the following is the best explanation for the irreversible nature of this reaction?
(A) The binding of pyruvate to the active site is very weak relative to the binding of phosphoenol pyruvate
(B) The reaction is coupled to the pyruvate dehydrogenase reaction
(C) The hydrolysis of ATP is highly favourble
(D) The change in free energy for the overall reaction is large and negative

694. Mitochondrial DNA does not code for:
(A) tRNA
(B) Enzyme for DNA replication
(C) mRNA
(D) rRNA

695. Chlorophyll consists of:
(A) Porphyrin head and phytol tail
(B) Phosphate head and porphyrin tail
(C) Phytol head and porphyrin tail
(D) Porphyrin head and phosphate tail

696. Most of the reactions of glyoxylate cycle occurs in:
(A) Glyoxysomes (B) Cytosol
(C) Mitochondria (D) All of these

697. A free living anaerobic bacterium capable of nitrogen fixation in the soil is:
(A) *Azotobacter* (B) *Clostridium*
(C) *Rhizobium* (D) *Streptococcus*

698. Element is essential for photolysis of water is:
(A) Nitrogen (B) Oxygen
(C) Chlorine (D) Carbon

699. Enzymes PEP carboxylase and RuBP carboxylase are located in chloroplasts of:
(A) C_3 plants (B) CAM plants
(C) C_4 plants (D) Both (B) and (C)

700. Constituents of pigment system I are located on:
(A) Granal thylakoids
(B) Stomal thylakoids
(C) Outer surface of granal and stomatal thylakoids
(D) Stroma

701. Which are stress hormones?
(A) Ethylene and IAA (B) IAA and ABA
(C) ABA and Ethylene (D) ABA and IBA

702. Light energy is converted into chemical energy in the presence of:
(A) Chloroplasts (B) Pyrenoids
(C) Ribosomes (D) Mesosomes

703. Calvin cycle occurs in:
(A) Cytoplasm (B) Mitochondria
(C) Glyxosomes (D) Chloroplasts

704. Inclusion of a cell extract containing all enzymes of glycolysis with [γ32P] ATP and unlabeled inorganic phosphate, results in the formation of which of the following labeled compounds (assume that pyruvate kinase is inactivated)?
1. Glucose-6 - ^{32}phosphate
2. (3-^{32}P)-phosphoglycerate
3. (1-phospho – 3 – ^{32} phospho)-bisphosphoglycerate
4. (1-phospho – 6 – ^{32} phospho)- fructose bisphosphate

(A) 1 and 3 (B) 1, 2 and 3
(C) 2 and 4 (D) Only 4

705. Plants purify air during:
(A) Photosynthesis (B) Respiration
(C) Transpiration (D) Desiccation

706. Which organelles is involved in ATP production by photophosphorylation?
(A) Mitochondria (B) Chloroplasts
(C) Leucoplasts (D) Chromoplasts

707. In case of palnts, cpDNA encodes which components of orgaeannellar ATPase?
(A) F1 ATPase
(B) F0 Factor
(C) Both (A) and (B)
(D) Some subunits of Both (A) and (B)

708. Nitrite reductase contains which prosthetic groups?
(A) Siroheme ($Fe_4 - S_4$) cluster
(B) FMN (Fe-S) cluster
(C) Mo cofactor (Fe-S) cluster
(D) All of these

709. Photorespiration occurs in:
(A) Green plants (B) All living cells
(C) Mitochondria (D) Root

710. Thylakoids possess photosynthetic units called:
(A) Quantasomes (B) Glyoxysomes
(C) Polysomes (D) Ribosomes

711. Noncyclic photophosphorylation is performed by:
(A) Dark reaction (B) PS I
(C) PS II (D) Both (B) and (C)

712. Photorespiration is affected by:
(A) Temperature (B) Light intensity
(C) CO_2 and O_2 (D) All the above

713. A free living anaerobic bacterium capable of nitrogen fixation in the soil:
(A) Azotobacter (B) Clostridium
(C) Rhizobium (D) Streptococcus

714. At carbon atom III, chlorophyll a has:
(A) Methyl group (B) Aldehyde group
(C) Carboxyl group (D) Magnesium

715. One of the carbon atoms of a glucose molecule is [14]-labeled. If $^{14}CO_2$ is released during the conversion of pyruvate to acetyl coenzyme-A, which carbon atom of glucose was radio labeled?
(A) C3 but not C 4 (B) C 3 or C 4
(C) C 1 or C 6 (D) C 1 but not C 6

716. Photorespiration is characteristic of:
(A) C_3 plants (B) C_4 plants
(C) CAM plants (D) All of these

717. Which organelles is involved in citric acid cycle?
(A) Mitochondria (B) Chloroplasts
(C) Leucoplasts (D) Chromoplasts

718. In case of yeast mitochondria, mt – DNA encodes which components of mitochondrial ATPase?
(A) F1 ATPase (B) F0 Factor
(C) Both (A) and (B)
(D) Some subunits of Both (A) and (B)

719. C_4 plants differ from C_3 plants with respect to:
(A) First product
(B) Substrate which accepts carbon dioxide
(C) Number of ATP molecules consumed
(D) All the above

720. *nif*HDK genes are organized in which ways?
(A) They are contigous
(B) D and K are separated by 11 kb DNA
(C) Separated by 91 kb DNA located between genes D and K
(D) Separated by 55 kb DNA between genes D & K

721. Photooxidation is:
(A) Photorespiration
(B) Photolysis
(C) Light and oxygen induced breakdown
(D) All the above

722. Core metal of chlorophyll is:
(A) Cu (B) Mg
(C) Fe (D) Ni

723. Which statement is true of glucokinase?
(A) It catalyzes the phosphorylation of fructose
(B) It has a lower K_m for glucose than does hexokinase
(C) It is found in liver
(D) It is inhibited by glucose-6-phosphate

724. Which of the following statement about photo-system II is incorrect?
(A) It is multimolecular transmembrane assembly containing several polypeptides, several chrolophyll molecules, a special pair of chlorophyll (P680), pheophytin, and plastoquinones
(B) It transfers electrons to phtosystem I via the cytochrome bf complex
(C) It uses light energy to create a separation of charge whose potential energy can be used to oxidize H_2O and to produce a reductant, plastoquinol
(D) It uses an Fe^{++} - Cu^{+} centre as a charge accumulator to form O_2 without genera-ting potentially harmful hydroxyl radials, superoxide anions, or H_2O_2

725. RUBISCO content of chloroplast is
(A) 20% (B) 5%
(C) 11% (D) 16%

726. In case of plants, cpDNA encodes which components of organellar ATPase?
(A) F1 ATPase (B) F0 factor
(C) Both (A) and (B)
(D) Some subunits of Both (A) and (B)

727. Main substance involved in transfer of electrons in photosynthesis is:
(A) Photochrome (B) Cytochrome
(C) Isosomes (D) Both (A) and (B)

728. Which one is required for photosynthetic evolution of oxygen?
(A) F-enzyme (B) Z-enzyme
(C) Co-enzyme (D) M-enzyme

729. The electron transport system and oxidative phosphorylation are concerned with?
(A) Mitochondria (B) Chloroplasts
(C) Leucoplasts (D) Chromoplasts

730. Azolla contains ab ebdosymbiont which fix nitrogen; this oprganism belongs to which genera?
(A) *Aulosira* (B) *Azosprillum*
(C) *Nostoc* (D) *Anabaena*

731. C_4 plants have higher net photosynthetic rate as they have:
(A) No photorespiration
(B) PEP as CO_2 acceptor
(C) Kranz anatomy
(D) Photosynthesis even at low light intensity

732. Azolla contains an endosymbiont which fixes nitrogen; this organism belong to which genera?
(A) *Aulosira* (B) *Azospirillum*
(C) *Nostoc* (D) *Ananaena*

733. Accessory pigments transfer their energy to chlorophyll molecules through:
(A) Fluorescence (B) Resonance
(C) Radiation (D) Transduction

734. C_4 cycle is connected with:
(A) Respiration (B) Photosynthesis
(C) Transpiration (D) Grwoth & development

735. ATP formation in chloroplast and mitochondrian is explained by:
(A) Cholodny-Went model
(B) Chemi-osmotic theory of Mitchell
(C) Munch's mass flow theory
(D) Relay pump theory of Godlewski

736. Primary carboxylation occurs in C_3 and C_4 plants with the help of:
(A) PEP carboxylase and pyrovate carboxylase
(B) PEP carboxylase and RuBP carboxylase
(C) RuBP carboxylase and PEPcarboxylase
(D) RuBP carboxylase and pyruvate carboxylase

737. Iodoacetamide irreversively inhibits few enzymes by reacting with the amino acid residue at the active site having the functional group:
(A) -SH (B) - COOH
(C) - OH (D) - NH_2

738. In case of yeast mitochndria, mtDNA encodes which components of mitochondrial ATPase?
(A) F1 ATPase (B) F0 factor
(C) Both (A) and (B)
(D) Some subunits of Both (A) and (B)

739. A specific function of light in the process of photosynthesis is:
(A) Reduction of CO_2
(B) Activate other molecules
(C) Split water molecules
(D) None of the above

740. 2,4-dinitrophenol uncouples electron transport from oxidative phosphorylaton in mitochondria by:
(A) Causing dissipation of the proton gradient generated by the electron transport
(B) Allowing the proton translocating function of ATP synthase while inhibiting its ATP synthesizing activity
(C) Activating a second proton pump that sends back the protons into the mitochondrial matrix
(D) Neutralizing the proton gradient by absorbing the protons generated during the electron transport

741. Which is an example of colourless plastid?
(A) Mitochondria (B) Chloroplasts
(C) Leucoplasts (D) Chromoplasts

742. C_4 plants differ from C_3 plants in:
(A) Number of ATP consumes per hexose molecule
(B) Type of end product
(C) Type of pigments
(D) Substrate for CO_2 in carbon assimilation

743. nifHDK genes are organized in which ways?
(A) They are contiguous
(B) D and K are separated by 11 kb DNA
(C) Separated by 91 kb DNA located between genes D and K
(D) Separated by 55 kb DNA between genes D & K

744. Warburg effect is decrease in the rate of photosynthesis at:
(A) Low carbon dioxide concentration
(B) High oxygen content
(C) High carbon dioxide concentration
(D) Both (A) and (B)

745. Which one of the following contain agranal chloroplasts?
(A) C_3 plants (B) Succulants
(C) C_4 plants (D) Hydrophytes

746. Which of the plants bears intermediate features between C_3 and C_4 plants?
(A) Maize (B) *Moricandia arvensis*
(C) *Amaranthus* (D) Both (A) and (C)

747. Which is not an example of C_4 plant?
(A) Sugarcane (B) Sorghum
(C) Potulaca (D) Wheat

748. Cyanobacteria differ from purple and green photosynthetic bacteria because they:
(A) Have a membrane-enclosed nucleus
(B) Use H_2S as an electron donor
(C) Do not require light
(D) Produce oxygen during photosynthesis

749. Which element is required for photosynthetic evolution of oxygen?
(A) Cu (B) Fe
(C) Zn (D) Mn

750. Thermogenin, the natural uncoupler in brown fat mitochondria, generates heat based on its ability to:
(A) Inhibits electron transport by binding to b-type cytochromes
(B) Allow protons to re-enter into the mitochondrial matrix
(C) Inhibit ATP production by binding to the ATP synthase
(D) Block electron transport to flavoproteins

751. The term carotenoid is related to:
(A) Carotene
(B) Chlorophyll
(C) Xanthophyll
(D) Both (A) and (C)

752. Nitrite reductase contains which prosthetic groups?
(A) Siroheme (Fe4-S4) cluster
(B) FMN (Fe-S) cluster
(C) Mo cofactor (Fe-S) cluster
(D) All of these

753. Protochlorophyll differs from chlorophyll in lacking:
(A) 4 hydrogen atoms in two pyrrole rings
(B) 2 hydrogen atoms in one of its pyrrole rings
(C) 2 hydrogen atoms in two pyrrole rings
(D) 1 hydrogen atoms in one pyrrole ring

754. Which is an example of C_4 plant?
(A) Maize (B) *Moricandia arvensis*
(C) *Amaranthus* (D) Both (A) and (C)

755. Which occurs during light reactions?
(A) Photolysis of water
(B) Photophosphorylation
(C) Fixation of CO_2 (D) Both (A) and (B)

756. In photosynthesis, protolysis of water is used in:
(A) Reduction of $NADP^+$ (B) Oxidation of NADP
(C) Oxidation of FAD (D) None of the above

757. Plant mitochondria were treated with antimycin A and sodium azide. This would lead to:
(A) Increased cyanide resistant respiration and reduced ATP synthesis
(B) Increased cyanide resistant respiration and increased ATP synthesis
(C) Increased cytochrome oxidation
(D) Reduced nuclear –mitochondria interaction

758. Photosynthesis occurs in:
(A) Leaves (B) Leaf cells
(C) Plants (D) Chloroplasts

759. In green cell the enzyme catalase is located in:
(A) Peroxisomes (B) Cytoplasts
(C) Lysosomes (D) Vacuoles

760. The wavelength of light most absorbed during photosynthesis is:
(A) 700 nm (B) 600 nm
(C) 550 nm (D) 400 nm

761. Maximum photosynthesis occurs in:
(A) Blue light (B) Red light
(C) Green light (D) White light

762. Which is the major pigment of chloroplasts?
(A) Chlorophyll (B) Xanthophyll
(C) Carotene (D) All of these

763. Which are stress hormones?
(A) Ethylene and IAA (B) IAA and ABA
(C) ABA and ethylene (D) ABA and IBA

764. Special feature of C_4 plants is:
(A) Thin cuticle
(B) Multi layeres epidermis
(C) Krantz anatomy
(D) Both (A) and (B)

765. Factors which is not normally limiting in photosynthesis is:
(A) Chlorophyll (B) Light
(C) Water (D) CO_2

766. Which of the following contains copper?
(A) Quinone (B) Plastoquinone
(C) Plastocyanin (D) Cyt b_5

767. Radioactive tracer studies with ^{14}C have shown that:
(A) Phloem transports organic nutrients
(B) Phloem transports inorganic nutrients
(C) Xylem transports organic nutrients
(D) Xylem transports inorganic nutrients

768. Piments of PS I occur in:
(A) Appressed part of granal thylakoids
(B) Stromal thylakoids and nonappressed parts of granal thylakoids
(C) Both appressed and nonappressed parts of granal thylakoids
(D) Stroma

769. Number of cell organelles involved in photorespiration is:
(A) One (B) Two
(C) Three (D) Four

770. Condition favouring cyclic photophosphorylation is:
(A) Anaerobic environment
(B) Aerobic and low light intensity
(C) Aerobic an optimum light
(D) Anaerobic and low light intensity

771. Which of the following statements about the enzyme complexes of the electron transport system is correct?
(A) They interact with one another via mobile electron carriers
(B) They are located in the mitochondria matrix
(C) They can not be separated from one another in a functional form
(D) They all have cytochromes

772. When sunlight is on the chloroplast, pH is lowest in the:
(A) Stroma
(B) Space enclosed by the inner & outer membranes
(C) Space enclosed by the thylakoid membranes
(D) Cytosol

773. Wave length of visible light is:
(A) 200-400 nm (B) 400-700 nm
(C) 700-900 nm (D) 100-200 nm

774. In photosynthesis synthesis of carbohydrates during CO_2 and:
(A) Water in the presence of sunlight with the help of chlorophyll
(B) Water in the presence of energy
(C) Water with the help of chlorophyll
(D) Presence of sunlight

775. Which is correct about C3 plants?
(A) During photosynthesis, RuBP carboxylase enzyme plays a key role in the fixation of CO_2
(B) 3- Phosphoglyceric acid (3-PGA) is the first stable products of photosynthesis
(C) Carbohydrates are formed through Calvin cycle
(D) All of these

776. During photosynthesis, dark reaction occurs in:
(A) Stroma (B) Thylakoids
(C) Oxysomes (D) Both (A) and (B)

777. Direction of conduction of food through phloem is:
(A) From below upwards
(B) From tip to bottom
(C) From leaves to roots
(D) Phloem never conducts food

778. Glycolysis uses, which substrate?
(A) Glucose (B) Sucrose
(C) Lipids (D) Fatty acids

779. RUBISCO has which activities?
(A) Carboxylase (B) Reductase
(C) Oxygenase (D) Both (A) and (C)

780. In photosynthesis oxygen is liberated during:
(A) Hydrolysis of carbohydrates
(B) Breakdown of proteins
(C) Reduction of CO_2
(D) Hydrolysis of water

781. DCMU inhibits photosynthesis at photosystem during:
(A) phe QA (B) QA Qb
(C) Cyt bfb P (D) Q8 ytbf6

782. The pigment present universally in all green oxygenic plants is:
(A) Chlorophyll a (B) Chlorophyll b
(C) Chlorophyll c (D) Chlorophyll m

783. Ferredoxin is:
(A) Phenol (B) Protein
(C) Fat (D) Carbohydrate

784. In which plants stomata are closed in day time and open during night?
(A) C_2 plants (B) C_4 plants
(C) CAM plants (D) All of these

785. Ethylene is mainly responsible for:
(A) Formation of flower
(B) Formation of root hair
(C) Ripening of fruits
(D) Formation of fruit

786. Green plants do not give out CO_2 during day time because they:
(A) Store the same
(B) Respire very slowly
(C) Do not respire
(D) Consume it in photosynthesis

787. The enzymes for the oxidation of amino acids, fatty acids and Kreb's cycle are present in:
(A) Lysosomes (B) Stroma
(C) Grana (D) Mitochondrial matrix

788. In response to stress, which amino acids is accumulated?
(A) Proline (B) Glycine
(C) Lysine (D) Phenylalanine

789. For fixing of one molecule of CO_2 in Calvin cycle, are required:
(A) 3 ATP + 1 $NADPH_2$ (B) 3 ATP + 2 $NADPH_2$
(C) 2 ATP + 3 $NADPH_2$ (D) 3 ATP + 3 $NADPH_2$

790. The rosette habit of cabbage can be changed drastically by the application of:
(A) GA (B) IAA
(C) ABA (D) CK

791. Enzyme required in early CO_2 fixation in C_4 cycle is:
(A) RuBP carboxylase (B) RuBP oxygenase
(C) PEP carboxylase (D) PGA dehydrogenase

792. During photosynthesis, light reaction occurs in:
(A) Stroma (B) Thylakoids
(C) Oxysomes (D) Both (A) and (B)

793. First reaction in photosynthesis is:
(A) Photolysis of water
(B) Excitation of chlorophyll molecules
(C) Formation of ATP (D) Fixation of CO_2

794. Carbon dioxide is fixed in:
(A) Light reaction (B) Dark reaction
(C) Aerobic respiration
(D) Anaerobic respiration

795. Three carbons of photophoglyceric acid formed during carbon fixation are derived from:
(A) PEP + CO_2 (B) RuBP
(C) CO_2 (D) RuBP + CO_2

796. Which of the following occurs in dark reaction of photosynthesis?
(A) Formation of ATP (B) Release of O_2
(C) Release of H_2 (D) Synthesis of PGAL

797. Photoperiod is the most relevant for:
(A) Abscission (B) Flowering
(C) Shoot bud formation (D) Root growth

798. Process of assimilation of CO_2 to produce carbohydrates is:
(A) Oxidative (B) Catabolic
(C) Reductive (D) All the above

799. Cytochrome b_6 and cytochrome *f* occur in:
(A) Ribosomes (B) Mitochondria
(C) Chloroplasts (D) Lysosomes

800. Light reaction of photosynthesis is also called:
(A) Calvin cycle (B) Hill reaction
(C) TCA cycle (D) All the above

801. Growth is a:
(A) Reversible increase in size
(B) Reversible increase in shape
(C) Irreversible increase in size
(D) None of the above

802. Starch is stored in Potato tuber as:
(A) Sugar is synthesized in leaves
(B) Sugar is transported from leaves to tubers
(C) In the tuber sugar is changed to starch
(D) All the above

803. Which is correct about C3 plants?
(A) During photosynthesis, PEP carboxylase plays a key role in the fixation of CO_2
(B) Malic is the first stable products of photosynthesis
(C) Carbohydrates are formed through Hatch Slack cycle
(D) All of these

804. Evidence for liberation of O_2 from water during photosynthesis comes from:
(A) Release of ^{18}O if water contains the same
(B) Isolated chloroplast supplied with reducing agent like - potassium, ferrocyanide evolved O_2 even in absence of CO_2
(C) Photosynthetic bacteria do not liberate O_2 and they use H_2S for obtaining reducing power
(D) All the above

805. Bakanae disease is caused by:
(A) Fungus (B) Alga
(C) Bacterium (D) Virus

806. Which is present in stroma?
(A) DNA (B) RNA
(C) Ribosomes (D) All of these

807. Chemical which absorbs light energy and changes it to chemical energy is:
(A) Xanthophyll (B) Chlorophyll a
(C) Chlorophyll b (D) Chlorophyll c

808. In maize, mesophyll cells perform photosynthesis cycle:
(A) C_4 (B) C_3
(C) C_2 (D) C_1

809. The rate of photosynthesis, can be measured by:
(A) Amount of carbohydrate accumulated
(B) The volume of CO_2 consumed
(C) The volume of oxygen librated
(D) All of these

810. Carotene protect plants from:
(A) Photooxidation (B) Desiccation
(C) Photorespiration (D) Photosynthesis

811. An excised leaf does not turn yellow if it is induced to root. This is attributed to the synthesis in root of:
(A) Ethylene (B) Cytokinins
(C) Gibberellins (D) Auxins

812. Which have circular DNA?
(A) Chloroplasts (B) Mitochondria
(C) Bacteria (D) All of these

813. A natural growth regulator is:
(A) NAA (B) 2, 4-D
(C) Benzaldehyde (D) Ethylene

814. The maximum growth phase is:
(A) Lag phase (B) Steady phase
(C) Log phase (D) Senescent phase

815. The main feature of light promoters of dicots is:
(A) TATA box
(B) An upstream AG or AGGA box
(C) Both TATA and AG/AGGA boxes
(D) CAAT box

816. The rate of growth is measured by:
(A) Auxanometer (B) Potometer
(C) Photometer (D) Hygrometer

817. During photosynthesis, which gases is evolved as byproduct?
(A) CO_2 (B) H_2
(C) O_2 (D) Both (B) and (C)

818. Anatomy of C_4 plant leaf shows:
(A) Presence of perroxisomes
(B) Presence of bundle sheath cells
(C) Absence of mitochondria
(D) Absence of bundle sheath cells

819. The growth of plants differs from growth of animals in being:
(A) Localised (B) Diffused
(C) Localised and Diffused (D) None of the above

820. Dark reaction is also called:
(A) Blackman reaction (B) Hill reaction
(C) Cyclic reaction (D) Both (A) and (B)

821. The term 'vernalization' was coined by:
(A) Garner and Allard (B) Darwin
(C) Geoffery (D) Lysenko

822. The alcohol dehydrogenase gene s induced by stress due to:
(A) Heat (B) Anaerobiosis
(C) Wound (D) Pathogene invasion

823. Gibberellins were first extracted from:
(A) Coleoptile tip (B) Root tip
(C) Fungus (D) Bacterium

824. Hill used a dye for his famous hill reaction:
(A) Sulphur green (B) Eosine
(C) Methylene blue
(D) Dichlorophenol indophenol

825. Etiolation in plants is caused when they are grown in:
(A) Red light (B) Blue light
(C) Dark light (D) Intense light

826. Translocation of carbohydrates mostly takes place in the form of :
(A) Maltose (B) Fructose
(C) Sucrose (D) Glucose

827. Substance which originated at the tip of stem and control growth elsewhere are:
(A) Enzymes (B) Hormones
(C) Vitamins (D) None of these

828. The short day plant is:
(A) Tomato (B) Sunflower
(C) Cotton (D) Potato

829. Which is not correct?
(A) Startpoint is typically a single base within the promoter
(B) The chief function of factor (of bacterial RNA polymerase) is to ensure that the holoenzyme binds stably at only promoter sequences
(C) Transcription factors are essential part of RNA polymerases
(D) Transcription factors can be divided into basal upstream and regulatory transcription factors

830. Which is correct about RNA polymerase I:
(A) It is present in nucleolus
(B) It constitutes 50-70% of total activity
(C) It transcribes genes for rRNA
(D) All of these

831. In case of eukaryotes, RNA processing is completed in:
(A) Cytoplasm (B) Nucleolus
(C) Nucleus (D) Vacuole

832. Which is correct about RNA polymerase III?
(A) It occurs in nucleoplasm
(B) It provides -10% of total polymerase activity
(C) It transcribes tRNA and 5SRNA genes
(D) All of these

833. The long day plant is:
(A) Tomato (B) Potato
(C) Cotton (D) Spinach

834. The day neutral plant is:
(A) Tobacco (B) Tomato
(C) Wheat (D) Oat

835. Light reaction is also called:
(A) Blackman reaction (B) Hill reaction
(C) Cyclic reaction (D) Both (A) and (B)

836. The synthesis of amylase in certain cereals is stimulated by:
(A) GA (B) IAA
(C) CK (D) ABA

837. Plant growth is:
(A) Irreversible (B) Increase in size
(C) Localised (D) All of these

838. Plant response to environment is mainly through:
(A) Induction of dormancy (B) Abscission of parts
(C) Synthesis of pigments (D) Growth

839. Darwin and Darwin discovered that canary grass coleoptile bends towards light only when:
(A) It is 5 cm long
(B) The seedling is growing in blue light
(C) Coleoptile tip is intact & exposed to light
(D) Nights are cooler

840. In case of C_4 plants, RuBISCO exists only in:
(A) Mesophyll cells (B) Bundle sheath
(C) Guard cells (D) All of these

841. Etiolation is characterized by:
(A) Slender yellowish stems
(B) Small yellowish leaves
(C) Subterminal hook
(D) All of these

842. Fe protein of nitrogenase is:
(A) Homomeric dimer (B) Hetermeric dimer
(C) homomeric tetramer (D) Hetermeric tetramer

843. How many ATP molecules are produced in anaerobic respiration of one glucose molecule?
(A) 2 (B) 4
(C) 6 (D) 8

844. On excess, CO_2:
(A) Accelerates growth
(B) Does not influence growth
(C) Inhibits growth
(D) Slow down growth

845. Which one regulates plant growth?
(A) Commensation (B) Climate
(C) Hormones (D) Both (B) and (C)

846. How many electrons are required for the reduction of one molecule of nitrogen to NH_4?
(A) 6 (B) 8
(C) 3 (D) 12

847. High C/N ratio produces:
(A) Solar tissues
(B) More mechanical tissues
(C) More growth hormones
(D) Growth inhibitors

848. Which is correct about C_3 plants?
(A) During photosynthesis, RuBP carboxylase enzyme plays a key role in the fixation of CO_2
(B) 3-Phosphogluceric acid (3-PGA) is the first stable product of photosynthesis
(C) Carbohydrates are formed through Calvin cycle
(D) All of these

849. HMP pathway is also known by:
(A) Pentose phosphate pathway
(B) Direct oxidase pathway
(C) EMP pathway
(D) Both (A) and (B)

850. Which one is not a growth promoter?
(A) ABA (B) GA
(C) IAA (D) CK

851. Which is not correct about transcription unit?
(A) It is that sequence of a DNA molecule that is transcribed into a single RNA molecule
(B) In case of all eukaryotes, each gene is a distinct transcription unit
(C) In case of prokaryotes, each operon functions as a single transcription unit
(D) Enhancer sequences are present within promoters

852. The GC-AG rule does not apply to introns of:
(A) Chloroplast introns
(B) Nuclear introns of yeast
(C) Mitochondrial inrtons
(D) All of these

853. Optimum growth occurs in :
(A) Blue light (B) Red light
(C) White light (D) Green light

854. DNP is:
(A) Cestrum (B) Cucumber
(C) Potato (D) Radish

855. Nitrogenase consists of:
(A) Mo protein (B) Fe-protein
(C) MoFe-protein (D) Both (B) and (C)

856. Major function of CK is to:
(A) Mobilise nutrients & prevent senescence
(B) Mobilise nutrients
(C) Prevent senescence
(D) Produce IAA

857. Growth promoter hormone are:
(A) IAA, ABA and CK (B) IAA, GA and ABA
(C) IAA, GA and CK (D) ABA, CK and GA

858. Glyoxysomes are involved in breakdown of
(A) Sugar (B) Protein
(C) Fatty acids (D) All of these

859. Pyruvic acid is the end product of:
(A) Hydrolysis of starch (B) Citric acid cycle
(C) Photolysis (D) Glycolysis

860. Hormone involved in inhibition of growth is:
(A) ABA (B) Ethylene
(C) None of these (D) Both (A) and (B)

861. Which is correct about C_4 plants?
(A) During photosynthesis, PEP carboxylase enzyme plays a key role in the fixation of CO_2
(B) Malic acidis the first stable product of photosynthesis
(C) Carbohydrates are formed through Hatch-Slack cycle
(D) All of these

862. Cyanobacteria protect nitrogenase from oxygen by:
(A) Nodule formation
(B) Heterocyst formation
(C) Phycocyanin (D) Leghaemoglobin

863. Cytokinins are mostly produced in:
(A) Shoot apex (B) Root apex
(C) Young leaves (D) Lateral buds

864. Nitrate reductase is found in:
(A) Vacuole (B) Mitochondria
(C) Cytoplasm (D) Chloroplasts

865. Cytokinins:
(A) Induce cell division and inhibit ageing
(B) Maintain dormancy
(C) Induced abscission
(D) Inhibit cell division

866. Removal of apical buds makes the plant:
(A) Bushy (B) Grow rapidly
(C) Grow slowly (D) Dormant

867. In germinating cereals, amylase synthesis is stimulated by:
(A) Cytokinins (B) Abscisic acid
(C) Gibberellins (D) Auxin

868. In organs of plants there is a variation in rate of growth. It is first slow, then accelerated upto maximum and slows down to stand still. The period of vigorous growth is known as:
(A) Lag phase (B) Period of growth
(C) Grand growth period (D) Log phase

869. Cyanobacteria protect nitrogenase from oxygen by:
(A) Nodule formation
(B) Heterocyst formation
(C) Phycocyanin (D) Leghaemoglobin

870. Bioassay for auxin is:
(A) Dwarf maize test
(B) Avena curvature test
(C) Cell division test (D) Green leaf test

871. Bioassay for gibberellin is:
(A) Avena curvature test (B) Seed dormancy test
(C) Dwarf maize test (D) Green leaf test

872. Site of oxidative election transport in cells is:
(A) Golgi bodies (B) Mitochondria
(C) Nucleus (D) Chloroplasts

873. Stomata plays a key role in:
(A) Photoperiodism (B) Transpiration
(C) Respiration (D) Vernalisation

874. Internodal elongation is promoted by the action of:
(A) Auxin (B) Gibberellic acid
(C) Cytokinin (D) Ascisic acid

875. Acetylene reduction to ethylene is a measure of:
(A) N_2 fixation (B) Nitrate reduction
(C) Nitrite reduction (D) CO_2 fixation

876. A green plant bends towards the source of light when exposed to the light on only one side. Which of the following is the best explanation of the phenomenon?
(A) The apices of the stems are attracted by light
(B) The plant needs light for photosynthesis
(C) Auxin accumulates on the shaded side to induce great cell elongation on that side
(D) Light reduces length

877. Which is plants bears intermediate features between C_3 and C_4 plants?
(A) Maize (B) *Moricandia arvensis*
(C) Amarathus (D) Both (A) and (C)

878. Nitrogenase consists of:
(A) Mo protein (B) Fe- protein
(C) MoFe – protein (D) Both (B) and (C)

879. Delay in senescence is caused by the spray of:
(A) IBA (B) GA
(C) Cytokinin (D) ABA

880. Cytokinins are generally:
(A) Acids (B) Aminopurines
(C) Phenols (D) Glucosides

881. HMP pathway is also known by:
(A) Pentose phosphate pathway
(B) Direct oxidative pathway
(C) EMP pathway (D) Both (A) and (B)

882. Apples are generally wrapped in waxed paper to:
(A) Prevent sunlight from changing its colour
(B) Prevent aerobic respiration by checking the entry of oxygen
(C) Prevent ethylene formation that hastens ripening
(D) Make the apples look costly

883. Seedless tomato are produced by:
(A) Supplying radioactive compounds
(B) Spraying hormones on flower
(C) Propagating tomato plants from cuttings
(D) Growing tomato plants in mineral solution

884. Which elements is related to cytochrome?
(A) P (B) Cu
(C) Fe (D) Zn

885. Avena coleoptile test is for which of the following hormone:
(A) GA (B) ABA
(C) IAA (D) NAA

886. Hormone involved in breaking dormancy of winter buds is:
(A) Ethylene (B) Gibberellin
(C) Auxin (D) Cytokinin

887. Fe protein of nitrogenase is:
(A) Homomeric dimmer
(B) Heteromeric dimmer
(C) Homomeric tetramer
(D) Heteromeric tetramer

888. Rapid growth of flowering axis in many plants is caused by:
(A) Lower level of absicisic acid
(B) Higher level of auxin
(C) Higher level of gibberellin
(D) Lower level of auxin

889. Which elements is related to cytochrome?
(A) P (B) Cu
(C) Fe (D) Zn

890. In Plants, respiration includes:
(A) Glycolysis (B) Kreb's cycle
(C) Electron transport system
(D) All of these

891. Differentiation of callus requires a specific ratio of two hormones:
(A) Auxin and gibberellin
(B) Auxin and abscisic acid
(C) Gibberellin and abscisic acid
(D) IAA and Cytokinin

892. Which is an example of C_4 plants?
(A) Maize (B) Moricandia arvensis
(C) Amarathus (D) Both (A) and (C)

893. In case of C_4 plants, RuBISCO exists only in:
(A) Mesophyll cells (B) Bundle sheath
(C) Guard cells (D) All of these

894. Ferulic acid, coumarin and hesperidin are:
(A) Secondary plant products
(B) Hormones
(C) Enzymes (D) Vitamins

895. Site of oxidative election transport in cells is:
(A) Golgi bodies (B) Mitochondria
(C) Nucleus (D) Chloroplasts

896. Gibberellins take part in:
(A) Bolting or rosette plants
(B) Replacing long day requirement
(C) Overcoming genetic dwarfism
(D) All the above

897. A decapitated plant exposed to unilateral light:
(A) Bends away source of illumination
(B) Bends towards the source of light
(C) Shows zigzag curvature
(D) Does not show bending movement

898. Correct sequence of different phases of growth:
(A) Cell division → Cell differentiation → Cell elongation
(B) Cell differentiation → Cell division → Cell elongation
(C) Cell elongation → Cell division → Cell differentiation
(D) Cell division → Cell elongation → Cell differentiation

899. Glyoxosomes are involve in breakdown of:
(A) Sugar (B) Protein
(C) Fatty acids (D) All of these

900. Correct sequence of different stages of development is:
(A) Fruiting → Flowering → Juvenility → Germination
(B) Germination → Juvenility → Flowering → Fruiting
(C) Flowering → Fruiting → Juvenility → Germination
(D) Juvenility → Flowering → Fruiting → Germination

901. Exponential growth is shown during:
(A) Lag phase (B) Log phase
(C) Maturation phase (D) Sebesence phase

902. During wall enlargement new wall materials are inserted amongst the older ones. The phenomenon is:
(A) Accretion (B) Integration
(C) Intussusception (D) Interpolation

903. Phosphorylation occurs in:
(A) Mitochondria (B) Glyoxysomes
(C) Chloroplasts (D) Both (A) and (C)

904. Temperature ranges for plant growth is:
(A) 25^0-35^0C (B) 10^0-35^0C
(C) 10^0-50^0C (D) 0^0-20^0C

905. Apical dominance is due to:
(A) C/N ratio (B) Photoperiodism
(C) Phototropism
(D) Auxin cytokinin balance

906. Acetylene reduction to ethylene is a measure of:
(A) N_2 fixation (B) Nitrate reduction
(C) Nitrite reduction (D) CO_2 fixation

907. Which is contains DNA?
(A) Peroxysomes (B) Mitochondria
(C) Centrosomes (D) Prions

908. Pineapple can be made to flower with the help of:
(A) Kinetin (B) Ethylene
(C) GA
(D) Exposure to low temperature

909. How many ATP molecules are produced in anaerobic respiration of one glucose molecule?
(A) 2 (B) 4
(C) 6 (D) 8

910. Growth of a shoot or a root involves:
(A) Cell enlargement (B) Cell division
(C) Cell differentiation (D) All of these

911. Agent Orange of Vietnam war was:
(A) 2,4,5-T (B) 2,4-D
(C) NAA (D) Mixture of (A) & (B)

912. In which plants stomata are closed in day time and open during night time?
(A) C_3 plants (B) C_4 plants
(C) CAM plants (D) All of these

913. During photosynthesis, oxygen is released from:
(A) CO_2 (B) H_2O
(C) SO_2 (D) O_3

914. Hormone involved in the formation of winter buds is:
(A) GA (B) CK
(C) ABA (D) Ethylene

915. Leaf abscission is prevented by:
(A) Gibberellin (B) Ethylene
(C) Abscisic acid (D) Auxin

916. Leaf abscisin is promoted by:
(A) Ethylene (B) Abscisic acid
(C) Cytokinin (D) Tryptophan

917. The main function of leghaemoglobin in root nodule is concerned with:
(A) N_2 supply (B) O_2 supply
(C) Mineral supply (D) CO_2 supply

918. Leaf fall occurs when:
(A) Auxin content is more in lamina than in the stem
(B) Auxin content is low in the lamina as compared to stem
(C) Abscisic acid is formed
(D) Morphactins are produced

919. It will not be any advantage to produce seedless fruits in:
(A) Mangoes (B) Watermelon
(C) Pomegranate (D) Oranges

920. Which is correct about transcription factors?
(A) These factors are required for transcription initiation at all the promoters of eukaryotes
(B) They join RNA polymerase II to form a complex around the startpoint
(C) They determine the site of transcription initiation
(D) All of these

921. If auxin concentration is increased in plant roots, the water absorption will:
(A) Increase (B) Decrease
(C) Remain the same (D) None

922. Pyruvic acid is the end product of:
(A) Hydrolysis of starch (B) Citric acid cycle
(C) Photolysis (D) Glycolysis

923. Cell mitosis is induced by:
(A) Colchicine (B) Gibberellic acid
(C) Nitrosoguanidine (D) Kinetin

924. Nitrate reductase is found in:
(A) Vacuole (B) Mitochondria
(C) Cytoplasm (D) Golgi body

925. Vernalisation is related to:
(A) Cold treatment
(B) Photoperiodism
(C) Development buffering
(D) Male sterility

926. Chrysanthemum flower in winter because:
(A) They are short day plant
(B) They required low temperature treatment
(C) They are long day plants
(D) Both (A) and (B)

927. Gibberellic acid has been successfully employed to induce flowering:
(A) In short day plants under long day conditions.
(B) In long day plants under short day conditions
(C) In all plants
(D) None of the above

928. Which enzymes is involved in conversion of ammonia to amino acid?
(A) Alanine transferase
(B) Ntrate reductase
(C) Glutamine synthetase
(D) Nitrite reductase

929. Plants which daily require shorter period of darkness for flowering are called:
(A) Short day plants (B) Long day plants
(C) Normal day plants (D) Ephemeral plants

930. Which have circular DNA?
(A) Chloroplasts (B) Mitochondria
(C) Bacteria (D) All of these

931. Which is not an example of C_4 plants?
(A) Sugarcane (B) Sorghum
(C) Portulaca (D) Wheat

932. Photoperiodism is associated with the formation of:
(A) Chlorophyll (B) Florigen
(C) Auxin (D) Gibberellin

933. Hormone is responsible for vernalization is:
(A) Florigen (B) Colchicine
(C) Abscisin (D) Vernalin

934. Stomata plays a key role in:
(A) Photoperiodism (B) Transpiration
(C) Respiration (D) Vernalization

935. Letuce seeds germinate only when:
(A) Red light is available
(B) Blue light is available
(C) Green light is available
(D) They are sown deep in the soil

936. Which of the following is required for the formation of flowers in the plants?
(A) Adequate oxygen
(B) Definite photoperiod
(C) Adequate manure
(D) Adequate water and mineral

937. Which is correct about RNA polymerase II?
(A) It is present in nucleoplasm
(B) It constitutes 20-40% of total activity
(C) It transcribes all the genes that produce mRNA
(D) All of these

938. Kinetin like properties are present in:
(A) Pineapple (B) Soybean
(C) Groundnut (D) Coconut milk

939. Low concentration of auxin inhibits growth in:
(A) Stem (B) Roots
(C) Leaves (D) Flowers

940. Apple growing on tree can have larger than normal size by:
(A) Thinning of blossoms (B) Defoliation
(C) Decapitation of stem apex
(D) Applying hormones

941. Which enzyme are responsible for fixation of nitrogen?
(A) Nitrite reductase (B) Nitrate reductase
(C) Carboxylase (D) Nitrogenase

942. An apple tree can be made to bear larger sized fruits by:
(A) Decapitation (B) Defoliation
(C) Dehydration (D) Thinning of blossom

943. Dark reaction is also called?
(A) Blackman reaction (B) Hill reaction
(C) Cyclic reaction (D) Both (A) and (B)

944. Which occurs during light reaction:
(A) Photolysis of water
(B) Photophosphorylation
(C) Fixation of CO_2 (D) Both (A) and (B)

945. Fruit drop occurs when fruit has:
(A) No auxin as in the stem
(B) Less auxin as compared to stem
(C) More auxin than the one present in the stem
(D) Auxin concentration equal to that of stem

946. How many electrons are required for the reduction of one molecules of nitrogen to NH_4-?
(A) 6 (B) 8
(C) 10 (D) 10

947. During photosynthesis, water splits in:
(A) PS I (B) PS II
(C) Cytoplasm (D) Stroma

948. Auxin does not increase:
(A) The plasticity of cell walls
(B) The rate of photosynthesis
(C) The uptake of water by cells
(D) Rate of respiration

949. In plants, respiration includes:
(A) Glycolysis (B) Kreb's cycle
(C) Electron transport system
(D) All of these

950. Induction of rooting in stem cutting would be most beneficial in:
(A) Marchantia (B) Wheat
(C) Cuscuta (D) Bougainvillea

951. Mitochondria and plastids present in eukaryotic cells contain RNA polymerases that are analogous to:
(A) Eukaryotic polymerase (B) Phage polymerases
(C) Bacterial polymerases (D) Both (B) and (C)

952. Weed killers have properties similar to:
(A) Hormones (B) Enzymes
(C) Insecticides (D) Vitamins

953. In Red (P_r) and Far-red (P_{fr}) (given alternatively) responses of plants, it is the:
(A) First treatment that is effective
(B) Middle treatment that is effective
(C) Last treatment that is important
(D) Sum total of all treatments that counts

954. During photosynthesis, dark reaction occurs in:
(A) Stroma (B) Thylakoids
(C) Oxysomes (D) Both (A) and (B)

955. Germination of some seeds is promoted by:
(A) Green light (B) Red light
(C) Far red light (D) Blue light

956. Flowering in some short day plants is promoted by:
(A) Auxins (B) Gibberellins
(C) Cytokinins/ABA (D) Ethylene

957. Mature root nodules mostly consists of:
(A) Diploid (B) Triploid
(C) Tetraploid (D) Pentaploid

958. Flowering in pineapple is promoted by:
(A) GA (B) IAA
(C) Ethylene (D) Both (B) and (C)

959. Reduction of nitrite to ammonium is catalyzed by nitrite reductase in:
(A) Chloroplasts (B) Mitochondria
(C) Nucleus (D) Golgi body

960. Mature root nodules mostly consist of which cells?
(A) Diploid (B) Triploid
(C) Tetraploid (D) Pentaploid

961. A decapitated plant does not bend towards light because:
(A) The root sends a message to shoot apex to do so
(B) There is accumulation of growth hormones at the cut end
(C) The hormone responsible for bending is not available
(D) Growth inhibitors are formed

962. During photosynthesis, light reaction occurs in:
(A) Stroma (B) Thylakoids
(C) Oxysomes (D) Both (A) and (B)

963. Low temperature treatment of some semigerminated seeds for some period often induces subsequently quicker flowering in plants. This process is called:
(A) Vernalization (B) Thermolysis
(C) Wintering (D) Devernalization

964. Which consensus sequences is recognized by more than one transcription factors?
(A) - 35 sequence (B) CAAT box
(C) Octamer sequence (D) GC box

965. Cytokinins have the ability to:
(A) Induce morphogenesis
(B) Substitute cold treatment
(C) Convert dwarf plants to tall plants
(D) None of the above

966. Which contains DNA?
(A) Pereroxysomes (B) Mitochondria
(C) Centrosomes (D) Prions

967. Which one of the following is gaseous growth regulator?
(A) IAA (B) NAA
(C) IBA (D) Ethylene

968. Sprouting of potato can be prevented in storage by:
(A) Maleic hydrazide (B) Gibbrellins
(C) Indole acetic acid (D) Cytokinins

969. A response of length of day in plants is called:
(A) Phototropism (B) Photorespiration
(C) Photoperiodism (D) Photo-oxidation

970. Which enzymes is responsible for fixation of nitrogen?
(A) Nitrite reductase (B) Nitrate reductase
(C) Carboxylase (D) Nitrogenase

971. The substance which accelerates the growth in stem is:
(A) Vitamin (B) Etiolin
(C) Enzyme (D) Auxin

972. Which of the following effects of auxins on plants is the basis for commercial application?
(A) Callus formation (B) Curvature of stem
(C) Induction of root formation in stem cuttings
(D) Stem growth

973. During photosynthesis, water splits in:
(A) PSI (B) PS II
(C) Cytoplasm (D) Stroma

974. Photomorphogenetic pigment in plants is:
(A) Phytochrome system (B) Chlorophyll
(C) Carotene (D) Chalcone pigment

975. The rate of photosynthesis can be measured by:
(A) Amount of carbohydrate accumulated
(B) The volume of CO_2 consumed
(C) The volume of oxygen librated
(D) All of these

976. Artificial ripening of fruits is accomplished by treatment with:
(A) Sodium chloride (B) IAA
(C) Ethylene gas (D) Kinetin

977. Plants bend towards light because:
(A) They need light for photosynthesis
(B) They need light for respiration
(C) Cells on the shaded side elongate more
(D) Lights attracts them

978. Which is usually surrounded by a GC rich sequence?
(A) TATA box (B) Pribnow box
(C) GC box (D) CAAT box

979. Which wavelength is more effective in photoperiodism:
(A) Green and yellow (B) Blue and red
(C) Blue and violet (D) Red and far-red

980. Vernalisation is related to:
(A) Cold treatment (B) Photoperiod
(C) Development buffering (D) Male sterility

981. Abscisic acid (ABA) promotes:
(A) Cell division (B) Leaf fall
(C) Germination (D) Budding

982. Phytotron is a device by which:
(A) Electrons are bombarded
(B) Protons are liberated
(C) Plants are grown in controlled environment
(D) Mutation are produced in plants

983. During photosynthesis, which of the gas is evolved as byproduct?
(A) CO_2 (B) H_2
(C) O_2 (D) Both A and

984. A green plant will turn yellow if placed in the dark. This is called:
(A) Phototropism (B) Dormancy
(C) Chlorosis (D) Etiolation

985. Which enzyme is involved in conversion of ammonia to amino acid?
(A) Alanine transferase
(B) Nitrate reductase
(C) Glutamine synthetase
(D) Nitrite reductase

986. Which of the following stimulates growth or cell enlargement?
(A) Gibberellin (B) Auxin
(C) Cytokinin (D) Ethylene

987. Which is occupies a fixed position?
(A) TATA box (B) Pribnow box
(C) GC box (D) CAAT box

988. Gibberellin was first extracted from:
(A) *Gibberella fujikuroi* (B) *Gelidium*
(C) *Aspergillus* (D) *Gracelaria*

989. Most of the reactions of glyxylate cycle occurs in:
(A) Glyoxysomes (B) Cytosol
(C) Mitochondria (D) All of these

990. Cytokinin:
(A) Is a hormone whose main function is the induction of cell division
(B) Is the process of cell division
(C) Refrs to cell movements
(D) Causes dormancy

991. In autumn leaf fall occurs because:
(A) Formation of abscission layer at the bases
(B) Leaf becomes heavy
(C) Leaf does not remain green
(D) Of low temperature

992. Higher concentration of some synthetic auxins is used for:
(A) Weed control
(B) Rooting of stem cuttings
(C) Parthenocarpy
(D) Increasing fruit yield

993. Translation of carbohydrates mostly takes place in the form of:
(A) Maltose (B) Fructose
(C) Sucrose (D) Glucose

994. Ethylene gas:
(A) Is saturated hydrocarbon
(B) Slows down ripening of apples
(C) Retards ripening of tomatoes
(D) Speeds up maturation of fruits

995. The main feature of light induced promoters of dicots is:
(A) TATA box
(B) An upstream AG or AGGA box
(C) Both TATA and AG/AGGA boxes
(D) CAAT box

996. Photochrome is involved in:
(A) Phototropism (B) Photorespiration
(C) Photoperiodism (D) Geotropism

997. Phosphorylation occurs in:
(A) Mitpchondria
(B) Glyoxysomes
(C) Chloroplasts
(D) Both (A) and (C)

998. Evergreen trees remain green throughout the year on account of:
(A) Absence of leaf fall
(B) Leaves falling in small number at intervals
(C) Supply of moisture throughout the year
(D) Cold climate

999. Hormone involved in phototropism is:
(A) IAA (B) Gibbrellin
(C) Kinetin (D) 2,4-D

1000. Which organelles is involved in ATP production by photophosphorylation?
(A) Mitochondria (B) Chloroplasts
(C) Leucoplasts (D) Chromoplasts

1001. Phytochrome is found in:
(A) Algae (B) Fungi
(C) Vascular cryptogams (D) Flowering plants

1002. In many plants the change over from vegetative to reproductive phase takes place in response to:
(A) The length of the day
(B) The severity of temperature
(C) Mainly the food material available in the soil
(D) The oxygen present in the air

1003. Light reaction is also called?
(A) Blackman reaction (B) Hill reaction
(C) Cyclic reaction (D) Both (A) and (B)

1004. When the dark period of short day plants is interrupted by a brief exposure of light the plant:
(A) Will not flower at all
(B) Flowers immediately
(C) Gives more flowers
(D) Turns into a long day plant

1005. Which is present in stroma?
(A) DNA (B) RNA
(C) Ribosomes (D) All of these

1006. Which is not affected by light:
(A) Photosynthesis (B) Flowering
(C) Fertilization (D) Transpiration

1007. In unisexual plants sex can be changed by application of:
(A) Ethanol (B) Auxin
(C) Cytokinin (D) ABA

1008. The main function of leghaemoglobin in root nodules is concenrned with:
(A) N_2 supply (B) O_2 supply
(C) Mineral supply (D) CO_2 supply

1009. Which is used as a weedicide in higher concentrations?
(A) Indole Acetic Acid
(B) Napthalene Acetic Acid
(C) Indole Butyric Acid (D) 2,4-D

1010. Exogenous GA induces formation of male flowers on genetically female plants of:
(A) Cucurbits (B) Luffa
(C) Cucumis (D) Carica

1011. Citric acid cycle occurs in:
(A) Mitochondria (B) Chloroplasts
(C) Leucoplasts (D) Chromoplasts

1012. For plant tissue culture which among the following is required:
(A) Trypsin (B) Kinetin
(C) Caffeine (D) Counarin

1013. Gibberelins cause:
(A) Elongation of internodes
(B) Curvature of coleoptile
(C) Cell division
(D) Initiation of lateral roots

1014. Which functional sequence/module is not found in eukaryotic promoters?
(A) TATA box (B) Pribnow box
(C) GC box (D) CAAT box

1015. Substances formed at stem tip and controlling growth elsewhere are:
(A) Auxins (B) Vitamins
(C) Enzymes (D) Metabolic intermediates

1016. Ethylene is :
(A) A gaseous metabolite (B) A gaseous enzyme
(C) A gaseous hormone (D) A solid hormone

1017. The enzymes for the oxidation of amino acids, fatty acids and Kreb's cycle are present in:
(A) Lysosomes (B) Stroma
(C) Grana
(D) Mitochondrial matrix

1018. Two climatic factors which affect growth are:
(A) Light and wind
(B) Light and temperature
(C) Rain and temperature
(D) Atmospheric humidity and temperature

1019. Herbicide DCMU kills plants by:
(A) Destroying chlorophyll
(B) Inhibiting PS II
(C) Inhibiting PS I
(D) Inhibiting oxidative phosphorylation

1020. Which one produces more ethylene?
(A) Green apple (B) Green Banana
(C) Ripening Banana (D) Fresh Potato tuber

1021. Which one shows red far-red interconversions?
(A) Carotenoids (B) Cytochromes
(C) Chlorophylls (D) Phytochrome

1022. The electron transport system and oxidative phosphorylation are concerned with:
(A) Mitochondria (B) Chloroplasts
(C) Leucoplasts (D) Chromoplasts

1023. Phytochrome is:
(A) Pigment that controls photomorphogen- esis and development
(B) Regulatory protein that controls dark dependent development processes
(C) Hormone
(D) Photosynthetic hormone

1024. Plant material used in tissue culture is:
(A) *Cycas revolute* (B) *Pinus roxburghii*
(C) *Cocos nucifera* (D) *Borassus flabellifer*

1025. Natural plant hormones were isolated for the first time from:
(A) Human urine and Rice plant
(B) Human urine and Corn germ oil
(C) Rice plants, Spinach leaves and cotton fruits
(D) *Gibberella,* Spinach leaves and *Avena* coleoptile

1026. Glycolysis uses as:
(A) Glucose (B) Sucrose
(C) Lipids (D) Fatty acids

1027. IAA was first isolated from:
(A) Corn germ oil (B) *Gibberella*
(C) Human urine (D) *Rhizopus*

1028. Auxin suppresses growth of:
(A) Lateral axillary buds (B) Apical buds
(C) Roots on stem cuttings (D) Parthenocarpy

1029. Which is often present ins several copies in a single promoter?
(A) TATA box (B) Pribnow box
(C) GC box (D) CAAT box

1030. Which one increases in the absence of light?
(A) Uptake of minerals (B) Uptake of water
(C) Elongation of internodes
(D) Ascent of sap

1031. Bioassay of IAA is:
(A) Avena curvature test (B) Callus test
(C) Leaf disc test (D) α-amylase test

1032. Which is the colourless plastid?
(A) Mitochondria (B) Chloroplasts
(C) Leucoplasts (D) Chromoplasts

1033. The hormone capable of replacing the requirement of long photoperiods for flowering is:
(A) Ethylene (B) Auxin
(C) Gibberellin (D) Cytokinin

1034. Which is false about auxin? It:
(A) Promotes hyper elongation
(B) Promotes cell division
(C) Promotes stem elongation
(D) Inhibits lateral bud growth

1035. Which can replace the requirement of vernalization?
(A) Cytokinin (B) Ethylene
(C) Gibbrellins (D) Auxin

1036. Mowing grass lawn facilities better maintenance because:
(A) Wounding stimulates regeneration
(B) Removal of apical dominance and stimulation of intercalary meristem
(C) Removal of apical dominance
(D) Removal of apical dominance and promotion of lateral meristem

1037. The term carotenoid is related to:
(A) Carotene (B) Chlorophyll
(C) Xanthophyll (D) Both (A) and (C)

1038. Which of the following is used in root formation on stem cuttings?
(A) Kinetin (B) GA_3
(C) ABA (D) IBA

1039. Abscission layer is formed when the concentration of:
(A) Auxin increases
(B) Auxin decreases
(C) Gibberellins decreases
(D) Gibberellins increases

1040. Hormone primarily connected with cell division is:
(A) IAA (B) NAA
(C) Cytokinin/Zeatin (D) Gibberellic acid

1041. Physiologically active form of phytochrome is:
(A) P_{730} (B) P_{660}
(C) P_{700} (D) P_{680}

1042. Which is the major pigment of chloroplasts?
(A) Chlorophyll (B) Xanthophyll
(C) Carotene (D) All of these

1043. Growth plotted against time gives a:
(A) Parabolic curve (B) Sigmoid curve
(C) Upright line (D) Horizontal line

1044. In plants, light is required for:
(A) Photosynthesis
(B) Controlling distribution of hormone
(C) Controlling growth and movement
(D) All the above

1045. Gibbrellins bring about:
(A) Yellowing of leaves
(B) Promotion of rooting
(C) Elongation of genetically dwarf plants
(D) Dwarfing of genetically tall plants

1046. Vernalization is:
(A) Growth curve related to light
(B) Effect of photoperiods on plant growth
(C) Speeding up ability to flower by low temperature treatment
(D) Diurnal photoperiodicity

1047. Phytochrome is involved in:
(A) Seed germination (B) Flowering
(C) Chloroplast orientation (D) All the above

1048. Flowering is influenced by:
(A) Soil water content
(B) Soil acidity
(C) Photoperiod
(D) Content of green pigment

1049. When seedling tip is cutoff, growth and light induced bending stop because it inhibits:
(A) Transpiration
(B) Photorespiration
(C) Respiration
(D) Perception of light stimulus

1050. Leaf fall occurs when content of:
(A) Abscisic acid decreases
(B) Auxin decreases
(C) Auxin increases
(D) Gibberellin decreases

1051. Gibbrellins promotes:
(A) Seed germination (B) Seed dormancy
(C) Leaf fall (D) Root elongation

1052. In case of prokaryotes, net gain of ATPs from complete oxidation of one molecule of glucose in respiration is:
(A) 40 (B) 38
(C) 36 (D) 30

1053. Apical dominance as found in higher plants is due to:
(A) Photoperiodism (B) Phototropism
(C) Hormones (D) Enzymes

1054. IAA precursor is:
(A) Tryptophan (B) Leucine
(C) Tyrosine (D) Phenylalanine

1055. Apical dominance is:
(A) Suppression of growth of apical bud by nearby lateral axillary buds
(B) Stimulation of growth of apical bud by removal of nearby axillary buds
(C) Suppression of growth of lateral axillary buds by removal of apical bud
(D) Suppression of growth of nearby lateral axillary buds by removal of apical bud

1056. Pigment taking part in photo-perception in flowering is:
(A) Carotene (B) Phytochrome
(C) Cytochrome (D) Lycopene

1057. Abscisic acid causes:
(A) Stomatal closure (B) Stem elongation
(C) Leaf expansion (D) Root elongation

1058. Abscisic acid controls:
(A) Cell division
(B) Leaf fall and dormancy
(C) Shoot elongation
(D) Cell elongation and wall formation

1059. Flowering dependent on low temperature exposure is:
(A) Vernalisation (B) Thermotrophy
(C) Cryoscopy (D) Cryostat

1060. Which is employed for artificial ripening of Banana fruits?
(A) Auxin (B) Coumarin
(C) Ethylene (D) Cytokinin

1061. Parthenocarpy can be achieved by:
(A) Zeatin (B) ABA
(C) Auxins (D) Kinetin

1062. Apical dominance is due to:
(A) Photoperiodism (B) Hormones
(C) Enzymes (D) Light demand

1063. Dwarfness can be controlled by treating the plant with:
(A) Cytokinin (B) Gibberellic acid
(C) Auxin (D) Antigibberellin

1064. Highest auxin concentration occurs:
(A) In growing tips
(B) In leaves
(C) At base of plant organs
(D) In xylem and phloem

1065. Cytokinins:
(A) Promote abscission
(B) Influence water movement
(C) Help retain chlorophyll
(D) Inhibit protoplasmic streaming

1066. Gibberellic acid induces flowering in:
(A) Some plants only
(B) Long day plants under short day conditions
(C) Short day plants under long day conditions
(D) Day neutral plants

1067. In short day plants, flowering is induced by:
(A) Photoperiod less than 12 hours
(B) Photoperiod below a critical length and unintruppted long night
(C) Long night
(D) Short photoperiod and interrupted long night

1068. Apical dominance is caused by:
(A) Abscisic acid in lateral bud
(B) Cytokinin in leaf tip
(C) Gibberellin in lateral buds
(D) Auxin in shoot tip

1069. How many cycles of Kreb's cycle is needed for complete oxidation of one glucose molecule?
(A) 1 (B) 3
(C) 2 (D) 4

1070. Which one is the test for gibbrellin?
(A) Bolting in cabbage
(B) Morphogenesis in tobacco callus
(C) Rapid division in carrot cells
(D) Elongation of oat coleoptile

1071. Which of the following hormone can replace vernalization?
(A) Auxin (B) Cytokinin
(C) Gibberellins (D) Ethylene

1072. Cut or excised leaves remain green for long if induced to root or dipped in:
(A) Gibberellins
(B) Cytokinins
(C) Auxins
(D) Ethylene

1073. Bananas can be prevented from over ripining by:
(A) Maintaining them at room temperature
(B) Refrigeration
(C) Dipping in ascorbic acid solution
(D) Storing in a freezer

1074. Which is induced flowering in short day plants?
(A) Auxin (B) Cytokinin
(C) Gibbrellin (D) Propylene

1075. At the time of seed germination, digestive enzymes are induced to be formed under the influence of:
(A) Cytokinin (B) Gibberellins
(C) Ethylene (D) Auxin

1076. Foolish seedling disease/bakane of rice helped in discovery of harmone:
(A) Gibberellin (B) IAA
(C) Ethylene (D) Cytokinin

1077. Which one can induce flowering in long day plants?
(A) IBA (B) IAA
(C) GA_3 (D) NAA

1078. Photoperiodism is probably due to synthesis of:
(A) Cytokinins (B) Gibberellins
(C) Auxin (D) Florigen

1079. Gibberellin application to a plant brings about :
(A) Leaf fall
(B) Diversification of root system
(C) Elongation of shoot system
(D) Delay in senescence

1080. In root, the region of fastest growth is:
(A) Root hair zone
(B) Behind the root tip
(C) Root tip
(D) Before the root tip

1081. What is a stress hormone? or The hormone produced during adverse environmental conditions is:
(A) Benzyl aminopurine
(B) Dichlorophenoxy acetic acid
(C) Ethylene (D) Abscisic acid

1082. Which organelles mainly synthesize amino acids in the cells?
(A) Mitochondria (B) Ribosomes
(C) Chloroplasts (D) Both (A) and (B)

1083. The regulator which retards ageing of plant parts is:
(A) Cytokinin (B) Auxin
(C) Gibberellin (D) Abscisic acid

1084. Richmond-Land effect is due to:
(A) Gibberellins (B) Auxin
(C) Ethylene (D) Kinetin

1085. Stomatal closure under stress conditions is due to:
(A) IAA (B) ABA
(C) IBA (D) 2,4-D

1086. Phytochrome is associated with:
(A) P_r and P_{fr} reactions
(B) Absorption of red light of 660 nm
(C) Absorption of far-red light of 740 nm
(D) Absorption of blue light by leaves

1087. In tissue culture, differentiation of shoot and root is controlled by:
(A) Light intensity
(B) Temperature shocks
(C) Modifying auxin and cytokinin ratios
(D) Using tissue of right size

1088. Removal of apical buds results in:
(A) Formation of new apical buds
(B) Elongation of main stem
(C) Death of plant
(D) Formation of lateral branching

1089. Maximum of growth occurs in
(A) Exponential phase (B) Decline phase
(C) Stationary phase (D) Lag phase

1090. Auxin synthesis occurs in :
(A) Root/shoot tips (B) Cortex
(C) Xylem (D) Phloem cells

1091. Phytotron is useful in:
(A) Protecting plants from excessive light
(B) Controlled plant growth
(C) Growth of drug plants
(D) Protection of plants

1092. Which one can absorb red and far red light?
(A) Carotene (B) Xanthophyll
(C) Chlorophyll (D) Phytochrome

1093. Which is correct about plastoquinones?
(A) They are special quinines present in plastids
(B) They carry two electrons from PS-II towards PS-I
(C) They transport H^+ from stroma into th thylakoid lumen
(D) All of these

1094. Movement of auxin is:
(A) Centripetal (B) Basipetal
(C) Acropetal (D) Both (B) and (C)

1095. Growth is maximum in zone of:
(A) Cell elongation (B) Cell division
(C) Cell maturation (D) All the above

1096. Which does not take place in short day plants?
(A) An interrupted critical dark period
(B) Critical period is interrupted by light
(C) Dark period is interrupted by red light followed by far-red light
(D) Critical period is not interrupted by white or red light

1097. Which is not a phytohormone:
(A) Phytochrome (B) Florigen
(C) GA (D) IAA

1098. Florigen is produced in the region of:
(A) Leaves (B) Fruit
(C) Root (D) Trunk

1099. Genetically dwarf plant can be made to grow taller by application of:
(A) Ethylene (B) Gibberellin
(C) Auxin (D) Cytokinin

1100. Abscisic acid causes:
(A) Seed dormancy (B) Apical dominance
(C) Root initiation (D) Phototropism

1101. Induction in cell division and delay in senescence are due to:
(A) Gibberellin (B) Auxin
(C) Cytokinin (D) Ethylene

1102. 2, 4-D causes:
(A) Delay in senescence
(B) Bolting
(C) Increase in branches
(D) Increased growth of all parts

1103. Lateral axillary buds are not allowed to grow by:
(A) Auxin (B) Ethylene
(C) Gibberellin (D) Cytokinin

1104. Which blue green algae does not form heterocysts?
(A) *Anabaena* (B) *Nostac*
(C) *Gleotrichia* (D) *Scytonema*

1105. Which is not a natural plant hormone:
(A) GA_3 (B) GA_2
(C) IAA (D) 2,4-D

1106. Cytokinins are mostly:
(A) Glucosides (B) Aminopurines
(C) Acidic (D) Phenolic

1107. Kinetin is:
(A) Indole butyric acid
(B) Indole acetic acid
(C) Butyric acid
(D) 6-furfuryl amino-purine

1108. One of the following is a growth retardant:
(A) Morphactin (B) Zeatin
(C) Adenine (D) Ascorbic acid

1109. Auxin takes part in:
(A) Promotion of cell division
(B) Promotion of cell enlargement
(C) Promotion of fruit ripening
(D) Inhibition of shoot growth

1110. Interfascicular cambium formation is induced by:
(A) Auxin (B) Cytokinin
(C) Gibberellin (D) Ethylene

1111. Hormone that has negative effect on apical dominance is:
(A) Cytokinin (B) Gibberellin
(C) Auxin (D) Both (B) and (C)

1112. Seasonality of plants is due to:
(A) Phototropism (B) Photosynthesis
(C) Photoperiodism (D) Photolysis

1113. Chemicals having profound effect on growth and development are:
(A) Enzymes (B) Phytohormones
(C) Catalytic agents (D) Manure

1114. Food products are kept in cold storage or refrigerator in order to:
(A) Make them tasty
(B) Use in off-season
(C) Maintenance of fressness, longevity and taste due to low respiration
(D) Keep them cool

1115. Which is not correct about starch?
(A) It is stored in photosynthetically inactive organs like roots, modified shoots and seeds
(B) The constituents of starch grains are amylase and amylopectin
(C) Both amylose and amylopectin are made up of D-glucose units
(D) Starch molecules have a definite molecular weight and are assembled into starch grains

1116. Food can be preserved by:
(A) Low temperature (B) High temperature
(C) Osmotic pressure (D) All the above

1117. Which one prevents photo-oxidation and pigment destruction?
(A) Phytochrome (B) Phytohormone
(C) Phycocyanin (D) Phyco-erythrin

1118. If a tree flowers thrice in a year in October, January and July in northern India, The plant is:
(A) Photo- and thermo sensitive
(B) Photo-and thermo insensitive
(C) Photo-sensitive and thermo-insensitive
(D) Thermo-sensitive but photo-insensitive

1119. Gibberellins induce:
(A) Cell division
(B) Leaf senescence
(C) Hydrolysing enzymes in germinating seeds
(D) Flowering

1120. Which is not caused by deficiency of mineral nutrition?
(A) Chlorosis
(B) Internode shortening
(C) Necrosis (D) Etiolation

1121. The hormone which promotes flowering in long day conditions and controls sex expression is:
(A) Auxin (B) Cytokinins
(C) Gibberellins (D) Ethylene

1122. In a plant subjected to continuous red light, phytochrome will show:
(A) Increased synthesis
(B) Decreased level
(C) Destruction
(D) Destruction and synthesis remain in balance

1123. Properties of starch molecules are mainly determined by:
(A) Degree of branching (including branch length)
(B) Types of bonds linking the glucose units
(C) Chain length
(D) All of these

1124. Development of shoot and root is determined by:
(A) Cytokinin and auxin ratio
(B) Enzymes
(C) Temporature
(D) Plant nutrients

1125. Phytochrome absorbs visible light maximum in red region at:
(A) 600 nm (B) 650 nm
(C) 660 nm (D) 730 nm

1126. Hormone discovered through tissue culture technique is:
(A) Auxin (B) Cytokinin
(C) Gibberellin (D) Abscisic acid

1127. Sometimes potato sprouts during storage due to formation of:
(A) Ethylene (B) Gibberellin
(C) Cytoinin (D) Auxin

1128. Flash of light in dark inhibits flowering in:
(A) SDP (B) LDP
(C) LSDP (D) DNP

1129. Which one controls flowering in long day plants and sex expression in certain species?
(A) Antiauxins (B) Auxins
(B) Cytokinins (D) Gibberellins

1130. Agent orange is:
(A) Weedicide with dioxin
(B) Chemical used in luminous paint
(C) Biodegradable insecticide
(D) Colour used in fluorescent lamp

1131. Food is preserved at low temperature because:
(A) Bacterial attack is minimized
(B) For easy cooking
(C) For easy digestion
(D) All the above

1132. Photosynthetic rate is the highest in:
(A) C_3 plants (B) C_4 plants
(C) CAM (D) Both (B) and (C)

1133. Which one is used for defoliation of forest trees?
(A) Phosphon D (B) Maleic hydrazide
(C) AMO 1618 (D) 2,4-D

1134. Which combination is used in fruit ripening?
(A) 80% CH_4 + 20% O_2
(B) 80% C_2H_4 + 20% CO_2
(C) 8% CO_2 + 20% CH_2
(D) 80% CH_4 + 20% CO_2

1135. Phytochrome is sensitive to:
(A) Blue light (B) Red light
(C) Red and far-red light (D) Green light

1136. Hormone produced during leaf fall is:
(A) Cytokinin (B) Florigen
(C) ABA (D) IAA

1137. Fruit drop is prevented by spraying:
(A) Auxin (B) Ethylene
(C) Gibberellins (D) Cytokinins

1138. Bakane disease of rice is due to:
(A) NAA (B) 2,4-D
(C) IAA (D) GA

1139. Which of the following ion is pulled out in apoplast by auxin during growth?
(A) K^+ (B) H^+
(C) Na^+ (D) Mg^{2+}

1140. Substances that accumulate around a site of infection in plants is:
(A) Phytolexins (B) Proteinome
(C) Saponins (D) All the above

1141. Plant growth is:
(A) Limited (B) Unlimited
(C) Unlocalised (D) Diffused

1142. Photoperiodic stimulus is perceived by:
(A) Leaves (B) Flowers
(C) Buds (D) Meristem

1143. Kreb's cycle starts with entry of:
(A) Acetyl-CoA (B) Oxaloacetate
(C) Both (A) and (B) (D) Pyruviic acid

1144. Parthenocarpic fruits can be produced by application of auxin:
(A) IAA (B) IBA
(C) NAA (D) All the above

1145. Hormone found in liquid endosperm of coconut is:
(A) Gibberellin (B) Auxin
(C) Ethylene (D) Cytokinin

1146. From which was kinetin isolated:
(A) Coconut milk (B) Pineapple
(C) Soyabean (D) Groundnut

1147. Cardinal temperature is related to:
(A) Minimum temperature
(B) Optimum temperature
(C) Maximum temperature
(D) All the above

1148. High concentration of auxin is present in:
(A) Root apex (B) Shoot apex
(C) Petiole (D) Node

1149. Hormone responsible for plant and seed dormancy during drought is:
(A) IBA (B) NAA
(C) ABA (D) Zeatin

1150. Growth regulator ethylene is used for:
(A) Retarding ripening of tomatoes
(B) Speeding up ripening of fruits
(C) Slowing down ripening of apples
(D) Both (B) and (C)

1151. In short day plants, flowering is inhibited by:
(A) Interruption of dark by white & red light
(B) Dark interruption by far red light
(C) Dark interruption by red light followed by far red light
(D) Not possible

1152. First chemical identification of auxin, indole 3-acetic acid, was carried out by:
(A) Went (B) Thimann
(C) Darwin (D) Paal

1153. Plant hormone causing abscission of leaves, senescence and inhibition of cell division is:
(A) IAA (B) Ethylene
(C) Cytokinins (D) ABA

1154. During photosynthesis, oxygen is released from:
(A) CO_2 (B) H_2O
(C) SO_2 (D) O_3

1155. Reduction of nitrite to ammonium is catalyzed by nitrite reductase in:
(A) Chloroplasts (B) Mitochondria
(C) Nucleus (D) Golgi body

1156. In which electron transport leading to phosphorylation is a common reaction?
(A) Chloroplasts (B) Mitochondria
(C) Both (A) and (B) (D) Golgi bodies

1157. The alcohol dehydrogenase gene is induced by stress due to:
(A) Heat
(B) Anaerobiosis
(C) Wound
(D) Pathogene invasion

1158. Which is correct?
(A) Flower senescence is promoted by ethylene
(B) Ethylene biosynthesis involves two enzymes, viz, ACC synthase and ACC oxidase
(C) ACC synthase enzyme is encoded by acs gene and ACC oxidase is encoded by aco gene
(D) All of these

1159. The metabolic activities of roots cells are the main cause of water absorption in:
(A) Active absorption
(B) Passive absorption
(C) Both passive and active absorption
(D) None of the above

1160. The movement of water is along:
(A) Turgor gradient (B) DPD gradient
(C) Diffusion gradient (D) Osmotic gradient

1161. The energy is required in absorption of water in:
(A) Active absorption
(B) Passive absorption
(C) Both passive and active absorption
(D) None of the above

1162. The most dominant role, in the absorption of water, is that of:
(A) Active absorption
(B) Passive absorption
(C) Both passive and active absorption
(D) None of the above

1163. No energy is required and roots play only the role of absorbing organs in:
(A) Active absorption
(B) Passive absorption
(C) Both passive and active absorption
(D) None of the above

1164. Which is influenced by opening and closing of stomata:
(A) Active water absorption
(B) Passive water absorption
(C) Both types of water absorption
(D) Rate of growth

1165. Root pressure is maximum at the time of:
(A) Water absorption as well as transpiration are low
(B) Both water absorption and transpiration are high
(C) Absorption is low and transpiration is high
(D) Absorption is high and transpiration is low

1166. The glyoxylate cycle is found in plants and bacteria but not in animals. The lack of this cycle in animals results in the inability to:
(A) Synthesize oxaloacetate from isocitrate
(B) Synthesize glutamate from malate
(C) Perform gluconeogenesis from amino acids
(D) Perform gluconeogenesis from fatty acids

1167. Which of the following statements about ribulose 1,5-bisphosphate carboxylase (Rubisco) is incorrect?
(A) It is present at low concentrations in the chlolroplast
(B) It is activated by the addition of CO_2 to the ε–amino group of a specific lysine to form a carba-mate that then binds a divalent metal action
(C) It catalyzes, as one part of its reaction sequence, an extremely exergonic reaction, the cleavage of a six–carbon diol derivate of carabinitol to form 2, 3-carbon compounds
(D) It catalyzes a reaction between ribulose 1, 5-bisphosphate and CO_2 that decreases the efficiency of photosynthesis

1168. The absorption of water is not affected when:
(A) Soil is water logged (poorly aerated)
(B) Soil temperature increases
(C) Soil solution is highly concentrated
(D) Soil water is available in maximum

1169. Guttation is loss of water:
(A) From root (B) Through stomata
(C) In liquid form (D) In vapour form

1170. For absorption of water, the root hairs act as:
(A) Sucking organs (B) Osmotic system
(C) Manometer (D) Anemometer

1171. Guttational drop comprises:
(A) Simple water
(B) Water and various salts dissolved in it
(C) Water and inorganic salts
(D) Water and organic salts

1172. A leafy shoot is enclosed air tightly in a flask. The flask becomes moist on the inner surface due to:
(A) Guttation (B) Evaporation
(C) Transpiration (D) Cooling effect

1173. Optimum temperature for water absorption is:
(A) 0^0-5^0C (B) 5^0-20^0C
(C) 20^0-35^0C (D) 40^0-50^0C

1174. Process of transpiration in plants helps in:
(A) Ascent of sap (B) Absorption of CO_2
(C) Opening of stomata (D) None of these

1175. At 0ºC soil temperature, the rate of water absorption:
(A) Increases (B) Not affected
(C) Decreases (D) Appreciable

1176. As absorbed water passes towards vascular cylinder, it must enter the cytoplasm of:
(A) Pericycle cells (B) Endodermal cells
(C) Cortical cells (D) Xylem parenchyma

1177. Which one of the following statement correctly describes the role of vernalisation?
(A) It reduces apical dominance and leads multiple branching
(B) It causes genetic modification that leads to early flowering
(C) It delays flowering and allows maximum vegetative growth leading to higher yield
(D) It causes epigenetic modification that pro-motes early flowering

1178. A living continuum of cells connected by plasmod-esmata is:
(A) Dermal tissue (B) Ground complex
(C) Donnan free space (D) Symplast

1179. Q_{10} is:
(A) Temperature Quotient
(B) Respiratory quotient
(C) Compensation point
(D) Time factor

1180. Water is absorbed by root hair when the external medium is:
(A) Hypotonic (B) Hypertonic
(C) Isotonic (D) Viscous

1181. Transpiration is least in:
(A) Good soil moisture
(B) High wind velocity
(C) Dry environment
(D) High atmospheric humidity

1182. In light phase of photosynthesis there is formation of:
(A) ATP
(B) $NADPH_2$
(C) Both ATP and $NADPH_2$
(D) Carbohydrates

1183. Absorption of water by root is increased with the:
(A) Increase in transpiration
(B) Increase in rate of photosynthesis
(C) Decrease in transpiration
(D) Decrease in salt uptake

1184. Cryptochromes:
(A) A class of blue light photoreceptors
(B) Occurs in both plants and animals
(C) Belongs to a family of flavoproteins
(D) All of these

1185. Contribution of passive water absorption to total water absorption is:
(A) 50% (B) 70%
(C) 80-90% (D) 96-100%

1186. Rate of water absorption generally follows closely the rate of:
(A) Transpiration (B) Photosynthesis
(C) Respiration (D) Growth

1187. Presently which view is considered best for turgor changes in guard cells:
(A) Photosynthesis of chloroplasts in guard cells
(B) Starch is converted into sugar in guard cells
(C) Starch is converted into glucose in guard cells
(D) Potassium is actively transported into guard cells

1188. Transpiration is reduced due to deposition of:
(A) Cutin (B) Pectin
(C) Mucilage (D) Cellulose

1189. Emersion effect indicates the existence of:
(A) Two pigment systems
(B) Two photosynthetic units
(C) Two photophosphorylations
(D) None of the above

1190. The phenomenon related to active water absorption is:
(A) Transpiration (B) Root pressure
(C) Osmotic pressure (D) Translocation

1191. In succulent plants the stomata open in night and close by day. Which among following would be best hypothesis to explain the mechanism of stomatal action in night only:
(A) CO_2 accumulates, reduces pH, stimulates enzy-mes, resulting in accumulation of sugars.
(B) Increase in CO_2 concentration, conversion of orga-nic acids into starch, resulting in the increased conversion into sugars, resulting in K^+ transport
(C) Low CO_2 concentration due to utilization, accumulation of organic acids resulting in the increased concentration of cell sap
(D) CO_2 used up, increases pH, results in accumulation of sugars

1192. Root pressure can be demonstrated by means of:
(A) Wilting (B) Guttation
(C) Transpiration (D) Exudation/bleeding

1193. The stomatal type of cereals which open only for a few hours during the day is:
(A) Barley type (B) Potato type
(C) Alfalfa type (D) Bean type

1194. Transpiring structure in plants is:
(A) Epidermis (B) Cortex
(C) Xylem (D) Phloem

1195. Dry weight of leaf is maximum during:
(A) Morning (B) Noon
(C) Afternoon (D) Night

1196. A soil is physiologically dry if it contains:
(A) No hygroscopic water
(B) No bound water
(C) No gravitational water
(D) High concentration of salt solution

1197. A water-logged soil is physiologically dry because of:
(A) Anaerobic conditions
(B) Nonmovement of water capillaries
(C) Increased viscosity of water
(D) Abundance of salts

1198. At field capacity the soil contains:
(A) Capillary and gravitational water
(B) Capillary and runaway water
(C) Capillary and hygroscopic water
(D) Capillary, hygroscopic and bound water

1199. Transpiration is high in:
(A) Rainy season/high humidity
(B) Winter season
(C) High temperature
(D) Low wind velocity

1200. Photosynthesis proceeds in sequence of:
(A) Dark phase and light phase
(B) Light phase alone
(C) Light phase and dark phase
(D) Dark phase alone

1201. The phenomenon which forces water upwards into tracheary elements of xylem in the root region is:
(A) Transpiration pull (B) Root pressure
(C) Turgor pressure (D) Imbibition pressure

1202. Root hairs absorb water when:
(A) They respire rapidly
(B) Soil solution is isotonic
(C) Salt concentration of cell sap is high
(D) Salt concentration of soil is high

1203. Transplanted seedlings often get killed because of:
(A) Damage to root hairs
(B) Damage to leaves
(C) Change in soil microorganisms
(D) Nonavailability of required minerals

1204. For process of photosynthesis all except one of the following items are essential point out the exception:
(A) Water, minerals
(B) Light, chlorophyll
(C) CO_2, optimum temperature
(D) O_2, sucrose

1205. Rate of transpiration is inversely related to:
(A) Humidity (B) Light
(C) Temperature (D) Water

1206. Which is produced during water stress that brings about stomatal closure:
(A) Ethylene (B) Abscisic acid
(C) Ferulic acid (D) Coumarin

1207. Wilting appears due to excessive:
(A) Respiration (B) Photosynthesis
(C) Absorption (D) Transpiration

1208. Stomata open and close due to:
(A) Circadian rhythm
(B) Genetic clock
(C) Pressure of gases inside the leaves
(D) Turgor pressure of guard cells

1209. Which colour of light gives maximum absorption peak of chlorophyll a:
(A) Red (B) Blue
(C) Orange (D) Yellow

1210. PS II performs:
(A) Reduction of CO_2
(B) Photolysis of water
(C) Liberation of energy
(D) Formation of water

1211. Stomata open when guard cells show:
(A) Increase in both osmotic and turgor pressure
(B) Decrease in both osmotic pressure and turgor pressure
(C) Increase in osmotic pressure but decrease in turgor pressure
(D) Decrease in osmotic pressure but increase in turgor pressure

1212. In guttation plants can:
(A) Excrete salts
(B) Compensate for loss of water
(C) Manufacture organic substances
(D) Get rid of excess water

1213. Light phase consists of:
(A) Two photochemical reactions
(B) Two photosynthetic units
(C) Two photorespiration reaction
(D) None of these

1214. An important factor in stomatal opening is:
(A) Hormone content of cells
(B) Protein content of cells
(C) Chlorophyll content of cells
(D) Shape of guard cells

1215. Which of the following shows guttation?
(A) *Pisum sativum* (B) *Ficus benghalensis*
(C) *Acacia nilotica* (D) *Tropaeolum*

1216. Scotoactive movement of stomata is that:
(A) Stomata open at night
(B) Stomata open during day
(C) Stomata close at night
(D) Stomata open both during day and night

1217. Which of the following does not affect light phase:
(A) Chlorophyll (B) Water
(C) Cytochrome (D) Temperature

1218. The most effective light for stomatal opening is:
(A) Yellow (B) Green
(C) Red (D) Blue

1219. The carbon di acceptor in CAM plants is:
(A) Malic acid (B) Oxalo-acetic acid
(C) Pyruvic acid
(D) Phosphoenol pyruvic acid

1220. During photosynthesis the oxygen in glucose comes from:
(A) H_2O (B) CO_2
(C) Both from H_2O and CO_2 (D) Oxygen in air

1221. Kranz anatomy occurs in:
(A) Leaves (B) Stem
(C) Flower (D) Seed

1222. Guttation is mainly due to:
(A) Root pressure (B) Osmosis
(C) Transpiration (D) Imbibition

1223. A specialised multicellular structure in leaves which excretes water droplets is:
(A) Lenticel (B) Stomata
(C) Hydathode (D) Bordered pit

1224. Who discovered purification of air in light by green plants:
(A) Ingenhouse (B) Priestley
(C) De Saussure (D) Mayer

1225. Phenyl mercuric acetate (PMA) results in:
(A) Reduced photosynthesis
(B) Reduced transpiration
(C) Reduced respiration
(D) Killing of plants

1226. Red drop occurs in wavelength of:
(A) 492 nm (B) 535 nm
(C) 586 nm (D) 680 nm

1227. An internal factor in transpiration is:
(A) CO_2 (B) O_2
(C) N_2 (D) Stomata

1228. The two pigment system theory of photosynthesis was proposed by:
(A) Hill (B) Blackman
(C) Emerson (D) Arnon

1229. How much energy is utilized in the synthesis of one molecule of glucose:
(A) 673 kcal (B) 686 kcal
(C) 666 kcal (D) 696 kcal

1230. The energy change in photosynthesis is from:
(A) Light energy to electric energy
(B) Light energy to molecular energy
(C) Light energy to chemical energy
(D) Light energy to activation energy

1231. Action of spectrum of photosynthesis was first studied by:
(A) Blackmann (B) Van Mayer
(C) Engelmann (D) Boussingault

1232. Plants with high light compensation point are:
(A) C_3 plants (B) Sun plants
(C) Shade plants (D) Mesophytes

1233. C_4 plants are adapted to:
(A) Temperate humid climate
(B) Tropical humid climate
(C) Temperate dry climate
(D) Tropical dry climate

1234. Cyclic photophosphorylation produces:
(A) NADPH (B) ATP and NADPH
(C) ATP, NADPH and O_2 (D) ATP only

1235. Photosynthate is translocated through:
(A) Sieve tubes/Phloem cells (B) Trachids
(C) Xylem vessels (D) Latex ducts

1236. In C_4 plants, synthesis of sugars occurs in:
(A) Mesophyll cells
(B) Bundle sheath cells
(C) Spongy parenchyma cells
(D) Palisade parenchyma cells

1237. A photosynthesizing plant is releasing ^{18}O more than the normal. The plant must have been supplied with:
(A) O_2 (B) H_2O with ^{18}O
(C) CO_2 with ^{15}O (D) $C_6H_{12}O_6$ with ^{15}O

1238. Chemistry of photosynthetic pigments was first studied by:
(A) Willstatter & Stoll (B) Pelletier & Caventou
(C) Priestley (D) Engelmann

1239. Which one is a C_4 plant?
(A) Papaya (B) Pea
(C) Potato (D) Maize

1240. During day time the green plants show an ratio of CO_2 consumption to CO_2 production of:
(A) 1 : 1 (B) 5 : 1
(C) 10 : 1 (D) 20 : 1

1241. In which wave length of light photosynthesis is maximum:
(A) Red light (B) Green light
(C) Blue light (D) Ultra-violet light

1242. In C_4 plants Calvin cycle enzymes are:
(A) Mesophyll chloroplasts
(B) Bundle sheath chloroplasts
(C) Guard cell chloroplasts
(D) Epidermal chloroplasts

1243. Which one of the following pigments does not occur in chloroplast:
(A) Chlorophyll (B) Carotene
(C) Anthocyanin (D) Xanthophyll

1244. Chlorophyll absorbs:
(A) Green light
(B) Red light only
(C) Blue light only
(D) Blue as well as red light

1245. The number of chlorophyll molecules in a quantasome is:
(A) 50-100 (B) 200-250
(C) 300-400 (D) 500-600

1246. The law of limiting factors for photosynthesis was enunciated by:
(A) R-Hill (B) Calvin
(C) Krebs (D) Blackman

1247. Carbon cycle occurs in:
(A) Chloroplasts (B) Cytoplasm
(C) Mitochondria (D) Glyoxysomes

1248. In bacterial photosynthesis, the hydrogen donor is:
(A) H_2O (B) H_2SO_4
(C) NH_2 (D) H_2S

1249. In photosynthesis:
(A) ATP is formed in light reaction and glucose in dark reaction
(B) Both ATP and glucose are produced in dark reaction
(C) Both ATP and glucose are produced in light reaction
(D) Both ATP and glucose are formed in light and dark

1250. C_4 cycle was discovered by:
(A) Hatch and Slack (B) Calvin
(C) Hill (D) Arnone

1251. During photosynthesis when PGA is changed into phosphogylceraldehyde, which of the following reactions occurs?
(A) Oxidation (B) Reduction
(C) Electrolysis (D) Hydrolysis

1252. During day-light hours the rate of photosynthesis is higher than that is respiration and the ratio of O_2 produced to that of consumed is:
(A) 1: 1 (B) 10 : 1
(C) 50 : 1 (D) 5 : 1

1253. Which one is most efficient converter of sunlight?
(A) Sugarcane (B) Rice
(C) Wheat (D) Papaya

1254. Conversion of ATP to ADP produces:
(A) Electricity (B) Energy
(C) Hormone (D) Enzyme

1255. Hydrilla is used for demonstrating photosynthesis because it shows:
(A) Little respiration
(B) Little transpiration
(C) Rapid photosynthesis
(D) Evolution of O_2 bubbles which can be collected once water

1256. Dark reaction of photosynthesis occurs in:
(A) Grana (B) Stroma
(C) Matrix (D) Cytoplasm

1257. The loss of which will harm the tree most:
(A) Bark (B) Half the leaves
(C) Half of branches (D) All the leaves

1258. Mass flow hypothesis explains:
(A) Transpiration (B) Translation
(C) Transcription (D) Translocation

1259. Synthesis of ADP + iP → ATP in grana/ photosynthesis is:
(A) Phosphorylation
(B) Photophosphorylation
(C) Oxidative phosphorylation
(D) Photolysis

1260. The number of light quanta required for evolution of one molecule of oxygen is called:
(A) Oxygen yield (B) Photosynthetic yield
(C) Quantum yield (D) Organic yield

1261. The photosynthetic unit having 250 chlorophyll molecules is called:
(A) Photon (B) Quantum
(C) Quantasome (D) Oxysomes

ANSWERSHEET

4. PLANT PHYSIOLOGY AND MOLECULAR BIOLOGY

1. (D)	2. (B)	3. (A)	4. (C)	5. (C)	6. (B)	7. (B)	8. (C)	9. (C)	10. (D)
11. (C)	12. (D)	13. (B)	14. (C)	15. (C)	16. (D)	17. (C)	18. (B)	19. (A)	20. (D)
21. (B)	22. (C)	23. (C)	24. (D)	25. (B)	26. (B)	27. (C)	28. (D)	29. (B)	30. (A)
31. (C)	32. (B)	33. (C)	34. (A)	35. (A)	36. (A)	37. (D)	38. (B)	39. (A)	40. (C)
41. (D)	42. (B)	43. (C)	44. (C)	45. (B)	46. (D)	47. (A)	48. (C)	49. (B)	50. (D)
51. (C)	52. (B)	53. (C)	54. (D)	55. (A)	56. (D)	57. (D)	58. (D)	59. (C)	60. (B)
61. (B)	62. (B)	63. (C)	64. (D)	65. (B)	66. (A)	67. (A)	68. (D)	69. (B)	70. (A)
71. (A)	72. (D)	73. (C)	74. (B)	75. (B)	76. (D)	77. (A)	78. (A)	79. (B)	80. (C)
81. (B)	82. (A)	83. (C)	84. (A)	85. (B)	86. (D)	87. (D)	88. (A)	89. (D)	90. (C)
91. (D)	92. (B)	93. (A)	94. (D)	95. (C)	96. (C)	97. (A)	98. (D)	99. (A)	100. (D)
101. (C)	102. (B)	103. (B)	104. (C)	105. (B)	106. (B)	107. (D)	108. (C)	109. (C)	110. (A)
111. (D)	112. (C)	113. (D)	114. (D)	115. (D)	116. (A)	117. (B)	118. (C)	119. (A)	120. (A)
121. (B)	122. (C)	123. (A)	124. (B)	125. (D)	126. (D)	127. (B)	128. (C)	129. (A)	130. (B)
131. (B)	132. (B)	133. (A)	134. (C)	135. (C)	136. (D)	137. (B)	138. (A)	139. (C)	140. (A)
141. (B)	142. (D)	143. (B)	144. (C)	145. (C)	146. (A)	147. (A)	148. (B)	149. (C)	150. (D)

151. (A)	152. (B)	153. (A)	154. (C)	155. (A)	156. (B)	157. (D)	158. (C)	159. (C)	160. (C)
161. (A)	162. (D)	163. (A)	164. (B)	165. (D)	166. (A)	167. (A)	168. (A)	169. (C)	170. (B)
171. (B)	172. (C)	173. (B)	174. (D)	175. (B)	176. (A)	177. (C)	178. (B)	179. (D)	180. (C)
181. (C)	182. (D)	183. (C)	184. (A)	185. (D)	186. (A)	187. (C)	188. (C)	189. (D)	190. (B)
191. (B)	192. (D)	193. (C)	194. (B)	195. (A)	196. (D)	197. (B)	198. (A)	199. (B)	200. (C)
201. (A)	202. (C)	203. (C)	204. (A)	205. (D)	206. (D)	207. (B)	208. (A)	209. (B)	210. (C)
211. (B)	212. (C)	213. (C)	214. (D)	215. (A)	216. (D)	217. (C)	218. (C)	219. (B)	220. (B)
221. (A)	222. (D)	223. (A)	224. (C)	225. (C)	226. (D)	227. (A)	228. (C)	229. (B)	230. (A)
231. (D)	232. (B)	233. (C)	234. (B)	235. (B)	236. (C)	237. (B)	238. (B)	239. (D)	240. (C)
241. (C)	242. (B)	243. (D)	244. (A)	245. (A)	246. (D)	247. (B)	248. (B)	249. (B)	250. (D)
251. (A)	252. (D)	253. (C)	254. (D)	255. (C)	256. (C)	257. (D)	258. (B)	259. (C)	260. (D)
261. (B)	262. (C)	263. (B)	264. (D)	265. (C)	266. (A)	267. (B)	268. (B)	269. (B)	270. (C)
271. (A)	272. (A)	273. (B)	274. (C)	275. (A)	276. (D)	277. (B)	278. (B)	279. (B)	280. (C)
281. (D)	282. (B)	283. (C)	284. (B)	285. (B)	286. (D)	287. (D)	288. (D)	289. (C)	290. (D)
291. (C)	292. (C)	293. (B)	294. (D)	295. (D)	296. (C)	297. (A)	298. (C)	299. (D)	300. (B)
301. (B)	302. (D)	303. (D)	304. (A)	305. (B)	306. (C)	307. (B)	308. (A)	309. (B)	310. (D)
311. (A)	312. (A)	313. (B)	314. (B)	315. (B)	316. (C)	317. (C)	318. (D)	319. (A)	320. (B)
321. (B)	322. (A)	323. (C)	324. (D)	325. (C)	326. (C)	327. (B)	328. (D)	329. (D)	330. (C)
331. (B)	332. (B)	333. (C)	334. (B)	335. (C)	336. (D)	337. (A)	338. (B)	339. (D)	340. (A)
341. (A)	342. (C)	343. (B)	344. (D)	345. (D)	346. (C)	347. (D)	348. (B)	349. (D)	350. (D)
351. (D)	352. (A)	353. (C)	354. (B)	355. (D)	356. (D)	357. (A)	358. (D)	359. (B)	360. (A)
361. (D)	362. (B)	363. (A)	364. (A)	365. (C)	366. (D)	367. (D)	368. (B)	369. (C)	370. (D)
371. (A)	372. (D)	373. (C)	374. (C)	375. (B)	376. (D)	377. (D)	378. (D)	379. (D)	380. (D)
381. (B)	382. (C)	383. (D)	384. (C)	385. (C)	386. (D)	387. (B)	388. (B)	389. (D)	390. (D)
391. (A)	392. (B)	393. (A)	394. (D)	395. (A)	396. (D)	397. (C)	398. (C)	399. (D)	400. (D)
401. (B)	402. (C)	403. (D)	404. (B)	405. (A)	406. (A)	407. (C)	408. (A)	409. (A)	410. (A)
411. (A)	412. (D)	413. (A)	414. (C)	415. (B)	416. (B)	417. (C)	418. (C)	419. (D)	420. (D)
421. (D)	422. (A)	423. (A)	424. (B)	425. (D)	426. (B)	427. (B)	428. (D)	429. (D)	430. (D)
431. (D)	432. (D)	433. (B)	434. (C)	435. (D)	436. (D)	437. (A)	438. (B)	439. (B)	440. (A)
441. I. (B) II. (A)	442. (A)	443. (A)	444. (A)	445. (B)	446. (A)	447. (A)	448. (C)	449. (A)	450. (D)
451. (C)	452. (A)	453. (A)	454. (D)	455. (C)	456. (A)	457. (B)	458. (A)	459. (D)	460. (B)
461. (B)	462. (D)	463. (B)	464. (B)	465. (C)	466. (B)	467. (D)	468. (D)	469. (A)	470. (A)
471. (C)	472. (B)	473. (A)	474. (C)	475. (A)	476. (B)	477. (D)	478. (B)	479. (A)	480. (D)
481. (A)	482. (C)	483. (D)	484. (D)	485. (D)	486. (A)	487. (D)	488. (D)	489. (B)	490. (D)
491. (B)	492. (B)	493. (A)	494. (A)	495. (A)	496. (A)	497. (B)	498. (B)	499. (A)	500. (C)
501. (D)	502. (D)	503. (B)	504. (A)	505. (D)	506. (B)	507. (C)	508. (A)	509. (B)	510. (B)
511. (C)	512. (D)	513. (B)	514. (A)	515. (B)	516. (D)	517. (B)	518. (B)	519. (C)	520. (B)
521. (A)	522. (C)	523. (D)	524. (C)	525. (A)	526. (B)	527. (C)	528. (B)	529. (A)	530. (C)
531. (C)	532. (D)	533. (C)	534. (C)	535. (C)	536. (D)	537. (D)	538. (C)	539. (B)	540. (C)
541. (A)	542. (B)	543. (D)	544. (A)	545. (C)	546. (D)	547. (B)	548. (B)	549. (D)	550. (B)
551. (B)	552. (A)	553. (D)	554. (C)	555. (D)	556. (D)	557. (B)	558. (D)	559. (C)	560. (B)
561. (A)	562. (C)	563. (B)	564. (C)	565. (C)	566. (B)	567. (B)	568. (A)	569. (C)	570. (C)

571. (B) 572. (C) 573. (C) 574. (D) 575. (C) 576. (B) 577. (A) 578. (A) 579. (D) 580. (A)
581. (A) 582. (D) 583. (B) 584. (B) 585. (B) 586. (B) 587. (C) 588. (B) 589. (D) 590. (D)
591. (B) 592. (A) 593. (A) 594. (A) 595. (C) 596. (C) 597. (C) 598. (A) 599. (A) 600. (A)
601. (B) 602. (B) 603. (A) 604. (C) 605. (A) 606. (C) 607. (B) 608. (D) 609. (C) 610. (D)
611. (B) 612. (D) 613. (A) 614. (D) 615. (B) 616. (C) 617. (B) 618. (A) 619. (D) 620. (A)
621. (B) 622. (D) 623. (D) 624. (A) 625. (C) 626. (D) 627. (D) 628. (B) 629. (A) 630. (D)
631. (B) 632. (B) 633. (A) 634. (D) 635. (C) 636. (D) 637. (A) 638. (A) 639. (C) 640. (C)
641. (B) 642. (D) 643. (D) 644. (C) 645. (C) 646. (D) 647. (C) 648. (A) 649. (A) 650. (C)
651. (C) 652. (C) 653. (A) 654. (D) 655. (C) 656. (A) 657. (D) 658. (A) 659. (A) 660. (A)
661. (B) 662. (D) 663. (B) 664. (B) 665. (D) 666. (D) 667. (C) 668. (B) 669. (A) 670. (C)
671. (B) 672. (A) 673. (A) 674. (A) 675. (C) 676. (B) 677. (A) 678. (D) 679. (C) 680. (D)
681. (B) 682. (B) 683. (A) 684. (C) 685. (B) 686. (D) 687. (B) 688. (C) 689. (B) 690. (C)
691. (B) 692. (C) 693. (D) 694. (B) 695. (A) 696. (A) 697. (B) 698. (C) 699. (C) 700. (C)
701. (C) 702. (A) 703. (D) 704. (B) 705. (A) 706. (B) 707. (D) 708. (A) 709. (A) 710. (A)
711. (D) 712. (D) 713. (B) 714. (A) 715. (B) 716. (A) 717. (A) 718. (B) 719. (D) 720. (B)
721. (D) 722. (B) 723. (C) 724. (D) 725. (D) 726. (D) 727. (A) 728. (B) 729. (A) 730. (D)
731. (A) 732. (D) 733. (B) 734. (C) 735. (B) 736. (C) 737. (A) 738. (B) 739. (C) 740. (A)
741. (C) 742. (A) 743. (B) 744. (B) 745. (C) 746. (B) 747. (D) 748. (D) 749. (D) 750. (B)
751. (D) 752. (A) 753. (B) 754. (D) 755. (D) 756. (A) 757. (A) 758. (D) 759. (A) 760. (D)
761. (D) 762. (A) 763. (C) 764. (C) 765. (A) 766. (C) 767. (A) 768. (B) 769. (C) 770. (B)
771. (A) 772. (C) 773. (B) 774. (A) 775. (D) 776. (A) 777. (C) 778. (A) 779. (D) 780. (D)
781. (D) 782. (A) 783. (B) 784. (C) 785. (C) 786. (D) 787. (D) 788. (A) 789. (B) 790. (A)
791. (C) 792. (B) 793. (B) 794. (B) 795. (B) 796. (B) 797. (B) 798. (C) 799. (C) 800. (B)
801. (C) 802. (D) 803. (D) 804. (D) 805. (A) 806. (D) 807. (B) 808. (A) 809. (D) 810. (A)
811. (B) 812. (D) 813. (D) 814. (C) 815. (C) 816. (A) 817. (C) 818. (B) 819. (A) 820. (A)
821. (D) 822. (B) 823. (C) 824. (D) 825. (C) 826. (C) 827. (B) 828. (D) 829. (C) 830. (D)
831. (C) 832. (D) 833. (D) 834. (B) 835. (B) 836. (A) 837. (D) 838. (D) 839. (C) 840. (B)
841. (D) 842. (A) 843. (A) 844. (A) 845. (D) 846. (B) 847. (B) 848. (D) 849. (D) 850. (A)
851. (D) 852. (D) 853. (C) 854. (B) 855. (D) 856. (A) 857. (C) 858. (C) 859. (D) 860. (A)
861. (D) 862. (B) 863. (A) 864. (C) 865. (A) 866. (A) 867. (C) 868. (D) 869. (B) 870. (B)
871. (C) 872. (B) 873. (B) 874. (B) 875. (A) 876. (C) 877. (B) 878. (C) 879. (C) 880. (B)
881. (D) 882. (C) 883. (B) 884. (C) 885. (C) 886. (B) 887. (A) 888. (C) 889. (C) 890. (D)
891. (D) 892. (D) 893. (B) 894. (A) 895. (B) 896. (D) 897. (D) 898. (D) 899. (C) 900. (B)
901. (B) 902. (C) 903. (D) 904. (B) 905. (D) 906. (A) 907. (B) 908. (B) 909. (A) 910. (D)
911. (C) 912. (C) 913. (B) 914. (C) 915. (D) 916. (A) 917. (B) 918. (B) 919. (C) 920. (D)
921. (A) 922. (D) 923. (D) 924. (C) 925. (A) 926. (D) 927. (B) 928. (C) 929. (B) 930. (D)
931. (D) 932. (B) 933. (D) 934. (B) 935. (A) 936. (B) 937. (D) 938. (D) 939. (A) 940. (D)
941. (D) 942. (D) 943. (A) 944. (D) 945. (B) 946. (B) 947. (B) 948. (B) 949. (D) 950. (D)
951. (D) 952. (A) 953. (C) 954. (A) 955. (D) 956. (C) 957. (C) 958. (B) 959. (A) 960. (C)
961. (C) 962. (B) 963. (A) 964. (C) 965. (A) 966. (B) 967. (D) 968. (A) 969. (C) 970. (D)
971. (B) 972. (C) 973. (B) 974. (A) 975. (D) 976. (C) 977. (C) 978. (A) 979. (C) 980. (A)
981. (B) 982. (C) 983. (C) 984. (D) 985. (C) 986. (B) 987. (A) 988. (A) 989. (A) 990. (A)

991. (A) 992. (A) 993. (C) 994. (D) 995. (C) 996. (C) 997. (D) 998. (B) 999. (A) 1000. (B)
1001. (D) 1002. (A) 1003. (B) 1004. (A) 1005. (D) 1006. (C) 1007. (B) 1008. (B) 1009. (D) 1010. (C)
1011. (A) 1012. (B) 1013. (A) 1014. (B) 1015. (A) 1016. (C) 1017. (D) 1018. (B) 1019. (B) 1020. (C)
1021. (D) 1022. (A) 1023. (A) 1024. (C) 1025. (B) 1026. (A) 1027. (C) 1028. (A) 1029. (C) 1030. (C)
1031. (A) 1032. (C) 1033. (C) 1034. (A) 1035. (C) 1036. (B) 1037. (D) 1038. (D) 1039. (B) 1040. (C)
1041. (A) 1042. (A) 1043. (B) 1044. (D) 1045. (C) 1046. (C) 1047. (D) 1048. (C) 1049. (D) 1050. (C)
1051. (A) 1052. (B) 1053. (C) 1054. (A) 1055. (D) 1056. (B) 1057. (A) 1058. (B) 1059. (A) 1060. (C)
1061. (C) 1062. (B) 1063. (B) 1064. (A) 1065. (B) 1066. (B) 1067. (B) 1068. (D) 1069. (C) 1070. (A)
1071. (C) 1072. (B) 1073. (C) 1074. (B) 1075. (B) 1076. (A) 1077. (C) 1078. (D) 1079. (C) 1080. (B)
1081. (D) 1082. (D) 1083. (A) 1084. (D) 1085. (B) 1086. (A) 1087. (C) 1088. (D) 1089. (A) 1090. (A)
1091. (B) 1092. (D) 1093. (D) 1094. (B) 1095. (A) 1096. (B) 1097. (A) 1098. (A) 1099. (C) 1100. (A)
1101. (C) 1102. (D) 1103. (A) 1104. (C) 1105. (D) 1106. (B) 1107. (D) 1108. (A) 1109. (D) 1110. (B)
1111. (A) 1112. (C) 1113. (B) 1114. (C) 1115. (D) 1116. (A) 1117. (A) 1118. (B) 1119. (C) 1120. (D)
1121. (C) 1122. (D) 1123. (D) 1124. (A) 1125. (C) 1126. (B) 1127. (B) 1128. (A) 1129. (D) 1130. (A)
1131. (A) 1132. (B) 1133. (D) 1134. (B) 1135. (C) 1136. (C) 1137. (A) 1138. (D) 1139. (B) 1140. (A)
1141. (A) 1142. (A) 1143. (A) 1144. (D) 1145. (D) 1146. (A) 1147. (D) 1148. (B) 1149. (D) 1150. (B)
1151. (A) 1152. (B) 1153. (D) 1154. (B) 1155. (A) 1156. (C) 1157. (B) 1158. (A) 1159. (A) 1160. (B)
1161. (A) 1162. (B) 1163. (B) 1164. (B) 1165. (D) 1166. (B) 1167. (A) 1168. (D) 1169. (C) 1170. (B)
1171. (B) 1172. (C) 1173. (C) 1174. (A) 1175. (B) 1176. (B) 1177. (D) 1178. (D) 1179. (A) 1180. (A)
1181. (D) 1182. (C) 1183. (A) 1184. (D) 1185. (D) 1186. (A) 1187. (D) 1188. (A) 1189. (A) 1190. (B)
1191. (C) 1192. (D) 1193. (A) 1194. (A) 1195. (C) 1196. (D) 1197. (A) 1198. (D) 1199. (C) 1200. (C)
1201. (B) 1202. (C) 1203. (A) 1204. (D) 1205. (A) 1206. (B) 1207. (D) 1208. (D) 1209. (B) 1210. (B)
1211. (A) 1212. (D) 1213. (A) 1214. (D) 1215. (D) 1216. (A) 1217. (D) 1218. (D) 1219. (D) 1220. (B)
1221. (A) 1222. (A) 1223. (C) 1224. (B) 1225. (B) 1226. (D) 1227. (D) 1228. (C) 1229. (B) 1230. (C)
1231. (C) 1232. (B) 1233. (D) 1234. (D) 1235. (A) 1236. (B) 1237. (B) 1238. (A) 1239. (D) 1240. (C)
1241. (A) 1242. (A) 1243. (C) 1244. (D) 1245. (B) 1246. (D) 1247. (A) 1248. (D) 1249. (A) 1250. (A)
1251. (B) 1252. (B) 1253. (A) 1254. (B) 1255. (D) 1256. (B) 1257. (A) 1258. (D) 1259. (B) 1260. (C)
1261. (C)

Chapter - 5

PLANT BIOCHEMISTRY

1. An antibody with enzymatic action, it is produced in response to stable analog of the transition state of the substrate for the given enzyme action is called:
(A) Abzyme (B) Enzyme
(C) Hormone (D) Activator

2. A phenolic compound produced by dicot plant cells at wound sites; act as signal molecule for the expression of vir regulation of *Agrobacterium tumefaciens* is:
(A) Acetosyringone (B) Abscisic acid
(C) Cytokinins (D) Auxin

3. The protein component of an enzyme that has a coenzyme/cofactor is:
(A) Abzyme (B) Enzyme
(C) Hormone (D) Apoenzyme

4. Enzyme that catalyze the transfer of amino groups from α – amino to α–keto acids is called:
(A) Transaminases (B) Amino transferases
(C) Both of these (D) None of these

5. ATP-dependent enzymatic esterification of the carboxyl group of an amino acid to the 3'-hydroxyl group of its corresponding tRNA is called:
(A) Transaminases
(B) Amino transferases
(C) Amino acid activation
(D) None of these

6. An enzyme having a regulatory function through its capacity to undergo a change in catalytic activity by allosteric mechanisms or by covalent modification is called:
(A) Regulatory enzyme
(B) Regulatory gene
(C) Regulatory sequence
(D) Reverse transcriptase

7. Enzymatic transfer of n amino group from an α – keto acid is called:
(A) Transmination (B) Transcription
(C) Transduction (D) Transformation

8. The enzymatic process whereby the genetic information contained is one strand of DNA is used to specify a complementary sequence of bases in an mRNA chain is called:
(A) Transmination (B) Transcription
(C) Transduction (D) Transformation

9. The most common reporter proteins are the:
(A) Chloramphenicol acetyl transferase
(B) β-glucuronidase
(C) Luciferase
(D) Green Fluorescent protein
(E) All of these

10. Genes for chloramphenicol acetyl transferase, β-galactosidase and β-glucuronidase are obtained from:
(A) *E. coli*
(B) Firefly-*Photinus pyralis*
(C) Jelly fish-*Aequaria victoria*
(D) None of these

11. Luciferase is obtained from:
(A) *E. coli*
(B) Firefly-*Photinus pyralis*
(C) Jelly fish-*Aequaria victoria*
(D) None of these

12. Green fluorescent proteins is obtained from:
(A) *E. coli*
(B) Firefly-*Photinus pyralis*
(C) Jelly fish-*Aequaria victoria*
(D) None of these

13. Proteins are clustered into families when they show close sequence homology, structural similarity and functional conservation is called:
(A) Protein family (B) Protein domain
(C) Protein fold (D) None of these

14. A functional unit of a protein in fold that is stable in solution can be defined as a:
(A) Protein family (B) Protein domain
(C) Protein fold (D) None of these

15. The number of major secondary structural elements, their arrangement in three-dimensional space and their topological connectivity is called:
(A) Protein family (B) Protein domain
(C) Protein fold (D) None of these

16. There are covalently linked carbohydrate (PG) containing proteins is called:
(A) Glycoproteins (B) Lipoproteins
(C) Nucleoproteins (D) Phosphoproteins

17. There are covalently linked lipid (PG) containing proteins is called:
(A) Glycoproteins (B) Lipoproteins
(C) Nucleoproteins (D) Phosphoproteins

18. **There are covalently linked nucleic acid (PG) containing proteins is called:**
(A) Glycoproteins (B) Lipoproteins
(C) Nucleoproteins (D) Phosphoproteins

19. **There are covalently linked phosphate (PG) containing proteins is called:**
(A) Glycoproteins (B) Lipoproteins
(C) Nucleoproteins (D) Phosphoproteins

20. **There are heme or iron porphyrin (PG) containing proteins is called:**
(A) Hemeproteins (B) phosphoproteins
(C) Hemoproteins (D) Glycoproteins

21. **Which type of protein is found in casein of milk is called:**
(A) Hemeproteins (B) Phosphoproteins
(C) Hemoproteins (D) Glycoproteins

22. **Which type of protein is found in hemoglobin is:**
(A) Hemeproteins (B) phosphoproteins
(C) Hemoproteins (D) Glycoproteins

23. **In some cases enzymes require additional organic and inorganic molecules is called:**
(A) Cofactors (B) Enzyme inhibitors
(C) Both (A) and (B) (D) None of these

24. **Many substances alter the activity of an enzyme and those which reduces the enzyme activity are called:**
(A) Cofactors (B) Enzyme inhibitors
(C) Both (A) and (B) (D) None of these

25. **The enzymatic rate constant (kcat/Km) of orotidine 5'-phosphate decarboxylase is $6 \times 10^{7}\ M^{-1}s^{-1}$ and the nonenzymatic rate constant (kn) is $3 \times 10^{-16} s^{-1}$. What is the value of the enzyme's catalytic proficiency?**
(A) $2 \times 10^{23} M^{-1}$ (B) $5 \times 10^{-24} M$
(C) $12 \times 10^{-9} M^{-1}s^{-2}$ (D) $8.3 \times 10^{7}\ Ms^{2}$

26. **Which one of the following is basic amino acid:**
(A) Lysine (B) Threonine
(C) Valine (D) Glycine

27. **Given the structure of DNA as described by Watson and Crick, all of the following must be true of DNA except:**
(A) The 3' end of one of the strands must be aligned with the 5' end of the other strand
(B) The total number of purine bases must equal to the total number of pyrimidine bases
(C) The sequence of bases on one strand must be identical to those on the other strand
(D) The sequence of bases in the genome of one organism must be different from the sequence of bases in the genome of a different organism

28. **DNA polymerase III requires a/an -------- for synthesis of DNA to occur:**
(A) DNA template (B) RNA primer
(C) Free 3'-OH end of a DNA strand
(D) All of these

29. **The difference between genomic DNA and cDNA is:**
(A) Genomic DNA is that which is expressed and cDNA is all the DNA
(B) Genomic DNA is found in the nucleus and cDNA is found in the cytoplasm
(C) Genomic DNA is all the DNA and cDNA is made only from genes that are expressed
(D) There is no difference between these two types of DNA

30. **The e subunit of DNA polymerase III is responsible for its --------- activity:**
(A) 5' → 3' polymerase
(B) Sliding clamp
(C) 3' → 5' exonuclease (proof reading)
(D) Ribosome assembly

31. **Which of the following contain sulphur:**
(A) Lipoic acid (B) Thiamin
(C) Riboflavin (D) A and B both

32. **The amino acids in polypeptide chains which contain sulphur(s) are:**
(A) Cysteine, cystine and methionine
(B) Cystine
(C) Methionine
(D) Cysteine
(E) Cysteine and methionine

33. **Which on of the following structure is a Zwitter ion:**
(A) H N – CH (R) – COOH
(B) H N + CH (R) – COO-
(C) H N – CH (R) – COO-
(D) None of these

34. **Acetylcholine on a smooth muscle cell is to phospholipase C as epinephrine on a liver cell is to?**
(A) Protein kinase C (B) cAMP
(C) Adenylyl cyclase (D) GTP

35. **A mixture of cytochrome-(MW-11.7 KD) and myoglobin (MW-17.2 KD) are to be separated by polyacrylamide gel electrophoresis. Their isoelectric pH (pI) values are 9.6 and 7.2 respectively. In which direction will each protein migrate at pH 8.5?**
(A) Myoglobin will migrate to anode and cytochrome - C will migrate to cathode
(B) Myoglobin will migrate to cathode and cytochrome - C will migrate to anode
(C) Both will migrate to anode
(D) Both will migrate to cathode

36. ***E. coli*** **DNA polymerase catalyzes polymerization in the --------- direction(s):**
(A) 5' → 3' (B) 3' → 5'
(C) 2' → 5' (D) Both 5' → 3' and 3' → 5'

37. **Enzymes that remain bound to their nascent chains through many polymerization steps are said to be--- ----------:**
(A) Constitutive (B) Distributive
(C) Conjugated (D) Processive

38. Which of the following is a monomeric unit of DNA:
(A) Nucleotide
(B) 2-Deoxy ribofuranose
(C) Nucleoside
(D) D-2-Deoxy-ribunucleotide

39. The pH at which the dipolar ion of amino acid do not migrate either to anode or cathode in an electric field is known as:
(A) Km Value (B) Vmax
(C) Turn over number (D) Isoelectric point

40. Fluorescent in situ hybridization is used to:
(A) Locate sequences of DNA on a chromosome
(B) Identify mini-and microsatellite DNAs
(C) Identify homologous pairs of chromosomes
(D) Identify the centromere portion of chromosomes

41. It is difficult to determine either Km or V_{max} from a graph of velocity vs. substrate concentration because:
(A) Too much substrate is required to determine them
(B) The graph is sigmoidal
(C) An asymptotic value must be determined from the graph
(D) The points on the graph are often not spread out on the hyperbola

42. Which of the following catalyzes the biosynthesis of DNA:
(A) Transcriptase (B) Nuclease
(C) Phosphorylase (D) DNA polymerase

43. The mechanism of action of lysozyme includes:
(A) Distortion of the substrat
(B) Acid catalysis
(C) Proximity effects
(D) Formation of a half – chair sugar form
(E) All of these

44. UV-spectrosopy is not used to quantitate the compounds based on absorbance:
(A) Nucleic acids have absorbance peak at 260 nm
(B) NAD (P) H have absorbance peak at 340 nm
(C) Aromatic amino acids have absorption maxima about 280 nm
(D) Sulphur containing amino acids have absorbance at 270 nm

45. Which of the following can not be digested by human:
(A) Glucose (B) Starch
(C) Cellulose (D) Dextrin

46. Serine and threonine are to? As tyrosine is to receptor tyrosine kinases:
(A) Phosphatases
(B) G proteins
(C) cAMP
(D) Protein kinases A & C

47. When Keq of a reaction = 1, then
(A) The forward reaction is faster than the reverse reaction
(B) The reverse reaction is faster than the forward reaction
(C) The forward and reverse reaction rates are constants and equal
(D) More products are formed than reactants
(E) Fewer products are formed than reactants

48. The processively of *E. coli* DNA polymerase III holoenzyme accounts for all of the following except:
(A) The relatively small number of enzymes needed to replicate the entire chromosome
(B) The directionality of polymerization
(C) The rapid rate of polymerization
(D) The ability of a single holoenzyme to add many nucleotides to a growing DNA chain

49. The 3' $\rightarrow$ 5' exonuclease activity of *E. coli* DNA polymerase III accounts for the -------------- of polymerization
(A) Low error rate (B) High speed
(C) Directionality (D) All of these

50. X-ray crystallographic examination of the active site of arginine kinase was possible because:
(A) It could be crystallized in the different structural forms
(B) Nitrate could be substituted for phosphate in the reaction
(C) It was readily purified (unlike most other enzymes)
(D) MgADP was involved in the reaction

51. Which of the following is a non-reducing sugar:
(A) Ribose (B) Galactose
(C) Sucrose (D) Fructose

52. Which of the following do not contain asymmetric carbon atom:
(A) Dihydroxy acetone (B) Fructose
(C) Glycine (D) Both (A) and (C)

53. Ultraviolet (UV) light can be used to estimate protein solution concentrations because:
(A) Phenylalanine absorbs at 260 nm
(B) All the amino acids absorb UV light
(C) Aromatic amino acids absorbs at 280 nm
(D) Tyrosine and tryptophan absorb at 280 nm
(E) All of these

54. The structure of different DNAs can vary in all the following ways except:
(A) The helix may be right-handed or left handed
(B) The sequence of nucleotide bases can be different
(C) The molecule can be fully hydrated as in the B form, or drier as in the A form
(D) The bases can be attached to the sugars or the phosphates of the backbone

55. Chymotrypsinogen in the native state and reduced unfolded states is loaded in wells A and B of a native polyacrylamide gel, respectively. After electrophoresis and staining, one of the following results is likely to be observed:

(A) Protein in well A would have moved longer distance than that in well B
(B) Protein in well A would have moved shorter distance than that in well B
(C) Protein in well A and B would have moved the same distance
(D) Protein in well A and B would have move at all

56. The dipeptide Lys-Glu on electrophoresis at pH 6.0 would:

(A) Move towards anode (+)
(B) Move towards cathode (-)
(C) Remain stationary
(D) Get degraded

57. Which of the following is not a reason that the actual number of human genes is still a mystery?

(A) The human genome is nowhere near completion
(B) It is not certain whether particular sequences of DNA actually contain genes
(C) Small genes with single exons can be missed by the analytical computers
(D) The genome's "dark matter" has yet to be accurately sequenced

58. Ryanodine and IP3 receptors are found on -------- membranes, whereas epinephrine and glucagons receptors are found on ------------- membranes:

(A) Plasma, intracellular
(B) Intracellular, plasma
(C) Mitochondrial, endoplasmic reticulum
(D) Outer, inner

59. Concentrations of some proteins cannot estimated by UV spectrophotometry because they are:

(A) High in non-absorbing amino acids
(B) Low in tryptophan and tyrosine
(C) Low in protein
(D) High in aromatic amino acids

60. What is the biological function of lysozyme?

(A) Converts trypsinogen to trysin
(B) Is a ligase that accelerates the polymerization of glycogen
(C) It regulates vascular constriction in chickens
(D) Hydrolyzes polysaccharides of bacterial cell walls

61. Fluorescence microscopy is based on the ability of certain molecules to:

(A) Absorb light of a constant wavelength
(B) Absorb light of many different wavelength
(C) Absorb light of a given wavelength and then emit light of a longer wavelength
(D) Absorb light of a given wavelength and then emit light of a shorter wavelength

62. Polymerization of the lagging strand is catalyzed by:

(A) A second core complex on the same polymerase holoenzyme that polymerizes the leading strand
(B) A completely separate replisome from that which polymerizes the leading strand
(C) DNA polymerase I holoenzyme
(D) The same polymerase holoenzyme that polymerized the leading strand, but with opposite directionality

63. The RNA primer at the beginning of each Okazaki fragment is removed by:

(A) DNA polymerase I
(B) DNA polymerase II
(C) DNA polymerase III
(D) 3' 5' exonuclease subunit of DNA polymerase III

64. Disulfide bridges can form in proteins:

(A) Only between cysteine residues side-by-side in the protein sequence
(B) Between cysteine residues that are close in three-dimentional space, but not necessarily close in the primary structure
(C) Between two cystine residues in proteins
(D) Between any two methionines or cysteines

65. The process of RNA biosynthesis is known as:

(A) Translation
(B) Transformation
(C) Replication
(D) Transcription

66. Glyoside hydrolases such as bacterial cellulase have been shown to differ in mechanisms from that of enzyme. They:

(A) Are alkaline catalysts
(B) Form a boat of the sugar
(C) Form a covalent sugar-enzyme intermediate
(D) Do not distort the substrate

67. Which of the following is not a reason that the actual number of human genes is still a mystery?

(A) The human genome is nowhere near completion
(B) It is not certain whether particular sequences of DNA actually contain genes
(C) Small genes with single exons can be missed by the analytical computers
(D) The genome's "dark matter" has yet to be accurately sequenced

68. An inhibitors binds to a site other than the active site of the enzyme. Which statement below correlated with this observation?

(A) It must be a competitive inhibitor
(B) The inhibition must be irreversible
(C) It could be noncompetitive or uncompetitive inhibition
(D) It could be irreversible competitive, noncompetitive or uncompetitive. The data do not relate to the type of inhibition

69. The overall shape of a protein is greatly influenced by:
(A) Amino acid R group properties
(B) Charged amino acids
(C) Hydrophobic amino acids
(D) pH
(E) Hydrophilic amino acids

70. Which of the following is a heterpolysaccharide:
(A) Starch (B) Insulin
(C) Cellulose (D) Gum

71. The surplus glucose units in the liver of animals are stored in the from of:
(A) Sucrose (B) Cellulose
(C) Starch (D) Glycogen

72. Nick translation is a process for:
(A) Filling in and sealing the gap in DNA after removal of the RNA primer
(B) Synthesis of the protein for primase from mRNA
(C) Removal of the RNA primer by DNA polymerase I
(D) Removal of okazaki fragments

73. Cultured hamster cells treated with methotrexate become resistant to the drug by amplifying the gene for DHFR. This is an example of:
(A) Selective gene amplification
(B) Random gene duplication and selective survival
(C) Evolution (D) All of these

74. Okazaki fragments are:
(A) The smallest subunits of DNA polymerase III
(B) Short stretches of DNA formed on the lagging strand
(C) Short RNA primers needed for initiation of polymerization
(D) Fragments of DNA polymerase I that lack 5' → 3' exonuclease activity

75. Time of action potential maximum for:
(A) Candiac muscle fibre
(B) Nerve fibre
(C) Skeleton muscle fibre
(D) Ganglionic fibre

76. Ramachandran determined the "allowed" values of the phi and psi angles primarily by considering:
(A) pKa values of the amino acids
(B) The hydropathy of amino acids
(C) Steric hindrance
(D) Hydrogen bonding effects

77. Induced fit studies by Koshland using the enzyme hexokinase showed that there are enzyme forms:
(A) Which react with the hydroxyl group of water
(B) Which react only with ATP and glucose present together
(C) Which are hydrophobic, excluding the competing hydroxyl group from water
(D) With and without glucose bound to each

78. One form of glycogen synthetase has a requirements for glucose-6-phosphate. What is the best explanation for this phenomenon?
(A) It is an example of negative feedback
(B) Glucose-6-phosphate is a substrate
(C) It is an example of allosteric activation
(D) It inhibits an enzyme that hydrolyses UDP-glucose

79. What is the function of the primosome?
(A) Removal of RNA from the lagging strand
(B) Synthesis of short RNA fragments
(C) Assembly of DNA polymerase III holoenzyme
(D) Detection of the origin of replication

80. Zymogens are inactive enzyme precursors which are made active by:
(A) Change in structure
(B) Selective proteolysis
(C) Secretion of new types of cells
(D) A, B and C

81. How many complete turns are there in an ideal α – helix that contains 15 amino acids and has a pith of 0.54 nm and a rise of 0.15 nm?
(A) 1 (B) 4
(C) 8 (D) 12

82. Insulin is entirely made up of:
(A) Fructose units
(B) Glucose and Fructose units
(C) Galactose units
(D) Glucose units

83. In eukaryotes, transposable elements arise by:
(A) Reverse transcription of RNA, followed by insertion into the genome
(B) Excision of a portion of the chromosome followed by insertion at a new place in the genome
(C) Duplication of transposable element's DNA followed by insertion into a new place in the genome
(D) All of these

84. Which is not a difference between B-DNA and A-DNA?
(A) A-DNA forms under more dehydrating conditions than B-DNA
(B) For polynucleotide strands containing the same number of nucleotide, the A-DNA strand will be shorter from end-to-end than the corresponding B-DNA
(C) Both are helical, but B-DNA is right –handed and A-DNA is left handed
(D) The planes of the nitrogenous bases are not perpendicular to the helix axis in A-DNA

85. What is a reasonable length for an RNA primer in *E. coli*?
(A) A few nucleotides (B) 50 nucleotides
(C) 100 nucleotides (D) 1000 nucleotides

86. The instant source of energy in human beings is:
(A) D-Glucose (B) Proteins
(C) Vitamins (D) Fats

87. In an enzyme reaction involving one enzyme and one substrate, the rate of the reaction depends on:
(A) Substrate concentration
(B) Enzyme concentration
(C) Both substrate and enzyme concentrations
(D) The enzyme concentration at first and the substrate concentration later on

88. Which one of the following bacteria contain T-DNA:
(A) *E. coli* (B) *Agrobacterium*
(C) *Rhizobium*
(D) *Pneumococci*

89. At the beginning of an enzyme-catalyzed reaction the -------- is negligible:
(A) Formation of ES
(B) Formation of E + P
(C) Conversion of ES to E + S
(D) Disappearance of ES

90. Which of the following is not correct about fatty acid degradation and biosynthesis

	Degradation	Biosynthesis
(A) Location of pathway in the cell	Mitochondrion	Cytosol
(B) Electron acceptor or donor used	FAD, NAD+	NADPH
(C) Participation of O_2	NO	Yes
(D) Carrier of intermediate acyl groups	$AP\text{-}(SH)_2$	CoA-SH

91. Fragments on the lagging strand are joined by the enzyme:
(A) DNA polymerase I (B) DNA synthase
(C) DNA primosome (D) DNA ligase

92. The concentrations of enzymes are determined by using pseudo first-order conditions in which:
(A) The concentration is kept the same and the rate is measured
(B) E concentration is kept the same and the rate is measured
(C) S concentration is constant, E concentration varies and the rate is measured
(D) E concentration is constant, S concentration varied and the rate is measured
(E) All of these

93. Which one of the following supports glycogen synthesis?
(A) High cyclic adenosine monophosphate (AMP) levels
(B) Inactive adenyl cyclase
(C) Active phosphorylase-alpha
(D) Epinephrine

94. Two strands of DNA double helix are held together tightly by:
(A) Electrostatic bonds
(B) Phosphodiester bonds
(C) Covalent bonds
(D) Hydrogen bonds

95. Which of the following is not a role that has been suggested for transposable elements in the process of evolution?
(A) Transposable elements bring previously unlinked segments together, possibly resulting in composite protein products
(B) Insertion of transposable elements into the genome results in gene duplication and the evolution of multigene families
(C) Transposable elements have no role in evolution; they are genetic parasites
(D) Natural selection always favours species with many transposable elements

96. Tachy candia means faster heart rate usually defined in adult person. It is ulases heart rate:
(A) By activating parasympathetic nervous system and blacking Na^+, Ca^{++} channels
(B) By activating sympathetic nervous system and blacking Na^+, Ca^{++} channels
(C) It deactivates sympathetic system and activates Na^+, Ca^{++} channels
(D) It activates sympathetic system, and generates action potential through the rapid entry of Na^+, Ca^{++} channels

97. When varying the substrate concentration at a fixed concentration of enzyme it is observed that at low concentrations of substrate the reaction is ----------, while at high concentrations of substrate the reaction is----------------:
(A) Maximal, initial
(B) Initial, maximal
(C) Second order; first order
(D) First order; second order
(E) First order; zero order

98. Site – directed cross-linking is a technique that gives insight into what property of integral membrane proteins?
(A) Number of transmembrane domains
(B) On which side of the membrane the N and C termini are found
(C) Proximity of various membrane-spanning domains to each other
(D) None of these

99. Why does E. coli need bth DNA polymerase III and DNA polymerase I?
(A) DNA polymerase III lacks the 5' $\rightarrow$ 3' exonuclease activity needed to remove RNA primers
(B) Each polymerase is specific for only one strand of DNA. DNA polymerase III acts only
(C) DNA polymerase I is needed to join the Okazaki fragments

(D) The DNA replication is bidirectional; one polymerase is used for each direction
(E) Only DNA polymerase I has proof reading abolity

100. **In candiac muscle fibress, generation of action potential through:**
(A) Rapid and slow entry of Na^+ and Ca^{++} ions, respectively
(B) Rapid entry of Na^+ only
(C) Slow entry of Ca^+ only
(D) Slow and rapid entry of Na^+ and Ca^{++}, respectively

101. **Induced fit of enzyme activation is a result of:**
(A) Contact (binding) of substrate
(B) Constant small and rapid motions of protein atoms
(C) Change to an active form
(D) A, B and C

102. **Active trypsin formation by the action of enteropeptidase can be viewed as the master activation step because:**
(A) Enteropeptidase can activate its own zymogen
(B) It is allosterically controlled
(C) Trypsin activates other pancreatic zymogens
(D) All of these

103. **Which one of the following correct about QRS complex in electrocardiogram:**
(A) It is ventricular depolarisation having voltage smv
(B) It alrial depolarization having voltage (1-1.5) m.v.
(C) It is ventricular cystole and has a voltage of. (1-1.5) m.v.
(D) It is ventricular depolarization, have a voltage of (1-1.5)

104. **In cellulose, different D-Glucose units are linked by:**
(A) a, D -1, 2-Glucosidic bonds
(B) b, D- 1, 4-Glucosidic bonds
(C) a, D- 1, 4- Glucosidic bonds
(D) b, D- 1, 4- Galactosidic bonds

105. **The structure of two enantiomers (enantiomorps) of glucose are different only in the configuration of carbon atom at:**

(A) C-1	(B) C- 2
(C) C-4	(D) C-5

106. **Patent ducters antriosus generally occurs in case of:**
(A) Tetralogy of fallot
(B) Blue baby syndrome
(C) When blood is purifies by lungs
(D) In newly born child when ducters does not lose properly

107. **Which might be a method used for distinguishing among several mechanistic possibilities in a multisubstrate reaction?**
(A) Measure Km in the presence of one substrate at a time
(B) Measure the rate with respect to one sub-strate while varying the concentration of another substrate
(C) Measure the rate while adding inactivators for all but one substrate
(D) Any of the above

108. **Which of the following statements about glycogen storage are incorrect (more than one is correct):**
P. Glycogen is stored in muscles and liver
Q. Glycogen is a major source of stored energy in brain
R. Glycogen reserves are less rapidly depleted than fat reserves during starvation
S. Glycogen storage ours in the form of dense granules in the cytoplasm of cells

(A) P and Q	(B) P and R
(C) Q and R	(D) Q and S

109. **In inspired air which one of the following gas shows no change in its partial pressure**

(A) O_2	(B) N_2
(C) CO_2	(D) H_2O vap

110. **Proteins with SH2 domains are to -------- as adenylyl cyclase and phospholipase C are to G protein-coupled receptors:**
(A) IRSs
(B) Receptor tyrosine kinases
(C) Insulin
(D) Autophosphorylation

111. **When the concentration of a substrate inside a cell falls below the substrate concentration at half maximal velocity:**
(A) ES is formed more easily
(B) ES dissociates to E + S
(C) ES is loser to the ground state than to the transition state
(D) E + S is in equilibrium with ES

112. **The initial velocity of an enzyme reaction (v) describes:**
(A) The concentration of the enzyme at maximal velocity
(B) The concentration of substrate at maximal velocity
(C) The concentration of both at the start of the reaction
(D) The rate of reaction at when the substrate and enzyme are first mixed

113. Which of the following is not true about ketogenesis:

(A) Acetyl – CoA acts as precursor

(B) HMG CoA synthase participates during reaction

(C) Occurs in the liver

(D) Acetone is the only product

114. Which of the following feature is not true about Z-DNA:

(A) Left handed helical structure

(B) Dinucleotide repeating unit

(C) 10.4 bp per helical turn

(D) Anti & syn conformation of glycosidic bond both

115. Ratio of end systolic volume stroke volume 2 end diasystolic volume:

(A) 1 : 1 : 2 (B) 3 : 7 : 9

(C) 1 : 1.5 : 3 (D) 2 : 6 : 8

116. Starting from a glucose residue in glycogen, how many net ATP molecules will be formed in the glycolysis of the residue to pyruvate?

(A) 1 (B) 2

(C) 3 (D) 4

117. Which mutant form of ras is likely to cause malignancy?

(A) ras that cannot hydrolyze GTP

(B) ras tht cannot bind to GTP

(C) ras that cannot bind to Grb2 or Sos

(D) ras that cannot bind to Raf

118. Match the following:

a. Renett cell 1. Round worms

b. Solenocyte 2. Arthropode

c. Metarephridia 3. Molluscass

d. Coxal gland 4. Fasciola

(A) a-1, b-4, c-3, d-2 (B) a-4, b-3, c-1, d-2

(C) a-1, b-2, c-3, d-4 (D) a-1, b-4, c-2, d-3

119. A membrane is examined using freeze-fracture and freeze-etch analysis. The surfaces are shown to have almost no pits or bumps. Which of the following is the most likely source of the membrane?

(A) Mitochondria, inner membrane

(B) Myelin (C) Red blood cell

(D) All of these

120. In which of the following respect A from of DNA differs from B-form of DNA

(A) Helic handedness (B) Repeating unit

(C) Conformation of glycosidic bond

(D) Base pair per helical turn

(E) All of these

121. Which of the following is common features of both glycogen synthesis and glycogen breakdown?

(A) Both require UDO-glucose

(B) Both involve gluose-1-phosphate

(C) Both are driven in part by the hydrolysis of pyrophosphate

(D) Both use the same enzyme for branching and debranching

122. The MAP kinase cascade stimulates transcription via specific transcription factors, and stimulates translation via:

(A) SREs (B) MAPK

(C) PHAS-I and eIF4E (D) All of these

123. Some antibody molecules are enzymatic if they are formed against:

(A) Enzyme

(B) Enzyme bound to other proteins

(C) Transition state analogs bound to other proteins

(D) Transition state analogs

124. The substrate specificity of serine proteases is primarily due to:

(A) A specificity pocket in the protein

(B) The positions of specific side chains of serine, histidine and aspartate

(C) Distinct backbone conformations of the individual proteins

(D) A, B and C

125. An update of Fisher's lock-and-key theory of enzyme specificity view the -------- as the lock and -------- as the key:

(A) Enzyme; substrate

(B) Substrate; enzyme

(C) Enzyme; transition state

(D) Transition state; enzyme

126. A helical wheel can be used to show:

(A) The pith of a helix structure

(B) The amphipathic nature of a helix

(C) DNA binding

(D) Pleated sheets

127. The sequence of events that occurs after insulin binds to its receptor tyrosine kinase is:

(A) Binding of IRS → Phosphorylation of proteins with SH2 domains → Effect

(B) Binding of proteins with SH2 domains → Phosphorylation of IRS → Effect

(C) Autophosphorylation and phosphory-lation of IRS → Phosphroylation of proteins with SH2 domains → Effect

(D) Autophosphorylation & binding of IRS → Phosphorylation of proteins with SH2 domains → Effects

128. What second messenger has been implicated in blood pressure regulation?

(A) cAMP (B) NO

(C) CO (D) IP3

129. Which one of the following organisms would you predict has the highest proportion of unsaturated fatty acids in their membranes?

(A) Antarctic fish

(B) Cactus

(C) Bacteria from thermal hot springs

(D) Humans

130. Which is not true about ß–sheets?
(A) The side-chains of all amino acids point to the same side of the sheet
(B) The polypeptide chains in the sheet are nearly fully extended
(C) The range of allowed phi and psi angles is broader than for those in the –helix
(D) In ant parallel sheets the hydrogen bonds between adjacent strands are nearly perpendicular to the backbones of the strands

131. After a meal, blood glucose enters cells and is stored as glycogen, particularly in the liver. Which of the following is the donor of new glucose molecules in glycogen?
(A) UDP-glucose-1-phosphate
(B) UDP-glucose
(C) UDP-glucose-6-phosphate
(D) Glucose-6-phosphate

132. Ab sandwich forms when:
(A) Two hydrophobic sides of b-sheets interact
(B) Two hydrophilic sides of b-sheets interact
(C) An a-helix separates two b-sheets
(D) Two amphipathic a-helices interact

133. Which is not an activity for DNA polymerase I?
(A) 5' → 3' polymerase activity
(B) Joining of Okazaki fragments
(C) 5' → 3' exonuclease activity
(D) Nick translation

134. What is a Klenow fragment?
(A) A fragment of DNA polymerase I that lacks 5'→3' exonuclease activity
(B) A topoisomerase responsible for unwinding DNA at the replication fork
(C) A portion of the replisome responsible for RNA primer synthesis
(D) Short stretches of DNA formed during polymerization of the lagging strand

135. Super secondary structures that contain recognizable combinations of α–helices, β-strands and loops (e.g. the Greek key) are called:
(A) Domains
(B) Folds
(C) Homologous regions
(D) Motifs

136. The beta-oxidation of a molecule of palmitic acid:
(A) Yields 8 molecules of acetyl-CoA
(B) Yields 16 molecules of acetyl-CoA only
(C) Yields carbon dioxide and water only
(D) Does both involve oxygen

137. A mutation in the FAK gene would likely result in:
(A) Abnormal cell shape
(B) Abnormal wound healing
(C) Abnormal cell growth
(D) All of these

138. The physiological roles of bile salts include which of the following:
P. They aid in the digestion of lipid
Q. They facilitate the absorption of sugars
R. They facilitate the absorption of lipid
S. They provide a means for excreting cholesterol

(A) P and Q (B) P and R
(C) Q and S (D) P, Q and S

139. Acid-base catalysis is estimated to accelerate a typical enzymatic reaction by what factor?
(A) 1 to 2 fold increase
(B) 10 to 100 fold increase
(C) 106 fold increase (D) 1023 fold increase

140. Which of the following statements about the hormonal regulation of glycogen synthesis and degradation are correct?
P. Insulin increases the capacity of the liver to synthesize glycogen
Q. Insulin in secreted in response to low levels of blood glucose
R. Glucagon and epinephrine have opposing effects on glycogen metabolism
S. Glucagon stimulate the breakdown of glycogen, particularly in the liver

(A) P and Q (B) Q and R
(C) P and R (D) P and S

141. What would you expect about the formation of an α–helix for a segment of a protein chain that contains lysine approximately every fourth residue with all other residues being mostly hydrophobic?
(A) Helix formation would be favoured at low pH
(B) Helix formation would be favoured at high pH
(C) Helix formation would be favoured at neutral pH
(D) Helix formation would never our regardless of pH

142. NA primers are synthesized about once every:
(A) Millisecond (B) Second
(C) Minute (D) 40 minutes

143. Which proteins are responsible for the unwinding of the double-stranded DNA during replication?
(A) Ligases (B) Helicases
(C) Topoisomerases (D) Primases

144. The sequence of genes and proteins used in the initiation and termination of DNA replication is:
(A) Tus → ter → origin → DnaA
(B) dnaA → DnaA → origin → Tus → ter
(C) Origin → tus → ter → DnaA
(D) Ter → Tus → dnaA → origin DnaA

145. The principle forces holding subunits of an oligomeric protein to each other are?
(A) Peptide bonds
(B) Hydrophobic interactions
(C) Covalent bonds
(D) Disulfide bonds

146. A key role of the hydroxyl group at position 6 in the purine ring in the formation of a transition state by the enzyme adenosine deaminase is obtained by comparing a ------ and a -----------:
(A) Competitive inhibitor; noncompetitive inhibitor
(B) Transition state analog; normal substrate
(C) Noncompetitive inhibitor; transition state analog
(D) Competitive inhibitor; transition state analog

147. The role of serine at the active site of serine proteases is to act as a (n) -------- catalyst, while the histidine residue serves as a(n) ------- catalyst:
(A) Strong; weak (B) Weak; strong
(C) Acid base; covalent (D) Covalent; acid base

148. A membrane-affiliated protein is isolated by detergent extraction, and found to be amphipathic and lacking any nonprotein components. To which class of membrane proteins does it belong?
(A) Integral proteins (B) Peripheral proteins
(C) Lipid-anchored proteins (D) None of these

149. Which of the following statements about the mobility of membrane components is true?
(A) Proteins and lipids move equally rapidly within the plane of the membrane
(B) Lipids generally move more rapidly than lipids within the plane of the membrane
(C) Proteins generally move more rapidly than lipids within the plane of the membrane
(D) Lipids and proteins "flip-flop" between membrane leaflets more readily than they move within the plane oi the membrane

150. Protein subunits in a multisubunit protein are held to each other primarily by:
(A) Covalent bonds
(B) Hydrophobic interactions exclusively
(C) Both strong and weak interactions
(D) Hydrophobic and other weak interactions

151. Which of the following is not about beta oxidation of fatty acid containing even number of carbons:
(A) End product – acetyl-CoA
(B) Cofactor required – NAD + and FAD
(C) Decreases during starvation
(D) Site – mitochondria

152. If phosphoglyceride A has a higher T_m than phosphoglyceride B, which of the following differences between A and B may exist? (In each case only one parameter-either chain length or double bonds is compared)
P. A has shorter fatty acid chains than B
Q. A has longer fatty acid chains than B
R. A has more unsaturated fatty acid chains B
S. A has more saturated fatty acid chains than B
(A) P and Q are correct
(B) Only Q is correct
(C) Q and S are correct
(D) R and S are correct

153. A tetramer protein contains:
(A) Four different subunits
(B) Four identical subunits
(C) Three subunits and one prosthetic group
(D) (A) or (B) only

154. Which of the following statements correctly describe what happens when acetyl CoA is abundant?
P. Pyruvate carboxylase is activated
Q. Phosphoenolpyruvate carboxykinase is activated
R. Phosphofrutokinase is activated
S. If ATP levels are high, oxaloacetate is diverted to gluconeogenesis
(A) P and Q (B) Q and R
(C) Q and S (D) P and S

155. Which of the following is mismatched?
(A) ter : Tus (B) dnaA : DnaA
(C) Ori : Primer (D) Tus : helicase inhibitor

156. The best enzyme to use in the Sanger method for DNA sequence determination is:
(A) DNA polymerase I
(B) Klenow fragment of DNA polymerase I
(C) DNA polymerase a high temperature bacterium
(D) SSB protein

157. Which enzyme below is fastest?
(A) Kinase, $K_{cat} = 10^3$
(B) Papain, $K_{cat} = 10$
(C) Carboxypeptidase, $K_{cat} = 10^2$
(D) Catalase, $K_{cat} = 10^7$

158. Which of the following glycosidic bond is responsible for branching in amylopectin portion of starch:
(A) a, D -1, 4-Glucosidic bonds
(B) b, D- 1, 4-Glucosidic bonds
(C) a, D- 1, 6- Glucosidic bonds
(D) a, D- 1, 2- Galactosidic bonds

159. The structure of two anomers of fructopyranose differ only in respect of orientation of -OH group at:
(A) C-2 (B) C-1
(C) C-5 (D) C-4

160. A hyperbolic binding curve differs from a sigmoidal binding curve in that the hyperbolic curve:
(A) Has a single equilibrium constant for oxygen binding
(B) Binds more oxygen after the initial proteins first bind oxygen
(C) Shows cooperativity
(D) Binds up to four molecules of oxygen

161. Transition state analogs should:
(A) Stabilize transition states
(B) Have a dissociation constant of 10^{-13} M or less
(C) Bind very tightly to the enzyme
(D) A and C

162. Which factors does not help to explain why many proteins exhibit quaternary structure?
(A) Oligomers are usually more stable than the free monomers
(B) Active sites can be formed when the protein chains associate
(C) The subunits always are able to maintain the same three-dimentional structure whether they are associated into an oligomer or not
(D) Increased efficiency by the sharing of the same subunit with the same function among different proteins

163. The catalytic triad of hymotrypsin and other serine proteases consists of:
(A) Three subunits of the enzyme
(B) Three amino acid residues adjacent in the primary structure which at to make serine a strong nucleophile
(C) Three amino acid residues close enough in space to make serine a strong nucleophile
(D) Three enzymes with very similar structural features

164. Regulation of fatty acid biosynthesis occurs at the enzymatic step catalyzed by:
(A) Carnitine acyl transferase I
(B) Acetyl – coA carboxylase
(C) Pyruvate carboxylase
(D) Citrate-malate translocase

165. All of the following statements about acetyl-CoA carboxylase are correct except:
(A) It undergoes protomer-polymer interconversion during its physiological regulation
(B) It requires biotin
(C) It is inhibited cAMP-mediated phosphorylation
(D) It is activated by both palmitoyl-CoA and cirate

166. Cooperative binding of oxygen by hemoglobin:
(A) Is induced by hemoglobin
(B) Is a result of different affinities for oxygen by each subunit protein
(C) Is induced by oxygenation
(D) Is a result of interaction with myoglobin

167. Which statement is false about the heme group?
(A) When oxygen binds to heme, the iron ion is oxidized from Fe^{2+} to Fe^{3+}
(B) If exposed to air, a free heme group (not associated with hemoglobin) is readily oxidized converting Fe^{2+} to Fe^{3+} and an no longer bind oxygen
(C) The heme group is tightly, but non-covalently, held in myoglobin molecule
(D) The chemical structure of the heme groups in myoglobin and hemoglobin are identical

168. Which one of the following set have a max different of pH?
(A) Saliva-uria
(B) Bile-blood
(C) Bile-gastric juice
(D) Saliva-gastric juice

169. The Michaelis constant, Km, is equal to the:
(A) Maximum velocity that any given enzyme reaction can achieve
(B) Substrate concentration which gives the best enzyme assay for an enzyme reaction
(C) Substrate concentration when the rate is equal to half its maximal value
(D) Maximum velocity divided by two

170. The ratio of kcat/km for each of two different substrates present in equal concentrations in an enzyme reaction will measure enzyme affinity because:
(A) The rate of P formation are given by the values found for each substrate
(B) The rate of P formation are the same for each substrate
(C) It is an indication of the formation of ES for each substrate
(D) All of these

171. The main difference between prokaryotic and eukaryotic DNA replication mechanisms is:
(A) No Okazaki fragments are made in eukaryotes
(B) Both leading and lagging strands are continuous in eukaryotes
(C) No primer synthesis is needed in eukaryotes
(D) The replication fork moves slower in eukaryotes

172. Functions of DNA polymerases and neccessory proteins in eukaryotic DNA replication include:
(A) PCNA protein, a sliding clamp unity for DNA
(B) β-DNA polymerase, for repair of damaged DNA
(C) RPC protein, and equivalent of SSB proteins in bacteria
(D) RPC protein, used for mitochondrial DNA synthesis

173. In packaging the eukaryotic DNA into nucleosomes, it is though that the requirement for --------- and their synthesis is a cause of the slower rate of movement of the replication fork:
(A) Okazaki fragments
(B) Larger DNA molecules
(C) Increase number of accessory proteins
(D) Histone proteins

174. Membrane carbohydrates occur as:
(A) Either glycoproteins or glycolipids, but are not free in the membrane
(B) Either glycoproteins, or free in the membrane, but are not attached to lipids
(C) Either glycolipids or free in the membrane, but are not attached to proteins
(D) None of these

175. In the Lineweaver-Burk plot of an enzyme reaction, the Km is given by the:
(A) x-intercept
(B) y-intercept
(C) Negative reciprocal of the x-intercept
(D) Reciprocal of the y-intercept

176. Substance X is a charged molecules of small molecular weight. It is found at higher concentrations outside of cells than inside. How might substance X enter cells?

(A) Diffusion through a channel
(B) Diffusion through the lipid bilayer
(C) Active transport
(D) None of these

177. An equimolar mixture of D-and L-alanine does not rotate the plane of polarized light is:

(A) True (B) False
(C) Unknown (D) Same

178. The proximity effect of speeding an enzyme-catalyzed reaction is explained by:

(A) A large loss of entropy when reactive groups are brought close to each other
(B) A large gain in entropy due to binding of the substrate
(C) A conversion of a reaction from endergonic to exergonic
(D) Increasing the flexibility of the substrate by increasing the degrees of freedom of rotation

179. The role of amino acid residues at the active site of enzymes can be determined by removing certain residues using the technique of:

(A) Specific hydrolysis
(B) Covalent binding
(C) Acylation of specific residues
(D) Site mutagenesis
(E) All of these

180. Which of the following disaccharide is made up of two D-Glucose units:

(A) Sucrose (B) Lactose
(C) Maltose (D) None of these

181. The C-2 epimer of D-Glucose is:

(A) D-Mannose (B) D-Galactose
(C) D-Fructose (D) D-Altrose

182. Aspartate and lysine are in the active site of an enzyme. They are both known to participate directly in catalysis. The pKa's of the residues are found to be 3.2 and 9.6, respectively for aspartate and lysine. The optimum pH for the enzyme is 6.4. Which forms of these two residues will predominate when the enzyme is most active?

(A) Aspartate is protonated; lysine is deprotonated
(B) Both residues are protonated
(C) Aspartate is deprotonated; lysine is protonated
(D) Both residues are dprotonated

183. The Lineweaver-Burk plot and other linear transformation of the Michaelis-Menten curve of kinetics are valuable for:

(A) Determination of Km
(B) Determination of Vmax
(C) Determination of kcat
(D) Determination of types of enzyme inhibition
(E) All of these

184. The Km values for enzyme reactions such as $A + B \rightarrow C + D$

(A) Cannot be determined using the Lineweaver-Burk plot analysis
(B) Can be determined by holding one (A or B) at high concentration, while varying the concentration of the other substrate
(C) Can be determined for one substrate and not the other
(D) Do not indicate the efficiency of the enzyme

185. Gorter and Grendel's earliest evidence that membranes were formed from a lipid bilayer included:

(A) Microscopic examination of red blood cell membranes
(B) Measuring the surface area of lipids extra-cted from red blood cells, and comparing that to the surface area of the cells
(C) Measuring the mobility rates of membrane proteins
(D) None of the above

186. The distribution of K+ across an artificial membrane was measured, and the concentrations were found to be equal on both sides. Which of the following statements is true about the distribution of K+?

(A) K+ must be at equilibrium across the cell membrane
(B) K+ cannot be at equilibrium across the membrane
(C) There cannot be a membrane potential under these conditions
(D) More information is needed to determine whether K+ is at equilibrium

187. Why is only one copy of DNA made per cell division cycle in eukaryotes despite the presence of multiple origins of replication?

(A) There is only one origin of replication per chromosome
(B) The rate of elongation proceeds at a rate that allows only one copy to be made by the time the cell divides
(C) The activity of an S-phase protein kinase controls the formation of the pre-replication complex
(D) Primase is activated by phosphorylation only at the onset of S-phase by S-phase protein kinase

188. The reaction catalyzed by DNA photolyase is an example of what type of repair?

(A) Recombination (B) Excision
(C) Mismatch (D) Direct

189. The assumptions made in calculating the Michaelis-Menten equation include:

(A) That the formation and decomposition of ES is the same for a period of time
(B) That the concentration of the substrate is much greater than the concentration of E
(C) That the value of K-2 can be ignored
(D) A, B and C (E) (A) and (B) only

190. Antibodies are:

(A) Globular proteins (B) Fibrous proteins
(C) Conjugated proteins (D) Metallo proteins

191. In which compound delocalisation is not possible?

(A) 2- butane (B) 1, 3-butadiene
(C) 1, 3, 5-hexatriene (D) Benzene

192. Stability of alkyl carbocations can be explained by:

(A) Induction effect (B) Hyperconjugation
(C) Both inductive (D) Electrometric effect

193. A mixture of three peptide A, B and C is loaded in a 18 reverse phase chromatography column. A mixture of water and acetonitrile is used as the solvent with increased concentration of the latter I a gradient mode. If the hydrophobicities of the peptides are in the order A < B < C, the order in which thoy will oluto from tho column would bo:

(A) A followed by B followed by C
(B) B followed by C followed by A
(C) C followed by B followed by A
(D) A followed by C followed by B

194. Which of the following is not true of membrane phosphoglycerides?

(A) They include phosphatidylcholine, phosphatidylethanolamine and cholesterol, among others
(B) They are amphipathic
(C) They have two fatty acids attached to a glycerol molecules, and a polar head group attached via a phosphate
(D) They have both saturated and unsaturated fatty acids

195. If a fresh water plant we put in sea water, its cells would:

(A) Undergo plasmolysis (B) Lyse
(C) Swell with turgor pressure (D) All of these

196. In photoreactivation, DNA damage caused by UV light is reversed by DNA photolyase which:

(A) Is activated by UV light to form a complex
(B) Removes nicks in the DNA backbone
(C) Forms a complex with thymine dimmers which absorbs visible light and reverses the dimerization
(D) Binds to the adenine bases opposite the thymine dimer

197. Why is DNA damage in skin cells from exposure to excessive UV light not completely reversed by photoreactivation?

(A) DNA photolyase in humans is activated only under conditions of heat shock
(B) Humans lack a photoreactivation repair mechanism
(C) The presence of histones greatly slows photoreactivation, thus humans depends primarily on other forms of DNA repair
(D) Photoreactivation is specific for the removal of thymine dimmers. Thymine dimmers do not form in human DNA

198. Which statement is true about the removal of reversible inhibitors from enzyme solutions?

(A) PAGE is the primary technique used for this purpose
(B) Dialysis and gel filtration are often used
(C) These inhibitors cannot be separated from the enzyme without treatment with Disopropylfluorophosphate
(D) Removal of reversible inhibitors is extremely difficult and can be achieved only after many purification steps

199. In a certain enzyme-catalyzed reaction the following steps occur:

1. A phosphate group on substrate A is transferred to a side chain of an active residue of the enzyme
2. The dephosphorylated form of substrate A dissociates from the enzyme
3. Substrate B enters the active site and is phosphorylated with simultaneous regeneration of the enzyme in its original form

What kind of kinetic mechanism is described?

(A) Random (B) Sequential
(C) Ordered (D) Ping-pong

200. For determining the K_m values of enzyme catalyzed reaction, it is customary to plot the kinetic data in a variety of ways. The plot of V versus V/S yields a straight line and the intercepts on the Y-axis gives:

(A) K_m/V_{max} (B) V_{max}/K_m
(C) $1/V_{max}$ (D) Hill coefficient, n

201. What distinguishes reversible inhibitors from irreversible inhibitors?

(A) Reversible inhibitors are not covalently bound to enzymes but irreversible inhibitors are
(B) There is an equilibrium between bound and unbound reversible inhibitor. There usually is little back reaction for the binding of an irreversible inhibitor
(C) Reversible inhibitors are easier to purify from solutions of enzymes than irreversible inhibitors
(D) All of these

202. The reason to rewrite the Michaelis-Menten equation (such as the Lineweaver-Burk plot) is to:

(A) Visualize reactions better
(B) Form enzyme kinetic data as a hyperbolic curve
(C) Calculate catalytic proficiency
(D) Calculate Vmax and Km

203. Two substances are used to produced a certain biological product in an enzyme-catalyzed react-ion. It is found that both substrates must bind to the enzyme, first one, then the other before the product is produced. This is an example of a/an:

(A) Linear reaction
(B) Ordered sequential reaction
(C) Random sequential reaction
(D) Ping-pong reaction

204. A thermodynamic pit occurs when:
(A) ES is not very stable
(B) ES forms faster than it dissociates
(C) ES is highly stable
(D) S is not bound tightly to an enzyme

205. Non classical competitive inhibition involves:
(A) Binding of the substrate at both the active site and at the inhibitor site
(B) Binding of either the substrate to the active site or the inhibitors to its own binding site thus preventing the other from binding
(C) Binding of the inhibitor to the substrate followed by binding of this complex to the active site
(D) Chemical removal of the substrate form the active site of reaction with the inhibitor

206. Which among the given compound is most stable?
(A) 1- butane
(B) 2, 3-dimethyl-2 butene
(C) cis-2-butene
(D) trans-2-butene

207. A mixture containing protein – 1, -2, -3, -4 and -5 with molecular weights 5000, 10000, 25000, 65000 and 100000, respectively, were separated as Sephadex G-50 column. The order of elution of these proteins from the column will be:
(A) Protein -1, protein – 2, protein – 3, protein – 4, and protein - 5
(B) Protein -5, protein – 4, protein – 3, protein – 2, and protein - 1
(C) Protein -1, -2, and – 3 elute first, followed by protein – 5 and - 4
(D) Protein - 4, and – 5 elute first, followed by protein – 3, - 2, and - 1

208. The general excision repair pathway for DNA repair has the following order:
(A) Endonuclease cuts → helicase or endonuclease removes → DNA polymerase → DNA ligase
(B) DNA ligase → Endonuclease cuts → helicase or endonuclease removes → DNA polymerase
(C) helicase or endonuclease removes → Endonuclease cuts → DNA polymerase → DNA ligase
(D) DNA polymerase → DNA ligase → helicase or endonuclease removes → Endonuclease cuts

209. In DNA repair, DNA glycosylases recognize deaminated bases within the DNA polymer and remove the deaminated:
(A) Nucleoside (B) Nucleotide
(C) Thymine dimmers (D) Bases

210. CH3 group of toluene is o, p-directing groups due to the:
(A) Hyperconjugation
(B) Resonance
(C) Inductive effect
(D) Electromeric effect

211. In a ping-pong reaction which does not occur?
(A) One product is released before a second substrate is bound
(B) The enzyme covalently binds a portion of the first substrate
(C) The enzyme is permanently converted to an altered form by the first substrate
(D) A group is transferred from one substrate to another

212. Which of the following statements about gluconeogenesis are true?
P. It occurs actively in the muscle during periods of exercise
Q. It occurs actively in the liver during periods of exercise or fasting
R. It occurs actively in adipose tissue during feeding
S. It occurs actively in the kidney during periods of fasting
(A) P and Q (B) Q and R
(C) Q and S (D) P and S

213. Which statement is correct for inductive effect?
(A) It is permanent effect
(B) It is the property of single bond
(C) It is cause permanent polarization in the mole
(D) All of these

214. Which statement is correct for toluene:
(A) It has six vinylic carbons
(B) It has one benzyl carbon
(C) Benzyl hydrogen is acidic in character due to resonance
(D) All are correct

215. Match the transition state or chemical entity of each enzyme that is responsible for their catalytic function:

Column I	Column II
P. Ribonuclease	1. Oxyanion
Q. Lysozyme	2. Pentacovalent phosphorus
R. Chymotrypsin	3. Carbonium ion
S. Carboxypeptidase	4. Mixed anhydride

(A) P-3, Q–2, R–4, S-1 (B) P-2, Q–3, R–1, S-4
(C) P-2, Q–1, R–3, S–4 (D) P-4, Q- 3, R–2, S–1

216. Which statement does not apply to transition states?
(A) Many have been detected experimentally
(B) Chemically bonds are in the process of being formed and broken
(C) Have lifetimes on the order of 10-14 to 10-13 seconds
(D) Differ in energy from the ground state by the activation energy

217. Two or more polypeptide chains present in the protein molecule are covalently cross linked by one or more:
(A) Diester bonds (B) Hydrogen bonds
(C) Disulphide bonds (D) Ionic bonds

218. Match the coefficients in group 1 with their corresponding downstream processing steps given in group 2:

Group 1	Group 2
P. Sedimentation coefficient	1. Aqueous two-phase extraction
Q. Partition coefficient	2. Ultrafiltration
R. Rejection coefficient	3. Dialysis
S. Activity coefficient	4. Centrifugation

(A) P-3, Q–1, R–4, S-2 (B) P-2, Q–1, R–4, S–3
(C) P-4, Q–3, R–1, S–2 (D) P-4, Q–1, R–2, S–3

219. Protamins and histones form conjugated proteins in association with:

(A) Carbohydrates (B) Glycerides
(C) Nucleic acids (D) Vitamins

220. Most Km values of enzymes for their substrates are on the order of ----------M:

(A) 10^{-2} (B) 10^{-3}
(C) 10^{-4} (D) 10^{-5}

221. A histidine was determined to be the critical residue involved in an enzyme catalyzed reaction. If the Ka of the histidine is known to be 6.5 in the active site and the pH of maximum catalytic activity is 7.2. What is likely the primary role of histidine in the catalytic reaction?

(A) Act as a proton donor
(B) Forms a covalent bond with the substrate
(C) Stabilizes a charged intermediate
(D) Reduces the entropy of the substrate

222. The number of elelocalised π–electrons in the given compound is:

(A) 4 (B) 6
(C) 8 (D) 10

223. Which statement is correct for CH_3- CH_2:

(A) It is paramagnetic is character
(B) It is neutral electrophile
(C) Formation take place hemolytic bond fission
(D) All of these

224. The T_m of phosphatidyl choline A is higher than T_m of phosphatidyl choline B because:

(A) A has shorter chain fatty acid and more unsaturated fatty acid than B
(B) A has longer chain fatty acid and more saturated fatty acid than B
(C) A has shorter chain fatty acid than B
(D) A has more is-unsaturated fatty acid than B

225. Deaminated bases must be removed from DNA because:

(A) They distort the molecule
(B) They form thymine dimmers
(C) They repair incorrectly during replication
(D) The mRNA formed may be irregular

226. In the holiday model describing recombination:

(A) Homologous sequences line up
(B) Strands of DN are nicked
(C) DNA strands associate with the homologous strand ("strand invasion")
(D) Strands rotate then are cleaved at the crossover point
(E) All of these

227. In the nonpolar environment of most enzyme active sites, which statement applies to charge-charge interaction between the enzyme and the substrate?

(A) They are rare due to the non polar environment
(B) They are frequent, but not very strong in the non polar environment
(C) They are stronger in the non polar environment
(D) The polarity of the active site has no effect on the strength of the charge-charge interactions

228. The Bohr effect in hemoglobin refers to the:

(A) Effect of pH on hemoglobin and myoglobin
(B) Higher pH found actively metabolizing tissues
(C) Increased affinity for O_2 at lower pH
(D) Reduced affinity for O_2 at lower pH

229. Primary structure of protein indicates:

(A) Sequence of amino acid present in it
(B) Helical and coiled shape of protein molecules
(C) Steric interactions of different amino acids present in it
(D) A and C both

230. The test is used to detect amino acids, peptides and proteins is:

(A) Polarimeter test (B) Ninhydrin test
(C) Bardfored test (D) Biuret test

231. Which one of the following is iron containing protein:

(A) Cytochrome 'C'
(B) Ascorbic acid anhydrase
(C) Alcohol dehydrogenase
(D) Nitrogenase

232. In toluene methyl group has neither -ve charge nor lp even then it is o, p-directing group for arse reaction due to:

(A) Induction effect (B) Electrometric effect
(C) Hyperconjugation (D) Mesomeric effect

233. How does hemoglobin carry carbon dioxide generated in tissues back to the lungs:

(A) By coordination with heme
(B) By forming N-terminal carbamate
(C) By forming – terminal carbamate
(D) By linking to the epsilon –amino group of lysine

234. The enzyme has an active site which:

(A) Fits the substrate exactly
(B) Fits the transition state
(C) May contain hydrogen bonds which are covalent
(D) Both (B) and (C)

235. The role of ser-195 in hymotrypsin cleavage of a peptide bond is that of a(n):
(A) Acid catalyst (B) Proximity effector
(C) Strong nucleophile (D) Weak nucleophile

236. Which one of the following reagent is used in the determination of amino acid sequences of protein at the N-terminal of the protein chain:
(A) 2,4-Dinitrofluorobenzene
(B) Carboxypeptidase
(C) 2,4-Dinitropheny hydrazine
(D) Hydroxyl amine

237. The formation of the anlinotiazolinone derivative (ATZ-amino acid) in anhydrous trifluoroaetic acid, thereby liberating the N-terminal amino acid residue in a cyclic form, and leaving the remaining peptide truncated by one residue is called:
(A) Coupling (B) Cyclization
(C) Conversion (D) None of these

238. Which of the following fatty acids has the lowest melting point?
(A) Fatty acids with sites of unsaturation with is double bonds
(B) Fatty acids with sites of unsaturation with trans double bonds
(C) Fatty acids with no sites of unsaturation
(D) Fatty acids with longer hydrophobic tails

239. During recombination in *E. coli* -------- recognize (s) regions of sequence similarity and promote(s) strand invasion, while ------- bind(s) to a Holiday junction promoting strand migration:
(A) RuvA and RuvB; RecBCD
(B) RecBCD: RecA
(C) RecA; RecBCD
(D) RecA; RuvA and RuvC

240. The sequence of proteins used in bacterial recombination is:
(A) RuvC $\rightarrow$ RuvB $\rightarrow$ RecA $\rightarrow$ RecBCD
(B) RecBCD $\rightarrow$ RecA $\rightarrow$ RuvAB $\rightarrow$ RuvC
(C) RecA $\rightarrow$ RecBCD $\rightarrow$ RuvAB $\rightarrow$ RuvC
(D) RuvAB $\rightarrow$ RuvC $\rightarrow$ RecBCD $\rightarrow$ RecA

241. Casein found in milk is an example of:
(A) Phosphoprotein (B) Fibrous protein
(C) Glycoprotein (D) Flavoprotein

242. Experiments on the bacterial serine protease subtilisin show that even when all three residues of the catalytic triad are mutated, the catalytic rate of the enzyme is still 3000 times the uncatalyzed reaction. Which mode of catalysis is likely responsible for this remaining catalytic activity?
(A) Acid-base catalysis
(B) Covalent catalysis
(C) Transition-state stabilization
(D) Hydrophobic effects

243. Which amino acid is least likely to participate in acid-base catalysis?
(A) Lysine (B) Tyrosine
(C) Aspartate (D) Histidine

244. Active holoenzymes are formed from ----------- in the presence of--------:
(A) Cofactors; proteins
(B) Proteins; cofactors
(C) Apoenzymes; cofactors
(D) Apoenzymes, proteins

245. Highly cooperative binding of a ligand to multiple binding sites on a macromolecule is best demonstrated by:
(A) Adair equation (B) Hill plot
(C) Lineweaver – Burk plot (D) Arrhenius plot

246. Because coenzymes are specific for the chemical groups that they accept and donate, they are referred to as:
(A) Cofactors (B) Reactive centres
(C) Activator ions
(D) Group-transfer reagents

247. Ca^{++} or Mg^{++} are most likely to be part of --------, while Zn^{++} or Fe^{++} are present in:
(A) Metal-activated enzymes; metalloenzymes
(B) Metalloenzymes, metal-activated enzymes
(C) Cofactors; coenzymes
(D) Coenzyme; cofactors

248. Ethanol (CH_3CH_2OH) is the alcohol found in beverages. It is oxidized in the body to acetaldehyde by the enzyme alcohol dehydrogenase. Methanol (CH_3OH) also known as wood alcohol, is converted to formaldehyde by the same enzyme. Acetaldehyde is toxic, but formaldehyde is far more toxic to humans, which is why the ingestion of relatively small amounts of methanol can cause blindness or death. One treatment for mild methanol poisoning is the administration of ethanol. Why might a doctor choose this treatment?
(A) Ethanol must at as a competitive inhibitor for the alcohol dehydrogenase and there fore, slows the formation of formaldehyde
(B) Ethanol likely irreversible binds to alcohol dehydrogenase which prevents the formation of formaldehyde
(C) The ethanol is likely an uncompetitive inhibitor and binds to a site other than the active site of the enzyme
(D) The doctor has given up on the patient and administers ethanol for sedation

249. A saturated characteristics common to lipids which allow them to function as good energy stores is:
(A) They are all hydrophobic
(B) They are all hydrophilic
(C) They have large numbers of carbon-phosphorus bonds
(D) They have large numbers of carbon-hydrogen bonds

250. Meselson and Stahl's experiments involved growing *E. coli* on a medium containing 15N (heavy nitrogen) which increased the density of the bacterial DNA. Cells were then grown on a medium containing only 14N. DNA samples were obtained for analysis after one, two and three rounds of replication (generations) and centrifuges on a density gradient. The double-helical DNA forms bands in the gradient depending on its density. How many bands did Meselson and Stahl see after two generation of cells were grown on the 14N medium?

(A) One with DNA containing both 15N and 14N
(B) Two, one with DNA containing only 15N and one with DNA containing only 14N
(C) Two, one with DNA containing both 15N and 14N and one with DNA containing only 14N
(D) Three, one with DNA containing only 15N, one with DNA containing both 15N and 14N and one with DNA containing only 14N

251. It takes about the same amount of time for eukaryotic chromosomes to be replicated in vivo as it does for *E. coli*. Why is this true?

(A) Eukaryotes have multiple origins of replication, *E. coli* has one unique origin
(B) Most chromosomes are approximately the same size
(C) Eukaryotic replication occurs at a faster rate, enabling replication of the larger chromosomes in about the same amount of time
(D) The replication of DNA in eukaryotes is faster due to a quicker, accurate repair of mismatches

252. The role of zinc in the mechanisms of carbonic anhydrase is to:

(A) Maintain the configuration of the holoenzyme
(B) Bind to three histidine residues
(C) Produce a nucleophilic attack on the substrate
(D) Promote ionization of bound water

253. Which of the following would produce a Hill plot with negative cooperativity:

(A) A protein has multiple subunits each with a single binding site. Binding a ligand to one site decreases affinity of other sites for ligand
(B) A protein is a single polypeptide chain with two ligand sites each with different ligand affinities
(C) A protein has a single binding site. But during purification the protein preparation is heterogeneous, hence there are different protein molecules with different ligand affinities
(D) A protein has multiple binding sites for different ligands. Binding of one type of ligand at a higher affinity site promotes binding of the other ligand at a lower affinity site

254. Insulin is a:

(A) Vitamin
(B) Hormone
(C) Protein
(D) Both (B) and (C)

255. Site directed mutagenesis is used to study enzymes by:

(A) Changing the location of the active site
(B) Producing enzymes with different amino acid residues
(C) Producing enzymes with modified R groups
(D) Changing the pH of the environment

256. What is the role of the magnesium ion in kinases that require the magnesium-ATP complex to donate phosphoryl groups?

(A) Maintain the configuration of the holoenzyme
(B) Shield the charged phosphate groups of ATP
(C) Produce an electrophilic attack on the substrate
(D) Promote ionization of bound water

257. Which one of the following enzymes is used to selectively break the peptide bond adjacent to free C-terminal of polypeptide:

(A) Co-carboxylase (B) Carboxypeptidase
(C) Carbonic anhydrase (D) Catalase

258. SDS is used while separating proteins by polyacrylamide gel electrophoresis because:

(A) It helps in solubilization of proteins thereby making it easier to separate
(B) It binds to proteins and confers uniform negative charge density thereby making them move during electrophoresis
(C) Decreases the surface tension of the buffer used for electrophoresis
(D) Stabilizes the proteins

259. An ion commonly found in metalloenzymes and which can undergo reversible oxidation and reduction is:

(A) Ca^{++} (B) Mg^{++}
(C) S^{-} (D) Fe^{++}

260. Replacement of the amino acid ----- at or near an active site of an enzyme is more likely to change enzyme activity than the replacement of ----- at or near the active site:

(A) Histidine; leucine
(B) Leucine; histidine
(C) Leucine; isoleucine
(D) Histidine; aspartate

261. In an alpha-helix hydrogen bonds are:

(A) Within a single chain
(B) Between chains that run side by side
(C) Between polar amino acids and water
(D) Only between amino acids of opposite charge

262. Hemoglobin and myoglobin both have all of the following characteristics except:

(A) Consist of subunits designed to provide hydrogen bonds and nonpolar interaction with other subunits
(B) Highly alpha helical
(C) Bind one molecules of heme per globin chain
(D) Bind heme in a hydrophobic pocket

263. S-adenosylmethionine (SAM) is formed from the amino acid methionine so that:
(A) Methyl groups can be formed
(B) It can react with nucleophilic acceptors
(C) It can donate methyl groups in many biosynthetic reactions
(D) Both (B) and (C)

264. Which of the following reagent is used to break cross linkage of polypeptide chains of a complex protein:
(A) Performic acid
(B) Periodic acid
(C) t-Butyl chromate
(D) Potassium permanganate

265. A nucleotide sugar coenzyme involved in carbohydrate metabolism (UDP – glucose) is formed when UTP reacts with a glucose molecules. More UTP is made available for additional reactions by the transfer of a phosphate group from:
(A) Another carbohydrate (B) ATP
(C) ADP (D) UTP

266. NAD and NADP dependent dehydrogenases catalysis substrates by transferring ------- to C- 4 of NAD^+ and $NADP^+$:
(A) One electron (B) Two electron
(C) One electron and one proton
(D) Two electron and one proton

267. The following data were obtained in the presence and absence of inhibitor. What type of inhibition is shown?

Substrate concentration (millimolar)	Rate without inhibitor (mmol/min)	Rate with inhibitor (mmol/min)
0.100	2.5	1.6
0.200	4.2	2.9
0.500	6.6	5.1
0.750	7.4	6.2
1.000	9.9	9.8
2.000	10.0	10.1

(A) Competitive (B) Uncompetitive
(C) Noncompetitive (D) Irreversible

268. Which statement is false about regulatory enzymes that are controlled allosterically?
(A) They are always less active when a modulator is bound to them
(B) They are often larger than other enzymes
(C) They have more than one binding site
(D) They often catalyze the first step in a reaction pathway

269. In aldol addition reaction product is always:
(A) β-hydroxy aldehyde
(B) β -hydroxyl ketoner
(C) α- β –unsaturated aldehyde
(D) α- β -unsaturated ketenes

270. Which one of the following methods is used for protein purification:
(A) Dialysis (B) Osmosis
(C) HPLC (D) GLC

271. One dimensional polyacrylamide gel electrophoresis in the absence of SDS (native gel electrophoresis) resolves proteins based on their mass and:
(A) Charge
(B) Hydrodynamic volume
(C) Secondary structure
(D) Hydrophobicity

272. Rank the following species in order of decreasing nucleophilicity in a polar solent:
1. $CH_3CH_2CH_2O$ 2. CH_3CH_2-CH_2S 3. $CH_3CH_2\overset{O}{C}$-O
(A) 3 > 1 > 2 (B) 2 > 3 >1
(C) 1 > 3 > 2 (D) 2 > 1 > 3

273. Which of the following statements is true:
(A) $CH_3CH_2S^-$ is both a stronger base and more than $CH_3CH_2O^-$
(B) $CH_3CH_2S^-$ is a both a stronger base but is less nucleophic than $CH_3CH_2O^-$
(C) $CH_3CH_2S^-$ is a weaker base and more nucleophilic than $CH_3CH_2O^-$
(D) $CH_3CH_3S^-$ is both a weaker base and less nucleophilic than $CH_3CH_2O^-$

274. Claisen-Shmidt reaction taken place between:
(A) Aldehyde & aldehyde
(B) Aromatic aldehyde & kelone
(C) Kelone & kelon
(D) Aromatic aldehyde & compound having active methyl group

275. The bonds in protein structure that are not broken on denaturation:
(A) Hydrogen bonds (B) Peptide bonds
(C) Ionic bonds (D) Disulphide bonds

276. The amino acids are said to be ketogenic when the carbon skeleton is finally degraded to:
(A) Succinyl –CoA (B) Fumarate
(C) Acetyl-CoA (D) Pyruvate

277. Which does not return to its original form after each catalysis?
(A) Prosthetic groups (B) Cosubstrates
(C) Metalloenzyme (D) None of these

278. Which pair will give claisen condensation reaction?
(A) $C_6H_5COOC_9H_5$ and $HCOOC_6H_5$
(B) $C_6H_5COOC_9H_5$ and CH_3CHO
(C) $CH_3COOC_9H_5$ and $HCOOC_9H_5$
(D) $HCOOC_9H_5$ and $HCOOC_9H_5$

279. Clemmenson reduction is used for the reduction of:
(A) Carbonyl compound (B) Ester
(C) Nitro group (D) All of these

280. Which statement is false about allosteric regulation?

(A) It is usually the mode of regulation for the last step in reaction pathways since this step produces the final product

(B) Cellular response is faster with allosteric control than by controlling enzyme concentration in the cell

(C) The regulation usually is important to the conservation of energy and materials in cells

(D) Allosteric modulators bind non-covalently at sites other than the active site and induce conformational changes in the enzyme

281. A solution of DNA has an absorbance at 260 nm of 0.212. What is the concentration of the DNA solution expressed as millimolarity:

(A) 0.03 Mm (B) 05 Mm
(C) 07 Mm (D) 01 Mm

282. Among the various interactions that help in the sustenance of the tertiary structure of proteins, the weakest one is:

(A) Hydrogen bonding
(B) Van der Waals interactions
(C) S-S bonding
(D) Hydrophobic interactions

283. All of the following are correct about ornithine except:

(A) It may be formed from or converted to glutamic semialdehyde
(B) It can be converted to proline
(C) It plays a major role in the urea cycle
(D) It is chemically spermidine

284. An oligonucleotide with the sequence GAACTACG-TTCGATCAAT is suspended in 750 μL of water. A 20 μL aliquot is diluted to a final volume of 1000 μL with water, and the absorbance at 260 nm is determined for the diluted sample. An OD of 0.242 is obtained. What is the concentr-ation of the oligonucleotide stock solution in pmol/μL?

(A) 59.46 (B) 61.35
(C) 04.35 (D) 50.24

285. You have 5 mL of bacteriophage λ stock having a concentration of 2.5 x 1011 λ phage/mL. You will purify λ DNA using the entire stock. Assuming 80% recovery, how many micrograms of λ DNA will you obtain? Phage genome size = 48502 bp:

(A) 53.4 μg (B) 60.04 μg
(C) 12.65 μg (D) 25.40 μg

286. Purines and pyrimidines exist in two tautomeric forms (lactam and lactim); nucleosides exist in two conformations (syn and anti). The most common forms found in cells are:

(A) Lactam and anti
(B) Lactum and syn
(C) Lactin and anti
(D) Lactim and syn

287. In DNA, phosphodiester bonds form with hydroxyl groups of carbons number ----- and --------- of each deoxyribose sugar:

(A) 2'; 3' (B) 2'; 4'
(C) 3'; 4' (D) 3'; 5'

288. Cells fed 5-methyl uracil (radioactive in the methyl group) will form radioactive:

(A) RNA (B) DNA
(C) Ribosomes (D) tRNA

289. Which is a true statement concerning the transition state of an SN2 reatcion?

(A) Closely resembles a carbocation intermediate
(B) The electrophile is responsible for the reaction
(C) Lower is energy than the starting materials
(D) Involves both the nucleophic and electrophile

290. Increasing the concentration of an electrophile in a typical SN2 reaction by a factor of 3 and the concentration of the nucleophic by the a factor of 3 will change the reaction rate to:

(A) Increase by a factor of 6
(B) Increase by a factor of 9
(C) Decrease by a factor of 3
(D) Remain about the same

291. Consider the following statements:

1. Bridgehead halices are inert towards both SN^1 and SN^2, reactions (till of the ring size is eight member ring)
2. The first step in both SN^1 and E1 reactions is the same
3. SN^2 reactions proceeds with total reaction of configuration
4. E2 elimination are by the use of a solvent of low polarity and high concentration of a strong base

Which of the above statement are correct?

(A) 1, 2 and 4 (B) 1 and 3
(C) 2, 3 and 4 (D) 1, 2, 3 and 4

292. Deoxythymidine is a ---- found mainly in DNA:

(A) Nucleotide (B) Deoxynucleotide
(C) Nucleotide (D) Deoxynucleoside

293. The difference between dGMP and GMP is:

(A) The type of phosphodiester linkage
(B) One hydroxyl group
(C) Syn versus anti conformation
(D) The presence of DNA or RNA

294. An enzyme is irreversibly inhibited by diisopropylfluorophosphate (DFP). What does this show?

(A) The enzyme has been denatured by DFP
(B) Serine is likely an important residue in the active site
(C) DFP is an allosteric modulator of the enzyme
(D) DFP is an analog of the enzyme's substrate

295. Adenylate and AMP each contain ---------- phosphate groups:

(A) 0 (B) 1
(C) 2 (D) 3

296. One of the characteristics of a peptide bond is that:
(A) There is no restriction in rotation around peptide bond
(B) There is a restricted rotation around peptide bond
(C) Its bond energy is approximately 5 Kcal/mole
(D) It is longer than 0.2 nm

297. Which of the following statements regarding reduced glutathione is not true?
(A) It contains one γ-carboxyglutomate, one cysteine, and one glycine residue
(B) It keeps the cystein residues of proteins in their reduced states
(C) It is regerenerated from oxidized glutathione by glutathione reductase
(D) It reacts with hydrogen peroxide and organic peroxides

298. The abbreviation dGp indicates:
(A) 5' deoxyguanylate
(B) 3' deoxyguanylate
(C) 3', 5' deoxyadenylate
(D) 5', 3' deoxyguanylate

299. The abbreviation pdAp indicates:
(A) 5', 3' deoxyadenylate
(B) 3', 5' deoxyadenylate
(C) 5' deoxyadenylate
(D) 3' deoxyadenylate

300. The abbreviation pppT indicates:
(A) TMP (B) TDP
(C) TTP (D) 3'5'TP

301. According to Chargaffs observations of nucleotide composition of DN samples:
(A) % of (G + C) + % of (A + T) = 100%
(B) A=T
(C) G = C
(D) % A + % G + % T = 100%

302. Much of the stability of the double-stranded DNA structure is the result of:
(A) Hydrogen bonding between purines
(B) The phosphodiester backbone
(C) The angle of the lanes of the bases with respect to the helix axis
(D) The stacking interactions between base pair

303. The rise and pitch of B-DNA are 0.33 nm and 3.40 nm, respectively. About how many helical turns are there in a fragment 1 mm in length?
(A) 3030 (B) 294
(C) 330 (D) 0.0034

304. Which one of the following is responsible for pigment formation in skin and hair:
(A) Collagen
(B) Elastin
(C) Fibroin
(D) Melanin

305. Regions of DNA that are most easily unwound have:
(A) About half G and half C
(B) Alternating A and G
(C) Greater G : C content
(D) Greater A : T content

306. Which is not true of the different conformations of DNA?
(A) Z-DNA is a left-handed spiral
(B) A-DNA and B-DNA are right handed spirals
(C) A-DNA and Z-DNA segments are limited to short regions of DNA
(D) Both A-DNA and B-DNA are dehydrated

307. In addition to knowing the chemical structures of the nucleotide, Watson and Crick used -------- of Franklin and Wilkins and the chemical equivalencies of Chargaff in order to propose their model of DNA structure:
(A) Sequence information
(B) UV spectra
(C) % (G + C) and % (A + T)
(D) X-ray diffraction data

308. In proteins, amino acids are linked by peptide bonds; in poynucleotides, nucleotides are linked by:
(A) Phosphoanhydride bonds
(B) 3'-5' phosphodiester bonds
(C) 5'-3' phosphodiester bonds
(D) B and C

309. The tatranucleotide AGTC (in DNA) has a free hydroxyl group on:
(A) A (B) A G, T and C
(C) C (D) G, T and C

310. The enzyme containing non-protein group in addition to a-amino acids are known as:
(A) Apo-enzyme (B) Co enzyme
(C) Holoenzyme (D) Metal activator

311. Five prime to three prime description of a DNA strand refers to:
(A) The phosphorylated 5' carbon at one, the 3' at the other end
(B) The location of the hydroxyl groups
(C) The phosphodiester backbone of DNA
(D) The phosphorylated 5 carbon on one sugar and the 3 carbon on the next sugar

312. Lysine is an amino acid with three ionizable groups. These are the α–COOH, 2α – amino and ε – amino groups with pKa values of 2.2, 9.2 and 10.8, respectively. The isoelectric point (pI) for lysine is:
(A) 5.7 (B) 6.5
(C) 9.2 (D) 10.0

313. Which one of the following possesses enzyme activity:
(A) Coenzyme (B) Apoenzyme
(C) Cofactor (D) Holoenzyme

314. DNA polymerase contains a lysine residues that is important for binding to DNA. Mutations were found that converted this lysine to either glutamate, glyine, valine or arginine. Which mutations would be predicted to be the most and least harmful to the ability of the enzyme to bind DNA:

	Most	Least
(A)	Glycine	Arginine
(B)	Arginine	Glycine
(C)	Glutamate	Valine
(D)	Glutamate	Arginine

315. The nitrogenous base present in lecithin:
(A) Choline (B) Ethanolamine
(C) Inositol (D) Serine

316. As B-DNA is gradually heated, the absorbance at 260 nm:
(A) Increases (B) Decreases
(C) Stays the same
(D) Is half way between that of poly (AT and poly, GC)

317. Adenine is a:
(A) Purine base (B) Pyrimidine base
(C) Pyridine base (D) Pteridene base

318. Which does not apply to most bacterial DNA?
(A) Circular (B) Square
(C) Not packed into nucleosomes
(D) Supercoiled

319. Which of the following is not true about alanine:
(A) At pH 1, the overall charge is +1
(B) At pH 1, it will move towards cathode
(C) An equimolar mixture of D and L alanine does not rotate the plane polarized light
(D) Contains branched side chain

320. Amino transferases:
(A) Usually require alpha-ketogluterate or glutamine as one of the reacting pair
(B) Catalyze reactions that results in a use or produ-ction of amino acids
(C) Catalyze irreversible reactions
(D) Require pyridoxal phosphate as an essential cofactor for the reaction

321. ------- plays and important role in storing the energy needed for the unwinding of DNA during replication, recombination, repair and transcription:
(A) A high percent of G : C base pairs
(B) The charge on basic side chains of histone proteins
(C) Negative supercoiling
(D) Overwinding of the helix

322. Scurvy disease is caused due to deficiency of:
(A) Retinol (B) Thiamine
(C) Nicotinic acid (D) Ascorbic acid

323. Which of the following is a molybdenum containing enzyme:
(A) Nitrogenase (B) Peroxidase
(C) Carbonic anhydrase (D) Catalase

324. Which type of RNA is the most abundant in living cells (by percent):
(A) Ribosomal (B) Messenger
(C) Small (D) Transfer

325. Nucleosides are condensation products of:
(A) Pentose sugar and phosphoric acid
(B) Nitrogen base and phosphoric acid
(C) Pentose sugar and nitrogen base
(D) A and B both

326. Histidine is often found at the active-site of enzymes because:
(A) It has a cyclic group
(B) It an form hydrogen bonds
(C) It has a pKa of 6.8
(D) It is an imino acid

327. Match the entries in column A against those in column B:

Column A	Column B
P. The amino acid with maximum number of codons	1. Cys
Q. The characteristic amino acid in lysozyme active site	2. Trp
R. The amino acid precursor in heme biosynthesis	3. Ser
S. The amino acid contributing the most to protein absorbance at 280 nm	4. Gly
	5. Glu

(A) P–2, Q–3, R–1, S-4 (B) P–3, Q–5, R–4, S-2
(C) P–5, Q–3, R–2, S-1 (D) P–4, Q–1, R–5, S-3

328. Which one of the following statements about prostaglandins is incorrect?
(A) Prostaglandins have a very short half-life
(B) Prostaglandins are synthesized only in the liver and the adrenal cortex
(C) Prostaglandins generally at locally on or near the tissue that produced them
(D) The common precursor of the prostaglandins us arachidonic acid

329. Which type of RNA carries activated amino acids to the ribosome for assembly into proteins?
(A) rRNA (B) tRNA
(C) mRNA (D) sRNA

330. What feature does a Ramachandran plot display?
(A) Allowed angles of phi and psi for a polypeptide backbone
(B) Preferred amino acids in an a-helix
(C) The hydropathy of amino acids
(D) The variation of pH versus volume of base added during titration to determine the pKa

331. Which is not an explanation of why site-directed mutagenesis is a more powerful tool than many other techniques for determining important residues in an enzyme?

(A) It is a more specific technique than irreversible inhibition experiments

(B) It can be used on enzymes for which no specific irreversible inhibitor is known

(C) It allows for testing of the specific function of amino acid side chains in an enzyme

(D) It is specially useful for enzymes whose sequence is not known

332. It is easier to melt DA richer in AT than GC because:

(A) It is more heat sensitive

(B) There is one less hydrogen bond

(C) The helix pitch is longer in AT rich regions

(D) All of these

333. A brown-eyed man has one allele for brown eyes and one for blue eyes. Half his sperm carry the allele for brown eyes, and half carry the allele for blue eyes. This is an illustration of:

(A) Mendel's law of independent segregation

(B) Mendel's law of independent assortment

(C) Mendel's law of segregation

(D) None of these

334. Both epinephrine and glucagons bind to G protein-coupled receptors and activate glycogen break-down. Therefore epinephrine and glucagons must:

(A) Have very similar structures and bind to the same receptor

(B) Bind to receptors with different ligand-binding sites but similar functions

(C) Bind to different receptors and activate different second messengers

(D) Bind to the same receptors, one being intracellular and the other, extracellular

335. Which of the following is not a characteristic of DNA?

(A) It is less susceptible to alkaline hydrolysis than RNA

(B) It has a hydroxyl group attached to the 3' carbon at the 3' end

(C) It is stabilized by hydrogen bonding, van der Waals forces and covalent interaction

(D) It an exist in four or more than four different helical conformations

336. The amino acid with pK, closest to physiological pH is:

(A) Serine (B) Histidine

(C) Threonine (D) Proline

337. What will be the molecular weight of poly (A) chain consisting of 100 residues, where the molecular weight of AMP is 300:

(A) 30000 (B) 29700

(C) 28218 (D) 29982

338. Which of the following enzyme catalyzes the hydrolysis of nucleic acid:

(A) Nuclease (B) Glycosidase

(C) Phosphorylase (D) Lipase

339. Glucagon binds to the glucagons receptor, which interacts with a membrane-bound G protein, whose a subunit releases GDP and binds GTP. The a subunit then dissociates from the G protein and diffuses to a membrane-bound adenylyl cyclase which produces cAMP. In this scenario, name the (1) transducer, (2) effector, (3) first messenger and (4) second messenger

(A) (1) Glucagon (2) cAMP (3) G Protein (4) GTP

(B) (1) G Protein (2) Adenylyl cyclase (3) Glucagon (4) cAMP

(C) (1) G protein (2) GTP (3) Glucagon (4) cAMP

(D) (1) Glucagon (2) G protein (3) adenylyl cyclase (4) cAMP

340. Which of the following contains magnesium:

(A) Nitrogenase (B) Chlorophyll

(C) b-carotene (D) Phytol

341. In RNA ribose is present in the form of:

(A) a, D-Ribofluranose (B) b, D-Ribofuranose

(C) b, D-Ribopyranose (D) a, D- Ribopyranose

342. Which is not a biological function of RNA?

(A) Participates in the modification of newly synthesized RNA

(B) It is the genome of some viruses

(C) Serves as a scaffold component for chromosomes

(D) Composes as integral part of the ribosome

343. Which one of the following statements about prostaglandin synthesis is incorrect?

(A) Prostaglandins synthesis is inhibited by glucocortioids and aspirin

(B) Prostagladins are derived from an essential fatty acid

(C) Prostaglandins are produced through the cation of lipoxygenase

(D) Arachidonic acid is a major precursor of prostaglandins

344. G proteins are said to be self-inactivating. What G – protein function accounts for this?

(A) GTPase activity of the a subunit

(B) Activation of the effector

(C) Binding to the receptor

(D) Dissociation of the subunits

345. Which one of the following amino acids has the largest number of rotatable bonds in the side chain:

(A) Tryptophan (B) Leucine

(C) Lysine (D) Cysteine

346. What is the number of hydrogen bonds in a double helical B-DNA structure of 100 base pairs with 20 adenines & 10 thymines in one of the two strands:

(A) 200 (B) 230

(C) 270 (D) 300

347. In nucleic acids, internucleotide linkage is known as:

(A) Glycosidic bond
(B) 3'5' phosphodiester bonds
(C) Peptide bond (D) Simple ester bond

348. People who suffer from thyroid adenomas have constitutively activated G proteins in some cells of the thyroid gland. Tumor cells not only secrete copious quantities of thyroid hormones but also divide and grow excessively. Based on these observations, in what kind(s) of signal transduction pathways do the mutant G proteins participate?

(A) cAMP–regulated pathways
(B) MAP kinase pathways
(C) Both cAMP and MAP kinase pathways
(D) Neither cAMP anor MAP kinase pathways

349. Which of the following is not true about fatty acid synthesis

(A) Cofactors – NADPH
(B) Precursor _ Two carbon unit
(C) Site – Cytosol
(D) Carrier – carnitine

350. Folate derivatives are required for the synthesis of each deoxynucleotides?

(A) Adenylate and guanylate
(B) Cytidylate and thymidylate
(C) Adenylate, guanylate and thymidylate
(D) Adenylate and Cytidylate

351. Depending on the cell, increase in intracellular Ca^{2+} can do all of the following except:

(A) Stimulate the increase of intracellular calcium
(B) Stimulate the removal of intracellular calcium
(C) Stimulate the breakdown of cAMP
(D) Stimulate the breakdown of calmodulin

352. Which one of the following vitamin contains isollaxazine ring in its structure:

(A) Vitamin A (B) Vitamin B_2
(C) Vitamin B_1 (D) Vitamin B_{12}

353. Genetic material in plant viruses is mostly:

(A) DNA (B) RNA
(C) Protein (D) None of these

354. Epinephrine binds to the same type of receptors in liver, fat and smooth muscle cells. Yet in liver, glycogen breaks down; in act, trisylglycerols break down; and smooth muscle cells relax. How can the same hormone produce three such different responses

(A) The same receptors activate three different second messengers
(B) The same second messengers activate three different intracellular cascades
(C) The same enzyme that breaks down glycogen also breaks down fat and relaxes muscle
(D) None of these

355. Which one of the following is a carrier of oxygen to muscles in vertebrates:

(A) Myoglobin (B) Hemoglobin
(C) Cytochromes (D) Immunoglobulin

356. A cell responds to an external stimulus by increasing the intracellular Ca^{2+} concentration, releasing IP3 from the membrane, and producing salicylic acid. What kind of cell does this describe?

(A) Bacterial (B) Viral
(C) Plant (D) Animal

357. Which of the following favours Z-DNA conformation?

(A) 2'-OH steric hindrance
(B) Anti-G and 2'-endo sugar
(C) Syn-G and 3'-endo sugar
(D) G4-DNA quartet

358. The distance along a helix axis for one complete turn is called:

(A) Axial distance (B) Rise
(C) Screw length (D) Pitch

359. Which of the following contain cobalt in its structure:

(A) Vitamin D (B) Vitamin K
(C) Vitamin B_5 (D) Vitamin B_{12}

360. An ideal α –helix has a pith of 0.54 nm and a rise of 0.15 nm. What is the length along the helix axis for a segment of an ideal α –helix that contains 30 amino acids?

(A) 200 nm (B) 4.50 nm
(C) 16.20 nm (D) 55.5 nm

361. Which one of the following is smallest in size:

(A) r-RNA (B) m-RNA
(C) t-RNA (D) DNA

362. The reaction catalyzed by a certain phosphatase enzyme is found to follow ping-pong kinetics and involves the transfer of a phosphate group from substrate A to substrate B. Which mode of catalysis is likely for this reaction?

(A) Sequential catalysis
(B) Acid-base catalysis
(C) Transfer catalysis
(D) Covalent catalysis

363. The molecular weight of a bacterial DNA molecule is 2.64 x 10^9. The average molecular weight of a nucleotide pair is 660. Assume that the average protein is made up of a chain of 400 amino acid residues. What is the maximum number of proteins that can be coded by the bacterial DNA molecule?

(A) 20000 (B) 3333
(C) 6667 (D) None

364. Thiamine is a constituent of:

(A) Co enzyme A (B) Co enzyme Q
(C) NADP
(D) Co-carboxylase coenzyme

365. **An unwanted DNA and a supercoiled DNA with the same linking number are:**

(A) Topologically and geometrically identical
(B) Topologically identical but geometrically different
(C) Topologically and geometrically different
(D) Topologically different by geometrically identical

366. **Ribosomal RNA modification such as methylation or pseudouridylation are guided by which of the following RNA in eukaryotic cells:**

(A) mi-RNA
(B) si-RNA
(C) sno-RNA
(D) sh-RNA

367. **Which was a major scientific accomplishment by Linus Pauling?**

(A) Determined the α-helix structure in keratin
(B) Discovered β – strands and described their assembly into sheets
(C) Determined the three-dimentional structure of hemoglobin and mycoglobin
(D) Determined the structure of penicillin

368. **A protein has a molecular weight of 5600 Daltons; its subunits are about 1960 Daltons. There are ------ protein chains per oligomer:**

(A) 4 (B) 3
(C) 2 (D) 1

369. **Which of the following is not true about RNA**

(A) RNA is chemically more reactive than DNA
(B) 2'-OH prevents RNA to adopt B-conformation
(C) RNA was discovered by Hoppe-Seyler
(D) Feulgen reaction is performed by both DNA and RNA

370. **Which of the following is not true? When RNA undergo for alkali-catalyzed cleavage:**

(A) Transesterification reaction occurs
(B) Cyclic 2'-3'-nucleoside monophosphate forms
(C) Cyclic 5'-3'-nucleoside monophosphate forms
(D) 2'-OH is converted into 2'-alkoxide ion

371. **Metal cations may do all of the following except:**

(A) Donate electron pairs to functional groups found in the primary structure of the enzyme protein
(B) Serve of Lewis acids in enzyme
(C) Participate in oxidation-reduction processes
(D) Stabilize the active conformation of an enzyme

372. **Which of the following answers completes the sentence correctly? The first product of purine nucleotide biosynthesis that contains a complete purine ring (hypoxanthine) is:**

(A) AMP
(B) IMP
(C) GMP
(D) Xynthylate

373. **The steady – state hypothesis in enzyme kinetics states that:**

(A) There is an equilibrium established between free enzyme (E) and substrate (S) versus the E-S complex
(B) There is an equilibrium established between the E-S complex and the products
(C) The rate of formation of the E-S complex in the same as the rate of breakdown of this complex either to products or back to reactants
(D) The rate of formation of products is the same as the rate of formation of E-S complex

374. **The void volume of a gel filtration column, V0=3-mL. A monomeric protein with a known molecular weight of 25 kDa elutes at a volume of 45 mL. The protein that you are trying to purify elutes at a volume of 35 mL. Which of the following is a valid conclusion?**

(A) Your protein is repelled by the gel filtration material
(B) Your protein Mr= 25 kDa
(C) Your protein Mr>25kDa
(D) Your protein Mr<25 kDa

375. **Which is not an example of an oligomeric protein?**

(A) Cytochrome
(B) Potassium channel protein
(C) MS2 capsid protein
(D) Hemoglobin

376. **You have mixture of proteins with the following properties:**

Protein 1 : Mr 12 kD pI = 10
Protein 2 : Mr 62 kD pI = 4
Protein 3 : Mr 28 kD pI = 8
Protein 4 : Mr 9 kD pI = 5

Predict the order of emergence of these proteins when a mixture of the four is chromatographed in DEAE-cellulose at pH 7.0

(A) 1, 3, 4, 2 (B) 2, 4, 3, 1
(C) Cannot be predicted (D) 5, 8, 4, 10

377. **The main property of myoglobin and hemoglobin that makes them an efficient system for oxygen delivery from lungs to muscles is:**

(A) Hydrophobicity
(B) Different binding affinities for oxygen
(C) Movement of the protein shapes
(D) Cooperativity

378. **Which one of the following statements is correct concerning the Michaelis constant, K_m?**

(A) Is measured in the same units as the rate of the enzyme-catalysed reaction
(B) Varies with the amount of enzyme used in measuring the reaction rate
(C) Can be considered to be a measure of the affinity an enzyme has for its substrate
(D) Has a low value to indicate a low affinity for substrate

379. Two proteins have the same molecular mass and have identical net charge at pH 7. The best way to separate them would be to use:

(A) SDS-polyacrylamide gel electrophoresis (SDS-PAGE)
(B) Native gel electrophoresis
(C) Cation-exchange chromatography
(D) Anion-exchange chromatography

380. The active site of a certain enzyme contains a serine residue. When the enzyme is incubated for a short time with its substrate, a form of the enzyme in which the active site serine is acetylated can be isolated and purified. In the native protein the serine is never found to be acetylated. This information supports:

(A) A covalent catalysis mode
(B) Intermediate state stabilization effects
(C) The acid base catalysis mode
(D) A polar group catalytic mechanism

381. Which statement is not true about enzyme inhibition?

(A) In competitive inhibition, the inhibitor binds to the active site of the enzyme
(B) In non-competitive inhibition, the inhibitor binds to the allosteric site of the substrate
(C) In irreversible inhibition, a poison binds to the enzyme so that it can never work again
(D) None of these

382. Three-dimensional images of the surface of the cells and tissues could be visualized through:

(A) Transmission electron microscope
(B) Scanning electron microscope
(C) Compound microscope
(D) Fluorescence microscope

383. From the data given below, identify the protein pair that would (P) give the least mobility band on a sodium dodecyl sulphate polyacrylamide gel electrophoresis (SDS-PAGE) experiment and (Q) elute last on an anion exchange e.g., DEAE column respectively:

Protein	pI	Subunit M.W.	Native M.W.
A	9	10000	20000
B	8	35000	35000
C	6	15000	90000
D	5	20000	80000
E	3	30000	30000

(A) (P) Protein C (Q) Protein A
(B) (P) Protein A (Q) Protein E
(C) (P) Protein B (Q) Protein E
(D) (P) Protein B (Q) Protein A

384. The cell in number (188) above would be in an environment that was:

(A) Hypertonic (B) Hypotonic
(C) Isotonic (D) Hypo-osmotic

385. The relative proportion (by weight) of lipid and protein in a membrane:

(A) Is always about equal
(B) Varies depending upon the source of the membrane
(C) Is always characterized by more protein than lipid
(D) Is always characterized by more lipid than protein

386. Glycogen is a branched polymer of glucose. It has:

(A) No reducing ends
(B) No non-reducing ends
(C) One reducing end and several non-reducing ends
(D) One non-reducing end & several reducing ends

387. Vitamin C is a vitamin but not a coenzyme because it:

(A) Does not bind to proteins
(B) Is a prosthetic group
(C) Is a reducing agent during hydroxylation of collagen
(D) Is not required human diets

388. Features of the fluid mosaic model of membrane structure include:

(A) A lipid bilayer
(B) Dynamic motion of both membrane lipids and proteins
(C) Proteins that may either rest on the surface or penetrate the membrane
(D) All of these

389. Which of the following is an example of an ion transport?

(A) Na + - K + ATPase (B) ABC transporter
(C) CFTR (D) All of these

390. Purified duplex DNA molecules cannot exist in which one of the following forms:

(A) Linear
(B) Circular and supercoiled
(C) Linear and supercoiled
(D) Circular and relaxed

391. Arrange the statement below in the proper sequence of events that occur in an action potential:

I. The resting potential is reestablished
II. A stimulus depolarizes the membrane to threshold
III. Na^+ gates open and Na^+ rushes into the cell
IV. K^+ gates open and K^+ rushes out of the cell

(A) II, III, IV, I (B) III, IV, II, I
(C) II, IV, III, I (D) I, II, IV, III

392. Of the four classes of cellular macromolecules, which one is not a component of cell membrane?

(A) Proteins (B) Lipids
(C) Nucleic acids (D) Carbohydrates

393. Which of the following answer complete the sentence incorrectly?

(A) Is a flavin nucleotide
(B) Is the major electron acceptor used metabolism
(C) Contains a nicotinamide ring that accepts a hydride ion during reduction
(D) Loses a plus charge upon reduction

394. Coenzymes and prosthetic groups are found in all the key metabolic pathways. Which one of the following statements is accurate about these key molecules?

(A) Carbonic anhydrase contains a tightly bound zinc ion that is essential for its structure but not for its activity
(B) Some enzymes contain organic molecules called prosthetic groups that are essential for activity; prosthetic groups are vitamins and cannot be synthesized in the body
(C) Coenzymes are small polypepetides that are involved in transfer reactions catalyzed by enzymes
(D) Coenzymes carry acyl or phosphate groups in many enzyme-catalyzed reactions that involve the transfer of these groups

395. A neurotransmitter that binds to a postsynaptic cell and opens K+ channels:

(A) Excites the postsynaptic cell
(B) Inhibits the postsynaptic cell
(C) Depolarizes the postsynaptic cell
(D) Must be acetylcholine

396. Prosthetic groups are:

(A) Required by all enzymes in the cell
(B) Loosely bound to enzymes via hydrogen bonds
(C) Sites on the enzyme molecule that permit allosteric modification of enzyme activity
(D) Tightly bound to enzymes and are required for their activity

397. The study of the energy changes during metabolic reactions is called:

(A) Bioinformatics (B) Metabodynamics
(C) Thermometrics (D) Bioenergetics
(E) Biological heat dynamics

398. An allosteric site on an enzyme is:

(A) The same as the active site
(B) Non-protein in nature
(C) Where ATP attaches and gives up its energy
(D) Often involved in feedback inhibition

399. How do membranes function as a locus of biochemical reactions?

(A) Membranes separate the reactants from the products of enzymatic reactions
(B) Membranes hold enzymes of sequential reactions in close proximity
(C) Membrane lipids catalize important cellular reactions
(D) Membrane increase the activation energy for catalysis

400. What region of polypeptide (backbone or side-chains) is involved in forming α–helices and β–sheets?

(A) The backbone (B) Side chain
(C) Both (A) and (B) (D) None of these

401. You put Protein X in methane, would you except Protein X to be:

1. You would expect that Protein X would be functional in methane
2. You would not expect that Protein X would be functional in methane
3. In a hydrophobic solution such as methane, it would be more energetically stable to have the hydrophobic amino acids on the outside of the structure and the polar amino acids on the inside of the structure.
4. The protein would be turned 'inside out' and no longer functional. Because some of these hydrophobic residues might also be involved in secondary structure, this level of protein structure would also be disrupted.

Correct answer is:

(A) 4 (B) 1
(C) 2 (D) 1, 3 and 4

402. What is the approximate diameter of a chromatin fiber?

(A) 3 cm (B) 30 nm
(C) 300 nm (D) 30 mm

403. Much of what we know about the nicotinic acetylcholine receptor has come from studies of:

(A) The torpedo marmarota electric organ
(B) The human heart
(C) Simple, single-celled organisms
(D) *E. coli* bacteria

404. Which one of the following characteristics best apples to an allosteric effector?

(A) Completes with substrate for the catalytic site
(B) Binds to a site on the enzyme molecule distinct from the catalytic site
(C) Changes the nature of the product formed
(D) Changes the substrate specificity of the enzyme

405. A spontaneous chemical reaction always has a:

(A) Positive Gibb's free energy
(B) Negative Gibb's free energy
(C) Positive enthalpy
(D) Negative enthalpy
(E) Positive entropy

406. In a chemical reaction, transition-state species have free energies:

(A) Lower than either the reactants or the products
(B) Higher than either the reactants or the products
(C) Lower than the reactants, but higher than the products
(D) Higher than the reactants, but lower than the products

407. More than 300 amino acids are present in cells but only 20 amino acid present in proteins called:
(A) Standard amino acids
(B) Nonstandard amino acids
(C) Both (A) and (B)
(D) None of these

408. Which is not a difference between RNA and DNA:
(A) The sugar ring of RNA is more oxidized than that in DNA
(B) RNA contains uracil; DNA usually does not
(C) RNA cannot form helices
(D) RNA is single-stranded; DNA is double-stranded

409. In a given reaction:

$$CH_3CH_2CHO + CH_3CHO \xrightarrow{dil} NaOH$$

(A) Three (B) Four
(C) Two (D) One

410. During the reduction of FAD:
(A) A flavin group is transferred
(B) An equivalent of a hydride ion is transferred
(C) The isoalloxazine become charged
(D) Two hydrogen atoms are added to the isoallexazine ring

411. If you break liver cells by homogenization, separate the broken cell membranes (particulate) from the cytoplasm (soluble), and add epinephrine to the soluble portion, the result will be:
(A) The production of cAMP
(B) Binding of the epinephrine to its receptor
(C) Activation of adenylyl cyclase
(D) None of these

412. Why it is important that the enzymes in lysosomes are more active at acidic pH than at neutral pH?
(A) Since lysosomes are primarily found in the stomach acid of mammals, their pH dependence allows for maximum efficiency for the digestion of food stuffs
(B) It prevents their diffusion out of the lysosomes
(C) It maximizes the interaction with their substrates which are always bases
(D) It prevents them from accidentally degrading the macromolecules in the cytosol
(E) It allows for regulation of their uptake by the mitochondria

413. Match the function of following cofactor:

Column I	Column II
P. Thiamine pyrophosphate	1. Acyl group transfer
Q. Coenzyme A	2. Transfer of one carbon component
R. Pyridoxal phosphate	3. Group transfer to/ or from amino acid
S. Tetrahydrofolate	4. Aldehyde transfer

(A) P-4, Q-3, R-1, S-2 (B) P-4, Q-3, R-2, S-1
(C) P-4, Q-1, R-3, S-2 (D) P-3, Q-1, R-4, S-2

414. Chromatin was first observed and named by:
(A) Walter Flemming (B) Francis Crick
(C) Oswald Avery (D) Friedrich Miescher

415. Why were the traits that Mendel chose to study in the pea plant considered lucky choices?
(A) They were the only traits governed by genes
(B) None of the traits he chose acted as if they were part of the same linkage group
(C) All the traits he chose could be found on the same plant
(D) They all obeyed the law of segregation

416. Acid base catalysis is estimated to accelerate a typical enzymatic reaction by what factor?
(A) 1 to 2 fold increase
(B) 10 to 100 fold increase
(C) 106 fold increase
(D) 1023 fold increase

417. Which of the following is mismatched?
(A) eRNA : 80% of cellular RNA
(B) tRNA : Stable RNA carrying the coded information from DNA
(C) mRNA: stable RNA carrying the coded information from DNA
(D) Small RNA: catalytic with or without proteins

418. Heterotropic enzymes:
(A) Are different isoenzymes
(B) Are modified by their substrate concentration
(C) Are stimulated or inhibited by an effector or modulator molecule other than their substrate
(D) Have same V_{max} but different k_m values

419. Which of the following statements about ribozymes is false?
(A) Ribozymes are capable of self-replication, but cannot catalyze other types of reactions
(B) Manfred Eigen found that RNA's could replicate themselves in solution without the aid of proteins
(C) In Tetrahymena thermophila, an intron was found that carries out its own excision and splicing
(D) A tRNA-processing enzyme containing RNA was found in which the RNA portion provided the catalysis

420. The observed optical rotation of an amino acid solution is directly proportional to the length of the light path through the polarimeter tube and the concentration of the amino acid solution is:
(A) True (B) False
(C) Unknown (D) Same

421. Stem-loop is a type of secondary structure not usually found in:
(A) rRNA
(B) tRNA
(C) mRNA
(D) Small RNA

422. According to the RNA-world theory:
(A) RNA molecules were the first organic molecules formed on earth
(B) Life evolved an another planet called the 'RNA world'
(C) All RNA molecules in cells are 'rhizymes'
(D) Primitive RNA molecules evolved before protein and DNA

423. You are an undergraduate researcher working in a lab that studies transport proteins. You have just identified a novel protein from human cells, Protein X, which you believe is is involved in intracellular transport. After careful investigation, you determine the amino acid sequence of Protein X. Below is a portion of the sequence that contains the part of the site where Protein X binds the molecule that it transports:
Val-Ala-Phe-Gly-Gly-Leu-Asp-Asp-Gly-Glu-Ala-Trp-Tyr-Met-Leu-Lys-Leu. Bold represents amino acids in the binding site. Next, you find mutant versions of Protein X that have single amino acid substitutions.
A mutant in which one of the Asp residues is mutated to a Glu residue. Even with this mutation, Protein X is still functional. Why?
1 The two Asp amino acids are both in the binding site, so they are important for biding the molecule to be transported.
2 Asp and Glu negatively charged amino acids. Asp and Glu are also fairly similar in size.
3 Assuming that the negative charge is important for binding, Glu can substitute for Asp because the charge is conserved.
4 Insufficient data
Correct answers are:
(A) 4 (B) 2
(C) 1 and 2 (D) 1, 2 and 3

424. Considering previous question, A mutant in which the Val residue is mutated to a Phe residue. This mutation causes Protein X to be nonfunctional. Yes or No?
1. Val in not in the binding site, so it is not directly involved in binding the transport molecule.
2. Val is changed to Phe. Both amino acids are hydrophobic or non polar
3. Phe, however, is a very large bulky amino acid compared to Val. Thus, it is possible that the change to Phe causes some disruption in the tertiary structure that prevents Protein X from being functional.
4. Protein will be functional
Correct answer is:
(A) 4 (B) 2
(C) 1 and 2 (D) 1, 2 and 3

425. Continuing to previous question, what level of protein structure do you know is changed in all of these mutant proteins?
(A) Primary (B) Secondary
(C) Tertiary (D) Quaternary

426. Continuing to previous question, after further analysis of your protein, you find that α–helices and β–sheets are extremely important for maintaining the structure of the binding site. What level of structure do α- helices and β– sheets represent?
(A) Primary (B) Secondary
(C) Tertiary (D) Quaternary

427. Proline distinct among the 20 commonly found amino acids because:
(A) It is a ring compound
(B) It is hydrophilic and ionic
(C) The nitrogen of the amino group is in a ring
(D) The carbon of the carboxyl group is in a ring

428. Involves a nucleophilic attack of the N-terminal – amino group on the thionate carbon of phenylisothiocyanate (PITC), to form the phenylthioarbamyl-peptide derivative (PTC- peptide) is called:
(A) Coupling (B) Cyclization
(C) Conversion (D) Repulsion

429. An enzyme's active site contains and arginine residue and a glutamate residue with pKa's of 2.9 and 9,1, respectively. Both residues are actively involved in the catalytic mechanism and they are the only two ionizable residues in the active site. What you expect for the optimum pH of the enzyme?
(A) 6.0 (B) 7.0 (C) 7.4
(D) No pH can be determined since the information is irrivalent to the optimum pH

430. Supercoiling of pure circular DNA is a result of:
(A) Underwinding or overwinding of the helix
(B) Entanglements because of the long length of the strands
(C) The lack of associated histones
(D) The activity of exonucleases

431. The atoms of the pyrimidine ring are derived from:
(A) Glutamine
(B) Carbamyol phosphate and aspertate
(C) Glutamine and N10 formyl tetrahydrofolate
(D) Aspartate ands N 10 formyl tetrahydrofolate

432. Nitrogen fixation is a:
(A) Oxidative
(B) Reductive
(C) Oxidative followed by reductive
(D) Reductive followed by oxidative

433. Stem-loop structures in RNA resemble the -------- form of double-helical DNA:
(A) A (B) B
(C) Z (D) Supercoiled circular

434. Alleles are:
(A) Different forms of the gene for the same trait
(B) Genes for different traits
(C) Always identical
(D) Inherited in pairs, both members of which come from the same parent

435. In a given reaction:

$CH_3CHO + C_6H_5CHO \xrightarrow{\text{dil}} NaOH$

(A) Three (B) Four
(C) Two (D) One

436. Which type of organisms convert nitrogen gas into ammonia:

(A) Prokaryotic (B) Eukaryotic
(C) Both (A) and (B) (D) None of these

437. Superoxide dismutase enzyme catalysis is faster than the rate of diffusion because it:

(A) Is an acid base catalysis
(B) Is a two step reaction
(C) Has an electric field around the active site
(D) Occurs in very high quantities in cells

438. The biological process of nitrogen fixation is catalyzed by an enzyme complex is called:

(A) Nitrogenase complex
(B) Dinitrogenase complex
(C) Both (A) and (B)
(D) None of these

439. When statements is true about aeration with an equilibrium constant, Keq, equal to 1000?

(A) The forward rate constant is 1000 times greater than the reverse rate constant
(B) The forward rate constant is 3 times greater than reverse rate constant
(C) The forward rate constant is 1000 times smaller than the reverse rate constant
(D) The forward rate constant is 3 times smaller than the reverse rate constant
(E) There is not enough information given to compare the forward and reverse rate constant

440. Tyrosine and tryptophan are less hydrophobic than phenylalanine because:

(A) Phenylalanine has an indole group
(B) Phenylalanine has no polar group in the side chain
(C) Phenylalanine is a phenol
(D) Tyrosine and tryptophan have smaller R groups
(E) All of these

441. Which of the following contributes nitrogen atoms to both purine and pyrimidine rings?

(A) Aspartate (B) Carbamoyl phosphate
(C) Carbon dioxide (D) Glutamine

442. A reaction that occurs with every collision between reactant molecules is called a/an:

(A) Saturation point
(B) Diffusion-controlled reaction
(C) Rate-determining step
(D) First order reaction

443. Zein is known as:

(A) Rice protein (B) Wheat protein
(C) Maize protein (D) Millet protein

444. The rearrangement of the liberated ATZ derivative to the corresponding phenylthiohydantoin derivative (PTH-amino acid) by treatment with aqueous acid is called:

(A) Coupling (B) Cyclization
(C) Conversion (D) Repulsion

445. What is the function of DNA topoisomerases?

(A) Cleaving nucleotides from the ends of DNA
(B) Adding or removing supercoils
(C) Methylation of the nitrogenous bases
(D) Packaging of DNA into nucleosomes

446. Maize protein is deficient in:

(A) Betadine (B) Tryptophane & Lysine
(C) Lysine (D) Tryptophane

447. Wheat protein is called as:

(A) Zein (B) Gluten
(C) Dhurin (D) Ricin

448. Essential oils are used in:

(A) Cosmetics & Toiletries (B) Food beverage
(C) Paint industry (D) All of these

449. First Face Pack was developed in:

(A) France (B) Rome
(C) Greece (D) None of these

450. Oil content in fresh herbage in menthol mint is:

(A) 0.1-0.5 % (B) 0.5-0.8 %
(C) 1.0-1.5 % (D) 1.5 – 2.0 %

451. Major constituent of *Occimum cannum* is:

(A) Methyl chavical (B) Thymal
(C) Linalool (D) Engenol

452. Protein content in rajmash is:

(A) 10.9 % (B) 22.9 %
(C) 25.9 % (D) 32.9 %

453. Carbohydrate content in rajmash is:

(A) 50.6 % (B) 60.6 %
(C) 70.6 % (D) 80.6 %

454. Protein content in pea is:

(A) 10 % (B) 22 %
(C) 25 % (D) 32 %

455. Carbohydrate content in potato is:

(A) 40.3 % (B) 70.3 %
(C) 81.3 % (D) 91.3 %

456. In a signal transduction pathway, what is passed from the upstream member of the pathway to the downstream member?

(A) Electrons (B) Protons
(C) Phosphates
(D) Information in the form of alterations in protein conformation

457. Which one of the following enzyme contains zinc:

(A) Peroxidase (B) Lipase
(C) Nitrogenase (D) Carbonix anhydrase

458. The protein molecules various amino acid residue are covalently linked in a head to tail arrangement by:
(A) Peptide bonds
(B) Phosphodiester bonds
(C) Glycosidic bonds
(D) Simple ester bonds

459. Given that alleles are located on chromosomes, and whole chromosomes are passed on to gametes, why don't offspring get all the alleles on a parental chromosome?
(A) Genetic recombination is a result of gene duplication during evolution
(B) Genetic recombination is a result of mutation
(C) Genetic recombination is a result of crossover between members of homologous pairs
(D) All traits on a given chromosomes are always found together in offspring

460. Cell signaling means all of the following except:
(A) Intercellular communication
(B) Export of cellular waste products
(C) Environmental monitoring
(D) Response to stimuli

461. Simple proteins on hydrolysis yield:
(A) L- Amino acids having – NH group on a- carbon
(B) D- Amino acids having – NH group on a- carbon
(C) L- Amino acids having – NH group on b- carbon
(D) D- Amino acids having – NH group on b- carbon

462. Rhodopsin is a protein complex the prosthetic group of which is:
(A) Retinol (B) Retinal
(C) Vitamin 'K' (D) Metal ion

463. The enzyme which exists in more than one structural forms in the same species is known as:
(A) Multienzyme complex (B) Zymogen
(C) Isozyme (D) Zymase

464. Changes in the topology and three-dimentional shapes of B-DNA can result from:
(A) Incubation with proteases
(B) Supercoiling
(C) Isomerization
(D) Hydration

465. Watson and Crick used all of the following information in deducting the structure of the genetic material except:
(A) Avery, MacLeod and McCarty's identification of DNA as the "transforming principle"
(B) X-ray diffraction studies by Astbury, Franklin, Pauling and Wilkins
(C) Chargaff's rules of DNA base composition
(D) Microscopic examination of DNA structure

466. Which one of the following contain b-ionine ring in its structure:
(A) Chlorophyll (B) Purine
(C) Vitamin 'A' (D) Inosine

467. Doing some more analysis of the function of Protein X reveals that in addition to binding the molecule it transports, it also binds ATP (at a different site). You have a transport assay that measures the activity of Protein X. You perform the transport assay under different conditions:

Transport Molecule	Nucleotide Added	Nucleotide Binding?	Transport Activity Level
Yes	ATP	+++	+++
Yes	ADP	+++	-
Yes	GTP	-	-
Yes	ATP-γS	+++	-

ATP – γS cannot substitute for ATP to transport the molecule because
(A) Impermeable to membrane
(B) Nonhydrolysable
(C) Both (A) and (B)
(D) It can substitute ATP

468. Continuing to question, GTP substitute for ATP to transport the molecule because:
1. This is most likely because ATP and GTP have different nitrogenous bases
2. Enzymes (and other proteins as well) have binding sites that are very specific. Due to the different structures of the nitrogenous bases, the ATP-binding site on Protein X cannot bind GTP.
3. Protein X could bind GTP; it probably sill could not substitute for ATP because GTP is not used as an energy source by the cell.
4. None

Correct answer is:
(A) 4 (B) 1
(C) 2 (D) 1, 2 and 3

469. Which one of the following enzymes contain copper as metal activator:
(A) Catalase (B) Carbonic anhydrase
(C) Cytochrome oxidase (D) Ascorbic acid oxidase

470. Individuals that share a haplotype are most likely to:
(A) Have all of the same genetic information
(B) Have the same SNPs in the region of the genome that contains that haplotype
(C) Be of different ethnic descent
(D) Be siblings

471. In DNA structure propeller twist means:
(A) Two bases orient antiparallel to each other
(B) Two bases flip with respect to each other
(C) Two flat bases counter rotate relative to each other
(D) Two bases stack with respect to each other with 3, 4 Å space

472. Select out sulphur containing amino acid:
(A) Arginine (B) Isoleucine
(C) Phenyl alanine (D) Methionine

473. You work in a lab that studies bacterial membrane proteins. You have isolated a protein, pMem, that spans the membrane and you determined the sequence of that protein:

1 KKLLPCTAL *VMCAGMACAQ* AEERND 50
WHFN *IGAMYEIENV* EGYGEDMDGL

51 AEPSVYFNAA *NGPWRIALAY* YQEGPV
DYSA *GKRGTWFDRP* ELEVHYQFLE 100

101 NDDFSFGLTG*GFRNYGYHYV* DEPG
KDTANM *QRWKIAPDWD *VKLTDDLRFN

151 GWLSMYKFAN* DLNTTGYADT* RVETE 200
TGLQY* TFNETVALRV* NYYLERGFNM

201 DDSRNNGEFS* TQEIRAYLPL* TLGNH 250
SVTPY *TRIGLDRWSN * WDWQDDIERE

251 GHDFNRVGLF *YGYDFQNGLS* VSLEY 301
AFEWQ* DHDEGDSDKF*
HYAGVGVNYS*F

Is the codon for phenylalanine (F) the last codon in the mRNA transcript?

(A) No. the last codon is always a stop codon
(B) It is F
(C) It can be any codon for amino acid
(D) None of these

474. Which one of the following is an essential amino acid:

(A) Alanine (B) Serine
(C) Lysine (D) Proline

475. Which of the following is mismatched?

(A) B-DNA : dehydrated
(B) Z-DNA : left-handed helix
(C) A-DNA : dehydrated
(D) Z-DNA : no major or minor grooves

476. Triple stranded DNA:

(A) Is a characteristic of SV 40 genome
(B) Involves Hoogsteen base pair
(C) Involves only intermolecular interactions
(D) Is characteristics by in the presence of a string of alternating purine-pyrimidine bases

477. Continue with above question, what us the best length (in nucleotides) of the mRNA that encodes for the above protein?

(A) 950 (B) 902
(C) 301 (D) 3303

478. Which one of the following yields two molecules of vitamin 'A' in intestine:

(A) a-carotene (B) Lycopene
(C) b-Carotene (D) g – Carotene

479. What factors are responsible for the specificity of G protein-coupled responses?

(A) Protein kinases can phosphorylate different proteins in different cells
(B) Receptors can exist as isoforms
(C) G-protein subunits can exist as isoforms
(D) All of these

480. How many monomers are there in an oligomeric protein designated αβ2γ2?

(A) 2 (B) 3
(C) 4 (D) 5

481. The term cistron, muton and recon were introduced by:

(A) Watson and crick (B) S. Benzer
(C) Meselson (D) Morgan

482. The nitrogen nases which pair with two hydrogen bonds are:

(A) Adenine and thymine
(B) Adenine and cytokinin
(C) Cytokinin and guanine
(D) Cytokinin and adenine

483. Extranuclear genetic material is found in:

(A) Plastid and nucleus
(B) Mitochondria and plastids
(C) Nucleus and cytoplasm
(D) Mitochondria and nucleus

484. The molecular formulae of deoxyribose sugar and ribose sugar respectively are:

(A) C5 H10 O4 and C5 H10 O6
(B) C5 H10 O4 and C5 H10 O5
(C) C5 H10 O5 and C5 H10 O4
(D) C5 H10 O5 and C6 H10 O4

485. Because DNA is a highly charged polyanion, its stability to heat denaturation:

(A) Does not depend on hydrophobic interactions
(B) Increases with increasing salt
(C) Is dependent of G – C content
(D) Decreases with increases salt

486. The most stabilizing force for nucleic acid is:

(A) Hydrogen bonds
(B) Electrostatic bond
(C) Van der Waals
(D) Conformational entropy

487. First evident that DNA is genetic material comes from the experiment of Griffith is:

(A) Transduction
(B) Transformation
(C) Replication
(D) Translation

488. In an agarose gel, DNA is moving from:

(A) Cathode to anode
(B) Left to right
(C) Anode to cathode
(D) A to B

489. Number of time the two DNA strands are interwined in circular DNA and is known as:

(A) Helix frequency
(B) Linking number
(C) Twisting number
(D) None of these

490. The higher order structure of DNA shows symmetry, whereas the higher order structures of most protein do not. Why is not protein shape more regular like DNA?

(A) DNA has one main function in cells whereas protein have many

(B) The many different amino acid R group on proteins confer many different shapes

(C) Some S amino acids cause proteins to bend; others cause protein to flatten

(D) All of these

491. In DNA, there are:

(A) Five bases known as adenine, guanine, thumine, tryptophan and cytosine

(B) Four bases known as adenine, guanine, thymine and cytosine

(C) Three bases known as adenine, guanine and cytosine

(D) Only two bases known as adenine and cytosine

492. In DNA, nucleotide are covalently joined together by:

(A) 3', 5' phosphodiester bonds to form a repetitive sugar – phosphate chain

(B) 2', 5' phosphodiester bonds to form a repetitive sugar – phosphate chain

(C) 2', 3' phosphodiester bonds to form a repetitive sugar – phosphate chain

(D) 3', 4' phosphodiester bonds to form a repetitive sugar – phosphate chain

493. Which of the following species lack a 'classical' histone H1?

(A) *Saccharomyces cerevisiae*

(B) *Strongylocentrotus purpuratus*

(C) *Xenopus laevis*

(D) *Arabidopsis thaliana*

494. In sperm heads, DNA is particularly highly condensed and the histones are replaced with small basic protein called:

(A) Protamines (B) Purines

(C) Oyrimidines (D) All of these

495. During conventional transformation experiments E. coli cells and plasmid DNA interacts in an environment of:

(A) High temperature and Ca^{++}

(B) Low temperature and Ca^{++}

(C) High temperature and Mg^{++}

(D) Low temperature and Mg^{++}

496. Which of the following is an active site on its single polypeptide chain for DNA polymerase I?

(A) 5' → 3' polymerase (B) 3' → 5' exonuclease

(C) 5' → 3' exonuclease (D) All of these

497. *E. coli* DNA polymerase II and III lack:

(A) 5'→ 3' exonuclease activity

(B) 5'→ 3' endonuclease activity

(C) Parially 5'→ 3' exonuclease activity

(D) Partially 5'→ 3' endonuclease activity

498. Automated DNA sequencing uses the chain termination method:

(A) But with an oligonucleotide primer labeled with a fluorescent dye

(B) But with a polynucleotide primer labeled with a fluorescent dye

(C) With no primer (D) Either (A) or (C)

499. Deoxyribonucleoside triphosphates, being need in the chain termination methods are:

(A) dATP (B) dGTP and dCTP

(C) dTTP (D) All of these

500. DNA helicase is used to:

(A) Unwind the double helix

(B) Interact the double helix closely

(C) Break a phosphodiester bond in DNA strand

(D) None of these

501. For the DNA replication in eukaryoted, the cell cycle consists of:

(A) G1, G2 and M phases

(B) S, G2 and M phases

(C) G1, S, G2 and M phases

(D) G2 and M phases

502. DNA replication in eucaryotes occurs only in:

(A) G1 phase (B) S phase

(C) G2 phase (D) M phase

503. The replication of a linear DNA molecule in a eukaryotic chromosome creates a problem which is reported to be sorted out by:

(A) Using enzyme called telomerasase

(B) Shortening of the double stranded replicated portion

(C) Using RNA as a template

(D) None of these

504. In DNA replication, one new strand is synthesized complementary to an old (parent) strand and is referred as:

(A) Mutational (B) Semiconservative

(C) Discontinuous (D) Anti parallel

505. Which of the following in DNA replication and transcription are common?

(A) Incorporation of deoxynucleotides

(B) Utilization the same enzyme

(C) Synthesis in the 5' – 3' direction

(D) None of these

506. The enzyme responsible for breaking the hydrogen bonds, and thus separating the DNA strands during DNA synthesis is:

(A) DNA polymerase (B) Strandase

(C) Helicase (D) None of these

507. An enzyme that recognizes foreign DNA, and results in a cut in that DNA is called:

(A) Transposase (B) Helicase

(C) Reverse transcriptase

(D) Restriction endonuclease

508. DNA segments capable of moving from one place in the genome to another is known as:
(A) Transposons (B) Retrovirus
(C) Introns (D) Moving elements

509. How does ethidium bromide interact with DNA?
(A) It binds to adenine
(B) It binds to phosphate
(C) It interactes between bases
(D) It does not interact

510. What is the only common methylation in the DNA of eucaryots?
(A) Anosine in Gp dinucleotides
(B) Guanosine in ApGpA trinucleotides
(C) Cytosine in CpG dinucleotides
(D) None of these

511. The accepted hypothesis for DNA replication is:
(A) Conservative theory
(B) Dispersive theory
(C) Semi-conservative theory
(D) Evolutionary theory

512. Enzymes, responsible for unraveling short segments of DNA is:
(A) DNA polymerase (B) Helicase
(C) DNA ligase (D) Primase

513. DNA replication takes place in which direction?
(A) 3' to 5'
(B) 5' to 3'
(C) Randomly
(D) Vary from organism to organism

514. Which enzyme catalyzes the elongation of DNA molecules?
(A) DNA polymerase (B) Helicase
(C) DNA ligase (D) Primase

515. Enzyme, responsible for proof reading base pairing is:
(A) DNA polymerase (B) Telomerase
(C) Primase (D) DNA ligase

516. Reverse trancriptase procudes:
(A) DNA from peptides
(B) RNA from DNA
(C) RNA from RNA
(D) DNA from RNA

517. The RNA primers used to initiate replication in E. coli:
(A) Result in Okazaki fragments on leading strand
(B) Are removed by Pol I
(C) Are joined together by DNA ligase
(D) Are removed by helicase + ATP

518. The nicks or gaps between Okazaki fragments are eventually sealed by:
(A) Gyrase (B) DNA ligase
(C) RNA ligase (D) Polymerase

519. Overwinding or overtightening of DNA during replication is caused and removed by:
(A) DNAb, DNA polymerase
(B) DNAa, gyrase
(C) DNAb, gyrase
(D) Single stranded protein, DNAa

520. The DNA polymerase in bacteria, responsible for DNA synthesis is:
(A) Polymerase I (B) Polymerase II
(C) Polymerase III (D) All of these

521. DNA replication results in:
(A) Two completely new DNA molecules
(B) Two DNA molecules such that each one contains a strand of the original
(C) One new DNA molecules 1 old molecule
(D) One new molecule of RNA

522. DNA gyrase in *E. coli*:
(A) Adds positive supercoils to chromosomal DNA
(B) Can be inhibited with antibiotics
(C) Is required only at the oric site
(D) Performs the same function as helicase in eucaryotes

523. Which polymerase is active in DNA repairing:
(A) Polymerase I (B) Polymerase II
(C) Polymerase III (D) None of these

524. Which DNA polymerase removes RNA primers in DNA synthesis?
(A) Polymerase I (B) Polymerase II
(C) Polymerase III (D) None of these

525. Proteins which help to open the double helix for DNA synthesis are:
(A) DNA helicases
(B) Single standed binding proteins
(C) Ligase
(D) DNA topoisomerase

526. Proteins involved in opening a replication bubble are:
(A) DNA helicases
(B) Single standed binding proteins
(C) Ligase
(D) DNA topoisomerase

527. An RNA primer is synthesized by:
(A) DNA helicases (B) Primase
(C) DNA topoisomerase (D) Ligase

528. The genetic code is degenerated. Which of the following codons represents the principle of degeneracy?
(A) UAA and UAC (B) AUG and AUA
(C) CAU and CAC (D) UUA and UUC

529. In prokaryotes, AUG encodes:
(A) Methionine
(B) N-formyl methionine
(C) A stop codon (D) Alanine

530. How many amino acids will be encoded by 5' GAU GGU UGA UGU 3' sequence?
(A) One (B) Two
(C) Three (D) Four

531. The sequence of one strand of DNA is: 5' ATTGCCA 3', what is the sequence of the other strand?
(A) 5' TAACGGT 3' (B) 5' TGGCAAT 3'
(C) 5' ATTGCCA 3' (D) 5' UAAGCCU 3'

532. The genetic code is:
(A) Universal
(B) Universal except for rare exceptions in mitochondria and some protozoa
(C) Species –specific
(D) Kingdom – specific

533. AUG codes for methionine act as a:
(A) Initiation code (B) Elongation code
(C) Termination code (D) Propagation code

534. The codons which do not specify an amino acid are called:
(A) Initiation code (B) Elongation code
(C) Termination code (D) Propagation code

535. In protein synthesis in prokaryotes:
(A) The initiating amino acid is N-formyl methionine
(B) The initiating amino acid is methionine
(C) The initiating amino acid is phenyl alanine
(D) None of these

536. in some organelles in eucaryotes, the genetic code for some codons:
(A) Differs from that used in prokaryotes
(B) Are same
(C) Are partially same (D) One of these

537. How many different codons are possible?
(A) 3 (B) 20
(C) 64 (D) An infinite number

538. Translation begins:
(A) At the replication fork
(B) On the lagging strand
(C) At the start codon
(D) In nucleus

539. Which of following RNA characterically contains unusual purines and pyrimidins?
(A) rRNA (B) nRNA
(C) mRNA (D) tRNA

540. A nickes RNA molecule can be ligated by:
(A) T4 RNA ligage (B) DNA polymerase
(C) T4 DNA ligase (D) All of these

541. In RNA, urasil pairs with:
(A) Adenine (B) Cytosine
(C) Thymine (D) Guanine

542. What is the function of messenger RNA?
(A) It carries amino acids
(B) It is a component of the ribosomes
(C) It is a direct copy of a gene
(D) It is the genetic material of some organisms

543. Amino acids required in the human diet and not synthesized by the body are called:
(A) Specialized (B) Trace
(C) Essential (D) Accessory

544. Protein fluorescence arises primarily from which residues?
(A) Arginine (B) Tryptophan
(C) Tyrosine (D) Phenylalanine

ANSWERSHEET

5. PLANT BIOCHEMISTRY

1. (A)	2. (A)	3. (D)	4. (C)	5. (C)	6. (A)	7. (A)	8. (B)	9. (E)	10. (A)
11. (B)	12. (C)	13. (A)	14. (B)	15. (C)	16. (A)	17. (B)	18. (C)	19. (D)	20. (A)
21. (B)	22. (C)	23. (A)	24. (B)	25. (A)	26. (A)	27. (C)	28. (D)	29. (C)	30. (C)
31. (D)	32. (E)	33. (B)	34. (C)	35. (A)	36. (A)	37. (D)	38. (D)	39. (D)	40. (A)
41. (C)	42. (D)	43. (E)	44. (D)	45. (C)	46. (D)	47. (C)	48. (B)	49. (A)	50. (B)
51. (C)	52. (D)	53. (E)	54. (D)	55. (B)	56. (C)	57. (A)	58. (B)	59. (A)	60. (D)
61. (C)	62. (A)	63. (A)	64. (D)	65. (D)	66. (C)	67. (A)	68. (C)	69. (D)	70. (D)
71. (D)	72. (C)	73. (D)	74. (B)	75. (C)	76. (C)	77. (D)	78. (C)	79. (B)	80. (D)
81. (B)	82. (A)	83. (D)	84. (C)	85. (B)	86. (A)	87. (D)	88. (B)	89. (A)	90. (D)
91. (D)	92. (D)	93. (B)	94. (C)	95. (D)	96. (D)	97. (D)	98. (C)	99. (A)	100. (B)
101. (D)	102. (C)	103. (A)	104. (A)	105. (D)	106. (D)	107. (B)	108. (C)	109. (B)	110. (B)
111. (B)	112. (D)	113. (D)	114. (C)	115. (C)	116. (C)	117. (A)	118. (A)	119. (B)	120. (E)

121. (B) 122. (C) 123. (C) 124. (A) 125. (C) 126. (B) 127. (C) 128. (B) 129. (A) 130. (A)
131. (B) 132. (A) 133. (B) 134. (A) 135. (D) 136. (A) 137. (D) 138. (B) 139. (B) 140. (D)
141. (B) 142. (A) 143. (B) 144. (B) 145. (B) 146. (D) 147. (A) 148. (A) 149. (B) 150. (D)
151. (C) 152. (C) 153. (D) 154. (D) 155. (D) 156. (B) 157. (D) 158. (C) 159. (A) 160. (A)
161. (A) 162. (C) 163. (C) 164. (B) 165. (D) 166. (C) 167. (A) 168. (C) 169. (C) 170. (A)
171. (D) 172. (C) 173. (D) 174. (A) 175. (C) 176. (A) 177. (A) 178. (A) 179. (E) 180. (C)
181. (A) 182. (A) 183. (E) 184. (B) 185. (B) 186. (D) 187. (C) 188. (D) 189. (E) 190. (A)
191. (A) 192. (C) 193. (A) 194. (A) 195. (A) 196. (C) 197. (B) 198. (B) 199. (B) 200. (B)
201. (D) 202. (D) 203. (B) 204. (C) 205. (B) 206. (B) 207. (D) 208. (A) 209. (D) 210. (A)
211. (C) 212. (C) 213. (D) 214. (D) 215. (B) 216. (A) 217. (C) 218. (D) 219. (C) 220. (C)
221. (A) 222. (B) 223. (D) 224. (B) 225. (C) 226. (E) 227. (C) 228. (D) 229. (A) 230. (B)
231. (A) 232. (C) 233. (B) 234. (D) 235. (C) 236. (A) 237. (B) 238. (A) 239. (C) 240. (C)
241. (A) 242. (C) 243. (B) 244. (C) 245. (B) 246. (D) 247. (A) 248. (A) 249. (D) 250. (C)
251. (A) 252. (D) 253. (A) 254. (D) 255. (B) 256. (B) 257. (B) 258. (B) 259. (D) 260. (A)
261. (A) 262. (A) 263. (C) 264. (A) 265. (B) 266. (D) 267. (A) 268. (A) 269. (A) 270. (A)
271. (A) 272. (D) 273. (C) 274. (B) 275. (B) 276. (C) 277. (D) 278. (C) 279. (A) 280. (A)
281. (A) 282. (B) 283. (D) 284. (C) 285. (A) 286. (A) 287. (D) 288. (B) 289. (D) 290. (B)
291. (A) 292. (D) 293. (B) 294. (B) 295. (B) 296. (B) 297. (A) 298. (B) 299. (A) 300. (C)
301. (D) 302. (D) 303. (B) 304. (D) 305. (D) 306. (D) 307. (D) 308. (B) 309. (C) 310. (C)
311. (A) 312. (D) 313. (D) 314. (D) 315. (A) 316. (A) 317. (A) 318. (C) 319. (D) 320. (D)
321. (C) 322. (D) 323. (A) 324. (C) 325. (C) 326. (C) 327. (B) 328. (B) 329. (B) 330. (A)
331. (D) 332. (B) 333. (C) 334. (B) 335. (C) 336. (B) 337. (C) 338. (A) 339. (B) 340. (B)
341. (B) 342. (C) 343. (C) 344. (A) 345. (C) 346. (C) 347. (B) 348. (C) 349. (D) 350. (C)
351. (D) 352. (B) 353. (B) 354. (B) 355. (A) 356. (C) 357. (C) 358. (D) 359. (A) 360. (B)
361. (C) 362. (D) 363. (B) 364. (D) 365. (B) 366. (C) 367. (C) 368. (B) 369. (D) 370. (C)
371. (A) 372. (B) 373. (C) 374. (C) 375. (A) 376. (A) 377. (B) 378. (C) 379. (B) 380. (A)
381. (D) 382. (B) 383. (C) 384. (A) 385. (B) 386. (C) 387. (C) 388. (D) 389. (D) 390. (D)
391. (A) 392. (C) 393. (A) 394. (D) 395. (B) 396. (D) 397. (D) 398. (D) 399. (B) 400. (A)
401. (D) 402. (B) 403. (A) 404. (B) 405. (B) 406. (B) 407. (A) 408. (C) 409. (B) 410. (D)
411. (D) 412. (A) 413. (C) 414. (A) 415. (B) 416. (B) 417. (C) 418. (C) 419. (A) 420. (A)
421. (C) 422. (D) 423. (D) 424. (D) 425. (A) 426. (B) 427. (C) 428. (A) 429. (A) 430. (A)
431. (B) 432. (B) 433. (A) 434. (A) 435. (C) 436. (A) 437. (C) 438. (A) 439. (A) 440. (D)
441. (A) 442. (B) 443. (C) 444. (C) 445. (B) 446. (B) 447. (B) 448. (D) 449. (B) 450. (B)
451. (C) 452. (B) 453. (B) 454. (B) 455. (B) 456. (D) 457. (D) 458. (A) 459. (B) 460. (B)
461. (A) 462. (B) 463. (A) 464. (B) 465. (D) 466. (C) 467. (B) 468. (D) 469. (D) 470. (B)
471. (C) 472. (D) 473. (A) 474. (C) 475. (A) 476. (B) 477. (A) 478. (C) 479. (D) 480. (D)
481. (B) 482. (A) 483. (B) 484. (B) 485. (B) 486. (C) 487. (B) 488. (A) 489. (B) 490. (D)
491. (B) 492. (A) 493. (A) 494. (A) 495. (B) 496. (A) 497. (A) 498. (A) 499. (D) 500. (A)
501. (C) 502. (B) 503. (A) 504. (B) 505. (C) 506. (C) 507. (D) 508. (A) 509. (C) 510. (C)
511. (C) 512. (B) 513. (B) 514. (A) 515. (A) 516(D) 517. (B) 518. (B) 519. (C) 520. (C)
521. (B) 522. (B) 523. (B) 524. (A) 525. (A) 526. (D) 527. (B) 528. (C) 529. (B) 530. (B)
531. (B) 532. (B) 533. (A) 534. (C) 535. (A) 536. (A) 537. (C) 538. (C) 539. (D) 540. (C)
541. (A) 542. (C) 543. (C) 544. (B)

Chapter - 6

AGRICULTURAL MICROBIOLOGY

1. Who is credited with discovery of Bacteria?
(A) Louis Pasteur (B) Leewenhoek
(C) Needham (D) Tyndall

2. Founder of modern bacteriology is:
(A) Leewenhoek (B) Pasteur
(C) Robert Koch (D) None of these

3. Father of bacteriological techniques is:
(A) Pasteur (B) Koch
(C) Gram (D) Buchner

4. An enzyme complex that forms ATP from ADP and phosphate during oxidative phosphorylation in the linear mitochondrial membrane or the bacterial plasma membrane, and during photophosphorylation in chloroplasts is called:
(A) ATP synthase (B) Pasteur efect
(C) Ionisation (D) DNA ligage

5. The immunized individual produced antibodies against the concerned pathogen is known as:
(A) Active immunity (B) Acquired immunity
(C) Innate immunity (D) None of these

6. The presence of a functional immune system that is capable of specifically recognizing and selectively eliminating foreign microorganisms and molecules are called:
(A) Active immunity
(B) Acquired immunity
(C) Innate immunity (D) None of these

7. The immunity can be visualized as comprising four types of defensive barriers: anatomic, physiologic, endocytic and phagocytic and inflammatory is called:
(A) Active immunity (B) Acquired immunity
(C) Innate immunity (D) None of these

8. The cDNA that codes for the human DNA polymerase ἡ has been cloned by hybridization using the sequence of the RAD 30 gene of:
(A) *S. cerevisiae* (B) *E. Coli*
(C) *Agrobacterium tumifaciens*
(D) None of these

9. The cDNA for human DNA polymerase K has been obtained similarly following with the gene for -------- DNA polymerase IV is:
(A) *S. cerevisiae* (B) *E. Coli*
(C) *Agrobacterium tumifaciens*
(D) None of these

10. The increase in cell numbers which involves increase in cell mass and cell size (elongate), duplication of the bacterial chromosomes, synthesis of new cell wall and plasma membrane, partitioning of the two chromosomes, septum fermentation and cell division is known as:
(A) Microbial growth (B) Biological space
(C) Binary fission (D) None of these

11. Population growth is limited by either the exhaustion of available nutrients or accumulation of inhibitory metabolites/end products or by exhaustion of space, in this case, called a lack of:
(A) Microbial growth (B) Biological space
(C) Binary fission (D) None of these

12. Immediately after inoculation of the cells into fresh medium, the population remains temporarily unchanged is called:
(A) Lag phase
(B) Exponential (Log) phase
(C) Stationary phase (D) Death phase

13. The growth is a pattern of balanced growth wherein all the cells are dividing regularly by binary fission, and are growing by geometric progression is called:
(A) Lag phase
(B) Exponential (Log) phase
(C) Stationary phase (D) Death phase

14. Exponential growth cannot be continued forever in a batch culture (e.g., a closed system such as a test tube or flask) is called:
(A) Lag phase
(B) Exponential (Log) phase
(C) Stationary phase (D) Death phase

15. The number of viable cells decreases geometrically (exponentially), essentially the everse of growth during the log phase is called:
(A) Lag phase
(B) Exponential (Log) phase
(C) Stationary phase (D) Death phase

16. Asexual process of reproduction is called:
(A) Microbial growth (B) Biological space
(C) Binary fission (D) None of these

17. The time interval required for the cells (or population) to divide is called:
(A) Microbial growth (B) Biological space
(C) Binary fission (D) Generation time

18. Phycoerythrin absorbs light maximally at:
(A) 550 nm (B) 620 nm
(C) 650 nm (D) 700 nm

19. Phycocyanin absorbs light maximally at:
(A) 550 nm (B) 620 nm
(C) 650 nm (D) 700 nm

20. Allophycocyanin absorbs light maximally at:
(A) 550 nm (B) 620 nm
(C) 650 nm (D) 700 nm

21. Anoxygenic phototrophs contain bacteriochlorophylls that absorbes maximally between :
(A) 550 nm (B) 620 nm
(C) 650 nm (D) 800-925 nm

22. Most algae and cyanobacteria show strong absorption of which type of light:
(A) Red (B) Blue
(C) Both (A) and (B) (D) None of these

23. The most widespread accessory pigments are the -------, which are always present in phototrophic organisms:
(A) Carotenoids (B) Phycobilins
(C) Phycobilisomes (D) None of these

24. Phycobiliproteins occur as high molecular weight aggregates which attached to the photosynthetic membranes is called:
(A) Carotenoids (B) Phycobilins
(C) Phycobilisomes (D) None of these

25. Light-harvesting pigments are also called:
(A) Accessary pigments (B) Antennae pigments
(C) Both (A) and (B) (D) None of these

26. In eukaryotes, photosynthesis is associated with special organelles called the:
(A) Chloroplasts (B) Thylakoids
(C) Chlorosomes (D) None of these

27. The chlorophyll pigments are attached to sheet-like membrane structures of the chloroplasts called the:
(A) Chloroplasts (B) Thylakoids
(C) Chlorosomes (D) None of these

28. In green bacteria, the photosynthetic apparatus is associated with both the cytoplasmic membrane and specialized non unit-membrane-encoded structures called:
(A) Chloroplasts (B) Thylakoids
(C) Chlorosomes (D) None of these

29. The process by which some bacteria use CO_2 as an electron sink in energy metabolism and reduce it to acetate as a fermentation end product using the acetyle CoA pathway is called:
(A) Acetogens (B) Methanogenes
(C) Methanogenesis (D) All of these

30. *Acetabacterium woodii* and *Clostridium aceticum* are strictly anaerobic and common examples of:
(A) Acetogens (B) Methanogenes
(C) Methanogenesis (D) All of these

31. *Methanobacterium, Methanobrevibacter, Methanococcus* and *Methanogenium* are important example of:
(A) Acetogens (B) Methanogenes
(C) Methanogenesis (D) All of these

32. The process by which some organisms gain energy by using H_2 to reduce CO_2 to CH_4 is called:
(A) Acetogens (B) Methanogenes
(C) Methanogenesis (D) All of these

33. The process by which many microorganisms synthesisc all their amino acids and other nitrogenous compounds such as purines and pyrimidines using ammonia and carbon chains derived from carbohydra mechanism is called:
(A) Acetogens
(B) Methanogenes
(C) Methanogenesis
(D) Nitrogen metabolism

34. The synthesis of the other amino acids requires that glutamate donates an amino group to an:
(A) α-ketocarboxylic acid
(B) α-ketoglutarate
(C) L-glutamate
(D) L-glutamine

35. Nitrogen fixation is found only in which type of microorganisms?
(A) Prokaryotic (B) Eukaryotic
(C) Both (A) and (B) (D) None of these

36. Nitrogen fixation requires the enzyme nitrogenase that consists of separate proteins called:
(A) Dinitrogenase
(B) Dinitrogenase reductase
(C) Both (A) and (B)
(D) None of these

37. The three genes coding of dinitrogenase and dinitrogenase reductase are part of a complex regulon called the:
(A) nif regulon (B) Prokaryotes
(C) Eukaryotes (D) None of these

38. Inorganic molecules such as derivatives of sulfur, nitrogen and iron are used by:
(A) nif regulon (B) Prokaryotes
(C) Eukaryotes (D) None of these

39. Inorganic nitrogen and sulfur compounds are reduced by many organisms are used as sources of cellular nitrogen and sulfur by:
(A) Assimilatory pathways
(B) Anaerobic respiration
(C) Chemolithotrophs
(D) Prokaryotes

40. Inorganic compounds are used instead of oxygen as terminal electron acceptors in a process called:
(A) Assimilatory pathways
(B) Anaerobic respiration
(C) Chemolithotrophs
(D) Prokaryotes

41. Oragnisms that obtain energy by the oxidation of inorganic compounds are called:
(A) Assimilatory pathways
(B) Anaerobic respiration
(C) Chemolithotrophs
(D) Prokaryotes

42. Most of the chemolithotrophs are autotrophs, that is they utilize CO_2 as the carbon source and are thus called:
(A) Chemolithoautotrophs
(B) Chemolithotrophs
(C) Both (A) and (B)
(D) Prokaryotes

43. Among the hydrogen-oxidizing bacteria are same that can also grow on carbon monoxide (CO) as the sole source of energy and carbon using oxygen, or in some cases nitrate (denitrifiers) as the electron acceptor. They are:
(A) Carboxydobacteria
(B) Hydrogen – oxidizing bacteria
(C) Sulfur-oxidizing bacteria
(D) All of these

44. Bacteria usually facultative chemolithotrophs and can live either autotrophically or heterotrophically; they also grow organotrophically. They are called:
(A) Carboxydobacteria
(B) Hydrogen – oxidizing bacteria
(C) Sulfur-oxidizing bacteria
(D) All of these

45. Bacteria include the photosynthetic sulfur oxidizers and the nonphotosynthetic sulfur oxidizers. They are called:
(A) Carboxydobacteria
(B) Hydrogen – oxidizing bacteria
(C) Sulfur-oxidizing bacteria
(D) All of these

46. A variety of anaerobic bacteria form propionate as the major end product of fermentation. They are:
(A) *Clostridium propionicum*
(B) *Megaspaera elsdenii*
(C) Both (A) and (B)
(D) *Saccharomyces cerevisiae*

47. Alcohol fermentation is the domain of yeasts, notably by:
(A) *Clostridium propionicum*
(B) *Megaspaera elsdenii*
(C) Both (A) and (B)
(D) *Saccharomyces cerevisiae*

48. Lactate is a very common end product of bacteria fermentations. The lactic acid bacteria is:
(A) *Lactobacillus lactis*
(B) *Streptococcus faecalis*
(C) *Pediococcus damnosum*
(D) All of these

49. Butyric acid fermentation is carried out by a group of strict anaerobic that belong to genera of:
(A) Clostridium (B) Butyrivibrio
(C) Eubacterium (D) Fusobacterium
(E) All of these

50. The bacterium is used in the fermentation of lactate to propionate in the manufacture of Swiss cheese:
(A) Propionibacterium (B) Enterobacteria
(C) Hematopoiesis (D) None of these

51. Low pH of media inhibits the growth of:
(A) Bacteria (B) Moulds
(C) Both of the above (D) None of these

52. Canning is a food Preservation technique first derived by:
(A) Appert (B) Pasteur
(C) Buchner (D) Menton

53. Which of the following genera is most likely to contain organisms capable of surviving high temperature?
(A) Clostridium (B) Penicillium
(C) Torula (D) Pseudomonas

54. In the microbial production of an amino acid, which form of the amino acid is synthesized?
(A) D-Stereoisomers (B) L-Stereoisomers
(C) L and D Forms (D) None of the above

55. Which of the following is/are Bio-insecticides?
(A) *Bacillus thuringiensis* (B) *Bacilllus popilliae*
(C) Both (a) and (B) (D) None of these

56. Funaric acid is produced by:
(A) *Aspergillus terrus*
(B) *Rhizopus nigricans*
(C) *Aspergillus niger*
(D) *Asergillus fumigatus*

57. Ionizing radiation to sterilize materials is called:
(A) Ionisation (B) Pasteurization
(C) Cold sterilization (D) Tyndallisation

58. Rhizosphere was coined by:
(A) Hiltner (B) Beijernick
(C) Winogradsky (D) None of these

59. ATP needed for N_2 fixation:
(A) 12 (B) 16
(C) 28 (D) 32

60. Factor which does not affect legume root nodulation?
(A) Temperature (B) Light
(C) Combined N_2 (D) H^+ ion conc.

61. **Microaerophillic N_2 fixing bacteria:**
(A) *Azotobacter* (B) *Azospirillium*
(C) *Clostridium* (D) *Enterobacter*

62. **Bacteria which can ferment sugar faster than yeast is:**
(A) *Clostridium* (B) *Zygomonas*
(C) *Thiobacillus* (D) *Bacillus*

63. **Father of soil microbiology is:**
(A) Winogradsky (B) Beijrenick
(C) Spermi (D) Hiltner

64. **Glycolysis inhibition in the presence of oxygen is called:**
(A) Warburg effect
(B) Hypoglycolytic effect
(C) Pasteur effect (D) None of these

65. **Wood sugar is:**
(A) Glucose (B) Mannose
(C) Xylose (D) Arabinose

66. **A mat of organism formed at the surface of a liquid culture is called:**
(A) Biofilm (B) Pellicle
(C) Scum (D) Foam

67. **Alginates are isolated from:**
(A) Rhodophyta (B) Bryophyta
(C) Chlorophyta (D) Phaeophyta

68. **The first antiseptic agent used was:**
(A) Mercuric chloride (B) Carbolic acid
(C) Alcohol (D) Hyposolution

69. **A fluorescent dye used in fluorescent microscopy is:**
(A) Acridine orange
(B) Phosphotungstic acid
(C) Ethidium bromide
(D) Luciferin

70. **What is the transformation of floral parts into leaf like structures?**
(A) Witche's broom (B) Faciation
(C) Phyllody (D) All of these

71. **All bacteria having a conjugative plasmid will have:**
(A) Drug resistance (B) Sex pili
(C) Flagella (D) All of these

72. **A mutation with multiple effect on the phenotype is called:**
(A) Multiple mutation
(B) Frame shift mutation
(C) Pleotrophic mutation
(D) None of these

73. **A DNA specific to plants and algae is:**
(A) Cytoplasmic DNA (B) Plastid DNA
(C) Heterocystic DNA
(D) Mitochondrial DNA

74. **The TCA cycle pathway is:**
(A) Anabolic (B) Catabolic
(C) Amphibolic (D) None of these

75. **Which among the following is used for insertion of foreign DNA into cells?**
(A) Agrobacterium (B) Vaccinia virus
(C) Hela cells (D) All of these

76. **Starting material for which wine making is:**
(A) Mollasses (B) Barley malt
(C) Grape juice (D) Beet root mash

77. **Virus quantification is a given sample is done by:**
(A) End point efflux
(B) End point dilution
(C) End point titration
(D) All of these

78. **The distance between the specimen and the nearest face of the objective lens when the specimen is in sharp focus in a microscope is:**
(A) Resolution distance (B) Numerical eperture
(C) Working distance (D) Nose length

79. **Plasmids with a similar or identical replion which cannot coexist in a cell are:**
(A) Compatible plasmids
(B) Iso replion plasmids
(C) Incompatible plasmids
(D) Strigent plasmids

80. **Which among the following is a flagellated cycanobacteria?**
(A) Chroococcum (B) Nostoc
(C) Both (a) and (B) (D) None of these

81. **Which among the following is a restriction modification in DNA?**
(A) Capping (B) Tailing
(C) Methylation (D) Phosphorylation

82. **Ethylene oxide exhibits its antimicrobial activity by ---- reaction:**
(A) Oxidation (B) Hydrogenation
(C) Reduction (D) Alkylation

83. **S^{38} incorporation helps in accumulation of radioactivity in the amino acid:**
(A) Proline (B) Arginine
(C) Valine (D) Methionine

84. **A micrometer is:**
(A) 10^{-3} mm (B) 10^{-9} mm
(C) 10^{-3} m (D) 10^{-9} m

85. **A medium in which different types of micro-organisms exhibit different growth forms so that they could be distinguished is called:**
(A) Selective medium
(B) Synthetic medium
(C) Differential medium
(D) Preferential medium

86. Cell-Cell interaction and differentiation is mostly studies on:
(A) *Arabidopsis thaliana*
(B) *Cenorabditis elegens*
(C) *Escherichia coli*
(D) *Dictyostlium discodeum*

87. Milky disease is caused by:
(A) *Bacillus thuringiensis* (B) *Lactobacillus lactis*
(C) *Bacillus popiliae* (D) None of these

88. mRNA which contains a transcript of more than one gene is said to be:
(A) Multigenic (B) Polygenic
(C) Polycistronic (D) Heterogenic

89. A bacterial cell which lacks a chromosome but contains all the components for transcription and translation:
(A) Maxi cell (B) Empty cell
(C) Mini cell (D) Ghost cell

90. Warburg's effect is due to:
(A) Increased photorespiration
(B) Reduced photorespiration
(C) Increased glycolysis
(D) decreased glycolysis

91. Fractional sterilization is:
(A) Appertization (B) Pasteurization
(C) Tyndalization (D) Cold sterilization

92. In gram staining the alcohol acts on:
(A) Teichoic acids (B) Peptidoglycan
(C) Periplasm (D) Membrane lipids

93. Partial diploid in bacteria is called:
(A) Merozygote (B) Heterokaryon
(C) False zygote (D) None of these

94. Specilised N-fixing cells in Filamentous Cyanobacterium symbiosis are:
(A) Alkinites (B) Endospores
(C) Cysts (D) Heterocysts

95. Cell to cell communication in legume Rhizobium symbiosis is carried out by:
(A) Lectins (B) Flavanoids
(C) Isoflavons (D) Haemoglobin

96. Bacterial endospores are characterized by the presence of:
(A) Diaminopimelic acid
(B) Polybetahydroxy butyrate
(C) Dipicolinic acid (D) All of these

97. In true photosynthesis the electron donor is:
(A) Oxygen (B) Water
(C) Photosystem I (D) ATP

98. The most predominant immunoglobulin is tears is:
(A) IgG (B) IgA
(C) IgE (D) IgD

99. In leghaemoglobin the heme portion is specified by:
(A) Plant genes (B) Bacterial genes
(C) Both of the above (D) None of these

100. Leghaemoglobin is structurally and functionally similar to:
(A) Haemoglobin (B) Actin
(C) Myosin (D) Myogolobin

101. The anticodon for ACG is:
(A) UGC (B) TGC
(C) CGU (D) CGT

102. Negative supercoiling in DNA is brought about by:
(A) DNA gyrase (B) DNA helicase
(C) DNA ligase (D) All of these

103. A substance that can elicit an immune response only when combined with another carrier:
(A) O antigen (B) Haptan
(C) Adjent (D) Immunogen

104. The non-cellulolytic polysaccharide fraction of plants are represented mainly by:
(A) Lignin (B) Pectin
(C) Hemicellulose (D) Suberin

105. Yeast are:
(A) Gram positive (B) Gram negative
(C) Gram variable (D) None of these

106. Giemsa stain are used as particularly applicable for staining:
(A) Rickettsias (B) Spores
(C) Protozoa (D) Both (A) and (C)

107. Name of organism which are predatory on bacteria:
(A) Virus (B) Viriod
(C) Bdellovibrios (D) Prion

108. In cocoa fermentation, the microbial inoculumn used is of:
(A) *Aspergillus niger* (B) *Candida krusei*
(C) *Rhizopus oryzae* (D) None of these

109. Riboflavin can be obtained by microbial fermentation using fungi:
(A) *Ashbya gossypii* (B) *Rhzopus oryzae*
(C) *Sacchromycetes* (D) None of these

110. Tetracycline is effective against:
(A) Gram +ve bacteria (B) Gram –ve bacteria
(C) Fungi (D) Broad spectrum

111. Bacitracin is effective against:
(A) Gram +ve bacteria (B) Gram –ve bacteria
(C) Fungi (D) Broad spectrum

112. Pili are filamentous hair like structures on the surface of only:
(A) Gram +ve bacteria
(B) Gram –ve bacteria
(C) Both of the above
(D) None of these

113. The phenomenon of inhibiting the growth of bacteria without killing them:
(A) Bactericidal (B) Bacteriostatic
(C) Both of the above (D) None of these

114. The morphological form of rhizobium cells within the root nodules of legumes is:
(A) Ovoid (B) Bacteroid
(C) Viorod (D) None of these

115. A cell wall component which anchors the outer membrane of enteric Gram negative bacteria to the peptidoglycan layer:
(A) Techioic acid (B) Lignic acid
(C) Braun lipoprotein (D) None of these

116. Protein coat of a virus:
(A) Capsule (B) Capsid
(C) Envolopo (D) Coat

117. An organism that uses inorganic compounds as electron donors and releies on chemical compounds for energy:
(A) Chemoautotroph (B) Photoautotroph
(C) Chemolithotroph (D) Photolithotroph

118. A relationship between members of different species living in proximity in which one organism benefits from the association but the other is not affected:
(A) Ammensalism (B) Communsalism
(C) Predation (D) Parasite

119. An agent which free from infection by killing the vegetative cells of microorganism:
(A) Surfactant (B) Disinfectant
(C) Antiseptic (D) None of these

120. The preservation of biological specimens by rapid freezing and rapid dehydration in a high vacuum is known as:
(A) Immobilisation (B) Cryopreservation
(C) Lyophilization (D) None of these

121. Periplasmic flagella is also called:
(A) Axial fibrils (B) Endoflagella
(C) Both of these (D) None of these

122. An antiviral substance produced by animal tissue:
(A) Virion (B) Interferon
(C) Antibody (D) Antigen

123. Temperature necessary for destroyed of all the microbes within 10 minutes is called:
(A) Decimal reduction time
(B) Thermal death point
(C) Thermal death time
(D) None of these

124. Nitrogenous enzyme first isolated from:
(A) *Clostridium pasteurianum*
(B) *Bacillus* sp.
(C) *Azotobacter* sp.
(D) *Penicillium*

125. Fungi used for bioassay of biotin:
(A) *Neurospora* (B) Yeast
(C) *Aspergillus* (D) *Penicillium*

126. Region around the plant root where materials released from the root increase microbial population activities is called:
(A) Rhizosphere (B) Rhizoplane
(C) Phyllosphere (D) None of these

127. ------------is a plant pathogen bacteria:
(A) Shigella (B) *E. coli*
(C) Salmonella (D) Erwinia

128. Anaerobic free living 'N' fixing bacterium:
(A) Azotobacter (B) Acetobacter
(C) Kelbseilla (D) Clostridium

129. Virus – 'Contangium viuum fluidum' was given by:
(A) A. Mayer (B) Beijerinck
(C) Iwanoski (D) None of these

130. Itaconic acid is produced by:
(A) *Aspergillus terrus*
(B) *Aspergillus itaconicus*
(C) Both of the above (D) None of these

131. Phagocytosis is discovered by:
(A) Robert Koch (B) Elie Metchnikoff
(C) Joseph Lister (D) Fenny Hesse

132. Lipopolysaccharide is found in cell wall:
(A) Gram +ve bacteria
(B) Gram –ve bacteria
(C) Both of the above (D) Fungi

133. In autoclave which form of heat is used?
(A) Dry heat (B) Moist heat
(C) Vacuum heat (D) Both (a) and (b)

134. Who developed cell free fermentation of modern science?
(A) Pasteur (B) Koch
(C) Buchner (D) None of these

135. Nitrogenase enzyme have how many component?
(A) 1 (B) 2
(C) 3 (D) 4

136. Antiseptic surgery is given by:
(A) Joseph Lister (B) Schcoedes
(C) Koch (D) Jenner

137. Five kingdom classification was developed by:
(A) Haeckel (B) Chatn
(C) Stainer (D) Whittaker

138. Number of chromosome in bacteria is:
(A) 0 (B) 1
(C) 2 (D) Many

139. Murein is present in:
(A) Cyanobacteria (B) Halobacterium
(C) Methanobacterium (D) None of these

140. Sterols are present in:
(A) Eubacteria (B) Fungi
(C) Mycoplasma (D) None of these

141. In -----, there is V-shaped structure due to snapping fission:
(A) Fusobacterium (B) Mycobacterium
(C) Spirochaetes (D) Cornyebacterium

142. Who proposed new grouping of living organisms into 3 kingdom i.e. Archaebacteria, Eubacteria and Eukaryotes?
(A) Eichels (B) Woose
(C) Haeckel (D) Chester

143. Father of bacteriology is:
(A) Leewenhock (B) Pasteur
(C) Koch (D) None of these

144. Microorganism which are associated with non legumes:
(A) Anabaena (B) Klebseilla
(C) Both of the above (D) None of these

145. Lipopolysaccharide do not contain:
(A) Lipid (B) D-Antigens
(C) Porins (D) Polysaccharides

146. In Archaebacteria which one is present in cell wall?
(A) Murein (B) Pseudomonas
(C) Both of them (D) None of these

147. The invaginations seen at localized areas specially near the site of cell division and participate in transverse septum formation is called:
(A) Lysosomes (B) Mesosomes
(C) Murein (D) Capsules

148. Smallest self-relicating prokaryotes capable of generating their own energy:
(A) Mycoplasma (B) Virus
(C) Chlamydia (D) Rickettsias

149. Species of mycoplasma are inhibited by:
(A) Penicilin (B) Tetracycline
(C) Both of them (D) None of these

150. Mesomes are well developed in:
(A) *E. coli* (B) *Proteus vulgaris*
(C) *Bacillus* (D) None of these

151. Principal sites for CO_2 fixation in autotrophic prokaryotes:
(A) Chromosomes (B) Carboxysomes
(C) Both of these (D) None of these

152. Causal organism of Q. fever:
(A) *Coexiella burnetti* (B) *Salmonella sp.*
(C) *E. coli* (D) None of these

153. In archaebacteria, first amino acid to initiate a new polypeptide chain is:
(A) Methionine (B) Cysteine
(C) Lysine (D) None of these

154. Which one is not correct about mesosomes?
(A) Prominent in gram +ve bacteria
(B) Prominent in gram –ve bacteria
(C) Helps in cell division
(D) Helps in DNA replication

155. In gram +ve bacteria Flagella:
(A) Only M ring is present
(B) Only S ring is present
(C) Only S and M rings are present
(D) All rings L, P, S and M are present

156. Bacteria are commonly seen in soils such as:
(A) Acidic (B) Neutral
(C) Alkaline (D) Saline

157. Diazotrophs are:
(A) Fix N_2 (B) Fix Phosphorus
(C) Fix O_2 (D) None of these

158. Amylases produced by:
(A) *Aspergillus* sp. (B) *Bacillus* sp
(C) Both of the above (D) None of these

159. Lipase produced by
(A) *Bacillus subtilis* (B) *Rhizopus* sp.
(C) *Aspergillus* sp. (D) None of these

160. Red pigment produced by:
(A) Serratia (B) Micrococcus
(C) Both of the above (D) None of these

161. Bacillus is a:
(A) Psychrophillic (B) Thermophilic
(C) Osmophilic (D) Gas former

162. Commercial (Now a days) production of penicillin is by:
(A) *Penicillium notatum*
(B) *Penicillium chrysogenium*
(C) *Aspergillus* sp.
(D) None of these

163. Mesophilic bacteria involved in composting:
1. Cellulomonas
2. Myxococcus
3. Bacillus
4. Clostridium
(A) 1, 2 (B) 1, 2, 3
(C) 2, 3 (D) All of these

164. Mesophilic fungi involved in composting:
1. Rhizopus
2. Fusarium
3. *Aspergillus niger*
4. Mucor
(A) 1, 2 (B) 2, 3
(C) 3, 4 (D) 1, 2, 3

165. Antibiotic streptomycin was isolated by:
(A) Buiknokles (B) Fleming
(C) Waksman (D) Duggar

166. An aminoglycoside antibiotic:
(A) Penicillin (B) Streptomycin
(C) Tetracycline (D) None of these

167. Which of the following have highest magnification?
(A) Simple microscope
(B) Compound microscope
(C) Phase contrast
(D) Oil immersion microscope

168. Mesophiles have temperature range of:
(A) 25-45^{0}C (B) 15-30^{0}C
(C) 20-45^{0}C (D) 15-45^{0}C

169. Root surface that can be colonized by the microbes:
(A) Rhizosphere (B) Rhizoplane
(C) Both of the above (D) None of these

170. Number of layers surrounding the heterocyst:
(A) One (B) Two
(C) Three (D) Four

171. Heterocyst lack:
(A) PSI (B) PSII
(C) Both of the above (D) None of these

172. Match the following:
(i) Acetic acid (A) *A. oryzae*
(ii) Lactic acid (B) *L. delbruckie*
(iii) Citric acid (C) *A. niger*
(iv) Gluconic acid (D) *Acetobacter sp.*

Codes:

	(i)	(ii)	(iii)	(iv)
(A)	a	b	c	d
(B)	c	b	d	a
(C)	d	b	c	a
(D)	b	a	d	c

173. Fungi used for bioassay of pantothenic acid:
(A) Yeast (B) Neurospora
(C) Mucor (D) Aspergillus

174. Fermenting organism involved in Yoghurt:
(A) Streptococcus (B) Aspergillus
(C) Podiococcus (D) None of these

175. Microbe used for Sarurkraut preparation:
(A) *A. niger* (B) *Streptococcus* sp.
(C) *Leuconostoc* sp. (D) None of these

176. Ropiness of unpacked bread is due to growth of:
(A) *Bacillus subtilis* (B) *Clostridium* sp
(C) *A. niger* (D) *Streptococcus* sp.

177. Example of chemolithotroph:
(A) *E. coli* (B) Bacillus
(C) Azotobacter (D) Nitrosomonas

178. Iron bacteria which oxidize ferrous into ferric compound is:
(A) Leptothrix (B) Beggiotoa
(C) Both the above (D) None of these

179. Match the following:
(i) Tetracyclime (A) *Bacillus subtilis*
(ii) Bacitrain (B) *Streptomyces aureofaciens*
(iii) Chloranphenicol (C) *S. grisens*
(iv) Cycloheximide (D) *S. venezuelae*

Codes:

	(i)	(ii)	(iii)	(iv)
(A)	a	b	c	d
(B)	b	a	d	c
(C)	a	b	d	c
(D)	d	b	c	a

180. Nitrate reductase enzyme present in:
(A) Cytoplasm (B) Chloroplast
(C) Mitochondria (D) Nucleus

181. Test organism for phenol coefficient:
(A) *Salmonella typhi*
(B) *Staphylococcus aureus*
(C) Both of the above
(D) None of these above

182. Microbes involved in fibre retting:
(A) *Micrococcus sp.*
(B) *Bacillus subtilis*
(C) *Clostridium butrycium*
(D) *Lactobacillus sp.*

183. Microbe involved in tobacco curing:
(A) *Micrococcus* sp. (B) *Bacillus subtilis*
(C) *Lactobacillus* sp. (D) *Aspergillus niger*

184. Urea degrading bacteria:
(A) *Bacillus pasturi*
(B) *Clostridium pasteurianum*
(C) *A. niger*
(D) *Micrococcus* sp.

185. 'P' solubilizing bacteria:
(A) *Bacillus megatherium*
(B) *Clostridium pasteurianum*
(C) *A. niger*
(D) *Micrococcus* sp.

186. Example of Eukaryotic inhibitor antibiotic:
(A) Cycloheximide (B) Penicillium
(C) Tetracycline (D) Streptomycin

187. Lactose sugar of milk is converted into Lactic acid by:
(A) *Streptococcus lactis* (B) *Lactobacillus sp.*
(C) *Clostridium sp.* (D) None of these

188. The size of microbe as:
(A) Viriod<Virus<Bacteria<Fungi
(B) Virus<Viriod<Bacteria<Fungi
(C) Viriod<Virus<Fungi<Bacteria
(D) Virus<Viriod<Fungi<Bacteria

189. Gram staining was discovered by Gram in year:
(A) 1762 (B) 1932
(C) 1884 (D) 1890

190. Match the following:

(i) Amylase (A) *Bacillus* sp.
(ii) Invertase (B) *Saccharromyces cersevisiae*
(iii) Proteases (C) *Aspergillus niger*
(iv) Lipases (D) *Rhizopus* sp.

Codes:

	(i)	(ii)	(iii)	(iv)
(A)	a	b	c	d
(B)	b	c	d	a
(C)	c	a	b	d
(D)	d	b	a	c

191. Which of the following is used to detect the presence of HIV?
(A) ELISA test (B) Benedict's test
(C) Widal test (D) Biuret's test

192. Which of the following does not produce any enzyme?
(A) Amoeba (B) Virus
(C) Bacteria (D) Fungi

193. A virus that may not destroy the host:
(A) Virulent phage
(B) Temperature phage
(C) Cyanophage
(D) Lytic cycle

194. Prions are:
(A) Organism containing only nucleic acid
(B) Proteins which are capable of replications in certain mammalian cells
(C) Small cells which are infectious
(D) Fungal toxins

195. A virion is a:
(A) Infectious nucleic acid
(B) Infectious virus particle
(C) A virus parasite on bacteria
(D) None of these

196. Which bacterium solubilizes tricalcium phosphate in soluble rock phosphatesin soils?
(A) *Bacillus polymyxa*
(B) *Pseudomonas striata*
(C) *Spirillum lipoferum*
(D) Both (A) and (B)

197. Number of flagella in Cyanobacteria are:
(A) 0 (B) 1
(C) 2 (D) 3

198. Type of relationship between *Acetobacter diazotrophicus* and sugarcane is:
(A) Symbiotic (B) Associative
(C) Endophytic (D) Free living

199. Example of VAM fungi:
(A) Pisoletecis (B) Sclerotenia
(C) Glomus (D) Trichoderma

200. Agar-Agar was developed by:
(A) Joseph Lister (B) Hoch
(C) Hesse (D) Pasteur

201. Match the following:

(i) Whisky (A) Grape wine
(ii) Rum (B) Potato
(iii) Brandy (C) Grain mash
(iv) Vodka (D) Molasses

Codes:

	(i)	(ii)	(iii)	(iv)
(A)	a	b	c	d
(B)	c	d	a	b
(C)	b	a	c	d
(D)	d	c	a	b

202. Match the following:

(i) Tryptophane (A) *Brevibacterium* sp.
(ii) Valine (B) *Enterobactericeae*
(ii) Glutamic acid (C) *Claviceps purpurea*
(iv) Lysine (D) *E. coli*

Codes:

	(i)	(ii)	(iii)	(iv)
(A)	a	b	c	d
(B)	c	a	d	b
(C)	b	c	a	d
(D)	d	c	a	b

203. The products of mixed acid fermentation are succinate, lactate, acetate, ethanol, formate, carbon dioxide and hydrogen gas is carried out by:
(A) Propionibacterium (B) Enterobacteria
(C) Hematopoiesis (D) None of these

204. The process by which blood cells are generated from a common precursor cell, known as the stem cell is called as:
(A) Propionibacterium (B) Enterobacteria
(C) Hematopoiesis (D) None of these

205. Gram stain is an example of:
(A) Simple stain (B) Differential stain
(C) Acid fast stain (D) None of these

206. Counter stain used in gram staining:
(A) Ethyl alcohol (B) Iodine solution
(C) Crystal violet (D) Safranine

207. Acid fast staining used for which bacteria:
(A) *Mycobacterium* (B) *Rhizobium*
(C) *Bacillus sp.* (D) *Clostridium*

208. Facultative free living 'N' fixing bacterium:
(A) Rhizobium (B) Azotobacter
(C) Klebsiella (D) None of these

209. In bacteria, cubical packets of 8 cells is called:
(A) Staphylococcus (B) Diplococcus
(C) Bacillus (D) Sarcina

210. Resolving power of electron microscope is ------ times that of light microscope.
(A) 10^1 (B) 10^2
(C) 10^3 (D) 10^4

211. In endospore staining, which one is used?
(A) Malachite green (B) Basic fuschin
(C) Indian Ink (D) Methylene blue

212. Braun's lipoprotein is present in :
(A) Gram +ve bacteria
(B) Gram –ve bacteria
(C) Bacteriophage
(D) Yeast

213. Pseudomorein is present in which organism?
(A) Archaebacteria (B) Eubacteria
(C) Eukaryote (D) Fungi

214. Mesosomes are well developed in:
(A) Gram +ve bacteria (B) *E. coli*
(C) Yeast (D) Mycoplasma

215. Energy parasite:
(A) Spirilla (B) Mycoplasma
(C) Chlaymidia (D) *Archaebacteria*

216. Example of pleomorphic bacteria is:
(A) *Acetobacter* (B) *Azotobacter*
(C) *Achromobacter* (D) *Arthrobacter*

217. Cyst formation is characteristics feature of :
(A) *Acetobacter* (B) *Arthrobacter*
(C) *Azotobacter* (D) None of these

218. N-reserve materials in cyanobacteria:
(A) Volutin granular
(B) PHB
(C) Polysaccharides
(D) Cyanophycin granules

219. Lysozyme treated cells of gram +ve bacteria are called:
(A) Protoplast (B) Sphaeroplast
(C) Cytoplasm (D) None of these

220. Lysozyme treated cells of gram -ve bacteria are called:
(A) Protoplast (B) Sphaeroplast
(C) Cytoplasm (D) Mesosomes

221. Bacteria having especially high rate of respiration:
(A) *Rhizobium* (B) *Azotobacter*
(C) *E. coli* (D) *Acetobacter*

222. Species of Mycoplasma are inhibited by:
(A) Penicillin (B) Tetramycin
(C) Both of the above (D) None of these

223. Xanthomonas is an example of:
(A) Monotrichous (B) Peritrichous
(C) Lophotrichous (D) Cephalotrichous

224. Test organism for pasteurization is:
(A) *Coxicella burnetti*
(B) *Clostridium pasteurizanum*
(C) *Bacillus subtilis*
(D) *Bacillus steareothermophillus*

225. "Ray Fungi" is the name given to:
(A) Rhizopus (B) Yeast
(C) Actinomycetes (D) Mycoplasma

226. Actinomycetes are:
(A) Gram +ve, aerobic
(B) Gram +ve, anaerobic
(C) Gram –ve, aerobic
(D) Gram –ve, anaerobic

227. Example for microaerophillic N_2 fixer:
(A) *Rhizobium* (B) *Azotobacter*
(C) *E. coli* (D) *Frankia*

228. Citric acid is produced by:
(A) *Aspergillus niger* (B) *Acetobacter*
(C) *Acetobutylicum* (D) None of these

229. Vitamin B2 is produced by:
(A) *Brevibacterium sp.* (B) *Pseudomonas*
(C) *Ashbya gossypii* (D) None of these

230. Bacitracin is produced by:
(A) *Bacillus subtilis* (B) *Aspergillus niger*
(C) *E. coli* (D) Yeast

231. Production of vinegar is by:
(A) *Bacillus subtilis* (B) *Glucanobacter*
(C) *Azotobacter* (D) None of these

232. First antifugal antibiotic:
(A) Cycloheximide (B) Aureofungin
(C) Neomycin (D) Nystatin

233. Neomycin is produced by:
(A) *S. nouesii* (B) *S. fradiae*
(C) *S. erthyreus* (D) *S. venezuelea*

234. Which one of the following is used for commercial production of vitamin B12?
(A) Brevibacterium (B) Mycobacterium
(C) Propionibacterium (D) None of these

235. Inhibitory action of penicillin is:
(A) Cell wall synthesis (B) Protein synthesis
(C) Both of the above (D) None of these

236. Organism involved in swiss cheese ripening is:
(A) *Penicillium* sp. (B) *Propionibacterium*
(C) *Streptococcus* sp. (D) None of these

237. Phosphorus solublizing microorganism:
(A) VAM (B) Mucor
(C) Rhizopus (D) *E. coli*

238. The term mycorrhiza is coined by:
(A) Frank (B) Hartig
(C) De bary (D) Winogradasky

239. Name the scientists involved in commercial production of Penicillin:
(A) Alexander Flemming
(B) Florey and Chain
(C) Tulsane Brothers
(D) Mouis Pasteur

240. Development of plants due to increase in number of cells is called:
(A) Hypertrophy (B) Hypotrophy
(C) Hyperplasia (D) Hypoplasia

241. Name of the bacterium producing endospore is:
(A) *Bacillus* (B) *Agrobacterium*
(C) *Xanthomonas* (D) *E. coli*

242. If 10% of the population of cells is killed by antimicrobial agent in the Ist minute of exposure, during IInd minute the percentage of killing will be:
(A) 9% (B) 10%
(C) 1% (D) 10.1%

243. The most resistant form of microbial life in the bacteria is:
(A) Cyst (B) Endospore
(C) Vegetative stage (D) None of these

244. If broth culture of following are heated in boiling after for 10 minutes which culture is most likely to be sterilized:
(A) *Bacillus sp.* (B) *Pseudomonas sp.*
(C) *Clostridium sp.* (D) None of these

245. Liquid N_2 has a temperature:
(A) -196^{0}C (B) -186^{0}C
(C) -176^{0}C (D) -172^{0}C

246. Ultraviolet light is most germicidal at the wavelength of:
(A) 245 nm (B) 255 nm
(C) 265 nm (D) None of these

247. UV light are microcidal due to the formation of:
(A) Pyrine dimers
(B) Pyrimidine dimers
(C) DNA damage (D) RNA damage

248. In general, an ideal antimicrobial chemical agent have a –spectrum activity:
(A) Limited (B) Broad
(C) Narrow (D) None of these

249. Azotobacter and bacillus polymyxa are:
(A) Decomposers
(B) Non-symbiotic nitrogen fixers
(C) Symbiotic nitrogen fixer
(D) Pathogenic bacteria

250. Methogenic bacteria are:
(A) Anaerobic in nature
(B) Aerobic in nature
(C) Microaerophillic in nature
(D) Saprophitic in nature

251. Urokinase enzyme is produced by:
(A) *Bacillus fastidious*
(B) *E. coli*
(C) *Bacillus thuringensis*
(D) *Pseudomonas putida*

252. For clearing oil pills which micro-organism is used?
(A) *Pseudomonas putida* (B) Nitrosomonas
(C) Rhizobium (D) *E. coli*

253. The antibiotic which inhibits bacterial growth are called?
(A) Bacteriocidal (B) Narrow spectrum
(C) Broad spectrum (D) Bacteriostatic

254. Most commonly used organism in SCP is:
(A) Bacillus (B) Lacto bacillus
(C) Spirulina (D) Pseudomonas

255. Biotechnology utilizes new organisms developed from present organisms utilizing:
(A) Mutation
(B) Recombinant DNA technology
(C) Both (A) and (B) (D) None of these

256. The method of growing microorganisms as a thin layer on nutrient medium is known as:
(A) Thin layer growth system
(B) Suspended growth system
(C) Support growth system
(D) None of these

257. Most of the organism which produce antibiotics line in the soil because:
(A) Darkness favours synthesis of antibiotics
(B) By the phenomenon antibiotics their growth, nutrition and survival value and enhanced in competitive world of microflora of the soil
(C) They cannot get nutrition out side the soil
(D) No one easily misuses their antibiotics

258. Which organism was used as bioweapon derived from:
(A) Clostridium (B) Yerstsinia pestis
(C) Fusarium species (D) Green algae

259. Genetically engineered bacteria are being used in commercial production of:
(A) Melatonin (B) Testoteron
(C) Human insulin (D) Thyroxin

260. The total number of 'nif' genes present in Klebsiella pneumonia is:
(A) 21 (B) 16
(C) 28 (D) 35

261. The bacterial genes responsible for nodule formation in Rhizobium --------- legume symbiosis are termed as:
(A) 'Sym' genes (B) 'nif' genes
(C) 'nod' genes (D) Nodulins

262. The nitrogen fixing potential of a nodulating plant can be assessed by;
(A) Nitrite reductase activity
(B) Acetylene reduction assay
(C) Nitrate reductase activity
(D) Estimation of ammonia formation

263. The endosymbiont responsible for nodulation in actinorhizal plants is an:
(A) Actinomycetes (B) Fungi
(C) Bacteria (D) Algae

264. The plant genes responsible for nodule formation are termed as:
(A) 'nod' genes (B) Nodulines
(C) 'ntr' genes (D) 'fix' genes

265. The site of nitrogen fixation in Frankia actinorhizal plants is:
(A) Spores (B) Bacteriodes
(C) Vesicles (D) Conidia

266. Ectendomycorrizas found in:
(A) Amaranthaceae (B) Ericaceae
(C) Monotrophaceae (D) Orchidaceae

267. One of the mechanisms by which plant growth promoting rhizobacteria improve plant growth:
(A) Siderophore (B) Drought resistance
(C) Heavy metal tolerance (D) salt resistance

268. The substances produced by microbes which can kill other microbes at low concentrations is known as:
(A) Antibiotics (B) Antibody
(C) Immunogen (D) Immunoglobulins

269. The covalently closed circular DNA present in bacterium is:
(A) Transposons (B) Insertional sequences
(C) Plasmids (D) Cosmids

270. The microbial process by which contaminants are removed from soil is termed as:
(A) Phytoremediation (B) Bioremediation
(C) Microbial remediation (D) Decomposition

271. The carrier material normally used for multiplication of VAM is:
(A) Lignite (B) Peat
(C) Vermiculite (D) Charcoal

ANSWERSHEET

6. AGRICULTURAL MICROBIOLOGY

1. (B)	2. (B)	3. (B)	4. (A)	5. (A)	6. (B)	7. (C)	8. (A)	9. (B)	10. (A)
11. (B)	12. (A)	13. (B)	14. (C)	15. (D)	16. (C)	17. (D)	18. (A)	19. (B)	20. (C)
21. (D)	22. (C)	23. (A)	24. (C)	25. (C)	26. (A)	27. (B)	28. (B)	29. (A)	30. (A)
31. (B)	32. (C)	33. (D)	34. (A)	35. (A)	36. (C)	37. (A)	38. (B)	39. (A)	40. (B)
41. (C)	42. (C)	43. (A)	44. (B)	45. (C)	46. (C)	47. (D)	48. (D)	49. (E)	50. (A)
51. (A)	52. (A)	53. (A)	54. (B)	55. (C)	56. (B)	57. (C)	58. (A)	59. (B)	60. (B)
61. (B)	62. (B)	63. (A)	64. (C)	65. (C)	66. (C)	67. (D)	68. (B)	69. (C)	70. (D)
71. (B)	72. (D)	73. (B)	74. (C)	75. (A)	76. (C)	77. (B)	78. (C)	79. (C)	80. (D)
81. (C)	82. (D)	83. (A)	84. (B)	85. (C)	86. (D)	87. (C)	88. (C)	89. (C)	90. (A)
91. (C)	92. (D)	93. (A)	94. (D)	95. (A)	96. (C)	97. (B)	98. (A)	99. (B)	100. (A)
101. (A)	102. (A)	103. (B)	104. (C)	105. (A)	106. (C)	107. (C)	108. (B)	109. (A)	110. (D)
111. (A)	112. (B)	113. (B)	114. (B)	115. (C)	116. (B)	117. (A)	118. (B)	119. (B)	120. (C)
121. (C)	122. (B)	123. (B)	124. (A)	125. (A)	126. (A)	127. (D)	128. (D)	129. (B)	130. (C)
131. (B)	132. (B)	133. (B)	134. (C)	135. (C)	136. (A)	137. (D)	138. (B)	139. (A)	140. (C)
141. (B)	142. (B)	143. (A)	144. (C)	145. (C)	146. (A)	147. (B)	148. (A)	149. (B)	150. (C)
151. (B)	152. (A)	153. (A)	154. (B)	155. (C)	156. (B)	157. (A)	158. (C)	159. (B)	160. (A)
161. (B)	162. (B)	163. (A)	164. (D)	165. (C)	166. (B)	167. (D)	168. (D)	169. (B)	170. (C)
171. (A)	172. (C)	173. (A)	174. (A)	175. (C)	176. (A)	177. (D)	178. (A)	179. (B)	180. (A)
181. (C)	182. (C)	183. (A)	184. (A)	185. (A)	186. (A)	187. (B)	188. (A)	189. (C)	190. (A)
191. (A)	192. (B)	193. (B)	194. (B)	195. (B)	196. (B)	197. (A)	198. (C)	199. (C)	200. (C)
201. (B)	202. (C)	203. (B)	204. (C)	205. (B)	206. (D)	207. (A)	208. (C)	209. (D)	210. (D)
211. (A)	212. (B)	213. (A)	214. (A)	215. (C)	216. (D)	217. (C)	218. (D)	219. (A)	220. (B)
221. (B)	222. (B)	223. (A)	224. (A)	225. (C)	226. (A)	227. (D)	228. (A)	229. (C)	230. (A)

231. (B)	**232. (D)**	**233. (B)**	**234. (C)**	**235. (A)**	**236. (C)**	**237. (A)**	**238. (A)**	**239. (B)**	**240. (C)**
241. (A)	**242. (B)**	**243. (B)**	**244. (B)**	**245. (A)**	**246. (C)**	**247. (B)**	**248. (B)**	**249. (B)**	**250. (A)**
251. (A)	**252. (A)**	**253. (D)**	**254. (C)**	**255. (C)**	**256. (C)**	**257. (B)**	**258. (C)**	**259. (C)**	**260. (B)**
261. (A)	**262. (B)**	**263. (A)**	**264. (B)**	**265. (C)**	**266. (C)**	**267. (A)**	**268. (A)**	**269. (C)**	**270. (C)**
271. (C)									

Chapter - 7

ECOLOGY AND ECOLOGICAL PRINCIPLES

1. Studies of the inter-relationships of organisms with their physical and biotic environments are termed as:
(A) Environmental biology (B) Ecology
(C) Both (A) and (B) (D) None of these

2. The term ecology is derived from which word?
(A) Greek (B) Latin
(C) Romen (D) English

3. Ecology term was first of all introduced by:
(A) Haeckel (B) Reiter
(C) Knight (D) None of these

4. Which of the following deals with the study of different habitats of the biosphere is:
(A) Habitat ecology (B) Ecosystem ecology
(C) Conservation ecology (D) Production ecology

5. Which of the following deals with the analysIs of ecosystem from structural and functional point of view including the interrelationship of physical (abiotic) and biological; (biotic) components of environment is:
(A) Habitat ecology
(B) Ecosystem ecology
(C) Conservation ecology (D) Production ecology

6. Which of the following deals with methods of proper management of natural resources such as land, water, forest, sea, mines, etc., for the benefit of human being is:
(A) Habitat ecology (B) Ecosystem ecology
(C) Conservation ecology
(D) Production ecology

7. Which of the following deals with the gross and net production of different ecosystems such as fresh water, sea water, crop fields, orchards, etc., and tries to do proper management of these ecosystems so that maximum yield can be obtained from them is:
(A) Habitat ecology
(B) Ecosystem ecology
(C) Conservation ecology
(D) Production ecology

8. Which of the following deals with the study of gross effects of radiations and radioactive substances over the environment and living organisms is:
(A) Radiation ecology (B) Taxonomic ecology
(C) Human ecology (D) Space ecology

9. Which of the following is concerned with the ecology of different taxonomic groups and eventually includes following subdivisions of ecology – plant ecology, insect ecology, invertebrate ecology, vertebrate ecology, microbial ecology etc. is:
(A) Radiation ecology (B) Taxonomic ecology
(C) Human ecology (D) Space ecology

10. Which of the following deals with the study of relationship of man with his environment is:
(A) Radiation ecology (B) Taxonomic ecology
(C) Human ecology (D) Space ecology

11. Which of the following concerned with the development of partially or completely regenerating ecosystems for supporting life of man during long space flights or during extended exploration of extra-terrestrial environments is:
(A) Radiation ecology (B) Taxonomic ecology
(C) Human ecology (D) Space ecology

12. Which of the following concerned with the analysis and understanding of the function and structure of ecosystem by the use of applied mathematics such as advanced statistical techniques, mathematical models and computer science is:
(A) Systems ecology (B) Taxonomic ecology
(C) Human ecology (D) Space ecology

13. The branch of ecology which deals with the ecology of populations is called:
(A) Demecology (B) Synecology
(C) Autoecology (D) Palaeoecology

14. Which of the following deals with the study of groups of organisms which are associated together as a unit (i.e. community) is called:
(A) Demecology (B) Synecology
(C) Autoecology (D) Palaeoecology

15. Which of the following deals with the study of the individual organism or an individual species is called:
(A) Demecology (B) Synecology
(C) Autoecology (D) Palaeoecology

16. Which of the following deals with the movements of biotic elements based on palaentological evidence, which provides information about ancestral organisms and environmental conditions existing in the past is called:
(A) Demecology (B) Synecology
(C) Autoecology (D) Palaeoecology

17. Which of the following deals with the adaptations of animals or preferences of particular organisms like insects to particular chemical substances is called:
(A) Chemical ecology (B) Synecology
(C) Autoecology (D) Palaeoecology

18. Which of the following deals with the study of the role of the environment in animal distribution is called:
(A) Demecology (B) Ecogeography
(C) Ecological energetics (D) Palaeoecology

19. Which of the following deals with energy conservation and its flow in the organisms within the ecosystem is called:
(A) Demecology (B) Ecogeography
(C) Ecological energetics (D) Palaeoecology

20. The sum total of physical and biotic conditions influencing the responses of the organisms is called:
(A) Atmosphere (B) Hydrosphere
(C) Lithosphere (D) Biosphere

21. The multilayered gaseous envelop surrounding the planet earth is called:
(A) Atmosphere (B) Hydrosphere
(C) Lithosphere (D) Biosphere

22. Which layers separates the stratosphere from the troposphere is:
(A) Tropopause (B) Stratpause
(C) Mesosphere (D) Ozonosphere

23. Which layer separates the mesosphere from stratosphere is:
(A) Tropopause (B) Stratopause
(C) Mesosphere (D) Ozonosphere

24. The lowest region of atmosphere which subjects to differential heating, temperature inversions and convection currents is:
(A) Troposphere (B) Stratosphere
(C) Mesosphere (D) Ionosphere

25. Which zone extends from the surface of the earth upto a height of 8 to 10 km at polar latitudes (poles), 10-12 km at moderate latitude and 16-20 km at the equator is:
(A) Troposphere (B) Stratosphere
(C) Mesosphere (D) Ionosphere

26. Many climatic events such as cloud formation, lightening, thunderstorm formation, thundering, etc., are held in the:
(A) Troposphere (B) Stratosphere
(C) Mesosphere (D) Ionosphere

27. The upper layer of troposphere, the temperature may decrease up to:
(A) 60°C (B) 30°C
(C) -60°C (D) -30°C

28. Upper region of the troposphere has a narrow boundary which has a constant temperature is called:
(A) Tropopause (B) Stratopause
(C) Mesopause (D) Ionopause

29. Upper layer of stratosphere form:
(A) Tropopause (B) Stratopause
(C) Mesopause (D) Ionopause

30. The percentage concentration of different gases in air does not vary with an increase in height is:
(A) Troposphere (B) Stratosphere
(C) Mesosphere (D) Ionosphere

31. The zone has about 30 km in height is called:
(A) Troposphere (B) Stratosphere
(C) Mesosphere (D) Ionosphere

32. The temperature of stratosphere increases up to:
(A) 60°C (B) 90°C
(C) -60°C (D) -90°C

33. The increase in temperature of stratosphere is due to ozone formation under the influence of ultra-violet component of sunlight. Such a layer of ozone is called:
(A) Troposphere (B) Stratosphere
(C) Mesosphere (D) Ozonosphere

34. Ozonosphere completely absorbs solar radiation, ultraviolet radiation from the sun and also a lot of the solar infra-red, thus, becoming ------- than adjacent layers above or below:
(A) Cooler (B) Warmer
(C) Moist and warmer (D) No change

35. The zone has about 40 km in height is called:
(A) Troposphere (B) Stratosphere
(C) Mesosphere (D) Ionosphere

36. In mesosphere, temperature shows:
(A) 60°C (B) 80°C
(C) -60°C (D) -80°C

37. The upper layer of mesosphere from the:
(A) Tropopause (B) Stratopause
(C) Mesopause (D) Ionopause

38. The zone has height about 300 km above the earth's surface is called:
(A) Troposphere (B) Stratosphere
(C) Mesosphere (D) Ionosphere

39. The zone contains several layers of ionized air. Thus most of the gaseous components which become ionized under the influence of radiant energy, remain as ions in:
(A) Troposphere (B) Stratosphere
(C) Mesosphere (D) Ionosphere

40. The outer fringe of earth's atmosphere and outer space begins after it is:
(A) Troposphere (B) Stratosphere
(C) Exosphere (D) Ionosphere

41. The gaseous mixture of troposphere, is utilized by most organisms in respiration to liberate energy from food during oxidation and is called:
(A) Air (B) Water
(C) Gas (D) Ice

42. In atmosphere, about 95 per cent of the total air is present up to the height of above the earth's surface about :
(A) 10 km (B) 20 km
(C) 30 km (D) 40 km

43. In atmosphere, about 5 per cent of the total air is present up to the height of above the earth's surface about :
(A) 100 km (B) 200 km
(C) 280 km (D) 400 km

44. Nitrogen gas is present in percent by volume in the atmosphere is:
(A) 77.00 (B) 78.00
(C) 78.09 (D) 79.00

45. Nitrogen gas is present in percent by weight in the atmosphere is:
(A) 77.00 (B) 78.09
(C) 75.00 (D) 75.54

46. Oxygen gas is present in percent by volume in the atmosphere is:
(A) 20.00 (B) 20.93
(C) 23.00 (D) 23.14

47. Oxygen gas is present in percent by weight in the atmosphere is:
(A) 20.00 (B) 20.93
(C) 23.00 (D) 23.14

48. Carbon di oxide gas is present in percent by volume in the atmosphere is:
(A) 0.340 (B) 0.034
(C) 0.0034 (D) 3.340

49. Carbon di oxide gas is present in percent by weight in the atmosphere is:
(A) 0.90 (B) 0.93
(C) 1.00 (D) 1.27

50. Miscellaneous gas is present in percent by volume in the atmosphere is:
(A) 0.01 (B) 0.02
(C) 0.03 (D) 0.04

51. Apparently the passage of oxygen across a membrane is accomplished by:
(A) Osmosis (B) Diffusion
(C) Both (A) and (B) (D) None of these

52. The rate of diffusion can be calculated by:
(A) Fick's law (B) Partial pressure
(C) Pressure plate (D) None of these

53. Carbon di oxide diffuses through water and animal tissues faster than oxygen is:
(A) 10 times (B) 20 times
(C) 25 times (D) 40 times

54. The oceans, rivers, streams, lakes, ponds, pools, polar ice caps, water vapour, etc. forms the:
(A) Atmosphere (B) Hydrosphere
(C) Lithosphere (D) Biosphere

55. About the earth's surface is covered with hydrosphere, the main component of which is water:
(A) One-third (B) One-half
(C) Three-fourth (D) Two-third

56. Certain animals such as, spider crab, *Maia*, etc., can tolerate only narrow fluctuations in salinity of water and are known as:
(A) Stenohaline (B) Euryhaline
(C) Both of these (D) None of these

57. Certain animals such as, *Mytilus, Aplysia,* etc., can withstand wider ranges of salinity of water and are known as:
(A) Stenohaline (B) Euryhaline
(C) Both of these (D) None of these

58. Certain animals such as, *Anguilla, Salmon,* etc., can tolerate both narrow and wider ranges of salinity of water and are known as:
(A) Stenohaline (B) Euryhaline
(C) Both of these (D) None of these

59. The terrestrial animals living under such conditions are called:
(A) Mesocoles (B) Hydrocoles
(C) Xerocoles (D) None of these

60. The animals that lives in the aquatic environment are called:
(A) Mesocoles (B) Hydrocoles
(C) Xerocoles (D) None of these

61. The animals living in such xeric conditions are called:
(A) Mesocoles (B) Hydrocoles
(C) Xerocoles (D) None of these

62. Plants which grow in areas where water is available in plenty are called:
(A) Mesophytes (B) Hydrophytes
(C) Xerophytes (D) None of these

63. Plants growing in water are called:
(A) Mesophytes (B) Hydrophytes
(C) Xerophytes (D) None of these

64. Plants can grow in ecosystem where water is scarce and where the day is very high are called:
(A) Mesophytes (B) Hydrophytes
(C) Xerophytes (D) None of these

65. Xerophytes living in physiologically dry soils, i.e., saline soils with high concentrations of salts such as NaCl, $MgCl_2$ and $MgSO_4$, are called:
(A) Halophytes (B) Hydrophytes
(C) Tenagophytes (D) None of these

66. The plants are amphibious in nature- they grow in water bodies as well as in water logged soil are called:
(A) Halophytes (B) Hydrophytes
(C) Tenagophytes (D) None of these

67. The plants grow in hydric conditions are called:
(A) Halophytes (B) Hydrophytes
(C) Tenagophytes (D) None of these

68. The solid component of earth is called:
(A) Atmosphere (B) Hydrosphere
(C) Lithosphere (D) Biosphere

69. The process takes place in cold humid climate under forest vegetation where leaf litter and other organic matter decompose slowly through fungal activity is called:
(A) Podsolization (B) Gleization
(C) Melanization (D) Calcification

70. The process take place in humid climates, due to high ground level or retention of surface water in the soil, waterlogging and consequent reducing conditions are fairly common is called:
(A) Podsolization (B) Gleization
(C) Melanization (D) Calcification

71. The process is very common in regions of low humidity where humus is formed from the organic matter is called:
(A) Podsolization (B) Gleization
(C) Melanization (D) Calcification

72. The process found in North India, where calcium in the form of calcium bicarbonate dissolves in water and leaches down escaping adsorption around dry particles are called:
(A) Podsolization (B) Gleization
(C) Melanization (D) Calcification

73. The process of iron and aluminium oxides which are resistant to decomposition, do not leach down but remain in the surface soil are called:
(A) Laterization (B) Gleization
(C) Melanization (D) Calcification

74. The soils are formed on poorly drained water filled regions with poor aeration, i.e., they are highly reducing soils in which ferric ions are reduced to ferrous are:
(A) Hydromorphic soils (B) Halomorphic soils
(C) Both (A) and (B) (D) None of these

75. The soils are characterized by the presence of high salt contents on the surface are:
(A) Hydromorphic soils
(B) Halomorphic soils
(C) Both (A) and (B) (D) None of these

76. The soils are found on mountain tops where parent rock is quite hard and resistant to weathering are:
(A) Lithosols (B) Regosols
(C) Both (A) and (B) (D) None of these

77. The soils are formed from volcanic deposits are called:
(A) Lithosols (B) Regosols
(C) Both (A) and (B) (D) None of these

78. In a well-saturated soil, the accessory (extra) amount of water displace air from the pore spaces between soil particles and percolates downwardly under gravitational force is called:
(A) Gravitational water (B) Capillary water
(C) Hygroscopic water (D) Combined water

79. The water which is held by capillary forces (i.e., surface tension and attraction forces of water molecules) in smaller soil channels. When the gravitational water and ground water have been drained, is called:
(A) Gravitational water (B) Capillary water
(C) Hygroscopic water (D) Combined water

80. Soil particles retain some water so firmly that the plants cannot absorb this. Such soil water is called:
(A) Gravitational water (B) Capillary water
(C) Hygroscopic water (D) Combined water

81. The water of chemical compounds held by chemical forces of molecules such as hydroxides of silicon, iron and aluminium is called:
(A) Gravitational water (B) Capillary water
(C) Hygroscopic water (D) Combined water

82. The water content (% oven dry weight) retained by the undisturbed soil is known as:
(A) Moisture equivalent
(B) Water holding capacity
(C) Soil water potential
(D) Field capacity

83. The extent to which a soil can held capillary water against gravity. It is equal to field capacity less hygroscopic water is called:
(A) Moisture equivalent
(B) Water holding capacity
(C) Soil water potential
(D) Field capacity

84. This is an expression of the total reduction of water potential in the soil, due to mineral matrix, solubility, external pressure and gravitation effects is called:
(A) Moisture equivalent
(B) Water holding capacity
(C) Soil water potential
(D) Field capacity

85. The water content of an undisturbed soil (% oven-dry weight) after it is saturated by rainfall and drainage of gravitational water has completely stopped is called:
(A) Moisture equivalent
(B) Water holding capacity
(C) Soil water potential
(D) Field capacity

86. The water vapour (%) absorbed by unit weight of dry soil, when placed in an atmosphere completely saturated with water vapour is called:

(A) Hygroscopic-coefficient
(B) Wilting coefficient
(C) Capillary potential
(D) Field capacity

87. The amount of water (% oven dry weight basis) that remains in a soil when permanent wilting is present in the plants growing in soil is called:

(A) Hygroscopic-coefficient
(B) Wilting coefficient
(C) Capillary potential
(D) Field capacity

88. The status of soil water is also represented in term of :

(A) Hygroscopic-coefficient
(B) Wilting coefficient
(C) Capillary potential
(D) Field capacity

89. Fungi of soil infect roots of plants and cause plant diseases such as cotton root rot and many kinds of wilts, rusts, blights and smuts is called:

(A) Parasitic fungi (B) Symbiotic fungi
(C) Saprophytic fungi (D) Actinomycetes fungi

90. Fungi of soil live on the roots of certain plants both fungus and plants are beneficial is called:

(A) Parasitic fungi (B) Symbiotic fungi
(C) Saprophytic fungi (D) Actinomycetes fungi

91. Fungi depend on dead organic matter of soil and derive the energy from decomposition of the latter is called:

(A) Parasitic fungi (B) Symbiotic fungi
(C) Saprophytic fungi (D) Actinomycetes fungi

92. Fungi prefer saline soils and bring about the decomposition of organic matter such as cellulose is called:

(A) Parasitic fungi
(B) Symbiotic fungi
(C) Saprophytic fungi
(D) Actinomycetes fungi

93. The "growth is dependent on the amount of food stuff that is present in minimum quantity". This statement has come to be known as:

(A) Liebig's law of minimum
(B) Factor interaction
(C) Factor productivity
(D) Law of tolerance

94. The work on limiting factors and substances has shown that a high level of one factor will modify the limiting effect of a second, a process described as:

(A) Liebig's law of minimum
(B) Factor interaction
(C) Factor productivity
(D) None of these

95. The idea that factors could be limiting at their maximum as well as minimum quantities was incorporated in:

(A) Liebig's law of minimum
(B) Factor interaction
(C) Factor productivity
(D) Law of tolerance

96. Plants bloom when the light duration is less than 12 hours per days are:

(A) Long day plants (B) Short day plants
(C) Day neutral plants (D) None of these

97. Plants bloom when the light duration is more than 12 hours per days are:

(A) Long day plants (B) Short day plants
(C) Day neutral plants (D) None of these

98. Plants show little response to length of day light are:

(A) Long day plants (B) Short day plants
(C) Day neutral plants (D) None of these

99. The effect of sunlight on the plant movement is called:

(A) Heliotropism (B) Phototropism
(C) Both (A) and (B) (D) None of these

100. The response of different organisms to environmental rhythms of light and darkness is termed:

(A) Photoperiodism (B) Phototropism
(C) Both (A) and (B) (D) None of these

101. The light directed growth mechanisms are called:

(A) Photoperiodism (B) Phototropism
(C) Both (A) and (B) (D) None of these

102. Each daily cycle inclusive of a period of illumination followed by a period of darkness is called:

(A) Photoperiodism (B) Phototropism
(C) Photoperiod (D) None of these

103. The biological rhythms synchronized with the low and high tides (i.e., the alternate rise and fall of the sea due to gravitational pull exerted by moon) in the sea are called:

(A) Circatidal rhythms (B) Circalunar rhythms
(C) Semilunar rhythms (D) Circannual rhythms

104. Biological rhythms which are synchronized with the phases of moon are called:

(A) Circatidal rhythms
(B) Circalunar rhythms
(C) Semilunar rhythms
(D) Circannual rhythms

105. The biological rhythms which are synchronized with the fortnightly cycle of spring tide (i.e., high tide occurring a day or two after the new of full moon) and neap tide (i.e., low tide which occurs in the middle of the second and fourth quarters of the moon) are called:

(A) Circatidal rhythms (B) Circalunar rhythms
(C) Semilunar rhythms (D) Circannual rhythms

106. The activity of some animals and plant as influenced by the seasons occurring once in a year are called:
(A) Circatidal rhythms (B) Circalunar rhythms
(C) Semilunar rhythms
(D) Circannual rhythms

107. The distance from the equator is called:
(A) Latitude (B) Longitude
(C) Altitude (D) None of these

108. Height above the sea level forms the:
(A) Latitude (B) Longitude
(C) Altitude (D) None of these

109. The mean temperature of air decreases for every degree of latitude north of equator and for every 300' (or 91.44 metres) of altitude:
(A) 10^0F (-12.2^{0}C) (B) 1^0C
(C) 10^0C (D) 20^0F

110. The fall in temperature in the Nilgiri hills is for every 100 metre rise on the Iceward side and 122 metres rise on the windward side is:
(A) 10^0F (-12.2^{0}C) (B) 1^0C
(C) 10^0C (D) 20^0F

111. At high altitudes, the velocity of wind remains high, temperature and air pressure decrease, humidity as well as intensity of light is:
(A) Increases (B) Decreases
(C) No change (D) Unknown

112. The density is number of individuals per unit area or volume is:
(A) Numerical density (B) Biomass density
(C) Ecological density (D) Crop density

113. The density is biomass per unit area or volume is:
(A) Numerical density (B) Biomass density
(C) Ecological density (D) Crop density

114. The density (number or biomass) per unit total space is:
(A) Numerical density (B) Biomass density
(C) Ecological density (D) Crop density

115. The density (number or biomass) per unit of habital space, i.e., available area or volume that can actually by colonized by the population is:
(A) Numerical density (B) Biomass density
(C) Ecological density (D) Crop density

116. In most types of populations, individuals are of different age. The proportion of individuals in each group is called:
(A) Age structure (B) Age distribution
(C) Both (A) and (B) (D) None of these

117. A model representing geometrically the proportion of different age groups in the population of any organism is called:
(A) Age pyramid (B) Age sex pyramid
(C) Both (A) and (B) (D) None of these

118. It indicates a rapidly expanding population with a high percentage of young individuals and only few old individuals is:
(A) Triangular structure (B) Bell-shaped polygon
(C) Urn-shaped structure (D) None of these

119. It indicates a stationary population having an equal number of young and middle – aged individuals is:
(A) Triangular structure
(B) Bell-shaped polygon
(C) Urn-shaped structure
(D) None of these

120. It indicates a low percentage of young individuals is:
(A) Triangular structure
(B) Bell-shaped polygon
(C) Urn-shaped structure
(D) None of these

121. Rapidly growing young population in:
(A) Triangular structure
(B) Bell-shaped polygon
(C) Urn-shaped structure
(D) None of these

122. The rate of growth become slow and stable in:
(A) Triangular structure
(B) Bell-shaped polygon
(C) Urn-shaped structure
(D) None of these

123. The declining in population in:
(A) Triangular structure
(B) Bell-shaped polygon
(C) Urn-shaped structure
(D) None of these

124. Covering the production of new individuals by birth, hatching, germination, or fission is called:
(A) Natality (B) Fecundity
(C) Fertility (D) None of these

125. The physiological notion which indicates that an organism is capable of breeding is:
(A) Natality (B) Fecundity
(C) Fertility (D) None of these

126. Ecological concept that is based on the numbers of offspring produced during a period of time is:
(A) Natality (B) Fecundity
(C) Fertility (D) None of these

127. The theoretical maximum production of new individuals under ideal conditions which simply means that there are no ecological limiting factors and that reproduction is limited only by physiological factors is:
(A) Physiological natality (B) Absolute natality
(C) Potential natality (D) All of these

128. Population increase under an actual, existing specific condition is:

(A) Ecological natality (B) Realized natality
(C) Both (A) and (B) (D) None of these

129. Mortality represents the theoretical minimum loss under ideal or non-limiting conditions. It may be constant for a population is:

(A) Minimum mortality (B) Specific mortality
(C) Potential mortality (D) All of these

130. The actual loss of individuals under a given environmental condition. It may not be constant for a population is:

(A) Ecological mortality (B) Realized mortality
(C) Both (A) and (B) (D) None of these

131. This type of curve (pattern I) in the figure is the characteristic of the species in which the mortality rate of the population is low until near the end of the life span is:

(A) Highly convex curve
(B) Highly concave curve
(C) Diagonal straight-line curve
(D) None of these

132. This type of curve (pattern III) is the characteristic of such species where mortality rate is high during the young stages is:

(A) Highly convex curve
(B) Highly concave curve
(C) Diagonal straight-line curve
(D) None of these

133. This type of curve (pattern II b) indicates an age-specific constant survivorship, i.e., a constant rate of mortality occurs at every age is:

(A) Highly convex curve
(B) Highly concave curve
(C) Diagonal straight-line curve
(D) None of these

134. Under natural conditions occurs when there is overcrowding in the migratory locust, lemming, grouse, snowy owl, snowshoe rabbit, arctic fox, gray squirrel and occasionally in other species is called:

(A) Emigration (B) Immigration
(C) Migration (D) None of these

135. It leads to a rise in population level, causing an overpopulation which may lead to an increase beyond the carrying capacity is called:

(A) Emigration (B) Immigration
(C) Migration (D) None of these

136. It is a peculiar kind of population dispersion which involves the mass movement of entire population is called:

(A) Emigration (B) Immigration
(C) Migration (D) None of these

137. Communities growing in condition of abundant light are called:

(A) Heliophyte (B) Mesophyte
(C) Xerophyte (D) Sciophytic

138. The extrinsic factors which tend to regulate the density of a population under different conditions, appearing to act on the population and inflict loss of individuals irrespective of the population density is called:

(A) Density independent factors
(B) Density dependent factors
(C) Density biotic factors
(D) Density abiotic factors

139. The intrinsic or biotic factors and they depend on coactions between individuals within the same population or between population of different species is called:

(A) Density independent factors
(B) Density dependent factors
(C) Density biotic factors
(D) Density abiotic factors

140. The communities are greatly influenced by inputs from adjacent communities is called:

(A) Minor community (B) Major community
(C) Global community (D) Concrete community

141. The communities are relatively independent and self-sufficient of their habitat is called:

(A) Minor community (B) Major community
(C) Global community (D) Concrete community

142. The community is an enormous mass of life, comprising all the plants and animals in the world is called:

(A) Minor community (B) Major community
(C) Global community (D) Concrete community

143. Since biotic province is an abstract community, so based on ecological criteria, the association are studied as:

(A) Minor community (B) Major community
(C) Global community (D) Concrete community

144. Among several species present in a community, a few exert a major controlling influence by virtue of their size, numbers or activities are called:

(A) Ecological dominance
(B) Relative dominance
(C) Both (A) and (B) (D) None of these

145. Different population in a community exist in relative proportions and this idea is called:

(A) Ecological dominance
(B) Relative dominance
(C) Both (A) and (B) (D) None of these

146. Communities growing in condition of moderately moist soils are called:

(A) Heliophyte (B) Mesophyte
(C) Xerophyte (D) Sciophytic

147. Communities growing in condition in arid or dry condition are called:

(A) Heliophyte (B) Mesophyte
(C) Xerophyte (D) Sciophytic

148. Communities growing in condition of shade are called:
(A) Heliophyte (B) Mesophyte
(C) Xerophyte (D) Sciophytic

149. The range of environmental conditions which a taxon can tolerate is called:
(A) Ecological dominance
(B) Relative dominance
(C) Ecological amplitude
(D) None of these

150. The growing buds of these plants are naked or covered with scale (i.e., they are not well protected), and situated in upright shoots much above the ground surface are called:
(A) Phanerophytes (B) Chamaephytes
(C) Hemicryptophytes (D) Cryptophytes

151. In these plants the buds are situated close to the ground surface and these buds get protection from fallen leaves and snow cover are called:
(A) Phanerophytes (B) Chamaephytes
(C) Hemicryptophytes (D) Cryptophytes

152. The plants are mostly found in cold temperate zone. Their buds are hidden under soil surface, protected in soil itself are called:
(A) Phanerophytes (B) Chamaephytes
(C) Hemicryptophytes (D) Cryptophytes

153. In these plants, the buds are usually buried in the soil or in bulbs and rhizomes where food is stored to withstand long periods of adverse climatic conditions are called:
(A) Phanerophytes (B) Chamaephytes
(C) Hemicryptophytes (D) Cryptophytes

154. These are seasonal plants, completing their life cycle in a single favourable season, and remain dominant throughout the rest unfavourable period of year in the form of seeds are called:
(A) Phanerophytes (B) Chamaephytes
(C) Therophytes (D) Cryptophytes

155. The transition zone between two or more diverse communities is called:
(A) Ecotone (B) Ecotypes
(C) Edge effect (D) None of these

156. Species with wide geographical ranges almost always develop locally adapted populations called:
(A) Ecotone (B) Ecotypes
(C) Edge effect (D) None of these

157. The potential for the ecotone to act as habitat for species found in neither major community is termed as the:
(A) Ecotone (B) Ecotypes
(C) Edge effect (D) None of these

158. The transitional series of communities which develop in a given area are called:
(A) Sere (B) Climax
(C) Succession (D) None of these

159. The final stable and mature community is called:
(A) Sere (B) Climax
(C) Succession (D) None of these

160. The development of the community by the action of vegetation on the environment leading to the establishment of new species is termed as:
(A) Sere (B) Climax
(C) Succession (D) None of these

161. If an area in any of the basic environments (such as terrestrial, fresh water or marine) is colonized by organisms for the first time, the succession is called:
(A) Primary succession
(B) Secondary succession
(C) Autogenic succession
(D) Allogenic succession

162. If the area under colonization has been cleared by whatsoever agency (such as burning, grazing, clearing, felling of trees, sudden change in climatic factors, etc.) of the previous plants, it is called:
(A) Primary succession
(B) Secondary succession
(C) Autogenic succession
(D) Allogenic succession

163. After the succession has began, in most of the cases, it is the community itself which, as a result of its reactions with the environment, modifies its own environment and, thus, causing its own replacement by new communities. This course of succession is known as:
(A) Primary succession
(B) Secondary succession
(C) Autogenic succession
(D) Allogenic succession

164. In some cases replacement of one community by another is largely due to forces other than the effects of communities on the environment is called:
(A) Primary succession
(B) Secondary succession
(C) Autogenic succession
(D) Allogenic succession

165. It is characterized by early and continued dominance of autotrophic organisms such as green plants. It begins in a predominantly inorganic environments and the energy flow is maintained indefinitely is called:
(A) Autotrophic succession
(B) Heterotrophic succession
(C) Induced succession
(D) Retrogressive succession

166. It is characterized by early dominance of heterotrophic organisms such as bacteria, actinomycetes, fungi and animals is called:
(A) Autotrophic succession
(B) Heterotrophic succession
(C) Induced succession
(D) Retrogressive succession

167. Activities such as overgrazing, frequent scraping, shifting cultivation or industrial pollution may cause deterioration of an ecosystem is called:
(A) Autotrophic succession
(B) Heterotrophic succession
(C) Induced succession
(D) Retrogressive succession

168. It means a return to simpler and less dense or even impoverished form of community from an advanced or climax community is called:
(A) Autotrophic succession
(B) Heterotrophic succession
(C) Induced succession
(D) Retrogressive succession

169. It is of local occurrence within a large community is called:
(A) Cyclic succession
(B) Heterotrophic succession
(C) Induced succession
(D) Retrogressive succession

170. The development of a new area without any form of life is known as:
(A) Nodation (B) Invasion
(C) Coaction (D) Reaction

171. The successful establishment of a species in a bare area is known as:
(A) Nodation (B) Invasion
(C) Coaction (D) Reaction

172. Due to aggregation of a large number of individuals of the species at the limited place, there develops competition for space and nutrition. Individuals of a species affect each other's life in various ways and this is called:
(A) Nodation (B) Invasion
(C) Coaction (D) Reaction

173. It includes mechanisms of the modification through the influence of living organisms on it is called:
(A) Nodation (B) Invasion
(C) Coaction (D) Reaction

174. After reaching to new area, the process of successful establishment of the species, as a result of adjustment with the conditions prevailing there, is known as:
(A) Ecesis (B) Segregation
(C) Pioneers (D) Sere

175. Colonization by successive offspring and new migrants helps to increase the population, a process called:
(A) Ecesis (B) Segregation
(C) Pioneers (D) Sere

176. Plants or autotrophic organisms which are the first to colonize and aggregate are called:
(A) Ecesis (B) Segregation
(C) Pioneers (D) Sere

177. The whole sequence of communities that replaces one another in the given area is called:
(A) Ecesis (B) Segregation
(C) Pioneers (D) Sere

178. According to the model each new communities in course of time prepares suitable ground to facilitate its own replacement by another better suited community is known as:
(A) Facilitation model (B) Tolerance model
(C) Inhibition model (D) None of these

179. According to the model, only such higher succession or climax species are able to join which can be tolerated by the early settlers is known as:
(A) Facilitation model (B) Tolerance model
(C) Inhibition model (D) None of these

180. According to the model, the early arrived species on a new habitat may develop counter – mechanism to normal replacement process is known as:
(A) Facilitation model (B) Tolerance model
(C) Inhibition model (D) None of these

181. The model is based on the concept of Initial Floritic Composition is:
(A) Facilitation model (B) Tolerance model
(C) Inhibition model (D) None of these

182. The microclimax theory was developed by:
(A) Fredrick Clements
(B) Tansley
(C) Whittaker, Macintosh and Sellack
(D) Leith, Odum and Golley

183. Polyclimax theory was developed by:
(A) Fredrick Clements
(B) Tansley
(C) Whittaker, Macintosh and Sellack
(D) Leith, Odum and Golley

184. Climax pattern hypothesis was developed by:
(A) Fredrick Clements
(B) Tansley
(C) Whittaker, Macintosh and Sellack
(D) Leith, Odum and Golley

185. Information theory was proposed by:
(A) Fredrick Clements
(B) Tansley
(C) Whittaker, Macintosh and Sellack
(D) Leith, Odum and Golley

186. Any ecological unit that includes all the organisms which interact among themselves and with the physical environment, so that a flow of energy leads to clearly defined trophic structure, biotic diversity and material cycle within the system, is known as:
(A) Ecological system
(B) Eco system
(C) Both (A) and (B)
(D) None of these

187. The total rate of photosynthesis including the organic matter used up in respiration during the measurement period id known as:
(A) Gross primary productivity
(B) Net primary productivity
(C) Net productivity
(D) None of these

188. The rate of storage of organic matter in plant tissues in excess of the respiratory utilization by plants during the measurement period is known as:
(A) Gross primary productivity
(B) Net primary productivity
(C) Net productivity
(D) None of these

189. The rate of storage of organic matter not used by the heterotrophic or consumers, i.e., equivalent to net primary production minus consumption by the heterotrophs during the unit period as a season or year, etc. is known as:
(A) Gross primary productivity
(B) Net primary productivity
(C) Net productivity
(D) None of these

190. The ratio of net to gross production is called:
(A) Gross production efficiency
(B) Net production efficiency
(D) Assimilation efficiency
(D) Ecological efficiency

191. The production of efficiencies with which organisms exploit their food resources and transform them into biomass available to the next higher trophic level is called:
(A) Gross production efficiency
(B) Net production efficiency
(D) Assimilation efficiency
(D) Ecological efficiency

192. The product of assimilation efficiency and net production efficiency is called:
(A) Gross production efficiency
(B) Net production efficiency
(D) Assimilation efficiency
(D) Ecological efficiency

193. Gross production efficiency of warm blooded terrestrial animals lie at:
(A) Exceed 5% (B) Below 1%
(C) 5-15 % (D) Exceed 30 %

194. Gross production efficiency of some birds and large mammals lie at:
(A) Exceed 5% (B) Below 1%
(C) 5-15 % (D) Exceed 30 %

195. Gross production efficiency of insects lie within the range of:
(A) Exceed 5% (B) Below 1%
(C) 5-15 % (D) Exceed 30 %

196. Gross production efficiency of some aquatic animals lie at:
(A) Exceed 5% (B) Below 1%
(C) 5-15 % (D) Exceed 30 %

197. Assimilatory efficiency of predatory species vary from:
(A) 60-90 % (B) 75-85%
(C) 5-15 % (D) Exceed 30 %

198. Old field herbs, crop species or aquatic plants have constantly net production efficiency of:
(A) 60-90 % (B) 75-85%
(C) 5-15 % (D) Exceed 30 %

199. Tropical vegetation exhibited net production efficiency of:
(A) Higher (B) Lower
(C) No change (D) Unknown

200. The cyclic movements of chemical elements of the biosphere between the organism and the environment are referred to as:
(A) Biogeochemical cycle
(B) Atmosphere reservoir cycle
(C) Sediment-reservoir cycle
(D) Hydrologic cycle

201. Biogeochemical cycles that have dominant atmospheric phases are often called:
(A) Biogeochemical cycle
(B) Atmosphere reservoir cycle
(C) Sediment-reservoir cycle
(D) Hydrologic cycle

202. Biogeochemical cycles those whose sedimentary phase is dominant are termed as:
(A) Biogeochemical cycle
(B) Atmosphere reservoir cycle
(C) Sediment-reservoir cycle
(D) Hydrologic cycle

203. Living organisms, atmosphere and earth maintain between them as circulation of water and moisture, which is referred to as :
(A) Biogeochemical cycle
(B) Atmosphere reservoir cycle
(C) Sediment-reservoir cycle
(D) Hydrologic cycle

204. Biotic and abiotic agents, both are driven by the flow of energy and both are tied to the:
(A) Biogeochemical cycle
(B) Atmosphere reservoir cycle
(C) Sediment-reservoir cycle
(D) Hydrologic cycle

205. Hydrologic cycle is driven by:
(A) Solar energy (B) Gravity
(C) Both (A) and (B) (D) None of these

206. Water is distributed throughout the earth:
(A) Evenly (B) Not evenly
(C) Unknown (D) Both evenly and not evenly

207. Which amount of the total water on earth is chemically bound into rocks and does not cycle is:
(A) 85 % (B) 90 %
(C) 95 % (D) 100 %

208. Almost 5 per cent of the total water, how much water is exist in the ocean?
(A) 95 % (B) 97.3 %
(C) 98.0% (D) 99 %

209. Almost 5 per cent of the total water, how much water is exist as ice in the polar caps and permanent glaciers?
(A) 2.1% (B) 97.3 %
(C) 5.2% (D) 0.60%

210. Almost 5 per cent of the total water, how much water is exist as the fresh water in the form of atmospheric water vapour, ground water, soil water, or inland surface water?
(A) 2.1% (B) 97.3 %
(C) 5.2% (D) 0.60%

211. Water that is evaporated directly from any surface other than a plant, such as a lake, soil surface or animal skin is called:
(A) Evaporation (B) Transpiration
(C) Evapo-transpiration (D) None of these

212. Water that evaporates from the surface of leaves of plant is called:
(A) Evaporation (B) Transpiration
(C) Evapo-transpiration (D) None of these

213. Water that losses in the form of evaporation as well as transpiration is called:
(A) Evaporation (B) Transpiration
(C) Evapo-transpiration (D) None of these

214. The acts to move the biogeochemical cycles for all mineral nutrients that enter the food chain via the roots of plants in:
(A) Evaporation (B) Transpiration
(C) Evapo-transpiration (D) None of these

215. If transpiration is related to the mechanism of nutrient uptake, the gross movement of soluble and solid particles in the ecosystem is accomplished largely by:
(A) Springs (B) Aquifer
(C) Artisen well (D) Runoff

216. The catchment area where ground water reaches the surface and run off is termed as:
(A) Springs (B) Aquifer
(C) Artisen well (D) Runoff

217. A rock body through which ground water flows is called an:
(A) Springs (B) Aquifer
(C) Artisen well (D) Runoff

218. A well drilled into an aquifer that has sufficient hydrostatic pressure to force water up into it is called:
(A) Springs (B) Aquifer
(C) Artisen well (D) Runoff

219. Which gas is the by-product of photosynthesis, is involved in the oxidation of carbohydrates with release of energy is:
(A) Oxygen (B) Sulphur
(C) Nitrogen (D) Carbon di-oxide

220. The protection from toxic effects of molecular O_2, cells posses the cellular organelles called:
(A) Oxygen (B) Sulphur
(C) Peroxisomes (D) Carbon di-oxide

221. Nitrogen is major constituents of atmosphere, comprising about:
(A) 79 % (B) 71 %
(C) 67 % (D) 60 %

222. Oxygen is second major constituents of atmosphere, comprising about:
(A) 20 % (B) 21 %
(C) 22 % (D) 23 %

223. The amino group ($-NH_2$) is liberated from organic molecules to form ammonia; this process is called:
(A) Deamination (B) Nitrification
(C) Denitrification (D) Ammonification

224. The ammonia is oxidized to nitrate and nitrite yielding energy; this process is called:
(A) Deamination (B) Nitrification
(C) Denitrification (D) Ammonification

225. Under certain circumstances, nitrate is either not produced in the nitrogen cycle or it is degraded before it can be utilized by autotrophs; this process is called:
(A) Deamination (B) Nitrification
(C) Denitrification (D) Ammonification

226. Fixation of nitrogen occurs by which types of microorganisms in soil are:
(A) Symbiotic (B) Nonsymbiotic
(C) Both (A) and (B) (D) None of these

227. The sulphur cycle is included:
(A) Gaseous phase (B) Sedimentary phase
(C) Both (A) and (B) (D) None of these

228. The sulphur is removed from the organic phase in the form of:
(A) Elemental sulphur (B) Sulphate
(C) Sulphite (D) Sulphide

229. Which cycle has no atmospheric phase:
(A) Nitrogen cycle (B) Sulphur cycle
(C) Phosphorus cycle (D) Mineral cycle

230. Elemental sulphur can also be utilized by other bacteria to form:
(A) Sulphate (B) Sulphite
(C) Organic sulphur (D) None of these

231. Under anaerobic conditions, elemental sulphur can still be oxidized to:
(A) Sulphite (B) Sulphide
(C) Sulphate (D) None of these

232. Freshwater ecosystem, the study of which is known as:
(A) Limnology (B) Ecology
(C) Aquaculture (D) Pisciculture

233. Ecosystem include all standing water (freshwater) habitats such as lakes, ponds, marshes, swamps, bogs, meadows, etc is called:
(A) Lentic ecosystem (B) Lotic ecosystem
(C) Both (A) and (B) (D) None of these

234. Ecosystem include running or moving water such as rivers, streams and related environment is called:
(A) Lentic ecosystem (B) Lotic ecosystem
(C) Both (A) and (B) (D) None of these

235. The oxygen level in a tropical pond exhibits diurnal variation-it remains at peak between:
(A) 06.00 – 10.00 hours (B) 12.00 – 14.00 hours
(C) 14.00 – 17.00 hours (D) 18.00 – 20.00 hours

236. Due to low salinity of freshwater, animals face the problem of:
(A) Osmoregulation (B) Marl
(C) Epilimnion (D) Hypolimnion

237. Deposition of calcium carbonate in water is called:
(A) Osmoregulation (B) Marl
(C) Epilimnion (D) Hypolimnion

238. The zone contains upper warm and oxygen rich circulating water layer which is called:
(A) Osmoregulation (B) Marl
(C) Epilimnion (D) Hypolimnion

239. The non circulating cold water with poor oxygen zone is called:
(A) Osmoregulation (B) Marl
(C) Epilimnion (D) Hypolimnion

240. The zone contain shallow water near the shore forms the:
(A) Littoral zone (B) Sublittoral zone
(C) Limnetic zone (D) Profundal zone

241. The zone extends from rooted vegetation to the non-circulating cold water in lakes with poor oxygen is:
(A) Littoral zone (B) Sublittoral zone
(C) Limnetic zone (D) Profundal zone

242. The open water zone away from the shore. It is the zone upto the depth of effective light penetration where rate of photosynthesis is equal to the rate of respiration is:
(A) Littoral zone (B) Sublittoral zone
(C) Limnetic zone (D) Profundal zone

243. The deep water area in lakes beneath limnetic zone and beyond the depth of effective light penetration is:
(A) Littoral zone (B) Sublittoral zone
(C) Limnetic zone (D) Profundal zone

244. It found only in deep lakes, since it begins at about 2000 meters from the surface is:
(A) Abyssal zone (B) Trophogenic zone
(C) Tropholytic zone (D) Profundal zone

245. The depth of light penetrating in lakes enabling photosynthesis includes littoral plus sublittoral zones is:
(A) Abyssal zone (B) Trophogenic zone
(C) Tropholytic zone (D) Profundal zone

246. The depth of light penetrating in lakes enabling photosynthesis into upper part of profundal zone is:
(A) Abyssal zone (B) Trophogenic zone
(C) Tropholytic zone (D) Profundal zone

247. The lakes contain high concentration of humic acid in water. The margins of these lakes have a low pH and develop into peat bogs is known:
(A) Dystrophic lakes (B) Desert salt lake
(C) Desert alkali lake (D) Volcanic lakes

248. The lakes exist in sedimentary drainages in arid climates where evaporation exceeds precipitation is:
(A) Dystrophic lakes (B) Desert salt lake
(C) Desert alkali lake (D) Volcanic lakes

249. The lakes occur in igneous drainage in arid climate, high pH and concentration of carbonates are the characteristic features of these lakes is:
(A) Dystrophic lakes (B) Desert salt lake
(C) Desert alkali lake (D) Volcanic lakes

250. Acid or alkali lakes associated with volcanic regions are called:
(A) Dystrophic lakes (B) Desert salt lake
(C) Desert alkali lake (D) Volcanic lakes

251. These are ponds which are found in stream flood plains. These are quite productive habitats due to accumulation of organic matter in them is:
(A) Flood plain ponds (B) Temporary ponds
(C) Artificial ponds (D) Beaver ponds

252. These ponds remain dry for most part of the year but support a unique community. Organisms in such ponds must be able to survive in a dormant stage during dry periods or be able to move in and out of pods is:
(A) Flood plain ponds (B) Temporary ponds
(C) Artificial ponds (D) Beaver ponds

253. These ponds are the result of damming of a stream or basin by man. They are artificially managed habitats is:
(A) Flood plain ponds (B) Temporary ponds
(C) Artificial ponds (D) Beaver ponds

254. Ponds are constructed by beavers (amphibious rodents) are called:
(A) Flood plain ponds (B) Temporary ponds
(C) Artificial ponds (D) Beaver ponds

255. The organisms depending on the substream are called:
(A) Pedonic forms (B) Limnetic forms
(C) Epineuston (D) Hyponeuston

256. The organisms that are free from it are the:
(A) Pedonic forms (B) Limnetic forms
(C) Epineuston (D) Hyponeuston

257. Animal that spend their lives on top of the air-water interface, such as water striders are termed:
(A) Pedonic forms (B) Limnetic forms
(C) Epineuston (D) Hyponeuston

258. Insects such as diving beetles and back swimmers, which spend most of their time on the underside of the air-water interface and obtain much of their food from within water are termed :
(A) Pedonic forms (B) Limnetic forms
(C) Epineuston (D) Hyponeuston

259. These are unattached organisms that live at the air-water interface is called:
(A) Neuston (B) Plankton
(C) Nekton (D) Benthos

260. These are forms which are found in all aquatic ecosystems except for fast-moving rivers are called:
(A) Neuston (B) Plankton
(C) Nekton (D) Benthos

261. The animals are swimmers and are found in all aquatic systems except for fastest moving rivers are called:
(A) Neuston (B) Plankton
(C) Nekton (D) Benthos

262. The organisms living at the bottom of the water mass are called:
(A) Neuston (B) Plankton
(C) Nekton (D) Benthos

263. Certain animal planktons or zooplanktons are exceedingly active and move relatively great distance considering their small size, but they are so small that their range is still controlled largely by currents and such plankton are also known as:
(A) Nektoplankton (B) Benthic epifauna
(C) Infauna (D) Hyponeuston

264. The benthos organisms living above the sediment-water interface are termed:
(A) Nektoplankton (B) Benthic epifauna
(C) Infauna (D) Hyponeuston

265. The benthos organisms living in the sediment itself are termed:
(A) Nektoplankton (B) Benthic epifauna
(C) Infauna (D) Hyponeuston

266. When silt and mud accumulation become high enough to be exposed at:
(A) Low tide (B) Deltas
(C) Tidal flats (D) Islands

267. Oligotrophic lake which is virtually sterile biologically has low production but its water remain well saturated with oxygen from top to bottom is:
(A) Orthograde oxygen curve
(B) Clinograde oxygen curve
(C) Both (A) and (B)
(D) None of these

268. Eutrophic lake has high productivity, but there is substantial oxygen depletion from top of the hypolimnion to the bottom, with depletion most extreme in the bottom layers is:
(A) Orthograde oxygen curve
(B) Clinograde oxygen curve
(C) Both (A) and (B)
(D) None of these

269. Some animals of rapidly flowing water streams live among the mosses and flowering plants is called:
(A) Plant inhabiting torrential forms
(B) Rocks inhabitant forms
(C) Inhabitant forms of spaces between rocks
(D) Inhabitants occurring beneath rocks

270. The animals that live on the tops of exposed rocks have an efficient mechanism for staying in one place, otherwise, they will be swept away is called:
(A) Plant inhabiting torrential forms
(B) Rocks inhabitant forms
(C) Inhabitant forms of spaces between rocks
(D) Inhabitants occurring beneath rocks

271. Numerous different kinds of animals live in the spaces between the rocks is called:
(A) Plant inhabiting torrential forms
(B) Rocks inhabitant forms
(C) Inhabitant forms of spaces between rocks
(D) Inhabitants occurring beneath rocks

272. Many species such as flatworms, annelid worms, other insect larvae, clams and some species of snails, live beneath the rocks is called:
(A) Plant inhabiting torrential forms
(B) Rocks inhabitant forms
(C) Inhabitant forms of spaces between rocks
(D) Inhabitants occurring beneath rocks

273. Waters of all streams and rivers ultimately drain into the sea; the place where this freshwater joins the salt water is called:
(A) Estuary (B) Deltas
(C) Tidal flats (D) Islands

274. As the river reaches the encroaching sea, the stream carried sediments are dropped in the quite water. These accumulate to form:
(A) Estuary (B) Deltas
(C) Tidal flats (D) Islands

275. --------- develop, which divide and braid the original channel of the estuary is:
(A) Estuary (B) Deltas
(C) Tidal flats (D) Islands

276. If more material is deposited than is carried away is called:
(A) Estuary (B) Deltas
(C) Tidal flats (D) Islands

277. Which sea is the largest (179.7 x 10^6 km^2), coldest (average 3.36°C), least saline (34.6 parts per thousand) and deepest (average depth 4028 m) is:
(A) Pacific ocean (B) Atlantic ocean
(C) Antarctic ocean (D) Indian ocean

278. Which ocean is warmer (average 3.73°C), more saline (average 34.9 parts per thousand) and relatively shallow (average depth 3310 m) is:
(A) Pacific ocean (B) Atlantic ocean
(C) Antarctic ocean (D) Indian ocean

279. Which ocean shows wide variations in temperature (-1.8 to 10°C) is:
(A) Pacific ocean (B) Atlantic ocean
(C) Antarctic ocean (D) Indian ocean

280. Which ocean covers 77 x 10^6 km^2 to an average depth of 3872 m is:
(A) Pacific ocean (B) Atlantic ocean
(C) Antarctic ocean (D) Indian ocean

281. Associated with a temperature gradient is a salinity gradient, especially at the higher latitudes is:
(A) Halocline (B) Pycnocline
(C) Both (A) and (B) (D) None of these

282. The density of sea water does increase in depth and does not reach its greatest density at 4°C as with freshwater, there is no seasonal overturn. This results in a normally stable stratification of density known as the:
(A) Halocline (B) Pycnocline
(C) Both (A) and (B) (D) None of these

283. The water is never static and waves, currents, and tides are the regular features in:
(A) Sea water (B) River water
(C) Canal water (D) Lake water

284. Most world wide water currents occur at the:
(A) Surface of sea water
(B) Great depth of sea water
(C) Both (A) and (B)
(D) None of these

285. Waves, currents and tides are regular features of sea water. All these phenomena are controlled by:
(A) Winds (B) Cosmic forces
(C) Water densities (D) All of these

286. There are certain regions in the world, notably at the west coasts of continents and the surface. These are called:
(A) Zone of upwelling (B) Tides
(C) Photic zone (D) Aphotic zone

287. The pelagic zone is divided from the surface to about 200 m is the:
(A) Zone of upwelling (B) Tides
(C) Photic zone (D) Aphotic zone

288. Water movements which are caused by some astronomical factors (i.e., due to a pull by moon and sun) is called:
(A) Zone of upwelling (B) Tides
(C) Photic zone (D) Aphotic zone

289. The pelagic zone in which there is sharp gradients of illumination, temperature and salinity. The region below this zone is called:
(A) Zone of upwelling (B) Tides
(C) Photic zone (D) Aphotic zone

290. Within the biogeographic realms and established by a complex interactions of climate, other physical factors and biotic factors, are large community units called:
(A) Biomes (B) Ecotone
(C) Permafrost (D) None of these

291. There is no sharp line of demarcation between adjacent biomes, instead each blends with the next through a fairly broad transition region termed as:
(A) Biomes (B) Ecotone
(C) Permafrost (D) None of these

292. The permanently frozen deeper soil layer is called:
(A) Biomes (B) Ecotone
(C) Permafrost (D) None of these

293. The region of mountain above the timber line contains a distinct flora and fauna and is referred to as the:
(A) Alpine zone (B) Sub-snow zone
(C) Meadow zone (D) Shrub zone

294. Immediately below the snow zone is:
(A) Alpine zone (B) Sub-snow zone
(C) Meadow zone (D) Shrub zone

295. Gradually merges into the timber zone is:
(A) Alpine zone (B) Sub-snow zone
(C) Meadow zone (D) Shrub zone

296. The zone lies in between the snow zone and timber zone is:
(A) Alpine zone (B) Sub-snow zone
(C) Meadow zone (D) Shrub zone

297. Forests dominated by trees of such species as *Abies*, *Pinus* and so on, occur at altitudes of 2900 to 3500 meters is called:
(A) Subalpine forests (B) Alpine forests
(C) Coniferous forests (D) Deciduous forests

298. Forests dominated by trees of such species as *Abies*, *Pinus* and so on, occur at altitudes of 3600 meters of more meters is called:
(A) Subalpine forests (B) Alpine scrubs
(C) Coniferous forests (D) Deciduous forests

299. Underground reservoirs of freshwater is called:
(A) Decreasers (B) Silviculture
(C) Aquifers (D) Soil erosion

300. Cold regions with high rainfall and strongly seasonal climates with long winters and fairly short summers are characterized by:
(A) Subalpine forests (B) Alpine forests
(C) Coniferous forests (D) Deciduous forests

301. The temperate forest biomes are characterized by a moderate climate and broad-leaved deciduous trees, which shed their leaves in fall, are bare over winter and grow new foliage in the spring is called:
(A) Subalpine forests (B) Alpine forests
(C) Coniferous forests (D) Deciduous forests

302. The term is applied to the northern range of coniferous forests is:
(A) Taiga (B) Ecotone
(C) Biomes (D) Tides

303. Tropical seasonal forests are found in southeast Asia, Central and South America, Northern Australia, Western Africa and the Tropical islands of the specific as well as India and Southeast Asia. In exceedingly wet tropical seasonal forests, commonly known as:
(A) Subalpine forests (B) Alpine forests
(C) Coniferous forests (D) Monsoon forests

304. Savannas of India are dominated by:
(A) Grasses (B) Sal tress
(C) Both (A) and (B) (D) None of these

305. The resources once gone, have very little chance of recovery or resynthesis is called:
(A) Non-renewable resources
(B) Renewable resources
(C) Both of these (D) None of these

306. The resources are constantly regenerated or renewed is called:
(A) Non-renewable resources
(B) Renewable resources
(C) Both of these (D) None of these

307. Uneaten food of whales is also found suitable for human consumption as food is called:
(A) Krill (B) Desertification
(C) Ecotone (D) Tides

308. The branch of agriculture which is mainly concerned with the breeding, feeding and caring of domestic animals is called:
(A) Animal husbandry
(B) Terrestrial agriculture
(C) Livestock management
(D) Shifting cultivation

309. When it incorporates the study of proper utilization of economically important domestic animals, it is called:
(A) Animal husbandry
(B) Terrestrial agriculture
(C) Livestock management
(D) Shifting cultivation

310. Destruction of biotic potential of land also leads to:
(A) Desertification (B) Krill
(C) Ecotone (D) Tides

311. The palatable plant species, which are sensitive to grazing is called:
(A) Decreasers (B) Silviculture
(C) Aquifers (D) Soil erosion

312. Rama Rao (1962) called creeping death of the soil is:
(A) Decreasers (B) Silviculture
(C) Aquifers (D) Soil erosion

313. The dust blows along with wind for flows in runoff water, usually after rains is called:
(A) Aquifers (B) Soil erosion
(C) Erosion (D) Drainage

314. The soil is removed in small but uniform amounts from all over and therefore, does not leave a mark behind is called
(A) Sheet erosion (B) Rill erosion
(C) Gully erosion (D) None of these

315. The runoff water moves rapidly and cuts small stream-like structure is called:
(A) Sheet erosion (B) Rill erosion
(C) Gully erosion (D) Riparion erosion

316. Several rills converge towards the steep slopes and join to form broad channels of water is called:
(A) Sheet erosion (B) Rill erosion
(C) Gully erosion (D) Riparion erosion

317. Humans need food as a source of energy and for tissue replacement, like any other animal. The source of much of the food consumed by man is called:
(A) Animal husbandry
(B) Terrestrial agriculture
(C) Livestock management
(D) Shifting cultivation

318. The energy is utilization is possible in volcanic regions or where hot springs and geysers occur is called:
(A) Geothermal energy (B) Tidal energy
(C) Dendrothermal energy (D) Wind energy

319. The power is a form of hydroelectricity produced by harnessing the ebb and flow of the tides is called:
(A) Geothermal energy (B) Tidal energy
(C) Dendrothermal energy (D) Wind energy

320. In Philippines, energy is experimentally generated from energy plantation is called:
(A) Geothermal energy (B) Tidal energy
(C) Dendrothermal energy
(D) Wind energy

321. Wind energy can be converted into:
(A) Mechanical energy (B) Electrical energy
(C) Both (A) and (B) (D) None of these

322. Gasification of biomass is an important means of harvest energy through thermo-chemical conservation. It yields:

(A) Biogas (B) Producer gas
(C) Pyrogas (D) All of these

323. Who provides a significant solution to present energy crisis, especially in rural areas is:

(A) Wind energy (B) Bio gas
(C) Geothermal energy (D) Petro plant

324. Plant production is harvested for use by man either directly or after processing is called:

(A) Animal husbandry (B) Crop agriculture
(C) Livestock management (D) Shifting cultivation

325. Crop from highly manipulated ecosystem is fed to domesticated animals is called:

(A) Animal agriculture
(B) Terrestrial agriculture
(C) Livestock management
(D) Shifting cultivation

326. Total manipulation of the natural system is practiced over a limited areas but for only a short (1-5 years) period of time is called:

(A) Animal husbandry
(B) Terrestrial agriculture
(C) Livestock management
(D) Shifting cultivation

327. Permanent manipulation of an ecosystem; the natural biota are removed and replaced with domesticated plants and animals are called:

(A) Animal husbandry
(B) Terrestrial agriculture
(C) Sedentary cultivation
(D) Shifting cultivation

328. The rivers during floods splash their water against the banks and thus, cuts through them is called:

(A) Stream bank erosion (B) Landslide erosion
(C) Sheet erosion (D) Rill erosion

329. The hydraulic pressure caused by heavy rains increases the weight of the rocks at cliffs which come under the gravitational force and finally slip or fall off is called:

(A) Stream bank erosion (B) Landslide erosion
(C) Sheet erosion (D) Rill erosion

330. Planting of crop in rows or strips under more slopy lands to check flow of water is called:

(A) Strip cropping
(B) Terrestrial agriculture
(C) Sedentary cultivation
(D) Shifting cultivation

331. Trees are planted in deserts which check the velocity of wind. Plantation of tees in short blocks are called:

(A) Wind breaks (B) Shelter belts
(C) Social forestry (D) Silviculture

332. Extensive plantation of trees are called:

(A) Wind breaks (B) Shelter belts
(C) Social forestry (D) Silviculture

333. The species or taxa, which are in danger of extinction and which may not survive if the adverse factors continue to operate is called:

(A) Endangered species (B) Vulnerable species
(C) Rare species (D) Threatened species

334. The species or taxa likely to move into the endangered category in the near future if the adverse factors continue to operate it is called:

(A) Endangered species
(B) Vulnerable species
(C) Rare species
(D) Threatened species

335. The species are not endangered and vulnerable but are at risk. These species are usually localized within restricted geographical areas or habitats or are thinly scattered over a more extensive range is called:

(A) Endangered species (B) Vulnerable species
(C) Rare species (D) Threatened species

336. The conservation of species which are in any one of the above three categories, viz. endangered, vulnerable or rare is called:

(A) Endangered species
(B) Vulnerable species
(C) Rare species
(D) Threatened species

337. Vertical distribution of organisms in marine (sea) habitat is called:

(A) Halobiotic (B) Limnobiotic
(C) Geobiotic (D) None of these

338. Vertical distribution of organisms in fresh water habitate is called:

(A) Halobiotic (B) Limnobiotic
(C) Geobiotic (D) None of these

339. Altitudinal distribution of organisms on land is called:

(A) Halobiotic (B) Limnobiotic
(C) Geobiotic (D) None of these

340. All the living organisms in a given region are termed as:

(A) Biota (B) Fauna
(C) Flora (D) None of these

341. The animals of a given region are collectively termed as:

(A) Biota (B) Fauna
(C) Flora (D) None of these

342. The plants of a given region are collectively termed as:

(A) Biota (B) Fauna
(C) Flora (D) None of these

343. The studies of the distribution of biota are collectively called:
(A) Biogeography
(B) Zoogeography
(C) Static biogeography
(D) Dynamic biogeography

344. The studies of the distribution of animals only are collectively called:
(A) Biogeography
(B) Zoogeography
(C) Static biogeography
(D) Dynamic biogeography

345. The studies which deals with the description of biota of different botanical and zoological areas of earth is called:
(A) Biogeography
(B) Zoogeography
(C) Static biogeography
(D) Dynamic biogeography

346. The studies which describes the forces which have brought about plant and animal distribution is called:
(A) Biogeography
(B) Zoogeography
(C) Static biogeography
(D) Dynamic biogeography

347. A taxon distributed throughout the world in all climatic zones or in all the continents in at least one climatic zone is said to:
(A) Continuous distribution
(B) Discontinuous distribution
(C) Both of these
(D) None of these

348. A taxon distributed in two or more widely separated geographical areas, is said to have:
(A) Continuous distribution
(B) Discontinuous distribution
(C) Both of these
(D) None of these

349. Such a taxon or a species occurs in all climatic zones is:
(A) Cosmopolitan distribution
(B) Circulation distribution
(C) Circumaustral distribution
(D) Pantropical distribution

350. Certain species are distributed in a belt around the north-pole is:
(A) Cosmopolitan distribution
(B) Circulation distribution
(C) Circumaustral distribution
(D) Pantropical distribution

351. Species which are restricted in their distribution to a small region is called:
(A) Cosmopolitan (B) Endemics
(C) Endemism (D) None of these

352. The species which are distributed in a near continuous belt in the temperate region of northern or southern hemisphere are said to have:
(A) Cosmopolitan distribution
(B) Circulation distribution
(C) Circumaustral distribution
(D) Pantropical distribution

353. Certain species are distributed throughout the tropical belt is:
(A) Cosmopolitan distribution
(B) Circulation distribution
(C) Circumaustral distribution
(D) Pantropical distribution

354. Species which are spread over a wide area or in different ecological conditions are said to be:
(A) Cosmopolitan (B) Endemics
(C) Endemism (D) None of these

355. The phenomenon of restricted distribution associated with some geographical or ecological factors is called:
(A) Cosmopolitan (B) Endemics
(C) Endemism (D) None of these

356. Some species exhibit very localized distribution and are called:
(A) Local endemics
(B) Pseudo endemics
(C) Progressive endemics
(D) Retrogressive endemics

357. Sometimes mutants appear and vanish without being able to compete with parental species, and are called:
(A) Local endemics
(B) Pseudo endemics
(C) Progressive endemics
(D) Retrogressive endemics

358. Some species may show a restricted distribution but cover large areas in course of time is called:
(A) Local endemics
(B) Pseudo endemics
(C) Progressive endemics
(D) Retrogressive endemics

359. Some old species may be restricted to small region because of a severe decline in their population, a phenomenon called:
(A) Local endemics
(B) Pseudo endemics
(C) Progressive endemics
(D) Retrogressive endemics

360. The endemics are supposed to have been the remnants of a once widely distributed taxon in the past is:
(A) Palaeoendemic (B) Neoendemics
(C) Both of these (D) None of these

361. The endemics taxa are supposed to have evolved only during the recent times and did not have sufficient time to extend their ranges of distribution is:
(A) Palaeoendemic (B) Neoendemics
(C) Both of these (D) None of these

362. Ecosystem may be defined as:
(A) A species alongwith environment
(B) Plants found in water
(C) Plants found on land
(D) All plants and animal species alongwith environment

363. Temperature changes in the environment affect most of the animals which are:
(A) Aquatic (B) Desert
(C) Poikilothermal (D) Homoiothermal

364. A pond/lake is:
(A) A biome
(B) A natural ecosystem
(C) An artificial ecosystem
(D) A community of plants and animals only

365. Biome is:
(A) The fauna of an ocean
(B) The flora of land
(C) Communities of organisms interacting with one another
(D) The part of the earth and its atmosphere which is inhabited by living organisms

366. Ecological niche of an organism denotes:
(A) Habitat of the organism
(B) On whom the organism feeds
(C) The status of an organism within a particular community
(D) All of these

367. The transition zone where two different types of communities meet is called:
(A) Ecotype (B) Ecocline
(C) Ecotone (D) Ecosystem

368. Grassland biome for its maintenance requires:
(A) Periodic fire (B) Intensive grazing
(C) Mowing (D) All of these

369. Which of the following is a biome:
(A) Wheat field (B) Rain water pond
(C) Grass field (D) Deciduous forest

370. Primary consumers in detritus food chain are:
(A) Herbivores
(B) Bacteria, fungi etc.
(C) Insect larvae, nematodes
(D) All of these

371. Total organic matter present in an ecosystem is called:
(A) Biomass (B) Biome
(C) Litter (D) Food

372. Detritus food chain in comparison to grazing food chain is:
(A) Generally longer (B) Generally equal
(C) Generally shorter (D) None of these

373. Purifying organisms are:
(A) Producer organisms
(B) Reducer organisms
(C) Consumer organisms (D) Parasitic organisms

374. The most important organisms for an ecosystem are:
(A) Herbivores (B) Carnivores
(C) Green plants (D) Protozoa

375. A peacock eats snakes which eat frogs. These eat hoppers which in turn thrive on leaves of the plants. The peacock is:
(A) Primary consumer
(B) Secondary consumer
(C) Decomposer
(D) Apex of the food pyramid

376. A food chain is:
(A) Group of organisms which eat the same type of food
(B) Animals eating the plants
(C) Series of plants/animals which are interrelated in the form of organism being eaten as food by the other
(D) All of these

377. A food chain consists of:
(A) Producers (B) Consumers
(C) Decomposers
(D) Producers and consumers

378. Trophic levels in a food chain is formed by:
(A) Producers (B) Consumers
(C) Decomposers (D) All of these

379. A food chain always starts with:
(A) Photosynthesis (B) Respiration
(C) Nitrogen fixation (D) Decay

380. Primay consumers are:
(A) Green plants (B) Herbivores
(C) Carnivores (D) All of these

381. Secondary consumers are:
(A) Green plants (B) Herbivores
(C) Carnivores (D) All of these

382. Which of the food chain directly depends on solar radiation:
(A) Predator (B) Grazing
(C) Detritus (D) None of these

383. Producers in any grazing food chain:
(A) Feed the herbivors
(B) Feed the carnivores
(C) Clean the atmosphere
(D) Capture solar energy

384. Converters or transducer organisms in the food chain are called:
(A) Herbivores (B) Carnivores
(C) Decomposers (D) Producers

385. Biological equilibrium is an equilibrium among the:
(A) Producers
(B) Producers and consumers
(C) Decomposers
(D) Producers, consumers and decomposers

386. Trophic level in a food chain which cycles the various materials between environment and living organisms is constituted by:
(A) Producers (B) Consumers
(C) Decomposers (D) All of these

387. Food web is constituted by:
(A) Various interlinked food chains in a community
(B) Relationship between animals and plants
(C) Relationship between organisms and their environment
(D) Relationship between animals, plants and microbes

388. Roots is feebly developed and stem is soft with a large number of air cavities in:
(A) Lithophytes (B) Halophytes
(C) Hydrophytes (D) Aerophytes

389. Vegetation of a geographic region with low rainfall, high temperature, loose and sandy soil is of the type called:
(A) Grassland (B) Scrub forest
(C) Xerophytic (D) Evergreen tropical forest

390. If phytoplankton are destroyed in the sea:
(A) It will affect food chain
(B) No effect will be seen
(C) Algae will get more space to grow
(D) Primary consumer will grow more

391. Ecological pyramid were given by:
(A) Elton (B) Haeckel
(C) Odum (D) Reiter

392. In terrestrial ecosystem, maximum energy is present in:
(A) T3 (B) T2
(C) T1 (D) T4

393. In pond ecosystem the shape of pyramid of number is:
(A) Linear (B) Irregular
(C) Upright (D) Inverted

394. In parasitic food chain the pyramid of number is:
(A) Linear (B) Upright
(C) Inverted (D) Inverted upright

395. In grassland ecosystem the pyramid of biomass is:
(A) Linear (B) Upright
(C) Inverted (D) Inverted upright

396. Pyramid of energy in ecosystem is:
(A) Linear (B) Upright
(C) Inverted (D) Inverted upright

397. In pyramid of food, the producers occupy:
(A) The base (B) Position near the base
(C) Apex (D) Position near apex

398. Energy flow in an ecosystem is:
(A) Unidirectional (B) Bidirectional
(C) Multidirectional (D) All of these

399. Ascending trophic level in grazing food chain pyramid represents:
(A) Gradual decrease in biomass from apex to base
(B) Gradual decrease in biomass from producers to the tertiary consumers
(C) Gradual increase of the biomass from producers to the tertiary consumers
(D) No change in biomass

400. Rate of storage of organic matter no used by heterotrophs is termed as:
(A) Net productivity
(B) Net primary productivity
(C) Gross primary productivity
(D) Secondary productivity

401. Graphic representation of biomass relationship between the producers and consumers in an ecosystem is called:
(A) Trophic level
(B) Ecological system
(C) Ecological niche
(D) Ecological pyramid

402. If all green plants of earth are destroyed:
(A) All plants shall die
(B) Only herbivorous animals shall die
(C) All animals shall die ultimately
(D) Nothing shall happen to animals

403. Animals is cold regions have fur for:
(A) Hair is a good insulating material
(B) Hair trap air which acts as insulator
(C) They like fur
(D) None of these

404. The increased productivity of lakes and streams brought about by nutrient enrichment is known as:
(A) Green house effect
(B) Eutrophication
(C) Biomagnification
(D) Biochemical oxygen demand

405. The typical biome of an area having 50 to 100 cm of rainfall is:
(A) Desert (B) Grassland
(C) Deciduous forest (D) Coniferous forest

406. An ecological niche is normally occupied by:
(A) A population (B) A small community
(C) Competing species (D) Competing genera

407. Grassland of USA is referred to as:
(A) Prairies (B) Steppes
(C) Pampas (D) Veldts

408. Grassland of Southern Russia, Siberia and Asia is referred to as:
(A) Prairies (B) Steppes
(C) Pampas (D) Veldts

409. Grassland of Africa is referred to as:
(A) Prairies (B) Steppes
(C) Pampas (D) Veldts

410. Grassland of south America (Argentina) is referred to as:
(A) Prairies (B) Steppes
(C) Pampas (D) Veldts

411. Plankton, Nekton and Benthos are not the components of one of ecosystems which is:
(A) Oceans (B) Fresh water rivers
(C) Grassland (D) Pond or lake

412. The food chain in which micro-organisms break down the energy rich compounds synthesized by producers:
(A) Predator food chain (B) Parasitic food chain
(C) Detritus food chain (D) None of these

413. When a big fish eats a small fish which eats water fleas supported by phytoplankton, water fleas are:
(A) Producers
(B) Primary consumers
(C) Secondary consumers
(D) Top consumers

414. Deciduous plants are those in which leaves fall:
(A) Once in a year (B) Twice in a year
(C) Many times in a year (D) None of these

415. Which forest have maximum diversity:
(A) Subtropical rain forests (B) Coniferous forests
(C) Tropical forests (D) Deciduous forests

416. Which one is not a factor of the abiotic environment:
(A) Sunlight (B) Decomposers
(C) Water (D) Temperature

417. A functional aspect of an ecosystem is:
(A) Producers, consumers and abiotic environment
(B) Regulation of population
(C) Light, temperature, oxygen and carbon di oxide
(D) All of these

418. Which is true of a food chain:
(A) Passage of food from one trophic level to the next higher one
(B) Passage of energy from one trophic level to the next higher one
(C) Dissipation of energy at successive trophic levels
(D) All of these

419. A food web:
(A) Increases variety of food at each trophic level
(B) Delicately balances the interrelations amongst organisms
(C) Decreases variety of food by increases quantity of food at each trophic level
(D) Increases variety as well as quantity of food at each trophic level

420. Which is true:
(A) Tigers prey upon herbivores
(B) Tigers prey upon all types of land animals
(C) Tigers prey upon most of land animals but can also eat aquatic animals
(D) Tigers prey upon medium-sized land animals

421. During winter Polar Bear:
(A) Hibernates
(B) Aestivates
(C) Remain active
(D) Migrates to warmer areas

422. How do they protect themselves from wave action:
(A) Nereis and tube-worms are fast swimmers
(B) Se Anemones and Limpets have burrowing habit
(C) Fucus and Laminaria are firmly attached to rocks
(D) All of these

423. Which is not true:
(A) Jelly Fishes are glassy
(B) Lions are brownish in colour
(C) Camels are sand coloured
(D) Sea cucumbers are olive-green

424. Niche represents:
(A) Habitat
(B) Microhabitat
(C) Habitat as well as interrelation
(D) Habitat as well as climate

425. Alpine tundra is found in:
(A) Siberia (B) Greenland
(C) Both (A) and (B) (D) Himalayas

426. Rainfall is absent by precipitation occur in:
(A) Arctic desert (B) Tropical desert
(C) Chapparal (D) Savannah

427. Taiga refers to:
(A) Temperate deciduous forest
(B) Subtropical semi-deciduous forest
(C) Evergreen forest
(D) North temperate coniferous forest

428. Chapparal occurs in area:
(A) Africa (B) Siberia
(C) Mediterranean (D) Arabia

429. Rainfall occurs in chapparal during:
(A) The whole year (B) Summer
(C) Winter (D) Spring-autumn

430. Deserts occur in areas of:
(A) Adverse human disturbance
(B) Underground saline water
(C) Little underground water
(D) Rain shadow

431. All types of latitudinal biomes are found telescoped on high hills found in:
(A) Tropical areas (B) Subtropical areas
(C) Temperate areas (D) None of these

432. Which ecosystem does not show variation dependent upon geographic location and rainfall:
(A) Marine ecosystem
(B) Fresh water ecosystem
(C) Desert ecosystem
(D) Tropical ecosystem

433. Photic zone is spread upto a depth of:
(A) 10 m (B) 10-100 m
(C) 100-200 m (D) 200 m

434. Plant life is restricted in sea to:
(A) Photic zone (B) Aphotic zone
(C) Abyssal area (D) Continental shelf

435. Ox-bow lakes occur in:
(A) Areas of previous glaciation
(B) Temperate regions
(C) Near sea shore
(D) Part of a river

436. Free floating plant found below the surface of water is:
(A) Pistia (B) Lemna
(C) Ranunculus (D) Ceratophyllum

437. Upper reaches of a river does not possess many living beings because of:
(A) Low temperature
(B) High speed
(C) Little depth
(D) Both low temperature and high speed

438. Agroecosystem is unstable because of:
(A) Lack of variety
(B) Lack of biological control
(C) Both (A) and (B)
(D) Being man-made

439. Biome represents:
(A) All the organisms of earth
(B) Flora of a land
(C) Flora and fauna of a land
(D) A major ecosystem delimited by climate or geography

440. Epiphytes are rare in Delhi because it has:
(A) Large number of automobiles
(B) Large amount of factories
(C) Semi-arid climate
(D) Moist climate

441. Annual rainfall in the area of a tropical deciduous forest is:
(A) Over 300 cm (B) 200-250 cm
(C) 100-150 cm (D) 50-75 cm

442. Vegetation of Delhi is predominantly:
(A) Halophytic (B) Xerophytic
(C) Mesophytic (D) Hygrophytic

443. Grassland with scattered trees is:
(A) Savannah (B) Alpine plants show
(C) Evergreen forest (D) Tropical rain forest

444. Alpine plants show:
(A) Mesophytism (B) Halophytism
(C) Xerophytism (D) Luxurian growth

445. A habitat with long severe winter and growing season of a few months of summer forms:
(A) Grasslands
(B) Tundra ecosystem
(C) Taiga
(D) Savannah

446. Alpine forests of Himalayas have:
(A) Tall evergreen coniferous trees
(B) Tall broad-leaved deciduous trees
(C) Tall broad-leaved deciduous trees
(D) Dwarf shrubby plants

447. *Rhododendron* is characteristic component of vegetation:
(A) Tropical forests (B) Mangroves
(C) Alpine zone (D) Gangetic plain

448. Temeperate forests occur in India in:
(A) Indo-Gangetic plains (B) Himalayas
(C) Eastern India (D) Southern Peninsula

449. In India, rain forests are found in:
(A) Uttar Pradesh and Bihar
(B) Madhya Pradesh and Orissa
(C) Tamilnadu and Andhra Pradesh
(D) Western Ghats and Eastern Himalayas

450. In India, temperate evergreen vegetation occurs mostly in:
(A) Rajasthan and South Punjab
(B) Eastern and Western Himalayas less than 3500 m
(C) Western Himalayas above 3500 m
(D) Western ghats and Assam

451. Diffirence between tropical rain forest and temperate forest is:
(A) Tropical forest have more of angiosperms while temperate forests have more of gymnosperms
(B) Tropical forest is homogeneous while temperate one is heterogeneous
(C) Plants of temperate forests are more mesophytic
(D) Trees of temperate forests are taller than those of tropical forests

452. The forest that colours in autumn is:
(A) Temperate evergreen forest
(B) Temperate deciduous forest
(C) Tropical evergreen forest
(D) Tropical deciduous forest

453. Ephemerals are a type of xerophytes:
(A) Drought escaping (B) Drought resisting
(C) Drought enduring (D) None of these

454. Succulents occur in:
(A) Deserts
(B) Tundra
(C) Temperate deciduous forests
(D) Tropical rain forests

455. Extreme xerophytic condition occurs in:
(A) *Cactus* (B) *Nerium*
(C) *Capparis* (D) *Brassica*

456. *Mesenbryanthemum* belongs to the category:
(A) Mesophyte
(B) Fleshy xerophytes
(C) Nonsucculent perennial xerophytes
(D) Epiphyte

457. Xerophytic plants have:
(A) Extensive root system, small & thick leaves
(B) Extensive shoot system
(C) Profuse and large flowers
(D) Extensive air space

458. Thick cuticle occurs on the leaves of plants belonging to:
(A) Cool habitats (B) Warm habitats
(C) Dry habitats (D) Wet habitats

459. Stomata are shunken in:
(A) Mesophytes (B) Xerophytes
(C) Hygrophytes (D) Hydrophytes

460. Relative to roots, the shoots are massive in plants of:
(A) Deserts
(B) Moist temperate
(C) Tundra
(D) Moist tropical forests

461. Tropical rain forests are found in:
(A) Andamans (B) Bihar
(C) Himanchal Pradesh
(D) Jammu and Kashmir

462. Submerged hydrophytes have a well developed:
(A) Vascular system (B) Aerenchyma
(C) Root system (D) Stomatal system

463. Stem of submerged hydrophytes is soft and week because:
(A) Absence of phloem
(B) Absence of stomata
(C) Absence of xylem
(D) Feebly developed supporting tissue & xylem

464. Aquatic plants generally have a:
(A) Well developed vascular system
(B) Reduced vascular system
(C) Well developed root system
(D) Well developed stomatal system

465. A plant devoid of roots is:
(A) *Wolffial / Ceratophyllum*
(B) *Vallisneria*
(C) *Hydrilla*
(D) *Azolla*

466. Typha is a:
(A) Submerged hydrophyte
(B) Floating hydrophyte
(C) Emerged hydrophyte
(D) Attached hydrophyte with floating leaves

467. In submerged hydrophytes the stomata occur:
(A) On lower surface
(B) On upper surface
(C) No where
(D) On both the surfaces

468. Which character is not present in the hydrophytes:
(A) Poor roots
(B) Plentiful xylem and sclerenchyma
(C) Leaves with stomata on upper surface or absent
(D) Abundant aerenchyma

469. Which character is not present in the xerophytes:
(A) Thick cuticle
(B) Well developed mechanical tissue
(C) Well developed conducting tissues
(D) Spongy parenchyma

470. The leaves of desert plants are not torn away by high wind velocity because of:
(A) Spines
(B) Bending towards opposite side
(C) Sclerenchymatous tissue that provides extra mechanical support
(D) Tearing of corners while the middle remains saved

471. Amongest hydrophytes finely dissecated leaves occur in:
(A) Rooted floating leaved plants
(B) Submerged plants
(C) Emerged plants
(D) Free floating plants

472. The term biocoenosis was given by:
(A) Warming (B) Carl Mobius
(C) Tansley (D) Odum

473. Who proposed the term ecosystem:
(A) Odum (B) Gardner
(C) Warming (D) Tanslay

474. The importance of ecosystem lies in:
(A) Cycling of materials (B) Flow of energy
(C) Both (A) and (B) (D) Its biomass

475. An ecosystem does not normally alter because it is in a state of:
(A) Homeostasis
(B) Imbalance
(C) Deficient light
(D) Deficient component

476. In India, tropical evergreen forest occur in:
(A) Himachal Pradesh (B) Madhya Pradesh
(C) Assam (D) Tamilnadu

477. The floating leaves of an aquatic plant has:
(A) Stomata on upper surface
(B) Stomata on lower surface
(C) Stomata on both the surface
(D) No stomata

478. Homeostasis is:
(A) Tendency of biological systems to change with change in environment
(B) Tendency of biological systems to resist change
(C) Disturbance of self regulatory system and natural controls
(D) Biotic materials used in homeopathic medicines

479. The two components of an ecosystem are:
(A) Plants and animals
(B) Weeds, trees, animals and man
(C) Energy flow and mineral cycling
(D) Biotic and abiotic

480. An ecosystem is:
(A) Different communities of plants, animals and microbes together with their physico-chemical environment
(B) Different communities of plants and microbes alongwith their physic-chemical environment
(C) A localized assemblage of several plants and animals
(D) An assemblage of plants, animals and their surroundings

481. The driving force for an ecosystem is:
(A) Biomass
(B) Producers
(C) Carbohydrates in producers
(D) Solar energy

482. Ecosystem is:
(A) Always open
(B) Always closed
(C) Both open and closed depending upon community
(D) Both open and closed depending upon biomass

483. Energy entering an organism from outside is 100% efficient in case of:
(A) Unicellular green algae
(B) All green plants
(C) Some photosynthetic bacteria
(D) No photoautotroph

484. The major components of an ecosystem are:
(A) Food chain and decomposers
(B) Mineral cycling and energy flow
(C) Food chain and energy flow
(D) Energy flow and decomposers

485. Biotic components of an ecosystem include:
(A) Producers, consumers and decomposers
(B) Producers and consumers
(C) Producers only
(D) Consumers only

486. A food chain can have trophic levels:
(A) Three or four (B) Three
(C) Two (D) One

487. In food chain, the initial organisms are:
(A) Top consumers (B) Secondary consumers
(C) Primary consumers (D) Photosynthates

488. In a food chain, herbivores/deer are:
(A) Primary producers
(B) Primary consumers
(C) Secondary consumers (D) Decomposers

489. Green plants constitute:
(A) Complete food chain
(B) Second trophic level
(C) First trophic level (D) Third trophic level

490. First link in food chain is green plant because:
(A) It alone can synthesise food
(B) It is fixed at one place
(C) It can pickup every thing
(D) It can eat every thing

491. What is true of ecosystem?
(A) Primary consumers are least dependent upon producers
(B) Primary consumers out-number producers
(C) Producers are more than primary consumers
(D) Secondary consumers are the largest and most powerful

492. Which is correct food chain in Grassland:
(A) Grass → Snake → Insect → Deer
(B) Grass → Wolf → Deer → Buffalo
(C) Bacteria → Grass → Rabbit → Wolf
(D) Grass → Insect → Frog → Snake

493. Food chain in which microorganisms breakdown the food formed by primary producers is:
(A) Parasitic food chain
(B) Detritus food chain
(C) Consumer food chain
(D) Predator food chain

494. A food chain comprises:
(A) Producers, herbivores and carnivores
(B) Producers, consumers and decomposers
(C) Producers and primary consumers
(D) Producers, carnivores and decomposers

495. Pickup the correct food chain:
(A) Grass → Chamelion → Insect → Bird
(B) Grass → Fox → Rabbit → Bird
(C) Phytoplankton → Zooplankton → Fish
(D) Fallen leaves → Bacteria → Insect larvae

496. Path of energy flow in an ecosystem is:
(A) Herbivores → Producers → Carnivores → Decomposers
(B) Herbivores → Carnivores → Producers → Decomposers
(C) Producers → Carnivores → Herbivores → Decomposers
(D) Producers → Herbivores → Carnivores → Decomposers

497. Biological equilibrium is equilibrium amongst:
(A) Producers and decomposers
(B) Producers and consumers
(C) Producers, consumers and decomposers
(D) Consumers and decomposers

498. An ecosystem is an interacting system of:
(A) Communities
(B) Communities and their physical environment
(C) Populations (D) Individuals

499. Upper part of Sea/aquatic ecosystem contains:
(A) Plankton (B) Nekton
(C) Plankton and nekton (D) Benthos

500. Energy transfer from organism to organism in a natural community develops:
(A) Biological control (B) Food chain
(C) Natural barriers (D) Food web

501. A group of interconnected food chains is:
(A) Complex of food chains (B) Food cycle
(C) Food web (D) Pyramid of energy

502. Which one is primary consumer:
(A) Carnivore (B) Scavenger
(C) Saprophyte (D) Herbivore

503. Decomposers of an ecosystem include:
(A) Microscopic animals
(B) Bacteria and fungi
(C) Both (A) and (B)
(D) Bacteria, fungi and macroscopic animals

504. Phytoplankton are destroyed. It will:
(A) Cause spurt in population of primary consumers
(B) Rapid growth of algae
(C) Affect the food chain
(D) Have no effect

505. In an aqueous environment, microscopic animals and plants are collectively known as:
(A) Plankton (B) Herbivore
(C) Commensals (D) Flora and fauna

506. Energy stored at the consumer level is:
(A) Gross primary productivity
(B) Net primary productivity
(C) Net productivity
(D) Secondary productivity

507. *Nepenthes* belongs to the category of:
(A) Primary consumer (B) Producer
(C) Secondary consumer
(D) Both (B) and (C)

508. Siddartha found a vegetarian field mouse being swallowed by snake which is turn formed the meal for a hawk. What is the position of snake in the food chain?
(A) Producer (B) Primary consumer
(C) Secondary consumer (D) Tertiary consumer

509. Which one occurs at the same trophic level:
(A) Crow and Cow (B) Snake & Earthworm
(C) Tiger and wild Boar (D) Deer and Bee

510. Where there is no difference:
(A) T1 and herbivores
(B) Primary consumers and herbivores
(C) Primary carnivores and T2
(D) Secondary consumers and herbivores

511. Abyssal zone in Ocean has:
(A) No sunlight but contain consumers and decomposers
(B) No sunlight but contains producers
(C) No sunlight but contains living beings
(D) Sunlight as well as producers

512. Flow of energy declines as it passes from lower to higher trophic level. This is explained by:
(A) First law of thermodynamics
(B) Second law of thermodynamics
(C) Newton's second law (D) Newton's third law

513. A young crocodile has eat a lamb. A hawk attacks the crocodile and consumes the same. In ecological terms, the hawk is:
(A) Producer (B) Primary consumer
(C) Secondary consumer (D) Tertiary consumer

514. If all the plants are to die, all the animals will also die due to deficiency of:
(A) Food (B) Carbon dioxide
(C) Oxygen (D) Timber

515. Pyramid of energy (in a forest/ponds ecosystem) is:
(A) Always inverted
(B) Always upright
(C) Spindle-shaped
(D) Both upright and inverted depending upon ecosystem

516. In an ecosystem, which one shows one-way passage:
(A) Free energy (B) Carbon
(C) Nitrogen (D) Potassium

517. Pyramid of numbers in a grassland/true ecosystem is:
(A) Always inverted (B) Always upright
(C) Both (A) and (B) (D) Spindle shaped

518. Suppose all the bacteria and fungi are destroyed. What will happen?
(A) There will be no disease and death
(B) No antibodies would become available
(C) Dead bodies and excretion will pile up
(D) Soil will become depleted of all nutrients

519. Maximum human population can be supported if all the humans are:
(A) Carnivores
(B) Pure vegetarian
(C) Vegetarian along with use of milk
(D) Vegetarian alongwith use of milk and eggs

520. The pyramid which cannot be inverted in a stable ecosystem is that of:
(A) Biomass (B) Number
(C) Energy (D) All of these

521. In a tree ecosystem, the pyramid of numbers is:
(A) Upright (B) Inverted
(C) Spindle like (D) Variable

522. Shallow lakes with abundant organic matter are:
(A) Saprotrophic (B) Oligotrophic
(C) Eutrophic (D) Heterotrophic

523. Total energy fixed by an ecosystem is:
(A) Primary production
(B) Gross production
(C) Net production
(D) Secondary production

524. Savannah are:
(A) Tropical rain forests
(B) Desert scrubs
(C) Grassland with scattered trees
(D) Dense forests with closed canopy

525. Which one has the maximum biomass:
(A) Temperate forest
(B) Tropical rain forests
(C) Alpine vegetation
(D) Taiga

526. Pyramid of numbers deals with number of:
(A) Species in an area
(B) Individuals in a community
(C) Individuals in a trophic level
(D) Subspecies in a community

527. Which ones lie at the level of primary consumers:
(A) Fishes (B) Eagles
(C) Cattle and insects (D) Snakes and frogs

528. Ecosystem creates:
(A) Food chain & food web (B) Food web
(C) Habitats (D) None of these

529. In lake ecosystem, the pyramid of biomass is generally:
(A) Spindle-shaped (B) Inverted
(C) Upright (D) Horizontal

530. Carnivores represent:
(A) Primary consumers
(B) Secondary and tertiary consumers
(C) Secondary consumers
(D) Reducers

531. In a food chain, man is:
(A) Primary and secondary consumer
(B) Primary consumer
(C) Secondary consumer
(D) Producer

532. Pyramid of numbers in a pond ecosystem is:
(A) Irregular (B) Inverted
(C) Upright (D) Spindle shaped

533. Pyramid of numbers in a forest ecosystem is:
(A) Upright (B) Inverted
(C) Irregular (D) Linear

534. Species diversity increases as one proceeds from:
(A) High altitude to low altitude and high latitude to low latitude
(B) Low altitude to high altitude and high latitude to low latitude
(C) Low altitude to high altitude and low latitude to high latitude
(D) High altitude to low altitude and low latitude to high latitude

535. In grass-deer-tiger food chain, grass biomass in one tone. The tiger biomass shall be:
(A) 100 kg (B) 10 kg
(C) 200 kg (D) 1 kg

536. In a biotic community, the most important factor for survival of an animal is:
(A) Day length (B) Soil moisture
(C) Green food (D) Predators

537. In a food chain, the largest population is that of:
(A) Producers (B) Decomposers
(C) Secondary consumers (D) Primary consumers

538. Xeric environment is characterized by:
(A) Precipitation
(B) Low atmospheric humidity
(C) Extremes of temperature
(D) High rate of vaporization

539. Second most important trophic level in a lake is:
(A) Zooplankton (B) Phytoplankton
(C) Benthos (D) Neuston

540. Which receives the maximum energy:
(A) Producers (B) Decomposers
(C) Primary consumers
(D) Secondary consumers

541. What is wrong about xerophytes:
(A) Sunken stomata
(B) Small spiny leaves
(C) Thick cuticle
(D) Larger number of stomata

542. Mechanical tissue is undeveloped in:
(A) Xerophytes (B) Hydrophytes
(C) Halophytes (D) Mesophytes

543. Tropical forests occur in India:
(A) Jammu and Kashmir (B) Rajasthan
(C) Kerala and Assam
(D) The forests do not occur in India

544. Which can be trophic level of a lion in forest ecosystem:
(A) T3 (B) T2
(C) T4 (D) T1

545. Tropical forests are denser due to :
(A) Wild animals
(B) High temperature and less rainfall
(C) Low temperature and excess rainfall
(D) High temperature and high rainfall

546. Which is correct:
(A) Least productive ecosystem is desert and deep lake
(B) Sugarcane is the most productive agroecosystem
(C) Most productive is coral reef system
(D) All of these

547. Bulk CO_2 fixation occurs in:
(A) Crop plants (B) Oceans
(C) Tropical rain forests (D) Temperature forests

548. Desert biome does not support much vegetation as it lacks:
(A) Favourable temperature (B) Sufficient light
(C) Sufficient water (D) Sufficient nutrients

549. Animals that can tolerate a narrow range of salinity are:
(A) Stenohaline (B) Euryhaline
(C) Anadromous (D) Catadromous

550. Hygrometer is employed to measure:
(A) Wind velocity (B) Temperature
(C) Relative humidity (D) Turbidity

551. Stratification is more common in:
(A) Tropical rain forest (B) Deciduous forest
(C) Temperate forests (D) Tropical savannah

552. Importance of ecosystem lies in:
(A) Cycling of materials (B) Flow of energy
(C) Both (A) and (B) (D) None of these

553. Number of primary producers per unit area would be maximum in:
(A) Forest ecosystem (B) Pond ecosystem
(C) Grassland ecosystem (D) Desert

554. A food chain begins with:
(A) Photosynthatic organisms
(B) Nitrogen fixing organisms
(C) Decomposers (D) Consumers

555. The most stable ecosystem is:
(A) Forest (B) Mountain
(C) Desert (D) Ocean

556. In the absence of decomposers, ecosystem functioning is adversely affected due to:
(A) Blocking of energy flow
(B) Blocking of mineral cycling
(C) Blocking of solar energy to herbivores
(D) Rate of decomposition of other components will increase

557. Maximum number in pond ecosystem is of:
(A) Producers (B) Consumers
(C) Top consumers (D) Decomposers

558. Which one is not a trait of xerophytes:
(A) Thick cuticle (B) Shunken stomata
(C) Aerenchyma
(D) Well developed mechanical tissue

559. In every ecosystem, the green plants are:
(A) Decomposers (B) Producers
(C) Consumers (D) None of these

560. The factors related to structure of earth surface is:
(A) Edaphic (B) Biotic
(C) Temperature (D) Topographic

561. Benthos of pond ecosystem are:
(A) Producers
(B) Primary consumers
(C) Secondary consumers
(D) Tertiary consumers

562. Lion is:
(A) Producer (B) Primary consumer
(C) Predator (D) Secondary consumer

563. Vegetation of Rajasthan is:
(A) Xerophytic (B) Alpine
(C) Arctic (D) Deciduous

564. During extreme aridity, Desert Rat:
(A) Stores water (B) Uses metabolic water
(C) Saves water (D) Does not use water

565. Rhizophora is an example of:
(A) Mesophyte (B) Hydrophyte
(C) Halophyte (D) Xerophyte

566. Which one lacks both roots and stomata:
(A) Hydrophytes (B) Mesophytes
(C) Hygrophytes (D) Halophytes

567. Which one is partially submerged and fixed in mud:
(A) *Marsilea* (B) *Cyperus*
(C) *Eichhornia* (D) *Typha*

568. 10% law for energy transfer in food chains was given by:
(A) Lindemann (B) Stanley
(C) Tansley (D) Weismann

569. Niche of a species is:
(A) Place of living
(B) Specific functions and competitive power
(C) Habitat and specific functions
(D) None of these

570. Benthos animals live:
(A) Deep in sea (B) Floating
(C) Submerged (D) Active swimmers

571. Major communities of the world are:
(A) Scrubs, jungles and gardens
(B) Ocean, fresh water and forest communities
(C) Grassland and desert communities
(D) Both (B) and (C)

572. Deciduous forests have:
(A) Variety of grasses
(B) Broad leaved trees
(C) Narrow leaved trees
(D) Variety of crocodiles

573. Food chain is:
(A) Number of human being forming a chain food
(B) Animals near a source of food
(C) Transfer of food energy from producers to consumers
(D) None of these

574. Which one is index of environmental conditions and changes:
(A) Individual (B) Communities
(C) Niche (D) Pollution

575. Increase in number of fauna and decrease in number of flora would be dangerous because of enhanced percentage of:
(A) Disease (B) CO_2
(C) O_2 (D) Radio-active pollution

576. In food chain, the largest population is that of:
(A) Producers (B) Primary consumers
(C) Secondary consumers (D) Tertiary consumers

577. Food levels of an ecosystem are known as:
(A) Producer levels (B) Herbivore levels
(C) Consumer levels (D) Trophic levels

578. Tropical dry deciduous forests occur in India in:
(A) Andamans (B) Eastern Himalayas
(C) Madhya Pradesh (D) Kerala

579. Plants of salty sea shore wet lands are:
(A) Heliophytes (B) Hydrophytes
(C) Halophytes (D) Saprophytes

580. *Eichhornia crassipes* is:
(A) Xerophyte (B) Hydrophyte
(C) Mesophyte (D) Parasite

581. Which one is the correct food chain:
(A) Eagle → Snake → Grasshopper → Grass → Frog
(B) Frog → Snake → Eagle → Grasshopper → Grass
(C) Grasshopper → Grass → Snake → Frog → Eagle
(D) Grass → Grasshopper → Frog → Snake → Eagle

582. Desert plants are able to tolerate water stress through:
(A) Root system reaching water table
(B) Stipular spines
(C) Absence of stomata
(D) Modification of stem into leaf like structures

583. Biome with broad-leaved resinous fire resistant drought enduring plants is:
(A) Savannah (B) Steppes
(C) Chapparal (D) Deciduous forest

584. Key stone species of an ecosystem are the ones which are:
(A) Most frequent
(B) Attain large biomass
(C) Present in maximum number
(D) Contribute to ecosystem properties

585. Which ecosystem has the highest gross primary productivity?
(A) Rain forests (B) Coral reefs
(C) Mangroves (D) Grasslands

586. During adverse season, therophytes survive through:
(A) Seeds (B) Bulbs
(C) Rhizomes (D) Corms

587. Which is dominant in desert:
(A) Hyla (B) Leopard
(C) Tiger (D) Lizard

588. Amount of energy transferred from one trophic level to next is:
(A) 1.5% (B) 10%
(C) 15% (D) 20%

589. Climate and edaphic factors controlling distribution of a community constitute:
(A) Ecotone (B) Ecophene
(C) Epharmony (D) Ecological niche

590. *Acacia arabica* (*A. nilotica*) is a:
(A) Hydrophyte (B) Mesophyte
(C) Xerophyte (D) None of these

591. Ecological pyramids are of types:
(A) Two (B) Three
(C) Four (D) Five

592. Species diversity is lowest in ecosystem:
(A) Desert (B) Tundra
(C) Grassland (D) Deciduous forest

593. In an ecosystem, bacteria are:
(A) Microconsumers (B) Macroconsumers
(C) Primary consumers
(D) Secondary consumers

594. Concept of life forms was given by:
(A) Weaver (B) Clements
(C) Raunkier (D) Turesson

595. Annual plant lacking perennating buds belongs to:
(A) Therophytes (B) Cryptophytes
(C) Phanerophytes (D) Chemaephytes

596. Who proposed the concept of ecological pyramids:
(A) Odum (B) Elements
(C) Tansley (D) Elton

597. Transfer of energy in different trophic levels of an ecosystem is:
(A) Bioenergetics (B) Biosystem
(C) Geobiocoenosis (D) Holocoenosis

598. Which one is the nature's cleaner:
(A) Consumers
(B) Producers
(C) Decomposers and scavengers
(D) Symbionts

599. Which one has maximum biomass:
(A) Pond ecosystem (B) Lake ecosystem
(C) Grassland ecosystem (D) Forest ecosystem

600. Submerged hydrophytes exchange gases through:
(A) Stomata (B) Hydathodes
(C) Lenticels (D) General surface

601. Biotic community alongwith its interacting physical environment comprises:
(A) Phytosociology (B) Phytogeography
(C) Ecosystem (D) Ecology

602. Number of individuals of a species in a particular ecosystem at a given time remains constant due to:
(A) Predators (B) Parasites
(C) Man (D) Food

603. Total amount of living material at various levels of a food chain is depicted by:
(A) Pyramid of energy (B) Pyramid of numbers
(C) Pyramid of biomass (D) All of these

604. Relatively more abundant animals of desert grassland are:
(A) Arboreal (B) Fossorial
(C) Diurnal (D) Aquatic

605. Tropical plants *Prosopis, Acacia* and *Capparis* belong to:
(A) Evergreen forest (B) Deciduous forest
(C) Grassland (D) Thorn forest

606. Energy enters an ecosystem through:
(A) Herbivores (B) Producers
(C) Carnivores (D) Decomposers

607. Rate of conversion of light energy into chemical energy of organic molecules in an ecosystem is:
(A) Gross primary productivity
(B) Net primary productivity
(C) Net secondary productivity
(D) Gross secondary productivity

608. Halophytes grow on physiologically dry soil due to:
(A) Excessive salt in water
(B) Excessive salt in plants
(C) Excessive humidity outside
(D) Dry soil

609. Tropical forests are dense due to:
(A) Wild animals
(B) High temperature and less rain
(C) Low temperature and excess rain
(D) High temperature and excess rain

610. An ecological niche is a:
(A) Habitat of a plant
(B) Habitat of a small community
(C) Habitat of nocturnal animals
(D) Kind of plant-animal association

611. Sal forest (*Shorea robusta*) is a:
(A) Deciduous forest (B) Evergreen forest
(C) Rain forest (D) None of these

612. Biotic factors refer to:
(A) Grass produced by industries
(B) Nutrient deficient soils
(C) Living organisms
(D) Fossil fuels

613. Decomposers are:
(A) Autotrophs (B) Heterotrophs
(C) Autoheterotrophs (D) Chemoautotrophs

614. Sun loving plants are:
(A) Halophytes (B) Sciophytes
(C) Heliophytes (D) Autotrophs

615. A submerged rooted hydrophyte is:
(A) *Vallisneria* (B) *Trapa*
(C) *Utricularia* (D) *Nymphaea*

616. Moderate rain during summer produces:
(A) Desert (B) Grassland
(C) Scrub forest (D) Deciduous forest

617. Relationships in an ecosystem can be depicted through:
(A) Pyramid of energy (B) Pyramid of biomass
(C) Pyramid of numbers (D) All of these

618. Water storage tissues has:
(A) Large-sized thin walled cells (B) Mucilage
(C) Large-sized vacuoles (D) All of these

619. Type of plants having adaptation to check transpiration is:
(A) Xerophytes (B) Lithophytes
(C) Halophytes (D) Epiphytes

620. Xerophytes possess:
(A) Sunken stomata (B) Deep roots
(C) Thick cuticle (D) All of these

621. Characteristic of hydrophytes is:
(A) Poorly developed roots
(B) Well developed roots
(C) Well developed xylem
(D) Stem with sclerenchyma

622. Viviparity and pneumatophores are features of:
(A) Hydrophytes (B) Halophytes
(C) Mesophytes (D) Xerophytes

623. Ecosystem has two components:
(A) Animals and humans (B) Plants and animals
(C) Biotic and abiotic (D) Weeds and trees

624. A habitat not suitable for primary productivity is:
(A) Cave (B) Pond
(C) Meadow (D) Forested river bank

625. Which part of the world has high density of organisms?
(A) Grasslands
(B) Savannah
(C) Deciduous forests
(D) Tropical rain forests

626. The most favoured habitat of a plant is called:
(A) Ecological habitat (B) Microhabitat
(C) Ecotype (D) Ecocline

627. Green plant constitute:
(A) First trophic level (B) Second trophic line
(C) Third trophic line (D) Fourth trophic line

628. Tip of an ecological pyramid is occupied by:
(A) Producers (B) Herbivores
(C) Carnivores (D) None of these

629. Zooplanktonic forms are:
(A) Primary producers
(B) Carnivores
(C) Primary consumers
(D) Secondary consumers

630. Which of the following occurs in abiotic component of ecosystem?
(A) Flow of energy
(B) Cycling of materials
(C) Consumers
(D) Flow of energy and cycling of materials

631. A man-made microecosystem is:
(A) Plants grown in a pond
(B) Crop field
(C) Tank formed naturally in your court yard
(D) Lake in a forest

632. A treeless biome is:
(A) Tundra (B) Grassland
(C) Desert (D) All of these

633. Correct sequence in a food chain is:
(A) Grass $\rightarrow$ Insects $\rightarrow$ Birds $\rightarrow$ Snake
(B) Grass $\rightarrow$ Snake $\rightarrow$ Insects $\rightarrow$ Deer
(C) Grass $\rightarrow$ Wolf $\rightarrow$ Deer $\rightarrow$ Buffalo
(D) Bacteria $\rightarrow$ Grass $\rightarrow$ Rabbit $\rightarrow$ Wolf

634. In an ecosystem, trophic level of man is:
(A) Producer (B) Herbivore
(C) Omnivore (D) Carnivore

635. Secondary producers are:
(A) Herbivores (B) Heterotrophs
(C) Carnivores (D) Green plants

636. Drought resistance of succulents is connected with:
(A) Higher frequency of stomata
(B) Regulation of stomatal movement
(C) Fleshy mesophyll
(D) Well developed xylem

637. Biome has organisms of various trophic levels:
(A) Reaching with abiotic environment
(B) Constitute a sociological unit
(C) Live a symbiotic life
(D) Inhabit a desert

638. A rat feeding on potato tuber is:
(A) Producer (B) Primary consumer
(C) Carnivore (D) Decomposer

639. Lentic ecosystem occurs in:
(A) Rain water (B) Running water
(C) Standing water (D) Gravitational water

640. A nonsucculent xerophytes with thick leathery leaves having white sticky waxy coating is:
(A) *Norium* (B) *Calotropis*
(C) *Bryophyllum* (D) *Ruscus*

641. An ecosystem contains:
(A) Producers (B) Consumers
(C) Decomposers (D) All of these

642. Climatology is science of:
(A) Edaphic factors (B) Topographic factors
(C) Climatic factors (D) Biotic factors

643. Ecosystem of a pond is:
(A) Lotic (B) Lentic
(C) Benthic (D) Xeric

644. Mangrove of marshy Sunderbans is characterized by:
(A) Pneumatophores (B) Prop roots
(C) Vivipary (D) All of these

645. Transition zone between two vegetational types is:
(A) Ecotone (B) Ecotype
(C) Ecocline (D) Ecosystem

646. What is true?
(A) Diaphragm helps in respiration of Frog
(B) Pericardial and cerebrospinal fluids are similar
(C) Some mammals hibernate
(D) Mesorchium connects ovary and kidney of frog

647. Organisms living at the bottom of a lake are:
(A) Nekton (B) Benthos
(C) Plankton (D) Pelagic

648. Maximum biomass of autotrophs is oceans is made of:
(A) Benthic brown algae, coastal red algae and daphnids
(B) Benthic diatoms and marine viruses
(C) Sea grasses and slime moulds
(D) Free floating microalgae, cyanobacteria and nanoplankton

649. Frog feeding of herbivorous insect is:
(A) Primary consumer
(B) Secondary consumer
(C) Tertiary consumer
(D) Top carnivore

650. A recently discovered ecosystem is:
(A) Crater (B) Tundra
(C) Floating iceberg (D) Vent

651. Controlling factor in ecosystem is:
(A) Soil moisture (B) Food
(C) Predation (D) Temperature

652. Vivipary occurs in:
(A) *Betula* (B) *Rhizophora*
(C) Mango (D) *Psidium*

653. Factor related to structure of earth's surface is:
(A) Edaphic (B) Topographic
(C) Biotic (D) Temperature

654. Which is an example of positive feedback?
(A) Seasonal growth
(B) Vessel diameter and blood pressure
(C) Allosteric inhibition of enzymes
(D) Both (A) and (B)

655. In upright pyramid of biomass herbivores occupy the position:
(A) 1 (B) 2
(C) 3 (D) 4

656. Which is amphibious?
(A) *Casuarina* (B) *Wolffia*
(C) *Polygorum* (D) *Hydrilla*

657. Hibernation occurring in certain animals is:
(A) Occasional (B) Intermittent
(C) Rhythmic (D) Periodic

658. No animal can live:
(A) Alone
(B) Without friend
(C) Without foes
(D) In area with high temperature

659. Animals have shorter and smaller extremities in colder areas. It is:
(A) Allen's law (B) Cope's law
(C) Dollo's law (D) Bergman's law

660. Terai forest is:
(A) Tropical (B) Coniferous
(C) Deciduous (D) Tropical deciduous

661. Savannah is found commonly in:
(A) USA (B) USSR
(C) Australia (D) India

662. Which one has always a steeper vertical gradient?
(A) Pyramid of mass
(B) Pyramid of energy
(C) Pyramid of numbers
(D) Pyramid of energy in aquatic ecosystem

663. Which one is exclusive xerophytic adaptation?
(A) Absence of stomata
(B) Long tap root system
(C) Stipular leaves
(D) Spines

664. Which one has evergreen vegetation and drought adapted animals?
(A) Chaparral (B) Savannah
(C) Tundra (D) Deciduous forest

665. Population means:
(A) All the animals of a place
(B) All the animals and plants found in an area
(C) Human beings living in an area
(D) Animals/plants of a species living in a specific area

666. Species ecology is:
(A) Autoecology (B) Synecology
(C) Community ecology (D) Palaeoecology

667. Population whose members reproduce asexually is:
(A) Panmictic (B) Species
(C) Amphimictic (D) Apomictic

668. Ecotype is:
(A) Genetically different but phenotypically similar
(B) Genetically similar but ecologically different individuals
(C) Genetically adapted ecological race
(D) Genetically and phenotypically dissimilar members of different species ecologically adapted to an area

669. The factors involved in formation of new species are:
(A) Competition and variation
(B) Isolation and competition
(C) Competition and mutations
(D) Isolation and mutations

670. An animal leading a solitary life is:
(A) Deer (B) Seal
(C) Fox (D) Hamster

671. Animal which lives in flocks is:
(A) Deer (B) Monkey
(C) Elephant (D) Parrot

672. An animal found in herds is:
(A) Deer (B) Swan
(C) Parrot (D) Monkey

673. A monogamous animal is:
(A) Seal (B) Deer
(C) Swan (D) Hamster

674. A polygamous animal is:
(A) Rat (B) Deer
(C) Fox (D) Wolf

675. The occurrence of two forms of individuals in the same species is known as:
(A) Differentiation (B) Dimorphism
(C) Trimorphism (D) Polymorphism

676. Presence of different types of flowers in male and female like mulberry/palm/papaya is:
(A) Dichogamy (B) Dimorphism
(C) Bisexuality (D) Monoecious nature

677. Cooperative interaction is essential for perpetuation of:
(A) Microorganisms (B) Plants
(C) Animals (D) Both (B) and (C)

678. Study of inter-relationship between organisms and their environment is:
(A) Ecology (B) Ecosystem
(C) Phytogeography (D) Ethology

679. Study of inter-relationships between a species/ individual and its environment in a forest is:
(A) Synecology (B) Forest ecology
(C) Autoecology (D) Ecology

680. Ecology is connected with the study of:
(A) Environmental factors
(B) Plant adaptations
(C) Effect of plants on environment
(D) All of these

681. Study of inter-relationships between an entire community and its environment is:
(A) Autoecology (B) Resource ecology
(C) Species ecology (D) Synecology

682. Synecology is study of inter-relationship between environment and:
(A) Individual plant (B) A population
(C) A community (D) Individual animal

683. Individuals of a species which occur in a particular area constitute:
(A) Flora (B) Fauna
(C) Population (D) Flora and fauna

684. Biotic potential is increase in population under:
(A) Natural conditions
(B) Climatic conditions
(C) Optimum conditions
(D) Given conditions

685. Proportion of young individuals is highest in case of:
(A) Declining population
(B) Static population
(C) Both (A) and (B)
(D) Expanding population

686. Large gap between birth and death rates will result in:
(A) More old persons
(B) Low rate of fertility
(C) More school going children
(D) Low dependency ratio

687. Competition is the most serve between two:
(A) Closely related species growing in different niches
(B) Closely related species growing in the same habitat
(C) Distantly related species growing in the same habitat
(D) Distantly related species growing in different niches

688. A population with equal number of births and deaths will show:
(A) Acceleration phase of growth
(B) Plateau phase or equilibrium phase
(C) Exponential growth phase
(D) Initial phase of growth

689. In case of scarcity of food or limitation of habitat, animal population is sure to starve and die. Human escape this disaster by:
(A) Emigration
(B) Immigration
(C) Transportation of food
(D) Both emigration & transportation of food

690. The word ecology was coined by:
(A) Linnaeus (B) Haeckel
(C) Harvey (D) Odum/Lamarck

691. The term of ecology was actually coined by:
(A) Aristotle (B) Reiter
(C) Linnaeus (D) Odum

692. Which can counteract biotic potential?
(A) Limitation of food
(B) Competition from other organisms
(C) Availability of lower trophic level
(D) None of these

693. The sum total of the population of the same kind of organisms constitute:
(A) Colony (B) Genus
(C) Community (D) Species

694. Ecology is the study of relationship of:
(A) Members of a family
(B) Man and environment
(C) Organisms and environment
(D) Soil and water

695. Competition is severe in a population that has a distribution :
(A) Irregular (B) Uniform
(C) Random (D) Nonrandom

696. Information about birth rate, death rate, sex ratio, age distribution, etc. of a population is obtained from:
(A) Mortality table (B) Life table
(C) Natality table (D) Both (A) and (C)

697. Which is correct about human races?
(A) All races can interbreed but most will produce infertile young ones
(B) All races can interbreed and produce fertile offspring
(C) Different races cannot interbreed
(D) Some human races can interbreed

698. Which is correct?
(A) Honey Bees communicate among themselves by secreting pheromones
(B) Last succession in series of biotic succession is called a sere
(C) Number of individuals of particular species present in a unit area at a given time is known as population size
(D) Home range of members of a species has larger area than that of territory

699. The aggregate of processes that determine the size and composition of any population is known as:
(A) Population dynamics
(B) Population dispersal
(C) Population explosion
(D) Population density

700. The abundance of a species population within its habitat is called:
(A) Absolute density (B) Regional density
(C) Relative density (D) Niche density

701. Synecology is study of:
(A) Human environment of a place
(B) Biotic environment of a place
(C) Abiotic component of a place
(D) Biotic community in relation to environment of a place

702. When a population reaches carrying capacity:
(A) Mortality rate = Birth rate
(B) Mortality rate > Birth rate
(C) Mortality rate < Birth rate
(D) None of these

703. Pheromones secreted by certain exocrine skin glands, affect:
(A) Genitalia
(B) Breast
(C) Skin colour
(D) Mutual behavior of members of a species

704. In a population, arte of addition of new members is higher than the rate of individuals lost. It shows:
(A) Zero population growth (B) Declined growth
(C) Exponential growth (D) None of these

705. For success of an animal population, the most important factor is?
(A) Adaptability (B) Natality
(C) Unlimited food (D) Inter-specific activity

706. Which one has the correct sequence of increasing organizational complexity?
(A) Population, species, community and ecosystem
(B) Population, community, species and ecosystem
(C) Population, variety, community and ecosystem
(D) Species, variety, ecosystem and population

707. Population of an area tends to increase when:
(A) Reproductivity decreases
(B) Immigration increases
(C) Emigration occurs
(D) Predation increases

708. A force that acts against maximum population growth is:
(A) Population pressure
(B) Carrying capacity
(C) Saturation point
(D) Environment resistance

709. The force operating in an ecosystem which oppose unchecked growth of population is:
(A) Fecundity
(B) Mortality
(C) Biotic control
(D) Environmental resistance

710. Gene flow is:
(A) Transfer of genes between genetically distinct but interbreeding population
(B) Transfer of genes from females to males of an organism
(C) Transfer of genes from outside to chromosomes
(D) Transfer of genes from sperm to egg

711. After exponential increase, population become stagnant. The growth curve is:
(A) J-shaped (B) S-shaped
(C) Fluctuating (D) Circular

712. Allopatric speciation is caused by:
(A) Temporal isolation
(B) Adaptive radiation
(C) Geographic isolation
(D) Reproductive isolation

713. Sympatric speciation is caused by:
(A) Genetic isolation
(B) Interbreeding
(C) Geographical isolation
(D) Reproductive isolation

714. Territoriality involves:
(A) Competition (B) Cooperation
(C) Predation (D) Parasitism

715. Pheromone is:
(A) Product of endocrine gland
(B) A protein
(C) mRNA
(D) Used for animal communication

716. Percentage ratio of natality over mortality is:
(A) Population dynamics (B) Vital index
(C) Population density (D) Total count

717. Human population shows:
(A) J-shaped growth curve
(B) Z-shaped growth curve
(C) S-shaped growth curve
(D) All of these

718. Scent producing glands are:
(A) Anal glands (B) Adrenal glands
(C) Bertholin's gland (D) Prostate gland

719. Number of immigration is more than emigration and death is lower than natality. Growth curve of population will show:
(A) Exponential phase (B) Lag phase
(C) Declining phase (D) Steady phase

720. Community is:
(A) A particular species
(B) A particular group of animals
(C) All the organisms in a given area
(D) None of these

721. Mutualism and proto-cooperation are:
(A) Positive interactions
(B) Negative interactions
(C) Both of these
(D) None of these

722. Association of nitrogen fixing bacteria with the leguminous roots is:
(A) Mutualism (B) Commensalism
(C) Neutralism (D) Parasitism

723. Many of the bacteria living in the human large intestine feeding on undigested food without harming the host are:
(A) Predators (B) Commensals
(C) Symbionts (D) Parasites

724. Mycorrhiza an association of fungus and angiospermic root is referred as:
(A) Amensalism (B) Mutualism
(C) Parasitism (D) Commensalism

725. Competition between two species:
(A) Results in dying of one species
(B) Forces one of the species to live elsewhere
(C) Switch over to another kind of food
(D) All of these

726. A group of several species living together with mutual tolerance or adjustment and beneficial interactions in a natural area is known as:
(A) Species (B) Population
(C) Biotic community (D) Family

727. Interaction between the two organisms in which one organism kills and feeds on the second organism is known as:
(A) Scavenging (B) Predation
(C) Parasitism (D) Symbiosis

728. The relationship between the two organisms where one receives benefit at the cost of the other is known as:
(A) Scavenging (B) Predation
(C) Parasitism (D) Symbiosis

729. The relationship between the two organisms or species in which both the partners derive benefit from each other is known as:
(A) Scavenging (B) Predation
(C) Parasitism (D) Symbiosis

730. Competiton among the plants in a community is for:
(A) Water (B) Nutrients
(C) Sunlight (D) All of these

731. Which is a protective device in animals?
(A) Competition (B) Commensalism
(C) Mimicry (D) Symbiosis

732. Mimicry is a device for:
(A) Concealment from predator
(B) Concealment from prey
(C) Both (A) and (B)
(D) Becoming conspicuous

733. Climax community is:
(A) First stable community
(B) Unstable
(C) Unstable that develops in the beginning of succession
(D) The last stable community

734. Succession is:
(A) Series of physical changes that occur in an area
(B) Development of biotic communities on a bare area
(C) Series of biotic communities that appear in a previously bare area
(D) Both (A) and (B)

735. Dominant species represents most abundant:
(A) Herb that binds the soil and provides organic matter to it
(B) Shrub that appears for the first time
(C) First tree
(D) Species having major effect on physical environment

736. Interaction of two different organisms in which one receives benefit without the cost of the other is:
(A) Parasitism (B) Scavenging
(C) Symbiosis (D) Commensalism

737. Series changes on previously barren area is:
(A) Sere (B) Climatic climax
(C) Primary succession
(D) Secondary succession

738. What is relationship between Egret and Cattle?
(A) Commensalism (B) Protocooperation
(C) Mutualism (D) Parasitism

739. What is true of commensalism?
(A) One organism is harmed while the other is unaffected
(B) One organism is benefitted while the other is unaffected
(C) One is benefitted while the other is harmed
(D) Both are benefitted

740. A climax community is recognized by:
(A) Grasses, herbs and shrubs
(B) Trees
(C) Uniform composition (D) High biomass

741. Lichens and mosses occur in:
(A) Psammosere (B) Hydrosere
(C) Xerosere (D) Hydrarch

742. Plants and animals are living in a particular area constitute:
(A) Flora and fauna (B) Community
(C) Ecosystem (D) Ecology

743. Community is:
(A) Species structure and diversity index of an area
(B) Biomass
(C) Autotrophic and hetertrophic components
(D) All of these

744. The phenomenon of one organism being benefitted without affecting the other is:
(A) Scavenging (B) Amensalism
(C) Commensalism (D) Symbiosis

745. Plankton is:
(A) Sea scum
(B) Floating microscopic plants and animals
(C) Floating plants
(D) Any layer of aquatic organisms

746. The spread of *Opuntia* was controlled by:
(A) Allowing cattle to graze over it
(B) Allowing sheep and goats to browse over it
(C) Introducing insect
(D) Spraying herbicide

747. The relationship between algae and fungi in a lichen is:
(A) Symbiosis (B) Parasitism
(C) Commensalism (D) Protocooperation

748. Association between Sea Anemone and Hermit Crab in gastropod shell is that of:
(A) Symbiosis (B) Parasitism
(C) Commensalism (D) Protocooperation

749. Association in which one is benefitted without affecting the other is:
(A) Symbiosis (B) Sapraphytism
(C) Parasitism (D) Commensalism

750. Zoochorellae in Hydra produce an association called:
(A) Predation (B) Parasitism
(C) Commensalism (D) Mutualism

751. Symbiosis is:
(A) Mutually benefitted association
(B) Mutually harmful association
(C) Harmful to one and beneficial to other
(D) Advantageous to one but neutral to other

752. Competition for light, nutrients and space is most severe between:
(A) Closely related organisms growing in different niche
(B) Closely related organisms growing in the same area/niche
(C) Distantly related organisms growing in the same habitat
(D) Distantly related organisms growing in different niches

753. Plant succession occurring in a sandy area is:
(A) Psammosere (B) Halosere
(C) Xerosere (D) Hydrosere

754. A community which starts succession is a habitat is:
(A) Pioneer community (B) Seral community
(C) Biotic community (D) Ecosere

755. The early settlers on a barren area are:
(A) Ferns (B) Mosses
(C) Lichens (D) Diatoms

756. What does the following sequence represent:
Blue-Green Algae → Crustose Lichens → Foliose Lichens → Mosses → Shrubs → Dicotyledonous trees
(A) A food pyramid
(B) Ecological succession
(C) Genetic drift (D) Phylogeneric trend

757. Succession in a 6 metre deep pond exhibits a sequence of:
(A) Submerged Plants → Reeds → Herbs → Floating plants → Shrubs → Trees
(B) Floating plants → Submerged Plants → Reeds → Herbs → Shrubs → Trees
(C) Shrubs → Trees → Submerged Plants → Floating plants → Reeds → Herbs
(D) Submerged Plants → Floating plants → Reeds → Herbs → Shrubs → Trees

758. Last established community in a plant succession is known as:
(A) Seral community
(B) Pioneer community
(C) Ecosere
(D) Climax community

759. Intermediate community between pioneer and climax communities is called:
(A) Seral community (B) Biotic community
(C) Temporary community (D) Ecosore

760. Climax vegetation of a region is generally:
(A) Xerophytic
(B) Hydrophytic
(C) Mesophytic
(D) Depends upon the climate

761. The first plants to reappear on a burnt forest area will be:
(A) Grasses (B) Ferns
(C) Mosses (D) Liverworts

762. A mutually beneficial association necessary for survival of both partners is:
(A) Mutualism/Symbiosis
(B) Commensalism
(C) Ammensalism
(D) Both (A) and (B)

763. Which one is true?
(A) Commensalism when none of the interacting populations affect each other
(B) Symbiosis when the interaction is useful to both the population
(C) Symbiosis when neither populations affects each other
(D) Commensalism when the interaction is useful to both the populations

764. In ecological succession from pioneer to climax community, the biomass shall:
(A) Decrease
(B) Increase and then decrease
(C) No relation
(D) Increase continuously

765. Association of animals when both partners are benefitted:
(A) Colony (B) Mutualism
(C) Commensalism (D) Ammensalism

766. An association of individuals of different species living in the same habitat and having functional interactions is:
(A) Population (B) Ecological niche
(C) Biotic community (D) Ecosystem

767. Insectivorous plants are adopted to soils:
(A) Rich in water
(B) Deficient in water
(C) Deficient in nitrogenous compounds
(D) Deficient in trace elements

768. Primary succession is development of communities on:
(A) Newly exposed habitat
(B) Cleared forest area
(C) Freshly harvested crop field
(D) Pond filled after a dry season

769. A parasite living within tissue of host is:
(A) Epiphyte (B) Endophyte
(C) Ectophyte (D) None of these

770. A bird introduced from another country became a serious pest due to:
(A) Better adaptation to new area
(B) More sexual reproduction
(C) Better nesting habitats
(D) Absence of natural competition

771. An example of mutualism is:
(A) *Riccia* (B) *Selaginella*
(C) Lichen (D) *Spirogyra*

772. Which take part in symbiosis of lichen?
(A) Alga-Fungus
(B) Alga –Alga
(C) Fungus – Fungus
(D) Fungus – Gymnosperm

773. Nature of climax community depends upon:
(A) Water (B) Temperature
(C) Soil fertility (D) Climate

774. Succession showing changes in communities at a place is called:
(A) Geographical succession
(B) Biotic succession
(C) Physiographic succession
(D) Climatic succession

775. The association between ants and some members of family rubiaceae is:
(A) Ornithophily (B) Entomophily
(C) Myrmecophily (D) Anemophily

776. Plants growing on sandy soils are:
(A) Oxylophytes (B) Psammophytes
(C) Psychrophytes (D) Lithophytes

777. Desert can be converted into green land by:
(A) Tropical trees (B) Oxylophytes
(C) Halophytes (D) Psammophytes

778. Which is not a primary colonizer?
(A) Viruses (B) Lichens
(C) Blue-green algae (D) Mosses

779. Which of the following is wrong?
(A) Lichen, an association of fungus and alga, is an example of mutualism
(B) Epiphytes, using other plants for support only and not for water or food, show commensalism
(C) Sea Anemone and Hermit Crab association is an example of protocooperation
(D) Mutualism, protocooperation & commensalism cannot be included under symbiosis

780. If green plants are removed from an aquarium, the fishes will:
(A) Die (B) Increase in number
(C) Decrease in size (D) Not be affected

781. Plants obtaining nourishment from other plants by haustoria are:
(A) Epiphytes (B) Parasites
(C) Xerophytes (D) Halophytes

782. Hyperparasite is parasite which:
(A) Kills its host
(B) Completes life cycle in one host
(C) Uses host machinery for reproduction
(D) None of these

783. Fungal association of roots of higher plants in mycorrhiza is:
(A) Parasitism (B) Hyperparasitism
(C) Mutualism (D) Commensalism

784. In commensalism:
(A) Both partners are benefitted
(B) Both partners are harmed
(C) Weaker is benefitted while stronger is unharmed
(D) None of these

785. Which is not characteristic of intestinal symbionts:
(A) Feeding on extra food of host
(B) Providing vitamins to host
(C) Anaerobic respiration
(D) Aerobic respiration

786. Ecological succession on sand is:
(A) Psammomere (B) Xerosere
(C) Halosere (D) Hydrosere

787. A successful parasite is the one which:
(A) Grows rapidly
(B) Reproduces fast
(C) Sticks to host for long
(D) Makes minimum demands from its host

788. Biosphere is defined as:
(A) Part of the earth water and atmosphere which inhabits living organisms
(B) Part of the soil where animals and plants are found
(C) Atmosphere is which life exists
(D) None of these

789. Biosphere consists of:
(A) Lithosphere (B) Atmosphere
(C) Hydrosphere (D) All of these

790. Component of biosphere related with soil is:
(A) Lithosphere (B) Hydrosphere
(C) Atmosphere (D) None of these

791. Source of energy in the biosphere is:
(A) Producer (B) Decomposer
(C) Sunlight (D) Heat of earth

792. Phenomenon of rain, dew, snow i.e., different forms of water coming to earth under force of gravity is called as:
(A) Geotropism (B) Climatic factors
(C) Precipitation (D) Calcification

793. Agents which make free atmospheric nitrogen available to plants are:
(A) Lighting
(B) Free living nitrogen fixing bacteria
(C) Symbiotic nitrogen fixing bacteria
(D) All of these

794. Maintenance of soil fertility without addition of nutrition is due to:
(A) Floods (B) Crop residue
(C) Favourable temperature
(D) Microbial activity

795. A logical sequence of carbon cycle is:
(A) Producer $\rightarrow$ Consumer $\rightarrow$ Decomposer
(B) Decomposer $\rightarrow$ Producer $\rightarrow$ Decomposer
(C) Consumer $\rightarrow$ Producer $\rightarrow$ Consumer
(D) Producer $\rightarrow$ Decomposer $\rightarrow$ Consumer

796. Rise of atmospheric temperature because of high concentration of carbon dioxide is known as:
(A) Green house effect (B) Biomagnification
(C) Pollution (D) Ecotone

797. Process of evapotranspiration and precipitation is called:
(A) Carbon cycle (B) Nitrogen cycle
(C) Hydrological cycle (D) All of these

798. Earth is an :
(A) Open system (B) Close system
(C) Both open and close system
(D) Neither open nor close system

799. Biogeochemical cycles are also known as:
(A) Sedimentary cycles (B) Gaseous cycles
(C) Material cycles (D) Cycles of water

800. In one year, one hectare of forest shall:
(A) Consume 10 tonnes of CO_2 & evolve 10 tonnes of O_2
(B) Consume 20 tonnes of CO_2 & evolve 10 tonnes of O_2
(C) Consume 30 tonnes of CO_2 & evolve 30 tonnes of O_2
(D) Consume 30 tonnes of CO_2 & evolve 10 tonnes of O_2

801. Amount of carbon fixed annually through photosynthesis is:
(A) 4.6×10^{10} kg (B) 7×10^{13} kg
(C) 7×10^{20} kg (D) 4.6×10^{21} kg

802. Nonbiological nitrogen fixation is:
(A) Electrochemical and photochemical
(B) Rhizobial
(C) Cyanobacterial (D) None of these

803. Climate of the world is threatened by:
(A) Increasing concentration of atmospheric oxygen
(B) Decreasing amount of atmospheric oxygen
(C) Increasing amount of atmospheric carbon dioxide
(D) Decreasing amount of atmospheric carbon dioxide

804. Green-house gases are:
(A) Absorbers of long wave radiation from earth
(B) Transparent to both solar radiation and long wave radiation from earth
(C) Absorbers of solar radiation for warming the atmosphere of earth
(D) Transparent to emissions from earth for passage into outer space

805. Pseudomonas is an important component of nitrogen cycle . It:
(A) Fixes elemental nitrogen
(B) Produces elemental nitrogen
(C) Transfers nitrogen
(D) Changes ammonium nitrogen to nitrate state

806. Water cycle is made up of two overlapping cycles:
(A) Ground water atmospheric cycle
(B) Surface water and atmospheric cycles
(C) Global and local cycles
(D) Oceanic and fresh water cycles

807. Sedimentation cycle is made up of two overlapping cycles:
(A) Phosphorus (B) Nitrogen
(C) Carbon (D) Sulphur

808. Earth is protected from ultra-violet radiations by means of:
(A) Ozone layer (B) Nitrogen
(C) Oxygen (D) Carbon dioxide

809. Ozone layer occurs in:
(A) Ionosphere (B) Stratosphere
(C) Thermosphere (D) Troposphere

810. Ecosystem obtained nitrogen from outside through the process of:
(A) Industrial fixation
(B) Electrochemical and photochemical fixation
(C) Biological fixation
(D) All of these

811. Earth is an open system for:
(A) Minerals (B) Energy
(C) Gases (D) Water

812. Lithosphere is reservoir of:
(A) Carbon (B) Oxygen
(C) Phosphorus (D) Nitrogen

813. Amount of nitrogen fixed by nitrogen fixing organisms is:
(A) 70% (B) 30%
(C) 50% (D) 90%

814. Carbon is found in lithosphere as:
(A) Carbohydrate (B) Carbonic acid
(C) Carbon (D) Carbonate

815. Which has a sedimentary cycle:
(A) Nitrogen (B) Carbon
(C) Water (D) Sulphur

816. Maximum control of hydrological cycle is exerted by:
(A) Epiphytes (B) Forests
(C) Plankton (D) Grassland

817. Ecological cycle directly driven by sunlight is:
(A) Carbon (B) Phosphorus
(C) Hydrologic (D) Nitrogen

818. Biosphere is:
(A) Area occupied by living beings
(B) Plants of the world
(C) Plants of a particular area
(D) Special plants

819. Source of carbon to plants is:
(A) Carbon dioxide of carbonate rocks
(B) Atmospheric carbon dioxide
(C) Fossil fuels (D) All of these

820. Biogeochemical cycling refers to cycling of:
(A) Energy in the ecosystem
(B) Nutrient in the ecosystem
(C) Water
(D) Gases between plants and animals

821. Gas being produced by paddy fields and involved in global warming is:
(A) Chlorine (B) Methane
(C) CO_2 (D) H_2S

822. Green house effect is warming due to:
(A) Infra-red rays reaching earth
(B) Moisture layer in atmosphere
(C) Increase in temperature due to increase in carbon dioxide concentration of atmosphere
(D) Ozone layer of atmosphere

823. Green house effect is related to:
(A) Increased growth of green algae
(B) Global warming
(C) Cultivation of vegetables in houses
(D) Development of terrace gardens

824. Green house gases include:
(A) CO_2, CFC, CH_4 and NO_2
(B) CO_2, O_2, N_2, NO_2 and NH_3
(C) CH_4, N_2, CO_2 and NH_3
(D) CFC, CO_2, NH_3 and N_2

825. The circulation or cycling of elements in an ecosystem is known as:
(A) Geological cycling
(B) Geo-chemical cycling
(C) Bio-geochemical cycling
(D) Chemical cycling

826. The sphere of living matter found on the surface of earth comprises:
(A) Lithosphere (B) Hydrosphere
(C) Atmosphere (D) Biosphere

827. The term biosphere is used for a zone containing living beings:
(A) In lithosphere, hydrosphere and atmosphere
(B) In lithosphere and hydrosphere
(C) In hydrosphere
(D) On lithosphere

828. Biosphere is made of:
(A) Living being and their remains
(B) Living beings + Lithosphere + Hydrosphere + Atmosphere
(C) Living beings + Lithosphere
(D) Living organisms + Lithosphere + Hydrosphere

829. Which is not a green house gas:
(A) Chlorofluorocarbon (B) Methane
(C) Carbon dioxide (D) Hydrogen

830. Concentration of green house gases is increasing due to:
(A) Deforestation
(B) Increased use of refrigerators
(C) Increased combustion of coal and petroleum
(D) All of these

831. Which gas contributes most to green house effect:
(A) CFC (B) Freon
(C) CO_2 (D) CH_4

832. Which gas of the atmosphere holds up ultraviolet rays:
(A) O_2 (B) O_3
(C) N_2 (D) CO_2

833. Checking of re-radiating heat by atmospheric dust, O_3, CO_2 and water vapours is:
(A) Ozone layer effect
(B) Solar effect
(C) Radioactive effect
(D) Green house effect

834. Liberation of H_2S/foul smell in water of tanks, ponds, etc, is due to:
(A) Anaerobiosis
(B) Aerobiosis
(C) Psammophytes
(D) Biological magnification

835. CO_2 and O_2 balance in atmosphere is due to:
(A) Photosynthesis (B) Respiration
(C) Leaf anatomy (D) Photorespiration

836. Biosphere is:
(A) A protected area
(B) Ecosystem
(C) In which individuals interact
(D) Earth crust

837. Fertilizer obtained from Sea Birds along the coast of Chile and Peru is:
(A) Guana (B) Bone meal
(C) Dung (D) Urea

838. Nitrogen fixation is:
(A) Nitrogen $\rightarrow$ Ammonia
(B) Nitrogen $\rightarrow$ Nitrates
(C) Nitrogen $\rightarrow$ Amino acid
(D) Both (A) and (B)

839. Ultraviolet radiation from sunlight cause a reaction which produces:
(A) O_3 (B) SO_2
(C) CO (D) CH_4

840. Nitrogen is a critical element of the ecosystem because it is:
(A) Essential element
(B) Abundant is atmosphere
(C) Labile
(D) Fixed by microbes

841. Carbon dioxide content of atmospheric has increased in the past 150 years from:
(A) 20-35 ppm (B) 270-340 ppm
(C) 0.027 – 0.034 ppm (D) 0.2 – 0.3 ppm

842. Carbon dioxide content of atmospheric is about:
(A) 0.0034% (B) 3.334%
(C) 0.34% (D) 0.034%

843. Zone of atmosphere near the ground is:
(A) Troposphere (B) Stratosphere
(C) Mesosphere (D) Homosphere

844. Role of bacteria in carbon cycle is:
(A) Photosynthesis
(B) Chemosynthesis
(C) Breakdown of organic compounds
(D) Assimilation of nitrogen compounds

845. Which green–house gas other than methane is being produced by agricultural fields is:
(A) Nitrous oxide (B) Ammonia
(C) Sulphure dioxide (D) Arsenic

846. In case CO_2 of earth's atmosphere disappears, the temperature of earth's surface would:
(A) Increase
(B) Decrease
(C) Depend on oxygen concentration
(D) Remain the same

847. Green house effect is mostly due to:
(A) CO_2 (B) CO
(C) NO_2 (D) H_2O

848. Old leguminous roots turn green due to:
(A) Conversion of leghaemoglobin to chl a
(B) Breakdown of leghaemoglobin
(C) Conversion of leghaemoglobin to chl b
(D) Conversion of leghaemoglobin to chl c

849. Most vital process for existence of life on earth is:
(A) Reproduction of plants and animals
(B) Respiration of animals
(C) Photosynthesis by plants
(D) Communication in animals

850. Which one depicts nitrogen fixation:
(A) $N_2 + 3H_2 \rightarrow 2NH_3$
(B) $2NH_3 \rightarrow N_2 + 3H_2$
(C) $2N_2$ + Glucose $\rightarrow$ 2 amino acids
(D) $2NH_4+ + 2O_2 + 8e- \rightarrow N_2 + 4H_2O$

851. Green house effect is due to:
(A) CO_2
(B) CH_4, SO_2 and NO_2
(C) Combinations of various gases
(D) Gases which absorb infra-red rays reflected from earth

852. Instrument used for measuring atmospheric humidity is:
(A) Potometer (B) Hygrometer
(C) Porometer (D) Auxanometer

853. Which one has the least green house effect?
(A) Carbon dioxide (B) Nitrogen gas
(C) Water vapours (D) Ozone

854. Belt on which life exists above and below the surface of earth is:
(A) Biotic belt (B) biome
(C) Biosphere (D) Ecosystem

855. The limiting factor in soil nitrification is:
(A) Soil pH (B) Light
(C) Temperature (D) Air

856. Green house effect is increasing due to:
(A) Increasing CO_2 concentration
(B) Increasing O_2 concentration
(C) Increasing concentration of chlorofluorocarbon
(D) Hole in ozone layer

857. Atmosphere consists of:
(A) Lithosphere + hydrosphere
(B) Lithosphere + Stratosphere + Hydrosphere
(C) Troposphere + Stratosphere + Ionosphere
(D) None of these

858. Environment can be improved by:
(A) Wind erosion (B) Excessive tree fall
(C) Conservation (D) None of these

859. Pollutants directly emanating from human activities are:
(A) Quantitative pollutants
(B) Qualitative pollutants
(C) Air pollutants
(D) Water pollutants

860. Which will not cause atmospheric pollution?
(A) SO_2 (B) CO_2
(C) CO (D) H_2

861. Acid rain is caused by:
(A) Excess production of coal gas
(B) Excess release of CO_2 due to increasing combustion and respiration
(C) Excess release of SO_2 and NO_2 from burning fossil fuels
(D) Excess production of gaseous hydrocarbons

862. SO_2 pollution is indicated by destruction of:
(A) Climbers (B) Lichens
(C) Mosses (D) Grasses

863. Sulphurdioxide affects:
(A) Cell wall
(B) Plasmamodesmata
(C) All membrane systems
(D) Nucleus

864. Addition of phosphorus and nitrates/fertilizers into water leads to:
(A) Increase growth of decomposers
(B) Reduced algal growth
(C) Increased algal growth
(D) Nutrient enrichment/eutrophication

865. Spraying of DDT produces pollution of:
(A) Air (B) Air & water
(C) Air & soil (D) Air, soil & water

866. Most abundant water pollutant is:
(A) Detergents (B) Pesticides
(C) Industrial wastes (D) Ammonia

867. Atmospheric pollutant is:
(A) CO_2 (B) CO
(C) O_2 (D) N_2

868. Domestic waste constitutes:
(A) Nonbiodegradable pollution
(B) Biodegradable pollution
(C) Effluents
(D) Air pollution

869. Effect of pollution is most marked on:
(A) Natural balance of nature
(B) Natural geochemical cycles
(C) Natural flora of place
(D) All of these

870. Chief source of water and soil pollution is:
(A) Mining
(B) Thermal power plant
(C) Agroindustry
(D) All of these

871. Fish die in water bodies polluted by sewage due to:
(A) Clogging of gills by silt (B) Pathogenes
(C) Reduction in oxygen (D) Foul smell

872. Pollution can bring about change in:
(A) Biogeochemical cycling (B) Abiotic environment
(C) Biotic environment (D) None of these

873. Which one is the major reason for pollution in big cities?
(A) Fossil fuel (B) Acid rain
(C) Heat dispersion (D) None of these

874. Heat islands are formed by:
(A) Water pollution (B) Air pollution
(C) Land pollution (D) All of these

875. Biodegradable pollutant is:
(A) Plastic (B) Asbestos
(C) Sewage (D) Mercury

876. Carbon dioxide is a major pollutant in:
(A) Air (B) Soil
(C) Water (D) Noise

877. Sudden mass death of fish is more likely to occur in:
(A) Eutrophic lake (B) Mesotrophic lake
(C) Oligotrophic lake (D) Oxalotrophic lake

878. Eutrophication causes reduction in:
(A) Dissolved hydrogen (B) Dissolved oxygen
(C) Dissolved salts (D) All of these

879. Continuous sewage flow into a stream would lead to:
(A) Eutrophication
(B) Algal bloom
(C) Depletion of oxygen
(D) Increase in temperature

880. Indiscriminate use of fertilizers causes:
(A) Air pollution (B) Water pollution
(C) Land pollution (D) All of these

881. Which pigment protects plants from ultra violet damage?
(A) Chlorophyll (B) Xanthophyll
(C) Carotenoids (D) Phycocyanin

882. Most harmful type of environmental pollutants are:
(A) Human organic waste
(B) Natural nutrients in excess
(C) Waste animal feed
(D) Nonbiodegradable chemicals

883. Environmental pollution affects:
(A) Biotic components
(B) Plants only
(C) Man only
(D) Biotic and abiotic components of environment

884. SO_2 pollution causes:
(A) Plasmolysis
(B) Destruction of cell wall
(C) Destruction of golgi bodies
(D) Destruction of chlorophyll

885. A non-conventional energy source is:
(A) Coal (B) Petroleum
(C) Solar radiation
(D) Electricity from nuclear plants

886. Sewage treatment in which a portion of decomposer bacteria present in waste is recycled into the beginning of the process:
(A) Cyclic treatment
(B) Activated sludge treatment
(C) Primary treatment
(D) Tertiary treatment

887. Use of lichens in case of pollution is:
(A) Treatment of pollutant water
(B) Bioindicators of pollution
(C) Promote pollution
(D) Lichens have no relation with pollution

888. When is a man a part of biotic community?
(A) Always
(B) Only when he lives in big cities
(C) Only when he lives in a forest
(D) Never

889. Metallic minerals and fossil fuels are:
(A) Non-renewable resources
(B) Inexhaustible resources
(C) Renewable
(D) None of these

890. Migratory birds are a resource that is:
(A) Inorganic (B) Multinational
(C) International (D) National

891. Conservation is:
(A) Judicious use of natural resources
(B) Protection of natural resources
(C) Management of natural resources
(D) All of these

892. Vegetation of a place is mainly determined by:
(A) Rainfall (B) Soil water
(C) Soil type (D) Amount of light

893. Which one is biotic in origin?
(A) Texture and porosity of soil
(B) Rainfall
(C) Carbon dioxide content of soil
(D) Light intensity

894. Best source of renewable energy is:
(A) Cattle (B) Petroleum
(C) Coal (D) Trees

895. A nonrenewable resources is:
(A) Forest (B) Coal/petroleum
(C) Water (D) Wild life

896. Main renewable source of energy is:
(A) Forest (B) Fuel gas
(C) Cattle (D) Coal

897. Geothermal energy is:
(A) Nonrenewable nonconventional energy source
(B) Nonrenewable conventional energy source
(C) Renewable nonconventional energy source
(D) Renewable conventional energy source

898. Minerals and metals are:
(A) Biodegradable resources
(B) Renewable
(C) Non-renewable
(D) Renewable and nonrenewable resources

899. L.P.G. is:
(A) Low Price Gas (B) Fossil fuel
(C) Low Pressure Gas (D) Biogas

900. Water is a resource:
(A) Nondegradable nonmaintainable
(B) Degradable maintainable
(C) Renewable
(D) Non-renewable

901. Sequence of humification and mineralization is:
(A) Dead organic matter → Litter → Duff → Humus → Minerals
(B) Humus → Minerals → Litter → Duff
(C) Minerals → Humus → Litter → Duff
(D) Dead organic matter → Duff → Litter → Minerals → Humus

902. Ecofriendly method is:
(A) Plantation of C_3 plants
(B) Plantation of sugarcane
(C) Energy plantation
(D) None of these

903. Edaphology is connected with:
(A) Plant and biosphere
(B) Soil and living microorganisms
(C) Animals and ecosystem
(D) Soil and biosphere

904. Which of the following means endemic?
(A) Indigenous (B) Exotic
(C) Ubiquitous (D) Naturalized

905. Those species whose populations have been seriously depleted and whose ultimate security is not assured are known as:
(A) Threatened species (B) Endangered species
(C) Vulnerable species (D) Rare species

906. First biosphere reserve was established in 1986 in:
(A) Nilgiri (B) Nanda Devi
(C) Little Rann of Kutchh (D) Sunderbans

907. A threatened species is:
(A) Endangered (B) Rare
(C) Vulnerable (D) All of these

908. MAB is:
(A) Man and Botany
(B) Man And Biosphere
(C) Man and Biotic community
(D) Man, Antibiotics and Bacteria

909. MAB programme of UNESCO was launched in:
(A) 1971 (B) 1981
(C) 1991 (D) 2001

910. Forests control drought through:
(A) Lot of water plants
(B) Increasing rainfall
(C) Retention of water and prevention of erosion
(D) Functioning as water shed

911. Which one is connected with conservation of forests?
(A) Gir (B) Silent valley
(C) Kaziranga (D) Bharatpur

912. New approach to conservation is establishment of:
(A) Reserve forests (B) Sanctuaries
(C) Biosphere reserves (D) National park

913. Social forestry is:
(A) Management of forests by cooperative societies
(B) Growing and management of useful plants on common lands
(C) Growing one type of trees
(D) Growing different types of plants together

914. Deforestation is the major causal agent of:
(A) Depletion of natural resources
(B) Environmental pollution
(C) Desertification of habitat
(D) Genetic erosion

915. In a National Park protection is provided to:
(A) Entire ecosystem (B) Flora and fauna
(C) Fauna only (D) Flora only

916. Which animals should be protected?
(A) Harmless (B) Economically useful
(C) Likely to perish (D) Ferocious

917. Animal species should be protected because they are:
(A) Lovely creatures
(B) Useful to humans
(C) To be studied by zoologists
(D) Man cannot recreate a lost species

918. Wild life is destroyed most when:
(A) There is lack of proper care
(B) Mass scale hunting for foreign trade
(C) Its natural habitat is destroyed
(D) Natural calamity

919. Deforestation doe not lead to:
(A) Quick nutrient cycling
(B) Soil erosion
(C) Alteration of local weather conditions
(D) Destruction of natural habitat of wild animals

920. Animals and plants are best protected in:
(A) Botanical gardens (B) Zoos
(C) National parks (D) Sanctuaries

921. Biosphere reserve project was started in India during:
(A) 1984 (B) 1985
(C) 1986 (D) 1989

922. If we remove half of the forest cover of earth, the crises that will occur:
(A) Many species would become extinct
(B) Population, pollution and ecological imbalance will rise
(C) Energy crises will commence
(D) The remaining in evaporating are near equator

923. Plant species on verge of extinction due to over-exploitation is:
(A) *Centella* (B) *Podophyllum*
(C) *Gloriosa* (D) All of these

924. WWF-N has logo for conservation of:
(A) Tiger (B) Giant Panda
(C) Red Panda (D) Polar Bear

925. Increased incidence of floods in the plains of North India are due to:
(A) Increased deforestation in catchment areas
(B) Increase in incidence of rainfall
(C) Silting of dams
(D) More area under cultivation

926. 'Red Data Book' provided data on:
(A) Biota of Red Sea
(B) Effect of red light or photosynthesis
(C) Red pigmented plants
(D) Threatened species

927. Which communities are more vulnerable to invasion by outside plants and animals?
(A) Tropical evergreen forests
(B) Temperate forests
(C) Mangroves
(D) Oceanic island communities

928. Adherence to two-children norm shall help on:
(A) Improving food web
(B) Checking pollution
(C) Stabilising ecosystem
(D) Fertility of soil

929. Jhoom cultivation refers to:
(A) Cultivation of Jamun trees
(B) Cultivation of medicinal plants by tribes
(C) Tribal method of shifting cultivation
(D) Cultivation of terraces

930. Species restricted to a given area is:
(A) Endemic species
(B) Allopatric species
(C) Sympatric species
(D) Sibling species

931. Which is mainly responsible for extinction of wildlife?
(A) Hunting of flesh
(B) Destruction of habitats
(C) Pollution of air and water
(D) All of these

932. Biosphere reserves are being threatened with:
(A) Population growth (B) Rains
(C) Population (D) All of these

933. Endangered plant species are conserved through:
(A) Herbarium (B) Gene library
(C) Gene Bank (D) Reducing pollution

934. Which one is endangered species?
(A) *Cuscuta* (B) *Acacia nilotica*
(C) *Nepenthes* (D) Both (B) and (C)

935. Afforestation in relation to urban development is component of:
(A) Social forestry
(B) Integrated forest development
(C) Agroforestry (D) Fuel forestry

936. Wildlife is conserved:
(A) In situ (B) Ex situ
(C) Both (A) and (B)
(D) Selective hunting of predators

937. The term succession was coined by:
(A) Buffon (1742) (B) King (1685)
(C) Halt (1885) (D) Cowles (1899)

938. Succession on sand dune is:
(A) Lithosphere (B) Psychrosere
(C) Psammosere (D) Halosere

939. Which relationship is beneficial to both interacting organisms but is not obligatory:
(A) Commensalism (B) Protocooperation
(C) Mutualism (D) Both (B) and (C)

940. The term homeostatsis was coined by:
(A) Whittaker (B) Watt
(C) Elton (D) Cannon

941. The amount of biogenic nutrient present in the abiotic environment is called:
(A) Standing state (B) Standing crop
(C) Detritus (D) Reserve nutrient

942. Phosphorus has:
(A) Imperfect nutrient cycle
(B) Perfect nutrient cycle
(C) Gaseous cycle (D) Mixed cycle

943. Carbosphere comprises:
(A) Atmosphere (B) Lithosphere
(C) Hydrosphere (D) All of these

944. Carbon dioxide is being added to the cycling pool by:
(A) Respiration and combustion
(B) Sedimentation of carbonates
(C) Precipitation, skeletons and fossilization
(D) Both (B) and (C)

945. A process that occurs without loss or gain of heat is:
(A) Coriolis effect (B) Eustatic effect
(C) Adiabatic (D) All of these

946. A source of nonbiological oxygen is:
(A) Photodissociation
(B) Photolysis
(C) Formation of fats from carbohydrates
(D) Both (B) and (C)

947. Adiabatic expansion resulting in cooling is called:
(A) Coriolis effect
(B) Joule-Kelvin effect
(C) Le Chatelier Principle
(D) Principle of Parsimony

948. Branch of biology dealing with distribution of plants and animals on earth surface and their inter-relationship is termed:
(A) Ecology (B) Phytology
(C) Zoogeography (D) Phytosociology

949. The most favoured habitat of plant is called:
(A) Ecological system (B) Ecological type
(C) Ecological habit (D) Microhabitat

950. Biosphere is:
(A) In which individuals interact with each other
(B) In which life originated
(C) The name of bird
(D) Organic compounds by which life diminishes

951. The main subdivision of global environment which includes all water termed as:
(A) Hydrosphere (B) Lithosphere
(C) Atmosphere (D) All of these

952. The entire inhabited part of the earth and its atmosphere including the living components forms:
(A) Biosphere (B) Atmosphere
(C) Lithosphere (D) Hydrosphere

953. Which of the following is a correct sequence of carbon cycle?
(A) Consumer → Photosynthesis → Decomposers
(B) Photosynthesis → Decomposers → Consumer
(C) Photosynthesis → Consumer→ Decomposers
(D) Decomposers →Consumer →Photosynthesis

954. Man occupies a unique position in biosphere:
(A) Large volume of brain (B) Skilled hand
(C) Superb memory (D) All of these

955. The cycling of elements in an ecosystem is called:
(A) Chemical cycle (B) Geo-chemical cycle
(C) Biogeochemical cycle (D) Geological cycle

956. Which of the following is a renewable source of energy?
(A) Petroleum (B) Coal
(C) Natural gas (D) Trees

957. Biosphere like a balanced aquarium is:
(A) Closed system
(B) An open system
(C) Partly open and partly closed system
(D) None of these

958. Which of the following is chemically inert?
(A) Sand particles (B) Clay particles
(C) Alluvial particles (D) All of these

959. Which of the following cycle would be affected if decomposers of an ecosystem vanish?
(A) Producers cycle
(B) Biogeochemical cycle
(C) Consumer cycle
(D) Decomposers cycle

960. Trowaway societies are:
(A) Todays's advanced industrialized countries
(B) Today's most polluting countries
(C) Asia and Australia
(D) Any of these

961. Biotic community can be grouped into:
(A) Producers (B) Consumers
(C) Decomposers (D) All of these

962. Which one of the following is an indirect ecological factor?
(A) Soil structure (B) Wind
(C) Light (D) Altitude

963. Types of forests found near equator are:
(A) Tropical rain forests (B) Deciduous forests
(C) Grasslands (D) Coniferous forests

964. Which of the following contribute to the carbon cycle?
(A) Respiration (B) Photosynthesis
(C) Fossil fuel combustion (D) All of these

965. A biogeochemical cycle without an atmospheric component is:
(A) Sulphur (B) Nitrogen
(C) Phosphorus (D) Oxygen

966. Green house effect was discovered by:
(A) Tyndall (B) Tansley
(C) Clements (D) Stanley

967. Which of the following is only source of energy which keeps natural ecosystems in a actual state?
(A) Nuclear power (B) Sun-light
(C) Thermal power (D) All of these

968. Organisms of an ecosystem are linked together in:
(A) Food chains (B) Food webs
(C) Biotic communities (D) All of these

969. Ecology of individual species and its population including the effect of other organisms and environment conditions on every stage of life cycle is known as:
(A) Autoecology (B) Genecology
(C) Synecology (D) None of these

970. An intrinsic property of all natural and complete ecosystem is:
(A) Self – sustainability (B) Tropical ecology
(C) Reasonable chains (D) None of these

971. Which one of the following is a direct ecological factor?
(A) Light (B) Temperature
(C) Atmosphere (D) All of these

972. The term biosphere is used for a zone containing living beings:
(A) In lithosphere, hydrosphere and atmosphere
(B) In lithosphere and hydrosphere
(C) In hydrosphere
(D) On lithosphere

973. Biosphere is made of:
(A) Living beings and their remains
(B) Living beings + lithosphere + hydrosphere + atmosphere
(C) Living beings + lithosphere
(D) Living organisms + lithosphere + hydrosphere

974. Biogeochemical cycle means:
(A) Cycling of energy in an ecosystem
(B) Cycling of gases between plants and the atmosphere
(C) Cycling of nutrients in an ecosystem
(D) Cycling of water

975. Green house effect is related to:
(A) Increased growth of green plants
(B) Global warming
(C) Cultivation of vegetables in houses
(D) Development of terrace gardens

976. Maximum control of hydrological cycle is exerted by:
(A) Epiphytes (B) Forests
(C) Plankton (D) Grassland

977. The circulation or cycling of elements in an ecosystem is known as:
(A) Geological cycling
(B) Geo-chemical cycling
(C) Bio-geochemical cycling
(D) Chemical cycling

978. Ecological cycle directly driven by sunlight is:
(A) Carbon (B) Phosphorus
(C) Hydrologic (D) Nitrogen

979. Biosphere constitutes:
(A) All living beings on earth
(B) Part of earth and its atmosphere inhabited by living organisms
(C) All parts on earth
(D) All animals on earth

980. Green houses effect is warming due to:
(A) Infra-red rays reaching earth
(B) Moisture layer in atmosphere
(C) Increase in temperature due to increase in carbon dioxide concentration of atmosphere
(D) Ozone layer of atmosphere

981. Biosphere is:
(A) Area occupied by living beings
(B) Plants of the world
(C) Plants of a particular area
(D) Special plants

982. Biogeochemical cycling refers to cycling of:
(A) Energy in the ecosystems
(B) Nutrients in the ecosystem
(C) Water
(D) Gases between plants and animals

983. Those species whose populations have been seriously depleted and whose ultimate security is not assured are known as:
(A) Threatened species (B) Endangered species
(C) Vulnerable species (D) Rare species

984. Silent valley (Kerala) is preserved because:
(A) It contains timber trees
(B) It is the best place for picnic
(C) It contains very rare species of plants and animals
(D) All of these

985. Those species whose populations have been seriously depleted and whose ultimate security:
(A) Threatened species (B) Endangered species
(C) Vulnerable species (D) Rare species

986. Conservation is:
(A) Protection of natural resources
(B) Management of natural resources
(C) Proper use of natural resources
(D) All of these

987. Which one is concerned with conservation of forests:
(A) Gir (B) Silent valley
(C) Kaziranga (D) Bharatpur

988. Animal species should be protected because they are:
(A) Lovely creatures
(B) Useful to humans
(C) To be studied by zoologists
(D) Man cannot recreate a lost species

989. The energy flow in an ecosystem is:
(A) Cyclic (B) Unidirectional
(C) Multidirectional (D) Square

990. Tiger project is a holistic approach for the conservation of:
(A) Ecosystem (B) Tiger
(C) Deer (D) Birds

991. The criteria for the selection of species under ex-situ conservation is/are:
(A) Vulnerability of the species to extinction
(B) Ecological importance of the species
(C) Economic & aesthetic importance of the species
(D) All of these

992. The establishment of national parks and sanctuaries come under:
(A) in-situ conservation of biodiversity
(B) ex-situ conservation of biodiversity
(C) Natural development plan of biodiversity
(D) All of these

993. Who coin the term biosphere?
(A) Odum (B) Haeckel
(C) Lindegren (D) Vernadsky

994. The main nitrogen reservoir in the biosphere is:
(A) Atmosphere (B) Ocean
(C) Organisms (D) Rocks

995. The largest reservoir of sulphur in the biosphere is the:
(A) Atmosphere (B) Ocean
(C) Rocks (D) Organisms

996. The largest reservoir of phosphorus in the biosphere is the:
(A) Atmosphere (B) Ocean
(C) Organisms (D) Rocks

997. Basic unit of ecological hierarchy is:
(A) Organisms (B) Biomes
(C) Microbes (D) Ecology

998. A distinct unit of life in nature:
(A) Organisms (B) Biomes
(C) Microbes (D) Ecology

999. The structural and functional unit of biosphere is:
(A) Mass (B) Environment
(C) Ecosystem (D) All of these

1000. The relationship between biotic and abiotic environment:
(A) Biogenesis (B) Symbiosis
(C) Holocoenosis (D) None of these

1001. The rate of biomass production is called:
(A) Productivity (B) Production
(C) Biodiversity (D) All of these

1002. The rate of total production of organic matter by the producer per unit area and time is:
(A) Primary productivity
(B) Secondary productivity
(C) Net primary productivity
(D) Gross primary productivity

1003. The rate at which organic matter is stored after the basic metabolic processes is:
(A) Primary productivity
(B) Secondary productivity
(C) Net primary productivity
(D) Gross primary productivity

1004. The productivity that is left after meeting the cost of respiration and maintenance of producers is:
(A) Primary productivity
(B) Secondary productivity
(C) Net primary productivity
(D) Gross primary productivity

1005. The rate of increase in the biomass of consumers per unit area and time is:
(A) Primary productivity
(B) Secondary productivity
(C) Net primary productivity
(D) Gross primary productivity

1006. Groups of species exploiting a common resources are called:
(A) Builds (B) Guilds
(C) Community (D) Population

1007. The geographical representation of ecological parameters is called:
(A) Ecological Pyramid (B) Ecology
(C) Ecological hill (D) Ecological mountain

1008. Ecological pyrmid are also called as:
(A) Eltonian pyramids (B) Charles pyramids
(C) hierarchical pyramids (D) All of these

1009. A population genetically adapted to its particular environment is known as:
(A) Ecotone (B) Ecotype
(C) Ecosystems (D) Ecology

1010. The environment which has been modified by human activities is called:
(A) Anthropogenic environment
(B) Natural environment
(C) Manmade environment
(D) Both (A) and (B)

1011. The green plants are called:
(A) Producers (B) Consumers
(C) Decomposers (D) All of these

1012. What is the components of an ecosystem are?
(A) Biotic (B) Abiotic
(C) Both (A) and (B) (D) None of these

1013. The largest unit of living organisms on earth is:
(A) Biosphere (B) Ecotone
(C) Ecosystem (D) Biome

1014. An eco-friendly slogan:
(A) 'Do or die'
(B) 'Birds of same feather fly together'
(C) 'Time and tide wait for none'
(D) 'Say no to crackers'

1015. An alternative source of energy to fuel energy:
(A) Tidal energy
(B) Petroleum
(C) Nuclear energy
(D) Compressed natural gas

1016. Which Biome is characterized by rapid nutrient cycling and high biomass?
(A) Deserts (B) Deciduous forests
(C) Tropical rain forests (D) All of these

1017. Biotic environment includes:
(A) Producers (B) Consumers
(C) Decomposers (D) All of these

1018. Which biomes is characterized by very large, mostly evergreen trees and ultisols?
(A) Equatorial (B) Tropical forests
(C) Both (A) and (B) (D) None of these

1019. Eco-labelling is governed by:
(A) ISO 14001 (B) ISO 14010
(C) ISO 14020 (D) ISO 14002

1020. All of the members of a particular species that live in one area are called a/an:
(A) Biome (B) Population
(C) Community (D) Ecosystem

1021. Eco-auditing focuses on:
(A) Assessment of actual effects of exist activities
(B) Pollution effects
(C) Climatic effects
(D) Water management

1022. What is the original source of almost all the energy in most ecosystems?
(A) Carbohydrates (B) Water
(C) Carbon (D) Sunlight

1023. What is an ecological model of the relationships that form a network of complex interactions among organisms in a community from producers to decomposers?
(A) Food web (B) An ecosystem
(C) food chain (D) A population

1024. The series of predictable changes that occurs in a community over time is called:
(A) Population growth
(B) Ecological succession
(C) Climax community
(D) Climate change

1025. Ecotone is:
(A) Ecosystem
(B) A zone of an ecosystem
(C) An ecological study
(D) A zone between two ecosystem

1026. A symbiosis in which both species benefit is:
(A) Commensalism (B) Mutualism
(C) Predation (D) Parasitism

1027. The repeated movement of water between Earth's surface and the atmosphere is called:
(A) Water cycle (B) Nitrogen cycle
(C) Carbon cycle (D) Phosphorus cycle

1028. Two species can occupy the same niche in the same habitat at the same time:
(A) Because of the interactions that shape the ecosystem
(B) Unless the species require different abiotic factors
(C) Because of the competitive exclusion principle
(D) Unless the species require different biotic factors

1029. A biotic or an abiotic resource in the environment that causes population size to decrease is a:
(A) Carrying capacity
(B) Limiting nutrient
(C) Limiting factor
(D) Growth factor

1030. As resources in a population become less available, population growth:
(A) Decline rapidly
(B) Increases slowly
(C) Reaches carrying capacity
(D) Enters a phase of exponential growth

1031. Which are two ways a population can decrease in size?
(A) Immigration and emigration
(B) Increased death rate and immigration
(C) Decrease birth rate and immigration
(D) Immigration and increased birthrate

1032. A biome is identified by its particular set of abiotic factors and its:
(A) Average precipitation and temperature
(B) Characteristics ecological community
(C) Distance from the equator
(D) Specific geographical location

1033. In a logistic growth curve, exponential growth is the phase in which the population:
(A) Reaches carrying capacity
(B) Grows quickly and few animals are dying
(C) Growth begins to slow down
(D) Growth stops

1034. When individuals in a population reproduce at a constant rate, it is called:
(A) Logistic growth (B) Growth density
(C) Exponential growth (D) Multiple growth

1035. Which two biomes have the least amount of precipitation?
(A) Tropical rain forest and temperate grassland
(B) Tropical savanna and tropical dry forest
(C) Tundra and desert
(D) Boreal forest and temperate woodland and shrub land

1036. One of the main characteristics of a population is its:
(A) Change over time
(B) Geographical distribution
(C) Dynamics
(D) Habitat

1037. Group of individuals of the same species that share common attributes are called:
(A) Community (B) Population
(C) Ecotype (D) Society

1038. Alpha diversity represents:
(A) Species richness (B) Species evenness
(C) Species richness and evenness (D) Species richness and dominance

1039. Energy flow in an ecosystem is:
(A) Cyclic (B) Unidiractional
(C) Multidirectional (D) Sequential

1040. The major type of flora and fauna found in aquatic ecosystem are determined by:
(A) Light (B) Temperature
(C) Soil (D) All of these

1041. The ocean zone that covers the continental shelf is the:
(A) Estuary (B) Coastal zone
(C) Littoral zone (D) Benthic zone

1042. The ecosystems with the world's highest net primary productivities per unit area are found in the:
(A) Estuary (B) Coastal zone
(C) Littoral zone (D) Benthic zone

1043. Organisms of the intertidal zone:
(A) Are constantly being swept away
(B) Are exposed to constant salinity levels
(C) Are exposed to both air and underwater conditions
(D) Must avoid drowning at high tide

1044. What is CITES?
(A) A treaty banning barrier island development
(B) A treaty banning trade in endangered species
(C) A treaty banning dumping plastic in the ocean
(D) A treaty banning over fishing

1045. In an aquatic ecosystem, the depth to which light penetrates is called:
(A) Aphotic (B) Photic
(C) Euphotic (D) Nonphotic

1046. The transfer of energy from organism to organism in a natural community establishes:
(A) Food habitat (B) Food chains
(C) Food nexus (D) All of these

1047. Driving force of ecosystem is:
(A) Solar energy (B) Water
(C) Wind (D) All of these

1048. Energy and nutrients enter a community by way of the:
(A) Producers (B) Consumers
(C) Detrivores (D) Both (A) and (B)

1049. The food chain that starts from plants and goes from smaller to larger animals is:
(A) Predator food chain
(B) Saprophytic food chain
(C) Detritus food chain
(D) Parasitic food chain

1050. Species ecology is:
(A) Autoecology (B) Synecology
(C) Community ecology (D) Palaeo ecology

1051. Biosphere consists of:
(A) Lithosphere (B) Atmosphere
(C) Hydrosphere (D) All of these

1052. Biogeochemical cycles are also known as:
(A) Sedimentary cycle (B) Gaseous cycle
(C) Material cycling (D) Cycle of water

1053. In grassland ecosystem the pyramid of biomass is:
(A) Upright (B) Inverted
(C) Spindle like (D) Variable

1054. In pond ecosystem the shape of pyramid of number is:
(A) Upright (B) Inverted
(C) Spindle like (D) Variable

1055. Controlling factor in ecosystem is:
(A) Precipitation (B) Light
(C) Predation (D) Nutrients

1056. Relationship between different organisms can best studied by:
(A) Charles pyramid (B) Eltonian pyramid
(C) Hierarchical pyramid (D) All of these

1057. Maximum oxygen comes from:
(A) Phytoplankton (B) Grassland
(C) Forests (D) Herbs and shrubs

1058. Biome is a:
(A) Community of plants
(B) Community of animals
(C) Small sized ecosystems
(D) Large sized ecosystem

1059. Total organic matter present in an ecosystem is called:
(A) Biomass (B) Biome
(C) Litter (D) Food

1060. Ecological niche of an organism denotes:
(A) Habitat of the organisms
(B) On whom the organisms feeds
(C) The status of an organism within a particular community
(D) All of these

1061. The most important organisms for an ecosystem are:
(A) Green plants (B) Herbivores
(C) Carnivores (D) Protozoa

1062. The transition zone where two different types of communities meet is called:
(A) Ecotone (B) Ecocline
(C) Ecotype (D) Ecosystem

1063. Ecosystem of a pond is:
(A) Lentic (B) Lotic
(C) Both (A) and (B) (D) None of these

1064. Which forests have maximum diversity?

(A) Tropical forests (B) Temperate forests
(C) Alpine forests (D) Thorn forests

1065. A treeless biome is:

(A) Tundra (B) Grassland
(C) Desert (D) All of these

1066. Lentic ecosystem occurs in:

(A) Standing water (B) Running water
(C) Tiver water (D) All of these

ANSWERSHEET

7. ECOLOGY & ECOLOGICAL PRINCIPLES

1. (C) 2. (A) 3. (B) 4. (A) 5. (B) 6. (C) 7. (D) 8. (A) 9. (B) 10. (C)
11. (D) 12. (A) 13. (A) 14. (B) 15. (C) 16. (D) 17. (A) 18. (B) 19. (C) 20. (A)
21. (A) 22. (A) 23. (B) 24. (A) 25. (A) 26. (A) 27. (C) 28. (A) 29. (B) 30. (A)
31. (B) 32. (B) 33. (D) 34. (B) 35. (C) 36. (D) 37. (C) 38. (D) 39. (D) 40. (C)
41. (A) 42. (B) 43. (C) 44. (C) 45. (D) 46. (B) 47. (D) 48. (B) 49. (D) 50. (B)
51. (B) 52. (A) 53. (C) 54. (B) 55. (C) 56. (A) 57. (B) 58. (C) 59. (A) 60. (B)
61. (C) 62. (A) 63. (B) 64. (C) 65. (A) 66. (C) 67. (B) 68. (C) 69. (A) 70. (B)
71. (C) 72. (D) 73. (A) 74. (A) 75. (B) 76. (A) 77. (B) 78. (A) 79. (B) 80. (C)
81. (D) 82. (A) 83. (B) 84. (C) 85. (D) 86. (A) 87. (B) 88. (C) 89. (A) 90. (B)
91. (C) 92. (D) 93. (A) 94. (B) 95. (D) 96. (B) 97. (A) 98. (C) 99. (C) 100. (A)
101. (B) 102. (C) 103. (A) 104. (B) 105. (C) 106. (D) 107. (A) 108. (C) 109. (A) 110. (B)
111. (A) 112. (A) 113. (B) 114. (D) 115. (C) 116. (C) 117. (C) 118. (A) 119. (B) 120. (C)
121. (A) 122. (B) 123. (C) 124. (A) 125. (C) 126. (B) 127. (D) 128. (C) 129. (D) 130. (C)
131. (A) 132. (B) 133. (C) 134. (A) 135. (B) 136. (C) 137. (A) 138. (A) 139. (B) 140. (A)
141. (B) 142. (C) 143. (D) 144. (A) 145. (B) 146. (B) 147. (C) 148. (D) 149. (C) 150. (A)
151. (B) 152. (C) 153. (D) 154. (C) 155. (A) 156. (B) 157. (C) 158. (A) 159. (B) 160. (C)
161. (A) 162. (B) 163. (C) 164. (D) 165. (A) 166. (B) 167. (C) 168. (D) 169. (A) 170. (A)
171. (B) 172. (C) 173. (D) 174. (A) 175. (B) 176. (C) 177. (D) 178. (A) 179. (B) 180. (C)
181. (B) 182. (A) 183. (B) 184. (C) 185. (D) 186. (C) 187. (A) 188. (B) 189. (C) 190. (B)
191. (D) 192. (A) 193. (A) 194. (B) 195. (C) 196. (D) 197. (A) 198. (B) 199. (B) 200. (A)
201. (B) 202. (C) 203. (D) 204. (D) 205. (C) 206. (B) 207. (C) 208. (B) 209. (A) 210. (D)
211. (A) 212. (B) 213. (C) 214. (B) 215. (D) 216. (A) 217. (B) 218. (C) 219. (A) 220. (C)
221. (A) 222. (B) 223. (A) 224. (B) 225. (C) 226. (C) 227. (C) 228. (A) 229. (C) 230. (A)
231. (C) 232. (A) 233. (A) 234. (B) 235. (C) 236. (A) 237. (B) 238. (C) 239. (D) 240. (A)
241. (B) 242. (C) 243. (D) 244. (A) 245. (B) 246. (C) 247. (A) 248. (B) 249. (C) 250. (D)
251. (A) 252. (B) 253. (C) 254. (D) 255. (A) 256. (B) 257. (C) 258. (D) 259. (A) 260. (B)
261. (C) 262. (D) 263. (A) 264. (B) 265. (C) 266. (A) 267. (A) 268. (B) 269. (A) 270. (B)
271. (C) 272. (D) 273. (A) 274. (B) 275. (C) 276. (D) 277. (A) 278. (B) 279. (C) 280. (D)
281. (A) 282. (B) 283. (A) 284. (C) 285. (D) 286. (A) 287. (C) 288. (B) 289. (D) 290. (A)
291. (B) 292. (C) 293. (A) 294. (B) 295. (D) 296. (C) 297. (A) 298. (B) 299. (C) 300. (C)
301. (D) 302. (A) 303. (D) 304. (C) 305. (A) 306. (B) 307. (A) 308. (A) 309. (C) 310. (A)
311. (A) 312. (D) 313. (C) 314. (A) 315. (B) 316. (C) 317. (B) 318. (A) 319. (B) 320. (C)

321. (C) 322. (D) 323. (B) 324. (B) 325. (A) 326. (D) 327. (C) 328. (A) 329. (B) 330. (A)
331. (A) 332. (B) 333. (A) 334. (B) 335. (C) 336. (D) 337. (A) 338. (B) 339. (C) 340. (A)
341. (B) 342. (C) 343. (A) 344. (B) 345. (C) 346. (D) 347. (A) 348. (B) 349. (A) 350. (B)
351. (B) 352. (C) 353. (D) 354. (A) 355. (C) 356. (A) 357. (B) 358. (C) 359. (D) 360. (A)
361. (B) 362. (D) 363. (C) 364. (B) 365. (C) 366. (C) 367. (C) 368. (D) 369. (D) 370. (B)
371. (A) 372. (C) 373. (B) 374. (C) 375. (D) 376. (C) 377. (D) 378. (D) 379. (A) 380. (B)
381. (C) 382. (B) 383. (D) 384. (D) 385. (D) 386. (C) 387. (A) 388. (C) 389. (C) 390. (A)
391. (A) 392. (C) 393. (C) 394. (C) 395. (B) 396. (B) 397. (A) 398. (A) 399. (B) 400. (B)
401. (D) 402. (C) 403. (B) 404. (B) 405. (B) 406. (A) 407. (A) 408. (B) 409. (D) 410. (C)
411. (C) 412. (C) 413. (B) 414. (A) 415. (C) 416. (B) 417. (B) 418. (D) 419. (D) 420. (C)
421. (A) 422. (C) 423. (D) 424. (C) 425. (D) 426. (D) 427. (A) 428. (C) 429. (C) 430. (D)
431. (A) 432. (A) 433. (C) 434. (A) 435. (D) 436. (D) 437. (D) 438. (C) 439. (D) 440. (C)
441. (C) 442. (B) 443. (A) 444. (C) 445. (B) 446. (D) 447. (C) 448. (B) 449. (D) 450. (B)
451. (A) 452. (B) 453. (A) 454. (A) 455. (C) 456. (B) 457. (A) 458. (C) 459. (B) 460. (D)
461. (A) 462. (B) 463. (D) 464. (B) 465. (B) 466. (C) 467. (C) 468. (B) 469. (D) 470. (C)
471. (B) 472. (C) 473. (D) 474. (C) 475. (A) 476. (C) 477. (A) 478. (B) 479. (D) 480. (A)
481. (D) 482. (C) 483. (D) 484. (B) 485. (A) 486. (A) 487. (D) 488. (B) 489. (C) 490. (A)
491. (C) 492. (D) 493. (B) 494. (A) 495. (C) 496. (D) 497. (C) 498. (B) 499. (A) 500. (B)
501. (C) 502. (D) 503. (B) 504. (C) 505. (A) 506. (D) 507. (D) 508. (C) 509. (D) 510. (B)
511. (A) 512. (B) 513. (D) 514. (C) 515. (B) 516. (A) 517. (B) 518. (C) 519. (B) 520. (C)
521. (B) 522. (C) 523. (B) 524. (C) 525. (B) 526. (C) 527. (C) 528. (A) 529. (B) 530. (B)
531. (A) 532. (C) 533. (C) 534. (A) 535. (B) 536. (C) 537. (A) 538. (B) 539. (A) 540. (A)
541. (D) 542. (B) 543. (C) 544. (C) 545. (D) 546. (D) 547. (B) 548. (C) 549. (A) 550. (C)
551. (A) 552. (B) 553. (D) 554. (A) 555. (D) 556. (B) 557. (A) 558. (C) 559. (B) 560. (D)
561. (B) 562. (C) 563. (A) 564. (C) 565. (C) 566. (A) 567. (D) 568. (A) 569. (C) 570. (A)
571. (D) 572. (B) 573. (C) 574. (C) 575. (B) 576. (A) 577. (D) 578. (C) 579. (C) 580. (B)
581. (D) 582. (D) 583. (C) 584. (D) 585. (B) 586. (A) 587. (D) 588. (B) 589. (D) 590. (C)
591. (B) 592. (B) 593. (A) 594. (C) 595. (A) 596. (D) 597. (A) 598. (C) 599. (D) 600. (D)
601. (C) 602. (A) 603. (C) 604. (B) 605. (D) 606. (B) 607. (A) 608. (A) 609. (D) 610. (B)
611. (A) 612. (C) 613. (A) 614. (C) 615. (A) 616. (D) 617. (D) 618. (D) 619. (A) 620. (D)
621. (A) 622. (B) 623. (C) 624. (A) 625. (D) 626. (B) 627. (A) 628. (C) 629. (C) 630. (B)
631. (B) 632. (D) 633. (A) 634. (C) 635. (B) 636. (C) 637. (A) 638. (B) 639. (C) 640. (B)
641. (D) 642. (C) 643. (B) 644. (D) 645. (A) 646. (C) 647. (B) 648. (D) 649. (B) 650. (A)
651. (C) 652. (B) 653. (B) 654. (A) 655. (B) 656. (C) 657. (D) 658. (D) 659. (A) 660. (A)
661. (C) 662. (B) 663. (D) 664. (A) 665. (D) 666. (A) 667. (C) 668. (C) 669. (D) 670. (D)
671. (D) 672. (A) 673. (C) 674. (B) 675. (B) 676. (B) 677. (D) 678. (A) 679. (C) 680. (D)
681. (D) 682. (C) 683. (C) 684. (C) 685. (D) 686. (A) 687. (B) 688. (B) 689. (D) 690. (B)
691. (B) 692. (A) 693. (D) 694. (C) 695. (A) 696. (D) 697. (B) 698. (D) 699. (A) 700. (D)
701. (D) 702. (A) 703. (D) 704. (C) 705. (A) 706. (A) 707. (B) 708. (D) 709. (D) 710. (A)
711. (B) 712. (C) 713. (D) 714. (A) 715. (D) 716. (B) 717. (C) 718. (A) 719. (A) 720. (C)
721. (A) 722. (A) 723. (B) 724. (B) 725. (D) 726. (C) 727. (B) 728. (C) 729. (D) 730. (D)
731. (C) 732. (C) 733. (D) 734. (D) 735. (D) 736. (C) 737. (C) 738. (B) 739. (B) 740. (B)

741. (C) 742. (B) 743. (D) 744. (C) 745. (C) 746. (C) 747. (A) 748. (C) 749. (D) 750. (D)
751. (A) 752. (B) 753. (A) 754. (A) 755. (C) 756. (B) 757. (D) 758. (D) 759. (A) 760. (C)
761. (B) 762. (A) 763. (B) 764. (D) 765. (B) 766. (C) 767. (C) 768. (A) 769. (B) 770. (D)
771. (C) 772. (A) 773. (D) 774. (B) 775. (C) 776. (B) 777. (D) 778. (A) 779. (D) 780. (A)
781. (B) 782. (D) 783. (C) 784. (C) 785. (D) 786. (A) 787. (D) 788. (A) 789. (D) 790. (A)
791. (C) 792. (C) 793. (D) 794. (D) 795. (A) 796. (A) 797. (C) 798. (C) 799. (C) 800. (D)
801. (B) 802. (A) 803. (C) 804. (A) 805. (B) 806. (C) 807. (D) 808. (A) 809. (B) 810. (D)
811. (B) 812. (C) 813. (A) 814. (D) 815. (D) 816. (B) 817. (C) 818. (A) 819. (A) 820. (B)
821. (B) 822. (B) 823. (C) 824. (A) 825. (C) 826. (D) 827. (A) 828. (B) 829. (D) 830. (D)
831. (C) 832. (B) 833. (D) 834. (A) 835. (A) 836. (D) 837. (A) 838. (D) 839. (A) 840. (C)
841. (B) 842. (D) 843. (A) 844. (C) 845. (A) 846. (B) 847. (A) 848. (B) 849. (C) 850. (A)
851. (D) 852. (B) 853. (B) 854. (C) 855. (A) 856. (A) 857. (C) 858. (C) 859. (B) 860. (D)
861. (C) 862. (B) 863. (C) 864. (D) 865. (D) 866. (C) 867. (B) 868. (B) 869. (A) 870. (D)
871. (C) 872. (C) 873. (A) 874. (B) 875. (C) 876. (B) 877. (A) 878. (B) 879. (C) 880. (B)
881. (C) 882. (D) 883. (C) 884. (D) 885. (C) 886. (B) 887. (B) 888. (A) 889. (A) 890. (B)
891. (D) 892. (A) 893. (C) 894. (D) 895. (B) 896. (A) 897. (A) 898. (D) 899. (B) 900. (B)
901. (A) 902. (D) 903. (B) 904. (A) 905. (C) 906. (A) 907. (D) 908. (B) 909. (A) 910. (D)
911. (B) 912. (C) 913. (B) 914. (C) 915. (B) 916. (C) 917. (D) 918. (C) 919. (A) 920. (C)
921. (C) 922. (B) 923. (B) 924. (C) 925. (A) 926. (D) 927. (D) 928. (C) 929. (C) 930. (A)
931. (D) 932. (C) 933. (C) 934. (C) 935. (B) 936. (C) 937. (C) 938. (C) 939. (B) 940. (D)
941. (A) 942. (A) 943. (D) 944. (A) 945. (C) 946. (A) 947. (B) 948. (A) 949. (D) 950. (A)
951. (A) 952. (A) 953. (C) 954. (D) 955. (C) 956. (D) 957. (C) 958. (A) 959. (B) 960. (D)
961. (B) 962. (A) 963. (A) 964. (D) 965. (C) 966. (A) 967. (A) 968. (A) 969. (A) 970. (A)
971. (B) 972. (A) 973. (B) 974. (C) 975. (B) 976. (B) 977. (C) 978. (C) 979. (B) 980. (C)
981. (A) 982. (B) 983. (C) 984. (C) 985. (C) 986. (D) 987. (B) 988. (D) 989. (C) 990. (A)
991. (A) 992. (B) 993. (D) 994. (A) 995. (C) 996. (D) 997. (A) 998. (A) 999. (C) 1000. (C)
1001. (A) 1002. (D) 1003. (C) 1004. (C) 1005. (B) 1006. (B) 1007. (A) 1008. (A) 1009. (B) 1010. (D)
1011. (A) 1012. (C) 1013. (A) 1014. (D) 1015. (A) 1016. (C) 1017. (D) 1018. (C) 1019. (A) 1020. (B)
1021. (A) 1022. (D) 1023. (A) 1024. (B) 1025. (D) 1026. (B) 1027. (A) 1028. (C) 1029. (C) 1030. (C)
1031. (C) 1032. (B) 1033. (B) 1034. (C) 1035. (C) 1036. (B) 1037. (C) 1038. (A) 1039. (B) 1040. (A)
1041. (C) 1042. (B) 1043. (C) 1044. (B) 1045. (B) 1046. (B) 1047. (A) 1048. (A) 1049. (A) 1050. (A)
1051. (D) 1052. (C) 1.53. (A) 1054. (A) 1055. (C) 1056. (B) 1057. (A) 1058. (D) 1059. (A) 1060. (C)
1061. (A) 1062. (A) 1063. (A) 1064. (A) 1065. (D) 1066. (A)

Chapter - 8

DIVERSITY OF LIFE

1. The Riccia is bryophyte because:
(A) It occurs mostly on land and has motile sperm
(B) It has heteromorphy alternation of generation and lacks leaves
(C) It has multicellular sex organs with a sterile Jacket and lacks vascular tissues
(D) Its sporophyte lacks differentiation and has a single –layered facket

2. Potato spindle – tuber causes potatoes to become gnarled and raked. This disease is caused by an infectious agent consisting of:
(A) Circular DNA molecules without a protein coat
(B) Circular RNA molecule with a protein coat
(C) DNA molecule with a protein coat
(D) Circular RNA molecule lacking a protein coat

3. Originally, organisms were classified according to a two-domain system that consisted of the bacteria and the eukaryotes. Recently, a three–domain system of classification has been proposed in which domain. Bacteria was divided into domain Archaea and domain Bacteria. this is because:
(A) There are so many different species of prokaryotes
(B) Of the large genetic differences between the archaebacteria and the bacteria
(C) The archaebacteria have a nucleus like that of eukaryotes
(D) The archaebacteria live in inhospitable environment, so they should be put in a different domain

4. Although they are vascular, and thus moderately advanced terrestrial plants, the ferns have not solved all of the problems of terrestrial life. How is this so?
(A) They require external water for sexual reproduction
(B) They are broad leafed
(C) They do not possess a gametophyte stage
(D) They lack advanced vascular tissue

5. The bryophytic gemetophytes are:
(A) Independent from sporophyte
(B) Partially dependent on sporrophytes
(C) Lack antheridia and archegonia
(D) Parasitically dependent of sporophytes

6. All of the following components of a retrovirus are encoded by the viral genome, except?
(A) Reverse transcriptase
(B) Viral RNAs
(C) Capsid proteins
(D) Envelope lipids

7. When a population of bacteria capable of conjugation transfers a specific chromosomal gene (say a gene coding for galactose metabolism) but no other genes, regardless of how long the bacteria are allowed to mate, these bacteria are said to be:
(A) F^+
(B) F^-
(C) Hfr
(D) F'

8. Which of the following enzyme has both DNA polymerase and Rnase H activities:
(A) Pancreatic ribonuclease A
(B) Reverse transcriptase
(C) DNAzyme
(D) RnaseP

9. Ecological equivalent describes:
(A) Groups of species with comparable roles
(B) Species that occupy the same niche in different geographical regions
(C) Diversity of habitats
(D) Social behaviors that enhance the fitness of other individuals in the population

10. Indicate which statement is not correct, T helper cells
(A) Stimulate migration of marophages
(B) Help B cells to produce antibodies
(C) Are cytotoxic to virus infected cells
(D) Help in generation of cytotoxic T cells

11. The body is normally able to distinguish its own self-antigens from foreign non-self antigens and does not mount an immunologic attack against itself. This phenomenon is called:
(A) Immune tolerance (B) Autoimmunity
(C) Transplantation (D) None of these

12. In haplontic life cycle:
(A) Only gametophyte generation is present
(B) Only sporophytic generation is present
(C) Predominent gametophyte generation is present
(D) Predominant sporophytic generation is present

13. Of the following ecological relationships, which one is the most different from the other three?
(A) Algae embedded in coral tissues
(B) Salmonella in human gastric tract
(C) Cellulolytic bacteria in a termite gut
(D) Pollen-colleting bees visiting flowers

14. The acceptable concept of ecological niche is that it represents the:
(A) Trophic status of a species in a community
(B) Unique microhabitat of a species
(C) Sum of total species use of biotic and abiotic sources of the environment
(D) Unique combination of the trophy status and the microhabitat of a species

15. How an organism is suited to live in a particular place is called?
(A) Competition (B) Adaptation
(C) Addition (D) Participation

16. Which of the following statements about energy flow is incorrect?
(A) Secondary productivity declines with each trophy level
(B) Only net primary productivity is available to consumers
(C) About 90% of the energy at one trophy level does not appear at the next
(D) Eating meat is probably the most economical way of acquiring the energy of photosynthetic productivity

17. The layer extends from about 90 km to between 500 and 1000 km is called:
(A) Troposphere (B) Stratosphere
(C) Mesosphere (D) Thermosphere

18. A pyramid of population indicating a moderately high proportion of young to old:
(A) Bell shaped
(B) Indicates stable population
(C) Urn shaped
(D) Diminising population

19. A fresh water wetland include(s):
(A) Riverine wetlands such as flood plains
(B) Lacustrine wetlands such as lake, ponds
(C) Palustrine wetland such as marshes, bogs, fens
(D) All of these

20. The layer starts at 50 km above earth's surface and goes upto 85 km high is called:
(A) Troposphere (B) Stratosphere
(C) Mesosphere (D) Thermosphere

21. The kingdom protista contains:
(A) Prokaryotic unicellular autotrophic organisms
(B) Eukaryotic unicellular photosynthestic/ non-photosynthetic organisms
(C) Prokaryotic multicellular heterotrophic organisms
(D) Eukaryotic multicellular heterotrophic organisms

22. Climax communities:
(A) Are more diverse than pioneer communities
(B) Are lass stable than pioneer communities
(C) Have greater entropy than pioneer communities
(D) Have a large number but fewer species of plants than pioneer communities

23. The main feature that distinguishes ferns from the other groups of seedless vascular plants is the presence of:
(A) A dominant sporophyte generation
(B) A leafy gametophyte
(C) Air-dispersed cuticle-covered spores
(D) Endospory

24. In the exponential phase of growth of a bacterial culture, 10 cfu/ml cells increased to 3200 cfu/ml cells in 2 hours. What is the generation time for this bacterium?
(A) 12 minutes (B) 15 minutes
(C) 24 minutes (D) 30 minutes

25. The tendency towards variety and density of species at community junctions is known as the:
(A) Edge effect (B) Variability
(C) Vital index (D) Biotic potential

26. Identify the wrong statement about dendrite cells:
(A) They take up antigen by endocytosis
(B) They express class – II molecules constitutively
(C) They cannot activate native T cells
(D) They express B, molecules all the time on the membrane

27. A technique for determining the concentration of a particular molecule in a sample, based on competitive binding between non-labeled and isotope labeled antigen for its specific antibody is:
(A) Immunofluorescene
(B) Immunuelectrophoresis
(C) Radioimmunoassay
(D) Rocket electrophoresis

28. The alpine biome refers to a distinct flora and fauna which characterizes the region:
(A) Below the timber line of the mountains
(B) Above the timber line of the mountains
(C) Between the sub-show zone and the meadow zone of the mountains
(D) Between the snow zone and sub-snow zone of the mountains

29. Which of the following is not an example of *in situ* conservation strategies?
(A) Biosphere reserve (B) Botanical garden
(C) National park (D) Sacred groove

30. Which of the following is not a characteristics of invasive species:
(A) Bentham and Hooker's system
(B) Engler and Pranti's system
(C) Huthinson's system
(D) Takhtajan system

31. The transition from swimming sperm to non-swimming sperm occurred in:
(A) Green algae (B) Bryophytes
(C) Ferns (D) Gymnosperms

32. Actin filaments are found in all of the following except the:
(A) Flagella of bacteria
(B) Sarcomeres of skeletal muscle cells
(C) Microvilli of the intestinal brush border
(D) Contractile rings of dividing animal cells

33. A community with low species diversity and high dominance is said to be:
(A) Unproductive but stable
(B) Productive but unstable
(C) Neither productive nor stable
(D) Both productive and stable

34. The layer extends upto 50 km and a thin layer of ozone is present at height of 15-30 km is called:
(A) Troposphere (B) Stratosphere
(C) Mesosphere (D) Thermosphere

35. Identify the wrong statement about B cells:
(A) They can present both exogenous and endogeneous antigens
(B) They can present only antigens for which they have surface immuniglobulin
(C) They can present only protein antigen
(D) They can present both protein and non-protein antigens

36. If Ab molecules are tagged with a fluorescent dye, or fluorochrome, immune complexes containing these fluorescently labeles antibodies can be detected by colored light emission when excited by light of the appropriate wavelength. Ab molecules bound to antigens in cells or tissue sections can similarly be visualized. This technique is known as:
(A) Immunofluorescene
(B) Immunuelectrophoresis
(C) Bioluminescene
(D) Rocket electrophoresis

37. Match list – I with list – II and select the correct answer using the codes given below the lists

List – I	List – II
P. Ecotone	1. Running water
Q. Benthic	2. Replacement of one community by another
R. Sere	3. Transitional zone between two communities
S. Lotic	4. Zone found at the bottom of the sea

(A) P–3, Q–4, R–2, S–1 (B) P–4, Q–3, R–2, S–1
(C) P–3, Q–4, R–1, S–2 (D) P–4, Q–3, R–1, S-2

38. All the energy in a food chain originates from:
(A) Plants (B) Animals
(C) Sun (D) Autotrophs

39. A certain place in India has (i) average annual temperature of 27°C, (ii) annual rainfall between 200 and 300 m (iii) number of rainy days ranging from 115 to 150. Which one of the following forest types would be found in such a place?
(A) Tropical wet evergreen forest
(B) Tropical dry deciduous forest
(C) Himalayan temperature forest
(D) Moist alpine scrub

40. When a plant chromosomal type aa pollinates a plant of type AA, what chromosome constitution of embryo and endosperm is expected in the resulting seeds?
(A) Diploid zygote of type Aa and triploid endosperm of type AAa
(B) Diploid zygote of type Aa and triploid endosperm of type Aaa
(C) Diploid zygote of type AA and triploid endosperm of type AAa
(D) Diploid zygote of type Aa and triploid endosperm of type aaa

41. Which of the following is not true bacterial flagella?
(A) Most of their length consists of a hollow, rigid tube
(B) They are constructed largely of a single protein called flagellin
(C) They spin like wheels, either clockwise or counterclockwise
(D) They use cytoplasmic ATP as their primary energy source

42. Biodiversity is described as:
(A) All genetic and species diversity
(B) The seasonal and daily changes in an environment
(C) The way species differ from one another
(D) The influence of physical factors on an environment

43. Which one of the following sets consists entirely of endangered wild life species of India?
(A) Himalayan musk, black buck, Indian lion, rhino, sambar
(B) Himalayan musk, deer, Indian lion, rhino, wild buffalo, golden cat
(C) Egret, Black bear, bison, spotted deer
(D) Himalayan musk deer, great Indian bustard, snow leopard, Kasmir stag, wild ass

44. Which of the following is not a typical characteristic of a r-shaped species?
(A) Large size
(B) Many, small offspring
(C) A large allocation of resources to reproduction
(D) Early reproductive maturity

45. Male gametophyte in Angiosperm is:
(A) Anther (B) Stamen
(C) Pollen sac (D) Pollen

46. Which of the following is a fact that argues against a symbiotic origin if eukaryotic flagella from bacterial flagella?
(A) Bacterial flagella typically contain flagellin, eukaryotic flagella contain tubulin
(B) Bacterial flagella rotate, eukaryotic flagella do not rotate
(C) Bacterial flagella are directly energized by electrons and protons from the respiratory system of the plasma membrane, eukaryotes are not
(D) More than one of the above argue against the symbiotic theory

47. Among the ecological relationship below, which is most different from the other three?
(A) Fungal mycilia and algal cells in lichen
(B) *Salmonella* in the human gut
(C) Cellulose digesting protozoa in termite gut
(D) N_2-fixing bacteria in root nodules

48. The layer extends at poles upto about 5-6 km, while at the equator it is about 18 km is called:
(A) Troposphere (B) Stratosphere
(C) Mesosphere (D) Thermosphere

49. Evidence indicating the chloroplasts were originally free-living prokaryotes that subsequently evolved a symbiotic relationship with a eukaryotic host includes all of the following except the:
(A) Similarities of rRNA sequences between chloroplasts and free-living prokaryotes
(B) Similarities of structures between chloroplasts and some contemporary free-living prokaryotes
(C) Presence of circular DNA in chloroplasts and in free-living prokaryotes
(D) Ability of chloroplasts to synthesize all their own protein

50. HIV replicated its genome using unique mechanisms. Which of the following statements about HIV is not correct:
(A) HIV is an enveloped RNA virus
(B) The virion contains an RNA dependent DNA polymerase
(C) A DNA copy of the HIV genome integrates into host cell DNA
(D) Virion contains an RNA dependent RNA polymerase

51. A type of competition when one species exploits a resource, such as food, space, in common with another species but without direct contact with that species:
(A) Interference competition
(B) Intraspecific competition
(C) Exploitation competition
(D) None of these

52. An IgG molecule after treatment with an enzyme generated a disulfide linked antigen binding fragment which could agglutinate cells. The enzyme is:
(A) Trypsin (B) Papain
(C) Pepsin (D) V_8 protease

53. When number of organisms and amounts of living material in successively higher trophy levels are compared, the values usually take the form of a pyramid, with the largest numbers and greatest biomass in the producer trophy level. However, in some marine ecosystems, the consumer trophic levels contain significantly greater amounts of living material than does the primary producer trophic level. The best explanation for this is which of the following?
(A) The main primary producers in marine ecosystems are microscopic algae with extremely high rates of population turnover
(B) Most consumers in marine ecosystems are filter feeders that must maintain large basket like structures for extracting food from the water
(C) The increased availability of solar radiation in marine ecosystems means that fewer primary producers are required to support marine food chains
(D) Marine zooplankton often produce large extensions of their bodies in order to increase buoyancy

54. Fertilizers can be washed into rivers by the rain. This can cause:
(A) Bioaccumulation
(B) Eutrophication
(C) Biodegradation
(D) Spontaneous combustion

55. Monoclonal antibodies are secreted by hybridomas which are generated by:
(A) Fusion of immune spleen cells with any type of cells capable of growing in tissue culture
(B) Fusion of immune spleen cells with plasmacytoma cells
(C) Growing of immune spleen cells in the presence of HAT
(D) Growing of immune spleen cells in the presence of B cell growth factors

56. --------- the terms that refer to the same physical process of light emission without heat is:
(A) Immunofluorescene
(B) Immunuelectrophoresis
(C) Chemiluminescene or Bioluminescene
(D) Rocket electrophoresis

57. Competition for resources is strongest among:
(A) Organisms belonging to single species
(B) Organisms belonging to two or more closely related species
(C) Organisms belonging to two or more distantly related species
(D) All species in an environment niche

58. Genetic races of a species with heritable variations that differentiated due to interplay with environmental conditions are called:
(A) Eads (B) Ecotypes
(C) Syntypes (D) Ecoclines

59. The most appropriate definition of seed in terms of homologous organs found in spore producing land plant is:
(A) A seed is an integumented sporangium with a single functional spore
(B) A seed is an integumented indehisent type
(C) A seed is an integumented gymnosporangium
(D) A seed is an integumented bud containing an egg

60. A bacterial culture contained 32 x 10^6 cells after 2.5 hours of exponential growth. If the doubling time was 30 min, what was the initial population number in this culture?
(A) 20 x 10^4 cells (B) 10 x 10^5 cells
(C) 40 x 10^5 cells (D) 16 x 10^6 cells

61. Acid rain damage depends on the buffering capacity of the soils in a given region. Damago has been greatest where the soil layer is:
(A) Thin and contains little calcium and magnesium
(B) Thin and contains abundant calcium and magnesium
(C) Thin and contains abundant but little magnesium
(D) Thick and contains abundant calcium and magnesium

62. An Hfr bacterium is one that contains:
(A) Many unusual plasmids
(B) Chromosomal material acquired from recipient cell
(C) The ability to undergo transduction
(D) A plasmid integrated into its chromosome

63. The maximum possible rate of increase of population of a given species under ideal conditions would depend upon its:
(A) Biotic potential (B) Potential natality
(C) Carrying capacity (D) Biomass

64. Which one of the following pairs is not correctly matched?
(A) Production of sulphur dioxide – burning of coal
(B) Depletion of ozone layer – release of chlorofluorocarbon
(C) Eutrophiation – increases in nitrogen and phosphorus contents in aquatic bodies
(D) Decreases in B.O.D. of pond water – increase in global temperature

65. Which of the following is an example of ritually endangered species in India?
(A) Black buck (B) Red panda
(C) Pigmy hog (D) Cheetah

66. A microspore will develop into:
(A) A gametophyte bearing archegonia
(B) A gametophyte bearing antheridia
(C) A sporophyte bearing archegonia
(D) A saprophyte bearing antheridia

67. Which statement is true of both prokaryotic and eukaryotic cells?
(A) Prokaryotic cells are generally much larger than eukaryotic cells
(B) Eukaryotic cells have ribosomes and prokaryotic cells do not
(C) Both have DNA as their primary genetic material
(D) Eukaryotic cells have plasma membranes and prokaryotic cells do not

68. Dikaryot hyphae are a kind of fungal tissue that:
(A) contain two chromosomes
(B) Form after plasmogamy and before karyogamy
(C) Are only a short stage of the life cycle of the Basidiomycota
(D) Are not found in the ascomycota

60. Photosynthetically active radiation is:
(A) 0.4-0.7 micrometer
(B) 0.04-0.07 micrometer
(C) 4-7 micrometer
(D) 0.004-0.007 micrometer

70. Identify the correct statements for the HAT medium:
P. Includes drug aminopterin to block major pathway for synthesis of deoxyribonucleotides
Q. Hypoxanthine is precursor for thymidine
R. Include drug aminopterin to block major pathway for synthesis of deoxyribonucleotides
S. Cells an grow in presence of aminopterin only if they have enzymes thymidine kinase and hypoxanthineguanine phosphoribosyl transferase
(A) P and Q (B) P and S
(C) R and S (D) Q and S

71. The technique that the antibody (rather than the antigen) is immobilized on a microtiter well by:
(A) Sandwich ELISA
(B) Immunuelectrophoresis
(C) Bioluminescene
(D) Rocket electrophoresis

72. A microscopic unicellular organism is observed to have the following characteristics: a food gullet, a flagellum, chloroplasts, mitochondria and a nucleus. This organism belongs to which kingdom?
(A) Protista (B) Plantae
(C) Fungi (D) Monera

73. All of the following may be considered as sites for secondary succession except:
(A) An abandoned agricultural field
(B) A bare rock shelf on a mountainside
(C) A forest gap resulting from a tree fall
(D) A hurricane-damaged forest

74. The pollen tube usually enters the embryo sac:
(A) Through one of the synergids
(B) By directly penetrating the egg
(C) Between one synergid and central cell
(D) By knocking off the antipodal cells

75. A endemic species is:
(A) A species found uniquely in one place
(B) A species carrying an epidemic disease
(C) A species at an early phase of its evolution
(D) A taxonomist's mistake

76. Which is the most modern method of classification of organisms?
(A) Morphology
(B) Gram staining
(C) Biochemical characterization
(D) 16S rRNA sequence

77. Which of the following is not a characteristic of invasive species?
(A) Small seed size
(B) Wide geographical range
(C) Phenotypic plasticity
(D) Slow reproduction rate

78. For most terrestrial ecosystem, pyramids of numbers, biomass and energy are essentially the same- they have a broad base and a narrow top. The primary reason for this pattern is that:
(A) Secondary consumers and top carnivores require less energy than producers
(B) At each step, energy is lost from the system as a result of keeping the organisms alive
(C) As matter passes through ecosystems, some of it is lost to the environment
(D) Bio-magnification of toxic materials limits the secondary consumers and top carnivores

79. Which one of the following statement is not correct:
(A) Almost all of the prokaryotes have a single circular chromosome
(B) Almost all of the prokaryotes have cell wall
(C) All prokaryotes have plasmids
(D) The genetic material in prokaryote is not enclosed in any structure surrounded by membranes

80. Which one of the following statements is true?
(A) All vertebrates contain amnion
(B) All chordates are vertebrates
(C) All tunicates are chordates
(D) Cephalochordates are characterized by the presence of a well defined skull

81. Numerical taxonomic results suggest:
(A) Evolutionary relationship
(B) Phylogenetic relationship
(C) Overall similarity
(D) Functional similarity

82. A pyramid of population indicating a moderately high proportion of young to old:
P. Bell shaped
Q. Indicates stable population
R. Urn shaped
S. Diminishing population
The correct sequence of the descending order in terms of the areas occupied by these zones is:
(A) P and Q (B) R and S
(C) P and S (D) R and Q

83. Which of the following is a false statement?
(A) All sponge are hermaphrodite
(B) Choanocytes are reproduced in function
(C) Sponges are multicellular organisms
(D) Porocyte allow incurrents of water

84. Which one of the following was the objective of signaling the 'Montreat protocol'?
(A) Protection of wild life
(B) Protection of ozone layer
(C) Control over the use of insecticides
(D) Control of noise pollution

85. Serum IgM cannot activate the complement by itself because:
(A) It does not have complement binding site
(B) It is planar in which complement binding sites in the F region are not accessible
(C) It gets degraded and hence unable to activate the complement
(D) It needs metal ions to activate complement

86. A negatively charged antigen is electrophoresed in a gel containing antibody can be detected by
(A) Immunofluorescene
(B) Immunuelectrophoresis
(C) Bioluminescene
(D) Rocket electrophoresis

87. Species having wide geographical ranges often develop locally adapted population called:
(A) Prototypes (B) Paratypes
(C) Syntype (D) Ecotypes

88. Upon germination, the typical moss spore develops into:
(A) The erect green stage
(B) A protonema
(C) The mature sporophyte
(D) An embryo

89. Which one of the following pair is not correctly matched:
(A) Early blight of potato : *Alternaria solani*
(B) Brown rot of potato : *Pseudomonas solanacearum*
(C) Leaf spot of potato : *Cercospora conconrs*
(D) Ring rot of potato : *Phytophthora infestans*

90. Which of these ecosystems has the lowest net primary production per square meter?
(A) A salt marsh (B) An open ocean
(C) A coral reef (D) A grassland

91. The Shannon-Weaver index (also known as Shannon-Wiener Index) is useful to measure:
(A) Wetland pollution levels
(B) Visibility in smog polluted areas
(C) Species diversity in an ecological community
(D) Rate of biomass transfer in a food chain

92. In the animal kingdom, several degree of inter-specific interactions can be recognized. Which one of the following is true for protocooperation interactions?
(A) Both partners gain but neither can survive without the other
(B) One of the species is harmed by the relationship
(C) Presence of one spccies appears to have no effect on a second species
(D) Both partners gain, though each is able to survive separately

93. A climatic zonation occurs with increasing altitude in the mountains:
(A) Increasing (B) Declining
(C) Similar (D) Uncertain

94. Biodiversity hot spots are recognized on the basis of:
(A) Their proximity to national parks and reserves
(B) The number of endemic species they contain
(C) The degree to which the included species are threatened with extinction
(D) B and C only

95. Interrupted mating process is used to determine order in which genes are present on chromosome. It is based on the kinetics of transfer of individual makes. In case of interrupted mating conjugation is mediated between:
P. Hfr donor and F^- recipient
Q. Antibiotic resistant strain of Hfr and F^-
R. F^+ and F^-
S. Prototropic Hfr donor &auxotropic F^- recipient
(A) P and S (B) P and Q
(C) R and S (D) Q and S

96. The total output of photosynthesis, including the organic matter used up in respiration during the period of measurement is called:
(A) Net productivity
(B) Gross primary productivity
(C) Net community productivity
(D) Secondary productivity

97. The amount of chemical energy in consumers' food that is converted to their own new biomass during a given time period is called:
(A) Biomass
(B) Standing crop
(C) Primary production
(D) Secondary production

98. The biomass of one trophy level getting incorporated into the biomass of the next trophy level is the:
(A) Relative ratio of energy flow
(B) Energy flow efficiency
(C) Ecological efficiency
(D) Ecological gradient

99. The process of electrophoresis as well as double immunidiffusion is called:
(A) Immunofluorescene
(B) Immunuelectrophoresis
(C) Bioluminescene
(D) Rocket electrophoresis

100. A food web:
(A) Only exist in pond ecosystem
(B) Shows how feeding relationship are interlinked
(C) Shows the number of organisms in a habitat
(D) Only shows important animals

101. Removal of which species will always result in great changes in an ecological system?
(A) Competitive species (B) Keystone species
(C) Carnivorous species (D) Rare species

102. Rabbits are important grazers, maintaining some important grassland habitats. If another catastrophe were to reduce rabbit populations what would happen?
(A) Important grassland butterflies would be lost
(B) There would be more grass for the surviving rabbits
(C) The grassland would be invaded by trees and shrubs
(D) All of these

103. Herd immunity can apply to all diseases:
(A) Yes (B) No
(C) Unknown (D) Indefinite

104. Which of the following pairs are correctly matched?
P. Amensalism — A population is inhibited whereas the other populations remain unaffected
Q. Predation — A population uses the other as its food
R. Mutualism — Interaction favourable to both and obligatory
(A) Q and R (B) P and Q
(C) P and R (D) P, Q and R

105. Find the incorrect statement
(A) Extinction is a natural process
(B) Habitat loss is primary reason for loss of biodiversity
(C) Hot spots are geographical regions that are rich in endemic species
(D) Black data books included name of endangered specks

106. What is the fundamental difference between matter and energy?

(A) Matter is cycled through ecosystems; energy is not
(B) Energy is cycled through ecosystems; matter is not
(C) Energy can be converted into matter; matter cannot be converted into energy
(D) Matter can be converted into energy; energy cannot be converted into matter

107. In India, tropical wet evergreen forests are restricted mainly to:

(A) Malabar coast and Assam
(B) Malabar coast and Western Himalaya
(C) Coromandal coast and Punjab
(D) Coromandal coast and Nilgiri

108. Two taxonomic species are distinguished from each other by:

(A) Their failure to interbreed
(B) Their ability to exchange gene freely
(C) Their similarity in morphological characters
(D) Discontinuity in a set of correlated characters

109. DDT has been banned from many developed countries because of its:

(A) High toxicity of mammals
(B) High degree of persistance in the environment
(C) Low toxicity to insects
(D) High solubility in water

110. The effect of mean temperature when we move from tropical to arctic region:

(A) Increasing (B) Declining
(C) Similar (D) Uncertain

111. The difference between an endangered species and a threatened one is that:

(A) An endangered species is closer to extinction
(B) A threatened species is closer to extinction
(C) Endangered species are mainly tropical
(D) There is no real difference

112. Somatic mutation of immunoglobulin genes accounts for:

(A) Allelic exclusion
(B) Class switching from IgM to IgG
(C) Affinity maturation
(D) Increased expression of the Ig gene

113. Genes that encode antigens isolated from a pathogen can be inserted into nonvirulent viruses or bacteria is called

(A) Attenuated vaccines
(B) Inactivated vaccines
(C) Recombinant vector vaccines
(D) Subunit vaccine

114. An ecosystem, such as an aquarium, is self-sustaining if it involves the interaction between organisms, the flow of energy , and the presence of:

(A) Equal numbers of plants and animals
(B) More animals than plants
(C) Material cycles
(D) Pioneer organisms

115. Which of the following statements regarding efficiency of energy transfer through food webs is false?

(A) A tropic level and support more herbivores than carnivores
(B) Birds and mammals have low production efficiency
(C) The efficiency of energy transfer depends on the total production at a particular tropic level
(D) Herbivores are less efficient than carnivores

116. A prophage is:

(A) An auxotropic mutant
(B) A gene
(C) A phage DNA incorporated into the host genome
(D) The DNA of lytic phage

117. The following activates the alternative pathway of complement:

(A) Lipopolysaccharides
(B) Some viruses and virus-infected cells
(C) Fungal and yeast cell walls
(D) All of these

118. Rate of storage of organic matter not used by heterotrophs is termed as:

(A) Net productivity
(B) Net community productivity
(C) Gross primary productivity
(D) Secondary productivity

119. Choose the mismatch:

(A) Littoral zone – rooted vegetation
(B) Limnetic zone – open water
(C) Benthic zone - plankton
(D) Profundal zone – only heterotrophs

120. A unicellular eukaryote that contains chlorophyll a (but no chlorophyll b), that functions as one of the major photosynthetic autotrophs in open – ocean ecosystem, and that has cell walls that contain silica is a:

(A) Red algae
(B) Green algae
(C) Dinoflagellate
(D) Diatom

121. Which of the following algal divisions is characterized by the pigment fucoxanthin and bodies composed of a holdfast, stipe, and blade?

(A) Phaeophyta
(B) Chlorophyta
(C) Pyrrophyta
(D) Rhodophyta

122. Which of the following statements about succession is correct?

(A) Secondary succession occurs where no soil exists

(B) Primary succession occurs in areas where soil remains after a disturbance

(C) Secondary succession an occur where a disturbance has left soil intact

(D) Some cases of succession involve facilitation, a phenomenon in which local species inhibit the growth of newcomers

123. Which of the following characterizes dicotyledonous plants?

(A) Their flowers always have superior ovaries

(B) Their seeds usually contain two embryos

(C) Their stems have scattered vascular bundles

(D) Their flowers usually have parts in multiples of four or five

124. A keystone species is one that:

(A) Preys heavily on a particular species

(B) Is especially vulnerable to extinction

(C) Is restricted to small geographic area

(D) Strongly influences the structure and functioning of its ecological community

125. An MHC class I molecule was run on an SDS-acrylamide gel. How many protein bands will be observed in the gel:

(A) 4 (B) 2

(C) 1 (D) 3

126. The vaccine fragments of pathogen's genome encoding antigenic proteins are injected directly into the host cells where they can integrate into the chromosomal DNA or exist as episomes is called:

(A) Attenuated vaccines (B) Inactivated vaccines

(C) DNA vaccines (D) Subunit vaccine

127. Life-history characteristics associated with K-selected organisms include:

(A) Rapid reproductive rates, short generation times and large body size

(B) Repeated reproduction, few progeny and large body size

(C) Inhabiting early successional state communities, rapid maturation rates and numerous large offspring

(D) Inhabiting climax communities, many small offspring and short life span

128. The phosphorus cycle differs from those of carbon and nitrogen in that:

(A) It lacks a gaseous phase

(B) It lacks a liquid phase

(C) Living organisms don't need phosphorus

(D) The phosphorus cycle does not differ importantly from the carbon and nitrogen cycles

129. Which of the following best describes a significant difference between vascular and nonvascular plants?

(A) Vascular plants have lignified cell walls to provide mechanical support, whereas nonvascular plants do not

(B) Non vascular plants are homosporous, whereas Vascular plants are heterosporous

(C) In nonvascular plants, the cells of the xylem and phloem are living, whereas in vascular plants they are not

(D) In vascular plants, the sporangia grow above the rest of the plant for easy disposal of spores, whereas in nonvascular plants they do not

130. In the angiosperms, the:

(A) Gametophyte is prominent, and the sporophyte is dependent upon the gametophyte

(B) Sporophyte is prominent, with the sporophyte and gametophyte living independently

(C) Sporophyte is prominent, and the gametophytes is dependent upon the sporophyte

(D) Gametophyte is prominent, and the sporophyte stage has disappeared

131. Which of the following may have an inverted biomass pyramid?

(A) Terrestrial and shallow water ecosystem

(B) Open or deep water situations

(C) Lakes and ponds

(D) All of these

132. The short term properties such as temperature, pressure, moisture of atmospheric conditions for a specific place and time is called:

(A) Macro climate (B) Micro climate

(C) Climate (D) Weather

133. Choose the correct set of matches between Group I and II:

Group I	Group II
P. Embryoid	1. An unorganized growth of plant cells in a culture medium
Q. Callus	2. A localized group of actively dividing cells from which permanent tissue systems such as root, shoot, leaf and flower are derived
R. Meristem	3. A process whereby specialized, non-dividing cells begin to proliferate by mitotic division, presumed to involve regression to an undifferentiated state
S. Scutellum	4. Mass of cells which has an external morphology resembling a proembryo
	5. The embryonic cotyledon of monocot plants

(A) P-4, Q-2, R-3, S-5 (B) P-4, Q-1, R-2, S-5

(C) P-3, Q-4, R-2, S-1 (D) P-2, Q-1, R-3, S-4

134. Match list – I (Name of species) with list – II (Categories) and select the correct answer using the codes given below the lists

List – I (Name of species)	List – II (Categories)
P. Leopard	1. Out of danger
Q. Asian lion	2. Rare
R. Golden cat	3. Vulnerable
S. Elephant	4. Endangered

(A) P–3, Q–4, R–1, S–2 (B) P–3, Q–4, R–2, S–1
(C) P–4, Q–3, R–2, S–1 (D) P–4, Q–1, R–1, S–2

135. Which of the following is not a positive interaction?
(A) Commensalism (B) Protocooperation
(C) Amensalism (D) Mutualism

136. Match the following

Column A	Column B
P. Retroviridase	1. Linear double stranded DNA
Q. Herpesviridae	2. Minus strand RNA
R. rhabdoviridae	3. Circular double stranded DNA
S. Baculoviridae	4. Plus strand RNA

(A) P–1, Q–2, R–3, S–4 (B) P–3, Q–4, R-1, S-2
(C) P–2, Q–3, R-4, S–1 (D) P–4, Q–1, R-2, S-3

137. Which of the following feature(s) about late successional stage is incorrect?
(A) Biomass Low
(B) Species High
(C) Resillience Low
(D) Resistance High

138. Anti-idiotypic antibodies refer to:
(A) Antibodies produced in vitro
(B) Antibodies that are highly cross-reactive
(C) Antibodies that are generated against the unique determinants in the variable regions of a specific antibody molecule
(D) Antibodies directed against the constant region of antibody only

139. The vaccine are produced by inactivation of pathogenic organisms by heat or chemical treatment so that organisms are unable to multiply in the host is called:
(A) Attenuated vaccines
(B) Inactivated vaccines
(C) DNA vaccines
(D) Subunit vaccine

140. Which one of the following is correct to measure ß–diversity?
(A) Species richness within an ecosystem
(B) Species evenness equitability
(C) Degree of change in species composition along an environmental gradient
(D) Species diversity of several habitats in a large geographical region

141. Match list – I with list – II and select the correct answer using the codes given below the lists

List – I
P. The amount of energy accumulation in green plants through the process of photosynthesis
Q. The total organic matter synthesized by the producers in the process of photosynthesis per unit time and area
R. The rate of storage of organic matter in plant tissues in excess of the respiratory utilization during the measurement period
S. Rate of energy storage at consumer level

List – II
1. Gross primary productivity
2. Primary productivity
3. Secondary productivity
4. Net primary productivity

(A) P–1, Q–2, R–3, S–4 (B) P–2, Q–1, R–3, S–4
(C) P–1, Q–3, R–4, S–2 (D) P–2, Q–1, R–4, S–3

142. The kingdom protista contains:
(A) Prokaryotic unicellular autotrophic organisms
(B) Eukaryotic unicellular photosynthetic/ non-photosynthetic organisms
(C) Prokarotic multicelular hetrotrophic organisms
(D) Eukaryotic multicelular hetrotrophic organisms

143. The sum total of an organism's interaction with the biotic and abiotic resources of its environment is called its :
(A) Habitat (B) Logistic growth
(C) Biotic potential (D) Ecological niche

144. Acoelomates are characterized by:
(A) The absence of cavity surrounding the internal organs
(B) The presence of huge body cavity, as in case of terrestrial animals
(C) The presence of air sacs, as in case of birds
(D) The absence of brain in a group of extinct species

145. Match list – I (Name of plant disease) with list – II (Name of pathogen) and select the correct answer using the codes given below the lists

List – I	List – II
P. Angular leaf spot of cotton	1. *Phytophthora infestans*
Q. Late blight of potato	2. *Colletotrichum falcatum*
R. Red rot of sugarcane	3. *Cercospora personata*
S. Tikka leaf spot of groundnut	4. *Xanthomonas malvacearum*

(A) P–3, Q–1, R–4, S–2 (B) P–4, Q–1, R–2, S–3
(C) P–3, Q–4, R–2, S–1 (D) P–4, Q–2, R–1, S–3

146. The long term pattern of weather in a locality, region, or even over the entire globe is called:
(A) Macro climate (B) Micro climate
(C) Climate (D) Weather

147. Nucleated cells tends to be more resistant to complement mediated lysis than RBC because:
(A) Many nucleated cells an endocytose the membrane attack complex
(B) Membrane attack complex cannot get inserted in the nucleated cell membrane
(C) Membrane attack complex can get inactivated by the nucleated cells
(D) Membrane attack complex get inactivated hence cannot get inserted in the nucleated cell membrane

148. The vaccines are prepared by attenuating pathogenic organisms by growing them in unfavourable conditions which result in gene mutations due to which organism looses pathogenicity but retains their capacity for transient growth is called:
(A) Attenuated vaccines (B) Inactivated vaccines
(C) DNA vaccines (D) Subunit vaccine

149. Match list – I with list – II and select the correct answer using the codes given below the lists:

List – I	List – II
P. Relatively fixed area defined by an animal because it contain a valuable resource	1. Habitat
Q. Clear orientation of communities along a steep environmental gradient	2. Territory
R. Place where an organism lives	3. Zonation
S. Complete account of how an organism uses its environment	4. Niche
	5. Ecotone

(A) P–5, Q–4, R–1, S–3 (B) P–2, Q–3, R–1, S–4
(C) P–2, Q–3, R–4, S–1 (D) P–3, Q–4, R–1, S–5

150. Ecotone is best described as the:
(A) State of equilibrium among various trophy levels in an ecosystem
(B) Potentiality of an animal to adjust to new circumstances
(C) Transition zone between two or more diverse communities
(D) Maximum biomass an ecosystem can support

151. In the statement Isotype is a duplicate specimen of x from the same collection of the same date and same locally, x stand for:
(A) Tototype (B) Holotype
(C) Paratype (D) Neotype

152. Barring the Himalayan region of India can be divided into following four major vegetational zones:
P. Tropical deciduous forests
Q. Tropical rain forests
R. Thorn and scrub forests
S. Tidal forests

The correct sequence of the descending order in terms of the areas occupied by these zones is:
(A) P, Q, R, S (B) P, Q, S, R
(C) Q, P, R, S (D) Q, P, S, R

153. The climatic conditions that prevail at a local scale, or in areas of limited size, such as the imme- diate surroundings of plants and animals is called:
(A) Macro climate (B) Microclimate
(C) Climate (D) Weather

154. For green house effect, which of the following statements are correct?
P. Intensive plantation can produce large amount of CO_2 which traps heat and produces the green house effect
Q. Extensive use of fossil fuel and intensive plantation result in the production of large amount of CO_2 and H_2O forming a thick layer that prevents the heat from being reradiated out
R. Excessive use of fossil fuel produces heat, which is absorbed by the CO_2 and water vapour in the atmosphere and causes green house effect
S. Increasing level of CO_2 tends to war air in the lower layer of the atmosphere of global sale resulting in global warming

The correct sequence of the descending order in terms of the areas occupied by these zones is:
(A) P and Q (B) Q and R
(C) R and S (D) P and S

155. A taxonomic system based on all phenotypic similarities, equally weighted and without regard to evolutionary relationship is called:
(A) Phylogeny (B) Cladistics
(C) Classical evolutionary taxonomy
(D) Phenetics

156. Choose the correct match from A, B, C and D:

Group I	Group II
P. IgA	1. Basophils
Q. IgE	2. δ heavy chain
R. IgG	3. Secretary component
S. IgM	4. Pentamer
	5. Crosses placenta

(A) P–3, Q–1, R–5, S–4 (B) P–3, Q–5, R-2, S–1
(C) P–2, Q–3, R-5, S–4 (D) P–2, Q–1, R-3, S–5

157. An individual may be exposed to an antigen to induce formation of antibodies a type of immunity known as:
(A) Native inactive immunity
(B) Natural active immunity
(C) Artificial active immunity
(D) Artificial inactive immunity

158. Which of the following is not correct about r-selected species?
P. High seed number
Q. Short life span
R. Type I survivorship curve
S. Small size
(A) R (B) P and R
(C) P, Q and S (D) P, Q, R and S

159. The presence of all of the following tend to increase species diversity except:
(A) Competitive exclusion
(B) Keystone predators
(C) Patchy environments
(D) Moderate disturbances

160. Of the three types of ecological pyramids, which pyramid gives the best overall picture of the functional nature of communities:
(A) Pyramid of numbers
(B) Pyramid of energy
(C) Pyramid of biomass
(D) Both (A) and (C)

161. The amount of energy entering a food chain depends on the:
(A) Direction of energy flow in the system
(B) Efficiency of energy recycling in the system
(C) Biomass of carnivores and their efficiency in locating and capturing animal prey
(D) Biomass of autotrophs and their efficie-ncy in transforming solar energy into chemical energy

162. Ecological dominants are those species that:
P. Exert major controlling influence
Q. Live in the community
R. Effect the environment of all other species
S. Control the energy flow
Which of the above statements is/are correct?
(A) P (B) Q
(C) P, R and S (D) P, Q, R and S

163. Food chains are usually short mainly because:
(A) Only a single herbivore species feeds on each plant species
(B) Local extinction of a species cause the extinction of other species in its food chain
(C) Most of the energy in a trophy level is lost as it passes to the next higher level
(D) Predators are less diverse and less abundant than prey

164. Choose the correct statement(s):
P. Groups of species that have similar or comparable roles in the community is termed guild
Q. No two species in the same general territory can occupy for long the same ecological niche
R. When one population produces a substance harmful or beneficial to a competing population then it is called allelopathy
(A) P and Q (B) Q and R
(C) P, Q and R (D) None of these

165. All of the following characteristics are typical of a selected population except:
(A) Occurrence in variable environments
(B) High intrinsic rate of growth
(C) Onset of reproduction at an early age
(D) Extensive parental are of offspring

166. Identify the correct statements
P. The second law of thermodynamics is also known as the law of conservation of energy
Q. Entropy is a measure of the available energy resulting from transformations
R. The transfer of energy through the food chain of an ecosystem is termed as 'energy flow'
S. The second law of thermodynamics deals with the transfer of energy towards more available state
(A) P, Q (B) P, R
(C) R, S (D) Q, S

167. With reference to ecological niche, consider the following statements:
P. Often within the same species, different stages in the organisms life history occupy different niches
Q. In some species, niche segregation may occur between sexes
R. When niches of two species overlap, it always results in the elimination of one species
S. Tidal forests
Which of the statement(s) given above is/are correct?
(A) Q (B) P and Q
(C) P and R (D) P, Q and R

168. The study of the relationship between living organisms and their environment, the interaction of organisms with each other, and the pattern and cause of the abundance and distribution of organism in nature is called:
(A) Ecology (B) Ecosystem
(C) Environment (D) None of these

169. The *ex situ* conservation of genetic resources can be done through:
P. Tissue culture practices
Q. Maintenane of santuaries
R. The establishment of germplasm banks
S. The stablishment of national parks
Which of these statement is/are correct?
(A) R (B) Q and S
(C) P and Q (D) P and R

170. The bryophytic gametophytes are:
(A) Independent from sporophyte
(B) Partially dependent on sporophytes
(C) Lack antheridia and archegonia
(D) Parasitically dependent on sporophytes

171. Choose the correct combination of statement from the following for MH II protein:
P. Has α_1, α_2, α_3 and β_2 domains
Q. α_1 and β_1 are the polymorphic domains
R. Involved in presenting antigen to cytotoxic T cells
S. Only α - chain is glycosylated
(A) Q and R (B) R and S
(C) R and S (D) P and Q

172. A taxonomic system based on all phenotypic similarities, equally weighted and without regard to evolutionary relationship is called:
(A) Physiology
(B) Cladistics
(C) Classical evolutionary taxonomy
(D) Phenetics

173. A type I survivorship curve is characteristic of the species with a rapid increase in mortality in old age. This type of curve is:
(A) Typical of many invertebrates that produce large number of offspring
(B) Typical of humans and other large mammals
(C) Almost never found in nature
(D) Typical of all species of birds

174. Which of the following statement identify the term biological magnification?
(A) Green house effect will have greater impact on tropical countries
(B) Toxins become concentrated in successive tropic levels of food webs
(C) Energy is lost at each level in a food chain
(D) Primary producers are at the bottom of the food chain

175. The correct sequence of various larvae in liver fluke is:
(A) Miracidium, sporocyst, cercaria, redia, metacercaria
(B) Miracidium, sporocyst, redia, cercaria, metacercaria
(C) Sporocyst, redia, miraidium, cercaria, metacercaria
(D) Cercaria, sporocyst, redia, miracidium, metacercaria

176. Match the following

Column – I	Column – II
P. Ctenidium	1. Termites
Q. Social insect	2. Nereis
R. Triploblastic metamerica-lly segmented	3. *Taenia solium*
S. Trochophore larva	4. Monopetinategill
T. Mehlis's gland	5. Sea mouse

(A) P – 2, Q – 5, R – 4, S – 1, T – 3
(B) P – 4, Q – 1, R – 5, S – 2, T – 3
(C) P – 3, Q – 4, R – 5, S – 1, T – 2
(D) P – 1, Q – 5, R – 2, S – 3, T – 4

177. Allopatric speciation is a speciation event that arises from:
(A) Temporal separation of two population
(B) Geographical separation of two populations
(C) An individual having parents of different species
(D) Reproductive isolation of two populations

178. Match list – I (Pollutants) with list – II (Methods of control/recovery) and select the correct answer using the codes given below the lists

List – I	List – II
P. Particulate matter from coal combustion	1. Incineration
Q. Sulphur dioxide fumes	2. Reverse osmosis
R. NO_2 from fumes	3. Combination modification
S. Municipal waste water	4. Scrubber installa-tion
	5. Electrostatic preci-pitators

(A) P–1, Q–3, R–4, S–2
(B) P–1, Q–3, R–2, S–4
(C) P–5, Q–4, R–3, S–2
(D) P–5, Q–4, R–2, S–3

179. Which of the following sequence about successional process?
(A) Nudation → Ecesis → Reaction → Competition
(B) Nudation → Reaction → Competition → Competition
(C) Nudation → Competition → Reaction → Ecesis
(D) Ecesis → Nudation → Competition → Reaction

180. Find out the correct statements for Linnaeus system of classification:
P. It is also known as artificial – sexual system of classification
Q. It was published in the name of 'Genera Plantarum'
R. In this system plants belonging to widely distant natural groups are placed under one order of a class
S. In this system gymnospermae and angiospe-rmae are placed under one order of a class
(A) P, R
(B) Q, R
(C) R, S
(D) P, R

181. Members of which tropic level would suffer the most from biological megnification of harmful substances:
(A) First tropic level
(B) Second tropic level
(C) Third tropic level
(D) Fourth tropic level

182. Supposing a genetic analysis was carried out on populations of salt water crocodiles along the east coast of India and freshwater crocodile in the rivers along the same coast. Theoretically, one would expect to find:
(A) Greater genetic differences between the saltwater crocodile populations
(B) Greater genetic differences the freshwater crocodile populations
(C) Lower mutation rates in freshwater populations
(D) Higher mutation rates in freshwater populations

183. Recently, a new species of frog was discovered in the Western Ghats whose closest relatives are in Seychelles and Madagascar. This proves that:

(A) India was connected to these islands in the past
(B) Frogs are good dispersing across oceans
(C) Frogs are living fossils
(D) India was a part of Gondwanaland

184. The deletion of a DNA sequence does not lead to any change in an organism's traits. This means that the sequence may have:

(A) Been non-functional
(B) Coded for RNA but not a protein
(C) Been part of an enhancer
(D) Been part of an exon

185. A forest patch has 100 deer on Sunday, May 1, 2007. Hunting season opens on May 8, 2007. Hunters kill 20 deer every Sunday. Each deer eats 1 kg of grass per day. If none of these populations are reproducing, how much grass would there need to be for the hunters to wipe out the deer population before the grass is wiped out?

(A) 700 kg (B) 1.4 tonnes
(C) 2.1 tonnes (D) 1.8 tonnes

186. Birds species A and B both nest in marshes. Species A is dominant over B when they are found together. Species B will occupy the entire marsh if species A is absent, but if species A is present, it will exclude Species B from the marsh center and restrict Species B to the edge of the marsh. For species B, the entire marsh would represent its:

(A) Realised niche (B) Habitat
(C) Range (D) Fundamental niche

187. Marsupials are found in:

(A) Australia and South America
(B) Australia alone
(C) Australia and Madagaskar
(D) Australia and Africa

188. Which of the following traits is not characteristic of all vascular plants?

(A) Presence of seeds
(B) Alteration of generations
(C) Presence of xylem and phloem
(D) Presence of lignin in cell walls

189. A genome consists of four distinct DNA sequences. Assuming that any subset of these sequences can determine a trait, the maximum number of possible traits is:

(A) 16 (B) 24
(C) 25 (D) 15

190. Two sources may be described as antagonistic when:

(A) Consuming both kills the animal
(B) One can replace the other in an animal's diet
(C) The animal requires proportionately more resources when both are consumed together
(D) The animal cannot consume both simultaneously

191. In a classical conditioning experiment, it was found that rats could trained to avoid food that made them sick a few hours after ingestion if the food was paired with a specific odour during presentation. The aversive conditioning did not, however, work if the food was paired with a sound rather than an odour. This experiment shows that:

(A) Rats are not as sensitive to sound stimuli as they are to odours
(B) Both sound and odours need to be presented simultaneously with the food for successful conditioning
(C) The efficacy of taste aversion condition varies for different kinds of stimuli
(D) Taste aversion is harder to condition than other behavior

192. Mongooses belong to the family :

(A) Mustelidae (B) Viverridae
(C) Herpestidae (D) Hysterreidae

193. Due to global warming, species of plants and animals can generally be expected to migrate from:

(A) Higher altitudes to lower altitudes
(B) Lower altitudes to higher altitudes
(C) West to East along a given latitude
(D) East to West along a given latitude

194. The following scientist is associated with both the field of sociobiology and the theory of island biogeography:

(A) Konrad Lorenz (B) Joseph Connell
(C) Stephen Hubbell (D) E.O. Wilson

195. Climate change, habitat loss, ultraviolet radiation, fungal disease, overexploitation and invasives have been implicated in the decline of:

(A) Frogs (B) Vultures
(C) Tigers (D) Snakes

196. Which of the following is not a source of genetic variation?

(A) Base substitution (B) Genetic drift
(C) Recombination (D) Base insertion

197. Chimpanzees are more closely related to humans than they are to gorillas, yet chimpanzees and gorillas appear morphologically very similar to each other. This is because:

(A) Morphological characters have evolved convergently in chimpanzees and gorillas
(B) Chimpanzees and gorillas have retained ancestral traits
(C) Chimpanzees and gorillas are apes
(D) Chimpanzees and gorillas hybridize

198. The cross between a mutant mother and a normal father gives rise to offspring that are all mutant; the reciprocal cross gives rise to offspring that are all normal. This means that the mutation is likely to be:

(A) In the mitochondrial genome
(B) In the nuclear genome
(C) Dominant (D) Recessive

199. Starting from the outside of a cell and moving towards the inside, which of the following components of a 'typical' plant cell are organizes correctly?

(A) Middle lamella ⇒ primary cell wall ⇒ secondary cell wall ⇒ plasma membrane ⇒ cytoplasm

(B) Plasma membrane ⇒ primary cell wall ⇒ secondary cell wall ⇒ middle lamella cytoplasm

(C) Cytoplasm ⇒ plasma membrane ⇒ middle lamella ⇒ secondary cell wall ⇒ primary cell wall

(D) Middle lamella ⇒ secondary cell wall ⇒ primary cell wall ⇒ plasma membrane ⇒ cytoplasm

200. Isolated island such as the Hawaiian and Galapagos islands have a higher proportion of closely related endemic species occupying different niches than the mainland. Which of the following statements is not a plausible explanation for the above observation?

(A) Very few species from the mainland colonize these islands

(B) Species that colonize these island eventually radiate into multiple species

(C) Isolated islands often have numerous unoccupied niches

(D) These islands have higher rates of species extinction

201. C4 plants and CAM plants are similar in their photosynthetic adaptations in which of these respects:

(A) In both cases, the stomata normally close during the day

(B) Both types of plants make their sugar without the C3 (Calvin) cycle

(C) In both cases, an enzyme other than Rubisco carries out the first step in carbon fixation

(D) Both types of plants make most of their sugar in the dark

202. In population A, each women bears exactly two female children, bom at maternal age 20 and 25. Population B is the same, except that the children are bom when the mother is 25 and 30. Assume that these are the only differences between A and B. When the sizes of the two population are monitored over time, we will find that:

(A) Both are stationary

(B) Both grow at the same rate

(C) B grows faster than A

(D) A grows faster than B

203. If the average molecular weight of one amino acid is 110, the molecular weight of a peptide made up of 10 amino acids is expected to be:

(A) 1100 (B) 938

(C) 876 (D) 744

204. The number of individuals or biomass at each trophie level is usually depicted as an ecological pyramid. Sometimes, inverted pyramids such as produced by a host parasite system, may be observed. These can be seen in:

(A) Pyramids of numbers and biomass, but never energy

(B) Pyramids of biomass energy, but never numbers

(C) Pyramids of energy numbers, but never biomass

(D) All kinds of pyramids

205. You wish to compare similarity in body form between four species of insects belonging to a particular group. To this end, you perform detailed morphometric measurements using 42 different characters on ten individuals of each species. An appropriate statistical test to quantitatively analyse morphological similarity would be:

(A) A principal components analysis

(B) An analyse of variance

(C) A multiple regression analysis

(D) A cluster analysis

206. Which of the following is least likely to explain greater bird species diversity in the terrestrial tropics along a latitudinal gradient?

(A) Higher plant diversity (B) Greater area

(C) Higher temperatures (D) Higher rainfall

207. The maximum absorbance of light for a visual pigment is at 420 nm. A point mutation in the gene encoding this pigment shifts the wavelength of maximum absorbance to 320 nm. This would result in a transition in colour vision for the organism from:

(A) Red to green sensitivity

(B) Green to blue sensitivity

(C) Blue to ultraviolet sensitivity

(D) Blue to green sensitivity

208. The total number of species of insects in two large fields is the same. Within each of these fields are two smaller patches. In field 1, the two patches share very few species. In the other, the two patches share most species. Which of the following statements is true?

(A) The individual patches in field 1 have more species than in field 2

(B) The individual patches in field 2 have more species than in field 1

(C) The individual patches in both fields have the same number of species

(D) There is high species turnover between field 1 and 2

209. Species that are most likely to have atrophied vision are those that:

(A) Live in caves or underground

(B) Live underwater

(C) Are nocturnal

(D) Live in the Arctic

210. Which statement about Mendel's cross of TT peas with tt peas is not true?

(A) Each parent can produce only one type of gamete
(B) F1 individuals produce two types of gametes
(C) Three genotypes are observed in the F2 generation
(D) Three phenotypes are observed in the F2 generation

211. The resolving power of the human eye is about:

(A) 1 metre (B) 0.1 m
(C) 1 millimetre (D) 0.1 millimetre

212. Let us suppose there are 10 red socks and 10 green socks in a drawer. Suppose we were to draw two socks from the drawer at random, one after the other, without replacement (i.e. without putting back the first sock). What is the probability of getting two green socks?

(A) 0.237 (B) 0.278
(C) 0.5 (D) 1

213. Which of the following is not a shared features of all chordates?

(A) Pharyngeal slits
(B) Notochord
(C) Dorsal nerve cord
(D) Four-chambered heart

214. In terrestrial vertebrates, sounds are sensed by:

(A) Ciliary cells
(B) Hair cells
(C) Pillar cells
(D) Basal epithelial cells

215. Consider a plant with dehiscent seed pods. At what angle to the ground should the seeds be expelled if they are to travel the largest possible horizontal distance?

(A) 30 degrees (B) 60 degrees
(C) 45 degrees (D) 90 degrees

216. An island in the Indian Ocean is first colonized by bird species A. Several decades later, bird species B, a close relative of species A with a similar sized beak, also arrives and establishes itself on this island. Subsequently, the beak sizes of the two species diverge. This is an example of:

(A) Adaptive radiation
(B) Character displacement
(C) Competitive exclusion
(D) Co-evolution

217. Oil starts to smell bad after being stored for a long time. This is because of:

(A) A process of adaptation on our sense of smell
(B) Glycolysis
(C) Sedimentation
(D) The conversion of hydrocarbon double bonds to single bonds

218. Individuals of a bird species that sings at dawn were captured and kept singly in complete darkness for two months and their singing activity was continuously monitored. Each bird typically produced song everyday but, with time, the singing period no longer coincided with the dawn. When exposed to normal light the birds, within a couple of days, synchronized their singing activity with the dawn. Which of the following statements is false?

(A) Light is not necessary for maintenance of a daily singing rhythm
(B) Singing is controlled by a system with a periodicity of exactly twenty four hours
(C) Singing is controlled by a system with a periodicity of approximately twenty four hours
(D) Light is necessary to reset the rhythm of the song control system

219. In the fruit fly, appoint mutation in one gene can cause the fly to develop four wings instead of two. This is likely to be because:

(A) The mutation removed a piece of the chromosome
(B) The mutation added an extra piece to the chromosome
(C) One gene can regulate the activities of many other genes
(D) There is a gene for four wings

220. The presence of living and extinct ratites (flightless birds such as the ostrich) in Australia, Africa, New Zealand, Madagaskar and South America can be best explained by:

(A) Continental drift
(B) Long – distance dispersal
(C) Convergent evolution
(D) Human mediated transport

221. Which statement about alleles is not true?

(A) They are different forms of the same gene
(B) There may be several alleles at a locus
(C) One allele might a dominant over the other
(D) They occupy different loci on the same chromosome

222. In honey bee colonies, the levels of mRNA of a specific gene called period (per) were found to be elevated in foragers as opposed in non-foraging members of the colony. Foragers are typically older than non-foragers in bee colonies. A few young 'precocious' foragers were found and their per mRNA levels were found to be as high as normal foragers. This experiment demonstrates that:

(A) Foraging leads to an increase in per mRNA level
(B) Increase in per mRNA level results in foraging behavior
(C) Per mRNA levels are higher in older bees, who also happen to be foragers
(D) There us a positive correlation between per mRNA level and foraging behavior

223. Ears of different moth species are found on different parts of the body, including the thorax, wings, abdomen and proboscis. This suggests that:

(A) Moths are evolving more rapidly than other insect groups
(B) Ears in moths arose as a single event in their phylogenetic history
(C) Ears in moths arose multiple times in their phylogenetic history
(D) The structure of moth ears is different from those of other insect groups

224. Imagine that we were interested in the surface are of a bird's wing, divided by the body length of the bird (call this "relative wing area" or RWA). From geometrical considerations alone, what would you expect when comparing the RWA of a small bird with that of a large bird? Assume that there is a strong positive correlation between wing length and body length:

(A) RWA of a small bird is expected to be larger than RWA of a large bird
(B) RWA of a small bird is expected to be smaller than RWA of a large bird
(C) RWA of a small bird is expected to be the same as the RWA of a large bird
(D) RWA does not depend on the size of the bird

225. The bacteria which cause septicaemia are fast becoming resistant to the antibiotics used for controlling them. Which one of the following explanations fits best?

(A) Each and every individual bacterium, in order to survive, develops a resistance to the antibiotic when exposed to it, and passes that resistance on to future
(B) Only some bacteria, in order to survive, develop a resistance to the antibiotics when exposed to it, and pass that resistance on to future generations
(C) A few bacteria have a natural heritable resistance to the antibiotics, so they survive when exposed to it, and pass that resistance on to future generations
(D) All the bacteria have a natural heritable resistance to the antibiotic, so they survive when exposed to it, and pass that resistance on to future generations

226. You carefully study populations of two very similar field mice, one from Karnataka and one from Kerala. You want to know whether the populations belong to the same species or to two different species. You could most confidently decide this if you could:

(A) Show that the ranges of the two mice overlap without hybridization occurring
(B) Bring the two types of mice into the laboratory to see if they will mate
(C) Demonstrate that the natural ranges of the types of mice are entirely allopatric
(D) Show that the Karnataka mice live in wetter habitats than the Kerala mice

227. When plants do not receive enough water their photosynthetic rate drops significantly. This is because:

(A) Water is a raw material needed for the light dependent reactions
(B) The stomata close and carbon dioxide is not available
(C) Sugar builds up and inhibits photosynthesis
(D) Not enough oxygen produced to keep glycolysis running

228. In an aversive conditioning paradigm, bees were trained to associate the colour yellow with bitter tasting KCl solution. Bees that once experienced the bitter taste for 30 seconds in the presence of a yellow card placed next to the solution refused to sample any solution in the presence of the yellow card on subsequent encounters when tested 10, 30, 60 min., one day and one week after training. If the bees were cooled to 4°C for 30 minutes immediately after encountering the stimuli, they no longer formed the association. If the cooling was started 10 min after the encounter however, they were able to make the association. This experiment shows that:

(A) Short term memory formation requires 30 minutes of exposure to the stimuli
(B) Long term memory formation requires 30 minutes of exposure to the stimuli
(C) Memory formation occurs within 10 minutes of encountering the stimuli
(D) Memory formation begins 10 minutes after encountering the stimuli

229. You can see the sal tree (*Shorea robusta*) growing naturally in which of these regions?

(A) Aravallis of Rajasthan
(B) Dooars of Bengal
(C) Western Ghats of Karnataka
(D) Andaman and Nicobar islands

230. There is a small population of tigers. A genetic study shows that there is relatively little genetic difference between individuals. The most likely cause for this is:

(A) They have a rare genetic disease
(B) Recently heavy mortality due to poaching
(C) The population was reduced to a few individuals in its recent history
(D) Low diversity of prey species

231. The lens of many vertebrate eyes is the crystallized form of a protein that also functions during digestion as a metabolic enzyme. This is an illustration for the fact that:

(A) Evolution is opportunistic
(B) Digestion co-evolved with vision
(C) Digestion evolved after vision
(D) Vision evolved after digestion

232. In humans, the genetic relatedness between first cousins is:

(A) 0.5 (B) 0.25
(C) 0.125 (D) 0.0625

233. An ordered data set consists of one million DNA sequences arranged according to length. If you were asked to find a sequence of desired length, choosing each sequence in your search at random, you should be able to find the desired sequence in about:

(A) 2 steps (B) 20 steps
(C) 200 steps (D) 2000 steps

234. During a study on fish in a pond, 200 fish are caught and marked on the first day. On the second day of sampling, 150 fish are caught, of which 50 are already found to be marked. What is the estimate of the total number of fish in the pond?

(A) 600 (B) 300
(C) 150 (D) 1200

235. A plant ecologist has to study the diversity of plants in a forest with a strong gradient of rainfall from North to South. A good sampling design would be:

(A) Random placement of quadrats in the forest
(B) Placement of quadrats from North to South
(C) Placement of quadrats from East to West
(D) Enumeration of all individuals in the forest

236. From our understanding of energy flow through various trophic levels of an ecosystem, which of the following statements is correct?

(A) About 50% of the energy in one trophic level is passed on to the next level
(B) Primary consumers have the highest ecological efficiency
(C) Only a negligible amount of energy is lost along each link in the food chain
(D) Producing meat is an inefficient way of obtaining energy trapped by photosynthesis

237. Which of the following is closest to being iso-osmoiic to its environment?

(A) A freshwater protozoan
(B) A marine jellyfish
(C) A marine reptile
(D) A freshwater bony fish

238. In flying insects such as dragonflies, butterflies and grasshoppers, the wing beat frequency is low and each nerve impulse results In wing muscle contraction. In flies and bees, however, the wing beat frequency is several times the nerve impulse frequency. Which of the following is not an adaptation to achieve this?

(A) Synchronous contraction of horizontal and vertical muscles
(B) An elastic thorax that acts as a mechanical oscillator
(C) Alternate contraction of horizontal and vertical flight muscles
(D) Fibrillar muscles that contract in response to being stretched

239. Which of the following is not an example of a fixed action pattern?

(A) Courtshin singing by a male fruitfly
(B) The headstand display of a fighting fish
(C) Gostings following their mother is single file
(D) Packing behavior of newly hatched chicks

240. Wild pigs regularly feed on sorghum crops but only within 1 m of the boundary of the crop field. Given this crop feeding behavior, if a farmer wished to minimize the area damaged by wild pigs, which of the following configurations of crop fields should he prefer?

(A) One 20 x 20 m field
(B) Four 10 x 10 m fields
(C) Hundred 2 x 2 m fields
(D) Twenty five 4 x 4 m fields

241. In a simple ecological system with two competitor species A and B competing over a single resource, thereby predicts that stable co-existence of the two species is most likely when:

(A) Interspecific competition is stronger than intraspecific competition in species A
(B) Intraspecific competition is weaker than interspecific competition in species B
(C) Intraspecific competition is stronger than interspecific competition in both species
(D) Interspecific competition is stronger than intraspecific competition in both species

242. Which of the following life history adaptation is least expected when predation pressure on a fish species that grows continuously throughout its lifespan is focused on large size classes?

(A) Allocate more resources preferentially to early reproductive effort than to growth
(B) Allocate more resources preferentially to growth than to early reproductive effort
(C) Early age at maturity
(D) Produce many small offspring in relatively few reproductive seasons

243. Which of the following is not an explanation for higher species richness in a community?

(A) Species are more specialized in their use of resources
(B) Species have more overlap in their use of resources
(C) A greater range of resources is available
(D) Some species are better at utilizing resources

244. Which of the following is likely to be heard at the furthest distance from source from the caller (given that all calls are produced with the same loudness)?

(A) The alarm call of a langur
(B) The chirping of a cricket
(C) The song of a magpierobin
(D) The clicks of a bat in a roost

245. Identical genomes can give rise to strikingly different phenotypes. Which of the following is not an example of this?

(A) Morphological castes within social insect species
(B) The larva and adult of a butterfly species
(C) Male and female peafowl
(D) Fraternal twins

246. The influx of which ion is critical for neurotransmitter release at the synapse?

(A) Sodium (B) Potassium
(C) Calcium (D) Magnesium

247. Which of the following is a wrong explanation for why temperate and Arctic lakes never freeze completely solid?

(A) The hydrogen bonds in ice make ice less dense than liquid water
(B) Water expands when it solidifies
(C) Water is denser as a solid than as a liquid
(D) Floating ice insulates the liquid water below

248. The axoloti, a species of salamander, does not metamorphose, continues to be aquatic, and retains juvenile traits such as external gills even as an adult. This developmental phenomenon where juvenile traits are retained in adults is known as:

(A) Embryogeny
(B) Neoteny
(C) Fluctuating asymmetry
(D) Developmental asynchrony

249. Which of the following is least likely as a general explanation for the evolution of aging?

(A) The force of selection on deleterious mutations acting late in the lifespan is weaker
(B) Because of predation and stochastic events organisms have a lower probability of living long and therefore should invest more in reproduction than in body maintenance and repair
(C) Because resources are typically limited, older individuals should avoid competing with the offspring for resources and therefore selection should reduce lifespan to prevent competition across generations
(D) Pleiotropic genes that increase fitness earlier in life may have negative effects later in life that result in ageing

250. The least likely explanation for the extinction of dinosaurs is:

(A) Competition with mammals
(B) Deccan trap volcanism
(C) Climate change
(D) Meteor impact

251. A type of immunity that occurs when the vaccination of a portion of the population (or herd) provides protection to unprotected individuals is termed:

(A) Hard immunity (B) Herd immunity
(C) Soft immunity (D) None of these

252. Which of the following is incorrectly paired with its description?

(A) Neritic zone → Shallow area over continental shelf
(B) Abyssal zone → Benthic region where light does not penetrate
(C) Littoral zone → Area of open water
(D) Intertidal zone → Shallow area at the edge of water

253. Group I lists relationships that an exist between two organisms. Their descriptions are given in group II. Find the correct set of matches between groups I and II:

Group I	Group II
P. Ammensalism	1. One population is benefited and the other is harmed
Q. Commensalism	2. Two dissimilar species living together in lose association
R. Symbiosis	3. Heterotrophic organisms that ingest other organisms
S. Parasitism	4. One population is inhibited but the other is not affected
	5. One population is benefited but the other is not affected

(A) P-4, Q–1, R–2, S-3 (B) P-3, Q–5, R–4, S-1
(C) P-5, Q–4, R–3, S-5 (D) P-4, Q–5, R–2, S–1

254. The main feature of the biological species concept is its emphasis on the

(A) Large morphological differences between different species
(B) Genetic variation within populations
(C) Role of sexual reproduction in maintaining diversity within a species
(D) Absence of gene flow between different species

255. Rearrange the following taxonomic terms in the correct hierarchical order:

Order – Family – Phylum – Class

(A) Family – Phylum – Class – Order
(B) Phylum – Order – Class - Family
(C) Class – Phylum – Family - Order
(D) Phylum – Class – Order – Family

256. A habitat is:

(A) Tropic level occupied by organisms
(B) The same as an ecosystem
(C) A particular are inhabitated by plants and animals
(D) The number of different organisms living in a specific area

257. Aggregate fruit is defined as the one which develops from:

(A) Multicarpellary, syncarpous gynoecium
(B) Multicarpellary, apocarpous gynoecium
(C) Complete inflorescence
(D) Multicarpellary, superior ovary

258. **Consider the following statements:**

P. In Bentham and Hooker's classification, gymnosperms are placed between dicots and monocots

Q. In Engler and Prantl's classification, monocots are regarded as more advanced than dicots

Which of these statement is/are correct?

(A) P (B) Q

(C) P and Q (D) None of these

259. **Early in their life cycle, amphibians and most invertebrates go through a larval stage that is distinctly different from the adult body form. Which of the following is often consequence of this phenomenon?**

(A) The young and adults of a species occupy different habitats or eat different foods, thereby reducing intraspecific competition

(B) Predators are less likely to prey on the young, thereby increasing the young animals survival rates

(C) The life cycles of the animals of a particular species are synchronized, thereby ensuring that all the animals become sexually mature at the same time

(D) The young are more sensitive than the adults to selection factors in the environment, thereby increasing the rate at which these species evolve

260. **Which of the following may have an inverted biomass pyramid?**

(A) Terrestrial and shallow water ecosystem

(B) Open or deep water situations

(C) Lakes and ponds

(D) All of these

261. **A condition in which structural or functional damage is produced by the action of immunologically competent cells or Ab against self antigen is called:**

(A) Immune tolerance (B) Autoimmunity

(C) Transplantation (D) None of these

262. **Match the correct combination of plasmid DNA to their properties**

Plasmid DNA	**Property**
P. Conjugated plasmid	1. Can integrate into the chromosome and replicate when the chromosome is coupled
Q. Cryptic plasmid	2. Capable of transferring itself between prokaryotes
R. Episome	3. Does not appear to have any function

(A) P – 1, Q – 3, R – 2 (B) P – 2, Q – 3, R - 1

(C) P – 2, Q – 1, R – 3 (D) P – 3, Q – 2, R – 1

263. **Increase in the emission of fossil fuels is through to result in global warming. This is possible because an increase in atmospheric CO_2 level would:**

(A) Increase the amount of sun light entering the earth atmosphere

(B) Increase the amount of infrared radiation entering the earth

(C) Absorb the infrared radiation reflected by the earth

(D) Insulate earth from old breeze coming from the space

264. **Which of the following statement is incorrect?**

(A) Enzyme activity associated with reverse transcriptase that digests the RNA template after DNA has been synthesized is RNase -H

(B) Viral envelop attaches the HIV to host cell is gp120

(C) Genome of retroviruses contain two copies of plus ssRNA

(D) HIV contains single copy of ssRNA

265. **Macrophages, which are also called monocytes, have the ability to:**

(A) Process and present antigens to T cells

(B) Produce antibodies

(C) Express IgM molecules on their cell surface

(D) Differentiate into dendrite cells when necessary

266. **Which one of the following places does not constitute in-situ conservation?**

(A) Biosphere reserves (B) Wildlife sanctuaries

(C) National parks (D) Botanical gardens

267. **When plant diversity is maintained in natural habitats, the conservation is called?**

(A) in-vitro (B) in-situ

(C) in-vivo (D) ex-situ

268. **Which of the following statements about water is correct?**

(A) It is critical for many of the chemical processes found in living systems

(B) Polar substances that can form hydrogen bonds will dissolve in it

(C) Nonpolar substances like oils are not soluble in it

(D) All of these

269. **Salt dissolves well in water as water molecules:**

(A) Form hydrogen bonds with the positively and negatively charged ions

(B) Make nonpolar covalent bonds with the positively charged ions only

(C) Surround the ions because of their charge but do not form hydrogen bonds

(D) Share electrons with the ions to make polar covalent bonds

ANSWERSHEET

8. DIVERSITY OF LIFE

1. (C)	**2. (D)**	**3. (B)**	**4. (A)**	**5. (A)**	**6. (D)**	**7. (D)**	**8. (B)**	**9. (B)**	**10. (B)**
11. (A)	**12. (C)**	**13. (B)**	**14. (D)**	**15. (B)**	**16. (D)**	**17. (D)**	**18. (A)**	**19. (D)**	**20. (C)**
21. (B)	**22. (D)**	**23. (B)**	**24. (C)**	**25. (A)**	**26. (C)**	**27. (C)**	**28. (D)**	**29. (B)**	**30. (A)**
31. (D)	**32. (A)**	**33. (A)**	**34. (B)**	**35. (D)**	**36. (A)**	**37. (A)**	**38. (C)**	**39. (B)**	**40. (A)**
41. (D)	**42. (A)**	**43. (A)**	**44. (A)**	**45. (D)**	**46. (D)**	**47. (B)**	**48. (A)**	**49. (D)**	**50. (D)**
51. (C)	**52. (C)**	**53. (A)**	**54. (B)**	**55. (B)**	**56. (C)**	**57. (A)**	**58. (B)**	**59. (A)**	**60. (B)**
61. (A)	**62. (D)**	**63. (C)**	**64. (D)**	**65. (C)**	**66. (B)**	**67. (C)**	**68. (B)**	**69. (A)**	**70. (B)**
71 (A)	**72. (C)**	**73. (B)**	**74. (A)**	**75. (A)**	**76. (D)**	**77. (D)**	**78. (B)**	**79. (C)**	**80. (C)**
81. (C)	**82. (A)**	**83. (B)**	**84. (B)**	**85. (B)**	**86. (D)**	**87. (D)**	**88. (B)**	**89. (D)**	**90. (B)**
91. (C)	**92. (D)**	**93. (C)**	**94. (D)**	**95. (A)**	**96. (B)**	**97. (D)**	**98. (C)**	**99. (B)**	**100. (B)**
101. (B)	**102. (D)**	**103. (B)**	**104. (D)**	**105. (D)**	**106. (A)**	**107. (A)**	**108. (A)**	**109. (B)**	**110. (B)**
111. (A)	**112. (C)**	**113. (C)**	**114. (C)**	**115. (C)**	**116. (C)**	**117. (D)**	**118. (B)**	**119. (C)**	**120. (D)**
121. (A)	**122. (C)**	**123. (D)**	**124. (D)**	**125. (B)**	**126. (C)**	**127. (B)**	**128. (A)**	**129. (A)**	**130. (C)**
131. (B)	**132. (D)**	**133. (B)**	**134. (C)**	**135. (C)**	**136. (D)**	**137. (A)**	**138. (C)**	**139. (B)**	**140. (D)**
141. (D)	**142. (B)**	**143. (D)**	**144. (A)**	**145. (B)**	**146. (C)**	**147. (A)**	**148. (A)**	**149. (B)**	**150. (C)**
151. (B)	**152. (A)**	**153. (B)**	**154. (C)**	**155. (D)**	**156. (A)**	**157. (C)**	**158. (A)**	**159. (A)**	**160. (B)**
161. (D)	**162. (D)**	**163. (C)**	**164. (C)**	**165. (D)**	**166. (C)**	**167. (D)**	**168. (A)**	**169. (D)**	**170. (A)**
171. (D)	**172. (D)**	**173. (B)**	**174. (B)**	**175. (B)**	**176. (B)**	**177. (B)**	**178. (C)**	**179. (A)**	**180. (D)**
181. (D)	**182. (B)**	**183. (D)**	**184. (A)**	**185. (C)**	**186. (C)**	**187. (A)**	**188. (A)**	**189. (A)**	**190. (D)**
191. (C)	**192. (B)**	**193. (A)**	**194. (D)**	**195. (C)**	**196. (D)**	**197. (A)**	**198. (A)**	**199. (A)**	**200. (C)**
201. (C)	**202. (D)**	**203. (B)**	**204. (A)**	**205. (A)**	**206. (B)**	**207. (C)**	**208. (B)**	**209. (A)**	**210. (D)**
211. (D)	**212. (B)**	**213. (D)**	**214. (A)**	**215. (C)**	**216. (B)**	**217. (D)**	**218. (B)**	**219. (B)**	**220. (A)**
221. (D)	**222. (B)**	**223. (C)**	**224. (C)**	**225. (B)**	**226. (B)**	**227. (B)**	**228. (C)**	**229. (A)**	**230. (C)**
231. (A)	**232. (B)**	**233. (A)**	**234. (B)**	**235. (B)**	**236. (D)**	**237. (A)**	**238. (A)**	**239. (C)**	**240. (C)**
241. (D)	**242. (B)**	**243. (C)**	**244. (B)**	**245. (A)**	**246. (C)**	**247. (C)**	**248. (B)**	**249. (D)**	**250. (D)**
251. (B)	**252. (C)**	**253. (D)**	**254. (D)**	**255. (D)**	**256. (C)**	**257. (B)**	**258. (C)**	**259. (A)**	**260. (B)**
261. (B)	**262. (B)**	**263. (C)**	**264. (D)**	**265. (A)**	**266. (D)**	**267. (B)**	**268. (D)**	**269. (C)**	

Chapter - 9

ENVIRONMENTAL BIOTECHNOLOGY

1. The pollutants which are already present in nature, but are released in substantial amounts by man are known as:
(A) Quantitative pollutants
(B) Qualitative pollutants
(C) Degradable pollutants
(D) Primary pollutants

2. The biological amplification of DDT in the various trophic levels is known as:
(A) Green house effect (B) Biomagnification
(C) Entrophication (D) Pollution

3. Environment can be improved by:
(A) Wind erosion (B) Excessive tree fall
(C) Conservation (D) None of these

4. Smog is produced due to :
(A) Air and metal dust
(B) Water and nitrogen oxide
(C) Smoke and fog
(D) None of these

5. The photochemical smog is formed by:
(A) Nitrogen oxides
(B) Hydrocarbons
(C) Nitrogen oxides and hydrocarbons
(D) Solar radiations upon nitrogen oxides and hydrocarbons

6. Photochemical smog brings about:
(A) Defoliation
(B) Bleaching and blazing of foliage
(C) Floral bud shedding
(D) Discolouration and curling of petals

7. In metropolitan cities the automobiles cause air pollution:
(A) 80% (B) 60%
(C) 50% (D) 100%

8. Huge quantity of sewage is dumped in a river. The BOD will:
(A) Increase (B) Decrease
(C) Slightly increase (D) Remain unchanged

9. Jet aeroplane releases pollutants in air is called:
(A) Smog
(B) Photo-chemical oxidants
(C) Aerosols
(D) Algae and bacteria

10. Fertilizer-pesticides and insecticides may cause:
(A) Air pollution (B) Water pollution
(C) Both (A) and (B) (D) None of these

11. Pollutants directly emanating from human activities are:
(A) Quantitative pollutants
(B) Qualitative pollutants
(C) Air pollutants
(D) Water pollutants

12. Atmospheric pollutants formed by human activities constitute of the total:
(A) 95% (B) 75%
(C) 25% (D) 0.05%

13. Most atmospheric pollutants do not rise above:
(A) 600 m (B) 1000 m
(C) 2000 m (D) 200 m

14. Aerosol is a system of colloidal particles having a size of:
(A) 5 μm (B) 1 – 2 μm
(C) Less than 1 μm (D) More than 10 μm

15. A secondary pollutants is:
(A) O_3 (B) CO
(C) CH_4 (D) Pb

16. A colourless, nonirritant, highly toxic gas that impairs respiration is:
(A) SO_2 (B) NO_2
(C) CO_2 (D) CO

17. Lead pollution in Indian cities is mainly due to:
(A) Lead pencils (B) Batteries
(C) Automobiles (D) Electronic gadgets

18. Pollutants likely to deplete ozone layer is:
(A) CO (B) Nitrogen oxides
(C) Chloroflurocarbons (D) Both (B) and (C)

19. Acid rain is caused by:
(A) Excess production of coal gas
(B) Excess release of CO_2 due to increasing combustion and respiration
(C) Excess release of SO_2 and NO_2 from burning fossil fuels
(D) Excess production of gaseous hydrocarbons

20. SO_2 pollution is indicated by destruction of:
(A) Climbers (B) Lichens
(C) Mosses (D) Grasses

21. Freom is:
(A) Ferrous sulphide pollutant
(B) Chloroflurocarbon
(C) Iron containing polycyclic hydrocarbons
(D) Metal fluoride

22. Maximum atmospheric NO_2 pollution is found in:
(A) Baroda (B) Chennai
(C) Nagpur (D) Mumbai

23. O_3 promotes:
(A) Oxidation of biochemicals
(B) Destruction of chlorenchyma
(C) Both (A) and (B)
(D) Growth of plants

24. Water soaked appearance of leaves due to sulphur dioxide pollution is caused by:
(A) Destruction of chlorophyll
(B) Damage to membranes
(C) Both (A) and (B)
(D) Corrrosive action over metals

25. In addition to SO_2, metals and textiles are damaged by:
(A) Nitrogen oxides (B) Fluorides
(C) Carbon monoxide (D) Aldehydes

26. Photochemical reaction of nitrogen oxides with unsaturation hydrocarbons produces:
(A) Methyl isocyanate
(B) Phosgene
(C) Benzpyrene
(D) Peroxy-acyl nitrates

27. Level of atmospheric CO_2 has increased from:
(A) 290 to 340 ppm (B) 250 to 300 ppm
(C) 300 to 350 ppm (D) 330 to 350 ppm

28. CO_2 content of the atmosphere is likely to double by the year:
(A) 2001 (B) 2010
(C) 2020 (D) 2050

29. Gravity setting tanks/porous filters are used to remove particles larger than:
(A) 2 μm (B) 50 μm
(C) 20 μm (D) 10 μm

30. Electronic precipitators are employed to remove particulate pollutants of the size of:
(A) 5-20 μm (B) 20-50 μm
(C) 50 μm and larger (D) 0.5-5.0 μm

31. B.O.D. is related to:
(A) Detergents (B) Putrescibility
(C) Inorganic pollutants
(D) Organic pollutants

32. A persistent insecticide is:
(A) Organophosphates (B) Carbamates
(C) Thiocarbamates
(D) Chlorinated hydrocarbons

33. An organoclorine insecticides is:
(A) DDT/aldrin (B) Malathion
(C) Tetraethyl pyrophosphate
(D) Thiocarbamate

34. Domestic water is:
(A) Effluent (B) Biodegradable
(C) Nonbiodgradable (D) Air pollutants

35. Inadequate drainage in a soil will lead to:
(A) Floods (B) Soil erosion
(C) Aridity (D) Salination

36. Major reason of soil erosion in India is:
(A) Tilling of land (B) Deforestation
(C) Overgrazing and deforestation
(D) Overgrazing

37. Jhuming occurs in India is:
(A) Hilly areas (B) Madhya Pradesh
(C) North – East (D) Malabar area

38. Ecological balance of a mining area is restored through:
(A) Conversion into agricultural area
(B) Revegetation of area
(C) Prevention of soil erosion
(D) Protection against grazing

39. Terrestrial life would become impossible with:
(A) Change in orbit of earth
(B) Change on atmosphere permitting all solar radiations to reach lithosphere
(C) Disappearance of moon
(D) Decrease in mean annual temperature by 10^0C

40. Disease aggravated by pollution is:
(A) Scurvy (B) Rheumation
(C) Haemophilia (D) Bronchitis

41. Beyond the lower atmosphere, extending out into space, is the stratosphere, in which temperature remains relatively constant from:
(A) -15 to -25 (B) -45 to -75
(C) -80 to -95 (D) -5 to -10

42. On of the characteristics of the stratosphere is the occurrence of a well-marked layer having an increased concentration of -------- at about 20 miles (32 km) above sea level:
(A) Carbon dioxide (B) Carbon monoxide
(C) Ozone (D) Hydrogen

43. In the ozone layer, the ozone does not exceed ------- ppm at 20 miles (32 km) above sea level (stratosphere):
(A) 5 (B) 10
(C) 15 (D) 20

44. The gaseous mixing in the atmosphere is, in general, similar at sea level and at heights upto ---- km.
(A) 20 (B) 40
(C) 60 (D) 80

45. Ozonated hexane led to distruption of ----- in plants:
(A) Membrane permeability
(B) Protein synthesis
(C) Lipid synthesis
(D) Stem elongation

46. Animals and nongreen plants continuously use oxygen and liberate ------ into the air:
(A) Sulfur dioxide (B) Carbon monoxide
(C) Carbon dioxide (D) Nitrous oxide

47. The rate of photosynthesis under the conditions of high light intensity may be controlled by:
(A) Light (B) Carbon dioxide
(C) Water (D) None of these

48. North of the equator, surface winds are known as:
(A) Northeast trade winds
(B) Southeast winds
(C) Westerlies
(D) Easterlies

49. In the humid tropics and subtropics the process of weathering (soil formation) is known as:
(A) Podzolization (B) Gleization
(C) Latosolization (D) None of these

50. The range of coffee production is limited approximately by the ------- ^{0}F of isotherms for the coldest month:
(A) 25 (B) 35
(C) 45 (D) 55

51. The diversity pertaining to a plant composition in several plant communities in one range of environment is known as:
(A) Alpha index of diversity
(B) Beta index of diversity
(C) Gamma index of diversity
(D) None of these

52. Among the types of natural succession, the one which begins in water is known as:
(A) Xerarch (xerosere)
(B) Halosere
(C) Hydrarch (hydrosere)
(D) Psammosere

53. In the earth's atmosphere at ----- cal/cm²/annum of solar energy is received:
(A) 15×10^6 (B) 15×10^8
(C) 15×10^{10} (D) 15×10^{12}

54. Which of the following crops utilizes even upto 10-12 per cent of the total solar energy:
(A) Sorghum (B) Rice
(C) Maize (D) Sugarcane

55. Fixation of each gramme of carbon from carbon dioxide to organic compounds during photosynthesis traps -------- of solar energy:
(A) 124 k cal (B) 134 k cal
(C) 144 k cal (D) 154 k cal

56. C_3 plants show a light saturation upto 40000 to 50000 lux, while C_4 plants upto full sun in tropics upto:
(A) 60000-70000 (B) 70000-80000
(C) 90000-100000 (D) 115000-120000

57. ----- pollutants are very harmful for human health and therefore, such polluted water should essentially be purified before their supply for drinking purposes:
(A) Biological (B) Physical
(C) Mechanical (D) Metallic

58. Ionising -------- are very important and harmful pollutants:
(A) Reflection (B) Deflection
(C) Radiation (D) Absorptions

59. The sound beyond --------dB are regarded as pollutant as it harms hearing ability:
(A) 60 (B) 80
(C) 100 (D) 120

60. If all the moisture present in the air at any time to converted into precipitation, it would be equal to 2.5 cm or about ------ of water supply by precipitation (average residence time):
(A) 5 days (B) 10 days
(C) 15 days (D) 20 days

61. About 2.5 cm rainfall sufficient to produce over 10 tonnes of stream flow water on a watershed area of about -------cm²:
(A) 3000 (B) 4000
(C) 5000 (D) 6000

62. Fresh water bodies like rivers and streams are known as:
(A) Palustrine (B) Lotic or riverine
(C) Lentic or lacustrice (D) None of these

63. On of the basis of temperature variation in the vertical column of water, or thermal stratification, lake ecosystem have been divided into the following zones wherein ----- comes in direct contact of sun and wind:
(A) Thermocline (B) Hypolimnion
(C) Epilimnion (D) None of these

64. Mercury level of ----- µg/g of fresh weight of fish is considered unsafe for human consumption:
(A) 0.5-1.0 (B) 1.6-2.0
(C) 2.1-2.5 (D) 2.6-3.0

65. Fish absorbs and may accumulate zinc in muscles amounting approximately upto how many time more than that in sea water:
(A) 3700 (B) 4700
(C) 5700 (D) 6700

66. Permissible limit of iron in drinking water is:
(A) 1.0 ppm (B) 1.5 ppm
(C) 2.0 ppm (D) 2.5 ppm

67. Potassium in natural waters never assume toxic proportions and remain less than ------ mg/litre:
(A) 10 (B) 15
(C) 20 (D) 25

68. When with chloride ion in drinking water, the dominant cation is potassium or magnesium, the salty taste is not felt even upto ---------- mg/litre:
(A) 500 (B) 750
(C) 1000 (D) 1250

69. The characteristic pungent odorous taste of onion is due to :
(A) Magnesium (B) Calcium
(C) Potassium (D) Sulphur

70. Many factory waters contain how much mg/litre of nitrate nitrogen:
(A) 20 (B) 30
(C) 40 (D) 50

71. Chloride added to drinking water supply when encountered with ------- gets neutralized and chloroamines are formed:
(A) Nitrous oxide (B) Carbon dioxide
(C) Mathane (D) Ammonia

72. The nitrate content say in the range of 30 to 40 mg/litre lead to poor oynation of blood haemoglobin and blood turns:
(A) Blackish (B) Whitish
(C) Bluish (D) Yellowish

73. In a study in USA, the earthworm inhabiting a foliarly fed DDT-tree, contained upto 80 ppm of DDT even after how months:
(A) 5 (B) 7.5
(C) 10 (D) 10.5

74. The coliform bacteria count in drinking water is regarded safe if it is:
(A) 5 or less (B) 10 or less
(C) 15 or less (D) 20 or less

75. The PSI value range ------ has been classified as hazardous (air) causing premature on set of certain diseases and aggravation of symptoms:
(A) 200-299 (B) 300-399
(C) 100-199 (D) 50-100

76. Methane gas once released in the atmosphere stays for how many years:
(A) 3-4 (B) 4-5
(C) 5-7 (D) 8-10

77. The earth during the last one century has warmed up by---------- ^{0}C as estimated:
(A) 0.3-0.7 (B) 0.8-1.0
(C) 1.2-1.3 (D) 1.4-1.5

78. The earth temperature during the past 20000 years has risen by --- ^{0}C:
(A) 2-3 (B) 4-5
(C) 6-7 (D) 8-9

79. Utraviolet radiation are of similar wavelengths than of visible light and those ranging from 320-400 nm are called:
(A) UV-C (B) UV-B
(C) UV-A (D) UV-D

80. CFCs methane and nitrogen oxide gases have the capacity to react with ozone to convert into:
(A) Nitrogen (B) Hydrogen
(C) Oxygen (D) None of these

81. The estimates reveal that the maximum NO_2 emission in European countries is in ----- in the range of 2Tg/annum (Tetagramme=10^{12} grammes):
(A) France (B) Germany
(C) England (D) None of these

82. The threshold concentration upto which no visible injury symptoms appear on plants due to NO_2 is 1 mg/m^2 on:
(A) Daily average (24 hrs) (B) Weekly average
(C) Monthly average (D) Annual average

83. The suggested threshold level of fluoride pollution in atmosphere is how much quantity (ppm)
(A) 10-40 (B) 50-80
(C) 90-100 (D) 110-115

84. Uranium contribute the radioactivity of about – pci/litre of ocean water:
(A) 2 (B) 3
(C) 4 (D) 5

85. The Department of Environment in independence India was formed in November:
(A) 1960 (B) 1970
(C) 1980 (D) 1990

86. Serious efforts on Environment protection in India began with the establishment of NCEPC by the Department of Science and Technology in February:
(A) 1952 (B) 1962
(C) 1972 (D) 1982

87. Environment Protection Act 1986 came into force in India on --------- 1968:
(A) October 19 (B) November 19
(C) September 19 (D) August 19

88. In December, 1989 standards were laid for the noise limits in day and night time in various area and silence zones upto ------- around educational institutions, hospitals and courts in India:
(A) 50 m (B) 100 m
(C) 150 m (D) 1200 m

89. The optimum rainfall range for growing maize is:
(A) 360-500 mm (B) 560-700 mm
(C) 300-400 mm (D) 460-600 mm

90. For urdbean, the suitable climate is:
(A) Hot and humid (B) Cool and humid
(C) Cool and dry (D) Hot and dry

91. The arhar (pigeonpea) is commonly grows well in:
(A) Tropical and subtropical climate
(B) Temperate climate
(C) Humid and temperate climate
(D) Sub-humid and temperate climate

92. High temperature above 41°C at flowering of castor results in:
(A) Rotting of pollens and poor fertilization
(B) Blasting of flowers and poor seed set
(C) Delayed pollination and poor fertilization
(D) Reduced pollination and rotting of seed

93. The crop of ---- is more liable to suffer from frost and cold:
(A) Toria (B) Pahari rai
(C) Sarson (D) Banarasi rai

94. Linseed is a which season crop:
(A) Warm (B) Cool
(C) Hot (D) Dry

95. Safflower is a which season crop:
(A) Dry (B) Summer
(C) Arid (D) Cool

96. The sunflower crop requires cool climate from:
(A) Germination to seedling stage
(B) Seedling to flowering stage
(C) Flowering to maturity stage
(D) Seedling to maturity stage

97. If temperature is high during maturation of sunflower, the linoleic acid:
(A) Remain unaffected
(B) Increases with slow rate
(C) Decreases
(D) Increases with rapid rate

98. Ozone layer is getting depleted because of:
(A) Excessive use of automobiles
(B) Excessive formation of industrial units
(C) Excessive use of man made compounds containing both fluorine and chlorine
(D) Excessive deforestation

99. Viruses, which cause hepatitis, are transmitted through:
(A) Air (B) Water
(C) Food (D) Personal contact

100. Vectors can be defined as:
(A) Animals carry the infecting agents from risk person to another healthy person
(B) Microorganisms which cause many diseases
(C) Infected person
(D) Diseased plants

101. Which of the following is a recently originated problem of environment?
(A) Ozone layer depletion (B) Green house effect
(C) Global warming (D) All of these

102. When we breathe in air, nitrogen also goes inside along with oxygen. What is the fate of this nitrogen?
(A) It moves alongwith oxygen into the cells
(B) It comes out with the CO_2 during exhalation
(C) It is absorbed only by the nasal cells
(D) Nitrogen concentration already move in the cells so it is not at all absorbed

103. Troposphere of the atmosphere extending upto a height from ground to:
(A) 8-18 km (B) 50 km
(C) 40 km (D) >50 km

104. Fog, dew, mist, rain etc occur in the layer of:
(A) Thermosphere (B) Mesosphere
(C) Stratosphere (D) Troposphere

105. Troposphere is the densest part of atmosphere and contains about:
(A) 55% of atmospheric mass
(B) 75% of atmospheric mass
(C) 85% of atmospheric mass
(D) 95% of atmospheric mass

106. The layer separated the stratosphere and mesosphere is called:
(A) Tropopause (B) Stratopause
(C) Mesopause (D) None of these

107. Inasphore is present in:
(A) Troposphere (B) Stratosphere
(C) Mesosphere (D) Thermosphere

108. Heat can flow between substance or within substance by:
(A) Conduction (B) Convection
(C) Radiation (D) All of these

109. The reason (s) for the decline in temperature with increase altitude is:
(A) Air is heated by the earth
(B) Water vapour content of air decreases with increase in altitude
(C) Expansion of air at higher elevation
(D) All of these

110. Vertical temperature decrease is called lapse rate and the normal lapse rate is:
(A) 4.5^0 C/km (B) 6.5^0 C/km
(C) 8.5^0 C/km (D) None of these

111. Heat transfer into the soil is mainly by:
(A) Conduction (B) Convection
(C) Radiation (D) None of these

112. The rate of heat flux into soil depends on:
(A) Temperature gradient
(B) Thermal conductivity
(C) A and B
(D) None of these

113. Thermal conductivity of soil decrease with:
(A) Increase porosity
(B) Increase organic matter
(C) Decrease soil moisture
(D) All of these

114. What will be the effect of tillage operation on thermal conductivity of soil?
(A) Remain unchanged
(B) Decrease
(C) Increase
(D) First increase then decrease

115. South-west monsoon contributes how much per cent of rainfall?
(A) 50 % (B) 75 %
(C) 80 % (D) 95 %

116. Average rainfall in India is around:
(A) 2120 mm (B) 1120 mm
(C) 850 mm (D) 500 mm

117. A source of nonbiological oxygen is:
(A) Photodissociation (B) Photolysis
(C) Formation of fats from carbohydrates
(D) Both (B) and (C)

118. A process that occurs without loss or gain of heat is:
(A) Coriolis effect (B) Eustatic effect
(C) Adiabatic (D) All of these

119. Adiabatic expansion resulting in cooling is called:
(A) Coriolis effect
(B) Joule-Kelvin effect
(C) Le Chatelier Principle
(D) Principle of Parsimony

120. Which area has four seasons:
(A) Tropics (B) Monsoonic
(C) Temperate (D) Arctic

121. Praying Mantis is green coloured predator insect that shows mimicry of the type:
(A) Concealing-protective (B) Warning-protective
(C) Alluring-aggressive
(D) Concealing-aggressive

122. Male Atlas Moth has huge antennae that can detect pheromone released by female moth from a distance of:
(A) 1 km (B) 2 km
(C) 5 km (D) 8 km

123. The biological amplication of DDT in the various trophic levels is known as:
(A) Green house effect (B) Biomagnification
(C) Entraphication (D) Pollution

124. Taj may be destroyed by:
(A) Flood in Yamuna
(B) Temperature mediated spoilage of marble
(C) Air pollutants from Mathura refinery
(D) All of these

125. Smog occurs in places of:
(A) Excess SO_2 (B) Low temperature
(C) High temperature (D) Excess NH_3

126. Acid rains is due to increase in atmospheric concentration of:
(A) Ozone and dust (B) CO_2 and CO
(C) SO_3 and CO (D) SO_2 and NO_2

127. Major pollution causing agents is:
(A) Hydrocarbons (B) Animals
(C) Human (D) Birds

128. Pollution is:
(A) Removal of top soil
(B) Release of toxic/undesirable materials in environment
(C) Conservation of energy
(D) All of these

129. Effect of pollution is most marked on:
(A) Natural balanced of nature
(B) Natural geochemical cycles
(C) Natural flora of place
(D) All of these

130. Chief sources of water and soil pollution is:
(A) Mining (B) Thermal power plant
(C) Agroindustry (D) All of these

131. Ultimate environment hazards to mankind is due to:
(A) Nuclear pollution (B) Water pollution
(C) Air pollution (D) Noise pollution

132. Which causes water pollution?
(A) Smoke/fly ash
(B) Automobile exhausts
(C) Aeroplanes
(D) 2, 4-D and pesticides

133. Air pollution is not caused by:
(A) Pollen grains (B) Hydroelectric power
(C) Industries (D) Automobiles

134. Pollution is controlled by:
(A) Banning atomic blasts
(B) Use of electrically operated automobiles
(C) Sewage treatment (D) All of these

135. Carbon monoxide is harmful to human being as it is:
(A) Carcinogenic
(B) Antagonistic to CO_2
(C) With higher affinity for haemoglobin as compared to oxygen
(D) Destructive to O_3

136. Disease caused by eating fish found in water contaminated with industrial waste having mercury is:
(A) Minamata disease (B) Bright's disease
(C) Hashimoto's disease (D) Osterosclerosis

137. Fish die in water bodies polluted by sewage due to:
(A) Pathogens
(B) Clogging of gills by silt
(C) Reduction in oxygen
(D) Foul smell

138. Lichens indicate pollution by:
(A) O_3 (B) SO_2
(C) NO_2 (D) CO

139. Drained sewage has B.O.D.:
(A) More than that of water
(B) Less than that of water
(C) Equal to that of water
(D) None of these

140. B.O.D. is connected with:
(A) Organic matter
(B) Microbes
(C) Microbes and organic matter
(D) None of these

141. Which one is the most important causative pollutants of soil?
(A) Glass junk (B) Iron junk
(C) Detergents (D) Plastics

142. BOD is:
(A) Biological oxygen deficit
(B) Biosphere oxygen demand
(C) Biological oxygen demand
(D) None of these

143. Pollution can bring about change in:
(A) Biogeochemical cycling (B) Abiotic environment
(C) Biotic environment (D) None of these

144. Which one is the major reason for pollution in big cities?
(A) Fossil fuel (B) Acid rain
(C) Heat dispersion (D) None of these

145. World environment day is:
(A) 28th Feb. (B) 5th June
(C) 7th Aug. (D) 10th April

146. Biodegradable pollutants is:
(A) Plastic (B) Asbestos
(C) Sewage (D) Mercury

147. Water pollution is due to:
(A) Sulphur dioxide (B) Carbon dioxide
(C) Oxygen (D) Industrial discharges

148. Major reason for soil erosion in India is:
(A) Tilling of land (B) Deforestation
(C) Overgrazing and deforestation
(D) Over grazing

149. SO_2 and NO_2 produce pollution by increasing:
(A) Alkalinity (B) Acidity
(C) Neutrality (D) Buffer action

150. Which one is found in higher concentration in the atmospheric?
(A) Ozone (B) Nitrous oxides
(C) Carbon dioxide, carbon monoxide and sulphur dioxide
(D) All of these

151. Carbon dioxide is a major pollutant in:
(A) Air (B) Soil
(C) Water (D) Noise

152. Phosphate pollution is caused by:
(A) Weathering of phosphate rocks only
(B) Agricultural fertilizers
(C) Phosphate rocks and sewage
(D) Sewage and agricultural fertilizers

153. Which pigment protects plants from UV damage?
(A) Chlorophyll (B) Xanthophyll
(C) Carotenoids (D) Phycocyanin

154. Indescriminate use of fertilizers causes:
(A) Air pollution (B) Water pollution
(C) Land pollution (D) All of these

155. SO_2 pollution causes:
(A) Plasmolysis
(B) Destruction of cell wall
(C) Destruction of Golgi bodies
(D) Destruction of chlorophyll

156. Depletion of ozone layer is due to:
(A) Oxides of nitrogen (B) Oxides of carbon
(C) Oxides of sulphur (D) None of these

157. Chipko movement originated in the district:
(A) Allahabad (B) Chamoli
(C) Hardwar (D) Dehradun

158. Chipko movement originated in Gopeshwar in:
(A) 1953 (B) 1963
(C) 1973 (D) 1083

159. A movement similar to Chipko which is spread in south is:
(A) Uliso (B) Appiko
(C) Belesu (D) Balasu

160. Chipko movement originated in:
(A) Silent valley of Kerala
(B) Kangra valley of Himachal Pradesh
(C) Tehri Garwal of UP
(D) Panchmari of MP

161. Extinction of numerous wild animals is due to:
(A) Hunting (B) Deforestation
(C) Overcrowding (D) Scarcity of food

162. Chipko movement is connected with:
(A) Conservation of natural resources
(B) Plant/forest conservation
(C) Plant Breeding
(D) Project tiger

163. Forest control drought through:
(A) Lot of water plants
(B) Increasing rainfall
(C) Retention of water and prevention of erosion
(D) Functioning as water shed

164. Overgrazing causes:
(A) Negative pollution
(B) Positive pollution
(C) Soil erosion
(D) Reduction in crop yields

165. Deforestation brings about:
(A) Increased sunlight
(B) Increased grazing area
(C) Weed control (D) Soil erosion

166. Deforestation will decrease:
(A) Soil erosion (B) Land slides
(C) Soil fertility (D) Rainfall

167. Deforestation is the major causal agent of:
(A) Depletion of natural resources
(B) Environmental pollution
(C) Desertification of habitat
(D) Genetic erosion

168. Deforestation does not lead to:
(A) Quick nutrient cycling
(B) Soil erosion
(C) Alteration of local weather conditions
(D) Destruction of natural habitat of wild animals

169. Forest destruction results in:
(A) Loss of wild life (B) Floods and drought
(C) Soil erosion (D) All of these

170. Afforestation should be with:
(A) Exotic species (B) Indigenous species
(C) Bamboos (D) Eucalyptus

171. If we remove half of the forest cover of earth, the crises that will occur:
(A) Many species would become extinct
(B) Population, pollution andecological imbalance will rise
(C) Energy crises will commence
(D) The remaining forest will correct the imbalance

172. Decreased rainfall in northern areas is due to:
(A) Change in sun
(B) Global phenomenon
(C) Deforestation in catchment areas
(D) Deforestation in evaporating areas near equator

173. Increased incidence of floods in the plains of northern India are due to:
(A) Increased deforestation in catchment areas
(B) Increase in incidence of rainfall
(C) Silting of dams
(D) More area under cultivation

174. Soil erosion can be prevented by:
(A) Restricted human activity
(B) Good plant cover
(C) Checking movement of animals
(D) Wind screen alone

175. Adherence to two-children norms shall help in:
(A) Improving food web
(B) Checking pollution
(C) Stabilising ecosystem
(D) Fertility of soil

176. Mist is fog horizontal visibility of:
(A) 100 m (B) 200 m
(C) 500 m (D) 1 km

177. Fog is formed by:
(A) Smoke
(B) Condensation of water vapours
(C) Condensation of water vapours over dust particles
(D) Both (A) and (C)

178. Succession on sand dune is:
(A) Lithosere (B) Psychrosere
(C) Psammosere (D) Halosere

179. The amount of biogenic nutrient present in the abiotic environment is called:
(A) Standing state (B) Standing crop
(C) Detritus (D) Reserve nutrient

180. Phosphorus has:
(A) Imperfect nutrient cycle
(B) Perfect nutrient cycle
(C) Gasseous cycle
(D) Mixed cycle

181. Carbosphere comprises:
(A) Atmosphere (B) Lithosphere
(C) Hydrosphere (D) All of these

182. Adiabatic expansion resulting in cooling is called:
(A) Coriolis effect
(B) Joule-Kelvin effect
(C) Le Chatelier Principle
(D) Principle of Parsimony

183. Cyclic/frontal rainfall generally occurs:
(A) All over the world
(B) Tropical areas
(C) Temperate areas
(D) Boundary of warm tropical air and cold polar air

184. Which area has four seasons:
(A) Tropics (B) Monsoonic
(C) Temperate (D) Arctic

185. The land of wind mills is:
(A) UK (B) USA
(C) Netherlands (D) Himachal Pradesh

186. The shining and thunder in clouds during rainy season:
(A) Start simulataneosly
(B) Occur thunder first and then shining starts
(C) Occur shining first and then thunder starts
(D) Occur mostly no thunder and no shining

187. Daily weather forecast is obtained in the country from:
(A) Pune (B) New Delhi
(C) Mumbai (D) Kolkata

188. The layer of atmosphere having uniform temperature is:
(A) Stratosphere (B) Exosphere
(C) Thermosphere (D) Troposphere

189. The physical state of the atmosphere at a given place and time is referred to as:
(A) Weather (B) Climate
(C) Both of these (D) None of these

190. The first meteorological department of India was started at Shimla in:
(A) 1863 (B) 1864
(C) 1865 (D) 1866

191. Climate word originated from:
(A) Greek (B) Latin
(C) English (D) Parsian

192. Climate word derived from:
(A) Klima+ Logos (B) Klima+Logo
(C) Clima + Logo (D) Clima+Logos

193. The basic meaning of klima is:
(A) Angle incidence of sun light
(B) Changes in temperature
(C) Incidence of humidity
(D) All of these

194. The science that seeks to describe and explain that nature of climate is called:
(A) Climatology (B) Meteorology
(C) Both of these (D) None of these

195. The study of physical status of climate is called:
(A) Physical climatology
(B) Statistical climatology
(C) Descriptive climatology
(D) Applied climatology

196. To study the physical atmospheric factors viz., temperature, relative humidity, clouds, wind velocity and direction is called:
(A) Physical climatology
(B) Statistical climatology
(C) Descriptive climatology
(D) Applied climatology

197. How many years of weather is optimum for climate:
(A) 30 years (B) 31 years
(C) 32 years (D) 33 years

198. To study the different factors of climate such as temperature an rainfall, are divided into state level is called:
(A) Physical climatology
(B) Statistical climatology
(C) Descriptive climatology
(D) Applied climatology

199. To study the relationship with other branched os climatology is called:
(A) Physical climatology
(B) Statistical climatology
(C) Descriptive climatology
(D) Applied climatology

200. The weather information collected from observatory are depicted in tables is called:
(A) Weather chart (B) Weather map
(C) Both of these (D) None of these

201. The weather information are depicted through map is called:
(A) Weather chart (B) Weather map
(C) Both of these (D) None of these

202. World Meteorological Organisation (WMO) was established at:
(A) Geneva (Switzerland) (B) Spain
(C) USA (D) Canada

203. World Meteorological Organisation (WMO) was established in:
(A) 1950 (B) 1951
(C) 1952 (D) 1953

204. International Meteorological Organisation was established in:
(A) 1877 (B) 1878
(C) 1879 (D) 1880

205. The science of climatological condition which have direct relation or relevance to agriculture is called:
(A) Agro-meteorology (B) Agro-climatology
(C) Meteorology (D) Climatology

206. The composite or generalization of the variety of day to day weather conditions is called:
(A) Climate (B) Weather
(C) Season (D) All of these

207. The observations recorded from open area of 2 meter height of soil surface is called:
(A) Macro climate (B) Micro climate
(C) Topo climate (D) Local climate

208. When weather data collected from sloppy land is called:
(A) Macro climate (B) Micro climate
(C) Topo climate (D) Local climate

209. When weather data collected from standing crop is called:
(A) Macro climate (B) Micro climate
(C) Topo climate (D) Local climate

210. The data recorded in macro climate is called:
(A) Synoptic Meteorological observation
(B) Asynoptic Meteorological observation
(C) Both of these (D) None of these

211. Upto 15 km above the earth is called:
(A) Troposphere (B) Mesosphere
(C) Stratosphere (D) None of these

212. 50 km above the earth is called:
(A) Troposphere (B) Mesosphere
(C) Stratosphere (D) None of these

213. In atmosphere mainly which gas is determined to atmospheric activities is:
(A) O_2 (B) Ozone
(C) Both of these (D) None of these

214. In atmosphere gasses found in:
(A) Mixed status
(B) Without mixed status
(C) Alone
(D) None of these

215. Four gasses- nitrogen, oxygen, argon and CO_2 accounts:
(A) 97 % (B) 98 %
(C) 99 % (D) 100 %

216. Nitrogen alone constituents nearly:
(A) One-third by volume
(B) Two-third by volume
(C) Three-fourth by volume
(D) Four-fifth by volume

217. Oxygen alone constituents nearly:
(A) One-third by volume
(B) Two-third by volume
(C) One-fifth by volume
(D) Four-fifth by volume

218. The blanket of gases and suspended liquids and solids the entirely envelops the earth is called:
(A) Atmosphere (B) Climate
(C) Clouds (D) Gasses

219. The powerful oxidizing agents and therefore a threat to life and property when concentrations become excessive is:
(A) Ozone (B) Nitrogen
(C) Oxygen (D) Carbon di oxide

220. The lower portion of the atmosphere, extending upto about 8 km at the poles and 16 km at the equator is called:
(A) Troposphere (B) Stratosphere
(C) Ionosphere (D) Exosphere

221. The upper portion of the toposphere, extending upto 50-60 km is called:
(A) Troposphere (B) Stratosphere
(C) Ionosphere (D) Exosphere

222. The upper portion of the stratosphere, extending upto 60-500 km is called:
(A) Troposphere (B) Stratosphere
(C) Ionosphere (D) Exosphere

223. The upper portion of the atmosphere, extending upto 500-1000 km is called:
(A) Troposphere (B) Stratosphere
(C) Ionosphere (D) Exosphere

224. Ozone gas mainly found in which region:
(A) Troposphere (B) Stratosphere
(C) Ionosphere (D) Exosphere

225. High air pressure and its affected area is called:
(A) Anticyclone (B) Cyclone
(C) Depression (D) Rainfall

226. Low air pressure and its affected area is called.
(A) Anticyclone
(B) Cyclone and Depression
(C) Fog (D) Rainfall

227. The unit of atmospheric pressure is:
(A) Millibar (B) Bar
(C) Pascal (D) Gram

228. What per cent of isolation is absorbed in the atmosphere by gasses, clouds and suspended solids?
(A) 19 % (B) 17 %
(C) 15 % (D) 13 %

229. Oxygen and ozone at high levels absorb most of which radiation:
(A) Ultra violet radiation (B) Gama radiation
(C) Blue radiation (D) All of these

230. The ratio of the amount of radiation reflected to the amount received on a surface is termed as:
(A) Albedo (B) Radiation
(C) Reflection (D) All of these

231. The ability to admit most of the insolation yet retard losses by reradiation from the earth's surface commonly is known as:
(A) Greenhouse effect (B) Atmosphere effect
(C) Both of these (D) None of these

232. The highest per cent of albedo is found:
(A) Desert (B) Tropical forest
(C) Deciduous forest (D) Coniferous forest

233. The instrument which measure sunshine and radiation is:
(A) Total radiation pyranometer
(B) Net radiometer
(C) Foster switch sunshine duration sensor
(D) All of these

234. The ratio of the mass of water vapour actually in the air to a unit of mass of air, including the water vapour is called:
(A) Specific humidity (B) Relative humidity
(C) Both of these (D) None of these

235. The ratio of the amount of water vapour actually in a volume occupied by air to the amount in the space could contain at saturation is called:
(A) Specific humidity
(B) Relative humidity
(C) Both of these (D) None of these

236. Early morning hours when temperatures are low the relative humidity will be:
(A) More (B) Less
(C) Not change (D) Unknown

237. The visible aggregates of water droplets, ice particles, or a mixture of both alongwith varying amounts of dust particles is called:
(A) Fog (B) Frost
(C) Clouds (D) All of these

238. The clouds are a cirriform layer or patch of small white flakes or tiny globules arranged in distinct group or lines is called:
(A) Cirrocumulus (B) Cirrostratus
(C) Altocumulus (D) Altostratus

239. The clouds is a thin white veil of cirrus is called:
(A) Cirrocumulus (B) Cirrostratus
(C) Altocumulus (D) Altostratus

240. The middle clouds predominate at height ranging from 2 to 8 km above ground levels is called:
(A) Altocumulus (B) Altostratus
(C) Both of these (D) None of these

241. The banner clouds above peaks are of which type clouds is called:
(A) Cirrocumulus (B) Cirrostratus
(C) Altocumulus (D) Altostratus

242. When the cloud layer lowers to become some what thicker and darker and falling rains or snow obscures its base it is called:
(A) Cirrocumulus (B) Cirrostratus
(C) Altocumulus (D) Nimbostratus

243. The clouds are broken into fragments by the wind it is called:
(A) Cirrocumulus (B) Cirrostratus
(C) Fractostratus (D) Altostratus

244. The clouds are dense, dome shaped and have flat bases is called:
(A) Cumulus (B) Cirrostratus
(C) Altocumulus (D) Altostratus

245. The heavy showers of rain, snow or hail, often accompanied by lighting and thunder occurs by:
(A) Cirrocumulus (B) Cumulonimbus
(C) Altocumulus (D) Altostratus

246. More than 60 per cent rains occurs by:
(A) South-west monsoon
(B) North-East monsoon
(C) Both of these
(D) None of these

247. Less than 30 per cent rains occurs by:
(A) South-west monsoon
(B) North-East monsoon
(C) Both of these
(D) None of these

248. North-East monsoon also called:
(A) Post monsoon period (B) Post rainy season
(C) Both of these (D) None of these

249. A counter clock wise rotation of wind in northern hemisphere is:
(A) Cyclone (B) Anticyclone
(C) Rainfall (D) Fog

250. A counter clock wise rotation of wind in southern hemisphere is:
(A) Cyclone (B) Anticyclone
(C) Rainfall (D) Fog

251. Wind speed measured by:
(A) Windvane (B) Anemometer
(C) Both of these (D) None of these

252. The barometric pressure is highest at its centre and decreased out ward is indication of:
(A) Cyclone (B) Anticyclone
(C) Rainfall (D) Fog

253. The common pressure that initiates thunder storms development are:
(A) Heating and convection in moist air over warm land surfaces
(B) Passage of cold, moist air over warm water
(C) Radiational cooling at upper levels
(D) Advection of cold air aloft
(E) All of these

254. The height of a thunderstorms is related to:
(A) Latitude (B) Season
(C) Both of these (D) None of these

255. Thunderstorms accused by surface heating over land are most common in summer in the:
(A) Afternoon (B) Early morning
(C) Both of these (D) None of these

256. Most violent storms of the lower troposphere is:
(A) Tornado (B) Watersprout
(C) Both of these (D) None of these

257. The moisture index is more in:
(A) Perhumid (B) Humid
(C) Semiarid (D) Arid

258. The moisture index is less in:
(A) Perhumid (B) Humid
(C) Semiarid (D) Arid

259. Human health and comfort suggest another possible approach to defining:
(A) Climatic types (B) Relative humidity
(C) Temperature (D) None of these

260. The summation of all atmospheric events an infinite time span does not changes is called:
(A) Climate (B) Weather
(C) Both of these (D) None of these

261. The periods of atmospheric changes on the order of millions of years is:
(A) Paleoclimates (B) Taiga climate
(C) Polar climate (D) Ocean climate

262. The mean monthly temperatures are all below 0^0C (snow, ice or barren rocks covers such areas) and vegetation is entirely lacking is called:
(A) Paleoclimates (B) Taiga climate
(C) Polar climate (D) Ocean climate

263. The key element in the water budget in which is also the link between moisture and energy exchange is.
(A) Evaporation (B) Transpiration
(C) Evapo-transpiration (D) None of these

264. Evapotraspiration can be measured by:
(A) Lysimeter
(B) Evapotranspirometer
(C) Both of these
(D) None of these

265. The frost occurs on clear nights with a temperature inversion and usually results in formation of ice crystals on bold objects is called:
(A) Radiation frost (B) Advection frost
(C) Black frost (D) None of these

266. Advection frost is also called:
(A) Radiation frost (B) Advection frost
(C) Black frost (D) None of these

267. The frost which results when the temperature at the surface in an air mass is below freezing is called:
(A) Radiation frost (B) Advection frost
(C) Black frost (D) None of these

268. The deficit that results when soil moisture is insufficient to meet the demands of potential evapotranspiration is:
(A) Drought (B) Rainfed
(C) Dryland (D) None of these

269. Drought may be classified as:
(A) Permanent drought may be associated with arid climate
(B) Seasonal drought, which occurs in climate with distinct amount periods of dry weather
(C) Drought due to precipitation variability
(D) All of these

270. Combating the drought effects are:
(A) Growing of less water requiring crops
(B) Decreasing the water need of crops
(C) Increasing the water supply
(D) All of these

271. When the annual rainfall is less than the normal over an area for a prolonged period is called:
(A) Meteorological drought
(B) Hydrological drought
(C) Agricultural drought (D) None of these

272. When a prolonged drought results in drying of the available water sources like streams, tanks, reservoir, well etc is called:
(A) Meteorological drought
(B) Hydrological drought
(C) Agricultural drought (D) None of these

273. Shortage of supply of water to meet the demand for crop maturity is called:
(A) Meteorological drought
(B) Hydrological drought
(C) Agricultural drought (D) None of these

274. During snowfall, the trees:
(A) Do not respire
(B) Do not photosynthesis
(C) Show minimum life function
(D) Show maximum transpiration

275. Chief air pollutant which is likely to deplete ozone layer is:
(A) Sulphur dioxide (B) Carbon dioxide
(C) Carbon monoxide
(D) Nitrogen oxides and fluorocarbons

276. Layer of ozone is present in:
(A) Stratosphere (B) Thermosphere
(C) Troposphere (D) None of these

277. The atmosphere stratum nearest to earth is:
(A) Thermosphere (B) Stratosphere
(C) Troposphere (D) None of these

278. Part of earth covered by water is about:
(A) 73% (B) 50%
(C) 92% (D) 87%

279. Floods can be checked by:
(A) Making dams (B) Deforestation
(C) Reforestation (D) None of these

280. Percentage of fresh water in the hydrosphere is:
(A) 2.7% (B) 30%
(C) 13% (D) 33%

281. Percentage of fresh water found in lakes, ponds, swamp, streams and rivers is:
(A) 22.4% (B) 77.2%
(C) 7.7% (D) 0.36%

282. Of the total fresh water, ground water constitutes:
(A) 90% (B) 10%
(C) 22.4% (D) 44.2%

283. Of the total water evaporated from oceans, precipitation returns:
(A) 100% (B) 90%
(C) 75% (D) 50%

284. Planning Commission agro-climatic region the whole country has been divided into:
(A) 15 (B) 20
(C) 25 (D) 30

285. The concept of agro-climatic zones were introduced during:
(A) NATP (B) NARP
(C) NREP (D) NAIP

286. The number of agro-climatic zones under NARP:
(A) 127 (B) 146
(C) 152 (D) 164

287. The number of agro-ecosystems identified under NATP:
(A) 6 (B) 9
(C) 14 (D) 21

288. Movement of air occurs in the atmosphere upto height of:
(A) 100km (B) 50 km
(C) 25 km (D) 10 km

289. Peak concentration of ozone in the stratosphere (ozonosphere) lies at a height of:
(A) 10 km (B) 25 km
(C) 40 km (D) 50 km

290. Planning Commission advice that the agro-climatic zones in which plan:
(A) Fifth Five Year Plan
(B) Sixth Five Year Plan
(C) Seventh Five Year Plan
(D) Eight Five Year Plan

291. Minamata disease is due to pollution of:
(A) Organic waste into drinking water
(B) Oil spill in water
(C) Industrial waste mercury into fishing water
(D) Arsenic into the atmosphere

292. The estimates postulate that the carbon dioxide content of atmosphere has increased from 280 parts per million to ---- during last three decades
(A) 300 ppm (B) 350 ppm
(C) 400 ppm (D) None of these

293. Nitrate levels in drinking water above ------- mg per litre are considered a human health hazard:
(A) 5 (B) 10
(C) 15 (D) 20

294. The discovery of oxygen was done by which scientist who proved the keystone of other discoveries that went far toward unlocking the mystery of plant life:
(A) John Mayow (B) John Woodward
(C) Benjamin Franklin (D) Priestly

295. There was not only one "Principle of vegetation" as reflected by agricultural writings during 17th and 18th centuries, was observed by:
(A) William Michael (B) Guy Baird
(C) Francis Home (D) Jone Cummings

296. More than which per cent of earth's surface is covered by water in the form of seas, lakes and rivers:
(A) 40 (B) 50
(C) 60 (D) 70

297. Depending upon the temperature factor, the upper warmer region of lacks is called:
(A) Epilimnion (B) Hypolimnion
(C) Thermocline (D) None of these

298. Secondary consequences of increasing elevation, both agronomically & economically significant are:
(A) Decreased temperature
(D) Decreased precipitation
(C) Decreased wind velocity
(D) Increased soil fertility

299. "Diversion of water from mountain streams by diversion dams to canals of different length and breadth and ditches of varying size, and finally directed to individual fields" is an important practice in:
(A) China (B) Egypt
(C) Philippines (D) Japan

300. The success or failure of the cropping season is determined by the intensity of:
(A) Climatic factors (B) Political factors
(C) Social factors (D) None of these

301. The ideal conditions of temperature for crop production are between:
(A) 12-18°C (B) 18-24°C
(C) 24-30°C (D) 30-36°C

302. The direct effect of strong winds on plants is of which nature:
(A) Chemical (B) Biological
(C) Mechanical (D) None of these

303. Altitude, because of it's ---- effects, induces a zonation of vegetation similar to that found through a wide range of latitude:
(A) Humidity (B) Radiation
(C) Temperature (D) None of these

304. Rice, maize and sugarcane are replaced by temperate zone species at elevations of possibly:
(A) 1000-3000' (B) 4000-6000'
(C) 7000-9000' (D) None of these

305. At 0°C water is in equilibrium with a water vapour pressure of ------- mm of mercury:
(A) 2.58 (B) 3.58
(C) 4.58 (D) 5.58

306. The atmosphere over one square foot of earth's surface at the Equator will contain approximately ----- pounds of water vapour:
(A) 20 (B) 30
(C) 40 (D) 50

307. It has been found that peas will remain in good harvest condition twice as long at ----------°F than at 80°F:
(A) 32 (B) 42
(C) 52 (D) 62

308. The changes that accompany hardening of plants include:
(A) Increased sugar concentration
(B) Decreased soluble proteins
(C) Decreased osmotic pressure
(D) Decreased bound water

309. Cacao, a tropical plant has been found to produce the maximum growth in approximately ------ of full sunlight:
(A) 5% (B) 10%
(C) 15% (D) 25%

310. When white light is passed through a prism, it is dispersed into different wavelengths of various colours and the orange colour appears at:
(A) 435-490 (B) 574-595
(C) 400-435 (D) 595-626

311. It has been demonstrated that the ------ light present in the white light prevents maximal growth rate:
(A) Blue (B) Red
(C) Green (D) Violet

312. Wavelengths longer than ----- m/μ are not visible to the eye, and are called infrared:
(A) 450 (B) 550
(C) 650 (D) 750

313. It has been observed that in a number of out breeding populations of rye grass, *Lolium perenne* could be brought about at low temperature (0-3°C) or ----:
(A) Long day (B) Short day
(C) Intermediate day (D) None of these

314. In -------, long days promote bulb formation whereas, short days promote flowering and seed production:
(A) Sweet potato (B) Artichoke
(C) Cassava (D) Onion

315. The ------- variety Maryland Mommoth in the latitude of Maryland (USA) produces only leaves and needs shorter days for seed production:
(A) Tomato (B) Carrots
(C) Radish (D) Tobacco

316. Cyclone collector is used for minimising:
(A) Radioactive pollution (B) Air pollution
(C) Noise pollution (D) Water pollution

317. Which abiotic components of environment contributes in various ways to fulfillment of human needs?
(A) Air (B) Water
(C) Soil (D) All of these

318. Environment consists of:
(A) Sum total of biotic factors
(B) Sum total of abiotic factors
(C) Both (A) and (B)
(D) Both (A) and (B) minus humans

319. Which constitutes environmental biotechnology?
(A) Waster treatment (B) Waste disposal
(C) Technologies generating less waste
(D) Both (A) and (B)

320. Which constitutes 'end of pipe' technology?
(A) Waster treatment (B) Waste disposal
(C) Technologies generating less waste
(D) Both (A) and (B)

321. Which constitutes 'front of pipe' technology?
(A) Waster treatment (B) Waste disposal
(C) Technologies generating less waste
(D) Both (A) and (B)

322. Recalcitrants xenobiotic compounds are highly stable due to the presence of which substitution groups?
(A) Halogens and carbamyl groups
(B) Nitro- and sulphonate groups
(C) Amino- and methoxy groups
(D) All of these

323. Bioaccumulation/biomagnifications may occurs due to:
(A) Xenobiotic compounds are continuously taken up from the environment and accumulated in the lipid deposits of the body
(B) Consumption of these organisms by other organisms leads to further accumulation of xenobiotic compounds in the bodies of latter
(C) The xenobiotic compounds are recalcitrant and lipid soluble
(D) All of these

324. Which diseases spread through contaminated water/food?
(A) Amoebiosis and typhoid
(B) Cholera and diarrhea
(C) Hepatitis (except B)
(D) All of these

325. Which is not an example of physical pollutants?
(A) Dust and mechanical stress
(B) Sound pollution and electrical storms
(C) Heavy metals
(D) Radiation and fires

326. Presence of coliform bacteria in water is indicative of contamination by mammalian excreta and suggestive of the presence of which pathogenic bacteria?
(A) *Vibrio cholera* (B) *Salmonella*
(C) *Shigella* (D) All of these

327. Which pollutants pose immediate threat to humans and is included in the black list?
(A) Oils and hydro carbons from crude oils
(B) Heavy metals, metaloides and compounds containing them
(C) Organophosphates, cyanides and asbestos
(D) All of these

328. Various approaches to waster treatment may be grouped into which categories?
(A) Anaerobic digestion
(B) Biofilters and aerobic digestion
(C) Landfill and burning
(D) All of these

329. Ladfil sites can be useful in which ways?
(A) Reclamation of derelict sites to develop landscaped gardens, etc
(B) As a source of biogas
(C) As a source of organic fertilizers
(D) Both (A) and (B)

330. Which microorganisms is the commonly found inaerobic digestion systems?
(A) Fungi (B) Bacteria
(C) Protozoa (D) Algae

331. Bioremeddiation of inorganic contaminants is primarily based on:
(A) Animal species (B) Microorganisms
(C) Plant species (D) All of these

332. The water (leachate) from landfills may contain :
(A) COD value of 1000-2000
(B) Total solids content upto 6000-7000 mg/l
(C) Sodium upto 800 mg/l
(D) All of these

333. Which statements is correct about landfill management?
(A) The risk due to methane can be removed by burning or tapping
(B) The risks of fires, offensive odours and increased vector populations may be circumvented by covering the waste with soil on daily basis
(C) The risk of damage from, landfill leachate may be avoided by lining the pit with an impermeable material
(D) All of these

334. The various parameters used to determine the level of pollution in water bodies, usually, determine:
(A) Amounts of organic matter present in the water
(B) Amounts of important inorganic nutrients
(C) Numbers of bacteria mililitre of waster water
(D) Both (A) and (B)

335. Treatment of the water proceeds through five stages; select the correct order of different stages of water treatment:
(A) Sludge treatment → preliminary treatment → primary treatment → secondary treatment → tertiary treatment
(B) Preliminary treatment→primary treatment → secondary treatment → tertiary treatment → Sludge treatment
(C) Preliminary treatment → Sludge treatment → primary treatment → secondary treatment → tertiary treatment
(D) Preliminary treatment → secondary treatment → tertiary treatment→ primary treatment → Sludge treatment

336. Which waste water treatments is an anaerobic process?
(A) Primary treatment
(B) Secondary treatment
(C) Tertiary treatment
(D) Sludge treatment

337. Preliminary treatment for waste water is necessary to prevent?
(A) Clogging of pipelines
(B) Damage to pump
(C) Interference with subsequent process
(D) All of these

338. Which is correct about aerobic digestion of waste water?
(A) Bacteria are the most common organisms
(B) Bacterial population may be more than 10^9 cells/ml
(C) Bacteria are responsible for the removal of about 80-90% of the BOD
(D) All of these

339. Practical application of microbes for xenobiotic degradation is facilitates by:
(A) Supply of sufficient nutrients or cometabolites
(B) Maintenance of the xenobiotic compounds at nontoxic levels
(C) Provision of microbial population or inoculums
(D) All of these

340. The quality of treated effluent from the digester depends mainly on :
(A) Sludge loading rate
(B) Organic loading rate
(C) Residence time
(D) All of these

341. NO_3^- is removed from water by the action of which bacteria?
(A) *Pseudomonas* (B) *Micrococcus*
(C) Alcaligenes (D) All of these

342. Sewage contains which pathogenic organisms?
(A) Bacteria causing typhoid, cholera, diarrhea
(B) Protozoa causing amoebic dysentery, giardiasis
(C) Viruses causing poliomyletis, hepatitis, gastroenteritis
(D) All of these

343. Protozoa are represented in waste waters by:
(A) Amoebal forms (B) Ciliates
(C) Flagellates (D) All of these

344. The most common type of protozoa found in waste waters:
(A) Amoebal forms (B) Ciliates
(C) Flagellates (D) All of these

345. Which form of protozoa is most diverse group?
(A) Amoebal forms (B) Ciliates
(C) Flagellates (D) All of these

346. Efficiency of anaerobic digestion can be enhanced by:
(A) By using solid support to retain biomass
(B) By recycling the active biomass
(C) By limiting the biomass loss to rate compatible with high population density in the digester
(D) Any of these

347. Bioremediation of organic contaminants is primarily based on:
(A) Microorganisms naturally present at the site
(B) Microbial inoculants developed in the laboratory and introduced at the site
(C) Suitable plants planted at the site
(D) All of these

348. Oil spills cause severe damage to the ecosystems and pose which threats?
(A) Ground water pollution due to percolation
(B) Air pollution due to exoporation
(C) Fire
(D) All of these

349. Which is not correct?
(A) The US Environment Protection Agency and the Exxon Company used microorganisms to clean up Alaskan beaches contaminated by the Valdez oil spill
(B) In India, a consortium of bacterial species has been developed to combat oilspills and oily sludge; the inoculants is aptly called 'oilzapper'
(C) Inoculation with Oilzapper reduced oilsludge contamination in soil to merely 0.5% in 360 days from the initial 13.41%
(D) Oilzapper has not been effective in field trials

350. The anaerobic digestion process involves a wide variety of organisms which digest the organic molecules, such as lipids, proteins, carbohydrates etc., into mainly :
(A) NO_3^- and PO_4^{--} (B) CH_4 and CO_2
(C) SO_4^{--} and CO (D) All of these

351. Which heterotrophic bacteria are responsible for oxidation of organic molecules?
(A) Pseudomonas and Escherichia
(B) Streptococcus and Salmonella
(C) Shigella and Aerobacter
(D) All of these

352. Which is not correct about halocarbons?
(A) They contain different numbers of halogen atoms in the place of He atoms
(B) They are used as solvents, in condenser units of cooling systems, etc
(C) The C_1-C_2 haloalkanes escape into the atmosphere where they destroys the protective ozone layer
(D) Pesticides leach into water bodies, where many of them are subjected to biomagnifications

353. Which is correct about alkylbenzyl sulphonates?
(A) These are surface-active detergents
(B) Sulphonate group present at one end resists microbial degradation
(C) At present, alkylbenzyl sulphonates are biodegraded by β–oxidation from their alkyl ends
(D) All of these

354. Phytoremediation may occur due to:
(A) Accumulation of the organic waste in plant tissues
(B) Translocation of the organic waste to leaf and volatilization from there
(C) The organic waste or in the rhizosphere itself
(D) Any of these

355. For phytoremediation of flyash, seedings of which of the following trees are established at abandoned flyash ponds/dams?
(A) Eucalyptus (B) Melia
(C) Poplar (D) All of these

356. Continued exposure of microorganisms to xenobiotic compounds can often lead to the evolution of metabolic processes needed to wholly or partly degrade the xenobiotic. These capabilities may arise due to:
(A) Mutation
(B) Transfer of plasmid –borne genes
(C) Transformation
(D) Both (A) and (B)

357. Continued exposure of microorganisms to xenobiotic compounds can often lead to the evolution of metabolic processes needed to wholly or partly degrade the xenobiotic. These can be expected to modify the active site of an enzyme so that it has an increased affinity for the xenobiotic compounds?
(A) Mutation
(B) Transfer of plasmid –borne genes
(C) Transformation (D) Both (A) and (B)

358. Alicyclic hydrocarbons are present naturally in:
(A) Wax from plants (B) Microbial lipids
(C) Crude oils (D) All of these

359. Which is not related to green-house gas?
(A) CH_2 (B) CO_2
(C) O_2 (D) CFC

360. Fluoride pollution mainly affects:
(A) Heart (B) Kidney
(C) Brain (D) Teeth

361. Biochemical oxygen demand measures:
(A) Industrial pollution
(B) Air pollution
(C) Dissolved O_2 needed by microbes to decompose organic waste
(D) Oxygen need to support microorganism growth

362. Smog is the product of:
(A) Smoke (B) Fog
(C) Soot (D) Both (A) and (B)

363. Green house effect is related to:
(A) Global warming (B) Global afforestation
(C) Deforestation (D) Both (A) and (C)

364. Increasing skin cancer and rate of mutation are the result of:
(A) Acid rain (B) Ozone depletion
(C) SO_2 pollution (D) CO_2 pollution

365. Which diseases is caused by eating fish inhabitating mercury contaminated water?
(A) Minamata disease (B) Osteosclerosis
(C) Hashimoto disease (D) Brain fevel

366. Depletion of ozone layer is caused by:
(A) CO_2 (B) SO_2
(C) CFC (D) Both (B) and (C)

367. The atmosphere of metro cities is polluted by:
(A) Pesticides
(B) Automobile exhausts
(C) House-hold wastes
(D) Radioactive fallout

368. Which associates with the blood haemoglobin more rapidaly than oxygen?
(A) CO (B) CO_2
(C) SO_4 (D) NO

369. Which decibles of sound becomes hazardous moise pollution?
(A) >20 (B) >80
(C) >120 (D) >140

370. Which plants can be used as indicators of SO2 pollution of air?
(A) Liverworts (B) Hornworts
(C) Epiphytic lichens (D) Fern

371. Attack of Asthama in certain persons may be due to:
(A) Inholation of some airborne pollution
(B) Exposure to cold temperature
(C) Exposure to certain pollen grains
(D) All of these

372. Which gases contributes the maximum to the 'green – house effect'?
(A) CO_2 (B) CFC
(C) CH_4 (D) N_2O

373. Which poses the greatest risk?
(A) Nuclear plants
(B) Thermal power plants
(C) Biogas plants
(D) Hydroelectric plants

374. Which is an example of nonrenewable resources?
(A) Petroleum (B) Biological agents
(C) Natural gas (D) Both (A) and (C)

375. Maximum deposition of DDT will occur in:
(A) Phytoplankton (B) Fish
(C) Crab (D) Sea gull

376. Which green-house gases is being generated by agricultural fields?
(A) Sulphur dioxide (B) SO3
(C) Nitrous oxide (D) Ammonia

377. Ozone day is observed on:
(A) September 16 (B) June 5
(C) May 1 (D) December 25

378. Which is an example of a renewable resource?
(A) Plants (B) Animals
(C) Microorganisms (D) All of these

379. Ozone depletion in the stratosphere will cause:
(A) Gobal warming (B) Increased sea level
(C) Increased incidence of skin cancer
(D) All of these

380. Ozone layer of upper atmosphere is being destroyed by:
(A) Smog (B) Photochemcial oxidation
(C) CFC (D) Soot

381. Bioremediation of organic wastes is generally based on:
(A) Bacteria (B) Animal species
(C) Plant species (D) Both (B) and (C)

382. Which is most ecofriendly?
(A) Bacteria (B) Animal species
(C) Plant species (D) Both (B) and (C)

383. Ultra-violet light causes:
(A) Formation of pyrimidines
(B) Sticky metaphases
(C) Photodynamics action
(D) Destruction of hydrogen bonds between complementary DNA strands

384. Sulphur dioxide affects:
(A) Cell wall (B) Plasmodesmata
(C) All membrane systems (D) Nucleus

385. Addition of phosphates and nitrates/fertilizers into water leads to:
(A) Reduced algal growth
(B) Increased growth of decomposers
(C) Increased algal growth
(D) Nutrient enrichment/eutrophication

386. Fluoride pollution produces necrosis or chlorosis of:
(A) Leaf margin and tip (B) Stem
(C) Petiole but not lamina (D) Midrib

387. Air pollutants photochemical oxidants include:
(A) Nitrous oxide, nitric acid and nitric oxide
(B) Oxygen, chlorine and nitric acid
(C) Ozone, peroxyacetyl nitrate and aldehydes
(D) Carbon monoxide and sulphur dioxide

388. Ultraviolet radiations are lethal because of inactivation of:
(A) Proteins, nucleic acids and pigments
(B) Carbohydrates, fats and vitamins
(C) Minerals, air and water
(D) Water, carbon dioxide and oxygen

389. Nonionising radiations with specific biological effects are:
(A) UV radiations (B) Beta-ray
(C) Gamma rays (D) X-rays

390. Carbon monoxide is pollutant as it:
(A) Inactibates nerves
(B) Inhibits glycolysis
(C) Combines with oxygen
(D) Combines with haemoglobin

391. Air pollutants that produce photochemical oxidants include:
(A) CO_2, CO and SO_2
(B) Nitrous oxide, nitric oxide and nitric acid
(C) Oxygen, clorine and nitric acid
(D) Ozone, chlorine and sulphur dioxide

392. Acid rains are produced by:
(A) Excess NO_2 and SO_2 from burning fossil fuels
(B) Excess production of NH_3 by industry and coal gas
(C) Excess release of carbon monoxide by incomplete combustion
(D) Excess formation of CO_2 by combustion and animal respiration

393. Spraying of DDT produces pollution of:
(A) Air (B) Air and water
(C) Air and soil (D) Air, water & soil

394. Most abundant water pollution is:
(A) Detergents (B) Pesticides
(C) Industrial waste (D) Ammonia

395. Atmospheric pollutants is:
(A) CO_2 (B) CO
(C) O_2 (D) N_2

396. Burning of fossil fuels is the main source of pollution:
(A) Nitrogen oxide
(B) Nitric oxide
(C) Nitrous oxide
(D) Sulphur dioxide

397. Modes of controlling pollution in large cities include:
(A) Cleanliness and less use of insecticides
(B) Proper disposal of organic wastes, sewage and industrial effluents
(C) Broader roads and shifting of factories out of the residential areas
(D) All of these

398. Domestic waste constitutes:
(A) Nonbiodegradable pollution
(B) Biodegradable pollution
(C) Effluents
(D) Air pollution

399. Ozone layer of upper atmosphere is being destroy by:
(A) Chlorofluorocarbons
(B) SO_2
(C) Photochemical oxidants/O_2 and CO_2
(D) Smog

400. SO_2 pollution is indicated by:
(A) Deschampsia (grasses)
(B) Sphagnum (mosses)
(C) Usnea (lichens)
(D) Cucurbita (climbers)

401. Increased asthmatic attacks in certain seasons are related to:
(A) Inhalation of seasonal pollen
(B) Eating of seasonal vegetables
(C) Low temperature
(D) Wet and dry environment

402. Ozone depletion in stratosphere shall results in:
(A) Forest fires
(B) Increased incidence of skin cancer
(C) Global warming (D) None of these

403. Which one of the following dissolves more rapidly in blood haemoglobin than oxygen?
(A) Ozone (B) Nitrous oxide
(C) Sulphur dioxide (D) Carbon monoxide

404. Which of the following forms a toxic substance in blood by combining with haemoglobin?
(A) Carbon dioxide
(B) Carbon monoxide
(C) Oxygen
(D) Methane

405. DDT is:
(A) Green house gas
(B) Degradable pollutant
(C) Nondegradable pollutant
(D) None of these

406. Ultraviolet radiation from sunlight causes a reaction that produces:
(A) Fluorides (B) Carbon monoxide
(C) Sulphur dioxide (D) Ozone

407. Noise pollution is measured in:
(A) Hertz (B) Fathoms
(C) Nanometres (D) Decibels

408. Most hazardous metal pollutant of automobile exhausts is:
(A) Mercury (B) Cadmium
(C) Lead (D) Copper

409. Pollution is rising due to:
(A) Rains
(B) Research institutes
(C) Population explosion
(D) Automobiles and industries

410. Pollution can be checked through:
(A) Reducing discharge of waste
(B) Increasing industrial development
(C) Throwing wastes in open spaces
(D) All of these

411. Continuous sewage flow into a stream would lead to:
(A) Eutrophication
(B) Algal bloom
(C) Depletion of oxygen
(D) Increase in temperature

412. Sewage infection of water will be demonstrated by cysts of:
(A) Entamoeba (B) *Escherichia coli*
(C) Pseudomonas (D) Leishmania

413. Measurement of rate of oxygen utilization by a unit volume of water over a period of time is to measure:
(A) Fermentation
(B) Biogas generation
(C) Biosynthetic pathway
(D) Biological oxygen demand

414. Carbon dioxide is a major pollutant in:
(A) Air (B) Soil
(C) Water (D) Noise

415. Increasing skin cancer and high mutation rate are due to:
(A) Acid rain (B) Ozone depletion
(C) CO pollution (D) CO_2 pollution

416. Sudden mass death of fish is more likely to occur in:
(A) Eutrophic lake (B) Mesotrophic lake
(C) Oligotrophic lake (D) Oxalotrophic lake

417. Which pollution is causes jaundice?
(A) Water (B) Air
(C) Land (D) Thermal

418. Which pollution causes leaf curling?
(A) CO (B) H_2S
(C) O_3 (D) SO_2

419. Water pollution causes:
(A) Increased photosynthesis
(B) Increased deoxygenation and turbidity
(C) Decreased turbidity
(D) Increased oxygenation

420. Natural sink of stratospheric ozone is:
(A) Sulphuric dioxide
(B) Sulphur fluxes of oceans
(C) Freons (D) HCFCs

421. Major source of methane in India is:
(A) Rice fields
(B) Sugarcane plantation
(C) Wheat fields (D) Fruit orchards

422. Ozone hole is maximum over:
(A) Europe (B) Antarctica
(C) India (D) Africa

423. In coming years, skin related disorders will become more common due to:
(A) Excessive use of detergents
(B) Water pollution
(C) Air pollution
(D) Depletion of ozone layer

424. Phosphate pollution is caused by:
(A) Agricultural fertilizers
(B) Weathering of phosphate rocks only
(C) Phosphate tocks ad sewage
(D) Sewage and agricultural fertilizers

425. Common indicator organism of water pollution is:
(A) *Escherichia coli* (B) *Cholera vibrio*
(C) *Salmonells typhi*
(D) *Entamoeba histolytica*

426. Which pigment protects plants from UV damage?
(A) Chlorophyll (B) Xanthophyll
(C) Carotenoids (D) Phycocyanin

427. Component of smog is:
(A) O_3 (B) PAN
(C) PPN and PBN (D) Both (A) and (B)

428. Indiscriminate use of fertilizers causes:
(A) Air pollution (B) Water pollution
(C) Land pollution (D) All of these

429. Most harmful type of environmental pollutants are:
(A) Human organic wastes
(B) Natural nutrients in excess
(C) Waste animal feed
(D) Nonbiodegradable chemicals

430. Which is not dangerous for life?
(A) Nuclear blast (B) Deforestation
(C) Ozone layer (D) Biopollutants

431. Spraying DDT on crops shall produce pollution of:
(A) Air and water (B) Air and soil
(C) Crops and air (D) Soil and water

432. Sewage water is purified by:
(A) Microorganisms (B) Light
(C) Fishes (D) Aquatic plants

433. Sewage treatment in which a portion of decomposer bacteria present in waste of recycled into the beginning of the process:
(A) Cyclic treatment
(B) Activated sludge treatment
(C) Primary treatment
(D) Tertiary treatment

434. Ozone hole refers to:
(A) Hole in ozone layers
(B) Reduction in thickness of ozone layer in stratosphere
(C) Reduction of thickness of ozone in troposphere
(D) Increase concentration of ozone

435. Photochemical smog is related to pollution of:
(A) Air (B) Water
(C) Soil (D) Nostoc

436. Among pollutants which has the most lasting effect:
(A) Carbon monoxide
(B) Pesticides
(C) Sulphur dioxide
(D) Smoke from chimneys

437. Pollution indicators plants:
(A) Are resistant to pollution
(B) Can purify the atmosphere
(C) Are very sensitive to pollutants
(D) Can indicate onset of rains

438. Depletion of ozone layer is due to:
(A) Oxides of nitrogen (B) Oxides of carbon
(C) Oxides of sulphur (D) None of these

439. Use of lichens in case of pollution is:
(A) Treatment of pollutant water
(B) Bioindicators of pollution
(C) Promote pollution
(D) Lichens have no relation with pollution

440. Peeling of ozone umbrella is due to:
(A) CFCs (B) PAN
(C) CO_2 (D) Coal burning

441. Ozone layer of stratosphere requires protection from indiscriminate use of:
(A) Fungicides, insecticides, bactericides and medicines
(B) Aerosols and high flying jets
(C) Atomic explosions and industrial wastes
(D) Baloons and turbo-prop aeroplanes

442. Environmental pollution affects:
(A) Biotic components
(B) Plants only
(C) Man only
(D) Biotic and abiotic components of environment

443. Anxiety and stress are caused by:
(A) Air pollution (B) Noise pollution
(C) Water pollution (D) Nuclear pollution

444. Water pollution is due to:
(A) Agricultural discharges
(B) Sewage and other wastes
(C) Industrial effluents
(D) All of these

445. Carbon monoxide kills because it destroys:
(A) Haemoglobin (B) Phytochrome
(C) Cytochrome (D) Both (A) and (B)

446. Water is often treated with chlorine to:
(A) Increase oxygen content
(B) Kill germs
(C) Remove hardness
(D) Remove suspended particles

447. Water blooms are formed by:
(A) Lemna (B) Hydrilla
(C) Water Hyacinth (D) Planktonic algae

448. DDT concentration recorded in human body has been:
(A) 0.3 – 0.7 ppm (B) 1.3 – 3.1 ppm
(C) 13 – 31 ppm (D) 30 - 70 ppm

449. Mottling of teeth is due to presence of an element in drinking water:
(A) Mercury (B) Fluorine
(C) Boron (D) Chlorine

450. Which one of the following is not pollutant?
(A) CO_2 (B) CO
(C) SO_2 (D) NO_2

451. As it passes into food chain, the concentration of DDT:
(A) Remains same (B) Decreases
(C) Increases (D) Unpredictable

452. NEERI is:
(A) National Environmental Engineering Research Institute
(B) National Ecological and Environmental Research Institute
(C) National Ethological and Ecological Research Institute
(D) National Eugenics and Ecological Research Institute

453. A person has impaired nervous system and signs of madness due to continued intake of metal contaminated water. The metal is:
(A) Mercury (B) Calcium
(C) Manganese (D) Lead

454. Wildsterben is an effect of:
(A) Ozone
(B) Nitrogen oxide and sulphur oxide
(C) Acid rain (D) All of these

455. Waldsterben is:
(A) Decline in forests
(B) Development of fog
(C) Harm to sculptures, status and buildings
(D) All of these

456. Ozone is:
(A) Pale blue (B) Highly reactive
(C) With penetrating odour
(D) All of these

457. An ozone depleting chemical which is used in fire-extinguishers is:
(A) Nitrogen oxide (B) Methyl bromide
(C) Halon (D) Chlorofluorocarbon

458. Smoked foods are not favoured now because of the possible toxicity of:
(A) Benzpyrene (B) Nitrogen oxides
(C) Sulphur dioxide (D) All of these

459. Benzpyrene is an ingredient of:
(A) Automobile exhausts
(B) Wood smoke and tobacco smoke
(C) Coal tar (D) All of these

460. Smoke or dust dispersed in air is called:
(A) Fog (B) Haze
(C) Mist (D) Smog

461. Mist is fog with horizontal visibility of:
(A) 100 m (B) 200 m
(C) 500 m (D) 1000 m

462. Fog is formed by:
(A) Smoke
(B) Condensation of water vapours
(C) Condensation of water vapour over dust particles
(D) Both (A) and (C)

463. The term green house effect was given by:
(A) Fourier (B) Tyndall
(C) Arrhenius (D) Linnaeus

464. Polychlorinated biphenyls are chemicals useful as:
(A) Fire – resistant (B) Fertilizers
(C) Pesticides (D) Germicides

465. Use of polychlorinated biphenyl is restricted because of their:
(A) Extreme toxicity (B) Persistent toxicity
(C) Biomagnification (D) Both (B) and (C)

466. Ecofriendly detergents do not contain:
(A) Phosphates and bleaches
(B) Alcohols
(C) Water softners
(D) Perborates

467. Sudden rattling of doors and windows is due to:
(A) Earthquake
(B) Sonic boom
(C) Movements of crustal layers of earth
(D) Slipping of glaciers on the nearby mountain

468. Plantation of rows of trees and shrubs along roads, rails, around residential complexes, educational institutes, etc is called:
(A) Green muffler (B) Shelter belt
(C) Wind break (D) Both (B) and (C)

469. Exposure of UV radiations causes:
(A) Damage of eye sight (B) Sunburn
(C) Skin cancer (D) All of these

470. Third pollution is:
(A) Controlled tipping
(B) Land or landscape pollution
(C) Noise pollution
(D) UV pollution

471. Sewage is broken down in:
(A) Primary treatment
(B) Secondary treatment
(C) Tertiary treatment (D) All of these

472. Tipping is a technique of:
(A) Sewage disposal (B) Waste disposal
(C) Smoke treatment (D) Effluent

473. Photocatalysis is useful in breakdown of:
(A) Acids and alkalies
(B) Chemicals containing heavy metals
(C) Toxic chemicals
(D) All of these

474. Chlorination of treated waste water required for:
(A) Killing all microbes
(B) Removing suspended materials
(C) Complete breakdown of organic matter
(D) All of these

475. Excess nitrate present in drinking water, vegetables and fruits shall result in:
(A) Methaemoglobinaemia
(B) Mottling of teeth
(C) Minamata disease
(D) Itai-itai

476. Bag filters are used for removal of pollutants from:
(A) Exhaust emissions
(B) Effluent water
(C) Sewage water
(D) Both (B) and (C)

477. Thermal plants have installed:
(A) Cyclone collector (B) Porous filters
(C) Electrostatic precipitators
(D) Adsorbents

478. Substitute of chlorofluorocarbon is:
(A) Hydrofluorocarbon
(B) Hydrochlorofluorocarbon
(C) Fluorocarbon
(D) Both (A) and (B)

479. Smog that occurs due to low temperature and has reducing environment is:
(A) Photochemical smog (B) Los Angeles smog
(C) Nonsulphurous smog (D) London smog

480. Green audit is:
(A) Estimation of crop yield and crop area
(B) Estimation of forested area and pasture land
(C) Checking of impact of an establishment over environment
(D) Effect of environment over vegetation of an area

481. Runoff from fields is:
(A) Point source pollution
(B) Line source pollution
(C) Area source pollution
(D) Diffuse source pollution

482. Stick and leaf insects resembles the surroundings. They undertake mimicry of the type:
(A) Concealing-protective
(B) Warning – protective
(C) Alluring-aggressive
(D) Concealing-aggressive

483. Nitrogen oxides released in the stratosphere cause ozone depletion due to:
(A) Catalysing conversion of ozone to oxygen
(B) Binding ozone
(C) Reacting with ozone
(D) All of these

484. In foggy and hazy environment, birds are able to locate position of sun and their path by means of:
(A) Infra-red wavelength
(B) Ultra-violet radiations
(C) Magnetic system
(D) Pheromones

485. The amount of biogenic nutrients present in the abiotic environment is called:
(A) Standing state (B) Standing crop
(C) Detritus (D) Reserve nutrient

486. Carbon dioxide is being added to the cycling pool by:
(A) Respiration and combustion
(B) Sedimentation of carbonates
(C) Precipitation, skeletons and fossilization
(D) Both (B) and (C)

487. A source of nonbiological carbon dioxide is:
(A) Volcanic eruptions
(B) Hot springs
(C) Weathering
(D) All of these

488. A succession that does not follow the natural course due to human activity is:
(A) Subsere (B) Plagiosere
(C) Prisere (D) Asere

489. Which of the following disorder in mango is caused by low temperature?
(A) Black heart (B) Crotch injury
(C) Collor injury (D) All of these

490. Use of green plants and their associated micro-organisms, soil amendments, and agronomic practices to remove, contain or render harmless environmental contaminants is known as:
(A) Phytoremediation
(B) Phytostabilization
(C) Both (A) and (B)
(D) None of these

491. Use of plants to sequester contaminants is cell wall lignins of root tissues, or to promote its binding to humus is known as:
(A) Phytoremediation
(B) Phytostabilization
(C) Both (A) and (B)
(D) None of these

492. The atmosphere of the earth is heated by radiations which are mainly:
(A) Radiated by the sun
(B) Re-radiated by land
(C) Re-radiated by water
(D) Re-radiated by land and water

493. If there were no atmosphere around the earth, the temperature of the earth will:
(A) Increase
(B) Go on decreasing
(C) Increase during day and decrease during night
(D) Be unaffected

494. What would happen, if all the oxygen present in the environment is converted to ozone :
(A) We will be protected more
(B) It will become poisonous and kill living forms
(C) Ozone is not stable, hence it will be toxic
(D) It will help harmful sun radiation to reach earth and damage many life forms

495. The two forms of oxygen found in the atmosphere are :
(A) Water and Ozone
(B) Water and Oxygen
(C) Ozone and Oxygen
(D) Water and Carbon dioxide

496. Rainfall pattern depend on:
(A) The underground watertable
(B) The density pattern of human population in an area
(C) The number of water bodies in an area
(D) The prevailing season in an area

497. Among the given options, which one is not correct for the use of large amount of fertilizers and pesticides?
(A) They are eco-friendly
(B) They fields barren after some time
(C) The adversely affect the useful component from the soil
(D) They destroy the soil fertility

498. The nitrogen molecules present in air can be converted into nitrates and nitrites by:
(A) A biological process of nitrogen fixing bacteria present in soil
(B) A biological process of carbon fixing factor present in soil
(C) Any of the industries manufacturing nitrogenous compounds
(D) The plants used as cereal crops in field

499. One of the following processes is not a step involved in the water cycle operating in nature:
(A) Evaporation (B) Temperature
(C) Precipitation (D) Photosynthesis

500. Which step is not involved in the carbon-cycle?
(A) Photosynthesis
(B) Transpiration
(C) Respiration
(D) Burning of fossil fuels

501. The term "water – pollution" can be defined in several ways, which of the following statements does not give the correct definitions?
(A) The addition of undesirable substances to water bodies
(B) The removal of desirable substances from water bodies
(C) A change in pressure of the water bodies
(D) A change in temperature of the water bodies

502. 'Ozone–hole' means:
(A) Large sized hole in the ozone layer
(B) Thinning of the ozone layer
(C) Small holes scattered in the ozone layer
(D) Thickening of ozone in the ozone layer

503. Environmental historians are:
(A) Paul Ehrlich
(B) Donella H. Meadows
(C) Rachel Carson
(D) All of these

504. Combustion of candle is a:
(A) Photochemical change
(B) Physical change
(C) Exothermic change
(D) Endothermic change

505. The heavily polluted zone of water reservoir is known as:
(A) Pleosaprophytic zone
(B) Mesosaprophytic zone
(C) Oligosaprophytic zone
(D) None of these

506. The rate of degradation and microbes resistance to toxic pollutants remain better when the:
(A) Mixed cell population is used
(B) Individual cell is used
(C) Mixed cell population along with metals is used
(D) Individual cell along with metal is used

507. Soil conservation is achieved through:
(A) Restricted human activities
(B) Good plant cover
(C) Wind screen
(D) Ploughing

508. Which of the following pairs is a non-renewable resource?
(A) Coal and minerals
(B) Plants and natural gas
(C) Water and natural gas
(D) Energy and water

509. The pollutants which are already present in nature but are released in substantial amounts by man are known as:
(A) Quantitative pollutants
(B) Qualitative pollutants
(C) Degradable pollutants
(D) Primary pollutants

510. Which of the following microorganisms leach metal out of rock ores and can accumulate silver?
(A) *Pseudomonas aeruginosa*
(B) *Thiobacillus*
(C) *Pseudomonas putida*
(D) *Zoogloea ramigera*

511. The photochemical smog is formed by:
(A) Nitrogen oxide
(B) Hydrocarbon
(C) Nitrogen oxide and hydrocarbon
(D) Solar radiation upon nitrogen oxides and hydrocarbon

512. Which of the following accumulates from factory waste?
(A) *Pseudomonas aeruginosa*
(B) *Thiobacillus*
(C) *Pseudomonas putida*
(D) *Zoogloea ramigera*

513. Which of the bacteria is considered for biological leaching?
(A) *T. thioxidans*
(B) *T. ferrooxidans*
(C) *Ferrobacillus terrooxidans*
(D) All of these

514. In what sphere above the earth, the satellites like INSAT are stationed?
(A) Troposphere (B) Ionosphere
(C) Microsphere (D) Stratosphere

515. Most primary productivity in the surface water of lakes and oceans is due to:
(A) Floating plants
(B) Red and brown algae
(C) Zooplankton
(D) Phytoplankton

516. Which of the following is most effective for the rice sheath light pathogen, *Pellicularia sasakii?*
(A) Polyoxin D (B) Polyoxin B
(C) Polyoxin L (D) Polyoxin C

517. Which of the following subgroup/(s) belong/(s) to the Baulo viruses?
(A) Nuclear polyhedrosis viruses
(B) Granulosis viruses
(C) Oryctestypes
(D) All of these

518. Which of the following have been used to oxidize ammonia and to prevent algal growth?
(A) Growth of hypomicrobium in presence of added methanol
(B) *Micrococcus denitrificans* cells encapsulated in liquid membranes
(C) Both (A) and (B)
(D) *Nitrosomonas europaen*

519. Ozone layer occurs in:
(A) Troposphere (B) Ionosphere
(C) Thermosphere (D) Stratosphere

520. Net productivity is:
(A) Gross production + Respiration
(B) Gross production – Respiration
(C) Gross production x Respiration
(D) Gross production ÷ Respiration

521. Source of methane pollution is:
(A) Rice fields
(B) Marshes
(C) Cattle and other herbivors
(D) All of these

522. *Streptomyces griseochromogenes* is reported to show a potent curative effect against:
(A) Rice blast disease
(B) Yellow blight of maize
(C) Milo of soybean
(D) Eye spot of sugarcane

523. Phenols degradation in waste water of hospitals, laboratories and coal processing to coke can be achieved by using:
(A) Fungus *Aureobasidium pollulans* adsorbed by the fibrous asbestos
(B) Cells of *Pseudomonas* spp. either adsorbed to anthracite coal or entrapped on alginінate gel
(C) Both (A) and (B)
(D) *Phanerochaete chrysosporium*

524. Halocarbons or halons are constituents of:
(A) Aerosols (B) Refrigerants
(C) Fire extinguishers (D) Pesticides

525. Ozone shield is components of:
(A) Troposphere (B) Stratosphere
(C) Mesosphere (D) Thermosphere

526. Aitken particles are:
(A) Activated in humid air
(B) Cloud's condensed nuclei
(C) Hygroscopic particles
(D) All of these

527. Carbon monoxide is harmful to human being as it is:
(A) Carcinogenic
(B) Antagonistic to CO2
(C) With higher affinity for haemoglobin as compared to oxygen
(D) Destruction to O3

528. Natural sink of stratospheric ozone is:
(A) Sulphure dioxide
(B) Sulphur fluxes of oceans
(C) Freons
(D) HCFCs

529. Common indicator organisms of water pollution is:
(A) *Escherichia coli*
(B) *Cholera vibrio*
(C) *Salmonella typhi*
(D) *Entamoeba histolytica*

530. Most poisonous pollutant in water is:
(A) Lead (B) Zinc
(C) Phosphate (D) Arsenic

531. Chemical conversion process are often employed for collection of:
(A) Particulate (B) Industrial wastes
(C) Gaseous pollutants (D) All of these

532. Most harmful type of environmental pollutants are:
(A) Human organic wastes
(B) Natural nutrients in excess
(C) Waste animal feed
(D) Non-degradable chemicals

533. Innundaton involves use of a/an:
(A) Large number of organisms over a short time to suppress/destroy a population
(B) Organism over a short time to suppress/destroy a population
(C) Large number of organisms over a time to suppress/destroy a population
(D) Organism over a long time to suppress/destroy a population

534. Thermal pollution is more prevalent near:
(A) Hot water springs
(B) Coal based power plants
(C) Temperate zones
(D) Tropical zones

535. The precipitation reaching the ground distribute in different ways as:
(1) Evaporate and reach the atmosphere
(2) Infiltrate the soil
(3) Detained in catchment area
(4) Become a form of ran-off
Select the correct answer from the codes:
(A) 1, 2 and 3 (B) 2, 3 and 4
(C) 1, 3 and 4 (D) All of these

536. The tropical cyclone is known as:
(A) Hurricane (B) Typhoon
(C) Catrina (D) Both (A) and (B)

537. Colourless, odourless and non-corrosive air pollutants are:
(A) Ozones (B) Sulphur dioxide
(C) Carbon dioxide (D) Carbon mononoxide

538. Which of the following herbicide is known to be effective against perennial shrubs, annual weeds and weeds of rice?
(A) Cycloheximide (B) Anisomycin
(C) Herbicide A and B (D) Biolaphos

539. Lead in water can cause:
(A) Kidney damage (B) Hair falling
(C) Eye disease (D) Minimata's disease

540. Deforestation is the major causal agent of:
(A) Depletion of natural resources
(B) Environmental pollution
(C) Desertification of habitat
(D) Genetic erosion

541. Milbemycin is known to exhibit a broad spectrum of activity against:
(A) Agricultural pests (B) Agricultural herbs
(C) Agricultural insects (D) All of these

542. The important categories of solid waste are:
(A) Municipal wastes (B) Industrial wastes
(C) Hazardous wastes (D) All of these

543. Which of the following/(s) can be used for the reduction of nitrate in waste water?
(A) growth of Hypomicrobium in presence of added methanol
(B) *Micrococcus denitrificans* cells encapsulated in liquid membranes
(C) Both (A) and (B)
(D) *Nitrosomonas europaen*

544. The important steps in solid waste management are:
(A) Waste generation and collection
(B) On-site handling, storage and processing
(C) Transport and disposal
(D) All of these

545. A waste is classified as hazardous it shows:
(A) Toxicity and ignitability
(B) Corrosivity
(C) Reactivity
(D) All of these

546. Which of the following is known to control Colorado potato beetle?
(A) *Baveria bassiana*
(B) *Metarhizium anisopliae*
(C) *Verticillium lecauli*
(D) *Nomuraea rileyi*

547. The factors that influence the quantity of municipal waste generated include:
(A) Collection frequency (B) Geographic location
(C) Salvaging & recycling (D) All of these

548. Major health effects due to depletion of ozone layer is/are:
(A) Increase in eye cataracts and mutation
(B) Increase in skin cancer and melonema
(C) Decrease in growth of ocean flora and fauna
(D) All of these

549. Overgrazing causes:
(A) Negative pollution
(B) Positive pollution
(C) Soil erosion
(D) Reduction in crop yield

550. Which of the following can be used for the removal of color from the effluents of pulp mill?
(A) *Coriolus versicolor* entrapped in calcium alginate beads
(B) *Phanerochaete chysosporium*
(C) Both (A) and (B)
(D) *Nitrosomonas europaen*

551. Environmental studies are very important for:
(A) Getting clean drinking water
(B) Hygienic living conditions
(C) Clean and fresh air
(D) All of these

ANSWERSHEET
9. ENVIRONMENTAL BIOTECHNOLOGY

1. (A)	2. (B)	3. (C)	4. (C)	5. (D)	6. (B)	7. (A)	8. (A)	9. (C)	10. (C)
11. (B)	12. (D)	13. (A)	14. (C)	15. (A)	16. (D)	17. (C)	18. (D)	19. (C)	20. (B)
21. (B)	22. (A)	23. (C)	24. (C)	25. (A)	26. (D)	27. (A)	28. (C)	29. (B)	30. (A)
31. (D)	32. (D)	33. (A)	34. (B)	35. (D)	36. (C)	37. (C)	38. (B)	39. (B)	40. (D)
41. (B)	42. (C)	43. (B)	44. (D)	45. (A)	46. (C)	47. (B)	48. (A)	49. (C)	50. (D)
51. (B)	52. (C)	53. (B)	54. (D)	55. (C)	56. (C)	57. (D)	58. (C)	59. (B)	60. (B)
61. (B)	62. (B)	63. (C)	64. (A)	65. (D)	66. (A)	67. (A)	68. (C)	69. (D)	70. (B)

71. (D)	72. (C)	73. (C)	74. (C)	75. (B)	76. (B)	77. (B)	78. (C)	79. (B)	80. (C)
81. (C)	82. (B)	83. (D)	84. (B)	85. (B)	86. (C)	87. (C)	88. (B)	89. (D)	90. (A)
91. (A)	92. (B)	93. (A)	94. (B)	95. (D)	96. (A)	97. (C)	98. (C)	99. (B)	100. (A)
101. (D)	102. (B)	103. (A)	104. (D)	105. (C)	106. (B)	107. (D)	108. (D)	109. (D)	110. (B)
111. (A)	112. (B)	113. (D)	114. (B)	115. (B)	116. (C)	117. (A)	118. (C)	119. (B)	120. (C)
121. (D)	122. (D)	123. (B)	124. (C)	125. (A)	126. (D)	127. (C)	128. (B)	129. (A)	130. (D)
131. (A)	132. (D)	133. (A)	134. (D)	135. (C)	136. (A)	137. (C)	138. (B)	139. (A)	140. (C)
141. (C)	142. (C)	143. (C)	144. (A)	145. (B)	146. (C)	147. (D)	148. (C)	149. (B)	150. (D)
151. (B)	152. (D)	153. (C)	154. (B)	155. (D)	156. (A)	157. (B)	158. (C)	159. (B)	160. (C)
161. (B)	162. (B)	163. (D)	164. (C)	165. (D)	166. (D)	167. (C)	168. (A)	169. (D)	170. (B)
171. (B)	172. (C)	173. (A)	174. (B)	175. (C)	176. (D)	177. (C)	178. (C)	179. (D)	180. (A)
181. (D)	182. (B)	183. (D)	184. (C)	185. (C)	186. (B)	187. (A)	188. (C)	189. (A)	190. (B)
191. (A)	192. (A)	193. (A)	194. (A)	195. (A)	196. (B)	197. (B)	198. (C)	199. (D)	200. (A)
201. (B)	202. (A)	203. (B)	204. (B)	205. (A)	206. (A)	207. (A)	208. (C)	209. (B)	210. (A)
211. (A)	212. (C)	213. (C)	214. (A)	215. (C)	216. (D)	217. (C)	218. (A)	219. (A)	220. (A)
221. (B)	222. (C)	223. (D)	224. (B)	225. (A)	226. (B)	227. (A)	228. (A)	229. (A)	230. (A)
231. (C)	232. (A)	233. (D)	234. (A)	235. (B)	236. (A)	237. (C)	238. (A)	239. (B)	240. (C)
241. (C)	242. (D)	243. (C)	244. (A)	245. (B)	246. (A)	247. (B)	248. (C)	249. (A)	250. (B)
251. (C)	252. (A)	253. (E)	254. (C)	255. (C)	256. (C)	257. (A)	258. (D)	259. (A)	260. (A)
261. (A)	262. (C)	263. (C)	264. (B)	265. (A)	266. (C)	267. (B)	268. (A)	269. (D)	270. (D)
271. (A)	272. (B)	273. (C)	274. (C)	275. (D)	276. (A)	277. (C)	278. (A)	279. (C)	280. (A)
281. (D)	282. (C)	283. (B)	284. (A)	285. (B)	286. (A)	287. (A)	288. (D)	289. (B)	290. (C)
291. (C)	292. (B)	293. (B)	294. (D)	295. (C)	296. (D)	297. (A)	298. (A)	299. (C)	300. (A)
301. (B)	302. (C)	303. (C)	304. (B)	305. (C)	306. (D)	307. (D)	308. (A)	309. (D)	310. (D)
311. (C)	312. (D)	313. (B)	314. (D)	315. (D)	316. (B)	317. (D)	318. (D)	319. (D)	320. (D)
321. (C)	322. (D)	323. (D)	324. (D)	325. (C)	326. (D)	327. (D)	328. (D)	329. (D)	330. (B)
331. (C)	332. (D)	333. (D)	334. (D)	335. (B)	336. (D)	337. (D)	338. (D)	339. (D)	340. (D)
341. (D)	342. (D)	343. (D)	344. (C)	345. (B)	346. (D)	347. (D)	348. (D)	349. (D)	350. (B)
351. (D)	352. (A)	353. (D)	354. (D)	355. (D)	356. (D)	357. (A)	358. (D)	359. (C)	360. (D)
361. (C)	362. (D)	363. (D)	364. (B)	365. (A)	366. (C)	367. (B)	368. (A)	369. (B)	370. (C)
371. (D)	372. (A)	373. (A)	374. (D)	375. (D)	376. (C)	377. (A)	378. (D)	379. (C)	380. (C)
381. (A)	382. (C)	383. (D)	384. (C)	385. (D)	386. (A)	387. (C)	388. (A)	389. (A)	390. (D)
391. (B)	392. (A)	393. (D)	394. (C)	395. (B)	396. (D)	397. (D)	398. (B)	399. (A)	400. (C)
401. (A)	402. (B)	403. (D)	404. (D)	405. (C)	406. (D)	407. (D)	408. (C)	409. (D)	410. (A)
411. (C)	412. (A)	413. (D)	414. (B)	415. (B)	416. (A)	417. (A)	418. (D)	419. (B)	420. (C)
421. (A)	422. (B)	423. (D)	424. (D)	425. (A)	426. (C)	427. (D)	428. (B)	429. (D)	430. (C)
431. (D)	432. (A)	433. (B)	434. (B)	435. (A)	436. (B)	437. (C)	438. (A)	439. (B)	440. (A)
441. (B)	442. (D)	443. (B)	444. (D)	445. (A)	446. (B)	447. (D)	448. (C)	449. (B)	450. (A)
451. (A)	452. (A)	453. (D)	454. (D)	455. (A)	456. (D)	457. (C)	458. (A)	459. (D)	460. (B)
461. (D)	462. (C)	463. (C)	464. (A)	465. (D)	466. (A)	467. (B)	468. (A)	469. (D)	470. (B)
471. (B)	472. (B)	473. (C)	474. (A)	475. (A)	476. (A)	477. (C)	478. (D)	479. (D)	480. (C)
481. (D)	482. (A)	483. (C)	484. (B)	485. (A)	486. (A)	487. (D)	488. (B)	489. (D)	490. (A)

491. (B)	**492. (D)**	**493. (C)**	**494. (B)**	**495. (C)**	**496. (C)**	**497. (A)**	**498. (A)**	**499. (D)**	**500. (B)**
501. (C)	**502. (B)**	**503. (D)**	**504. (C)**	**505. (A)**	**506. (A)**	**507. (B)**	**508. (A)**	**509. (A)**	**510. (B)**
511. (D)	**512. (D)**	**513. (D)**	**514. (B)**	**515. (D)**	**516. (A)**	**517. (D)**	**518. (D)**	**519. (D)**	**520. (B)**
521. (D)	**522. (A)**	**523. (C)**	**525. (C)**	**525. (B)**	**526. (D)**	**527. (C)**	**528. (C)**	**529. (A)**	**530. (C)**
531. (A)	**532. (D)**	**533. (A)**	**534. (B)**	**535. (D)**	**536. (B)**	**537. (B)**	**538. (A)**	**539. (C)**	**540. (C)**
541. (A)	**542. (C)**	**543. (C)**	**544. (D)**	**545. (D)**	**546. (A)**	**547. (D)**	**548. (D)**	**549. (C)**	**550. (C)**
551. (D)									

Chapter - 10

NATURAL AND FOREST CONSERVATION

1. **The species introduced from outside (often require specific environment) produced ecological imbalance due to removal of biological control is known as:**
(A) Exotic species (B) Endemic species
(C) Endangered species (D) Vulnerable species

2. **The species which is found in a particular natural habitat beyond which is unknown, is known as:**
(A) Exotic species (B) Endemic species
(C) Endangered species (D) Vulnerable species

3. **The species which is liable to become extinct if not allowed to realize full biotic potential by providing protection from exotic species/human exploitation/ habitat determination/depletion of food is called:**
(A) Exotic species (B) Endemic species
(C) Threatened species (D) Endangered species

4. **The species is facing danger of imminent extinction due to reduction in habitat, abundance of predator/pollutant, very few in number is called:**
(A) Exotic species (B) Endemic species
(C) Endangered species (D) Vulnerable species

5. **Abundant population at present but threatened with depletion in number due to some factors is called:**
(A) Exotic species (B) Endemic species
(C) Endangered species
(D) Vulnerable species

6. **The species with naturally small populations, either localized or thinly scattered, which are always at risk from pests/pathogens/predators/ exotic species is called:**
(A) Rare species (B) Endemic species
(C) Endangered species
(D) Indeterminate species

7. **The species are in danger of extinction but the reason is not known is called:**
(A) Exotic species
(B) Indeterminate species
(C) Rare species (D) Vulnerable species

8. **Formation of organic matter takes place by:**
(A) Chemical weathering
(B) Soil micro-organisms
(C) Herbivores (D) Soil water

9. **Layer of ozone is present in:**
(A) Stratosphere (B) Thermosphere
(C) Troposphere (D) None of these

10. **The atmospheric stratum nearest to earth is:**
(A) Thermosphere (B) Stratosphere
(C) Troposphere (D) None of these

11. **Which one of the following is renewable resource?**
(A) Water (B) Metals
(C) Fossil fuel (D) All of these

12. **Metallic minerals and fossil fuels are:**
(A) Non-renewable resources
(B) Inexhaustible resources
(C) Renewable (D) None of these

13. **The common source of energy is our villages is:**
(A) Coal
(B) Wood and animal dung
(C) Electricity (D) Sunlight

14. **The natural source of energy which is most important is:**
(A) Atomic energy (B) Biogas
(C) Sunlight (D) Fossil fuels

15. **Floods can be checked by:**
(A) Making dams (B) Deforestation
(C) Reforestation (D) None of these

16. **Extinction of many wild animals is due to:**
(A) Scarcity of food (B) Over-crowding
(C) Hunting (D) Deforestation

17. **Which one is not a resource for Onge tribe?**
(A) Water (B) Food
(C) Gold (D) Wood

18. **Oil, coal and natural gas constitute:**
(A) Metals (B) Renewable resource
(C) Fossil fuels (D) None of these

19. **Underground water is:**
(A) International Resource
(B) Non-renewable Resource
(C) Renewable Resource
(D) mixed Resource

20. **Per capita daily energy consumption in nineteenth century industrialized society was:**
(A) 6000 kcal (B) 10000 kcal
(C) 30000 kcal (D) 70000 kcal

21. **Per capita daily energy consumption of USA is:**
(A) 100000 kcal (B) 30000 kcal
(C) 70000 kcal (D) 250000 kcal

22. Movement of air occurs in the atmosphere upto height of:
(A) 100 km (B) 50 km
(C) 25 km (D) 10 km

23. Of the total water evaporated from oceans, precipitation returns:
(A) 100% (B) 90%
(C) 75% (D) 60%

24. Conservation is:
(A) Judicious use of natural resources
(B) Protection of natural resources
(C) Management of natural resources
(D) All of these

25. Vegetation of a place is mainly determined by:
(A) Rainfall (B) Soil water
(C) Soil type (D) Amount of light

26. American water plant that has become a troublesome water weed in India is:
(A) *Cyperus rotundus*
(B) *Eichhornia crassipes*
(C) *Trapa latifolia* (D) *Trapa bispinosa*

27. River water deposits:
(A) Loamy soil (B) Alluvial soil
(C) Laterite soil (D) Sandy soil

28. Alluvial soils are found in:
(A) Northern India (B) Southern India
(C) Gangetic-Jamuna plains
(D) Eastern India

29. Development of oil from rocky matter is:
(A) Pedogenesis (B) Pedology
(C) Edaphic factors (D) Edaphic climax

30. Good soil is the one which:
(A) Allows water to percolate slowly
(B) Allows water to pass quickly
(C) Allows limited amount of water into it
(D) Holds whole of water entering it

31. Humus is good for plant growth because:
(A) It improves physical condition of soil
(B) It makes the soil porous
(C) It increases water holding and aeration of soil
(D) All of these

32. Humus is important for plant growth as it is:
(A) Made of dead organic matter
(B) Derived from leaves
(C) Rich in nutrients and increases water holding capacity of soil
(D) Partially decomposed

33. Humus is example of:
(A) A fertilizer
(B) Component of soil structure
(C) Organic colloids
(D) Crystalloids

34. Soil fertility can be maintained without addition of nutrients due to:
(A) Floods (B) Crop residues
(C) Activity of microorganisms
(D) Favourable temperature and water

35. Fertility of soil is measured by its ability of:
(A) Retain nutrients
(B) Hold organic materials
(C) Hold water (D) Support life

36. A fertile agricultural soil appears deep coloured at the surface as compared to soil one metre down. The reason for colour of top soil is:
(A) More moisture
(B) Rich in organic matter
(C) Rich in iron, calcium and magnesium
(D) Recent formation

37. Water holding capacity is highest in:
(A) Sandy soil (B) Loamy soil
(C) Silt soil (D) Clay soil

38. Soil water available to roots is:
(A) Free water (D) Gravitational water
(C) Capillary water (D) Hygroscopic water

39. Water held tightly by soil particles as a thin film over their surface is:
(A) Hygroscopic water (B) Chemical water
(C) Gravitational water (D) Capillary water

40. Which governs the solubility and availability of nutrients in the soil?
(A) Aeration (B) pH
(C) Temperature (D) Porosity

41. Waterlogging commonly occurs in:
(A) Sandy soil (B) Gravelly soil
(C) Clayey soil (D) Loamy soil

42. Greatest problem in water conservation is to reduce the amount is:
(A) Ground water (B) Precipitation
(C) Runoff water (D) Evaporation

43. Soil conservation is:
(A) Conservation of sterile soil into fertile one
(B) Aeration of soil
(C) Erosion of soil
(D) Protection against loss

44. Removal of top soil by wind or water is called:
(A) Siltation (B) Soil erosion
(C) Weathering of soil (D) Leaching

45. Sheet erosion is due to:
(A) Fast running rivers (B) Wind
(C) Heavy rains (D) Occasional rains

46. Maximum production (on global basis) is shown by:
(A) Forests (B) Aquatic ecosystem
(C) Deserts (D) Grassland

47. Which one is biotic in origin?
(A) Rainfall
(B) Texture and porosity of soil
(C) Carbon dioxide content of soil
(D) Light intensity

48. Best source of renewable energy is:
(A) Cattle (B) Petroleum
(C) Coal (D) Trees

49. A nonconventional energy source is:
(A) Coal
(B) Petroleum
(C) Solar radiations
(D) Electricity from nuclear plants

50. A nonrenewable resources is:
(A) Forest
(B) Coal/Petroleum
(C) Water
(D) Wild life

51. Geothermal energy is:
(A) Nonrenewable nonconventional energy source
(B) Nonrenewable conventional energy source
(C) Renewable nonconventional energy source
(D) Renewable conventional energy source

52. Recurrence of floods in the plains of North India is due to excessive:
(A) Siltation of dams
(B) Rainfall
(C) Agriculture
(D) Deforestation in catchment area

53. Mulching is helpful in:
(A) Increasing soil fertility
(B) Improvement of soil structure
(C) Moisture conservation
(D) Better crops

54. Eroded soils are:
(A) Devoid of plant nutrients
(B) Richer in plant nutrients
(C) Unaltered in plant nutrients
(D) Fit for agriculture

55. A depleting source of energy is:
(A) Sunlight (B) Water
(C) Wind (D) Fossil fuel

56. Renewable source of energy is:
(A) Biomass (B) Coal
(C) Petroleum (D) Kerosene

57. Main renewable source of energy is:
(A) Forest (B) Fuel gas
(C) Cattle (D) Coal

58. Biomass produced by plants in ocean accounts for:
(A) 85% (B) 75%
(C) 65% (D) 55%

59. Soil conservation achieved through:
(A) Restricted human activities
(B) Good plant cover
(C) Wind screen (D) Ploughing

60. Soil fertility is reduced by:
(A) Crop rotation
(B) Nitrogen fixing bacteria
(C) Decaying organic matter
(D) Intensive agriculture

61. Water is a resource:
(A) Nondegadable nonmaintainable
(B) Degradable maintainable
(C) Renewable
(D) Non-renewable

62. Largest amount of fresh water is found in:
(A) lakes and streams (B) Underground
(C) Polar ice and glaciers (D) Rivers

63. Bulk fixation of carbon through photosynthesis takes place in:
(A) Tropical rain forests
(B) Tropical rain forest and crop plants
(C) Crop plants (D) Oceans

64. Guano is the major source of:
(A) Nitrogen (B) Phosphorus
(C) Sulphur (D) Zinc

65. Sequence of humification and mineralization is:
(A) Dead organic mater → Litter → Duff→ Humus → Minerals
(B) Duff→ Litter → Humus →Minerals → Dead organic mater
(C) Litter →Dead organic mater → Minerals → Duff → Humus
(D) Dead organic mater → Litter → Humus → Duff → Minerals

66. Soil is composed of:
(A) Mineral + Water + Air
(B) Mineral + Organic matter + Water
(C) Mineral + Organic matter + Air + Water
(D) Organic matter + Water

67. Source of energy which does not evolve CO_2 is:
(A) Coal (B) Oil
(C) Organic compounds (D) Nuclear energy

68. Fossil fuels have been formed by a:
(A) Conventional process (B) Natural process
(C) None – conventional process
(D) All of these

69. Plants die in waterlogged soils due to:
(A) Nutrient leaching
(B) Stoppage of root respiration
(C) Dilution of soil nutrients
(D) Dilution of cell sap

70. Ecofriendly methods is:
(A) Plantation of C3 plants
(B) Plantation of sugarcane
(C) Energy plantation
(D) None of these

71. Fuel wood crisis can be overcome by:
(A) Better wooden stoves
(B) Efficient wood extraction
(C) Greater afforestation
(D) Using alternate energy source

72. Edaphology is connected with:
(A) Plant and biosphere
(B) Soil and living microorganisms
(C) Animals and ecosystem
(D) Soil and biosphere

73. Top soil is darker and:
(A) Contains more Na and Mg
(B) It drier than subsoil
(C) Contains more organic matter
(D) Is wetter than subsoil

74. When is a man a part of biotic community?
(A) Always
(B) Only when he lives in big cities
(C) Only when he lives in a forest
(D) Never

75. Amount of water available to the plants in the soil is called as:
(A) Holard (E + C) (B) Chresard
(C) Echard (D) None of these

76. Humus is defined as:
(A) Physical texture of soil
(B) Chemical composition of soil
(C) Decomposed organic matter in soil
(D) None of these

77. Almost decomposed organic matter in which original form is lost is in:
(A) Duff (B) Litter
(C) Leaf mould (D) None of these

78. Edaphic factors are related to:
(A) Soil (B) Man
(C) Animals (D) temperature

79. The kind of soil water most useful to plant is:
(A) Hygroscopic water (B) Gravitational water
(C) Capillary water (D) None of these

80. Major constitutes of soil is:
(A) Organic matter (B) Inorganic matter
(C) Soil air (D) Soil water

81. Which of the following means endemic:
(A) Indigenous (B) Exotic
(C) Ubiquitous (D) Naturalized

82. Those species whose population have been seriously depleted and whose ultimate security is not assured are known as:
(A) Threatened species (B) Endangered species
(C) Vulnerable species (D) Rare species

83. Chipko movement is related to:
(A) Plant/Forest conservation
(B) Preventing soil pollution
(C) Preserving threatened species
(D) None of these

84. The average life of a crop variety is:
(A) 1-5 years (B) 5-15 years
(C) 15-20 years (D) 20-25 years

85. Gene for resistance to brown plant hopper (*Nilaparvata lugens*) of modern rice has been obtained from:
(A) Wild rice
(B) African rice
(C) Old rice varieties from Kerala
(D) IR-8

86. Which one is most troublesome weeds of agricultural fields?
(A) *Chenopodium album*
(B) *Avena fatua*
(C) *Parthenium hysterophorus*
(D) *Eichhornia crassipus*

87. Valley of flowers is found in:
(A) Nilgiri Hills
(B) North Eastern Himalyan
(C) North Western Himalyan
(D) Satpura Hills

88. Congress grass is:
(A) *Cynadon dactylon*
(B) *Cyperus rotundus*
(C) *Parhenium hysterophorus*
(D) *Digitaria purpurea*

89. Poor man's timber is:
(A) Mango wood (B) Bamboo
(C) Neem timber (D) Shisham wood

90. Tendu/Kendu leaves used as wrappers for 'bidis' are got from:
(A) *Diospyros obenum*
(B) *Dipterpcarpus indicus*
(C) *Dipterpcarpus tuberculatus*
(D) *Diospyros exsculpta*

91. The plant yielding Sola Pith (*Aeschynomene aspara*) is:
(A) Large sized tree
(B) Herb
(C) Palm-like monocot
(D) Perennial shrub with spongy floating stem

92. Major forest types in India are:
(A) Tropical evergreen and tropical thorn
(B) Temperate mixed and tropical deciduous
(C) Tropical dry deciduous and tropical moist deciduous
(D) Tropical evergreen and tropical moist deciduous

93. Chipko movement originated in:
(A) Silent valley of Kerala
(B) Kangra valley of Himachal Pradesh
(C) Tehri Garwal of UP
(D) Pachmari of MP

94. First biosphere reserve was established in 1986 at:
(A) Nilgiri (B) Nanda Devi
(C) Little Rann of Kutch (D) Sunderbans

95. Extinction of numerous wild animals is due to:
(A) Hunting (B) Deforestation
(C) Overcrowding (D) Scarcity of food

96. Limited private operations are allowed in:
(A) Zoos (B) Botanical gardens
(C) National Parks (D) Sanctuaries

97. A threatened species is:
(A) Endangered (B) Rare
(C) Vulnerable (D) All of these

98. National Wildlife Protection Act was enacted in India in:
(A) 1950 (B) 1962
(C) 1972 (D) 1986

99. MAB programme of UNESCO was launched in:
(A) 1971 (B) 1981
(C) 1991 (D) 2001

100. MAB is:
(A) Man and Botany
(B) Man And Biosphere
(C) Man and biotic community
(D) Man, Antibiotics and Bacteria

101. Conservation is:
(A) Protection of natural resources
(B) Management of natural resources
(C) Proper use of natural resources
(D) All of these

102. Overgrazing causes:
(A) Negative pollution (B) Positive pollution
(C) Soil erosion
(D) Reduction in crop yield

103. Forest management is related to :
(A) Silviculture (B) Sericulture
(C) Apiculture (D) Olericulture

104. Forest controls drought through:
(A) Lot of water plants (B) Increasing rainfall
(C) Retention of water and prevention of erosion
(D) Functioning as water shed

105. Deforestation brings about:
(A) Increased sunlight
(B) Increased grazing area
(C) Weed control (D) Soil erosion

106. Silent Valley has been preserved as it has:
(A) Recreational value
(B) Valuable timber plants
(C) Natural forest
(D) Rare plants and animals

107. Silent valley having rare plants and animals is located at:
(A) Kerala (B) Karnataka
(C) UP (D) Maharashtra

108. Which one is connected with conservation of forests?
(A) Gir (B) Silent valley
(C) Kaziranga (D) Bharatpur

109. Deforestation will decrease:
(A) Soil erosion (B) Land slides
(C) Soil fertility (D) Rainfall

110. New approach to conservation is establishment of:
(A) Reserve forests (B) Sanctuaries
(C) Biosphere reserves (D) National parks

111. Social forestry is:
(A) Management of forests by cooperative societies
(B) Growing and management of useful plants on common lands
(C) Growing one type of trees
(D) Growing different types of plants together

112. Deforestation is the major causal agent of:
(A) Depletion of natural resources
(B) Environmental pollution
(C) Desertification of habitat
(D) Genetic erosion

113. In a National Park protection is provided to:
(A) Entire ecosystem (B) Flora and fauna
(C) Fauna only (D) Flora only

114. Ranganathittoo Sanctuary (Mysore) is known for population of:
(A) Bison (B) Tiger
(C) Lion (D) Birds

115. Whale washed ashore will die because of:
(A) Incapability of breathe
(B) Increase in body temperature
(C) Crushing of bones under body weight
(D) Cracking skin

116. Which animal has become extinct from India?
(A) Snow leopard (B) Hippopotamus
(C) Wolf (D) Cheetah

117. Which animal should be protected?
(A) Harmless (B) Economically useful
(C) Likely to perish (D) Ferocious

118. Animal species should be protected because they are:
(A) Lovely creatures
(B) Useful to humans
(C) To be studied by zoologists
(D) Man cannot recreate a lost species

119. Wild life is destroyed most when:
(A) There is lack of proper care
(B) Mass scale hunting for foreign trade
(C) Its natural habitat is destroyed
(D) Natural calamity

120. Deforestation does not lead to:
(A) Quick nutrient cycling
(B) Soil erosion
(C) Alteration of local weather condition
(D) Destruction of natural habitat of wild animals

121. Forests takes part in:
(A) Control of atmospheric pollution
(B) Maintenance of natural balance
(C) Prevention of soil erosion
(D) All of these

122. Forest destruction results in:
(A) Loss of wild life (B) Floods and drought
(C) Soil erosion (D) All of these

123. Afforestation should be with:
(A) Exotic species
(B) Indigenous species
(C) Bamboos
(D) Eucalyptus

124. Animals and plants are best protected in:
(A) Botanical gardens (B) Zoos
(C) National Parks (D) Sanctuaries

125. Troublesome Americal water weeds found in India is:
(A) *Eichhornia* (B) *Trapa*
(C) *Cyperus* (D) *Typha*

126. First national park of India is:
(A) Kanha National Park
(B) Periyar National Park
(C) Corbett National Park
(D) Bandipur National Park

127. Breeding place of Flamingo (Hansawar) in India is:
(A) Chilka Lake (B) Sambar Lake
(C) Rann of Kutch (D) Ghana Vihar

128. If we remove half of the forest cover of earth, the crisis that will occur:
(A) Many species would become extinct
(B) Population, pollution and ecological imbalance will rise
(C) Energy crisis will commence
(D) The remaining forest will correct the imbalance

129. Decreased rainfall in northern areas is due to:
(A) Changes in sun
(B) Global phenomenon
(C) Deforestation in catchment areas
(D) Deforestation in evaporating are near equator

130. Biosphere reserve project was started in India during:
(A) 1984 (B) 1985
(C) 1986 (D) 1987

131. Decrease in species diversity in tropical countries is mainly due to:
(A) Urbanisation (B) Pollution
(C) Deforestation (D) Soil erosion

132. Plant species on verge of extinction due to over-exploitation is:
(A) *Centella* (B) *Podophyllum*
(C) *Gloriosa* (D) All of these

133. WWF-N has logo for conservation of:
(A) Tiger (B) Giant Panda
(C) Red Panda (D) Polar Bear

134. Increased incidence of floods in the plains of North India are due to:
(A) Increased deforestation in catchment areas
(B) Increase in incidence of rainfall
(C) Silting of dams
(D) More area under cultivation

135. Red Data Book provides data on:
(A) Biota of Red Sea
(B) Effect of red light or photosynthesis
(C) Red pigmented plants
(D) Threatened species

136. Soil erosion can be prevented by:
(A) Restricted human activity
(B) Good plant cover
(C) Checking movement of animals
(D) Wind screen alone

137. Which communities are more vulnerable to invasion by outside plants and animals?
(A) Tropical evergreen forests
(B) Temperate forests
(C) Mangroves
(D) Oceanic island communities

138. Adherence to two-children norm shall help in:
(A) Improving food web
(B) Checking pollution
(C) Stabilizing ecosystem
(D) Fertility of soil

139. Which of the following has high extinction prospects?
(A) Himalayan Bear and Musk Deer
(B) Lion and Leopard
(C) Tiger and Bustard
(D) Crocodile and Elephant

140. Which of the following will help most in conservation of wildlife?
(A) Making stringent laws
(B) Making numerous zoos
(C) Making numerous sanctuaries
(D) All of these

141. Jhoom cultivation refers to:
(A) Cultivation of Jamun trees
(B) Cultivation of medicinal plants by tribes
(C) Tribal method of shifting cultivation
(D) Cultivation of terraces

142. A species restricted to a given area is:
(A) Endemic species (B) Allopatric species
(C) Sympatric species (D) Sibling species

143. Which is mainly responsible for extinction of wildlife?
(A) Hunting of flesh
(B) Destruction of habitats
(C) Pollution of air and water
(D) All of these

144. Biosphere reserves are being threatened with:
(A) Population growth (B) Rains
(C) Pollution (D) All of these

145. *Idri idri* occurs in:
(A) India (B) Mauritius
(C) Fizi (D) Madagascar

146. Endangered plant species are conserved trough:
(A) Herbarium (B) Gene library
(C) Gene bank (D) Reducing pollution

147. Animal species protected in Kaziranga National Park is:
(A) *Panthera leo* (B) *Rhinoceros unicornis*
(C) *Panthera tigris* (D) *Macaca mulatta*

148. Which one is endangered species is?
(A) *Cuscuta* (B) *Acacia nilotica*
(C) *Nepenthes* (D) Both (B) and (C)

149. Afforestation in relation to urban development is component of:
(A) Social forestry
(B) Integrated forest development
(C) Agroforestry (D) Fuel forestry

150. Wildlife is conserved:
(A) *In situ* (B) *Ex situ*
(C) Both (A) and (B)
(D) Selective hunting of predators

151. The term wildlife was coined by:
(A) Hornday (B) Augus
(C) Montreal Protocol (D) Wiener

152. Chipko movement originated in the district:
(A) Allahabad (B) Chamoli
(C) Hardwar (D) Dehradun

153. Chipko movement originated in Gopeswar in:
(A) 1953 (B) 1963
(C) 1973 (D) 1983

154. A movement similar to Chipko which is spread in South is:
(A) Uliso (B) Appiko
(C) Belesu (D) Balasu

155. Dachigam sanctuary is associated with conservation of:
(A) Hangul (B) Rhino
(C) Barking Deer (D) Both (B) and (C)

156. The practice of raising plants and animals is:
(A) Agriculture (B) Horticulture
(C) Agronomy (D) Silviculture

157. Agriculture originated about:
(A) 2300-3000 years back
(B) 20000-30000 years back
(C) 7000-13000 years back
(D) 90000-100000 years back

158. Kung tribe is:
(A) Primitive agriculturist
(B) Hunter-gatherer
(C) Advanced agriculturist
(D) Completely dependent on horticulture

159. In various parts of the world, agriculture:
(A) Came from Asia Minor
(B) Has spread from Yellow Rive Valley
(C) Originated independently
(D) Has developed to various degree

160. Fire helped in the cultivation of:
(A) Fruit plants
(B) Fibre Plants
(C) Drug yielding plants
(D) Grain yielding plants

161. Early human civilization was based on:
(A) Development of fibre crops (B) Fire
(C) Agriculture
(D) Fire and Agriculture

162. A nonagricultural society living today is:
(A) Kung tribe (B) Urban Society
(C) Hintza tribe (D) Kuntizha tribe

163. Early man preferred plants:
(A) Grain yielding (B) Fruit yielding
(C) Multipurpose (D) Fibre yielding

164. Peruvian Andes are believed to be centre of origin of:
(A) Tomato (B) Potato
(C) Tomato and Potato
(D) Tomato, Potato and Chillies

165. Potato famine of Ireland occurred in:
(A) 1645 (B) 1745
(C) 1845 (D) 1854

166. Potato famine of Ireland was caused by the attack of fungus
(A) *Phytophthora palmivora*
(B) *Plasmopora viticola*
(C) *Puccinia graminis*
(D) *Phytophthora infestans*

167. Natural resistance to late blight of potato is present in:
(A) *Solanum tuberosum* (B) *Solanum acaule*
(C) *S. demissum* (D) *S stoloniferum*

168. Gene bank are part of:
(A) Ex situ conservation
(B) In situ conservation
(C) Both ex situ and in situ conservation
(D) Tribal diet

169. *In situ* conservation are carried out through:
(A) Natural parks (B) Wildlife santuaries
(C) Biosphere reserves (D) All the above

170. *Ex situ* conservation is carried out through:
(A) Funneling the threatened species into trade
(B) Offsite collection
(C) Gene banks
(D) All the above

171. Gene banks comprise:
(A) Seed banks, orchards, tissue culture and cryopreservation
(B) Seed banks, tissue culture and cryopreservation
(C) Seed banks and cryopreservation
(D) Tissue culture and cryopreservation

172. Orthodox seeds can be preserved for long periods at a temperature of:
(A) -196 ^{0}C (B) -10 to -20 ^{0}C
(C) 0 – 4 ^{0} C (D) 10 to 15 ^{0}C

173. Dwarfing gene of wheat is:
(A) Pal -1 (B) Atomita – 1
(C) Norin – 10 (D) Pelita -2

174. Jojoba yields wax for cosmetics from its:
(A) Stem (B) Leaves
(C) Seeds (D) Fruit pulp

175. IR 36 was developed through breeding of:
(A) Six rice varieties and *Oryza nivara*
(B) 13 rice varieties and *Oryza nivara*
(C) *Oryza sativa* and *O. indica*
(D) *Oryza indica* and *O. nivara*

176. A social forest tree is:
(A) Teak (B) Sandal wood
(C) Subabul (D) Rosewood

177. A plant species threatened with extinction can be preserved in:
(A) Herbarium (B) National park
(C) Botanical Garden (D) Gene Bank

178. A living fossil is:
(A) Gnetum (B) Ginkgo
(C) Riccia (D) Pinus

179. Jojoba is:
(A) Crotolaria (B) Leucaena
(C) Simmondsia (D) Parthenium

180. A place where endangered genetic materials of plants are kept is:
(A) Gene library (B) Genebank
(C) Gene pool (D) Herbarium

181. Wild varieties of plants must be conserved to:
(A) Feeding wild animals
(B) Maintain ecosystem
(C) Future evolution
(D) Incorporate useful traits in future crop varieties

182. "The dominant species with maximum canopy coverage of two or more species" represents a community character known as:
(A) Society (B) Location
(C) Consociation (D) Association

183. Agricultural growth is very soon likely to be hampered by scarcity of:
(A) Natural resources
(B) Man-made resources
(C) International cooperation
(D) Local coordination

184. A set of shared, similar characteristics (as applied to logically deduced and symbolically expressed models and observations) may be called as:
(A) Negative isomorphism
(B) Positive isomorphism
(C) Neutral isomorphism
(D) None of these

185. Choose the correct sequences:
(A) CO_2 in atmosphere → decomposers → organic carbon in animals → organic carbon in plants
(B) CO_2 in atmosphere → organic carbon in plants → organic carbon in animals→inorganic carbon in soil
(C) Inorganic carbonate in water → organic carbon in plants → organic carbon in animals → scavengers
(D) Organic carbon in animals → decomposers → CO_2 in atmosphere → organic carbon in plants

186. Around 50% world's population lives in:
(A) America (B) Africa
(C) Europe (D) Asia

187. The most prominent pastoral community in Africa is:
(A) Maasai (B) Gujjar
(C) Ahir (D) Pardhis

188. The Serengeti National park is in:
(A) Somali (B) Kenya
(C) Tswana (D) Tanzania

189. The word 'pastoral' refers to:
(A) Hunting
(B) Cattle herding
(C) Earning money by business
(D) Earning money by handicrafts

190. The vast meadow in the high mountains are known as:
(A) Bugyal (B) Bhabar
(C) Gaddar (D) Platean

191. The agricultural revolution came first of all in:
(A) France (B) Germany
(C) England (D) Austria

192. Growth rate refers to:
(A) Birth rate-death rate
(B) Number of people added in a year
(C) Number of males and females added in a year
(D) Total population –people who added have gone abroad

193. Sex ratio is:
(A) Distribution of population in the three sectors
(B) Distribution of population into three age groups
(C) Number of females per 1000 males
(D) Educated people in the country

194. The virgin vegetation, which are purely Indian are known as:
(A) Endemic (B) Exotic
(C) Flora (D) Cultivated crops

195. A very large ecosystem on land having distinct types of vegetation and animal life is called as:
(A) Biodiversity
(B) Natural vegetation
(C) Virgin vegetation
(D) Biome

196. Identify the type of forest; they are also called monsoon forest, trees shed their leaves for about 6 to 8 weeks in dry summer:
(A) Tropical evergreen forest
(B) Tropical deciduous forest
(C) Mountain forest
(D) Mangrove forest

197. Acacias, palms, euphorbias and cacti are found in the areas of rainfall less than:
(A) 70 cm (B) 80 cm
(C) 100 cm (D) Less than 50 cm

198. Identify the vegetation found in the coastal areas influenced by tides:
(A) Mountain forest
(B) Tropical dry deciduous forest
(C) Mangrove forest
(D) Moist tropical forest

199. One horned rhinocerous are found in:
(A) Swampy and marshy lands of Assam and West Bengal
(B) Arid areas of the Rann of Kachchh
(C) Thar desert
(D) Gir forest of Gujarat

200. Animal which can survive in extreme cold, i.e., in Laddakh's freezing high altitude are:
(A) Polar bear (B) Panda
(C) The Bharal (Blue sheep) (D) Leopard

201. Which of the following bio reserves of India is not included in the world net work of bio reserves?
(A) Manas (B) Gulf of mannar
(C) Sunderbans (D) Nanda Devi

202. Which bio-reserve is situated at the tri-junction of Kerala, Karnataka and Tamilnadu?
(A) Panchmarhi
(B) Nokrek
(C) Achanakmar-Amarkntak
(D) Nilgiris

203. Name the vegetation that grow at high altitudes, generally more than 3600 m above sea level:
(A) Temperate grasslands (B) Alpine grasslands
(C) Tundra vegetation (D) Alpine vegetation

204. Gir forest is the last remaining habitat of:
(A) Nilgai (B) Lions
(C) Wild ass and Camels (D) Asiatic lion

205. A very large ecosystem on land having various types of vegetation and animal life is called:
(A) Biosphere (B) Biome
(C) Biodiversity (D) Ecosystem

206. A forest area where flora and fauna are protected in their natural environment is called:
(A) Biosphere reserve (B) Biome
(C) Biodiversity (D) Ecosystem

207. It is a narrow zone of contact between lithosphere, hydrosphere and atmosphere where all living being live is known as:
(A) Biosphere (B) Biome
(C) Biodiversity (D) Ecosystem

208. A system which consists of the physical environment living within is called:
(A) Biosphere (B) Biome
(C) Biodiversity (D) Ecosystem

209. How many bioreserve are found in India to protect flora and fauna is?
(A) 12 (B) 13
(C) 14 (D) 15

210. Tropical deciduous forests are also called:
(A) Mountain forest
(B) Tropical dry deciduous forest
(C) Monsoon forest
(D) Moist tropical forest

211. The variety and abundance of species of plants and animals in a habitation, terrestrial or aquatic (fresh and marine water) with a large biosphere is called:
(A) Biosphere (B) Biome
(C) Biodiversity (D) Ecosystem

212. Which among the following produces seeds?
(A) Tallophyta (B) Bryophyta
(C) Pteridophyta (D) Gymnosperms

213. Which one is a true fish?
(A) Jelly fish (B) Star fish
(C) Dog fish (D) Silver fish

214. Well defined nucleus is absent in:
(A) Blue green algae (B) Diatoms
(C) Algae (D) Yeast

215. Which one of the following is not a viral disease?
(A) Dengue (B) AIDS
(C) Typhoid (D) Influenza

216. Which one of the following is not a bacterial disease?
(A) Cholera (B) Tuberculosis
(C) Anthrax (D) Influenza

217. Which one of the following disease is caused by protozoans?
(A) Dengue (B) AIDS
(C) Malaria (D) Influenza

218. Which of the following can make you ill if you come in contact with an infected person?
(A) High blood pressure
(B) Genetic abnormalities
(C) Sneezing (D) Blood cancer

219. AIDS cannot be transmitted by:
(A) Sexual contact (B) Hugs
(C) Breast feeding (D) Blood transfusion

220. Biotic component of biosphere is not constituted by:
(A) Producers (B) Consumers
(C) Decomposer (D) Air

221. Oxygen is returned to the atmosphere mainly by:
(A) Burning of fossil fuel (B) Respiration
(C) Photosynthesis (D) Fungi

222. Low visibility during cold weather is due to:
(A) Formation of fossil fuel
(B) Unburnt carbon particles or hydrocarbons suspended in air
(C) lack of adequate power supply
(D) None of these

223. Growth of Lichens on barren rocks is followed by the growth of:
(A) Moss (B) Ferns
(C) Gymnosperms (D) Algae

224. Marked temperature changes in aquatic environment can affect:
(A) Breeding of animals
(B) More growth of aquatic plants
(C) Process of digestion in animal
(D) Availability of nutrients

225. Oxygen is harmful for:
(A) Ferns (B) Nitrogen fixing bacteria
(C) Chara (D) mango tree

ANSWERSHEET

10. NATURAL AND FOREST CONSERVATION

1. (A)	2. (B)	3. (C)	4. (C)	5. (D)	6. (A)	7. (B)	8. (B)	9. (A)	10. (C)
11. (A)	12. (A)	13. (B)	14. (D)	15. (C)	16. (D)	17. (C)	18. (C)	19. (C)	20. (D)
21. (D)	22. (A)	23. (B)	24. (D)	25. (A)	26. (B)	27. (B)	28. (C)	29. (A)	30. (A)
31. (D)	32. (C)	33. (C)	34. (C)	35. (D)	36. (B)	37. (D)	38. (C)	39. (A)	40. (B)
41. (C)	42. (D)	43. (D)	44. (B)	45. (C)	46. (B)	47. (C)	48. (D)	49. (B)	50. (B)
51. (A)	52. (D)	53. (C)	54. (A)	55. (D)	56. (A)	57. (A)	58. (A)	59. (B)	60. (D)
61. (B)	62. (C)	63. (D)	64. (B)	65. (A)	66. (C)	67. (D)	68. (C)	69. (B)	70. (C)
71. (D)	72. (B)	73. (C)	74. (A)	75. (B)	76. (C)	77. (C)	78. (A)	79. (C)	80. (B)
81. (A)	82. (C)	83. (A)	84. (B)	85. (C)	86. (C)	87. (C)	88. (C)	89. (B)	90. (D)
91. (D)	92. (C)	93. (C)	94. (A)	95. (B)	96. (D)	97. (D)	98. (C)	99. (A)	100. (B)
101. (D)	102. (C)	103. (A)	104. (D)	105. (D)	106. (C)	107. (A)	108. (B)	109. (D)	110. (C)
111. (B)	112. (C)	113. (A)	114. (D)	115. (C)	116. (D)	117. (C)	118. (D)	119. (D)	120. (A)

121. (D) 122. (D) 123. (B) 124. (C) 125. (A) 126. (C) 127. (A) 128. (B) 129. (C) 130. (C)
131. (C) 132. (B) 133. (C) 134. (A) 135. (D) 136. (B) 137. (D) 138. (C) 139. (A) 140. (D)
141. (C) 142. (A) 143. (D) 144. (C) 145. (D) 146. (C) 147. (B) 148. (C) 149. (B) 150. (C)
151. (A) 152. (B) 153. (C) 154. (B) 155. (A) 156. (A) 157. (C) 158. (B) 159. (C) 160. (D)
161. (D) 162. (A) 163. (B) 164. (D) 165. (C) 166. (D) 167. (C) 168. (A) 169. (D) 170. (D)
171. (A) 172. (B) 173. (C) 174. (C) 175. (B) 176. (C) 177. (C) 178. (B) 179. (C) 180. (B)
181. (D) 182. (D) 183. (A) 184. (B) 185. (D) 186. (B) 187. (A) 188. (D) 189. (B) 190. (A)
191. (C) 192. (A) 193. (C) 194. (A) 195. (D) 196. (D) 197. (A) 198. (C) 199. (A) 200. (C)
201. (A) 202. (D) 203. (D) 204. (D) 205. (B) 206. (A) 207. (A) 208. (D) 209. (C) 210. (C)
211. (C) 212. (D) 213. (C) 214. (A) 215. (B) 216. (D) 217. (C) 218. (C) 219. (B) 220. (D)
221. (C) 222. (B) 223. (A) 224. (A) 225. (B)

Chapter - 11

GENOMICS AND RECOMBINANT DNA TECHNOLOGY

1. Introduction of foreign genes for improving genotype is:
(A) Tissue culture
(B) Immunisation
(C) Biotechnology
(D) Genetic engineering

2. Genetic engineering means:
(A) Meiotic division of cells
(B) Nucleotide transfer
(C) Deletion and repair mechanism
(D) None of these

3. The set of techniques that enables DNA from different sources to be identified, isolated and recombined so that new characteristics can be introduced into an organism is called:
(A) Recombinant DNA technology
(B) Genetic engineering
(C) Molecular biology technology
(D) All of these

4. The recombinant DNA technique was engineered by:
(A) Stanley Norman Cohen (B) Herbert Boyer
(C) Both of these (D) None of these

5. Phosphatediester bond of DNA and RNA involves:
(A) 2'C and 1'C (B) 5'C and 3'C
(C) 5'C and 2'C (D) 3'C and 1'C

6. Production of recombinant DNA and its transfer into a suitable host for multiplication/expression of the recombinant DNA. The recombinant DNA may or may not become integrated into the host genome is called:
(A) Genetic engineering
(B) Genetic transformation
(C) Genetic material (D) Genetic resource

7. The chemical that contains the genetic information, i.e., of which genes are composed; in most of the organisms, DNA is the genetic material; in only some viruses, RNA serves this purpose is called:
(A) Genetic engineering
(B) Genetic transformation
(C) Genetic material (D) Genetic resource

8. Genetic engineering was born on:
(A) 1972 (B) 1973
(C) 1974 (D) 1975

9. Sum total of all the genes and their alleles present in a species and its wild relatives is called:
(A) Genetic engineering
(B) Genetic transformation
(C) Genetic material
(D) Genetic resource

10. Introduction of DNA or recombinant DNA in to a suitable cell/organism is called:
(A) Genetic engineering
(B) Genetic transformation
(C) Genetic material
(D) Genetic resource

11. A DNA molecule created in the laboratory by ligation of two or more different pieces of DNA is called:
(A) Recombinant DNA (B) Recombinant clone
(C) Genetic counseling (D) Genetic diversity

12. A bacterial clone containing a recombinant DNA molecules is called:
(A) Recombinant DNA (B) Recombinant clone
(C) Genetic counseling (D) Genetic diversity

13. Educating perspective parents, either suffering from or suspected to be heterogygous for some genetic disease, on the risk of their children suffering from the same disease is called:
(A) Recombinant DNA (B) Recombinant clone
(C) Genetic counseling (D) Genetic diversity

14. The variability present within a species, between individuals of a population or different population groups of the same species for one or a group of traits is called:
(A) Recombinant DNA (B) Recombinant clone
(C) Genetic counseling (D) Genetic diversity

15. Origin of genetic variation in a relatively high frequency, usually, much higher than that ascribable to spontaneous mutations is called:
(A) Genetic instability
(B) Recombinant clone
(C) Genetic counseling
(D) Genetic diversity

16. The DNA segment to be cloned is called:
(A) DNA ligase (B) DNA insert
(C) DNA array (D) DNA methylation

17. A gene construct in which the gene is located in antisense orientation so that the natural sense strand of the gene is now transcribed to produce antisense RNA; used is antisense RNA Technology is called:
(A) Antisense gene construct
(B) Antisense RNA technology
(C) Both (A) and (B)
(D) None of these

18. The process of an antisense construct of the target gene is integrated into the genome to suppress the concerned endogenous genes is called:
(A) Antisense gene construct
(B) Antisense RNA technology
(C) Both (A) and (B)
(D) None of these

19. The antigenic peptides bind within the cleft of MHC I (Major Histrocompatibility Complex I) or MHC II molecules and are displayed on the surface of antigen processing cells (APCs) is called:
(A) Antigen presentation
(B) Antisense orientation
(C) Antisense strand
(D) None of these

20. Reverse orientation of a gene in relation to its promoter; the promoter is now located at the 5'-end of the antisense strand so that the natural sense is now transcribed is called:
(A) Antigen presentation
(B) Antisense orientation
(C) Antisense strand
(D) None of these

21. The strand of a native gene, which is used by RNA polymerase as template to genetrate its complementary copy of RNA; this RNA ultimately functions as messenger RNA; the promoter sequence is, as a rule, located at the 3'-end of the antisense strand is called:
(A) Antigen presentation
(B) Antisense orientation
(C) Antisense strand
(D) None of these

22. Colonies from a master plates are either lifted or replica plates onto a nitrocellulose filter, cells are lysed, DNA is denatured & hybridized with radioactive probe to detect the colonies having DNA/RNA sequences complementary to the probe is called:
(A) Colony hybridization
(B) Northern hybridization
(C) Southern hybridization
(D) None of these

23. An enzyme that seals nicks in DNA strands by catalyzing the formation of phosphodiester bond between the contiguous nucleotides is called:
(A) DNA ligase (B) DNA insert
(C) DNA array (D) DNA methylation

24. RNA sample is subjected to gel electrophoresis, RNA transferred from the gel onto a nylon membrane, immobilized and hybridized with a single stranded radioactive probe; autoradiography reveals the RNA band hybridized to the probe, used for detection of transcription of transgenes is called:
(A) Colony hybridization
(B) Northern hybridization
(C) Southern hybridization
(D) Somatic hybridization

25. Restriction digested DNA is subjected to get electrophoresis, denatured, blotted onto a nitrocellulose/nylon membrane fixed, hybridized with a single-stranded DNA/RNA probe and hybridization of probe with DNA is detected by autoradiography; used for detection of transgene integration into host genome, RFLP detection, DNA fingerprinting, etc., is called:
(A) Colony hybridization
(B) Northern hybridization
(C) Southern hybridization
(D) None of these

26. Production of hybrid plants via protoplast fussion; feasible in sexuality incompatible combinations, is sterile/non-flowering species, etc by:
(A) Colony hybridization
(B) Northern hybridization
(C) Southern hybridization
(D) Somatic hybridization

27. A procedure for transferring denaturated RNA from an agarose gel onto a nitrocellulose filter to allow its hybridization with a probe by:
(A) Southern blotting (B) Northern blotting
(C) Both (A) and (B) (D) None of these

28. A procedure includes electrophoresis of RNA sample and transfer of RNA from the gel onto a nylon membrane by:
(A) Southern blotting (B) Northern blotting
(C) Both (A) and (B) (D) None of these

29. The characteristics two dimensional pattern (on paper or gel) formed by the separation of a mixture of peptides resulting from partial hydrolysis of a protein is known as:
(A) Genetic map (B) Peptide mapping
(C) Peptide map (D) None of these

30. A series of DNA sequence fixed as distinct spots on a suitable solid support, like glass chip; basically two types, spotted DNA array and printed oligonucleotide chips is called:
(A) DNA ligase (B) DNA insert
(C) DNA array (D) DNA methylation

31. Addition of methyl (-CH3) residues, chiefly to C residues, usually, located in CG or CNG (N means any one of the four DNA bases) sequences, after they are incorporated in DNA is called:
(A) DNA ligase (B) DNA insert
(C) DNA array (D) DNA methylation

32. The biotin-labelled mRNA from non-stressed cell is called:
(A) Driver mRNA (B) cDNA
(C) Both (A) and (B)
(D) None of these

33. The copy of DNA or complementary DNA produced by using RNA (Usually mRNA) as tempelate; produced by reverse transcriptase enzyme is called:
(A) Driver mRNA (B) cDNA
(C) Both (A) and (B) (D) None of these

34. RNA produced by transcription of the antisense strand of a gene; in case of structural genes encoding proteins, it usually means mRNA is:
(A) Driver mRNA (B) cDNA
(C) Sense RNA (D) None of these

35. A diagram showing the relative sequence and position of specific genes along a chromosomes is called:
(A) Genetic map
(B) Genetic information
(C) Genetic code (D) Gene expression

36. The hereditary information contained in a sequence of nucleotide bases in chromosomal DNA or RNA is known as:
(A) Genetic map
(B) Genetic information
(C) Genetic code (D) Gene expression

37. The set of triplet code words in DNA (or mRNA) coding for the amino acids of proteins is known as:
(A) Genetic map
(B) Genetic information
(C) Genetic code (D) Gene expression

38. Transciption, and in the case of proteins, translation, to yield the product of a gene, a gene is expressed when its biological product is present and active is known as:
(A) Genetic map
(B) Genetic information
(C) Genetic code (D) Gene expression

39. Recombination between two DNA molecules of similar sequence, occurring in all cells; occurs during meiosis and mitosis in eukaryotes is called:
(A) Homologous Genetic Recombination
(B) Recombinant DNA
(C) Recombinational DNA repair
(D) Recombination

40. Peptide mapping is also known as:
(A) Peptide fingerprinting
(B) DNA fingerprinting
(C) DNA footprinting (D) None of these

41. The transfer of genetic information from one cell to another by means of a viral vector is called:
(A) Transmination (B) Transcription
(C) Transduction (D) Transformation

42. Introduction of an exogenous DNA into a cell, causing the cell to require a new phenotype is called:
(A) Transmination (B) Transcription
(C) Transduction (D) Transformation

43. The process in which the genetic information present in an mRNA molecule specifies the sequence of amino acids during protein synthesis is called:
(A) Translation (B) Transcription
(C) Transduction (D) Transformation

44. The movement of gene or set of genes from one site in the genome to another is called:
(A) Transmination (B) Transposition
(C) Transduction (D) Transformation

45. A DNA sequence at which RNA polymerase may bind, leading to initiation of transcription is called:
(A) Transmination (B) Transcription
(C) Promoter (D) Transformation

46. DNA formed by the joining of genes in to new combinations is called:
(A) Recombinant DNA (B) Recombination
(C) Recombinational DNA repair
(D) None of these

47. Which of the variations of PCR aims to enhance specificity?
(A) Anchored PCR (B) Hotstart PCR
(C) Touchdown PCR (D) Both (B) and (C)

48. Which *E. coli* vectors yields single-stranded copies of cloned DNA?
(A) pUC18/19 (B) λ - phage vectors
(C) Phage M13 vectors (D) Cosmid vectors

49. Which bacteria has at least two circular chromosomes in its genome?
(A) *Bacillus* (B) *Pseudomonas*
(C) *Vibrio chlerae* (D) *Mycoplasma genitalium*

50. Which is not correct about DNA?
(A) Double helix model of DNA was produced by Watson and Crix
(B) DNA molecules ordinarily consist of two polynucleotide chains/strands
(C) DNA double helix is stabilized by hydrogen bonds
(D) Two strands of a DNA molecule are oriented antiparallel to each other

51. Which is not correct?
(A) If the base sequence of one DNA strand is known, the base sequence of its other strand can be easily deduced
(B) The formation of hydrogen bonds between adenine and thymine, and between guanine & cytosine is known as complementary base pairing
(C) Usually, B-DNA molecule has right-handed coiling
(D) Hydrophobic interactions are important in transcription

52. Match the column A and B and select the correct option from those listed below them.

Column A	Column B
A. Genomics (functional)	1. Single nucleotide polymorphism
B. Marker	2. Proteins
C. Proteomics	3. Gene isolation
D. Probe	4.Function of all genes

Options:
(A) A–1, B–3, C–2, D-4 (B) A–4, B–1, C–2, D-3
(C) A–2, B–1, C–3, D-4 (D) A–1, B–4, C–3, D-2

53. Multiple copies of a gene can be obtained by :
(A) Gene cloning
(B) Polymerase chain reaction
(C) Combination of PCR with chemical synthesis
(D) All of these

54. DNA polymerase III serves which type of functions?
(A) DNA repair
(B) DNA replication
(C) Initiation of DNA replication
(D) All of these

55. Which is not correct about DNA replication?
(A) The presence of replication fork in E. coli chromosome was first demonstrated by Cairns in 1963
(B) DNA replication always proceeds in the 5'→ 3' direction
(C) Both strands of DNA replicate continuously
(D) DNA replication of DNA replication synthesis of a small RNA primer

56. Which enzyme is called molecular scissor?
(A) DNA ligase (B) DNA gyrase
(C) Restriction enzyme (D) Topoisomerase

57. A class of enzymes, which cleaves DNA only within or near those sites that have specific base sequences is called:
(A) Type I endonuclease
(B) Type III endonuclease
(C) Type II endonuclease (D) Exonuclease

58. Which scientists was awarded Nobel Prize in 1978 for the discovery of endonucleases?
(A) Smith (B) Nathans
(C) Arber (D) All of these

59. Near-isogenic lines can be developed by:
(A) Single seed descent
(B) Mutagenesis
(C) Backcross method
(D) Continued inbreeding

60. Which is an example of type I restriction endonuclease?
(A) *Eco*K (B) *Hinf*III
(C) *Hind*II (D) *Bgl*II

61. Which of the following nucleotide sequences is recognized by EcoRI?
(A) G/ATTC
CTTAA/G (B) T/GA
AGC/T
(C) A/GATCT
TCGAG/A (D) A/AGCTT
TTCGA/A

62. IRGSP stand for:
(A) International Rice Genome Sequencing Project
(B) International Rice and Groundnut Sequencing Project
(C) International Research on Groundnut, Sorghum and Pigeonpea
(D) Indian Rice Genetic Resources Programme

63. Most of the type II restriction endonucleases have recognition sites of how much length?
(A) 4 bp (B) 5 bp
(C) 6 bp (D) 4-6 bp

64. Which restriction enzymes produces blunt ends?
(A) *Bam*H1 (B) *Alu*I
(C) *Pst*I (D) *Hind*III

65. Which one has the largest number of genes?
(A) *Arabidopsis* (B) Yeast
(C) *Caenorhabditis* (D) *Drosophila*

66. Identification of protein using antibodies may be done by:
(A) Enzyme-linked immunosorbant assay (ELISA)
(B) Western blotting
(C) Precipitation and electrophoresis
(D) All of these

67. Which scientists isolated ribosomal RNA genes in 1965?
(A) Maxam and Gilbert
(B) Wallace and Birnstiel
(C) Sanger and Southern
(D) Brown and Gilbert

68. Which has the smallest genome size?
(A) *Mycoplasma genitalium*
(B) *Saccharomyces cerevisiae*
(C) *E. coli*
(D) *Caenorhabditis elegans*

69. The lambda genome contains:
(A) Origin of replication
(B) Genes for head and tail proteins
(C) Proteins involved in lysogeny
(D) All of these

70. Which population are used for linkage mapping?
(A) F_2 population
(B) Backcross population
(C) Recombinant inbred lines
(D) All of these

71. Restrictive polyacrylamide gels are used for:
(A) Viroid test
(B) Analysis of point mutations
(C) DNA sequencing
(D) All of these

72. During DNA isolation, if bacterial cell extract has a heavy load of proteins, it is treated with?
(A) Proteinase K (B) Trypsin
(C) Pronase (D) Both (A) and (C)

73. Which is the least likely to accumulate property folded proteins encoded by an eukaryotic transgene?
(A) Animal cells (B) *Saccharomyces*
(C) *E. coli* (D) *Pichia*

74. Which one has the largest genome size?
(A) *Arabidopsis* (B) Yeast
(C) *Caenorhabditis* (D) *Drosophila*

75. In general, which is used for initial cloning?
(A) *B. subtilis* (B) *Agrobacterium*
(C) *E. coli* (D) *B. thuringiensis*

76. Use of an antibiotic resistance and the *lac*Za in a vector, eliminates the need for:
(A) Plating on the antibiotic containing
(B) Replica plating
(C) Plating on a second selection medium
(D) Both (B) and (C)

77. AFLP may arise due to:
(A) A difference in restriction site
(B) Deletions and insertions within the amplified restriction fragments
(C) Mutations beyond the restriction sites that affect complementarity with the selection nucleotides of the AFLP primers
(D) All of these

78. In case of linkage maps, the distance, between the markers are measured in terms of:
(A) Recombination frequency
(B) Mutation frequency
(C) Angstroms
(D) Nucleotide number

79. Which of the following cleaves both the DNA strands at random sites?
(A) DNase I (B) S1 nuclease
(C) *Eco*R 1 (D) *Hind*III

80. Which one has smallest genome size?
(A) *Arabidopsis* (B) Yeast
(C) *Caenorhabditis* (D) *Drosophila*

81. Which of the following enzymes is used for converting protruding ends into blunt ends?
(A) Polynucleotide kinase
(B) S1 nuclease
(C) Klenow fragment of *E. coli* DNA polynuclease I
(D) Both (B) and (C)

82. Which is the preferred strategy for converting protruding ends into blunts ends?
(A) Polynucleotide kinase
(B) S1 nuclease
(C) Klenow fragment of *E. coli* DNA polynuclease I
(D) Both (B) and (C)

83. DNA fingerprinting is used for:
(A) Identification of varieties
(B) Resolving disputed parentages
(C) Identification of criminals
(D) All of these

84. ARMS PCR achieves differential amplification of allele due to:
(A) Primer sequence
(B) Annealing temperature
(C) Ion concentration
(D) Specific PCR protocol

85. In most PCR products, a single nucleotide overhang is of:
(A) A at the 3' - end (B) A at the 5'- end
(C) A at the 3' – end (D) T at the 5' - end

86. The genome of eukaryotes was first to be sequenced:
(A) *Arabidopsis* (B) Yeast
(C) *Caenorhabditis* (D) *Drosophila*

87. Which is not an example of insertion vector?
(A) λgt10 (B) λEMBL4
(C) λgt11 (D) λZAPII

88. Identification of protein using antibodies may be done by:
(A) Enzyme-linked immunosorbant assay (ELISA)
(B) Precipitation and electrophoresis
(C) Western blotting (D) All of these

89. Replacement vector, λ ENBL4 can be used to clone DNA of upto which size?
(A) 10 kb (B) 20 kb
(C) 40 kb (D) 45 kb

90. Bacterial cells are usually multiply in:
(A) Jensson medium
(B) Luria-Bertani (LB) medium
(C) YE medium (D) Ganborg medium

91. Pulsed-field gel electrophoresis is able to separate chromosomes of:
(A) Yeast (B) Bacteria
(C) Protozoa (D) All of these

92. Who developed pBR322:
(A) Boliver Rodriguez (B) Berg Rosenberg
(C) Benfy Ris (D) Bonner Riggs

93. Which vectors has been constructed for use in two different hosts?
(A) Shuttle vectors (B) Cosmids
(C) Phasmids (D) Phagemids

94. DNA markers can be used for:
(A) Preparation of linkage map
(B) Map-based cloning of genes
(C) Mapping of quantitative trait loci (QTL)
(D) All of these

95. Which is not correct about DNA replication?
(A) Mode of DNA replication is semi-conservative
(B) Mode of semi-conservative DNA replication was proposed by Watson and Crick
(C) Evidence for semi-conservative replication of DNA was first reported by Meselson and Stahl in 1958
(D) Semi-conservative mode of DNA replication was first demonstrated in T2 phage of *E. coli*

96. Which is not correct about RNA?
(A) RNA molecules ordinarily consist of a single polyribonucleotide strand
(B) RNA molecules contain uracil in the place of cytosine
(C) Phosphodiester bonds are formed between ribonucleotides in the same manner as in the case of deoxyribonucleotides
(D) RNA was, most likely, the first genetic material, it was much later replaced by DNA

97. Isozyme markers behave like:
(A) Recessive (B) Dominant
(C) Codominant (D) Overdominant

98. During 1957, Frankel-Conrat and Singer proved that RNA is the genetic material in which organisms?
(A) Cauliflower mosaic virus
(B) Tobacco mosaic virus
(C) Cucumber mosaic virus
(D) Tomato mosaic virus

99. Eukaryotic chromosomes consist of:
(A) Histone and non-histone proteins
(B) Nucleoproteins
(C) DNA
(D) Nucleic acids

100. Lod score method was developed by:
(A) Paterson (B) Morton
(C) Peltier (D) Rigby

101. Which is not correct?
(A) Double helix RNA generally follows the same rule of base-pairing as does DNA
(B) The smallest RNA molecule is tRNA
(C) Antisense RNA is also called micRNA
(D) Ribosomal RNA is processed from larger primary transcripts

102. Who was appointed as the first Director of Human Genome Project:
(A) A. H. Paterson (B) P.M Gresshoff
(C) K.E. Davis (D) J.D. Watson

103. Direct repeat means:
(A) Repetition of certain length of nucleotides in reverse order
(B) Repetition of certain length of nucleotides in the same order
(C) Repetition of certain length of nucleotides in either the same or the reverse order
(D) Repetition of certain length of nucleotide to form a palindrome with rotational symmetry

104. Which is not correct about *Taq* DNA polymerase?
(A) It is isolated from *Thermus aquaticus*
(B) It is stable at higher temperatures
(C) It has proof-reading property
(D) It is used in PCR for DNA amplification

105. Among the following statements about B- and Z-DNA forms, which one is not common to both forms?
(A) Double helix
(B) Length of one full turn of helix is 34 Å
(C) Two strands of double helix are antiparallel
(D) Complementary base pairing

106. Which is not correct about *E. coli*?
(A) *E coli* is rarely used for gene cloning
(B) Single generation of *E. coli* is completed within 20-30 minutes, under optimum conditions
(C) Chromosome length of *E. coli* is approximately 1 mm
(D) Rate of chromosome replication is 30-40 μ per minute under optimum conditions

107. Biotin labeled probes are generally prepared by:
(A) Primer labeling (B) Nick translation
(C) End labeling (D) All of these

108. Which is not correct?
(A) A mixture of fragments of genomic DNA is used to produce a genomic library
(B) Genomic library contains only those genes that are expressed in the concerned tissue/organism
(C) cDNA obtained from mRNA may be cloned to give rise to a cDNA library
(D) cDNA library contains DNA fragments that represent genes

109. A phagemid can be used as:
(A) A sequencing vector
(B) An expression vector
(C) A cloning vector
(D) All of these

110. Who isolated ribosomal RNA genes in 1965?
(A) Maxam and Gilbert
(B) Wallace and Birnstiel
(C) Sanger and Southern
(D) Brown and Gilbert

111. Which is used in colony hybridization?
(A) Antibody (B) Enzyme
(C) Probe (D) Substrate

112. Enzyme procedure of DNA sequencing was developed by:
(A) Maxam and coworkers
(B) Gilbert and coworkers
(C) Sanger and coworkers
(D) Collins and coworkers

113. Which is correct?
(A) Genome is the basic set of chromosomes, which is equivalent to the gametic chromosome complement number of a diploid species
(B) Genome map is the schematic depiction of structural and functional organization of all chromosomes in the genome of an organism
(C) Genome project was started to develop genome maps, genome sequence and its annotation in man and other experimental organisms
(D) All of these

114. Which is used for non-radioactive labeling of probe/nucleic acids?
(A) Digoxigenic (B) Fluorescent molecules
(C) Biotin (D) All of these

115. Which techniques can be used for radioactive labeling of probes?
(A) Nick translation and direct labeling
(B) End labeling and primer extension
(C) Method based on RNA polymerases
(D) All of these

116. Transgene becomes integrated chiefly by homologous recombination in:
(A) Yeast (B) Animals
(C) Plants (D) Both (B) and (C)

117. Which enters *E. coli* cell in linear form?
(A) A cosmid vector (B) A λ phage vector
(C) pBR322 (D) Both (A) and (B)

118. Which affect the level of recombinant protein accumulation?
(A) Promoter
(B) Context of AUG codon (initiator codon)
(C) Codon usage (D) All of these

119. RAPD is generated by using of:
(A) A single arbitrary primer
(B) Single-sequence specific primer
(C) Two arbitrary primers
(D) A pair of sequence-specific primers

120. The term 'recombinant DNA' is specifically used to denote:
(A) A DNA molecules having sequences from two or more sources
(B) A vector into which a DNA insert is integrated
(C) A transgene having sequences from two different organisms
(D) A chromosome into which is transgene is integrated

121. A cosmid vector can achieve which of the following that a plasmid vector can not achieve?
(A) Cloning of 40 kb DNA insert
(B) Packaging in empty λ phage heads
(C) Transfection of *E. coli* cells
(D) All of these

122. Which restriction endonucleases is not used in gene cloning?
(A) Type I (B) Type II
(C) Type III (D) Both (A) and (C)

123. Which markers are uses PCR as well as restriction enzymes?
(A) AFLP (B) RFLP
(C) RAPD (D) STMS (STR loci)

124. Which sequences forms a palindrome with rotational summary?
(A) GAAAAG (B) GAAGAA
(C) GAATTC (D) Both (A) and (B)

125. Which is a simple palindromic sequence?
(A) GAAAAG (B) GAAGAA
(C) GAATTC (D) Both (A) and (B)

126. Which is a tandem direct repeat sequence?
(A) GAAAAG (B) GAAGAA
(C) GAATTC (D) Both (A) and (B)

127. The method of determining the activity of the promoter is also known as:
(A) Reporter assay (B) Primer extension
(C) Expression vector (D) Codon bias

128. Which is not correct about amplified genomic library?
(A) Particular recombinant DNAs increase in frequency
(B) Particular recombinant DNAs decrease in frequency
(C) Some recombinant DNAs may be eliminated
(D) Specified recombinant DNAs are amplified

129. Translational fusion offers:
(A) Enhanced translation efficiency
(B) Protection of transgene product
(C) May facilitate purification
(D) All of these

130. Minisatellite variant PCR is an example of:
(A) Hotstart PCR (B) Touchdown PCR
(C) Nested PCR (D) ARMS PCR

131. Which is not correct about minisatellite DNAs?
(A) They are usually 200-2000 bp (0.2-2.0 kb) long DNA fragments
(B) They consist of tandem repeats of 15-60 bp
(C) In human genome, they are concentrated in the proterminal regions of chromosomes
(D) They are a good marker system for human genome

132. A phagemid vector has:
(A) Phage origin of replication
(B) Plasmid origin of replication
(C) Both (A) and (B)
(D) *cos* sequence from λ phage

133. ARMS PCR achieve:
(A) High efficiency amplification
(B) Highly specific amplification
(C) Differential amplification of alleles
(D) Both (A) and (B)

134. Which aims at increasing the specificity of PCR without lowering the efficiency?
(A) Overlap extension PCR (B) Inverse PCR
(C) Touchdown PCR (D) Anchored PCR

135. Which can be used when sequence of only one end of the desired segment is known?
(A) Overlap extension PCR (B) Inverse PCR
(C) Touchdown PCR (D) Anchored PCR

136. Selection nucleotides are used in which of the marker systems?
(A) RAPD (B) RFLP
(C) AFLP (D) STMS (STR loci)

137. A probe for colony hybridization may be obtained from:
(A) Sequence of the concerned gene
(B) Amino acid sequence of the concerned protein product
(C) Same gene of another organism
(D) All of these

138. Which of the following observations provided evidence that Mendel's "units" were located on the chromosomes?
(A) Chromosomes occur as homologous pairs
(B) The number of linkage groups and chromosomes is always equal
(C) One member of each homologous pair goes to each gamete during meiosis
(D) All of these

139. Technique of Southern blotting was originally based on:
(A) Capillary action (B) Electromotive force
(C) Vacuum (D) Both (A) and (C)

140. Which is called molecular glue?
(A) DNA polymerase (B) DNA ligase
(C) Exonuclease (D) Endonuclease

141. Theoretically, a recognition sequence of bases, e.g., 5'GAATTC3' (for EcoRI), is expected to occur at, on an average, a distance of how many nucleotides?
(A) 8188 bp (B) 1024 bp
(C) 4096 bp (D) 2048 bp

142. In which of the PCR variants are used two pairs of primers?
(A) Asymmetric (B) Nested
(C) Anchored (D) Inverse

143. Southern hybridization can be used for:
(A) Preparation of RLFP maps
(B) DNA fingerprinting
(C) Confirmation of transgene integration
(D) All of these

144. At the end of n cycles of PCR, how many copies of the amplified segment (amplicon) are ideally expected?
(A) n^2 (B) 2^n
(C) $n^2/2$ (D) 2n

145. If 8 bases primer is used in the PCR, a sequence complementary to it may be expected to occur once every base pair of:
(A) 8^4 bp (B) 4^8 bp
(C) 8^8 bp (D) 4^4 bp

146. Complementation assay is based on:
(A) Protein function
(B) DNA-sequence homology
(C) Protein-protein interaction
(D) DNA-protein interaction

147. Which of the following is the basis of Southern-Western assay?
(A) Protein function
(B) DNA-sequence homology
(C) Protein-protein interaction
(D) DNA-protein interaction

148. Which of the PCR variants have flanking regions of the target gene are amplified?
(A) Asymmetric (B) Nested
(C) Anchored (D) Inverse

149. When two selectable markers are used in a vector, where the DNA insert has to be integrated?
(A) In one of the selectable markers
(B) In both the selectable markers
(C) Anywhere in the vector
(D) Either A or B

150. Which enzymes removes nucleotides from the 5'-ends?
(A) S1 nuclease (B) Exonuclease Bal31
(C) *E. coli* exonuclease III (D) λ exonuclease

151. Which enzymes converts protruding ends into blunt ends?
(A) S1 nuclease (B) Exonuclease Bal31
(C) *E. coli* exonuclease III (D) λ exonuclease

152. DNA microarrays are high throughout version of:
(A) Southern analysis
(B) Differential screening
(C) Gene tagging
(D) Colony hybridization

153. The process of identification of genetic material began in 1928 with the experiments by:
(A) Griffith (B) Hershey and Chase
(C) Watson (D) Avery & coworkers

154. Prokaryotic DNA polymerase I is known to function in:
(A) DNA replication (B) DNA repair
(C) Recombination (D) Both (A) and (B)

155. Genome of which of the following organisms was the first to be sequenced?
(A) *Pseudomonas putida*
(B) *Mycoplasma genitalium*
(C) *Saccharomyces cerevisiae*
(D) *Haemophilus influenzae*

156. Length of lambda genome is:
(A) 39 kb (B) 49 kb
(C) 59 kb (D) 29 kb

157. A bacterial artificial chromosome vector can be used to clone DNA inserts of upto?
(A) 100 kb (B) 300 kb
(C) 800 kb (D) 600 kb

158. Which enzymes is used in PCR?
(A) *Vent* polymerase (B) *Taq* polymerase
(C) *Pfu* polymerase (D) All of these

159. PCR can be used for:
(A) Study of DNA polymorphism
(B) Detection of presence of transgene in transformed cells
(C) Production of cDNA
(D) All of these

160. Meltinng temperature ™ of DNA can be calculated using by:
(A) [2(G + C)] + [4(A + T)]
(B) [4(G + C)] + [2(A + T)]
(C) [3(G + C)] + [2(A + T)]
(D) (4G + C) + (2A + T)

161. A plasmid vector can not achieve:
(A) Gene expression
(B) Propagation of recombinant DNA
(C) Propagation in two different hosts
(D) Packaging in empty λ particles

162. Which endonucleases introduces single-strand nicks in DNA?
(A) DNase I (B) S1 nuclease
(C) *Eco*R 1 (D) *Hind*III

163. Which of the following has the recognition sequence 5'GAATTC3'?
(A) DNase I (B) S1 nuclease
(C) *Eco*R 1 (D) *Hind*III

164. Treatment with which of the following prevents vector circulation?
(A) Polynucleotide kinase
(B) S1 nuclease
(C) Alkaline phosphatase
(D) λ exonuclease

165. Which vectors contains λ phage *cos* sequence?
(A) Cosmid (B) Phagemid
(C) Phasmid (D) Both (A) and (C)

166. The λ exonuclease digests from:
(A) 3'- end having –OH
(B) 5'- end having P residue
(C) 5'- end having –OH
(D) Both (B) and (C)

167. A good vector has at least how many marker genes?
(A) One (B) Two
(C) Three (D) Four

168. Which is a type II restriction endonuclease?
(A) *Eco*PI (B) *Hinf*III
(C) *Eco*P15 (D) *Pst*I

169. Hershey and Chase used which organisms for their experiments to show that DNA is the genetic material?
(A) T_2 phage and *E. coli* (B) T_4 phage and *E. coli*
(C) λ phase and *E. coli* (D) T_7 phage and *E. coli*

170. Germplasm is contained in:
(A) Somatic cells (B) Growing points
(C) Reproductive cells (D) Both (A) and (C)

171. Which one will be best for germplasm collection:
(A) Britain (B) France
(C) Portugal (D) Peru

172. A gram-negative soil bacterium that produces hairy-root disease of plants due to its Ri (Root inducing) plasmid; the natural genetic engineer is:
(A) *Agrobacterium rhizogenes*
(B) *E. coli*
(C) *Agrobacterium tumifaciens*
(D) None of these

173. A gram negative soil bacterium that produces crown gall in dicots by transferring the T-DNA of its Ti (Tumour inducing) plasmid; the natural genetic engineer is:
(A) *Agrobacterium rhizogenes*
(B) *E. coli*
(C) *Agrobacterium tumifaciens*
(D) None of these

174. Graft obtained from genetically different individuals of the same species is:
(A) Allograft (B) Pseudograft
(C) Nanograft (D) All of these

175. A haploid obtained from an allo polyploidy is:
(A) Autohaploid (B) Allo haploid
(C) Androgenesis haploid (D) None of these

176. Haploid obtained from the activity of male gamete is:
(A) Autohaploid
(B) Allo haploid
(C) Androgenesis haploid
(D) None of these

177. Molecular maps have become increasingly useful in:
(A) Genetic analysis of animals and plants
(B) Location of functional genes within the genome
(C) Physical mapping of eukaryotic genomes
(D) All of these

178. DNA ligase is used for:
(A) Joining of more than one fragment of DNA
(B) Synthesis of DNA
(C) Replication of DNA
(D) Cleavage of DNA

179. Prokaryotic DNA polymerase I is known to function in:
(A) DNA replication (B) DNA repair
(C) Recombination (D) Both (A) and (B)

180. Who proposed the concept of central dogma?
(A) Crick (B) Watson
(C) Meselson (D) Cairns

181. Which have been involved in *Arabidopsis thaliana* genome evolution?
(A) Polyploidization
(B) Genome reduction
(C) Single cycle of (A) followed by (B)
(D) More than one cycle of (A) followed by (B)

182. Who got Nobel Prize in 1980 for the development for recombinant DNA technology?
(A) W. Gilbert (B) F. Sanger
(C) D. Baltimore (D) P. Berg

183. The S1 nuclease is isolated from:
(A) *Aspergillus oryzae* (B) *Escherichia coli*
(C) *Neisseria lactamica* (D) *Proteus vulgaris*

184. Which is an example of virulence plasmid?
(A) T1 Plasmid (B) Col plasmid
(C) Ri plasmid (D) Both (A) and (C)

185. The phenomenon of transformation was discovered by:
(A) N. Zinder (B) F. Griffith
(C) J. Leaderberg (D) E.L. Tatum

186. Which is not correct about microsatellite DNAs?
(A) They are usually, larger than 2 kb long DNA fragments
(B) They consist of repeats of 2-3 bp sequences
(C) They are randomly and frequently distributed in all eukaryotic genomes, except yeast
(D) They are almost ideal markers for genome mapping

187. Probes can be used for:
(A) To confirm the integration of a DNA insert into the host genome
(B) To prepare genome maps of eukaryotes
(C) To identify the desired recombinant clone
(D) All of these

188. Which is an example of artificial chromosome vector?
(A) pBeloBAC11 (B) pBACe3.6
(C) pBAC108L (D) pBeloBAC1

189. Which approaches is the most successful for chemical synthesis of DNA?
(A) Phosphodiester
(B) Phosphite trimester
(C) Phosphotriester
(D) Phosphite diester

190. How many alleles are possible at an SNP locus?
(A) 1 (B) 2
(C) 4 (D) 3

191. Which marker systems is corrently used for human DNA fingerprinting?
(A) RAPD (B) RFLP
(C) AFLP (D) STMS (STR loci)

192. Synthetic oligonucleotides can be used to fulfill the:
(A) To produce a complete gene
(B) As linkers and adaptors for gene cloning
(C) For enzymatic DNA sequencing
(D) All of these

193. Contigs facilitate:
(A) Rapid sequencing
(B) Sequence alignment
(C) Sequence annotation
(D) Sequence interpretation

194. PCR amplification is sensitive to:
(A) Haem breakdown product (B) Humic acid
(C) Phenolics (D) All of these

195. Which if not correct about PCR?
(A) Only intact genomic DNA can be used for PCR
(B) PCR does not use radioactivity
(C) Allele-specific primers can be used to detect specific alleles
(D) PCR requires nanogram quantities of DNA

196. Which is not PCR based molecular marker?
(A) RAPD (B) RFLP
(C) STS (D) AFLP

197. In 1970, which scientists synthesized the gene for alanine-tRNA of yeast?
(A) Maxam and coworkers
(B) Sanger and coworker
(C) Khorana and coworkers
(D) Smith and coworkers

198. The genome of which plants was the first to be sequenced?
(A) *Zea mays*
(B) *Oryza sativa*
(C) *Hordeum vulgare*
(D) *Arabidopsis thaliana*

199. Which of the following is the basis for colony hybridization?
(A) Protein function
(B) DNA-sequence homology
(C) Protein-protein interaction
(D) DNA-protein interaction

200. Which of the following is the basis of ELISA?
(A) Protein function
(B) DNA-sequence homology
(C) Protein-protein interaction
(D) DNA-protein interaction

201. Initially, which marker systems was used for human DNA fingerprinting?
(A) RAPD (B) RFLP
(C) AFLP (D) STMS (STR loci)

202. Transgene becomes integrated chiefly by homologous recombination in:
(A) Yeast (B) Animals
(C) Plants (D) Both (B) and (C)

203. Which transgene integration generally occurs at random sites?
(A) Yeast (B) Animals
(C) Plants (D) Both (B) and (C)

204. Which PCR variants generates single-stranded products?
(A) Asymmetric (B) Nested
(C) Anchored (D) Inverse

205. Which plants has the smallest genome size?
(A) *Zea mays* (B) *Oryza sativa*
(C) *Hordeum vulgare*
(D) *Arabidopsis thaliana*

206. In which of the PCR variants is target DNA cleavage and ligation is involved?
(A) Asymmetric (B) Nested
(C) Anchored (D) Inverse

207. In which of the PCR variants have the concentrations of one primer is much higher than that of the other?
(A) Asymmetric (B) Nested
(C) Anchored (D) Inverse

208. Human genome is estimated to contain about how many genes?
(A) 35000 (B) 25000
(C) 50000 (D) 40000

209. In which of the PCR variants are primers annealing to the 5'-ends of the target gene are used?
(A) Asymmetric (B) Nested
(C) Anchored (D) Inverse

210. Which plants has the largest genome size?
(A) *Zea mays* (B) *Oryza sativa*
(C) *Hordeum vulgare*
(D) *Arabidopsis thaliana*

211. A cosmid vector has :
(A) A selectable marker (B) λ Phage *cos* site
(C) Terminase binding site (D) All of these

212. Tailing uses which enzymes?
(A) Polynucleotide kinase
(B) Klenow fragments of *E. coli* DNA polymerase I
(C) Terminal deoxynucleotidyl transferase
(D) Both (B) and (C)

213. Which enzymes removes nucleotide from 3'-ends?
(A) S1 nuclease
(B) Exonuclease Bal31
(C) *E. coli* exonuclease III (D) λ exonuclease

214. Which one has the smallest number of genes?
(A) *Arabidopsis* (B) Yeast
(C) *Caenorhabditis* (D) *Drosophila*

215. Which enzymes is used to digest either the 3'- or the 5'- protruding ends from DNA fragments?
(A) S1 nuclease (B) Exonuclease Bal31
(C) *E. coli* exonuclease III (D) λ exonuclease

216. Match the column A and B and select the correct option from those listed below them.

Column A	Column B
A. *Arabidopsis thaliana*	1. 20000
B. *Drosophila melanohaster*	2. 6000
C. *Caenorhabditis elegans*	3. 26000
D. *Saccharomyces cerevisiae*	4. 16000

Options:
(A) A–1, B–3, C–2, D-4 (B) A–4, B–1, C–2, D-3
(C) A–3, B–4, C–1, D-2 (D) A–1, B–4, C–3, D-2

217. Which enzymes is used to shorten DNA fragments?
(A) S1 nuclease
(B) Exonuclease Bal31
(C) *E. coli* exonuclease III
(D) λ exonuclease

218. Which is not correct about PCR?
(A) It does not use radioactivity
(B) It does not require specific skills
(C) DNA purity is not critical
(D) It always uses a pair of primers

219. Which version of PCR is used for site-directed mutagenesis?
(A) Overlap extension PCR
(B) Inverse PCR (C) Touchdown PCR
(D) Anchored PCR

220. Most of the PCR products have which type of ends?
(A) Single nucleotide overhang
(B) Blunt (C) Cohesive
(D) Either (B) or (C)

221. Two-hybrid yeast system studies in:
(A) Protein function (B) Heterosis
(C) Protein-protein interaction
(D) Protein expression

222. An enzymatic process by which the linear arrangement of nucleic acid sequences in a chromosome is altered by cleavage and rejoining is called:
(A) Recombinant DNA
(B) Recombination
(C) Recombinational DNA repair
(D) None of these

223. Recombinational processes that are directed at the repair of DNA strand breaks or cross-links, especially at inactivated replication forks is called:
(A) Recombinant DNA
(B) Recombination
(C) Recombinational DNA repair
(D) None of these

224. To study the patterns of expression of all the different genes present in the genome of organisms that leads to the growth and development of the organism is called:
(A) Functional genome (B) Functional proteomic
(C) Gene regulation (D) DNA transfection

225. Use of proteomics techniques to analyze the characteristics of molecular protein networked involved in a living cell is called:
(A) Functional genome
(B) Functional proteomics
(C) Gene regulation
(D) DNA transfection

226. To study all the transcriptional level; that is how a given gene produces a transcript by:
(A) Functional genome (B) Functional proteomics
(C) Gene regulation (D) DNA transfection

227. The process of reintroduction of a cloned DNA into a cell is known as:
(A) Functional genome (B) Functional proteomics
(C) DNA transfection (D) Gene regulation

228. The construction of expression vectors is called:
(A) Vector engineering (B) Host engineering
(C) In vitro transcription (D) Codon bias

229. Alanine has four codons: CGA, GCC, GCG and GCT. Humans use GCC four times more than GCG for codon alanine. This is known as:
(A) Vector engineering (B) Host engineering
(C) In vitro transcription (D) Codon bias

230. Introgression of genes from improved varieties of crop plants into their wild relatives is called:
(A) Gene flow (B) Gene stacking
(C) Gene replacement (D) Gene correction

231. Correcting the illness by substituting an abnormal gene by a normal one in the basis of:
(A) T-DNA (B) Gene tagging
(C) Knockout organism (D) Gene therapy

232. To obtain optimal expression levels, such host factors have to be minimized by choosing an appropriate mutant of the host in which to express the heterologous gene. Such manipulation of the host for optimal expression of the introduced heterologous gene is known as:
(A) Vector engineering (B) Host engineering
(C) In vitro transcription (D) Codon bias

233. In the cell, transcription takes place inside the nucleus or close to the nucleotide in bacteria by the action of the RNA polymerase on DNA. This process can also be mimicked outside the cell, that is closed DNA can be mixed with RNA polymerase and the four nucleotides in tube and under appropriate conditions, RNA transcripts can be formed at is does inside the cell. This is known as:
(A) Vector engineering (B) Host engineering
(C) In vitro transcription (D) Codon bias

234. The *Agrobacterium tumefaciens* bacteria causes a cancerous growth in the infected plant due to the introduction of this DNA fragment derived from the plasmid. The plasmid is called the Ti plasmid and the introduced fragment of DNA is called the:
(A) T-DNA (B) Gene tagging
(C) Knockout organism (D) Gene therapy

235. The method of insering a recognizable DNA fragment within a gene such that the function of the gene is distrupted and the gene can be easily recognized by virtue of the inserted fragment is known as:
(A) T-DNA (B) Gene tagging
(C) Knockout organism (D) Gene therapy

236. Transgenic organisms can also be produced to inactivate a particular gene in which case it is known as:
(A) T-DNA (B) Gene tagging
(C) Knockout organism (D) Gene therapy

237. Many transgenes are transferred into a single variety through a single transformation event is called:
(A) Gene flow (B) Gene stacking
(C) Gene replacement (D) Gene correction

238. Depending upon the type of change one targets in the defective gene if the purpose is to increase its activity it is known as:
(A) Gene flow (B) Gene stacking
(C) Gene replacement (D) Gene correction

239. The purpose is to replace a defective gene by a functional one is called:
(A) Gene flow (B) Gene stacking
(C) Gene replacement (D) Gene correction

240. The deliberate modifications of an organism's genetic information by directly changing its nucleic acid content is a subject matter of:
(A) Genetic engineering
(B) Population genetics
(C) Microbiology
(D) Protein engineering

241. Vectors are:
(A) Molecules that degrade nucleic acids
(B) Molecules that help in replication
(C) Molecules that are able to covalently bond to and carry foreign DNA into cells
(D) Molecules that protect host cells from invasion by foreign DNA

242. Electroporation is:
(A) The process of combining foreign DNA to an electrically charged vector molecules
(B) The application of high voltage pulse
(C) The process of multiplication of the cells
(D) The process of separating charged molecules through a gel maintained in an electric field

243. Which of the following enzyme is used to covalently bond foreign DNA to a vector plasmid?
(A) DNA helicase
(B) DNA polymerase
(C) Restriction endonuclease
(D) DNA ligase

244. Charged molecules are separated based on varying rates of migration through a solid matrix when subjected to an electric field. This technique is known as:
(A) Gel electrophoresis (B) Blotting
(C) Photoreaction (D) Autoradiography

245. Which of the following is obtained using processed mRNA molecules as a template?
(A) rDNA (B) cDNA
(C) mDNA (D) tDNA

246. Agrobacterium tumifaciens is:
(A) A bacterium that can be used to introduce DNA into fungai
(B) A bacterium that can be used to introduce DNA into plants
(C) A disease in humans that causes loss of sight
(D) A fungi that is used to produce antibiotics in large amounts

247. Problem in obtaining large amounts of proteins encoded by recombinant genes can often be overcome by using:
(A) YACS (B) BACS
(C) Expression vectors (D) All of these

248. What is the normal role of restriction endonucleases in bacterial cells?
(A) To degrade invading phage DNA
(B) To produce RNA primers for replication
(C) To degrade the bacterial chromosome into small pieces during replication
(D) All of these

249. Bacterial cells protect their own DNA from degradation by restriction endonucleases by:
(A) Not producing any restriction endonucleases
(B) Having anti restriction endonucleases
(C) Deleting all recognition sites from the genome
(D) Methylation the DNA at the site that the enzyme recognizes

250. Which of the following is not commonly used as vector?
(A) Fungi
(B) Artificial chromosome
(C) Plasmid
(D) Cosmid

251. A genomic library is:
(A) A data base where the sequence of an organism's genome is stored
(B) A place where the information of the genetic organization of organisms are kept
(C) A collection of many clones possing different DNA fragments from the same organisms bound to vectors
(D) A book that describes how to isolate DNA from a particular organism

252. A molecular technique in which DNA sequences between two oligonucleotide primers can be amplified is known as:
(A) Southern blotting
(B) Polymerase chain reaction
(C) DNA replication
(D) Northern blotting

253. An animal, that has gained new genetic information from the acquisition of foreign DNA, is considered as:
(A) A transgenic animal
(B) A vector
(C) An enzyme that links DNA molecules
(D) A chimera

254. The piece of equipment, that introduces DNA into cells via DNA-coated microprojectiles is known as:
(A) Laser
(B) DNA probe
(C) Gene gun
(D) Inoculating needle

255. A southern blotting technique depends on:
(A) Similarities between the sequences of probe protein and experimental protein
(B) Similarities between the sequences of probe DNA and experimental DNA
(C) The molecular mass of proteins
(D) Similarities between the sequences of probe RNA and experimental RNA

256. Which of the following vector can maintain the largest fragment of foreign DNA?
(A) YAC (B) Cosmid
(C) Plasmid (D) Phage

257. A short molecule containing 2-20 nucleotide is:
(A) Plasmid (B) Mononucleotide
(C) Oligonucleotide (D) Vector

258. **The advantage of using DNA polymerases from thermophilic organisms in PCR is that:**
(A) The DNA polymerases of these bacteria never make mistakes while replicating DNA
(B) The DNA polymerases of these bacteria can withstand the high temperatures needed to denature the DNA strands
(C) The DNA polymerases of these bacteria are much faster than from other organisms
(D) All of these

259. **For gene probes to be useful they must:**
(A) Be labeled in some manner to allow detection
(B) Be large enough to contain gene-specific sequences
(C) Both (A) and (B)
(D) None of these

260. **In genetic engineering, a chimera is:**
(A) A plasmid that contains foreign DNA
(B) An enzyme that links DNA molecules
(C) A virus that infects bacteria
(D) A fungi

261. **Which type of restriction endonuclease cuts the DNA within the recognition site?**
(A) Type I (B) Type II
(C) Type III (D) All of these

262. **Enzymes that recognize and cleave specific 4 to 8 base pair sequences of DNA are:**
(A) DNA ligase
(B) DNA gyrase
(C) Restriction endonucleases
(D) Helicases

263. **Cos site of the cosmids:**
(A) Consists of 12 bases
(B) Helps whole genome in circularization and ligation
(C) Both (A) and (B)
(D) Contains cleavage site

264. **Conjugative plasmids:**
(A) Exhibit antibiotic resistance
(B) Do not exhibit antibiotic resistance
(C) Carry transfer genes called the tra genes
(D) Do not carry transfer genes

265. **Maximum size of foreign DNA that can be inserted into an insertion vector is:**
(A) 35 kb (B) 18 kb
(C) 50 kb (D) 27 kb

266. **Maximum size of foreign DNA that can be inserted into a replacement vector is:**
(A) 25-30 kb (B) 18-20 kb
(C) 20-25 kb (D) 40-50 kb

267. **Cosmid lack:**
(A) Genes coding for viral proteins
(B) Origin of replication
(C) Marker genes coding for replication
(D) Cleavage site for the insertion of foreign DNA

268. **Charon 34 and Charon 35 can clone DNA upto:**
(A) 1-2 kb (B) 6-7 kb
(C) 9-20 kb (D) 30-35 kb

269. **M 13 is an example of:**
(A) Filamentous phage
(B) Single stranded DNA vector
(C) Both (A) and (B)
(D) Plasmid

270. **For cloning of the gene, which among following is not important:**
(A) Vector (B) Host
(C) Restriction enzymes (D) DNase

271. **The isolation and production of indefinite number of copies of a single gene is called:**
(A) Gene bank
(B) Genetic engineering
(C) Gene pool
(D) Gene cloning

272. **Which enzyme plays an important role in reverse transcription?**
(A) DNA polymerase (B) RNA polymerase
(C) Reverse transcriptase (D) Isomerase

273. **During the process of reverse transcription:**
(A) mRNA is synthesized from DNA
(B) RNA is synthesized from DNA
(C) Proteins are formed from mRNA
(D) DNA is synthesized from mRNA

274. **The group of enzyme that cut DNA molecules at specific sites are known as:**
(A) Recombinant DNA
(B) DNA ligase
(C) Restriction endonuclease
(D) Plasmid DNA

275. **The technique to clone the first gene was developed in:**
(A) 1972 (B) 1975
(C) 1976 (D) 1980

276. **Complete set of cloned DNA fragments of an organism is known as:**
(A) Genomic library (B) cDNA library
(C) mRNA library (D) Gene bank

277. **Horizontal gene flow can be reduced b expressing genes in:**
(A) Nucleus (B) Chloroplast
(C) Both (A) and (B) (D) None of these

278. **Transfer of genes (gene flow) in between different species is known as:**
(A) Verticular gene flow
(B) Horizontal gene flow
(C) Lateral gene flow
(D) None of these

279. Which among the following is not used as vector for gene cloning?
(A) Cosmid (B) Phagemid
(C) PAC (D) MAC

280. Bacterial genes lack:
(A) Operator (B) Introns
(C) Promoters (D) None of these

281. DNA replication is:
(A) Conservative (B) Semi-conservative
(C) Both (A) and (B) (D) None of these

282. The bond between sugar and nitrogenous base in case of DNA is known as:
(A) Glycosidic bond
(B) Phosphodiester bond
(C) Peptide bond
(D) None of these

283. Which bond is the DNA provides it the backbone?
(A) Glycosidic bond (B) Phosphodiester bond
(C) H-bond (D) Peptide bond

284. DNA is fit for making tools in case of nanotechnology. This is due to:
(A) Branching nature of DNA
(B) Flexibility in DNA conformation
(C) Small size of DNA
(D) All of these

285. Which of the following DNA structure forms left hand helix?
(A) DNA A (B) DNA B
(C) DNA C (D) DNA Z

286. Procaryotic DNA polymerase comprises five subunits, which are:
(A) $\alpha\beta\mu\delta\varepsilon$ (B) $\alpha_2\beta\beta'\delta$
(C) $\alpha\varepsilon\cdot\omega\omega$ (D) None of these

287. Eucaryotic DNA replication is carried out by:
(A) DNA POLδ (B) DNA POLμ
(C) DNA POLβ (D) None of these

288. DNA replication occur in which stage of cell division?
(A) M Phase (B) G_2 Phase
(C) S Phase (D) None of these

289. Genomic library can be prepared by:
(A) Colony hybridization (B) Shotgun experiment
(C) PCR technique (D) All of these

290. Genetically engineered bacteria is used for the production of:
(A) Ephedrine (B) Cortisoles
(C) Human insulin (D) Thyroxine

291. Manipulation of the genetic material towards a desired end in a directed and predetermined way is called:
(A) Genetic engineering
(B) Gene cloning
(C) Recombinant DNA technology
(D) All of these

292. Which gene is used for development of salt tolerant crops?
(A) LEA genes
(B) Genes involved in proline biosynthesis
(C) Heat shock proteins
(D) All of these

293. For cloning eukaryotic genes in prokaryotes, genes should be isolated from:
(A) cDNA library (B) Genomic library
(C) Eukaryotic host (D) Prokaryotic host

294. Essential constituents of a cloning vector are:
(A) Replicon, promoter and selectable marker
(B) Replicon, unique restriction site and selectable marker
(C) Promotor, operator and restriction site
(D) Replicon, unique restriction site and promoter

295. Essential constituents of expression vector are:
(A) Replicon, promoter and selectable marker
(B) Replicon, unique restriction site, selectable marker and promotor
(C) Promotor, operator and restriction site
(D) Replicon, unique restriction site and promoter

296. Restriction site for cloning should be:
(A) Tetranucleotide
(B) Palindromic in nature
(C) Repeated sequence
(D) Hexa nucleotide

297. GAATTC is restriction sequence of CTTAAG:
(A) Eco R1 (B) NCo I
(C) Sau 3 (D) None of these

298. Commonly used host in gene cloning is:
(A) *E. coli* (B) Pseudomonas
(C) Nitrosomonas (D) Bacillus

299. Which of the following vector can be used for cloning bigger DNA fragments?
(A) E. coli (B) Cosmid
(C) YACs (D) Plasmid

300. A nucleoside is formed of:
(A) Pentose sugar, phosphate and nitrogen base
(B) Phosphate and nitrogen base
(C) Pentose sugar and phosphate
(D) Pentose sugar and nitrogen base

301. A nucleotide is formed of:
(A) Purine, pyrimidine and phosphate
(B) Purine, sugar and phosphate
(C) Nitrogen base, sugar and phosphate
(D) Pyrimidine, sugar and phosphate

302. Amino acid binding site of tRNA is:
(A) 5' end (B) Anticodon loop
(C) DHU loop (D) -CCA 3' end

303. Anticodon occurs in:
(A) tRNA (B) mRNA
(C) mtRNA (D) rRNA

304. Bacterial plasmid contains:
(A) RNA (B) RNA + protein
(C) DNA
(D) Photosynthetic structures

305. Base pair present in one turn of DNA are:
(A) 12 (B) 11
(C) 10 (D) 9

306. Distance between two pairs of DNA is:
(A) 34 nm (B) 3.4 nm
(C) 0.68 nm (D) 0.34 nm

307. Distance between two strands of DNA is:
(A) 34 Å (B) 20 Å
(C) 3.4 Å (D) 340. Å

308. Distance between two successive nitrogenous bases or base pairs of DNA is:
(A) 34. Å (B) 3.4 Å
(C) 10 Å (D) 5 Å

309. DNA and RNA show similarity in having:
(A) polymer of nucleotides (B) similar pyrimidines
(C) Double strands (D) Similar sugars

310. DNA is composed of repeating units of:
(A) Ribonucleosides
(B) Deoxyribonucleosides
(C) Ribonucleosides
(D) Deoxyribonucleotides

311. DNA resembles RNA as both have:
(A) Polymers of nucleotides
(B) Similar sugars
(C) Similar pyrimidine bases
(D) Ability to replicate

312. Double chained DNA strand is made radioactive in both its chain. It is allowed to replicate twice in nonradioactive medium. The result would be:
(A) All strands have radioactivity
(B) Half the strands have radioactivity
(C) Three strands have radioactivity
(D) Radioactivity is absent in all strands

313. Gene is segment of:
(A) RNA (B) DNA
(C) Either RNA or DNA (D) Both DNA and RNA

314. Haploid plant cultures are got from:
(A) Leaves (B) Root tip
(C) Pollen grain (D) Buds

315. Hydrogen bonds present between cytosine and guanosine are:
(A) 2 (B) 3
(C) 1 (D) 4

316. A riboside is:
(A) Base + phosphate
(B) Ribose + phosphate
(C) Ribose + phosphate + base
(D) Ribose + base

317. Adenine is:
(A) Purine (B) Pyrimidine
(C) Nucleoside (D) Nucleotide

318. Chemofusion and electrofusion are:
(A) Eugenics (B) Protoplast fusion
(C) Cloning (D) Mutations

319. DNA and RNA are similar in having:
(A) Similar nucleotides (B) Similar pyrimidines
(C) Similar purines (D) Similar sugars

320. DNA replication is:
(A) Conservative and discontinuous
(B) Semi conservative and semi discontinuous
(C) Semi conservative and discontinuous
(D) Conservative

321. DNA does not occur in:
(A) Nucleus (B) Ribosomes
(C) Mitochondria (D) Plastids

322. Functional unit of gene that specifies synthesis of one polypeptide is:
(A) Codon (B) Cistron
(C) Recon (D) Muton

323. In DNA, adenine pairs with:
(A) Guanine (B) Thymine
(C) Cytosine (D) Uracil

324. In DNA, guanine lies opposite:
(A) Cytosine (B) Thymine
(C) Uracil (D) Adenine

325. Most abundant RNA of the cell of:
(A) tRNA (B) rRNA
(C) mRNA (D) mnRNA

326. Nucleosides are:
(A) Sugar + phosphoric acid
(B) Purine/pyrimidine + sugar + phosphate
(C) Purine/pyrimides + phosphoric acid
(D) Purine/pyrimidine + sugar

327. Protein helping in opening of DNA double helix in front of replication fork is:
(A) DNA gyrase
(B) DNA polymerase I
(C) DNA gyrase
(D) DNA topoisomerase

328. Restriction endonuclease is employed for cutting:
(A) A single stranded DNA
(B) Double stranded DNA
(C) RNA fragment
(D) mRNA

329. Reverse transcriptase is:
(A) RNA dependent RNA polymerase
(B) DNA dependent RNA polymerase
(C) DNA dependent DNA polymerase
(D) RNA dependent DNA polymerase

330. RNA does not possess:
(A) Uracil (B) Thymine
(C) Adenine (D) Cytosine

331. Semiconservative DNA/chromosome replication using 14N was demonstrated by:
(A) Messelson (B) Tylor
(C) Messelson and Stahl (D) Hershey and Chase

332. Genome is:
(A) Genes on nuclear DNA
(B) Nuclear DNA + Mitochondrial DNA
(C) Nuclear DNA + chloroplast DNA
(D) Nuclear DNA + Mitochondrial DNA + Chloroplast DNA

333. The complete set of chromosomal and extrachromosomal genes of an organisms is called:
(A) Genome (B) Gene pool
(C) Gene bank (D) Gene library

334. The study of all the proteins coded by the genome is called:
(A) Proteome (B) Proteomics
(C) Genome (D) Protein formation

335. Sequencing of genomic DNA is included under:
(A) Structural genomics
(B) Functional genomics
(C) Proteomics
(D) Transgenesis

336. Gene expression, regulation and phenotype production are studied in second phase of genome analysis called:
(A) Structural genomics
(B) Functional genomics
(C) Proteomics
(D) Transmeiosis

337. In forensic science which of the following is used?
(A) Bacterial cloning (B) DNA foot printing
(C) DNA finger printing (D) DNA cloning

338. DNA finger printing is based on:
(A) Occurrence of VNTR's
(B) Knowledge of human karyotype
(C) Cloned DNA
(D) Recombinant DNA

339. VNTR's represents:
(A) New terminal regions in DNA
(B) Functional genes in the DNA
(C) Split genes in the sample DNA
(D) Specific non-coding sequences with unique tandem repeats

340. Restriction enzymes are used in genetic engineering because they:
(A) Can join DNA fragment
(B) Cut DNA at specific base sequence
(C) Cut DNA at variable sites
(D) Are proteolytic enzymes which degrade harmful proteins

341. DNA strand with nitrogen base sequence ATTGCC will have sequence in mRNA:
(A) UAACGC (B) ATCGCC
(C) STTGCA (D) AGGACC

342. DNA sequence of ATTCGATG is transcribed as:
(A) AUUCGAUG (B) UAAGCUAC
(C) CAUCGAAU (D) GUAGCUUA

343. DNA sequence is TAG. What shall be the sequence in anticodon of tRNA:
(A) UAG (B) ATC
(C) ATG (D) UAC

344. DNA sequence is ATG. What would be the sequence of bases in anticodon of tRNA:
(A) ATG (B) AUG
(C) UAC (D) TAC

345. A DNA nucleotide chain has AGCTTCGA sequence of other chain would be:
(A) TCGAAGCT (B) GCTAAGCT
(C) TAGCATAT (D) GATCCTAG

346. A strand of DNA has base sequence CATGACTAG. The base sequence on the other strand would be:
(A) CATTAG GAC (B) GTA CTG ATC
(C) GAT GTC ATC (D) TAC ACT GCT

347. Pyrimidine base present in RNA in place is thymine of DNA is:
(A) Uracil (B) Adenine
(C) Cytosine (D) Gaunine

348. Restriction endonucleases are:
(A) Used in genetic engineering for uniting two DNA molecules
(B) Used for in vitro DNA synthesis
(C) Present in mammalian cells for degeneration of DNA of dead cells
(D) Synthesized by bacteria for their defence

349. RNA that picks up specific amino acid from amino acid pool of cytoplasm to carry it to ribosome during protein synthesis is:
(A) tRNA (B) mRNA
(C) rRNA (D) gRNA

350. Successive nucleotides are covalently linked through:
(A) Glycosidic bonds
(B) Phosphodiester bonds
(C) hydrogen bonds
(D) Nitrogen bonds

351. The codon for anticodon 3' UUUA – 5' is:
(A) 5'AAAU – 3' (B) 5' UUUA – 3'
(C) 3' UAAD – 5' (D) 3' AUUU – 5'

352. A flowering plant lily have ---------- more DNA than humans:
(A) 10 times (B) 15 times
(C) 18 times (D) 20 times

353. African plant Pentadiplandra is used as:
(A) Low calories sweetner
(B) 2000 times sweeter agent
(C) Sweetner for diabetic patients
(D) All of these

354. Three dimentional shape of tRNA is:
(A) L – shaped (B) Clover leaf like
(C) X –shaped (D) Y-shaped

355. A segment of DNA has 120 adenine and 120 cytosine bases. The total number of nucleotides present in the segment is:
(A) 120 (B) 240
(C) 320 (D) 480

356. Callus is:
(A) Tissue that forms embryo
(B) An insoluble carbohydrate
(C) tissue that grows to form embryoid
(D) Unorganised actively dividing mass of cells maintained in culture

357. Choose the correct statement:
(A) DNA is hereditary material
(B) RNA is hereditary material
(C) DNA is hereditary material but where it is absent RNA can function as hereditary material
(D) Both DNA and RNA are hereditary materials

358. DNA duplex shows:
(A) Right handed coiling and parallel
(B) Right handed coiling and antiparallel
(C) left handed coiling and antiparallel
(D) Left handed and parallel

359. Double helical structure of DNA was proposed by:
(A) Kornberg
(B) Nirenberg
(C) Watson and Crick
(D) Wilkins and Franklin

360. In AGCT of DNA hydrogen bonds and base pairing occur between:
(A) A-U, C-G (B) A-C, G-T
(C) A-G, C-T (D) A-T, C-G

361. In double helix of DNA, the two DNA strands are:
(A) Coiled around a common axis
(B) Coiled around each other
(C) Coiled differently
(D) Coiled over protein sheath

362. Initiation codon of protein synthesis (in eucaryotes) is:
(A) GUA (B) GCA
(C) CCA (D) AUG

363. Nitrogen bases of DNA are:
(A) ATUC (B) UTGC
(C) ATGC (D) AUGC

364. Plasmids are vectors for gene cloning because they:
(A) Self replicate in bacterial cells
(B) Replicate freely outside bacterial cells
(C) Can be multiplied in culture
(D) Can be multiplied in laboratories using enzymes

365. RNA contains:
(A) Hexose (B) Ribose
(C) Fructose (D) Glucose

366. Termination of polypeptide chain is brought about by:
(A) UUG, UAG and UCG
(B) UAA, UAG and UGA
(C) UUG, UGC and UCA
(D) UCG, GCG and ACC

367. The common feature amongst nucleus, chloroplast and mitochondria is:
(A) Lamellae (B) Cristae
(C) DNA (D) All of these

368. Similarity between DNA and RNA is that both have;
(A) Similar sugars
(B) Similar mode of replication
(C) Similar pyrimidines
(D) Polymers of nucleotides

369. In a DNA molecule cytosine is 18%. Percentage of adenine would be:
(A) 32 % (B) 36 %
(C) 56 % (D) 64 %

370. In callus culture, roots can be induced by the supply of:
(A) Auxin and no-cytokinin
(B) Higher concentration of auxin and lower concentration of cytokinin
(C) Higher concentration of cytokinin and lower concentration of auxin
(D) Both auxin and cytokinin in equal proportion

371. In RNA, thymine is replaced by:
(A) Adenine (B) Guanine
(C) Cytosine (D) Uracil

372. Nucleotides present in one turn of DNA helix:
(A) 4 (B) 8
(C) 9 (D) 10

373. Nonsense codon takes part in:
(A) Terminating message of gene controlled protein synthesis
(B) Formation of unspecified amino acids
(C) Conversion of sense DNA into non-sense one
(D) Releasing tRNA from polypeptide chain

374. Preserving germplasm in frozen state is:
(A) Cryopreservation (B) Cold storage
(C) in-situ preservation (D) Vernalisation

375. The basic unit of a nucleic acid is:
(A) Pentose sugar (B) Nucleotide
(C) Nucleoside (D) Nucleotide

376. Variation observed during tissue culture of some plants are known as:
(A) Clonal variation
(B) Somatic variation
(C) Somaclonal variation
(D) Tissue culture variation

377. DNA differs from RNA in:
(A) Presence of deoxyribose sugar
(B) Presence of thymine base
(C) Property of replication
(D) All of these

378. In nucleoside, nitrogen base is attached to pentose sugar at:
(A) Carbon – 1 of pentose sugar
(B) Carbon – 2 of pentose sugar
(C) Carbon – 4 of pentose sugar
(D) Carbon – 5 of pentose sugar

379. The enzyme DNA polymerase can work only in:
(A) 3□® 5 - direction (B) 5□® 3 - direction
(C) Both the direction (D) 5□® 5 - direction

380. During DNA replication, the reunion or recoiling of separated DNA strand is prevented by:
(A) Helix destabilizing protein
(B) Single strand binding protein
(C) Rep protein
(D) Both (A) and (B)

381. Which of the following enzyme is required to release the tension imposed by uncoiling of strands:
(A) Endonuclease (B) DNA ligase
(C) DNA gyrase (D) DNA helicase

382. Formation of mRNA from DNA is called:
(A) Transformation (B) Translocation
(C) Translation (D) Transcription

383. The codon which may present at 3-end of mRNA:
(A) UAA (B) UAG
(C) UGA (D) Any of these

384. In recombinant DNA technology a plasmid vector is cleaved by:
(A) Modified DNA ligage
(B) A heated alkaline solution
(C) The same enzyme that cleave the donor DNA
(D) The different enzyme other thn that cleave the donor DNA

385. 'Nif gene' for nitrogen fixation is ceral crops like wheat, jowar etc. is introduced by cloning:
(A) Rhizobium meliloti
(B) Bacillus thuringiensis
(C) Rhizopus (D) Rhizophora

386. DNA molecules makes a complete turn after every:
(A) 20Å (B) 34 Å
(C) 3.4 Å (D) 10 base pairs

387. The distance between two successive nitrogenous base pairs is:
(A) 34 Å (B) 36 Å
(C) 20 Å (D) 3.4 Å

388. Enzyme required for removing RNA primer during DNA replication is:
(A) DNA primase (B) DNA ligage
(C) DNA polymerase I (D) DNA polymerase III

389. The cellular composition of mRNA is:
(A) 5-10% (B) 3-5 %
(C) 10-20 % (D) 70-80 %

390. The ratio of purines and pyrimidinesin mRNA is not 1:1 because the nitrogenous bases are:
(A) Unpaired (B) Paired
(C) Paired only in loop (D) Paired in stems

391. The most common plasmid vector used in genetic engineering is:
(A) PBR 328 (B) PBR 322
(C) PBR 325 (D) PBR 330

392. Eco R1 is an:
(A) Ligage (B) Polymerase
(C) Restriction enzyme (D) Gyrase

393. If the strand of DNA has 35 nucleotide how many phosphodiester bonds would exist:
(A) 34 (B) 35
(C) 36 (D) 37

394. In eukaryotic DNA replication, lagging stand is formed by:
(A) RNA fragments
(B) Okazaki fragments
(C) DNA fragments
(D) Nucleotide fragments

395. The enzyme that cuts the bonds of DNA molecule at the origin of replication is:
(A) Endonuclease (B) DNA polymerase
(C) DNA gyrase (D) DNA ligage

396. Which of the following is not tool of genetic engineering?
(A) Vectors (B) Enzymes
(C) Foreign DNA (D) GMO

ANSWERSHEET

11. GENOMICS & RECOMBINANT DNA TECHNOLOGY

1. (D) **2. (C)** **3. (D)** **4. (C)** **5. (B)** **6. (A)** **7. (C)** **8. (B)** **9. (D)** **10. (B)**
11. (A) **12. (B)** **13. (C)** **14. (D)** **15. (A)** **16. (B)** **17. (A)** **18. (B)** **19. (A)** **20. (B)**
21. (C) **22. (A)** **23. (A)** **24. (B)** **25. (C)** **26. (D)** **27. (A)** **28. (B)** **29. (B)** **30. (C)**
31. (D) **32. (A)** **33. (B)** **34. (C)** **35. (A)** **36. (B)** **37. (C)** **38. (D)** **39. (A)** **40. (A)**
41. (C) **42. (D)** **43. (A)** **44. (B)** **45. (C)** **46. (A)** **47. (B)** **48. (C)** **49. (C)** **50. (C)**
51. (D) **52. (B)** **53. (D)** **54. (B)** **55. (C)** **56. (C)** **57. (C)** **58. (D)** **59. (C)** **60. (A)**
61. (A) **62. (A)** **63. (D)** **64. (B)** **65. (A)** **66. (D)** **67. (B)** **68. (A)** **69. (D)** **70. (D)**
71. (D) **72. (D)** **73. (C)** **74. (D)** **75. (C)** **76. (D)** **77. (D)** **78. (A)** **79. (A)** **80. (B)**
81. (D) **82. (C)** **83. (D)** **84. (A)** **85. (A)** **86. (B)** **87. (B)** **88. (D)** **89. (B)** **90. (B)**
91. (D) **92. (A)** **93. (A)** **94. (D)** **95. (D)** **96. (B)** **97. (C)** **98. (B)** **99. (B)** **100. (B)**
101. (B) **102. (D)** **103. (B)** **104. (C)** **105. (B)** **106. (A)** **107. (B)** **108. (B)** **109. (D)** **110. (B)**
111. (C) **112. (C)** **113. (D)** **114. (D)** **115. (D)** **116. (A)** **117. (D)** **118. (D)** **119. (A)** **120. (B)**
121. (D) **122. (D)** **123. (A)** **124. (C)** **125. (A)** **126. (B)** **127. (A)** **128. (D)** **129. (D)** **130. (D)**
131. (D) **132. (C)** **133. (C)** **134. (C)** **135. (D)** **136. (C)** **137. (D)** **138. (D)** **139. (A)** **140. (B)**
141. (C) **142. (B)** **143. (D)** **144. (B)** **145. (B)** **146. (A)** **147. (D)** **148. (D)** **149. (A)** **150. (D)**
151. (A) **152. (B)** **153. (A)** **154. (D)** **155. (D)** **156. (B)** **157. (B)** **158. (D)** **159. (D)** **160. (B)**
161. (D) **162. (B)** **163. (C)** **164. (C)** **165. (D)** **166. (B)** **167. (B)** **168. (D)** **169. (A)** **170. (C)**
171. (D) **172. (A)** **173. (C)** **174. (A)** **175. (B)** **176. (C)** **177. (D)** **178. (A)** **179. (D)** **180. (A)**
181. (D) **182. (D)** **183. (A)** **184. (D)** **185. (B)** **186. (A)** **187. (D)** **188. (C)** **189. (B)** **190. (C)**
191. (D) **192. (D)** **193. (D)** **194. (D)** **195. (A)** **196. (B)** **197. (C)** **198. (D)** **199. (B)** **200. (C)**
201. (B) **202. (A)** **203. (D)** **204. (A)** **205. (D)** **206. (D)** **207. (D)** **208. (B)** **209. (D)** **210. (C)**
211. (D) **212. (C)** **213. (C)** **214. (B)** **215. (A)** **216. (C)** **217. (B)** **218. (D)** **219. (A)** **220. (A)**
221. (C) **222. (B)** **223. (C)** **224. (A)** **225. (B)** **226. (C)** **227. (C)** **228. (A)** **229. (D)** **230. (A)**
231. (D) **232. (B)** **233. (C)** **234. (A)** **235. (B)** **236. (C)** **237. (B)** **238. (C)** **239. (D)** **240. (A)**
241. (C) **242. (B)** **243. (D)** **244. (A)** **245. (B)** **246. (B)** **247. (C)** **248. (A)** **249. (D)** **250. (A)**
251. (C) **252. (B)** **253. (A)** **254. (C)** **255. (B)** **256. (A)** **257. (C)** **258. (B)** **259. (C)** **260. (A)**
261. (B) **262. (C)** **263. (C)** **264. (C)** **265. (B)** **266. (C)** **267. (A)** **268. (C)** **269. (C)** **270. (D)**
271. (D) **272. (C)** **273. (D)** **274. (C)** **275. (A)** **276. (A)** **277. (B)** **278. (B)** **279. (D)** **280. (B)**
281. (B) **282. (A)** **283. (B)** **284. (D)** **285. (D)** **286. (B)** **287. (A)** **288. (C)** **289. (B)** **290. (C)**
291. (D) **292. (D)** **293. (A)** **294. (B)** **295. (B)** **296. (B)** **297. (A)** **298. (A)** **299. (C)** **300. (D)**
301. (C) **302. (D)** **303. (A)** **304. (C)** **305. (C)** **306. (D)** **307. (B)** **308. (B)** **309. (A)** **310. (D)**
311. (A) **312. (B)** **313. (B)** **314. (C)** **315. (B)** **316. (D)** **317. (A)** **318. (B)** **319. (C)** **320. (C)**
321. (B) **322. (B)** **323. (B)** **324. (A)** **325. (B)** **326. (D)** **327. (A)** **328. (B)** **329. (D)** **330. (B)**
331. (C) **332. (D)** **333. (A)** **334. (B)** **335. (A)** **336. (B)** **337. (C)** **338. (A)** **339. (D)** **340. (B)**
341. (A) **342. (B)** **343. (A)** **344. (B)** **345. (A)** **346. (B)** **347. (A)** **348. (D)** **349. (A)** **350. (B)**
351. (A) **352. (C)** **353. (D)** **354. (B)** **355. (D)** **356. (D)** **357. (C)** **358 (B)** **359. (C)** **360. (D)**
361. (A) **362. (D)** **363. (C)** **364. (A)** **365. (B)** **366. (B)** **367. (C)** **368. (B)** **369. (A)** **370. (B)**

371. (D)	**372. (D)**	**373. (A)**	**374. (B)**	**375. (D)**	**376. (C)**	**377. (D)**	**378. (A)**	**379. (B)**	**380. (D)**
381. (C)	**382. (D)**	**383. (D)**	**384. (C)**	**385. (A)**	**386. (D)**	**387. (D)**	**388. (C)**	**389. (B)**	**390. (A)**
391. (B)	**392. (C)**	**393. (A)**	**394. (B)**	**395. (A)**	**396. (D)**				

Chapter - 12

TRANSGENICS

1. The plants with specific genes or traits obtained through genetic engineering or DNA recombinant technology is called:
(A) Transgenic plant (B) Transgenic animals
(C) Transgenic microbes (D) None of these

2. The animals having specific genes obtained from outside is known as:
(A) Transgenic plant
(B) Transgenic animals
(C) Transgenic microbes
(D) None of these

3. The microbes used in industry for producing different biochemical and various functions is called:
(A) Transgenic plant (B) Transgenic animals
(C) Transgenic microbes
(D) None of these

4. Transgene may become deleted or modified by recombination, following their integration into the host genome is called:
(A) Transgene instability
(B) Transgene silencing
(C) Transgene flow (D) Gene splicing

5. A decline in the level of transgene expression with the passage of generation of the transgenic line is called:
(A) Transgene instability
(B) Transgene silencing
(C) Transgene flow
(D) Gene splicing

6. Transfer of gene from a transgenic variety to non-transgenic varieties, wild species and wild relatives through hybridization is called:
(A) Transgene instability
(B) Transgene silencing
(C) Transgene flow
(D) Gene splicing

7. The enzymatic attachment of one gene, or part of a gene, to another is called:
(A) Transgene instability (B) Transgene silencing
(C) Transgene flow (D) Gene splicing

8. A transgenic rice that accumulates ß-carotene in its seeds; expected to combat vitamin A deficiency areas where rice is the staple food is:
(A) Golden rice (B) Hybrid rice
(C) Boiled rice (D) Desi rice

9. A DNA sequence that appears at the end of a transcriptional unit and signals the end of transcription is called:
(A) Termination sequence
(B) Terminal transferase
(C) Supercoiled DNA
(D) Suppressor mutation

10. An enzyme that catalyzes the addition of nucleotide residues of a single kind to tho 3' ond of DNA chains is called:
(A) Termination sequence
(B) Terminal transferase
(C) Supercoiled DNA
(D) Suppressor mutation

11. DNA that twists upon itself because it is under-or overwound (and thereby strained) relative to B-form DNA is called:
(A) Termination sequence
(B) Terminal transferase
(C) Supercoiled DNA
(D) Suppressor mutation

12. A mutation that totally or partially restores a function lost by a primary mutation; located at a site different from the site of the primary mutation is called:
(A) Termination sequence
(B) Terminal transferase
(C) Supercoiled DNA
(D) Suppressor mutation

13. A mutation is a gene that cause no detectable change in the biological characteristics of the gene product is called:
(A) Termination sequence
(B) Terminal transferase
(C) Silent mutation
(D) Suppressor mutation

14. An amino acid sequence, often at the amino terminus, that signals the cellular fate or destination of a newly synthesized protein is called:
(A) Termination sequence
(B) Terminal transferase
(C) Signal sequence
(D) Suppressor mutation

15. The process by which an extracellular signal (chemical, mechanical or electrical) is amplified and converted to a cellular response is called:
(A) Signal transduction
(B) Terminal transferase
(C) Signal sequence
(D) Suppressor mutation

16. Highly repeated, nontranslated segments of DNA in eukaryotic chromosomes; most often associated with the centromeric region is called:
(A) Satellite – DNA (B) Ribosomal RNA
(C) Small nuclear RNA (D) Transfer RNA

17. A class of RNA molecules serving as components of ribosomes is called:
(A) Satellite – DNA (B) Ribosomal RNA
(C) Small nuclear RNA (D) Transfer RNA

18. Any of several small RNA molecules in the nucleus; most have a role in the splicing reactions that remove introns from mRNA, tTRNA and rRNA molecules is called:
(A) Satellite – DNA (B) Ribosomal RNA
(C) Small nuclear RNA (D) Transfer RNA

19. A class of RNA molecules, each of which combines covalently with a specific amino acid as the first step in protein synthesis is called:
(A) Satellite – DNA (B) Ribosomal RNA
(C) Small nuclear RNA (D) Transfer RNA

20. A mutation caused by insertion of gene or more extra bases, or a mutagen, between successive bases in DNA is called:
(A) Insertion mutation (B) Foot printing
(C) DNA chimera (D) DNA chip

21. A technique for identifying the nucleic acid sequence bound by a DNA or RNA-binding protein is called:
(A) Insertion mutation (B) Foot printing
(C) DNA chimera (D) DNA chip

22. DNA containing genetic information derived from two different species is called:
(A) Insertion mutation (B) Foot printing
(C) DNA chimera (D) DNA chip

23. An informal term for a DNA microarray, referring to the small size of typical microarrays is called:
(A) Insertion mutation (B) Foot printing
(C) DNA chimera (D) DNA chip

24. A collection of cloned DNA fragments is called:
(A) DNA library (B) DNA ligage
(C) DNA looping (D) DNA microarray

25. An enzyme that creates a phosphodiester bond between the 3'-end of one DNA segment and the 5' end of another is called:
(A) DNA library (B) DNA ligage
(C) DNA looping (D) DNA microarray

26. The interaction of proteins bound at distant sites on a DNA molecule so that the intervening DNA forms a loop is called:
(A) DNA library (B) DNA ligage
(C) DNA looping (D) DNA microarray

27. A collection of DNA sequences immobilized on a solid surface, with individual sequence laid out in patterned arrays that can be probed by hybridization is called:
(A) DNA library (B) DNA ligage
(C) DNA looping (D) DNA microarray

28. An enzyme that catalyzes template – dependent synthesis of DNA from its deoxyribonucleoside 5'-triphosphate precursors:
(A) DNA polymerase (B) DNA ligage
(C) DNA looping (D) DNA microarray

29. The entire complex of enzymes and specializes proteins required in biological DNA replication is called:
(A) DNA replicase system
(B) DNA supercoiling
(C) cDNA library
(D) Complementary DNA

30. The coiling of DNA upon itself, generally as a result of bending, underwinding, or over writing of the DNA helix is called:
(A) DNA replicase system
(B) DNA supercoiling
(C) cDNA library
(D) Complementary DNA

31. DNA library consisting entirely of cloned cDNAs from a particular organism or cell type is called:
(A) DNA replicase system
(B) DNA supercoiling
(C) cDNA library
(D) Complementary DNA

32. A DNA used in DNA cloning, usually made by reverse transcriptase, complementary to a given mRNA is called:
(A) DNA replicase system
(B) DNA supercoiling
(C) cDNA library
(D) Complementary DNA

33. Who described the method of chemical sequencing in 1977 is:
(A) Maxam and Gilbert
(B) Sanger and coworkers
(C) Arthur Komberg and coworker
(D) Phillip Leder

34. Who devised the enzymatic method of DNA base sequencing in 1977 is:
(A) Maxam and Gilbert
(B) Sanger and coworkers
(C) Arthur Komberg and coworker
(D) Phillip Leder

35. Enzymatic synthesis *in vitro* of DNA complementary to a given DNA template is possible by using:
(A) DNA polymerase I (B) DNA polymerase
(C) DNA looping (D) DNA library

36. In 1958, Arthur Komberg and coworker purified an enzyme from the soluble fraction of the lysate of *E. coli* that could synthesize DNA in vitro and it was named:
(A) DNA polymerase I (B) DNA polymerase
(C) DNA looping (D) DNA library

37. With a natural DNA tempelate, addition of all the four deoxyribonucleoside 5'-triphosphates was found essential for the poly-merization reaction catalyzed by the:
(A) DNA polymerase I (B) DNA polymerase
(C) DNA looping (D) DNA library

38. A lambda cloning vecter can therefore be constructed by deletion (*in vivo* or *in vitro* by restriction deletion) of a part of the nonessential region that the remainder is not less than 35 kb is:
(A) Insertion-type lambda vectors
(B) Substitution type lambda vectors
(C) Both (A) and (B)
(D) None of these

39. The lambda vectors have two EcoR, sites or two Bam HI sites in the nonessential; region is:
(A) Insertion-type lambda vectors
(B) Substitution type lambda vectors
(C) Both (A) and (B)
(D) None of these

40. A pBLUESCRIPT phagmid contains the genes for:
(A) Ampicillin resistance
(B) Lac repressor (Lac I)
(C) β-galactosidase (Lac Z)
(D) All of these

41. Cosmid vectors were first developed by Collins and coworkers in:
(A) 1977 (B) 1978
(C) 1979 (D) 1980

42. Yeast artificial chromosomes was developed by D.T. Burke and C. F. Carle in:
(A) 1987 (B) 1988
(C) 1989 (D) 1990

43. Bacterial artificial chromosomes was developed by Melsimon and coworkers in:
(A) 1990 (B) 1991
(C) 1992 (D) 1993

44. Which polypeptide sequences are used as tags to facilitate recombinant protein purification?
(A) Glutathione-S-transfication
(B) Maltose-binding protein
(C) *myc* epitope
(D) All of these

45. Which generates high level transient expression in some specific mouse cell lines?
(A) SV40 replicons (B) BPV replicons
(C) Murine-polymavirus (D) EBV replicons

46. Agrobacterium proteins functions as an endonuclease?
(A) VirA (B) VirG
(C) VirD1 (D) VirD2

47. Which Cry proteins is specific to nematodes?
(A) Cry II (B) Cry V
(C) Cry IV (D) Both (B) and (C)

48. Which of the following is/are true?
P. Physical map unit is centimorgan
Q. Physical mapping is done using restriction enzymes
R. Physical maps help during genome sequencing projects
S. Physical maps are also called genetic maps
(A) P (B) R
(C) Q and R (D) P, R and S

49. A mutant strain of Neurospora was crossed with the wild type. A total of 132 asci were analyzed:
First division segregation : 104
Second division segregation: 28
Findout the map distance (M) of the gene from the centromere.
(A) 5.3 (B) 8.3
(C) 10.6 (D) 21.2

50. When a population of bacteria capable of conjugation transfers a specific chromosomal gene (say a gene coding for galactose metabolism) but no other genes, regardless of how long the bacteria are allowed to mate, these bacteria are said to be:
(A) F^+ (B) F^-
(C) Hfr (D) F'

51. Which leads to stable episomal transformation in various luman cell lines?
(A) SV40 replicons (B) BPV replicons
(C) Murine-polymavirus (D) EBV replicons

52. Which statement is not correct?
(A) The enzyme thymidine kinase belongs to the salvage pathway
(B) The F-protein of Sendal Virus envelope mediates the fusion of viral envelope with cell plasma membrane
(C) Lipofection is the method of choice for transfection of mammalian cells in vitro
(D) Virosomes contain viral genome

53. The slow ripening transgenic tomato Flavr Savr was developed in USA using which approaches/ technologies?
(A) Gene disruption
(B) Antisense RNA technology
(C) Ribozyme technology
(D) Cosupression approach

54. Which insects shows 7 –fold (lowest) increase in resistance against Cry protein after 21 generations of laboratory selection?
(A)*Helicoverpa virescence* (B) *Cadra cutella*
(C) *Plodia interpunctella* (D) *Plutella xylostella*

55. Drastic reduction in ethylene production can be achieved by:
(A) Expression of antisense constructs of ACC synthase or ACC oxidase
(B) Cosuppression of either of these genes
(C) Expression of such enzymes that metabolize S-adenosyl methionine (SAM), e.g., SAM hydrolase gene from bacteriophage
(D) All of these

56. Production of recombinant human protein in insect larvae poses which problems?
(A) Protein folding
(B) Downstream processing
(C) Both (A) and (B) (D) Glycosylation

57. Tranforming principle was discovered by:
(A) Hershey and Chase
(B) Avery, MacLeod and McCarty
(C) Zinder and Lederberg
(D) Lederberg and Tatun

58. Intein sequence facilitates which steps in recombinant protein production?
(A) Separation from other proteins
(B) Removal of tag
(C) Over expression of the proteins
(D) Both (A) and (B)

59. Anti-freeze protein promoter was isolated from:
(A) Frog (B) Fish
(C) Snake (D) Drosophila

60. Which countries has the largest area under cultivation of transgenic crops?
(A) Canada (B) Brazil
(C) USA (D) Argentina

61. Which Cry proteins is specific to diptera?
(A) Cry I (B) Cry II
(C) Cry IV (D) Cry V

62. SV40 transducing vectors must have:
(A) SV40 origin including the surrounding region containing the transcriptional regulatory signals
(B) Genes encoding large-T, VP1, VP2 and VP3
(C) Total size (including, that of the DNA insert) between 3900 bp and 5300 bp for packaging into virions
(D) All of these

63. Which Agrobacterium proteins undergoes dimerization in response to acetosyringone?
(A) VirA (B) VirG
(C) VirD1 (D) VirD2

64. Which is not correct?
(A) MAHYCO collaborated with Ciba Geigy, UA, for exploiting potential of Bt technology in cotton in India
(B) The Bt transgenic cotton showed significantly higher level of protection against *Helicoverpa armigera* with reduced number of insecticidal sprays
(C) Based on the extended biosafety analysis and agronomic trials three hybrids, viz., Bt MECH12, Bt MECH 162 and Bt MECH 184 were approved for release
(D) Cotton hybrids Bt MECH 12, Bt MECH 162 and Bt MECH 184 contain cry1Ac gene

65. Which is correct?
(A) The sequence flanking the intein are exeins
(B) Intein sequence is derived from the VAM1 gene of *Saccharomyces cerevisiae*
(C) The chitin-binding domain (CBD) from *Bacillus circulans* has a high affinity for chitin column
(D) All of these

66. A mechanism of genetic exchange that does not occur in a culture medium containing the enzyme DNase is:
(A) Specialized transduction
(B) Conjugation
(C) Transformation
(D) Generalized transduction

67. A culture of tetracycline-sensitive bacteria was infected by a phage that is delivered from the lysis of a tetracycline-resistant bacterial strain. This results in developments of tetracycline-resistance in the original culture. What phenomenon has occurred?
(A) Conjugation (B) Recombination
(C) Transformation (D) Transduction

68. Which is not correct about fish vectors?
(A) A typical fish vector is s plasmid
(B) These vectors usually contain a selectable marker
(C) The origin of replication (ori) from *E. coli* plasmid pBR322
(D) An enhancer/promoter sequence from CaMV

69. Which techniques is used to obtain information on the relative levels of mRNA present in the tissues of different transgenic individuals?
(A) Reverse transcription PCR
(B) Nuclease protection assay
(C) In situ hybridization
(D) All of these

70. Which of the enzyme enables substitution of the carbon source in the medium?
(A) ß-gluconidase
(B) Xylose isomerase
(C) Mannose-6-phosphate isomerase
(D) Both (B) and (C)

71. **Which constitutes gene silencing?**
(A) Low level of transgene expression in T0
(B) Low level of transgene expression in T0
(C) High level of transgene expression in to, but reduced levels of expression in sub-sequent generation
(D) Both (A) and (B)

72. **Which Cry proteins is specific to coleoptera?**
(A) Cry I (B) Cry II
(C) Cry IV (D) Cry V

73. **F-virosomes are prepared from Sendai Virus; they contain:**
(A) Viral envelop (B) Viral F-protein
(C) Recombinant DNA (D) All of these

74. **Gene barnase is expressed in which tissues in order to produce male sterility?**
(A) Pollen mother cell (B) Stamen
(C) Tapetum (D) Pollen grain

75. **Which insects shows resistance against *Bacillus thuringiensis* formulations?**
(A) *Helicoverpa virescence*
(B) *Cadra cutella*
(C) *Plodia interpunctella*
(D) *Plutella xylostella*

76. **Which transgenes can be used for development of insect resistant transgenics?**
(A) Lectin gene
(B) Cowpea trypsin inhibitor
(C) Serine protease inhibitor
(D) All of these

77. **Which is not correct about Knockout mice?**
(A) In a knockout mouse, one gene is distrupted by the technique of site directed mutagenesis
(B) The name Knockout signifies the fact that a specific gene in the genomes of these animals has been made non-functional
(C) Production of Knockout mice employs the embryonic stem cell technology
(D) Knockout mice are invaluable for various kinds of basic and applied research

78. **Which Agrobacterium proteins guides the T-DNA into the plant cell nucleus?**
(A) VirA (B) VirG
(C) VirD1 (D) VirD2

79. **RNAi suffers from which limitations?**
(A) Exogenous siRNA must be delivered into the cells, which is not highly efficient
(B) RNAi based on exogenous siRNA is rather short-lived, e.g., for 4-6 days only
(C) Longer dsRNA molecules trigger antiviral; repose and a general suppression of gene expression
(D) All of these

80. **Which is the chief advantages of both late and early region replacement SV40 vectors?**
(A) The recombinant DNA replicates to a high copy number
(B) The recombinant DNA is packaged into virions
(C) Both (A) and (B)
(D) The maximum size of DNA insert is -2.5 kb

81. **Virus mediated transfer of cellular genetic material from one bacterial cell to another is called:**
(A) Transfection (B) Transformation
(C) Transposition (D) Transduction

82. **Which promoters specified anaerobic expression?**
(A) Vicilin (B) Extensin
(C) 35S (D) Adh1

83. **Which is not correct about edible vaccines?**
(A) They are edible parts of transgenic plants
(B) The transgenic encoded an antigen from plants
(C) They are consumed in large quantities
(D) They are in clinical trial stage

84. **The 'intein' sequence function in:**
(A) RNA splicing (B) Protein splicing
(C) Both (A) and (B) (D) Recombination

85. **Which is not correct about *Agrobacterium tumefaciens*?**
(A) It is sometimes called natural genetic engineer
(B) It is generally used for gene transfer in plants
(C) It is a Gram positive soil bacterium
(D) Optimum temperature for its growth is 28^0C

86. **Human diseases caused by mutation in mitochondrial genome:**
(A) Are inherited from both parents
(B) Are inherited from the father
(C) Are inherited from the mother
(D) Do not exist because the mutation is always complemented by the normal

87. **Which of the vector can be used for high level transient expression in insect larvae?**
(A) Adeno-associated virus (B) Baculovirus
(C) Both (A) and (B) (D) Vaccinia virus

88. **Which poses problems in gene transfer in plants through genetic engineering?**
(A) Low frequency of transformation
(B) Low level of transgene expression
(C) Gene silencing
(D) All of these

89. **In mouse cell lines, plasmid vectors having which replicons yield stable episomal transformation?**
(A) SV40 replicons (B) BPV replicons
(C) Murine-polymavirus (D) EBV replicons

90. **How many type of gametes will be produced by individuals of AABb genotype?**
(A) Two (B) Four
(C) Six (D) Nine

91. Which is not obtained by Southern hybridization?
(A) Number of sites of transgene integration per genome
(B) Presence of mRNA/protein produced by transgene
(C) Both (A) and (B)
(D) Confirmation of integration of transgene

92. Genes encoding ultimately yields a substitute for cytokinin in the medium?
(A) ß-glucuronidase
(B) Mannose-6-phosphate isomerase
(C) Xylose isomerase
(D) All of these

93. Gene silencing may cause:
(A) Additional expenditure in evaluation
(B) Delayed release of transgenics as varieties
(C) Problems in registration of the transgenic varieties
(D) All of these

94. In Mendel's experiments with Garden Pea, round seeds shape (RR) were dominant over wrinkled seeds (rr), yellow cotyledons (YY) were dominant over green cotyledons (yy). What are expected phenotypes if F2 generation RRYY x rryy?
(A) Only wrinkled seeds with green cotyledons
(B) Only wrinkled seeds with yellow cotyledons
(C) Only round seeds wit green cotyledons
(D) Round seeds with yellow cotyledons and wrinkled seeds with yellow cotyledons

95. The methotrexate (Mtx) resistant cell lines developed through cell selection show:
(A) Reduced Mtx uptake
(B) Overproduction of DHFR
(C) Reduced affinity of DHFR for Mtx
(D) All of these

96. Which countries has the smallest area under cultivation of transgenic crops?
(A) Canada (B) Brazil
(C) USA (D) Argentina

97. Which Cry proteins is specific to Lepidoptera?
(A) Cry I (B) Cry II
(C) Cry IV (D) Cry V

98. Independent assortment of genes occurs due to the orientation of chromosomes at:
(A) Metaphase of mitosis
(B) Metaphase I of meiosis
(C) Metaphase II of meiosis
(D) Any phase of the cell division

99. Which Agrobacterium proteins are phorphorylatedin response to acetosyringone?
(A) VirA (B) VirB11
(C) VirG (D) Both (A) and (C)

100. Which is not correct about small interfering RNA (siRNA)?
(A) The two strands of an siRNA molecule are either synthesized chemically or they are produced in vivo by transcription
(B) Chemically synthesized siRNA molecules have to be introduced into the cells, and they inhibit target gene expression for only 4-6 days
(C) There are two common mechanisms, viz., RNase H-dependent antisense mechanism, and siRNA mechanism by which short synthetic oligonucleotides can be used to modulate gene expression in mammalian cells
(D) The siRNA and RNase H-dependent mechanisms are not similar in any respect

101. Sex linked inheritance:
(A) Occurs due to gene present on x-chromosomes only
(B) Occurs due to presence of gene in only one sex
(C) Occurs as a result of individual sex hormone
(D) Not occurs in animals

102. A women with normal vision but with colourblind father marries a colourblind man. The fourth child of the couple is a boy. This boy
(A) May or may not be colourblind
(B) Will be partially colourblind due to being heterozygous
(C) Must have normal vision
(D) Must be colourblind

103. Tomato transgenic variety 'Endless summer' has which features?
(A) Delayed ripening
(B) Bacterial wilt resistance
(C) Delayed fruit softening
(D) Early ripening

104. Which products is being produced in transgenic plants at commercial scale?
(A) Avidin
(B) Hirudin
(C) ß-glucuronidase (GUS) enzyme
(D) All of these

105. To linked genes a and b show 20% recombination. The individuals of a dihybrid cross between + + /+ + x ab/ab shall show gametes:
(A) + + 80 : ab 20
(B) + + 50 : ab 50
(C) + + 40 : ab 40 : + a 10 : + b 10
(D) + + 30 : ab 30 : + a 20 : + b 20

106. Whhich advantages is not offered by chloroplast transformation?
(A) Reduced risk of evolution of resistance in insect pests
(B) Ideal for eukaryotic gene expression
(C) Very high levels of transgene expression
(D) Elimination of the risk of transgene dispersal through pollen

107. Which of the following is not correct?
(A) Huntington disease is autosomal dominant disease
(B) Myotonic dystrophy is X-linked dominant disease
(C) Gene involved in Kennedy disease contain trinucleotide repeat
(D) None of these

108. Corn transgenic, 'KnockOut' has been developed by:
(A) Monsanto (B) Novartis
(C) DuPont (D) Anentis

109. Which of the following is not used as a genetic marker?
(A) Single nucleotide polymorphisms (SNPs)
(B) RFLP
(C) SINE (D) VNTR

110. GFP fluoresces bright green when exposed to which?
(A) UV light (B) Blue light
(C) White light (D) Both (A) and (B)

111. In case of fatty acid biosynthesis, which is sequentially added during chain elongation?
(A) Acetyl – CoA (B) Aceto-acetyl-CoA
(C) Malonyl-CoA (D) Butyryl-CoA

112. Consider the following terms:
Protein, DNA, RNA, trait
Select the correct option that depicts their biological relationship from among those given below?
(A) DNA RNA Protein Trait
(B) DNA RNA Protein Trait
(C) DNA RNA Protein Trait
(D) DNA RNA Protein Trait

113. An individual with d genes was crossed with would + +. On test crossing F_1, the progeny was + c105, + d 115, cd 880 and + + 900. Distance cd genes is:
(A) 11 maps units (B) 5.5 maps units
(C) 44 maps units (D) 88 maps units

114. Which Agrobacterium proteins has ATPase activity?
(A) VirA (B) VirB11
(C) VirG (D) All of these

115. Which is not correct?
(A) Useful vectors are created by joining together segments performing specific functions
(B) The uptake of foreign DNA by plant cells is called transfection
(C) When viral genome is introduced into plant cells with the help of Agrobacterium, it is called agroinfection
(D) A degenerate primer is a set of primers having the same sequence, except for one or more position where variation occurs

116. Which of the following types of DNA modification results in silencing of a region of the genome in such a way that the silencing can be passed on to offspring?
(A) Acetylation (B) Methylation
(C) Phosphorylation (D) Ubiquitination

117. Homologous repeats are involved in which of the following types of mutation?
(A) Frameshifts (B) Nonsense mutations
(C) Transition (D) Splicing mutations

118. Linkage map of x-chromosome of fruitfly has 66 map units with yellow body gene (y) at one end and bobbed hair (b) at the other. The recombin-ation frequency between Y and b gene would be:
(A) 66 % (B) >50 %
(C) ≤50 % (D) 100 %

119. Genes encoding which can be used as a selectable marker when combined with a compound added to the medium?
(A) β-glucuronidase
(B) Mannose-6-phosphate isomerase
(C) Xylose isomerase
(D) All of these

120. Gene silencing leads to:
(A) Reduced transgene expression
(B) Variable transgene expression
(C) Increased transgene expression
(D) Both (A) and (B)

121. If a transgene has undesirable pleiotropic effect on the transgenic plants, it may be corrected by:
(A) Targetting the transgene product into the appropriate cellular compartment
(B) Changing the genetic background of transgene through hybridization
(C) Mutagenesis in the transgene
(D) Both (A) and (B)

122. Which of the following would occur through specialized transduction?
(A) Acquisition of Hfr plasmid
(B) Transfer of genes for toxin production
(C) Transfer of genes for capsule formation
(D) Transfer of a plasmid with genes for degrading pesticides

123. A recessive mutation was discovered in a small island population (1000 people). The mutation was observed in the heterozygous state, but never in the homozygous state. What is a possible explanation for this observation?
(A) Only individuals with opposite alleles at this locus reproduce
(B) The recessive allele is lethal in a homozygous state
(C) The heterozygous state is fixed in this population
(D) The population size is too small to give an explanation for this phenomenon

124. Which produces high level transient expression in COS cell lines?
(A) SV40 replicons (B) BPV replicons
(C) Murine-polymavirus (D) EBV replicons

125. Genetic use restriction technology (GURT) is also called:
(A) Terminator technology
(B) Control of plant gene expression
(C) Antifarmer technology
(D) All of these

126. A disorder which is linked to the Y chromosome in humans (holandric) will:
(A) Only be expressed in male whose mothers were a carrier of the gene
(B) Never be passes from father to child
(C) Show a pattern of skipping generations in a family
(D) Be passed only from father to son

127. Glyphosate resistance in plants can be generated by expressing in them transgenes that serve which purpose?
(A) Produce an enzyme that inactivates glycophosate
(B) Over-production of EPSPS enzyme
(C) Encode an EPSPS enzyme that is tolerant to glyphosate
(D) Any of these

128. Which is not correct?
(A) Granule-bound starch synthase I (GBSS I) is the main enzyme of amylase biosynthesis
(B) Suppression of GBSS I activity in potato (by antisense RNA technology) yielded an amylase-free or waxy starch potato
(C) Production of waxy starch in potato is the first successful genetic modification of starch
(D) All of these

129. Prophage, i.e., phase genome integrated into a bacterial genome, is selfish DNA because:
(A) It always propagates together with the genome of the lysogen
(B) It replicates independently of the genome of the lysogen
(C) It provides immunity against infection by further virions
(D) All of these

130. Which Agrobacterium proteins are most likely involved in the tube formation between bacterial and plant cells?
(A) VirB (B) VirD4
(C) VirC (D) Both (A) and (B)

131. A tryptophan auxotrophic Hfr strain of *E. coli* is mixed with F- histidine auxotroph and plated on to minimal medium to select for prototrophs. The prototrophs are formed due to:
(A) Conjugation (B) Transduction
(C) Transformation (D) Transfection

132. Which is not correct about genetic transformation of chloroplasts?
(A) The first successful chloroplast transformation in higher plants was achieved in 1989, when spectinomycin resistance was transferred into tobacco
(B) pZS148 vector can be used for plastid transformation
(C) Particle gun is widely used and effective method of transforming chloroplasts
(D) Transgenes integrated into the chloroplast genome generally show low levels of expression

133. Enzyme aminoglycoside 3'-phosphotransferase II (APH (3') II) is commonly called:
(A) Luciferase
(B) Neomycin phosphotranferase II
(C) ß-glucuronidase
(D) Acyltransferase

134. A prophage is:
(A) An auxotrophic mutant
(B) A gene
(C) A phage DNA incorporated into the host genome
(D) Host DNA packed into viral heads

135. Which is correct about SV40 vectors?
(A) It is a spherical virus with circular, double-stranded 5243 bp chromosome, which encodes 5 proteins
(B) Large-T is essential for viral replication
(C) VP1 (virion protein 1), VP2 and vP3 form viral capsid
(D) All of these

136. Which of the following is not true about the human Y chromosome (in a male)?
(A) It is a single linear DNA molecule
(B) Its genes are known as holandric genes
(C) It is packaged in nucleosomes
(D) During mitosis, prior to metaphase, it contains only one chromatid

137. The IV (intermediate vector) component of Ti plasmid cointegrate vector system does not have:
(A) Origin for replication in *E.coli*
(B) Origin for replication in Agroo bacterium
(C) T-DNA sequence lacking borders
(D) A selectable marker

138. A transition mutation:
(A) Occurs when a purine is substituted for a pyrimidine, or vice-versa
(B) Results from the insertion of one or two bases into the DNA chain
(C) Decreases in frequency in the presence of base analogs
(D) Results from substitution of one purine for another or of one pyrimidine for another

139. Which is not correct about lectins?
(A) Lectin gene was isolated from snowdrop (*Galanthus nivalis*)
(B) Lectins are plant glycoproteins which act against piercing and sucking insects like aphids
(C) They are effective only when ingested in microgram quantities
(D) As a result of item 'C', it has strong insecticidal effectiveness in transgenic plants to be of practical usefulness

140. Which is not correct?
(A) Introduction of a transgene encoding AFPase may increase the activity of this enzyme and starch yield
(B) Lauric acid occurs naturally in Brassica species
(C) The rate limiting step in fatty acid biosynthesis is the carboxylation of acetyl-CoA to malonyl-CoA
(D) Most successful example of fatty acid modification is the increased lauric acid content of *B. napus* transgenic, Laurical

141. The karyotype designation 47, xx, +13 designates which of the following:
(A) Female with trisomy 13
(B) Female with monosomy 13
(C) Female with 13 extra chromosomes
(D) Female with extra genetic material on chromosomes 13

142. Which is used as processor for PHB synthesis?
(A) Acetyl – CoA (B) Aceto-acetyl-CoA
(C) Malonyl-CoA (D) Butyryl-CoA

143. Choose the mismatch:
(A) Cri-du-chat syndrome – 5p
(B) Patan syndrome – 47, +13
(C) Retinoblstoma – 13q, 14
(D) Edwards syndrome – 47, +21

144. What is the form of chromosomal aberration that leads to the fusion of two formerly separation non-homologous chromosomes into a single chromosome or to the separation of formerly joined sections of a single chromosome into two chromosomes?
(A) Reciprocal translocation
(B) Isochromosome
(C) Robertsonian translocation
(D) Centromeric fusion

145. In bacteria, which enzyme binds single-stranded DNA, denatures double-stranded DNA and matches the single stranded DNA with complementary denatured DNA?
(A) RecA (B) RecBcD
(C) UvrABc (D) UvrD

146. Which Agrobacterium proteins are DNA binding protein?
(A) VirG (B) VirD1
(C) VirE (D) All of these

147. In many plants, certain traits related to quality can be improved by reducing or eliminating the expression of specific genes. The level of endogenous gene expression can be reduced by using which approaches?
(A) Antisense RNA
(B) Small interfering RNA (siRNA)
(C) Gene disruption
(D) Any of these

148. Which transgenes shows protection from salt stress in tobacco?
(A) *mtl*1D (B) *bet*A
(C) P5CS (D) Any of these

149. Which transgenes leads protection against heat stress in tobacco?
(A) *sac*B (B) *hsf*
(C) *bet*B (D) *ala*3

150. Galatosemia is a recessive single gene genetic disorder, caused due to the mutation in any one of the three genes involved in galactose catabolism. A family consists of 10 normal children with both parents suffering from galatosemia. This is most likely because of:
(A) Epistatsis (B) Reversion
(C) Suppression (D) Complementation

151. Which is not correct about Xenopus oocytes?
(A) Each mature oocyte has a large nucleus called germinal vesicle
(B) Nucleus is located in the darkly pigmented hemisphere of the oocytes
(C) Oocyte nucleus is arrested at first meiotic metaphase
(D) Oocytes serves as a valuable expression system of a wide variety of genes from both animals and plants

152. The presence of transgene/DNA segment in plant cells can be detected by:
(A) DT blot assay (B) PCR amplification
(C) Southern hybridization (D) All of these

153. Which of the following statements is not true for transposable element system?
(A) It consists of both autonomous and non-autonomous elements
(B) Transposases is transcribed by the central region of autonomous elements
(C) Dissociation elements are autonomous in nature
(D) Certain repeats in the genome remain fixed even after the element transposes out

154. The genes, M, P and S are linked. crossover map distance determined by two-point crosses are:
P-C = 7, S–M = 10, C-M = 8, S-C= 2 and P – S = 5.
The relative positions of these four linked loci are:
(A) PSCM (B) SCPM
(C) PCSM (D) SCMP

155. Freedom II is a transgenic variety of:
(A) Squash (B) Maize
(C) Cotton (D) Tobacco

156. P element:
(A) Is a transposon
(B) Has a role in hybrid dysgenesis
(C) Occurs in *D. melanogaster*
(D) All of these

157. Agroinfection achieves:
(A) Genetic transformation
(B) Introduction of viral genome into plant cells
(C) Crown gall formation
(D) Both (B) and (C)

158. Temperature-sensitive mutant for a specific gene does not allow growth at the restrictive temperature by:
(A) Inhibiting the expression of the specific gene
(B) Expressing an unfolded form of the specific gene
(C) Degrading the protein as soon as it is expressed
(D) Inhibiting translation

159. A mechanism that an cause a gene to move from one linkage group to another is:
(A) Translocation (B) Inversion
(C) Crossing over (D) Duplication

160. Which transgenes generates protection against freeze stress in tobacco?
(A) *sac*B (B) *hsf*
(C) *bet*B (D) *ala*3

161. A normal couple has seven children (2 daughters and 5 sons). Three of the sons suffer from a hereditary disorder but none of the daughter is affected. Which is the inheritance type?
(A) Sex-limited recessive
(B) Autosomal dominant
(C) Sex-linked dominant
(D) Sex-linked recessive

162. According to classical genetic, which of the following statements is true?
(A) Recessive alleles are detected by the phenotype of the F_1 generation
(B) The closer two genes are, the more frequently they recombine
(C) Genes on different autosomes segregate independently
(D) Gene on sex chromosomes segregate with the same pattern as autosomal genes

163. Which of the following is not correct about genomic imprinting?
(A) It is an epigenetic effect
(B) It occurs as a result of acetylation of histone protein
(C) DNA methylation is responsible for imprinting
(D) Paternal and maternal alleles different pattern of methylation

164. In Drosophila sex is determined by:
(A) X and Y chromosomes
(B) Ratio of pairs of X-chromosomes to the pairs of autosomes
(C) Ratio of number of x-chromosomes to the sets of autosomes
(D) Whether the egg is fertilized or develops parthenogenetically

165. The most common cause of the pleiotropic effect of a gene is due to:
(A) The same product of the given gene being involved in different metabolic pathways
(B) The gene making very different products in different cell types
(C) The DNA sequence of the gene getting changed in cell specific manner
(D) The gene not functioning in some cells

166. Which transgenes shows protection against drought in tobacco?
(A) *sac*B (B) *hsf*
(C) *bet*B (D) *ala*3

167. Which one the following statement is not correct?
(A) Almost all of the prokaryotes have a single circular chromosome
(B) Almost all of the prokaryotes have cell wall
(C) All prokaryotes have plasmids
(D) The genetic material in prokaryote is not enclosed in any structure surrounded by membrane

168. In case of ribozyme approach of suppression of endogenous gene expression, the ribozyme encoding sequence is used as:
(A) Independent of the target gene
(B) Fused with a sequence of the target gene in sense orientation
(C) Fused with a sequence of the target gene in sense orientation
(D) Either B and C

169. Soybean transgenic variety Roundup Ready is resistant to:
(A) Phosphinothricin (B) Atrazine
(C) Yellow mosaic (D) Glyphosate

170. The bar gene conferring resistance to phosphinothricin was isolated from:
(A) *Streptomyces* sp.
(B) *Azotobactor vinelandii*
(C) *Bacillus subtilis*
(D) *E. coli*

171. The *barnase* and *barstar* genes were isolated from:
(A) *Bacillus thuringiensis*
(B) *Bacillus subtilis*
(C) *Agrobacterium rhizogenes*
(D) *Bacillus amyloliquefaciens*

172. Which of the following is not related to fragile X syndrome?

(A) Trinucleotide repeat
(B) Inherited mental retardation
(C) Robertsonian translocation
(D) Defect in FMR1 gene

173. Which is not correct about marker genes?

(A) Scorable markers provide valuable tools for identification of and tracking genetic modification in plants
(B) They are not useful in quantifying both transformation efficiency and gene expression in transformants
(C) They are particularly important where genetically modified plants can not be regenerated from single cells and direct selection is feasible or effective
(D) They identify and tag transformed cells

174. Which presents problems in gene transfer in plants?

(A) Undesirable features of transonic plants
(B) Contamination by Agrobacterium
(C) Low frequency of regeneration from transformed cells
(D) All of these

175. If a mutation in a gene makes the product of the gene defective in binding to its ligand, then the best what can be predicted about the gene and the mutation is:

(A) The gene codes for a receptor and the mutation is a gain – of function mutation
(B) The gene codes for a kinase and the mutation is a gain – of – function mutation
(C) The gene codes for a receptor and the mutation is a loss – of – function mutation
(D) The gene codes for a kinase and the mutation is a loss – of – function mutation

176. If Drosophila males of genotype AaBb are crossed with females of genotype AABB, what is the percentage of white-eyed wrinkled-winged offspring in the first generation (a-white eye, A-red eye; B-smooth wings, b-wrinkled wings)?

(A) 0 % (B) 10 %
(C) 20 % (D) 25 %

177. Which transgenes provides protection against osmotic stress in tobacco?

(A) *sac*B (B) *hsf*
(C) *bet*B (D) *ala*3

178. In a dihybrid cross AABB x aabb F_2 progeny of AABB, AABb, AaBB and AaBb occurs in the ratio of:

(A) 1 : 1 : 1 : 1 (B) 9 : 3 : 3 : 1
(C) 1 : 2 : 2 : 1 (D) 1 : 2 : 2 : 4

179. In a particle gun, acceleration of particles is achieved by:

(A) Blank cartridges (B) Pressurised helium
(C) Electrostatic energy (D) All of these

180. Which of the following feature is not associated with material inheritance?

(A) It is mostly uniparental
(B) It occurs due to nuclear genes
(C) Organelles genome is responsible for maternal inheritance
(D) Variegation in four O-clock plants is an example of maternal inheritance

181. Which of the following does not participate during site specific recombination between genome of λ–phage and bacterial chromosome?

(A) λ integrase protein (λ Int)
(B) Integration host factor
(C) attB and attP
(D) RecBCD

182. The high frequency of recombinant obtained from a cross between Hfr and F- strains are because :

(A) The F for integrated into the chromosome mediates transfer of large regons of chromosomes into the recipient
(B) The strain bears high number of F pilli on its surface
(C) There are high number of copies of the F plasmid per cell
(D) It includes recombination at high rate

183. In case of agrobacterium – mediated genetic transformation, 'wound response' ha a reference to:

(A) Release of a signal molecule from plant wounds
(B) Release of nutrients from plant wound sites
(C) Release of antibiotic compounds from plant wounds
(D) Both (A) and (B)

184. Which is not correct about Golden Rice-2?

(A) During 2005, a team of researchers at biotechnology company Zeneca (now Syngenta) produced a variety of golden rice called Golden Rice-2
(B) The team of researchers combined a phytene synthase gene from maize with a *lyc* and *cry*1 from the original golden rice
(C) Golden rice-2 produces 23 times more betacarotene (37 micrograms per gram) than golden rice (1.6 micrograms per gram)
(D) Consumption of golden rice, helps in removal of diseases associated with deficiencies of vitamin A and iron

185. A silent mutation in a gene results in:

(A) No change in the nucleotide sequence of the mRNA encoded by the gene
(B) No change in the amino acid sequence of the encoded by the gene
(C) No expression of the protein encoded by the gene
(D) An amino acid substitution that has a significant effect on the functional activity of the protein encoded by the gene

186. Male child with blood group AB is colorblind. His parents could be:

(A) Father normal vision with blood group A, mother colour bind with group O

(B) Father colour blind with group O, mother colour blind with blood group AB

(C) Father normal vision with blood group A, mother colour blind with blood group B

(D) Father colour blind with blood group O, mother normal vision with blood group O

187. A temperature phage differs from a virulant phage in that, temperature phage:

(A) Can exhibit only lytic cycle

(B) Can choose between lytic and lysogenic cycle

(C) Can exhibit only lysogenic cycle

(D) Engages in lysogenic cycle only at high temperature

188. The cross between a mutant female mouse and a normal male mouse gives rise to offspring that are all mutant; the reciprocal cross gives rise to offspring that are all normal. This means that the mutation is likely to be:

(A) In the mitochondrial genome

(B) In the nuclear genome

(C) Dominant

(D) Recessive

189. In a cross involving polygenic inheritance, only 2/125 of the offspring (F_2) were as extreme as one of the P_1 parents. How many gene pairs are involved?

(A) 3 (B) 1

(C) 4 (D) 2

190. Which is the most successful approach for gene transfer in plants?

(A) Electroporation

(B) PEG induced DNA uptake

(C) Microinjection

(D) *Agrobacterium*-mediated transformation

191. Match list-I with list II and select the correct answer using the codes given below the lists:

List –I (Character of man)	List – II (Example)
P. Sex-linked	1. Baldness
Q. Sex-influenced	2. Aquired immuno deficiency syndrome
R. Sex –limited	3. Klinefelter's syndrome
	4. Haemophilia
	5. Tuft of hairs (hypertrichosis) on pinna

(A) P – 4, Q – 1, R – 5 (B) P – 5, Q – 3, R – 2

(C) P – 5, Q – 1, R – 3 (D) P – 4, Q – 3, R – 2

192. *B. napus* transgenic variety 'Laurical' has a transgene that encoded:

(A) Acyl-ACP desaturase

(B) Acyl-ACP thioesterase

(C) Acyl transferase

(D) Both (A) and (C)

193. In Drosophila melanogaster, cinnabar eye (n) & vestigial wing (vg) are simple recessive traits. A female, heterozygous for both genes, was crossed with a male with cinnabar eyes & vestigial wings. The offspring resulting from this cross are listed below:

Class	Phenotype	Number
I	Wild type	445
II	Cinnabar eye, wild type wing	51
III	Wild type eye, vestigial wing	49
IV	Cinnabar eye, vestigial wing	455

From the data shown, what is the distance between the n and vg genes and if n and vg were unlinked what percent of the offspring would be expected to have either cinnabar or vestigial wings or both?

(A) 500 map units and 100%

(B) 100 map units and 25 %

(C) 50 map units and 50%

(D) 10 map units and 75%

194. The terminator technology uses:

(A) A lea promoter

(B) A recombine gene

(C) A ribosome inactivating protein gene

(D) All of these

195. Which transgenic varieties could not be commercialized due to IPR problems?

(A) Endless Summer tomato

(B) Flavr Savr tomato

(C) Rainbow papaya (D) Laurical *B. napus*

196. Consider the following statements:

P. B-chromosomes are indispensable and their inheritance is non-Mendelian

Q. B-chromosomes may increase in number due to a drive involving non-disjunction during pollen mitosis

R. When B – chromosomes are present in high number, they suppress vigour and fertility

S. B-chromosome do not show homology with the basic A chromosomes

Which of the statements given above are correct?

(A) P, Q and R (B) Q, R and S

(C) P, R and S (D) P, Q and S

197. A woman trying to get pregnant has a sister whose child has an autosomal recessive disease characteristic by dysfunction of mucus-secreting cells. As a result, this child has abnormally thick mucus that tends to precipitate into dense plugs that obstruct the pancreatic ducts, bronchic and bronchioles, as well as the bile ducts. Which of the following tests could be performed to determine if this woman and her husband are carriers of this disease?

(A) Northern blot (B) PR and sequencing

(C) Southern blot (D) Western blot

198. Which is based on homologous recombination?

(A) Targetted gene transfer

(B) Gene disruption

(C) Gene replacement

(D) All of these

199. Use of antibiotic resistance markers in production of transgenic plants may pose which risks?
(A) Toxicity
(B) Allergenicity
(C) Resistance in gut microflora
(D) All of these

200. Which of the following is not true?
P. RecBCD has both helicase and nuclease properties
Q. RecBCD activity is ATP dependent
R. RuVC resolves the Holiday junction
S. RecA protein play role in homologous recombination in bacteria
Which of the statements given above are correct?
(A) P and S (B) P R and S
(C) P, Q and S (D) P, Q, R and S

201. 5-methylytosines are common sites for mutations because they:
(A) Can mispair with adenine
(B) Can determinate to thymidine
(C) Prevent discrimination between the daughter and parental strand
(D) Can deaminate to uracil

202. Choose the wrong statement(s):
P. QTLs (Qualitative trait loci) control polygonic traits
Q. RFLP is used as marker in QTL mapping
R. QTLs control a quantitative trait
S. QTLs are always present in single chromosome
(A) S (B) Q and R
(C) P and S (D) P and Q

203. A suitable expression vector ordinarily contains which of the following?
(A) Promoter
(B) Transcription termination sequence
(C) Usually, a polyadenylation site
(D) All of these

204. Which is required for the function of modified intein used in recombinant protein production?
(A) Low (4^0C) temperature
(B) ß- mercaptoethanol
(C) High (40^0C) temperature
(D) Both (A) and (B)

205. Genetic mosaics have:
P. Highly methylated chromosome
Q. An extra chromosome
R. An inactivated chromosome
S. Nucleic containing Barr body
(A) Q and R (B) P, Q, R and S
(C) P, Q and S (D) P, R and S

206. Which type of chemical mutagens are incorporated into the genome by the DNA polymerase during genome replication?
(A) Alkylation agents (B) Base analogos
(C) Deaminating agents (D) Intercalating agents

207. In a test cross of an individual heterozygous for each of the three lined genes, the most frequent classes of progeny were ABc/abc and abC/abc, and the least frequent classes were ABC/abc and abc/abc. What is the order of the genes?
(A) A - C – B (B) B – A – C
(C) A – B – C (D) None of these

208. Which is the most commonly used vector for gene transfers in plants?
(A) Ri plasmid vectors
(B) Ti plasmid vectors
(C) 2 µ plasmid vectors
(D) Cauliflower mosaic virus

209. Match the following

Column – I	Column– II
A. Down's syndrome	P. An additional sex chromosome
B. Cri-dy-chat Syndrome	Q. Loss of a part is chromosome
C. Klinefelter's Syndrome	R. Absence of an extra chromosome
D. Turner's syndrome	S. Presence of an extra chromosome
	T. Presence of two extra chromosomes

(A) A–S, B–Q, C-P, D–R (B) A-T, B-S, C-P, D–Q
(C) A–S, B–Q, C-Q, D–R (D) A–S, B–Q, C-R, D–P

210. Choose the correct statements:
P. Is element is a transposon
Q. Direct repeat acts as hallmark for transposable element
R. Transposons are only present in eukaryotes
S. Yeast Ty element is a retrotransposon
(A) Q and S (B) P and Q
(C) P, Q and S (D) P, Q, R and S

211. In a particle gun, is the most commonly used
(A) Blank cartridges
(B) Pressurised helium
(C) Electrostatic energy
(D) All of these

212. Match the following

Column –I (Type of interaction)	Column– II (F_2 phenotypic ratio)
P. Recessive epistasis	A. 12 : 3 : 1
Q. Dominant epistatsis	B. 13 : 3
R. Duplicate recessive epistasis	C. 9 : 6 : 1
S. Dominant & recessive epistasis	D. 9 : 3 : 4
	E. 9 : 7
	F. 15 : 1

(A) P-B, Q-A, R-B, S-E (B) P-D, Q-A, R-E, S-B
(C) P-F, Q-C, R-B, S-A (D) P-A, Q-E, R-C, S-D

213. Which of the following is true?

P. Transversional mutations are more harmful than transitional mutations

Q. Frameshift mutations may lead to premature termination

R. Missense mutations are more deleterious than nonsense mutations

S. GC mutation is a transversional mutation

Which of the statements given above are correct?

(A) P, Q, and R (B) P , Q, R and R
(C) P, Q and R (D) P, Q and S

214. The controversial terminator technology seeks to provide :

(A) Transgenic varieties that can be grown by farmers generation after generations
(B) Protection from piracy of transgenes
(C) Seeds fit for consumption but not for use as seed
(D) Both (B) and (C)

215. If only two segregating loci contribute to cob length and we represent the parental cross as AABB (9-in average cob length) x aabb (3-in average), the fraction of the F_2 expected to be 4.5 in is:

(A) 1/8 (B) 1/4
(C) 1/16 (D) 3/16

216. Cells take in calcium phosphate-DNA precipitate particles by:

(A) Pincocytosis (B) Phagocytosis
(C) Both (A) and (B) (D) Lipofection

217. Homologous recombination is exploited by:

(A) Specific gene function 'Knockout' mice
(B) Gene replacement (in gene therapy)
(C) Transgenic mice
(D) Both (A) and (C)

218. A colour blind man and a phenotypically normal woman have four children (each has 46 chromosomes)- a boy with haemophilla but no other traits; a boy with colour blindness but no other traits; a colourblind girl but no other traits; and a phenotypically normal girl (all traits mentioned are inherited as x-linked recessives).What is the probability that the mother is heterozygous for colour blindness)?

(A) ¼ (B) 1/2
(C) ¾ (D) 1

219. Mendel could not find recombination and crossing over as:

(A) Traits he chosen were either present on different chromosomes or were far apart
(B) Traits chosen by him were not influenced by genes
(C) He did not have a high power microscope
(D) He selected only pure types

220. In 1983, the first transgenic plant was produced in:

(A) Tobacco (B) Tomato
(C) Cotton (D) Maize

221. A 12 year old boy develops a disorder also present in his father. No one else in the family is known to be affected. Which of the following modes of inheritance is least likely?

(A) Autosomal recessive
(B) Autosomal dominant
(C) X-linked recessive
(D) Y-linked

222. Which is the most difficult problem in case of *Agrobacterium*-mediated transformation?

(A) Inability to transform apical meristems
(B) Inability to transform monocot cells/tissues
(C) High frequency of transient transformations
(D) Lack of 'wounds response' in certain dicots as well

223. Among the given scorable markers, which one is generally used in plants?

(A) Nopaline synthase (nos)
(B) ß-glucuronidase (gus)
(C) Luciferase (lux)
(D) Octopine synthase (ocs)

224. Albinism is condition that results from the lack of normal pigmentation. Individuals possessing two recessive alleles (aa) are albino. Similarly, attached result from two recessive alleles (ee). In a family, the husband is an albino with attached earlobes and the wife if pigmented with non-attached earlobes. All their 20 children are pigmented with non-attached earlobes. What is the genotype of the mother?

(A) aaee (B) aaEE
(C) AaEe (D) AAEE

225. The cryIIB protein is specific to:

(A) Coleoptera (B) Diptera
(C) Orthoptera (D) Lepidoptera

226. The patent entitled 'Control of Plant Gene Expression' was granted in which country?

(A) USA (B) Japan
(C) European Union (D) India

227. Consider the terms maternal inheritance and maternal effect and choose the correct statement below:

(A) There is no difference between the terms
(B) Maternal effect is the genetic information carried by the mRNAs deposited in the egg
(C) Maternal inheritance is the genetic information carried by the mRNAs deposited in the egg
(D) Chloroplasts only provide the genetic information that constitute the maternal inheritance

228. The techniques for gene transfer and transgenic animal production were developed by using which animals as models?

(A) Sheep (B) Rabbits
(C) Mice (D) Pigs

229. Match list-I with list II and select the correct answer using the codes given below the lists:

List –I	**List – II**
P. Incomplete dominance	1. Human skin colour
Q. Codominance	2. Purple colour in maize due to anthrocyanin
R. Polygenes	3. Human being belonging to AB blood group
S. Compelementary genes	4. Pink flower in 4 o' clock plant

(A) P–4, Q–3, R–1, S–2 (B) P–3, Q–4, R–2, S–1
(C) P–4, Q–3, R–2, S–1 (D) P–3, Q–4, R–1, S–2

230. A protein that is normally localized to nucleus has been mutated in such a way that its localization is no longer nuclear, but cytosolic instead. The function of the protein, however, is 99 of the wild type protein. Which of the followings could be said about the mutation?
(A) It is a frameshift mutation
(B) It is a nonsense mutation
(C) It is a silent mutation
(D) It is a small deletion

231. Which is correct about CAD (bacterial origin) protein:
(A) It is encoded by Cad gene
(B) It is a multifunctional enzyme and catalyzes the first three reaction of endogenous pyrimidine synthesis
(C) PALA-resistant cell lines developed through selection overproduce CAD protein due to a high amplification of the Cad gene
(D) All of these

232. In a cross in Neurospora involving two alleles B and b, the following tetrad patterns were observed:

Tetrad pattern	Number
BBbb	36
bbBB	44
BbBb	4
bBbB	6
BbbB	3
bBBb	7

What will be the distance between the gene and the centromere
(A) 10 map unit (B) 15 map unit
(C) 40 map unit (D) None of these

233. What could be the best term to describe effect of one gene on another in a way that one would hide the effect of another on a phenotype?
(A) Pleitropy (B) Homeostasis
(C) Epistasis (D) Hyperstasis

234. Which of the following is not an example of allopolyploid?
(A) Rephanobrassia (B) Evening prime rose
(C) Triticale (D) *Tritichum aestivum*

235. Which is correct about development of insect resistance against Cry proteins?
(A) Resistance to Cry proteins may be due to reduced binding of the toxin to midgut epithelium (e.g., *P. xylostella*) or complete degradation of Cry proteins (e.g., *S. frugiperda*)
(B) The inheritance of resistance to Cry proteins in insect populations is usually recessive, and is governed by one or few major genes
(C) It appears that insects may evolve resistance to some groups of toxins which faster than no others
(D) All of these

236. Phenylketonuria (PKU) is inherited disease that is characterized by:
(A) Elimination of genetisic acid in urine
(B) Increased occurrence of phenylalanine in blood and tissues
(C) Elimination of sugar in urine
(D) Decrease in phenylalanine in blood and tissues

237. A recombination process occurring in bacteria always refers to the:
(A) Affect of a physical nature changing the nature of the genes
(B) Reversal of a microbial mutation
(C) Loss of genes from a cell
(D) Acquisition of DNA by an organism

238. Which is an example of a dominant selectable marker?
(A) CAD protein
(B) DHFR and XGPRT
(C) Neomycin phosphotranferase
(D) All of these

239. Which is not correct about methotexate (Mtx)?
(A) Normal nonmutant cultured cells are very sensitive to mtx
(B) It is a weak inhibitor of dihydrofolate reductase (DHFR)
(C) Enzyme DHFR is involved in endogenous synthesis of nucleotides
(D) The Chinese hamster ovary cell line 'A29' overproduces an altered DHFR so that it is extremely resistant to mtx

240. Which of the following feature is to related to inversion?
(A) Position effect
(B) Crossover suppression
(C) Formation of dicentric bridge
(D) Pseudodominance

241. Consider the following pairs

P. Acridine compounds	:	Frame-shift mutation
Q. Ultraviolet light	:	Conversion of one base into another
R. Bromouracil	:	Thymidine dimer

Which of the above is/are correctly matched?
(A) P (B) P and Q
(C) P and R (D) Q and R

242. A gene can be obtained from:
(A) Genomic library (B) Chemical synthesis
(C) cDNA library (D) All of these

243. Which transgenic varieties was the first to reach the market?
(A) Liberty Link (B) Endless Summer
(C) Flavr Savr (D) Bollgard

244. In studying an *E. coli* strain containing a known frameshift mutation, you isolate a mutant in which this frameshift is suppressed. You therefore, expect to find which of the following associated with the suppression? (Assume the frameshift has not been corrected in the coding for the peptide).
(A) Suppressor tRNAs encodes an amino acid for a stop signal
(B) Suppressor tRNAs encode a two-nucleotide anticodon
(C) Suppressor tRNAs encode a four-nucleotide anticodon
(D) Both (B) and (C)

245. In tomato the following genes are located on chromosome 3:
+ tall plant + normal leaves + smooth fruit
d dwarf plant m mottled leaves p pubescent fruit
Results of the cross : +++//dmp X dmp/dmp were

+++	430
dmp	452
+mp	45
d++	38
++p	16
dm+	17
+m+	1
d+p	1

Which one of the following progeny groups represents double crossovers?
(A) +mp/d++ (B) ++p/dm+
(C) +m+/d+p (D) +++/dmp

246. TA29 promoter has which features?
(A) Light-responsiveness
(B) Wound responsiveness
(C) Tissue specificity
(D) Development stage-specificity

247. The term transformation is used to denote:
(A) Delivery of DNA into bacterial/plant cell by a virus
(B) Uptake of DNA by a cell
(C) Conversion of a normal vertebrate cells into a tumourigenic cells
(D) Both (B) and (C)

248. The barnase gene was first used for producing male sterility in 1990 by:
(A) Thorpe and coworkers
(B) Paterson and coworkers
(C) Mariani and coworkers
(D) Scowcroft and coworkers

249. Transformation through electroporation has been reported in:
(A) Rice (B) Tobacco
(C) Wheat (D) All of these

250. An x-linked recessive gene produces red-green colour blindness in humans. A women with normal colour vision whose father was colour-blind marries a colour – blind man. What is the probability that their son will be colour-blind?
(A) 0 (B) 1/4
(C) ½ (D) ¾

251. RecA is a protein involved in:
(A) Recombinational repair
(B) Mismatch repair
(C) Nucleotide excision repair
(D) Base excision repair

252. Which has been the most often used promoter for gene transfers in plants?
(A) ShI early promoter (B) nos promoter
(C) 35S promoter (D) Adh1 promoter

253. Which is not correct definition of transgene?
(A) A gene from a related species
(B) A gene from an unrelated species
(C) A gene transferred into an organism using recombinant DNA technology
(D) A synthetic gene

254. The bacterial XCPRT gene serves as dominant selectable marker on a culture medium containing shih of the following?
(A) Xanthine and Adenine
(B) Mycophenolic acid
(C) Aminopterin
(D) All of these

255. Which is correct about insertional mutagenesis?
(A) It often produces aberrant phenotypes
(B) It occurs by chance and can not be controlled
(C) Occurrence of insertional mutagenesis leads to accidental discovery of unsuspected genes and gene functions
(D) All of these

256. 'Nature Gard' is a transgenic variety of:
(A) Brinjal (B) Potato
(C) Cotton (D) Corn

257. DEAE-dextran brings about DNA uptake by :
(A) Pincocytosis (B) Endocytosis
(C) Active transport (D) Osmosis

258. The concept of genetically engineered cross protection can be exploited using which approaches?
(A) Expression of satellite RNA and replicase sequences
(B) Expression of viral protease and movement proteins
(C) Use of satellite RNA and antisense RNA
(D) All of these

259. A silent mutation is one that:
(A) Result in a truncated polypepetide
(B) Replaces an amino acid with an equivalent amino acid in a polypepetide
(C) Does not change the amino acid sequence of the polypeptide
(D) Changes the leading frame of the mRNA leading to an altered polypeptide

260. Which of the following conditions is likely to interfere with the transfer of genetic material by conjugation in bacteria?
(A) Pre-treatment of the recipient cells with DNase
(B) Pre-treatment of the recipient cells by application of strong shearing forces
(C) Treatment of the recipient cells with cycloheximide
(D) Treatment of the mating cell pairs by application of strong shearing forces

261. Neomycin phosphotransferase II (antII) inactivates by which phosphorylation:?
(A) Kanamycin (B) Neomycin
(C) G418 (D) All of these

262. Which of the edible vaccines for humans?
(A) Honey (B) Banana
(C) Apple (D) All of these

263. Which is not correct about SV40 plasmid vectors?
(A) Plasmid vectors are generally used for stable transfection
(B) They integrate into the host genome with a frequency of 10-5 to 10-3
(C) Plasmid vectors have been mainly used to investigate sequences involved in transcriptional and post-transcriptional regulation
(D) pRSV series of vectors are derivatives of pSV vectors

264. Among the given proteins, which has been evaluated for insect control?
(A) Lectin and aprotinin
(B) Cowpea trypsin inhibitor and serine protease inhibitor
(C) Cystein protease inhibitor and proteinase inhibitor II
(D) All of these

265. The first transgenic plant was produced in 1983, when a tobacco line expressing which genes was produced?
(A) Ampicillin resistance
(B) Kanamycin resistance
(C) Hygromycin resistance
(D) Phosphinothricin resistance

266. Stability/expression of transgenes in plants can be affected by:
(A) Gene silencing
(B) Chromosomal rearrangements
(C) Pleiotropy
(D) All of these

267. Which has been used in chrysanthemum & pelargonium to make these plants dwarf, delayed their leaf senescence and increased chlorophyll pigmentation:
(A) Suppression of endogenous genes of chalcone synthase
(B) Alteration of the cytokinins: auxin ratio by introducing and expressing specific transgenes
(C) Over expression of the gene encoding phytochrome A
(D) Gene disruption technology

268. In Drosophila, XXY is female. In humans it represents an abnormal male beause:
(A) Y – chromosome induces male traits in humans
(B) Y – chromosome is essential for female sex in Drosophila
(C) Y – chromosome is not essential for male sex in humans
(D) None of these

269. Crystal proteins from *B. thuringiensis* are solubilized in the alkaline environment of insect midgut and then proteolytically processed to yield a toxic core fragment of which size?
(A) 50 kDa (B) 60 kDa
(C) 80 kDa (D) 70 kDa

270. *Petite* mutants in yeast stand for slow growing and small colony forming yeast cells. Genetic crosses between different haploid yeast strains show that the *Petite* mutations do not segregate with any known chromosome. The defect could be in the:
(A) Nucleus (B) Mitochondria
(C) Golgi body (D) ER

271. The first animal vector was devised for :
(A) Simian Virus 40 (SV 40)
(B) Primate Papova Virus
(C) Papolloma Virus
(D) Both (A) and (B)

272. Among the given tests which one is the most rapid and convenient?
(A) Immunoblotting
(B) ELISA
(C) Radioimmunoprecipitation
(D) Immunnohistochemical staining of tissue sections

273. CryI proteins are insecticidal to:
(A) Coleoptera (B) Lepidoptera
(C) Orthoptera (D) Diptera

274. Gene is in T-DNA determines the synthesis of:
(A) Cytokinin (B) Indole acetic acid
(C) Opine (D) Isopentenyl adenine

275. How many different types of gametes could be produced by an individual with the arbitrary genotype of AAbbCCDdEe?
(A) Two (B) Four
(C) Six (D) Eight

276. The TA29 promoter sequence was isolated from:
(A) Cotton (B) Tomato
(C) Tobacco (D) Cauliflower

277. 'Control of Plant gene Expression' was dubbed as 'Terminator Technology' by:
(A) Delta and Pine Land Company
(B) United States Department of Agriculture, USA
(C) Rural Advancement Foundation International, Canada
(D) Department of Biotechnology, New Delhi

278. T-DNA is flanked on both the sides by direct repeat sequences of how many base-pairs?
(A) 24 bp (B) 44 bp
(C) 34 bp (D) 14 bp

279. Modification of flower colour intensity in petunia, rose, carnation has been achieved by:
(A) Suppression of endogenous genes of chalcone synthase
(B) Alteration of the cytokinins: auxin ratio by introducing and expressing specific transgenes
(C) Over expression of the gene encoding phytochrome A
(D) Gene disruption technology

280. Who said that we need to have a National Biodiversity Act, a Plant Variety Act, Farmers Right's Act and a Geographic Appellation Act to deal with the problems created by terminator technology?
(A) M.S. Swaminathan (B) B.P. Pal
(C) R.S. Paroda (D) Panjab Singh

281. SV 40 genome has been used to develop by:
I. Plasmid vectors
II. Transforming vectors
III. Tranducing vectors
IV. Early replacement vectors
(A) I, II, III, IV (B) I, II, III
(C) I, III, IV (D) II, III, IV

282. Which is a more general approach for identification of transgenic animals?
(A) Dot blot techniques
(B) PC amplification
(C) Both (A) and (B)
(D) Southern hybridization

283. Particle gun is suitable for transformation of:
(A) Embryogenic calli
(B) Leaf discs
(C) Immature zygotic embryos
(D) All of these

284. Assume that two pairs of genes with two alleles each, Aa and Bb determine plant height additively in a population. The homozygote AABB is 50 m high and aabb 30 m. What is the height of the F_1 of a cross between them?
(A) 50 cm (B) 40 cm
(C) 30 cm (D) 20 cm

285. Stable integration of T-DNA has been reported in:
(A) Sunflower and Brassica
(B) Cotton and tomato
(C) Petunia and tobacco
(D) All of these

286. Which sequences has been used to produce a virus resistant variety of squash?
(A) Defective virus genome
(B) Ribozyme sequence
(C) cDNA of satellite RNA
(D) Virus coat protein gene

287. Which can be used to modify size and shape of ornamental plants?
(A) Suppression of endogenous genes of chalcone synthase
(B) Alteration of the cytokinins: auxin ratio by introducing and expressing specific transgenes
(C) Over expression of the gene encoding phytochrome A
(D) Gene disruption technology

288. Potato transgenic variety `New Leaf is resistant to:
(A) Potato virus
(B) Colorado potato beetle
(C) Late blight
(D) Frost

289. Teminism refers to:
(A) DNA-RNA hybridization
(B) DNA denaturation
(C) Reverse transcription
(D) Transcription

290. Among The given gene transfer techniques, which is the most commonly used in case of mice embryos?
(A) Retroviral infection
(B) Microinjection
(C) Embryonic stem cell technology
(D) Electroporation

291. Which of the following aims at transformation of cells in the developing inflorescence?
(A) Macroinjection (B) Electroporation
(C) Microinjection (D) Partcle gun

292. If inducible gene expression is driven by metallothionein promoter, transcription is induced by which metal ions?
(A) Cadmium (B) Zinc
(C) Iron (D) Both (A) and (B)

293. The term transfection is used to denote:
(A) Genetic transformation of bacteria via virus infection
(B) Genetic transformation of animal cells
(C) Delivery of recombinent DNA into animal cell via a virus
(D) All of these

294. Liposime-mediated DNA uptake has been reported in protoplasts of:
(A) Petunia (B) Tobacco
(C) Carrot (D) All of these

295. Which approaches can be used to increase the toxicity of Cry proteins?
(A) Increasing copy number of the cry genes
(B) Site directed mutagenesis within the toxin encoding region
(C) Production of hybrid cry genes by fusing selected toxin domains of two or more different cry genes
(D) Both (B) and (C)

296. TA29 promoter is specific to:
(A) Tapetum (B) Pollen
(C) Ovule (D) Pollen mother cells

297. If a chiasm forms between the loci of genes A and B in 20% of the tetrads of an individual of genotype AB/ab, the percentage of gametes expected to be Ab is:
(A) 40 (B) 20
(C) 10 (D) 5

298. Transgenics have been produced in which animals?
(A) Cow and goat (B) Sheep and Rabbits
(C) Mice, fish and pigs (D) All of these

299. Transgenic animal has:
(A) Foreign DNA in all its cells
(B) Foreign RNA in all its cells
(C) Foreign DNA in some of the cells
(D) Both (B) and (C)

300. Which viral transduction expression vectors is used for in vivo gene transfer in mammalian hosts?
(A) Adeno-associated virus (B) Baculovirus
(C) Both (A) and (B) (D) Vaccinia virus

301. The chromatography column used to separate recombinant protein having polyhistidine tags has :
(A) Divalent nickel (B) Ca^{++}
(C) K^{+} (D) Both (A) and (B)

302. Which is correct about toxic fragments of cry proteins?
(A) They have three domains, domain I functions in pore and iron channel formation, domain II is involved in binding of the receptor, and domain III is involved in receptor recognition
(B) They are -80000 times more toxic to the target insects than organophosphate insecticides
(C) They are highly selective in their action
(D) All of these

303. Which states of India has the lowest area under cultivation of transgenics cotton varieties?
(A) Uttar Pradesh (B) Madhya Pradesh
(C) Maharastra (D) Andhra Pradesh

304. Which contributes to high level transgene expression?
(A) Inclusion of an intron in the expression cassette
(B) Removal of unnecessary untranslated sequences
(C) Codon optimization
(D) All of these

305. Cloning is a mean to:
(A) Replace original genotype
(B) Preserve genotype
(C) Production of HGH gene in *Escherichia coli*
(D) None of these

306. Which Agrobacterium proteins functions as a receptor?
(A) VirA (B) VirG
(C) VirD1 (D) VirD2

307. 'Golden rice' has transgenes that provide which enzyme activities?
(A) Phytoene synthesase
(B) Phytoene desaturase
(C) Lycopene ß–cyclase
(D) All of these

308. Mammalian artificial chromosomes & human artificial chromosomes are linear vectors that contain :
(A) Centromeres and telomeres
(B) Origin of replication recognition sequences
(C) Transcriptionally active chromosomal domains
(D) All of these

309. Which is not correct about promoter/enhancer?
(A) A promoter sequence is the site to which RNA polymerase first binds during the initiation of transcription
(B) Promotor must be located at the 5' –end of the antisense trand of the gene
(C) The promoter sequence determines the elevel of transcription of the gene and tissue or cell type where the gene will be expressed
(D) A suitable enhancer sequence may be used alongwith the promoter to enhance the promoter activity

310. Which promoters has been the most wisely used in dicots?
(A) Vicilin (B) Extensin
(C) 35 S (D) Adh1

311. Which is correct about edible vaccines?
(A) They can be used as regular component of food
(B) Honey could be used as edible vaccine
(C) Their production cost is comparable to that of conventional vaccines
(D) They are useful only when the native antigen is orally inactive

312. Which Agrobacterium proteins functions as transcription activator?
(A) VirA (B) VirG
(C) VirD1 (D) VirD2

313. All the transfection techniques are applicable to cultured animal cells, but microinjection is ordinarily not used due to :

(A) Limited number of cells can be handled
(B) Tediousness of the technique
(C) Avilability less demanding alternative technique
(D) All of these

314. Which Cry proteins is specific to Lepidoptera and Diptera?

(A) Cry II (B) Cry V
(C) Cry IV (D) Both (B) and (C)

315. Maximize and yield gard are transgenic varieties of:

(A) Cotton (B) Maize
(C) Cotton (D) Tobacco

316. Early region replacement and late region replacement vectors of SV40 have:

(A) Transient expression
(B) Produce virions, which infect host cells
(C) Mammalian cells are hosts of SV40-derived vectors
(D) All of these

317. Which statement is not correct?

(A) In case of electroporation, circular super-coiled DNA is far more efficient in transfection than linearized DNA
(B) The male pronucleus is chosen for microinjection because it is much larger than the female pronucleus
(C) The use of cationic lipids for construction of liposomes offers a distinct advantage
(D) Recombinant retroviruses produce virions

318. In which insects, a >250- fold (highest) increase in the resistance against Cry proteins was observed after 36 generation of laboratory selection?

(A)*Helicoverpa virescence*
(B) *Cadra cutella*
(C) *Plodia interpunctella*
(D) *Plutella xylostella*

319. Which states of India has the highest area under cultivation of transgenics cotton varieties?

(A) Uttar Pradesh (B) Madhya Pradesh
(C) Maharastra (D) Andhra Pradesh

320. The DNA of bacteriophage T4 contains approximately 200,000 nuclotide pairs. The a11 region of the T4 genome coupies about 1% of its total genetic length. Benzer has found that 300 sites are separable by recombination within the r11 region. What will be average number of nucleotides in each recon?

(A) About 20 (B) About 300
(C) About 7 (D) Cann't be determined

321. Which Agrobacterium proteins topoisomerase functions?

(A) VirA (B) VirG
(C) VirD1 (D) VirD2

322. Xenopus oocytes can be used for expression of :

(A) mRNAs obtained by in vitro transcription of eukaryotic genes using T7RNA polymerase
(B) Eukaryotic mRNAs isolated from various tissues
(C) Genomic eukaryotic DNA micro-injected into the nucleus
(D) All of these

323. Which is not correct?

(A) MAHYCO developed Bt cotton hybrids by transferring Bt (Cry 1Ac) gene from cotton transgenic variety Coker 312 into bred lines
(B) Cotton hybrids Bt ECH 12, Bt MECH 162 and Bt MECH 184 were approved Bt Genetic Engineering Approval Committee of the Ministry of Environment, Govt. of India, for commercial release
(C) Till March 2005, a total of 20 Bt-transgenic cotton hybrids has been released for commercial cultivation in India
(D) In 2004, of the 81.0 million hectares under transgenics, developing countries share was 53.4 mha 966%) as compared to 27.6 mha (34%) in the industrial countries

324. Which is not correct about transgene integration in animal genomes?

(A) Usually, in a given cells, integration occurs at a single location
(B) In general, multiple copies are integrated when small amounts of DNA are used for transfection
(C) When multiple copies are integrated, they are mostly integrated at one site joined to each other head-to-tail
(D) Transgene integration frequently leads to various forms of rearrangements

325. A species of plant is discovered in which individual plants produce either white or purple flowers. True breeding plants that have short stems with white flowers are crossed with true breeding plants that have tall stems and purple flowers. The resulting offspring all have tall with purple flowers. When one of these tall, purple-flowered offspring is crossed with a short, white – flowered plant, plants with the following characteristics were obtained in equal proportions. White flowers on tall stems and purple flowers on tall stems. White flowers on short stems, purple flowers on short stems. Given the information above, which of the following is most likely true?

(A) The white-flower allele and the tall-stem allele occur at the same locus
(B) The genes for flower colour and stem length are loosely linked
(C) Extensive crossing-over between the purple-flower allele the white-flower allele has occurred
(D) The genes for flower colour and stem length are not linked

326. Which promoters is seed specific?
(A) Vicilin (B) Extensin
(C) 35S (D) Adh1

327. Which can reduce transgene silencing?
(A) Same orientation of the transgenes in the recombinant DNA
(B) A.T composition of transgene simiar to that of the host chromosome
(C) Single copy integration of transgene
(D) All of these

328. Papaya transgenic Rainberg is resistant to:
(A) Papaya Mosaic Virus
(B) Leaf Curl Virus
(C) Papaya Ring Spot Virus
(D) Distorted Ring Spot Virus

329. An Hfr strain of *E. coli* contains:
(A) A bacterial chromosome with a human gene inserted
(B) A bacterial chromosome with the F factor inserted
(C) A human chromosome with a transposable element inserted
(D) A bacterial chromosome with a phage inserted

330. The yeast artificial chromosomes have found regular use for cloning large genome DNA fragments in various genome sequencing projects is:
(A) BACs (B) PACs
(C) Both (A) and (B) (D) None of these

331. For gene therapy, we need to have gene-sized human DNA fragments including their:
(A) Promoters (B) Control elements
(C) Both (A) and (B) (D) None of these

332. The restriction maps provide the basic knowledge of the physical organization of a genomic region or a whole genome with respect to the cleavage sites of:
(A) Restriction endonuclease
(B) Detritylation
(C) Coupling
(D) Capping

333. The chosen derivatized solid support, with the required 3' end nucleotide hooked onto it, is treated with trichloreacetic acid is called:
(A) Restriction endonuclease
(B) Detritylation
(C) Coupling
(D) Capping

334. The nest nucleotide is chosen as the phosphoramide derivative whose 5' – OH group is also protected with dimethoxytrityl group is called:
(A) Restriction endonuclease
(B) Detritylation
(C) Coupling
(D) Capping

335. There is always a possibility that some of the 3'-nucleotide (the first nucleotide from the 3' end in this case) may have escaped the coupling reaction and still have a reactive free 5'-OH group is known as:
(A) Restriction endonuclease
(B) Detritylation
(C) Coupling (D) Capping

336. The linkage between the first and the second nucleotide is then converted from the phosphate to the more stable phsphotriester by using iodine as the oxidizing agent and water as the oxygen donor is called:
(A) Oxidation (B) Detritylation
(C) Coupling (D) Capping

337. The deletion of small phenolic compounds released by wounded plants resulting in autophosphorylation and then phophorylation of vir G by which vir gene?
(A) Vir A (B) Vir G
(C) Vir D1/D2 (D) Vir D2

338. The activation of other vir gene by which gene:
(A) Vir A (B) Vir G
(C) Vir D1/D2 (D) Vir D2

339. Recognize the 25 bp border sequence and produce a single-stranded endonucleal ytic cleavage in the bottom strand of each border by which vir gene:
(A) Vir A (B) Vir G
(C) Vir D1/D2 (D) Vir D2

340. Which vir gene is remains tightly associated with the 5' end of the T strand:
(A) Vir A (B) Vir G
(C) Vir D1/D2 (D) Vir D2

341. The Ss binding protein rendering it resistant to nucleolytic degradation by which vir gene:
(A) Vir A (B) Vir G
(C) Vir E2 (D) Vir B

342. Which vir gene provide energy for the translocation of the T-complex:
(A) Vir A (B) Vir G
(C) Vir E2 (D) Vir B

343. All genetic transformations in monocot plants had been performed using direct:
(A) DNA delivery system (B) Cis-inactivation
(C) Cosuppression (D) Gene silencing

344. When silencing takes place at the same place due to multiple insertion it can be referred to as:
(A) DNA delivery system (B) Cis-inactivation
(C) Cosuppression (D) Gene silencing

345. The post-transcriptional gene silencing had been reported in multiple homologous transgenes and the endogeneous genes that share common coding sequences, the process referred to as:
(A) DNA delivery system (B) Cis-inactivation
(C) Cosuppression (D) Gene silencing

346. Single copy transgene is also reported to be inactivated due to:
(A) DNA delivery system (B) Cis-inactivation
(C) Cosuppression (D) Gene silencing

347. To demarcate those plants which does not bear a transgene i.e. gene from other organism, rather it carries gene from itself but in a modified version is:
(A) Cis genic (B) Multi genic
(C) Uni genic (D) None of these

348. The first genes available for genetic engineering crop plants for pest resistant were:
(A) Cis genic (B) Multi genic
(C) Uni genic (D) Cry genes

349. The first genes available for genetic engineering crop plants for pest resistance were Cry genes a bacterium *Bacillus thuringiensis*, popularly known as:
(A) Bt genes (B) Multi genic
(C) Uni genic (D) Cry genes

350. Bt cotton is resistant to a notorious insect pest of:
(A) *Helicoverpa armigera*
(B) Termites
(C) Spodoptera
(D) Aphids

351. Bt cotton was adopted in India in:
(A) 2000 (B) 2001
(C) 2002 (D) 2003

352. The transgenic technology provides an alternative and innovative method to improve pest-control management which is:
(A) Eco-friendly
(B) Effective
(C) Sustainable and increase yield
(D) All of these

353. Bt cotton contains:
(A) Bt gene (B) Multi genic
(C) Cry IAC gene (D) Cry genes

354. Genetic engineering can also be used to develop transgenic fruits that show:
(A) Delayed fruit softness (B) Improving nutrient
(C) Processing qualities (D) All of these

355. Genetically engineered rice (golden rice), which is enriched in:
(A) Provitamin A (B) Vitamin B
(C) Mineral (D) Vitamin K

356. The production of pathogen-free plants with uniform quality for agronomic cultured and also for germplasm conservation is:
(A) Micro propagation
(B) In-vitro culture
(C) Dedifferentiated cultures
(D) Differentiated cultures

357. The production of secondary metabolites under controlled cultures conditions is:
(A) Micro propagation (B) In-vitro culture
(C) Dedifferentiated cultures
(D) Differentiated cultures

358. In vitro cultures are currently being used for secondary metabolite production of callus and suspension cultures is:
(A) Micro propagation (B) In-vitro culture
(C) Dedifferentiated cultures
(D) Differentiated cultures

359. In vitro cultures are currently being used for secondary metabolite production of transformed roots and shoots is:
(A) Micro propagation (B) In-vitro culture
(C) Dedifferentiated cultures
(D) Differentiated cultures

360. A transgene expression can achieve which of the following?
(A) Prevent expression of native gene
(B) Modify an existing biosynthetic pathway
(C) Produce protein that itself produces the phenotype of interest or is the product of interest
(D) All of these

361. Transgenic study can be done by:
(A) Transcription profiling
(B) RT-PCR
(C) Western hybridization
(D) All of these

362. The study to create transgenic organisms allows genetic characteristics to be transferred beyond:
(A) Genus (B) Species lines
(C) Across kingdom boundaries
(D) All of these

363. Which of the following statement is true regarding the production of transgenic cassava?
(A) A non- toxic cyanide derivative was production by metabolic engineering
(B) The release of cyanide was inhibited by antisense gene regulation
(C) The release of cyanide was inhibited by immunolocalization
(D) The production of cyanide was enhanced by overexpression of an enzyme

364. Which of the following transgenic methods were used (in separate experiments) in an attempt to increase the prevalence of certain amino acids is seeds?
(A) Altered tRNA gene; over expression of sunflower seed albumin (SSA)
(B) Altered tRNA gene; over expression of human serum albumin (HSA)
(C) Altered rRNA gene; over expression of sunflower seed albumin (SSA)
(D) Altered tRNA gene; repression of sunflower seed albumin (SSA) expression

365. Bt cotton is resistant to boll worm, it is due to:
(A) Bt is repellent to boll worm
(B) Bt is contact poison for boll worm
(C) Bt is stomach poison; it change the permeability of stomach due to generation of pores in alkaline environment of stomach
(D) Bt is causing death due to asphagsiation of insect

366. In Bt cotton expression of crystal protein occurs:
(A) On surface of leaves (B) Systemic expression
(C) Only in stem part (D) Stem and leaves

367. Hirudin is obtained from the transgenic plant:
(A) *Brassica napus* (B) *Hibiscus rosasinensis*
(C) *Raphanus sativus* (D) *Phseolus vulgare*

368. Bt cotton is:
(A) Cloned plant (B) Transgenic plant
(C) Hybrid plant (D) Mutated plant

369. A technique of using very small metal particles coated with desired gene in the gene transfer is called:
(A) Electroporation (B) Microinjection
(C) Macroinjection (D) Biolistics

370. Which ones produce androgenic haploids in anther cultures?
(A) Anther wall
(B) Tapetal layer of anther wall
(C) Connective tissue
(D) Young pollen grains

371. Callus is:
(A) Tissue that forms embryo
(B) An insoluble carbohydrate
(C) Unorganised actively dividing mass of cells maintained in culture
(D) Tissue that growth to form embryoid

372. To raising of plants from a small tissue in culture is known as:
(A) Macroproduction (B) Micropropagation
(C) Tissue culture (D) Mass production

373. The transgenic plant flavor savr tomato carries an artificial gene for:
(A) Delay ripening process (B) Longer shell life
(C) Added flavours (D) All of these

374. The term transgenic was coined by:
(A) Dale (B) Dale and Irwin
(C) Gordon and Ruddle (D) Meyer

375. Gurdon and Ruddle coined the term transgenic in:
(A) 1980 (B) 1981
(C) 1982 (D) 1983

376. Herbicide resistance for glufosinate from streptomyces has been transferred into:
(A) Wheat (B) Tomato
(C) Potato (D) All of these

377. In potato, for improving protein quality gene has been utilized from:
(A) Wheat (B) Human
(C) Chicken (D) Barley

378. In rapeseed, resistance to seedling blight has been transferred from:
(A) Cowpea (B) Common bean
(C) Arabidopsis (D) Bt

379. In maize, resistance to European corn Borer has been transferred from:
(A) Cowpea (B) Common bean
(C) Arabidopsis (D) Bt

380. In wheat, herbicide resistance to glufosinate has been transferred from:
(A) Bt (B) Arabidopsis
(C) E. coli (D) Streptomyces

381. In tobacco, cold resistance has been transferred from:
(A) Cowpea (B) Common bean
(C) Arabidopsis (D) Tomato

382. In tomato, antifreezing gene has been transferred from:
(A) Fish (B) Arabidopsis
(C) Chicken (D) None of these

383. Transgenic may be:
(A) A plant (B) An animal
(C) Amicrobe (D) All of these

384. Maximum transgenic plants have been released in:
(A) Rapeseed (B) Potato
(C) Tobacco (D) Tomato

385. Plasmid method exhibits:
(A) Host specificity
(B) Somaclonal variation
(C) Slow regeneration
(D) All of these

386. In alfalfa protein quality has been improved by transferring ovalbumin gene from:
(A) Fish (B) Chicken
(C) Duck (D) All of these

387. In cotton, resistance to Helicoverpa has been transferred from:
(A) Common bean
(B) *Bacillus thuringiensis*
(C) Fish (D) Cowpea

388. Which of the following methods of foreign gene transfer exhibits severe host specificity:
(A) Particle bombardment (B) Plasmid method
(C) Micro injection (D) Direct DNA uptake

389. Foreign genes (transgenes) can be transferred by:
(A) Plasmid methods
(B) Particle bombardment
(C) Micro-injection (D) All of these

390. In tobacco, protein quality has been improved by transferring glutenin gene from:

(A) Barley (B) Oat

(C) Wheat (D) Rice

391. In rapeseed, male sterility has been transferred from:

(A) *Bacillus amyloliquefaciens*

(B) Cowpea

(C) Bt

(D) Arabidopsis

ANSWERSHEET

12. TRANSGENICS

1. (A)	2. (B)	3. (C)	4. (A)	5. (B)	6. (C)	7. (D)	8. (A)	9. (A)	10. (B)
11. (C)	12. (D)	13. (C)	14. (C)	15. (A)	16. (A)	17. (B)	18. (C)	19. (D)	20. (A)
21. (B)	22. (C)	23. (D)	24. (A)	25. (B)	26. (C)	27. (D)	28. (A)	29. (A)	30. (B)
31. (C)	32. (D)	33. (A)	34. (B)	35. (A)	36. (B)	37. (B)	38. (A)	39. (B)	40. (D)
41. (B)	42. (A)	43. (A)	44. (D)	45. (C)	46. (D)	47. (D)	48. (C)	49. (C)	50. (D)
51. (D)	52. (D)	53. (B)	54. (B)	55. (D)	56. (D)	57. (B)	58. (B)	59. (B)	60. (C)
61. (C)	62. (D)	63. (B)	64. (A)	65. (D)	66. (C)	67. (D)	68. (D)	69. (D)	70. (D)
71. (C)	72. (B)	73. (D)	74. (C)	75. (D)	76. (D)	77. (A)	78. (D)	79. (D)	80. (C)
81. (D)	82. (D)	83. (C)	84. (B)	85. (C)	86. (C)	87. (B)	88. (D)	89. (B)	90. (A)
91. (B)	92. (A)	93. (D)	94. (D)	95. (D)	96. (B)	97. (A)	98. (B)	99. (D)	100. (D)
101. (C)	102. (A)	103. (A)	104. (D)	105. (C)	106. (B)	107. (B)	108. (B)	109. (C)	110. (D)
111. (C)	112. (A)	113. (A)	114. (B)	115. (B)	116. (B)	117. (A)	118. (C)	119. (D)	120. (D)
121. (D)	122. (C)	123. (D)	124. (A)	125. (D)	126. (D)	127. (D)	128. (D)	129. (C)	130. (D)
131. (A)	132. (D)	133. (B)	134. (C)	135. (D)	136. (D)	137. (B)	138. (D)	139. (D)	140. (B)
141. (A)	142. (A)	143. (D)	144. (A)	145. (A)	146. (D)	147. (D)	148. (D)	149. (B)	150. (D)
151. (C)	152. (D)	153. (C)	154. (A)	155. (A)	156. (D)	157. (B)	158. (A)	159. (A)	160. (D)
161. (D)	162. (C)	163. (B)	164. (C)	165. (A)	166. (A)	167. (C)	168. (C)	169. (D)	170. (A)
171. (D)	172. (C)	173. (B)	174. (D)	175. (C)	176. (A)	177. (C)	178. (D)	179. (D)	180. (B)
181. (D)	182. (B)	183. (A)	184. (D)	185. (B)	186. (C)	187. (B)	188. (A)	189. (A)	190. (D)
191. (A)	192. (B)	193. (D)	194. (D)	195. (A)	196. (C)	197. (B)	198. (D)	199. (D)	200. (D)
201. (B)	202. (A)	203. (D)	204. (D)	205. (D)	206. (B)	207. (A)	208. (B)	209. (A)	210. (C)
211. (B)	212. (B)	213. (D)	214. (D)	215. (B)	216. (B)	217. (D)	218. (D)	219. (A)	220. (A)
221. (C)	222. (A)	223. (B)	224. (D)	225. (B)	226. (A)	227. (B)	228. (C)	229. (A)	230. (D)
231. (D)	232. (A)	233. (C)	234. (B)	235. (D)	236. (B)	237. (D)	238. (D)	239. (B)	240. (D)
241. (A)	242. (D)	243. (C)	244. (D)	245. (C)	246. (C)	247. (D)	248. (C)	249. (D)	250. (C)
251. (A)	252. (C)	253. (C)	254. (D)	255. (D)	256. (D)	257. (B)	258. (D)	259. (B)	260. (D)
261. (D)	262. (D)	263. (A)	264. (D)	265. (B)	266. (D)	267. (C)	268. (A)	269. (B)	270. (B)
271. (D)	272. (B)	273. (B)	274. (C)	275. (B)	276. (C)	277. (C)	278. (A)	279. (A)	280. (A)
281. (C)	282. (C)	283. (D)	284. (B)	285. (D)	286. (D)	287. (B)	288. (B)	289. (C)	290. (B)
291. (A)	292. (D)	293. (D)	294. (D)	295. (D)	296. (A)	297. (C)	298. (D)	299. (A)	300. (C)
301. (A)	302. (D)	303. (A)	304. (D)	305. (B)	306. (A)	307. (D)	308. (D)	309. (B)	310. (C)
311. (B)	312. (B)	313. (D)	314. (A)	315. (B)	316. (D)	317. (A)	318. (C)	319. (C)	320. (C)

321. (C)	**322. (D)**	**323. (D)**	**324. (B)**	**325. (D)**	**326. (A)**	**327. (D)**	**328. (C)**	**329. (B)**	**330. (C)**
331. (C)	**332. (A)**	**333. (B)**	**334. (C)**	**335. (D)**	**336. (A)**	**337. (A)**	**338. (B)**	**339. (C)**	**340. (D)**
341. (C)	**342. (D)**	**343. (A)**	**344. (B)**	**345. (C)**	**346. (D)**	**347. (A)**	**348. (D)**	**349. (A)**	**350. (A)**
351. (C)	**352. (D)**	**353. (C)**	**354. (D)**	**355. (A)**	**356. (A)**	**357. (B)**	**358. (C)**	**359. (D)**	**360. (D)**
361. (D)	**362. (D)**	**363. (D)**	**364. (A)**	**365. (C)**	**366. (B)**	**367. (A)**	**368. (B)**	**369. (D)**	**370. (D)**
371. (C)	**372. (B)**	**373. (D)**	**374. (C)**	**375. (B)**	**376. (D)**	**377. (B)**	**378. (B)**	**379. (D)**	**380. (D)**
381. (C)	**382. (A)**	**383. (D)**	**384. (A)**	**385. (D)**	**386. (B)**	**387. (B)**	**388. (B)**	**389. (D)**	**390. (C)**
391. (A)									

Chapter - 13

MOLECULAR MARKERS

1. A genetic marker that is detected as differential mobility of a protein/DNA g fragment or as differential amplification of a DNA sequence present in different individuals is:
(A) Molecular marker (B) Molecular biology
(C) Molecular breeding (D) Molecular farming

2. A specific location on a chromosome that is defined by naked eye polymorphism, protein or DNA sequences; the inheritance of a genetic marker can be followed in a mapping population is known as:
(A) Molecular marker (B) Genetic marker
(C) Molecular breeding (D) Molecular farming

3. The plant breeding programme that are supported by the used of DNA-based markers and recombinant DNA technology is called:
(A) Molecular marker (B) Molecular biology
(C) Molecular breeding (D) Molecular farming

4. Production of proteins encoded by transgenes in animal/plants; the protein is recovered from milk, urine, blood (in animals) seeds etc. (in plant) is called:
(A) Molecular marker (B) Molecular biology
(C) Molecular breeding
(D) Molecular farming

5. A branch of biology concerned with the study of molecular aspects of biology phenomenon is called:
(A) Molecular marker (B) Molecular biology
(C) Molecular breeding (D) Molecular farming

6. It depicts the order of genetic markers and the relatives distances between them; measured by recombination frequency between the markers is called:
(A) Linkage map (B) Genetic map
(C) Genome map (D) Restriction map

7. A schematic representation of various genes and molecular markers located in different chromosomes; different strategies are used to construct linkage maps, cytological or cytogenetic maps and physical maps is called:
(A) Linkage map (B) Genetic map
(C) Genome map (D) Restriction map

8. A detailed schematic description of the structural and functional organization of the complete genome of an organism is called:
(A) Linkage map (B) Genetic map
(C) Genome map (D) Restriction map

9. A linear map of a DNA molecule showing the relative positions of the recognition sites for a number of different restriction enzyme is known as:
(A) Linkage map (B) Genetic map
(C) Genome map (D) Restriction map

10. A single protein species (interferons) produced by both leucocytes and fibroblasts in response to virus infection is:
(A) Interferon α (B) Interferon ß
(C) Interferon γ (D) Interferon ω

11. A single protein species produced by T-lymphocytes in response to antigen-activation is:
(A) Interferon α (B) Interferon ß
(C) Interferon γ (D) Interferon ω

12. A single protein species produced by virus infected cells is:
(A) Interferon α (B) Interferon ß
(C) Interferon γ (D) Interferon ω

13. It generally implies analysis of chromosome pairing relationship among genomes of different species is called:
(A) Genome analysis (B) Genome map
(C) Genome compaction (D) Genomic library

14. Reduction in genome size achieved by reducing sequences not involved in genetic functions is called:
(A) Genome analysis (B) Genome map
(C) Genome compaction (D) Genomic library

15. A collection of plasmid clones or phage lysates containing recombinant DNA molecule so that the sum total of DNA inserts in this collection, ideally, represents the entire genome of the concerned organisms is called:
(A) Genome analysis (B) Genome map
(C) Genome compaction (D) Genomic library

16. Development of vectors permitting the synthesis of pure proteins encoded exclusively the DNA inserts is called:
(A) Transcriptional fusion
(B) Transcriptional silencing
(C) Transcription
(D) Transcription units

17. Genes, which have major effect on the expression of the concerned traits is called:
(A) Oligogenes (B) Oncogene
(C) Luxury gene (D) Gene

18. **Denotes a suppressions of transcription of the transgenes, it is the result of methylation of promoter region of transgene is called:**
(A) Transcriptional fusion
(B) Transcriptional silencing
(C) Transcription
(D) Transcription units

19. **Production of complementary RNA copy of a DNA strand, catalyzed by RNA polymerase is called:**
(A) Transcriptional fusion
(B) Transcriptional silencing
(C) Transcription
(D) Transcription units

20. **The segment of genome, which is transcribed into a single primary mRNA transcript; operon in case of prokaryotes, and one gene in eukaryotes is called:**
(A) Transcriptional fusion
(B) Transcriptional silencing
(C) Transcription
(D) Transcription units

21. **An enhancer that makes the promoter activity stimulus specific is called:**
(A) Stimulus-specific enhancer
(B) Stimulus-specific gene
(C) Stimulus-specific promoter
(D) None of these

22. **Gene expressed in response to only a specific stimulus is called:**
(A) Stimulus-specific enhancer
(B) Stimulus-specific gene
(C) Stimulus-specific promoter
(D) None of these

23. **A promoter that becomes active/inactive only in response to a specific stimulus usually, due to enhancer/silencer sequences is called:**
(A) Stimulus-specific enhancer
(B) Stimulus-specific gene
(C) Stimulus-specific promoter
(D) None of these

24. **A gene involved in tumour development, in case of Agrobacterium, genes specifying auxin and cytokinin biosynthesis is called:**
(A) Oligogenes (B) Oncogene
(C) Luxury gene (D) Gene

25. **Those genes, which are expressed either in specialized cells or in response to an specific stimuli is called:**
(A) Oligogenes (B) Oncogene
(C) Luxury gene (D) Gene

26. **Molecular marker pattern of an individual obtained by using highly polymorphic markers; used for unequivocal identification of strains, individuals, criminals etc., is called:**
(A) DNA fingerprint (B) DNA finger printing
(C) DNA footprinting (D) DNA sequencing

27. **Developing the DNA fingerprint of an individual is called:**
(A) DNA fingerprint
(B) DNA finger printing
(C) DNA footprinting (D) DNA sequencing

28. **It enables identification of the site on a DNA molecule to which a protein, such as RNA polymerase, binds is called:**
(A) DNA fingerprint (B) DNA finger printing
(C) DNA footprinting (D) DNA sequencing

29. **Determination of the base sequence of a DNA fragment is called:**
(A) DNA polymerase (B) DNA microarray
(C) DNA footprinting (D) DNA sequencing

30. **The enzyme that catalyses synthesis of DNA copies of template DNA, separate enzymes involved in DNA replication and DNA repair is called:**
(A) DNA polymerase (B) DNA microarray
(C) DNA footprinting (D) DNA sequencing

31. **An assembly of a very large number of small does onto a solid support, each dot containing a short sequence of a different gene of the organism concerned is called:**
(A) DNA polymerase (B) DNA microarray
(C) DNA footprinting (D) DNA sequencing

32. **A gene, which permits easy and rapid selection for the cell/organism in which it is present and expressed is:**
(A) Molecular marker (B) Genetic marker
(C) Selectable marker (D) Scorable marker

33. **A gene that permits easy and rapid identification of the cell/organism in which it is present and expressed in:**
(A) Molecular marker (B) Genetic marker
(C) Selectable marker (D) Scorable marker

34. **A gene that gives rise to a product involved in the regulation of the expression of another gene; for example, a gene coding for a repressor protein is called:**
(A) Structural gene
(B) Regulatory gene
(C) Regulatory sequence
(D) Reverse transcriptase

35. **A gene coding for a protein or RNA molecule; as distinct from a regulatory gene is called:**
(A) Structural gene (B) Regulatory gene
(C) Regulatory sequence
(D) Reverse transcriptase

36. **A DNA sequence involved in regulating the expression of a gene, for example, a promoter or operator is called:**
(A) Regulatory enzyme
(B) Regulatory gene
(C) Regulatory sequence
(D) Reverse transcriptase

37. An RNA-directed DNA polymerase in retroviruses; capable of making DNA complementary to an RNA is called:
(A) RNA polymerase
(B) Regulatory gene
(C) Regulatory sequence
(D) Reverse transcriptase

38. An enzyme that catalyzed the formation of RNA from ribonucleoside 5'-triphosphates, using a strand of DNA or RNA as a template is called:
(A) RNA polymerase
(B) Regulatory gene
(C) Regulatory sequence
(D) Reverse transcriptase

39. Removal of introns and joining of exons in a primary transcript is called:
(A) RNA polymerase
(B) RNA splicing
(C) Regulatory sequence
(D) Reverse transcriptase

40. The rate of sedimentation of a macromolecule can be enhanced many fold by applying:
(A) DNA polymerase I (B) DNA polymerase
(C) DNAlibrary (D) Centrifugal force

41. Which varieties is the first variety in Indian developed by MAS?
(A) Pusa Basmati 1
(B) Pusa 1401
(C) Pusa 1121
(D) Improved Pusa Basmati 1

42. In eukaryotic DNA replication:
(A) Only one replisome forms because there is a single origin of replication
(B) The Okazaki fragments are 1000 to 2000 nucleotides in length
(C) Helicase dissociates from DNA as soon as the initiation bubble forms
(D) FEN1 (flap endonuclease 1) is involved in removing the primer

43. A strain of *E. coli* has the following genotype with respect to the lac operon: $I^-OZ^+Y^-A^+$. This strain will
(A) Produce permease only in the presence of lactose
(B) Not produce beta galactosidase if lactose is present in the medium
(C) Not produce permease irrespective of the presence or absence of lactose in the medium
(D) Produce trans-acetylase only when induced

44. What is incorrect about tRNA?
(A) It binds with an amino acid at its 3' end
(B) It has five double stranded regions
(C) It has a anticodon at one end which recognises codon of mRNA
(D) It looks like clover leaf in 3-dimentional structure

45. Pusa 1406 has which of the following genes transferred through MAS?
(A) Xa21 (B) Xa13
(C) Both (A) and (B) (D) Xa14

46. Which of the following statement about restriction endonucleases is not true?
(A) The enzymes are natural products of bacteria
(B) The enzymes cut DNA at defined sequences
(C) The enzymes cut only single-stranded DNA molecules
(D) The enzymes are named because they restrict the growth of bacteriophage

47. SDS-PAGE does not involve:
(A) Separation of proteins based on their molecular weights
(B) Protein with heat and chemicals
(C) Application of an electric field to proteins
(D) Creating a temperature gradient for denaturation of proteins

48. Primer designing for sequence-tagged sites (STS) is based on:
(A) Random genomic DNA
(B) Ends of large genomic DNA fragments
(C) Sequences of cDNA (expressed sequence tags)
(D) All of these

49. Improved Pusa Basmati 1 (Pusa 1406), the first variety in India developed through MAS, has been improved by transferring :
(A) Vitamin A (B) QTLs for grain size
(C) Disease resistance (D) Cooking quality

50. Polynucleotide ligase joins two DNA molecules together by forming a covalent bond between :
(A) Two OH groups
(B) To phosphate groups
(C) Complementary nucleotides on opposite strands
(D) A 3'-OH group and a $5'-PO_4^{-2}$ group

51. To test whether a new fertilizer is effective, rice is grown in some fields without fertilizer and in the other fields with fertilizer and the average yields from each of these fields is calculated in terms of its mean and variance. The statistical test to be used to find the effectiveness of the fertilizer would be:
(A) F – test (B) t -test
(C) Chi-square test (D) ANOVA

52. Which markers systems has been used for developing Pusa 1460?
(A) RFLP (B) STMS
(C) ISSR (D) SCAR

53. What would stop DNA replication in bacteria?
P. Addition of a DNA polymerase inhibitor
Q. Addition of dideoxyribonucleaotides (ddNTP)
R. Addition of lamin inhibitor
S. Addition of actin inhibitor
(A) P and Q (B) Q and R
(C) R and S (D) P and S

54. Protein binding regions of DNA are identified by one of the following techniques:
(A) Finger printing (B) Footprinting
(C) Southern printing (D) Western printing

55. Development of which of the following populations requires the greatest effort?
(A) F_2 (B) Backcross
(C) RILs (D) NILs

56. Once an RAPD marker linked to a trait of interest is found, it can be converted into a more reproducible PCR type assay by use of :
(A) Allele-specific ligation
(B) Allele-specific PCR (AS-PCR)
(C) Sequence characterized amplified region (SCAR) assay
(D) All of these

57. If 5-terminal of a linear DNA molecule with sticky ends are being modified in a way that the ends cannot be ligated with each other anymore (to form a circular molecule), what enzyme was used to modify the molecule?
(A) Methylase (B) Kinase
(C) Phosphatase (D) Dehydrogenase

58. For a nucleic acid probe to hybridize, which of the following must be true?
(A) Both the probe and the sequence to which it hybridizes must be RNA
(B) Both the probe and the sequence to which it hybridizes must be DNA
(C) The sequence to which it hybridizes must be identical to that of the probe
(D) The sequence to which it hybridizes must be complementary to that of the probe

59. Electrocardiogram (ECG) is:
(A) Measures electrical activity generated by the brain
(B) The hormonal activity of the heart over time
(C) Recording of the electrical activity of the heart over time
(D) None of these

60. A major problem with the computational assembly of DNA sequences of complex eukaryotic genomes is the presence of:
(A) Multiple chromosomes
(B) Mitochondrial DNA
(C) Introns within the genome
(D) Repetitive sequence

61. Positional cloning refers to:
(A) Using a selection procedure to lone a cDNA
(B) Cloning a portion of a gene using PCR
(C) Isolating a gene by PCR using primers from another species
(D) Mapping a gene to a chromosomal region and then identifying and cloning the genomic copy of the gene from the region

62. Which step of translation does not consume high energy phosphate bond?
(A) Translocation
(B) Peptidyl transferase reaction
(C) Amino acid activation
(D) Aminoacyl tRNA binding to A-site

63. The first case of linkage between a quantitative trait and a qualitative character was reported in *Phaseolus vulgaris* between:
(A) Seed size and seed colour
(B) Seed size and flower colour
(C) Seed size and pod colour
(D) Seed yield and seed colour

64. All of the following statements about type I topoisomerase (Topo I) are true except:
(A) Topo I removes DNA supercoils in an ATP-dependent reaction
(B) Topo I from eukaryotic cells can remove both positive and negative supercoils
(C) Topo I from *E coli* an remove only negative superscoils
(D) Topo I is essential for viability in *E. coli*

65. When a growing replication fork encounters an RNA polymerase complex engaged in transcription:
(A) Both DNA replication and transcription stops
(B) Replication stops but transcription continues
(C) Both replication and transcription continues but slowly
(D) Transcription pauses to allow replication fork to progress

66. Spilicing of GU-AG intron:
(A) Involves transesterification reaction
(B) Occurs in prokaryotic mRNA
(C) Form Y-shaped branch structure
(D) Requires maturase activity

67. The segregation ratios for dominant and codominant markers will differ in which of the following mapping populations:
(A) F_2 (B) Backcross
(C) RILs (D) NIls

68. What electronic transitions are the most important in UV spectroscopy of organic molecules because they result in the absorbance of longer wavelength radiation outside the UV region?
(A) $\sigma \rightarrow \pi$, $n \rightarrow \pi$ (B) $\sigma \rightarrow \sigma^*$, $n \rightarrow \sigma^*$
(C) $\pi \rightarrow \pi^*$, $n \rightarrow n^*$ (D) $\pi \rightarrow \pi^*$, $n \rightarrow \pi^*$

69. Which is not correct about molecular markers?
(A) The sequence tagged sites (STS) marker system has all the advantages of PCR technique
(B) RFLP is the first molecular marker that was developed in 1980s
(C) RFLP markers are highly reproducible
(D) A major limitation of RAPD markers in their non-reproducibility due to low annealing temperature

70. What is the best method to identify the cellular location of a protein?
(A) Place a reporter gene next to the promoter of the gene encoding the protein, and identify the cellular location of the reporter protein
(B) Use of labeled antibody to identify the cellular location of the protein
(C) Separate the cellular compartments by centrifugation and screen the different compartments with an antibody
(D) Tag the protein with fluorescent amino acids and identify the cellular location of fluorescence microscopy

71. To increase the stringency of a wash buffer so as to allow the detection of only completely matched hybrids you would?
(A) Lower the temperature and raise the salt concentration
(B) Lower the temperature and lower the salt concentration
(C) Raise the temperature and raise the salt concentration
(D) Raise the temperature and lower the salt concentration

72. Transgenic plants are easier to produce the transgenic animals because:
(A) Plants can more easily be grown from single cultured cell into which foreign DNA has been introduced
(B) Plant DNA is easier to clone
(C) Plant cells can be transformed by bacterial infection
(D) DNA passes more readily through the plant cell wall than through the animal cell membrane

73. Okazaki fragments are formed during DNA synthesis, because
P. DNA synthesis extends from 5' to 3' direction
Q. Their synthesis are opposite to the direction of replication fork
R. DNA ahead of replication fork is positively supercoiled
S. DNA synthesis is semi-conservative
(A) Q and R (B) P and Q
(C) R and S (D) S and P

74. A recombination-dependent mechanism for repairing damaged DNA is:
(A) SOS repair
(B) Excision repair
(C) Mismatch repair
(D) Photo-reactivation repair

75. Which mapping populations is used for linkage mapping of molecular markers?
(A) Near-isogenic lines (NILs)
(B) Recombinant inbred lines (RILs)
(C) Backcross progeny
(D) Any of these

76. A reporter gene is used to:
(A) Identity regulatory sequences from the upstream regions of other genes
(B) Determine if a protein binds to a given sequence element
(C) Determine if a gene contains introns
(D) Determine the stability of a protein

77. Plasmid vectors for cloning:
(A) Can generally accommodate larger insects than phage vectors
(B) Grow within bacteria, and are present in bacterial colonies on an agar plate
(C) Can accommodate inserts of over 100 kilobases
(D) Include centromeres to allow propagation in yeast

78. Eukaryotes differ from prokaryotes in DNA replication:
(A) Different enzyme for leading and lagging strands
(B) Use of DNA primer instead of RNA primer
(C) Discontinuous rather than semi-discontinous replication
(D) Unidirectional rather than bidirectional replication

79. Which is not correct about recombinant inbred lines (RILs)?
(A) RILs are also called inbred-derived lines
(B) RILs are homozygous lines derived from a suitable cross
(C) RILs are created by single seed descent (SSD) through atleast five generations of continued selfing
(D) At the end of SSD, seeds from each plant are harvested separately to obtain as many RILs as there are individual plants

80. UV - B induced damage of DNA for formation of pyrimidine dimmers are repaired by photolyase on activation by:
(A) UV - C light (B) Green light
(C) Blue light (D) Infrared light

81. Which of the following is an example of RNA editing?
(A) Removal of introns from an RNA transcript
(B) Degradation of an RNA molecule by nucleases
(C) Alternation of the nucleotide sequence of an RNA molecule
(D) Caping of the 5' end of an RNA transcript

82. Why were the first genetically engineered human insulin genes split into separate A and B chain sequences and cloned in separate *E. coli* strains?
(A) To control the proportions of A and B chains produced
(B) To prevent an *E. coli* strain producing an active insulin
(C) Because *E coli* cannot fold the resultant peptide with the chain present
(D) To prevent loss of insulin by secretion

83. Sodium Dodecyl Sulphate (SDS) is used while separating proteins by PAGE because:
(A) It helps in solubilization of proteins thereby making it easier to separate
(B) It binds to proteins and confers uniform negative charge density thereby making move during electrophoresis
(C) Decreases the surface tension of the buffer used for electrophoresis
(D) Stabilizes the proteins

84. Which of the following types of vector would be most suitable for introducing DNA into a human cell?
(A) Plasmid (B) Bacteriophage
(C) Cosmid (D) Adenovirus

85. Which of the following activities is most closely associated with alkaline phosphatase?
(A) Cutting DNA
(B) Preventing vector dimerization
(C) Joining DNA fragments with cohesive ends
(D) Homopolymer tailing

86. Group I introns are remarkable because:
(A) They are spliced by external RNA molecules with the help of protein
(B) They are spliced by protein molecules in the absence of external RNA molecules
(C) They are autocatalytic
(D) They are only present in mitochondrial and chloroplast genomes

87. Which is not correct about the applications of molecular markers?
(A) The molecular markers that are uniformly distributed throughout the genome can be used to estimate the genetic contribution of each parent to each individual of a segregating population
(B) MAS can not be used to improve complex traits like yield
(C) Molecular markers would enable an effective selection for horizontal resistance even in the presence of vertical resistance genes
(D) Two rice varieties 'Angke' and 'Coonde', developed by MAS for bacterial leaf blight resistance, have been released in Indonesia

88. Molecular breeding requires which:
(A) Genetic maps and molecular markers linked to agronomic traits
(B) Automated and high throughout diagnostic techniques
(C) A modification if breeding practice that takes full advantages of the information provided by such diagnositic assays
(D) All of these

89. In 1923, the theory of QTL mapping was described by:
(A) Stern (B) Sax
(C) Mather (D) Galton

90. Northern blotting is difficult from Southern blotting in that:
P. The target is DNA
Q. The probe should be selected taking introns into consideration
R. The label is not radioactive
S. The targets form hairpins
(A) Q (B) P and Q
(C) Q and S (D) P, Q and R

91. Which of the following statements(s) concerning the Edman degradation method is/are true?
P. Phenyl isothiocyanate reacts with the amino terminal residue
Q. If a protein has a blocked amino-terminal residue (as does N-formyl methionine, for example), it cannot react with phenyl isothiocyanate
(A) P (B) Q
(C) P and Q (D) None of these

92. ATP-independent chromatin remodeling enzymes mediate:
(A) Nucleosome sliding
(B) Transfer of a histone octamer from one DNA strand to another
(C) Weakening of the interaction between DNA and histone octamer
(D) Replacement of a histone with a histone variant

93. Which combination of the following statements with regard to gene expression is true?
P. Heteronuclear RNA represents contiguous segment of genomic DNA
Q. UTRs are part of heteronuclear RNA but not of mRNA
R. UTRs are part of exons
S. Translation start codon must be in the first exon
(A) P – True, Q – False, R – False, S - False
(B) P – True, Q – False, R – True, S - False
(C) P – True, Q – True, R – False, S - True
(D) P – False, Q – True, R – True, S - True

94. Double chained DNA strand is made radioactive in both its chains. It is allowed to replicate twice in non-radioactive medium. The result would be:
(A) All strands have radioactivity
(B) Half the strands have radioactivity
(C) Three strands have radioactivity
(D) Radioactivity is absent in all strands

95. In an experiment designed to determine the number of phage particles in a sample 56 plaques were recorded when 100 l of a 10^{-4} dilution was plated. How many phage are there per ml in the original?
(A) 5.6×10^3
(B) 5.6×10^4
(C) 5.6×10^5
(D) 5.6×10^6

96. The polymerase chain reaction or PR is a technique that:

(A) Was used to demonstrate DNA as the genetic material

(B) Is used to determine the content of minerals in a soil sample

(C) DNA primers and a thermostable DNA polymerase to replicate specific DNA sequences in vitro

(D) Measures theriosome transfer rate during translation

97. Which is not correct about application of molecular markers?

(A) Molecular markers do not permit the identification of putative resistant plants in the absence of disease test

(B) In resistance breeding, molecular markers can be very useful in gene pyramiding

(C) Molecular markers may be useful in identifying heterotic combinations

(D) MAS can be successfully used to select for such traits whose phenotypic evaluation is either problematic or costly

98. Why is the gene inactivation is a useful technique to determine the function of a gene?

(A) Gene activation provides information about the expression of a gene

(B) Gene inactivation provides information about the cellular location of a gene product

(C) Gene inactivation provides an opportunity to identify phenotypic changes associated with the loss of the functional gene

(D) Gene activation provides information about the structure of the gene product

99. The lineal use of recombinant DNA technology is to have rapidly replicating bacteria produce large amounts of specific proteins (e.g., hormone). Expression of a eukaryotic gene in prokaryotes involves:

(A) A shine-Dalgarno (SD) sequence in mRNA

(B) Introns

(C) Regulatory elements upstream of the gene

(D) All of these

100. Begining with 600 template DNA molecules, after 25 cycles of PCR, how many amplicons will be produced?

(A) 2×10^{10} (B) 2^{25}

(C) 600 X 25 (D) 2×10^{25}

101. Which is not correct about Cleaved Amplified Polymorphic Sequences (CAPS)?

(A) This marker captures some of the advantages of RFLP assay

(B) It avoids the disadvantages of Southern blot analysis

(C) CAPS was initially developed for use in pea

(D) CAPS detects DNA sequence variation in terms of the lengths of DNA fragments generated by the restriction digestion of PCR products

102. DNA helicase in *E. coli*:

(A) Moves in direction opposite of replication fork

(B) Binds with template of the leading strand

(C) Is a hexameric protein with ATPase activity

(D) Catalyzes formation of primer

103. Hybridization between mature mRNA and DNA an identify introns in eukaryotic genes. When mature mRNA produced by the insulin gene is hybridized with denatured chromosomal DNA, which of the following will most likely be observed?

(A) No hybridization would occur under any conditions

(B) Hybridization of mRNA would occur with random sections of chromosomal DNA

(C) Hybridization of mRNA with DNA would occur in a continuous stretch that is equal to the length of the mRNA

(D) Hybridization of mRNA with DNA would occur but with many single – stranded loops of DNA

104. Amplified Fragment Length Polymorphism (AFLP)

P. A PCR – based tool

Q. Uses restriction enzymes to cut genomic DNA

R. Involves ligation of adaptors to the sticky ends of the restriction fragments

S. Involves amplification using primers complementary to the adaptor

(A) P and Q (B) R and S

(C) P, Q and S (D) P, Q, R and S

105. Which of the following statements are correct?

P. The complement fixation test is a relatively simple procedure used in serological microbiology

Q. In an agglutination test, a visible lump of particles appears to the observer

R. In the precipitation test, cells and large particles enter the reaction and clump together

S. The double-diffusion Oudin test utilizes the precipitation reaction as its basis

(A) P and Q (B) Q and S

(C) P, Q and R (D) P, Q, R and S

106. Degeneracy of genetic code implies that:

(A) The codons degenerate after the synthesis of polypeptide chain

(B) More than one codon can code for one amino acid

(C) Some codons degenerate as they are not involved in coding for any amino acid

(D) One codon can code for more than one amino acid

107. DNA replication in prokaryotes requires:

(A) Primase to make small primers containing deoxyribonucleotides

(B) DNA polymerase III to join okazaki fragments

(C) DNA polymerase alpha to synthesize DNA

(D) DNA polymerase I to cut out primers and fill in gaps in the lagging strand

108. Mapping of human chromosomes:
(A) Has been restricted to the sex chromosomes because of small family sizes
(B) Proceed much more successfully as large number of DNA markers became available
(C) Has determined that the number of linkage groups is about twice the number of chromosomes
(D) Has demonstrated that almost all of the DNA is involved in coding for genes

109. Two purified DNA samples A and B contain equal number of basepairs. Each of these DNA samples has one site each for EcoR1 and BamHI restriction enzymes. Complete digestion with both the enzymes yielded 3 DNA bands and 2 DNA bands respectively for A and B upon electrophoresis of the digestion products. Which one of the following explains the observation?
(A) A is circular DNA and B is linear
(B) B is circular DNA and A is linear
(C) A is circular DNA &B could be linear or circular
(D) B is circular DNA & A could be linear or circular

110. Which of the following statements about DNA ligage is/are incorrect?
P. It forms a phosphodiester bond between a 5'-hydroxyl and a 3' –phosphate in duplex DNA
Q. It requires a cofactor, either NAD^+ or ATP, depending on the source of the enzyme, to provide the energy to form the phosphodiester bond
R. It catalyzes its reaction by a mechanism that involves the formation of a covalently lined enzyme adenylate
S. It catalyzes its reaction by a mechanism that involves the activation of a DNA phosphate through the formation of a phosphoanhydride bond with AMP
(A) P and S (B) P and R
(C) S (D) P

111. Which problems is associated with RAPD process?
(A) It produces difficulties in reproducing RAPD patterns in different laboratories
(B) RAPD patterns are not reproducible due to impurities in DNA preparation
(C) RAPD patterns are not reproducible due to the use of small amount of DNA
(D) Both (A) and (B)

112. Which approaches can be used for the mapping of genes for monogenic traits in crop plants?
(A) Bulked-segregant analysis of F_2 population
(B) Use of near isogenic lines
(C) Bulked segregant analysis of RILs
(D) All of these

113. Which populations, au contain released varieties as well?
(A) F_2 (B) Backcross
(C) NIls (D) RILs

114. What would happen if the concentration of dideoxynucleotides was too high in a chain termination sequencing reaction?
(A) The reactions would yield very long molecules and there would be little sequence data close to the primer
(B) The reactions would yield very short molecules
(C) The reactions would not proceed as the high concentrations of the dideoxynucleotides would inhibit the DNA polymerase
(D) The fluorescence of the sequencing product would be too high and difficult to read

115. The function of the 3' to 5' exonuclease activity of a DNA polymerase is to:
(A) Remove the 5' end of the polynucleotide strand that is attached to the template strand that is being copied
(B) Remove damaged nucleotide from the template strand during DNA synthesis
(C) Remove nucleotides from the ends of DNA molecules to ensure the generation of blunt ends
(D) Remove incorrect nucleotide from the newly synthesized strand of DNA

116. Mapping of disease resistance genes has been based on:
(A) ISSR (B) RAPD
(C) Microsatellite (D) All of these

117. Which of the following statement about the 3' poly (A) tall of mRNA is false?
(A) It is transcribed for DNA sequence
(B) It is added to the primary transcript in the nucleus
(C) It is essential for protein synthesis
(D) It contributes to the stability and lifespan of the mRNA

118. Telomerase :
P. Is a ribonucleoprotein
Q. Participates in replication of linear dsDNA
R. Is species specific
S. Is absent in somatic cells
(A) P, Q and R (B) P, R and S
(C) P, Q, R and S (D) P

119. T4 polynucleotide kinase is used for:
(A) Remove 5'-phosphates from DNA and RNA
(B) Transfers terminal phosphate groups from ATP to 5'-OH groups
(C) Adds homopolymer tails to the 3'end of DNA
(D) Used in mapping studies

120. A microorganism when viewed under a compound microscope with objective of 40X and eye piece of 10X magnification measured 400 μ in length. The same microorganism when viewed under a dissecting microscope with 10X lens, would measure:
(A) 100 μ (B) 10 μ
(C) 40 μ (D) 400 μ

121. In bacterial RNA synthesis, which of the following is the function of factor rho?
(A) Allow proper initiation of transcription
(B) Eliminate the binding of RNA polymerase to the promoter
(C) Participate in the proper termination of transcription
(D) Increase the rate of RNA synthesis

122. The tertiary structure of protein is detected by:
(A) X-ray diffraction/crystallography
(B) Spectrophotometry
(C) Electrophoresis
(D) Chromatography

123. Which of the following sets of factors are needed for translation termination at an amber codon (UAG) and release of the finished polypeptide chain?
(A) RF2, RF3, EF-G and RRF
(B) RF1, RF3, EF-Tu and RRF
(C) RF1, RF3, EF-G and RRF
(D) RF2, RF3, EF-G and RRF

124. DNA sequencing by the Sanger method takes advantage of which property of DNA synthesis to generate a sequencing ladder?
(A) Addition of nucleotides requires a free OH group on the 3' of the DNA strand
(B) DNA has a free 5' phosphoryl group
(C) DNA polymerase has proof reading capabilities
(D) U is a base only found in RNA

125. Which of the following technique is used to inactivate a gene by altering the DNA encoding it?
(A) Homologous recombination
(B) Acid blocks
(C) Antibody microinjection
(D) Introduction of dominant inhibitory mutants

126. Which of the following types of DNA replication or repair systems is dysfunctional in individuals with xeroderma pigmentosum?
(A) Mismatch repair
(B) Base excision repair
(C) Nucleotide excision repair
(D) DNA helicase

127. Which of the following statements are correct? Very long DNA fragments (>100 kb) from eukaryotic genomes:
P. Can be efficiently packaged in a λ phage vector
Q. Can be propagated in yeast artificial chromosomes
R. Must have cohesive ends for cloning
S. Can be separated by pulsed field gel electrophoresis techniques
(A) P and Q (B) P and S
(C) Q and R (D) Q and S

128. What is the advantage of probing a transcriptome with mRNAs bound to ribosomes?
(A) Eukaryotic mRNA molecules are difficult to isolate unless complexed to a ribosome
(B) mRNA molecules isolated from ribosomes represent those that are actively being translated into proteins
(C) mRNA molecules isolated from ribosomes are more stable than other mRNA molecules
(D) mRNA molecules that are not being translated by ribosome still contain their intron sequences

129. In selecting a DNA sequence to be used as a template for construction of PCR primers, many investigators look for sequences rich in codons for tryptophan or methionine because:
(A) Tryptophan and methionine have only one codon each (i.e., there is no wobble)
(B) Tryptophan and methionine have three codon each
(C) Codon for tryptophan and methionine abundantly present in mRNA
(D) None of these

130. In methyl-directed mismatch repair in *E. coli*, the daughter strand containing the mismatched base nicked by:
(A) MutH-endonuclease
(B) UvrAB-endonuclease
(C) AP-endonuclease (D) 3' to 5' exonuclease

131. Reverse transcriptase requires the following for the conversion of a single-strand RNA into a double-strand DNA:
P. An RNA template
Q. A primer
R. A DNA template
S. All four dNTPs
(A) P and Q (B) Q and R
(C) P and S (D) P, Q and S

132. Among the following which statement is correct regarding the promoter of the gene:
(A) It is always located upstream of transcription start site
(B) It is always located downstream of translation start site
(C) Different promoter elements are recognized by different RNA polymerase
(D) Promoters are recognized due to their different secondary structures

133. Humen DNA (3×10^6 Kb) is replicated in 5 hrs at a rate of 1 Kb/min. The number off origins of replication utilized are:
(A) 1 (B) 3
(C) 300 (D) 10000

134. Hydroxyurea blocks DNA replication by:
(A) Inhibiting DNA polymerase
(B) Inhibiting ribonucleotide reductase
(C) Inhibiting lagging strand DNA synthesis
(D) Inhibiting replicative helicase

135. **Which of the following is true about RFLP?**
(A) Acts as genetic marker
(B) RLFPs have more than two alleles
(C) Used in case of FISH
(D) Both A and B

136. **Two double stranded DNA samples that are identical with respect to the number of base pairs, but differ significantly in their GC content, can be separated by:**
(A) Density gradient centrifugation
(B) Electrophoresis
(C) Dialysis
(D) Oligo-dT column chromatography

137. **In which of the following splicing events, the intermediates do not form the lariat or branched structures?**
(A) Trans splicing
(B) Group I intron splicing
(C) Group II intron splicing
(D) Spliceosome mediated splicing

138. **The product of gene X is a cytosolic protein. A repressor of gene X expression is overexpressed in a cell line using recombinant DNA technology. After the over expression, the cell line is observed to be susceptible to UV lights and higher mortality rates are observed upon short exposures to the UV. What could be the function of the gene X related to?**
(A) Photosynthesis
(B) DNA repair
(C) DNA replication
(D) X- chromosome assembly

139. **Which of the following is not part of the pathway for nuclear pre-mRNA splicing?**
(A) Transesterification where the 5' OH of the downstream exon attacks the donor junction
(B) Transesterification where the 2' OH of the branched point attacks the donor junction
(C) Recognition of the donor junction by the UI snRNP
(D) Dissociation of U4 from U6

140. **The minimum components of an artificial yeast chromosomes include all of the following except:**
(A) All the histones except H1
(B) An autonomously replicating sequence (ARS)
(C) A centromere (CEN) sequence
(D) A telomere (TEL) sequence

141. **A DNA sample when it is diluted by a factor of 100 gives an optical density reading at 260 nm of 0.369. Assuming 50 μg/ml of DNA gives an absorbance reading of 1, the concentration of the original DNA sample is:**
(A) 36.9 μg/ml
(B) 1845 μg/ml
(C) 36900 μg/ml
(D) 184.5 μg/ml

142. **In a eukaryotic gene, one of the introns us mutated at the 3' end as shown:**
Normal : 5' ... TTTCCCACCCTTAG 3'
Normal : 5'.... TTTCCCACCCTTCG 3'
Which of the following processes is this mutation most likely to affect?
(A) Capping (B) Hybridization
(C) Polyadenylation (D) Splicing

143. **Which of the following is/are true?**
P. **RNA polymerase needs general transcription factors to elongate the message**
Q. **Tissue–specific transcription factors bind to TATA box**
R. **Generalized transcription factors bind to proximal upstream of the promoters**
S. **C-terminus of RNA polymerase II is an important control point**
(A) P only (B) R only
(C) Q and R (D) R and S

144. **Sequence tagged sites have which of the following properties?**
(A) They are present only one within a genome and possesses an RFLP site
(B) They are present only one within a genome and their sequence is known
(C) Their sequence is known and they must contain repetitive DNA sequences
(D) They must contain the sequences of a gene, and no repetitive DNA sequences can be present

145. **An alpha-helical conformation of a globular protein in solution is best determined by which of the following?**
(A) Ultraviolet-visible absorbance spectroscopy
(B) Fluorescence spectroscopy
(C) Electron microscopy
(D) Circular dichroism

146. **Which of the following reagents would be useful for labeling the oligodeoxyribo-nucleotide d(GGATATCC)?**
P. **[γ- ^{32}P] ATP**
Q. **Polynucleotide kinase**
R. **DNA-dependent RNA polymerase**
S. **DNA ligase**
(A) P and Q (B) R and S
(C) P and R (D) Q and S

147. **What is the principal function of reporter molecules in a DNA hybridization assay?**
(A) Enhances the stringency of the hybridization reaction
(B) Binds the target DNA to the solid support
(C) Aids in base pairing
(D) Aids in the detection of robe-target hybrids

148. Standard deviation of a data is 6. When observation is increased by 1, then the standard deviation of new data is:

(A) 5 (B) 7
(C) 6 (D) 8

149. Which of the following statements is incorrect?

(A) Most of the eukaryotic mRNAs have a 7-methylguanosine cap at their 5' end
(B) TATA binding protein is involved in the synthesis of mRNA, tRNA and rRNA
(C) Histones have no other function except in chromatin organization
(D) Eukaryotic RNA polymerase II consists of more than 4 subunits

150. One of the significant discoveries which allowed the development of recombinant DNA technology was:

(A) The discovery of antibiotics used for selecting transformed bacteria
(B) The identification and isolation of restriction endonucleases permitting specific DNA cutting
(C) The discovery of DNA and RNA polymerase allowing workers to synthesize any DNA sequence
(D) The development of the polymerase chain reaction

151. Choose the correct statement(s):

P. GrI intron show homing
Q. GrII intron show retrohoming
R. Many GrI and GrII introns have open reading frame
S. GrI intron acts as riboswitch

(A) P and Q (B) Only S
(C) P, Q and R (D) P and S

152. UV-spectroscopy is not used to quantitative the following compounds based on absorbance:

(A) Nucleic acids have absorbance peak at 260 nm
(B) NAD(P)H have absorbance peak at 340 nm
(C) Aromatic amino acids have absorption maxima at about 280 nm
(D) Sulfur containing amino acids have absorbance at 270 nm

153. A leader sequence in an mRNA of eukaryotes can be found:

(A) After the 'stop' codon
(B) Between transcriptional start site and translational start site
(C) Within the first exon
(D) In the exon-intron boundaries

154. If you were designing a method to specifically inhibits prokaryotic transcription, but not eukaryotic transcription, interfering with which of the following would work best?

(A) DNase activity
(B) Recognition of the promoter region
(C) An interalating agent
(D) Ribosomal binding

155. Khorana *et al.* synthesized RNA with copolymer of UGUGUGUGUG ...... It produced a peptide with alternatives cysteine and valine. The codons for the two are:

(A) UGU and GUG (B) UUG and GGU
(C) GUG and UGU (D) UGG and GUU

156. Electrophoretic separation of RNA molecules based on their molecular weight, requires a denaturant such as formaldehyde to be present in the gel system, because:

(A) RNAs are sensitive to nucleases
(B) RNAs are single stranded
(C) RNAs are usually small
(D) RNAs are usually present in high amounts

157. Large quantities of useful products can be produced through genetic engineering involving:

(A) Bacteria containing recombinant plasmids
(B) Yeast carrying foreign genes
(C) Transgenic plants
(D) All of these

158. Match items in group 1 with correct options from those given in group 2:

Group 1	Group 2
P. VNTR sequence	1. Gene regulation on the same chromosome
Q. Leader sequence	2. Ribosome binding site
R. SD sequence	3. DNA finger printing
S. Cis-acting sequence	4. Functions in attenuation

(A) P–3, Q–1, R–4, S–2 (B) P-2, Q–3, R–1, S–4
(C) P-3, Q–4, R–2, S–1 (D) P–3, Q–1, R–2, S–4

159. Four identical PCR reactions were carried out in tubes named I, II, III and IV. Besides the usual mix of dNTPs, each of the tubes respectively contained γ-^{32}P dATP, β-^{32}P dATP, α-^{32}P dATP and α -^{32}P rNTP. Which one of the tubes will have radiolabeled PCR product?

(A) Tube I (B) Tube II
(C) Tube III (D) Tube IV

160. What is the minimum length of time required for the synthesis of a 110 KDa protein in *E. coli*?

(A) About 96 sec (B) About 55 sec
(C) About 5 sec (D) More than 200 sec

161. With reference to protein synthesis, consider the following events:

P. Activation of amino acid
Q. Peptide bond formation
R. Formation of amino acid tRNA complex
S. Binding of aminoacyl tRNA complex with mRNA

What is the correct sequence of the above events in protein synthesis?

(A) P, R, Q, S (B) R, P, S, Q
(C) R, P, Q, S (D) P, R, S, D

162. Some compounds called anti-inducers bind to repressors such as the *lac* repressor and inhibit the action of inducers that is, transcription is repressed and higher concentrations of inducer are required to induce transcription. Anti-inducers
(A) Compete for binding to the repressor
(B) Compete for binding to the operator
(C) Compete for binding to the RNA pol
(D) Compete for binding to the activator

163. A viral genomic DNA that has two HindIII sites and two EcoR1 sites, was cut with both of the enzymes simultaneously for 5 hrs at 37^{0}C. The analysis of the restriction enzyme digestion was made of a 1% agarose gel. How many bands would one expect to see on the gel?
(A) 3 (B) 4
(C) 5 (D) 16

164. What is genomic library?
(A) A collection of recombinant molecules with inserts that contain all of the genes of an organism
(B) A collection of recombinant molecules with inserts that contain all an organism's genome
(C) A collection of recombinant molecules that express all of the genes of an organism
(D) A collection of recombinant molecules that have been sequenced

165. An enhancer element shows all of the following properties except:
(A) Acts as a binding site for transcription factors
(B) Acts as a binding site for RNA polymerase
(C) Is functional when located upsteam or downstream of the transcription initiation site
(D) Is functional when located in an intron

166. Choose the false statement:
(A) Peptidyl-tRNA binds in the P-site
(B) Shine-Dalgarno is a polypurine sequence that helps in ribosome binding
(C) Eukaryotic mRNAs are mostly monocistronic
(D) Translocation of ribosome requires ATP hydrolysis

167. Metaphase chromosomes were initially used for flourescent *in situ* hybridization, but the results were somewhat limiting because:
(A) Many regions of a chromosome are condensed and cannot hybridize to probes
(B) The probes will hybridize preferentially to repeated sequences present multiple chromosome
(C) The chromosomes are not stable in the condensed stable, and the signal diffuses when the chromosome are relaxed
(D) Only low-resolution mapping is possible, as the chromosomes are condensed

168. The coefficient of correlation r satisfies:
(A) $0 < r < 1$ (B) I Γ I >1
(C) I Γ I ≤1 (D) $-1 < r < 0$

169. A restriction enzyme with a four-base recognition site would leave DNA with a statistical frequency of one every:
(A) 256 base pairs (B) 4 base pairs
(C) 4096 base pairs (D) 65.5 kb

170. Simple tendem repeat, polymorphisms in humans are most useful for:
(A) Solving criminal and paternity cases
(B) Reconstructing the relationships of humans and chimps
(C) Estimating relationships of humans and Neanderthals
(D) Transferring disease resistance factors into bone marrow cells

171. The DNA mixture containing unlabelled and N^{15} labelled DNA from phage T4 was denatured and allowed to reanneal. How many bands one would observe upon $CsCl_2$ density gradient centrifugation of the above mixture?
(A) 0 (B) 1
(C) 2 (D) 3

172. Suppressor tRNA differ from normal tRNA in that:
(A) They are similar in size
(B) They do not carry an amino acid
(C) They have to be charged with a special aminoacyl-tRNA synthetase
(D) They do not obey regular genetic code

173. How is the latiat structure formed during splicing of a GU-AG intron?
(A) After cleavage of the 5' splice site, a new phosphodiester bond is formed between the 5' nucleotide and the 2' carbon of the nucleotide at the 3' splice site
(B) After cleavage of the 5' splice site, a new phosphodiester bond is formed between the 5' nucleotide and the 2' carbon of an internal adenosine
(C) After cleavage of the 3' splice site, a new phosphodiester bond is formed between the 5' nucleotide and the 2' carbon of the nucleotide at the 5' splice site
(D) After cleavage of the 3' splice site, a new phosphodiester bond is formed between the 5' nucleotide and the 2' carbon of an internal adenosine

174. Choose the mismatch:
(A) Pol I – Primer removal and DNA repair
(B) Pol III- DNA replication
(C) Pol V – Translesion replication
(D) Pol γ – Nuclear DNA replication

175. In *E. coli*, attenuation and antitermination utilize which structure?
(A) Stem loop structures in RNA
(B) Stem loop structures in DNA
(C) RNA/DNA hybrids
(D) Differential protein folding

176. A mixture containing two proteins having similar molecular mass but different oligomeric properties can be separated by:

(A) SDS-PAGE analysis
(B) Native PAGE analysis
(C) Isoelectric focusing
(D) Both (B) and (C)

177. During mismatch repair in *E. coli*, the parental strand is recognized by:

(A) Single-stranded breaks
(B) Glycosylated adenines
(C) Methylated adenines
(D) Methylation of the 6th position of guanine residues

178. Match the item in group I with correct options in group II

Group I	Group II
P. DNA footprinting	1. Protein-protein interaction
Q. Yeast two-hybrid system	2. VNTR
R. DNA fingerprinting	3. DNA binding protein
S. SAGE	4. Transcriptome analysis

(A) P-1, Q–2, R–4, S–3 (B) P-3, Q–1, R–2, S–4
(C) P-3, Q–4, R–1, S–2 (D) P-4, Q–2, R–1, S–3

179. Selenocystein:

(A) Only presents in prokaryotes
(B) Coded by UGA (stop codon)
(C) Derives from cysteine
(D) Forms post translationally

180. Which of the following could not possibly give rise to a restriction fragment length polymorphism (RFLP)?

(A) A missense mutation within the protein coding region of a gene
(B) A silent mutation within the protein coding region of a gene
(C) A single base change within the intron sequence of a gene
(D) An error in RNA splicing that mistakenly removes an exon during RNA processing

181. Histones have very high percentage of basic amino acids (15-30%). For this class of proteins which of the following reagents would be a suitable choice for generating peptide in determination of amino acid sequence of the protein:

(A) Cyanogen bromide
(B) Chymotrypsin
(C) Trypsin
(D) ß- bromosuccinamide

182. Restriction endonucleases from two different organisms that recognize the same DNA sequence for cleavage are called:

(A) Isoschizomers (B) Polymorphisms
(C) Concatemers (D) Palindromes

183. RNA interference works by which of the following methods?

(A) Using antisensce RNA molecules to block the translation of mRNA molecules
(B) Using RNA polymerase inhibitors to block the transcription of specific genes
(C) Using short, double-stranded RNA molecules that will cause the degradation of an mRNA molecule
(D) Using modified tRNA molecules to block the translation of mRNA molecules

184. In Scanning Electron Microscope (SEM), to form an image of the specimen?

(A) Electron should pass through the specimen
(B) Electrons are scattered from the surface of the specimen
(C) A thin film of heavy metal is evaporated
(D) Specimens are stained

185. siRNA interfere at:

(A) Transcriptional level
(B) Post-transcriptional level
(C) DNA replication level (D) Translational level

186. Which of the following techniques yields the most information about the positional and spatial characteristics of all atoms in a protein?

(A) Automated Edman analysis
(B) Circular dichroism spectroscopy
(C) Magnetic resonance spectroscopy
(D) Reversible unfolding/refolding

187. Which of the following is a role of gRNA?

(A) RNA editing (B) Self splicing
(C) Chemical modification of rRNA
(D) Polydenylation

188. In a bacterial cell, a mutation in an aminoacyl-tRNA synthetase leads to charging of the entire $tRNA^{Ser}$ population with alanine. Which of the following describes the result of using these aminoacyl tRNAs for protein synthesis in the cell?

(A) Protein synthesized using the alanyl-$tRNA^{Ser}$ will contain neither alanine nor serine
(B) Protein synthesized using the alanyl-$tRNA^{Ser}$ will contain only serine where alanine would normally occur
(C) Protein synthesized using the alanyl-$tRNA^{Ser}$ will contain only alanine where alanine would normally occur
(D) Protein synthesized using the alanyl-$tRNA^{Ser}$ will randomly contain either alanine or serine where serine would normally occur

189. Which is not correct about RILs?

(A) RILs are created by SSD through at least five generations of continued selfing
(B) Continued selfing yields a set of lines each of which contains a different combinations of linkage blocks from the original parents
(C) The different linkage blocks in RILs provide a basis from linkage analysis
(D) RILs are, theoretically, mortal

190. PCR based DNA amplification is an essential feature of which of the following combination of molecular markers?
(A) RLFP, AFLP and SSR
(B) AFLP, SSR and RAPD
(C) RFLP, RAPD and SSR
(D) RAPD, RFLP and SSR

191. Dominant markers supply as much information as codominant markers in :
(A) Doubled haploids
(B) One of the two backcross populatons in coupling phase
(C) RILs
(D) All of these

192. Which of the following characteristics of eukaryotic genes is also commonly true of most prokaryotic genes?
(A) Genes are packaged into nucleosomes in chromatin
(B) Many genes contain introns
(C) Usually only one gene is contained on a given mRNA
(D) The initiation codon for protein synthesis is usually AUG

193. An open reading frame is:
(A) All of the nucleotides of a gene that are transcribed into mRNA
(B) All nucleotides of a gene that make up the codons specifying amino acids
(C) The nucleotides of an mRNA molecule before the introns have been removed
(D) The amino acid sequence of a polypeptide

194. What is the order of the following steps in Western blotting?
P. Protein denaturation
Q. Hydrogen peroxide reduction
R. Primary antibody binding
S. Transfer onto membrane
(A) P, Q, R, S (B) Q, P, R, S
(C) P, Q, S, R (D) P, S, R, Q

195. Which of the following elements must be present in a plasmid cloning vector?
P. Origin of replication
Q. Selectable marker
R. Unique restriction sites
S. Promoter element upstream of the unique restriction site
(A) S and P (B) Q and R
(C) R and S (D) P, Q and R

196. Aminoacyl-tRNA synthetases must be capable of recognizing which of the following:
(A) A specific tRNA and a specific amino acid
(B) A specific rRNA and a specific amino acid
(C) A specific tRNA and the 40S ribosomal subunit
(D) A specific amino acid and the 40S ribosomal subunit

197. Which of the following statement are incorrect?
P. The frequency of recombination is a measure of linkage between genes on the same chromosome
Q. DNA polymerase I is the true DNA replicase in *E. coli*
R. The conserved element closest to the transcription initiation site is called the CAAT box
S. The introns in the nuclear pre-mRNAs are excised by spliceosomes
(A) P and Q (B) Q and R
(C) P and R (D) P and S

198. RFLP analysis is a technique that:
(A) Uses hybridization to detect specific DNA restriction fragments in genomic DNA
(B) Is used to determine whether a gene is transcribed in specific cells
(C) Measures the transfer frequency of genes during conjugation
(D) Is used to detect genetic variation at the protein level

199. In *E. coli*, which of the following codons are reorganized by the release factor RF1?
(A) UAG and UGA (B) UAA and UGG
(C) UAG and UAA (D) UAG and UUA

200. Which of the following statements is/are correct about leucine zipper motif?
P. A DNA binding motif
Q. Homodimer or heterodimer
R. Present in transcription factors-Fos and Jun
S. Present in steroid recepters
(A) P only (B) P, Q and R
(C) Q, R and S and R (D) P and S

201. Which one of the following transcription factor is involved in promoter recognition in eukaryotes?
(A) TFIIF (B) TFIIB
(C) TFIID (D) TFIIE

202. Which of the following is not correct about freeze-fracture technique?
(A) It involves physical breaking of frozen biological samples
(B) Fixation in glutaraldehyde
(C) Involves vacuum sublimation of ice
(D) Cryoprotection with glycerol

203. The chronological order in which the following landmark discoveries in molecular biology were made is:
(A) Recombinant DNA technology, DNA sequencing, polymerase chain reaction
(B) DNA sequencing, recombinant DNA technology, polymerase chain reaction
(C) Polymerase chain reaction, DNA sequencing, Recombinant DNA technology,
(D) Polymerase chain reaction, Recombinant DNA technology, DNA sequencing,

204. When change in the phenotype or gene expression occur without changes in the underlying DNA sequence, the phenomenon is called:

(A) Mutation (B) Eugenics
(C) Epigenetics (D) Epistasis

205. Klenow fragment is a key enzyme in molecular biology research. Which one of the following statements is not correct for the Klenow fragment isolated from *E. coli?*

(A) Klenow fragment is a constituent of DNA polymerase I
(B) Klenow fragment has DNA polymerase activity
(C) Klenow fragment has 3' to 5' exonuclease activity but it devoid the 5' to 3' exonulease activity
(D) While Klenow fragment is primarily asso-ciated with DNA repair function, DNA polymerase I is essential for DNA replication

206. Edmann degradation is used for the:

(A) Determination of amino acid sequence from the N-terminus of a protein
(B) Nucleotide sequence of DNA
(C) Determination of amino acid sequence from the C–terminus of a protein
(D) Determination of RNA structure

207. On average, how many fragments would a restriction enzyme which recognizes a specific 4 base sequence in DNA be expected to generate by leaving a double-stranded bacteriophag DNA with a genome size of 5000 bp into?

(A) About 2 (B) About 4
(C) About 20 (D) About 50

208. Match the terms in group I with terms in group II

Group I	Group II
P. RNase-P	1. Lac operon
Q. Leucine zipper	2. rRNA gene transcription
R. RNA pol-I	3. tRNA gene transcription
S. Attenuation	4. Transcription factors
	5. Ribozymes
	6. trpoperon
	7. mRNA splicing

(A) P-7, Q–5, R–3, S-1 (B) P-4, Q–5, R–2, S-1
(C) P-5, Q–4, R–2, S-6 (D) P-4, Q–5, R–3, S–6

209. The large (Klenow) fragment of *E. coli* DNA polymerase I contains which of the following activities?

(A) Reverse transcriptase activity and nick-translation activity
(B) Polymerase activity and 3'-5'- exonuclease activity
(C) Polymerase activity and nick-translation activity
(D) 5' – 3'- exonuclease activity

210. Approximate molecular weight (KDa) of the product after translation of a 390 bases mRNA will be:

(A) 48 (B) 26
(C) 39 (D) 14

211. The catabolite activating protein (CAP):

(A) Binds to allolactose
(B) Interacts with β–subunits of RNA polymerase
(C) Binds the operator sequence of the la operon
(D) Causes a bending in the DNA near the start site of transcription

212. Which of the following statements is false?

(A) β-mercapteothanol acts as reducing agent and used to break disulfide bonds
(B) Bromophenol blue used as tracking dye during electrophoresis
(C) Coomassie brilliant blue used for staining proteins
(D) Sodium dodecyl sulphate is a cationic detergent

213. The yeast two-hybrid system is designated to identify which of the following?

(A) All of the components of a multiprotein complex
(B) Human proteins that are required for binding RNA polymerase
(C) Two proteins that directly interact with one another
(D) Two proteins that are involved in the same metabolic pathway

214. What phenotype would you except for a mutation that eliminates the tetramerization domain of *lac* repressor:

(A) Recessive, inducible
(B) Recessive, constitutive
(C) Dominant, inducible
(D) Dominant, constitutive

215. Wobble between a codon and an anticodon occurs between:

(A) The first nucleotide of the codon and the first nucleotide of the anticodon
(B) The first nucleotide of the codon and the third nucleotide of the anticodon
(C) The third nucleotide of the codon and the first nucleotide of the anticodon
(D) The third nucleotide of the codon and the third nucleotide of the anticodon

216. A homogeneous protein of native molecular weight 100000 gave a single band of molecular weight 50000 on SDS-PAGE in presence of β–mercapto-ethanol. N-terminal analysis gave two amino acids alanine and leucine in equal proportions. Hence the:

(A) Protein is a homodimer
(B) Protein is contaminated with another protein
(C) Protein has two polypeptides linked by disulfide bridges
(D) None of these

217. Antibiotic inhibiting interaction between tRNA and mRNA during protein synthesis in bacteria is:

(A) Tetracycline (B) Neomycin
(C) Erythromycin (D) Streptomycin

218. Which of the following is characteristics of eukaryotic transcription but not prokaryotic transcriptions?
(A) Transcription and translation are coupled
(B) There is one form of RNA polymerase
(C) RNA polymerase II is responsible for synthesis of mRNA
(D) RNA can be polycistronic

219. SDS-polyacrylamide gel electrophoresis and the isoelectric-focusing method for the separation of protein have which of the following characteristics in common? Both
(A) Separate native proteins
(B) Separate proteins according to their mass
(C) Required a pH gradient
(D) None of these

220. Microsatellites are used more commonly than minisatellite as DNA markers because:
(A) Minisatellites are present in too many locations within the genomes
(B) Restriction enzymes can be used to type microsatellites but not minisatellite
(C) There are very few microsatellites in eukaryotic genome so they are easily identified and analyzed
(D) Microsatellites are present throughout eukaryotic genomes and are easily amplified using PCR

221. In eukaryotic transcription by RNA polymerase II, formation of a pre-initiation complex:
(A) Begins with the binding of TBP to the TATA box of the promoter
(B) Involves the ordered addition of several transcription factors and the RNA polymerase
(C) Allows an ATP-dependent opening of the two strands of DNA
(D) All of these

222. If the *E. coli* cells are grown in a medium containing both glucose and lactose, what is likely to happen?
(A) Both the sugars would be utilized simultaneously
(B) The culture will exhibit synchronous growth
(C) Lactose will be utilized first followed by glucose
(D) Glucose will be utilized first followed by lactose

223. To join together two Okazaki fragments during DNA replication in *E. coli*, which of the following sets of enzymes are used?
(A)Pol I (5' → 3' exonuclease), Pol I (polymerase), ligase
(B) Pol III (5'→ 3' exonuclease), Pol III (polymerase), ligase
(C) DNA gyrase, Pol III (polymerase), ligase
(D) SSB, Pol I, ligase

224. Which of the following is true about poly (A) tail?
P. It is synthesized post-transcriptionally
Q. It is usually shorter than 300 nt
R. One of its functions is to protect the mRNA from cytoplasmic RNases
S. One of its functions is to stimulate translation
(A) P only (B) Q and R
(C) Q and S (D) P, Q, R and S

225. In vertebrate genes, transcription regulatory regions that contain CpG islands are inactivated by which CpG modification?
(A) Methylation (B) Myristylation
(C) Phosphorylation (D) Acetylation

226. A primer with the sequence ATCGGTAACGATTACATTC is to be used in a PCR reaction. Using the formula: $T_m = 2^0 (A + T) + 4^0 (G + C)$, what is the T_m of this oligonucleotide?
(A) 42ºC (B) 52ºC
(C) 20ºC (D) 62ºC

227. What is the purpose of performing a homology search with a DNA sequence
(A) To determine if any genes with similar sequences are present in the DNA databases
(B) If the sequence is already in the DNA database
(C) To search for consensus exon-intron boundaries
(D) To determine the codon bias for a specific gene

228. A specific aminoacyl-tRNA always carries the same amino acid that is attached by a specific amioacyl-tRNA synthetase. Regions on tRNA that are required for the specificity of the enzyme is/are the:
(A) Anticodon and aceptor region
(B) D loop and acceptor region
(C) Anticodon alone
(D) The TΨC loop and anticodon

229. Which of the following is required for the initiation of translation in prokaryotes?
(A) The mRNA base pairing with the 3' end of 23S rRNA
(B) The Kozak sequence positioning
(C) Met –tRNA binding to the A site on the ribosome
(D) GTP bound to initiation fator 2 (IF 2)

230. Gene conservation is observed when:
(A) During meiosis, one allele is replaced by another allele
(B) During meiosis, the expected 2:2 ratio of alleles is replaced with a 4:0 ratio
(C) During meiosis, the expected 2:2 ratio of alleles is replaced with a 3 : 1 ratio
(D) All of these

231. Which of the following reagents would be useful for visualizing DNA restriction fragments that have been separated by electrophoresis in an agarose gel and remain in the wet gel?
(A) Ethidium bromide (B) [α -^{32}P] ATP
(C) Diphenylamine (D) $^{32}P_i$

232. If a mutation occurred in the phosphorylation site in the CTD domain of RNA polymerase II, which of the following outcomes would be observed?
(A) Transcription of tRNA would occur normally
(B) Transcription of mRNA would not occur
(C) Transcription of precursor rRNA would not occur
(D) All of these

233. How do eukaryotic ribosomes identify the start sites of proteins?
(A) Via their Shine-Dalgarno sequences
(B) By scanning from the 5' end of the mRNA for the first AUG
(C) The 'cap' structure of a eukaryotic mRNA overlaps with the first AUG
(D) The transcription start site of a eukaryotic mRNA overlaps with the translational start site

234. After isolating and purifying to homogeneity a small enzyme (110 amino acids long) from a culture of bacteria, you are confused as to whether you grew wild-type bacteria or a mutant strain that produced the enzyme with a valine residue at position 66 instead of the glycine found in the wild-type strain. For quick determination of nature of protein you will use:
(A) Mass spectroscopy
(B) Ion exchange chromatography
(C) SDS-PAGE (D) HPLC

235. A mixture of proteins (W, X, Y, Z) elute from Sephadex G-200 column in the order W, X, Y, Z. The protein with maximum electrophoretic mobility on SDS-PAGE will be:
(A) W (B) X
(C) Y (D) Z

236. Which of the following characteristics is undesirable in cloning vectors?
(A) Readily isolated from cells
(B) Higher copy number
(C) Vulnerable at several sites to a restriction enzyme
(D) Small in size

237. SmaI and XmaI are neoshizomers. A circular DNA has two SmaI and one XmaI sites. How many fragments will be generated by digestion with XmaI?
(A) 1 (B) 2
(C) 3 (D) 4

238. Which of the following techniques, based on migration of DNA fragments in a gel in the presence or absence of proteins, is used to identify proteins that bind to DNA?
(A) Nuclear magnetic resonance spectroscopy
(B) Gel retardation
(C) Nuclease protection
(D) DNA footprinting

239. Which is not correct about marker-assisted selection?
(A) Use of linked molecular markers would allow selection for desirable traits in early segregating generations at the seedling stage independent of environment influences
(B) Marker-assisted selection saves time and other resources needed for raising large segregating populations for several generations
(C) Indirect selection is expected to be less effective than the direct selection due to the absence of confounding effects of the environment in the case of latter
(D) Indirect selection allows pyramiding (accumulation) of genes for characters like disease resistance, which is difficult through the use of conventional methods of plant breeding

240. A set of genes from *Bacillus subtillis* that encode the proteins required for sporulation have conserved DNA sequenes-35 and -10 nucleotides before the site of transcript initiation, although the sequence at -35 is different from that seen in most other genes from that species. Which of the following best explains this differences?
(A) A noval sigma factor is required for transcription initiation at these genes
(B) The -35 sequene is the binding site for a repressor of transcription
(C) The replication of these genes requires a specifically modified DNA polymerase
(D) Translation of the mRNAs transcribed from these genes requires specific ribosomes that recognize a modified shine-Dalgarno sequence

241. It is the early 1960s. In an attempt to elucidate the genetic code you have performed a series of experiments similar to those of Khorana. You have prepared synthetic mRNA & used *E. coli* cell extracts to translate them into corresponding poly-pepetides. You have then characterized the amino acid composition of the protein. Your results are as follows:

mRNA	Amino acid components
5'UUUUUUUUU....3'	Phenylalanine
5'AUAUAUAUA....3'	Isoleucine, tyrosine
5'UAUUAUUA ....3'	Asparagine, Isoleucine
5' AAAAAAAAA...3'	Lysine
5' GUAGUAGUA ...3'	Serine, Valine
5'AUUAUUAUU ...3'	Isoleucine, Leucine, Tyrosine

Based on the above information, the codon for isoleucine is:
(A) AUA (B) UAA
(C) AUU (D) UAU

242. All of the following cell types contain the enzyme telomerase which protects the length of telomeres at the end of chromosomes except:
(A) Germinal (B) Somatic
(C) Haemopoetic (D) Tumour

243. Enhancer elements are known to increase the rate of transcription when present at the upstream side of the promoter sequences. If the same enhancer element is placed in reverse orientation:
(A) The rate of transcription increases as compared to upstream side
(B) The rate of transcription decreases as compared to upstream side
(C) The rate of transcription remains same as compared to upstream side
(D) No transcription is observed

244. How many loci can be detected from a single cross through AFLPs?
(A) >25 (B) > 400
(C) >70 (D) > 1000

245. Codon bias refers to which of the following?
(A) Some codons for an amino acids are more frequently used in all species
(B) Acids are rarely used in the proteins of some organisms
(C) Some codons for an amino acids are more frequently used and the bias varies in different species
(D) Some codons code for rare amino acids such as salenocysteine in some species

246. Which of the statement are correct? Spliceosomes
P. Are composed of RNA and protein
Q. Recognize RNA sequences that signal for removal of introns
R. Can produce different mRNA molecules by splicing at alternate site
(A) P and Q (B) P and R
(C) Q and R (D) P, Q and R

247. The single primer amplification reaction (SPAR) is also called as:
(A) MP-PCR (B) RAMP
(C) SSLP (D) All of these

248. The Shine-Dalgarno sequence is responsible for:
(A) Binding of RNA polymerase to gene during transcription
(B) Binding of DNA polymerases to origin of replication during DNA replication
(C) Binding of ribosomes to mRNA during initiation of translation
(D) Binding of snurps during splicing

249. Which is not correct about sequence tagged microsatellite sites (STMS)?
(A) STMS markers are dominant
(B) These markers are extremely polymorphic and highly informative
(C) The most important uses for STMS markers are tagging genes for economic traits and for the study of gentic diversity
(D) STMS describes such microsatellite based markers that are analysed by PCR amplification of a short genomic region containing the repeated sequence

250. A plasmid that will express a eukaryotic protein x in *E. coli* should contain:
P. Origin of replication
Q. Bacterial promoter
R. Mammalian promoter
S. Genomic sequence of gene
T. Genomic sequence of gene x
U. DNA sequence of gene x
(A) Q, S and T (B) P, R, S and T
(C) P, R, S and U (D) P, Q, S and U

251. X-ray crystallography are:
P. Based on the elastic scattering of X-rays.
Q. Non-destructive analytical techniques
R. Used in determining the arrangement of atoms within a crystal
S. Based on Bragg's law
(A) P and Q (B) R and S
(C) P, Q and R (D) P, Q, R and S

252. Sequence Tagged Microsatellite Sites (STMS) can be used as markers for:
(A) Identification of germplasm and analysis of gene pool varieties
(B) Cultivar identification and determination of hybrid purity
(C) Linkage mapping and diagnostic markers for traits of economic value
(D) All of these

253. Which of the following event does not occur during pre-tRNA processing?
(A) 5' end cleavage by the RNaseP
(B) Addition of poly(A) site
(C) Addition of A sequence at 3' end
(D) Chemical modification of bases

254. Proteins are separated in SDS-electrophoresis on the basis of their:
(A) Molecular size (B) Charge
(C) Amino acid composition
(D) N-terminal sequence

255. An RNA-dependent RNA polymerase is likely to be present in the virion of a:
(A) DNA virus that multiplies in the cytoplasm
(B) DNA virus that multiplies in the nucleus
(C) Minus-strand RNA virus
(D) Transforming virus

256. Which is not correct about application of molecular markers?
(A) Molecular markers are very useful in unequivocal identification of plant varieties
(B) Molecular markers can not be used for elimination of undesirable genes from segregating population
(C) Molecular markers are used for mapping of QTLs
(D) They allow construction of high density linkage maps since the number of markers available per cross is very large

257. The predictive value of molecular markers used in marker-assisted selection depends on :

(A) Their inherent repeatability
(B) Map position
(C) Linkage with economically important traits
(D) All of these

258. During DNA replication significant proportions of newly synthesized DNA in the lagging strand exists as small Okazaki fragments. The size of these units in bacteria are approximately:

(A) 100 nucleotides
(B) 1000 nucleotides
(C) 100 base pairs
(D) 1000 base pairs

259. What is the correct order of the following steps in protein synthesis?

1. A peptide bond is formed
2. The small ribosomal subunit is loaded with initiation factors, mRNA and initiation aminoacyl-tRNA
3. The intact ribosome slides forward three bases to read a new codon
4. The primed small ribosomal subunit binds with the large ribosomal subunit
5. Elongation factors deliver aminoacyl-tRNA to bind to the A site

(A) 1, 2, 5, 4, 3 (B) 2, 3, 4, 5, 1
(C) 4, 5, 1, 2, 3 (D) 2, 4, 1, 3, 5

260. Three of the four eukaryotic rRNAs are synthesized from a single transcription unit consisting of the rDNA which one of the following does not belong to this group?

(A) 5.8S (B) 5S
(C) 18S (D) 28S

261. A solution contains DNA polymerase I, Mg^{++} salts of dATP, dGTP, dCTP and dTTP, and an appropriate buffer. Which of the following DNA molecules would serve as a template for DNA synthesis when added to this solution?

(A) A single –stranded closed circle
(B) A single-stranded closed circle base-paired to a shorter strand with a 3-terminal hydroxyl
(C) A single-stranded closed circle base-paired to a shorter strand with a 3-terminal phosphate
(D) A double-stranded closed circle

262. The suitability of RFLP for elucidation of QTLs has been shown in :

(A) Tomato (B) Maize
(C) Potato (D) All of these

263. Which of the most critical in Marker-Assisted Selection (MAS)?

(A) Their inherent repeatability
(B) Map position
(C) Linkage with economically important traits
(D) All of these

264. Which of the following is not a difference between a genomic library and a cDNA library?

(A) The presence or absence of introns in cloned genes
(B) The presence or absence of promoters in cloned genes
(C) The presence or absence of a prokaryotic origin of replication on the vector
(D) The presence or absence of DNA from intergenic regions

265. A multiple cloning site:

(A) Contains many copies of a cloned gene
(B) Allows flexibility in the choice of restriction enzymes for cloning
(C) Allows flexibility in the choice of organism for cloning
(D) Contains many copies of the same restriction enzyme site

266. Apart from the 20 naturally occurring amino acids, 700 unnatural amino acids have been reported in the literature. It all the 720 were to be accommodated in the genetic code, what is the minimum number of base needed in a codon:

(A) 4 (B) 5
(C) 6 (D) 7

267. Codominant markers, such as RFLPs, are useful for:

(A) Marker-assisted selection
(B) Evolutionary studies
(C) Linkage mapping
(D) All of these

268. Saturated molecular linkage maps have been developed for :

(A) Potato (B) Tomato
(C) Maize (D) All of these

269. What of the following are required to be present in a plasmid, for it to be useful in a cloning experiment?

P. Multiple cloning site (polylinker)
Q. Antibiotic- resistance gene
R. Beta-galactosidase gene
S. Origin of replication

(A) P and Q (B) P, Q and R
(C) P, Q and S (D) P, Q, R and S

270. Plasmid A and plasmid B were digested with BamHI and analysed by agarose gel electrophoresis. If plasmid A gave two fragments and plasmid B gave three fragments, then which of the following inferences are correct?

P. Plasmid A has three sites and is circular
Q. Plasmid B has three sites and is circular
R. Plasmid A has two sites and is linear
S. Plasmid B has two sites and is linear

(A) P and Q (B) Q and R
(C) P and S (D) Q and S

271. The saturated molecular linkage maps initially developed in tomato. Potato and maize used which of the following molecular marker systems?
(A) AFLP (B) RFLP
(C) SNP (D) DAF

272. Which of the following is false about DNA pol V of *E. coli*?
(A) Belongs to Y-family of DNA polymerase
(B) Catalyzes translesion of DNA synthesis
(C) Has 5' → 3' exonuclease activity
(D) Catalyzes error-prone replication

273. The first strategy used for molecular mapping was based on:
(A) Recombinant inbred lines
(B) Bulked segregant analysis
(C) Near-isogenic lines
(D) F2 population

274. A researcher is purifying a mixture of 60 kDa (P60) and 40 kDa (P40) polypeptides through a Sephadex G-100 column. What will be the fate of these two polypeptides when they pass through the column?
(A) P40 will run faster
(B) P60 will run faster
(C) Always a 50% mixture of p40 and P60 will come out of the column
(D) P40 and P60 will come out in a ratio of 3 : 2 at any given time.

275. Development of which of the following populations requires the least amount of effort?
(A) F_2 (B) Backcross
(C) RILs (D) NILs

276. A family of closely related proteins (interferons) produced by leucocytes in response to virus infection is:
(A) Interfron α (B) Interferon β
(C) Interferon γ (D) Interferon ω

277. When DNA molecules of same density but different masses are to be separated by:
(A) Velocity sedimentation
(B) Equilibrium sedimentation
(C) Gel Electrophoretic separation
(D) Centrifugal force

278. When DNA molecules of different density but same or different masses are to be separated by:
(A) Velocity sedimentation
(B) Equilibrium sedimentation
(C) Gel Electrophoretic separation
(D) Centrifugal force

279. Separation of charged molecules under an electric field by:
(A) Velocity sedimentation
(B) Equilibrium sedimentation
(C) Gel Electrophoretic separation
(D) Centrifugal force

280. The first cloning of a gene and its expression in a foreign organism were achieved in:
(A) 1971 (B) 1972
(C) 1973 (D) 1974

281. The markers are available for comparative determination of sizes of single stranded RNA molecules is
(A) RNA markers (B) DNA markers
(C) Genetic markers (D) DNA size markers

282. Since many of the viral DNA genomes can easily be isolated in intact form, these molecules cleaved with a specific restriction endonuclease have been popular as:
(A) RNA markers (B) DNA markers
(C) Genetic markers (D) DNA size markers

283. A specific location on a chromosome that is defined by naked eye polymorphism, protein or DNA sequence; the inheritance of a genetic marker can be followed in a mapping population is known as:
(A) RNA markers (B) DNA markers
(C) Genetic markers (D) DNA size markers

284. Restriction endonuclease digest of plasmid DNA is also used as:
(A) RNA markers
(B) DNA markers
(C) Genetic markers
(D) DNA size markers

285. The DNA fragments obtained by cleaving the genome of the bacterial virus lambda (λ) with the restriction endonuclease Hind III are very often used for determining the sizes of DNA fragments between:
(A) 2 and 20 kb (B) 700 base to 8 kb
(C) 200 bases to 8 kb (D) None of these

286. The same genome of the virus lambda (λ) when cleaved with the restriction endonuclease BstE II can be used as DNA markers between the size of:
(A) 2 and 20 kb (B) 700 base to 8 kb
(C) 200 bases to 8 kb (D) None of these

287. Fragments obtained by cleaving the genome of the bacteria; virus Ø x 174 with the restriction endonuclease Hae III are used as marker for DNA fragments between:
(A) 2 and 20 kb (B) 700 base to 8 kb
(C) 200 bases to 8 kb (D) None of these

288. The first reported purification of chromosomal DNA was in:
(A) 1943 (B) 1944
(C) 1945 (D) 1946

289. Plasmid DNA can be purified by -------- in a cesium chloride-ethidium bromide gradient is:
(A) Equilibrium centrifugation
(B) Centrifugal force
(C) Both (A) and (B)
(D) None of these

290. There are many genetic engineering operations in which very pure plasmid DNA is:
(A) Required (B) Not required
(C) Unknown (D) Known but doubtful

291. Once the cell lysate is obtained, RNA is isolated through steps similar to that os:
(A) DNA isolation (B) DNA purification
(C) DNA looping (D) DNA library

292. Ribosomal RNA constitutes 80-85% of total cellular RNA, transfer RNA and small nuclear RNA (in case of eukaryotes) constitute about 10-15%, while messenger RNA constitutes only about:
(A) 1-5% (B) 5-10%
(C) 10-15% (D) 15-20%

293. For most genetic engineering experiments which is most important:
(A) Messenger RNA (B) Messenger DNA
(C) D-RNA (D) cDNA

294. Bacterial messenger RNA (mRNA) can be separated from ribosomal and transfer RNAs by :
(A) Velocity sucrose density-gradient centrifugation
(B) Gel electrophoresis
(C) Both (A) and (B)
(D) None of these

295. Method for purification of eukaryotic mRNA was developed by Phillip Leder in:
(A) 1971 (B) 1972
(C) 1973 (D) 1974

296. An important requirement in gene cloning is:
(A) Vecter (B) Vehicle
(C) Both of these (D) None of these

297. For gene cloning which following has been used :
(A) Plasmids (B) Viral DNA
(C) Both (A) and (B) (D) None of these

298. The foreign DNA is then spliced to the DNA from the plasmid or virus by using the enzyme is:
(A) Plasmids (B) Viral DNA
(C) DNA ligage (D) Cloning vector

299. The vecter containing the ligated foreign DNA is then introduced into the recipient organism by:
(A) Transformation (B) Transfection
(C) Both (A) and (B) (D) None of these

300. Derivatives of the genome of bacteriophage lambda have been constricted to serve as:
(A) Plasmids (B) Viral DNA
(C) DNA ligage (D) Cloning vector

301. Yeast one-hybrid system is used for cloning genes encoding a DNA-binding protein specific for a particular:
(A) DNA sequence
(B) DNA – binding Domain
(C) Yeast-two hybrid system
(D) Yeast three-hybrid system

302. Yeast two hybrid system is based on the observation that most proteins involved in promoting transcription of a gene (transcription factors) contain two distinct parts or:
(A) DNA sequence
(B) DNA – binding Domain
(C) Yeast-two hybrid system
(D) Yeast three-hybrid system

303. A protein interacting with a specific DNA sequence, commonly known as DNA binding protein, can be studied and the DNA encoding it can be cloned using a method known as:
(A) Yeast one hybrid system
(B) DNA – binding Domain
(C) Yeast two- hybrid system
(D) Yeast three-hybrid system

304. When two proteins interact with each other, their corresponding genes are known as interacting genes and can be cloned using a method known as the:
(A) DNA sequence
(B) DNA – binding Domain
(C) Yeast two- hybrid system
(D) Yeast three-hybrid system

305. Interaction between protein and RNA molecules can be investigated using a method known as:
(A) DNA sequence
(B) DNA – binding Domain
(C) Yeast-two hybrid system
(D) Yeast three-hybrid system

306. The genomic genes are often much larger that the:
(A) cDNA (B) mRNA
(C) rRNA (D) cRNA

307. To identify particular locations in the genome by:
(A) Molecular markers (B) Genomes
(C) Selectable marker (D) Scorable marker

308. Molecular markers rare:
(A) Based on nucleic acid hybridization (mostly DNA-DNA)
(B) Amplification based (PCR based)
(C) Both (A) and (B)
(D) None of these

309. The human genome contains:
(A) 10000-20000 genes
(B) 20000-30000 genes
(C) 30000-40000 genes
(D) More than 40000 genes

310. Each base in the genome I, therefore, a potential mutable unit. These alterations if present is less than 1% of the population are called:
(A) Mutations (B) Polymorphisms
(C) Gene (D) None of these

311. Structural unit of inheritance in living organisms is:
(A) Mutations (B) Polymorphisms
(C) Gene (D) None of these

312. Each base in the genome is, therefore, a potential mutable unit. These alterations of the frequency is higher, they are called:
(A) Mutations (B) Polymorphisms
(C) Gene (D) None of these

313. The sequence of the -------- billion chemical base pairs that make up human DNA:
(A) 2 (B) 3
(C) 4 (D) 5

314. Genes govern the functions of the cells, and thus of the entire organism:
(A) Structure (B) Metabolic
(C) Both (A) and (B) (D) None of these

315. Molecular markers identify particular location in the:
(A) Genome (B) Gene
(C) DNA (D) All of these

316. Molecular markers based on:
(A) Nucleic acid hybridization (mostly DNA-DNA)
(B) Amplification based (PCR based)
(C) Both (A) and (B)
(D) None of these

317. The molecular markers are based on the hybridization of a cloned piece of DNA to total genomic is:
(A) Genome (B) Gene
(C) DNA (D) All of these

318. Which of the following is a dominant marker?
(A) RAPD (B) RFLP
(C) SSR (D) AFLP

319. High density molecular map can be developed by:
(A) RFLP (B) RAPD
(C) AFLP (D) SNPs

320. For molecular characterization of plant species which of the following molecular markers are used?
(A) RAPD (B) AFLP
(C) SSR/STMS (D) All of these

321. Microsatellites are also known as:
(A) VNTRs (B) RFLPs
(C) STRs (D) RAPDs

322. Minisatellites are also known as:
(A) VNTRs (B) RFLPs
(C) STRs (D) RAPDs

323. RAPD molecular markers are of which nature?
(A) Dominant (B) Co-dominant
(C) Recessive (D) Neutral

324. AFLP molecular markers are of which nature?
(A) Dominant (B) Co-dominant
(C) Recessive (D) Neutral

325. Which of the following technique is used for DNA finger printing?
(A) HPLC (B) RFLP
(C) TLC (D) ELISA

326. Which of the following technique is used for amplification of DNA?
(A) HPLC (B) PCR
(C) RFLP (D) ELISA

327. Marker free selection can be done by:
(A) Co-transformation (B) Deletion
(C) Insertion (D) None of these

328. The molecular scissors used to cut DNA into specific genes of interest are called:
(A) Polymerases (B) Restriction endonucleases
(C) Ligases (D) None of these

329. Antibiotic resistance is used as a selectable marker for:
(A) Selection of vector
(B) Selection of recombinant vectors
(C) Selection of transformed cells
(D) Selection of resistant bacteria

330. Taq polymerase used in PCR is:
(A) Thermotolerant
(B) High progressive enzyme
(C) Osmotolerant
(D) All of these

331. Development of high-density molecular linkage map of a particular mapping population can be used for:
(A) MAS (B) Map based cloning
(C) Species characterization
(D) All of these

332. The central dogma of molecular genetics was proposed by:
(A) Mendel (B) Mmonad
(C) Khorana (D) Crick

333. The following amongst the molecular markers in codominant in nature:
(A) RAPD (B) SSR and RFLP
(C) VNTR (D) None of these

334. Genomic imprinting occurs in:
(A) Plants (B) Animals
(C) Humans (D) All of these

335. If the DNA of a species has the mole fraction of G + C = 0.48, the mole fraction of A will be:
(A) 0.26 (B) 0.20
(C) 0.48 (D) 0.52

336. A virus free clone from a virus infected plant can be obtained by:
(A) Tissue (B) Leaf
(C) Stem (D) Meristem culture

337. Chemical knives/molecular scissors of DNA are:
(A) Restriction endonucleases
(B) Polymerases
(C) Ligases
(D) Transcriptases

ANSWERSHEET

13. MOLECULAR MARKERS

1. (A)	2. (B)	3. (C)	4. (D)	5. (B)	6. (A)	7. (B)	8. (C)	9. (D)	10. (B)
11. (C)	12. (D)	13. (A)	14. (C)	15. (D)	16. (A)	17. (A)	18. (B)	19. (C)	20. (D)
21. (A)	22. (B)	23. (C)	24. (B)	25. (C)	26. (A)	27. (B)	28. (C)	29. (D)	30. (A)
31. (B)	32. (C)	33. (D)	34. (B)	35. (A)	36. (C)	37. (D)	38. (A)	39. (B)	40. (D)
41. (D)	42. (D)	43. (C)	44. (D)	45. (C)	46. (C)	47. (D)	48. (D)	49. (C)	50. (D)
51. (C)	52. (B)	53. (A)	54. (B)	55. (D)	56. (D)	57. (C)	58. (D)	59. (C)	60. (D)
61. (D)	62. (B)	63. (A)	64. (A)	65. (D)	66. (A)	67. (A)	68. (D)	69. (B)	70. (A)
71. (D)	72. (A)	73. (B)	74. (A)	75. (D)	76. (A)	77. (B)	78. (A)	79. (A)	80. (C)
81. (C)	82. (C)	83. (B)	84. (D)	85. (B)	86. (C)	87. (B)	88. (D)	89. (B)	90. (C)
91. (C)	92. (D)	93. (B)	94. (B)	95. (D)	96. (C)	97. (A)	98. (C)	99. (D)	100. (A)
101. (C)	102. (C)	103. (D)	104. (D)	105. (B)	106. (B)	107. (D)	108. (B)	109. (B)	110. (D)
111. (D)	112. (D)	113. (C)	114. (B)	115. (D)	116. (D)	117. (A)	118. (C)	119. (B)	120. (B)
121. (C)	122. (A)	123. (C)	124. (A)	125. (A)	126. (C)	127. (D)	128. (B)	129. (A)	130. (A)
131. (D)	132. (C)	133. (D)	134. (B)	135. (A)	136. (A)	137. (B)	138. (B)	139. (A)	140. (A)
141. (D)	142. (D)	143. (D)	144. (B)	145. (D)	146. (A)	147. (D)	148. (C)	149. (C)	150. (B)
151. (C)	152. (D)	153. (B)	154. (B)	155. (A)	156. (B)	157. (D)	158. (C)	159. (C)	160. (B)
161. (D)	162. (A)	163. (C)	164. (B)	165. (B)	166. (D)	167. (D)	168. (C)	169. (A)	170. (A)
171. (D)	172. (D)	173. (B)	174. (D)	175. (A)	176. (D)	177. (C)	178. (B)	179. (B)	180. (D)
181. (C)	182. (A)	183. (A)	184. (B)	185. (B)	186. (C)	187. (A)	188. (C)	189. (D)	190. (B)
191. (D)	192. (D)	193. (B)	194. (D)	195. (D)	196. (A)	197. (B)	198. (A)	199. (C)	200. (B)
201. (C)	202. (C)	203. (A)	204. (C)	205. (D)	206. (A)	207. (C)	208. (C)	209. (B)	210. (D)
211. (D)	212. (D)	213. (C)	214. (D)	215. (C)	216. (A)	217. (B)	218. (C)	219. (D)	220. (D)
221. (D)	222. (D)	223. (A)	224. (D)	225. (A)	226. (B)	227. (A)	228. (D)	229. (D)	230. (C)
231. (A)	232. (B)	233. (B)	234. (A)	235. (D)	236. (C)	237. (C)	238. (B)	239. (C)	240. (A)
241. (D)	242. (B)	243. (C)	244. (C)	245. (C)	246. (D)	247. (D)	248. (C)	249. (A)	250. (D)
251. (D)	252. (D)	253. (B)	254. (A)	255. (C)	256. (B)	257. (D)	258. (B)	259. (D)	260. (B)
261. (B)	262. (D)	263. (C)	264. (C)	265. (B)	266. (B)	267. (D)	268. (D)	269. (C)	270. (D)
271. (B)	272. (C)	273. (C)	274. (B)	275. (A)	276. (A)	277. (A)	278. (B)	279. (C)	280. (C)
281. (A)	282. (B)	283. (C)	284. (D)	285. (A)	286. (B)	287. (C)	288. (B)	289. (A)	290. (B)
291. (A)	292. (A)	293. (A)	294. (C)	295. (B)	296. (C)	297. (C)	298. (C)	299. (C)	300. (D)
301. (A)	302. (B)	303. (A)	304. (C)	305. (D)	306. (A)	307. (A)	308. (C)	309. (C)	310. (C)
311. (C)	312. (B)	313. (B)	314. (C)	315. (A)	316. (C)	317. (C)	318. (A)	319. (D)	320. (D)
321. (C)	322. (A)	323. (A)	324. (B)	325. (B)	326. (B)	327. (B)	328. (B)	329. (C)	330. (A)
331. (D)	332. (D)	333. (B)	334. (D)	335. (A)	336. (D)	337. (A)			

Chapter - 14

PLANT TISSUE CULTURE

1. An increase in the number of copies of gene per genome of the organism in comparison to that naturally present is called:
(A) Gene amplification (B) Gene tagging
(C) Gene silencing (D) Gene expression

2. Use of Agrobacterium T-DNA, transposons etc., for identification and/or isolation of a gene by first integrating it into the concerned gene and isolating mutants so produced is known as:
(A)Gene amplification (B) Gene tagging
(C) Gene silencing (D) Gene expression

3. Suppression of transgene expression in later generations of transgenic plants; usually, due to transgene (including promoter) methylation; more common when the transgene is integrated in higher copy number is called:
(A) Gene amplification (B) Gene tagging
(C) Gene silencing (D) Gene expression

4. When a gene performs the specific function, which it is capable of, it is called:
(A) Gene amplification (B) Gene tagging
(C) Gene silencing (D) Gene expression

5. B cells marked by negative selection die by apoptosis is called :
(A) Clonal deletion (B) Clonal propagation
(C) Clonal selection (D) None of these

6. Multiplication of plants via asexual reproduction, e.g., via micropropagation, vegetative propagation, etc. is called
(A) Clonal deletion (B) Clonal propagation
(C) Clonal selection (D) None of these

7. Phenomenon of selective prolifiration of lymphocyetes, e.g., B-cells, in response to their interaction with an antigen is called:
(A) Clonal deletion (B) Clonal propagation
(C) Clonal selection (D) None of these

8. A base sequence occurring in majority of the DNA segments serving the same function is called:
(A) Consensus sequence (B) Conserved sequence
(C) Both (A) and (B) (D) None of these

9. A base sequence present without any change in ll DNA sequences having the same function is called:
(A) Consensus sequence
(B) Conserved sequence
(C) Both (A) and (B) (D) None of these

10. It is a method of breeding crops to increase their nutritional value; it can be achieved through conventional selective breeding or through genetic engineering is called:
(A) Consensus sequence (B) Conserved sequence
(C) Biofortification (D) None of these

11. A crop variety expressing a transgene, produced by genetic engineering is called:
(A) Genetically modified crops
(B) Genetically modified food
(C) Genetically modified organism
(D) Genetically modifies plants

12. Food and additives obtained from genetically modifies organisms is called:
(A) Genetically modified crops
(B) Genetically modified food
(C) Genetically modified organism
(D) Genetically modifies plants

13. An organism with a unique gene combination that does not occur naturally, and is produced either by using genetic engineering or protoplast fusion is called:
(A) Genetically modified crops
(B) Genetically modified food
(C) Genetically modified organism
(D) Genetically modifies plants

14. Plant with unique gene combinations that do not occur naturally and are produced either by using genetic engineering or protoplast fusion is called:
(A) Genetically modified crops
(B) Genetically modified food
(C) Genetically modified organism
(D) Genetically modifies plants

15. In vitro derived virus-free shoots are grafted onto healthy root-stocks seedlings grown in vitro to obtain virus free plants is called:
(A) Micrografting (B) Micropropagation
(C) Micro injection (D) None of these

16. In vitro production of a large number of small plantlets from cultured cells/tissues; through axillary branching, adventitious shoot bud regeneration or somatic embryogenesis is called:
(A) Micrografting
(B) Micropropagation
(C) Micro injection
(D) None of these

17. Delivery of DNA directly into the cytoplasm/ nucleus of cells/male pronucleus of fertilized egg/ young embryos; micro-manipulations are used to hold the eggs and microinject DNA is called:
(A) Micrografting (B) Micropropagation
(C) Micro injection (D) None of these

18. In prokaryotic gene regulation; a gene that encodes a:
(A) Regulator gene (B) Regulator protein
(C) Reporter gene (D) All of these

19. A protein that either represses (= repressor protein) or activates (= activator protein) transcription of a transcription unit by binding to a DNA sequence in a highly specific manner is called:
(A) Regulator gene (B) Regulator protein
(C) Reporter gene (D) All of these

20. A gene, which allows either easy selection (selectable markers) or identification (scorable markers) of the cells/organisms in which it is present and expressed is called:
(A) Regulator gene (B) Regulator protein
(C) Reporter gene (D) All of these

21. Microbial culture used for the assay of biological activities of microbial products is:
(A) Assay culture (B) Batch culture
(C) Cell culture (D) Root culture

22. Growth in such cultures shows a typical sigmoid curve, and they require subculture/termination after a specific period of time in:
(A) Assay culture (B) Batch culture
(C) Cell culture (D) Root culture

23. In vitro culture of single or relatively small groups of cells, e.g., suspension cultures, this term is often used for callus culture as well is:
(A) Assay culture (B) Batch culture
(C) Cell culture (D) Root culture

24. Culture of roots in vitro; initiated from piece of root containing root-tip; can be maintained indefinitly is:
(A) Assay culture (B) Batch culture
(C) Cell culture (D) Root culture

25. Use of jumping and linking libraries to determine the correct order in a chromosome of the DNA fragments generated by the restriction enzyme used to prepared these libraries is called:
(A) Chromosome jumping
(B) Chromosome walking
(C) Cybridization
(D) Hybridization

26. Use of restriction maps to identify overlapping DNA fragments in a small region of a chromosome is called:
(A) Chromosome jumping
(B) Chromosome walking
(C) Cybridization (D) Hybridization

27. The process of cybrid production; usually, to transfer cytoplasm from one species in to other is called:
(A) Chromosome jumping
(B) Chromosome walking
(C) Cybridization
(D) Hybridization

28. The process of producing, usually, by crossing; also by protoplast fusion. In case of nucleic acid, association of two complementary DNA sequence/RNA sequences or one RNA and one DNA sequence by complementary base pairing is:
(A) Chromosome jumping
(B) Chromosome walking
(C) Cybridization
(D) Hybridization

29. A hybrid cell/individual derived from the fusion of a cytoplast from one parent and protoplast of the other patent, contains nucleus from one parent but cytoplasm from both the parents is:
(A) Cybrid (B) Hybrid
(C) Cybridization (D) Hybridization

30. Individual obtained by crossing/mating individuals /strains/varieties of dissimilar genotypes; produced by sexual or protoplast fusion routes is:
(A) Cybrid (B) Hybrid
(C) Cybridization (D) Hybridization

31. A science devoted broadly to the understanding of cellular and organism genomes is called:
(A) Hybrid (B) Genotype
(C) Genomes (D) Cybrid

32. A genetic constitution of an organism, as distinct from its physical characteristics, or phenotype called:
(A) Genotype (B) Genome
(C) Cybrid (D) Hybrid

33. "In glass"; that is, in the test tube is:
(A) In vitro
(B) In vivo
(C) Either (A) or (B)
(D) Neither (A) nor (B)

34. "In life"; that is, in the living cell or organism is:
(A) In vitro (B) In vivo
(C) Either (A) or (B) (D) Neither (A) nor (B)

35. For most genetic engineering experiments which RNA is important:
(A) mRNA (B) tRNA
(C) rRNA (D) nRNA

36. Bacterial messenger RNA (mRNA) can be separated from ribosomal and transfer RNA by:
(A) Velocity sucrose density-gradient centrifugation
(B) Gel electrophoresis
(C) Both (A) and (B)
(D) None of these

37. The PAC cloning system combines the features of the vectors derived from the bacteriophage P1 and the :

(A) BAC vectors (B) YAC vectors
(C) PAC vectors (D) None of these

38. The process of reintroduction of a cloned DNA into a cell is known as:

(A) DNA transfection (B) Gene regulation
(C) Southern blot (D) Northern blot

39. To study at the transcriptional level, that is how a given gene produces a transcript is called:

(A) DNA transfection (B) Gene regulation
(C) Southern blot (D) Northern blot

40. To detect the presence of a known piece of DNA in any given DNA preparation from an organism by:

(A) DNA transfection (B) Gene regulation
(C) Southern blot (D) Northern blot

41. The method of estimating the size and abundance of a particular transcript by:

(A) DNA transfection (B) Gene regulation
(C) Southern blot (D) Northern blot

42. To determins the 5' end transcripts, the most reliable method is :

(A) Primer extension (B) Gene tagging
(C) Gene therapy
(D) Chromosome engineering

43. The method of inserting a recognizable DNA fragment within a gene such that the function of the gene is distrupted and the gene can be easily recognized by virtue of the inserted fragment is:

(A) Primer extension
(B) Gene tagging
(C) Gene therapy
(D) Chromosome engineering

44. Correcting the illness by substituting an abnormal gene by a normal one is the basis of:

(A) Primer extension
(B) Gene tagging
(C) Gene therapy
(D) Chromosome engineering

45. Manipulation of individual chromosomes or their parts is called:

(A) Primer extension
(B) Gene tagging
(C) Gene therapy
(D) Chromosome engineering

46. Genes can be delivered during gene therapy in:

(A) in vivo (B) ex vivo
(C) Both (A) and (B) (D) None of these

47. The procedures like electroporation, lipofection, etc, can all be used for gene therapy in the:

(A) in vivo (B) ex vivo
(C) Both (A) and (B) (D) None of these

48. More precisely the regulation of gene expression can take place at different stages is called:

(A) Gene regulation (B) Gene silencing
(C) DNA (D) RNA levels

49. When gene expression falls, resulting in:

(A) Gene regulation (B) Gene silencing
(C) DNA (D) RNA levels

50. Transcriptional gene silencing takes place due to modification of the:

(A) Gene regulation (B) Gene silencing
(C) DNA (D) RNA levels

51. Among the given supplements/additives which is an example of complex organic supplements/ additive?

(A) Coconut milk and tomato juice
(B) Corn milk and malt extract
(C) Yeast extract and casein hydrolysate
(D) All of these

52. Tobacco cultivar A.C. Change is resistant to:

(A) Leaf curl (B) Black rot
(C) Damping off (D) TMV

53. Evergreen is a somaclone of:

(A) Blackberry (B) Potato
(C) Cucumber (D) Carrot

54. Among the given surface which is best sterilizing agents is?

(A) Silver nitrate (B) Cetrimide
(C) Calcium or sodium hypochlorite
(D) Bromine water

55. Slow-growth shoot culture protocols have been developed for:

(A) Coconut and citrus
(B) Sweet potato and garlic
(C) Ginger and banana (D) Both (B) and (C)

56. Somaclone Scarlet has improvement in:

(A) Skin colour (B) Insect resistance
(C) Yield (D) Disease resistance

57. Hyperhydracity of cultures may not be reduced by:

(A) Agar hydrolysate (B) Growth retardants
(C) Higher agar concentration
(D) Gibberillic acid

58. The cultivar A.C. Change and Delgold were released for commercial cultivation in:

(A) Japan (B) China
(C) Canada (D) Germany

59. Which variety of rice was developed through haploid phase?

(A) IR 36 (B) Shin Shu
(C) Pragati (D) IR 8

60. Andro is a somaclone of:

(A) Flax (B) Apple
(C) Maize (D) Cotton

61. Somaclone Adro represents improvement in:
(A) Salt tolerance (B) Oil quality
(C) Disease resistance (D) Yield

62. In protoplast isolation and culture, ionic osmoticum is not preferred because of:
(A) It reduced protoplast survival
(B) It reduces protoplast yield
(C) It reduces protoplast fusion
(D) It inhibits cell wall regeneration

63. For long-term in vitro storage of plant germplasm, which is in practical use?
(A) Cryopreservation (B) Lyophilization
(C) Slow-growth cultures (D) Both (A) and (C)

64. PEG-induced protoplast fusion is enhanced by:
(A) High pH PEG treatment medium
(B) High pH high Ca^{++} washing medium
(C) High Ca^{++} culture medium
(D) Low pH PEG treatment medium

65. Distant crosses may fail due to:
(A) Lack of pollen germination
(B) Failure of pollen tube growth
(C) Degeneration of endosperm
(D) All of these

66. Among the given wheat varieties, which one was developed through haploid phase?
(A) Jinghua 1 (B) Janak
(C) Arjun (D) Pb 543

67. Somaclone Andro has improvement in which traits?
(A) Rust immunity
(B) Powdery mildew resistance
(C) Wilt resistance (D) Leaf spot resistance

68. Bio-13 is a somaclone of:
(A) *Mentha arvensis* (B) *B. juncea*
(C) *Zea mays* (D) *Cironella java*

69. Intergeneric hybrids from which crosses have been recovered through embryo culture?
(A) *Horedeum vulgare x Triticum aestivum*
(B) *Triticum aestivum x Agropyron repens*
(C) *Hordeum vulgare x Secale cereal*
(D) All of these

70. Anther culture ordinarily yields:
(A) Haploid plants (B) Disomic plants
(C) Polyploid plants
(D) Doubled haploid plants

71. Somaclonal variation arises due to:
(A) Change in chromosome number
(B) Deletion/duplication
(C) Gene mutation/gene amplification
(D) All of these

72. Sugarcane somaclone Ono was isolated from:
(A) CO 1148 (B) Pindar
(C) BO 54 (D) COS 771

73. 'New Alleles' of known genes and even 'new mutations' of genes not known so far have been recovered through :
(A) Embryo rescue
(B) Somaclonal variation
(C) Somatic hybridization
(D) Cybridization

74. Which is not correct about micropropation?
(A) Vitrification may be a problem in some species
(B) Somaclonal variation may arise during in vitro culture
(C) Hugh production cost has limited the use of this technique to more valuable ornamental crops and some fruit trees
(D) Suitable techniques of micro-propagation are available for all valuable species

75. Which is not correct about taxol?
(A) It acts on spindle like colchicines
(B) It promotes dissolution of microtubules into tubulin molecules
(C) It is used for treatment of breast and overian cancer
(D) It is obtained from *Trichosanthes sp.*

76. Secondary metabolites include which compounds?
(A) Terpenoids (B) Alkaloids
(C) Phenylpronoids (D) All of these

77. Biomass yields from cell cultures can generally be enhanced by which approaches?
(A) Use of larger inoculums size
(B) Modification of nutrient medium
(C) Selection of high producing cell clones
(D) Both (A) and (B)

78. Biochemical production by cultural cells can be increased by using which approaches?
(A) Development of high producing clones
(B) Optimization of medium and culture conditions
(C) Use of elicitors
(D) Use of organ cultures and two –phase systems
Options:
(A) I, II, III (B) IIII, IV
(C) I, II, IV (D) I, II, III, IV

79. Which is not an example of alkaloid?
(A) Morphine (B) Codeine
(C) Camphor (D) Berberine

80. Somatic hybridization produces:
(A) Cybrids (B) Symmetric hybrids
(C) Asymmetric hybrids (D) All of these

81. Somatic hybrids are also called:
(A) Parasexual hybrids (B) Cytoplasmic hybrids
(C) Vybrids (D) Both (A) and (B)

82. Which elements is not essential for plant cell and tissue culture?
(A) Ni (B) K
(C) Ca (D) P

83. Which is not correct about asymmetric hybrids?
(A) Asymmetric hybrids provide an opportunity for introgression of few desirable genes from one fusion partner into the other
(B) Degree of chromosome elimination does not appear to depend on the irradiation dose; it seems to depend mainly on the species combination used in the fusion
(C) Asymmetric hybrids can be produced by fusion of cytoplasts of a species with irradiated protoplasts of another species
(D) Several asymmetric hybrids have been produced in *Nicotiana, Brassica*, etc.

84. Which is not correct about biotransformation?
(A) Biotransformation reactions are catalyzed by enzymes present in cells
(B) Products generated by biotransformation are more useful or valuable than the precursors used
(C) Low rate of biotransformation prevents commercially exploitation of the very large number of bioconversions known for plant cells
(D) Arbutin is produced commercially by biotransformation

85. Which plant species, biotransforms hydroquinone into β–D-glucoside (Arbutin)?
(A) *Rauwolfia serpentine* (B) *Datura innoxia*
(C) *Catharanthus roseus* (D) All of these

86. A spin-filter bioreactor is a good example of:
(A) Fluidized bed reactor
(B) Continuous-flow bioreactor
(C) Air lift bioreactor
(D) All of these

87. Micropropagation of woody tress presents which problems?
(A) Difficulties in rooting
(B) Browning in medium
(C) Poor growth in vitro
(D) All of these

88. Protoplasts can be used for genetic transformation by:
(A) PEF-mediated DNA uptake
(B) Liposome-mediated delivery
(C) Electroporation
(D) All of these

89. Which limitations is associated with biochemical production from cultured plant cells?
(A) Often cultured plant cells do not produce compounds of high commercial value
(B) High production cost is the major disadvantage
(C) Due to lack of basic knowledge of the biosynthetic pathways, many systems are difficult to improve
(D) All of these

90. Which is not a feature of somaclonal variation?
(A) A very effective selection can be practiced at the cell level for several traits
(B) Some new alleles or even 'new' mutations may be isolated, which were not available in the germplasm or through mutagenesis
(C) Somaclonal variation occurs in very low frequency
(D) Use of somaclonal variation may reduce the time required for the release of new variety as compared to mutation breeding

91. For rapid development of homozygous lines haploids can be produced by which approaches?
(A) Anther culture
(B) Ovary culture
(C) Interspecific hybridization
(D) All of these

92. Which is not used as growth retardant?
(A) ABA and Chlomequat (CCC)
(B) TIBA and paclobutrazole
(C) Daminazide
(D) All of these

93. In case of slow growth cultures, high osmotic concentrations as created by:
(A) Mannitol (B) Sorbitol
(C) Sucrose (D) All of these

94. Which chemicals interferes with virus multiplication?
(A) Cycloheximide (B) Actinomycin D
(C) Virazole (ribavirin) (D) All of these

95. Cryoprotectants reduce cryoinjury by:
(A) Reduction in cellular dehydration
(B) Reduced ice crystal formation
(C) Reduced dislocation of structural water
(D) All of these

96. Cryoinjury results from:
(A) Ice crystal formation
(B) Toxic solution effect
(C) Dislocation of structural water
(D) All of these

97. Cryoprotectants is done at which temperatures?
(A) 0^0C (B) $5\text{-}10^0C$
(C) $10\text{-}15^0C$ (D) $15\text{-}20^0C$

98. Vitrification/encapsulation eliminated the need during cryopreservation:
(A) Controlled freezing (B) Cryoproetctant
(C) LN refrigerator (D) Both (A) and (B)

99. Partial desiccation step can be combined with cryopreservation of:
(A) Somatic embryos
(B) Vitrified somatic embryos
(C) Encapsulated somatic embryos
(D) Both (A) and (C)

100. Which is not correct about cryoprotectants?
(A) DMSO is highly permeable
(B) Sorbitol is nonpermeable
(C) Sucrose is slowly permeable
(D) Glycerol is highly permeable

101. Which is not correct?
(A) Meristematic cells suffers from less cryoinjury than differentiated cells
(B) Dividing cells are more likely to survive cryopreservation than nondividing cells
(C) Cryoprotectants minimize is of structural water
(D) Partial desiccation of tissues reduces cryoinjury

102. Which vitamins is essential for plant tissue cultures?
(A) Thiamine (B) Pyridoxine
(C) m-Inositol (D) Nicotinic acid

103. Which was the first to be achieved?
(A) Embryo culture (B) Anther culture
(C) Somatic hybridization (D) Embryogenesis

104. Which was the first to be applied to solve a practical problems?
(A) Embryo culture (B) Anther culture
(C) Somatic hybridization (D) Embryogenesis

105. Which was the last to be achieved?
(A) Embryo culture (B) Anther culture
(C) Somatic hybridization
(D) Embryogenesis

106. A batch culture has following features:
(A) Typical sigmoidal growth curve
(B) Progressive nutrient depletion
(C) Limited culture duration
(D) All of these

107. Which following techniques used for cell cloning are used for proplast culture as well?
(A) Microchamber technique
(B) Filter paper raft-nurse tissue
(C) Microdrop method
(D) Bergman's plating technique

108. A high auxin concentration is required during:
(A) SE induction (B) SE development
(C) SE maturation (D) SE conversion

109. GA3 has a beneficial effect during:
(A) SE induction (B) SE development
(C) SE maturation (D) SE conversion

110. ABA is generally promotive during:
(A) SE induction (B) SE development
(C) SE maturation (D) SE conversion

111. Rejuvenation may be brought about by:
(A) Repeated sprays of a cytokining
(B) Repeated grafting onto seedlings
(C) Repeated subculture on cytokinin containing media
(D) All of these

112. GRs are not used during which stages of micropropagation?
(A) Preparation of mother plants
(B) Culture establishment
(C) Multiplication
(D) Rooting

113. Auxin is used :
(A) Preparation of mother plants
(B) Culture establishment
(C) Multiplication
(D) Rooting

114. In micropropagation, in vitro hardening is achieved by:
(A) PFG in medium
(B) Use of a Pseudomonas culture
(C) Cooling of culture vessels at bottoms
(D) All of these

115. Which chemicals has been used to eliminate viruses from plants?
(A) Ribovirin (B) Actinomycin D
(C) Malachite green (D) All of these

116. Virus free plants produced by various in vitro techniques can be used for:
(A) Breeding programmes
(B) For germplasm exchange
(C) Commercial cultivation
(D) Both (A) and (B)

117. 'Genetic erosion' is the consequence of:
(A) Reduction in genetic variability among cultivated varieties
(B) Reduction in genetic variability within cultivated varieties
(C) Extinction of wild relatives of crop species
(D) All of these

118. Which is a sweetening agent?
(A) Thaumatic (B) Shikonin
(C) Dioxgenin (D) Pyrethrin

119. Which is a steroid?
(A) Thaumatic (B) Shikonin
(C) Dioxgenin (D) Pyrethrin

120. Which has insecticidal activity?
(A) Thaumatic (B) Shikonin
(C) Dioxgenin (D) Pyrethrin

121. Biochemical accumulation by cultured plant cells is affected by:
(A) Chemical composition of medium
(B) Light quality
(C) Inoculum size
(D) All of these

122. Which can serve as elicitors?
(A) Chitosan (B) Cell wall pectins
(C) Some glycoproteins (D) All of these

123. **Certain biochemicals are produced by specialized cells present in organized structures. Which culture systems will be useful for production of such compounds?**

(A) Immobilized cell cultures
(B) Organ cultures
(C) Callus cultures
(D) Suspension cultures

124. **Arbutin is commercially produced from hydroquinone by:**

(A) Chemical reactions
(B) Biotransformation by Digitalis cells
(C) Biotransformation by Datura cells
(D) A combination of B and A

125. **In which of the following, dihaploids are used for breeding, while tetraploids are used as commercial varieties?**

(A) Wheat (B) Potato
(C) Sugarcane (D) Tobacco

126. **In which plant species, YY males are sued to produce all male populations?**

(A) *Silene latifolia* (B) *Coccinia*
(C) *Asparagus* (D) *Trichosanthes*

127. **Which breeding procedures is applicable to only autopolyploid crop species?**

(A) Hybrid sorting
(B) Analytical breeding
(C) Somaclonal breeding
(D) Both (A) and (B)

128. **Which involved in haploid production?**

(A) Hybrid sorting (B) Analytical breeding
(C) Somaclonal breeding (D) Both (A) and (B)

129. **In most species, cells become committed to gametophytic development after which stages?**

(A) Pollen mother cell (B) Uninucleate pollen
(C) Binucleate pollen (D) Trinucleate pollen

130. **Which have been shown to prevent precocious germination of cultured embryos?**

(A) High sucrose levels (B) ABA
(C) Casein hydrolysate (D) All of these

131. **Asymmetric hybrids can be produced by:**

(A) Spontaneous chromosome elimination
(B) Irradiation-incuded chromosome elimination
(C) Fusion of protoplasts of one species with microplats of the other species
(D) All of these

132. **Somatic hybridization can be resorted to only when which criteria are satisfied?**

(A) Isolation of protoplasts in large quantities
(B) Totipotency of the isolated protoplasts
(C) Availability of effective strategy for selection of hybrid cells
(D) Both (A) and (B)

133. **It is desirable to preserve which of the following tissues to minimize the risk of genetic instability?**

(A) Zygotic embryos (B) Shoot-tips
(C) Suspension cultures (D) Either A or B

134. **Which one is hybrid variety of mango:**

(A) Neelam (B) Langra
(C) Sumarr Bahist (D) Chausa

135. **Amrapali is a variety of mango developed by cross of:**

(A) Dashhari x Neelam (B) Neelam x Dashhari
(C) Chausa x Langra (D) Dashhari x Langra

136. **Mallika is a variety of mango developed by cross of:**

(A) Dashhari x Neelam
(B) Neelam x Dashhari
(C) Chausa x Langra
(D) Dashhari x Langra

137. **Gaurav is a variety of mango developed by cross of:**

(A) Dashhari x Totapari (B) Neelam x Dashhari
(C) Chausa x Langra (D) Dashhari x Langra

138. **Saurabh is a variety of mango developed by cross of:**

(A) Dashhari x Fazali jafrani
(B) Neelam x Dashhari
(C) Chausa x Langra
(D) Dashhari x Langra

139. **Rajiv is a variety of mango developed by cross of:**

(A) Dashhari x Romani
(B) Neelam x Dashhari
(C) Chausa x Langra
(D) Dashhari x Langra

140. **Which of the following practice is very much required for higher fruiting in bottle guard?**

(A) Blanching (B) Chlorosis
(C) Staking (D) Digging

141. **Flower dropping of chilli can be checked by:**

(A) By spraying of planofix
(B) By applying of nitrogen
(C) By spraying of clorine
(D) By spraying of urea solution

142. **Which is suitable method for seed production of cabbage:**

(A) Head to seed method (B) Head intact method
(C) Stump method (D) All of these

143. **Alternate bearing of mango can be checked by:**

(A) Deblossoming (B) Ringing
(C) Root prunnig (D) All of these

144. **Heading back pruning refers to:**

(A) To remove extra and overlapping branches
(B) To remove terminal portion of all branches
(C) To remove the flowers only
(D) To remove flowers and top portion of main trunk

145. Teonic in compatibility refers to:
(A) Scion and root stock are adversely affected by each other
(B) Scion and root stock are positively affected by each other
(C) No effect of scion and rootstock on each other
(D) None of these

146. The regularity in mango bearing can be achieved by soil application of:
(A) Cultar @ 0.5 g/plant (B) 15 ppm NAA
(C) 50 ppm Ethylene (D) All of these

147. Fruit thining and pre harvest of fruit can be controlled in mango and citrus by:
(A) 100-150 ppm NAA (B) 200 ppm NAA
(C) 10 ppm GA (D) 20 ppm IAA

148. Which of the following is a evergreen temperate plant?
(A) Apricot (B) Chilgoza
(C) Plum (D) Almond

149. Fruiting of apple occurs on:
(A) New fruiting spurs (B) Old branches
(C) Whole plant (D) All of these

150. Fruit cracking in litchi can be controlled by:
(A) Spraying of water
(B) Spraying of hormones
(C) Spraying of 2,4-D
(D) Spraying of Bordeaux mixture

151. Poor fruiting in loquat is mainly due to:
(A) Poor growth (B) Low fertility
(C) Self incompatibility (D) None of these

152. Pusa Rasraj is :
(A) A pureline variety of muskmelon
(B) A hybrid variety of muskmelon
(C) A Seedless variety of watermelon
(D) A hybrid variety of watermelon

153. The primary objective of green house cultivation of vegetables is:
(A) To get early crop
(B) To get off season crop
(C) To protect the crop from disease
(D) To protect the crop from excessive rains

154. Which of the following is a stem vegetable?
(A) Carrot (B) Sweet potato
(C) Knoll khol (D) Radish

155. Which of the following is a ripening hormone?
(A) Ethylene (B) Cytokinin
(C) Auxin (D) Gibberellin

156. Which of the following root stock of grape is useful under high salt content and nematodes?
(A) Rumani (B) Olour
(C) Dogridge (D) Rangpur lime

157. The bud dormancy in grapes can be broken by use of:
(A) Cultar (B) Hydrogen cynamide
(C) IAA (D) NAA

158. Whipe tail is caused by deficiency of?
(A) Zn (B) P
(C) Mo (D) B

159. Buttoning is caused by:
(A) N deficiency (B) P deficiency
(C) K deficiency (D) Ca deficiency

160. Browning is caused by:
(A) N deficiency (B) B deficiency
(C) K deficiency (D) Ca deficiency

161. Which of the following factor is responsible for higher yield under protected cultivation?
(A) High photosynthetic efficiency
(B) High O_2 availability
(C) High temperature (D) All of these

162. Standard root stock of apple against woolly aphid is:
(A) Malling-Merton (B) DX-1
(C) PT-1 (D) None of these

163. Which of the following is a very vigorous root stock is used for propagation of apple?
(A) Malling -12 (B) Malling-13
(C) Malling-16 (D) All of these

164. Which of the following is a very dwarf root stock for propagation of apple?
(A) M-27 (B) M-12
(C) M-13 (D) None of these

165. How much percentage of male plants should be planted in papaya orchard for proper pollination?
(A) 5% (B) 10%
(C) 15% (D) 20%

166. Stem of banana is a:
(A) Pseudomonas (B) Main stem
(C) Pseudostem (D) All of these

167. Propping refers to:
(A) Staking of banana stem at the time of frui-ting to avoid bending or breaking of stem due to heavy weight of fruits
(B) To remove extra suckers
(C) To provide shade to the fruits
(D) None of these

168. Fruiting of guava during November–January is known as:
(A) Mrig bahar (B) Hast bahar
(C) Ambe bahar (D) Neno bahar

169. Fruiting of guava during February-April is known as:
(A) Mrig bahar (B) Hast bahar
(C) Ambe bahar (D) Neno bahar

170. Fruiting of guava during Rainy season is known as:
(A) Mrig bahar (B) Hast bahar
(C) Ambe bahar (D) Neno bahar

171. Which of the root stock is used in citrus fruits?
(A) Karna Khatta (B) Jumberi
(C) Sour orange (D) All of these

172. Frame working refers to:
(A) Change into old plant into high grade plants
(B) All side branches are grafted with high quality scions
(C) Only top part of plant is removed
(D) All of these

173. Pruning is an essential cultural activity in fruit plants because:
(A) To provide proper shape
(B) To provide strong structure
(C) To beautify the plant
(D) All of these

174. Notching of fruit plants refers :
(A) To make a V shaped notch just below N above the bud in fruit plants
(B) It helps in vegetative growth
(C) It is common in Fig
(D) None of these

175. Which of the following plant hormone is used to control pre harvest fruit drop?
(A) 2, 4-D (B) NAA
(C) 2, 4-D and NAA (D) None of these

176. The cuttings are prepared by removing of cuttings from mother plant is:
(A) 3 cm below the bud at basal or lower part and 1.23 cm above the bud at upper part
(B) Removing of cuttings from upper part of the twig of mother plant
(C) Take centre part of the mother plant
(D) All of these

177. Top working of fruit plants refers to:
(A) Modify the form of old plants
(B) Convert old plants into new plants
(C) Old plants are modified into new species
(D) All of these

178. What is blanching?
(A) It is a method of fruit ripening
(B) It is a method of bunding
(C) It is a method of protecting fruits from light and hot
(D) It is a method of plant protecting from insects

179. Blindness in cauliflower means:
(A) Full development of flower/fruit
(B) Full development of plant without flower
(C) Started plant growth
(D) All of these

180. Training of fruit trees means:
(A) Remove few unwanted branches to make plant more beautiful and strong
(B) Remove few flowers
(C) Remove and cut few roots
(D) None of these

181. Temperate fruit plants can be trained by:
(A) Leader method
(B) Open centered method
(C) Modified leader system
(D) All of these

182. Which of the training is considered more suitable?
(A) Closed centered system
(B) Open centered system
(C) Modified leader system
(D) All of these

183. In quincunx system, the total plants are nearly double than:
(A) Square method (B) Rectangular method
(C) Contour method (D) Hexagonal method

184. In quincunx system, the filler plants are generally:
(A) Fast growing and early fruiting
(B) Slow growing but early fruiting
(C) Late fruiting (D) All of these

185. Loping of root stock refers to:
(A) To remove the upper part of root stock after setting of buds
(B) To remove the lower part of root stock after setting of buds
(C) To remove extra new branches from the root stock
(D) All of these

186. Which of the following inhances rootings in cuttings?
(A) Auxin (B) Phenolic substances
(C) Auxin-Phenolic complex
(D) All of these

187. Which of the following chemical is used to enhances rootings in cuttings?
(A) IAA (B) IBA
(C) NAA (D) All of these

188. Which of the following hormone is used for fast rootings in cuttings?
(A) Seredix-A (B) Vitamin A
(C) Vitamin E (D) None of these

189. The rooting of cuttings can be induced by cuttings before planting in:
(A) IAA (B) GA
(C) ABA (D) Zn solution

190. In above case, the concentration of IAA should be used:
(A) 100-500 ppm (B) 1000-5000 ppm
(C) 10-50 ppm (D) All of these

191. IAA, belongs to which group of plant hormone:
(A) Auxin (B) Gibberellin
(C) Cytokinin (D) None of these

192. Diploid variety of apple are?
(A) Self fruitful (B) Self unfruitful
(C) Both (A) and (B) (D) All of these

193. Triploid varieties of apple are?
(A) Self fruitful (B) Self unfruitful
(C) Both (A) and (B) (D) All of these

194. Which of the following is a diploid variety of apple?
(A) Jonathan (B) Rome beauty
(C) Newton wonder (D) All of these

195. Which of the following is a triploid variety of apple?
(A) Baldvin (B) Blenhenn orange
(C) Cox's orange pippin (D) All of these

196. Mahmud bahar is a cross of:
(A) Bombay green x Kallapady
(B) Neelam x Chausa
(C) Neelam x Langra
(D) Neelam x Dashhari

197. Which of the following is a polyembryonic variety?
(A) Mazagaon (B) Chandra karan
(C) Neleswar (D) All of these

198. Which of the root stock has dwarfing effect on Dashhari?
(A) Rumani (B) Olour
(C) Dogridge (D) Rangur lime

199. Olor is a polyembryonic variety of:
(A) Guava (B) Citrus
(C) Mango (D) Strawberry

200. Which of the following dwarfing root stock is used for mango?
(A) Olour (B) Totapari Red small
(C) Dashhari (D) Alfanso

201. Dehorning refers to:
(A) Removal of branching from fruit trees
(B) Removal of terminal portion of branches
(C) Removal of flowers
(D) All of these

202. Pinching refers to:
(A) To remove top portion of plant with few leaves
(B) It is like a head back pruning
(C) It is done in ornamental plants
(D) All of these

203. Stripping refers to:
(A) Removal of barks in stripes of 1.25 – 2.5 cm length
(B) Removal of extra flowers
(C) Removal of terminal portion of branches
(D) None of these

204. Pollinizers are:
(A) Fruit plants that provide pollens to other plants
(B) Helps in fertilization and fruiting
(C) They are planted middle of fruiting plants
(D) All of these

205. Fruiting can be enhanced by:
(A) Root pruning (B) Ringing
(C) Notching (D) All of these

206. Fruiting can be induced by:
(A) NAA (B) 2,4-D
(C) IAA (D) All of these

207. Spliting of fruits can be checked by spraying of:
(A) 1 ppm NAA (B) Orange rind oil
(C) Bordeaux mixture (D) All of these

208. Periclinal chimeras in fruit plants can be produced by:
(A) Colchicine (B) X-rays
(C) High temperature (D) All of these

209. Polygamous plants are those plants that produce:
(A) Male, female and hermaphrodite flowers
(B) Male flowers only
(C) Female flowers only
(D) Hermaphrodite flowers

210. Which one is the hybrid variety of mango is:
(A) Alphanso (B) Dashahari
(C) Mallika (D) Langra

211. Browning of cauliflower is caused by the deficiency of:
(A) Zinc (B) Manganese
(C) Iron (D) Boron

212. The citrus diseases namely 'Reclamation Disease' and 'Die back' are caused by the deficiency of:
(A) Zinc (B) Iron
(C) Copper (D) Molybdenum

213. Fastening of vines to a upright support of a tree called panagara (*Erithrina indica*) is known as:
(A) Kniffen system (B) Head system
(C) Single stake system (D) Cordon system

214. Maximum genetic diversity of crop plants occurs where agriculture is:
(A) Run commercial line (B) Advanced
(C) Rainfed (D) Primitive

215. Which is more suitable for germplasm collection:
(A) Centre of production (B) Homelands
(C) Tropical areas (D) Temperate areas

216. Which one will be scarched for obtaining maximum genetic diversity:
(A) Recent varieties (B) Old varieties
(C) Wild relatives (D) All the above

217. In a gene bank, genetic material is stored as:
(A) Genes (B) Seeds
(C) Vegetative material/tissue culture
(D) Both (B) and (C)

218. Genetic erosion is due to:
(A) Deforestation
(B) Shifting cultivation
(C) Adoption of genetically uniform varieties
(D) All the above

219. Exchange of germplasm is carried mostly through shoot tips because it is:
(A) Cheap (B) Small and handy
(C) Virus free (D) Genetically stable

220. The gynophore of groundnut is commonly referred to as the:
(A) Perinth (B) Peg
(C) Ovate (D) Monadelph

221. The rice variety containing 'Dee-geo-woo-gen' gene is:
(A) Indrasan (B) Basmati
(C) Tilak (D) IR 8

222. Use of diploids in breeding of crops like potato offers:
(A) Much easier hybridization with diploid wild species, which offers many genes of great value
(B) Much simpler segregation ratios & as a result, the need for similar segregating populations
(C) A greater efficiency opf selection
(D) All of these

223. Gametoclonal may occur due to :
(A) Residual mutation (B) Gene mutation
(C) Gene multiplication (D) All of these

224. In 1922, Blakeslee and coworkers, for the first time, reported occurrence of haploids in Datura due to:
(A) Androgenesis (B) Gynogenesis
(C) Parthenogenesis (D) Mutagenesis

225. During distant hybridization, problem of endosperm abortion can be overcome by:
(A) Embryo culture (B) Ovary culture
(C) Ovule culture (D) All of these

226. In plant tissue culture, ABA may effect:
(A) Enhance shoot bud differentiation
(B) Promote somatic embryo maturation
(C) Induced somatic embryo conversion
(D) Both (A) and (B)

227. Plants with gametic chromosome number are termed as:
(A) Diploid (B) Dihaploid
(C) Haploid (D) Monoploid

228. Micropropagation can be achieved through:
(A) Enhanced axillary branching
(B) Somatic embryogenesis
(C) Production of adventitious shoot buds
(D) All of these

229. Among the given hydrogels, which is most preferred?
(A) Agarose (B) Sodium alginate
(C) Agar (D) Gel-Rite

230. Somatic embryo can be used for:
(A) Genetic transformation
(B) Production of uniform plant populations
(C) *In vitro* mutagenesis
(D) All of these

231. Progress in haploid and dihaploid production is still slow due to:
(A) High levels of albinism, especially in monocots
(B) Low frequency of plantlet regeneration
(C) Genetic instability
(D) All of these

232. Somaclonal variation occurs due to:
(A) Methylation and demethylation in the promoter region of genes
(B) Activation of silent genes in multigene families
(C) Non-reciprocal mitotic recombination and activation of transposable elements
(D) All of these

233. GA3 is known to effect:
(A) Suppress callus formation
(B) Promote somatic embryo conversion
(C) Promote shoot regeneration in some species
(D) All of these

234. Survival of encapsulated somatic embryo has often been reported to be lower than that of the non-encapsulated ones probably due to:
(A) Poor respiration under the almost anaerobic condition within the capsule
(B) Poor photosynthesis
(C) Poor storage of food material
(D) All of these

235. Ethylene promotes:
(A) Activity of cell wall destroying enzymes
(B) Ageing of leaves
(C) Activity of chlorophyllase
(D) All of these

236. Guha and Maheshwari produced haploids from anther culture of:
(A) Nicotiana (B) Datura
(C) Ginkgo (D) Hordeum

237. Somaclonal variation generally serves the same purpose as:
(A) Analytical breeding
(B) Mutation breeding
(C) Backcross breeding
(D) Hybrid sorting

238. Which is not correct about abscissic acid?
(A) It is a naturally occurring growth inhibitor
(B) It destroys the effect of GA and promotes dormancy in seeds
(C) It is synthesized in leaves
(D) Precursor of ABA is malic acid

239. How first reported interspecific hybridization through protoplast fusion in two species of Nicotiana?
(A) Powars and coworkers
(B) Carlson and coworkers
(C) Miller and coworkers
(D) Cocking and coworkers

240. Which minimizes the risk of variation during in vitro conservation of germplasm?
(A) Shoot spices (B) Zygote embryos
(C) Somatic embryos (D) Both (A) and (B)

241. Success in germplasm storage is determined by which factors?
(A) Inherent longevity and physiological storage behavior of the species
(B) Initial quality (e.g., moisture content) of material stored
(C) Storage methods and conditions used
(D) All of these

242. Pusa Jaikisan (Bio 902) is a somaclonal variety of:
(A) Rapeseed (B) Indian mustard
(C) Groundnut (D) Sunflower

243. Which material is conserved as germplasm?
(A) Wild taxa related to crop species, weedy forms and land races
(B) Obsolate varieties and advanced cultivars
(C) Parents of hybrids, mutants and cytogenetic stocks
(D) All of these

244. Which varieties developed through somatic hybridization followed by back cross?
(A) Delgold (B) Sigma
(C) A.C. Chang (D) Both (A) and (C)

245. Which variety that was developed from somaclonal variants?
(A) Delgold (B) Sigma
(C) A.C. Chang (D) Both (A) and (C)

246. Why conservation of genetic resources is so important?
(A) Due to piracy of genetic resource
(B) Due to regular use in plant breeding
(C) Due to genetic erosion and epidermics
(D) All of these

247. Cryopreservation is based on:
(A) Liquid CO_2 (B) Liquid helium
(C) Liquid nitrogen (D) Both (A) and (C)

248. Which is thermolabile?
(A) Zeatin (B) Pectinase
(C) ABA (D) All of these

249. Which scientists is regarded as father of plant tissue culture?
(A) Herbert E. Street
(B) Gottlieb Haberlandt
(C) Jean P. Nitsch (D) Philip R. White

250. The technique of embryo culture can be used to obtain:
(A) Interspecific hybrids
(B) Propagation of orchids
(C) Dormancy and to shorten breeding cycle
(D) All of these

251. For log-term storage of germplasm, which is used?
(A) Carbon dioxide (B) Liquid oxygen
(C) Liquid nitrogen (D) Mineral oil

252. Which enzymes is used for obtaining protoplasts?
(A) Macerozyme (B) Cellulase
(C) Pectinase (D) All of these

253. In general, callus cultures are subcultured after:
(A) 4-6 days (B) 4-6 weeks
(C) 8-10 weeks (D) 2-3 months

254. Somatic embryo maturation is not promoted by:
(A) Cytokinin (B) High sucrose
(C) Partial desiccation (D) Abscissic acid

255. The technique of cryopreservation involves:
(A) Freezing and storage
(B) Thawing and reculture
(C) Freezing and reculture
(D) Both (A) and (B)

256. The term somaclonal variation was coined by:
(A) Vasil and Vasil
(B) Larkin and Scowcroft
(C) Skirvin and Karp
(D) Evans and Sharp

257. Protoplast fusion is induced by:
(A) High voltage electric pulse
(B) High pH (10.5) and high (50 m mol/l) Ca^{++} concentration
(C) Polyethylene glycol (PEG)
(D) All of these

258. In general, suspension cultures are sub-cultured after which periods:
(A) 1-2 weeks (B) 10-15 days
(C) 3-10 days (D) 15-18 days

259. Among the given auxins, which one is the most frequently used in plant tissue culture?
(A) 2, 4-D (B) IBA
(C) NAA (D) IAA

260. Pusa Jai kisan (Bio 902) somaclone of Indian mustard was isolated from:
(A) Pusa Bold (B) Kranti
(C) Varuna (D) Vaibhav

261. Delgold and A.C. Chang, developed through somatic hybridization, are cultivars of:
(A) *Nicotiana rustica*
(B) *Nicotiana tabacum*
(C) *Lycopersicon esculentums*
(D) *Brassica napus*

262. **Among the given compounds, which one is the most commonly used as cryoprotectant?**
(A) Proline (B) Sorbitol
(C) DMSO (D) Glycerol

263. **In which tree species, embryo culture is used to overcome dormancy?**
(A) Taxus (B) Iris
(C) Prunus (D) All of these

264. **Among the given varieties to tobacco, which one was developed through haploid phase?**
(A) F 211 (B) DD 413
(C) GSH3 (D) All of these

265. **Sugarcane somaclone Ono, isolated from the parental variety Pindar, is resistant to a disease caused by:**
(A) Fungus (B) Virus
(C) Mycoplasma (D) Bacterium

266. **Scarlet is somaclone of:**
(A) Wheat (B) Sorghum
(C) Sweet potato (D) Potato

267. **Among the given agents, which one is used as cryoprotectants?**
(A) DMSO, proline and valine
(B) DMSO, glycerol and proline
(C) DMSO, glycerol and acetic acid
(D) DMSO, arginine and acetic acid

268. **Delgold is a variety of:**
(A) Cheroot tobacco
(B) Flue-cured tobacco
(C) Chewing tobacco
(D) Cigar and wrapper tobacco

269. **The first plant from a mature plant cell was regenerated by:**
(A) Braun (B) Haberlandt
(C) Cocking (D) White

270. **During gynogenesis, haploid plants generally originated from:**
(A) Synergids (B) Antipodal cells
(C) Egg cell (D) Polar nuclei

271. **Somatic hybridization has been the most successful in:**
(A) Graminae (B) Leguminosae
(C) Malvaceae
(D) Solanaceae and Crucifereae

272. **Post transcriptional gene silencing takes place at the:**
(A) Gene regulation (B) Gene silencing
(C) DNA (D) RNA levels

273. **Cross breeding is also a very important method for getting useful:**
(A) Hybrids
(B) Conventional breeding
(C) Selection breeding (D) None of these

274. **The green revolution was a product of:**
(A) Hybrids
(B) Conventional plant breeding
(C) Selection breeding (D) None of these

275. **The process of choosing plants with desired characters followed by breeding is called:**
(A) Hybrids
(B) Conventional breeding
(C) Selection breeding (D) None of these

276. **Development of alleles from differentiated explants is termed as:**
(A) Dedifferentiation phase
(B) Cybrids
(C) In vitro method (D) Ex-vitro method

277. **The variable cell resulting from the fusion of a cytoplast with a whole cell, thus creating a cytoplasmic hybrid is known as:**
(A) Dedifferentiation phase (B) Cybrids
(C) In vitro method (D) Ex-vitro method

278. **The material could be stored in the form of suspension cultures, meristem, callus cultures and embryoids by:**
(A) Dedifferentiation phase (B) Cybrids
(C) In vitro method (D) Ex-vitro method

279. **The method for preservation of genetic resources for cultivated as also the wild materials by:**
(A) Dedifferentiation phase (B) Cybrids
(C) In vitro method (D) Ex-situ method

280. **The method to preserve land races and wild relatives which could be used for future breeding programme by:**
(A) In-situ method (B) Cybrids
(C) In vitro method (D) Ex-vitro method

281. **A place where stores DNA for future analysis is:**
(A) DNA bank (B) Gene bank
(C) Both (A) and (B) (D) None of these

282. **The Royal Botanic Garden (at Kew) DNA Bank contains over 13000 samples of plant genomic DNA (as at the beginning of 2002), all stored at:**
(A) 0^0C (B) -20^0C
(C) -40^0C (D) -80^0C

283. **The most common, straight forward and speedy method for the isolation of gene is:**
(A) Restriction of gene
(B) Polymerase Chain Reaction (PCR)
(C) Recombinant DNA Technology
(D) None of these

284. **The speedy but little tricky method of gene isolation is:**
(A) Restriction of gene
(B) Polymerase Chain Reaction (PCR)
(C) Recombinant DNA Technology
(D) None of these

285. Which has been used extensively with microbial system, is also an important tool for the direct genetic manipulation of plant is called:
(A) Restriction of gene
(B) Polymerase Chain Reaction (PCR)
(C) Recombinant DNA Technology
(D) None of these

286. *Agrobacterium tumefaciens* are capable of producing which disease in almost all dicotyledonous plants through wounded tissues:
(A) Crown galls (Plant tumors)
(B) Hairy root (extra root)
(C) Both (A) and (B)
(D) None of these

287. *Agrobacterium rhizogens* are capable of producing which disease in almost all dicotyledonous plants through wounded tissues:
(A) Crown galls (Plant tumors)
(B) Hairy root (extra root)
(C) Both (A) and (B) (D) None of these

288. The T-DNA region includes important genes, which encode enzymes that synthesize the plant hormone auxin (IAA):
(A) iaaM (tms 1) (B) iaaH (tms2)
(C) Both (A) and (B) (D) None of these

289. Vir genes are located on the Ti-plasmid itself beyond the region of:
(A) TDNA (B) RDNA
(C) CDNA (D) MDNA

290. In tissue culture disease resistance can be obtained by:
(A) Somaclonal variation (B) Embryoculture
(C) Somatic hybridization (D) Anther culture

291. Selection of protoplast fussion hybrids can be done by:
(A) Visual selection (B) Pollen culture
(C) Ovary culture (D) Any of these

292. Micropropagation can be done by:
(A) through shoot and proliferation
(B) Through adventitious buds
(C) Meristem tip culture
(D) All of these

293. Most commonly used method for transcription of plants is:
(A) Agrobacterium mediated transformation
(B) Protoplast method
(C) Microinjection
(D) None of these

294. Most commonly used method for transcription of animals is:
(A) Agrobacterium mediated transformation
(B) Protoplast method
(C) Microinjection
(D) Electro transformation

295. In tissue culture regeneration of shoot and root occurs by manipulation the balance of:
(A) Auxin and ABA (B) Auxin and Cytolinin
(C) Cytokinin and ABA (D) Auxin and Ethylene

296. Best method to determine paternity is:
(A) Protein analysis
(B) Chromosome counting
(C) Gene counting
(D) DNA finger printing

297. Development of shoot and root in tissue culture is determined by:
(A) Cytokinin and auxin ratio
(B) Enzymes
(C) Temperature (D) Plant nutrients

298. In tissue/bacterial culture glassware and nutrients are sterilized through.
(A) Water bath at 200^{0}C
(B) Dry air oven at 200^{0}C
(C) Dehumidifire (D) Autoclave

299. In tissue culture, callus can be induced to form shoot or root by altering the ratio of:
(A) Auxin to cytokinin
(B) Cytokinin to ethylene
(C) Auxin to gibberellins
(D) Gibberellin to cytokinin

300. Introduction of foreign genes for improving genotype is:
(A) Tissue culture (B) Immunisation
(C) Biotechnology (D) Genetic engineering

301. It is now possible to bred plants and animals with desired characters through:
(A) Genetic engineering
(B) Chromosome engineering
(C) Ikebana technique
(D) Bonsia technique

302. Okazaki segments are formed during:
(A) Transduction (B) Transcription
(C) Replication (D) Translation

303. Plants developed in vitro culture from pollen grains are:
(A) Androgenic plants (B) Pollen grains
(C) Male plants (D) Sterile plants

304. Pomato is somatic:
(A) Poppy and potato (B) Potato and tomato
(C) Poppy and tamarind (D) Poppy and tomato

305. The plant part which is used for regeneration is called:
(A) Clone (B) Callus
(C) Explant (D) None of these

306. A mass of unorganized regenerated cells in culture medium is called:
(A) Embryoid (B) Somatic bud
(C) Callus (D) All of these

307. **The ability of a plant cell to develop into a whole plant in culture medium is known as:**
(A) Viability (B) Longetivity
(C) Totipotency (D) Prepotency

308. **The process of differentiation of root and shoot from somatic embryo in culture medium is called:**
(A) Gametogenesis (B) Sporogenesis
(C) Organogenesis (D) Embrygenesis

309. **Somatic hybridization leads to development of:**
(A) Symmetrical hybrids
(B) Symmetrical hybrids
(C) Cybrids (D) All of these

310. **Hybrid cell involving protoplast of two different species is called:**
(A) Homokaryon (B) Heterokaryon
(C) Eukaryon (D) None of these

311. **Somatic hybrid may be:**
(A) Interspecific (B) Intergeneric
(C) Intertribal (D) All of these

312. **Organ culture refers to:**
(A) Anther culture (B) Ovule culture
(C) Embryo culture (D) All of these

313. **Micropropagation is widely used in:**
(A) Horticulture (B) Floriculture
(C) Forestry (D) All of these

314. **In vegtatively propagated species, somaclonal variation is:**
(A) Low (B) Moderate
(C) High (D) Very high

315. **Multiple identical copies of a gene can be obtained by:**
(A) Gene cloning (B) Gene sequencing
(C) Gene splicing (D) Micro cloning

316. **Callus refers to a mass of:**
(A) Cells (B) Unorganized cells
(C) Organized cells (D) ll of these

317. **Somatic hybridization is useful in producing:**
(A) Allotetraploids (B) Transgenics
(C) Distant hybrids (D) All of these

318. **Allotetraploids can be produced in a single step by:**
(A) Anther culture (B) Microcloning
(C) Somatic hybridization (D) All of these

319. **The sexual process is bypassed in:**
(A) Budding (B) Grafting
(C) Somatic hybridization (D) All of these

ANSWERSHEET

14. PLANT TISSUE CULTURE

1. (A)	2. (B)	3. (C)	4. (D)	5. (A)	6. (B)	7. (C)	8. (A)	9. (B)	10. (C)
11. (A)	12. (B)	13. (C)	14. (D)	15. (A)	16. (B)	17. (C)	18. (A)	19. (B)	20. (C)
21. (A)	22. (B)	23. (C)	24. (D)	25. (A)	26. (B)	27. (C)	28. (D)	29. (A)	30. (B)
31. (C)	32. (A)	33. (A)	34. (B)	35. (A)	36. (C)	37. (A)	38. (A)	39. (B)	40. (C)
41. (D)	42. (A)	43. (B)	44. (C)	45. (D)	46. (C)	47. (C)	48. (A)	49. (B)	50. (C)
51. (D)	52. (B)	53. (A)	54. (C)	55. (D)	56. (A)	57. (D)	58. (C)	59. (B)	60. (A)
61. (A)	62. (D)	63. (D)	64. (B)	65. (D)	66. (A)	67. (A)	68. (D)	69. (D)	70. (A)
71. (D)	72. (B)	73. (B)	74. (B)	75. (D)	76. (D)	77. (D)	78. (D)	79. (C)	80. (D)
81. (A)	82. (A)	83. (C)	84. (D)	85. (D)	86. (B)	87. (D)	88. (D)	89. (D)	90. (C)
91. (D)	92. (D)	93. (D)	94. (D)	95. (D)	96. (D)	97. (A)	98. (A)	99. (D)	100. (D)
101. (B)	102. (A)	103. (A)	104. (A)	105. (C)	106. (D)	107. (C)	108. (A)	109. (A)	110. (C)
111. (D)	112. (B)	113. (D)	114. (D)	115. (D)	116. (D)	117. (D)	118. (A)	119. (C)	120. (D)
121. (C)	122. (C)	123. (B)	124. (A)	125. (B)	126. (C)	127. (B)	128. (D)	129. (C)	130. (D)
131. (D)	132. (D)	133. (D)	134. (C)	135. (A)	136. (B)	137. (A)	138. (A)	139. (A)	140. (A)
141. (A)	142. (D)	143. (D)	144. (B)	145. (A)	146. (A)	147. (A)	148. (A)	149. (A)	150. (A)
151. (C)	152. (D)	153. (B)	154. (C)	155. (A)	156. (C)	157. (B)	158. (C)	159. (A)	160. (B)
161. (D)	162. (A)	163. (D)	164. (A)	165. (B)	166. (C)	167. (A)	168. (A)	169. (B)	170. (C)

171. (D) **172. (D)** **173. (D)** **174. (A)** **175. (C)** **176. (A)** **177. (D)** **178. (C)** **179. (B)** **180. (A)**
181. (D) **182. (C)** **183. (A)** **184. (A)** **185. (A)** **186. (C)** **187. (D)** **188. (A)** **189. (A)** **190. (B)**
191. (A) **192. (A)** **193. (B)** **194. (D)** **195. (D)** **196. (A)** **197. (D)** **198. (A)** **199. (C)** **200. (A)**
201. (A) **202. (D)** **203. (A)** **204. (D)** **205. (D)** **206. (D)** **207. (D)** **208. (D)** **209. (A)** **210. (C)**
211. (D) **212. (C)** **213. (C)** **214. (D)** **215. (B)** **216. (D)** **217. (D)** **218. (D)** **219. (C)** **220. (B)**
221. (D) **222. (D)** **223. (D)** **224. (C)** **225. (D)** **226. (D)** **227. (C)** **228. (D)** **229. (B)** **230. (D)**
231. (D) **232. (D)** **233. (D)** **234. (A)** **235. (D)** **236. (B)** **237. (B)** **238. (D)** **239. (B)** **240. (D)**
241. (D) **242. (B)** **243. (D)** **244. (D)** **245. (B)** **246. (D)** **247. (C)** **248. (C)** **249. (B)** **250. (D)**
251. (C) **252. (D)** **253. (B)** **254. (A)** **255. (D)** **256. (B)** **257. (D)** **258. (C)** **259. (A)** **260. (C)**
261. (B) **262. (C)** **263. (D)** **264. (A)** **265. (B)** **266. (C)** **267. (B)** **268. (B)** **269. (A)** **270. (C)**
271. (D) **272. (D)** **273. (A)** **274. (B)** **275. (C)** **276. (A)** **277. (B)** **278. (C)** **279. (D)** **280. (A)**
281. (A) **282. (D)** **283. (A)** **284. (B)** **285. (C)** **286. (A)** **287. (B)** **288. (C)** **289. (A)** **290. (A)**
291. (D) **292. (D)** **293. (A)** **294. (C)** **295. (B)** **296. (D)** **297. (A)** **298. (D)** **299. (A)** **300. (B)**
301. (A) **302. (C)** **303. (A)** **304. (B)** **305. (C)** **306. (C)** **307. (C)** **308. (C)** **309. (D)** **310. (B)**
311. (D) **312. (D)** **313. (D)** **314. (D)** **315. (A)** **316. (B)** **317. (D)** **318. (C)** **319. (D)**

Chapter - 15

BIOINFORMATICS

1. LCR using reverse transcriptase; specific for detection of RNA by:
(A) AG-LCR (B) Asymetrical PCR
(C) Both (A) and (B) (D) None of these

2. The two PCR primers used in a 100:1 ratio so that one primer is finished 10 cycles before the other; yields predominantly single-strand copies that can be used for DNA seque4ncing in:
(A) AG-LCR (B) Asymetrical PCR
(C) Both (A) and (B) (D) None of these

3. The application of computational techniques for the analysis of biological data (computational biology) that provides not just sequence analysis, but also structural bioinformatics and the stimulation of molecular interactions like metabolic pathways and gene networks is called:
(A) Bioinformatics (B) Bioethics
(C) Computer hardware (D) None of these

4. The information sciences, computer programming skills especially with scripting languages are useful:
(A) PERL (B) EMBL
(C) EST (D) GenBank

5. The important data bases in bioinformatics are used:
(A) SWISS-PROT (B) EMBL
(C) GenBank (D) Pfam
(E) All of these

6. The primary nucleotide sequence databases are:
(A) EMBL (B) GenBank
(C) DDBI (D) All of these

7. KRAF_HUMAN is the entry name for the raf-1 oncogene from:
(A) *Homo sapiens*
(B) *Avena sativa*
(C) *Pisum sativum*
(D) *Hoemophilus influenza*

8. The database has sequences deposited normally after verification is:
(A) GenBank (B) DDBI
(C) EST (D) EMBL

9. The data base have cDNA single-pass sequences and contain a high degree of error is:
(A) GenBank (B) DDBI
(C) EST (D) EMBL

10. Deletions and insertions are collectively known as:
(A) Indels (B) Intels
(C) EST (D) None of these

11. Length of the DNA sequence affected by the:
(A) Mutational event (B) Point mutation
(C) Segmoidal mutation (D) KEGG database

12. To affect a single nucleotide are called:
(A) Mutational event (B) Point mutation
(C) Segmoidal mutation (D) KEGG database

13. The mutations that affect several adjacent nucleotides is:
(A) Mutational event (B) Point mutation
(C) Segmoidal mutation (D) KEGG database

14. The databases are those of metabolic pathways and resources is:
(A) Mutational event (B) Point mutation
(C) Segmoidal mutation (D) KEGG database

15. The KEGG database is an effort to computerize current knowledge of:
(A) Molecular (B) Cellular biology
(C) Both (A) and (B) (D) None of these

16. The protein sequence database are:
(A) SWISS-PROT (B) PIR
(C) Both (A) and (B) (D) None of these

17. The translated DNA databases are:
(A) PROTEIN (NCBI) (B) TREMBL
(C) Both (A) and (B) (D) None of these

18. The sequence derived databases are:
(A) PROTEIN (NCBI) (B) TREMBL
(C) Pfam (D) SWISS-PROT

19. The database for 3D structures of biological macromolecules determined by X-ray crystallography and NMR is:
(A) PDB (B) SCOP
(C) Pfam (D) PAM

20. Secondary data base derived from PDB are the:
(A) PDB (B) SCOP
(C) Pfam (D) PAM

21. The dot plot was first described by:
(A) Gibbs and McIntyre
(B) Margaret Dayhoff
(C) Senger etal
(D) None of these

22. The steps in implementing the algorithm for alignment using dynamic programming is:
(A) Matrix initialization (B) Matrixfill
(C) Trace back (D) All of these

23. The programme is used to compare a protein sequence with a nucleotide (translated into all six frames) sequence database is:
(A) TBLASTN (B) TBLASTX
(C) PROSITE (D) PROMOT

24. The programme is used to compare a nucleotide (translated into all six frames) sequence with a nucleotide (translated into all six frames) sequence database is:
(A) TBLASTN (B) TBLASTX
(C) PROSITE (D) PROMOT

25. A collection of functional sites and sequences patterns found in proteins, it also has search tools for matching patterns/motifs is:
(A) TBLASTN (B) TBLASTX
(C) PROSITE (D) PROMOT

26. The search tool can be used to match a sequence against the PROSITE database is:
(A) TBLASTN (B) TBLASTX
(C) PROSITE (D) PROMOT

27. The PAM matrix was developed by:
(A) Gibbs and McIntyre (B) Margaret Dayhoff
(C) Senger etal (D) None of these

28. The ancestral relationship of a set of species which determined by the comparison of specific features (characters) of a species with the natural assumption that similar species-with similar features are genetically close is:
(A) Phylogeny (B) Microarray
(C) LigBuilder (D) None of these

29. Phylogeny usually described:
(A) Plant (B) Tree
(C) Ferns (D) Algae

30. The function of genes identified using an ab initio method is identified by comparing the gene sequence against known proteins using:
(A) BLAST (B) TBLASTX
(C) PROSITE (D) PROMOT

31. The availability of the complete set of genes from an organism, with their sequences, has enabled the study of expression profiles using:
(A) PROSITE (B) BLAST
(C) Microarray (D) PROMOT

32. To introduce a mathematical formulation to determine the potential energy of a molecular called:
(A) Microarray
(B) Molecular force field
(C) Shot gun sequencing strategy
(D) Shot gun approach of sequence

33. Genome sequencing became possible on a large scale with the discovery of the:
(A) Microarray (B) Molecular force field
(C) Shot gun sequencing strategy
(D) Shot gun approach of sequencing

34. The random selected clones are sequenced until all clones in the genomic library are analyzed, the sequences are aligned using computers by:
(A) Microarray
(B) Molecular force field
(C) Shot gun sequencing strategy
(D) Shot gun approach of sequencing

35. The partially digested genomic DNA is subjected to agarose gel electrophoresis or sucrose gradient configuration for separation from the mixture of fragments of appropriate size. The fragments are then inserted into a suitable vector for cloning by:
(A) Shot gun approach for gene cloning
(B) Molecular force field
(C) Shot gun sequencing strategy
(D) Shot gun approach of sequence

36. A de novo ligand design program which buildup ligands from a library of organic fragments taking into consideration the structural consist of the target protein is:
(A) LigBuilder (B) POCKET module
(C) GROW module (D) LINK module

37. LigBuilder progarmme basically consists of:
(A) POCKET (B) GROW
(C) LINK (D) None of these

38. The binding site of the target protein and derives the key interaction sites within this site is:
(A) LigBuilder (B) POCKET module
(C) GROW module (D) LINK module

39. Implemenats the sequential growing strategy of building up the molecular from a single starting block known as the seed by:
(A) LigBuilder (B) POCKET module
(C) GROW module (D) LINK module

40. Buildup the molecule from several seed structures which have been placed in the binding site so as to maximize interactions with the protein by:
(A) LigBuilder (B) POCKET module
(C) GROW module (D) LINK module

41. Which is concerned with determination of the function of a gene product?
(A) Functional proteomics
(B) Structural proteomics
(C) Functional genomics
(D) Structural genomics

42. The sequence of a human genome has how many nucleotide pairs?
(A) 4.2×10^6 (B) 4.6×10^9
(C) 3.2×10^9 (D) 3.8×10^9

43. One megabase pairs (Mb) is equal to how many base pairs?
(A) 10^5 bp (B) 10^6 bp
(C) 10^8 bp (D) 10^9 bp

44. Databases on DNA sequences contain which types of sequences?
(A) EST and GST sequences
(B) Genomic DNA and cDNA sequences
(C) Organellar DNA sequences and sequences of other molecules such as tRNA, small RNAs
(D) All of these

45. Which is an example of major nucleotide sequence database?
(A) DNA Databank of Japan
(B) Nucleotide Sequence Database maintained by EMBL
(C) Gene Bank held by NCBI, USA
(D) All of these

46. In which risk groups, agents are not associated with diseases in healthy humans?
(A) Risk group 3 (B) Risk group 1
(C) Risk group 2 (D) Risk group 4

47. Functional genomics includes answer to:
(A) How is a gene expressed?
(B) How is a gene product related in sequence and structure to products of other genes of the same organisms
(C) How does a gene interacts with the proteins?
(D) All of these

48. The α–amylase inhibitor from *Phaseolus vulgaris* (rajma) caused allergy when expressed in:
(A) Wheat (B) Maize
(C) Peas (D) Tomato

49. Which eukaryotic genome was sequenced in 1999?
(A) Caenorhabditis elegans (worm)
(B) *Drosophila melanogaster*
(C) *Saccharomyces cerevisiae*
(D) Both (A) and (C)

50. Non-infectious illnesses is associated with production of steroids:
(A) Increased body weight
(B) Feminization of males
(C) Hypertension (D) All of these

51. The potential risks from novel DNA sequences used during transgenic development may be summed up as:
(A) Protein products of transgenes or products generated by them may be toxic
(B) The transgene product may be allergenic
(C) Antibiotic resistance genes may be taken up by human intestinal microflora, which, as a consequence, become resistant to these antibiotics
(D) All of these

52. Strategy of temporal and tissue-specific expression for preventing transgene transfer from transgenic varieties of crops is based on which techniques?
(A) Terminator technology
(B) Tapetum-specific promoters allow expression only when the suppressor is present
(C) Recoverable block function
(D) Inducible promoters allow expression only when the inducer is present

53. Which is not correct about printed oligonucleotide chips?
(A) These are produced by a light-directed printing technology
(B) These chips have double-stranded oligonucleotides of 20-25 bases
(C) Oligonucleotides are synthesized directly onto the chips
(D) Each sequence is represented by a set of ~ 20 non-overlapping oligonucleotides to reduce false positive results

54. DNA molecule used for synthesis of complementary sequences either in vivo or in PCR is called:
(A) T-DNA (B) Template DNA
(C) Z-DNA (D) Carrier DNA

55. The two strands of a DNA molecule are coiled together in a left-handed helix having diameter of 18Å, and the length of one turn of the helix is Ca 69Å; it contains 12 base pairs per turn in the helix is called:
(A) T-DNA (B) Template DNA
(C) Z-DNA (D) Carrier DNA

56. Which scientists proposed that amino acid sequences of proteins may be used to study evolutionary relationships among organisms?
(A) Zuckerland and Pauling
(B) Sensen and Buchler
(C) Dayhoff and Rashidi
(D) Tolin and Vidaver

57. The vector under host-vector system includes:
(A) Nonconjugative plasmids
(B) Derivative of nonconjugative plasmids
(C) Variants of bacteriophages
(D) All of these

58. Most of the recombinant microorganism released in the environment of genes by:
(A) Transposon insertion
(B) Deletion of genes by molecular genetic procedure *in vitro*
(C) Genetic tagging of bacteria with specific
(D) Any of these

59. Insertional mutagenesis has been reported in:
(A) Mouse and Drosophila melanogaster
(B) Yeast and roundworm
(C) Zebrafish and Arabidopsis
(D) All of these

60. In genetic transformation; unrelated DNA added to the solution of DNA used for transformation, enhances transformation frequency, rearrangements and multicopy integration is called:
(A) T-DNA (B) Template DNA
(C) Z-DNA (D) Carrier DNA

61. Which is used for comparing the submitted nucleotide sequence with nucleotide database?
(A) BLASTn (B) BLASTx
(C) tBLASTx (D) tBLASTn

62. The rough draft of human genome was announced on:
(A) June 26, 2000 (B) June 26, 2001
(C) June 26, 1999 (D) June 26, 2003

63. The *Mycoplasma genitalium* complete genome was chemically synthesized and assembled in which year?
(A) 2005 (B) 2006
(C) 2007 (D) 2008

64. Which is not correct about homologous recombination?
(A) Exchange of identical segments occurs between two heterologous DNA molecules/chromosomes
(B) To achive transgene integration by homologous recombination, the marker gene is flanked by appropriate sequences from or around the target gene
(C) Mouse embryonic stem cells, may show upto 0.1% integration by homologous recombination
(D) In prokaryotes, integration by homologous recombination may occur upto 90% of the time when suitable targeting vectors are used

65. Which is not correct about Gene trap vector?
(A) In this vector, the insertion element contains a reporter gene, e.g., lacZ or gusA, downstream of a splice acceptor site
(B) Reporter gene will be expressed only when the vector inserts within the transcribed region of a gene
(C) It is used for detection and isolation of genes
(D) This vector was isolated from *Bacillus thuringiensis*

66. Which programmes can be used for the identifications/detection of genes from genome sequence of prokaryotes?
(A) Glimmer (B) GENMARK
(C) GeneFinder (D) Both (A) and (B)

67. India has not been involved in sequencing the genomes of which organisms as:
(A) Pigeon (B) Rat
(C) Pigeonpea (D) Humans

68. Which database is used by ENTREZ for bibliographic or citation search?
(A) SWISS-PROT (B) PubMed
(C) TrEMBL (D) PDB

69. Databases on DNA sequences contain which types of sequences?
(A) EST and GST sequences
(B) Genomic DNA and cDNA sequences
(C) Organellar DNA sequences and sequences of other molecular such as tRNA, small RNAs
(D) All of these

70. Effective dissemination of plant-associated micro-organisms beyond the green house can be prevented by using:
(A) Organisms that have been genetically disabled to survive in nature
(B) Organisms whose natural mode of transmission requires injury to the host
(C) Organisms which require another organisms for their survival
(D) Both (A) and (B)

71. The first recombinant microorganism (Tn5-containing strain of *Pseudomonas fluorescens)* was released in which countries?
(A) USA (B) Canada
(C) Netherlands (D) Germany

72. Which advantages is offered by Real-Time PCR?
(A) Multiplex PCR reactions can be used by employing molecular beacons/TagMan probes
(B) It eliminates the post – PCR processing of PCR products needed for quantitative measurements of PCR products
(C) Real-Time PCR offers a much wider dynamic range than conventional RT-PCR
(D) All of these

73. In case of experiments with animals, which procedures is followed at all biosafety levels?
(A) Animal carecass is disposed off to avoid it use as food
(B) A permanent record is maintained of the experimental use and disposal of each animal
(C) The animals are confined to security-fenced areas
(D) All of these

74. Which is not correct about NIH guidelines?
(A) A major revision of the guidelines was effected in 1982
(B) In the revision, the containment levels were made more stringent
(C) Experiments that were previously prohibited, were changed to category requiring review and approval by NIH
(D) In USA, the NIH guidelines are followed by all federal agencies that fund research on recombinant DNA

75. The short gun sequencing approach would be most suited for which genomes?
(A) *E. coli* (B) Yeast
(C) Humans (D) *Drosophila*

76. A Ca 23 kb segment of Ti/Ri plasmid that has 24 bp direct repeat border sequence and it transferred into the host plant genome during pathogenesis; has genes for auxin, cytokinin and opion biosynthesis is called:
(A) T-DNA (B) Template DNA
(C) Z-DNA (D) Carrier DNA

77. Which is not correct about TaqMan probes?
(A) They are oligonucleotides longer than the primers
(B) They contain a fluorescent dye usually on the 5'-base, and a quenching dye typically on the 3'-base
(C) When Taq polymerase replicates a tempelate to which a TaqMan probe is paired, its 5'-exonuclease activity cleavage the TaqMan probe
(D) The fluorescence decreases proportionately to PCR amplification

78. Which is correct about genome annotation?
(A) GAME is a programme for describing experimental evidence to support annotation
(B) DAS is useful in indexing and visualization
(C) BioPerl 2001 is used for storing, manipulating and visualizing the genome annotation
(D) All of these

79. The type of regulation necessary during field trials should depend on:
(A) Ability of the modified plants to survive in nature
(B) Their ability for dispersal and reproduction
(C) Their ability to hybridize with crop and weed plants
(D) All of these

80. Which is not correct?
(A) RecA- strains lack recombination
(B) Transposon vectors lacking transposase gene are not able to transpose to other bacteria
(C) Auxotrophic mutants can survive in nature
(D) Transgenes integrated into plasmids may be transferred to other bacteria

81. Which terms is not used as synonyms of planned introduction?
(A) Field testing
(B) Deliberate release
(C) Contained use
(D) Deliberate introduction

82. The first comprehensive collection of amino acid sequences was compiled in the Atlas of Protein Sequence and Structure by:
(A) European Bioinformatics Institute
(B) National Biomedical Research Foundation
(C) National Centre for Bioinformatics Information
(D) European Molecular Biology Laboratory

83. Which is used for comparing the submitted nucleotide sequence with nucleotide database?
(A) BLASTn (B) BLASTx
(C) rBLASTx (D) tBLASTn

84. Transgenic plants could serve as vectors for transgene dissemination through:
(A) Seed
(B) DNA released from debris of transgenic plants
(C) Pollen (D) All of these

85. For minimizing pollen-mediated gene flow from transgenic varieties of *Solanum tuberosum*, how many meters of isolation distance is recommended?
(A) 10 (B) 20
(C) 30 (D) 40

86. Which of the search tools is used for comparing the submitted nucleotide sequence with amino acid sequence database?
(A) BLASTn (B) BLASTx
(C) rBLASTx (D) tBLASTn

87. Which is correct about transmission of transgenes from one crop species to another?
(A) In case of cross-pollinated species isolation distance is rarely sufficient to prevent inter-population mating
(B) If male sterility is introduced into transgenic cross - pollinated crops, it can check the movement of transgenes through pollen
(C) It transgenes are integrated in the chloroplast genome, it would almost eliminated the transmission of transgene through pollen grains
(D) All of these

88. Which is not correct?
(A) European Molecular Biology laboratory endeavours to collect, organise and distribute nucleotide sequence data and related information
(B) National Centre of Bioinformatics was established in USA and serves as primary information databank and provider of information
(C) RNA data bank was established by Japan
(D) National Biomedical Research Foundation established the protein Information System in 1984

89. Which of the search tools is used to compare submitted amino acid sequence with nucleotide sequence database?
(A) BLASTn (B) BLASTx
(C) rBLASTx (D) tBLASTn

90. Physical containment relates to:
(A) Containment equipment
(B) Laboratory design
(C) Special procedures
(D) All of these

91. Which of the tools compares submitted amino acid sequence with amino acid database?
(A) BLASTn (B) BLASTx
(C) rBLASTx (D) tBLASTn

92. Which is not correct?
(A) Oral exposure to transgenic tomato containing cry protein poses no additional risk to human and animal health
(B) In general, the quality of produce from cry gene carrying transgenic crops is regarded different from that the nontransgenic plants of the same variety
(C) Cowpea trypsin inhibitors do show some adverse effects on rats in nutritional studies
(D) Significant hypertrophy of small intestine and altered enzyme activities in brush border occur due to wheat germ agglutinin

93. Single nucleotide polymorphisms (SNPs) may be located in:
(A) In the recognition sequence of an endonuclease
(B) Within the concerned gene
(C) Very close to the gene to show a tight linkage
(D) All of these

94. Which approaches can be used for gene identification?
(A) Exon trapping (B) cDNA capture
(C) cDNA selction (D) All of these

95. The biosafety guidelines are developed to ensure an adequate level of protection during which activities of modern biotechnology?
(A) Safe transfer (B) Handling
(C) Use of GMOs (D) All of these

96. In case of prokaryotes, which represents the frequency of unique ORFs?
(A) 50% (B) 100%
(C) 25% (D) 75%

97. The NIH, USA guidelines specify the practices for constructing and handling are:
(A) Recombinant DNA molecules
(B) Organisms and viruses containing recombinant DNA molecules
(C) Products obtained from B
(D) Both (A) and (B)

98. Which is not correct about TaqMan probes?
(A) They are oligonucleotides longer than the primers
(B) They contain a fluorescent dye usually on the 5'-base, and a quenching dye typically on the 3'-base
(C) When Taq polymerase replicates a template to which a TaqMan probe is paired, its 5'-exonuclease activity cleavage the TaqMan probe
(D) The fluorescence decreases proportionately to PCR amplification

99. Sequencing of the 24 human chromosomes was completed in which years?
(A) 2004 (B) 2005
(C) 2006 (D) 2007

100. Only one primer is used in combination with dideoxynucleotides; the primer is either radio or fluorescence labeled. The product of PCR reaction is subjected to PAGE for determining the base sequence of the template DNA is :
(A) Representational Difference Analysis
(B) Thermal Cycle Sequencing PCR
(C) Restriction Fragment Length Polymorphism (RFLP)
(D) Reverse Transcription PCR (RT-PCR)

101. It is feared that genetically engineered microorganisms may disturb the ecosystem and its processes, in which ways?
(A) They may rapidly multiply and out complete the native microbes
(B) They may transfer genes related to virulence or pathogenesis into the native bacterial populations
(C) Both (A) and (B)
(D) They may cause unexpected health hazards

102. DNA sequences/molecules used for preparing spotted DNA arrays may be obtained from:
(A) PCR amplification (B) Genomic DNA
(C) cDNA clones (D) All of these

103. Which eukaryotic genome was sequenced in 1999?
(A) *Caenohabditis elegans* (worm)
(B) *Drosophila melanogaster*
(C) *Saccharomyces cerevstae*
(D) Both (A) and (C)

104. Which is not correct about Cry proteins?
(A) Cry proteins are rapidly degraded by the stomach juices of vertebrates
(B) They could have harmful effects on non target insect species
(C) It has been reported that Cry proteins leak into soil from transgenic maize roots, where they persist for over 3 weeks
(D) The Cry proteins in the soil become associated with soil clay/humus and become prone to microbial degradation

105. When PCR is used to achieve substractive cloning, this is called:
(A) Representational Difference Analysis
(B) Pulsed-Field Gel Electrophoresis (PFGE)
(C) Restriction Fragment Length Polymorphism (RFLP)
(D) Reverse Transcription PCR (RT-PCR)

106. Which advantages is offered by Real-Time PCR
(A) Multiplex PCR reactions can be used by employing molecular beacons/TaqMan probes
(B) It eliminates the post-PCR processing of PCR products needed for quantitative measurements of PCR products
(C) Real-time PCR offers a much wider dyanamic range than conventional RT-PCR
(D) All of these

107. The sequence of a human genome has how many nucleotide pairs?

(A) 4.2 x 10^6 (B) 4.6 x 10^9
(C) 3.2 x 10^9 (D) 3.8 x 10^9

108. Which is not correct?

(A) The public funded Human Genome Project employed the clone-by-clone Celera Genomics followed whole genome shot-gun approach
(B) Clone-by-clone approach is based on contigs
(C) In the whole genome shot-gun approach, genome DNA is sheared to obtain 2 to 10 kb fragments, which are then cloned and sequenced from both the ends
(D) Assigning biological functions to the DNA sequences using a variety of tools & techniques is described as structural genomics

109. Which is not correct about Real-time PCR?

(A) Real-time PCR can detect the size of the amplicon
(B) Design of primers, TaqMan probes and molecular beacons requires almost complete information of the target sequence
(C) With SYBR green method, primer dimer artifact is a problem
(D) The one-step RT_PCR can not be used for multiplex real-time RT-PCR

110. Difference between individuals in terms of the size of fragments generated by restriction enzyme digestion of genomic DNA and detected by hybridization with specific DNA sequences used as probe is:

(A) Random Amplified Polymorphic DNA (RAPD)
(B) Pulsed-Field Gel Electrophoresis (PFGE)
(C) Restriction Fragment Length Polymorphism (RFLP)
(D) Reverse Transcription PCR (RT-PCR)

111. Which is not correct about bromoxynil?

(A) It is a non-biodegradable herbicide
(B) It causes birth defects in laboratory animals
(C) It is toxic to fish and may cause cancer in humans
(D) Bromoxynil is absorbed through the skin

112. India has initiated genome sequencing project of:

(A) Rice (B) Tomato
(C) Pigeon (D) Human

113. The α–amylase inhibitor gene was transferred to modify which traits?

(A) Insect resistance (B) Amino acid balance
(C) Food digestibility (D) Both (A) and (C)

114. Which is considered as secondary physical containment?

(A) Laboratory design
(B) Containment equipment
(C) Special procedures
(D) Both (A) and (B)

115. Which is not correct?

(A) Both TaqMan probes and molecule beacons allow multiplex reactions, and thereby, detection of multiple DNA sequences in a single PCR reaction
(B) In multiplexing, different probes/beacons are labelled with different reporter dyes
(C) Multiplexing does not allow amplification in a single PCR reaction of both target and endogeneous control sequences
(D) The use of probe/beacons avoids problems due to mispriming, primer-dimer artifacts, etc.

116. Which is not correct about genomics?

(A) In 1980, David Botstein and coworkers working at the Whitehead Institute of Biomedical Research, USA proposed that one could construct a genome map
(B) In 1995, the first bacterial genome to be completely sequenced was that of *Haemophilus influenza*
(C) The sequencing of whole genome of first multicellular organism, round worm (*Caenorhabdits elegans*), was published in 1999
(D) The sequencing of rice genome by International Rice Genome Sequencing Project was based on whole genome shot-gun approach

117. Which statements is correct about comprehensive risk assessment?

(A) It is after the initial risk assessment
(B) The level of containment appropriate for an experiment is decided on the basis of comprehensive risk assessment
(C) The containment level for the experiment may, in the end, be comparable to that of the risk groups (RG) of the agent or it may be raised or lowered as a result of the comprehensive risk assessment
(D) All of these

118. First of all cDNA of the mRNA is produced using the enzyme reverse transcriptase; the cDNA single – strand is then used for amplification to obtain cDNA deplexes:

(A) Random Amplified Polymorphic DNA (RAPD)
(B) Pulsed-Field Gel Electrophoresis (PFGE)
(C) Restriction Fragment Length Polymorphism (RFLP)
(D) Reverse Transcription PCR (RT-PCR)

119. In which of the following is found the lowest density of transposable elements?

(A) *Arabidopsis* (B) *Drosophila*
(C) *C. elegans* (D) Humans

120. It is feared that transgenic plants, may lead to:

(A) Transfer of their transgenic feature to weeds
(B) May themselves become weeds
(C) Degrade the environment
(D) All of these

121. Which is not correct?
(A) The public funded Human Genome Project employed the clone-by-clone approach, while the private company Celera Genomics followed whole genome shot-gun approach
(B) Clone-by-clone approach is based on contigs
(C) In the whole genome shot-gun approach, genomic DNA is sheared to obtain 2 to 10 kb fragments, which are then cloned and sequenced from both the ends
(D) Assigning biological functions to the DNA sequences suing a variety of tools & techniques is described as structural genomics

122. Which is not correct?
(A) Transgene isolated from a food crop guarantees the food safety of the transgene product
(B) Acute toxic effect depends on the concentration of the compounds in question
(C) Psoralen is known to be toxic, carcinogenic and mutagenic
(D) Any outstanding insect resistance whose bio-chemical basis is not known should be conside-red as the prime candidate for toxicological evaluation

123. Which is not correct?
(A) In databases, base sequence of only one strand is listed
(B) Base sequence runs from the 3' to the 5'-directions
(C) In single letter codes, R denoted guanine or adenine, while Y denotes thymine or cytosine
(D) The nomenclature system adapted in bioinform-atics is based on he recommendations of Intern-ational Union of Pure and Applied Chemistry

124. DNA is driven through the gel using alternating pulses of electrical current that are at different an-gles to each other; it can achieve a good separation of DNA fragments that vary from 50 kb to 10 mb is:
(A) Random Amplified Polymorphic DNA (RAPD)
(B) Pulsed-Field Gel Electrophoresis (PFGE)
(C) Restriction Fragment Length Polymorphism (RFLP)
(D) Reverse Transcription PCR (RT-PCR)

125. Which is not correct about biosafety Guidelines in India?
(A) Initially, the Indian Recombinant DNA safety Guidelines and Regulation were prepared in 1990 b Recombinant DNA Advisory Committee, Department of Biotechnology, New Delhi
(B) The Indian Recombinant DNA safety Guide lines and Regulation were revised in 1994
(C) Field trials, in India using transgenic plants began in 1995
(D) Canada become the first country to begin commercial use of virus resistant transgenic tobacco and tomato in the early 1990

126. In case of *E. coli*, the host for any HV 1 system is:
(A) Strain K -12
(B) A derivative of strain K-12
(C) Either (A) or (B) (D) Strain chi 1776

127. In which of the following is found the highest density of transposable elements?
(A) *Arabidopsis* (B) *Drosophila*
(C) *C. elegans* (D) Humans

128. The strategy uses nucleic acid hybridization in colony hybridization or chromosome walking/ jumping or is based on PCR is:
(A) Sequence – dependent screening
(B) Sequence–Tagged Microsatellites Sites (STMS)
(C) Sequence Tagged Site (STS)
(D) Single Locus Probe Analysis (SLP)

129. Which is concerns with determination of the function of a gene product?
(A) Expression proteomics
(B) Structural genomics
(C) Functional genomics
(D) Structural proteomics

130. Which is correct about wheat germ agglutinin?
(A) It is effective against insects at concentration of 10 mg/kg body weight
(B) It produces cytotoxic effects in rats at a lower concentration (7 mg/kg)
(C) Safe food has it only at 0.3 mg/kg
(D) All of these

131. DNA sequences/molecules used for preparing spotted DNA arrays may be obtained from:
(A) PCR amplification (B) Genomic DNA
(C) cDNA clones (D) All of these

132. The *Mycoplasma genitalium* complete genome was chemically synthesized and assembled in which year?
(A) 2005 (B) 2006
(C) 2007 (D) 2008

133. Shot-gum sequencing was conceptualized by:
(A) Craig Venter (B) Michelmore
(C) Giovannoni (D) Young

134. Which phenomenon comes under post-translat-ional modifications?
(A) Phosphorylation (B) Proteolysis
(C) Glycosylation (D) All of these

135. In the early 1970s which were considered as the possible potential health hazards of cloning recombinant DNA molecules?
(A) Molecular biologists may not be well-versed with the laboratory practices needed to handle serious human and animal pathogens
(B) 'Hybrid organisms' could be created with biological activities of an unpredictable nature
(C) 'Hybrid organisms' may escape from the laboratory with unpredictable consequences
(D) All of these

136. The chief objectives of risk assessment during laboratory research are to decide about which of the proposed research?
(A) Laboratory procedures
(B) Biological containment
(C) Physical containment
(D) All of these

137. Which approaches can be used for gene identification?
(A) Exon trapping (B) cDNA capture
(C) cDNA selection (D) All of these

138. Which is not correct about homologous recombination?
(A) Exchange of identical segments occurs between two heterologous DNA molecules chromosomes
(B) To achieve transgene integration by homologous recombination, the marker gene is flanked by appropriate sequences from or around the target gene
(C) In prokaryotes, integration by homologous recombination may occur upto 90% of the time when suitable targeting vectors are used
(D) Mouse embryonic stem cells may show upto 0.1% integration by homologous recombination

139. In the assessment of allergenicity, the source of transgene is the first consideration; the transgene may be obtained from:
(A) Commonly allergenic food
(B) Less commonly allergenic food
(C) No documented history of food allergenicity
(D) Any of these

140. Which software have been developed for the prediction/detection of genes from genome sequence of eukaryotes?
I. HMM Gene and GRAIL
II. GeneFinder and GENESCAN
III. GENIE
IV. GENMARK
Options:
(A) I, II, III (B) II, III, IV
(C) I, II (D) I, II, III, IV

141. In yeast chromosomes, it is a 20-70 tandem repeat of the 6-base sequence 5'CCCCAA3' (its complementary sequence, 5'TTGGGG3', occurs in the other strand), there is a hairpin loop formation at the terminus, which makes the DNA duplex resistant to endonuclease action is called:
(A) Target sequence
(B) Telomeric sequence
(C) Recognition sequence (D) None of these

142. Which is used to comparing the submitted nucleotide sequence with amino acid sequence data base?
(A) BLASTn (B) BLASTx
(C) tBLASTx (D) tBLASTn

143. A transgenic variety of which crop is to withdrawn from cultivation following due to allergy?
(A) Maize (B) Soybean
(C) Tomato (D) *Brassica napus*

144. The assessment of risk during laboratory research is usually done in which steps?
(A) Initial risk assessment
(B) Comprehensive risk assessment
(C) Advanced risk assessment
(D) Both (A) and (B)

145. Which is used to compare submitted amino acid sequence with nucleotide sequence data base?
(A) BLASTn (B) BLASTx
(C) tBLASTx (D) tBLASTn

146. Which is used to compares submitted amino acid sequence with amino acid database?
(A) BLASTn (B) BLASTx
(C) tBLASTx (D) tBLASTn

147. Which is not correct about lectins?
(A) They are polysaccharides
(B) They are involved in defence mechanism of plants
(C) Several lectins have been identified as food allergens
(D) α-Amylase/tripsin inhibitor family proteins are potential allergens

148. Which PCR strategies monitors the fluorescence emitted during reaction to measure the amount of amplification product generated during the PCR cycles?
(A) Anchored PCR (B) Touchdown PCR
(C) Real-time PCR (D) Nested PCR

149. Which data base is used by ENTREZ for bibliographic or citation search?
(A) SWISS-PROT (B) PubMed
(C) TrEMBL (D) PDB

150. Which is not correct about NIH guidelines?
(A) The first NIH guidelines were prepared in 1975
(B) NIH guidelines were more liberal than the recommendations of the Asilomar Conference
(C) The guidelines were revised after two years
(D) In 1977, NIH prepared an Environmental Impact Statement

151. In risk assessment, a lack of scientific knowledge or scientific consensus should not necessarily be interpreted as indicating:
(A) A particular level of risk
(B) Absence of risk
(C) Acceptable risk
(D) Any of these

152. The rate of gene-flow is affected by :
(A) Mode of seed dispersal
(B) Mating system
(C) Mode of pollination (D) All of these

153. Which is not correct about genomics?
(A) Sequencing, sequence assembly and annotation of the whole genome of an organism constitutes functional genomics
(B) The first plant genome to be completely sequenced is that of the weed, *Arabidopsis thaliana*
(C) *Arabidopsis thaliana* is a wild relative of mustards
(D) The *Arabidopsis* Genome Initiatives followed the same methods of sequencing as the Human Genome Project

154. The sequences recognized by restriction endonucleases for generating cleavage is called:
(A) Target sequence
(B) Telomeric sequence
(C) Recognition sequence
(D) None of these

155. Difference between individuals in terms of DNA regions either being or not being amplified in a polymerase hair reaction primed by random ~10 nucleotide long sequence by:
(A) Random Amplified Polymorphic DNA (RAPD)
(B) Pulsed-Field Gel Electrophoresis (PFGE)
(C) Restriction Fragment Length Polymorphism (RFLP)
(D) Reverse Transcription PCR (RT-PCR)

156. Insertional mutagenesis has been reported in:
(A) Mouse and *Drosophila melanogaster*
(B) Yeast and roundworm
(C) Zebrafish and *Arabidopsis*
(D) All of these

157. Occurence of different nucleotides at a specific site in the genome of different individuals of a species, human genome shows SNPs at about every 1000 bp is called:
(A) Single Nucleotide Polymorphism (SNP)
(B) Single Primer Amplification Reaction (SPAR)
(C) Single Strand Conformational Polymerase (SSCP)
(D) SSR-Anchored PCR (SSR-APCR)

158. DNA marker system that uses primers based on microsatellites and amplifies inter-SSR and DNA sequences, produces multiple fragments per reaction is called:
(A) Single Nucleotide Polymorphism (SNP)
(B) Single Primer Amplification Reaction (SPAR)
(C) Single Strand Conformational Polymerase (SSCP)
(D) SSR-Anchored PCR (SSR-APCR)

159. Development of practices/procedures, assays/tests, biological agents, equipment and working design and structures to minimize the risks is called:
(A) Risk regulation
(B) Risk management
(C) Risk perception
(D) Risk determination

160. The marker relies on secondary and tertiary structure difference between denaturated and rapidly cooled amplified DNA fragments that differ slightly in their base sequence. Different SSCP alleles are resolved on non-denaturing acrylamide gel, usually at low temperature by:
(A) Single Nucleotide Polymorphism (SNP)
(B) Single Primer Amplification Reaction (SPAR)
(C) Single Strand Conformational Polymerase (SSCP)
(D) SSR-Anchored PCR (SSR-APCR)

161. This is essentially single primer amplification reaction, except that the primer is either anchored at the 3'-or 5' – terminus with two to four arbitrary nucleotides by:
(A) Single Nucleotide Polymorphism (SNP)
(B) Single Primer Amplification Reaction (SPAR)
(C) Single Strand Conformational Polymerase (SSCP)
(D) SSR-Anchored PCR (SSR-APCR)

162. In case of which products, an empirical approach based on data obtained through experimentation about the source of potential hazards has proved satisfactory?
(A) Pesticides (B) Drugs
(C) Vaccines (D) All of these

163. India has initiated genome sequencing project of:
(A) Rice (B) Tomato
(C) Pigeonpea (D) Humans

164. Which progamme can be used for the identification/detection of genes from genome sequence of prokaryotes?
(A) Glimmer (B) GENMARK
(C) Gene-finder (D) Both (A) and (B)

165. The desired segment that is to be amplified by PCR is called:
(A) Target sequence (B) Telomeric sequence
(C) Recognition sequence (D) None of these

166. Which is found the highest density of transposable elements?
(A) *Arabidopsis* (B) *Drosophila*
(C) *C. elegans* (D) Humans

167. Which is concerned with the development and application of computer hardware and software for acquisition, storage, analysis and visualization of biological information?
(A) Computer science
(B) Bio-infotechnology
(C) Bioinformatics
(D) Information technology

168. Non experts intuitive judgment about risks, which often are not in agreement with the level of risk as judged by experts is known as:
(A) Risk regulation (B) Risk management
(C) Risk perception (D) Risk determination

169. Task of regulators to review risk data and make regulatory decisions based on the data with a view to minimize risk. The process often is influenced by consideration of costs and benefits as well as by public perceptions:

(A) Risk regulation (B) Risk management
(C) Risk perception (D) Risk determination

170. It is the estimation of the overall risk posed by a GMO based on the evaluation of the likelihood and consequences of the identified adverse effects being realized is called:

(A) Risk regulation (B) Risk management
(C) Risk perception
(D) Risk determination

171. Programme is used to convert protein sequence into nucleotide sequence and to compare it with a nucleotide sequence database is:

(A) tBLASTn (B) tBLASTx
(C) BLASTn (D) tBLAST

172. Programme is used to translate the submitted nucleotide sequence as well as the nucleotide sequence database into amino acid sequence and search for homology between the two is:

(A) tBLASTn (B) tBLASTx
(C) BLASTn (D) tBLAST

173. The programme is used to compare a nucleotide sequence with a nucleotide sequence database is:

(A) BLASTn (B) BLASTx
(C) BLASTp (D) tBLAST

174. It compare the submitted protein sequence against a protein database is:

(A) BLASTn (B) BLASTx
(C) BLASTp (D) tBLAST

175. The programme translated the submitted nucleotide sequence into amino acid sequence and compares the letter with a protein database is:

(A) BLASTn (B) BLASTx
(C) BLASTp (D) tBLAST

176. Sequencing of the 24 human chromosomes was completed in which year?

(A) 2004 (B) 2005
(C) 2006 (D) 2007

177. The term genomics was coined by:

(A) Thomas H. Roderick (B) H. Winkler
(C) A.H. Paterson (D) Benzer

178. According to biosafety guidelines of India, physical containment envisages to limit the spread of dangerous microorganisms by:

(A) Laboratory practices (B) Safety equipment
(C) Laboratory design and facilities
(D) All of these

179. Shot-gun sequencing was conceptualized by:

(A) Craig Venter (B) Michelmore
(C) Giovannoni (D) Young

180. The principal purpose of transgenic plant containment is to avoid the unintentional transmission of a recombinant DNA containing plant genome including:

(A) Organellar hereditary material
(B) Nuclear hereditary material
(C) Release of recombinant DNA-derived organisms associated with plants
(D) Any of these

181. In which risk groups, is concerned with agents which are associated with such serious or lethal human diseases for which preventive or therapeutic interventions may be available?

(A) Risk group 3 (B) Risk group 2
(C) Risk group 1 (D) Risk group 4

182. Non-infectious illness like allergic asthama and dermatitis are associated with production of:

(A) Brewing (B) Single cell protein
(C) Antibiotic (D) All of these

183. The rough draft of human genome was announced on:

(A) June 26, 2000 (B) June 26, 2001
(C) June 26, 2002 (D) June 26, 2003

184. Which can be used to identify genes in bacterial genomes?

(A) Protein-coding genes are composed of open reading frames
(B) Coding region begins with an initiation codon (usually ATG)
(C) Coding region ends with a termination codon (TAA, TAG or TGA)
(D) All of these

185. Which scientists proposed that amino acid sequences of proteins may be used to study evolutionary relationships among organisms?

(A) Zuckerland and Pauling
(B) Sensen and Buchler
(C) Dayhoff and Rashidi (D) Tolin and Vidaver

186. Which PCR strategies monitors the fluorescence emitted during reaction to measure the amount of amplification product generated during the PCR cycle?

(A) Anchored PCR (B) Touchdown PCR
(C) Real-Time PCR (D) Nested PCR

187. Which is not correct?

(A) Both TaqMan probes and molecular beacons allow multiplex reactions, and thereby, detection of multiple DNA sequences in a single PCR reaction
(B) In multiplexing, different probes/beacons are labeled with different reporter dyes
(C) Multiplexing does not allow amplification in a single PCR reaction of both target and endogeneous control sequences
(D) The use of probes/beacons avoids problems due to mispriming, primer dimmer artifacts, etc.

188. *E. coli* genome has how many operons?
(A) 200 (B) 600
(C) 1000 (D) 2000

189. The shot-gun sequencing approach would be the mostly suited for which genome?
(A) *E. coli* (B) Yeast
(C) Humans (D) Drosophila

190. Which is concerned with organisms which are associated with such human diseases, which are rarely serious and for which preventive or therapeutic intervention are often available?
(A) Risk group 4 (B) Risk group 3
(C) Risk group 2 (D) Risk group 1

191. Which hazards is associated with operation of bioreactor fermentation?
(A) Spillage
(B) Effluent contamination of gases
(C) Aerosols from reactor
(D) All of these

192. Lateral gene transfer is also called:
(A) Horizontal gene transfer
(B) Vertical gene transfer
(C) Parallel gene transfer
(D) Both (B) and (C)

193. Which is not correct about TaqMan probes?
(A) They are oligonucleotides longer than the primers
(B) They contain a fluorescent dye usually on the 5'-base , and a quenching dye typically on the 3' - base
(C) When Taq polymerase replicates a tempelate to which a TaqMan probe is paired, its 5'-exonuclease activity cleaves the TaqMan probe
(D) The fluorescence decreases proportionately to PCR amplification

194. In which years, a proposal for sequencing of human genome was prepared:
(A) 1984 (B) 1985
(C) 1986 (D) 1987

195. Which is not correct about printed oligonucleotide chips?
(A) These are produced by a light-directed printing technology
(B) These chips have double-stranded oligonucleotides of 20-25 bases
(C) Oligonucleotides are synthesized directly onto the chips
(D) Each sequence is represented by a set of ~20 non-overlapping oligonucleotides to reduce false positive results

196. In case of Real-time PCR which methods can be used for the detection of amplification products?
(A) Molecular beacons (B) Scorpions
(C) TaqMan probes (D) All of these

197. Which is not correct?
(A) National Biomedical Research Foundation established the Protein Information System in 1984
(B) National centre for Bioinformatics Institute was established in USA and serves as primary information databank and provider of information
(C) RNA Data Bank was established by Japan
(D) European Molecular Biology Laboratory endeavours to collect, organize and distribute nucleotide sequence data and related information

198. Which was the first prokaryotic genome to be sequenced?
(A) *E. coli*
(B) *Agrobacterium tumefaciens*
(C) *Haemophilus influenza*
(D) *Proteus vulgaris*

199. Which phenomenon comes under post-translational modification?
(A) Phosphorylation (B) Proteolysis
(C) Glycosylation (D) All of these

200. The amino acids were conventionally represented by three-letter symbols, while inbioinformatics, each of them is denoted by how many letters?
(A) Single letter (B) Two letters
(C) Three letters (D) Four letters

201. Which is concerned with organisms, which cause serious or lethal disease for which preventive or therapeutic invention are not usually available?
(A) Risk group 1 (B) Risk group 2
(C) Risk group 3 (D) Risk group 4

202. Non-infectious illness is associated with endotoxin production:
(A) Flu symptoms (B) Dermatitis
(C) Asthama (D) Cancer

203. Which procedures can be used for the identification/discovery of SNPs?
(A) DNA chips
(B) Microarrays
(C) Analyzing the sequence data
(D) All of these

204. Single nucleotide polymorphisms (SNPs) may be located in:
(A) In the recognition sequence of an endonuclease
(B) Within the concerned gene
(C) Very close to the gene to show a tight linkage
(D) All of these

205. Crop yields can be stabilized by introducing which genes in transgenic varieties?
(A) Genes for improvement nutritional quality
(B) Genes for better tolerance/resistance to biotic and/or abiotic stresses
(C) Genes for herbicide resistance
(D) All of these

206. The microsatellite based markers are analysed by PCR amplification using a pair of primers specific to the unique sequences flanking a SSR sequence, alleles differ for the number of repeats of the short repeating unit is:
(A) Sequence – dependent screening
(B) Sequence – Tagged Microsatellites Sites (STMS)
(C) Sequence Tagged Site (STS)
(D) Single Locus Probe Analysis (SLP)

207. Which is not correct about prokaryotic genomes?
(A) The genome size varies considerably
(B) Pseudogenes do not occur in prokaryotic genome
(C) The frequency of paralogues increases with the genome size
(D) The gene density is however, relatively constant at one per kilo base pairs

208. Which statement is not correct about eukaryotic genomes?
(A) As the genome size increases, gene density decreases
(B) A high percentage of proteins encoded by human genome have orthologs in other eukaryotic genomes
(C) About 75% of repetitive DNA in human genome is due to LINEs ad SINEs sequences
(D) Intron sequences are as variable as exon sequences

209. When pair of PCR primers amplifies only a single sequence within a haploid genome, the amplified region of the genome is called:
(A) Sequence – dependent screening
(B) Sequence – Tagged Microsatellites Sites (STMS)
(C) Sequence Tagged Site (STS)
(D) Single Locus Probe Analysis (SLP)

210. Which features can be used to identify genes in bacterial genomes?
(A) Protein coding genes are composed of open reading frames
(B) Coding region begins with an initiation codon (usually ATG)
(C) Coding region ends with a termination codon (TAA, TAG or TGA)
(D) All of these

211. PCR based markers may be:
(A) Locus specific (SSR or SNP)
(B) Locus unspecific (RAPD or AFLP)
(C) Both (A) and (B)
(D) None of these

212. Which has operons similar to those of bacteria?
(A) *Arabidopsis* (B) *Drosophila*
(C) *C. elegans* (D) Humans

213. Use of a probe that identifies a single DNA fragment for fingerprinting is called:
(A) Sequence – dependent screening
(B) Sequence–Tagged Microsatellites Sites (STMS)
(C) Sequence - Tagged Site (STS)
(D) Single Locus Probe Analysis (SLP)

214. Which is not correct about genomics?
(A) In 1980, David Botstein and coworkers working at the Whitehead Institute of Biomedical Research, USA proposed that one could construct a genome map
(B) In 1995, the first bacterial was that of *Haemophilis influenzae*
(C) The sequencing of whole genome of first multicellular organism, round worm (*Caenorhabdits elegans*), was published in 1999
(D) The sequencing of rice genome by International Rice Genome Sequencing Project was based on whole genome shot-gun approach

215. The fragment connecting and sequential-grow approaches of building up a ligand and the whole building process is governed by a:
(A) Genetic algorithm (B) Taxonomy Browser
(C) Phylogeny (D) None of these

216. To tool provides taxonomy information on various species by:
(A) Genetic algorithm
(B) Taxonomy Browser
(C) Phylogeny
(D) None of these

217. The basic tasks of any decision, the following procedure must accomplish is:
(A) Characterization of the binding sites
(B) Positioning of the ligand into the binding site
(C) Evaluating the strength of interaction for a specific ligand receptor complex
(D) All of these

218. In 1997 Lipiski and coworkers analyzed 2245 compounds from the World Drug Index, and found that in a high percentage of compounds, poor absorption or permeation occurred when:
(A) There are more than five H-bond donors (expressed as the sum of OHs and NHs)
(B) The molecular weight is over 500
(C) The CLogP (calculated octanol/water partition coefficient) is over 5
(D) There are more than 10H-bond acceptors (expressed as the sum of Ns and Os)
(E) All of these

219. A network of protein-protein interaction generated from yeast two-hybrid data is:
(A) *Saccharomyces cereviseae*
(B) *E. coli*
(C) *Agrobacterium*
(D) None of these

220. The scoring matrix for protein sequences is:
(A) PAM matrix (B) BLOSUM matrix
(C) RNA sequences (D) Thure Etzold

221. The matrix for protein sequences is:
(A) PAM matrix (B) BLOSUM matrix
(C) RNA sequences (D) Thure Etzold

222. Dot plot are still used as quick and easy methods to identify regions of repeating sequences, where there are a series of parallel diagonal lines and potential hairpins is:
(A) PAM matrix (B) BLOSUM matrix
(C) RNA sequences (D) Thure Etzold

223. Sequence retrival system (SRS) developed by:
(A) PAM matrix (B) BLOSUM matrix
(C) RNA sequences (D) Thure Etzold

224. Bioinformatics encompasses which of the following:
(A) Biology (B) Computer Science
(C) Information Technology (D) All of these

225. Whole genome sequencing can be done by:
(A) Whole Genome Shot Gun Method
(B) Hierarchical Clone by Clone Method
(C) Both (A) and (B) (D) None of these

226. Which of the following chemical is used for Methylation of guanine at N7 position in Maxam Gilbert Method?:
(A) Dimethyl sulphate (B) Hydrazine
(C) NaOH (D) Acid

227. In Sangers method of sequencing, which of the following is used in Chain termination?
(A) TAMRA (B) RGG
(C) RIIO (D) ddNTP

228. Pyrosequencing was developed by:
(A) Maxam and Gilbert (1977)
(B) Ronaghi and Nyren (1990)
(C) Watson and Crick (1953)
(D) Jones and Sanger (1977)

229. The sequencing data is recorded in the form of:
(A) PDF (B) RTF
(C) SCF (D) DOC

230. Which of the following refers to Next Gen Sequencing?
(A) 454 Pyro sequencing
(B) Illumina (Solexa) sequencing
(C) Solid sequencing (D) All of these

231. The freeware commonly used for genome assembly is/are:
(A) Phred (B) Phrap
(C) Consed (D) All of these

232. Algorithms used for pair wise alignment of sequence:
(A) Needlemann & Wunsch
(B) Smith and Waterman
(C) Both of these
(D) None of these

233. The first Bioinformatics database was created by:
(A) Richard Durbin (B) Dayhoff
(C) Dunn (D) Pearson

234. SWISSPORT Protein seqeunces Database was created in:
(A) 1985 (B) 1987
(C) 1990 (D) 1991

235. An example of homology search tool is:
(A) PROSPECT (B) EMBOSS
(C) RASMOL (D) BLAST

236. The commonly used tool for motif identification is:
(A) COPIA (B) PATTERN HUNT
(C) BLAST (D) PROSPECT

237. ExPASy, the first molecular biology server was established in the year:
(A) 1991 (B) 1992
(C) 1993 (D) 1994

238. Deposition of cDNA into inert structure is:
(A) DNA fingerprinting (B) DNA Microarray
(C) Epigenomics (D) DNA toe-printing

239. The identification of drugs through genomic study is:
(A) Genomics (B) Cheminformatics
(C) Pharmagenetics (D) Eugenics

240. National Centre of Bioinformation was established in:
(A) USA (B) Japan
(C) England (D) China

241. National Centre of Bioinformation serves as:
(A) Primary Information data bank
(B) Provides of information
(C) Both (A) and (B)
(D) None of these

242. European Molecular Biology Laboratory endeavours to:
(A) Collect, Organise and distribute nucleotide sequence data
(B) Primary information data Bank
(C) Protein information system
(D) All of these

243. National Biomedical Research Foundation was established in:
(A) 1983 (B) 1984
(C) 1985 (D) 1986

244. National Biomedical Research Foundation serves as:
(A) Primary information data Bank
(B) Protein information system
(C) Both (A) and (B) (D) All of these

245. Margaret Dayhoff developed the first protein sequence database called:
(A) SWISS PROT (B) PDB
(C) Atlas of protein sequence and structure
(D) Protein sequence databank

246. Each record in a database is called an:
(A) Entry (B) File
(C) Record (D) Ticket

247. Literature database include:
(A) MEDLINE and PubMED
(B) MEDLINE and PDB
(C) PubMED and PDB
(D) MEDLINE and PDS

248. FlyBase is a:
(A) Biodiversity database
(B) Model organism database
(C) Literature database
(D) Biomolecular database

249. Which of the following is an *E. coli* model organism database:
(A) EcoGene (B) EcoBase
(C) EcoSeq (D) ColGene

250. Which of the following is a protein sequence database:
(A) DDBJ (B) EMBL
(C) GenBank (D) PIR

251. GenBank, the nucleic acid sequence database is maintained by:
(A) Brookhaven laboratory
(B) DNA database of Japan (DDBJ)
(C) European Molecular Biology Laboratory (EMBL)
(D) National Centre for Biotechnology Information (NCBI)

252. Submission to GenBank are made using:
(A) Bankit and Sequin
(B) Benkit and Bankin
(C) Sequin and Bankin
(D) Entrez

253. STAG is maintained by:
(A) Brookhaven laboratory
(B) DNA database of Japan (DDBJ)
(C) European Molecular Biology Laboratory (EMBL)
(D) National Centre for Biotechnology Information (NCBI)

254. A comprehensive database for the study of human genetics and molecular biology is:
(A) PDB (B) STAG
(C) OMM (D) PSD

255. All the following are protein sequence database except:
(A) PIR (B) PSD
(C) SWISS PROT (D) EMBL

256. The information retrieval tool of NCBI GenBank is:
(A) Entrez (B) STAG
(C) Seqin (D) Text search

257. The first secondary database developed was:
(A) PRINTS (B) PROSITE
(C) PDB (D) PIR

258. Which of the following is a sequence alignment tool:
(A) BLAST (B) PRINT
(C) PROSITE (D) PIR

259. MEtDB is a model organism database for:
(A) Mouse (B) Human
(C) E. coli (D) Arabidopsis

260. The combination of biology and IT is:
(A) Cell biology (B) Biotechnology
(C) Bioinformatics (D) Nanotechnology

261. Bioinformatics is useful in the study of:
(A) Genomics (B) Proteomics
(C) Metabolomics (D) All of these

ANSWERSHEET
15. BIOINFORMATICS

---

1. (A)	2. (B)	3. (A)	4. (A)	5. (E)	6. (D)	7. (A)	8. (A)	9. (C)	10. (A)
11. (A)	12. (B)	13. (C)	14. (D)	15. (C)	16. (D)	17. (C)	18. (C)	19. (A)	20. (B)
21. (A)	22. (D)	23. (A)	24. (B)	25. (C)	26. (D)	27. (B)	28. (A)	29. (B)	30. (A)
31. (C)	32. (B)	33. (C)	34. (D)	35. (A)	36. (A)	37. (A)	38. (B)	39. (C)	40. (D)
41. (C)	42. (C)	43. (B)	44. (D)	45. (D)	46. (B)	47. (D)	48. (C)	49. (D)	50. (D)
51. (D)	52. (D)	53. (B)	54. (B)	55. (C)	56. (A)	57. (D)	58. (B)	59. (D)	60. (D)
61. (A)	62. (B)	63. (D)	64. (A)	65. (D)	66. (D)	67. (D)	68. (B)	69. (D)	70. (D)
71. (A)	72. (D)	73. (D)	74. (B)	75. (A)	76. (A)	77. (D)	78. (D)	79. (D)	80. (C)
81. (C)	82. (B)	83. (A)	84. (D)	85. (B)	86. (C)	87. (D)	88. (C)	89. (D)	90. (D)

91. (B) **92. (B)** **93. (D)** **94. (D)** **95. (D)** **96. (B)** **97. (D)** **98. (D)** **99. (C)** **100. (B)**

101. (C) **102. (D)** **103. (D)** **104. (D)** **105. (A)** **106. (D)** **107. (C)** **108. (D)** **109. (A)** **110. (C)**

111. (A) **112. (A)** **113. (A)** **114. (C)** **115. (C)** **116. (D)** **117. (D)** **118. (D)** **119. (B)** **120. (D)**

121. (D) **122. (A)** **123. (B)** **124. (B)** **125. (D)** **126. (C)** **127. (D)** **128. (A)** **129. (C)** **130. (D)**

131. (D) **132. (D)** **133. (A)** **134. (D)** **135. (D)** **136. (D)** **137. (D)** **138. (A)** **139. (D)** **140. (A)**

141. (B) **142. (C)** **143. (B)** **144. (D)** **145. (D)** **146. (B)** **147. (A)** **148. (C)** **149. (B)** **150. (B)**

151. (D) **152. (D)** **153. (A)** **154. (C)** **155. (A)** **156. (D)** **157. (A)** **158. (B)** **159. (A)** **160. (C)**

161. (D) **162. (D)** **163. (C)** **164. (D)** **165. (A)** **166. (D)** **167. (C)** **168. (C)** **169. (B)** **170. (D)**

171. (A) **172. (B)** **173. (A)** **174. (C)** **175. (B)** **176. (C)** **177. (A)** **178. (D)** **179. (A)** **180. (D)**

181. (A) **182. (B)** **183. (B)** **184. (D)** **185. (A)** **186. (C)** **187. (C)** **188. (B)** **189. (A)** **190. (C)**

191. (D) **192. (A)** **193. (D)** **194. (C)** **195. (B)** **196. (D)** **197. (C)** **198. (C)** **199. (D)** **200. (A)**

201. (D) **202. (A)** **203. (D)** **204. (D)** **205. (B)** **206. (B)** **207. (B)** **208. (D)** **209. (C)** **210. (D)**

211. (C) **212. (C)** **213. (D)** **214. (D)** **215. (A)** **216. (B)** **217. (D)** **218. (E)** **219. (A)** **220. (A)**

221. (B) **222. (C)** **223. (D)** **224 (D)** **225 (C)** **226 (A)** **227 (D)** **228 (B)** **229 (C)** **230 (D)**

231 (D) **232 (C)** **233 (B)** **234 (B)** **235 (D)** **236 (A)** **237 (C)** **238 (B)** **239 (C)** **240 (A)**

241 (C) **242 (A)** **243 (B)** **244 (B)** **245. (C)** **246. (A)** **247. (A)** **248. (B)** **249. (A)** **250. (D)**

251. (D) **252. (A)** **253. (B)** **254. (C)** **255. (D)** **256. (A)** **257. (B)** **258. (A)** **259. (D)** **260. (C)**

261. (D)

Chapter - 16

SEED BIOTECHNOLOGY

1. Seed is a:
(A) Developed ovule
(B) Fertilised and developed ovule
(C) Developed ovary
(D) Fertilised and developed ovary

2. The smallest and the lightest seed is that of:
(A) *Lemma* (B) *Morus*
(C) *Orchis* (D) *Wolfia*

3. The largest and heaviest seed is that of:
(A) *Lodoicea* (B) Coconut
(C) *Rafflesia* (D) Mango

4. The seed bearing plants belong to the group of:
(A) Pteridophyta (B) Bryophyta
(C) Vascular cryptogams (D) Phancrogams

5. Outer seed coat is:
(A) Testa (B) Tegman
(C) Hilum (D) Funiculus

6. Seed bud is:
(A) Plumule (B) Tegmen
(C) Ovule (D) Mesocotyl

7. Tegmen is the name of:
(A) Embryo shoot (B) Embryonic node
(C) Cotyledon (D) Inner seed coat

8. Hilum of seed is:
(A) Scar of micropyle
(B) Scar of funiculus
(C) Area through which pollen tube enters the ovule
(D) Glandular and attractant region of ovule

9. Tigellum is:
(A) Embryonic axis (B) Embryonic root
(C) Embryonic shoot (D) Embryonic leaf

10. Seed leaf is:
(A) Leaf present over the plumule
(B) Cotyledon
(C) Coleoptile
(D) Mesocotyl

11. Which one is the reproductive unit having an embryo, reserve food and protective covering:
(A) Spore (B) Fruit
(C) Seed (D) Fruitlet

12. In exalbuminous seeds, there is:
(A) Nonformation of endosperm
(B) Abundant formation of endosperm
(C) Abundnt protein
(D) Consumption of endosperm during development of seed

13. In exalbuminous seed the food is generally stored in:
(A) Endosperm initially but hypocotyls later on
(B) Cotyledons
(C) Endosperm from the beginning
(D) Testa

14. The seed in which reserve food is present in the perisperm:
(A) *Nymphaea* (B) Coconut
(C) Onion (D) Litchi

15. An endospermic seed is:
(A) Pea (B) Gram
(C) Castor (D) Bean

16. Endosperm of Areca/Betel Nut is:
(A) Tetraploid (B) Soft
(C) Smooth (D) Ruminate

17. Integmentary outgrowth present over the micropyle of Castor is:
(A) Chalaza (B) Caruncle
(C) Aril (D) Wing

18. Hair of cotton are:
(A) Outgrowth of testa
(B) Outgrowth of inner layer of perisperm
(C) Constituents of aril
(D) Formed from caruncle

19. Which is an albuminous seed:
(A) Gram (B) Castor
(C) Maize (D) Both (B) and (C)

20. The remnants of nucleus present in the seed constitute:
(A) Endosperm (B) Endocarp
(C) Perisperm (D) Aril

21. Cotyledons is absent in:
(A) *Polygonum* (B) *Cuscuta*
(C) *Santalum* (D) *Shorea*

22. The plant in which hypocotyls stores food is:
(A) *Bertholettia* (B) *Nymphaea*
(C) Cotton (D) Rice

23. A dicot exalbuminous seed is:
(A) Pea (B) Castor
(C) Rice (D) Wheat

24. A dicot albuminous seed is:
(A) Pea (B) Castor
(C) Bean (D) Gram

25. In endospermic seed the cotyledons are:
(A) Fleshy (B) Leathery
(C) Papery (D) Green

26. The single shield shaped cotyledon of maize is called:
(A) Tigellum (B) Scutellum
(C) Obturator (D) Coleoptile

27. In Maize, a tubular sheath covers the plumule. It is:
(A) Coleoptile (B) Coleorhiza
(C) Mesocotyl (D) Scutellum

28. Aleurone layer is:
(A) Outer layer of scutellum in contact with endosperm
(B) Layer of pericarp specialized in absorption of water
(C) Layer present in the ovule that guides pollen tube
(D) Layer present on the outside of endosperm and having protein grains

29. Seed of Black Pepper is:
(A) Endospermic (B) Exalbuminous
(C) Acotyledonous (D) Perispermic

30. Coleorhiza is:
(A) Covering of radical
(B) Covering of plumule
(C) Plumule and rudimentary sheath
(D) Radicle and root cap

31. Aleurone layer takes part in:
(A) Protection of delicate embryo
(B) Enzyme synthesis
(C) Transfer of food to cotyledons
(D) Transfer of food from cotyledons to embryo tips

32. Which one is a monocotyledonous seed:
(A) *Pisum sativum*
(B) *Cicer arietinum*
(C) *Dolichos lablab*
(D) *Triticum aestivum*

33. A spongy white bilobed structure present at the narrow tip of Castor seed is:
(A) Strophile (B) Caruncle
(C) Raphe (D) Remains of chalaza

34. The function of caruncle is to:
(A) Protect the radical end of embryo
(B) Check entry of pathogens through micropyle
(C) Protect against mechanical injury
(D) Absorb water and its inward transfer through micropyle

35. Seeds are required for fruit growth:
(A) Throughout (B) Early phase
(C) Late phase (D) Mid phase

36. A seed which does not possess micropyle and helium is:
(A) Pea (B) Gram
(C) Castor (D) Maize

37. In *Ricinus*, the outer white papery covering of endosperm is:
(A) Perisperm/tegmen (B) Tigellum
(C) Strophiole (D) Exo-endosperm

38. Cotyledonary node of embryo lies between:
(A) Hypocotyl and radical
(B) Epicotyle and plumule
(C) Epicotyle and hypocotyle
(D) Cotyledons and radicle

39. Seed having the longest viability is:
(A) *Chenopodium* (B) *Quercus*
(C) *Nelumbo* (D) *Eucalyptus*

40. Viability of the seed is tested with:
(A) Triphenyl chloride (B) Indole acetic acid
(C) Mercuric chloride (D) 2,4-D

41. A seed cut into two halves and immersed in 0.1% triphenyl chloride solution. Its viability is indicated by development of colouration:
(A) Yellow (B) Red
(C) Blue (D) Pink

42. Fleslhy liberated seeds of *Erianthus hiemalis* do not germinate due to:
(A) Occurrence of impermeable seed coat
(B) Immaturity of embryo
(C) Occurrence of inhibitors
(D) Absence of growth hormones

43. Fleshy shed seeds can germinate immediately under favourable conditions in:
(A) *Xanthium* (B) *Pisum sativum*
(C) *Zea mays* (D) Both (B) and (C)

44. Common cause of seed and bud dormancy is the presence of:
(A) Ethylene (B) Cytokinins
(C) Abscisic acid (D) Both (B) and (C)

45. Seeds of Tomato do not germinate in its pulp due to:
(A) Presence of ferulic acid
(B) Presence of excess salts
(C) Absence of oxygen
(D) Presence of ABA

46. Seed dormancy may be due to:
(A) Permeable seed coat
(B) Hard impermeable seed coat
(C) Tin seed coat
(D) Lack of reserve food

47. Wheat germ is:
(A) Cotyledon (B) Endosperm
(C) Embryo (D) Ovule

48. The hormone which can break seed dormancy is:
(A) Coumarin (B) Ferulic acid
(C) ABA (D) GA

49. Mechanical injuring of seed coat to break dormancy is called:
(A) Scarification (B) Stratification
(C) Impaction (D) Compaction

50. Thiourea is used in overcoming seed dormancy by:
(A) Chemical scarification
(B) Counteracting inhibitors
(C) Inducing cell division
(D) Develop osmotic pressure

51. The most important external factor for seed germination is:
(A) Light (B) Soil
(C) Oxygen (D) Water

52. The first process which occurs when the seed is placed in the soil is:
(A) Photosynthesis (B) Respiration
(C) Imbibition (D) Solubilisation of food

53. Seeds placed deep in the soil do not germinate because they are:
(A) Unable to get sufficient oxygen
(B) Without sufficient food to bring the seedling to the surface
(C) Under pressure of overlying soil layers
(D) Unable to get light

54. During germination, micropyle of seed takes part in:
(A) Forming weak point for emergence of radical
(B) Absorption of water
(C) Passage of gases
(D) Leaching of inhibitors

55. Which are the external conditions required for seed germination:
(A) Oxygen, carbon dioxide & suitable temperature
(B) Oxygen, light and suitable temperature
(C) Light, moisture and suitable temperature
(D) Oxygen, moisture & suitable temperature

56. A seed which does not require oxygen for germination is:
(A) Pea (B) Rice
(C) *Typha* (D) Both (B) and (C)

57. A seed which is unable to germinate in the presence of light is:
(A) *Viscum* (B) Onion
(C) Bean (D) Maize

58. The seed in which dormancy can be broken by red light is:
(A) Pea (B) Gram
(C) Lettuce (D) Castor

59. Part of the embryo which comes out first during seed germination is:
(A) Radicle (B) Plumule
(C) Epicotyl (D) Hypocotyl

60. Which is not essential for seed germination in most cases:
(A) Oxygen (B) Light
(C) Suitable temperature (D) Moisture

61. Part of seed which forms the shoot at the time of germination is:
(A) Radicle (B) Cotyledons
(C) Epicotyl (D) Plumule

62. Epigeal germination occurs in:
(A) Pea (B) Gram
(C) Castor (D) Maize

63. Germination is hypogeal in:
(A) Cotton (B) Pea
(C) Castor (D) Bean

64. Vivipary is seed germination:
(A) In strong light
(B) Without involving cotyledons
(C) With cotyledons coming above ground
(D) While contained inside the fruit

65. Cotyledons constitute the first pair of leaves in:
(A) Castor (B) Maize
(C) Pea (D) Gram

66. Vivipary occurs in:
(A) Mangrove plants
(B) Alpine plants/Orchids
(C) Tropical plants/Sea Weeds
(D) Desert plants/Vallisneria

67. Seed coats develop from:
(A) Ovule (B) Embryo sac
(C) Integuments (D) Chalaza

68. Seed develops from:
(A) Ovary (B) Embryo
(C) Embryo sac (D) Ovule

69. Caruncle develops from:
(A) Outer integument (B) Funiculus
(C) Cotyledon (D) Inner integument

70. Seeds are products of sexual reproduction because they:
(A) Give rise to new plants
(B) Have viability
(C) Are formed by fusion of gametes
(D) Are formed by fusion of pollen tube

71. Milky water in green coconut is:
(A) Liquid female gametophyte
(B) Liquid endosperm
(C) Liquid nucellus
(D) Liquid chalaza

72. **Perisperm is:**
(A) Remnant of endosperm
(B) Persistent nucellus
(C) Peripheral part of endosperm
(D) Disintegrated secondary nucleus

73. **Food is stored in albuminous seed in:**
(A) Testa (B) Cotyledon
(C) Ensdosperm (D) Persperm

74. **Unused nucellus constitutes:**
(A) Tapetum (B) Tegmen
(C) Endosperm (D) Perisperm

75. **Aleurone layer occurs in the peripheral region of:**
(A) Ensdosperm (B) Coleoptile
(C) Cotyledon (D) Coleorhiza

76. **Proteinaceous part of Maize endosperm is:**
(A) Apophysis (B) Aleurone layer
(C) Peripheral layer (D) Scutellum

77. **Micropyle of seed is involved in the passage of:**
(A) Male gamete (B) Pollen tube
(C) Water (D) Gases

78. **Ruminate endosperm occurs in:**
(A) Annonaceae/Areca Nut
(B) Compositae
(C) Cruciferae (D) Euphorbiacee

79. **The plants in which germination of embryo occurs when the fruit is still attached to the plant:**
(A) Mangrove plant (B) Mango
(C) Epiphytes (D) Artacarpus

80. **Scutellum is:**
(A) Single fleshy cotyledon of Trapa
(B) Single shield-shaped cotyledon of cereals
(C) Covering of plumule
(D) Covering of radical

81. **Cotyledon represents:**
(A) Covering of radical
(B) Covering of cotyledon
(C) Covering of plumule
(D) Alternate name of plumule

82. **Cotyledons form the first leaves in:**
(A) Maize (B) Gram
(C) Pea (D) Castor

83. **Seed of pea is:**
(A) Epigeal and exalbuminous
(B) Hypogeal and exalbuminous
(C) Hypogeal and albuminous
(D) EEpigeal and albuminous

84. **Vivipary is:**
(A) Seed germination with subterranean cotyledons
(B) Seed germination with epiterranean cotyledons
(C) Fruit development without pollination
(D) Seed germination inside the fruit while attached to the plant

85. **In hypogeal germination, the cotyledons do not turn green because they:**
(A) Lack chloroplasts (B) Contain inhibitors
(C) Remain below soil (D) Abscise early

86. **Dry seeds can endure high temperature and other adverse conditions better than germinating seeds due to:**
(A) Hydration makes the enzymes sensitive to temperature
(B) Seeds take part in perennation
(C) Dry seeds possess more reserve food
(D) Dry seeds possess thick seed coats

87. **Tegmen develops from:**
(A) Funiculus (B) Chalaza
(C) Inner integument (D) Outer integument

88. **Outermost layer of endosperm of Maize grain is:**
(A) Epidermis (B) Pericarp
(C) Tunica (D) Aleurone

89. **Plant which buries its seeds in the soil in:**
(A) Peanut (B) Potato
(C) Sweet potato (D) Cashewnut

90. **Oil reserve of groundnut is present in:**
(A) Embryo (B) Cotyledons
(C) Endosperm (D) Underground tubers

91. **Vivipary is characteristic of:**
(A) Mesophytes (B) Xerophytes
(C) Hygrophytes (D) Halophytes

92. **The storage tissue of rice and other cereal grains is:**
(A) Embryo (B) Seed
(C) Endoperm (D) Fruit

93. **The blackish brown covering around the endosperm containing kernel of coconut is:**
(A) Seed coat (B) Epicarp
(C) Mesocarp (D) Endocarp

94. **An albuminous seed showing hypogeal germination is:**
(A) Castor (B) Bean
(C) Gram (D) Maize

95. **Oil is stored in the endosperm of:**
(A) Groundnut (B) Soybean
(C) Coconut (D) Cashewnut

96. **Germination of seed within fruit is:**
(A) Ovipary (B) Vivipary
(C) Hypogeal (D) Epigeal

97. **Alerurone layer helps in:**
(A) Storage of food in endosperm
(B) Protection of embryo
(C) Utilization of stored food
(D) All of these

98. **In the legume seed, food is stored in:**
(A) Cotyledons (B) Endosperm
(C) Perisperm (D) Seed coats

99. Epithelial layer of maize scutellum secretes the hormone:
(A) Auxin (B) Cytokinin
(C) Gibberellin (D) Ethylene

100. Microphyle occur in:
(A) Ovary (B) Seeds
(C) Ovule (D) Both (B) and (C)

101. Shield-shaped cotyledon/scutellum occurs in:
(A) Maize (B) Gram
(C) Pea (D) Cucumber

102. Outgrowth developing alongwith helium of the seed is:
(A) Plumule (B) Radicle
(C) Strophiole (D) Perisperm

103. A gas required for germination of pea seed is:
(A) Nitrogen (B) Oxygen
(C) Hydrogen (D) Water vapours

104. In groundnut seeds are:
(A) Geocarpic (B) Amphicarpic
(C) Photocarpic (D) Epicarpic

105. Which is active in Maize?
(A) Maltase (B) Zymase
(C) Diastase (D) Urease

106. Part of Castor seed that yields oil is:
(A) Caruncle (B) Endosperm
(C) Nucellus (D) Cotyledon

107. Which one is essential for seed germination in pea?
(A) O_2 (B) H_2
(C) N_2 (D) CO_2

108. Protective covering over radical during seed germination is:
(A) Coleorhiza (B) Coleoptile
(C) Suspensor (D) Epithelium

109. After removal of covering in pea, the seed consists of:
(A) Cotyledons
(B) Embryo
(C) Cotyledons + Endosperm
(D) Cotyledons + Endosperm + Pericarp

110. The character found only in halophytes is:
(A) Vivipary (B) Velamen
(C) Heterophylly (D) Sunken stomata

111. Endosperm is formed in angiosperms due to double fertilization. It is however, absent in certain seeds due to lack of:
(A) Certain enzymes
(B) Growth hormone
(C) Dicotyledonous hormone
(D) Nutrients

112. Germination is epigeal in:
(A) *Zea mays* (B) *Helianthus*
(C) *Mangifera* (D) *Pisum*

113. Coleorhiza is cap-like covering over:
(A) Radicle in dicots
(B) Plumule in dicots
(C) Plumule in monocots
(D) Radicle in monocots

114. Seed dormancy allows the plants to:
(A) Overcome unfavourable climatic conditions
(B) Develop healthy seeds
(C) Reduce viability
(D) Prevent deterioration of seeds

115. Which of the following is associated with dormancy of seeds?
(A) IBA (B) GA
(C) Ethylene (D) Ferulic acid

116. Among the following which compound can induce seed dormancy:
(A) Gibberellins (B) Caffeine
(C) ABA (D) Potassium nitrate

117. Fruit of peanut (*Arachis hypogeal*) is:
(A) Pod (B) Achene
(C) Caryopsis (D) Drupe

118. A method of breaking dormancy and allowing ample absorption of water is:
(A) Stratification (B) Scarification
(C) Vernalisation (D) Devernalisation

119. During seed germination, seed coat ruptured due to:
(A) Differentiation of cotyledons
(B) Massive glycolysis in endosperm and cotyledons
(C) Massive imbibitions of water
(D) Sudden increase in cell division

120. Seed dispersal:
(A) Avoids competition
(B) Produces mixed population
(C) Promotes cross population
(D) All of these

121. Jaculator mechanism of seed dispersal is found in:
(A) *Impatiens* (B) *Ruellia*
(C) *Abrus* (D) *Ecballium*

122. Dispersal of seeds through explosive mechanism occurs in:
(A) *Tecoma* (B) *Sonchus*
(C) Squirting cucumber (D) Poppy

123. The fruit which breaks into five valves on touching and results in explosive dispersal of seeds is:
(A) *Ecballium* (B) *Balsam*
(C) *Entada gigas* (D) *Ruekkia*

124. In *Abrus* and pea the fruit disperses its seeds through:
(A) Explosive mechanism
(B) Winds
(C) Water (D) Animals

125. Autochory occurs in:
(A) *Geranium* (B) Lotus
(C) *Plumbago* (D) Mango

126. Hairy styles help in the dispersal of fruits in:
(A) *Aristolochia* (B) *Clematis*
(C) *Naravelia* (D) Both (B) and (C)

127. In censer mechanism, seed dispersal occurs through:
(A) Birds
(B) Entangling of seeds to skin of roaming animals
(C) Shaking by wind
(D) Floating on water

128. Very small and light seeds which are dispersed by air are:
(A) Dust seeds (B) Winged seeds
(C) Baloon seeds (D) Plumed seeds

129. Censer mechanism of seed dispersal occurs in:
(A) *Bombax* (B) *Alstonia*
(C) *Moringa* (D) *Poppy*

130. Dust seeds occur in:
(A) *Ulmus* (B) *Orchis*
(C) *Cinchona* (D) *Ailanthus*

131. Fruits are flattened to be easily dispersed by air in:
(A) *Taraxacum* (B) *Nymphaea*
(C) *Pupalia* (D) *Dalbergia*

132. Propellar mechanism of anemochory occurs in:
(A) *Tecoma* (B) *Hiptage*
(C) *Ailanthus* (D) *Shorea*

133. Winged fruits occur in:
(A) *Lagerstroemia* (B) *Moringa*
(C) *Colutea* (D) *Shorea*

134. Winged seed occur in:
(A) *Moringa* (B) *Ulmus*
(C) *Shorea* (D) *Hiptage*

135. In Hiptage, anemochory occurs because of:
(A) Winged seeds (B) Baloon fruits
(C) Winged fruits (D) Plumed seeds

136. In anemochorous *Tecoma*, the modification present is:
(A) Dust seeds (B) Winged seeds
(C) Winged fruits (D) Plumed seeds

137. A fleshy fruit dispersed by inflated persistent calyx is:
(A) *Bombax* (B) *Albizia*
(C) *Physalis* (D) *Naravelia*

138. Inflated fruits which are dispersed by air occur in:
(A) *Aristolochia* (B) *Heritiera*
(C) *Nymphaea* (D) *Cardiospermum*

139. Anemochorous seed bearing hair at two ends is:
(A) *Alstonia* (B) *Calotropis*
(C) *Bombax* (D) Cotton

140. Hygroscopic hair for fruit dispersed to moist areas occurs in:
(A) *Sonchus* (B) Cotton
(C) *Mucuna gigantea* (D) *Entada gigas*

141. Plumed seeds occur in:
(A) *Naravelia* (B) *Taraxacum*
(C) *Calotropis* (D) *Colutea*

142. Parachute mechanism of fruit dispersal is found in:
(A) *Bombax* (B) *Taraxacum*
(C) *Holoptelia* (D) *Polygonum*

143. Coconut fruit is dispersed by:
(A) Air (B) Animals
(C) Autochory (D) Water

144. In coconut, hydrochory occurs due to presence of:
(A) Papery epicarp (B) Fibrous mesocarp
(C) Stony endocarp (D) Liquid endosperm

145. Water dispersal of coconut is not efficient because:
(A) It is restricted
(B) Coconut fruit decays within a couple of days
(C) The fruit cannot remain afloat for than one day
(D) The statement is incorrect

146. Sticky zoochorous fruits occur in:
(A) *Cinchona* (B) *Polygonum*
(C) *Plumbago* (D) *Mulberry*

147. In Boerhaavia, fruit dispersal occurs due to presence of:
(A) Barbs (B) Hooks
(C) Sticky glands (D) Hygroscopic hair

148. Hooks help in fruit dispersal of:
(A) *Martynia* (B) *Hentiera*
(C) *Viscum* (D) *Colutea*

149. Which of the following is dispersed by animals due to presence of barbs and hooks:
(A) *Pupalia* (B) *Boerhaavia*
(C) *Ecballium* (D) *Cordia*

150. Spear grass (*Aristida*) is dispersed by:
(A) Air
(B) Animals
(C) Water
(D) Explosive mechanism

151. Forced zoochory occurs in:
(A) *Mulberry* (B) *Viscum*
(C) *Xanthium* (D) *Cordia*

152. *Tribulus* and *medicago* are dispersed by animals because they possess:
(A) Sticky glands (B) Sweet pulpy fruits
(C) Bright colour (D) Hooks

153. Seeds dispersed by animals and passing through their alimentary canal are of plant:
(A) *Cordia* (B) *Viscum*
(C) *Loranthus* (D) Peepal

154. Fruits having sticky pulp for zoochorous dispersal are of:
(A) Guava (B) Banyan
(C) *Cordia* (D) Cleome viscose

155. A turf of hair present at one or both ends of the seed (for dispersal) is called:
(A) Propellar (B) Coma
(C) Pappus (D) Wing

156. Which one bears coma for anemochory:
(A) *Cinchona* (B) *Calotropis*
(C) *Alstonia* (D) Both (B) and (C)

157. Pappus for parachute mechanism of anemochory develops from:
(A) Calyx (B) Fruit wall
(C) Seed wall (D) Style

158. Seeds of *Ruellia* tuberose are disseminated by:
(A) Censar mechanism
(B) Parachute mechanism
(C) Jaculator mechanism
(D) Explosive mechanism

159. The fruits of *Tribulus* are dispersed by:
(A) Water (B) Wind
(C) Animals (D) Explosive mechanism

160. Seeds of Cinchona are dispersed by:
(A) Wind because they are very small and light
(B) Wind because they are winged
(C) Water because they possess aerenchyma
(D) Animals as they have barbs

161. Hooks and barbs are helpful in dispersal of:
(A) *Viscum* (B) *Boerhaavia*
(C) *Plumbago* (D) *Bidens*

162. Wide distribution of coconut in coastal areas of different countries is due to:
(A) Its fruits can float on water
(B) It is economically very important
(C) It requires saline soil
(D) It yields oil

163. Fruit of *Xanthium* is zoochorous as it has:
(A) Sticky surface (B) Hooks
(C) Edible pulp (D) Sticky pulp

164. Cotton and Madar (*Calotropis*) seeds are dispersed by air as they have:
(A) Pappus (B) Wings
(C) Hair (D) Flattened surface

165. Anemochorous seeds and fruits are:
(A) Round in shape
(B) Flattened
(C) Light in weight
(D) Without any bright colour

166. Orchid seeds are usually:
(A) Small, light and dry (B) Small and sticky
(C) Large and sticky (D) Large and heavy

167. Which would have wider distribution, those distributed by:
(A) Spores (B) Vegetative means
(C) Seeds (D) Fruits

168. *Clematis* and *Naravelia* are dispersed by air with the help of:
(A) Persistent inflated calyx
(B) Persistent hairy styles
(C) Hair (D) Wings

169. Seeds dispersed by winds are:
(A) Heavy (B) Round
(C) Irregular (D) Light

170. Seeds of *Orchis*/Orchids are:
(A) Large and heavy (B) Small and hairy
(C) Small and light (D) Large and sticky

171. A fruit dispersed by grazing goats and cows is:
(A) *Xanthium* (B) *Cucurbita*
(C) *Helianthus annus* (D) *Brassica compestris*

172. Seeds are dispersed by wind in:
(A) *Petunia* (B) *Calotropis*
(C) *Oryza* (D) *Tribulus*

173. Censar mechanism of seed dispersal is found in:
(A) Papaveraceae (B) Liliaceae
(C) Leguminoceae (D) Rosaceae

174. Parachute type of mechanism of seed dispersal is found in:
(A) Pea/Leguminoceae
(B) Mustard/Crucifereae
(C) Cotton/Malvaceae
(D) *Taraxacum* / Compositeae

175. Winged seeds occur in:
(A) *Shorea* (B) *Moringa*
(C) *Hiptage* (D) *Calotropis*

176. Parachute mechanism of fruit dispersal as found in compositeae is due to structure named:
(A) Bract (B) Pappus
(C) Coma (D) Barbs

177. Dispersal of seed in Opium Poppy/Argemone occurs by:
(A) Explosive mechanism
(B) Parachute mechanism
(C) Censar mechanism
(D) Jaculator mechanism

178. Which of the following traits helps in wind dispersal:
(A) Barbs and hooks
(B) Hair and wings
(C) Sticky glands
(D) Round and compact seeds

179. Winged fruits occur in:
(A) *Papaver* (B) *Moringa*
(C) *Hiptage* (D) *Cinchona*

180. Which of the following is dispersed by wind:
(A) *Calotropis* (B) Coconut
(C) *Nelumbium* (D) *Nymphaea*

181. Dispersal of coconut fruit occurs by:
(A) Wind
(B) Explosive mechanism
(C) Animals
(D) Water

182. Persistent calyx produces a dry bladder around an edible berry in:
(A) *Capsicum* (B) *Solanum*
(C) *Nicotiana* (D) *Physalis*

183. Seeds are beset with tiger's claw in:
(A) Martynia (B) *Xanthium*
(C) Spear grass (D) Punture vine

184. Parachute type dispersal occurs in:
(A) Tomato (B) Mustard
(C) Pea (D) Cotton

185. Winged seeds are:
(A) Rough and sticky (B) Smooth and sticky
(C) Dry and spinning (D) None of these

186. Pappus helps in seed dispersal in:
(A) Asteraceae (B) Brassicaceae
(C) Malvaceae (D) Solanaceae

187. In which of the following fruits dispersal occurs by grazing animals?
(A) *Xanthium* (B) Mango
(C) Coconut (D) All of these

188. Calyx is modified into wings in:
(A) *Shorea* (B) *Hopea*
(C) *Dioscorea* (D) *Acer*

189. Seed lot certificate colour is:
(A) Orange (B) Green
(C) Purple (D) Orange or green

190. Seed sample certificate colour is:
(A) Blue (B) Orange
(C) Purple (D) White

191. Hot water treatment is used to break dormancy in:
(A) Lentil (B) Cherry
(C) Cotton (D) None

192. Smaller plant virus is:
(A) SBYV (B) TMV
(C) Satellite virus (D) PVX

193. Fungi and Bacteria usually enters through:
(A) Stomata (B) Wounds
(C) Hydathodes (D) Insect puntyre

194. How many primary samples will be taken from 3 containers of 100 kg?
(A) 3 (B) 5
(C) 15 (D) 9

195. Difference between seed and grain is:
(A) Seed processing (B) Genetic purity
(C) Seed quality (D) Cost

196. Embryo exised test ----- used for surface disinfestation:
(A) NaOCl (B) Lactophenol
(C) $AlK(SO_4)_2$ (D) All of these

197. Fanning mill is:
(A) Air screen cleaner (B) Hammer mill
(C) Blower (D) Rice mill

198. Haryali yojna is related to:
(A) Soil management
(B) Water management
(C) Crop management (D) Air management

199. Mak-12 is a variety of :
(A) Maize (B) Medikago
(C) Bt Cotton (D) Tomato

200. National Nematode collection of India is located at
(A) New Delhi (B) Dharwad
(C) Anand (D) Bangalore

201. Leverage ratio is:
(A) Owner's equity/Total assets
(B) Owner's equity/Current liabilities
(C) Current liabilities/Owner's equity
(D) Total debt/Owner's equity

202. Size of saltation is;
(A) 0.1 – 0.5 mm (B) <0.1 mm
(C) 0.55 – 1 mm (D) >0.1 mm

203. Monocarpic flower is:
(A) Mango (B) Agave
(C) Amaranthus (D) Mulberry

204. Seed coat perodidase test is used for:
(A) Sorghum (B) Wheat
(C) Soybean (D) Groundnut

205. Tarai Development Corporation (TDC) was established in:
(A) 1965 (B) 1969
(C) 1970 (D) 1971

206. Alkaloid test is used for:
(A) Lucern (B) Lupins
(C) Crysanthemum (D) Cereals

207. Average diameter of bulbet of onion (seed standard) shall not be less than:
(A) 2.5 cm (B) 1 cm
(C) 3 cm (D) 2 cm

208. Pure seed fraction in *Abelmosschus esculentus*:
(A) 98 (B) 95
(C) 99 (D) 96

209. Harvesting index is maximum in:
(A) Carrot (B) Pigeonpea
(C) Rice (D) Wheat

210. ---------- sugar found in germinating seed in large amount:
(A) Maltose (B) Sucrose
(C) Cellulose (D) Lactose

211. Match the following:

List I	List II
A. Accelerating Ageing test	a. Maize
B. Cold test	b. Soybean
C. A_6 Test	c. Virus Y
D. M_1 Test	d. Leaching Test
E. Cool Germination Test	e. Groundnut
F. Mitochondrial Activity	f. Cotton

	A	B	C	D	E	F
(A)	a	b	c	d	e	f
(B)	a	b	c	d	f	e
(C)	b	a	c	d	e	f
(D)	b	a	c	d	f	e

212. In maize, a foundation single cross production field must be isolated by not less than ------ mts from maize with icernal colour or texture different from that of seed part:
(A) 400 mts (B) 600 mts
(C) 200 mts (D) 250 mts

213. A group of similar plants in structural features and performance is known as:
(A) Variety (B) Clone
(C) Pure line (D) Inbred line

214. The blotter method of seed health testing detects primarily:
(A) Virus (B) Fungi
(C) Bacteria (D) Nematodes

215. Possible reason for seed dormancy is:
(A) Presence of pathogens (B) Cracking of hulls
(C) Immature embryo (D) Green distoration

216. A sound seed certification programme requires:
(A) Direct participation of breeder
(B) Use of sophisticated equipment
(C) Classification of seed
(D) Support of law

217. The seed which farms generally used are called:
(A) Breeder Seed (B) Foundation seed
(C) Certified seed (D) Nucleus seed

218. The rudimentary root of the seed or seedling that forms the primary root of the young plant is known as:
(A) Rachis (B) Radicle
(C) Rachilla (D) Raceme

219. New gene are created by:
(A) Mutation (B) Recombination
(C) Inversion (D) None

220. Mode of pollination in maize is
(A) Self –pollination (B) Vegetative propagation
(C) Cross-pollination (D) None

221. The genome contain in bread wheat are:
(A) ABC (B) XYZ
(C) ABD (D) None

222. Seed drying is very important to maintain its:
(A) Viability and vigour
(B) Protein content
(C) Oil content
(D) Chemical composition

223. The two most important factors influencing the life span of seeds under storage are:
(A) Relative humidity and temperature
(B) Rainfall and Relative humidity
(C) Climate and temperature
(D) Seed size and moisture content

224. The chemical seed treatment is given to control:
(A) Fungi and bacteria (B) Soil insects
(C) Storage insects (D) All of these

225. Maximum moisture content for safe storage in wheat seed is:
(A) 12 (B) 15
(C) 7 (D) None of these

226. The factors affecting the final marketing price of crops a seed are:
(A) Inputs costs
(B) Supply and demand
(C) Price of other farm produce and time trend
(D) All of these

227. The equipment used to apply chemicals that involves suspension and wettable powder treatment material in water is:
(A) Slurry treaters (B) Direct treaters
(C) Panogen treaters treaters (D) Mist-o-matic

228. Mutalism are shown by:
(A) Pollination by animals
(B) Dispersal of fruits and seed
(C) Lichen
(D) All of these

229. In India, normally how many generation system seed are produced?
(A) 1 (B) 2
(C) 3 (D) 4

230. Off types are removed by:
(A) +ve selection (B) -ve selection
(C) Pure line selection (D) Mass selection

231. Dicot endospermic seed is:
(A) Castor (B) Fenugreek
(C) Both (A) and (B) (D) Bean

232. Basic input of agriculture is:
(A) Seed (B) Fertilizer
(C) Soil (D) Irrigation

233. When the pathogen is loosely mixed with seed in the form of sclerotia, galls etc are called:
(A) Infection (B) Infestation
(C) Contamination
(D) Concomitant contamination

234. Seed certification involves:
(A) 1 phase (B) 3 phase
(C) 5 phase (D) 7 phase

235. Slow drying seed is:
(A) Cereals (B) Rapeseed and mustard
(C) Grass (D) Pulses

236. Mite don't survive below:
(A) 5^0C (B) 7^0C
(C) 10^0C (D) 15^0C

237. Seed dormancy are shown by:
1 Nigella
2. Xanthium
3. Tomato
4. Lettuce
(A) Only 1 and 2 (B) Only 1, 2 and 3
(C) Only 1, 2 and 4 (D) Only 2

238. +ve photoblastic seeds are:
(A) Nigella (B) Silene
(C) Nemophills (D) Arraranutum

239. Epigeal germination is found in:
(A) Pea (B) Gram
(C) Cucumber (D) Mango

240. Germination is inhibited by:
(A) Red light (B) Blue light
(C) IR light (D) UV light

241. Pyrenoids are:
(A) Protein surrounded by starch grain
(B) Starch surrounded by protein
(C) Fat surrounded by starch grain
(D) Starch surrounded by fat

242. Tetrazolium test determine the level of activities of enzyme:
(A) ATPase (B) Dehydrogenase
(C) Carboxylase (D) Isomerase

243. Production of seedless grapes require:
(A) Gibberellins (B) NAA
(C) Ethylene (D) IAA

244. Which of the following vigour test method estimates decarboxylation activities?
(A) Leachate test
(B) GADA test
(C) ATPase test
(D) Exhaustion test

245. ISTA was established in the year:
(A) 1871 (B) 1876
(C) 1921 (D) 1924

246. While sampling from a seed lot of 50 bags, the minimum number of primary sample should be:
(A) 5 (B) 10
(C) 15 (D) 20

247. Certified seed of cotton should have minimum germination of:
(A) 50% (B) 60%
(C) 70% (D) 80%

248. Half gram is working sample is required for parity analysis of:
(A) Tomato (B) Onion
(C) Berseem (D) Tobacco

249. For seed samples kept in an incubator for germination test, light is:
(A) Always essential (B) Never essential
(C) Not harmful (D) Harmful

250. Development of embryo without fertilization is:
(A) Apomixis (B) Amghimixis
(C) Parthenocarpy (D) None of these

251. Male sterility is used to develop:
(A) Inbreds (B) Hybrids
(C) Synthetic (D) None

252. Occurrence of more than one embryo in seed is known as:
(A) Polysomy (B) Polyembryony
(C) Apogamy (D) None of these

253. The moisture content for safe storage of cereals is:
(A) 12-14% (B) 14-16%
(C) 16-18% (D) 18-20%

254. The GoT is done for verification of:
(A) Physical purity (B) Genetic purity
(C) Germination % (D) None of these

255. T_2 test colour of living tissues of seed changed to:
(A) Red (B) Blue
(C) Yellow (D) Green

256. The minimum number of counts to be taken for an area upto 2 ha in wheat design any field inspection are:
(A) 7 (B) 5
(C) 3 (D) 10

257. The standard for germination % (min.) for certified seed of gram is:
(A) 85 (B) 90
(C) 80 (D) 75

258. The recommended isolation distance (meters) for foundation seed production of radish, turnip and cauliflower is:
(A) 800 (B) 1200
(C) 1600 (D) 2000

Match the following

259. **List – I** / **List – II**

List – I	List – II
a. Stratification	1. Peas
b. Bleached seed	2. Thermodormancy
c. Hybrex	3. Father of seed technology
d. Nobbe	4. Wheat

Codes:

	a	b	c	d
(A)	1	2	3	4
(B)	2	1	4	3
(C)	2	3	4	1
(D)	4	2	3	1

260.

List – I	List – II
a. Cotton	1. Bolting
b. Reddish	2. Heterostyly
c. Tobacco	3. Delinting
d. Brinjal	4. Steckling

Codes:

	a	b	c	d
(A)	1	4	3	2
(B)	1	2	3	4
(C)	3	4	1	2
(D)	3	2	4	1

261.

List – I	List – II
a. Genetic purity	1. Cross polination
b. Scarification	2. Offtype
c. Dormancy	3. Seed dormancy
d. Roguing	4. Somatic mutation

Codes:

	a	b	c	d
(A)	4	3	2	1
(B)	1	3	4	2
(C)	4	1	3	2
(D)	2	1	3	4

262.

List – I	List – II
a. Seed research	1. Certification
b. Poor storage	2. Polyembryony
c. Notified variety	3. ISST
d. Citrus	4. Low germination

Codes:

	a	b	c	d
(A)	3	2	1	4
(B)	3	1	2	4
(C)	2	1	4	3
(D)	3	4	1	2

263.

List – I	List – II
a. Orthodex seed	1. Legume
b. Recalcitrant seed	2. Sulphuric acid
c. Hard seed	3. Wheat
d. Scarification	4. Cocca

Codes:

	a	b	c	d
(A)	1	2	3	4
(B)	3	4	1	2
(C)	1	4	2	3
(D)	4	3	2	1

264.

List – I	List – II
a. Germination test	1. Other crop seed
b. Moisture test	2. Formazan
c. Purity test	3. PUG
d. T2 test	4. Pre drying

Codes:

	a	b	c	d
(A)	2	4	3	1
(B)	3	4	1	2
(C)	1	3	2	3
(D)	3	1	2	4

265.

List – I	List – II
a. Vigour test	1. Trier
b. Nobbe	2. Hiltners test
c. Ganga 5	3. Hybrid maize
d. Quick viability	4. T2 test

Codes:

	a	b	c	d
(A)	4	1	2	3
(B)	1	2	3	4
(C)	2	1	3	4
(D)	3	4	2	1

266.

List – I	List – II
a. Seed Act	1. 1968
b. NSC	2. 1961
c. Seed rules	3. 1966
d. World seed day	4. 1963

Codes:

	a	b	c	d
(A)	2	4	3	1
(B)	3	4	1	2
(C)	4	1	3	2
(D)	1	2	3	4

267.

List – I	List – II
a. NSP	1. 1961
b. UPOV	2. 1983
c. Seed control order	3. 1976
d. New Seed Act	4. 2006

Codes:

	a	b	c	d
(A)	2	4	3	1
(B)	1	2	3	4
(C)	4	2	1	3
(D)	3	1	2	4

268.

List – I	List – II
a. Breeder seed	1. Blue
b. Foundation seed	2. Buff
c. Nucleous seed	3. Golden yellow
d. Certified seed	4. White

Codes:

	a	b	c	d
(A)	1	3	2	4
(B)	3	4	2	1
(C)	2	3	4	1
(D)	1	4	2	1

269. There are ------ seed certification agencies in India at present:
(A) 11 (B) 15
(C) 19 (D) 23

270. The isolation distance required to produce the certified seed of chickpea, lentil, mungbean and urdbean is ------ meters:
(A) 25 (B) 15
(C) 10 (D) 5

271. The number of plants per hectare to be observed in taking one count in seed production of pearlmillet, sorghum, rice and wheat is ----as recommended:
(A) 500 (B) 1000
(C) 1500 (D) 2000

272. The sampling lot in agricultural seed of large size should not exceed ----- kg except that of maze (where 40000 kg is allowed):
(A) 5000 (B) 10000
(C) 15000 (D) 20000

273. Use of linear measurements of plumule growth as a vigour test for cereals and sugarbeet was first suggested by:
(A) S. Mathews in 1929 (B) J.S. Burris in 1939
(C) Germ in 1949 (D) Hegarty in 1959

274. The seed, the progeny of which comes from foundation seed or registered seed, is handled to maintain genetic purity, approved and certified by certifying agency is called:
(A) Certified seed (B) Foundation seed
(C) Nucleus seed (D) Registered seed

275. The seed of oil palm require as high as --------^{O}C temperature to break the dormancy:
(A) 30-40 (B) 40-50
(C) 50-60 (D) 60-70

276. The seeds of ---- are positively photoblastic:
(A) Tomato (B) Tobacco
(C) Chillies (D) Cucurbits

277. Heteromorphic seeds of the plant families Asteraceae and Chenopodiaceae do not permit ------- exchange:
(A) Water (B) Chemical
(C) Gaseous (D) Light

278. Loose smut is:
(A) Externally seed borne
(B) Internally seed borne
(C) Air borne
(D) None of these

279. Solar treatment is used for:
(A) Loose smut (B) Stem rust
(C) Powdery mildew (D) Blast

True/False

280. Sunflower is an example of protandry.

281. Detasseling is done in sorghum hybrid seed production.

282. More than half of the size of a broken seed is considered as a pure seed.

283. Tip sterility is common in bajra.

284. Hard seed are FUS.

285. In light sensitive seed dormancy is controlled by phytochromes.

286. All varieties can be certified.

287. In storage the loss of seed vigour presses loss of viability.

288. For long storage seed with recommended moisture, should be packed in moisture vapour proof containers.

289. Foundation stage II seed shall be the progeny of breeder seed.

290. Field inspection is made to verify seed source.

291. Seed analysis is done to verify prescribed seed standards.

292. Basic or general seed certification standards are same for all classes of seed.

293. Isolation distance is maintained to protect the crop from insect pest.

294. Inspection made at flowering stage gives more realistic picture.

295. Cultivar purity of a seed lot is evaluated through a standard laboratory test.

296. The inspection at flowering stage in case of cross-pollinated crop should be in the presence of grower.

297. Pollen shadder are found in maize.

298. Only those variety are eligible for seed production and certification which are already notified.

299. Recocitrant seeds follow Harrigton's thump rule.

Fill in the blanks

300. Under seed act -------- is voluntary and --- is compulsory.

301. ------day and ----------continuous induce stimulative flowers in cucurbits.

302. Isolation distance for soybean-----------------------.

303. ---------------------is used for delinting of cotton seed.

304. --------is used for separation of discoloured seed.

305. Harvest index----------------------------.

306. T2 test is used for evaluating the -----of seed.

307. Maintainer line hybrid seed production is also known as ------------------------.

308. Pre-chilling treatment to seed is given for overcoming -----------------------of seed.

309. The first generation hybrid between two single crosses is known as ----------------------.

310. The immature seed is also classified as-------- seed.

311. Sexual function of the sperm and egg cell are called ----------------------

312. The outer most wall of the ovary is called the ------- ------------------

313. The species which has male and female flowers on the different plants, are known as ---------------------

314. A flower containing both male and female reproductive organs is called ------------------ flower.

315. The tissue of seeds that develops from sexual fusion of the polar nuclei of the ovule and the second male sperm cell, the term used as-------------------.

316. The portion of the embryo or seedling above the cotyledons is known as -------------------

317. Trueness to type or variety, usually referring to seed is known as-----------------

318. The hereditary make up of a plant or variety, which determine its inheritance is called-------------------.

319. Competition between individual pollen tubes belonging to different genotypes and growing at different rates is called --------------------.

320. Method of emasculation in maize-------------------

321. Change in genetic make up of varieties if grown over a long period in areas outside their adaptation due to natural selection, specially in composites, synthetic & open pollinated varieties is called -----

322. A machine used to separate seeds on the basis of seed shape used for seed cleaning----------------

323. In hybrid seed production based on male sterility, a plant belonging to A line that produces pollen is termed as ----------------

324. --------refers to the percentage of seed sample that would produce seedlings of the variety under certification.

325. Restorer line restores fertility to ------------------

326. --------refers to a specified quantity of seed which is physically identifiable from other such quantities.

327. -----------------technique is used for the commercial production of virus free seed potatoes.

328. A machine which devides seeds on the basis of their weight and size which is used for seed cleaning grading ----------------------

329. Small sugar beet roots, stored over winter and planted for the production of seed is called -------

330. Seed produced by a breeder of an entry that has been identified for pre-release multiplication but has not been released and notified-------------------.

331. ------------------refers to the weight in grams of 1000 seeds or grains.

332. A machine used to separate seeds on the basis of their surface smoothness, used for seed cleaning----- --------------.

333. Minimum germination percentage of certified seed of hybrid maize is--------------------.

334. Minimum germination percentage of certified seed of cauliflower and Bhindi is-------------------------.

335. Minimum germination percentage of certified seed of carrot and chilli is--------------------.

336. Minimum germination percentage of sorghum and rice is-------------------.

337. The maximum permissible level of moisture in soybean certified seed is ------------------

338. Minimum pure seed in certified seed of carrot ------

339. Weight of working sample is usually-----------------

340. The Tarai Development Corporation was established at Pant Nagar on ---------

341. Fine grinding is usually done for moisture determination for --------------------, ---------------- and coarse grinding for ----------------

342. For moisture determination at low temperature oven method what is the temperature required ------ ----------------

343. For moisture determination at high temperature oven method what is the temperature required ------ ---------------

344. Phosphorus pentroride method is a method used for ---------------determination.

345. Seed testing rules are revised every ---------------- years.

346. Enzyme responsible for pollen release ----------------

347. Basic unit in seed sampling ----------------

348. Seed value (1968) has ---------------- number of rules.

349. For moisture determination how much is the sample size required for submitted sample -------.

350. In which stage seed dry weight and fresh weight is maximum ----------------.

351. Minimum germination percentage of Hybrid jowar- ----------------

352. Minimum germination percentage of wheat--------

353. Minimum germination percentage of rice--------

354. Minimum germination percentage of water melon--- -----------------

355. Minimum germination percentage of hybrid bajra--- -------------

356. Minimum germination percentage of hybrid maize-- ---------------

357. Outer most layer of endosperms ------------------

358. --------refers to the process used to test vegetatively reproduced plants (e.g. potato) for freedom from virus diseases before multiplying them.

359. Top of the paper method for germination is usually done for --------------------sized seeds.

360. Tobacco and lettuce are examples of ---- seeds.

361. Minimum genetic purity is required for foundation seed-----------------

362. Minimum genetic purity is required for breeder seed----------------------

363. Minimum genetic purity is required for certified seed ---------------------

364. Minimum genetic purity is required for composite varieties ----------------------

365. Minimum genetic purity is required for hybrids synthetics---------

366. Naked seeds are generally found in ---------------

367. A legislation passed in western countries that entitles the breeder of a variety, a prescribed loyality for using the variety ----------------------.

368. Cold test is a type of vigour test used for -------------.

369. Artificial seed refers to:
(A) Seeds of a plant with desired characteristics synthesized with recombinant DNA technology
(B) Somatic embryos encapsulated in a gel
(C) Seeds of synthetic plants
(D) All of these

370. The most appropriate definition of seed in terms of homologous organs found in spore producing land plant is:
(A) A seed of an integumented sporangium with a single functional spore
(B) A seed of an integumented indehisent sporangium containing an egg
(C) A seed of an integumented gymnosporangium
(D) A seed of an integumented bud containing an egg

371. Terminator seeds produced by seed companies using techniques of genetic engineering are in fact:
(A) Hybrid seeds which germinate in the farmer's field but do not produce seeds in the next generation
(B) Hybrid seeds which give a very high yield
(C) Seeds which germinate and produce incompatible gametes
(D) Seeds that develop into fertile plants which produce non germinable seeds

372. Synthetic seeds are:
(A) Protoplasts (B) Somaclones
(C) Cybris (D) Somatic embryos

373. A transgenic rice that accumulates ß-carotene in its seeds; expected to combat vitamin A deficiency areas where rice is the staple food is:
(A) Golden rice (B) Hybrid rice
(C) Boiled rice (D) Desi rice

374. Label denotes:
(A) History of seeds
(B) Quality of seeds
(C) Both (A) and (B)
(D) None of these

375. Tag denotes:
(A) History of seeds (B) Quality of seeds
(C) Both (A) and (B) (D) None of these

376. Genetic purity of breeder seed in:
(A) 100% (B) 99.5%
(C) 99% (D) 95%

377. Genetic purity of foundation seed is:
(A) 100% (B) 99.5%
(C) 99% (D) 95%

378. Genetic purity of certified seed is:
(A) 100% (B) 99.5%
(C) 99% (D) 98%

379. Genetic purity of truthfull seed is:
(A) 100% (B) 99.5%
(C) 99% (D) 90%

380. Synthetic seeds are:
(A) Protoplasts (B) Somaclones
(C) Cybrid (D) Somatic embryos

381. Cytoplasmic hybrids are known as:
(A) Hybrid (B) Cybrids
(C) Both (A) and (B) (D) None of these

382. Which phytohormone is used for development of synthetic seeds?
(A) IAA (B) GA
(C) ABA (D) Ethylene

383. Which of the following is involved in seedless watermelon production?
(A) Apomixes (B) Interspecific crosses
(C) Tetraploidy (D) None of these

384. Improved seed distributed to farmers for commercial cultivation is:
(A) Nucleus (B) Breeder
(C) Foundation (D) Certified

385. Synthetic seeds are:
(A) Artificially synthesized seeds
(B) Somatic embryos encapsulated in suitable matrix
(C) Seeds of plants modified genetically
(D) None of these

386. Which of the following is not true about synthetic seeds?
(A) Can be stored for a year without the loss of valuables
(B) Easy to handle
(C) Can be directly sown in the soil like natural seeds
(D) Need hardening in the green house

387. Recalcitrant seeds are:
(A) Resistant to drying and freezing temperature
(B) Killed by drying and freezing temperature
(C) Both (A) and (B)
(D) None of these

388. Encapsulation is necessary to produce and protect synthetic seeds. The encapsulation is carried out by various types of hydrogels, which are:
(A) Soluble in water
(B) Soluble in organic solvents
(C) Insoluble in water
(D) Insoluble in organic solvents

389. The production of high quality and uniform embryos has been limited to only;
(A) Carrot (B) Alfalfa
(C) Both (A) and (B) (D) Sandalwood

390. Somatic embryoids are termed as:
(A) Identical with zygotic embryos and without seed coats
(B) Identical with zygotic embryos and with seed coat
(C) Non-identical with zygotic embryos and without seed coats
(D) Non-identical with zygotic embryos and with seed coats

391. The encapsulation of somatic embryos can be carried out by:
(A) Automatic encapsulation process
(B) Gel complexation
(C) Both (A) and (B)
(D) Coating proteins

392. The preserved embryoids are termed as:
(A) Synthetic seeds
(B) Semi-synthetic seeds
(C) Natural seeds
(D) Fermented seeds

393. The seeds Act 2004 includes:
(A) Agriculture (B) Horticulture
(C) Forestry (D) All of these

394. The seeds Act 2004 includes:
(A) Plantation crops (B) Medicinal plants
(C) Aromatic plants (D) All of these

395. The seed Act 2004 permits sale of:
(A) Non-branded seeds
(B) Truthfully labeled seeds
(C) Certified seeds
(D) All of these

396. As per the seeds Act 2004, the seed bag should contain:
(A) Name of variety
(B) Physical purity
(C) Germination percentage
(D) All of these

397. Compulsory licence has to be obtained by:
(A) Seed dealers (B) Seed sellers
(C) Seed growers (D) All of these

398. As per the seeds Act 2004, which of the following is to be registered?
(A) Seed producer (B) Seed dealer
(C) Horticulture nursery (D) All of these

399. Basic requirement for registration under PPV & FR Act are:
(A) Novelty (B) Distinctveness
(C) Uniformity (D) Stability
(E) All of these

400. Farmers can sell seeds matching to certified seeds in terms of:
(A) Germination (B) Physical purity
(C) Genetic purity (D) All of these

401. As per the Seeds Act 2004, the variety of field crops can be registered for:
(A) 5 years (B) 10 years
(C) 15 years (D) 20 years

402. The variety of perennial crops can be registered for:
(A) 10 years (B) 15 years
(C) 18 years (D) 20 years

403. Seed meant for general distribution to the farmers for commercial crop production refers to:
(A) Foundation seed (B) Breeder seed
(C) Certified seed (D) Nucleus seeds

404. Initial seed of an improved variety is called:
(A) Nucleus seed (B) Breeder seed
(C) Foundation seed (D) Certified seed

405. Seed certification requires:
(A) An improved variety (B) Genetic purity
(C) Physical purity
(D) Prescribed germination

406. Improved seed includes:
(A) Nucleus seed (B) Breeder seed
(C) Foundation seed (D) Certified seed

407. Certification is not required for:
(A) Breeder seed (B) Foundation seed
(C) Certified seed (D) All of these

ANSWERSHEET
16. SEED BIOTECHNOLOGY

1. (B)	2. (C)	3. (A)	4. (D)	5. (A)	6. (C)	7. (D)	8. (B)	9. (A)	10. (B)
11. (C)	12. (D)	13. (B)	14. (A)	15. (C)	16. (D)	17. (B)	18. (A)	19. (D)	20. (C)
21. (B)	22. (A)	23. (A)	24. (B)	25. (C)	26. (B)	27. (A)	28. (D)	29. (D)	30. (A)
31. (B)	32. (D)	33. (B)	34. (D)	35. (A)	36. (D)	37. (A)	38. (C)	39. (C)	40. (A)

41. (D)	42. (B)	43. (D)	44. (C)	45. (A)	46. (B)	47. (C)	48. (D)	49. (A)	50. (B)
51. (D)	52. (C)	53. (A)	54. (B)	55. (D)	56. (D)	57. (B)	58. (C)	59. (A)	60. (B)
61. (D)	62. (B)	63. (B)	64. (D)	65. (A)	66. (A)	67. (C)	68. (D)	69. (A)	70. (C)
71. (B)	72. (B)	73. (C)	74. (D)	75. (A)	76. (B)	77. (C)	78. (A)	79. (A)	80. (B)
81. (C)	82. (D)	83. (B)	84. (D)	85. (C)	86. (C)	87. (C)	88. (D)	89. (A)	90. (B)
91. (D)	92. (C)	93. (A)	94. (D)	95. (C)	96. (B)	97. (C)	98. (A)	99. (C)	100. (D)
101. (A)	102. (C)	103. (B)	104. (A)	105. (C)	106. (B)	107. (A)	108. (A)	109. (B)	110. (A)
111. (B)	112. (B)	113. (D)	114. (A)	115. (D)	116. (C)	117. (A)	118. (B)	119. (C)	120. (D)
121. (B)	122. (C)	123. (B)	124. (A)	125. (A)	126. (D)	127. (C)	128. (A)	129. (D)	130. (B)
131. (D)	132. (C)	133. (D)	134. (A)	135. (C)	136. (B)	137. (C)	138. (D)	139. (A)	140. (A)
141. (C)	142. (B)	143. (D)	144. (B)	145. (D)	146. (C)	147. (C)	148. (A)	149. (A)	150. (B)
151. (C)	152. (D)	153. (D)	154. (C)	155. (B)	156. (D)	157. (A)	158. (C)	159. (C)	160. (B)
161. (D)	162. (A)	163. (B)	164. (C)	165. (C)	166. (A)	167. (D)	168. (B)	169. (D)	170. (C)
171. (A)	172. (B)	173. (A)	174. (D)	175. (B)	176. (B)	177. (C)	178. (B)	179. (C)	180. (A)
181. (D)	182. (D)	183. (A)	184. (D)	185. (C)	186. (A)	187. (A)	188. (A)	189. (D)	190. (A)
191. (A)	192. (C)	193. (A)	194. (D)	195. (B)	196. (A)	197. (A)	198. (B)	199. (C)	200. (A)
201. (D)	202. (A)	203. (B)	204. (C)	205. (B)	206. (B)	207. (D)	208. (C)	209. (A)	210. (A)
211. (D)	212. (A)	213. (A)	214. (B)	215. (C)	216. (A)	217. (C)	218. (B)	219. (A)	220. (C)
221. (C)	222. (A)	223. (A)	224. (D)	225. (A)	226. (D)	227. (A)	228. (D)	229. (C)	230. (B)
231. (C)	232. (A)	233. (D)	234. (C)	235. (D)	236. (A)	237. (A)	238. (D)	239. (C)	240. (C)
241. (A)	242. (B)	243. (A)	244. (C)	245. (D)	246. (B)	247. (B)	248. (D)	249. (C)	250. (A)
251. (B)	252. (B)	253. (A)	254. (B)	255. (A)	256. (B)	257. (A)	258. (C)		

Match the following

259. (B)	260. (C)	261. (C)	262. (D)	263. (B)	264. (B)	265. (C)	266. (B)	267. (D)	268. (B)
269. (C)	270. (D)	271. (B)	272. (D)	273. (C)	274. (D)	275. (C)	276. (B)	277. (C)	278. (B)
279. (A)									

True/False

280. T	281. F	282. F	283. T	284. F	285. T	286. T	287. T	288. T	289. F
290. F	291. T	292. F	293. F	294. T	295. F	296. T	297. T	298. T	299. F

Fill in the blanks

300. certification, labeling	301. long, dry	302. 3 m	303. H_2SO_4	304. electric colour sorter
305. Economic yield x 100/biological yield	306. viability	307. B. line	308. Dormancy	309. Double hybrid
310. pure	311. syngamy	312. pericarp	313. dioecious	314. perfect
315. endosperm	316. epicotyl	317. genetic purity	318. genotype	319. certation
320. detassling	321. genetic shift	322. spiral	323. seddy tassel	324. RLV
325. A. line	326. seed lot	327. seed plot	328. specific gravity	329. stackling
330. stock seed	331. test weight	332. vulvet roll separator	333. 90%	334. 65%

335. 60%	**336. 60%**	**337. 12**	**338. 95**	**339. 25 gms**
340. 27 Feb. 1969	**341. cereals, cotton and legume**	**342. 103 ± 20C**	**343. 130-135°C**	**344. moisture content**
345. 3	**346. Callase**	**347. g**	**348. 39 rules**	**349. 100 gm and 50 gm**
350. physiological maturity	**351. 80**	**352. 85**	**353. 80**	**354. 60**
355. 75	**356. 90**	**357. aleurone layer**	**358. indexis**	**359. small**
360. photoblastic	**361. 99%**	**362. 100 or 99.9%**	**363. 98%**	**364. 98%**
365. 98%	**366. Gymnosperm**	**367. PBR**	**368. Maize**	**369. (B)**
370. (A)	**371. (D)**	**372. (D)**	**373. (A)**	**374 (A)**
375 (B)	**376 (A)**	**377 (C)**	**378 (D)**	**379 (D).**

380. (D)	**381. (B)**	**382. (C)**	**383. (A)**	**384. (D)**	**385. (B)**	**386. (D)**	**387. (B)**	**388. (A)**	**389. (C)**
390. (A)	**391. (C)**	**392. (A)**	**393. (D)**	**394. (D)**	**395. (C)**	**396. (D)**	**397. (D)**	**398. (D)**	**399. (E)**
400. (D)	**401. (C)**	**402. (C)**	**403. (C)**	**404. (A)**	**405. (C)**	**406. (A)**	**407. (A)**		

Chapter - 17

PLANT PROTECTION BIOTECHNOLOGY

1. Studies of plant diseases is called:
(A) Virology (B) Plant Pathology
(C) Mycology (D) None of these

2. Who studies the plant diseases:
(A) Virologist (B) Pathologist
(C) Plant Pathologist (D) Mycologist

3. Plant Pathology word derived from three Greek words. Which of the following is not related to it?
(A) Phyton (B) Pathos
(C) Logus (D) Logy

4. Who is known as father of plant pathology?
(A) Anton de Bery (C) Luis Pasteur
(C) R.L. Tulasne (D) Julious Coon

5. Who laid the foundation of experimental plant pathology?
(A) Anton de Bery (C) Luis Pasteur
(C) Morren (D) Tillet

6. Downey mildew fungus was discovered by:
(A) Anton de Bery (C) Luis Pasteur
(C) Morren (D) Tillet

7. Virus is found in cells in form of:
(A) Crystalline groups (B) Liquid from
(C) Suspension form (D) None of these

8. Indian Phytopathological Society was established by:
(A) E.T. Butler (B) S.N. Dasgupta
(C) R. Verma (D) B.B. Mundkur

9. Who is the father of Indian plant pathology?
(A) E.T. Butler (B) K.C. Mehta
(C) R.S. Singh (D) R. Verma

10. Who told, 'Virus is a contagium vivum fluidum'?
(A) Beijerinck (B) Ivanowski
(C) Meyor (D) R. Kotch

11. Who is associated with the study of wheat rust in India:
(A) B.B. Mundkur (B) E.J. Butler
(C) K.C. Mehta (D) J.F. Dastur

12. Which type of rust does not occur on wheat:
(A) White rust (B) Black rust
(C) Brown rust (D) Yellow rust

13. On germination, the Teliospore produces:
(A) Germ tube (B) Germ sporangium
(C) Promycelium (D) None of these

14. In a rust fungus, how many types of spores are produced:
(A) One (B) Two
(C) Four (D) Five

15. Which one is considered as the perfect spore:
(A) Pycniospore (B) Aeciospore
(C) Teliospore (D) Uredospore

16. Heteroecious rust completes its life cycle on:
(A) One host (B) Two unrelated hosts
(C) Three hosts (D) None of these

17. Pustules are often arranged in rows in:
(A) Black rust (B) Brown rust
(C) Yellow rust (D) None of these

18. *Puccinia recondita* is the causal organism of:
(A) Black rust (B) Brown rust
(C) Yellow rust (D) White rust

19. Appearance of wheat rusts in India takes place through the dissemination of:
(A) Pycniospores (B) Aeciospores
(C) Uredospores (D) Teliospores

20. Erysiphe graminis tritici is:
(A) Obligate saprophyte
(B) Obligate parasite
(C) Facultative parasite
(D) Facultative saprophyte

21. Ear cockle of wheat is a :
(A) Fungal disease (B) Bacterial disease
(C) Viral disease (D) Nematode disease

22. The pathogen of loose smut of wheat survives:
(A) With the seeds (B) On the seeds
(C) Within the seeds (D) None of these

23. Which one is the abiotic disease of rice:
(A) Blast (B) False smut
(C) Khaira (D) Brown spot

24. Only the naked rachis remains in the last in case of:
(A) Covered smut of barley
(B) Loose smut of wheat
(C) Grain smut of Jowar (D) Bajra smut

25. Solar energy treatment against loose smut of wheat was evolved by:
(A) Mehta (B) Mundkur
(C) Butler (D) Luthra & Sattar

26. The causal organism of ear cockle of wheat is:
(A) *Ustilago tritici* (B) *Puccinia triticina*
(C) *Anguina tritici* (D) *Alternaria triticina*

27. Whip smut occurs in:
(A) Maize (B) Sorghum
(C) Bajra (D) Sugarcane

28. Karnal bunt is the name because:
(A) Symptoms appear in Kernal
(B) Reported from Karnal
(C) Reported by Karnal (D) None of these

29. False smut is the name because its causal organism does not belong to:
(A) Masticomycotina (B) Zygomycotina
(C) Ascomycotina (D) Basidiomyotina

30. Disease development is rapid in Karnal bunt due to:
(A) Frequent light rains (B) Cloudy weather
(C) Low temperature
(D) Combination of above three factors

31. Which disease has caused the famous 'Bengal famine':
(A) Rust of wheat (B) Red rot of sugarcane
(C) Tikka disease of groundnut
(D) Brown spot of paddy

32. The toxin 'Cochliobolin' is produced by the causative fungus of:
(A) Chickpea grey mould (B) Sugarcane wilt
(C) Rice brown spot (D) Rice blast

33. In blast of rice, the symptoms appear on:
(A) Leaves only (B) Leaf-sheath only
(C) Panicle only (D) On all above parts

34. A disease which occurs widely but periodically is:
(A) Epidemic disease (B) Endemic disease
(C) Sporadic disease (D) Pandemic disease

35. A disease which occurs at a short interval and its severity is more called:
(A) Epidemic disease (B) Endemic disease
(C) Sporadic disease (D) Pandemic disease

36. The pathogen of brown spot of rice is:
(A) Soil borne (B) Seed borne
(C) Soil & seed borne (D) None of these

37. Smuts, in general are considered as:
(A) Obligate parasites
(B) Obligate saprophytes
(C) Facultative parasites
(D) Facultative saprophytes

38. The conidia of *Pyricularia oryzae* are:
(A) One celled (B) Two celled
(C) Three celled (D) Many celled

39. Mycellium of powdery mildew pathogen is:
(A) Ectophytic (B) Endophytic
(C) Intercellular (D) Intracellular

40. In red rot of sugarcane, the reddening is found in:
(A) Parenchyma (B) Pith
(C) Phloem (D) Xylem

41. Setae are produced in:
(A) *Fusarium* (B) *Pythium*
(C) *Colletotrichum* (D) *Cercospora*

42. In *Pyricularia oryzae*, the spots on leaves are mostly:
(A) Reniform (B) Pyriform
(C) Doliform (D) Spherical

43. In Ascochyta blight, the spots on leaves are mostly:
(A) Angular (B) Circular
(C) Oblong (D) Irregular

44. Fusaric acid plays an important role in pathogenesis of:
(A) Wilt (B) Blight
(C) Rust (D) Smut

45. The perfect size of *Colletotrichum falcatum* is:
(A) *Pyrenophora graminae*
(B) *Mycosphaerella arachidicola*
(C) *Glumerellatu cumanensis*
(D) *Cochliobolus miyabeanus*

46. In red rot of sugarcane, symptoms also appear on:
(A) Tip of the leaves (B) Margin of the leaf
(C) Midrib of the leaves (D) None of these

47. In *Colletrotrichum falcatum*, the conidia are produced in:
(A) Pycnidia (B) Acervuli
(C) Synnemata (D) Sporodochia

48. Karathane is basically used for the control of:
(A) Downey mildew (B) Powdery mildew
(C) Wilt (D) Blast

49. Bordeaux mixture is a:
(A) Copper fungicide (B) Sulphur fungicide
(C) Dithiocarbamate (D) Systemic fungicide

50. Bacterial blight of cotton is also called as:
(A) Tar leaf spot (B) Angular leaf spot
(C) Zonate leaf spot (D) Red leaf spot

51. Bacterial blight of cotton does not produce the symptoms on:
(A) Seedlings (B) Stems
(C) Bolls (D) Roots

52. The disease is a typical vascular wilt in rice:
(A) Blast (B) Brown leaf spot
(C) Bacterial leaf streak
(D) Bacterial leaf blight

53. The main source of primary inoculum of bacterial blight of cotton is:
(A) Seed (B) Soil
(C) Air (D) Water

54. Kresek phase is found in the rice disease:
(A) Bacterial leaf streak (B) Bacterial blight
(C) Blast (D) Sheath blight

55. Which factor does not favour the rapid spread of bacterial blight of rice:
(A) Rainy weather (B) Strong wind
(C) High humidity
(D) Temperature below 20^{0}C

56. The symptoms only appear on leaf in case of:
(A) Angular leaf spot of cotton
(B) Bacterial leaf streak of rice
(C) Blast of rice
(D) Brown spot of rice

57. In the genus *Fusarium*, conidia are produced in:
(A) Pycnidia (B) Acervuli
(C) Sporodocia (D) Synnemata

58. Wilt disease can be controlled by eliminating soil borne infection through:
(A) Crop rotation (B) Seed treatment
(C) Spraying of fungicide (D) None of these

59. Wilt pathogen is mainly:
(A) Soil borne (B) Seed borne
(C) Air borne (D) None of these

60. The size of pesticide molecules, if all other conditions are constant, greatly influences it is:
(A) Absorption (B) Solubilization
(C) Adsorption (D) Volatilization

61. Some pesticides with positively charged groups are also absorbed by:
(A) Illite clays (B) Kaolinite clays
(C) Monmorillonite clays (D) Silicate clays

62. In general, herbicides are --------- than either fungicides or insecticides:
(A) More mobile (B) Less mobile
(C) Equally mobile (D) More immobile

63. Some degradation products of pesticides are even -------- than the original pesticides.
(A) Less toxic (B) More toxic
(C) Non-toxic (D) Equally toxic

64. The total number of bacteria frequently -------- after fumigation:
(A) Increases (B) Decreases
(C) Remains constant (D) None of these

65. The process of ammonification is often --------- by the use of pesticides:
(A) Harmful (B) Benefitted
(C) Not affected (D) None of these

66. The principle that in modern times "the distribution of plant species on the earth is not uniform" was established by:
(A) Charles Darwin (B) De Candolle
(C) Nikolai Vavilov (D) Carl Von Linnaeus

67. Vavilov recognized primary centres of probable origin of cultivated plants numbering:
(A) 4 (B) 5
(C) 8 (D) 10

68. Species and subdivisions in plants were designated respectively at Linneons and Jordanons by:
(A) W.F. Libby (B) L.P. Lotsy
(C) R.E. Reeves (D) M.Y. Nuttonson

69. The settled village community farming based on small grain and animal domestication occurred in Near East about ---------- years ago:
(A) 5000 (B) 7500
(C) 10000 (D) 10500

70. The causal organism of black rust in wheat is:
(A) *Pucinia recondite tritici*
(B) *Puccinia streiformis*
(C) *Puccinia graminis tritici*
(D) *Ustilago nuda tritici*

71. Karnal bunt or partial bunt was first observed in experimental seed plots at:
(A) Kernal (B) Pantnagar
(C) New Delhi (D) Hisar

72. Brown stripe downy mildew of maize can be controlled by use of:
(A) Bordeaux mixture (4:450)
(B) Bavistin
(C) Karathane-45
(D) Dithane M-45

73. The causal agent of anthracnose diseases in sorghum is:
(A) *Puccinia purpurea*
(B) *Helminthosporium turcicum*
(C) *Colletotrichum graminicola*
(D) *Gleourcospora sorghi*

74. The chickpea wilt is caused by:
(A) *Fusarium oxysporum*
(B) *Ascochyta rabi*
(C) *Bortrylis cineraria*
(D) *Scelerotinia sclerotiorum*

75. Chickpea pod borer can be controlled by the spray of:
(A) Monocrotophos @ 1 ml/litre of water
(B) Aldrin @ 15 kg/ha
(C) Seed treatment by captan (0.25%) @ 4g/ha of seed
(D) BHC @ 10 kg/ha

76. The seed of safflower should be treated with Agrosan GN
(A) 30 g/kg seed
(B) 30mg/kg seed
(C) 30 decigram/kg seed
(D) 100 decigram/kg seed

77. Safflower rust can be controlled by:
(A) Spray of Zineb
(B) Seed reatment with thiram
(C) Destroying the infected material
(D) Following the 3 to 4 years crop rotation

78. Treatment of cotton seed with organo-mercurial compound (s) helps to minimize the disease:
(A) Soil borne (B) Air borne
(C) Water borne (D) Seed borne

79. The most destructive disease of sugar cane is:
(A) Red rot (B) Wilt
(C) Smut (D) Albino

80. To avoid rotting in early planting of potato, seeds should be treated with:
(A) 0.5 per cent Dithane M-45
(B) 2.0 per cent Dithane M-45
(C) 1.5 per cent Aretan
(D) 2.0 per cent Tafasan

81. Treat the sugarcane seed setts with--------- against fungaldise:
(A) Diathane M-45 (B) Zineb
(C) Thiram (D) Aretan

82. The ------- botanist Micheli (1679-1737) was the first scientist who in 1729 studied fungi and show their spores:
(A) German (B) Dutch
(C) Italian (D) Austrian

83. In ---------, the French botanist Tillet published a paper on bunt or stinking smut of wheat:
(A) 1555 (B) 1655
(C) 1755 (D) 1855

84. By 1857, Berkley and Schacht discovered root knot nematodes and cyst nematode in---------:
(A) Sweet sorghum (B) Sugarcane
(C) Potato (D) Sugarbeet

85. The discovery of --------- mosaic by Meyer in 1886 was the beginning of studies on viruses as incitants of diseases in plants:
(A) Tomato (B) Mungbean
(C) Tobacco (D) Soybean

86. ---------, before 1910 initiated exhaustive study of Indian fungi and diseases caused by them:
(A) E.J. Butler (B) J.G. Moseman
(C) R.E. Nelson (D) A. Livney

87. Dr. K.C. Mehta of Agra College made an outstanding contribution to plant pathology by discovering the disease cycle of cereal in India:
(A) Smuts (B) Rusts
(C) Wilts (D) Blights

88. In bacterial brown rot and wilt of potato the amount of -------- produced by pathogens are proportional to severity of symptoms:
(A) Gibberellins (B) Lipids
(C) Polysaccharides (D) Proteins

89. Some ------varieties of apple exude toxic wax on leaf surface to prevent germination of conidia of *Podosphaera leucotricha*:
(A) Scab resistant (B) Leaf blight resistant
(C) Powdery mildew resistant
(D) Rust resistant

90. Red scales of ---------- (red varieties) contain protocatechuic acid and catechol which may exude in drops to resist the attack of *Colletrotrichum circinans*:
(A) Garlic (B) Onion
(C) Leek (D) Celery

91. Some varieties of ---------- are resistant to wilt due to the presence of hydrocyanides (HCN) in root exudates:
(A) Sunflower (B) Safflower
(C) Castor (D) Linseed (flax)

92. Oil from American elm seed contains 55 per cent capric acid toxic to *Ceratocystis ulmi*, the pathogen causing the famous disease:
(A) English elm (B) Dutch elm
(C) American elm (D) French elm

93. The epidermal cells of --------- plant become filled with toxic terpenoid aldehyde after they are few days old and these cells are rarely penetrated by pathogens:
(A) Jute (B) Flax
(C) Cotton (D) Okra

94. In *P. coffeae* (nematode), a --------- species, the optimum temperature for multiplication is reported to be 29.5ºC on rough lemon:
(A) Temperate (B) Tropical
(C) Humid (D) Sub-humid

95. Strawberry plants infected with *P. penetrans* (nematode) have been successfully treated with exposure in hot water at:
(A) 49.4ºC for 17.5 min. (B) 52.8ºC for 6 min.
(C) 54.4ºC for 4 min. (D) 56ºC for 2 min.

96. Leaf curl is a serious disease of:
(A) Moong (B) Pea
(C) Gram (D) Lentil

97. Ear cockle disease is found in:
(A) Sugarcane (B) Wheat
(C) Cotton (D) Rice

98. Blight in pulses can be controlled by:
(A) Spray of manganese carbamate @ 2.0 kg /ha
(B) Ziram 80% @ 2.0 kg/ha before flowering
(C) Seed treatment with Benlate and Thiram
(D) All of these

99. Downey mildew in pulses can be controlled by spraying of:
(A) Mencozeb @ 0.2 %
(B) Thiofinate methyl 70% @ 0.15%
(C) Grow resistant variety (D) All of these

100. Powdery mildew in pea can be controlled by spraying of:
(A)Dusting of Pendimethalin @ 2.0-3.0 l/ha
(B) Karathan @ 0.06% spray
(C) Stomp @ 2.0 l/ha (D) None of these

101. Khaira disease is caused due to deficiency of:
(A) Iron (B) Zinc
(C) Sulphur (D) Nitrogen

102. Wilt is an important disease of:
(A) Pea (B) Arhar
(C) Mustard (D) Wheat

103. The wilt resistant variety of chickpea is:
(A) K-850 (B) KWR-108
(C) Sadabahar (D) Radhey

104. Seed treatment of sugarcane is done by:
(A) Agalol (B) Thiram
(C) Vitavax (D) Zinc sulphate

105. Which one the important disease of groundnut is:
(A) Phyllody (B) Wilt
(C) Tikka (D) Cercospora leaf spot

106. Which one the important disease of sesame is:
(A) Phyllody (B) Wilt
(C) Tikka (D) Cercospora leaf spot

107. Sterility mosaic is a disease related to:
(A) Fungal (B) Bacterial
(C) Viral (D) Physiological

108. Black tip is a common disease of:
(A) Banana (B) Grape
(C) Orange (D) Mango

109. Which wheat rust normally appears earlier in plains of Uttar Pradesh:
(A) Black rust (B) Yellow rust
(C) Brown rust (D) None of these

110. Teliospores are bicelled and stalked in:
(A) *Uromyces* (B) *Melampsora*
(C) *Ustilago* (D) *Puccinia*

111. Heterokaryosis is the phenomenon of:
(A) Fission of two dissimilar nuclei in fungus cell
(B) Breaking of fungus nuclei into two nuclei
(C) Production of spores from nucleus
(D) All of these

112. Dr. K.C. Mehta was associated with:
(A) Cereal rust in India (B) Smut research
(C) Bacterial research (D) Virus research

113. Viroid is a particle with:
(A) Naked nucleic acid (B) Covered nucleic acid
(C) Protein capsid (D) None of these

114. Potato spindle tuber disease is caused by:
(A) Virus (B) Viroid
(C) Mycoplasma (D) Fungus

115. All plant pathogenic bacteria are:
(A) Gram positive
(B) Gram negative
(C) Gram positive and gram negative
(D) None of these

116. Multiplication of bacteria is caused by:
(A) Mitosis (B) Meiosis
(C) Binary fussion (D) All of these

117. The cell wall of oomycetes is made of:
(A) Chitin
(B) Cellulose
(C) Both chitin and cellulose
(D) None of these

118. Cell wall of fungi is made of:
(A) Chitin (B) Cellulose
(C) Both chitin and cellulose
(D) All of these

119. Infectious disease is caused by:
(A) Parasites (B) Non parasites
(C) Virus (D) All of these

120. Non infectious disease is caused by:
(A) Nutrient deficiency (B) Mineral deficiency
(C) Pollutants (D) All of these

121. Saprophytic pathogens feed on:
(A) Only dead matter
(B) Only living organisms
(C) Both (A) and (B) (D) None of these

122. Parasites feed on:
(A) Only dead matter
(B) Only living organisms
(C) Both (A) and (B) (D) None of these

123. Which of the following has its own metabolic activity?
(A) Fungus (B) Virus
(C) Viroid (D) Plasmids

124. Litchen is a example of:
(A) Mutalistic association
(B) Symbiotic association
(C) Parasitism
(D) All of these

125. A plant disease spreads on a larger area with great loss is called:
(A) Endemic disease (B) Epidemic disease
(C) Sporadic disease (D) All of these

126. A plant disease spreads after irregular intervals at different locations in big way, is called:
(A) Endemic disease (B) Epidemic disease
(C) Sporadic disease (D) All of these

127. A plant disease spreads on a regular intervals is called:
(A) Endemic disease (B) Epidemic disease
(C) Sporadic disease (D) All of these

128. An obligate pathogen can:
(A) Grow only on living organism
(B) Grow on dead matters
(C) Grow on both living and non living matters
(D) None of these

129. Fungi is devoid of:
(A) Chlorophyll (B) Cell wall
(C) Nucleus capsid (D) None of these

130. Plasmids are:
(A) Extra chromosomal DNA fragments found in bacterial cells
(B) Extra chromosomal RNA fragments found in bacterial cells
(C) DNA particles found in viral cells
(D) All of these

131. Homothallic fungus is:
(A) Dioecious (B) Monoecious
(C) Both of these (D) None of these

132. Clamp connection helps in:
(A) Communication of two cells of same hyphae
(B) Communication of two cells of different hyphae
(C) To provide support (D) None of these

133. The fungi grown on dung of herbivorous animals are called as:
(A) Coprophilous fungi (B) Keratinaceous fungi
(C) Amphibious fungi (D) All of these

134. The fungi grown on dead bodies are called as:
(A) Coprophilous fungi
(B) Keratinaceous fungi
(C) Amphibious fungi (D) All of these

135. The root like structure of fungi is called as:
(A) Rhizoid (B) Stroma
(C) Rhizomorph (D) Sclerotium

136. The hyphae aggregation that bear more than one fortification is called as:
(A) Rhizoid (B) Stroma
(C) Rhizomorph (D) Sclerotium

137. Hyphae compact resting body is known as:
(A) Rhizoid (B) Stroma
(C) Rhizomorph (D) Sclerotium

138. Rope like strands of hyphae is known as:
(A) Rhizoid (B) Stroma
(C) Rhizomorph (D) Sclerotium

139. Spermatia are formed in:
(A) Oogamy (B) Ascus
(C) Spermogonia (D) None of these

140. Ascospores are formed in:
(A) Oogamy (B) Ascus
(C) Spermogonia (D) None of these

141. The process in which two gamete mix each other and make diploid zygote is known as:
(A) Oogamy (B) Karyogamy
(C) Spermogonia (D) Plasmogamy

142. Which of the following spore is formed in sporangia?
(A) Conidia (B) Zoospore
(C) Oidia (D) Chlamydospore

143. Isogamy refers to:
(A) Copulation between two similar gametes
(B) Copulation between two dissimilar gametes
(C) Copulation of between two gametes
(D) All of these

144. Which of the following is a facultative parasite?
(A) Fusarium (B) Rhizoctonia
(C) Alternaria (D) All of these

145. Which of the following is a facultative saprophyte?
(A) Smut (B) Helminthosporium
(C) Leaf curl fungus (D) All of these

146. In which of the following group of fungi has similar sporangiophore to its vegetative hyphae?
(A) Pythium (B) Albugo
(C) Permospora (D) Sclerospora

147. Aflatoxin poison is produced by:
(A) *Aspergillus flavus* (B) *Claviceps purpurea*
(C) *Pencillium spp* (D) All of these

148. Gibberellic acid is formed by:
(A) *Gibberella fujikuroi* (B) *G. mosshata*
(C) *Rhizopus niger* (D) *Aspergillus flavus*

149. The member of oomycetes are:
(A) Uniflagellate (B) Biflagellate
(C) Triflagellate (D) None of these

150. Amphigynous antheridium is found in:
(A) Pythium (B) Phytophthora
(C) Both (A) and (B) (D) None of these

151. Which of the following is a obligate parasite?
(A) Puccinia (B) Pernospora
(C) Albugo (D) All of these

152. Which of the following is not saprophyte?
(A) Erysiphe (B) Aspergillus
(C) Mucor (D) Penicillium

153. Indole acetic acid is obtained from:
(A) *Gibberella spp.* (B) *Rhizopus niger*
(C) *Aspergillus flavus* (D) *Claviceps purpurea*

154. Alkaloid can be obtained from:
(A) *Gibberella* spp. (B) *Rhizopus niger*
(C) *Aspergillus flavus*
(D) *Claviceps purpurea*

155. Pin moulds are also known as:
(A) Mucorales (B) Algal fungi
(C) Yeast fungi (D) None of these

156. Which of the following fruiting bodies produces asexual spores?
(A) Pycnidia (B) Sporodochia
(C) Acervuli (D) All of these

157. Sexual reproduction is not found in:
(A) Phycomycetes (B) Ascomycetes
(C) Basidiomycetes (D) Imperfect fungi

158. Pycnidia formation is very common in :
(A) Rust fungi
(B) Downy mildew fungi
(C) White blister fungi (D) All of these

159. Synnema formation is a common feature of:
(A) Ascomycetes (B) Phycomycetes
(C) Basidiomycetes (D) Imperfect fungi

160. The cushion like aggregation of fungal tissues and bearing conidiophores is known as:
(A) Sporodochia (B) Aecium
(C) Pycnidia (D) All of these

161. Basidiospores are borne on spore mother sac is known as:
(A) Basidiophore (B) Basidium
(C) Sporangia (D) None of these

162. Which of the spore is produced by perethecium?
(A) Zygospore (B) Zoospore
(C) Ascospore (D) Chlamydospore

163. Which of the following is the most important feature of Phycomycetes?
(A) Asexual spores are either motile or non motile
(B) Unicellular and halocarpic
(C) Mycellium is coenocytic
(D) All of these

164. Which of the following is known as fruiting bodies?
(A) Pycnidia (B) Sporodochia
(C) Acervuli (D) All of these

165. Acervuli produce:
(A) Conidiospores (B) Spores
(C) Acervuli (D) Aecium

166. Which of the following fungi produces infertile spores?
(A) Phytomycetes (B) Ascomycetes
(C) Basidiomycetes (D) Deuteromycetes

167. Flask shape fruiting body is known as:
(A) Pycnidia or spermogonia
(B) Acervuli (C) Sori
(D) None of these

168. Mucors may be sexually differentiated by:
(A) ++ strains (B) -- strains
(C) + - strains (D) All of these

169. Incomplete reduction division is also known as:
(A) Meiosis (B) Brachymeiosis
(C) Mitosis (D) None of these

170. Brachymeiosis was found in:
(A) *Erysiphe polygoni*
(B) *Lycopersium esculentum*
(C) *Rhizopus* spp
(D) All of these

171. Food is spoiled by fungus:
(A) *Aspergillus niger* (B) *Aspergillus repens*
(C) *Aspergillus glaucas* (D) All of these

172. Concentrated packed milk is spoiled by fungus:
(A) *Aspergillus niger*
(B) *Aspergillus repens*
(C) *Aspergillus glaucas* (D) All of these

173. Meat is spoiled by fungus:
(A) *Aspergillus niger* (B) *Aspergillus repens*
(C) *Aspergillus glaucas* (D) All of these

174. Yeast complete its life cycle known as:
(A) Haplobiontic type (B) Diplobiontic type
(C) Haplo diplobiontic (D) All of these

175. The sexual structure in ascomycetes is known as:
(A) Pycnidia (B) Acervuli
(C) Ascocarp (D) Synnemata

176. Mucorales are generally:
(A) Saprophyte (B) Parasite
(C) Predators (D) All of these

177. Which of the fungus does not produce mycelium?
(A) Yeast (B) Rhizopus
(C) Mucor (D) Algal fungi

178. Sexual reproduction in yeast occurs under:
(A) Favourable condition
(B) Unfavourable condition
(C) Both (A) and (B) (D) None of these

179. Asperigillosis is caused by:
(A) Aspergillus fungus (B) Bacteria
(C) Virus (D) None of these

180. Wood is deteriorated by the species of Panicillium:
(A) *Penicillium divaricatum*
(B) *Penicillium expensum*
(C) *Penicillium glacuum* (D) None of these

181. Rust fungi mainly a:
(A) Saprophyte (B) Parasite
(C) Facultative parasite (D) All of these

182. Which of the alkaloid is produced by ergot fungi?
(A) Ergotin (B) Penicillin
(C) Cutrinin (D) All of these

183. The initial stage of ergot infection is known as:
(A) Honey dew stage (B) Scleotial stage
(C) Spore formation stage (D) All of these

184. The sexual fruiting body of ergot fungi is:
(A) Apothecium (B) Cleistothecium
(C) Synnemata (D) Perithecium

185. Primary mycelium is also known as:
(A) Monocaryon mycelium
(B) Dicaryon mycelium
(C) Haploid mycelium (D) All of these

186. Tertiary mycelium is found only in:
(A) Phycomycetes (B) Ascomycetes
(C) Basidiomycetes (D) All of these

187. Which of the following fungi have well developed dicaryophase?
(A) Phycomycetes (B) Ascomycetes
(C) Basidiomycetes (D) All of these

188. Smut fungi are:
(A) Obligate parasite (B) Obligate saprophyte
(C) Facultative saprophyte
(D) Facultative parasite

189. Rust fungi are:
(A) Polymorphic (B) Homothallic
(C) Monoecious (D) All of these

190. Smut means:
(A) Sooty powder (B) Wet powder
(C) Grey powder (D) All of these

191. Basidiocarps are absent in:
(A) Teliomycetes (B) Hymenoycetes
(C) Gasteromycetes (D) None of these

192. In basidiomycetes, conidia are always produced on:
(A) Primary mycelium
(B) Secondary mycelium
(C) Tertiary mycelium (D) All of these

193. Basidiomycetes are:
(A) Homothallic (B) Heterothallic
(C) Both (A) and (B) (D) None of these

194. Basidiomycetes are produced on:
(A) Hymen layer (B) Sterigmata
(C) Basidiophores (D) None of these

195. Eu-form rust are also known as:
(A) Long cycled rust (B) Short cycled rust
(C) Demi cycled rust (D) None of these

196. Which is not an example of long cycled rust?
(A) Wheat rust (B) Sunflower rust
(C) Lucerne rust
(D) White rust of Brassica

197. The life cycle of rust fungi, in which uredinial stage is not found is known as:
(A) Macro cycle (B) Demi cycle
(C) Micro cycle (D) None of these

198. The life cycle of rust fungi, in which all five stages are not found is known as:
(A) Macro cycle or long cycle
(B) Demi cycle
(C) Micro cycle
(D) None of these

199. The life cycle of rust fungi where only dicoryotic telial stage is not found is known as:
(A) Macro cycle (B) Demi cycle
(C) Micro cycle (D) None of these

200. Autocious fungi complete their life cycle on:
(A) Single host (B) Two host
(C) Dead matter (D) None of these

201. Heterocious fungi complete their life cycle on:
(A) Single host (B) More than one host
(C) Dead matter (D) None of these

202. Chlamydospores are produced always on:
(A) Primary mycelium
(B) Secondary mycelium
(C) Both (A) and (B)
(D) None of these

203. Most appropriate time of black rust attack in Northern Plains is:
(A) Late February – Early March
(B) December- January
(C) October-November (D) All of these

204. Most appropriate time of black rust attack in southern Plains is:
(A) Late February – March
(B) December- February
(C) October-November (D) All of these

205. Optimum temperature for black rust spread is:
(A) 17-20^{0}C (B) 25-30^{0}C
(C) 30-35^{0}C (D) 35-40^{0}C

206. Which is the alternate host of yellow rust?
(A) *Tholictrum spp* (B) *Berberis spp*
(C) Wild wheat (D) Not known

207. Which of the following is the collateral host of yellow rust?
(A) *Agropyron semicostatum*
(B) *Bromus catharticus*
(C) *Bromus japonicus*
(D) All of these

208. Which of the following urediniospores of rust can require the highest temperature for its germination?
(A) Black rust (B) Yellow rust
(C) Brown rust (D) All of these

209. Telospores are:
(A) Unicellullete (B) Bicellilate
(C) Sessile (D) Round shaped

210. Urediniospores are:
(A) Uninucleate (B) Binucleate
(C) Both (A) and (B) (D) None of these

211. Which of the following stages is the alternate host of black rust?
(A) 0 and I stage (B) II and III stages
(C) IV and 0 stages (D) All of these

212. Which of the following is the collateral host of stem rust?
(A) *Birza minor* (B) *Bromes patulus*
(C) *Avena fatua* (D) All of these

213. Mature teliospore has a:
(A) Heploid nucleus (B) Diploid nucleus
(C) Dikaryotic (D) None of these

214. Which of the following sequence of rust occurrence in UP:
(A) Leaf rust > Stripe rust > Stem rust
(B) Stripe rust > Leaf rust > Black rust
(C) Black rust > Brown rust > Yellow rust
(D) Yellow rust > Leaf rust > Black rust

215. Teliospore of smut fungi are called as:
(A) Terminal
(B) Intercalary
(C) Intercalary and endogenous
(D) None of these

216. Which of the mycelium of smut fungi is ineffective?
(A) Primary (B) Secondary
(C) Tertiary (D) All of these

217. Hypomycetes has:
(A) Well developed mycelium
(B) Budding absence
(C) Sterile mycelium
(D) All of these

218. Bajra smut is a:
(A) Air borne (B) Seed borne
(C) Soil borne (D) Soil and seed borne

219. The colour of bajra smut spores is:
(A) Black (B) Yellow
(C) Brownish green (D) Red

220. The sorus of bajra smut is formed usually on:
(A) Leaf (B) Stem
(C) Ovary (D) All aerial parts

221. The main symptoms of early blight of potato is:
(A) Formation of concentric ringes on infected spots
(B) Formation of cottony growth
(C) Formation of chlorosis
(D) All of these

222. Optimum temperature for development of early blight of potato is:
(A) 12-19^{0}C (B) 25-26^{0}C
(C) 30-36^{0}C (D) 38-40^{0}C

223. Which of the following spores are not formed in Fusarium wilt?
(A) Micro conidia (B) Macro conidia
(C) Chlamydospores (D) Onidia

224. The symptoms of false smut of rice appears on:
(A) Leaf (B) Stem
(C) Panicle (D) Root

225. The main symptoms of false smut of rice is:
(A) Apearance of round bluish spots on leaves
(B) Conversion of few seeds into scleratial bodies covering by green colour layer
(C) Formation of black powder in whole panicle
(D) All of these

226. The main symptoms of rice blast is:
(A) Round black spots on leaves
(B) Oval shaped round spots on leaves
(C) Boat shaped spots with grey red borders and grey ash colour centre
(D) None of these

227. Rice blast is also known as neck rot because of:
(A) Neck of panicle becomes shriveled due to severity of disease
(B) Neck becomes stronger as fungus produces pyricularin toxin
(C) It produces hyaline spores
(D) None of these

228. The cell wall of bacteria is made of:
(A) Pectin (B) Chitin
(C) Mycopeptide (D) None of these

229. Fission fungi is also known as:
(A) Bacteria (B) Virus
(C) Nematodes (D) None of these

230. The coma (,) shaped bacteria are called :
(A) Bacillus (B) Vibrio
(C) Streptococcus (D) None of these

231. Bacteria without flagella is known as:
(A) Atrichous (B) Monotrichous
(C) Lophotrichous (D) Petritrichous

232. Which type of bacteria can be killed by Penicillin?
(A) Gram +ve (B) Gram – ve
(C) Both (A) and (B) (D) None of these

233. The pathogenic bacteria are generally:
(A) Gram +ve (B) Gram –ve
(C) Both (A) and (B) (D) None of these

234. The main function of mesosomes in bacterial cell is:
(A) To help in cell division
(B) To help in respiration
(C) To assist in photosynthesis
(D) All of these

235. Bacteria having flagella at its whole body is known as:
(A) Monotrichous (B) Amphitrichous
(C) Petritrichous (D) None of these

236. Bacteria having a bunch of flagella at one end is known as:
(A) Monotrichous (B) Lophotrichous
(C) Petritrichous (D) None of these

237. Which of the following disease is caused by bacteria?
(A) Brown rot of potato
(B) Angular leaf spot of cotton
(C) Fire blight of apple and pear
(D) All of these

238. Which of the following disease is caused by bacteria?
(A) Red stripe of sugarcane
(B) Soft rot of cotton
(C) Leaf roll of potato
(D) Stalk rot of maize

239. Which of the following is a autotrophic bacteria?
(A) *Chlorobium* (B) *Thiospirillum*
(C) *Nitrobacter* (D) All of these

240. Which of the following is a heterotrophic bacteria?
(A) *Chlorobium* (B) *Pseudomonas*
(C) *Xantthomonas* (D) All of these

241. Sex pili is found only in:
(A) Male bacteria (B) Female bacteria
(C) Both male and female (D) None of these

242. Sex factor of bacteria is known as:
(A) H factor (B) F factor
(C) S factor (D) None of these

243. Cyst formation is very common in which of the following bacteria?
(A) Azotobacter (B) Myxococcus
(C) Chondromyces (D) All of these

244. Nucleus of bacteria cell is called as:
(A) Nucleoplasm (B) Nucleoid
(C) Both (A) and (B) (D) None of these

245. Which of the following feature of gram positive bacteria?
(A) Homogenous cell wall that contains >85% mucopeptides and its polymers
(B) Teichoic acids, ribitol, glutamic acid and alanine are found in more quantity
(C) Violet or blue colour after staining with solution of crystal violet and iodine
(D) All of these

246. Which of the following feature of gram negative bacteria?
(A) Cell wall is thin and made of mainly lipoprotein and its polymers
(B) They react with counter stain safranine and become red to pink in colour
(C) Teichoic acid is not formed
(D) All of these

247. Generally plant virus posses a:
(A) Single stranded RNA
(B) Double stranded RNA
(C) Doubled stranded DNA
(D) None of these

248. Viroid is a:
(A) Naked RNA particles
(B) Naked DNA particles
(C) Naked RNA and DNA particles
(D) None of these

249. Prions are:
(A) Infectious protein particles
(B) Infectious DNA particles
(C) Both (A) and (B)
(D) None of these

250. The virus enzymes are collectively known as:
(A) Transcriptase (B) Polymerase
(C) Peplomers (D) None of these

251. Virus can be classified under the Phylum of:
(A) Procaryota (B) Vira
(C) Eucaryota (D) Virophyta

252. Viruses are resistant to:
(A) Acids (B) Bases
(C) Salts (D) All of these

253. Virus is a:
(A) Obligate parasite (B) Obligate saprophyte
(C) Facultative parasite
(D) Facultative saprophyte

254. Chemically virus is a:
(A) Nucleoprotein (B) Proteir
(C) Nucleic acid (D) All of these

255. Virus has mainly two parts, one is nucleic acid core and second part is:
(A) Cell wall (B) Nucleic membrane
(C) Protein core (D) None of these

256. Which of the following disease is caused by virus?
(A) Small pox (B) Measles
(C) Common cold (D) All of these

257. Ultra microscopic particle capable of infecting live cells is called:
(A) Plasmic (B) Nucleiod
(C) Virion (D) None of these

258. The size of virus particle is:
(A) 10-300 μ (B) 10-20 μ
(C) 50-80 μ (D) 25-35 μ

259. Virus can be seen only with:
(A) Naked eyes (B) Microscope
(C) Electron microscope (D) All of these

260. Yellow Jack is a:
(A) Yellow fever (B) Small pox
(C) Swine fever (D) None of these

261. Cuscuta is a:
(A) Holoparasite on stem
(B) Semi-parasite on stem
(C) Holoparasite on root
(D) Semi-parasite on root

262. Loranthus is a:
(A) Stem Holoparasite
(B) Stem semi-parasite
(C) Root Holoparasite
(D) Root semi-parasite

263. Orobanche is a:
(A) Stem Holoparasite (B) Stem semi-parasite
(C) Root Holoparasite (D) Root semi-parasite

264. Striga is a:
(A) Stem Holoparasite (B) Stem semi-parasite
(C) Root Holoparasite
(D) Root semi-parasite

265. Broom rape is also known as:
(A) Orobanche (B) Striga
(C) Loranthus (D) Cuscuta

266. Witch weed is also known as:
(A) Orobanche (B) Striga
(C) Loranthus (D) Cuscuta

267. Vrikshabhaksha is known as:
(A) Loranthus (B) Banda
(C) Giant mistletoe (D) All of these

268. Nematodes are:
(A) Saprophyte (B) Predator
(C) Pathogenic (D) All of these

269. Nematodes may cause following symptoms?
(A) Cellular hypertrophy/hyperplasia
(B) Suppression of mitosis
(C) Cell necrosis (D) All of these

270. Plant nematodes always have:
(A) Stylet (B) Respiratory organs
(C) Circulatory system (D) None of these

271. Which of the following system is not present in plant nematodes?
(A) Respiratory system (B) Circulatory system
(C) Reproductive system (D) None of these

272. Which of the following nematicides are used to control of nematodes?
(A) Nemagon (B) Nemaphos
(C) Hexanema (D) All of these

273. Which stage of root knot nematode attacks to plant?
(A) First larvae (B) Second larvae
(C) Third larvae (D) Fourth Larvae

274. Male nematodes differ with female nematodes in:
(A) Male nematodes are cylindrical and filliform while female are bigger in size with round and oval shaped
(B) Male nematodes have respiratory system
(C) Females do not have stylets
(D) None of these

275. Main symptoms of root knot of tomato is:
(A) Knot formation in roots (B) Yellowing of leaves
(C) Stunting growth (D) All of these

276. The nematodes take their food from host with the help of:
(A) Mouth (B) Stylet
(C) Haustoria (D) None of these

277. The important feature of plant parasitic nematodes is:
(A) Body cavity absent
(B) Transparent segmentless body
(C) Respiratory and circulation systems absent
(D) All of these

278. Which of the following nematodes acts as carrier of viruses?
(A) *Xiphineiva* spp. (B) *Longidorus* spp.
(C) *Trix chodorous* spp. (D) All of these

279. Cyst forming nematodes species is:
(A) *Meloidogyne* spp. (B) *Heterodera* spp.
(C) *Aphelenchoides* spp. (D) *Pratylenchus* spp.

280. Leaf nematodes species is:
(A) *Meloidogyne* spp. (B) *Heterodera* spp.
(C) *Aphelenchoides* spp. (D) *Pratylenchus* spp.

281. Root injury creating nematodes species is:
(A) *Meloidogyne* spp. (B) *Heterodera* spp.
(C) *Aphelenchoides* spp. (D) *Pratylenchus* spp.

282. Which of the following nematode species is the causal agent for root knot disease in plants?
(A) *Meloidogyne* spp. (B) *Heterodera* spp.
(C) *Aphelenchoides* spp. (D) *Pratylenchus* spp.

283. Reverse transcription is found only in:
(A) Bacteria (B) Virus
(C) Protozoa (D) Fungus

284. Which of the following disease is caused by MLO's?
(A) Papaya bunchy top
(B) Sugarcane grassy shoot
(C) Witch's broom of groundnut
(D) All of these

285. Stinking smut is also known as:
(A) Hill bunt of wheat (B) Karnal bunt
(C) Covered smut of jowar (D) None of these

286. Which of the following is used for disinfestations of ware houses?
(A) $CuSO_4$
(B) Formaldehyde
(C) Fumigation with chloropicrin tear gas
(D) All of these

287. Exclusion includes:
(A) Compulsory quarantines and inspection
(B) Embargoes
(C) Certification
(D) All of these

288. Which of the following gas is used for sterilization?
(A) Chlorine (B) Sulphur dioxide
(C) Formadehyde (D) All of these

289. Tyndalization is a:
(A) Fractional sterilization
(B) Complete sterilization
(C) Incomplete sterilization
(D) None of these

290. Tyndalization is used for:
(A) Sterilization of medium
(B) Sterilization of inoculation materials
(C) Sterilization of inoculating pathogens
(D) None of these

291. Vericides are used to control:
(A) Bacteria (B) Virus
(C) Nematodes (D) Fungus

292. Damping off of tobacco is caused by:
(A) *Pythium aphanidermatum*
(B) *Pythium infestans*
(C) *Phytophthora infestans*
(D) None of these

293. White Blister disease is caused by:
(A) *Pythiaceae* (B) *Albuginaceae*
(C) *Perouosporaceae* (D) None of these

294. Downy mildew disease is caused by:
(A) *Pythiaceae* (B) *Albuginaceae*
(C) *Perouosporaceae* (D) None of these

295. Late blight of potato is caused by:
(A) *Phytophthora infestans*
(B) *Albugo candida*
(C) *Peronospora pisi*
(D) None of these

296. Downy mildew of pea disease is caused by:
(A) *Phytophthora pisi* (B) *Peronospora pisi*
(C) *Albigo candida* (D) None of these

297. Green ear disease is caused by:
(A) *Albugo candida*
(B) *Sclerospora graminicola*
(C) *Erysiphe graminis* (D) None of these

298. White blister of crucifers disease is caused by:
(A) *Albugo candida* (B) *Erysiphe graminis*
(C) *Peronospora pisi* (D) None of these

299. Downy mildew of brassica is caused by:
(A) *Albugo candida*
(B) *Peronospora brassicae*
(C) *Phytophthora infestans* (D) None of these

300. Downey mildew of maize is caused by:
(A) *Sclerospora sacchari*
(B) *Sclerospora rayssiae* var. *zeae*
(C) *Sclerospora philippinensis*
(D) All of these

301. Stem gall of coriander is caused by:
(A) *Phycomycetes* (B) *Ascomycetes*
(C) *Basidiomycetes* (D) All of these

302. Causal organism of stem gall of coriander is:
(A) *Protomyces macrosporus*
(B) *Taphrina deformans*
(C) *Raphrina maculans*
(D) *Penicillium expensum*

303. Spot rot of apple is caused by:
(A) *Protomycese macrosporus*
(B) *Penicillium expensum*
(C) *Raphrina mauclans*
(D) *Taphrina deformans*

304. Leaf spot of turmeric is caused by:
(A) *Taphrina maculans* (B) *Erasiphe polygoni*
(C) *Erysiphe graminis* (D) All of these

305. Peach leaf curl is caused by:
(A) *Taphrina maculans* (B) *Erasiphe polygoni*
(C) *Erysiphe graminis* (D) All of these

306. Powdery mildew of pea is caused by:
(A) *Erysiphe graminis* (B) *Erysiphi polygoni*
(C) *Uncinula necator* (D) All of these

307. Powdery mildew of cereals is caused by:
(A) *Erysiphe graminis* (B) *Erysiphi polygoni*
(C) *Taphrina deformans* (D) All of these

308. Powdery mildew of apple is caused by:
(A) *Uncinula necator*
(B) *Podosphaera leuxotricha*
(C) *Sclerotium oryzae* (D) All of these

309. Stem rot of paddy is caused by:
(A) *Sclerotium oryzae* (B) *Claviceps purpurea*
(C) *Ustilaginoidea virens* (D) All of these

310. False smut of paddy is caused by:
(A) *Ustilaginoidea virens* (B) *Claviceps purpurea*
(C) *Erysiphe graminis* (D) None of these

311. Ergot of bajra is caused by:
(A) *Claviceps microcephala*
(B) *Claviceps purpurea*
(C) *Sclerotium oryzae* (D) None of these

312. Black rust of wheat is caused by:
(A) *Puccinia graminis* var. *tritici*
(B) *Puccinia recondita*
(C) *Puccinia striformis*
(D) *Melampsora lini*

313. Leaf rust of wheat is caused by:
(A) *Puccinia graminis*
(B) *Puccinia recondita*
(C) *Puccinia striformis* (D) *Melampsora lini*

314. Stripe rust of wheat is caused by:
(A) *Puccinia graminis* (B) *Puccinia recondita*
(C) *Puccinia striformis* (D) *Melampsora lini*

315. Stem rust of wheat is also known as:
(A) Black rust (B) Brown rust
(C) Yellow rust (D) White rust

316. Leaf rust of wheat is also known as:
(A) Black rust (B) Brown rust
(C) Yellow rust (D) White rust

317. Stripe rust of wheat is also known as:
(A) Black rust (B) Brown rust
(C) Yellow rust (D) White rust

318. Causal organism of pea rust is:
(A) *Uromyces pisi* (B) *Uromyces aestivum*
(C) *Puccinia recondita* (D) None of these

319. Causal organism of linseed rust is:
(A) *Uromycese pisi* (B) *Uromyces fabae*
(C) *Melampsora lini* (D) None of these

320. Causal organism of loose smut of wheat is:
(A) *Ustilago nuda*
(B) *Ustilago nuda var tritici*
(C) *Ustilago hordei*
(D) *Tolyposporium penicillariae*

321. Causal organism of grain smut of bajra is:
(A) *Sphacelotheca sorghi*
(B) *Sphacelotheca hordi*
(C) *Tolyposporium penicillariae*
(D) None of these

322. Causal organism of smut of bajra is:
(A) *Fusarium oxysporum*
(B) *Tolyposorium penicillariae*
(C) *Ustilago nuda* (D) None of these

323. Causal organism of early blight of potato is:
(A) *Alternaria solani* (B) *Alternaria infestans*
(A) *Pyricularia oryzae*
(D) *Cercospora personata*

324. Causal organism of Alternaria leaf blight of wheat is:
(A) *Alternaria tritici* (B) *Alternaria aestivum*
(C) *Pyricularia triticina* (D) None of these

325. Causal organism of lentil wilt is:
(A) *Fusarium oxysporum* (B) *Fusarium udum*
(C) *Fusarium solani* (D) *Fusarium pisi*

326. Causal organism of arhar wilt is:
(A) *Fusarium oxysporum* (B) *Fusarium udum*
(C) *Fusarium solani* (D) *Fusarium pisi*

327. Causal organism of linseed wilt is:
(A) *Fusarium oxysporum* (B) *Fusarium udum*
(C) *Fusarium solani* (D) *Fusarium pisi*

328. Causal organism of tikka disease of groundnut is:
(A) *Cercospora personata*
(B) *Cercospora arachidicola*
(C) Both (A) and (B)
(D) None of these

329. Which of the following causal organism of tikka disease of groundnut is more fatal?
(A) *Cercospora personata*
(B) *Cercospora arachidicola*
(C) Both (A) and (B) (D) None of these

330. Which of the following disease is caused by mycoplasma like organisms (MLO)?
(A) Rice yellow dwarf
(B) Sugarcane grassy shoot
(C) Potato witch's broom
(D) All of these

331. Which of the following disease is caused by Rickettsia like microorganisms (RLO's)?
(A) Ratoon stunting (B) Aster yellow
(C) Khaira disease (D) All of these

332. Which of the following characteristics is not associated with MLO's?
(A) Unicellular (B) Usually immobile
(C) Cell wall absent (D) Cell wall present

333. Khaira disease is a example of:
(A) Endemic disease (B) Epidemic disease
(C) Non infectious disease
(D) Infectious disease

334. Green ear disease is a example of:
(A) Endemic disease (B) Epidemic disease
(C) Sporodic disease (D) Infectious disease

335. Late blight of potato disease is a example of:
(A) Endemic disease (B) Epidemic disease
(C) Sporodic disease (D) Infectious disease

336. Necrotrophs are pathogens that:
(A) Kill host plants
(B) Do not kill host plant
(C) Provide nutrition to host plants
(D) None of these

337. Which of the following is a example of necrotroph pathogen?
(A) Pythium (B) Rhizoctonia
(C) Soft rot bacteria (D) All of these

338. Facultative parasites are:
(A) Primarily saprophytes but can attack on living organisms under favourable conditions
(B) Primarily saprophytes and do not attack on living organisms
(C) Primarily obligate parasites
(D) All of these

339. Dolipore septa are found in:
(A) Oomycetes (B) Ascomycetes
(C) Basidiomycetes (D) All of these

340. The process in which two dissimilar and motile gametes copulate is called as:
(A) Isogamy (B) Anisogamy
(C) Heterogamy (D) All of these

341. The process in which one non motile female gamete copulates with small motile gamete is known as:
(A) Isogamy (B) Anisogamy
(C) Heterogamy (D) All of these

342. The process in which two gamete mix each other and make diploid zygote is known as:
(A) Isogamy (B) Anisogamy
(C) Keryogamy (D) Plasmogamy

343. Oospore is a:
(A) Resting spore (B) Active spore
(C) Both (A) and (B) (D) None of these

344. The thallus that is partially converted into a reproductive structure is known as:
(A) Holocarpic (B) Eucarpic
(C) Miecious (D) Dioeicious

345. LSD is a:
(A) Stimulant (B) Pain killer
(C) Antibiotic (D) None of these

346. LSD is abbreviated as:
(A) Least Stimulant Dose
(B) d-Lysergic Acid diethylamide
(C) Least Statistical Deviation
(D) None of these

347. Irish blight is also known as:
(A) Late blight of potato
(B) White blight of mustard
(C) Yellow rust
(D) Loose smut of wheat

348. Which of the following difference between Phytophthora and Albugo fungi?
(A) Albugo fungi are obligate parasite while Phytophthora are mainly facultative saprophyte
(B) Sporangiophores in phytophthora are simple and branched but in albugo they are unbranched and clavate type
(C) Chlamydospores are formed in phytophthora only
(D) All of these

349. Which of the fruiting body is related to sexual reproduction?
(A) Cleistothecia (B) Perithecia
(C) Apothecia (D) All of these

350. Which of the following is the most important difference between Oomycetes and Zygomycetes?
(A) Asexual spores are always non motile in zygomycetes while motile and non motile in Oomycetes
(B) Oomycetes produce laterally biflagellate zoospores but zygomycetes produce aplanospores
(C) Sexual zygote in oospore is oomycetes while zygospore is zygomycetes
(D) All of these

351. Sterile filaments of perethecium are called as:
(A) Paraphyses (B) Periphyses
(C) Appendages (D) All of these

352. Hymenium is a:
(A) Layer containing asci in the fruiting body
(B) Type of spore
(C) Fruiting body (D) None of these

353. Stromata is a:
(A) Filament like structure associated with perethecia
(B) Fruiting body
(C) Type of spore (D) None of these

354. Which is difference between Mucor and Rhizopus:
(A) Rhizopus are absent in mucor but mucor are present in Rhizopus
(B) Rhizopus have stolon but mucor do not have
(C) Single sporangiospore in mucor but incluster form in rhizopus
(D) All of these

355. Dikaryophase is found in:
(A) Phycomycetes (B) Ascomycetes
(C) Deuteromycocetes (D) All of these

356. The sexual fruiting body of plectomycetes is:
(A) Cleistothecium (B) Apothecium
(C) Perithecium (D) All of these

357. Penicillin antibiotic is prepared by:
(A) *Penicillium chrysogenum*
(B) *Penicillium notatum*
(C) *Penicillium albidum*
(D) *Penicillium citrinum*

358. Citrinin is prepared by:
(A) *Penicillium citrinum*
(B) *Penicillium flavus*
(C) *Penicillium albidum*
(D) *Penicillium citrinum*

359. Penicillin B is prepared by:
(A) *Penicillium flavum*
(B) *Penicillium notatum*
(C) *Penicillium albidum*
(D) *Penicillium citrinum*

360. The fruit like apple, pear and grapes are spoiled by:
(A) *Penicillium expensum*
(B) *Penicillium notatum*
(C) *Penicillium flavum*
(D) *Penicillium citrinum*

361. Which of the *penicillium* spp is harmful for food products?
(A) *Penicillium expensum*
(B) *Penicillium glaccusm*
(C) *Penicillium digitatum*
(D) All of these

362. Aspergillin antibiotic was prepared by which fungus?
(A) *Aspergillus niger*
(B) *Aspergillus fumigatus*
(C) *Aspergillus flavus* (D) All of these

363. The causal organism of powdery mildew of grape vine is:
(A) Unicellular nector
(B) Podosphaera leucotricha
(C) Sclerotium oryzae (D) All of these

364. Most important symptoms of ergot disease is:
(A) Transformation of kernels into black structure
(B) Dispersion of spores
(C) Both (A) and (B) (D) None of these

365. The symptoms of powdery mildew of pea appear first of all on:
(A) Leaves (B) Stem
(C) Whole plant (D) Roots

366. Nature of mycelium growth of Basidiomycetes is mainly:
(A) Ectophytic (B) Endophytic
(C) Both (A) and (B) (D) None of these

367. Powdery mildew differs with downy mildew in which of the following ways?
(A) Powdery mildew on both surface of leaf but downy mildew growth appears only on lower surface of leaf
(B) Powdery mildew causes necrosis while downy growth causes chlorosis
(C) Powdery mycelium ectophytic while downy mycelium endophytic
(D) All of these

368. Which of the following difference between Erysiphe and Aspergillus is correct?
(A) Erysiphe is obligate ectophytic parasite while Aspergillus is saprophyte
(B) Erysiphe has appressoria and haustoria while these are absent in Aspergillus
(C) In Rhizopus, conidiophore produces sterigmeta, while sterigmata are absent in Erysiphe
(D) All of these

369. A Septate basidium is known as:
(A) Probasidium (B) Phragmobasidium
(C) Holobasidium (D) None of these

370. Dicaryotization is carried out in basidiomycetes with the help of:
(A) Dolipore septa
(B) Clamp connection
(C) Cell division (D) None of hese

371. Which of the following stage is not found in micro cyclic rust?
(A) Aerial (B) Uredinal
(C) Aerial and Uredinial (D) None of these

372. Which of the following genera has stalked teliospores?
(A) Puccinia (B) Uromuces
(C) Melampsora
(D) Puccinia and Uromyces

373. Pycnial and aerial stage of rust fungi are always found on:
(A) Primary host (B) Alternate host
(C) Main host (D) All of these

374. Which of the order belong to rust fungi?
(A) Uredinales (B) Ustilaginales
(C) Teliomycetes (D) None of these

375. Which of the following stages are found in rust fungi?
(A) Pycnial, Aerial, Uredinal, Telial and Basidial stages
(B) Pycnial, Aerial, Apothecial, Telial and Basidial stages
(C) Conidial, Aerial, Uredinal, Telial and Basidial stages
(D) None of these

376. Conidia are produced in basidiomycetes on:
(A) Primary mycelium
(B) Secondary mycelium
(C) Both (A) and (B) (D) None of these

377. Aseptic basidium is known as:
(A) Probasidium (B) Phragmobasidium
(C) Holobasidium (D) None of these

378. Telial stage of rust fungi is always found on:
(A) Primary host (B) Secondary host
(C) Dead matter (D) All of these

379. How many physiological strains of black rust are found?
(A) 100 (B) 200
(C) 250 (D) 300

380. Which of the following is not the collateral host of yellow rust?
(A) *Agropyron semicostatum*
(B) *Bromes cathartics*
(C) *Puccinia striiformis* (D) All of these

381. Which of the following is the alternate host of black rust?
(A) Wheat (B) Barbary
(C) Calotropis (D) *Phalaris minor*

382. Basidiospores of smut fungi are called as:
(A) Sporidia (B) Pycniospores
(C) Oidiospores (D) None of these

383. Basidiospores of stem rust of wheat attack on:
(A) Wheat plant
(B) Barbary plant
(D) *Phalaris minor*
(D) *Chenopodium album*

384. Which of the following plant is the natural alternate host of *Puccinia recondita*?
(A) *Berberis vulgaris*
(B) *Isopyron fumarioides*
(C) *Avena fatua* (D) *Triticum aestivum*

385. Which of the following rust requires the lowest temperature for its over summering?
(A) Black rust (B) Leaf rust
(C) Strip rust (D) All of these

386. Linseed rust is a:
(A) Obligate parasite
(B) Obligate saprophyte
(C) Facultative parasite
(D) Facultative saprophyte

387. Linseed rust fungus is a:
(A) Autocious (B) Heterocious
(C) Both (A) and (B) (D) None of these

388. In *Alternaria solani*, the conidia are formed at:
(A) Top end of the condiophores
(B) Intercalary formation
(C) Both (A) and (B) (D) None of these

389. The infected seed of bajra smut are converted into:
(A) Sori (B) Nod
(C) Mycelium (D) None of these

390. Grain smut of jowar may cause following disease in humen beings:
(A) Hey fever (B) Cholera
(C) Cold (D) None of these

391. Loose smut is a:
(A) Internally seed borne disease
(B) Air borne disease
(C) Soil borne disease
(D) Externally seed borne disease

392. The spores of covered smut of barley are covered by:
(A) Yellow membrane (B) Black membrane
(C) Shining silvery membrane
(D) None of these

393. Which of the following feature is found in deuteromycetes?
(A) Absence of sexual stages
(B) Well developed septate mycelium
(C) Conidia is a sexual spore
(D) All of these

394. Alternaria solani is a:
(A) Soil borne disease (B) Seed borne disease
(C) Air borne disease (D) None of these

395. Which of the perfect stage of red rot of sugarcane:
(A) *Glomerella tucumanensis*
(B) *Physalospora tucumansis*
(C) Both (A) and (B)
(D) None of these

396. The perfect stage of tikka disease of groundnut is:
(A) *Mycosphaerella berkeleyii*
(B) *Mycosphaerella arachidicola*
(C) Both (A) and (B)
(D) None of these

397. Rotten neck disease is known as:
(A) Blast of rice (B) Red rot of sugarcane
(C) Leaf rust of wheat (D) None of these

398. What is the main symptoms of red rot of sugarcane?
(A) Dull red pith (B) Stunted growth
(C) Shorter internodes (D) None of these

399. Which of the main symptoms of leaf blight of wheat?
(A) Small, oval, discolored, irregular lesions scattered on leaves
(B) Leaf starts drying from tip
(C) Black powdery conidia of the fungus cover the lesion under moist conditions
(D) All of these

400. The bacteria arranged like grapes bunches are known as:
(A) Streptococcus (B) Staphylococcus
(C) Sarcinae (D) None of these

401. Which of the following RNA is found in bacterial cell?
(A) m RNA (B) t RNA
(C) r RNA (D) All of these

402. Photosynthetic bacteria have photosynthetic pigments made of:
(A) Lamellae and chromatophores
(B) Ribosomes
(C) Endoplasmic reticulum
(D) All of these

403. Which of the following plant disease is caused by MLO's?
(A) Witch brooms of potato (B) Mulbery dwarf
(C) Aster yellows (D) All of these

404. Which of the following feature is related to MLO.s?
(A) Can be grown on cell free media
(B) They are resistant to Penicillin and susceptible to tetracycline
(C) They can pass through anti bacterial filters
(D) All of these

405. Which of the following does not possess the cell wall?
(A) Bacteria (B) MLO's
(C) Fungi (D) All of these

406. When bacterial cells are arranged in a long series, the bacteria known as:
(A) Diplococcus (B) Streptococcus
(C) Sercinae (D) Vibrio

407. When bacterial cells from structure like cubical packet, the bacteria are known as:
(A) Diplococcus (B) Streptococcus
(C) Sercinae (D) Vibrio

408. Slime layer around the bacterial cells is important for:
(A) Protection against phagocytes
(B) Store food products
(C) Help in infection (D) All of these

409. Literary meaning of virus is:
(A) Morbid poison (B) Acute poison
(C) Poison (D) All of these

410. Antibiotics are prepared mainly from:
(A) Actinomycetes (B) Rhizobium
(C) Pseudomonas (D) Azotobacter

411. What is lysogany?
(A) Fussion of the genetic material of bacteria and virus
(B) Conjugation of one viral cells
(C) Conjugation of two viral cells
(D) None of these

412. *Escherichia coli* is a:
(A) Monomorphic (B) Dimorphic
(C) Both (A) and (B) (D) None of these

413. Which of the following does not posses mitochondria?
(A) Fungi (B) Bacteria
(C) Virus (D) MLO's

414. The virus that attacks on bacteria is called:
(A) Bacteria parasite (B) Bacteria phages
(C) Virophages (D) None of these

415. Viroids have only:
(A) Naked nucleic acid particles
(B) Nucleic acid protein core
(C) Only protein core (D) None of these

416. The nucleic code of plant virus is made of:
(A) RNA (B) DNA
(C) RNA and DNA (D) None of these

417. Which of the following does not possess metabolic ability?
(A) Fungi (B) Bacteria
(C) Virus (D) All of these

418. Protein core of virion is also known as:
(A) Capsid (B) Capsomer
(C) Structural unit (D) All of these

419. Ear cockle of wheat or Sehun disease is caused by:
(A) *Anguina tritici* (B) *Heterodera avenae*
(C) *Meloidogyne javanica* (D) All of these

420. Molya disease of barley is caused by:
(A) Bacteria (B) Virus
(C) Nematodes (D) None of these

421. The causal organism of molya diseases is:
(A) *Heterodera avenae* (B) *Melodogne spp.*
(C) *Anguina spp.* (D) None of these

422. Molya disease of barley first reported at:
(A) Gujarat (B) UP
(C) Punjab (D) Rajasthan

423. The bacteria phages that combine its genome with bacterial genome is called:
(A) Lytic bacteriophages
(B) Lysogenic bacteriophages
(C) Transduction
(D) None of these

424. Which of the following crops are affected by orobanche?
(A) Tobacco (B) Brinjal
(C) Tomato (D) All of these

425. Ear cockle of wheat is often associated with:
(A) Yellow ear rot or Tundu disease
(B) Loose smut
(C) Rust
(D) None of these

426. Which of the following striga species are found in India?
(A) *Striga lutea* (B) *Striga densiflora*
(C) *Striga euphrasiodes* (D) All of these

427. Loranthus is parasite of:
(A) Mangoes (B) Guava
(C) Pipal (D) All of these

428. Prophylaxis includes:
(A) Exclusion (B) Eradication
(C) Direct protection (D) All of these

429. The medium or culture is generally sterilized by:
(A) Autoclave (B) Hot air oven
(C) Light (D) None of these

430. Oogonia filled with oospore are known as:
(A) Plerotic (B) Aplerotic
(C) Filled (D) Unfilled

431. Oogonia partially filled with oospore are known as:
(A) Plerotic (B) Aplerotic
(C) Filled (D) Unfilled

432. Mycoses disease is caused by:
(A) Fungi (B) Bacteria
(C) Virus (D) Fly

433. Heterothallic fungus is:
(A) Dioecious (B) Monoecious
(C) Both (A) and (B) (D) None of these

434. The conjugation of oogonium and antheridium is known as:
(A) Oogamy (B) Karyogamy
(C) Plasmogamy (D) All of these

435. Zygospore is formed after conjugation of:
(A) Two similar gametangia only
(B) Two gametangia
(C) Oogonium and anthesidium
(D) None of these

436. Oospore is formed after conjugation of:
(A) Two similar gametangia
(B) Two similar spores
(C) Oogonium and anthesidium
(D) Two dissimilar gametangia

437. The sucking part of fungi, that draws their neourishment is known as:
(A) Hypha (B) Haustoria
(C) Septa (D) Spore

438. Biotrophs are pathogens that:
(A) Complete their life cycle on living cells
(B) Produce haustoria
(C) Rarely kill to host plants
(D) All of these

439. Facultative parasites are those pathogen that are:
(A) Basically saprophytes but can attack on living organism under favourable conditions
(B) Basically attack on living organism but can survive on dead materials under unfavourable conditions
(C) Grow equally on dead and living organism
(D) All of these

440. The stage in which the nucleoli are scattered in cytoplasm of aseptate fungi is called as:
(A) Coenocytic (B) Uninucleate
(C) Dolipore (D) None of these

441. A cells having two distinct gametes, is called:
(A) Sporangia (B) Gametangia
(C) Conidiosporangia (D) All of these

442. Sexual reproduction in mastigomycotina occurs by which of the following method:
(A) Oogamy
(B) Somatogamy
(C) Planogametic conjugation
(D) All of these

443. Which of the following is the difference between Pythium and Phytophthora genus?
(A) Pythium complete its life cycle on both dead matters and host while phytophthora require living host to complete its life cycle
(B) Pythium are homothallic while Phytophthora are heterothallic fungi
(C) Pythium do not make haustoria but phytophthora make haustoria in the host plants
(D) All of these

444. The dense fascicles made by aggregation of erect conidiophores are known as:
(A) Synnemata (B) Sori
(C) Acervuli (D) None of these

445. Which is the main difference between Sclerospora and Pernospora fungi?
(A) Sclerospora attack mainly on members of monocot plants while Pernospora on dicot plants and very rarely on monocot plant
(B) Both are obligate parasite
(C) Sclerospora is more harmful than pernospora
(D) None of these

446. Which of the spore is produced by cleistothecia?
(A) Ascospore (B) Basidiospore
(C) Oospore (D) Conidia

447. Which of the class is not a true fungi?
(A) Myxomycetes (B) Phycomycetes
(C) Ascomycetes (D) Deuteromycetes

448. The stage under which the production of aeciospores in liquid medium occurs in mucor is known as:
(A) Torula stage (B) Sexual stage
(C) Production stage (D) All of these

449. Vegatative life cycle in yeast mainly occurs by:
(A) Budding (B) Fission
(C) Spermatization (D) Conjugation

450. Vegetative life cycle in yeast occurs under:
(A) Favourable condition
(B) Unfavourable condition
(C) Both (A) and (B)
(D) None of these

451. Asexual reproduction in Aspergillus fungi occurs through:
(A) Conidia (B) Aplanospores
(C) Oospores (D) Ascospores

452. Ascus are naked in which of the following class:
(A) Hemiascomycetes (B) Plectosmycetes
(C) Pyrenomycetes (D) Discomycetes

453. Basidiospores are produced as:
(A) Endogenous (B) Exogenous
(C) Both (A) and (B) (D) None of these

454. How many basidiospores are produced by each basidium?
(A) 2 (B) 4
(C) 6 (D) 8

455. Polymorphism refers to the process having:
(A) > 2 life cycle stages (B) Only one life cycle
(C) < 2 life cycle stages (D) All of these

456. The different spore stages of rust fungi are known as:
(A) 0, I, II, III and IV stages
(B) 1, 2, 3, 4 and 5 stages
(C) 0, I, II, III, IV and IV stages
(D) None of these

457. What is causal organism of yellow rust of wheat?
(A) *Puccinia striiformis* (B) *Puccinia recondita*
(C) *Puccinia tritici* (D) All of these

458. What is optimum temperature for germination of Urediniospores?
(A) 3-5^{0}C (B) 15-16^{0}C
(C) 30-35^{0}C (D) 40-45^{0}C

459. What is optimum temperature for spread of loose smut disease?
(A) 18^{0}C (B) 23^{0}C
(C) 28^{0}C (D) 33^{0}C

460. Grain smut disease of jowar is a disease of:
(A) Internally seed borne
(B) Externally seed borne
(C) Air borne
(D) Soil borne

461. Which of the following genus comes under moniliales order of imperiti fungi?
(A) Alternaria (B) Helminthosporium
(C) Cercospora (D) All of these

462. Which of the following class of deuteromycetes has false mycelium and reproduction by budding?
(A) Blastomycetes (B) Hypomycetes
(C) Coelomycetes (D) All of these

463. Infected hyphae of grain smut of fungi is a:
(A) Uninucleate (B) Dikaryotic
(C) Diploid (D) None of these

464. The optimum temperature for development of leaf blight of wheat is:
(A) 15^{0}C (B) 20^{0}C
(C) 25^{0}C (D) 30^{0}C

465. Which of the following fruiting body is formed in red rot of sugarcane?
(A) Acervuli (B) Pycnidia
(C) Apothecia (D) All of these

466. Mechanism of wilting can be understood by:
(A) Plugging of vessels by fungus hyphae
(B) Toxin theory-production of toxins like Fusaric acids
(C) Enzyme theory-Production of enzymes like Pectolytic enzyme
(D) All of these

467. Silica content in leaf helps rice plant to resist against:
(A) Blast (B) False smut
(C) Bacterial blight (D) None of these

468. Mycoplasma was first discovered from:
(A) Pleuropneumonia disease of sheep
(B) *Bacillis vibrio*
(C) Anthrax of cattle
(D) None of these

469. The sedimentation coefficient of bacterial ribosome is:
(A) 60 s (B) 70 s
(C) 80 s (D) 90 s

470. The pilli or Fimbria are hair like structures found on bacterial cell and its main function is related to:
(A) Conjugation (B) Movement
(C) Cell division (D) None of these

471. The unit of sedimentation is:
(A) Pascal (B) Svedberg
(C) Virus (D) MLO's

472. Which of the following asexual method is very common in bacteria?
(A) Conidia formation
(B) Endospore formation
(C) Binary fission (D) None of these

473. The method through which the genetic material of bacteria is transferred from one cell to other by bacteriophages, is known as:
(A) Conjugation (B) Transformation
(C) Transduction (D) None of these

474. Ghost light is a result of:
(A) Bacterial luminescence
(B) Phosphorescent glow
(C) Bioluminescence
(D) All of these

475. The main symptoms of black leg or soft rot of potato is:
(A) Brown – black or jet black colour of stem at soil level
(B) Black leaf coloration
(C) Black crust on tubers
(D) None of these

476. Virus can be passed through:
(A) Brekfeld and chamberlain filters
(B) Only brekfeld filter
(C) Only chamberlain filter
(D) None of these

477. Plasmid are found in:
(A) Bacterial cells (B) Viral cells
(C) Fungal cells (D) None of these

478. Plasmid are mainly:
(A) Extra chromosomal particles of DNA
(B) Particles of proteins
(C) Micro organisms (D) None of these

479. Virus can be precipitated with the help of:
(A) Acetone (B) Safranin
(C) Ethyl alcohol (D) All of these

480. If the virus multiplies in host bacterium which can undergoes complete lysis to release the daughter phages, this type of virus is called:
(A) Lytic bacterio phages
(B) Lysogenic bacteriophages
(C) Transformation
(D) None of these

481. Root knot nematodes have higher growth at :
(A) 15-30^{0}C (B) 20-25^{0}C
(C) 25-28^{0}C (D) 30-35^{0}C

482. **Which of the following disease is caused by MLO's?**
(A) Little leaf of egg plant (B) Citrus greening
(C) Purple top roll of potato (D) All of these

483. **Which of the following disease is caused by virus?**
(A) Fiji disease of sugarcane
(B) Leaf curl of papaya
(C) Ratoon stunting of sugarcane
(D) None of these

484. **The process in which bacterial genome is multiplied with bacteriophages is called?**
(A) Transformation (B) Transduction
(C) Reproduction (D) None of these

485. **Which of the crop is affected mostly by striga?**
(A) Sugarcane (B) Maize
(C) Millets (D) All of these

486. **Which of the following gas can stimulate the seeds of striga?**
(A) Ethylene (B) Nitrous oxide
(C) Carbon mono oxide (D) Oxygen

487. **Strigol is used to control of:**
(A) Striga (B) Cuscuta
(C) Orobanche (D) Loranthus

488. **Which of the following is a synthetic culture medium?**
(A) Oat agar medium (B) Plant extracts
(C) Rechard's solution (D) None of these

489. **Hill bunt of wheat is caused by:**
(A) *Tiletia foetida* (B) *Neovossia indica*
(C) *Uromystis tritici* (D) None of these

490. **Which of the following is used for tree surgery?**
(A) Bordeaux mixture (B) Chaubatia paste
(C) Santar A (D) All of these

491. **Main symptoms of karnal bunt is:**
(A) Formation of black powder below the pericarp of seed
(B) Produces rotten fish odour
(C) Bad odour is due to Trimethylamine
(D) All of these

492. **Hill bunt can be controlled by:**
(A) Seed treatment with 0.2% vitavex
(B) Use of resistant varieties
(C) Clean the field (D) All of these

493. **Which of the following culture is a natural medium for isolation of pathogen?**
(A) Potato dextrose agar medium
(B) Milk
(C) Egg (D) All of these

494. **Trychoderma can be used as:**
(A) Soil (B) Seed
(C) Both of these (D) None of these

495. **Which of the filters are used to sterilize the medium?**
(A) Brekfeld filter (B) Seitz filter
(C) Chamber lain filter (D) None of these

496. **Root knot disease of tomato is caused by:**
(A) *Meloidogyne javanica*
(B) *Heterodera avenaei*
(C) *Anguina tritici* (D) None of these

497. **The spores of grain smut of jowar are attached with long solid structure known as:**
(A) Columella (B) Pith
(C) Paraphysis (D) None of these

498. **Coelomycetes class of imperfect fungi is characterized by:**
(A) Well developed mycelium
(B) Budding absence
(C) Pycnidium & acervulus fruiting bodies present
(D) All of these

499. **The collateral host of linseed rust is:**
(A) *Linum usitatissimum* (B) *Linum mysorense*
(C) *Linum pisi*
(D) All of these

500. **Covered smut of barley is a disease of:**
(A) Externally seed borne
(B) Internally seed borne
(C) Air borne
(D) None of these

501. **In melampsora, the teliospores are:**
(A) Sessile and unicellular
(B) Strigmated and unicellular
(C) Sessile and multinucleate
(D) None of these

502. **The causal organism of powdery mildew of wheat is:**
(A) *Erysiphe graminis var tritici*
(B) *Erysiphe polygoni*
(C) *Erysiphe aestivum*
(D) None of these

503. **Pin moulds are also known as sugar fungi because of:**
(A) They use simple carbohydrates
(B) They do not use carbohydrate
(C) They use polysaccharides
(D) All of these

504. **Ascogonium produces one tubular structure that receives the nuclei of antheridium. This structure s known as:**
(A) Trichogyne (B) Crosier
(C) Fertilization tube (D) All of these

505. **How many ascospores are produced by an ascus?**
(A) Two (B) Four
(C) Eight (D) Twelve

506. Some filaments protrude out through ostiole of perethecium are known as:
(A) Paraphyses (B) Periphyses
(C) Appendages (D) All of these

507. The spores and spore bearing hyphae are some times grouped into small or large masses or clusters, known as:
(A) Synnemata (B) Sori
(C) Pycnidia (D) None of these

508. Disc like, saucer shaped or cup like fruiting body raised on long or short stalk is known as:
(A) Aecium (B) Sporodochium
(C) Apothecium (D) All of these

509. Aeciospores are formed in:
(A) Pycnidia (B) Synnemata
(C) Acervuli (D) Aecium

510. Setae is present in which of the following fruiting body:
(A) Pycnidia (B) Acervuli
(C) Aecium (D) All of these

511. The thallus that is converted fully into a reproductive structure is known as:
(A) Holocarpic (B) Eucarpic
(C) Monoecious (D) Dioecious

512. In gametangial contact, ascogonia contact with antheridia with the help of a special structure known as:
(A) Trichogyne (B) Dikaryon hyphae
(C) Stroma (D) All of these

513. Which of the following is motyle spore?
(A) Chlamydospore (B) Conidia
(C) Odia (D) Zoospore

514. Which of the following is a biotroph pathogen?
(A) Viruses
(B) Rust fungi
(C) Downy and Powdery mildew
(D) All of these

515. The pathogen that can grow only under controlled environment is called:
(A) Ecologically obligate
(B) Bio-chemically obligate
(C) Facultative parasite
(D) All of these

516. Apisomes are also known as:
(A) Plasmids (B) RNA particles
(C) Golgi bodies (D) All of these

517. Fungi have been originated from:
(A) Polygeneric line (B) Phytogeneric line
(C) Polygendric line (D) None of these

518. The carrier of leaf curl tomato virus is:
(A) White fly (B) Beetle
(C) Bacteria (D) Fungus

519. Leaf curl of tomato is caused by:
(A) Tobacco virus 16 (B) Tomato virus 16
(C) Virus No. 10 (D) Virus No.16

520. Which of the following sexual reproduction method is found in bacteria?
(A) Conjugation (B) Transformation
(C) Transduction (D) All of these

521. Hemaglutination is the process in which bacteria cells can:
(A) Attach with RBC cells
(B) Attach with water surface
(C) Attach with virons (D) None of these

522. Bacterial nucleus is characterized by:
(A) Undeveloped nucleus
(B) Not covered by nucleus membrane
(C) Multiplied by binary fission
(D) All of these

523. Which of the nucleic acid is found in the nucleus of cauliflower mosaic virus?
(A) RNA (B) DNA
(C) Both RNA and DNA (D) All of these

524. England famine in 1846 was caused mainly by:
(A) Late blight of potato
(B) Helminthosporium leaf spot of rice
(C) Rust of wheat
(D) Leaf curl virus in tomato

525. One of the most important reason of Bengal famine in 1943 was:
(A) Late blight of potato
(B) Helminthosporium leaf spot of rice
(C) Rust of wheat
(D) Leaf curl virus in tomato

526. Cuscuta is a:
(A) Complete stem parasite
(B) Complete root parasite
(C) Semi stem parasite
(D) Semi root parasite

527. Loranthus is a:
(A) Complete stem parasite
(B) Complete root parasite
(C) Semi stem parasite
(D) Semi root parasite

528. Striga is a:
(A) Complete stem parasite
(B) Complete root parasite
(C) Semi stem parasite
(D) Semi root parasite

529. Orobanche is a:
(A) Complete stem parasite
(B) Complete root parasite
(C) Semi stem parasite
(D) Semi root parasite

530. Seed treatment with vitavex is the mainly control :
(A) Loose smut (B) Rust
(C) Downy mildew (D) Powdery mildew

531. Loose smut of wheat can be controlled by:
(A) Seed treatment by vitavex before sowing
(B) Seed treatment with hot water
(C) Use of disease resistant varieties
(D) All of these

532. The important symptoms of black rot in cauliflower is:
(A) Chlorosis (B) Drying of leaves
(C) Burning of leaves
(D) Black spots on petioles and veins

533. Purple black of onion is caused by:
(A) Fungus (B) Bacteria
(C) Virus (D) None of these

534. Onion smut can be avoided by:
(A) Spray of Dithane M-45
(B) Seed treatment by captan/Thiram
(C) Weed control (D) Irrigation

535. Carrot yellow is caused by:
(A) Fungus (B) Algae
(C) Bacteria (D) Virus

536. Bacterial soft rot is the result of:
(A) High temperature
(B) Very low temperature
(C) Waterlogging (D) None of these

537. Black rot is main disease of sweet potato and it can be controlled by:
(A) Selection of healthy cuttings
(B) Proper drainage facilities
(C) Cuttings should be treated with 6.25% Areton solution for 8-10 minutes
(D) All of these

538. Wilt is the main disease of nursery of sweet potato and it is caused by:
(A) Bacteria (B) Fungus
(C) Virus (D) Nematodes

539. Why waterlogging area is not suitable for chilli cultivation:
(A) Damping off (B) Bacterial spot
(C) Leaf curl (D) All of these

540. Leaf curl of chilli is mainly spread by:
(A) Air (B) White fly
(C) Water (D) Nematode

541. Mango black tip is caused by:
(A) Bacteria (B) Virus
(C) Smoke of Brick kline (D) Fungi

542. Which of the following disease is the most harmful to mango orchard?
(A) Mango wither tip (B) Anthracnose
(C) Bunchy top (D) Black tip

543. Which of the following disease is not found in papaya orchard?
(A) Anthracnose (B) Collar rot
(C) Papaya mosaic (D) Black tip

544. Panama disease of banana is caused by:
(A) *Fusarium oxysporum* (B) *Bacillus spp.*
(C) Actinomycetes (D) Virus

545. Bunchy top of banana is caused by:
(A) Bacteria (B) Fungi
(C) Virus (D) Actinomycetes

546. Which of the following disease is most harmful to citrus plants?
(A) Citrus canker (B) Gummosis
(C) Anthracnose (D) All of these

547. Citrus canker is caused by:
(A) *Phytophthora infestans*
(B) *Xanthomonas citri*
(C) *Collectotrichum* spp.
(D) *Bacillus moniliformis*

548. Citrus canker can be controlled by :
(A) Blitox-50 @ 0.25% solution
(B) Endosulphan @ 0.2% solution
(C) 2,4-D @ 1.5 kg a.i./ha
(D) None of these

549. Powdery mildew of apple can be controlled by:
(A) Spray of Krathane @ 0.1% solution
(B) Spray of Endosulphan @ 0.2% solution
(C) Spray of Diathane -78 @ 0.1% solution
(D) All of these

550. Which of the following is an important disease of peach?
(A) Blight (B) Leaf curl
(C) Gummosis (D) All of these

551. Suitable bioagents for controlling of wilt disease of chickpea, lentil and pigeonpea is:
(A) *Trychoderma harianum* (B) *Trichoderma viridi*
(C) Both of these (D) None of these

552. Karnal bunt is a disease of:
(A) Viral (B) Nematodes
(C) Fungal (D) Bacterial

553. The pigeonpea wilt is caused by:
(A) *Fusarium oxysporum* (B) *Fusarium udum*
(C) *Ascochyta rabi* (D) *Bortrytis cineraria*

554. Which viruses is likely to be commercialized in India for control of *Helicoverpa (Heliothis) armigera*?
(A) Nucleopolyhedrosis viruses (NPV)
(B) Granulosis viruses (GV)
(C) Advenoviruses (D) All of these

555. Pusa 1406 has resistance for :
(A) Bacterial leaf blight (B) Blast
(C) Tungro (D) Grassy stunt

556. Annual loss due to pests, diseases and weeds annually is estimated nearly:
(A) >50000 crores (B) <50000 crores
(C) > 1.0 crores (D) < 1.0 crores

557. India is using biopesticides presently:
(A) 1% of total pesticide
(B) 5 % to total pesticide
(C) 10 % of total pesticide
(D) 15 % of total pesticide

558. Average biopesticide use globally is:
(A) 5 % of total pesticide
(B) 10 % of total pesticide
(C) 12 % of total pesticide
(D) 15 % of total pesticide

559. Total pesticide consumption in India during 2006-07 is :
(A) 35.65 million tonnes
(B) 37.95 million tonnes
(C) 39.26 million tonnes
(D) 42.56 million tonnes

560. Directorate of Plant Protection Quarantine and Storage was started in:
(A) 1907 (B) 1914
(C) 1921 (D) 1946

561. Destructive Insects and Pest Act was launched in India in:
(A) 1907 (B) 1914
(C) 1921 (D) 1946

562. Which of the airport does not have plant quarantine station?
(A) Lucknow (B) Kolkata
(C) New Delhi (D) Madras

563. Browning of cauliflower is caused by the deficiency of:
(A) Zinc (B) Manganese
(C) Iron (D) Boron

564. The citrus diseases namely 'Reclamation Disease' and 'Die back' are caused by the deficiency of:
(A) Zinc (B) Iron
(C) Copper (D) Molybdenum

565. Which Cry proteins is specific to Lepidoptera?
(A) Cry I (B) Cry II
(C) Cry IV (D) Cry V

566. Which Cry proteins is specific to coleoptera?
(A) Cry I (B) Cry II
(C) Cry IV (D) Cry V

567. Which Cry proteins is specific to diptera?
(A) Cry I (B) Cry II
(C) Cry IV (D) Cry V

568. Which Cry proteins is specific to nematodes?
(A) Cry II (B) Cry V
(C) Cry IV (D) Both (B) and (C)

569. Which Cry proteins is specific to Lepidoptera and diptera?
(A) Cry II (B) Cry V
(C) Cry IV (D) Both (B) and (C)

570. Papaya transgenic Rainberg is resistant to:
(A) Papaya Mosaic Virus
(B) Leaf Curl Virus
(C) Papaya Ring Spot Virus
(D) Distorted Ring Spot Virus

571. Which Cry proteins is specific to Lepidoptera?
(A) Cry I (B) Cry II
(C) Cry IV (D) Cry V

572. Tobacco cultivar A.C. Change is resistant to:
(A) Leaf curl (B) Black rot
(C) Damping off (D) TMV

573. Which of the following disorder in mango is caused by low temperature?
(A) Black heart (B) Crotch injury
(C) Collor injury (D) All of these

574. At present the most serious disease of wheat in Northern plains is:
(A) Powdery mildew (B) Foliage blight
(C) Stem rust (D) Karnal bunt

575. Primary inoculum of loose smut of wheat survives as:
(A) Spore on seed surface (B) Spore in the seed
(C) Spore in soil (D) Mycellium in seed

576. Curved rod shaped bacterium is:
(A) Xanthomonas (B) Erwinia
(C) Streptomyces (D) Pseudomonas

577. Cancer disease of citrus in caused by :
(A) Bacteria (B) Fungas
(C) Virus (D) Nematodes

578. Tungru is a viral disease of:
(A) Rice (B) Citrus
(C) Wheat (D) Tobacco

579. Systemic insecticide used to control pests is:
(A) Endosulphan (B) Chlorpyriphos
(C) Phosphamidon (D) Quinolphos

580. Ear cockel disease of wheat is caused by:
(A) Fungi (B) Bacteria
(C) Virus (D) Nematodes

581. Saw fly damage toria crop at:
(A) Early stage of crop grown
(B) Filling of silique
(C) Maturity stage
(D) Harvesting stage

582. Spongy tissue in a physiological disorder in the fruit of:
(A) Mango (B) Litchi
(C) Guava (D) Pomegranate

583. Loose smut of wheat is controlled by:
(A) Seed treatment
(B) Soil sterilization
(C) Crop rotation
(D) Thiram spray on crop

584. Sulphur fungicides are very effective for:
(A) Leaf blight (B) Downey mildew
(C) Powdery mildew (D) Wilt disease

585. Most effective and economical method of plant disease control is:
(A) Protecting the crop by chemical sprays
(B) Use of resistant varieties
(C) Crop rotation
(D) Seed treatment with systemic fungicide

586. `Ratoon stunting' of sugarcane is caused by:
(A) A virus (B) A fungus
(C) Mycoplasma like organism
(D) A bacterium

587. A complete virus particle is called as:
(A) Viroid (B) Virion
(C) Virus (D) Naked virus

588. Insecticidal activity of various neem products is due to the process of toxic agents are:
(A) Nimbin (B) Karanjin
(C) Terpenes (D) Azadiractin

589. `Economics threshold' level for insect damage is always:
(A) Below economic injury level (EIL)
(B) Equal to economic injury level
(C) More than economic injury level
(D) None of these

590. Yellow mosaic in soybean is spread by:
(A) Seeds (B) Soils
(C) Air (D) White fly

591. Internally seed born disease of wheat is:
(A) Covered smut (B) Loose smut
(C) Kernal bunt (D) Black nest

592. Black arm disease of cotton is caused by:
(A) Bacteria (B) Virus
(C) Fungus (D) Nematodes

593. Uzi-fly is the paraiste by:
(A) Silkworm (B) Honey-bee
(C) Lac insect (D) Grass hopper

594. Jikka disease is related to:
(A) Groundnut (B) Gram
(C) Pea (D) Barley

595. The main objective of pest management to:
(A) Not to kill the pest
(B) To kill the pest
(C) To maintain the pest below economic injury level
(D) Not to minimize the pest level

596. Bacterial disease are controlled by:
(A) Antibiotics (B) Viricides
(C) Fungicides (D) Kelthane

597. Horizontal resistance is:
(A) Resistance against one race of pathogen
(B) Resistance against all the races of pathogen
(C) Resistance against lodging
(D) None of these

598. Root know is caused by which of the following nematodes:
(A) Aphlencoides (B) Melidogyne species
(C) Trichodorous (D) Both (B) and (C)

599. What is the chemical control of potato blights:
(A) Spraying of endosulfan
(B) Soil and Seed treatment
(C) Spraying of Gama BHC
(D) Spraying of Dithan M45

600. The causal organism of `red rot' a fungal disease of sugarcane is present in:
(A) Soil (B) Climate
(C) Seed (Planting material) (D) Virus

601. The most suitable chemical for seed treatment of wheat against `Kernel Bunt' is:
(A) BHC (B) Agrosan GN
(C) Vita vax (D) Captan

602. Who was the first plant protection advisor to Govt. of India:
(A) Dr. Sardar Singh (B) Dr. S. Pradhan
(C) Dr. M.L. Roonwal (D) Dr. H.S. Pruthi

603. White grub in the pest of:
(A) Rice (B) Wheat
(C) Groundnut (D) Mustard

604. Which of the following insecticide in safest for use:
(A) Methyl parathion (B) Carbaryl
(C) Phosphomidan (D) DDT

605. Dissemination of plant pathogens take place only by:
(A) Wind (B) Wind and water
(C) Wind, water, soil
(D) Wind, water, soil and insects

606. One of the factors causing malformation of mango in:
(A) Fungus (B) Nematodes
(C) Virus (D) Bacteria

607. The insecticides act was passed by parliament in the year:
(A) 1973 (B) 1965
(C) 1968 (D) 1972

608. `Yellow Vein Mosaic' is a disease of:
(A) Citurs (B) Okra
(C) Litchi (D) Loquat

609. Insects with sucking type of mouth parts need following group of insecticides for effective control:
(A) Stomach poisons (B) Systemic poisons
(C) Contact poisons (D) Fumigants

610. Insecticides under integrated control system are applied at which one of the following stage:
(A) General equilibrium level
(B) Economic threshold level
(C) Economic injury level
(D) Total loss level

611. The most convenient and safes chemical to control stored grain insects in rural areas is:
(A) EDB (B) Celphos
(C) DDT (D) Nephthaline ball

612. Cynogas pump is a :
(A) Sprayer (B) Duster
(C) Flame thrower (D) Fumigator

613. Which one is the insect trap crop is:
(A) Arhar (B) Paddy
(C) Bhindi (D) Potato

614. The main objective of pest management in:
(A) To reduce pest status
(B) To kill pests
(C) To prevent pest entry
(D) Not to apply insecticides

615. Thiram is used as a:
(A) In seed treatment
(B) At the time of sowing in furrow
(C) As top dressing
(D) As insecticides

616. Severe problem in gram cultivation is:
(A) High fertilizer requirement
(B) Weed
(C) Rust
(D) Pod borer

617. Virus free plants can be obtained by:
(A) Meristem culture (B) Pollen culture
(C) Overy culture (D) Tissue culture

618. Plant viral diseases can be controlled by:
(A) Antisense RNA technology
(B) By using ribosomes
(C) RNAi technology (D) All of these

619. Disease causing organisms are identified by using:
(A) MAB's (B) Gene transfer
(C) DNA probes (D) Micro projectiles

620. Which one of the following cultures is stables, easier to regenerate and produces virus-free clonal plants?
(A) Shoot tip (B) Cells from pith
(C) Root tip (D) Phloem

621. A set of disease resistance mechanisms that are not specific to a particular pathogene comes under:
(A) Active immunity (B) Innate immunity
(C) Passive immunity (D) None of these

622. Which of these is/are the characteristic/s of the specific immunity?
(A) Antigenic specificity
(B) Diversity and self/non self recognition
(C) immunologic memory
(D) All of these

623. Introduction of a tobacco mosaic virus transgene into plants in 1986 was used o demonstrate:
(A) Cross-inoculation (B) Co-inoculation
(C) Co-transformation (D) Cross protection

624. When disease spreads to the whole of plants, it is termed as:
(A) Localised (B) Systemic
(C) Endemic (D) Sporadic

625. Insecticides are specific inhibitors of:
(A) Nervous system (B) Digestive system
(C) Excretory system (D) Blood vascular system

626. Micrografting is used to produce plants free from:
(A) Virus (B) RLO's
(C) MLO's (D) Bacteria

627. Black heart is a physiological disorder of:
(A) Tomato (B) Chilli
(C) Potato (D) Cauliflower

628. Which of the disease(s) induce floral abnormalities:
(A) Downey mildew of mustard
(B) Green ear of bajra
(C) White rust of crucifers (D) Both (B) and (C)

629. Which one of the following is/are entomopathogenic fungi:
(A) *Beaveria bassina*
(B) *Metarrhizium anisopliae*
(C) *Trichoderma harzianum*
(D) Both (A) and (B)

630. Application of potash increases:
(A) Resistance to water logging
(B) Frost resistance in plants
(C) Disease resistance in plants
(D) None of these

631. J.C. Luthra and his associates developed the solar heat treatment of wheat seeds for the control of:
(A) Black rust (B) Loose smut
(C) Covered smut (D) Brown rust

632. The root-knot nematode generally have interaction with:
(A) *Pseudomonas solanacearum*
(B) *Xanthomonas citri*
(C) *Bacillus subtilis*
(D) *Streptomyces scabies*

633. Viruses contain:
(A) RNA (B) DNA
(C) Both RNA and DNA
(D) Either RNA or DNA, never both

634. The mature particle of a plant virus is generally known as:
(A) Capsid (B) Viriod
(C) Virion (D) None of these

635. In plants buckling, puckering and blistering symptoms are produced by:
(A) Bacteria (B) Fungi
(C) Viruses (D) Nematodes

636. The adjacent cells of the same fungal hypha brought into communication with each other by means of loops is known as:
(A) Transduction (B) Transformation
(C) Clamp connection (D) Conjugation

637. Suidal germination takes place in:
(A) Dodder (B) Striga
(C) Loranthus
(D) Dendrophthoe falcata

638. Yellow mosaic of legumes was first reported in India from:
(A) Faridabad (B) Solan
(C) Delhi (D) Kanpur

639. For quick and accurate detection of viruses can be done by:
(A) ELISA (B) HADAS
(C) IEM (D) All of these

640. A virus that infects bacteria and destroys them is called as:
(A) Bactericide (B) Bacteriophase
(C) Appressocium (D) Agglutination

641. The science which deals with cause of the plant diseases in nature of causal agent:
(A) Fertilization (B) Etiology
(C) Aetiology (D) Both (B) and (C)

642. Loose smut of wheat and barley are:
(A) Internally seed borne
(B) Externally seed borne
(C) Both (A) and (B) (D) None of these

643. MLO and spiroplasma are mostly:
(A) Xylem inhibiting (B) Phloem inhibiting
(C) Both (A) and (B) (D) Stomata inhibiting

644. Which one of the following can not be detected by ELISA technique:
(A) Virus (B) Bacteria
(C) Viriod (D) Nematodes

645. The kernel bunt of wheat first time reported in Karnal by Mitra in:
(A) 1931 (B) 1940
(C) 1950 (D) 1951

646. The Karnal bunt of wheat is caused by:
(A) *Neossia indica* (B) *Tilletia horrid*
(C) *Melamsora lini* (D) *Ustilago maydis*

647. Rice blast pathogen perfect stage is:
(A) *Pyricularia oryzae* (B) *Magnaporthe grisea*
(C) Helminthosporium (D) Rhizoctonia

648. The nematodes are:
(A) Monoblastic (B) Duoblastic
(C) Triploblastic (D) All of these

649. Cirus greening caused by the:
(A) Fastidious bacteria (B) Fungi
(C) Virus (D) Algae

650. Shape of MLO is:
(A) Rigid (B) Cuboid
(C) Circular (D) Polymorphic

651. Cell wall lack microorganism:
(A) Bacteria (B) Virus
(C) Fungi (D) MLO

652. A pesticide of lipophylic in nature:
(A) 2,4-D (B) DDT
(C) BHC (D) Carbamates

653. Repeated application of pesticides produce:
(A) Immune strains (B) Bioconcentration
(C) Biomagnification (D) All of these

654. How do the common weedicides act?
(A) Interfere with respiration
(B) Interfere with photosynthesis
(C) Defence mechanism is destroyed
(D) Causes wilting

655. Use of light traps for the management of pest is an example of:
(A) Biological control
(B) Mechanical method of control
(C) Legal method of control
(D) Cultural control

656. Which energy is utilized for making fog, smoke and vapour sprays:
(A) Hydraulic (B) Gaseous
(C) Centrifugal/electrical (D) Thermal

657. In India Genetic Engineering Approval Committee (GEAC) is responsible for monitoring of GM crops come under ministry of :
(A) Agriculture
(B) Environment and Forests
(C) Information and Broadcasting
(D) Science and Technology

658. The higher LD 50 value indicates the following:
(A) Neutral (B) Less toxic
(C) Highly toxic (D) None of these

659. Work on GM mustard is carried out by:
(A) Monsanto (B) Syngenta
(C) Mahyco (D) Proagro

660. I.P.M. relies heavily on:
(A) Economic threshold level
(B) Total loss of crops
(C) Resistant pest
(D) Susceptible pest

661. Bt cabbage is developed against:
(A) Diamond Back Moth (B) Cabbage butter fly
(C) Mealy bug (D) Green bug

662. Virus free plants can be obtained through:
(A) Antibiotic treatment (B) Bordeaux mixture
(C) Root tip culture (D) Shoot tip culture

663. Which of the following are NOT possible uses for fructans as presented in class?
(A) Natural insecticides (B) Fat substitute
(C) Amino acids (D) Sweetners

664. Which of the following is true of glyphosate?
(A) It has an animal DL 50 of 0.4 g/kg
(B) It is more selective than 2, 4-D
(C) It can be manufactured without a license from the patent holder
(D) None of these

665. Which of the following attempts to increase the utility of fungal herbicides succeeded?
(A) Expression of a phytotoxic protein at abnormally high levels in Fusarium
(B) Expression of a phytotoxic protein from Fusarium in another fungal species
(C) Expression of a phytotoxic protein from Fusarium in a crop plant
(D) Expression of a phytotoxic protein from Fusarium in a weed

666. What is a disadvantage of the current use of transgenic antibody fragments to induce resistance to the herbicide Picloram?
(A) Each antibody mmolecule can neutralize only one herbicide molecule
(B) Picloram is an auxinic herbicide, thus the antibodies will neutralize endogenous auxin
(C) The antibodies induce morphological defects (e.g. epinasty)
(D) The antibodies are quickly degraded by host enzymes, so they have little effect on picloram

ANSWERSHEET

17. PLANT PROTECTION BIOTECHNOLOGY

1. (B)	2. (C)	3. (D)	4. (A)	5. (A)	6. (A)	7. (A)	8. (D)	9. (A)	10. (A)
11. (C)	12. (A)	13. (C)	14. (D)	15. (C)	16. (B)	17. (C)	18. (B)	19. (C)	20. (B)
21. (D)	22. (C)	23. (C)	24. (B)	25. (D)	26. (C)	27. (D)	28. (B)	29. (D)	30. (D)
31. (D)	32. (C)	33. (D)	34. (A)	35. (B)	36. (C)	37. (D)	38. (C)	39. (A)	40. (B)
41. (C)	42. (B)	43. (B)	44. (A)	45. (C)	46. (C)	47. (B)	48. (B)	49. (A)	50. (B)
51. (D)	52. (D)	53. (A)	54. (B)	55. (D)	56. (B)	57. (C)	58. (A)	59. (A)	60. (C)
61. (D)	62. (A)	63. (B)	64. (A)	65. (B)	66. (C)	67. (C)	68. (B)	69. (D)	70. (A)
71. (A)	72. (A)	73. (C)	74. (A)	75. (A)	76. (C)	77. (B)	78. (D)	79. (A)	80. (A)
81. (D)	82. (C)	83. (C)	84. (D)	85. (C)	86. (A)	87. (C)	88. (C)	89. (C)	90. (B)
91. (D)	92. (B)	93. (C)	94. (B)	95. (A)	96. (A)	97. (B)	98. (D)	99. (D)	100. (B)
101. (B)	102. (B)	103. (B)	104. (A)	105. (C)	106. (A)	107. (C)	108. (D)	109. (B)	110. (D)
111. (A)	112. (A)	113. (A)	114. (B)	115. (C)	116. (C)	117. (B)	118. (D)	119. (A)	120. (D)
121. (A)	122. (B)	123. (A)	124. (A)	125. (B)	126. (C)	127. (A)	128. (A)	129. (A)	130. (A)
131. (B)	132. (A)	133. (A)	134. (B)	135. (A)	136. (B)	137. (D)	138. (C)	139. (C)	140. (B)
141. (B)	142. (B)	143. (A)	144. (D)	145. (D)	146. (A)	147. (A)	148. (A)	149. (B)	150. (B)
151. (A)	152. (A)	153. (B)	154. (D)	155. (A)	156. (D)	157. (D)	158. (A)	159. (D)	160. (A)
161. (B)	162. (C)	163. (D)	164. (D)	165. (A)	166. (A)	167. (A)	168. (D)	169. (B)	170. (A)
171. (A)	172. (B)	173. (D)	174. (D)	175. (C)	176. (A)	177. (A)	178. (B)	179. (A)	180. (A)
181. (B)	182. (A)	183. (A)	184. (D)	185. (A)	186. (C)	187. (C)	188. (C)	189. (A)	190. (A)
191. (A)	192. (C)	193. (B)	194. (B)	195. (A)	196. (D)	197. (B)	198. (A)	199. (C)	200. (A)

201. (B)	202. (B)	203. (A)	204. (C)	205. (A)	206. (D)	207. (D)	208. (B)	209. (B)	210. (B)
211. (A)	212. (D)	213. (B)	214. (A)	215. (C)	216. (B)	217. (D)	218. (A)	219. (C)	220. (C)
221. (A)	222. (B)	223. (D)	224. (C)	225. (B)	226. (C)	227. (D)	228. (C)	229. (A)	230. (B)
231. (A)	232. (A)	233. (B)	234. (A)	235. (C)	236. (B)	237. (D)	238. (C)	239. (D)	240. (D)
241. (A)	242. (B)	243. (D)	244. (A)	245. (D)	246. (D)	247. (A)	248. (A)	249. (A)	250. (A)
251. (B)	252. (D)	253. (A)	254. (A)	255. (C)	256. (D)	257. (C)	258. (A)	259. (C)	260. (A)
261. (A)	262. (B)	263. (C)	264. (D)	265. (A)	266. (B)	267. (D)	268. (D)	269. (D)	270. (A)
271. (A)	272. (D)	273. (B)	274. (A)	275. (D)	276. (B)	277. (D)	278. (D)	279. (B)	280. (C)
281. (D)	282. (A)	283. (B)	284. (D)	285. (A)	286. (D)	287. (D)	288. (D)	289. (A)	290. (A)
291. (B)	292. (A)	293. (B)	294. (C)	295. (A)	296. (B)	297. (B)	298. (A)	299. (B)	300. (B)
301. (B)	302. (A)	303. (B)	304. (A)	305. (A)	306. (B)	307. (A)	308. (B)	309. (A)	310. (A)
311. (A)	312. (A)	313. (B)	314. (C)	315. (A)	316. (B)	317. (C)	318. (A)	319. (C)	320. (B)
321. (A)	322. (B)	323. (A)	324. (A)	325. (A)	326. (B)	327. (A)	328. (C)	329. (A)	330. (D)
331. (A)	332. (D)	333. (C)	334. (C)	335. (B)	336. (A)	337. (D)	338. (A)	339. (B)	340. (B)
341. (C)	342. (C)	343. (A)	344. (B)	345. (A)	346. (B)	347. (A)	348. (D)	349. (D)	350. (D)
351. (A)	352. (A)	353. (A)	354. (D)	355. (B)	356. (A)	357. (A)	358. (A)	359. (B)	360. (A)
361. (D)	362. (A)	363. (A)	364. (A)	365. (A)	366. (B)	367. (D)	368. (D)	369. (B)	370. (B)
371. (C)	372. (D)	373. (B)	374. (A)	375. (A)	376. (C)	377. (C)	378. (A)	379. (C)	380. (C)
381. (B)	382. (A)	383. (B)	384. (B)	385. (C)	386. (A)	387. (A)	388. (A)	389. (A)	390. (A)
391. (A)	392. (C)	393. (D)	394. (A)	395. (C)	396. (C)	397. (A)	398. (A)	399. (D)	400. (B)
401. (B)	402. (A)	403. (D)	404. (D)	405. (B)	406. (B)	407. (C)	408. (D)	409. (A)	410. (A)
411. (A)	412. (B)	413. (A)	414. (B)	415. (A)	416. (A)	417. (C)	418. (D)	419. (A)	420. (C)
421. (A)	422. (D)	423. (B)	424. (D)	425. (A)	426. (D)	427. (D)	428. (D)	429. (A)	430. (A)
431. (B)	432. (A)	433. (A)	434. (A)	435. (B)	436. (C)	437. (B)	438. (D)	439. (A)	440. (A)
441. (B)	442. (D)	443. (D)	444. (A)	445. (A)	446. (A)	447. (A)	448. (A)	449. (A)	450. (A)
451. (A)	452. (A)	453. (B)	454. (B)	455. (A)	456. (A)	457. (A)	458. (B)	459. (B)	460. (B)
461. (D)	462. (A)	463. (B)	464. (C)	465. (A)	466. (D)	467. (A)	468. (A)	469. (B)	470. (A)
471. (B)	472. (C)	473. (C)	474. (D)	475. (A)	476. (A)	477. (A)	478. (A)	479. (D)	480. (A)
481. (C)	482. (D)	483. (D)	484. (B)	485. (D)	486. (A)	487. (A)	488. (C)	489. (A)	490. (D)
491. (D)	492. (D)	493. (D)	494. (C)	495. (D)	496. (A)	497. (A)	498. (D)	499. (B)	500. (A)
501. (A)	502. (A)	503. (A)	504. (A)	505. (C)	506. (B)	507. (B)	508. (C)	509. (D)	510. (D)
511. (A)	512. (A)	513. (D)	514. (D)	515. (A)	516. (A)	517. (B)	518. (A)	519. (A)	520. (D)
521. (A)	522. (D)	523. (B)	524. (A)	525. (B)	526. (A)	527. (C)	528. (D)	529. (B)	530. (A)
531. (D)	532. (D)	533. (A)	534. (B)	535. (D)	536. (C)	537. (D)	538. (B)	539. (A)	540. (B)
541. (C)	542. (C)	543. (D)	544. (A)	545. (C)	546. (A)	547. (B)	548. (A)	549. (A)	550. (D)
551. (C)	552. (C)	553. (B)	554. (A)	555. (A)	556. (A)	557. (A)	558. (C)	559. (B)	560. (D)
561. (B)	562. (A)	563. (D)	564. (C)	565. (A)	566. (B)	567. (C)	568. (D)	569. (A)	570. (C)
571. (A)	572. (B)	573. (D)	574 (D)	575 (B)	576 (C)	577 (A)	578 (A)	579 (C)	580 (D)
581 (A)	582 (A)	583 (A)	584 (C)	585 (C)	586 (A)	587 (B)	588 (D)	589 (A)	590 (D)
591 (B)	592 (A)	593 (A)	594 (A)	595 (C)	596 (A)	597 (B)	598 (B)	599 (D)	600 (C)
601 (C)	602 (D)	603 (C)	604 (B)	605 (D)	606 (A)	607 (C)	608 (B)	609 (C)	610 (B)
611 (A)	612 (D)	613 (C)	614 (D)	615 (A)	616 (D).	617. (A)	618. (D)	619. (C)	620. (A)

621. (B)	622. (D)	623. (D)	624. (B)	625. (A)	626. (A)	627. (C)	628. (D)	629. (D)	630. (C)
631. (B)	632. (A)	633. (D)	634. (C)	635. (C)	636. (C)	637. (B)	638. (C)	639. (D)	640. (B)
641. (D)	642. (A)	643. (B)	644. (C)	645. (A)	646. (A)	647. (B)	648. (C)	649. (A)	650. (D)
651. (D)	652. (B)	653. (A)	654. (B)	655. (A)	656. (C)	657. (B)	658. (B)	659. (D)	660. (A)
661. (A)	662. (D)	663. (A)	664. (C)	665. (B)	666. (A)				

Chapter - 18

INDUSTRIAL BIOTECHNOLOGY

1. Bacterium commonly used in plant genetic engineering is:
   (A) Agrobacterium (B) Corynebacterium
   (C) Bacillus subtilis (D) Salmonella typhisae

2. Antibiotics are drugs commonly used to cure disease of:
   (A) Fungi (B) Viruses
   (C) Protozoan (D) Bacteria

3. Genetically engineered bacteria are become employed for production of:
   (A) Human insulin (B) Thyroxine
   (C) Cortisol (D) Epinephrine

4. In 1928, a scientist discovered the first effective antibiotics. Scientist and antibiotics are:
   (A) Flemmin – Streptomycin
   (B) Flemming - Penicillin
   (C) Waksman Penicillin
   (D) Waksman - Streptomycin

5. The nature of biomass obtained from energy crops is:
   (A) Wood (B) Sugar and starch
   (C) Hydrocarbon (D) All of these

6. Biotechnological production of antibiotics require:
   (A) Search for antibiotic producing microorganisms
   (B) Isolation of gene for antibiotics production
   (C) Join the gene for antibiotic with plasmid of *Escherichea coli*
   (D) All of these

7. Which is not correct about the drug zidovudin (AZT)?
   (A) The first successful drug used for the treatment of AIDS
   (B) It is a nucleoside analogue and targeted at reverse transcriptase
   (C) Incorporation of this base analogue into a cDNA chain enhances its extension
   (D) Long term use produces several side effects and mutant viruses develop in treated patients

8. In case of AIDS, highly active antiretroviral therapy (HAART) has:
   (A) It reduces viral load in blood below detection limits
   (B) It lowers death rates due to AIDS
   (C) This therapy is very expensive (upto US$ 15000 per day)
   (D) All of these

9. Biohydrogen is produced by which routes?
   (A) Bt photolysis of water
   (B) During anaerobic fermentation
   (C) An *in vitro* system created by coupling together the photosynthetic unit and hydrogenase
   (D) All of these

10. The current processes of production of ethanol from cellulose are based on which of the following microorganisms?
    (A) *Trichoderma reesei*
    (B) *Saccharomyces cerevisiae*
    (C) Recombinant *E. coli*
    (D) All of these

11. The important reason for limiting the use of bioethanol as fuel?
    (A) Ethanol run engines use about 10% more fuel than petrol
    (B) Due to highly hydrophilic nature, ethanol can absorb moisturc from atmosphere, which interferes with combustion and also causes corrosion
    (C) The downstream processing for ethanol recovery is costly
    (D) Ethanol produces engine starting problems when air is cool

12. Blood flows from the heart to the lugs in the pulmonary artery and returns from the lungs to the heart in the pulmonary vein. The blood in the pulmonary artery is:
    (A) Higher in O_2 and lower in CO_2 content than the blood in the pulmonary vein
    (B) Higher in both O_2 and CO_2 content than the blood in the pulmonary vein
    (C) Lower in O_2 and higher in CO_2 content than the blood in the pulmonary vein
    (D) Lower in both O_2 and CO_2 content than the blood in the pulmonary vein

13. Which is correct about conventional vaccines?
    (A) Conventional vaccines consist of whole pathogenic organisms
    (B) Pathogens may either be killed or live, but attenuated
    (C) Conventional vaccines are produced in whole animals or cultured cells
    (D) All of these

14. Bioethanole can be obtained from:
(A) Cereals (B) Tuber crops
(C) Root crops (D) All of these

15. Oxygen is carried in blood:
(A) Bound to hemoglobin
(B) Dissolved in the plasma
(C) Dissolved in the cytosol of erythrocytes
(D) All of these

16. Cardia output is the
(A) Volume of blood pumped per minute by both ventricles
(B) Volume of blood flowing through the systemic circulation each minute
(C) Product of the number of heartbeats per minute and the volume pumped per beat
(D) Both (B) and (C)

17. Which statements are correct about hydrolytic and fermentative bacteria?
I. They include both obligate and facultative anaerobes
II. They may occur upto $10^8 - 10^9$ cells/ml of sewage sludge digesters
III. They remove O_2 and create anaerobic conditions
IV. Usually, only 50% of the polysaccharides present in the waster may be digested

Options:
(A) I, II (B) I, II, III, IV
(C) I, II, IV (D) I, II, III

18. Which is organic acids?
(A) Long chain fatty acids
(B) Formate and butyrate
(C) Acetate and propionate
(D) All of these

19. Which produces hydrogen gas which can be used as a pollution free fuel?
(A) Chlamydomonas (B) Clostridium
(C) Both (A) and (B) (D) Green plants

20. The cellulose enzyme systems (produced by both bacteria and fungi) carry by:
(A) Binding to cellulose
(B) Hydrolysis at the end of a cellular chain
(C) Cleavage of cellulose chain at an internal site
(D) All of these

21. Consider the following with reference to the mechanism of action of a steroidal hormone on target cell:
P. Bonding of activated complex with specific genomic site
Q. Activation of transcriptional apparatus
R. Attachment with cytosolic receptor protein
S. Transportation of complex to the nucleus
T. Formation of new proteins in the cytoplasm
What is the correct sequence of the above in said mechanism?
(A) P, Q, R, S, T (B) S, P, R, Q,T
(C) Q, R, P, S, T (D) R, S, P, Q, T

22. During chemical transmission, the following events occur at synaptic junction:
P. Transmitters cross synaptic left and attach to receptors on the postsynaptic neurons
Q. Postsynaptic membrane is depolarized, initiating the impulse
R. Chemical excitation at the presynaptic terminal and release of neurotransmitters
S. Immediately after impulse is triggered, the recovery of pre and postsynaptic membranes follows and neurotransmitter is destroyed
Which of the statements given above is/are correct?
(A) P, R, S, Q (B) R, P, S, Q
(C) R, P, Q, S (D) P, R, Q, S

23. Therapeutic agents from non-recombinant organisms as originate from:
(A) Plant cell cultures (B) Animal cell cultures
(C) Microorganisms (D) All of these

24. Column A lists some tumour-associuated antigens, while column B lists different types of cancers in which antigen are found. Match both the Columns carefully and select the correct option from those given below them:

Column A	**Column B**
A. CEA	1. Some head, lung, etc. tumours
B. MUCI	2. Breast and ovarian tumours
C. HER 2	3. Breast cancer
D. EGFR	4. Colon cancer

Options:
(A) A–3, B–4, C–1, D–2 (B) A–4, B–3, C–2, D–1
(C) A–3, B–1, C–4, D–2 (D) A–1, B–3, C–2, D–4

25. Which diagnosed with anti-dsDNA:
(A) Polymyositis
(B) Scleroderma
(C) Systemic lupus erythematus
(D) Myasthenia gravis

26. The recombinant *E. coli* used for pentose fermentation has transgenes encoding by which enzymes:
(A) Pyruvate decarboxylase
(B) Alcohol dehydrogenase
(C) Acetaldehyde dehydrogenase
(D) Both (A) and (C)

27. Methanogenic bacteria are:
(A) *Methanobacterium omelianskii*
(B) *Methanosarcina barkeri*
(C) Both (A) and (B)
(D) *Methylococcus capsulatus*

28. Monoclonal antibodies can be used for:
(A) Disease diagnosis (B) Disease treatment
(C) Immunopurification (D) All of these

29. Which enzyme is obtained commercially from Aspergillus?
(A) Protease and pectinase
(B) α-Amylase and glucoamylase
(C) Both (A) and (B) (D) Papain

30. Hydrolytic and fermentative bacteria hydrolyse and ferent the organic materials and produce by:
(A) Organic acids (B) CO_2
(C) H_2 (D) All of these

31. Ethanol has which of the following advantages over petrol?
I. Ethanol has a lower latent heat of vaporization than petrol
II. Ethanol has a higher octane number (99) than petrol (80-100) as a result; preignition does not occur when ethanol is used in engines set for petrol
III. Higher octane rating of ethanol allows the compression ratio of the engines to be increased; this results in increased production of power
IV. Ethanol is burnt more completely so that hydrocarbon emission is drastically lower as compared to that in case of petrol

Options:
(A) I, II (B) II, III, IV
(C) I, II, III, IV (D) I, II, III

32. Which is the examples of hydrophobic amino acids?
(A) Leucine and isoleucine
(B) Methionine and vaine
(C) Phenylalanine and often tyrosine
(D) All of these

33. Rennet enzyme is produced by:
(A) Mucor miehei (B) Calf stomach
(C) Calf liver (D) Both (A) and (B)

34. Anti-RNA polymerase I is used to diagnose which autoimmune diseases?
(A) Polymyositis
(B) Scleroderma
(C) Systemic lupus erythematus
(D) Myasthenia gravis

35. Enzyme 'diastase' was identified by:
(A) Dubrunfaut (B) Reamur
(C) Payen and Persoz (D) Hansen

36. In 1986, which of the scientists discovered the catalytic action of some RNA molecules?
(A) Baltimore (B) Smith
(C) Cech (D) Berg

37. The first antibiotic to be isolated was:
(A) Terramycin (B) Streptomycin
(C) Neomycin (D) Penicillin

38. Penicillin acylases can be used as:
(A) Penicillin G (B) Penicillin V
(C) Penicillins (D) Both (A) and (B)

39. Penicillin is obtained from:
(A) *Streptomyces griseus*
(B) *Penicillin chrysogenum*
(C) *Penicillum griseofulvum*
(D) *Aspergillus fumigatus*

40. Humilin is:
(A) Antibiotic (B) Human insulin
(C) Digestive enzyme (D) A form of chitin

41. Which yields highly purified enzymes in a single step?
(A) Gel filteration
(B) Affinity chromatography
(C) Ion exchange chromatography
(D) Both (B) and (C)

42. Which is a multienzyme complex?
(A) Urease
(B) Tryptophan synthase from *E.coli*
(C) Aldoase
(D) Both (B) and (C)

43. Which enzymes are used in the largest quantity?
(A) Textile industry (B) Alcohol production
(C) Detergents (D) Leather industry

44. Which is not an example of an energy crop that yields hydrocarbons?
(A) *Asclepia speciosa*
(B) *Copaifera multijuga*
(C) *Euphorbia lathyrus*
(D) *Eucalyptus globules*

45. Which metal ion is associated with urase enzyme?
(A) Fe (B) Ni
(C) Mg (D) Cu

46. Thiamine pyrophosphate (TPP) coenzyme is found in:
(A) Glutamin synthase
(B) Glyceraldehyde 3-phosphate dehydrogenase
(C) Pyruvate dehydrogenase complex
(D) DNA ligase

47. Dextran is employed in:
(A) Bleeding (B) Plasma extending
(C) Blood clotting (D) Blood preservation

48. Lyase class of enzyme does not contain :
(A) Decarboxylases (B) Dehydratases
(C) Aldolases (D) Proteases

49. Erythromycin and chloramphenicol are obtained from:
(A) *Penicillium species* (B) *Aspergillus species*
(C) *Streptomyces species* (D) *Bacillus species*

50. Which enzymes catalysis an irreversible reaction of glycolysis?
(A) Phosphofructokinase
(B) Glucose-6-isomerase
(C) Aldolase
(D) Phosphoglycerate mutase

51. All chemical reaction possess major properties:
(A) Kinetic (B) Thermodynamic
(C) Both (A) and (B) (D) Dynamic

52. Which diagnosed by anti-acetylcholine receptor?
(A) Polymyositis
(B) Scleroderma
(C) Systemic lupus erythematus
(D) Myasthenia gravis

53. Smith and Nathans are famous for discovery of:
(A) Gene therapy
(B) Restriction enzyme
(C) Humulin
(D) Second generation vaccine

54. Isomerases include:
(A) Intramolecular transferases
(B) Epimerases
(C) Racemases
(D) All of these

55. A broad spectrum antibiotic is effective:
(A) In low concentration
(B) On a variety of microorganisms
(C) Against all bacteria and viruses
(D) Against both pathogens and hosts

56. α-amylase is produced from:
(A) Malted barley (B) Bacillus
(C) Aspergillus (D) All of these

57. Antibiotics belong to the category of:
(A) Steroids (B) Toxins
(C) Medicines (D) Tonics

58. Flemming is famous for his discovery of:
(A) Actinomycetin (B) Penicillin
(C) Actinomycine (D) Neomycin

59. Streptomycin is prepared from:
(A) *Streptomyces antibioticus*
(B) *Streptomyces nodasus*
(C) *Streptomyces griscus*
(D) *Streptomyces rimosus*

60. Antibiotics are mostly obtained from:
(A) Fungi
(B) Actinomycetes/Bacteria
(C) Both (A) and (B)
(D) Cyanobacteria

61. *Streptomycin griscus* produces the antibiotic:
(A) Terramycin (B) Chloramphenicol
(C) Neomycin (D) Streptomycin

62. Antibiotics destroy or inhibit:
(A) Bacteria, fungi and viruses
(B) Bacteria, algae and viruses
(C) Bacteria and fungi
(D) Bacteria and viruses

63. Proteases are used to digest away proteins present in:
(A) Blood stains (B) Dirt stains
(C) Grass stains (D) All of these

64. Who established the scientific basis of vaccination?
(A) Jenner (B) Pasteur
(C) Waksman (D) Kohler and Milstein

65. Which is not correct about aspartame?
(A) It is a dipeptide
(B) It is synthesized from an equal mixture of L-aspartic acid and methyl ester of L-phenylalanine
(C) It is 180 times sweeter than sucrose
(D) It is used as high caloric sweetner

66. Granulocyte colony stimulating factor can be used for the treatment of?
(A) AIDS
(B) Cancer
(C) Bone marrow transplantation
(D) All of these

67. An 'indel' defines a site in the genome that has ?
(A) Insertion of a sequence
(B) Deletion of a sequence
(C) Either A or B
(D) Inversion of a sequence

68. Most enzymes are used extra cellularily for which medical purposes?
(A) Removal of toxic substances
(B) Treatment of disorders within blood circulatory system
(C) Topical application
(D) All of these

69. Current hepatitis B vaccine is:
(A) An interferon
(B) First generation vaccine
(C) Second generation vaccine
(D) Third generation vaccine

70. Kohler and Milstein developed biotechnology for:
(A) Monoclonal antibiotics
(B) Myclomas
(C) Immonilisation of enzymes
(D) Steroid conversion

71. Synzymes must possess:
(A) Substrate binding site
(B) Catalytically effective site
(C) Allosteric site
(D) Both (A) and (B)

72. The term antibiotic was coined by:
(A) Flemming (B) Waksman
(C) Funk (D) Jenner

73. Which steps are involved in downstream processing?
(A) Separation of particles & disintegration of cells
(B) Extraction and concentration
(C) Purification and drying
(D) All of these

74. Which indicates the correct order of ascorbic acid production?

(A) Glucose → Ascorbic acid → D-sorbitol → L-sorbose

(B) Glucose → L-sorbose → D-sorbitol → Ascorbic acid

(C) Glucose → D-sorbitol → L-sorbose → Ascorbic acid

(D) D-sorbitol → L-sorbose → Glucose → Ascorbic acid

75. Which enzyme is isolated from hypertherophiles?

(A) DNA polymerase (B) α - amylase

(C) α - Glucosidase (D) All of these

76. Waksman got nobel prize for his discovery of:

(A) Penicillin (B) Neomycin

(C) Chloromycetin (D) Streptomycin

77. Continuous flow reactors offer which advantages?

(A) Greater productivity per unit amount of enzyme

(B) Uniform product quality

(C) They can be used for substrates having low solubility

(D) All of these

78. Cells obtained from cancerous tissues are:

(A) Myelomas (B) Hybridomas

(C) Monoclonal (D) Lymphocytes

79. Hybridoma are used for:

(A) Killing cancer cells

(B) Formation of monoclonal antibiotics

(C) Antibiotic synthesis

(D) Formation of somatic hybrids

80. Biosensor for cocaine detection uses which biological agents?

(A) Whole cell (B) Antibody

(C) Enzyme (D) Either (A) or (C)

81. The first biotransformation to be commercialized was:

(A) Deoxycorticosterone to contocosterone

(B) Ethanol to acetic acid

(C) Glucose to gluconic acid

(D) Fructose to mannitol

82. Biological component of biosensors perform which function?

(A) Recognise the analyte

(B) Interact with the analyte in such a manner, which produces some physical change detectable by the transducer

(C) Both (A) and (B)

(D) Inactivate toxic analytes

83. Ovulation may be stimulated by the administration of :

(A) Clomiphene

(B) Human chorionic gonadrotrophin (hCG)

(C) Human menopausal gonadotrophin (hMG)

(D) Any of these

84. Most of the eubacterial antibiotics are obtained from:

(A) Streptomyces (B) Pseudomonas

(C) Streptococcus (D) Bacillus

85. Which products generated by genetically engineered microbes is approved for marketing in India?

(A) Granulocyte macrophage colony stimulating factor and interferons

(B) Insulin and hepatitis-B surface antigen

(C) Granulocyte colony stimulating factor & insulin

(D) All of these

86. During 'invert syrup' production, small amounts of xylene are added to achieve to:

(A) To prevent microbial contamination

(B) To facilitate product recovery

(C) Substrate activation

(D) To prevent product inhibition

87. The enzymology, abbrebiation EC stands for:

(A) Exotic Collection

(B) Enzyme Commission

(C) Enzyme Collection (D) Enzyme Cofactor

88. Which enzymes is obtained from bacteria?

(A) α- and ß-amylases (B) Glucose isomerase

(C) Penicillin amidase (D) All of these

89. In case of glucose isomerase used for production of high fructose syrup, which is used as cofactor?

(A) Ca^{++} (B) Co^{++}

(C) Mg^{++} (D) Both (B) and (C)

90. Which statements is most correct about ribozymes?

(A) Enzyme activity associated with ribosomes

(B) Enzymes which catalyse both synthesis and degradation of small oligonucleotides

(C) Enzyme that use RNA as substrate

(D) RNA molecules having enzymatic properties

91. Streptomycin is useful for treatment of pathogens:

(A) Gram (+) (B) Gram (-)

(C) Both (A) and (B) (D) Gram neutral

92. Nuclease enzyme which begins its attack from free end of polynucleotide is:

(A) Endonuclease (B) Exonuclease

(C) Polymerase (D) Kinase

93. *Bacillus thuringiensis* can be used for the control of:

(A) Heliothis and other Lepidoptera

(B) Coleoptera pests

(C) *Spodoptera litura*

(D) Both (A) and (B)

94. Biotransformations are performed by microbes is far more efficient and economical due to:

(A) High metabolic rates of microorganisms

(B) Rapid microbial growth

(C) Both (A) and (B)

(D) More efficient enzymes

95. **Which options lists the correct order of the mechanism of enzyme reaction?**
(A) $E + S \rightarrow EP \rightarrow ES \rightarrow E + P$
(B) $E + S \rightarrow ES \rightarrow EP \rightarrow E + P$
(C) $E + S \rightarrow EP \rightarrow ES \rightarrow E + P$
(D) $E + P \rightarrow ES \rightarrow ES \rightarrow E + P$

96. **The bulk of enzymes, both in terms of quantity and variety, are produced from which of the following sources? The correct option lists these sources in describing order.**
(A) Animals → Plants → Microorganisms
(B) Microorganisms → Animals → Plants
(C) Microorganisms → Plants → Animals
(D) Plants → Microorganisms → Animals

97. **Which of the following features is correct about asymmetric gap ligase chain reaction (AG-LCR)?**
(A) It utilize reverse transcriptase
(B) It is used for detection of RNA
(C) It is used for detection of antibodies
(D) Both (A) and (B)

98. **Erythropoietin (EPO) is used to stimulate red blood production in:**
(A) Kidney dialysis (B) AIDS patients
(C) Cancer patients (D) Both (A) and (C)

99. **Antibodies can be used as biological agents in which types of biosensors?**
(A) Optical (B) Acoustic wave
(C) Potentiometric (D) Both (A) and (B)

100. **Aspartate is an artificial sweetner, it is chemically:**
(A) Fatty acid (B) Carbohydrate
(C) Dipeptide (D) Oligopeptide

101. **Modified antibiotics are got through:**
(A) Vernalisation (B) Ultrafiltration
(C) Ultracentrifuge (D) Genetic engineering

102. **Interferon is:**
(A) Antibacterial (B) Antiviral
(C) Antianimal cell (D) Bacterium

103. **Waksman and Woodruff (1941) is responsible for the isolation of:**
(A) Antiseptic (B) Antibiotic
(C) Actinomycin (D) Antibodies

104. **A packed bed reactor can take care of:**
(A) Product inhibition (B) Substrate activation
(C) Reaction reversibility (D) All of these

105. **Thermodynamic predictions are based on:**
(A) Stabilities of reactants
(B) Stabilities of products
(C) Both (A) and (B)
(D) Amounts of enzymes

106. **Hydrolase class of enzymes does not include:**
(A) Lipases (B) Esterases
(C) Oxygenases (D) Glycosidases

107. **Consider the following parts:**
P. Reissner's membrane
Q. *Tapetum lucidum*
R. *Fenestra rotunda*
S. *Forea entrails*
Which of these are associated with auditory organs?
(A) P and Q (B) R and S
(C) P and R (D) Q and S

108. **Which enzymes is not obtained commercially for malted barley?**
(A) β - Glucanase (B) α - Amylase
(C) Glucoamylase (D) β – Amylase

109. **Which techniques is used for obtaining information on three dimentional structure of enzymes?**
(A) SDS-PAGE
(B) Nuclear magnetic resonance (NMR)
(C) X-ray differaction
(D) Both (B) and (C)

110. **Enzymes are similar to catalysts in respects to:**
(A) They lower the activation energy of reaction
(B) They do not participate in the reaction
(C) They only increase the reaction rate
(D) All of these

111. **Consider the following statements:**
P. Grey matter surrounds the white matter in the brain
Q. Grey matter surrounds the white matter in the spinal cord
R. White matter surrounds the grey matter in the spinal cord
S. White matter surrounds the grey matter in the brain
Of these statements:
(A) Q (B) S
(C) P and S (D) P and R

112. **Which enzymes is not obtained commercially for Bacillus?**
(A) Protease and glucose isomerase
(B) Pectinase
(C) α and β amylase (D) *Penicillin amidase*

113. **Which is false about antibiotics?**
(A) Coined by Waksman (1942)
(B) Produced by microorganisms
(C) Produce allergy in some persons
(D) Cure all diseases

114. **First hormone produced artificially by culturing bacteria is:**
(A) Insulin (B) Thyroxine
(C) Testosterone (D) Adrenaline

115. **Antiviral substance produced by vertebrates in case of viral infection is:**
(A) Virion (B) Interferon
(C) Antigen (D) Antivirin

116. **Which enzymes is not obtained from animals?**

(A) Ficin (B) Lipase
(C) Trypsin (D) Catalase

117. **Which enzymes is not obtained from plants?**

(A) α - and β - amylase
(B) β - glucanase and papain
(C) Pectinase
(D) Ficin

118. **Site- directed mutagenesis can be effected by which techniques?**

(A) Oligonucleotide-directed mutagenesis
(B) Casette mutagenesis
(C) Cut-paste mutagensis
(D) Both (A) and (B)

119. **As a general rule of thumb, hydrophobic residues in a protein tend to be situated at:**

(A) Extending out of the molecule
(B) Half exposed on the surface
(C) Interior of the molecule
(D) Exposed on the surface

120. **Which is not affected by enzymes?**

(A) Reaction rate (B) ΔG
(C) ΔG^{+}_{+} (D) Both (B) and (C)

121. **Which biosensor is most sensitive?**

(A) Potentiometric (B) Conductimetric
(C) Piezoelectric (D) Thermometric

122. **Which statement is the most correct about enzymes?**

(A) They are always proteins
(B) They are heat stable
(C) All the enzymes are highly specific for their substrates
(D) They can be inhibited by substrate analogues

123. **Km is the substrate concentratuion at wgcug reaction rate if equal to:**

(A) V_{max} (B) ½ V_{max}
(C) $2V_{max}$ (D) ¼ V_{max}

124. **In an electrocardiogram, the QRS complex represents the:**

(A) Depolarisation of the atria
(B) Repolarisation of the atria
(C) Depolarisation of the ventricles
(D) Repolarization of the ventricles

125. **An enzyme used in detergents must have:**

(A) Safe to use
(B) Cost-effective
(C) Able to function at pH between 8 to 10.5
(D) All of these

126. ***Streptomyces venezuelae* produces antibiotics:**

(A) Chloromycetin (B) Aureomycin
(C) Tetracycline (D) Streptomycin

127. **Column A lists of some enzymes, while column B lists their use. Match both the columns carefully & select the correct option from these listed below the.**

Column A	Column B
A. Lysozyme	1. Gout
B. Uricase	2. Skin ulcer
C. Streptokinase	3. Blood clots
D. Rhodonase	4. Cyanide poising
	5. Antibiotic

Options:

(A) A–3, B–4, C–1, D–2 (B) A–3, B–1, C–2, D–4
(C) A–5, B–4, C–3, D–4 (D) A–4, B–3, C–1, D–2

128. **The glucose syrup composition is typically:**

(A) 90-95% glucose, 10-15% maltose and 3-5% isomaltose
(B) 95-97% glucose, 1-2% maltose and 0.5-2.0% isomaltose
(C) 80-90% glucose, 5-10% maltose and 8-10% isomaltose
(D) 85-90% glucose, 5-10% maltose and 5-10% isomaltose

129. **Assertion (A): Immobilized cells are particulary vulnerable to contaminations**
Reason (R): An antibiotic may have to be used in such cases

Options:

(A) Both A and R are correct and R is the correct remedy of A
(B) Both A and R c rrect, but R does not explain A
(C) Only A is correct
(D) Only R is correct

130. **Na^{+}-K^{+} ATPase is related to:**

(A) Active transport (B) Passive transport
(C) Group translocation (D) Both (A) and (C)

131. **Unit of Km is:**

(A) cm (B) moles/litre
(C) s-1 (D) It has no unit

132. **A person has developed interferons in his body. He seems to carry infection of:**

(A) Tetanus (B) Malaria
(C) Measles (D) Typhoid

133. ***Rhizobium leguminosarus* does not nodulate:**

(A) Clover (*Trifolium spp*)
(B) Pea (*Pisum sativum*)
(C) Khesari (*Lathyrus sativus*)
(D) Lentil (*Lens culinaris*)

134. **β- oxidation of palmitic acid produces how many net ATPs?**

(A) 96 (B) 146
(C) 130 (D) 123

135. **At which of the following sites is the partial pressure of carbon dioxide highest?**

(A) Exhaled gas (B) Alveolar gas
(C) Systemic arterial blood
(D) Systemic venous blood

136. Which of the following growth factors is used for the treatment of osteoporosis?
(A) Fibroblast growth factor
(B) Thymosin α
(C) Insulin like growth factor
(D) Human growth hormone

137. Consider the following statements:
Sympathetic nervous system is a characterized by
P. Cetylcholine a neuro secretion
Q. Fight and flight activities
R. Longer preganglionic fibres
S. Non-medullated postganglionic fibres
T. Arising from thoracic-lumbar portion
Which of these are associated with auditory organs?
(A) Q and S (B) Q, S and T
(C) P, Q and R (D) P, R and T

138. Cerebellum of brain is concerned with:
(A) Static balance
(B) Initiation of muscular contraction
(C) Regulation of body posture and equilibrium
(D) Co-ordination of muscular movements

139. In case of covalent bonding, the most commonly employed matrix is:
(A) Cellulose (B) Polyacrylamides
(C) Agarose (D) All of these

140. In 1953, Sanger determined the amino acid sequence of which proteins?
(A) Casein (B) Insulin
(C) Haemoglobin (D) Heat shock protein

141. What is one of the main ethical questions involved with germ-lie therapy?
(A) Effects on future generations
(B) Effects on humans s a species
(C) Psychological effects
(D) Prejudice against those not able to conceive

142. Which of the following drugs for AIDS is not a protease inhibitor?
(A) Delavirdine (B) Indinavir
(C) Nelfinavir (D) Seqionavir

143. Due to which of the following features, enzymes are attractive for commercial application?
(A) They operate under mild conditions of temperature, pressure and pH, which saves energy
(B) Undesirable by-products are not produced
(C) Certain stereochemical reactions are impractical with chemical methods
(D) All of these

144. Which is biological materials is most commonly used in biosensors?
(A) Antibodies (B) Nucleic acids
(C) Enzymes (D) Lectins

145. Which is highly toxic?
(A) Carbodimides (B) Chloroformates
(C) Cyanogen bromide (D) Glutaraldehyde

146. Assertion (A): Initially, plant and animal enzymes were preferred over microbial enzymes
Reason (R): Considerations of safety and the fear of contaminations by microorganisms or toxins limited the use of microbial sources
Options:
(A) Both A and R are correct and R roughly explains A
(B) Both A and R correct, but R does not explain A
(C) Only A is correct
(D) Only R is correct

147. Which enzymes is obtained commercially from animal pancreas?
(A) Trypsin (B) Rennet
(C) Lipase (D) All of these

148. Which of the following statements best describes the effect of increased carbon dioxide pressure in the blood?
(A) It affects oxygen affinity and oxygen carrying capacity
(B) It increases the oxygen affinity but has no effect on oxygen-carrying capacity
(C) It reduces the oxygen affinity and the oxygen-carrying capacity
(D) It reduces oxygen affinity but has no effect on oxygen-carrying capacity

149. Rhodopsin, - adrenergic receptor, and musarinic acetylcholine receptors share which of the following features?
(A) Each causes inhibitory intracellular response
(B) Each activates a tyrosine kinase cascade
(C) Each is composed of an dimer
(D) Each functions through a heterotrimeric G-protein

150. Which is the basis of different VNTR loci?
(A) Sequence of repeating units
(B) Number of repeating units
(C) Length of repeating units
(D) Both (A) and (C)

151. Recombinant vaccines include:
(A) DNA vaccines
(B) Recombinant polypeptide vaccines
(C) Purified antigen vaccines
(D) Both (A) and (B)

152. Subtilisin inactivation by oxygen produced by bleaches is due to:
(A) Oxidation of the methionine residue at position 222
(B) Reduction of the methionine residue at position 222
(C) Oxidation of Ser residue in the active site
(D) All of these

153. Stealth liposomes are made up of ?
(A) Lipid (B) Monoclonal antibodies
(C) PEG (D) All of these

154. Enzyme kinetics (i.e., activity) may be affected due to:
(A) Matrix used for immobilization
(B) Conformational changes in the enzyme molecules induced by the procedure of immobilization
(C) Both (A) and (B)
(D) Altered charge of enzyme molecules

155. Whole cells have been used to develop which type of biosensor devices?
(A) Potentiometric (B) Thermometric
(C) Amperomteric (D) All of these

156. Which is not is used for treatment of thrombosis?
(A) Interferon
(B) Granulocyte macrophage colony stimulating factor
(C) Streptokinase
(D) Didanosine

157. The nature of interaction between analyte and the biological material used in bio-sensors may be of which types?
(A) The analyte may be converted into a new chemical molecule by enzymes
(B) The analyte may simply bind to the biological material present on biosensor
(C) The analyte may alter the biological agent used in the biosensor
(D) Both (A) and (B)

158. Packed bed enzyme reactors suffer from:
(A) Difficulty in temperature control
(B) Clogging by colloids
(C) Channel formation
(D) All of these

159. Enzyme ability in the presence of heavy metals can be improved by:
(A) Replacing Cys residue
(B) Replacing Met residue
(C) Removing surface carboxyl groups
(D) All of these

160. Which enzyme is regarded as catalytically perfect enzyme?
(A) Catalase
(B) Triphosphate isomerase
(C) Carbonic anhydrase
(D) All of these

161. Which of the following statement(s) concerning the rate of action potential propagation is/are true?
(A) It is faster in large-diameter axons than in small – diameter ones
(B) It is faster for a strong stimulus than for a weak one
(C) It s faster in myelinated nerve fibres than in non-myelinated ones
(D) Both (A) and (C)

162. Lowered pH results in:
(A) Increase in O_2 affinity of respiratory pigments
(B) Decrease in O_2 affinity of respiratory pigments
(C) Decrease in the number of red blood corpuscles
(D) Increase in the number of red blood corpuscles

163. At a VNTR locus, which of the following is the basis for different alleles?
(A) Sequence of repeating units
(B) Number of repeating units
(C) Length of repeating units
(D) Both (A) and (C)

164. Name of drug produced through biotechnology for cancer treatment:
(A) Insulin (B) Interferon
(C) HGH (D) TSH

165. Which of the following pairs are correctly matched?

P. Stenochaline	Limited tolerance to change in the osmotic concentrations of external environment
Q. Euryhaline	Wider range tolerance of osmotic concentrations
R. Osmoconformers	Animals which are incapable of osmoregulation
S. Osmoregulators	Animals which change in volume ar the external osmotic concentration changes

Select the correct answer using the codes given below
(A) Q, R and S (B) P, Q and S
(C) P, R and S (D) P, Q and R

166. Which one of the following is not a neurotransmitter?
(A) Glutamine (B) Glutamate
(C) Glycine (D) Acetylcholine

167. The term vaccine is derived from:
(A) Vacca (B) Vacular
(C) Vaccination (D) Vascular

168. Penicillin inhibits bacterial multiplication because it:
(A) Checks RNA synthesis
(B) Checks DNA synthesis
(C) Destroys chromatin
(D) Inhibits cell wall formation

169. Which of the following diseases is diagnosed on the basis of probes?
(A) Kala-azar and filarial
(B) Malaria and sleeping sickness
(C) Herpex Virus type I, type 2
(D) All of these

170. Hybridoma cells are:
(A) Nerve cells
(B) Hybrid cells formed from myeloma cells
(C) Cells having oncogenes
(D) Product of spore formation in bacteria

171. Antibiotic cephalosporin is got from:
(A) Bacterium (B) Alga
(C) Fungus (D) Mycoplasma

172. In case of downstream processing, purification is achieved through :
(A) Crystallization
(B) Chromatographic methods
(C) Evaporation (D) Both (A) and (B)

173. Which is correct about propteases?
(A) They help in removal of dirt
(B) Serine proteases are produced by *Bacillus licheniformis* and *B. amyloliquefaciens*
(C) They are packed inside dust-free granules for use in detergents
(D) All of these

174. In human beings, interferons are:
(A) Leucocytic (B) Immune
(C) Fibroblastic (D) All of these

175. Which is not antibiotic:
(A) Blastidin (B) Ceresin
(C) Penicillin (D) Cyclohexamide

176. Plasmids are used in genetic engineering because they are:
(A) Easily available (B) Able to replicate
(C) Able to integrate with host chromosome
(D) Inert

177. The choice of reactor type depends on:
(A) Form (free or immobilized) of enzyme to be used
(B) Kinetics of reaction
(C) Scale of operation (D) All of these

178. The chief limitations of membrane reactors are:
(A) Regular replacement of membrane adds to cost
(B) Need for substrate diffusion through membrane
(C) Useful for production of small quantities of products
(D) All of these

179. Consider the following statements:
Constriction of the renal artery would result in an increase in mean arterial pressure due to
P. Increased rennin secretion
Q. Increased angitensin – II formation
R. Decreased plasma bradykinin levels
S. Increased aldosterone secretion
Which of the statements given above is/are correct?
(A) P and S (B) Q, R and S
(C) P, Q and R (D) P, Q, R and S

180. The sequence of flow of blood in the hepatic portal system of a mammal is:
(A) Caudal vein, Kidney, liver, posterior venacava
(B) Intestine, liver, hepatic vein, posterior venacava
(C) Kidney, liver, intestine, sinus, venosus
(D) Posterior venacava, intestine, liver, hepatic vein, sinus, venosus

181. Glucose isomerase enzyme is obtained commercially from:
(A) Bacillus (B) Streptomyces
(C) Aspergillus (D) Both (A) and (B)

182. Which is the most likely first event in colon cancer development?
(A) Loss of APC function
(B) Loss of DCC function
(C) Activation of k-ras (D) Loss of p53 function

183. Which play the major role in hisyocompatibility?
(A) MHC loci (B) ABO blood group
(C) Rh blood group (D) MN blood group

184. Which is false about antibiotics?
(A) The term was coined by Waksman in 1942
(B) Antibiotics are capable of curing all diseases
(C) Some persons develop allergy by antibiotics
(D) Antibiotics are produced by microorganisms

185. K_m may be influenced by :
(A) Temperature (B) Ionic strength
(C) pH (D) All of these

186. Which is correct about BRCAI gene?
(A) It causes human breast cancer
(B) It is -100 kb long
(C) It encodes a protein having over 1800 amino acids
(D) All of these

187. Typically, fermentation of 100 g glucose by selected strains of *Saccharomyces cerevisidae* and *Saccharpmyces carlsbergensis* yields how many grams of ethanol?
(A) 25-29 (B) 45-49
(C) 60-70 (D) 70-89

188. Monoclonal antibiotics are produced by:
(A) Sarcoma cells (B) Germ cells
(C) Somatic cells
(D) Hybrid of carcinogen cell and bone cell

189. Homologues of BRCAI gene were not found in:
(A) Mice (B) Chickens
(C) Sheep (D) Pig

190. Which is not an example of immunosuppressive drugs?
(A) Methotrexate (B) Zidovudine
(C) Corticosteroids (D) Azathioprine

191. Which is not used in asymmetric gap-LCR?
(A) NAD (B) DNA ligase
(C) DNA polymerase (D) Reverse transcriptase

192. In case of histocompatibility, which is recognized directly (without processing) by TH cells?
(A) Major histocompatibility complex antigens
(B) Minor histocompatibility antigens
(C) A and B antigens (ABO blood group)
(D) Both (A) and (B)

193. Recombinant protein vaccines suffer from:
(A) Their cost is very high and often prohibitive
(B) They have to be stored at low temperature
(C) Storage and transportation, especially in developing countries are problematic
(D) All of these

194. Candidate disorder for gene therapy are selected on the basis of which criteria?
I. Disease should be life-threatening
II. Gene responsible for disease should have been cloned
III. A precise regulation of the gene should be required
IV. A suitable delivery system should be available
Options:
(A) I, II, IV (B) II, III, IV
(C) I, II, III, IV (D) I, II, III

195. Which is not used in gap-LCR?
(A) Reverse transcriptase (B) DNA ligase
(C) DNA polymerase (D) NAD

196. Which statements is not correct about AIDS?
(A) In 1981, it was first reported in India
(B) About 30% of infants from women having AIDS get infected unless the mothers are treated with antiviral agents before delivery
(C) In general, antibodies against HIV proteins appear in the patients 3 months after infection
(D) The decrease in CD_4^+ T cells is the hallmark of AIDS

197. The lead for development of the antihypertensive drug Captopril came from:
(A) South American pit viper
(B) Banded Krait
(C) King Cobra (D) Rattle snake

198. Protooncogenes (and their oncogenic counterparts) encode proteins that are involved in:
(A) Induction of cellular proliferation
(B) Regulation of programmed cell death
(C) Inhibition of cellular proliferation
(D) All of these

199. Generally, the waste are of which types?
(A) Biological (B) Physical
(C) Chemical (D) All of these

200. Which is correct about purifies antigen vaccines?
I. These vaccines are based on purified antigens of the concerned pathologens isolated from GMOs
II. They do not contain the organism so that the risk of pathogenicity is avoided
III. Successful examples of such vaccines are mostly from bacteria
IV. Cost of purified antigen vaccines is higher than that of conventional vaccines
Options:
(A) I, II, III (B) II, III, IV
(C) I, II, III, IV (D) I, II, IV

201. Immuno-PCR detects the presence of:
(A) DNA (B) RNA
(C) Antigen (D) Immunogen

202. SCID may result from a deficiency of:
(A) RAG-1/RAG-2 (B) ADA
(C) JAK-3 (D) All of these

203. Which of the following is most likely the last is\to occur in colon cancer development?
(A) Loss of APC function
(B) Loss of DCC function
(C) Loss of p53 function
(D) Activation of k-ras

204. Biogas produced by bacterial digestion of organic matter through which processes?
(A) Aerobic digestion
(B) Anaerobic digestion
(C) Both (A) and (B)
(D) Partially aerobic digestion

205. Consider the following statements:
P. Chymotrypsinogen is converted into chymotrypsin by trypsin
Q. Cholecystokinin is secreted by stomach
R. Pepsin acts on peptide linkage in which tyrosine and phenylalanine are involved
Which of the statements given above is/are correct?
(A) P and Q (B) Q and R
(C) P and R (D) P, Q and R

206. The term renal autoregulation refers in part to the fat that:
(A) The kidney does not require blood flow to sustain its active transport
(B) The kidney contains baroreceptors (pressure receptors that contribute to the regulation of ardia output)
(C) Renal blood flow is relatively constant over a wide range of systemic arterial pressures
(D) Renal blood flow is not affected by activation of the sympathetic nerves that innervate the kidney

207. Which of the following is not a function of the liver?
(A) Production of bile
(B) Detoxication of drugs
(C) Storage of glucose
(D) Storage of vitamin

208. When antidiuretic hormone (ADH) is released by the pituitary gland it acts on the distal tubules and collecting ducts of the nephrons to increase their permeability to water and promote water retention by the body. This action of ADH on the kidney has which of the following effects on the body?
(A) Lowering blood osmolarity
(B) Raising blood pH
(C) Lowering blood pressure
(D) Raising blood glucose

209. Which of the following diseases is not known to be caused by over-expression of trinucleotide repeat?
(A) Huntington's disease
(B) Fragile X syndrome
(C) Alzheimer's disease
(D) Kennedy's diseases

210. Syntrophic H_2 producing bacteria are also called :
(A) Obligate H_2 producing bacteria
(B) Obligate proton educing bacteria
(C) Facultative O_2 reducing bacteria
(D) Both (A) and (B)

211. Steroid biotransformation yields which adrenocortical hormones?
(A) Cortisone (B) Corticosterone
(C) Hydrocortisone (D) All of these

212. Rhizobium can be used in:
(A) Pulses (B) Forage crops
(C) Cereals (D) Both (A) and (B)

213. Bioethanol production on commercial scale is based on :
(A) Oil seed crops (B) Starch crops
(C) Sugar crops (D) Both (B) and (C)

214. The correct sequence for sperm migration after its production in testes is:
(A) Seminiferous tubule → Epididymis → Vas deferens → Urethra
(B) Urethra → Vas deferens → Epididymis → Seminiferous tubule
(C) Epididymis → Vas deferens → Urethra → Seminiferous tubule
(D) Seminiferous tubule → Vas deferens → Epididymis → Urethra

215. Calcium is important in skeletal muscles contraction because it:
(A) Activates the myosin ATPase by binding to it
(B) Binds to troponin to remove a constant inhibition of cross-bridge attachment
(C) Prevents the formation of bonds between the myosin cross bridges and the actin filament
(D) Is required to detach the myosin head from the actin filament

216. Salivary α–amylase becomes inactive in the stomach primarily due to:
(A) Inactivation by low pH
(B) Degradation by gastric pepsin
(C) Inhibition by chloride ion
(D) Inhibition by peptides

217. Cerebrospinal fluid:
(A) Has a composition almost identical to that of the rain extracellular fluid
(B) Is produced by the choroid plexus
(C) Circulates from the subarachnoid space into the cerebral ventricles
(D) Is absorbed through the arachnoid villi

218. If communication between the SA node and the AV node become blocked, which will most likely occur?
(A) The rate of ventricular contraction will decrease
(B) Afterload will increase
(C) The rate of atrial contraction will decrease
(D) Stroke volume will increase to 5L/beat

219. Biomass may be used in :
(A) Direct burning to generate heat
(B) Thermochemical conversion by pyrolysis/ gasification
(C) Biological conversion
(D) All of these

220. Rennet was purified by:
(A) Waksman (B) Flemming
(C) Hansen (D) Dubrunfaut

221. Contraction of the abdominal muscles is important in:
(A) Normal (quiet) inspiration
(B) Forced (maximum) inspiration
(C) Normal (quiet) expiration
(D) Forced (maximum) expiration

222. Biodiseal can be obtained from :
(A) Lipids accumulated in plants and algae
(B) Hydrocarbons produced by some plants & algae
(C) Both (A) and (B)
(D) Biologically conversion of diesel

223. Which of the following statements is not true of the endocrine system?
(A) It is one of two major regulatory systems of the body
(B) It is composed of glands that secrete chemical messengers into the blood
(C) It is an important regulator of homeostatic mechanisms
(D) None of these

224. Which is represents a gain of function?
(A) Loss of APC function
(B) Loss of DCC function
(C) Activation of k-ras
(D) Loss of p53 function

225. In case of downstream processing, commonly used approach of drying is:
(A) Spray drying (B) Freeze drying
(C) Vacuum drying (D) All of these

226. Tumour necrosis factor can be used for the treatment of:
I. Thrombolysis
II. Sepsis
III. Tumours
IV. Viral diseases
Options:
(A) I, II, III (B) III, IV
(C) I, II, III, IV (D) I, II, IV

227. A Graffian follicle is:
(A) An immature developing follicle
(B) A mature follicle ready to ovulate
(C) A follicle undergoing apoptosis
(D) Ovulated follicle

228. A shift in the oxygen-hemoglobin dissociation curve to the right occurs in:
(A) Hypothermia
(B) Carboxyhemoglobin
(C) Fetal hemoglobin S
(D) An increase in the 2-3-bisphosphoglycer-ate (2,3-BPG)

229. High doses of antibiotics can destroy the bacterial flora of the large intestine. This can result in impaired:
(A) Absorption of protein (D) Blood coagulation
(C) Bone resorption (D) Respiratory control

230. During the rising phase of an action potential:
(A) Voltage – gated Na^+ channels open
(B) Voltage – gated K^+ channels open
(C) Voltage-gated Na^+ channels lose
(D) Voltage-gated K^+ channels lose

231. Uric acid is the chief end product of nitrogenous excretion in adult reptiles and birds. It is often excreted in concentrated crystalline form. This pattern of excretion is mainly an adaptation for:
(A) Conservation of urea producing enzymes
(B) Raising osmotic concentration of blood
(C) Conservation of water
(D) Storage of waste materials

232. The function of the ileum is:
(A) Absorb nutrients
(B) Absorb vitamin B_{12} and bile salts
(C) To introduce bile and pancreatic juices
(D) Absorb alcohol and aspirin

233. Which is correct about ideal vaccine?
(A) It should not contaminate the environment
(B) It should not be tumorigenic or toxic or pathogenic
(C) It should have very low levels of side-effects in normal individuals
(D) The technique of vaccination should be complex

234. Polyesters of lactic acid and glycolic acid are used for achieving slow release of :
(A) Proteins
(B) Oligonucleotides
(C) Polysaccharides
(D) All of these

235. The resting membrane potential
(A) Occurs only in nerve and muscle cells
(B) Is the same in all cells
(C) Is oriented so that the cell's interior is positive with respect to the extracellular fluid
(D) None of these

236. Which statement are correct about cancer?
I. It refers specifically to malignant tumours
II. Cancers invade healthy tissues by a process called metastasis
III. They are classified as carcinomas, leukaemias/ lymphomas and sarcomas according to the embryonic origin of the tissue from which the tumour is derived
IV. It is induced by a variety of agents, such as, chemical carcinogens, irradiation and certain viruses

Options:
(A) I, II, III (B) II, III, IV
(C) I, II, III, IV (D) I, II, IV

237. The rate of water reabsorption from the proximal tubule is determined primarily by the:
(A) Rate of dissolved particle (solute) reabsor-ption from the proximal tubule
(B) Concentration of ADH (antidiuretic hormone) in the blood
(C) Osmotic pressure developed by plasma proteins in the proximal tubule
(D) Active transport of water molecules by the proximal tubule cells

238. An ECG would be useful for determining a patient's:
(A) Heart murmur
(B) Stroke volume
(C) Cardia output
(D) Blockage of condition of electrial signals between the atria and the ventricles

239. The internal pacemaker that sets biological rhythms:
(A) Is located in the brain
(B) Is located in the heart
(C) Does not function in the absence of light or other environmental uses
(D) B and C

240. Hormones:
(A) Are chemical regulators that are conveyed from one organ to another via the blood stream
(B) May be secreted by endocrine cells
(C) May be secreted by nerve cells
(D) All of these

241. Piercidin has insecticidal activity; it is isolated from which microorganisms?
(A) *Streptomyces hygrosopicus*
(B) *Streptomyces sagamonensis*
(C) *Streptomyces monaraensis*
(D) *Streptomyces avermitilis*

242. Which one of the following is not a heme protein?
(A) Catalase
(B) Cytochrome oxidase
(C) Microsomal P_{450} drug oxidizing system
(D) Fe-S protein complex

243. Which of the following applies uniquely to secondary lamphoid organs?
(A) Presence of precursor B and T- cells
(B) Circulation of lymphocytes
(C) Terminal differentiation
(D) Cellular prolifiration

244. For an action potential to occur:
(A) The stimulus must reach or exceed threshold
(B) Na^+ influx must exceed K^+ efflux
(C) The membrane must be out of the relative refractory period
(D) Both (A) and (B)

245. As blood passes along the glomerular capillaries from the afferent to efferent arteriole, the net filtration pressure:
(A) Increases
(B) Decreases
(C) First decreases, reached a minimum about half way along the capillary, then increases
(D) First increases, reaches a maximum about half way along the capillary, then decreases

246. Propagation of a regenerative action potential along an axon can be accelerated by which of the following?
(A) A decrease in the transmembrane resistance
(B) A decrease in the axoplasmic resistance
(C) Reduced myelin wrapping
(D) Shortened internodal lengths

247. Commercial preparations of *B. thuringiensis* usually contain?
(A) Crystal proteins (B) Spores
(C) Inert carriers (D) All of these

248. Which one of the following groups of enzymes is involved in the digestion of proteins?
(A) Pepsin, amylopsin and trypsin
(B) Amylopsin, trypsin and chymotrypsin
(C) Trypsin, chymotrypsin and steapsin
(D) Pepsin, trypsin and chymotrypsin

249. A patient undergoes a total gastrectomy because of a proximally located gastric cancer. After the surgery, which of the following digestive enzymes will be produced in inadequate amounts?
(A) Amylase (B) Chymotrypsin
(C) Lipase (D) Pepsin

250. Most of the CO_2 that is transported in blood:
(A) Is dissolved in the plasma
(B) Is bound to hemoglobin
(C) Is in carbonic acid
(D) Is in bicarbonate ion

251. Which moiety of hemoglobin molecule binds 2,3-BPG at pH 7.0?
(A) Amino terminal (B) Carboxylic terminal
(C) Sulphadryl group (D) Alcoholic terminal

252. Lipophillic hormones:
(A) Include peptide hormones such as EGF
(B) Are hormonally active in their original secreted form
(C) Must bind to membrane receptors to exert effects
(D) Pass through cell membranes into the cytoplasm

253. All of the following are known to involve a calcium activated, vescile-mediated secretory event except:
(A) Synaptic transmissions
(B) Release of histamine from mast cells
(C) Sperm acrosomal reactions
(D) Constitutive secretion of collagens

254. White matter of the brain is composed of:
(A) Glial cells
(B) Non-mylenated nerve fiber
(C) Mylenated nerve fiber
(D) Dendrites

255. Which bacteria increases the solubility/availability of phosphate?
(A) *Glomus* (B) *Thiobacillus*
(C) *Bacillus* (D) Both (B) and (C)

256. Which hydrolytic enzymes is produced in bulk quantity from plants?
(A) α- and β-amylase (B) Papain
(C) β-Glucanase (D) All of these

257. When a person dives deep under water and resurfaces rapidly, which one of the following sequences of events causes decompression sickness?
(A) Increase in pressure - N_2 dissolves in blood - Rapid fall in presence - N_2 bubbles in blood
(B) Increase in pressure - Pressure on ear ossicles - Loss of orientation -Imbalance
(C) Fall in pressure – Loss of O_2 from blood – deoxygenation - Fatique
(D) Increased pressure – enhanced blood pressure – Headache – Loss of memory

258. Complex media contain:
(A) Molasses (B) Cornsteep liquor
(C) Soybean meal (D) Any of these

259. Biocontrol agents have which features?
(A) They should be specific to the target organism
(B) They should have a high multiplication rate
(C) They should not damage the environment (including human beings)
(D) All of these

260. Some neurons in the vagus nerve terminate on sinoatrial (pacemaker) cells in the heart. These neurons secrete acetylcholine, which ultimately results in a decreased heart rate. this is an example of:
(A) Endorine control (B) Exorine control
(C) Hormonal control (D) Neural control

261. In the statement "there are several ventricles in the brain and some of these ventricles are connected by foramen of Monro. This foraman connects X which are located in the Y of the brain", X and Y stand respectively for:

(A) Lateral ventricles and cerebral hemisphere
(B) Third and fourth ventricles in cerebral hemisphere and medulla oblongata
(C) Lateral ventricles and third ventricle in cerebral hemispheres and diencephalon
(D) Lateral ventricles and third ventricle in cerebral hemisphere and mesencephalon

262. Which techniques are used to study of foetal cells for detection of genetic disorders?

I. Hybridization with oligonucleotide probes
II. RFLP analysis
III. Enzyme assay
IV. Karyotype analysis

Options:

(A) I, II, III (B) II, III, IV
(C) I, II, III, IV (D) I, II, IV

263. Which one of the following events is not triggered by the hormone glucagon in liver?

(A) Activation of glycogen synthase
(B) Stimulation of cAMP production
(C) Activation of protein kinase
(D) Activation of phosphorylase kinase

264. Oxidoreductase class of enzymes does not include:

(A) Aldolases (B) Dehydrogenases
(C) Oxidases (D) Peroxidases

265. Which protective agents reduces the detrimental effects of lyophilization?

(A) Skimmed milk (B) Salt
(C) Sucrose (D) Both (A) and (C)

266. What is the name and number of the nerve that transmits information from the basilar and tutorial membranes of the internal ear to the brain?

(A) Olfactory and I cranial nerve
(B) Optic and II cranial nerve
(C) Auditory and VIII cranial nerve
(D) Trigeminal and V nerve

267. In which of the following combinations is the name of the hormone, its chemical type and its tissue of origin correctly matched?

(A) Aldosterone-peptide-pancrease
(B) Glucagon-peptide-adrenal cortex
(C) ACTH-polypeptide-adrenal cortex
(D) Vasopession – peptide-posterior pituitary

268. Chorionic gonadotropin hormone, synthesized in the placenta, plays an important role in the establishment and maintenance of pregnancy in:

(A) Cows
(B) Pigs
(C) Rodents
(D) Humans

269. During the cardiac cycle:

(A) The volume of blood leaving the left side of the heart is greater than that leaving the right side
(B) The pressure of blood leaving the right side of the heart is greater than that leaving the left side
(C) The duration of systole is greater than that of diastole
(D) The duration of diastole is greater than that of systole

270. Trypsin differs from pepsin in that:

(A) Trypsin digests protein in an acidic medium while pepsin does so in an alkaline medium
(B) Trypsin digests protein in an alkaline medium while pepsin does so in an acidic medium
(C) Trypsin is secreted from the gastric glands while pepsin is erected from the pancreas
(D) Trypsin production is influenced by peptidergic neurohormones while pepsin is influenced by steroids

271. Which of the following is the primary factor regulating normal coronary blood flow?

(A) Aortic diastolic pressure
(B) Coronary perfusion pressure
(C) Systolic wall tension
(D) Myocardial oxygen consumption

272. The ability of a desert rodent's kidneys to form highly concentrated urine is largely attributable to:

(A) A high glomerular filteration rate
(B) The production of a very salty ultrafilterate in the glomerulus
(C) The ability of cells lining the proximal convoluted tubule to absorb water
(D) Long loops of Henle that pass through tissues of increasing salt concentrations

273. Which one of the following statements is correct?

(A) Chloride shift of Hamburger phenomenon occurs from the blood to liver in a fresh fish to maintain its osmotic balance
(B) Chloride shift takes place from stomach to the blood vascular system of man to produce HCl
(C) Chloride shift involves the passage of chloride ions from the plasma to RBC to balance the bicarbonate ions which have passed out from RBC to plasma
(D) Chloride shift occurs to balance chloride ions and sodium ions in a muscle cell during muscle contraction

274. Which microbial culumns is used for studies on product isolation, identification and strain improvement?

(A) Assay culture
(B) Development culture
(C) Research culture
(D) Both (B) and (C)

275. Which one of the following sequences is the correct route of circulation in the human heart?
(A) Right auricle – right ventricle – pulmonary artery – lungs – pulmonary veins – left auricle – left ventricle - aorta
(B) Left auricle - Right auricle – right ventricle – pulmonary artery – lungs – pulmonary veins — left ventricle - aorta
(C) Right auricle – left auricle - right ventricle – pulmonary artery – lungs – pulmonary veins — left ventricle - aorta
(D) Right auricle – right ventricle – pulmonary veins– lungs — pulmonary artery - left auricle – left ventricle - aorta

276. The rates of forward and reverse reactions depend on:
(A) Rate constants for the forward and backward reactions
(B) Concentrations of the products
(C) Concentrations of the reactants
(D) All of these

277. Which is regarded as the third revolution in vaccines?
(A) Recombinant protein vaccines
(B) Recombinant polypeptide vaccines
(C) DNA vaccines
(D) All of these

278. In human eyes, light exposure to the retinal photoreceptors:
(A) Causes its depolarization
(B) Causes its hyperpolarization
(C) Opens Na^+ channels of the photoreceptors
(D) Opens K^+ channels of the photoreceptors

279. Which biosensor has the highest cost?
(A) Potentiometric (B) Conductimetric
(C) Piezoelectric (D) Thermometric

280. In organisms with closed circulatory systems, fluid leaves the blood of capillary networks at the arterial end and returns to blood at the venous end for which of the following reasons?
(A) Osmotic pressure increases prior to dehydration
(B) Osmotic pressure difference dominates at the arterial end; hydrostatic pressure difference dominates at the venous end
(C) Hydrostatic pressure difference dominates at the arterial end; osmotic pressure difference dominates at the venous end
(D) Hydrostatic pressure drops during diastole

281. Cutting the posterior root of a spinal nerve would:
(A) Impair motor control of skeletal muscle
(B) Interfere with the flow of sensory impulses
(C) Interfere with the ability of brain to transmit impulses
(D) Interfere with the circulation of CSF

282. Insulin and thyroxine arrive at an organ at the same time. Thyroxine causes can effect on the organ but insulin does not. Why?
(A) Thyroxine is a lipid-soluble hormone and insulin is not
(B) The target cells in the organ are up-regulated by thyroxine
(C) The organ's cells have receptors for thyroxine but not for insulin
(D) Thyroxine is a local hormone and insulin is a circulating hormone

283. Consider the following water conservation mechanisms:
P. Absence of sweat glands
Q. Dependence on metabolic water
R. Greater ADH level in the blood
S. Lying more on protein-rich diet
The kangaroo (*Dipodiomys*) living in deserts can survive without drinking water because of:
(A) P, Q and R (B) P, Q and S
(C) Q, R and S (D) P, R and S

284. Release of which of the following peptide leads to an increase in the secretion of pancreatic-enzymes into the small intestine?
(A) Cholecystokinins (B) Gastrin
(C) Motilin (D) Secretin

285. Drugs for AIDS are generally designed to block:
(A) Integrase
(B) Reverse transcriptase (C) Viral protease
(D) All of these

286. Compared with systemic arterial blood, pulmonary arterial blood has a higher:
(A) Oxygen content
(B) pH
(C) Bicarbonate ion concentration
(D) Hb concentration

287. Consider the following statements:
For osmoregulation in aquatic medium
P. A freshwater fish has to drink water continuously
Q. A marine bony fish has to drink water continuously
R. A fresh water fish produces copious urine, which is hyposmotic to blood
S. A marine fish produces scanty urine, which is hyper osmotic to blood
Which of the statements given above is/are correct?
(A) P (B) S
(C) P and S (D) Q and R

288. Which of the following would cause a decrease in the binding affinity of haemoglobin for oxygen:
(A) Increased pH of the blood
(B) Increased temperature of the blood
(C) Decreased BPG levels in erythrocytes
(D) Both (A) and (B)

289. Polycythemia is a condition of blood in which the:

(A) Viscosity of blood is reduced due to deficiency in the blood count

(B) Surface tension of the blood is reduced due to the subnormal concentration of the haemoglobin in blood

(C) Viscosity of the blood is enhanced due to the solidification of the plasma

(D) Viscosity of the blood is increased due to the presence of excessive blood cells

290. Which of the following are functions of sertolic cells?

1. Protection of developing spermatogenic cells
2. Nourishment of spermatocytes, spermatids and sperm
3. Phagocytosis of excess sperm cytoplasm as development proceeds
4. Mediation of the effects of testosterone and FSH
5. Control of movement of spermatogenic cells and release of sperm into the lumen of seminiferous tubules

(A) 1, 2, 4 and 5 (B) 1, 2, 3 and 5
(C) 2, 3, 4 and 5 (D) 1, 2, 3, 4 and 5

291. Match list – I with list – II and select the correct answer using the codes given below the lists

List – I	List – II
P. Dehydration	1. Aldosterone
Q. Antidiuretic hormone	2. Osmoreceptor
R. Reabsorption of Na^+ in the kidney	3. Blood hyperosmotic urine
S. Water diuresis	4. Vasopressin
	5. Blood hyposmotic urine

(A) P–4, Q–1, R–2, S–3 (B) P–4, Q–4, R–5, S–1
(C) P–3, Q–1, R–2, S–5 (D) P–3, Q–4, R–1, S–5

292. Which of the following is not true about cortisol?

(A) Increases the circulating lamphoytes

(B) Increases the circulating eosinophils

(C) Decreases the production of prostaglandins

(D) Inhibits the production of fibroblasts

293. Mesonephric kidney in which tubules extend posteriority throughout the length of coelom, such kidney is referred as:

(A) Pronephros (B) Opisthonephros
(C) Holonephros (D) Metanephros

294. Which of the following statements accurately describes sex hormones?

(A) They bind specific membrane receptors

(B) They interact with DNA directly

(C) They cause release of a proteinaeous second messenger from the cell membrane

(D) They enhance transcription when bound to receptors

295. Most of the enzymes are used extracellularily for which medical purposes?

(A) Removal of toxic substances

(B) Treatment of disorders within blood circulatory system

(C) Topical applications

(D) All of these

296. Colorimetric test strips for glucose contain:

(A) Horse-radish peroxidase

(B) A weekly coloured chromogen

(C) Glucose oxidase

(D) All of these

297. Which yields antibiotics?

(A) *Cladonia* (Lichens) (B) *Streptomyces*
(C) *Usnea* (D) All of these

298. Which polymer matrices is commonly used for cell immobilization?

(A) Polyglycol oligomers (B) Polyacrylamide gel
(C) Calcium alginate (D) *K*-Carageenan

299. Cell immobilization may be achieved in:

(A) Cells directly bound to water insoluble carriers

(B) They can be cross-linked to bi-or multifunctional reagents

(C) Polymer matrices may be used for entrapping

(D) All of these

300. Sterilization can be achieved by:

(A) Heating (B) Filtration
(C) Radiation (D) All of these

301. Glucose isomerase is obtained from:

(A) *Streptomyces sp.*

(B) *Actinoplanes missouriensis*

(C) *Bacillus coagulans*

(D) All of these

302. Increased calcium influx across the membranes of presynaptic neurons has which of the following effects on synaptic transmission?

(A) Increases release of neurotransmitter

(B) Increases in voltage-dependent potassium efflux

(C) Decreases release of neurotransmitter

(D) Decrease in voltage-dependent potassium efflux

303. Dissociation curve of hemoglobin shifts to the right when:

(A) CO_2 concentration decreases

(B) O_2 concentration decreases

(C) CO_2 concentration increases

(D) Chloride concentration increases

304. Which is based on chemical interaction between the carrier and the product molecules?

(A) Covalent chromatography

(B) Gel Filtration

(C) Group specific affinity chromatography

(D) Both (A) and (C)

305. In 1929, Fleming discovered which antibiotics?
(A) Streptomycin (B) Tetracyclin
(C) Penicillin (D) Gentamycin

306. Which biosensor type has the lowest cost?
(A) Potentiometric (B) Conductimetric
(C) Piezoelectric (D) Thermometric

307. Biotransformation of progesterone to 11 α–hydroxyprogesterone is achieved by which microorganisms?
(A) *Rhizopus arrhizus*
(B) *Thiobacillus thiooxidans*
(C) *Rhizopus nigricans*
(D) Both (A) and (C)

308. Which biosensor has the fastest response?
(A) Potentiometric (B) Conductimetric
(C) Piezoelectric (D) Thermometric

309. Luteinising hormone (LH) in female:
(A) Helps in the appearance of secondary sexual characters
(B) Stimulates ovary to secrete estradiol
(C) Helps in release of the ovum from the ovary
(D) Control the blood pressure

310. Who is recognized as father of microbiology:
(A) Robert Koch (B) Edward Jenner
(C) Louis Pasteur (D) Aristotle

311. N-butanol is used for making:
(A) Formaldehyde and urea
(B) Petrol
(C) Resin
(D) All of these

312. Which acts as an agent of bioinformation?
(A) Resting and dried microbial cells
(B) Vegetative cells and spores of microbes
(C) Immobilized cells (D) All of these

313. Microbiology is widely consisted to have begun in 1857 when Pasteur reported his studies on :
(A) Lactic acid fermentation
(B) Microscopic features of the microorganisms
(C) A suitable medium for lactic acid fermentation
(D) All of these

314. In industrial microorganisms, which one is commonly used?
(A) River mud (B) Soil
(C) Lake (D) All of these

315. Interferon can be used for the treatment of:
(A) AIDS (B) Cancer
(C) Viral diseases (D) All of these

316. Bestatin is produced by which organisms?
(A) *Trichoderma polysporum*
(B) *Rhizoctonia leguminicola*
(C) *Streptomyces olivoreticuli*
(D) *Streptomyces avermitilis*

317. Gluconic acid is produced from:
(A) *Aspergillus niger*
(B) *Penicillum purpurogenous*
(C) *Rhizopus stolonifer*
(D) *Staphylococcus aureus*

318. Which is not correct about Rhizobium?
(A) These are Gram-positive soil bacteria capable of forming root nodules in most leguminous plants
(B) *Azorhizobium caulinodans* nodulates the stems of *Sesbania rostrata*
(C) Root nodulating rhizobia are of common occurrence
(D) Rhizobium cells contain genes for nitrogen fixation (nif genes)

319. Which products are obtained from yeast fermentation?
(A) Acetic acid, caproic acid and lactic acid
(B) Glycerol, ethanol, butanol and n-propanol
(C) Amyl alcohol and ethyl acetate
(D) All of these

320. A germplasm bank of AMF (arbuscular mycorrhizal fungi) cultures was established at:
(A) National Bureau of Plant Genetic Resources, New Delhi
(B) The Energy and Resources Institute, New Delhi
(C) Indian Institute Pulses Research, Kanpur
(D) Indian Agricultural Research Institute, Delhi

321. Correct sequence of hormone secretion from beginning of menstruation is
(A) FSH, progesterone, estrogen
(B) Estrogen, FSH, progesterone
(C) FSH, estrogen, progesterone
(D) Estrogen, progesterone, FSH

322. Fungus *Ampelomyces quisqualis* is used for the control of which powdery mildew?
(A) Grapes (B) Tomato
(C) Straw berries (D) All of these

323. Which is an example of broad spectrum antibiotics?
(A) Chloramphenicol
(B) Tetracyclin
(C) Penicillin
(D) Both (A) and (B)

324. Which of the following is not true for adrenaline?
(A) Is synthesized by demethylation of noradrenaline
(B) Increases coronary blood flow
(C) Increases free fatty acids in the blood
(D) Mobilizes glycogen stores from the liver

325. *Trichoderma parasitoid* can be used for the control of:
(A) Bollworms of cotton
(B) Stem borer (Sorghum)
(C) Sugarcane internode borer
(D) All of these

326. Which bacteria induces leaching of copper as copper sulphate from ores?
(A) *Thiobacillus organoparus*
(B) *T. thiooxidans* (C) *T. acidophilus*
(D) All of these

327. Acid protease is produced by which microorganisms?
(A) *Mucor pusillus* (B) *Endothica parasitica*
(C) *Mucor miehei* (D) All of these

328. Lactic acid is produced by activity of which micro-organisms using molasses/starch as substrate?
(A) *Rhizopus*
(B) *Lactobacillus bulgaricus*
(C) *Streptococcus lactis* (D) All of these

329. In case of downstream processing separation of cells from liquid medium is achieved by:
(A) Centrifugation (B) Filtration
(C) Flocculation and floating (D) All of these

330. Which is the commonly used techniques of filtration?
(A) Depth filtration
(B) Cross-flow filtration
(C) Rotary drum vacuum filtration
(D) Surface filtration

331. Which diseases occurs due to a single gene mutation?
(A) Tay Sach's disease and sickle cell anaemia
(B) Cystic fibrosis and Huntington's chorea
(C) Haemophilia and Lesch-Nyhan syndrome
(D) All of these

332. Bestatin shows which activities?
(A) Antitumour (B) Insecticidal
(C) Herbicide (D) Antifungal

333. Disruption of microbial cells is usually difficult due to:
(A) Small cell size (B) Strong cell wall
(C) High osmotic pressure inside cells
(D) All of these

334. The high glucagons/insulin ratio seen in starvation:
(A) Promotes mobilization of fatty acids from adipose stores
(B) Stimulates beta oxidation by inhibiting the production of malonyl CoA
(C) Leads to increased concentrations of ketone bodies in the blood
(D) All of these

335. Which of the following statement is true?
(A) All veins carry deoxygenated blood
(B) All arteries carry deoxygenated blood
(C) All veins except one carry oxygenated blood
(D) All arteries except one carry oxygenated blood

336. Bioreactors are extensively used for:
(A) Fermentation (B) Food processing
(C) Waste management (D) All of these

337. Azotobacter can be used in:
(A) Wheat (B) Vegetables
(C) Rice (D) All of these

338. Which is correct about azathioprine?
(A) A mitotic inhibitor
(B) Acts on S phase cells
(C) Blocks synthesis of inosinic acid
(D) All of these

339. The process involving *Thiobacillus ferrooxidans* is used to recover:
(A) Uranium (B) Gold
(C) Copper (D) All of these

340. Typically, corn medium yields butanol, acetone and ethanol in which ratios?
(A) 3 : 3 : 1 (B) 6 : 3 : 1
(C) 3 : 6 : 1 (D) 1 : 3 : 6

341. Which one of the following comparisons between oogenesis and spermatogenesis in human is not correct?
(A) FSH promotes development in both eggs as well as sperms
(B) LH triggers ovulation in ovary and androgen production in testes
(C) Primary oocytes and primary spermatocytes follow similar pattern of development through meiosis
(D) An ovum is not produced until it fuses with the sperm whereas sperm is produced even in the absence of the ovum

342. Fungus (*Vertticillium lecanii*) can be used for the control of which insects?
(A) Aphids (B) Leaf hopper
(C) White flies (D) Both (A) and (C)

343. Diuring penicillin G production which precursors is used?
(A) Phenylacetic acid (B) Acetadehyde
(C) Phenoxyacetic acid (D) Both (A) and (C)

344. Which is used as antifoaming agent during antibiotic production?
(A) Mineral oil (B) Vegetable oil
(C) Silicones (D) Both (B) and (C)

345. Aabomycin A shows which activities?
(A) Antifungal (B) Antibacterial
(C) Antiviral (D) Antiprotozoal

346. Which indicates the correct order of steps involved in production of ethanol from starch substrate?
(A) Starch → Pyruvate → Glucose → Acetaldehyde → Ethanol
(B) Starch → Glucose → Pyruvate → Acetaldehyde → Ethanol
(C) Starch → Acetaldehyde → Glucose → Pyruvate → Ethanol
(D) Starch → Pyruvate → Acetaldehyde → Glucose → Ethanol

347. Long chain immunosuppressive therapy increases the risk of:
(A) Hypertension (B) Cancer
(C) Metabolic bone disease (D) All of these

348. Bioinformations are performed by:
(A) Plant cells (B) Animal cells
(C) Microbes (D) All of these

349. Which is the commonly used to achieve cell disruption:
(A) Drying (B) Lysis
(C) Mechanical cell disruption
(D) All of these

350. Which is not correct about α–amylase?
(A) This enzyme is produced by bacillus amyloliquefaciens
(B) This is an exohydrolase
(C) It hydrolyses starch into components, which have 3 or more linear α -1, 4 glucan units
(D) The end products of starch hydrolysis are dextrins

351. Recombinant insulin is produced in:
(A) *E. coli*
(B) Continuous cell lines
(C) Yeast
(D) Both (A) and (C)

352. Which options indicates the correct order for bioinformation of penicillin G and penicillin V?
(A) Penicillin G/penicillin V → 6-aminopenicillanic acid → semisynthetic penicillins and cephalosporins
(B) Penicillin G/penicillin V → semisynthetic penicillins and cephalosporins → 6-aminopenicillanic acid
(C) Penicillin G/penicillin V → natural penicillin → semisynthetic penicillins
(D) Penicillin G/penicillin V → semisynthetic penicillins and cephalosporins → synthetic antibiotics

353. Pernicious anemia is due to:
(A) Blockage of vitamin B_{12} absorption
(B) Blockage of vitamin A absorption
(C) Deficiency of Vitamin C
(D) Deficiency of vitamin B_2

354. Which bacteria induces leaching of copper as copper sulphate from ores?
(A) *Arthrobacter simplex*
(B) *Thiobacillus ferroxidans*
(C) *Bacillus subtilis*
(D) All of these

355. Which vectors has been designated for targeted gene transfer on humans?
(A) Insertion vectors (B) Suicide vectors
(C) Replacement vectors (D) Both (A) and (C)

356. Which options indicates the correct order of 11α – hydroxyprogesterone production?
(A) Diosgenin/stigmasterol → progesterone → 11 - α Hydroxyprogesterone
(B) Diosgenin → stigmasterol → 11 α –Hydroxyprogesterone
(C) Progesterone → diosgenic → 11 α –Hydroxyprogesterone
(D) Stigmasterol → 11 α -Hydroxyprogesterone → Diosgenic

357. In ruminants the correct order of the four chambers is:
(A) Rumen, reticulum, abomasums, omasum
(B) Rumen, reticulum, omasum, abomasums
(C) Omasum, abomasums, reticulum, rumen,
(D) Omasum, reticulum, rumen, abomasums

358. Chemical fertilizers suffers from which limitations?
(A) They damage the environment
(B) They are costly
(C) They utilize petroleum (specially nitrogenous fertilizers)
(D) All of these

359. Application of the process based on *Thiobacillus* for copper recovery has resulted in:
(A) Reduction in price of copper by about 2-3 fold
(B) Elimination of the release of SO_2 into the atmosphere
(C) Both (A) and (B)
(D) Increased purity of the product

360. LCR can detect which molecules?
(A) DNA (B) Protein
(C) RNA (D) Both (A) and (C)

361. Which bacteria can absorb metals on their cell surfaces?
(A) Proteus (B) Bacillus
(C) Streptococcus (D) All of these

362. Which enzyme is used for manufacturing of detergents?
(A) Lactases (B) Amylases
(C) Glucoamylases (D) Proteases

363. Bioreactors are used for:
(A) Production of ethanol
(B) Production of enzymes on large scale
(C) Production of cell culture on large scale
(D) All of these

364. Which of the following organism is not used in any of the traditional industrial fermentation?
(A) *E. coli* (B) Yeast
(C) Bacillus (D) Pseudomonas

365. Yeast is used in the production of:
(A) Ethyl alcohol (B) Acetic acid
(C) Cheese (D) Curd

366. Which of the following microorganisms help in the ethanol production?
(A) *Saccharomyces cerevisae*
(B) *Zygomonas mobilis*
(C) *Saccharomyces uvarum*
(D) All of these

367. Which of the following is used to check vortex and to improve aeration efficiency in a fermentor?
(A) Impeller (B) Baffles
(C) Sparger (D) All of these

368. What are different substrates used for ethanol production?
(A) Starch containing substrate
(B) Juices from sugarcane or molasses or sugar beet
(C) Waste product from wood or processed wood
(D) All of these

369. High ethanol concentration:
(A) Promotes yeast growth
(B) Inhibits yeast growth
(C) Promotes bacterial growth
(D) Inhibits bacterial growth

370. The dried malt is the source of:
(A) Amylase (B) Proteinases
(C) Both (A) and (B) (D) Cellulase

371. After the fermentation is over, ethanol is recovered by:
(A) Centrifugation (B) Distillation
(C) Filtration (D) Cell disintegration

372. Ethanol is produced by:
(A) Continuous fermentation
(B) Batch fermentation
(C) Both (A) and (B)
(D) None of these

373. Which enzymes are used for converting corn starch into high fructose syrup?
(A) Glucoisomerase (B) Amylases
(C) Glucoamylases (D) All of these

374. The enzyme diastase was identified by:
(A) Payon and Persoz (B) Alenxander Fleming
(C) Christian Hansen (D) S.A. Waksman

375. Vitamin B2 is commercially obtained from:
(A) *Escherichia coli*
(B) *Leuconostoc mesenteroids*
(C) *Closteridium butyricum*
(D) *Streptomyces alivacaeus*

376. Dextran is used in:
(A) Bleeding (B) Blood transfusion
(C) Blood clotting (D) Blood Pressure

ANSWERSHEET

18. INDUSTRIAL BIOTECHNOLOGY

1. (A)	2. (D)	3. (A)	4. (B)	5. (D)	6. (D)	7. (C)	8. (D)	9. (D)	10. (D)
11. (C)	12. (C)	13. (D)	14. (D)	15. (D)	16. (D)	17. (C)	18. (D)	19. (C)	20. (D)
21. (D)	22. (C)	23. (D)	24. (B)	25. (C)	26. (D)	27. (C)	28. (D)	29. (C)	30. (D)
31. (B)	32. (D)	33. (D)	34. (B)	35. (C)	36. (C)	37. (D)	38. (D)	39. (B)	40. (B)
41. (B)	42. (B)	43. (C)	44. (D)	45. (B)	46. (C)	47. (B)	48. (D)	49. (C)	50. (A)
51. (C)	52. (D)	53. (B)	54. (D)	55. (B)	56. (D)	57. (C)	58. (B)	59. (C)	60. (C)
61. (D)	62. (A)	63. (D)	64. (B)	65. (D)	66. (D)	67. (C)	68. (D)	69. (C)	70. (A)
71. (D)	72. (B)	73. (D)	74. (C)	75. (D)	76. (D)	77. (D)	78. (A)	79. (B)	80. (D)
81. (A)	82. (C)	83. (D)	84. (D)	85. (D)	86. (A)	87. (B)	88. (D)	89. (D)	90. (D)
91. (C)	92. (B)	93. (D)	94. (C)	95. (B)	96. (C)	97. (D)	98. (D)	99. (D)	100. (C)
101. (D)	102. (B)	103. (C)	104. (D)	105. (C)	106. (C)	107. (C)	108. (C)	109. (D)	110. (D)
111. (D)	112. (B)	113. (D)	114. (A)	115. (B)	116. (A)	117. (C)	118. (D)	119. (C)	120. (B)
121. (D)	122. (D)	123. (B)	124. (C)	125. (D)	126. (A)	127. (C)	128. (B)	129. (A)	130. (A)
131. (B)	132. (C)	133. (A)	134. (C)	135. (A)	136. (C)	137. (B)	138. (D)	139. (D)	140. (B)
141. (A)	142. (A)	143. (D)	144. (C)	145. (C)	146. (A)	147. (D)	148. (D)	149. (D)	150. (D)
151. (D)	152. (A)	153. (D)	154. (C)	155. (D)	156. (C)	157. (D)	158. (D)	159. (D)	160. (D)
161. (D)	162. (B)	163. (B)	164. (B)	165. (D)	166. (A)	167. (A)	168. (D)	169. (D)	170. (B)

171. (C) **172. (D)** **173. (D)** **174. (D)** **175. (D)** **176. (B)** **177. (D)** **178. (D)** **179. (D)** **180. (B)**

181. (D) **182. (B)** **183. (A)** **184. (B)** **185. (D)** **186. (D)** **187. (B)** **188. (D)** **189. (B)** **190. (B)**

191. (C) **192. (A)** **193. (D)** **194. (A)** **195. (A)** **196. (A)** **197. (A)** **198. (D)** **199. (D)** **200. (B)**

201. (C) **202. (D)** **203. (C)** **204. (B)** **205. (C)** **206. (C)** **207. (D)** **208. (A)** **209. (C)** **210. (D)**

211. (D) **212. (A)** **213. (D)** **214. (A)** **215. (B)** **216. (A)** **217. (B)** **218. (A)** **219. (D)** **220. (C)**

221. (D) **222. (C)** **223. (D)** **224. (C)** **225. (C)** **226. (B)** **227. (B)** **228. (D)** **229. (B)** **230. (A)**

231. (C) **232. (B)** **233. (D)** **234. (D)** **235. (D)** **236. (C)** **237. (A)** **238. (D)** **239. (B)** **240. (D)**

241. (C) **242. (D)** **243. (C)** **244. (D)** **245. (B)** **246. (B)** **247. (D)** **248. (D)** **249. (D)** **250. (D)**

251. (A) **252. (D)** **253. (D)** **254. (C)** **255. (D)** **256. (D)** **257. (B)** **258. (D)** **259. (D)** **260. (D)**

261. (C) **262. (C)** **263. (A)** **264. (A)** **265. (D)** **266. (C)** **267. (B)** **268. (D)** **269. (D)** **270. (B)**

271. (D) **272. (D)** **273. (C)** **274. (C)** **275. (A)** **276. (D)** **277. (C)** **278. (B)** **279. (D)** **280. (C)**

281. (B) **282. (C)** **283. (C)** **284. (A)** **285. (D)** **286. (B)** **287. (D)** **288. (B)** **289. (D)** **290. (D)**

291. (D) **292. (B)** **293. (B)** **294. (D)** **295. (D)** **296. (D)** **297. (D)** **298. (C)** **299. (D)** **300. (D)**

301. (D) **302. (A)** **303. (C)** **304. (D)** **305. (C)** **306. (B)** **307. (D)** **308. (C)** **309. (C)** **310. (C)**

311. (D) **312. (D)** **313. (D)** **314. (D)** **315. (D)** **316. (C)** **317. (B)** **318. (A)** **319. (D)** **320. (B)**

321. (C) **322. (D)** **323. (D)** **324. (A)** **325. (D)** **326. (D)** **327. (D)** **328. (D)** **329. (D)** **330. (C)**

331. (D) **332. (A)** **333. (D)** **334. (D)** **335. (D)** **336. (D)** **337. (D)** **338. (D)** **339. (D)** **340. (B)**

341. (C) **342. (D)** **343. (D)** **344. (D)** **345. (C)** **346. (B)** **347. (D)** **348. (D)** **349. (A)** **350. (B)**

351. (D) **352. (A)** **353. (A)** **354. (B)** **355. (D)** **356. (A)** **357. (B)** **358. (D)** **359. (C)** **360. (D)**

361. (D) **362. (D)** **363. (D)** **364. (A)** **365. (A)** **366. (D)** **367. (B)** **368. (D)** **369. (B)** **370. (C)**

371. (B) **372. (C)** **373. (C)** **374. (A)** **375. (D)** **376. (B)**

Chapter - 19

FOOD AND BEVERAGE BIOTECHNOLOGY

1. Fleming is famous for his discovery is:
(A) Actinomycetes (B) Penicillin
(C) Chloromycetin (D) Streptomycin

2. Vitamin B 1 was first of all isolated by:
(A) Mc Collum (B) Mellanbby
(C) Funk (D) Szent-Gyorgy

3. Starch rich material like wheat, rice, maize, etc., that are added to barley malt during beer production is called:
(A) Adjuncts (B) Adjuvant
(C) Adsorption (D) None of these

4. Molecules adhere to the surface of carrier matrix due to a combination of hydrophobic effects and the formation of several salt links per molecules is known as:
(A) Adjuncts (B) Adjuvant
(C) Adsorption (D) None of these

5. A non-antigenic precipitation known to stimulate immune response is:
(A) Adjuncts (B) Adjuvant
(C) Adsorption (D) None of these

6. Citric acid is used as an acidulant is:
(A) Food (B) Confectionary
(C) Beverages (D) All of these

7. Production of citric acid from sugar solution was first realized by:
(A) *Aspergillus niger* (B) Penicillium
(C) Candida (D) Acetobacter

8. Now-a-days the most widely used organism for production of citric acid is:
(A) *Aspergillus niger* (B) Penicillium
(C) Candida (D) Acetobacter

9. To oxidize ethanol solely to acetic acid by :
(A) *Aspergillus niger* (B) Penicillium
(C) Gluconobacteria (D) Acetobacter

10. The most common species used in fermentation of acetic acid is:
(A) *Gluconobacter oxydans* (B) *Acetobacter aceti*
(C) Both (A) and (B) (D) None of these

11. Gluconic acid is commonly used in the manufacture of:
(A) Metal (B) Leather
(C) Food (D) All of these

12. Now-a-days the fungus are used for production of gluconic acid:
(A) *Aspergillus niger*
(B) *Acetobacter suboxydans*
(C) Both (A) and (B) (D) None of these

13. The process by which metals are dissolved from ore-bearing rocks by the use of microorganisms is called:
(A) Bioleaching (B) Bioprocess technology
(C) Bioreactors (D) Food Processing

14. The application of the principles of process technology (which are essentially derived from chemical engineering) to biological systems is called:
(A) Bioleaching
(B) Bioprocess technology
(C) Bioreactors (D) Food Processing

15. In any fermentation process -------- plays a central role in determining the process efficiency is:
(A) Bioleaching
(B) Bioprocess technology
(C) Bioreactors (D) Food Processing

16. Food preservation is a most important part of:
(A) Bioleaching
(B) Bioprocess technology
(C) Bioreactors
(D) Food Processing

17. The microorganism is most widely used in bio leaching is:
(A) *Thiobacillus ferroxidasns*
(B) *Pseudomonas species*
(C) Both (A) and (B)
(D) None of these

18. Food processing is done in order to achieve the:
(A) Increased safety and shelf life
(B) Improve the ease and reliability of preparation by consumers
(C) Protect from contamination and thus help storage and transportation
(D) All of these

19. For bioremediation have involved natural microorganisms such as the:
(A) *Thiobacillus ferroxidasns*
(B) *Pseudomonas species*
(C) Both (A) and (B)
(D) None of these

20. Any vessel with facilities for aeration and agitation can be used as a:
(A) Bioreactor (B) Batch culture
(C) Fed-batch culture (D) Sterilization

21. A system where either a simple shake flask or a bioreactor is used to grow cells is:
(A) Bioreactor (B) Batch culture
(C) Fed-batch culture (D) Sterilization

22. One of the most favoured cultivation techniques used in the industry is the use of:
(A) Bioreactor (B) Batch culture
(C) Fed-batch culture (D) Sterilization

23. The process of getting rid of contaminating agents like bacteria, viruses, etc from the system so that we can use it for growing the desired microorganisms is called:
(A) Bioreactor (B) Batch culture
(C) Fed-batch culture (D) Sterilization

24. Making antiviral drugs is more difficult than making anti-bacterial medicines because:
(A) Viruses make use of host machinery
(B) Viruses are on the border line of living and non-living
(C) Viruses have very few biochemical mechanisms of their own
(D) Viruses have a protein coat

25. You are aware of Polio Eradication Programme is your city. Children are vaccinated because:
(A) Vaccination kills the polio causing microorganisms
(B) Prevents the entry of polio causing organism
(C) It creates immunity in the body
(D) All of these

26. Which organisms used for SCP production shows the lowest growth rate?
(A) Bacteria (B) Algae
(C) Yeast (D) Filamentous fungi

27. Germinating barley seeds are used for preparation of:
(A) Lactic acid (B) Wine
(C) Cheese (D) Beer

28. Rennel is employed in:
(A) Synthesis of antibiotics
(B) Preparation of cheese
(C) Fermentation (D) Bread making

29. Yeast is an important source of:
(A) Vitamin (B) Riboflavin
(C) Sugars (D) Proteins

30. Growing microorganisms in thin layer of nutrient medium is:
(A) Suspended growth (B) Thin layer growth
(C) Support growth system
(D) None of these

31. Due to which features, microorganisms are generally considered promising source of lipids?
(A) They show high growth rate in simple media
(B) They are amenable to numerous biotransformation reaction
(C) They produce value-added types of fats or oils
(D) All of these

32. Organic acid first produced through fermentation is:
(A) Propionic acid (B) Lactic acid
(C) Citric acid (D) Oxalic acid

33. Vinegar is prepared from alcohol with the help of:
(A) *Lactobacillus* (B) *Acetobacter*
(C) *Azotobacter* (D) *Rhizobium*

34. Vitamin B12 is formed during fermentation of:
(A) *Ashbya gossipii*
(B) *Rhizopus stolonifer*
(C) *Propionibacteria*
(D) *Saccharomyces cervisiae*

35. Which ones are involved in preparation of alcohol?
(A) Yeasts (B) Slime Moulds
(C) Bacteria (D) Water moulds

36. Which proteases is used to recover meat (in form of slurry) adhering to bones?
(A) Papain
(B) Heat labile fungal protease
(C) Subtilism
(D) Neutral or alkaline proteases

37. Pasteurization frees foodstuffs from:
(A) All living bacteria
(B) All living microorganisms
(C) All vegetative forms of bacteria
(D) Vegetative forms of all pathogenic bacteria

38. Fermentated food Gari is produced from which substrates?
(A) Cassava (B) Rice
(C) Milk (D) Urad

39. Which organisms used for SCP production shows the highest growth rate?
(A) Bacteria (B) Algae
(C) Yeast (D) Filamentous fungi

40. Yeast yields:
(A) Vitamin B_{12} (B) Tocopherol
(C) Riboflavin (D) Ascorbic acid

41. Fermenattion of sugar to yield alcohol is carried out by:
(A) Microorganisms
(B) Zymase
(C) Raised temperature
(D) Decomposition of sugar

42. *Torulopsis utilis* is:
(A) Food yeast
(B) An important intestinal commensal
(C) Microorganism that yields third generation vaccines
(D) Employed for synthesis of citric acid

43. In cheese microorganisms are required for:
(A) Ripening
(B) Souring of milk
(C) Souring and ripening
(D) Development of resistance to spoilage

44. Which bacteria have been used for production of SCO in Japan?
(A) *Chromobacterium viscosum*
(B) *Mycobacterium sp.*
(C) *Agrobacterium sp.*
(D) All of these

45. Glucose fermentation by yeast yields:
(A) Ethanol + CO_2 (B) Ethanol + H_2O
(C) Methanol + CO_2 (D) H_2O + CO_2

46. Yeast is used for the preparation of:
(A) Acetic acid (B) Curd
(C) Ethyl alcohol (D) Cheese

47. Cheese and yoghurt are products of:
(A) Pasteurization (B) Fermentation
(C) Dehydration (D) Distillation

48. When substrate is rich in fermentable sugars, which enzymes is used for saccharification?
(A) Fungal α–amylase
(B) Heat stable bacterial α-amylase
(C) α-Amylase and glucanase
(D) Glucanase

49. Curd, milk, cheese and butter are produced with the use of:
(A) Yeast (B) *Penicillin*
(C) *Streptococus* (D) None of these

50. Which enzymes is used during post-fermentation stages?
(A) Papain (B) Neutral proteases
(C) Cellulases (D) All of these

51. Which proteases is used for meat tenderization?
(A) Papain
(B) Heat labile fungal protease
(C) Subtilism
(D) Neutral or alkaline proteases

52. Which one is not used in the production of Yoghurt?
(A) *Streptococcus lactis*
(B) *Streptococcus thermophilus*
(C) *Lactobacillus bulgaricus*
(D) *Acetobacter aceti*

53. For the production of Ang-Kaka red rice, rice is inoculated with which microorganisms?
(A) *Leuconostoc mesenteroides*
(B) *Monascus purpureus*
(C) *Endomycopsis burtonii*
(D) *Geotrichum candidum*

54. Cells of which of the organisms are used for dough fermentation:
(A) *Candida utilis*
(B) *Saccharomyces cerevisiae*
(C) *Fusarium gaminearum*
(D) *Kluyveromyces fragilis*

55. Distillation of wine is necessary as it:
(A) Increases duality
(B) Prevents further fermentation and spoiling of wine
(C) Decreases toxicity
(D) Has no action on nature of wine

56. Distribtion of clean and quality milk throughout the world has been made possible through work of:
(A) Leeuwenhoek (B) Koch
(C) Pasteur (D) Blackman

57. Which enzymes is used for production of high gravity beer?
(A) Glucoamylase
(B) Fungal α–amylase
(C) Either or Both (A) and (B)
(D) Neutral proteases

58. Anaerobic respiration of yeast produces:
(A) Alcohol
(B) Carbon dioxide
(C) Alcohol, carbon dioxide, beer, wine and other beverages
(D) None of these

59. Production of low calorie light bears specifically uses which enzymes?
(A) Glucoamylase
(B) Fungal–amylase
(C) Either or Both (A) and (B)
(D) Neutral proteases

60. Yoghurt is produced with the help of:
(A) *Lactrobacillus bulgaricus*
(B) *Lactobacillus thermophilus*
(C) *Streptococcus thermophilus*
(D) Both (A) and (C)

61. Which one is used in production of alcohol?
(A) *Saccharomyces cerevisiae*
(B) *Tonilopsis utilis*
(C) *Clostridium botulinum*
(D) *Leuxonostoc citrovorum*

62. Cheeses are classified on the basis of:
(A) Flavour (B) Colour
(C) Texture (D) All of these

63. For the fermentation of brew, which species of yeast is used?
(A) *Saccharomyces carisbergensis*
(B) *Kluyveromyces fragilis*
(C) *Sacchromyces cerevisiae*
(D) Either A or C

64. Renin employed in cheese industry is:
(A) Inhibitor (B) Alkaloid
(C) Enzyme (D) Vitamin

65. Which organisms has been used for commercial SCP production from starch hydrolysates?
(A) *Candida utilis*
(B) *Saccharomyces cerevisiae*
(C) *Fusarium gaminearum*
(D) *Kluyveromyces fragilis*

66. During fermentation by yeast, alcohol is formed from:
(A) Proteins (B) Sugars
(C) Mucopolysaccharides (D) Fats

67. Enzyme that changes glucose to ethyl alcohol is:
(A) Zymase (B) Diastase
(C) Invertase (D) Maltase

68. A new strain of bacteria for alcohol industry produced through biotechnology belongs to:
(A) *Escherichia coli*
(B) *Bacillus subtilis*
(C) *Pseudomonas putida*
(D) *Saccharomyces cerevisiae*

69. Caric acid is got from:
(A) *Aspergillus niger*
(B) *Rhizobium nigricans*
(C) *Penicillium citrinum*
(D) *Lactobacillus vulgaricus*

70. Microorganisms grown on molasses and sold as flavouring substances is:
(A) *Saccharomyces* (B) *Rhizopus*
(C) *Acetonacter* (D) *Lactobacillus*

71. Product of fermentation of glucose by yeast is:
(A) $(C_6H_{10}O_5)_n$ (B) C_2H_5OH
(C) $C_6H_{12}O_6$ (D) CH_3OH

72. When lactose is hydrolysed by lactase, which monosaccharides is produced?
(A) D-glucose + D-glucose
(B) D-glucose +ß -D-galactose
(C) ß-D-galactose + ß-D-galactose
(D) D-glucose + D-fluctose

73. In France, SCP production using *Kluyveromyces fragilis* uses which as a substrate?
(A) Whey (B) Molasses
(C) Starch hydrolysate
(D) Confectionary effluent

74. Pomato is somatic hybrid between:
(A) Poppy and Potato
(B) Potato and Tomato
(C) Poppy and Tamarind
(D) Poppy and Tomato

75. Which mushrooms can be grown on straw?
(A) *Coprinus sp.* (B) *Pleurotus ostreatus*
(C) *Volvariella volvaceae* (D) All of these

76. Which proteases is used for partial hydrolysis of soy protein?
(A) Papain
(B) Heat labile fungal protease
(C) Subtilism
(D) Neutral or alkaline proteases

77. Beer production from fermentation of barley malt uses which species of yeast?
(A) *Kluyveromyces fragils*
(B) *Saccharomyces cerevisiae*
(C) *Saccharomyces carisbergensis*
(D) Both (B) and (C)

78. In USA, Torula yeast (*Candida utilis)* has been used for commercial SCP production using which as a substrate?
(A) Methanol
(B) Ethanol (from fermentation)
(C) Ethanol (from ethylene)
(D) Dairy effluent

79. Which proteases is used for gluten hydrolysis in dough from high gluten wheat varieties?
(A) Papain
(B) Heat labile fungal protease
(C) Subtilism
(D) Neutral or alkaline proteases

80. Which enzymes is used particularly when wheat is added as an adjunct?
(A) Papain (B) Neutral proteases
(C) Cellulases (D) All of these

81. Ordinarily, saccharification during beer production is stopped by boiling the wort when about how much of the starch has been converted into fermentable sugars?
(A) 25% (B) 50%
(C) 75% (D) 100%

82. Methane is utilized by which bacterium?
(A) *Methaomonas methanica*
(B) *Pseudomonas methanica*
(C) *Methylococcus capsulatus*
(D) All of these

83. Which enzymes enhances fermentation rates?
(A) Papain
(B) Neutral proteases
(C) Cellulases
(D) All of these

84. In cheese ripening, which generates the strongest bitter flavor?
(A) Medium-sized peptides
(B) Small peptides
(C) Large peptides
(D) Both (B) and (C)

85. Which distilled alcoholic drink is produced by fermentation of a large number of substrates rich in carbohydrates?
(A) Rum (B) Whisky
(C) Vodka (D) All of these

86. Which mushrooms is commonly known as Chinese or straw mushrooms?
(A) *Coprinus sp.*
(B) *Pleurotus ostreatus*
(C) *Volvariella volvaceae*
(D) All of these

87. Among the given bacteria, which one utilizes methane as well as fixes atmospheric nitrogen?
(A) *Methanomonas methanica*
(B) *Methylococcus capsulatus*
(C) *Pseudomonas methanitrificans*
(D) *Pseudompnas methanica*

88. Which mushrooms can be cultivated during summer and rainy seasons?
(A) *Coprinus sp.*
(B) *Pleurotus ostreatus*
(C) *Volvariella volvaceae*
(D) All of these

89. Brix percentage of sauce is:
(A) 15 (B) 20
(C) 25 (D) 30

90. Brix percentage of ketchup is:
(A) 15-16 (B) 20-22
(C) 25-26 (D) 28-30

91. Which of the following is not a fruit preservation method?
(A) Drying (B) Canning
(C) Freezing (D) Grafting

92. Which of the following fruit is used for Jam preparation?
(A) Apple (B) Grape
(C) Papaya (D) All of these

93. Which of the following is a natural colouring agent of chilli?
(A) Oleoresin (B) Turmeric
(C) Resinocine (D) None of these

94. Yellow coloured fruits and vegetables are rich source of:
(A) Vitamin A (B) Vitamin B
(C) Vitamin C (D) Vitamin E

95. What is main ingredient of pungency in chilli?
(A) Alkaloid (B) Acids
(C) Capsicin (D) Sulphur

96. Which of the following method is used for temporary preservation of vegetables?
(A) Drying
(B) Use of low concentration of antiseptic
(C) Pasteurization (D) All of these

97. Which of the following method is used for permanent preservation of vegetables?
(A) Cold storage (B) Bottling
(C) Drying (D) All of these

98. Which of the following vegetables can be stored in salt solution?
(A) Potato (B) Cauliflower
(C) Carrot (D) All of these

99. The main gradient of garlic oil is:
(A) Alicin
(B) Alilo propel disulphide
(C) Alicin and Alicin propel disulphide
(D) Acetelin

100. Which of the garlic varieties are more pungent:
(A) White varieties (B) Green varieties
(C) Red varieties (D) All of these

101. Bitter guard is very useful for patients of:
(A) AIDS (B) Rickets
(C) Diabetes (D) Malaria

102. Garlic is useful medicine for patients having problems of:
(A) High blood pressure (B) Low blood pressure
(C) AIDS (D) All of these

103. Which is the main ingredients of special smell in onion?
(A) Sulphur dioxide (B) Methane
(C) Allyl propile disulphide
(D) All of these

104. Onion is very useful against:
(A) Hot air and cholera (B) Malaria
(C) Diabetes (D) All of these

105. Jelly is a:
(A) Transparent (B) Opaque
(C) Semi-transparent (D) All of these

106. Fruits are the source of mainly:
(A) Minerals (B) Vitamins
(C) Carbohydrates (D) All of these

107. Vitamin C is found in:
(A) Grape (B) Straw berry
(C) Papaya (D) All of these

108. Vitamin B_2 is found in:
(A) Paapaya (B) Banana
(C) Litchi (D) All of these

109. Vitamin B_1 is found in:
(A) Ber (B) Apple
(C) Cherry (D) All of these

110. Folic acid is found in:
(A) Grape (B) Banana
(C) Lemon (D) All of these

111. Fruits are not the source of:
(A) Vitamis (B) Fat
(C) Minerals (D) Carbohydrates

112. Cultar is a name of:
(A) Paclobutrazol (B) Methyl oxide
(C) Calcium carbonate (D) None of these

113. Sodium benzoate is added for the preservation of fruit juice at the rate of:
(A) 0.01-0.5% (B) 0.06-0.10%
(C) 1.10-1.5% (D) 1.6-2.0%

114. Papain is obtained from which tree :
(A) Mango (B) Papaya
(C) Banana (D) Guava

115. Which of the following preservative is used in the preparation of squash?
(A) Formaldehyde (B) NaCl
(C) Potassium-meta bi-sulphide
(D) Na benzoate

116. The bitterness in colocasia corms is due to the presence of:
(A) Calcium oxalate (B) Calcium carbonate
(C) Calcium chloride (D) Potassium oxalate

117. Mango is a rich source of:
(A) Vitamin A (B) Vitamin C
(C) Vitamin A and Vitamin C
(D) Vitamin E

118. For temporary fruit preservation the fruits should be pasteurized at:
(A) 160-180^{0}F for 20-30 minutes
(B) 120-130^{0}F for 20-30 minutes
(C) 160-180^{0}F for 5 minutes
(D) 140-150^{0}F for 15-20 minutes

119. For temporary preservation, fruits and vegetables can be stored at temperature:
(A) 10-20^{0}C (B) 15-20^{0}C
(C) 30-35^{0}C (D) 39-45^{0}C

120. Fruits and vegetables should be sterilized for permanent preservation at:
(A) 160^{0}F for 10-15 minutes
(B) 180^{0}F for 12-35 minutes
(C) 212^{0}F for 12-35 minutes
(D) 225^{0}F for 30 minutes

121. Which of the following hormone is used for blossom thinning?
(A) 250-500 ppm NAA (B) 1000 ppm NAA
(C) 1500 ppm NAA (D) All of these

122. Which of the following plant hormone is used for ripening of apples?
(A) 4 ppm 2,4,5-T (B) 10 ppm 2,4,5-T
(C) 15 ppm 2,4,5 –T (D) 20 ppm 2,4,5-T

123. Which of the following chemical is used to control the moisture in the storage plants?
(A) $CaCl_2$ (B) Na Cl
(C) $CaSO_4$ (D) All of these

124. Marmalade can be prepared generally from:
(A) Apple (B) Mango
(C) Citrus fruits (D) Banana

125. Jelly can be prepared from the fruits that has high amount of:
(A) Sugar (B) Water
(C) Pectin (D) Carbohydrte

126. Which of the following solution can be used for preservation of fruits and their products?
(A) 6-8% salt (B) 42-50% sugar
(C) 0.5-1.5% acetic acid (D) All of these

127. Vapour heat treatment is required for disinfestations of:
(A) Fresh fruits (B) Dry fruits
(C) Grains and pulses (D) Preserved fruits

128. Sugar content in grape is:
(A) 5% (B) 10%
(C) 20% (D) 25%

129. The sweetness in muskmelon shall be higher at:
(A) Low temperature and high rainfall
(B) High temperature and high rainfall
(C) High temperature and dry climate
(D) Low temperature and dry climate

130. If the potato is to be stored for 30 weeks, the optimum temperature and relative humidity of storage should be:
(A) 2-3^{0}C and 85-90 % (B) 4-5^{0}C and 85-90 %
(C) 2-3^{0}C and 75-80 % (D) 2-3^{0}C and 65-70 %

131. Which of the following antiseptic solutions can be used for the preservation of vegetables?
(A) 2% Acetic acid (B) 2-3 % salt
(C) 60-70 % sugar (D) All of these

132. The most of the vegetables can be stored safely at temperature is:
(A) < 10^{0}C (B) 15^{0}C
(C) 20^{0}C (D) 25^{0}C

133. Wax coating treatment enhances the self life of fruits because it blocks:
(A) Transpiration (B) Respiration
(C) Ripening (D) Photosynthesis

134. Optimum temperature for cold storage of garlic is:
(A) 0-2^{0}C (B) 2-5^{0}C
(C) 5-8^{0}C (D) 8-10^{0}C

135. Cabbage head can be stored long time at:
(A) 0^{0}C and 90-95% RH
(B) 5-6^{0}C and 90-95% RH
(C) 10^{0}C and 90-95% RH
(D) 10^{0}C and 70-75% RH

136. Cabbage is a rich source of:
(A) Vitamin B (B) Vitamin C
(C) Vitamin D (D) Vitamin E

137. Litchi is a rich source of:
(A) Fe and P (B) Vitamin C
(C) Vitamin D (D) All of these

138. Which of the following fruit is not suitable for jam making:
(A) Mango (B) Banana
(C) Aonla (D) Lemon

139. Guarana is:
(A) Fodder plant (B) Oil plant
(C) Beverage plant (D) Food plant

140. Cola, the common beverage, is obtained from:
(A) Seeds (B) Flowers
(C) Leaves (D) Stems

141. Which one is beverage plant:
(A) *Ilex paraguriensis* (B) *Caltha edulis*
(C) *Poullinia cupana* (D) All of these

142. Potential food plant is:
(A) Winged bean (Psophocarpus tetragonolobus)
(B) Jajoba (*Simmondsia chinensis*)
(C) Guayule (*Parthenium argentatum)*
(D) *Leucaena*

143. Cryopreservation is carried out at:
(A) -10 to -20 ^{0}C (B) -50 to -60 ^{0}C
(C) -196 ^{0}C (D) -100 ^{0} C

144. Kaffir beer is produced from which substrate?
(A) *Eleusine coracana* (B) Cassava
(C) *Sorghum caffrorum* (D) Both (A) and (C)

145. Which organisms used for SCP may contain significant amounts of nitrogen in an unavailable form?
(A) Bacteria (B) Algae
(C) Yeast (D) Filamentous fungi

146. Propionibacteria yield:
(A) Vitamin B12 (B) Lactic acid
(C) Citric acid (D) Vitamin B6

147. Rennel is used in:
(A) Bread making
(B) Fermentation
(C) Cheese making
(D) Synthesis of antibiotics

148. The term antibiotic was coined by:
(A) Flemming (B) Waksman
(C) Funk (D) Jenner

149. Which are common contaminants of milk?
(A) *Streptococci* (B) *Lactobacilli*
(C) *Micrococci* (D) All of these

150. Which enzymes is used to prevent 'chill-haze' of beer?
(A) Papain (B) Neutral proteases
(C) Cellulases (D) All of these

151. Germinating barley is employed for preparation of:
(A) Wine (B) Beer
(C) Lactic acid (D) Cheese

152. Pasteurization is heating at:
(A) 120^{0}C for 60 minutes
(B) 60^0-70^{0}C for thirty minutes
(C) 70^{0}C for 60 minutes
(D) 80^{0}C for 30 minutes

153. B2 is got from:
(A) *Acetobacter*
(B) *Bacilus megatherium*
(C) *Ashbya gossypii*
(D) *Pseudomonas*

154. Which fermented food is produced from milk?
(A) Cheese (B) Yoghurt
(C) Kaffir (D) All of these

155. Which organisms used for SCP production shows the highest protein content?
(A) Bacteria (B) Algae
(C) Yeast (D) Filamentous fungi

156. Which microorganism is employed for production of dextran?
(A) *Leuconostoc* (B) *Torulopsis*
(C) *Rhizopus* (D) *Lactobacillus*

157. Which leads to curd formation during cheese production?
(A) Cleavage of k-casein
(B) Cleavage of α–casein
(C) Lactic acid production
(D) Both (A) and (B)

158. A bioindicator is:
(A) Fermented tank
(B) Culture containing radioactive isotopes
(C) Culture for synthesis of new chemicals
(D) Hybridoma

159. Heat kills all organisms in the process called:
(A) Immunity (B) Pasteurization
(C) Sterilization (D) None of these

160. Yeast cell can be immobilized in:
(A) Silica gel (B) Calcium alginate
(C) Porcelain column (D) Encapsulation

161. High-value (having high nutritive value) microbial lipids that are under current consideration are those used for production of a food product similar to:
(A) Cocoa butter
(B) Cocoa polyunsaturated fatty acids
(C) Cocoa fat (D) All of these

162. Milk is changed into curd by:
(A) *Acetobacter aceti*
(B) *Bacillus megatherium*
(C) *Xanthomonas citri*
(D) None of these

163. Rennet is purified by:
(A) Waksman (B) Flemming
(C) Hansen (D) Dubrunfaut

164. Leaved bread is produced with the help of:
(A) Yeast (B) *Rhizopus*
(C) *Neurospora* (D) Lactobacillus

165. Heating milk and wines to 66^{0}C and sudden cooling is:
(A) Pasteurization (B) Sterilisation
(C) Fermentation (D) Preservation

166. Yoghurt is produced by which microorganisms?
(A) *Streptococcus thermophilus*
(B) *Lactobacillus bulgaricus*
(C) *Lactobacillus delbrueckii*
(D) Both (A) and (B)

167. To produce citric acid from carbohydrates or alkalines with is:
(A) *Aspergillus niger* (B) Penicillium
(C) Candida (D) Acetobacter

168. To oxidize ethanol first to acetic acid then further to CO_2 and H_2O by:
(A) *Aspergillus niger* (B) Penicillium
(C) Candida (D) Acetobacter

169. Agar-agar is obtained from:
(A) Moulds (B) Sea weeds
(C) Algae (D) Egg protein

170. The vinegar means:
(A) Bitter wine (B) Sweet wine
(C) Sour wine (D) None of these

171. Germinating barley seeds are employed in the preparation of:
(A) Cheese (B) Beer
(C) Wine (D) Lactic acid

172. Which of the following enzyme is used in brewing?
(A) ß amylase (B) Cellulose
(C) Lactase (D) Pectinase

173. Which organism is used to accumulate copper from factory waste?
(A) *Bacillus subtillis*
(B) *Pseudomonas putida*
(C) *Zaogloca ramigera*
(D) *Bacillus thuringensis*

174. Which yeast train is naturally used for ethanol production?
(A) *Sacchromyces cereviciae* (B) *S. diastaticus*
(C) Both (A) and (B) (D) None of these

175. Who of the following scientists showed that *Sacchromyces cerevisiae* caused fermentation products such as beer and butter milk?
(A) Louis Pasteur (B) Sach
(C) Alexander Fleming (D) Salman Waksman

176. Ales are the beer in which fermentation is carried out using:
(A) Top yeast (B) Bottom yeast
(C) Either of the above (D) Middle yeast

177. The process of making malt as soluble as possible by using enzymes adjuncts etc is known as:
(A) Brewing (B) Malting
(C) Mashing (D) Pitching

178. Lagers are the beer in which fermentation is carried out using:
(A) Top yeast (B) Bottom yeast
(C) Either of the above (D) Middle yeast

179. The fermentator can be sterilized by:
(A) Keeping it in the oven
(B) Steam under pressure
(C) Boiling
(D) All of these

180. Which of the following is used for agitation in a fermentor?
(A) Impeller (B) Baffles
(C) Sparger (D) All of these

181. Which of the following is used for aeration in a fermentor?
(A) Impeller (B) Baffles
(C) Sparger (D) All of these

182. Block beer is prepared from:
(A) Roasted germinated barley seeds
(B) Rice
(C) Wheat (D) Grapes

183. Maturation of beers is carried out at:
(A) 14ºC (B) 10ºC
(C) 6ºC (D) 2ºC

184. Sonti is:
(A) Rice beer or wine of India
(B) Wheat beer or wine
(C) Barley beer or wine
(D) Rice beer or wine of Japan

185. Sake is:
(A) Rice beer or wine of India
(B) Wheat beer or wine
(C) Barley beer or wine
(D) Rice beer or wine of Japan

186. The crusted grapes are known as:
(A) Malt (B) Must
(C) Wort (D) Sonti

187. The yeast generated during the fermentation of beer is generally separated by:
(A) Centrifugation (B) Filtration
(C) Cell disruption (D) All of these

188. After storage and recarbonation, left over yeast is separated by:
(A) Centrifugation (B) Filtration
(C) Cell disruption
(D) Need to be separated

189. Wine yeast:
(A) Is sensitive to sulphur dioxide
(B) Is resistant to sulphur dioxide
(C) Shows enhanced enzyme production in presence of sulphur dioxide
(D) None of these

190. Punching down is done:
(A) For the same reasons as pumping over
(B) To extract tannin
(C) Both (A) and (B)
(D) To keep the dry cap

191. Which is not the purpose of punching down:
(A) To oxidize the wine
(B) To extract tannin
(C) To reduce possible microbial spoilage
(D) To keep the cap moist

192. Cidar vinegar is produced from:
(A) Fruit juices (B) Malted grain
(C) Ethanol (D) Ale

193. Alegar vinegar is produced from:
(A) Fruit juices (B) Malted grain
(C) Ethanol (D) Ale

194. Malt vinegar is produced from:
(A) Fruit juices (B) Malted grain
(C) Ethanol (D) Ale

195. Sprit vinegar is produced from:
(A) Fruit juices (B) Malted grain
(C) Ethanol (D) Ale

196. Which substrate is used in the fermentation of citric acid?
(A) Beet molasses (B) Sucrose
(C) Starch hydrolysate (D) All of these

197. Shery is a type of :
(A) Wine (B) Beer
(C) Brandy (D) None of these

198. Vinegar production consists of:
(A) Aerobic fermentation
(B) Anaerobic fermentation
(C) Aerobic fermentation followed by anaerobic fermentation
(D) Anaerobic fermentation followed by aerobic fermentation

199. Ethanol is oxidized to acetic acid aerobically using:
(A) Acetobacter (B) Gluconobacter
(C) Both (A) and (B) (D) lactobacillus

200. Fining a wine is defined as:
(A) Adding one substance to remove another
(B) Removing small particles
(C) Adding acid (D) Removing tannin

201. In sparkling wine production, the cuvee is:
(A) The wire cage that holds the cork in place
(B) The sugar addition just before corking
(C) The sugar/yeast addition to the base wine
(D) The base wine

202. Carbonic maceration wines:
(A) Are made from crushed grapes
(B) Are very fruity white wines
(C) Are produced in California
(D) Are low tannin red wines

203. Chaptalization is:
(A) Adding sucrose to must
(B) Adding grape juice to must
(C) Adding sugar to wine before bottling
(D) Adding grape juice to wine before bottling

204. Malolactic fermentation:
(A) Produces lactic acid
(B) Reduces total acidity
(C) Produces diacetyl (D) All of these

205. Wine is obtained after alcoholic fermentation of:
(A) Rice (B) Wheat
(C) Grapes (D) Barley

206. Which of the following is true for the ale beer?
(A) Top yeast of *S. cerevisiae* is used
(B) Fermentation is quick and completes in 5-7 days
(C) Quantity of hops used is more than in lager beer
(D) All of these

207. Wort is:
(A) An aqueous extract of malt
(B) Malted barley
(C) Coagulated protein obtained during boiling
(D) None of these

208. Grist is:
(A) Coarse powder obtained after milling of malted barley
(B) An aqueous extract of malt
(C) Malted barley
(D) Coagulated protein obtained during boiling

209. As per fermentation for the ethanol production is over, the cells are separated to get the biomass of yeast cells that is used:
(A) For next fermentation
(B) As single cell protein fro animal feed
(C) As manure (D) All of these

210. How is air supply to the fermentor sterilized:
(A) By heat filtration
(B) Filtration through fibrous material
(C) Filtration through granular material
(D) All of these

211. *Candida utilis* is used for the fermentation of waste sulphate liquor since it ferments:
(A) Hexoses only (B) Pentoses only
(C) Both (A) and (B) (D) None of these

212. The zygomonas has a merit over yeast because:
(A) It has more osmotic tolerance to higher sugar concentration
(B) It has high tolerance to ethanol
(C) It has more specific growth rate
(D) All of these

213. To maintain aseptic conditions during fermentation which of the following is needed?
(A) Sterilization of fermentor
(B) Sterilization of air supply
(C) Aeration and agitation
(D) All of these

214. If starch containing substrates are used for ethanol production, yeast strain cannot be used directly because:
(A) It does not contain amylases to hydrolyze starch
(B) Starch is not a suitable substrate for the production of ethanol
(C) It is converted to pentose sugar
(D) None of these

215. *S. cerevisiae* start producing ethanol with in:
(A) 10 h (B) 12 h
(C) 15 h (D) 24 h

216. Malting process allows malt amylase and proteinases to degrade starch and protein to:
(A) Glucose and peptone as well as peptides
(B) Glucose and amino acids
(C) Maltose and amino acids
(D) Maltose and peptone as well as paptides

217. Grist is:
(A) Coarse powder obtained after milling of malted barley
(B) An aqueous extract of malt
(C) Malted barley
(D) Coagulated protein obtained during boiling

218. Wort is boiled with the hops:
(A) To help in protein coagulation
(B) To provide bitter taste and flavor to the beer
(C) To help in the coagulation of tannins
(D) All of these

219. The organism used for the preparation of sake is:
(A) *Rhizopus sonti*
(B) *Aspergillus oryzea*
(C) *Lactobacillus vermiformis*
(D) *Sacchromyces pyriformiss*

220. The organism used for the preparation of sonti is:
(A) *Rhizopus sonti*
(B) *Aspergillus oryzea*
(C) *Lactobacillus vermiformis*
(D) *Sacchromyces pyriformis*

221. Which type of press produces the most high quality juice or wine?
(A) Membrane press
(B) Continuous press
(C) Vertical basket press
(D) Horizontal basket press

222. Which of the following organism is used for the fermentation of grapes?
(A) *Rhizopus sonti*
(B) *Aspergillus oryzea*
(C) *Lactobacillus vermiformis*
(D) *Sacchromyces cerevisiae*

223. If the malolactic fermentation is not carried out:
(A) The wine will have an odd odor
(B) Malic acid may precipitate, leaving small crystals
(C) The wine may be microbially unstable
(D) The wine will be too low in acid

224. In a hot climate, grapes lose acidity because:
(A) Enough tartaric acid is not made
(B) Enough malic acid is not made
(C) Of break down of malic acid
(D) Of breakdown of tartaric acid

225. If a wine maker wanted to make a deeply coloured red wine, which of the following should not be done?
(A) Bleed off juice prior to fermenting on skins
(B) Ferment at a high temperature
(C) Use whole clusters
(D) Punch down at least once a day

226. How long does it take to complete fermentation process in case of red wine?
(A) 1 – 3 days (B) 3 – 5 days
(C) 5 – 7 days (D) 7 – 11 days

227. How long does it take to complete fermentation process in case of white wine?
(A) 1 – 3 days (B) 3 – 5 days
(C) 5 – 7 days (D) 7 – 11 days

228. The crystal like material some times found in wine bottles is:
(A) Residual diatomaceous earth
(B) Tannin
(C) Prevented by cold stabilizing
(D) Caused by excessive malic acid

229. Which is not correct about blush wines?
(A) They are stored in barrels
(B) They are bottled within a year
(C) They are usually slightly sweet
(D) They are fermented in stainless steel

230. Vinegar is typically produced in fed batch reactors because:

(A) A fed batch reactor can be used to maintain low acetic acid concentrations

(B) A fed batch can be used to maintain low ethanol concentrations

(C) Acetic acid bacteria tend to ferment at high ethanol concentrations

(D) All of these

231. Which of the following nitrogen sources is added in the fermentation medium of citric acid production?

(A) Ammoniun salt (B) Urea

(C) Both (A) and (B) (D) Pepton

232. In koji process for the fermentation to produce citric acid:

(A) *A. niger* is used with solid substrate as sweet potato

(B) *A. niger* floats on the surface of solution

(C) Fungal mycelium grows through out the solution in a deep tank

(D) Both (A) and (B)

233. In liquid surface culture process for the fermentation to produce citric acid:

(A) *A. niger* is used with solid substrate as sweet potato

(B) *A. niger* floats on the surface of solution

(C) Fungal mycelium grows through out the solution in a deep tank

(D) Both (A) and (B)

234. In submerged process for the fermentation to produce citric acid:

(A) *A. niger* is used with solid substrate as sweet potato

(B) *A. niger* floats on the surface of solution

(C) Fungal mycelium grows through out the solution in a deep tank

(D) Both (A) and (B)

235. What temperature is maintained during anaerobic fermentation of red wine?

(A) 20-24^{0}C (B) 24-27^{0}C

(C) 27-31^{0}C (D) 31-34^{0}C

236. What temperature is maintained during anaerobic fermentation of white wine?

(A) 10-21^{0}C (B) 24-27^{0}C

(C) 27-31^{0}C (D) 31-34^{0}C

ANSWERSHEET

19. FOOD AND BEVERAGE BIOTECHNOLOGY

1. (B)	2. (C)	3. (A)	4. (C)	5. (B)	6. (D)	7. (B)	8. (A)	9. (C)	10. (C)
11. (D)	12. (C)	13. (A)	14. (B)	15. (C)	16. (D)	17. (A)	18. (D)	19. (B)	20. (A)
21. (B)	22. (C)	23. (D)	24. (C)	25. (C)	26. (B)	27. (D)	28. (B)	29. (D)	30. (C)
31. (D)	32. (B)	33. (B)	34. (C)	35. (A)	36. (D)	37. (D)	38. (A)	39. (A)	40. (C)
41. (B)	42. (A)	43. (C)	44. (D)	45. (A)	46. (C)	47. (B)	48. (B)	49. (C)	50. (A)
51. (A)	52. (D)	53. (B)	54. (B)	55. (B)	56. (C)	57. (D)	58. (C)	59. (C)	60. (D)
61. (A)	62. (A)	63. (D)	64. (C)	65. (C)	66. (B)	67. (A)	68. (C)	69. (A)	70. (A)
71. (B)	72. (B)	73. (A)	74. (B)	75. (D)	76. (C)	77. (D)	78. (C)	79. (B)	80. (C)
81. (C)	82. (D)	83. (B)	84. (A)	85. (D)	86. (C)	87. (C)	88. (C)	89. (C)	90. (D)
91. (D)	92. (D)	93. (A)	94. (A)	95. (C)	96. (D)	97. (D)	98. (D)	99. (C)	100. (C)
101. (C)	102. (A)	103. (C)	104. (A)	105. (A)	106. (D)	107. (D)	108. (D)	109. (D)	110. (D)
111. (B)	112. (A)	113. (B)	114. (B)	115. (C)	116. (A)	117. (C)	118. (A)	119. (D)	120. (C)
121. (A)	122. (A)	123. (C)	124. (D)	125. (C)	126. (D)	127. (A)	128. (D)	129. (C)	130. (A)
131. (D)	132. (A)	133. (A)	134. (A)	135. (A)	136. (B)	137. (D)	138. (D)	139. (C)	140. (A)
141. (D)	142. (A)	143. (C)	144. (D)	145. (D)	146. (A)	147. (C)	148. (B)	149. (D)	150. (A)
151. (B)	152. (B)	153. (C)	154. (D)	155. (A)	156. (A)	157. (D)	158. (A)	159. (C)	160. (B)
161. (D)	162. (D)	163. (C)	164. (A)	165. (A)	166. (D)	167. (C)	168. (D)	169. (C)	170. (C)
171. (B)	172. (A)	173. (C)	174. (B)	175. (A)	176. (A)	177. (C)	178. (B)	179. (B)	180. (A)

181. (C) **182. (A)** **183. (D)** **184. (A)** **185. (D)** **186. (B)** **187. (A)** **188. (B)** **189. (B)** **190. (D)**

191. (A) **192. (A)** **193. (D)** **194. (B)** **195. (C)** **196. (D)** **197. (A)** **198. (D)** **199. (C)** **200. (A)**

201. (D) **202. (D)** **203. (A)** **204. (D)** **205. (C)** **206. (D)** **207. (A)** **208. (A)** **209. (B)** **210. (D)**

211. (C) **212. (D)** **213. (D)** **214. (A)** **215. (B)** **216. (D)** **217. (A)** **218. (B)** **219. (B)** **220. (A)**

221. (A) **222. (D)** **223. (C)** **224. (C)** **225. (C)** **226. (B)** **227. (D)** **228. (C)** **229. (A)** **230. (B)**

231. (C) **232(A)** **233. (B)** **234. (C)** **235. (B)** **236. (A)**

Chapter - 20

INTELLECTUAL PROPERTY RIGHT, PATENTING AND BIOSAFETY BIOTECHNOLOGY

1. A form of IPR protection, they cover such qualities, reputation or other characteristics of a product that can be essentially attributable to the geographical origin of the product, and that could be used to recognize that a product had originated from a particular terriority is called:
(A) Geographical Indications
(B) Trademark
(C) Copyright
(D) None of these

2. It is an idea, a design, an invention, a manuscript etc., which can ultimately give rise to useful product/application is called:
(A) Intellectual Property
(B) Intellectual Property Right
(C) Both (A) and (B)
(D) None of these

3. The right of an inventor to derive economic benefits from his invention, i.e., intellectual property, and to exclude others from doing so is called:
(A) Intellectual Property
(B) Intellectual Property Right
(C) Both (A) and (B)
(D) None of these

4. The European Patent Convention (EPC) excludes from patent protection such inventions that are contrary to the public order of morality is under the act of:
(A) Article 53 (A) (B) Article 53 (b)
(C) Article 54 (A) (D) Article 54 (b)

5. The European Patent Convention excludes from patent protection (1) plant and animal varieties and (2) essential biological processes for the production of plants and animals under the act of:
(A) Article 53 (A) (B) Article 53 (b)
(C) Article 54 (A) (D) Article 54 (b)

6. A collection of right that are granted to an individual or groups of individuals for any work or creation that involved intellectual inputs is called:
(A) Intellectual Property
(B) Intellectual Property Right
(C) Both (A) and (B)
(D) None of these

7. Intellectual Property is categories into:
(A) Industrial property
(B) Artistic and Literary property
(C) Both (A) and (B)
(D) None of these

8. The protection to any kind of confidential information from the industry that has a commercial value or industrial significance is called:
(A) Trade secrete
(B) Industrial design
(C) Geographical indications
(D) Trade mark

9. In plant-related inventions, the parent lines employed to produce a hybrid plant can be kept as a:
(A) Trade secrete
(B) Industrial design
(C) Geographical indications
(D) Trade mark

10. Industrial property are:
(A) Trade secrete
(B) Industrial design
(C) Geographical indications
(D) Trade mark
(E) All of these

11. Artistic and Literary property are:
(A) Copy right
(B) Industrial design
(C) Geographical indications
(D) Trade mark

12. The deals with protection of the expression of an idea; however, it does not protect the idea itself is called:
(A) Copy right
(B) Industrial design
(C) Geographical indications
(D) Trade mark

13. The exclusive rights that are granted to the inventor to prevent third parties from making, using, selling, importing and offering for sale the invention, unless an authorization or license is obtained from the patentee is called:
(A) Patents (B) Industrial design
(C) Copy right (D) Trade mark

14. In biotechnology, the significance of copyright would be:
(A) Possibility of copyrighting sequences
(B) Methods or manuals of exponential procedures
(C) Both (A) and (B)
(D) None of these

15. The intellectual property that have protected by an alternative type of protection termed as:
(A) Sui Generic system
(B) Trademark
(C) Copyright
(D) None of these

16. Sui Generic protection are:
(A) Plant variety protection
(B) Integrated computer circuits
(C) Data base protection
(D) All of these

17. The TRIPs Agreement came is:
(A) 1994 (B) 1995
(C) 1996 (D) 1997

18. The TRIPs Agreement came in to force in January 1, 1995 with immediate effect but with certain exceptions for which provisions were made which were through to come into force by:
(A) 1998 (B) 1999
(C) 2000 (D) 2001

19. The patent Act, 1970 was amended first time in the year of:
(A) 1999 (B) 2002
(C) 2005 (D) 2006

20. The patent Act, 1970 was amended second time in the year of:
(A) 1999 (B) 2002
(C) 2005 (D) 2006

21. The patent Act, 1970 was amended third time in the year of:
(A) 1999 (B) 2002
(C) 2005 (D) 2006

22. The term of protection shall not end before the expiration of a period of twenty years counted from the date of filing in:
(A) Article 33 (B) Article 34
(C) Article 35 (D) Article 36

23. The time period accorded for protection of the invention has been set uniformly for all categories of inventions at:
(A) 10 years (B) 15 years
(C) 20 years (D) 25 years

24. An 'invention' as 'a new product or process involving an inventive step and capable of industrial application' under:
(A) Section 2(1) (j) (B) Section 2(1) (ja)
(C) Section 3 (j) (D) Section 3 (ja)

25. In 'inventive step' as 'inventive step means a feature of an invention that involves technical advance as compared to the existing knowledge of having economic significance or both and that makes the invention not obvious to a person skilled in the art' under:
(A) Section 2(1) (j) (B) Section 2(1) (ja)
(C) Section 3 (j) (D) Section 3 (ja)

26. Essentially biological process falls under the exclusion of patentability under the section of the Indian patent act :
(A) Section 2(1) (j) (B) Section 2(1) (ja)
(C) Section 3 (j) (D) Section 3 (ja)

27. Patentability requirements as per the Indian Patents Act are:
(A) It should be new
(B) It should involve an inventive step (nonobviousness of the invention)
(C) It should be capable of industrial application
(D) All of these

28. Under Article 4 related to patents in TRIPs and ZEU Directives the following shall not be patentable:
(A) Plant and animal varieties
(B) Essentially biological processes of the production of plants and animals
(C) Both (A) and (B)
(D) None of these

29. Essentially biological process falls under the exclusion of patentability under the section of the Indian Patent Act:
(A) Section 2(1) (j) (B) Section 2(1) (ja)
(C) Section 3 (j) (D) Section 3 (ja)

30. The EU Directive Article 4.1 (b) appears to strongly exclude plant and animal varieties, but Article ---- makes clear that – transgenic plants and animals are patentable, as they are produced by 'microbiological' or other technical process:
(A) Article 33 (B) Article 4.3
(C) Article 35 (D) Article 36

31. The process by which used in making transgenic plants and animals are biological process, and so transgenic plants and animals should not be patentable:
(A) Transformation (B) Transfection
(C) Both (A) and (B) (D) None of these

32. Whether or not a (non microbiological) process is to be considered as 'essentially biological' within the meaning of Article -------- EPC has to be judged on the basis of the essence of the invention, taking into account the totality of human intervention and its impact on the results to be achieved:
(A) Article 33 (B) Article 53(b)
(C) Article 35 (D) Article 36

33. Gene splicing technique is one of the important techniques of:
(A) R-DNA technology (B) C-DNA technology
(C) m-DNA technology (D) D-DNA technology

34. Patentable processes involving R-DNA technology are:
(A) Method of transferring cells or complete biological entity such as microbes, plants or animals
(B) Method of cloning and expression of desired gene in the same or different cell
(C) Method of optimizing gene expressions to produce useful therapeutic proteins
(D) Method of producing a transgenic plant or animal
(E) Method of serving and detection of specific DNA sequences
(F) Method of designing probes, primers, recombinant vectors
(G) Method of modifying genomic sequences
(H) All of these

35. Patentable products involving R-DNA technology:
(A) Different cloning vehicle (vectors) such as plasmids, cosmids, phagemids, etc
(B) Probes, primers and oligonucleotides
(C) Vaccines and other useful proteins
(D) Transgenic/genetically modified organisms
(E) All of these

36. For patenting of DNA/RNA sequences, ESTs, SNPs, the following steps should be undertaken:
(A) Identify novel genetic sequences
(B) Specify the resulting product from the sequence
(C) Specify how the product functions in nature (use)
(D) Enable one skilled in the field to use the sequence for its stated purpose
(E) All of these

37. The genes are not patentable when they are in the:
(A) In-situ (B) Ex-situ
(C) Both (A) and (B) (D) None of these

38. The major differences between the 1995 guidelines and the 2001 guidelines was the requirement that the claimed utility has to be:
(A) Substantial (B) Specific
(C) Credible (D) All of these

39. The 2001 utility Examination Guidelines follow the 1999 Revised Interim Utility Guidelines in adopting the structure of:
(A) Specific, substantial and credible utility test
(B) Well established utility test
(C) Both (A) and (B) (D) None of these

40. Under both the 1999 and 2001 guidelines, a utility is 'specific' when it particularly concerns the:
(A) Subject matter claimed
(B) Real world use
(C) DUS test (D) Novelty

41. A European patent (EP 1417338) was granted in 2004 for genes and SNPs associated with eating disorders like:
(A) Anorexia nervosa (B) Bulimia
(C) Both (A) and (B) (D) None of these

42. Under both 1999 and 2001 guidelines establish that a 'substantial utility' is one that defines a:
(A) Subject matter claimed (B) Real world use
(C) DUS test (D) Novelty

43. A new variety shall be registered under PPVFR Act 2001 if it confirms to the criteria of novelty and fulfils the:
(A) Subject matter claimed (B) Real world use
(C) DUS test (D) Novelty

44. The variety should not have been sold (offered for sale) or marketed earlier than one year for the country and four years for other countries where protection I desired is:
(A) Subject matter claimed (B) Real world use
(C) DUS test (D) Novelty

45. A US patent (US5,972,609) has been awarded in 1999 for a utrophin gene promoter that they may be used to control of:
(A) Transcription of heterologous sequences
(B) PPVFR Act 2001
(C) Distinctiveness (D) Uniformity

46. Indian chose to opt for a Sui Generis System for protection of plant and plant varieties by enacting the:
(A) Transcription of heterologous sequences
(B) PPVFR Act 2001
(C) Distinctiveness (D) Uniformity

47. The variety should be clearly distinguishable from any other variety of common knowledge is:
(A) Transcription of heterologous sequences
(B) PPVFR Act 2001
(C) Distinctiveness (D) Uniformity

48. The variety is uniform/homogeneous if, subject to the variation that may be expected from the particular features of it propagation, it is sufficiently uniform in it relevant characteristics is:
(A) Transcription of heterologous sequences
(B) PPVFR Act 2001
(C) Distinctiveness (D) Uniformity

49. The variety is stable if its relevant (essential) characteristics remain unchanged after repeated propagation is:
(A) Stability (B) Microorganisms
(C) Protection (D) None of these

50. TRIPs allows members to exclude from patentability, plants and animals other than:
(A) Stability (B) Microorganisms
(C) Protection (D) None of these

51. Plants found in the wild and other land races cannot be protected under an IPR system as they do not meet the requirements for:
(A) Stability (B) Microorganisms
(C) Protection (D) None of these

52. All types of genes can be patented as long as an inventive step has been involved to:
(A) Identify (B) Isolate
(C) Modify and/or transfer (D) All of these

53. There has been a decreasing intellectual property activity in the area of biotechnology since:
(A) 2000 (B) 2001
(C) 2002 (D) 2003

54. Which Indian organisation is more active in the area of patenting:
(A) ICAR (B) CSIR
(C) ICMR (D) BARC

55. Some of the Indian companies active in the area of intellectual property right are:
(A) Ranbaxy (B) Dabur
(C) Ayur (D) All of these

56. The institution of CSIR active in the area of intellectual property right are:
(A) NBRI (B) IICT
(C) CIMAP (D) All of these

57. Patenting of stem cells may involved:
(A) Product patents (B) Process patents
(C) Both (A) and (B) (D) All of these

58. There are different forms of protection for plant-related inventions are:
(A) Utility patents
(B) Plant Patent Act
(C) Plant Variety Protection Act
(D) All of these

59. The utility patents and the plant protection patents are granted by:
(A) USPTO (B) PVTO
(C) Both (A) and (B) (D) None of these

60. The plant variety protection act is granted by:
(A) Plant Variety Protection Officer
(B) Plant Breeder
(C) Plant Protection Officer
(D) All of these

61. The patenting can also be done in:
(A) Transformations vectors
(B) Selectable marker genes
(C) Reporter genes
(D) All of these

62. In India, animals in whole and any part thereof are not patentable under section of Patent Act:
(A) Section 2(1) (j) (B) Section 2(1) (ja)
(C) Section 3 (j) (D) Section 3 (ja)

63. Which amendment of the Patent Act also also makes it mandatory to deposit the biological materials mentioned in the specification with IDA is:
(A) 2000 amendment
(B) 2001 amendment
(C) 2002 amendment
(D) 2003 amendment

64. Plant varieties may be protected in India under:
(A) PVPFRA (B) PVRA
(C) Both (A) and (B) (D) None of these

65. Under Section 3(j) of Indian Patents Act, 1970 plants are not considered patentable subject matter under section 3(h) of the Act are:
(A) Method of agriculture
(B) Method of horticulture
(C) Both (A) and (B)
(D) None of these

66. The first plant patent was issued in 1931 to Henry Rosenberg of New Brunswick NJ for:
(A) Climbing rose (B) Trailing rose
(C) Either (A) or (B) (D) None of these

67. The plant patent for a rose plant named -------- granted to Ezio Bree in 2003 is:
(A) Breapum (B) Zembla
(C) Both (A) and (B) (D) None of these

68. The patent for chrysanthemum plant named is:
(A) Breapum (B) Zembla
(C) Both (A) and (B) (D) None of these

69. A patent for chrysanthemum plant named 'Zembla' was granted mainly for:
(A) Decorative inflorescence
(B) Excellent Past harvest longevity
(C) Both of these
(D) None of these

70. The invention for climbing roses bearing true:
(A) Blooming characters
(B) Vegetative characters
(C) Growth characters
(D) Transplanting characters

71. In biotechnological components that are also patentable:
(A) Transformation vectors
(B) Selectable marker genes
(C) Reporter genes
(D) Promoter and enhancers
(E) All of these

72. The act covers varieties of asexually propagated crops, e.g., ornamental and fruit trees by:
(A) The US Plant Patent Act (1930)
(B) The US Plant Variety Protection Act (1970)
(C) The Utility Patents Act (1985) of USA
(D) All of these

73. The US version of the plant breeders right system followed by European Union and several other countries by:
(A) The US Plant Patent Act (1030)
(B)The US Plant Variety Protection Act (1970)
(C) The Utility Patents Act (1985) of USA
(D) All of these

74. The act was originally meant to cover man-made industrial inventions and processes. Non-obviousness is the main criterion of utility patent; patent for plant varieties are now being granted under the provisions of this act is:
(A) The US Plant Patent Act (1030)
(B) The US Plant Variety Protection Act (1970)
(C) The Utility Patents Act (1985) of USA
(D) All of these

75. The first patent of a life form was issued on:
(A) March 31, 1980 (B) March 31, 1981
(C) March 31, 1982 (D) March 31, 1983

76. Trade secret relates to:
(A) Certain materials, e.g., bacterial strains
(B) Processes (C) Formulae
(D) Any one of these

77. Which is not correct about patent?
(A) It is the right of an individual or a group of individuals to gain profit from a particular invention or unique manufacturing process
(B) It is granted by the Government of a country to the applicant for an unlimited period of time
(C) When a patent is granted, the inventor becomes the owner of the patent like any other form of property
(D) Patents are granted for an invention (including a product), innovation/improvement in an invention, process/product of invention and concept

78. In the area of biotechnology, which materials can be kept as trade secret?
(A) Strains of microorganism
(B) Cell lines
(C) Production processes
(D) All of these

79. Trade secret can not be applied for:
(A) Plant varieties (B) Equipment design
(C) Books (D) All of these

80. Which peoples are considered under farmer's rights for sharing the profit earned from the development of high yielding varieties evolved from the use of raw materials/plant genetic resources?
(A) Tribal people
(B) Rural communities
(C) Traditional farming families
(D) All of these

81. Indian has created a National Biodiversity Authority with headquarters at:
(A) New Delhi (B) Kolkata
(C) Chennai (D) Hyderabad

82. Which is not correct about breeder's exemption?
(A) The use of a protected variety for the development of new varieties is exempted from protecton
(B) The PBR for a new variety evolved from the initial variety will be of the breeder who developed the new variety and of the holder of PBR title of the initial variety as well
(C) Under the UPOV 1978 Act, all new varieties evolved using a protected variety were exempted from protection under this provision
(D) The UPOV 1991 Act has limited breeders exemption in that 'essentially-derived varieties' are covered by the PBR granted to the initially variety

83. An otherwise excellent transgenic variety of which crops developed during 1990a could not be released due to IPR problems?
(A) Potato (B) Tomato
(C) Tobacco (D) All of these

84. The copyright of a book may be held by:
(A) Author(s) (B) Editor(s)
(C) Publisher (D) Any of these

85. In USA, which systems are available for protection of IPRs related to plants?
(A) The Plant Patents Act (1930)
(B) The Plant Variety Protection Act (1970)
(C) The Utility Patents Act (1985)
(D) All of these

86. In USA, which systems are available for protection of IPRs is considered to be the most powerful and the most expansive in scope of coverage?
(A) The Plant Patents Act (1930)
(B) The Plant Variety Protection Act (1970)
(C) The Utility Patents Act (1985)
(D) All of these

87. The copyright suffers from:
(A) It provides protection for a specific period and only from reproduction as such of the copyright material
(B) It does not prevent another person from using either the idea or he information contained in a copyright material
(C) In case of DNA sequence, one may get around this protection by designing an alternative sequence to encode the same protein
(D) All of these

88. PPVFR Act (2001) is comparable to UPOV Act (1991) with respect to:
(A) Requirement of novelty, distinctiveness uniformity and stability for registration of a variety
(B) Protection extended to commercial use of all the material of the protected variety
(C) Essentially-derived varieties being subjected to PBR protection granted to the concerned initial varieties
(D) Both (A) and (C)

89. Which countries has 30 years protection for inbred lines of maize and for clovers and a few grasses?
(A) France (B) Germany
(C) USA (D) UK

90. Which is the main considerations for the development of PBR systems?
(A) It allows breeders to benefit from the varieties developed by them
(B) It encourages plant breeding activities
(C) Private sector is encouraged to invest in plant breeding and seed industry
(D) All of these

91. Government of India has now established the PPVFR Authority (PPVFRA) with headquarters at:
(A) Chennai (B) Kolkata
(C) New Delhi (D) Hyderabad

92. Which biotechnological inventions can be patented?
(A) Microorganisms, cell lines, plant/animal lies obtained through biotechnological approaches
(B) DNA sequences/genes and the proteins encoded, if any, by them
(C) Biotechnological process/technologies for modification of the properties of various organisms
(D) All of these

93. Which features is not concerned with the PPVFR Act (2001)?
(A) The Central Government is to constitute a National Gene Fund
(B) Citizens of convention countries will have the same rights as citizens of India under the Act
(C) Application for registration of a variety may be made in India within 2 years from the date of application for registration of the same plant variety made in the convention country
(D) The rights of PBR holder 'shall not be deemed infringed by a farmer who was not aware of the existence of such right'

94. Which biotechnology related maters can be patented?
(A) Method/process of production
(B) Products, etc.
(C) Applications of the various processes/products
(D) Any of these

95. In case of PPVFR Act (2001), the genen fund shall be used fro:
(A) Paying compensation to communities for their contributions to the development of a variety
(B) Conservation of genetic resources
(C) Sustainable use of genetic resources
(D) All of these

96. Methods for which can not be patented?
(A) Treatment of disease
(B) Diagnosis of disease
(C) Prevention of disease
(D) All of these

97. Which life forms can not be patented in India under the provisions of Indian Patents Act (1970)
(A) Microorganisms (B) Animals
(C) Plants (D) All of these

98. Patent activity is the lowest in:
(A) USA (B) Japan
(C) Europe (D) India

99. Which technologies and/or raw materials were used to develop Endless Summer tomato by DNA Plant Technology?
(A) Selectable marker *npt*II
(B) CaMV 35S promoter & the ACC synthase gene
(C) Antisense RNA technology and *Agrobacterium tumerfaciens* mediated genetic transformation
(D) All of these

100. DNA Plant Technology faced difficulties during release of Endless Summer due to the use of:
(A) Selectable marker *npt*II
(B) CaMV 35S peromoter & the ACC synthase gene
(C) Antisense RNA technology and *Agrobacterium tumerfaciens* mediated genetic transformation
(D) All of these

101. The PPVFR Authority (PPVFRA) began registration of plant varieties in which years?
(A) 2005 (B) 2006
(C) 2007 (D) 2008

102. Which is an example of geographical indications?
(A) Champagne wine (B) Scotch whiskey
(C) California wine (D) All of these

103. Geographical indications includes:
(A) Natural products (B) Agricultural goods
(C) Goods of handicraft (D) All of these

104. Which is not correct about geographical indications (GIs):
(A) Once enforced by legislation, GIs exclude others from using a GI as trademark
(B) The chief requirement for GI protection is that a given quality, reputation or some other characteristics of the product be essentially attributable to the locality of its origin
(C) The GIs do not cover manufactured products and even food products
(D) Indian has enacted the Geographical Indication Act (1999) to claim GI for a variety of goods, including 'Basmati rice'

105. Monitoring and tackling the IPR aspects of inventions has:
(A) Demand time (B) Enhances cost
(C) May act as a distinctive for R & D efforts
(D) All of these

106. PPVFR Act (2001) allows registration of:
(A) Farmer's varieties (B) Extant varieties
(C) New Varieties (D) All of these

107. In the first instance, PPVFR began registration of varieties in how many of the crops?
(A) 6 (B) 12
(C) 2 (D) 20

108. Physical containment relates to:
(A) Containment equipment
(B) Laboratory design
(C) Special procedures (D) All of these

109. Which is considered as primary physical containment?
(A) Laboratory design
(B) Containment equipment
(C) Special procedures
(D) Both (A) and (B)

110. Which is considered as secondary physical containment?
(A) Laboratory design
(B) Containment equipment
(C) Special procedures
(D) Both (A) and (B)

111. It is feared that transgenic plants, may lead to:
(A) Transfer of their transgenic feature to weeds
(B) May themselves become weeds
(C) Degrade the environment
(D) All of these

112. In the early 1970s which were considered as the possible potential health hazards of cloning recombinant DNA molecules?
(A) Molecular biologists may not be well-versed with the laboratory practices needed to handle serious human and animal pathogens
(B) 'Hybrid organisms' could be created with biological activities of an unpredictable nature
(C) 'Hybrid organisms' may escape from the laboratory with unpredictable consequences
(D) All of these

113. A transgenic variety of which crops had to be withdrawn from cultivation following due to allergy?
(A) Maize (B) Soybean
(C) Tomato (D) *Brassica napus*

114. Which is not correct about NIH guidelines?
(A) The first NIH guidelines were prepared in 1975
(B) NIH guidelines were more liberal than the recommendations of the Asilomar Conference
(C) The guidelines were revised after two years
(D) In 1977, NIH prepared an Environmental Impact Statement

115. Which terms is not used as synonyms of planned introduction?
(A) Field testing
(B) Deliberate release
(C) Contained use
(D) Deliberate introduction

116. Which is not correct about NIH guidelines?
(A) A major revision of the guidelines was effected in 1982
(B) In the revision, the containment levels were made more stringent
(C) Experiments that were previously prohibited, were changed to category requiring review and approval by NIH
(D) In USA, the NIH guidelines are followed by all federal agencies that fund research on recombinant DNA

117. The biosafety guidelines are developed to ensure an adequate level of protection during which activities of modern biotechnology?
(A) Safe transfer (B) Handling
(C) Use of GMOs (D) All of these

118. The NIH, USA guidelines specify the practices for constructing and handling are:
(A) Recombinant DNA molecules
(B) Organisms and viruses containing recombinant DNA molecules
(C) Products obtained from B
(D) Both (A) and (B)

119. In case of which products, an empirical approach based on data obtained through experimentation about the source of potential hazards has proved satisfactory?
(A) Pesticides (B) Drugs
(C) Vaccines (D) All of these

120. In risk assessment, a lack of scientific knowledge or scientific consensus should not necessarily be interpreted as indicating:
(A) A particular level of risk
(B) Absence of risk
(C) Acceptable risk
(D) Any of these

121. The assessment of risk during laboratory research is usually done in which steps?
(A) Initial risk assessment
(B) Comprehensive risk assessment
(C) Advanced risk assessment
(D) Both (A) and (B)

122. The chief objectives of risk assessment during laboratory research are to decide about which of the proposed research?
(A) Laboratory procedures
(B) Biological containment
(C) Physical containment
(D) All of these

123. In which risk groups, agents are not associated with diseases in healthy humans?
(A) Risk group 3
(B) Risk group 1
(C) Risk group 2
(D) Risk group 4

124. Which is not correct about biosafety Guidelines in India?

(A) Initially, the Indian Recombinant DNA safety Guidelines and Regulation were prepared in 1990 by Recombinant DNA Advisory Committee, Department of Biotechnology, New Delhi

(B) The Indian Recombinant DNA safety Guide lines and Regulation were revised in 1994

(C) Field trials, in India using transgenic plants began in 1995

(D) Canada become the first country to begin commercial use of virus resistant transgenic tobacco and tomato in the early 1990

125. Which statements is correct about comprehensive risk assessment?

(A) It is after the initial risk assessment

(B) The level of containment appropriate for an experiment is decided on the basis of comprehensive risk assessment

(C) The containment level for the experiment may, in the end, be comparable to that of the risk groups (RG) of the agent or it may be raised or lowered as a result of the comprehensive risk assessment

(D) All of these

126. It is feared that genetically engineered microorganisms may disturb the ecosystem and its processes, in which ways?

(A) They may rapidly multiply and out complete the native microbes

(B) They may transfer genes related to virulence or pathogenesis into the native bacterial populations

(C) Both (A) and (B)

(D) They may cause unexpected health hazards

127. Which is not correct?

(A) Oral exposure to transgenic tomato containing cry protein poses no additional risk to human and animal health

(B) In general, the quality of produce from cry gene carrying transgenic crops is regarded different from that from the nontransgenic plants of the same variety

(C) Cowpea trypsin inhibitors do show some adverse effects on rats in nutritional studies

(D) Significant hypertrophy of small intestine and altered enzyme activities in brush border occur due to wheat germ agglutinin

128. In which risk groups, is concerned with agents which are associated with such serious or lethal human diseases for which preventive or therapeutic interventions may be available?

(A) Risk group 3

(B) Risk group 1

(C) Risk group 2

(D) Risk group 4

129. Which is concerned with organisms which are associated with such human diseases, which are rarely serious and for which preventive or therapeutic intervention are often available?

(A) Risk group 3 (B) Risk group 1

(C) Risk group 2 (D) Risk group 4

130. Which is concerned with organisms, which cause serious or lethal disease for which preventive or therapeutic invention are not usually available?

(A) Risk group 3 (B) Risk group 1

(C) Risk group 2 (D) Risk group 4

131. The α–amylase inhibitor from *Phaseolus vulgaris* (rajma) caused allergy when expressed in:

(A) Soybean (B) Tomato

(C) Peas (D) Maize

132. The α–amylase inhibitor gene was transferred to modify which traits?

(A) Insect resistance (B) Amino acid balance

(C) Food digestibility (D) Both (A) and (C)

133. In case of *E. coli*, the host for any HV 1 system is:

(A) Strain K -12

(B) A derivative of strain K-12

(C) Either A or B (D) Strain chi 1776

134. The vector under host-vector I system includes:

(A) Nonconjugative plasmids

(B) Derivative of nonconjugative plasmids

(C) Variants of bacteriophages

(D) All of these

135. Effective dissemination of plant-associated microorganisms beyond the green house can be prevented by using:

(A) Organisms that have been genetically disabled to survive in nature

(B) Organisms whose natural mode of transmission requires injury to the host

(C) Organisms which require another organisms for their survival

(D) Both (A) and (B)

136. The type of regulation necessary during field trials should depend on:

(A) Ability of the modified plants to survive in nature

(B) Their ability for dispersal and reproduction

(C) Their ability to hybridize with crop and weed plants

(D) All of these

137. Which is not correct?

(A) RecA- strains lack recombination

(B) Transposon vectors lacking transposase gene are not able to transpose to other bacteria

(C) Auxotrophic mutants can survive in nature

(D) Transgenes integrated into plasmids may be transferred to other bacteria

138. Which is correct about transmission of transgenes from one crop species to another?

(A) In case of cross-pollinated species isolation by distance is rarely sufficient to prevent inter-population mating

(B) If male sterility is introduced into transgenic cross - pollinated crops, it can check the movement of transgenes through pollen

(C) It transgenes are integrated in the chloroplast genome, it would almost eliminated the transmission of transgene through pollen grains

(D) All of these

139. Which is not correct about Cry proteins?

(A) Cry proteins are rapidly degraded by the stomach juices of vertebrates

(B) They could have harmful effects on non target insect species

(C) It has been reported that Cry proteins leak into soil from transgenic maize roots, where they persist for over 3 weeks

(D) The Cry proteins in the soil become associated with soil clay/humus and become prone to microbial degradation

140. The first recombinant microorganism (Tn5-containing strain of *Pseudomonas fluorescens*) was released in which countries?

(A) USA (B) Canada

(C) Netherlands (D) Germany

141. Most of the recombinant microorganism released in the environment of genes by:

(A) Transposon insertion

(B) Deletion of genes by molecular genetic procedure in vitro

(C) Genetic tagging of bacteria with specific

(D) Any of these

142. Non-infectious illnesses is associated with production of steroids:

(A) Increased body weight

(B) Feminization of males

(C) Hypertension

(D) All of these

143. Non-infectious illness is associated with endotoxin production:

(A) Flu symptoms (B) Dermatitis

(C) Asthama (D) All of these

144. Which hazards is associated with operation of bioreactor fermentation?

(A) Spillage

(B) Effluent contamination of gases

(C) Aerosols from reactor

(D) All of these

145. Non-infectious illness like allergic asthama and dermatitis are associated with production of:

(A) Brewing (B) Single cell protein

(C) Antibiotic (D) All of these

146. In case of experiments with animals, which procedures is followed at all biosafety levels?

(A) Animal carecass is disposed off to avoid it use as food

(B) A permanent record is maintained of the experimental use and disposal of each animal

(C) The animals are confined to security-fenced areas

(D) All of these

147. According to biosafety guidelines of India, physical containment envisages to limit the spread of dangerous microorganisms by:

(A) Laboratory design and facilities

(B) Laboratory practices

(C) Safety equipment

(D) All of these

148. Crop yields can be stabilized by introducing which genes in transgenic varieties?

(A) Genes for improvement nutritional quality

(B) Genes for better tolerance/resistance to biotic and/or abiotic stresses

(C) Genes for herbicide resistance

(D) All of these

149. The potential risks from novel DNA sequences used during transgenic development may be summed up as:

(A) Protein products of transgenes or products generated by them may be toxic

(B) The transgene product may be allergenic

(C) Antibiotic resistance genes may be taken up by human intestinal microflora, which, as a consequence, become resistant to these antibiotics

(D) All of these

150. Trangenic plants could serve as vectors for transgene dissemination through :

(A) Seed

(B) DNA released from debris of transgenic plants

(C) Pollen

(D) All of these

151. Which is not correct?

(A) In European Union, the maximum permissible limit for GMO content in conventional food products and food ingredients is 1% at the marketing stage

(B) All foods that contain more than 1% GMOs related to the different ingredients have to be labeled

(C) In case of organic farming, utilization and presence of GMOs is excluded

(D) Volunteers are not likely to pose problems in crops like rapeseed and mustard

152. The rate of gene-flow is affected by :

(A) Mode of seed dispersal

(B) Mating system

(C) Mode of pollination

(D) All of these

153. Which is not correct about lectins?
(A) They are polysaccharides
(B) They are involved in defence mechanism of plants
(C) Several lectins have been identified as food allergens
(D) α-Amylase/tripsin inhibitor family proteins are potential allergens

154. In the assessment of allergenicity, the source of transgene is the first consideration; the transgene may be obtained from:
(A) Commonly allergenic food
(B) Less commonly allergenic food
(C) No documented history of food allergenicity
(D) Any of these

155. Which is correct about wheat germ agglutinin?
(A) It is effective against insects at concentration of 10 mg/kg body weight
(B) It produces cytotoxic effects in rats at a lower concentration (7 mg/kg)
(C) Safe food has it only at 0.3 mg/kg
(D) All of these

156. Which is not correct?
(A) Transgene isolated from a food crop guarantees the food safety of the transgene product
(B) Acute toxic effect depends on the concentration of the compounds in question
(C) Psoralen is known to be toxic, carcinogenic and mutagenic
(D) Any outstanding insect resistance whose biochemical basis is not known should be considered as the prime candidate for toxicological evaluation

157. Which is not correct about bromoxynil?
(A) It is a non-biodegradable herbicide
(B) It causes birth defects in laboratory animals
(C) It is toxic to fish and may cause cancer in humans
(D) Bromoxynil is absorbed through the skin

158. For minimizing pollen-mediated gene flow from transgenic varieties of *Solanum tuberosum*, how many meters of isolation distance is recommended?
(A) 10 (B) 20
(C) 30 (D) 40

159. Strategy of temporal and tissue-specific expression for preventing transgene transfer from transgenic varieties of crops is based on which techniques?
(A) Tapetum-specific promoters allow expression only when the suppressor is present
(B) Terminator technology
(C) Recoverable block function
(D) Inducible promoters allow expression only when the inducer is present

160. Roundup Ready transgenic variety of soybean shoed, on an average, how much yield decrease as compared to the non-transgenic parent varieties in USA, in a comparison based on field trials at 8200 different locations?
(A) 15% (B) 10%
(C) 5% (D) -5%

161. The principal purpose of transgenic plant containment is to avoid the unintentional transmission of a recombinant DNA containing plant genome including:
(A) Organellar hereditary material
(B) Nuclear hereditary material
(C) Release of recombinant DNA-derived organisms associated with plants
(D) Any of these

162. Biopatents are ----------:
(A) Right to use invention
(B) Right to use biological entities
(C) Right to use products
(D) Right to use process

163. A set standards used to regulate own or community activity in relation to biological world is:
(A) Biopotency (B) Biopiracy
(C) Biowar (D) Bioethics

164. Biopiracy means:
(A) Use of biopatents
(B) Thefts of plant and animals
(C) Stealing of bioresources
(D) Exploitation of bioresources without authentic permission

165. Bioethics is related to:
(A) Preventing bioparacy
(B) Regulation of unethical activities like gene cloning in animals
(C) Preventing theft of living materials
(D) Moral guidance to the problem in biology

166. For patenting a plant variety should be:
(A) DUS (B) DUD
(C) DDS (D) Any of these

167. Products will be applicable for patenting only if they are:
(A) Novel (B) Inventive
(C) Useful (D) All of these

ANSWERSHEET

20. INTELLECTUAL PROPERTY RIGHT, PATENTING AND BIOSAFETY BIOTECHNOLOGY

1. (A)	2. (A)	3. (B)	4. (A)	5. (B)	6. (B)	7. (C)	8. (A)	9. (A)	10. (E)
11. (A)	12. (A)	13. (A)	14. (C)	15. (A)	16. (D)	17. (B)	18. (C)	19. (A)	20. (B)
21. (C)	22. (A)	23. (C)	24. (A)	25. (B)	26. (C)	27. (D)	28. (C)	29. (C)	30. (B)
31. (C)	32. (B)	33. (A)	34. (H);	35. (E)	36. (E)	37. (A)	38. (D)	39. (C)	40. (A)
41. (C)	42. (B)	43. (C)	44. (D)	45. (A)	46. (B)	47. (C)	48. (D)	49. (A)	50. (B)
51. (C)	52. (D)	53. (C)	54. (B)	55. (D)	56. (D)	57. (C)	58. (D)	59. (A)	60. (A)
61. (D)	62. (C)	63. (C)	64. (A)	65. (C)	66. (C)	67. (A)	68. (B)	69. (C)	70. (A)
71. (E)	72. (A)	73. (B)	74. (C)	75. (C)	76. (D)	77. (B)	78. (D)	79. (D)	80. (D)
81. (C)	82. (B)	83. (B)	84. (D)	85. (D)	86. (C)	87. (D)	88. (D)	89. (A)	90. (D)
91. (C)	92. (D)	93. (C)	94. (D)	95. (D)	96. (D)	97. (D)	98. (D)	99. (D)	100. (D)
101. (C)	102. (D)	103. (D)	104. (C)	105. (D)	106. (D)	107. (B)	108. (D)	109. (D)	110. (C)
111. (D)	112. (D)	113. (B)	114. (B)	115. (C)	116. (B)	117. (D)	118. (D)	119. (D)	120. (D)
121. (D)	122. (D)	123. (B)	124. (D)	125. (D)	126. (C)	127. (B)	128. (A)	129. (C)	130. (D)
131. (C)	132. (A)	133. (C)	134. (D)	135. (D)	136. (D)	137. (C)	138. (D)	139. (D)	140. (A)
141. (B)	142. (D)	143. (A)	144. (D)	145. (B)	146. (D)	147. (D)	148. (B)	149. (D)	150. (D)
151. (D)	152. (D)	153. (A)	154. (D)	155. (D)	156. (A)	157. (A)	158. (B)	159. (D)	160. (C)
161. (D)	162. (A)	163. (D)	164. (D)	165. (D)	166. (A)	167. (D)			

Chapter - 21

ANIMAL BIOTECHNOLOGY

1. The most used domesticated animal by Eskimo is:
(A) Cow (B) Sheep
(C) Goat (D) Husky

2. Light chain of antibody consists of how many amino acids?
(A) 440 (B) 220
(C) 320 (D) 360

3. Exotic breeds is:
(A) Require specific environment
(B) Hardy
(C) Sturdy (D) Take less food

4. Which is not correct about human blood clotting factor VIII?
(A) Recombinant factors VIII is inferior to natural factor VIII
(B) The complete cDNA has been driven by the promoter of whey acidic protein gene of pig
(C) Factor VIII is now produced in pig mammary gland and is secreted in the milk
(D) Recombinant factor VIII appears to be identical to the native human protein

5. Exotic breeds are:
(A) Used for cross breeding
(B) Allowed to multiply and replace local breeds
(C) Cheaper
(D) Resistant to local pests and pathogens

6. Number of breeds of Zebu cattle is:
(A) 16 (B) 26
(C) 6 (D) 13

7. Number of breeds of Indian Buffalo is:
(A) 7 (B) 13
(C) 16 (D) 17

8. A milch breed of cow is:
(A) Haryana (B) Malvi
(C) Kankrej (D) Sahiwal

9. A draught breed of cattle is:
(A) Red Sindhi (B) Gir
(C) Malvi (D) Haryana

10. An exotic breed of cow is:
(A) Ongole (B) Friesian
(C) Halliker (D) Deoni

11. Hybrid breed of cattle is:
(A) Sunandini (B) Holstein
(C) Brown swiss (D) Kankrej

12. Which is not correct?
(A) A prolonged administration of human menopausal gonadotrophin (hMG) for stimulation of ovulation produces some side effects like abdominal discomfort and pleural effusion
(B) Human chorionic gonadotrophin (hCG) is administered to prevent the inhibition of LH (luteinizing hormone) surge by hMG
(C) Lapsroscopy is usually done after ovulation
(D) The time of natural ovulation is determined by monitoring the rise in the level of LH either in urine or in blood

13. Zebu cattle is:
(A) Water Buffalo (B) Indian Buffalo
(C) Cow (D) Sheep

14. Infertility of local breeds of cattle can be overcome by use of:
(A) Cross breeding with exotic breeds
(B) Good nourishment
(C) Stilbesterol
(D) Gonadotropin

15. Stilbesterol is used for:
(A) Induction of lactation
(B) Artificial insemination
(C) Super-ovulation
(D) Cryopreservation

16. Surrogate mother is:
(A) Mother without lactation
(B) Future mother with embryo implanted from another
(C) Carrying several embryos
(D) Artificially inseminated female

17. Animal cell cultures are used to produce:
(A) Enzymes and hormones
(B) Interferons and monoclonal antibodies
(C) Virus vaccines and hybrid antibodies
(D) All of these

18. Artificial insemination is better than natural insemination in cattle because:
(A) Semen of good bulls can be provided everywhere
(B) There is no likelihood of contagious diseases
(C) It is economical
(D) All of these

19. Cryopreservation is:
(A) Preservation of living being in chemicals
(B) Preservation at very low temperature
(C) Preservation through exposure to irradiation
(D) Preservation through use of gases

20. Cryopreservation is useful for:
(A) Preservation of semen (B) Very young fetuses
(C) Living cells (D) All of these

21. Which one is most recent domestication:
(A) Buffalo (B) Sheep
(C) Turkey (D) Silkworm

22. Monoclonal antibodies produced for carcinoembryonic antigen and some tumour associated antigens can be used in histochemical assays that permit the identification of:
(A) Tumour cell type
(B) Benign and malignant nature of tumours
(C) Early cases of metastasis
(D) All of these

23. Sheep are reared mainly for:
(A) Meat (B) Mutton
(C) Wool (D) Milk

24. Pashmina is obtained from a variety of:
(A) Sheep (B) Goat
(C) Rabbit (D) Yak

25. Famous angora wool is got from a:
(A) Sheep (B) Rabbit
(C) Goat (D) Yak

26. Lohi is:
(A) A breed of sheep
(B) A breed of goat
(C) A breed of Domestic Fowl
(D) A breed of Geese

27. Merino sheep has been imported for:
(A) Increasing wool output through merino farms
(B) Increasing mutton availability
(C) Improving indigenous breeds
(D) Developing hybrid between sheep and goat

28. Complement system activation can occur in:
(A) Lectin pathway (B) Alternate pathway
(C) Classical pathway (D) All of these

29. Which one is poor men's cow:
(A) Sheep (B) Yak
(C) Goat (D) Llama

30. Goats feed on:
(A) Low herbs (B) Grass
(C) Low shrubs (D) Shrubs and low trees

31. Earliest animal to the domesticated by primitive men was:
(A) Goat (B) Dog
(C) Horse (D) Cat

32. Domesticated livestock of high mountains is:
(A) Yak (B) Goat
(C) Sheep (D) Horse

33. Most economic meat making animal is:
(A) Yak (B) Pig
(C) Goat (D) Sheep

34. Polo game played with the help of:
(A) Donkey (B) Horse
(C) Camel (D) Cock

35. Bhutia is a breed of:
(A) Chicken (B) Goat
(C) Sheep (D) Horse

36. Water buffalo is:
(A) European breed of buffalo that prefers living in water for most of the day
(B) Buffalo-like animal living in rivers
(C) Llama (D) Buffalo

37. Ship of desert is:
(A) Elephant (B) Camel
(C) Sheep (D) Goat

38. Cytotoxic T lymphocytes eliminate :
(A) Tumour cells (B) Cells of tissue grafts
(C) Virus infected cells (D) All of these

39. Silk industry is related to:
(A) Sericulture (B) Apiculture
(C) Pisciculture (D) Horticulture

40. The animal most useful on difficult terrains is:
(A) Mule (B) Yak
(C) Camel (D) Elephant

41. Bacterian camel is characterized by:
(A) Two humps and long neck
(B) Two humps and long limbs
(C) Two humps and thick coat
(D) Single humps and thick coat

42. A breast of burden which needs little care is:
(A) Pig (B) Donkey
(C) Mule (D) Yak

43. The largest land animal is:
(A) Camel (B) Elephant
(C) Rhino (D) Python

44. A queen honey bee lays eggs of:
(A) One type from which all castes develop
(B) Two types, one forming queen and workers and second forming drones
(C) Three types forming queen, drone and workers
(D) Unfertilised die while fertilized ones form all castes

45. Lac is a product of:
(A) Faecal matter
(B) Secretion from body
(C) Excretion from body
(D) Excess food oozing out of the body

46. Large-scale in vitro culture of hybridoma cells has been done in:
(A) In vessels based on immobilized cells
(B) Stirred bioreactors
(C) Airlift fermentors
(D) All of these

47. Silk is produced by:
(A) Larva (B) Adult moth
(C) Both (A) and (B) (D) Cocoon/Pupa

48. Which among the following is real product of Honey Bee:
(A) Honey (B) Pollen
(C) Bee wax (D) Propolis

49. Super ovulation and embryo transplantation are meant for improving:
(A) Human race (B) Livestock
(C) Poultry (D) Plants

50. Worker bees are:
(A) Fertile males (B) Fertile females
(C) Sterile females (D) Sterile males

51. The young chicken raised specifically for meat are called:
(A) Broilers (B) Cockerels
(C) Pellets (D) Hen

52. Ranikhet disease is connected with:
(A) Honey Bee (B) Hens
(C) Fishes (D) Pigs

53. Induced breeding is carried out in case of:
(A) Pisciculture (B) Apiculture
(C) Sericulture (D) Lac culture

54. Silk is the product of:
(A) Cuticle of larva (B) Cuticle of adult
(C) Salivary gland of larva
(D) Salivary gland of adult

55. Rearing of bees is:
(A) Horticulture (B) Apiary
(C) Apiculture (D) Poultry

56. Kappa light chain gene has:
(A) 300 different V (variable) segments
(B) Five different J (joining) segments
(C) A single C (constant) segment
(D) All of these

57. Which is a product of insect body:
(A) Honey, wax and silk
(B) Silk, lac and wax
(C) Royal jelly, wax and lac
(D) All of these

58. Lac is produced by:
(A) Males
(B) Females
(C) More by females than males
(D) More by males than females

59. Silkworm larva spins silk from:
(A) Inside to outside
(B) Outside to inside
(C) Random fashion
(D) Anterior to posterior side

60. Three carp fishes, *Catla labeo* and *Cirrhina*, can be grown together in the same pond more economically as they have:
(A) Positive interactions
(B) Commensalism
(C) Symbiosis
(D) No competition for food

61. Which is not correct about interferon?
(A) Interferons induce in the cells production of 2,5-adenosine polymerase
(B) Interferons enhance the cytotoxic activity of natural killer cells
(C) Interferons are known to enhance growth of some types of tumours
(D) Large-scale (1000 l to 10000 l) cell cultures are infected with Sendai Virus and incubated for 24 hours after which the supernatant is collected , centrifuged and used for isolation of interferon

62. Fish introduced in India by foreigners is:
(A) *Labeo rohita* (B) *Mysnis singhala*
(C) *Pomphret* (D) *Clasius batrachits*

63. Best aquarium is located at:
(A) Vishakhapatnam (B) Chennai
(C) Tarapur, Mumbai (D) Z.S.I., Calcutta

64. To increase milk yield, cow is given:
(A) Sorbitol (B) Stilbesterol
(C) Prolactin (D) Gonadotropin

65. In silkworm, if juvenile hormone is absent at the time of larval moulting, the worm will:
(A) Die (B) Moult into larval stage
(C) Moult into pupa (D) Moult into adult

66. Name of sheep cloned for the first time is:
(A) Dolly (B) Polly
(C) Molly (D) Holly

67. Nagpuri buffalo is:
(A) Milker (B) Draught cattle
(C) Dual purpose (D) Grazer

68. Which amongst the following is used in raising super milk cows:
(A) Artificial insemination with pedigree bull
(B) Embryo transplantation
(C) Superovulation of high yielding cow
(D) All of these

69. Inland fisheries are:
(A) Deep sea fishes
(B) Capturing fishes from sea coast
(C) Raising and capturing fishes in fresh water
(D) Oil extraction from fish

70. Silkworm is a:
(A) Beetle (B) Worm
(C) Fly (D) Moth

71. Which of the following is not a plant product:
(A) Hemp (B) Silk
(C) Cotton (D) Flax

72. Pisciculture is rearing and production of:
(A) Fishes (B) Birds
(C) Reptiles (D) Wool yielding animals

73. Which is not a plant fibre:
(A) Coir (B) Silk
(C) Sunnhemp (D) Manila Hemp

74. Hormone used for inducing lactation in sterile cows is:
(A) Estorogen (B) Progesterone
(C) Relaxin (D) Stilbestrol

75. The bulls of which bred of buffalo are extensively used for upgrading inferior stock:
(A) Ravi (B) Bhadawari
(C) Murrah (D) Jaffarabadi

76. There are about ------breeds of cattle in India besides hybrids:
(A) 14 (B) 20
(C) 26 (D) 32

77. There are ---- breeds of buffaloes in India, in addition to a large number of non-descript:
(A) 4 (B) 6
(C) 10 (D) 12

78. Which vaccine was the first to be produced from cell cultures?
(A) Polio (B) Rabbies
(C) Measles (D) Hepatitis-B

79. --------is a bred of cattle:
(A) Bhadawari (B) Tharparkar
(C) Kashmiri (D) Landrace

80. ---------- is a breed of buffalo:
(A) Berkshire (B) Beetal
(C) Jaffarabadi (D) Sindhi

81. --------- is a breed of goat:
(A) Berbari (B) Large White Yorkshire
(C) Sahiwal (D) Nili

82. ------- is a breed of swine:
(A) Malwi (B) Ravi
(C) Gaddi (D) Chester white

83. Hereford is a breed of:
(A) Cattle (B) Goat
(C) Swine (D) Buffalo

84. Durock is a breed of:
(A) Swine (B) Cattle
(C) Buffalo (D) Goat

85. Tamworth and middle White Yorkshire are breeds of:
(A) Buffalo (B) Goat
(C) Cattle (D) Swine

86. Jamanapari and Barabari are breeds of:
(A) Cattle (B) Buffalo
(C) Swine (D) Goat

87. Surti and Beetal are breeds of:
(A) Swine (B) Goat
(C) Cattle (D) Buffalo

88. The medium used for in vitro culture of hybridoma cells must serve:
(A) Support adequate cell proliferation
(B) Give high antibody yield
(C) Both (A) and (B)
(D) Promote cell differentiation

89. ------ of England applied the term "cell" to the cavities he saw in sections of cork:
(A) Robert Hooke (B) William Winter
(C) Ralph Turner (D) Kleiner Baird

90. Marcello Malpighi in -------- published "Anatomy of Plants", the first systematic study of cell structure:
(A) 1575 (B) 1675
(C) 1775 (D) 1875

91. In 1839, ---------, an animal anatomist formulated the cell theory which set forth the concept that "elementary parts of all tissues are formed of cells though highly diversified in manner":
(A) Robert Hooke (B) Mercello Malphighi
(C) Jheodor Schwann (D) Alred White

92. The nucleus contains ------ per cent of the RNA in the cell:
(A) 4-10 (B) 6-12
(C) 7-17 (D) 8-20

93. The ribosome is composed of two sub-units and has an approximate size of :
(A) 200 Å (B) 210 Å
(C) 200 Å (D) 195 Å

94. Jersey calves suffering from ------ have broad skull, large eye-sockets, short and broad nasal bones and forehead broader than normal:
(A) Aguatha (B) Cerebral hernia
(C) Umblical hernia
(D) Bulldog head (prognathism)

95. Sex linkage has been indicated for more than 200 traits in:
(A) Cattle (B) Sheep
(C) Horse (D) Humans

96. In case of ---------, the ovaries are enclosed by membranous covering which is continuous with the wall of the tube to ensure that the discharged ova passes into the tubes and is not lost into the body:
(A) Dog (B) Cow
(C) Goat (D) Rabbit

97. The sac of the scrotium in bulk is thermoregulatory in function, that is, it provides protection to the tests and maintains them at a temperature lower by ------- ^{0}F than the abdomen which is necessary for active spermatogenesis:

(A) 2-4 (B) 5-7
(C) 8-10 (D) 11-13

98. An examination of the rainfall map of India how that it is invariably in the higher elevations, and in humid areas that the cattle are:

(A) Large and non-descript
(B) Large and well-descript
(C) Small and well descript
(D) Small and non-descript

99. In each ovary (of farm animals) there may be as many -------- follicles:

(A) 25000 (B) 50000
(C) 75000 (D) 100000

100. In case of cow, the vagina is about ------- cm long:

(A) 7.5-10.0 (B) 12.5-15.0
(C) 16.5-19.0 (D) 20.0-25.0

101. The process of spermatogenesis in the male and oogenesis in the female finally gives rise to --------- daughter cells:

(A) 2 (B) 4
(C) 6 (D) 8

102. The egg after the formation of fertilization membrane decreases in diameter by------- per cent due to the secretion of some fluid:

(A) 4 (B) 8
(C) 11 (D) 14

103. The parathyroid glands were discovered by sandstorm in:

(A) 1840 (B) 1860
(C) 1880 (D) 1900

104. --------- compounds have been isolated from the adrenal cortex which are all sterol derivatives:

(A) Twenty-eight (B) Twenty-six
(C) Twenty-four (D) Twenty-three

105. The critical hormones control electrolytes as well as water metabolism which helps in retention of -------- in the body:

(A) Iodine (B) Sulphur
(C) Sodium (D) Phosphorus

106. The ovary can be examined upto ------------months of pregnancy after which it sinks deeper into abdominal cavity:

(A) One (B) Two
(C) Three (D) Four

107. Dystokia, among the domesticate species is mostly met with in :

(A) Dogs (B) Cats
(C) Monkeys (D) Dairy cattle

108. The amount of --------- needed by an adult cow for efficient reproduction is about 10-12 g per day:

(A) Potassium (B) Phosphorus
(C) Calcium (D) Sodium

109. The discovery of living cells would have been difficult, if most impossible, before the compound microscope was invented by Zachaias Jansen of -------- in 1590:

(A) England (B) Poland
(C) France (D) Holland

110. Vitamins --------- are synthesized by the ruminants in their body:

(A) A and D (B) B and K
(C) E and D (D) C and D

111. ----------- per cent of animals contracting bang's diseases become either temporarily or permanently sterile:

(A) 10-15 (B) 15-20
(C) 18-23 (D) 25-30

112. The milk exudes from ------------ separate gland tubes that open at the base of the "mammary hairs":

(A) 50-95 (B) 100-150
(C) 40-80 (D) 155-195

113. The udder usually increases about -------- in size between milking, depending upon the quality of the udder and the amount of milk secreted:

(A) One-fourth (B) Half
(C) One-third (D) Two-thirds

114. In Holstein cow, between two months and the time of puberty (7-10 months), the mammary gland grows three and half times faster than the body weight and reaches the peak pubertal development at -------- months of age:

(A) 6-8 (B) 10-12
(C) 11-13 (D) 14-16

115. It is estimated that ------- volumes of blood flow through the mammary gland for each volume of milk synthesized:

(A) 600 (B) 500
(C) 400 (D) 300

116. An acceptable record is 70 per cent non-returns upto 60 days after breeding or 60 per cent non-returns upto ------- days after breeding:

(A) 70 (B) 80
(C) 90 (D) 100

117. A fertile herd takes less than ---------- months between calving:

(A) 13 (B) 14
(C) 15 (D) 16

118. A cow during her productive life must remain pregnant only for ------ months in every year:

(A) 6 (B) 7
(C) 8 (D) 9

119. In unipara as the cow, mare etc., usually single ovum is released, in sheep one or two and occasionally three, whereas in swine several ova upto -------- are released in a single ovulation:

(A) 15 (B) 20
(C) 25 (D) 30

120. According to legend artificial insemination had its origin in --------- at which time and Arab Chieftain used it to impregnate a prized mare with semen stealthily collected by night from the sheath of an enemy's stallion:

(A) 1310 (B) 1315
(C) 1319 (D) 1322

121. The first scientific research on artificial insemination of domestic animals was conducted with --------- by the Italian psychologist, Lazarro Spallanzani in 1780:

(A) Dogs (B) Sheep
(C) Goats (D) Cows

122. The discovery of spermatozoa dates back to ---------- when the compound microscope was invented by Leewenhock, the Dutch Naturalist:

(A) 1627 (B) 1677
(C) 1697 (D) 1707

123. Mikhail Fyalorovich Ivanoy, a Russian Physiologist commenced his investigation on artificial insemination in farm animals as early as :

(A) 1879 (B) 1889
(C) 1897 (D) 1899

124. In Denmark, Edward Sorensen and Jens Gylling Ham organized the first Cooperative Artificial Breeding Association in:

(A) 1916 (B) 1926
(C) 1936 (D) 1946

125. Dr. Sampath Kumaron employed artificial insemination method for the first time in India in ------- at the Palace Dairy Farm, Mysore:

(A) August, 1939 (B) September, 1939
(C) August, 1938 (D) September, 1938

126. "Broad helix" of "Spiral body" or mitochondrial sheath is ------- rich structure believed to be derived from mitochondria:

(A) Carbohydratte (B) Lipid
(C) Protein (D) Minerals

127. The human prostatic secretion contains fibrinolysin that is so powerful that 2 cc of the same can liquidify 100 cc of clotted blood in -------- hours at 37^{0}C:

(A) 18 (B) 16
(C) 14 (D) 12

128. The prostatic secretion represents the main source of citric acid and acid phosphate for whom ------ semen:

(A) Stallion (B) Bull
(C) Human (D) Dog

129. The identification of the seminal sugars as -------- by Mann (1956) opened up the way for detailed studies on its generation capacity of the accessory tissues:

(A) Glucose (B) Fructose
(C) Sucrose (D Ribulose

130. In species viz., ---- semen after ejaculation remains liquid unlike in others wherein it tends to gelate or coagulate shortly after ejaculation:

(A) Horse and Ass (B) Sheep and Goat
(C) Bull and Dog (D) Cat and Camel

131. Haemacytometer contains in the centre 16 big squares divided into -------- small squares:

(A) 128 (B) 256
(C) 384 (D) 512

132. The percentage of abnormal semen in a normal bull should not exceed:

(A) 21-26 (B) 5-12
(C) 15-20 (D) 26-30

133. Which is not heat labile?

(A) Growth factors and serum
(B) Trypsin and proteins
(C) Peptone and salts (D) Both (A) and (B)

134. In practice, the food value of ether extract has been taken as yielding -------- k cal of gross energy per gramme of the product while carbohydrates and proteins yield about five K cal per gramme:

(A) 6.35 (B) 7.35
(C) 8.35 (D) 9.35

135. The term protein (from word Protrus) was suggested in 1840 by:

(A) Mulder (B) Shudder
(C) Coleure (D) Waternail

136. There may be about -------- different enzymes in a single cell:

(A) 250 (B) 500
(C) 750 (D) 1000

137. Wool is rich in cystine and contains ------ per cent of sulphur:

(A) 1.5 (B) 2.0
(C) 3.0 (D) 4.0

138. About -------- per cent of the plasma copper is firmly bound to an alpha-2 globulin as ceruloplasmin, the form in which it is transported in the body:

(A) 96 (B) 94
(C) 92 (D) 90

139. As early as ----------, a Swiss physician, J. Francois Coindet, first recommended iodine as a remedy for goiter:

(A) 1800 (B) 1820
(C) 1840 (D) 1860

140. By late ----- synthetic vitamin A became available on the commercial market:

(A) 1925 (B) 1930
(C) 1935 (D) 1940

141. Vitamin K was identified by ------------ in 1935 as a factor present in green leaves:
(A) Dam (B) Dixon
(C) Dukes (D) Diggins

142. Jansen and Donath in ------------ succeeded in crystaling Vitamin B_1 in pure form:
(A) 1906 (B) 1916
(C) 1926 (D) 1936

143. The intestine represents a long tube approximately -------- feet long in average dairy cattle:
(A) 180 (B) 160
(C) 200 (D) 220

144. Trypsin is preferred over other enzymes for tissue disaggregation because of:
(A) Effective for many tissues
(B) Tolerated by most cells
(C) Nneutralized by serum
(D) All of these

145. The large intestine of ------- is the shortest and simplest of all the domestic animals:
(A) Cat (B) Dog
(C) Cow (D) Horse

146. The rate of gas production in the rumen is most rapid immediately after meal and in the cow may exceed ------- per hour:
(A) 15 (B) 20
(C) 25 (D) 30

147. It has been estimated that the human stomach wall is replete with -------glands:
(A) 100000000 (B) 150000000
(C) 300000000 (D) 350000000

148. In bovine --- per cent of the total rumen bacteria have a proteolytic activity:
(A) 32 (B) 34
(C) 36 (D) 38

149. Serum contains which growth factors?
(A) Epidermal and fibroblast growth factors
(B) Transforming growth factor ß
(C) Endoethelial & platelet –derived growth factors
(D) All of these

150. Which is the preferred enzyme for disaggregation of animals tissues?
(A) Collagenase (B) Trypsin
(C) Elastase (D) Mucase

151. The manure gutter should have a gradient of one inch for every ---- feet length:
(A) 15 (B) 14
(C) 12 (D) 10

152. A system of grazing known as Henbenhein system was developed in:
(A) Holland (B) Poland
(C) England (D) Germany

153. Diphtheria is caused by:
(A) *S. lactis*
(B) *Diplitheria bacillis*
(C) *B. dysenteriae* (D) All of these

154. Abortion is caused in cattle mainly by
(A) *Brucella abortus* (B) *Brucella melitensis*
(C) *Salmonella spp* (D) All of these

155. Paratyphoid fever is caused by:
(A) *Salmonella typhosa*
(B) *Salmonda paratyphi*
(C) *Brucella abortus* (D) None of these

156. The colour of colostrums is yellow due to presence of:
(A) Carotene (B) Blood particles
(C) Immune globulin (D) All of these

157. Which of the following element has lower content in colostrums than milk:
(A) Clorine (B) Iron
(C) Potassium (D) All of these

158. Colostrum is more concentrated than milk due to presence of:
(A) Total solids (B) Clorine ions
(C) Globulin (D) All of these

159. Vitamin A is stored in:
(A) Liver (B) Pancrease
(C) Gall bladder (D) Lungs

160. Which of the following is rich in coline?
(A) Ghee (B) Egg
(C) Liver (D) All of these

161. The nutritive value of a fodder can be evaluated by which method:
(A) Digestibility trials
(B) Metabolic experiments
(C) Calorimetric measurement
(D) All of these

162. Gastro-enteritis is caused mainly by:
(A) *Staphylococcus aureus* (B) *Proteus vulgaris*
(C) *Clostridium welchii* (D) All of these

163. Septic sore throat is caused by:
(A) *Vibrio sp.* (B) *S. scarlatine*
(C) *Diphtheria bacilli* (D) *S. epidermicus*

164. Undulant fever is very common in:
(A) Goat (B) Cow
(C) Sheep (D) All of these

165. On attainment of maturity, between first and second year a bull can serve ------ cows:
(A) 18 (B) 20
(C) 23 (D) 25

166. Ordinarily, lambs are not used for servicing until ------- months of age:
(A) 20 (B) 18
(C) 16 (D) 14

167. Barned variety of Plymoth Rock was developed in about ------ from the cross of the Dominique and Black Cochin breeds:
(A) 1835 (B) 1845
(C) 1855 (D) 1865

168. The poultry breed New Hampshire was developed both for meat and egg production in New Hampshire from single comb Rhode Island Red and Red Stock during:
(A) 1910 (B) 1920
(C) 1930 (D) 1940

169. The poultry breed New Hampshire developed in ------ about 200 years ago, primarily as a table bird:
(A) Poland (B) Holland
(C) New Zealand (D) England

170. The commercial egg farms in India maintain Austro White, a cross between:
(A) Australorp male x White Leghorn female
(B) Australorp male x leg horn female
(C) Australorp male x New Hampshire female
(D) Australorp mole x Wyandotte female

171. Artificial hatching was practiced some ------ years ago in China and Egypt:
(A) 2500 (B) 3000
(C) 3500 (D) 4000

172. In 1770, an Englishman named ------, hatched chicks by passing hot air through the room where the eggs were located:
(A) John Watchman (B) John William
(C) John Champion (D) John Laggard

173. Infectious coryza is a:
(A) Serious disease of poultry in India
(B) Caused by *Hemophilus gallinarum* bacterium
(C) Nasal discharge & sneezing are common symptoms
(D) All of these

174. Bronchitis in poultry refers to:
(A) Viral disease but death loss is highest among young
(B) Raspy cough, gasp and discharge from nose and eyes
(C) Egg production suddenly drops
(D) All of these

175. C- ELISA kit is useful for:
(A) Diagnosis and treatment of rinderpest disease
(B) It is a vaccine against T.B.
(C) It is a technique to test milk fat in a adulterated milk
(D) It is a machine used for quick milking

176. A calf starter refers to:
(A) High energy ration for calves
(B) A kind or liquid milk fed to calves
(C) A kind of dried milk to calves
(D) All of these

177. Generally calf starter contains:
(A) Ground grain of cereals
(B) Iodized salt and aureomycin
(C) Steamed bone meal
(D) All of these

178. How much aureomycin hydrochloride (Aureomycin) can be given to calf per day?
(A) 50 mg (B) 80 mg
(C) 100 mg (D) 120 mg

179. PETA refers to:
(A) People for Ethical Treatment of Animals
(B) People for Exotic Treated Animals
(C) Problems in Exotic Treated Animals
(D) None of these

180. The main function of camel hump is:
(A) Acts as an energy food store in form of fatty acids
(B) It stores water
(C) It helps in riding (D) All of these

181. Camel can tolerate the water loss upto:
(A) 20 % of body weight
(B) 30% of body weight
(C) 40% of body weight
(D) 50% of body weight

182. The camel can sustain without water for longer period because of:
(A) It has higher water in the blood plasma
(B) It releases out meager water in urine and faeces
(C) Both (A) and (B) (D) None of these

183. Flushing refers to:
(A) Proper feeding of sows/gilts before two weeks of breeding
(B) Measuring of fat % in pig's carcasses
(C) Removing extra teats of boar
(D) None of these

184. Milk and milk product order, 1992 is related to:
(A) Registration of dairy man if he handles >10000 litres or 500 MT milk solids/year
(B) Production of clean milk
(C) Production of butter
(D) Production of ghee

185. Turkey production is not popular in India because of:
(A) High cost (B) Low growth
(C) Poor laying capacity in summer
(D) All of these

186. The true stomach in ruminants is known as:
(A) Rumen (B) Abomasum
(C) Reticulum (D) Omasum

187. The main function of abomasums is:
(A) Storage of feed
(B) Squeezing of water from feed
(C) Cheating of feeds
(D) Secretion of gastric juice

188. Which of the following animal has not a simple digestive tract?
(A) Swine (B) Poultry
(C) Goat (D) Men

189. How much amino acids are essential for poultry rations?
(A) 7 (B) 10
(C) 11 (D) 15

190. How much amino acids are essential for simple stomach animals?
(A) 7 (B) 10
(C) 11 (D) 15

191. Which of the following antibiotic is added in the animal feed?
(A) Aureomycin (B) Terramycin
(C) Penicillin (D) All of these

192. Which of the following is a major mineral required for animal's growth and development?
(A) Salt (B) P
(C) Ca (D) All of these

193. Stilbestrel refers to:
(A) An artificial hormone
(B) Improves growth rate of animals
(C) Resembles with natural female hormone estrogen
(D) All of these

194. Transquilizers have:
(A) Growth stimulating effect
(B) Calming effect upon livestock
(C) They also reduces shrink in cattle when given prior to shipping
(D) All of these

195. Iodinated casein is a:
(A) Similar to thryroxine hormone
(B) It induces greater milk production
(C) It is more useful to brood sows
(D) All of these

196. Main function of Gizzard is:
(A) Grind and crush the coarse feed
(B) Digestive juices are secreted and mixed with feed
(C) Chewing of feed
(D) All of these

197. The function of reticulum is:
(A) To hold foreign harmful bodies
(B) To furnish additional storage space
(C) Both (A) and (B)
(D) Squeezing out the water from feed

198. The causal organism of T.B. is:
(A) *Mycobacterium tuberculosis*
(B) *Mycobacterium typhii*
(C) *Streptococcus typhii*
(D) *Bacillus tuberculosis*

199. Causal organism of undulant fever is:
(A) *Brucella melitensis* (B) *B. abortus*
(C) *Salmonella typhosa* (D) All of these

200. Scalet fever is caused by:
(A) *S. scartaline* (B) *S. lactis*
(C) *B. dysenteriae* (D) *Salmonelle sp.*

201. It has been estimated that half a dozen average-size pigs will throw off the energy at the rate of about 1 kwh which should be actively dispersed to maintain normal body temperature at:
(A) 39 ^{0}C (B) 38 ^{0}C
(C) 37 ^{0}C (D) 40 ^{0}C

202. Albumine carries which molecules into cells?
(A) Vitamins (B) Hormones
(C) Lipids (D) All of these

203. Typhoid fever is caused by:
(A) *S. lactis* (B) *Brucella abortus*
(C) *Salmonella typhosa* (D) All of these

204. The quantitative production of pig response to a change in feed dose lies at the sore of a computer model developed at the university of -------- in 1970:
(A) Michigan (B) Kansas
(C) Edinburgh (D) California

205. --------- poisoning is common in areas of USA where it is known as alkali disease or blind staggers:
(A) Cobalt (B) Chlorine
(C) Magnesium (D) Selenium

206. The finest woolled sheep of the Duwn breeds and the South down (of Britain), has a count of ------- microns (1000000 microns = 1 meter):
(A) 27-28 (B) 25-26
(C) 23-24 (D) 29-30

207. Wool yields vary from about 1.3 kg for a Welsh ewe to 2.6 kg for a Suffolk and -------- kg for an English Leicester:
(A) 3 (B) 4
(C) 5 (D) 6

208. Match the somatic chromosome number with the animal:

Somatic chromosome number	Animal species
(A) 78	(i) Sheep
(B) 54	(ii) Cattle
(C) 60	(iii) Human
(D) 46	(iv) Hens

Answer:
(A) (A) i (B) ii (C) iii (D) iv
(B) (A) i (B) iii (C) ii (D) iv
(C) (A) iv (B) i (C) ii (D) iii
(D) (A) iv (B) ii (C) iii (D) i

209. Plasticware can be made wettable and negatively charged by which treatments?
(A) Oxidizing agents (B) UV irradiation
(C) Strong acids (D) All of these

210. Match the sex chromosome number with the animals embodying it

Chromosomes number	Animal
(A) 23 pairs	(i) Cattle
(B) 30 pairs	(ii) Sheep
(C) 27 pairs	(iii) Wheat
(D) 21 pairs	(iv) Human

Answer:

(A) (A)i (B) ii (C) iii (D) iv
(B) (A)i (B) iii (C) ii (D) iv
(C) (A)iv (B) i (C) ii (D) iii
(D) (A)iv (B) ii (C) iii (D) i

211. Development of serum-free formulations of culture media are based on:
(A) Limiting factor (B) Analytical
(C) Synthetic (D) All of these

212. Camel pox is generally found in:
(A) Young camel (B) Adult camel
(C) Old camel (D) None of these

213. Camel pox is a:
(A) Contagious disease
(B) Non contagious disease
(C) Infectious disease
(D) Non-infectious disease

214. The common symptoms of camel pox is:
(A) Lesion on skin of upper and lower lips
(B) Lesions on other body parts too at acute form
(C) High temperature
(D) All of these

215. Which of the following disease of pigs is caused by virus?
(A) FMD (B) Hog Cholera
(C) Swine pox (D) All of these

216. Diamond skin disease of pigs is known as:
(A) Anthrax (B) Swine crysipelas
(C) Swine pox (D) FMD

217. Which of the following disease is found in newly born pigs?
(A) Dysentery (B) Piglet influenza
(C) Novel ill (D) None of these

218. Which of the following internal parasite causes damage to pigs?
(A) Roundworm (B) Mange
(C) Lice (D) None of these

219. Mange disease of pigs is caused by:
(A) Lice (B) Mites
(C) Roundworm (D) Virus

220. New castle disease (Ranikhet) in chicks refers to:
(A) Viral disease first discovered in Ranikhet
(B) Highly infections and caused death from 25 to 100%
(C) Chicks sneeze, gasp and have difficult in breathing
(D) All of these

221. Coccidiosis disease is very serious in poultry because of:
(A) Usually occurs in 3-9 weeks old chickens and caused by protozons
(B) Blood diarrhea leads to quick death
(C) Outbreaks occur most often during warm and humid region
(D) All of these

222. Fowl pox (chicken pox or Sore-Head) refers to:
(A) It is a viral infections poultry disease
(B) It has two forms either affects skin and lambs or mucous membranes
(C) Small blister like spots on face, comb and wattles are major symptoms
(D) All of these

223. Larygotracheitis refers to:
(A) Highly communicable and fetal viral poultry disease
(B) Labored breathing, gasping, rattling with head extended and beak open are main symptoms
(C) Vaccinate all birds at the age of 6 weeks to avoid this disease
(D) All of these

224. Which of the following external parasite is a serious threat to poultry?
(A) Round worms (B) Tapeworms
(C) Ticks (D) All of these

225. Infectious Bovine Rhinotrachitis (IBR) is a:
(A) Disease of bovines caused by virus mainly excretes through semen
(B) Bacterial disease
(C) Protozoan disease
(D) Nutritional disorder

226. Which of the following disease is found in camel?
(A) Surra (B) Camel pox
(C) Anthrax (D) All of these

227. The main symptoms of rabies in camel is:
(A) Aggressiveness
(B) Glassy look
(C) Dribbling of saliva from mouth
(D) All of these

228. The main symptoms of Anthrax of camel is:
(A) High body temperature
(B) Cessation of rumination
(C) Rapid breathing
(D) All of these

229. Which of the following animal is not a ruminant?
(A) Cow (B) Goat
(C) Pig (D) Sheep

230. The stomach of ruminant is divided into:
(A) 2 parts (B) 3 parts
(C) 4 parts (D) 5 parts

231. Which of the following is not a compartment of ruminant stomach?
(A) Rumen (B) Reticulum & Omasum
(C) Abomasum (D) Abdomen

232. Which of the part of ruminant stomach is the largest one?
(A) Rumen (B) Reticulum
(C) Abomasum (D) Omasum

233. Regurgitation takes place in:
(A) Rumen (B) Reticulum
(C) Abomasum (D) Omasum

234. Which of the following has not a simple digestive tract?
(A) Swine (B) Poultry
(C) Goat (D) Men

235. The main function of proventriculus in poultry is:
(A) Digestive juices are secreted and mixed with feed
(B) Grinding of feed into small prices
(C) Chewing of feed
(D) All of these

236. The main function of Gizzard is:
(A) Grind and crush the coarse food
(B) Digestive juices are secreted and mixed with food
(C) Chewing of feed
(D) All of these

237. Which of the following part is not found in the digestive systems of poultry?
(A) Crop (B) Proventriculus
(C) Gizzard (D) Abomasum

238. The main function of abomasums is:
(A) Storage of feed
(B) Squeezing of water from feed
(C) Cheaning of feeds
(D) Secretion of gastric juice

239. Which is essential for maintenance of heterogeneous animal cell cultures?
(A) Precursor cells (B) Stem cells
(C) Differentiated cells (D) Both (A) and (C)

240. The function of reticulum is:
(A) To hold foreign harmful bodies
(B) To furnish additional storage space
(C) Both (A) and (B)
(D) Squeezing out the water from feed

241. What is true for omasum?
(A) Third compartment of ruminant stomach
(B) It has strong muscular walls
(C) It helps in squeezing out water from feed before it enters to abomasums
(D) All of these

242. American buffalo is known as:
(A) Bison (B) Toda
(C) Mehsana (D) Jaffrabadi

243. Which of the following breed has highest fat in milk:
(A) Surti (B) Mehsana
(C) Bhadawari (D) Murrah

244. Which of the breed of buffalo gives highest milk production?
(A) Bhadawari (B) Murrah
(C) Jaffrabadi (D) Surti

245. How many breeds of cattle are found in India?
(A) 10 (B) 15
(C) 21 (D) 27

246. Deoni is a:
(A) Draft breed (B) Dairy breed
(C) Dual purpose breed (D) All of these

247. Which of the following is not a dairy breed of cattle?
(A) Sahiwal (B) Sindhi
(C) Nagore (D) Gir

248. Which enzymes is the most commonly used for disaggregation of explants?
(A) Elastase and pappain
(B) Mucase and elastase
(C) Trypsin and collagenase
(D) All of these

249. Which of the Indian cattle breed was exported to Brazil?
(A) Sahiwal (B) Jersey
(C) Kankrej (D) All of these

250. Zebu is a name of Indian breeds of cattle in:
(A) Europe (B) South America
(C) Both (A) and (B) (D) None of these

251. Which of the following is not a exotic breed of cattle?
(A) Jersey (B) Hallikar
(C) Holstein –Friesian (D) Brown swiss

252. Which cell types present in heterogeneous animal cell cultures are the least likely to divide?
(A) Precursor cells (B) Stem cells
(C) Differentiated cells (D) Both (A) and (C)

253. Huskies are:
(A) Yaks (B) Thick coated dog
(C) Donkeys (D) Water buffaloes

254. Best source for dietary protein for a vegetarian is:
(A) Soybean (B) Gram
(C) Groundnut (D) Milk

255. Cow Jill is a cross of:
(A) Ayreshire x Haryana
(B) Ayrshire x Holstein Friesian
(C) Jersey x Sindhi
(D) Sahiwal x Sindhi

256. Hisasale is a cross of:
(A) Bikaneri ewes x Merino rams
(B) Merino ewes x Bikaneri rams
(C) Rambouillet ewes x Merino rams
(D) Merino ewes x Rambonillet rams

257. Cell lines having specific cell types may be obtained by:
(A) Selective culture (B) Cloning
(C) Physical cell separation (D) All of these

258. Which of the following is a sheep breed of southern region of India?
(A) Deccani (B) Nellore
(C) Mandy (D) All of these

259. Which of the following sheep breed is found in arid western region?
(A) Kathiawari (B) Marwari
(C) Bikaneri (D) All of these

260. Which of the following sheep breed is most popular for its fine wool?
(A) Rambouillet (B) Corriedale
(C) Merino (D) Gurej

261. Which of the following sheep breed of temperate region?
(A) Gurej (B) Bhakarwal
(C) Gaddi (D) Lohi

262. Which is not correct about production of artificial skin?
(A) Intact skin explants is cultured on a suitable medium
(B) Cell division occurs in keratinocytes
(C) Corneocytes present in the explants do not contribute to the artificial skin
(D) Irradiated 3T3 fibroblast cells are used as feeder layer

263. What is correct for Bharat merino sheep:
(A) Fine wool sheep breed
(B) Indigenously developed cross breed
(C) Cross of Rambouillet and Merino with indigenous breeds
(D) All of these

264. *In vitro* transformation of animal cells may occur due to:
(A) Spontaneously (B) Chemically induced
(C) Induced by virus (D) All of these

265. Which of the following goat breed is the highest milk producer?
(A) Barbari (B) Bengil black
(C) Jamanapari (D) Assam local

266. The most striking feature of Jamanapari goat is:
(A) Romen nose coupled with a shortened upper lip
(B) High milk production
(C) High reproductive rate
(D) All of these

267. Which of the following is a horse breed?
(A) Kathiawari (B) Marwari
(C) Bhutia (D) All of these

268. Which of the following horse breed is found in Pakistan and India both?
(A) Baluchi (B) Hizzai
(C) Unmol (D) All of these

269. Which of the following horse breed is generally known as a race horse?
(A) Arabi (B) Thoroughbred
(C) Spiti (D) None of these

270. Which of the following horse breed is popular in hilly areas of India?
(A) Manipuri (B) Spiti
(C) Bhutia (D) All of these

271. Which is not correct about animal organ cultures?
(A) They can be maintained only for few months
(B) Hormone-dependent organs remain hormone-dependent
(C) Results from them are comparable to those from whole animals
(D) The development of foetal organs in vitro is comparable to that in vitro

272. Which of the following breed of donkey is the smallest one?
(A) Indian ass (B) Poitou
(C) Egyptian white (D) Arab

273. Which of the following breed of donkey is the largest one in the world?
(A) Poitou (B) White Egyptian
(C) Arab (D) Persian

274. Which of the following donkey breed is found in France?
(A) Poitou (B) White Egyptian
(C) Arab (D) Persian

275. Which of the following is a breed of a donkey?
(A) White Egyptian (B) Damascus
(C) Persian (D) All of these

276. Which of the following is considered as a parent of a present day donkey?
(A) Nubian ass (B) Ponies
(C) Mare (D) Mules

277. Mules are most suitable for:
(A) Hilly tracts (B) Military operations
(C) Stress climate (D) All of these

278. Which of the following is a baggage type of camel breed in India?
(A) Hill camel (B) Reverine camel
(C) Both (A) and (B) (D) None of these

279. Most of cell lines will grow for a period of about how many weeks?
(A) 5 (B) 12
(C) 15 (D) 18

280. Which of the following breed of camel is found in India?
(A) Riverine camel (B) Desert camel
(C) Hill camel (D) All of these

281. Which of the following camel breed is used for riding purposes?
(A) Hill camel (B) Riverine camel
(C) Desert camel (D) All of these

282. Which of the following camel breed is not found in India?
(A) Bacterian (B) Arabian
(C) Both (A) and (B) (D) None of these

283. Riverine camel is mostly found in:
(A) UP and Punjab (B) Rajasthan and TN
(C) Assam & Uttaranchal (D) Kerala and AP

284. Which is not correct?
(A) Cells derived from a single cell through meiosis constitute a clone
(B) Plating efficiency of continuous cell lines is generally 10% or higher
(C) A feeder layer is a layer of cells, which has been treated to prevent their growth
(D) During colony isolation, desired colony may be shielded and the remaining colonies irradiated by a lethal dose of X-rays

285. Which of the following is a exotic pig breed?
(A) Hampshire (B) Poland China
(C) Tamworth (D) All of these

286. Which of the following is a exotic pig breed?
(A) Chester white (B) Duroc
(C) Danish Landrace (D) All of these

287. Which of the following pig breed is known as bacon hog?
(A) Wessex Saddleback (B) Hampshire
(C) Duroc (D) Bershire

288. Which of the pair is correct?
(A) Tamworth – Red colour
(B) Hampshire – Black but white belt encircling
(C) Berkshire – Black
(D) All of these

289. Which of the following is a American fowl?
(A) Plymoth Rock (B) Rhode Island Red
(C) New Hampshire (D) All of these

290. Continuous cell lines provide which advantages?
(A) Lower serum requirement
(B) Ability to grow in suspension
(C) Faster rate of growth
(D) All of these

291. Which of the following poultry breed is suitable for wet and heavy rainfall regions?
(A) Australorp (B) White leghorn
(C) White Rock (D) White Cornish

292. Which of the following is a most popular egg breed of fowl in all over world?
(A) White Leghorn (B) Plymoth Rock
(C) New Hampshire (D) White Rock

293. Which of the following breed is known as best round breed of all poultry?
(A) White Leghorn (B) Plymoth Rock
(C) Rhode Island Red (D) White Rock

294. Which of the following is the largest breed of poultry?
(A) Black Minorca (B) White Leghorn
(C) Light Sussex (D) New Hampshire

295. Which of the following belongs to duck breeds?
(A) Sylhet met (B) Nageswari
(C) Indian Runner (D) All of these

296. Cell cloning is used for which purposes?
(A) To obtain homogeneous cell lines from heterogeneous cell cultures
(B) To develop hybridoma clones
(C) To isolate biochemical mutants and cell strains with marker chromosomes
(D) All of these

297. Which of the duck breed produces the largest size of egg?
(A) Khaki Campbell (B) Sylhet Mete
(C) Nagaswari (D) Indian runner

298. Most popular duck breed in India is:
(A) Khaki Campbell (B) Sylhet Mete
(C) Nagaswari (D) Indian runner

299. Which of the following is a geese breed?
(A) Indian goose (B) Farn chinese
(C) Emden (D) All of these

300. The heaviest breed of geese is:
(A) Emden (B) English white
(C) Romen (D) White Chinese

301. The lightest breed of geese is:
(A) Emden (B) Indian goose
(C) English grey (D) Toulouse

302. Which normal cells do not give rise to continuous cell lines?
(A) Human glia cells
(B) Human fibroblast cells
(C) Chick fibroblasts
(D) All of these

303. Which of the following a not a breed of Turkey?
(A) Cambridge (B) British white
(C) New Hampshire (D) Norfolks

304. Which of the following breed of Turkey is the most popular and heaviest one?
(A) Cambridge
(B) Broad Breasted Bronze
(C) Western white
(D) Beltsville white

305. Indubrazil is a cattle breed:
(A) Cross between Indian breed x Local of Brazil cow
(B) New cross breed of Brazil
(C) Suitable for beef and milk
(D) All of these

306. Cell culture are used for obtaining which viral vaccines?
(A) Foot and mouth (B) Polio and mumps
(C) Measles and rabies (D) All of these

307. Which of the following pair is not matched correctly?
(A) Amritmahal – Mysore (B) Malvi – Malva tract
(C) Nagore – Nagore (D) Gir – South India

308. What is correct to Tharparkar breed of Cattle?
(A) Dual purpose breed
(B) Originated from Tharparkar district of Western Pakistan
(C) Main breed of Jaisalmer and Jodhpur in Rajasthan
(D) All of these

309. Which of the cattle breed of south India?
(A) Rath (B) Mewati
(C) Tharparkar (D) Ongole

310. Which of the following is not matched correctly?
(A) Toda - Nilgiri hills (B) Bhadawari – UP
(C) Parlakmedi – Orissa (D) Surti – Punjab

311. Which of the following group belongs to buffalo breeds?
(A) Murrah, Jaffarabadi, Surti and Mehsana
(B) Gir, Nagore, Rath and Knakrej
(C) Pashmina, Beetal, Barbari and Malabari
(D) Lohi, Merino, Rambouillet and Corriedale

312. Which of the following group belongs to sheep breeds?
(A) Murrah, Jaffarabadi, Surti and Mehsana
(B) Gir, Nagore, Rath and Knakrej
(C) Pashmina, Beetal, Barbari and Malabari
(D) Lohi, Merino, Rambouillet and Corriedale

313. Which of the following group belongs to goat breeds?
(A) Murrah, Jaffarabadi, Surti and Mehsana
(B) Gir, Nagore, Rath and Knakrej
(C) Pashmina, Beetal, Barbari and Malabari
(D) Lohi, Merino, Rambouillet and Corriedale

314. The highest milk producer cattle breed is:
(A) Holstein Friesian
(B) The Trademarks sable Fashion
(C) Jersey (D) Brown swiss

315. Cell cultures are used for obtaining:
(A) Interleukins and interferons
(B) Hormones, enzymes and antibodies
(C) Viral vaccines (D) All of these

316. Which of the following is not a draft cattle breed?
(A) Nagore (B) Amritmahal
(C) Gir (D) Halikar

317. Which of the following criteria is used for selecting the dairy milking animals?
(A) Breed and pedigree
(B) Production and records
(C) Physical appearance
(D) All of these

318. Which of the most important criteria is used for selecting the dairy animals?
(A) General appearance (B) Dairy character
(C) Body appearance
(D) Mammary appearance

319. Which of the following criteria is used for selecting the dairy bull?
(A) Dairy character (B) Body capacity
(C) Physical appearance (D) All of these

320. Which of the following method is used to dry up cows:
(A) Intermittent milking (B) Incomplete milking
(C) Abrupt milking (D) All of these

321. Which of the following method of drying off is most suitable for high yielder animals?
(A) Intermittent milking (B) Incomplete milking
(C) Abrupt milking (D) All of these

322. Strip cup is used for:
(A) Removing two –three streams of milk from each quarter before milking
(B) Checking the normalcy of milk
(C) Cleaning of treat canals
(D) All of these

323. Milk replacer refers to:
(A) Readymade dry feed mixture
(B) Substitute of milk
(C) It is also fortified with vitamin A, D and antibiotics
(D) All of these

324. Which of the following pair is correct:
(A) Holstein Friesian – Holland
(B) Ayrshire – Scotland
(C) Brown swiss – Switzerland
(D) Gurney – USA

325. Most common system for housing dairy cattle is:
(A) Stanchion barn
(B) Open barn
(C) Half open and half closed system
(D) All of these

326. Which of the following method is used for market of calves:
(A) Ear notches (B) Tattooing
(C) Metal ear tags (D) All of these

327. EDTA readily disaggregates in which tissues?
(A) Fibraous tissues (B) Epithelium
(C) Skin (D) Muscles

328. Which of the following marketing method of calves is considered the best one?
(A) Ear notches (B) Tattooing inside
(C) Metal ear tags (D) All of these

329. Adult cattle are generally marked successfully with:
(A) Ear notches
(B) Hot iron branding on hind flanks
(C) Neck chains (D) All of these

330. Which of the following marking method is used for marketing of buffaloes:
(A) Horn branding
(B) Hot iron branding
(C) Totooning the inside of the root of the tail
(D) All of these

331. Under the feeding rule, the roughage should provide the total dry matter to dairy animal is:
(A) 1/3 (B) 2/3
(C) >2/3 (D) ½

332. Which of the following statement is correct?
(A) If roughage contains no legumes, the protein in concentrate ration should be 17-20 %
(B) If cow is receiving all high quality legume hay or legume silage, concentrate ration should have 9-10 % protein
(C) If animal is fed on green fodder, there is no need to provide concentrate ration
(D) All of these

333. An antibody molecule consists of:
(A) Two identical light chains
(B) Two identical heavy chains
(C) Two α and 2δ chains
(D) Both (A) and (B)

334. Which of the following goat breeds births twice in a year?
(A) Jamanapari (B) Barbari
(C) Beetal (D) Saanen

335. Which of the following pair is correct:
(A) Cattle – Calving (B) Horse – Faoling
(C) Pig – Farrowing (D) All of these

336. Which of the chemical is used for blocking the heat cycle in sows?
(A) ICI-33828 (B) IC – 35
(C) IAA (D) All of these

337. Shearing in sheep is done during:
(A) Winter season
(B) Summer season
(C) After winter and rainy season
(D) All of these

338. Sandiness of ice cream is due to:
(A) High fat %
(B) Low lactose %
(C) Lactose crystallization
(D) None of these

339. Sandiness of ice cream can be reduced by:
(A) Increasing fat % (B) Increasing lactose %
(C) Reducing fat % (D) None of these

340. What is the role of ice cream ageing?
(A) It solidifies the ice cream
(B) It increases whipping ability
(C) It also improves flavour of ice cream
(D) All of these

341. 100 g ice cream is equal to:
(A) 206 calories (B) 218 calories
(C) 306 calories (D) 350 calories

342. What is main difference between cow and buffalo khoa?
(A) Cow khoa contains higher per cent of water, protein and lactose
(B) Buffalo khoa contains higher per cent of fat and ash
(C) Cow khoa contains higher Fe than buffalo khoa
(D) All of these

343. At what temperature, citric acid should be mixed in milk for curding?
(A) 50^{0}C (B) 60^{0}C
(C) 70^{0}C (D) 82^{0}C

344. The quality parameters of judging chhana are given below in decreasing importance. Which of the following order is correct?
(A) Flavour>Texture>Colour and body> Modified product
(B) Texture>Flavour>Colour and body>Modified product
(C) Flavour>Colour and body>Texture>Modified product
(D) All of these

345. What is colouring agent is mixed in milk for paneer preparation?
(A) 0.003-0.012% (B) 0.04-0.05%
(C) 0.5-0.7 (D) 1.0-1.2%

346. Which is not correct about plastic substrates?
(A) Most plasticware are nonautoclavable and are supplied sterile for a single use
(B) Polystyrene is the most commonly used but polyethylene, Perspex, PVC, polycarbonate, Teflon, cellophane and cellulose acetate are all suitable substrates
(C) All the organic substrates must be suitably treated to make them wettable and positively charged
(D) Thin Teflon films are available as Petridishes or as membranes to be used as rafts; these films are permeable to O_2 and CO_2

347. Which of the following parameter is most important for judging quality of cheese?
(A) Flavour (B) Body
(C) Texture (D) Colour

348. Nurse cow refers to:
(A) A cow that feeds the calves of another cow
(B) A feeding cow of her own calves
(C) A diseased cow
(D) None of these

349. Which of the following method is considered best, feeding system of calves?
(A) Liberal milk feeding
(B) Limited milk feeding plus a dry calf starter
(C) Using milk replacers
(D) None of these

350. Which of the following is a calf feeding system?
(A) Liberal milk feeding
(B) Using milk replacers
(C) Limited milk feeding plus a dry calf starter
(D) All of these

351. Which of the following is a external parasite of cattle?
(A) Screw worm (B) Live fluke
(C) Protozoa (D) All of these

352. Which of the following is a infectious disease of cattle?
(A) Rinderpest (B) FMD
(C) Brucellosis (D) All of these

353. Which of the following disease is not a infectious disease?
(A) FMD (B) Black quarter
(C) Haemorrhagic Septicaemia
(D) Mastitis

354. High temperature and bloody discharge from natural opening are the major symptoms of:
(A) FMD (B) Mastitis
(C) Anthrex (D) Black quarter

355. Which of the following is a serious threat to young calves?
(A) Pneumonia (B) Milk fever
(C) Warts (D) Calf scours

356. Milk fever is caused by:
(A) Ca deficiency in blood
(B) P deficiency
(C) Virus (D) Bacteria

357. Milk fever disease is found in:
(A) Cows (B) High producing cows
(C) Calves (D) Camel

358. Which of the following is a internal parasite of cattle?
(A) Screw worm (B) Ticks
(C) Liver fluke (D) Ring worm

359. Which of the following disease of cattle is not caused by virus?
(A) Rinderpest (B) Cowpox
(C) Black quarter (D) FMD

360. Tympanitis is a:
(A) Disorder caused by over feeding resulting more gas production known as a bloat
(B) Disease of cattle caused by bacteria
(C) Instrument used for castrating male animals
(D) It is highly infectious disease of cattle

361. What is correct for mastitis?
(A) Caused by bacteria
(B) It is a disease of lactating dairy cattle
(C) Milk will be stringy and sometimes bloody
(D) All of these

362. What is correct for ketosis?
(A) It is a disease caused by shortage of sugar in body
(B) High producing cows are most affected
(C) Affected animals are nervous and loss their appetite
(D) All of these

363. Which of the following is known as Bang's disease?
(A) Brucellosts (B) FMD
(C) Rinderpest (D) Black quarter

364. Which of the following method is used for making of flock/ewes/rams?
(A) Paint brands (B) Metal tags
(C) Ear notches (D) All of these

365. Which of the external parasite of sheep?
(A) Blow fly (B) Lice
(C) Ticks (D) All of these

366. Bloody severe diarrhea is a main symptoms of:
(A) Coccidiosis (B) Black quarter
(C) Rinderpest (D) Mastitis

367. Shooting diarrhea with fetid smell and high temperature are the symptoms appear in sheep due to the disease of:
(A) Sheep pox (B) Rinderpest
(C) Black quarter (D) Anthrax

368. Sheep scab is caused by:
(A) Mites (B) Lice
(C) Fungus (D) Bacteria

369. The continuous sniffing and nose run in sheep are caused by:
(A) Liver fluke (B) Sheep bots
(C) Ticks (D) Lice

370. Which of the bacterial disease of goat?
(A) Anthrax (B) Brucellosis
(C) Vibriosis (D) All of these

371. Which of the following goat's disease is caused by virus?
(A) Goat pox (B) Rinderpest
(C) Pleuropnemonia (D) All of these

372. Dystocia is a:
(A) Condition in which goat has difficulty in kidding due to malposition of kids
(B) Disease caused by internal parasites
(C) Bacterial disease
(D) Viral disease

373. Trichomoniasis is:
(A) Contagious protozoan veneral disease of both sexes
(B) Infectious bacterial disease of goats
(C) Infectious viral disease of goats
(D) None of these

374. Ketosis is a serious disease of goat herd caused by:
(A) Metabolic disorder of carbohydrates and fats
(B) Bacteria
(C) Virus
(D) External parasites

375. Which of the following parasite causes disease in goat herds?
(A) Flukes (B) Tapeworms
(C) Protozoan (D) All of these

376. Florosis in goats is caused by:
(A) Excess florine (B) Excess P
(C) P deficiency (D) None of these

377. Mycotic dermetitis is caused by:
(A) Fungus (B) Bacteria
(C) Roundworm (D) Liver flukes

378. Which of the following disease is not found in equines?
(A) Epizoatic Lymphangitis
(B) Tetanus
(C) Colic
(D) FMD

379. Which of the most important disease of camel?
(A) Camel pox (B) Rabies
(C) Surra (D) All of these

380. Surra disease is transmitted with the help of:
(A) Water
(B) Housefly
(C) Biting and blood sucking flies
(D) Mosquitoes

381. The important symptoms of surra is:
(A) High body temperature
(B) Anemia (C) Emaciation
(D) All of these

382. Rabies is transmitted to camels by:
(A) Bites of rabid dogs (B) Jackal
(C) Wolves (D) All of these

383. The true stomach in ruminants is known as:
(A) Rumen (B) Reticulum
(C) Abomasum (D) Omasum

384. Santa Gertrudis is a cross of:
(A) Bull of Indian breed x Shorthorm cow of USA
(B) Jersey x Local breed
(C) Kankry x Dargi
(D) None of these

385. Which of the following cattle breed in the smallest one?
(A) Jersey (B) Ayrshire
(C) Brown swiss (D) Holstein – Friesians

386. Which of the following breed considered the most beautiful?
(A) Red deoni (B) Ayreshire
(C) Jersey (D) Tharparkar

387. Which of the following is considered best draft breed of buffalo?
(A) Surti (B) Nagpuri
(C) Nili (D) Murrah

388. Which of the following exotic cattle breed is known as dual purpose type?
(A) Jersey (B) Red Dane
(C) Brown swiss (D) Holstein - Friesian

389. Which of the following is not a dual purpose cattle breed?
(A) Kankrej (B) Haryana
(C) Gir (D) Ongole

390. Which of the following is a oldest cross bred wool breed of sheep?
(A) Corriedale (B) Hissardale
(C) Bharat merino (D) None of these

391. Indian sheep produce mostly:
(A) Carpet wool (B) Fur wool
(C) Fine wool (D) Long wool

392. How many recognized breeds of goat is found in India?
(A) 15 (B) 18
(C) 21 (D) 23

393. Which of the following horse breed is in great demand for polo and racing?
(A) Manipuri (B) Spiti
(C) Kathiawari (D) Bhutia

394. Which of the camel breed has two humps?
(A) Bacterial camel (B) Arabian camel
(C) Both (A) and (B) (D) None of these

395. Which of the following pig breed produces high quality pork?
(A) Middle white x Yorkshire
(B) Berkshire
(C) Danish Landrace
(D) All of these

396. Which of the following poultry breed is suitable for drier areas?
(A) White Leghorn (B) Australorp
(C) White Rock (D) White Cornish

397. Khaki Campbell is a breed of:
(A) Hen (B) Pigeon
(C) Duck (D) Pea cock

398. Which of the following is a breed of Turkey?
(A) Beltville white (B) Breasted Bronze
(C) Norfolks (D) All of these

399. Cell fusion is enhanced by:
(A) Polyethylene glycol
(B) Ultraviolet inactivated Sendai Virus
(C) Colchicine
(D) Both (A) and (B)

400. Better quality chhana has:
(A) Rose flavour (B) Rancid flavour
(C) Nutty flavour (D) None of these

401. Pasty chhana can be prepared by:
(A) Cow milk (B) Sheep milk
(C) Colostrum (D) Goat milk

402. Which of the following is main reason of paraffining of cheese?
(A) The holes of cheese are closed
(B) The cheese moisture is not changed
(C) The cheese is protected from fungal growth
(D) All of these

403. When should colour be added in milk in the process of cheddar preparation?
(A) After adding of lactic acid or before rennet adding
(B) After rennet adding or before lactic acid adding
(C) Before adding of lactic acid or rennet
(D) All of these

404. Evaporated milk means:
(A) Sugar free milk is condensed in vaccum
(B) Sweet milk is condensed in open pan
(C) Milk is condensed at 40°C
(D) None of these

405. In animal cell cultures, differentiation and cell proliferation (cell division) are affected by:
(A) Cell density and serum
(B) Ca^{++} ions and hormones
(C) Cell to cell interactions
(D) All of these

406. Milk powder means:
(A) Evaporated milk in powder form with <3% moisture
(B) Milk powder has 30% water
(C) It is like a condensed milk
(D) None of these

407. What is sponification value of milk?
(A) 210-235 (B) 250-260
(C) 261-265 (D) 266-270

408. Relative density of milk can be calculated with the help of following formula:
(A) $\frac{\text{Lactometer reading} + 1}{1000}$
(B) $\frac{1 - \text{Lactometer reading}}{1000}$
(C) $\frac{\text{Lactometer reading} - 1}{100}$
(D) None of these

409. Pure ghee has a special odour due to presence of the following:
(A) Lactose (B) Diacetyle
(C) Vit. B complex (D) Minerals

410. Colestrol is a:
(A) Milk fat sterole (B) Phytosterole
(C) Phosphonized fat (D) All of these

411. Which is the reason for production of antibodies of a single specificity by a hybridoma clone?
(A) Fusion with a single B cell
(B) Production of antibodies of a single specificity by a single B cell
(C) Production of no antibodies by the other fusion partner of the hybridoma clone
(D) All of these

412. Which of the following two elements are not found in proper amount in normal milk?
(A) Fe and Cu (B) Ca and Mg
(C) K and Ca (D) All of these

413. Which of the following essential amino acid is not found in optimum dose in normal milk?
(A) Riboflavin (B) Nicotinic acid
(C) Thymine (D) None of these

414. What is the disadvantages of deshi method of butter preparation?
(A) Butter may have cooked flavour
(B) Loss of Vitamin B_2 and C due to more heating of milk
(C) More water in butter
(D) All of these

415. What will happen if cream of more acidity is churned for butter?
(A) To reduce fat recovery
(B) Bad odour developed
(C) Reduced keeping quality of butter
(D) All of these

416. Generally, cell proliferation is promoted by:
(A) Low Ca^{++} ions
(B) High growth factor levels
(C) Low cell density
(D) All of these

417. Keeping quality of dahi in comparison to milk is:
(A) More (B) Less
(C) Same (D) Unknown

418. Cheesy butter is caused due to:
(A) Breaking of casein by proteolytic bacteria
(B) Low salt in butter
(C) Pasteurization at high temperature
(D) All of these

419. What is the reason for leaky butter?
(A) Churning of cream at higher temperature
(B) Low working
(C) Storage at high temperature
(D) All of these

420. Which of the following affects the quality of dahi?
(A) Flavour (B) Texture and body
(C) Acidity (D) All of these

421. Which is not correct about suspension cultures?
(A) Cells of certain type are grown in liquid media as suspension cultures
(B) Cells harvested from late log phase cultures are inoculated into fresh liquid medium maintained at 20^{0}C
(C) The initial cell density is usually between 5 x 10 ^{4} to 2 x 10^5 cells/ml
(D) The cultures are stirred at 200-350 rpm

422. Which of the following has highest marks in judging the quality of ghee?
(A) Flavour (B) Body and texture
(C) Colour (D) Acidity

423. B.T. test is used for identification of:
(A) Vegetable oil in ghee
(B) Cotton seed oil in ghee
(C) Pig fat in ghee (D) None of these

424. Fat per cent in ice cream is important for:
(A) Softness (B) Smoothness
(C) Flavour (D) All of these

425. Which is not correct about artificial skin?
(A) The skin explants used for obtaining artificial skin may be either obtained from the patient concerned or from the foreskin of new born babies
(B) Skin cells of newborns grow as fast as those from adult skin
(C) Use of PGA (a synthetic polymer) allows the newborn skin to grow without scars
(D) Artificial skin from newborn skin explants is used to cover the wound till artificial skin is obtained from patient's own skin

426. Ice cream may be pasteurized at:
(A) 60^{0}C for 30 minutes (B) 68^{0}C at 30 minutes
(C) 71^{0}C for 30 minutes (D) All of these

427. Ice cream should be stored at:
(A) -2 to -3^{0}C (B) -5 to -10^{0}C
(C) 2-3 ^{0}C (D) None of these

428. Egg solids in ice cream increases:
(A) Whipping capacity (B) Body and texture
(C) Flavour (D) All of these

429. Colour of cow khoa is:
(A) White (B) Light yellow
(C) Yellow (D) Colourless

430. Colour of buffalo khoa is:
(A) White (B) Light yellow
(C) Yellow (D) Colourless

431. Chhana refers to:
(A) It is a milk product
(B) It is prepared after curding of milk
(C) The solid particles are separated from paneer water
(D) All of these

432. The goats are reared under the housing system of:
(A) Loose housing system
(B) Closed housing system
(C) Stanchion system
(D) All of these

433. Which of the housing system is considered the best one for goats?
(A) Loose housing system
(B) Closed housing system
(C) Stanchion system
(D) All of these

434. 10% solution of ammonia is used mainly for:
(A) To kill the coccidice spores on goat floors
(B) To kill all type of bacteria and viruses
(C) To kill protozoa (D) None of these

435. Animal cell cultures may be recommended by:
(A) Viruses (B) Other cell lines
(C) Bacteria (D) All of these

436. Kerosol is used for:
(A) To disinfect the goat floor by killing most the bacteria and its spores
(B) To kill the eternal parasites of goats
(C) To kill internal parasites of goats
(D) None of these

437. Which of the following is a common type of goat keeping systems?
(A) Subsistence (B) Extensive
(C) Intensive (D) All of these

438. The housing place of a horse is known as:
(A) Ward (B) Stable
(C) Brood (D) None of these

439. Memory cells lack in:
(A) IgD (B) IgG
(C) IgM (D) IgA

440. A house of 6 m x 12 m may accommodate:
(A) 200 hens of heavy breeds
(B) 200 hens of light breeds
(C) 300 hens of light breeds or 260 hens of heavy breeds
(D) 260 hens of light breeds or 300 hens of heavy breeds

441. Cell fusion leads to the production of?
(A) Homokaryons (B) Heterokaryons
(C) Hybrid cells (D) All of these

442. Which of the following method is used for dehorning?
(A) Chemical (B) Hot Iron
(C) Clippers and saws (D) All of these

443. Which chemical is used for dehorning of calves?
(A) Caustic potash (B) Potassium chloride
(C) Benzene (D) All of these

444. Which is not correct?
(A) Proteins produced by genes transferred by genetic engineering into selected hosts are called recombinant proteins
(B) In general, antigens are immunogenic
(C) Antibodies belong to a special category of proteins called immunoglobulins
(D) Monoclonal antibodies are produced from hybrid cells obtained by fusing a memory B cell with a myeloma cell

445. The old cattle may be dehorned with the help of:
(A) Electric dehorner (B) Chemicals
(C) Clippers and saws (D) All of these

446. The age of cattle can be determined with the help of:
(A) Tail (B) Teeth
(C) Health (D) Pregnancy

447. Castration means:
(A) Feed properly to cattle
(B) Removal of male parts from male calves
(C) Complete milking of cow
(D) None of these

448. Age of male calves for castration is:
(A) 6 months (B) 12 months
(C) 15 months (D) 20 months

449. Best season for castration is:
(A) Rainy season (B) Spring season
(C) Winter season (D) All of these

450. Which of the following method is recommended for castration the male calves?
(A) Removal of testicles by making incision
(B) Crush and destroy spermatic cord & blood vessels with the help of Burdizzo castrator
(C) Removal of whole male organs
(D) All of these

451. A hybrid cell obtained by cell fusion has:
(A) Two genetically different nuclei
(B) A single nucleus obtained by fusion of two genetically different nuclei
(C) Cytoplasm derived from the two fusion partners
(D) Both (B) and (C)

452. Castrated male goat is called as:
(A) Ram (B) Buck
(C) Wether (D) Chhanga

453. Male goats should be castrated at a age of:
(A) Just after birth
(B) 5 days after birth
(C) 10 days to 3 weeks of age
(D) All of these

454. What are the following bacteria present in starter?
(A) *Lactobacillus acidophilus*
(B) *Streptococcus lactis*
(C) *L. bulgaricus*
(D) All of these

455. Dahi has more nutritive value than milk because of:
(A) Dahi has more Ca and P
(B) Dahi has bacteria that helps to prepare Vit. B and C
(C) Dahi casein is more digestible than milk casein
(D) All of these

456. Dahi like product refers to:
(A) Dahi made of *Lactobacillus acidophilus*
(B) Dahi made of *Streptococcus lactis*
(C) Dahi made of artificial milk
(D) All of these

457. What is the main component of ghee adulteration?
(A) Vegetable fat (B) Pig fat
(C) Edible oil (D) None of these

458. Which of the following fur wool is most valuable?
(A) Persian lamb (B) Broadtail
(C) Karakul (D) All of these

459. Persian lamb fur is collected from:
(A) <10 days old lambs (B) Newly born lambs
(C) >2 weeks old lambs (D) > 3 weeks old lambs

460. The sheep of southern region are suited mainly for:
(A) Fine wool (B) Carpet wool
(C) Meat (D) Fine wool and meat

461. Which of the following pair is matched correctly?
(A) Merino – Spain
(B) Rambouillet – France
(C) Corriedale – New Zealand
(D) All of these

462. Which of the following pair is matched correctly?
(A) Cheviot – Scotland (B) Leicester – England
(C) Karakul – USA (D) All of these

463. Wethers refer to:
(A) Male lambs castrated at an early age before they reach sexual maturity
(B) Male lambs ready for slaughter
(C) Lambs bearing a full year's growth of wool
(D) Mating male sheep

464. Shearer lambs refers to:
(A) One year full growth of wool
(B) Fattened male ready for slaughter
(C) Mating male sheep
(D) Castrated male sheep

465. Feeder sheep refers to:
(A) Over one year old fattened sheep for marketing purposes
(B) < 1 year old fattened sheep
(C) >2 years old fattened sheep
(D) All of these

466. Flushing refers to:
(A) Good feeding to sheep before mating
(B) Good feeding to sheep before marketing
(C) Good feeding to sheep before shearing
(D) All of these

467. Flushing is useful for:
(A) It brings ewes into heat early
(B) Higher conception rate
(C) Possibility of more twins
(D) All of these

468. Grafting in case of sheep refers to:
(A) To put dead skin of a lamb over the lamb of another
(B) To remove fleece
(C) To remove tails (D) None of these

469. Docking refers to:
(A) Removal of testicles (B) Removal of tails
(C) Removal of wool (D) Removal of fleece

470. Docking can be performed with the help of:
(A) Knife and Emasculator (B) Hot iron
(C) Rubber band (D) All of these

471. Which of the following method is used for marketing of flock/ewes/rams?
(A) Paint brands (B) Metal tags
(C) Ear notches (D) All of these

472. Tagging in sheep refers to:
(A) Removal of wool from around the dock before breeding season starts
(B) Putting tags on necks for their identification
(C) Making colour marks for their identification
(D) All of these

473. Eyeing in sheep refers to:
(A) Clipping of wool from around the eyes
(B) Eye operation in case of eye problems
(C) Tearing of eyes (D) All of these

474. Shearing in sheep refers to:
(A) Clipping of wool from whole body of rams and ewes before breeding season
(B) Removal of wool from eyes only
(C) Removal of wool from dock only
(D) None of these

475. Ringing refers to:
(A) Removal of wool from the neck & around the penis of rams before breeding season
(B) Making of colour rings in necks fur identification
(C) Putting belt around the neck of sheep
(D) All of these

476. Conventional linkage maps suffer from:
(A) The number of genes producing distinct morphological effects is rather limited
(B) Their mapping is tedious and time-taking
(C) Genes governing quantitative traits can not be mapped
(D) All of these

477. Which of the following is a recognizes quality of sheep wool for export?
(A) Joria (B) Harnai
(C) Rajpootana (D) All of these

478. The sheep wool is divided on the basis of length into:
(A) Carpel and apparel wool
(B) Course and rough wool
(C) Klemp and fine wool
(D) All of these

479. Souring of wool refers to:
(A) Removal of grease dirt and vegetable matters from wool
(B) Removal of thick and short hairs from fine wool
(C) Removal of clotted fleece from the wool
(D) All of these

480. Klemp is a:
(A) Fine hair (B) Course hair
(C) Very fine hairs (D) All of these

481. Shrinkage of sheep wool refers to:
(A) Weight of grease, dirt etc expressed as a percentage of the grease fleece weight
(B) Shrinking of hairs
(C) Loss of wool weight
(D) None of these

482. Nitric acid is used for detection of:
(A) Adulteration of wax, vegetable ghee and animal fat in ghee
(B) Adulteration of vegetable oil in ghee
(C) Adulteration of cotton oil in ghee
(D) All of these

483. Why does goat rearing is beneficial profession?
(A) Quick onset of reproduction and high fecundity
(B) High survival rate under adverse conditions
(C) Production of meat, milk, skins and fibre
(D) All of these

484. Among the given biological fluids used as culture medium, which one is the most commonly used in animal cell cultures?
(A) Serum (B) Amniotic fluid
(C) Insect haemolymph
(D) Aqueous humour from eye

485. How are fat per cent and cream acidity related for cream churning?
(A) Positive (B) Linear
(C) Inverse (D) Negative

486. Fishy flavour in butter is developed due to:
(A) Hydrolysis of lecithin
(B) Oxidation of butter due to high acidity
(C) Oxidation of Fe and Cu
(D) All of these

487. Mottled butter is caused by:
(A) Improper washing of butter
(B) Improper salt mixing
(C) Improper working of butter
(D) All of these

488. Rancidity of butter is caused by:
(A) Hydrolysis of fat in cream by lipase enzyme
(B) High acidity of cream
(C) Storage at low temperature
(D) All of these

489. Dahi refers to:
(A) Fermented milk products
(B) Lactose is converted into lactic acid
(C) Starter is mixed at 21^{0}C milk temperature
(D) All of these

490. Which of the following bacteria is responsible for good flavour in dahi?
(A) *Streptococcus lactis*
(B) *Streptococcus diacetyl aromaticus*
(C) *Lactonostoc citrovorum*
(D) All of these

491. Which of the following animal is known as ship of the desert?
(A) Donkey (B) Horse
(C) Camel (D) Zebra

492. Foal-heat refers to:
(A) Mares come into heat just after 4-11 days of foaling
(B) Mating of mares with stallion
(C) First heat of a mare
(D) None of these

493. Mule is a cross of:
(A) She ass x Stallion pony
(B) Male donkey x Mare
(C) Male horse x Female donkey
(D) Zebra x Horse

494. Mules are:
(A) Fertile (B) Infertile
(C) Both of these (D) None of these

495. Which of the following method is commonly used in pigs identification?
(A) Iron branding (B) Ear notching
(C) Neck strips (D) All of these

496. Which techniques was used for development of 'Dolly"?
(A) Nucleus removal (C) Cell culture
(C) Cell fusion (D) All of these

497. Ringing pigs refer to:
(A) To put ring or iron ring in the nose of pig
(B) To make a ring by removing hair from the neck
(C) To brand the hogs with round hot iron ring
(D) All of these

498. Cannibalism in hens refers to:
(A) Picking of each other by hens
(B) Laying of eggs
(C) Breaking of eggs by other hens
(D) None of these

499. Brooder house refers to:
(A) House for chicks (B) House of pullets
(C) House of layers (D) House of cocks

500. Caponette refers to:
(A) Female chicks produced after implanting of synthetic female hormones in young cockerels
(B) Chicks without male organs
(C) Both (A) and (B)
(D) None of these

501. Capon refers to:
(A) Female chicks produced after implanting of synthetic female hormones in young cockerels
(B) Chicks without male organs
(C) Both (A) and (B)
(D) None of these

502. Which is not correct about T cells?
(A) During T cell maturation, rearrangement of genes encoding T cell receptor (TCR) takes place
(B) Each TCR molecule is a heteromer consisting of an α and β chain or a δ and γ chain
(C) Each T cell expresses TCR of a single antigenic specificity
(D) TCR gene rearrangements follows the same general scheme as that for antibody genes

503. Curdy of icecream is due to:
(A) High acidity (B) Hydrolysis of fat
(C) High over run (D) None of these

504. Yellow level ghee is known as:
(A) Cow ghee (B) Buffalo ghee
(C) Normal ghee (D) Special ghee

505. Blue level ghee is known as:
(A) Cow ghee (B) Buffalo ghee
(C) Normal ghee (D) Special ghee

506. Red level ghee is known as:
(A) Cow ghee (B) Buffalo ghee
(C) Normal ghee (D) Special ghee

507. Grey level ghee is known as:
(A) Cow ghee
(B) Buffalo ghee
(C) Normal ghee
(D) Special ghee

508. Tollowy flavour in ghee is caused by:
(A) Oxidation (B) Reduction
(C) Hydrolysis (D) Protolysis

509. The techniques of cyclic exposure to medium and gas phase is successful in long-term cultures of which human adult tissues?
(A) Mammary epithelium (B) Oesophagus
(C) Uterine endocervix (D) All of these

510. Which of the following milk has higher amount of lower fatty acid (i.e. Capric, Caprylic and Caproic acids)?
(A) Cow milk (B) Goat milk
(C) Buffalo milk (D) Camel milk

511. The fat per cent of diarrhea infected animal will be:
(A) Low (B) More
(C) Same (D) None of these

512. The fat per cent of mastitis infected animal will be:
(A) Low (B) More
(C) Same (D) None of these

513. The milk pH of mastitis infected animal will be:
(A) Increased (B) Decrease
(C) Same (D) Unknown

514. Which one is the mostly used as culture medium?
(A) Liver extract
(B) Chick embryo extract
(C) Plasma (D) Bone morrow

515. Which of the protein is not found in milk?
(A) Casein (B) Albumin
(C) Globulin (D) All of these

516. Who reported first time the germicidal property in goat milk?
(A) Fokker (B) R. Hook
(C) Mc Millon (D) Michael

517. Milk pasteurization refers to:
(A) To kill bacteria before its use
(B) To heat milk at 60^{0}C for 20 minutes
(C) It does not affect the quality of milk and cream layer
(D) All of these

518. What is benefit of pasteurization of milk?
(A) No effect on Ca, P and Vit. A, D and B complex
(B) No or little effect on flavour
(C) No effect on cream layer
(D) All of these

519. What are the benefit of homogenization of milk?
(A) No creamy layer at milk surface
(B) High digestibility
(C) Better colour and flavour
(D) All of these

520. What is used for cream standardization?
(A) Water (B) Milk
(C) Separata (D) Whey

521. Separator slime is a:
(A) Concentrated and sticky product of milk that is found in the inner walls of bowl hood
(B) The colour is grey white to ash
(C) It is a mixture of impurities in milk
(D) All of these

522. The surface available for the attachment of cultured cells is called:
(A) Surface attachment site(B) Platform
(C) Substrate (D) Both (A) and (B)

523. Which country used centrifugal force first time to separate the cream from milk?
(A) Germany (B) Sweden
(C) Denmark (D) Pakistan

524. Which country used centrifugal force first time to separate the cream from milk?
(A) India (B) Sweden
(C) Denmark (D) Germany

525. Butter refers to:
(A) It is manufactured from either cream or milk
(B) It should not have <80% fat and >2% other ingredients
(C) It should not have >16% water
(D) All of these

526. Which one of the hardiest breed of poultry?
(A) Rhode Island Red (B) White leghorn
(C) Australorp (D) New Hempshire

527. Gummar refers to:
(A) Sheep having no teeth in her mouth
(B) Sheep that has broken teeth
(C) Sheep with only 2 teeth
(D) None of these

528. The condition, Broken mouth in case of sheep, refers to:
(A) All teeth disappeared
(B) Few teeth broken
(C) All teeth have become red and brown
(D) All of these

529. Feeder lambs refers to:
(A) Under one year age used for slaughter
(B) Under 2 year age used for slaughter
(C) Lambs are fed properly&make ready for mating
(D) Castrated lambs

530. Cashgora refers to:
(A) Goat breed
(B) It is a cross between Changthangi x Angora
(C) Both (A) and (B)
(D) None of these

531. Which techniques is used for the separation of X- and Y-chromosomes bearing sperms?
(A) Gel electrophoresis
(B) Sephadex gel column
(C) Zoobloting (D) Both (A) and (B)

532. Which events occurs during inflammatory response?
(A) An increase in blood vessel diameter
(B) An increase in capillary permeability
(C) Influx of phagocytes from the capillaries into the tissues
(D) All of these

533. Which of the following goat breed is found in Peninsular India?
(A) Osmunabadi (B) Sangamneri
(C) Malabari (D) All of these

534. Which of the following breed is found in Himalayan region of India?
(A) Angora (B) Kashmiri
(C) Guddi and Chamba (D) All of these

535. Barbari goat is more useful because of:
(A) Good milk and meat producer
(B) More prolific as it may kid twice with two kids
(C) Higher butter fat (5%)
(D) All of these

536. Billy refers to:
(A) A goat breed
(B) It is found in Barren Island in Andamans
(C) It survive on sea water
(D) All of these

537. Which is correct for Angora breed of goat?
(A) It is a goat breed and its hair are known as Mohair
(B) It is found in the Himalayan mountain region
(C) It originated in Asia-minor and Turkey
(D) All of these

538. The minimum standards of milk and milk product quality is maintained by the rules/Acts of:
(A) Prevention of Food Adulteration Rules, 1955 and amended in 1976
(B) Food Adulteration Act 1954
(C) Indian Dairy Corporation.
(D) None of these

539. Goat milk is easily digested because of:
(A) Small size fat globules and soft curd
(B) Low fat per cent
(C) Fat & protein per cent are equal to human milk
(D) All of these

540. Lymph nodes are located abundantly in:
(A) Groin (B) Arm pit
(C) Neck (D) All of these

541. Which of the following is the highest producer of milk?
(A) Alpine (B) Anglo-nubian
(C) Saanen (D) Toggenburg

542. Most beautiful goat breed is:
(A) Barbari (B) Jamanapari
(C) Beetal (D) Surti

543. Pashmina is a:
(A) It is a goat breed
(B) It is found in Ladakh, Lahaul and Spiti valleys
(C) It produces fine and silky hair
(D) All of these

544. Which of the following is a dual purpose goat breed?
(A) Barbari (B) Jamanapari
(C) Mehsana (D) All of these

545. Which of the following goat breeds are suitable for meat purpose?
(A) Marwari (B) Bengal goat
(C) Barbari (D) All of these

546. Unripened paneer is known as:
(A) Cheddar (B) Swiss
(C) Cottage (D) All of these

547. Which of the following is a soft paneer?
(A) Limberger (B) Camembert
(C) Cottahe (D) All of these

548. Which of the following paneer is without gas holes?
(A) Cheddar (B) Swiss
(C) Limberger (D) All of these

549. Which of the following paneer is with gas holes?
(A) Cheddar (B) Swiss
(C) Limberger (D) All of these

550. Paneer is mainly source of:
(A) Milk fat (B) Milk protein
(C) Milk lactose (D) All of these

551. Which type of paneer is most popular in the world?
(A) Swiss (B) Cheddar
(C) Limberger (D) Roquefort

552. Which is an example of natural medium?
(A) Tissue extract (B) Plasma clot
(C) Biological fluids (D) All of these

553. Cheddar paneer is prepared by which milk:
(A) Buffalo milk (B) Cow milk
(C) Both milk (D) None of these

554. Which of the following is considered the most important parameter for judging the quality of khoa?
(A) Flavour (B) Body and texture
(C) Colour (D) Packing

555. Which of the following has higher protein per cent?
(A) Milk (B) Ice cream
(C) Cream (D) Butter

556. Rancidity of ice cream is caused by:
(A) High acidity (B) Hydrolysis of fat
(C) High over run (D) None of these

557. Foamy of ice cream is caused by:
(A) High acidity (B) Hydrolysis of fat
(C) High over run (D) None of these

558. Serum can be obtained from:
(A) Placental cord blood (B) Adult human blood
(C) Horse blood (D) All of these

559. The colour of milk is:
(A) Colourless (B) White
(C) Grey (D) None of these

560. The main reason of milk fat rancidity is:
(A) Conversion of fat into methyl ketones
(B) Conversion of fat into butyric acid
(C) Conversion of fat into hydroperoxide
(D) All of these

561. Evaporated milk means:
(A) Sugar free milk is condensed in vaccum
(B) Sweet milk is condensed in open pan
(C) Milk is condensed at 40^{0}C
(D) None of these

562. Paraffining of cheese means:
(A) To cover cheese with thin layer of wax
(B) To cover cheese with thin layer of salt
(C) To cover cheese with thin layer of colour
(D) All of these

563. Which one is the commonly used in animal cell culture media?
(A) Placental cord blood (B) Cattle serum
(C) Foetal calf serum (D) Horse serum

564. Which of the following protein is available in curd?
(A) Lactose (B) Casein
(C) Glycerol (D) All of these

565. Which of the following cow has the highest fat per cent in the milk?
(A) Sindhi (B) Haryana
(C) Tharparkar (D) Shahiwal

566. Embryo transfer technology suffers from:
(A) The donor females are removed from production for the period they are used as donor of young embryos
(B) A high degree of expertise is required for an efficient and successful operation
(C) The cost of producing each progeny is several-fold higher than that from natural events
(D) All of these

567. Which of the following casein is highest available in milk?
(A) α Casein (B) β Casein
(C) γ Casein (D) All of these

568. Lacto globulin is known mainly for:
(A) It provides immunity to newly born babies
(B) It saves people from heat strokes
(C) It increases body weight
(D) All of these

569. Lactose in milk is a:
(A) Carbohydrate (B) Fat
(C) Protein (D) All of these

570. Which of the following amylase is present in milk?
(A) α amylase (B) β amylase
(C) γ amylase (D) All of these

571. Which is not correct about T-cells?
(A) T lymphocytes originate from myeloid stem cells
(B) Lymphoid stem cells that migrate into thymus ultimately mature into T cells
(C) During T cell maturation, rearrangement of the genes encoding T cell receptor (TCR) takes place
(D) T cells are primarily of two types, viz., T helper cells and T cytotoxic cells

572. Pasty in cheese is caused by:
(A) High moisture and acidity
(B) Low moisture and acidity
(C) High salt per cent
(D) All of these

573. If the humidity is less than 70%, the cheese may get:
(A) Dry (B) Wet
(C) Fungus growth (D) All of these

574. If the humidity is more than 70%, the cheese may get:
(A) Dry (B) Wet
(C) Fungus growth (D) All of these

575. Which of the following cheese has the highest protein per cent:
(A) Cheddar (B) Swiss
(C) Limberger (D) Camembert

576. In comparison to cow and goat milk has fat globules of:
(A) Larger size (B) Smaller size
(C) Same (D) All of these

577. Which of the following is considered TB free milk?
(A) Goat milk (B) Cow milk
(C) Sheep milk (D) Camel milk

578. Lactometer is used to measure the:
(A) Relative density of milk
(B) Relative density of water
(C) Relative density of fat
(D) Relative density of protein

579. What will the effect of mixing of SNF in milk:
(A) Relative density will increase
(B) Relative density will decrease
(C) Relative density will not change
(D) All of these

580. Which of the following will increase with advancement of lactation period?
(A) Ca and P (B) Cl
(C) Fat (D) All of these

581. What will be the impact of high temperature on milk fat?
(A) Increase (B) Decrease
(C) Same (D) All of these

582. If the interval of milking increases, the fat per cent in milk will:
(A) Increase (B) Decrease
(C) No change (D) Not known

583. The fat per cent in evening milk is found higher because of:
(A) Less interval in milking
(B) Cow produces more concentrated milk in evening than morning
(C) Cow eats more feed in day and gives higher fat per ent in evening
(D) None of these

584. Which of the following milk has higher per cent of fat?
(A) Fore milk (B) Middle milk
(C) Stripped milk (D) All of these

585. Lymphoid stem cells give rise to:
(A) Lymphocyte progenitor cells
(B) Pro-B and Pro-T cells
(C) Pre B and Pre –T cells
(D) All of these

586. During heat period, the milk will:
(A) Increase (B) Decrease
(C) No change (D) Not known

587. How much energy in available from 1 gm milk?
(A) 8 calorie (B) 9 calorie
(C) 10 calorie (D) 12 calorie

588. What is the temperature range for survival of psychrophilic bacteria?
(A) 3-20^{0}C (B) 20-50^{0}C
(C) 50-70^{0}C (D) None of these

589. What is the temperature range for survival of mesophilic bacteria?
(A) 3-20^{0}C (B) 20-50^{0}C
(C) 50-70^{0}C (D) 75-80^{0}C

590. What is the temperature range for survival of thermophilic bacteria?
(A) 3-20^{0}C (B) 20-50^{0}C
(C) 50-70^{0}C (D) 75-80^{0}C

591. *Pseudomonas fragi* and *Pseudomonas fluorescens* belong to which of the following category?
(A) Psychrophilic bacteria
(B) Mesophilic bacteria
(C) Thermophilic bacteria
(D) All of these

592. Sliminess is caused due to:
(A) Fermentation of milk cream
(B) Fermentation of lactose
(C) Fermentation of protein
(D) None of these

593. *Streptococcus lactis* and *S. cremoris* belong to which of the following group?
(A) Psychrophilic bacteria
(B) Mesophilic bacteria
(C) Thermophilic bacteria
(D) Thermoduric bacteria

594. Which of the following is the thermophilic bacteria?
(A) *Micrococcus luteus*
(B) *Micrococcus condidus*
(C) *Streptococcus thermophilus*
(D) All of these

595. *Lactobacillus, thermophilus* and *Bacillus calidolactis* are the names of:
(A) Psychrophilic bacteria
(B) Mesophilic bacteria
(C) Thermophilic bacteria
(D) Thermoduric bacteria

596. Sweet curding of milk is caused by:
(A) *Streptococcus subtilis* (B) *Bacillus mycoides*
(C) *B. cereus* (D) All of these

597. The oiliness or cardboard flavour or sourness in milk is caused by:
(A) *Streptococcus liquefaciens*
(B) *Micrococcus pitiutoporus*
(C) *Pseudomonas flourescens*
(D) All of these

598. Oiliness of milk is caused mainly due to:
(A) Oxidation of fat
(B) Reduction of fat
(C) Fermentation of lactose
(D) Fermentation of fat

599. Sweet curdling is caused mainly due to:
(A) Fermentation of casein
(B) Fermentation of fat
(C) Oxidation of ash
(D) Fermentation of ash

600. Normal fermentation of milk is caused by:
(A) *Streptococcus lactis*
(B) *Bacillus subtilis*
(C) *Streptococcus liquefaciens*
(D) All of these

601. Rancid flavour in milk is caused due to:
(A) Fermentation of fat
(B) Fermentation of lactose
(C) Fermentation of protein
(D) Fermentation of ash

602. Sliminess or ropy fermentation is caused by:
(A) *Achromobactor aerogans*
(B) *Bacterium cloacae*
(C) *Alkaligenes viscosus*
(D) All of these

603. Frothiness in milk is caused due to:
(A) *Aerobacter aerogens* (B) *Torula cremoris*
(C) *Torula spherical* (D) All of these

604. Blue colour of milk is developed due to:
(A) *Pseudomonas syncyanea*
(B) *Streptococcus lutea*
(C) *Streptococcus marcescens*
(D) All of these

605. Red colour of milk is developed due to:
(A) *Pseudomonas syncyanea*
(B) *Streptococcus lutea*
(C) *Streptococcus marcescens*
(D) All of these

606. Yellow colour of milk is developed due to:
(A) *Pseudomonas syncyanea*
(B) *Streptococcus lutea*
(C) *Streptococcus marcescens*
(D) All of these

607. Which of the following yeast is found in milk?
(A) *Torula cremoris* (B) *Torula spherical*
(C) *Torula lactis* (D) All of these

608. Which of the following fungus involved in milk industry?
(A) *Penicillium glaucum*
(B) *Penicillium rogueforti*
(C) *Oidium spp.*
(D) All of these

609. Roquefort and Gorgonzola paneer are made with the help of:
(A) Bacteria (B) Yeast
(C) Fungi (D) All of these

610. Which of the following bacteria acts first on milk?
(A) *Lactobacillus*
(B) *Streptococcus lactis*
(C) *Bacillus subtilis* (D) None of these

611. Germicidal property of milk is due to presence of:
(A) Lactenin (B) Fatty acids
(C) Amino acids (D) All of these

612. The lymphocytes protect from:
(A) Viral pathogens (B) Bacterial pathogens
(C) Cancer (D) All of these

613. Milking pot may be cleaned with:
(A) Clorine water (B) Hot water
(C) Soda water (D) All of these

614. Milking pot may be sterilized by:
(A) Keep pot in water at 82^{0}C for 15 minutes
(B) Keep pot in water at 101.6^{0}C for 15 minutes
(C) Keep pot in chlorine water for 15 minutes
(D) None of these

615. Which work is performed by lymph nodes?
(A) Preparation of blood
(B) Elimination of CO_2
(C) Destruction of bacteria
(D) All of these

616. At which temperature the milk is heated for pasteurization at temperature long time (L.T.L.T.) method:
(A) 62.8^{0}C for 30 minutes (B) 60^{0}C for 15 minutes
(C) 62.8^{0}C for 15 minutes (D) 70^{0}C for 30 minutes

617. Batch pasteurization method refers to:
(A) Milk is heated at 62.8 to 65.6^{0}C for 30 minutes
(B) Milk is heated with the help of either stream or hot water
(C) Milk is filled in big containers
(D) All of these

618. Antibodies fight against:
(A) Starvation (B) Ageing
(C) Infection (D) All of these

619. What is the benefit of bottle pasteurization:
(A) High keeping quality
(B) Free from pathogenic bacteria
(C) Free from *Bacillus coli*
(D) All of these

620. High temperature short time (HTST) method refers to:
(A) Continuous pasteurization
(B) Flash method
(C) Heating milk at 71.7^{0}C for 15 minutes
(D) All of these

621. Vacuum pasteurization refers to:
(A) Milk is heated at 72.85^{0}C for few seconds
(B) Milk is cooled at 43.5^{0}C
(C) It is a continuous process
(D) All of these

622. Stassanization refers to:
(A) Discovered by Dr. Stassano for milk pasteurization
(B) Milk is heated at 74^{0}C for 15-16 seconds
(C) Milk is cooled at 15^{0}C temperature
(D) All of these

623. Uperization of milk means:
(A) Milk is heated at 148.8^{0}C for 1 second
(B) Milk is depressed to remove bad flavour
(C) Milk is aerated just to avoid loss of vitamins and Vitamin C
(D) All of these

624. What is the benefit of uperization of milk?
(A) Milk is completely sterilized
(B) No effect on quality of milk
(C) Bad flavour is removed
(D) All of these

625. Bactofugation refers to:
(A) A method developed by Prof S. Monart to remove bacteria from milk instead of killing bacteria
(B) Milk is heated at 75^{0}C
(C) Milk is run under speed centrifuge section to remove 99.99% bacteria
(D) All of these

626. What is the benefit of bactofugation of milk?
(A) Bacteria killed up to 99.99%
(B) Milk flavour remains like fresh milk
(C) Higher keeping quality than pasteurizaton milk
(D) All of these

627. Which of the thermoduric bacteria is found in pasteurized milk?
(A) *Micrococcus epidermidis*
(B) *Microbacterium liquefaciens*
(C) *Streptococcus faecalis*
(D) All of these

628. Which of the thermophilic bacteria is found in pasteurized milk?
(A) *Micrococcus epidermidis*
(B) *Lactobacillus thermophilus*
(C) *Streptococcus lactis*
(D) All of these

629. Pasteurized milk may be infected with:
(A) Mesophilic bacteria
(B) Psychrophilic bacteria
(C) Thermophilic bacteria
(D) Thermoduric bacteria

630. Sterilization of milk refers:
(A) Milk is heated at 93.3 to 99.4^{0}C for 30 minutes
(B) Milk is heated at 80^{0}C for 20 minutes
(C) Milk is heated at 90^{0}C for 20 minutes
(D) All of these

631. Bottle milk is sterilized at:
(A) 100-110^0 C for one hour
(B) 98^{0}C for 40 minutes
(C) 100^{0}C for 20 minutes
(D) All of these

632. Which of the following is the main benefit of sterilization of milk?
(A) Increase keeping quality
(B) Free from all bacteria
(C) Increase digestibility
(D) All of these

633. Homogenization of milk refers to:
(A) To break fat globules and mix uniformly in the milk
(B) To avoid deposition of creamy layer at surface
(C) To break fat globules and allow to deposit at surface
(D) All of these

634. Homogenization of milk requires:
(A) 60-65.5^{0}C temperature and 2000-2500 lbs per square inch pressure
(B) 40^{0}C temperature and 2000 lbs per square inch pressure
(C) 60-65.5^{0}C temperature and 1500 lbs per square inch pressure
(D) All of these

635. During heat period, the fat per cent in milk will:
(A) Increase (B) Decrease
(C) No change (D) None of these

636. What are the cooling agent is used for cooling the milk?
(A) Salt solution (B) Ammonia solution
(C) Ice (D) All of these

637. Cream is a:
(A) Part of milk which has higher fat per cent than milk
(B) Generally it has >25% fat
(C) It is collected from milk through gravity or centrifugal forces
(D) All of these

638. Cream has more dark yellow colour than milk is due to:
(A) Lower carotene and xanthophylls in milk
(B) Higher carotene & xanthophylls in cream
(C) Cream has higher density than milk
(D) Milk has higher density than cream

639. Food poisoning is caused by:
(A) *Staphylococcus aureus*
(B) *Bacillus* spp.
(C) *Streptococcus lactis*
(D) All of these

640. The pathogenic and non-pathogenic bacteria may be killed if the milk is heated at the temperature:
(A) 62.8^{0}C for 30 minutes (B) 60^{0}C for 15 minutes
(C) 62.8^{0}C for 15 minutes (D) 80^{0}C for 45 minutes

641. Bottle pasteurization refers to:
(A) Hot milk is filled in hot bottles before pasteurization
(B) The bottle are then heated in hot water at 62.8 to 65.6^{0}C for 30 minutes
(C) The filled bottles may be heated in hot air chamber at 64^{0}C for 30 minutes
(D) All of these

642. Which of the following is responsible for better flavour in cream?
(A) Diacetyl
(B) Acidic and Propionic acid
(C) Acetyl methyl carbinol
(D) All of these

643. Separata refers to:
(A) Milk without cream
(B) Milk after separated butter
(C) Milk with cream
(D) All of these

644. Butter milk refers to:
(A) The part of milk and cream that left after separating butter
(B) Milk without cream
(C) Part of milk left after separating butter
(D) All of these

645. Whey refers to:
(A) Part of milk left after separating paneer
(B) Part of milk left after separating cream
(C) Part of milk after separating butter
(D) None of these

646. Who showed that leucocytes can divide outside the body?
(A) Meyers (B) Walker
(C) Arnold (D) Adams

647. Which of the following has higher SNF?
(A) Whey (B) Milk
(C) Butter milk (D) Separata

648. Which of the following has higher mineral matters?
(A) Ghee (B) Separata
(C) Butter milk (D) Milk

649. What is the suitable temperature for cream pasteurization?
(A) 44^{0}C for 30 minutes (B) 58^{0}C for 30 minutes
(C) 62^{0}C for 30 minutes (D) 70^{0}C for 20 minutes

650. Why cream pasteurization is required?
(A) To get higher fat per cent in butter
(B) To improve taste, texture and flavour of butter
(C) To get homogenous butter
(D) All of these

651. What is the breaking stage of cream churning?
(A) Separation of fat globules from lassi/butter milk/separata
(B) Formation of fat globules likesorghum seed size
(C) Churning glass become clear
(D) All of these

652. Which is not provided by serum?
(A) Carbohydrated and lipids
(B) Plasma proteins
(C) Alkaloids (D) Some enzymes

653. Which of the following is used for colour development in butter?
(A) Annatto (B) Yellow
(C) Blue (D) Red

654. Milk fat can be tested with the help of:
(A) Butyrometer (B) Lactometer
(C) Thermometer (D) All of these

655. Which of the following are tested under organolaptic tests:
(A) Milk appearance (B) Milk taste
(C) Milk flavour (D) All of these

656. Hansa test is famous for:
(A) Detection of buffalo milk in mixture
(B) Detection of *Lactobacillus lactis* in cream
(C) Detection of antiserum in cow milk
(D) None of these

657. Pure ghee should have Reichert Meissal value:
(A) >28 (B) <24
(C) >15 (D) None of these

658. Halphen test is used to detect:
(A) Adulteration of vegetable oil in ghee
(B) Adulteration of animal fat in ghee
(C) Adulteration of cotton oil in ghee
(D) All of these

659. Which cell cultures is an example of continuous cell lines?
(A) HeLa cell line (B) MDCK dog kidney
(C) 3T3 fibroblasts (D) All of these

660. Iodine test is used to detect the milk adulterated with:
(A) Starch (B) Sucrose
(C) Water (D) All of these

661. Valenta test is used to detect:
(A) Adulteration of vegetable oil in pure ghee
(B) Adulteration of butter in pure ghee
(C) Adulteration of animal fat in ghee
(D) None of these

662. Nitric acid test detects the milk adulteration with:
(A) Starch (B) Separata
(C) Milk powder (D) All of these

663. Bomer's phytosterol test is used to detect:
(A) Adulteration of vegetable oil in pure ghee
(B) Adulteration of butter in pure ghee
(C) Adulteration of tallow in ghee
(D) None of these

664. Which is commonly used for promoting cell growth in animal cell cultures?
(A) Ethanolamine (B) Insulin
(C) Transferrin (D) All of these

665. Which of the following is used as milk preservative?
(A) Formalin (B) Borex & Boric acid
(C) Salicylic acid (D) All of these

666. Nitrate test is used for detection of:
(A) Nitrate in milk (B) Formalin in milk
(C) Borex in milk (D) None of these

667. Brine solution is used as:
(A) Cleaning agent (B) Refrigerant
(C) Heating agent (D) None of these

668. Which of the following can be mixed in ghee?
(A) Vegetable ghee (B) Lard and tallow
(C) Safflower oil (D) All of these

669. Vieth ratio may be used for:
(A) Adulteration of milk (B) Urea in milk
(C) Fat estimation in milk (D) None of these

670. Milk powder may be prepared from:
(A) Milk (B) Separata
(C) Cream (D) All of these

671. Condensed milk may be prepared from:
(A) Normal milk (B) Separata milk
(C) Both milk & separata (D) All of these

672. Evaporated milk may be made of:
(A) Sugar free milk (B) Sweet milk
(C) Both type (D) None of these

673. Bitterness in cheese is caused by:
(A) High acidity (B) Poor quality of milk
(C) Low salt per cent (D) All of these

674. Rancidity in cheese is caused by:
(A) Lipolytic organisms (B) High acidity
(C) Low salt per cent (D) None of these

675. Which is not correct about animal organ cultures?
(A) The first attempt was made by Loeb in 1887
(B) Loeb maintained adult rabbit liver, kidney, thyroid & ovary on small plasma clots in test tubes
(C) Loeb that cultured organs retained their histological features for 13 days
(D) In 1919, Loeb and Fleischer reported that culture tubes must be filled with O_2 to prevent central necrosis of explants

676. Pearson method is used for:
(A) Standardization of fat in milk and cream
(B) Milk adulteration
(B) Judging the quality of cheese
(D) All of these

677. Red colour formation after mixing milk and resorcinol indicates the adulteration of milk with:
(A) Water (B) Sucrose
(C) Starch (D) All of these

678. Boudovin test is used for detection of:
(A) Adulteration of sesame oil in ghee
(B) Adulteration of tallow in ghee
(C) Adulteration of cotton oil in ghee
(D) All of these

679. Alezarine test is useful for testing the:
(A) Milk pH (B) Milk fat
(C) Milk flavour (D) All of these

680. Soda ash test is used to detect:
(A) Adulteration of vegetable fat and animal fat in ghee
(B) Adulteration of preservatives in ghee
(C) Adulteration of potato in ghee
(D) All of these

681. Milk pasteurization can be detected by:
(A) Phosphate test (B) Boudovine test
(C) Nickel test (D) All of these

682. First clone Dolly was made in?
(A) Sheep (B) Goat
(C) Cow (D) Buffalo

683. Dolly created by:
(A) B.P.Pal (B) William Godd
(C) Wilmont (D) Kurean

684. Mechanism of heat tolerance is found in:
(A) Goat (B) Buffalo
(C) Cow (D) Fish

685. Buffalo meat contains white fat as:
(A) Carotene converted to vitamin A
(B) RBC
(C) Xanthophyll dominant
(D) None of these

686. Flavouring compound for butter is:
(A) Lactic acid (B) Diacetyl
(C) Acetyl methyl carbinol (D) Lactose

687. Paneer is a:
(A) Concentrated milk product
(B) Dried milk product
(C) Coagulated milk product
(D) None of these

688. Stiff lamb disease caused by deficiency of:
(A) Vitamin E (B) Vitamin K
(C) Vitamin B2 (D) Vitamin B 12

689. Morocco leather produced from skin of:
(A) Goat (B) Horse
(C) Sheep (D) Cow

690. Methylene Blue Reduction Test (MBRT) is conducted to know the efficiency of:
(A) Toned milk (B) Double toned milk
(C) Sterilized milk (D) Pasteurized milk

691. Polenske value of milk fat is due to:
(A) Water soluble fatty acids
(B) Water insoluble fatty acids
(C) Protein nitrogen
(D) Fats

692. Which of the following vitamins remains most resistant on heat treatment of milk:
(A) Vitamin A (B) Vitamin C
(C) Vitamin K (D) Vitamin E

693. Pasteurization kills:
(A) Bacteria along spores
(B) Bacteria spores alive
(C) Bacteria + fungi kill
(D) Bacteria + virus

694. Which one of the hormone contains carbohydrates?
(A) FSH (B) Progestrone
(C) Oxytocin (D) Androgen

695. Oogenesis and spermatogenesis are a major physiological action of (hormone released from pituitary gland):
(A) Androgen
(B) Prolactin
(C) Follicle stimulating hormone
(D) Oxytocin

696. Mating of related individual with 4-6 generation is known as:
(A) Inbreeding (B) Line breeding
(C) Close breeding (D) Out crossing

697. Mating of unrelatedpure breed animal within the same breed is:
(A) Inbreeding (B) Line breeding
(C) Close breeding (D) Out crossing

698. Mating of animals of different breeds is known as:
(A) Inbreeding (B) Line breeding
(C) Cross breeding (D) Out crossing

699. When two breeds are crossed alternatively is known as:
(A) Criss crossing (B) Triple crossing
(C) Back crossing (D) None of these

700. When three breeds are crossed in a rotational manner:
(A) Triple crossing (B) Back crossing
(C) Criss crossing (D) None of these

701. Mating of a cross breed animal back to one of the pure parent is:
(A) Inbreeding (B) Triple crossing
(C) Back crossing (D) Line breeding

702. Meat, milk, egg, soybean and fish powder have maximum amount of vitamin:
(A) Vitamin K (B) Vitamin C
(C) Vitamin D (D) Vitamin B

703. The major group of organisms which are Psychotropic in raw milk are:
(A) Bacillus spp. (B) Streptococcus spp.
(C) Pseudomonas spp. (D) Staphylococcus spp.

704. Diseases transmitted from animal to man and vice-versa are known as:
(A) Contagious diseases (B) Zoonotic diseases
(C) Infectious diseases (D) Protozoan diseases

705. Which part of a cell carries the information that is passed from one generation to the next?
(A) Cell (B) Nucleus
(C) Cytoplasm (D) Chloroplast

706. Which of these carries the thousands of genes needed to allow animal and plant cells to pass on inherited characteristic?
(A) Chromosomes (B) Chromatogram
(C) Chromatography (D) Chromium

707. Which of the following best describes artificial insemination?
(A) Transplanting an embryo into the uterus
(B) Fertilization of an egg in a test tube
(C) Selectively breeding health animals
(D) Taking the sperm and placing it directly

708. The technique, mainly used for the diagnosing birth defects in the fetus by means of needle, is called:
(A) Amniocentesis (B) Ectogensis
(C) Transplantation (D) All of these

709. In which year, Dolly the first mammalian clone was boron?
(A) 1995 (B) 1996
(C) 1997 (D) 1998

710. Superovulation is an:
(A) Increased ovulatory response by external hormonal therapy
(B) Decreased ovulatory response by external hormonal therapy
(C) Decreased ovulatory response by internal hormonal therapy
(D) Increased ovulatory response by internal hormonal therapy

711. Genetic change can occur only in:
(A) One way (B) Two ways
(C) Three ways (D) Four ways

712. The nucleous of mature unfertilized ovum may be removed by:
(A) Irradiation
(B) Surgery
(C) Both (A) and (B)
(D) Neutralization and homogenization

713. When fertilized egg cell develops into an embryo, the entropy of the living system:
(A) Decreases (B) Increases
(C) Remains constant (D) None of these

714. Multiple ovulation and embryo transfer (MOET):
(A) Can increase the rate of progress in dairy cattle
(B) Allows progeny testing of males
(C) Allows progeny testing of females
(D) All of these

715. Transfection refers to which of the following:
(A) Synthesis of mRNA from DNA template
(B) Synthesis of protein based on mRNA sequence
(C) Introduction of foreign gene in to a cell
(D) The process by which a cell become malignant

716. For the transfer of whole individual chromosomes, they are isolated from the cells at:
(A) Prpphase (B) Metaphase
(C) Telophase (D) Anaphase

717. DNA microinjection into the egg has been used to produce which of the following transgenic animals?
(A) Mice (B) Chicken
(C) Pigs (D) All of these

718. DNA into fish is injected into:
(A) Pronuclei (B) Cytoplasm
(C) Both (A) and (B) (D) None of these

719. Transgenic goats have been used to produce which of the following protein that is used for dissolving blood clots?
(A) Amyloid precursor protein
(B) α_1-anti trypsin (C) Casein
(D) A variant of human tissue-type plasminogen activator

720. In transgenic fish, the genes are introduced by:
(A) Microinjection in fish
(B) Viruses
(C) Transfer of whole nuclei
(D) All of these

721. DNA is microinfected into the fertilized egg:
(A) After the fusion of male and female nuclei
(B) Before the fusion of male and female nuclei
(C) At the time of fusion of male and female nuclei
(D) Any time, it can be infected

722. Which of the following is the word used to describe what happens when the nucleus of a sperm joins with the nucleus of an egg cell?
(A) Sex (B) Fertilization
(C) Intercourse (D) Reproduction

723. Some types of variation are due to changes in the genetic material. What is this type of change called?
(A) Fertilization (B) Mutation
(C) Radiation (D) Sterilisation

724. Which of the following is the process of choosing parent organisms for the characteristics that is wanted in their offspring?
(A) Active selection
(B) Reproductive selection
(C) Selective breeding
(D) Breeding selection

725. Cloning is a method by which numbers of genetically identical organisms are derived from a single organism by:
(A) Vegetative propagation
(B) Vegetative initiation
(C) Vegetative termination
(D) None of these

726. For the rapid genetic progress, embryo transfer (ET):
(A) Will be as potent tool as AI (artificial insemination)
(B) Will not be as potent a toll as AI (artificial insemination)
(C) ET cannot be used at all
(D) Will be a superior tool than AI

727. Enucleation of the cells can be achieved by treating the cells with:
(A) Polyethylene glycol (B) Cytochalasin B
(C) Both (A) and (B) (D) Alcohol

728. Karyoplast is:
(A) Cells devoid of cell wall
(B) Nuclei
(C) Nuclei with only some residual plasma membrane
(D) Cell with nucleus

729. Superovulation is primarily the result of:
(A) Decreased atresia in medium and large follicles (>1.7 mm)
(B) Increased atresia in medium and large follicles (>1.7 mm)
(C) Increased atresia in the large follicles (>1.7 mm)
(D) Decreased atresia in small follicles (>1.7 mm)

730. Which of these established cell lines originate from a mouse embryo?
(A) BHK (B) 3T3
(C) HeLa (D) BTK

731. The success of embryo transplantation depends upon the ability to obtain an increased number of embryos from animals of superior genetic merit, which can be achieved by he use of:
(A) Exogenous gonadaatrophic hormones
(B) Endogenous gonadatrophic hormones
(C) Both (A) and (B)
(D) FSH (Follicle stimulating Hormone)

732. Production of transgenic animals require transfections of:
(A) Red blood cells (B) Stem cells
(C) Eggs or embryos (D) All of these

733. Embryo transfer (ET) refers to the technique by which fertilized ova are collected from the reproductive tract of a genetically:
(A) Superior female and transferred to the inferior
(B) Inferior female and transferred to the superior
(C) Balanced female
(D) None of these

734. When the cells are enucleated and centrifuged, nuclei:
(A) Detach from cell and remain in the supernatant
(B) Detach from cell and settle at the bottom
(C) Nuclei remain attached to the cell
(D) None of these

735. The number of follicles at any particular stage is governed by:
(A) The rate of entry of dormant follicle
(B) The rate of growth of follicles
(C) The rate of loss of follicles
(D) All of these

736. Which of the following is/are the method of transfection for making transgenic animals?
(A) Transfer of whole nuclei
(B) Transfer of whole individual chromosomes or fragment
(C) Transfer of DNA (D) All of these

737. Which of the following gene have been introduced into the transgenic fish?
(A) Human or rat gene for growth hormone
(B) Chicken gene for delta crystalline protein
(C) *E. coli* gene for β-galactosidase
(D) All of these

738. Chromosomes may be isolated from metaphase cells by:
(A) Hypertonic lysis (B) Hypotonic lysis
(C) Either (A) and (B) (D) Isotonic lysis

739. Transgenic goats produce a variant of human tissue type plasminogen activator protein in:
(A) Blood (B) Urine
(C) Milk (D) Muscles

740. Which protein has been produced generating a transgenic sheep that is used for replacement therapy for individuals at risk from emphysema?
(A) α-anti trypsin (AAT)
(B) Plasminogen activator
(C) Casein
(D) Amyloid precursor proteins

741. Fusion of karyoplast with the enucleated cell is achieved in presence of:
(A) Cytohalasin B (B) Polyethylene glycol
(C) Both (A) and (B) (D) Alcohol

742. Dolly sheep was genetically similar to:
(A) The mother from which nucleated fertilized egg was taken
(B) The mother from which nuclear DNA of udder cell was taken
(C) The surrogate mother
(D) Both surrogate mother and nuclear donor mother

ANSWERSHEET

21. ANIMAL BIOTECHNOLOGY

1. (D) 2. (B) 3. (A) 4. (A) 5. (A) 6. (B) 7. (A) 8. (D) 9. (C) 10. (B)
11. (A) 12. (C) 13. (C) 14. (D) 15. (A) 16. (B) 17. (D) 18. (D) 19. (B) 20. (D)
21. (C) 22. (D) 23. (C) 24. (B) 25. (B) 26. (A) 27. (C) 28. (D) 29. (C) 30. (D)
31. (B) 32. (A) 33. (B) 34. (B) 35. (D) 36. (D) 37. (B) 38. (D) 39. (A) 40. (A)
41. (C) 42. (B) 43. (B) 44. (B) 45. (B) 46. (D) 47. (D) 48. (C) 49. (B) 50. (C)
51. (A) 52. (B) 53. (A) 54. (C) 55. (C) 56. (D) 57. (B) 58. (C) 59. (B) 60. (D)
61. (C) 62. (C) 63. (C) 64. (B) 65. (C) 66. (A) 67. (A) 68. (D) 69. (C) 70. (D)
71. (B) 72. (A) 73. (B) 74. (D) 75. (C) 76. (C) 77. (C) 78. (A) 79. (B) 80. (C)
81. (A) 82. (D) 83. (C) 84. (A) 85. (D) 86. (D) 87. (B) 88. (C) 89. (A) 90. (B)
91. (C) 92. (D) 93. (A) 94. (D) 95. (D) 96. (A) 97. (B) 98. (D) 99. (C) 100. (D)
101. (B) 102. (D) 103. (C) 104. (A) 105. (C) 106. (C) 107. (D) 108. (D) 109. (D) 110. (B)
111. (D) 112. (B) 113. (C) 114. (B) 115. (B) 116. (C) 117. (A) 118. (D) 119. (C) 120. (D)
121. (A) 122. (B) 123. (D) 124. (C) 125. (A) 126. (B) 127. (A) 128. (C) 129. (B) 130. (C)
131. (B) 132. (C) 133. (C) 134. (D) 135. (A) 136. (D) 137. (D) 138. (A) 139. (B) 140. (D)
141. (A) 142. (C) 143. (A) 144. (D) 145. (B) 146. (D) 147. (D) 148. (D) 149. (D) 150. (D)
151. (D) 152. (D) 153. (B) 154. (A) 155. (B) 156. (A) 157. (A) 158. (C) 159. (A) 160. (D)
161. (D) 162. (D) 163. (D) 164. (A) 165. (D) 166. (B) 167. (D) 168. (C) 169. (D) 170. (A)
171. (D) 172. (C) 173. (D) 174. (D) 175. (A) 176. (A) 177. (D) 178. (B) 179. (A) 180. (A)
181. (B) 182. (C) 183. (A) 184. (A) 185. (D) 186. (B) 187. (D) 188. (C) 189. (C) 190. (B)
191. (D) 192. (D) 193. (D) 194. (D) 195. (D) 196. (A) 197. (C) 198. (A) 199. (A) 200. (A)
201. (A) 202. (D) 203. (C) 204. (C) 205. (D) 206. (A) 207. (C) 208. (C) 209. (D) 210. (C)
211. (D) 212. (A) 213. (A) 214. (D) 215. (D) 216. (B) 217. (C) 218. (A) 219. (B) 220. (D)
221. (D) 222. (D) 223. (D) 224. (D) 225. (A) 226. (D) 227. (D) 228. (D) 229. (C) 230. (C)
231. (D) 232. (A) 233. (A) 234. (C) 235. (A) 236. (A) 237. (D) 238. (D) 239. (B) 240. (C)

241. (D) 242. (A) 243. (C) 244. (B) 245. (C) 246. (B) 247. (C) 248. (C) 249. (A) 250. (C)
251. (B) 252. (C) 253. (B) 254. (D) 255. (A) 256. (A) 257. (D) 258. (D) 259. (D) 260. (C)
261. (D) 262. (A) 263. (D) 264. (D) 265. (C) 266. (D) 267. (D) 268. (D) 269. (B) 270. (D)
271. (C) 272. (A) 273. (A) 274. (A) 275. (D) 276. (A) 277. (D) 278. (C) 279. (B) 280. (D)
281. (C) 282. (C) 283. (A) 284. (A) 285. (D) 286. (D) 287. (A) 288. (D) 289. (D) 290. (D)
291. (A) 292. (A) 293. (C) 294. (A) 295. (D) 296. (D) 297. (A) 298. (A) 299. (D) 300. (A)
301. (B) 302. (D) 303. (C) 304. (B) 305. (D) 306. (D) 307. (D) 308. (D) 309. (D) 310. (D)
311. (A) 312. (D) 313. (C) 314. (A) 315. (D) 316. (C) 317. (D) 318. (A) 319. (C) 320. (D)
321. (C) 322. (D) 323. (D) 324. (D) 325. (A) 326. (C) 327. (B) 328. (B) 329. (B) 330. (C)
331. (C) 332. (D) 333. (D) 334. (B) 335. (D) 336. (A) 337. (C) 338. (C) 339. (A) 340. (D)
341. (A) 342. (D) 343. (D) 344. (A) 345. (A) 346. (C) 347. (A) 348. (A) 349. (A) 350. (D)
351. (A) 352. (D) 353. (D) 354. (C) 355. (A) 356. (A) 357. (B) 358. (C) 359. (C) 360. (A)
361. (D) 362. (D) 363. (A) 364. (D) 365. (D) 366. (A) 367. (B) 368. (A) 369. (B) 370. (D)
371. (D) 372. (A) 373. (A) 374. (A) 375. (D) 376. (A) 377. (A) 378. (D) 379. (C) 380. (C)
381. (D) 382. (D) 383. (C) 384. (A) 385. (A) 386. (B) 387. (B) 388. (B) 389. (C) 390. (A)
391. (A) 392. (C) 393. (A) 394. (A) 395. (A) 396. (A) 397. (C) 398. (D) 399. (D) 400. (C)
401. (C) 402. (D) 403. (A) 404. (A) 405. (D) 406. (A) 407. (A) 408. (A) 409. (B) 410. (A)
411. (D) 412. (A) 413. (B) 414. (D) 415. (D) 416. (D) 417. (A) 418. (A) 419. (D) 420. (D)
421. (B) 422. (A) 423. (A) 424. (D) 425. (B) 426. (D) 427. (A) 428. (D) 429. (B) 430. (A)
431. (D) 432. (D) 433. (A) 434. (A) 435. (D) 436. (A) 437. (D) 438. (B) 439. (A) 440. (C)
441. (D) 442. (D) 443. (A) 444. (D) 445. (C) 446. (B) 447. (B) 448. (B) 449. (C) 450. (B)
451. (D) 452. (C) 453. (C) 454. (D) 455. (D) 456. (A) 457. (A) 458. (A) 459. (A) 460. (C)
461. (D) 462. (D) 463. (A) 464. (A) 465. (A) 466. (A) 467. (D) 468. (A) 469. (B) 470. (D)
471. (D) 472. (A) 473. (A) 474. (A) 475. (A) 476. (D) 477. (D) 478. (A) 479. (A) 480. (B)
481. (A) 482. (D) 483. (D) 484. (A) 485. (C) 486. (D) 487. (D) 488. (D) 489. (D) 490. (D)
491. (C) 492. (A) 493. (B) 494. (B) 495. (B) 496. (D) 497. (A) 498. (A) 499. (A) 500. (A)
501. (B) 502. (D) 503. (A) 504. (A) 505. (B) 506. (D) 507. (C) 508. (A) 509. (D) 510. (B)
511. (B) 512. (A) 513. (A) 514. (B) 515. (B) 516. (A) 517. (D) 518. (D) 519. (D) 520. (A)
521. (D) 522. (C) 523. (A) 524. (D) 525. (D) 526. (A) 527. (A) 528. (A) 529. (A) 530. (C)
531. (B) 532. (D) 533. (D) 534. (D) 535. (D) 536. (D) 537. (D) 538. (A) 539. (D) 540. (D)
541. (D) 542. (B) 543. (D) 544. (D) 545. (D) 546. (C) 547. (D) 548. (A) 549. (B) 550. (B)
551. (B) 552. (D) 553. (C) 554. (A) 555. (D) 556. (B) 557. (C) 558. (D) 559. (A) 560. (D)
561. (A) 562. (A) 563. (C) 564. (B) 565. (A) 566. (D) 567. (A) 568. (A) 569. (A) 570. (A)
571. (A) 572. (A) 573. (A) 574. (C) 575. (B) 576. (B) 577. (A) 578. (B) 579. (A) 580. (D)
581. (A) 582. (B) 583. (A) 584. (C) 585. (D) 586. (B) 587. (B) 588. (A) 589. (B) 590. (C)
591. (A) 592. (A) 593. (B) 594. (D) 595. (C) 596. (D) 597. (D) 598. (A) 599. (A) 600. (A)
601. (A) 602. (D) 603. (D) 604. (A) 605. (C) 606. (B) 607. (D) 608. (D) 609. (C) 610. (B)
611. (A) 612. (D) 613. (D) 614. (A) 615. (C) 616. (A) 617. (D) 618. (C) 619. (D) 620. (D)
621. (D) 622. (D) 623. (D) 624. (D) 625. (D) 626. (D) 627. (D) 628. (B) 629. (C) 630. (A)
631. (A) 632. (D) 633. (A) 634. (A) 635. (A) 636. (D) 637. (D) 638. (B) 639. (A) 640. (A)
641. (D) 642. (D) 643. (A) 644. (A) 645. (A) 646. (C) 647. (D) 648. (B) 649. (C) 650. (D)
651. (A) 652. (C) 653. (A) 654. (A) 655. (D) 656. (A) 657. (A) 658. (C) 659. (D) 660. (A)

661. (C)	**662. (C)**	**663. (A)**	**664. (D)**	**665. (D)**	**666. (A)**	**667. (B)**	**668. (B)**	**669. (A)**	**670. (D)**
671. (D)	**672. (A)**	**673. (D)**	**674. (A)**	**675. (C)**	**676. (A)**	**677. (B)**	**678. (A)**	**679. (A)**	**680. (D)**
681. (A)	**682. (A)**	**683. (C)**	**684. (B)**	**685. (A)**	**686. (B)**	**687. (C)**	**688. (A)**	**689. (A)**	**690. (D)**
691. (B)	**692. (A)**	**693. (B)**	**694. (A)**	**695. (C)**	**696. (A)**	**697. (D)**	**698. (C)**	**699. (A)**	**700. (A)**
701. (C)	**702. (D)**	**703. (C)**	**704. (B)**	**705. (B)**	**706. (A)**	**707. (D)**	**708. (A)**	**709. (B)**	**710. (A)**
711. (D)	**712. (C)**	**713. (A)**	**714. (D)**	**715. (C)**	**716. (B)**	**717. (D)**	**718. (B)**	**719. (D)**	**720. (A)**
721. (B)	**722. (B)**	**723. (B)**	**724. (C)**	**725. (A)**	**726. (B)**	**727. (B)**	**728. (C)**	**729. (A)**	**730. (B)**
731. (A)	**732. (C)**	**733. (A)**	**734. (B)**	**735. (D)**	**736. (D)**	**737. (D)**	**738. (B)**	**739. (C)**	**740. (A)**
741. (B)	**742. (B)**								

Chapter - 22

MODEL TEST PAPERS

1. Which is often called molecular glue?
(A) DNA ligase (B) Synthetase
(C) DNA polymerase I (D) RNA polymerase

2. Which assays involves antigen-antibody interaction?
(A) Southern hybridization (B) ELISA
(C) Colony hybridization (D) Dot-blot assay

3. Organarily, the recognition sites of restriction enzymes are:
(A) Palindromes
(B) Nonpalindromic
(C) Asymmetrical
(D) Palindromes with rotational symmetry

4. A RAPD fragment is generated when sequences complementary to the single arbitrary primer used in the PCR reaction are located at two sites in the target DNA, having:
(A) In the same direction
(B) In the opposite directions
(C) Within 5 kb distance
(D) Both (B) and (C)

5. Totipotency is reflected by:
(A) Somatic embryogenesis (B) Shoot regeneration
(C) Pollen embryogenesis (D) All of these

6. A vaccines contains :
(A) Inactivated pathogen
(B) Attenuated pathogen
(C) Purified antigen from pathogen
(D) Any of these

7. Which vectors can be used to clone DNA inserts of about 40kb?
(A) Plasmid (B) Cosmid
(C) Phagemid (D) M13

8. Which human chromosomes has the highest extent of segmental duplications?
(A) Autosome I (B) X chromosome
(C) Ychromosome (D) Autosome 13

9. Which involves restriction mapping of DNA fragments?
(A) Chromosome jumping
(B) Primer walking
(C) Chromosome walking
(D) Both (B) and (C)

10. Which uses two pairs of PCR primers?
(A) Nested PCR (B) Asymetrical PCR
(C) inverse PCR (D) Anchored PCR

11. Monoclonal antibodies are used for:
(A) Disease diagnosis (B) ELISA
(C) Immunopurification (D) All of these

12. MHC molecules are present on the surface of:
(A) Helper T cells (B) Tc cells
(C) Antigen presenting cells
(D) Pathogens

13. Which markers behave as dominant marker?
(A) RFLP (B) RAPD
(C) AFLP (D) Both (B) and (C)

14. Which molecular markers is expected to produce a single band?
(A) Sequence-tagged microsatellite marker
(B) RAPD
(C) Amplified fragment length polymorphism
(D) ISSR (Inter-simple sequence repeats)

15. In case of single cell proteins, high content is problematic?
(A) Nucleic acids (B) Lipids
(C) Carbohydrates (D) Proteins

16. Which molecules functions as a protein kinase?
(A) VirA (B) VirB
(C) VirD (D) VirG

17. Which proteins is active against nematodes?
(A) CryII (B) CryIA
(C) CryIII (D) CryVI

18. 'Laurical'variety of *B. napus* has which transgenes?
(A) Lauroyl-ACP desaturase
(B) Lauroyl-ACP thioesterase
(C) Lauroyl-ACP saturase
(D) Stearoyl-ACP desaturase

19. Which research reagents is being produced from transgenic plants on a commercial scale?
(A) Avidin (B) ß-glucuronidase
(C) Both (A) and (B) (D) Polygalacturonidase

20. Tissue culture raised plantlets require hardening due to:
(A) Poor cuticle development
(B) Poor vascular bundle development
(C) Improper stomatal function
(D) All of these

21. Which crop has heat shock induces androgenesis?
(A) Rice (B) *Brassica sp*
(C) Barley (D) Datura

22. Somatic hybridization has been the most successful in:
(A) Solanaceae (B) Papilionaceae
(C) Graminae (D) Malvaceae

23. Which is the least reliable strategy for virus indexing?
(A) Indicator plants (B) Serological tests
(C) Visible disease symptoms
(D) Both (B) and (C)

24. In cheese production, Rennet leads to curd formation by digesting:
(A) α-casein (B) ß-casein
(C) κ-casein (D) All of these

25. *Trichoderma viridae* is used for the control of:
(A) Fungal pathogen (B) Bacterial pathogen
(C) Viral pathogen (D) All of these

26. Unfertilized ovaries/ovules are cultured for:
(A) Gynogenesis (B) In vitro pollination
(C) Both (A) and (B) (D) Embryo rescue

27. Selection of hybrid cells based on which of the following is the most widely applicable?
(A) Visual markers
(B) Complementation
(C) Transgenic selectable markers
(D) Both (A) and (B)

28. Which is the strongest auxin?
(A) IAA (B) 2,4-D
(C) NAA (D) 2,4,5-T

29. A test cross is made mainly to demonstrate:
(A) 1 : 1 ratio of gametes in monohybrids
(B) Random union of gametes
(C) Equal survival of zygotes
(D) Both (A) and (B)

30. In case of species propagated by asexual reproduction, germplasm is represented by:
(A) Pollen grains (B) Seed
(C) Vegetative propagules
(D) Zygotic embryos

31. Which is a transgene?
(A) A human gene transferred into *E. coli*
(B) A human gene used for gene therapy
(C) Both (A) and (B)
(D) An *Aegilops sp.* gene transferred into *Triticum sp.* by backcrossing

32. A biotechnological production process may use:
(A) A microbe
(B) An animals/a plant cell
(C) An enzyme
(D) Any of these

33. S1 nuclease is used for:
(A) DNA footprinting
(B) Production of blunt ends from protruding ends
(C) DNA fingerprinting
(D) DNA sequencing

34. Which of the following is used for positional cloning of genes?
(A) Chromosome walking
(B) Chromosome jumping
(C) Chromosome sorting
(D) Both (A) and (B)

35. Orthologue means 'homologous' genes within:
(A) Different tissues of an organism
(B) Same tissue of an organism
(C) Individuals of different species
(D) Different individuals of the same species

36. In case of retrovirus vectors, sequence is required for:
(A) Transcription (B) Replication
(C) Reverse transcription
(D) Packaging into virions

37. Restriction site recognition is affected by:
(A) Base substitution
(B) Base deletion/addition
(C) Base methylation (D) All of these

38. Which has enabled successful transformation of monocots using Agrobacterium?
(A) Exogenous acetosyringone
(B) Agrobacterium strain
(C) Explant tissue (D) Plant variety

39. In case of mice, DNA is microinjected into:
(A) Male pronucleus (B) Female pronucleus
(C) Ovum cytoplasm (D) Both (A) and (C)

40. Which marker systems would be preferred for testing the purity of hybrid seeds?
(A) RAPD (B) RFLP
(C) STMS (D) AFLP

41. In which of the following types of cultures, a constant volume of spent medium is regularly replaced by equal volume of fresh medium?
(A) Perfusion cultures (B) Fed-batch cultures
(C) Agar cultures (D) Batch cultures

42. In which of the following types of cultures, fresh medium is added gradually?
(A) Perfusion cultures
(B) Fed-batch cultures
(C) Agar cultures (D) Batch cultures

43. In which of the following types of cultures, there is an increase during the culture period?
(A) Perfusion cultures
(B) Fed-batch cultures
(C) Agar cultures
(D) Batch cultures

44. Conversion of palm oil to cocoa butter is based on which enzymes?
(A) Lipase
(B) Fatty acid synthase I
(B) Fatty acid desaturase
(D) Fatty acid thioesterase

45. Chitin synthase inhibition is used as screen for detection of microorganisms producing:
(A) Antibiotics (B) Antivirals
(C) Antifungals (D) Antitumour agents

46. The biological component of biosensors is responsible for:
(A) Detection of analyte
(B) Specificity of biosensor
(C) Sensitivity of biosensor
(D) All of these

47. Lipases are used in biphasic systems during the production of:
(A) Wax (B) Isomyl acetate
(C) Fatty acids (from lipid) (D) Both (A) and (C)

48. Chromosome walking is used in case of eukaryotes to generate:
(A) Linkage map (B) Restriction map
(C) Physical map (D) Both (B) and (C)

49. Anther culture is related to:
(A) Gynogenesis (B) Androgenesis
(C) Parthenogenesis (D) Embryogenesis

50. Which is a Zn containing enzyme:
(A) Ureas
(B) Alkaline phosphatase
(C) Hexokinase
(D) Phosphoglycerate kinase

51. Which techniques yields high purification of molecules in a single step?
(A) Affinity chromatography
(B) Gel filteration
(C) SDS-PAGE
(D) Column chromatography

52. The terms 'single cell protein' refers to:
(A) Protein recovered from microbes
(B) Protein recovered from single cells
(C) Protein recovered from cultured plant and animal cells
(D) Microbial biomass used as food/feed supplement

53. Alcohol dehydrogenase gene is induced by:
(A) Anaerobiosis (B) Wound
(C) Oxygen (D) Pathogen invasion

54. RNA acts as the genetic material:
(A) Tobacco mosaic virus
(B) Rhizobium
(C) Fungi
(D) Archaebacterium

55. Plant obtained by chromosome doubling of haploids are known as:
(A) Amphidiploids (B) Dihaploids
(C) Double haploids (D) Diploids

56. Dot blot assay detects the presence of:
(A) Polypepetide (B) DNA sequence
(C) mRNA (D) Both (B) and (C)

57. *Bacillus thuringiensis* strains are active against which group of insects?
(A) Diptera (B) Nematodes
(C) Coleoptera (D) All of these

58. Genes whose products are necessary for the maintenance of all types of cells of a species are called:
(A) Smart genes (B) House-keeping genes
(C) Luxury genes (D) Both (A) and (B)

59. Which uses as restriction enzymes?
(A) Chromosome jumping (B) Restriction mapping
(C) AFLP (D) All of these

60. T-DNA is transferred into the plant cells most likely as:
(A) Single-stranded circular molecule
(B) Double-stranded circular molecule
(C) Single-stranded linear molecule
(D) Double-stranded linear molecule

61. *Bacillus thuringiensis* Cry protein is inactivated by:
(A) UV light (B) Humidity
(C) Red light (D) Both (A) and (B)

62. RuBISCO functions is:
(A) Photorespiration in C_3 plants
(B) Fixation of CO_2 in C_3 plants
(C) Respiration
(D) Both (A) and (B)

63. In response to stress, which amino acids accumulates?
(A) Phenylalanine (B) Lysine
(C) Proline (D) Histidine

64. Monoclonal antibodies are being used for diagnosis of which diseases?
(A) Hepatitis B (B) Cancer
(C) Venereal diseases (D) All of these

65. Plant cell cultures are being used for the production of:
(A) Berrberine (B) Shikonin
(C) Ginseng (as biomass) (D) All of these

66. The main features of light induced promoters of dicots relates to:
(A) An upstream AG or AGGA box
(B) TATA box
(C) Both TATA and AG/AGGA boxes
(D) CAAT box

67. The action of chloranphenicol involves:
(A) Inhibition of DNA replication
(B) Inhibition of transcription
(C) Inhibition of peptidyl transferase
(D) Prevention of initiator tRNA from entering the P site

68. The two complementary strands of a DNA molecule are held together chiefly by:
(A) Salt bridges (B) Ionic interactions
(C) Hydrogen bonds
(D) Hydrophobic interactions

69. Map based cloning requires:
(A) Saturated RFLP map (B) Genetic map
(C) Karyotype (D) Integrated A & B

70. Molecular markers can be used for:
(A) Assessment of genetic variability
(B) Creation of genetic variability
(C) Selection of desirable genotype
(D) Both (A) and (C)

71. Affinity chromatography techniques can be best used for:
(A) Enzyme identification
(B) Enzyme classification
(C) Enzyme purification
(D) Quantitative assays of enzyme activity

72. The marker enzyme for mitochondria is:
(A) Cytochrome C (B) Catalase
(C) Amino acid permease (D) 5'-Nucleotidase

73. A eukaryotic nuclear gene with a bacterial promoter can not be expressed in bacteria because of which differences between prokaryotes and eukaryotes?
(A) Translation (B) RNA processing
(C) DNA replication (D) Genetic code

74. Rabies virus affects which organs?
(A) Thymus (B) Liver
(C) Kidney (D) Spinal cord

75. Which is a prokaryotes:
(A) Yeast (B) Green algae
(C) Blue-green algae (D) Mushroom

76. RNA is not involved in:
(A) Enzymatic activity (B) Protein synthesis
(C) Chromosome structure (eukaryotes)
(D) Heredity

77. Genetic material of tobacco mosaic virus will show:
(A) $A \neq T$ (B) $G = C$
(C) $A = U$ (D) $A \neq U$

78. Which is considered as primary containment?
(A) Containment equipment
(B) Laboratory practices
(C) Both (A) and (B)
(D) Special laboratory design

79. Which is considered as secondary containment?
(A) Containment equipment
(B) Laboratory practices
(C) Both (A) and (B)
(D) Special laboratory design

80. Synthetic vaccines have been developed for:
(A) Typhoid and rabies
(B) Cancer (feline leukaemia virus)
(C) Cholera and foot and mouth disease of cattle
(D) All of these

81. Which is the most suitable transfection technique for animal ova?
(A) Lipofection
(B) Calcium phosphate precipitation
(C) Microinjection (D) Particle gun

82. Technology of haploid production leads to:
(A) Genetic variation (B) Hybrid sorting
(C) Homozygous lines (D) All of these

83. Which vitamins is commercially produced by microbes?
(A) Riboflavin (B) B12
(C) Carotene (D) All of these

84. Which one became the first country to begin commercial use of virus resistant transgenic tobacco and tomato in early 1990s?
(A) India (B) USA
(C) China (D) Canada

85. Which is responsible for the loss of biodiversity?
(A) Deforestation and overgrazing
(B) Biological invasion and hunting
(C) Destruction of habitats and overexploitation
(D) All of these

86. Which viruses gas acquired a gene from its host genome?
(A) Adenoviruses
(B) Acute transforming retroviruses
(C) Epstein-Barr Virus
(D) Replication competent tumour retroviruses

87. Which reason has already been responsible for extinction of biodiversity?
(A) Introduction of foreign species
(B) Habitat destruction
(C) Infrastructure development
(D) All of these

88. Biological diversity, called biodiversity in short, refers to:
(A) Diversity among species
(B) Diversity among ecosystems
(C) Genetic diversity within a given species
(D) All of these

89. Who said, "the cost of conserving biodiversity is far less than the penalty for allowing its degradation"
(A) M.S. Swaminathan (B) Edouard Sauma
(C) Suman Sahai (D) Panjab Singh

90. **The term 'biological diversity' was for the first time introduced by Norse and McManus in 1980. This term was shortened as 'biodiversity' in 1985 by which scientists?**
(A) W.G. Rosen (B) M.S. Swaminathan
(C) N.I.Vavilov (D) J.R.Harlan

91. **Plant tissue culture is regarded to have began when Haberlandt attempted to culture in vitro:**
(A) Leaf discs (B) Isolated leaf cells
(C) Cambium tissue (D) Zygotic embryos

92. **Production of which of the following does not constitute biotechnology?**
(A) Plastic (B) Antibiotic
(C) Idli (D) α-amylase

93. **Viruses from plants can be eliminated by:**
(A) Thermotherapy (B) Cryotherapy
(C) Chemotherapy (D) Any of these

94. **A single-nucleotide polymorphic site has how many alleles?**
(A) 1 (B) 4
(C) 2 (D) 3

95. **Nick translation uses which enzymes?**
(A) *E. coli* DNA polymerase III
(B) *E. coli* Klenow fragment
(C) *E. coli* DNA polymerase II
(D) DNA polymerase I of E. coli

96. **Which marker is uses restriction of enzymes?**
(A) RFLP
(B) Microsatellite markers
(C) AFLP (D) Both (A) and (C)

97. **Trade secret can be used to protect:**
(A) A formula (B) A design
(C) A manuscript (D) A transgene

98. **Single cell proteins are obtained from:**
(A) Bacteria (B) Fungi
(C) Algae (D) All of these

99. **'Edible vaccines' belong to which category?**
(A) Conventional vaccines
(B) Purified antigen vaccines
(C) Recombinant vaccines
(D) Both (A) and (B)

100. **The human Genome is estimated to contain how many genes?**
(A) 25000 (B) 35000
(C) 45000 (D) 55000

101. **Embryo culture is routinely used for propagation of:**
(A) Lily (B) Orchid
(C) Rose (D) Apple

102. **Negative cell selection allows which type of cells to divide?**
(A) Wild type cells (B) Variant cells
(D) Both (A) and (B) (D) Neither A nor B

103. **Which contains nuclear genome from a single fusion parent?**
(A) Asymetric hybrids (B) Cybrids
(C) Symmetric hybrids (D) Both (A) and (B)

104. **Haploid plants can be obtained by:**
(A) Anther culture
(B) Ovary culture
(C) Interspecific hybridization
(D) All of these

105. **Which Vir proteins functions as activator of vir operon?**
(A) VirE (B) VirG
(C) VirH (D) VirD

106. **Which approaches for virus resistance has been the most successful?**
(A) Coat protein gene
(B) cDNA of satellite RNA
(C) Antisense RNA approach
(D) Ribozyme approach

107. **Phylate is used for phosphate storage in seeds, and it interfere with absorption in human intestine:**
(A) ß-Carotene (B) Zinc
(C) Iron (D) Phosphate

108. **Which would reduce transgene silencing?**
(A) Single copy integration in genome
(B) Transgene AT composition similar to host genome
(C) Each transgene having a different promoter and polydenylation signal
(D) All of these

109. **Affinity biosensors are based on:**
(A) Enzyme (B) Antibody
(C) Nucleic acid (D) Either (B) or (C)

110. **Which enzymes is produced in largest quantity?**
(A) ß-Amylase (B) Lipase
(C) Lactase (D) α-Amylase

111. ***In vitro* regeneration potential is affected by:**
(A) Oligogenes (B) Polygenes
(C) Plasma genes (D) All of these

112. **Which serves as cryoprotectant?**
(A) DMSO (B) Proline
(C) Glycerol (D) All of these

113. **Frame-shift mutations may result from:**
(A) Base substitution
(B) Trinucleotide expansion
(C) Tautomeric shift
(D) Transposition

114. **The term 'wound reponse' refers to the production of which of the following by the plant cells at wound sites?**
(A) Protein VirA (B) Acetosyringone
(C) Protein VirG (D) Lentins

115. Golgi bodies function in:
(A) Protein synthesis (B) Protein modification
(C) Protein packaging (D) Both (B) and (C)

116. Which is the chief reason for the reduction in chromosome number from 2n to n as a result of meiosis?
(A) Chromosome pairing during MI
(B) Separation of chromosomes during AL
(C) The A above, followed by B
(D) Two nuclear divisions without an intervening S phase

117. Which processes involves DNA-protein interaction?
(A) Replication initiation
(B) Transcription activation
(C) Transcription repression
(D) All of these

118. North-western screening is based on:
(A) Antigen-antibody interaction
(B) RNA-protein interaction
(C) DNA: protein interaction
(D) RNA : DNA interaction

119. Which has more than one chromosome in its genome?
(A) *E. coli* (B) *B. subtilis*
(C) *Agrobacterium* (D) *Vibrio cholera*

120. Paralogues means 'homologous' genes within :
(A) Same cell (B) Same organism
(C) Same genus (D) Both (A) and (B)

121. Which techniques does not use antigen-antibody interaction for screening of expression libraries?
(A) Complementation (B) ELISA
(C) FACS (D) Both (A) and (C)

122. 'Nick translation' is used for:
(A) DNA labeling
(B) Recombinant protein production
(C) DNA replication
(D) Expression – based screening

123. Beer production is estimated to have began about how many years ago:
(A) 4000 (B) 6000
(C) 400 (D) 600

124. Which enzymes is used to prevent circulation of opened vectors during recombinant DNA production?
(A) Acid phosphatase
(B) Polynucleotide kinase
(C) Alkaline phosphatase
(D) Klenow fragment

125. Which techniques identifies the DNA fragment to which a given protein binds?
(A) DNA footprinting (B) DNA fingerprinting
(C) Gel retardation assay
(D) Restriction mapping

126. Which techniques identifies the region of the given DNA fragment to which the concerned protein binds?
(A) DNA footprinting (B) DNA fingerprinting
(C) Gel retardation assay (D) Restriction mapping

127. Which identifies is used for physical mapping?
(A) DNA footprinting (B) DNA fingerprinting
(C) Gel retardation assay
(D) Restriction mapping

128. Which techniques may used PCR?
(A) DNA footprinting
(B) DNA fingerprinting
(C) Gel retardation assay (D) Restriction mapping

129. Which of the following techniques identifies, DNase in digestion of DNA is involved?
(A) DNA footprinting (B) DNA fingerprinting
(C) Gel retardation assay (D) Restriction mapping

130. Which generally produces transient transfection?
(A) DEAE-Dextran mediated transfection
(B) Calcium-phosphate precipitation
(C) Microinjection (D) retroviral infection

131. Which proteins is specific to Diptera?
(A) CryI (B) CryII
(C) CryIII (D) CryIV

132. Which *Agrobacterium* proteins has helicase activity?
(A) VirC (B) VirG
(C) $VirD_2$ (D) VirE

133. Which of the proteins is single-strand binding protein?
(A) VirC (B) VirG
(C) $VirD_2$ (D) VirE

134. Which genetic transformation techniques can be used in monocots?
(A) Particle gun
(B) Agrobacterium-mediated
(C) Electroporation
(D) All of these

135. Change in medium pH is used as screen for detection of microbial strains producing:
(A) Protease (B) Urease
(C) Lipase (D) Amylase

136. Which is used as the main nitrogen source for riboflavin production?
(A) Pancreatic digest of gelatin
(B) Peanut granules
(C) Corn-steep liquor (D) Soybean meal

137. Which is used as the nitrogen source for pencillin production?
(A) Pancreatic digest of gelatin
(B) Peanut granules
(C) Corn-steep liquor
(D) Soybean meal

138. Lectins are used as ligands in affinity chromatography for the separation of:
(A) Antigens (B) Glycoproteins
(C) Polysaccharides (D) Both (B) and (C)

139. Solid state fermentation is used for?
(A) Food processing
(B) Enzyme production
(C) Organic acid production
(D) All of these

140. Which vectors is the most promosing for development of DNA vaccines?
(A) Vaccinia virus (B) Retrovirus
(C) Herpes virus (D) Adenovirus

141. Glucose biosensor based on glucose oxidase detects which changes?
(A) Evolution of a gas
(B) Change in temperature
(C) Redox reaction (D) Fluorescence

142. Which can be used for in situ recovery of products?
(A) Vacuum (B) Two-phase systems
(C) Dialysis (D) All of these

143. In plants, mitosis occurs during:
(A) Embryogenesis (B) Parthenogenesis
(C) Gametogenesis (D) All of these

144. Nucleic acids are transformed from a gel onto a membrane filter in which techniques?
(A) Northern hybridization
(B) Dot-blot hybridization
(C) Calony hybridization
(D) Zoo blot hybridization

145. Which is not a operation in a biotechnological process?
(A) Strain development
(B) Downstream processing
(C) Process engineering
(D) Safety evaluation

146. Anzymes are:
(A) Antigen-antibody complexes
(B) Antigens with enzyme activity
(C) Antibodies with enzyme activity
(D) RNA molecules with enzyme activity

147. The least used growth regulator in plant tissue cultures is:
(A) Kinetin (B) GA3
(C) BAP (D) 2,4-D

148. Enzyme technology concerns with:
(A) Modification of the amino acid sequence of the enzymes
(B) Enzyme immobilization
(C) Both (A) and (B)
(D) Modification of glycosylation pattern of enzymes

149. Gel filtration separates macromolecules on the basis of:
(A) Colour (B) Mass
(C) Charge (D) Specific binding

150. Penicillin acylase (amidase) is being used industrially for production of:
(A) Detergents
(B) Semisynthetic penicillins
(C) Secondary metabolites
(D) Mitigation of penicillin allergy

151. An endonuclease like DNase I is used in which techniques?
(A) Nick-translation (B) DNA foot-printing
(C) Both (A) and (B)
(D) Chromosome walking

152. An endonuclease like DNase II is used in which techniques?
(A) Nick-translation (B) DNA foot-printing
(C) Both (A) and (B)
(D) Chromosome walking

153. Immobilization of enzymes may involve:
(A) Covalent bonds (B) Cross-linking
(C) Ionic bonds (D) Any of these

154. Nick-translation utilizes which enzymes?
(A) DNase I (B) DNA polymerase I
(C) Klenow fragment (D) Both (A) and (B)

155. Stable integration of agene into plant genome can be conclusively shown by:
(A) PCR
(B) Northern hybridization
(C) Dot blot assay
(D) Southern hybridization

156. Virus resistant transgenic plants can be produced by transferring into them:
(A) cDNA of the satellite RNA of the RNA virus
(B) Coat protein gene of the virus
(C) Defective genome of the virus
(D) Any of these

157. Virus resistant transgenic plants has been exploited commercially:
(A) cDNA of the satellite RNA of the RNA virus
(B) Coat protein gene of the virus
(C) Defective genome of the virus
(D) Any of these

158. A protein has isoelectric pH of 5.7. What will be the net charge of this protein at pH 7.0?
(A) Negative (B) Positive
(C) Neutral (D) Either A or B

159. Introduction of purified phage DNA molecules into bacterial cells is called:
(A) Conjugation (B) Transfection
(C) Transduction (D) Transformation

160. Which of the following shall be used when naked recombinant DNA is introduced into a bacterial cell?
(A) Conjugation (B) Transfection
(C) Transduction (D) Transformation

161. Some introns perform which type of functions?
(A) Enhance promoter action in expression vectors
(B) Encode enzymes
(C) Perform self-splicing
(D) Any of these

162. The recognition sites for restriction enzymes used in gene cloning are:
(A) Random sequences
(B) Palindromes with rotational symmetry
(C) Simple palindromes
(D) Nonpalindromic sequences

163. The polysaccharides necessary for attachment of *A. tumefaciens* cells to plant cells is encoded by genes present in:
(A) Vir region
(B) Bacterial chromosome
(C) T-DNA
(D) Both (A) and (C)

164. In a disarmed Ti plasmid, which genes are deleted?
(A) *iaa*M and *iaa*H (B) *ipt*
(C) *nos* (D) Both (A) and (B)

165. Selection of toxin resistant cells for the developm-ent of disease resistant plant is the most successful in the case of pathogens producing is:
(A) Toxin (B) Specific toxin
(C) Nonspecific toxin (D) No toxin

166. Isoenzyme are variants of a single enzyme differing in :
(A) Catalysis function
(B) Mobility during electrophoresis
(C) Substrate preference
(D) All of these

167. Nitrate reductase consists of:
(A) Mo cofactor and one polypeptide
(B) One polypeptide
(C) Two polypeptides
(D) Three polyppetides

168. Crystal formation in plant cells preserved in liquid nitrogen would take place if their temperature rises above:
(A) -130^{0}C (B) -5^{0}C
(C) -190^{0}C (D) -160^{0}C

169. Southern hybridization differs from northern hybridization in respect to:
(A) Membrane filter
(B) Need for denaturation
(C) Type of nucleic acid involved
(D) All of these

170. In India, molecular markers can not be used to resolve issues related to:
(A) Biodiversity
(B) Plant breeding rights
(C) Plant breeder's right
(D) Farmer's rights

171. An extremozyme is an enzyme that has which features?
(A) Extremely active
(B) Extremely long-lived
(C) Extremely mobile during electrophoresis
(D) Stable and active under extreme environments

172. Chloramphenicol acetyltransferase gene has been used in gene transfer experiments as:
(A) Bacterial resistance gene
(B) Selectable marker
(C) Scorbale marker
(D) Both (B) and (C)

173. A mutant of *Agrobacterium tumefaciens* defective in VirG shows which features?
(A) Can be used to transform dicots
(B) Can be used to transform monocots
(C) Can not be used for plant transformation
(D) Both (A) and (B)

174. Acid phosphatase acts as a marker for:
(A) Mitochondria (B) Lysosome
(C) Golgi complex (D) Endoplamic reticulum

175. For double-starnded DNA, which base ratios always equals one?
(A) (A+T)/(G+C) (B) (A+G)/(C+T)
(C) C/T (D) A/G

176. Many of the *Bacillus species* are pathogenic to Lepidopteran pests due to:
(A) Endotoxin production
(B) Exotoxin production
(C) Antibiotic production
(D) Chitinase production

177. AFLP utilize:
(A) Chromosome walking
(B) Southern hybridization
(C) PCR
(D) In situ hybridization

178. Na+ - K+ ATPase is related to:
(A) Group translation (B) Active transport
(C) Passive transport (D) All of these

179. Which has eukaryotic regulatory sequences?
(A) T-DNA (B) vir regulon
(C) ΦX174 (D) λphage

180. In which organelles, phosphorylation is a common reaction?
(A) Chloroplast (B) Mitochondria
(C) Golgi bodies (D) Both (A) and (B)

181. Which causes oncogenesis?
(A) DNA viruses (B) Retroviruses
(C) Mutation (D) All of these

182. Which is known as Drosophilla of plant kingdom?
(A) *Nicotiana tabaccum*
(B) *Arabidopsis thaliana*
(C) *Daucus carota* (D) *Capsicum annum*

183. Which is correct about vinegar?
(A) Acetic acid is commonly called vinegar
(B) Used a souring agent, preservative & condiment
(C) It is 10-13% ripened acetic acid
(D) All of these

184. Which strain of bacteria is used for the degradation of petroleum and management of oil spills?
(A) *Clostridium acetobutylicum*
(B) *Methylophilus methylotrophus*
(C) *Pseudomonas putida*
(D) *Pseudomonas fluorescens*

185. In which transgene integration by homologous recombination is almost the rule?
(A) Yeast (B) Drosophila
(C) Mice (D) Humans

186. It is feared that transgenic plants may pose which risks?
(A) Transfer their transgenic features to weeds
(B) Degrade the environment
(C) Theme selves become weeds
(D) All of these

187. The gene encoding which proteins serves as a selectable marker for animal cells?
(A) Neomycin phosphotransferase
(B) Dihydrofolate reductase
(C) Thymidine kinase
(D) All of these

188. In case of which vectors, selection for recombinant clones is based on replica plating technique?
(A) pBR322 (B) pUC18
(C) pUC8 (D) Both (B) and (C)

189. Which of the marker genes leads to an amplification of the DNA insert linked to it?
(A) Neomycin phosphotransferase
(B) Dihydrofolate reductase
(C) Thymidine kinase
(D) All of these

190. Risks assessment consists of:
(A) Hazard identification and hazard analysis
(B) Consequence analysis and determination
(C) Risks evaluation
(D) All of these

191. Functional food provides the body with the required amount needed for healthy survival:
(A) Carbohydrates (B) Proteins and fats
(C) Vitamins (D) All of these

192. Which features is correct about BABAC2 and TAC plant vectors?
(A) They can transfer multiple genes
(B) They can be used for positional cloning of genes identified by mutation
(C) Genomic libraries have been constructed using these vectors
(D) All of these

193. Which can be used for induction of mutations in plants and also to serve as molecular tags for the genes so mutated?
(A) T-DNA (B) Transposable elements
(C) EMS (D) Both (A) and (B)

194. Which approaches are used for development of fungal disease resistant transgenic crops?
(A) Transgene encoding antifungal compounds
(B) Transgenes which encodes a product that leads to cell death upon pathogen infection
(C) The R gene may be transferred to initiate resistance response
(D) All of these

195. Which features is correct about ribosome-inactivating proteins (RIPs)?
(A) An effective antifungal protein from barley seeds called RIP has been isolated
(B) It acts by inhibiting protein synthesis via N-glucosidase modification of fungal 28SrDNA
(C) Transgenic crop plants expressing RIP proteins are now being tested against many fungi
(D) All of these

196. Molecular markers offer which advantages over morphological/other conventional markers?
(A) They show high degree of polymorphism
(B) They can be established at cellular and tissue levels
(C) They can be influenced by the environmental fluctuation
(D) All of these

197. The ability to design and execute molecular strategies for generating male sterility has heen possible due to:
(A) Isolation, cloning and characterization
(B) Anther/pollen specific promoters
(C) Chemically inducible promoters
(D) All of these

198. For the recovery of virus-free stocks of clonal crops, meristem culture is combined with:
(A) Thermotherapy (B) Chemotherapy
(C) Cryotherapy (D) Any of these

199. Expressed sequence tags have been used for:
(A) As physical markers for genome maps
(B) Discovery of genes
(C) Differential – display PCR
(D) All of these

200. Which is not correct about micropropagation?
(A) Mass clonal ropagation of desired genotype
(B) Plantlets are usually pathogen free
(C) Shoot cultures of some species are maintained as 'slow growth cultured' for germplasm conservation
(D) There is no chance of somaclonal variation

201. Which is correct about fluorescencein diacetate (FDA)?
(A) During testing of viability, cells are treated with 0.01% solution of FDA
(B) Live cells cleave FDA by esterase activity and produce fluorescencein
(C) When stained cells are exposed to UV rays, living cells appear green
(D) All of these

202. Artificial seed is a bead of gel that contains :
(A) Pesticides and antibiotics
(B) Nutrients and growth regulators
(C) Shoot bud/somatic embryo
(D) All of these

203. Which steps is necessary for the production of artificial seeds by desiccated system?
(A) Somatic embryo (SEs) are first hardened to withstand desiccation
(B) Hardened SEs are encapsulated in a suitable coating material
(C) SEs may be hardened neither by coating/ treating mature SEs with a suitable polymer followed by drying during their maturation phase
(D) All of these

204. Which is not correct about production of hydrated artificial seed?
(A) Somatic embryos are enclosed in gels, which are then desiccated
(B) Calcium alginate is the most suitable gel
(C) The artificial seed are sticky and difficult to handle
(D) The seeds are difficult to store ffor long periods

205. Somatic embryos can be stored for prolonged periods through:
(A) Cryopreservation (B) Desiccation
(C) Cold storage (D) All of these

206. The problem of endosperm abortion can be overcome by excising and culturing:
(A) Hybrid endosperms (B) Hybrid fruits
(C) Hybrid embryos (D) Hybrid zygote

207. In cryopreservation, which favours storage of plant cells, tissues organs in liquid nitrogen at -196°C?
(A) All metabolic processes and growth are suppressed
(B) Occurrence of genetic, karyotypic morphological and biochemical changes are prevented
(C) It is the most reliable method for long term prevention of cell cultures
(D) All of these

208. In case of plant variety is a chimera, which route of micropropagation can be used to conserve the chimaeric state?
(A) Somatic embryogenesis
(B) Adventitious shoot buds
(C) Shoot – tip culture (D) Both (A) and (B)

209. Which disrups chimaeras and produces nonchimeric progeny plants?
(A) Somatic embryogenesis
(B) Adventitious shoot buds
(C) Shoot – tip culture
(D) Both (A) and (B)

210. Eukaryotic DNA segments are usually cloned initially in:
(A) Bacillus (B) Yeast
(C) *E. coli* (D) Both (A) and (B)

211. Air – lift fermenters do not have provision for:
(A) Aeration (B) Stirring
(C) Temperature control (D) Sterilization

212. Which vectors can not be used to efficiently clone DNA inserts longer than 5 kb?
(A) Cosmid (B) Plasmid
(C) Lambda phage (D) M13

213. Which growth regulators suppresses callus formation?
(A) GA_3 (B) Auxin
(C) Cytokinin (D) Both (A) and (B)

214. In which PCR procedures, target DNA is cut and circulated, then cut open to produce linear molecules?
(A) Anchored PCR (B) Asymetric PCR
(C) Inverse PCR (D) Nested PCR

215. The melting temperature for a DNA sequence having 6G : C and 4 A : T base pairs will be:
(A) 32°C (B) 28°C
(C) 36°C (D) 30°C

216. In India, a transgenic variety can be protected by:
(A) Patent
(B) Plant Breeder's Rights
(C) Both (A) and (B) (D) Trade Secret

217. Which generated more number of different proteins from the same number of genes?
(A) Autosplicing (B) Trans-splicing
(C) Alternative splicing (D) RNA editing

218. The term clonal selection signifies which in response to an antigen?
(A) B cell proliferation (B) T cell proliferation
(C) B/T cell maturation (D) Both (A) and (B)

219. A helper Ti of binary vector system des have:
(A) vir region
(B) DNA insert
(C) Origin of replication for *E. coli*
(D) T-DNA

220. Which can be used to suppress the expression of endogenous genes?
(A) Consuppression (B) Antisense RNA
(C) RNAi (D) All of these

221. In the case of barnase – barstar male sterility system, phosphinothricin resistance gene bar is used to achieve:
(A) Restoration of male fertility
(B) Reinforcing male sterility
(C) Enhancing Barstar action
(D) Eliminating male fertile plants

222. The 'slow growth' cultures of tropical plants would suffer cold damage at which temperature?
(A) 4^0C (B) 8^0C
(C) Both (A) and (B) (D) 15^0C

223. Androgenesis would yield a dihaploid:
(A) Rice (B) Potato
(C) Tobacco (D) Both (B) and (C)

224. Somatic embryo maturation is promoted by:
(A) Auxin (B) ABA
(C) Cytokinin (D) GA3

225. Disease resistant plants can be developed by cell selection using which as selection agents?
(A) Specific toxins (B) Nonspecific toxins
(C) Both (A) and (B) (D) Neither A and B

226. A somatic hybrid plant may have chloroplasts from:
(A) Both the parents (B) First fusion parent
(C) Second fusion parent (D) Any of these

227. Plant vitrification solutions are used for:
(A) Slow-growth cultures (B) Cryopreservation
(C) Micropropagation
(D) Prevention of hyperhydration

228. Which is the most commonly used for root regeneration?
(A) IBA (B) IAA
(C) 2,4-D (D) ABA

229. In production of glucose syrup, -amylase brings about:
(A) Gelatinization (B) Liquefaction
(C) Saccharification (D) All of these

230. Which solubilizes phosphate?
(A) *Thiobacillus* sp. (B) *Glomus* sp.
(C) Both (A) and (B) (D) *Azospirillum*

231. In hypersensitive response, the avr gene of pathogen is responsible for generating:
(A) Toxin (B) Receptor
(C) Elicitor (D) Repressor

232. Which is uses insertional mutagenesis?
(A) Gene tagging
(B) Exon trapping
(C) Differential screening
(D) Chromosomal jumping

233. Which is the most correct?
(A) Each normal chromosomes has, as a rule, a single centromere
(B) The cnetromeric activity is always confined to a single locus in the chromosomes
(C) In most species, the centromeric activity is confined to a single locus in the chromosomes
(D) Polycentric chromosomes represent structural aberrations

234. Klenow fragment is used fro:
(A) Filling up of protruding ends
(B) Nick translation
(C) DNA sequencing
(D) Both (A) and (C)

235. Which is the first recombinant protein to be approved for use in humans?
(A) Insulin (B) Interferon
(C) Hirudin (D) Erythroprotein

236. A restriction enzyme has six-base recognition sequence. This sequence may be expected to occur at which intervals?
(A) 4^6bp (B) 6^4 bp
(C) 4 x 6 bp (D) $4^6/6$ bp

237. Paralogues means 'homologous' genes within:
(A) Same cell (B) Same organism
(C) Same genus (D) Both (A) and (B)

238. Which has more than one chromosome in its genome?
(A) *E. coli* (B) *B. subtilis*
(C) *Agrobacterium* (D) *Vibrio cholera*

239. Improved Pusa Basmati I (Pusa 1460), the first variety developed in India through MAS, has been improved for which traits?
(A) VitA content
(B) Cooking quality
(C) Bacterial leaf blight resistance
(D) Blast resistance

240. F-virosomes derived from Sendai virus do not contain:
(A) Viral envelope (B) Viral HN protein
(C) Recombinant DNA (D) Viral F protein

241. The satellite RNA present in some RNA viruses may produce which effects on disease severity?
(A) Enhance disease severity
(B) Decrease disease severity
(C) No effect on disease severity
(D) All of these

242. Enhanced metabolite production may be achieved by selection for:
(A) Auxotropihic mutants
(B) Analogue resistant mutants
(C) Revertant of a nonreproducing mutant
(D) All of these

243. Which serves as inducer of pollulenase production?
(A) Maltose (B) Sucrose
(C) Dextrin (D) Starch

244. In 1860 the first synthetic medium for microorganisms was used to study which fermentations?
(A) Lactic acid (B) Alcohol
(C) Acetic acid (D) Citric acid

245. Which is used in gap –ligase chain reaction?
(A) Thermostable DNA ligase
(B) Thermostable DNA polymerase
(C) NAD (D) All of these

246. Which elements is not essential for plant cells?
(A) Ni (B) Na
(C) K (D) P

247. The Km of the substrate for any enzyme is increased by:
(A) Competitive inhibitor
(B) Noncompetitive inhibitor
(C) Substrate concentration
(D) Both (A) and (C)

248. Somaclonal variation is the least likely to occur in:
(A) Protoplast culture (B) Suspension culture
(C) Meristem culture (D) Callus culture

249. Which statements is correct?
(A) All enzymes are heat stable
(B) Enzymes are always proteins
(C) Enzymes are always specific for a single substrate
(D) Enzymes can be inhibited by substrate analogues

250. Antisen RNA technology may be used to achieve:
(A) Minimise the expression of a gene
(B) Enhance the expression of a gene
(C) Stabilize the expression of a gene
(D) Change the expression pattern of a gene

251. Inactive precursors of enzymes are called:
(A) Isozymes (B) Zymogenes
(C) Allozymes (D) Pre-enzymes

252. Which techniques does not use radioactivity?
(A) RFLP (B) Western hybridization
(C) RAPD (D) Southern blotting

253. In thermotherapy for virus elimination, the best results are obtained when high temperature treatment is applied to:
(A) Shoot-tip cultures (B) Seeds
(C) Whole plants, followed by culture of their shoot-tips
(D) Whole plants

254. Biological extraction of copper an duranium utilizes:
(A) *Thiobacillus thiooxidans*
(B) *T. ferroxidans*
(C) Both (A) and (B) (D) *T. organoparus*

255. Production of biochemicals by cultured plant cells is markedly improved by:
(A) Treatment with elicitors
(B) Selection of high producing clones
(C) Optimisation of culture conditions
(D) All of these

256. Which agents is used for virus elimination?
(A) Ribavirin (B) Actinomycin D
(C) 2-Thiouracil (D) All of these

257. Phenotypic variability having nongenetic basis is called:
(A) Epigenetic variation
(B) Soamclonal variation
(C) Gametoclonal variation
(D) Protoclonal variation

258. The virus titre in shoot-tip merism cells is much lower than that in the differentiated cells; this may be due to:
(A) Absence of plasmodesmata
(B) Rapid cell division
(C) Inhibitory action of growth regulators
(D) All of these

259. Lectin are proteins that have:
(A) DNA binding proteins
(B) Bind to glycoproteins
(C) Bind to lipoproteins
(D) Hormone binding proteins

260. The most effective fusion treatment for plant protoplasts is:
(A) High pH and high Ca^{++}
(B) PEG
(C) PEG combined with high pH & high Ca^{++}
(D) $NaNO_3$

261. Which *Agrobacterium tumefaciens* proteins is activated by acetosyringone?
(A) VirA (B) VirC
(C) VirG (D) VirD2

262. Which is an approach of germplasm conservation?
(A) Slow-growth shoot cultures
(B) DNA banks
(C) Cryopreservation
(D) All of these

263. ABA is related to:
(A) Flowering in plants
(B) Dormancy
(C) Abscission of leaves
(D) Both (B) and (C)

264. A polyprotein is a protein that has:
(A) Several functions
(B) Homopolymer
(C) Heteropolymer
(D) Cleaved into two or more functional proteins

265. WHO has launched Global Immunization Progarmme for which disease?
(A) Pertusis and Tetanus
(B) Diphtheria and Measles
(C) Polio and T.B.
(D) All of these

266. A yeast minichromosome vector does not contain which molecules?
(A) ARS sequence
(B) Telomere sequence
(C) CEN sequence
(D) A functional yeast gene

267. Cosmid contain which sequences from the λ phage genome?
(A) Terminase biding site
(B) cos site
(C) cos and terminase binding sites
(D) cos, terminase binding and cleavage sites

268. A conditioned medium permits:
(A) Subculture at low cell density
(B) Elimination of lag phase
(C) Production of greater biomass
(D) Both (A) and (B)

269. Hydrogen peroxide is degraded by the enzyme catalase in which organelles?
(A) Peroxisomes (B) Lysosomes
(C) Mitochondria (D) Glyoxysomes

270. A transgene stably integrated into the plant genome shows:
(A) Cytoplasmic inheritance
(B) Segregation distortion
(C) Medelian inheritance
(D) All of these

271. Restriction endonucleases are sometimes unable to digest the DNA even in the presence of the correct recognition sequence; this may be due to:
(A) Acetylation of DNA
(B) Methylation of DNA
(C) Phosphorylation of DNA
(D) All of these

272. A circular plasmid DNA gives two bands of 6.5 kb and 5.5 kb with EcoRI and only one band of 12 kb with BamHI. How many bands will be expected when it is digested with bth the enzymes?
(A) 2 (B) 4
(C) 3 (D) 5

273. A double – stranded DNA molecule of 100000 bp length has about how many complete turns?
(A) 8000 (B) 10000
(C) 20000 (D) 12000

274. Which plant is used as a model system for transgenic plant production?
(A) Potato (B) Arabidopsis
(C) Tobacco (D) Datura

275. Which is gram positive bacterium?
(A) *Bacillus subtilis* (B) *Agrobacterium sp.*
(C) *Rhizobium sp.* (D) *E. coli*

276. As a general rule of thumb, hydrophobic residues in a protein tend to be situated at:
(A)C-terminal (B) N-terminal
(C) Interior of the molecules
(D) Surface of the molecules

277. In which random transgene integration is far more frequent than that by homologous recombination?
(A) Animals (B) Plants
(C) Yeast (D) Both (A) and (B)

278. Which can be used for DNA delivery into *E. coli* cells?
(A) $CaCl_2$ treatment (B) Microinjection
(C) Electroporation (D) Both (A) and (C)

279. During gynogenesis, haploid plants generally originate from:
(A) Polar nuclei (B) Egg cell
(C) Antipodals (D) Synergids

280. It is feared that transgenic plants may pose which risks?
(A) Transfer their transgenic features to weeds
(B) Degrade the environment
(C) Themselves become weeds
(D) All of these

281. Which is not correct about in situ conservation of biodiversity?
(A) Plants are conserved in their natural habitats
(B) Only the wild relatives are candidates for in situ conservation
(C) Plants can continue to evolve
(D) In situ conservation is rather simple

282. Which is not correct about BIBAC2 and TAC plant vectors?
(A) Both have ampicilin resistance marker for selection in bacteria
(B) Both contain hygromycin resistance marker for selection of transformed plant cells
(C) Both have the sacB marker (containing a polylinker for DNA insert integration) for negative selection in bacteria
(D) Both have the Ri origin of maintenance in Rhizobium

283. Nutraceticals differ from dietary supplements in respect to:
(A) They must not only supplement the diet, but should also aid in prevention or treatment of diseases or disorders
(B) They are represented for use as the conventional food or as the sole item of meal or diet
(C) They include some products from nonfood sources
(D) Both (A) and (B)

284. The use of GM crops relates to:
(A) Biodiversity, ecology and biosafety
(B) Environmental and social concerns
(C) Ethical concerns
(D) All of these

285. Which features is relevant to an ideal herbicide?
(A) No toxicity towards human, animals or soil organisms
(B) Kills selectively weeds
(C) Biodegradability
(D) All of these

286. Selection of appropriate molecular marker for applied plant breeding would depend on:
(A) Trait(s) under consideration
(B) Crop in question
(C) Reproducibility of molecular marker
(D) All of these

287. Which statement is correct about molecular markers?
(A) The ultimate goal of molecular marker research is to devlop plant varieties more efficiently and effectively
(B) Molecular markers tightly linked to male sterility gene can be used to identify male fertile sister plants at very early stage
(C) RAPD is cost effective and simple, but its reproducibility is not high
(D) All of these

288. Terminator seed technology is seen to result in:
(A) Encourage monopoly of multinational seed companies
(B) Force the farmers to purchase seeds of even self-pollinated crops every year
(C) Jeoparadise food security
(D) All of these

289. For toxicity or allergenicity is not reported:
(A) Cry protein
(B) Trypsnin inhibitor
(C) Virus coat protein
(D) Both (A) and (C)

290. Expressed sequence tags (ESTs) have been used for:
(A) Identification of genes in genomic clones
(B) Designing of PCR primers to generate large numbers of target sequences for cDNA microarrays
(C) Gene expression analysis
(D) All of these

291. The protoplast fusion products recovered after a treatment with a fusion inducing agent consist of:
(A) Homokaryons
(B) Heterokaryons
(C) Unsafed protoplasts of the two species/strains
(D) All of these

292. Which is not correct about meristem culture?
(A) Cultivation of axillary or apical shoot meristems
(B) The shoot apical meristem is the portion lying distal to the youngest leaf primordium and is 250μ m in length
(C) The shoot apex/shoot tip consists of the apical meristem plus 1-3 young leaf primordial
(D) Meristem culture usually involves the regeneration of a new shoot meristem

293. Somatic embryos are used for micropropagation in:
(A) Coconut (B) Oil palm
(C) Orchids (D) Date palm

294. Gene integration occurs generally by homologous recombination in:
(A) Yeast (B) Mammals
(C) Tobacco (D) Tomato

295. 'Single cell protein' is represented by :
(A) Protein from microbes
(B) Protein from cell cultures
(C) Microbial biomass
(D) Both (A) and (B)

296. The site where a protein binds a DNA molecule is determined by:
(A) DNA footprinting (B) DNA fingerprinting
(C) Gel retardation assay (D) FISH

297. Somatic embryo maturation is promoted by:
(A) Auxin (B) ABA
(C) Cytokinin (D) GA_3

298. GA_3 promotes shoot regeneration:
(A) Tobacco (B) Arabidopsis
(C) Carrot (D) Citrus

299. Eukaryotic enzymes essentially functions as reverse transcriptase:
(A) DNA polymerase (B) Primase
(C) Telomerase (D) RNA polymerase

300. Enhanced metabolite production may be achieved by selection for:
(A) Auxotrophic mutants
(B) Analogue resistant mutants
(C) Revertant of a nonproducing mutant
(D) All of these

ANSWERSHEET

22. MODEL TEST PAPERS

1. (A)	2. (B)	3. (D)	4. (D)	5. (D)	6. (D)	7. (B)	8. (C)	9. (C)	10. (A)
11. (D)	12. (C)	13. (D)	14. (A)	15. (A)	16. (A)	17. (C)	18. (B)	19. (C)	20. (D)
21. (B)	22. (A)	23. (C)	24. (C)	25. (A)	26. (C)	27. (C)	28. (B)	29. (A)	30. (C)
31. (C)	32. (D)	33. (B)	34. (D)	35. (C)	36. (D)	37. (D)	38. (A)	39. (A)	40. (C)
41. (A)	42. (B)	43. (B)	44. (A)	45. (C)	46. (D)	47. (D)	48. (D)	49. (B)	50. (B)
51. (A)	52. (D)	53. (C)	54. (A)	55. (C)	56. (D)	57. (D)	58. (B)	59. (D)	60. (C)
61. (D)	62. (D)	63. (C)	64. (D)	65. (D)	66. (C)	67. (B)	68. (D)	69. (D)	70. (D)
71. (C)	72. (A)	73. (B)	74. (D)	75. (C)	76. (C)	77. (D)	78. (C)	79. (D)	80. (D)
81. (C)	82. (D)	83. (D)	84. (C)	85. (D)	86. (B)	87. (D)	88. (D)	89. (B)	90. (A)
91. (B)	92. (A)	93. (D)	94. (B)	95. (D)	96. (D)	97. (A)	98. (D)	99. (C)	100. (A)
101. (B)	102. (A)	103. (B)	104. (D)	105. (B)	106. (A)	107. (C)	108. (D)	109. (D)	110. (D)
111. (D)	112. (D)	113. (D)	114. (B)	115. (D)	116. (D)	117. (D)	118. (B)	119. (D)	120. (D)
121. (A)	122. (A)	123. (B)	124. (C)	125. (C)	126. (A)	127. (D)	128. (B)	129. (A)	130. (A)
131. (D)	132. (A)	133. (D)	134. (D)	135. (B)	136. (A)	137. (C)	138. (D)	139. (D)	140. (A)
141. (C)	142. (D)	143. (D)	144. (A)	145. (D)	146. (C)	147. (B)	148. (C)	149. (B)	150. (B)
151. (C)	152. (D)	153. (D)	154. (D)	155. (D)	156. (D)	157. (B)	158. (A)	159. (B)	160. (D)
161. (D)	162. (B)	163. (B)	164. (D)	165. (B)	166. (B)	167. (A)	168. (A)	169. (D)	170. (D)
171. (D)	172. (C)	173. (C)	174. (B)	175. (B)	176. (A)	177. (C)	178. (B)	179. (A)	180. (D)
181. (D)	182. (B)	183. (D)	184. (C)	185. (A)	186. (D)	187. (D)	188. (A)	189. (B)	190. (D)
191. (D)	192. (D)	193. (D)	194. (D)	195. (D)	196. (D)	197. (D)	198. (D)	199. (D)	200. (D)
201. (D)	202. (D)	203. (D)	204. (A)	205. (D)	206. (C)	207. (D)	208. (C)	209. (D)	210. (C)
211. (B)	212. (D)	213. (A)	214. (C)	215. (A)	216. (B)	217. (C)	218. (D)	219. (A)	220. (D)
221. (D)	222. (C)	223. (D)	224. (B)	225. (A)	226. (D)	227. (B)	228. (A)	229. (B)	230. (C)
231. (C)	232. (A)	233. (C)	234. (D)	235. (A)	236. (A)	237. (D)	238. (D)	239. (C)	240. (C)
241. (D)	242. (D)	243. (A)	244. (B)	245. (D)	246. (A)	247. (A)	248. (C)	249. (D)	250. (A)
251. (B)	252. (C)	253. (C)	254. (C)	255. (D)	256. (D)	257. (A)	258. (D)	259. (B)	260. (C)
261. (A)	262. (D)	263. (D)	264. (D)	265. (D)	266. (B)	267. (D)	268. (D)	269. (A)	270. (C)
271. (D)	272. (C)	273. (B)	274. (C)	275. (A)	276. (C)	277. (D)	278. (D)	279. (B)	280. (D)
281. (D)	282. (A)	283. (D)	284. (D)	285. (D)	286. (D)	287. (D)	288. (D)	289. (D)	290. (D)
291. (D)	292. (D)	293. (B)	294. (A)	295. (C)	296. (A)	297. (B)	298. (B)	299. (C)	300. (D)

Chapter - 23

PREVIOUS YEARS SOLVED PAPERS

BASED ON MEMORY

CSIR-JRF-NET examination on June 2009

1. RNA interference is used to control different viral diseases in case of plants and animals. The active fragment involving in interference is known as:

(A) Si RNA (B) Micro RNA
(C) Both (A) and (B) (D) None of these

2. If all the parameters related with cockroach are doubled such as height, width and length, it will not survive because of:

(A)Low surface are of volume ratio
(B) High surface area of volume ratio
(C) Exchange of gases
(D) Problem of excretion

3. Disorder, which occur due to single nucleotide change:

(A) Sickle cell anemia (B) Thalisemia
(C) Leishmenia (D) Anemia

4. Relative mean kinetic energy for helium atom (atomic weight 4) and argon (atomic weight 40) would be:

(A) 1 : 10 (B) 1 : 4
(C) 1 : 100 (D) 1 : 16

5. For constructing recombinant plasmid and DNA to be inserted are digested with same restriction enzyme and kept in same reaction solution. To prevent self sealing of plasmid, which of the following enzyme is utilized?

(A) Alkaline phosphatase
(B) Polynucleotide kinase
(C) Terminal transferase (D) Ligase

6. Perennial habit among trees would be more preferred under conditions:

(A) Low survival during sapling stage and high during adult
(B) High survival during sapling stage and high during adult
(C) Low survival during sapling stage and low during adult
(D) All of these

7. At ground state of hydrogen atom, its Bohr radius is 5.3×10^{-11} m and mean velocity is 2.1×10^{6} m/s. What would be the value of fundamental time unit?

(A) 2.52×10^{-17} sec. (B) 2.52×10^{-5} sec
(C) 1.2×10^{-17} sec (D) 1.52×10^{-17} sec

8. Angle between two vectors 2i + 3j and 3i + 2j will be:

(A) 30^0 (B) 45^0
(C) 60^0 (D) 90^0

9. Maximum absorption of UV light at wavelength 280 nm by a protein is due to:

(A) Aromatic amino acids
(B) Aromatic amino acids and peptide bond
(C) Aliphatic amino acids
(D) Aromatic and aliphatic amino acids

10. Immunotoxins are:

(A) Bacterial toxins
(B) Antibody for specific antigen tagged with toxin
(C) Low immunigenic toxin
(D) Anti-toxin

11. What would be effect on time period of pendulum one placed on equator and other on pole?

(A) No effect
(B) Time period would be greater at pole
(C) Time period would be greater at equator
(D) Pendulum will stop at poles

12. In hydra if any part is lost remaining portion repattern itself and give rise to compete organism. Such a pattern of development is termed as:

(A) Epimorphosis (B) Morphallaxis
(C) Regeneration (D) Healing

13. Which of the following is not a major green house gas in stratosphere?

(A) CO_2 (B) Methane
(C) Ozone (D) Water vapors

14. Mostly inner material of earth remains in solid state. Seldom has it melted and do not remain inside and expelled to the surface of each because:

(A) It is just beneath the earth because
(B) Due to buoyancy
(C) Due to high pressure
(D) More density of surrounding rocks

15. Transduction has been used extensively for genome mapping of bacteria. Which of the following process is useful for gene mapping?

(A) Generalized transduction
(B) Specialized translation
(C) Site specific recombination
(D) Bacterial lysis

16. A aminopurine is attached to ribose sugar by N_9-C' glycosidic bond would be termed as:
(A) Nucleotide (B) Deoxyadenosine
(C) Adenosine (D) Adenosine monophosphate

17. If two sides of isosceles triangle are 7 m respectively. What would be the length of its third side?
(A) 7 cm (B) 16 cm
(C) 23 cm (D) $(16^2 - 7^2)^{-1/2}$

18. Elevated level of RBC and low affinity of hemoglobin for oxygen is an adaptation for:
(A) High altitudes (B) 2 poles
(C) Low altitudes (D) Marine

19. During combustion of carbon in presence of oxygen CO_2 is formed. What will be effect of release of CO_2 if availability of oxygen is doubled?
(A) No change (B) Will double
(C) Will half (D) Will increase four times

20. Half life of any radioactive material is 50 days. How many half life it will take to become 12.5% of the original amount?
(A) 1 (B) 2
(C) 3 (D) 4

21. Terminal electron acceptor for metabolic reactions in organisms is:
(A) CO_2 (B) H_2
(C) O_2 (D) H_2O

22. Among the following which ocean receives maximum sediments?
(A) Arabian ocean (B) Indian Ocean
(C) Bay of Bengal (D) Dead sea

23. What percentage of photoactive radiation are actually utilized for photosynthesis by plants?
(A) Lesser than 1 per cent (B) 1 – 3 per cent
(C) 10 – 20 per cent (D) > 20 per cent

24. The among of rainfall in summer at any place is shown in Table:

January	March	June	December
550	30	10	450

The probable place would be:
(A) India (B) North America
(C) Australia (D) Srilanka

25. Molecular marker can not be utilized for:
(A) Mapping of genes
(B) Identifying the clones
(C) Identifying the locus of gene on chromosome
(D) Identifying the expressed product

26. Type 2 diabetes mellitus occur due to
(A) Deficiency of insulin
(B) Inactivation of insulin recepters
(C) Both of the above
(D) None of the above

27. A committee of two members has to be selected out of 3 men and 2 women. In how many probable ways it can be done?
(A) 10 (B) 25
(C) 110 (D) 120

28. If parents with AABBccddeeFF and aabbCCDDEEff are crossed, the genotype of resulting progeny will be:
(A)AABBccDDeeff (B) AaBbCcDdEeFf
(C) aaBBccDDeeFF (D) AbBbCCddEeFf

29. Boiling point of water at sea level is 100^{0}C. What would be its boiling point at top of the Mount Everest?
(A) 100^{0}C (B) 104^{0}C
(C) 114^{0}C (D) 74^{0}C

30. Which of the following organism excrete uric acid?
(A) Human (B) Fish
(C) Frog (D) Bird

31. If bird is kept in a closed age such that all external clues are blocked then what would be effect on its biological clock:
(A) No effect
(B) It will lag behind to small level
(C) It will be random
(D) It will stop functioning

32. Major weight of human body is due to:
(A) C (B) P
(C) N (D) O

33. What is the probability of getting first three female pups out of a litter of 7?
(A) 1/8 (B) 7/8
(C) 3/27 (D) 3/64

34. The display of a heterologous target protein under study on phage particle is known as:
(A) Phage display (B) Chimeric protein
(C) Both of these (D) None of these

35. Area is required to store fats in seed as compared to carbohydrate would be:
(A) Equal (B) More
(C) Less (D) Slightly more

36. The essential mineral required for cell adhesion protein adherin is:
(A) Calcium (B) Magnesium
(C) Iron (D) Sodium

37. Sxl genes of Drosophila regulate expression at:
(A) Transcriptional level
(B) Post transcriptional level
(C) Translational level
(D) Post translational level

38. Which kinase activity is associated with phytochrome photoreceptors responsible for Red/Far red response?
(A) Histidine (B) Tyrosine
(C) Aspartate (D) Ser/Thr kinase

39. Which molecules has property of self replication?
(A) Protein (B) Carbohydrate
(C) Lipids (D) Nucleic acid

40. A crystalline sphere of radius 1 m is broken into pieces of 0.01 m radius each. What would be change in surface area?
(A) 0 (B) 10
(C) 100 (D) 1000

41. The common reporter genes obtained from *E coli* is:
(A) Chloramphenicol acetyl transferase (CAT)
(B) ß -galactosidase
(C) ß-glucuronidase (GUS)
(D) All of these

42. A non-malignant tumor with radius 'r' shrinks at constant rate with time 't'. It can be represented by equation:
(A) $r = r_0 + k/t$ (B) $r = r_0 - k/t$
(C) $r = r_0 - kt$ (D) None of these

43. Fas protein involved in cell mediated immune response:
(A) Have death domain (B) Act as inducer
(C) Generated G protein (D) Inhibit apoptosis

44. A population of 200 is in Hardy-Weinberg equilibrium with allele frequency of 'A' = 0.7 and 'a' = 0.3. The number of carriers in population will be:
(A) 18 (B) 42
(C) 84 (D) 94

45. In an early embryonic transplantation experiment prospective skin cells were transferred near future muscle cell but then also it differentiates into skin cell. The cell would be termed:
(A) Determined (B) Committed
(C) Totipotent (D) Differentiated

46. Bubonic plaque caused by *Yersinia.pestis* cannot be eradicated completely because:
(A) Causal organism cannot be culture in invitro
(B) Antibodies are not generated by causal organism
(C) Causal organism do not express surface antigen
(D) *Yersinia pestis* have broad host range

47. The Mendelian law of independent of assortment is due to arrangement of chromosome during:
(A) Anaphase I (B) Anaphase II
(C) S phase (D) Cytokinesis

48. The model organism to study cell lineage is:
(A) *Xenopus* (B) Yeast
(C) *Caenorhabditis elegans*
(D) *Drosophila*

49. Main function of sweating is:
(A) Thermoregulation of body
(B) Excrete salt
(C) Maintenance of blood volume
(D) Osmoregulation of body

50. IPR stands for:
(A) Intellectual Property Rights
(B) Indigenous Product Right
(C) Indian Property Rights
(D) International Product Rights

51. Which of the following is a component of MAP kinase signal transduction pathway?
(A) IP3 (B) ERK
(C) Protein kinase B (D) JAK kinase

52. Rolling of sheet of cell over other cells during gasturation is termed as:
(A) Epiboly (B) Ingression
(C) Involulation (D) Delamination

53. Cysteine aspartate protein kinase involved in process of apoptosis function as:
(A) Initiator and executioner:
(B) Initiator and inflammatory
(C) Initiator, inflammatory and executioner
(D) Inflammatory and executioner

54. Serum contains:
(A) Non-fibrinogen proteins, minerals and glucose
(B) Cells corpuscles, mineral and glucose
(C) Cell corpuscles, minerals and non-fibrinogen proteins
(D) Minerals and glucose

55. Recombinants are hybrid DNA molecules consisting of autonomously replicating DNA segment plus inserted elements. Such hybrid molecules are also called:
(A) Chimera (B) Ligation
(C) Transformation (D) Cloning

56. Red wine and red grapes are important source of which anti-tunour agent:
(A) Taxol (B) Vincristine
(C) Resveratrol (D) Bradystanin

57. Which era is characterized by dramatic diversification among eukaryotes?
(A) Cambrian (B) Devonian
(C) Carboniferous (D) Triassic

58. During germination of seeds, after imbibitions of water first step would be:
(A) Mobilization of reserve food
(B) Transcription of specific genes
(C) Cell division
(D) Embryo differentiation

59. Sendai virus enters host cell by:
(A) Endocytosis
(B) Phagocytosis
(C) Cell fusion
(D) Receptor mediated endocytosis

60. Among the following most variable stage of cell cycle is:
(A) G1 (B) S
(C) G2 (D) M

61. Glucose is never seen in urine because it is mainly absorbed by:
(A) Proximal tubule
(B) Collecting duct
(C) Ascending loop of Henle
(D) Glucose never enter into Bowman's capsule

62. Mycobacteria tuberculosis is able to cause disease because as it enters host cell it does not allow endosome to mature into:
(A) Lysosomes (B) Peroxisomes
(C) ER (D) Golgi

63. An organism influence the evolutionary pace of the other organism in:
(A) Co-evolution (B) Parallel evolution
(C) Convergent evolution (D) Divergent evolution

64. In tissue culture experiment to initiate shoots from undifferentiated mass of cell the medium must contain:
(A) Low auxin and high cytokinin
(B) High auxin and high cytokinin
(C) High auxin and low cytokinin
(D) Low auxin and low cytokinin

65. What would be effect on photosynthesis in C_3 and C_4 plants by elevating the concentration of CO_2 under light saturated conditions?
(A) No effect on both type of plants
(B) C_3 plant will saturate fast and C_4 plant remain unaffected
(C) C_4 plants saturate fast and C_3 plants remain unaffected
(D) Both type plants will saturate fast

66. The techniques to clone the first gene were developed in:
(A) 1972 (B) 1987
(C) 1976 (D) 1977

67. Air inhaled during breathing contains principal gases in order $N_2>O_2>CO_2>H_2$. The gases exhaled in air would be in order:
(A) $N_2>O_2>CO_2>H_2$ (B) $N_2> O_2>CO_2 >H_2$
(C) $N_2> CO_2>H_2>O_2$ (D) $N_2> H_2>CO_2> CO_2$

68. During evolution increased ornamentation in male is a result of:
(A) Directional selection (B) Co-evolution
(C) Sexual selection (D) Natural selection

69. If there are only 20 individuals in a population then as per IUCN it would be kept under category:
(A) Extinct (B) Rare
(C) Endangered (D) Critically endangered

70. Which function is not related with TH1 cells?
(A) Secretion of IL-2
(B) Promoting antibody binding to soluble entigens
(C) IFN-γ
(D) Induce phagocytosis

71. During evolution increased ornamentation in male is a result of:
(A) Directional selection (B) Co-evolution
(C) Sexual selection (D) Natural selection

72. Type of mutation which is most suitable for study of regulation of cell like DNA replication is:
(A) Gain of function (B) Loss of function
(C) Suppressor mutation
(D) Condition mutation

73. In aroid plant the temperature of inflorescence rise around 8-10^{0}C as compare to plant during maturation. It is due to activity of:
(A) Dehydrogenase (B) Cytochrome oxidase
(C) Alternative oxidase (D) Peroxidase

74. Common metabolites in nucleotide biosynthesis from glucose by pentose phosphate pathway is:
(A) Phospho ribosyl pyrophosphate (PRPP)
(B) Glyceraldehyde - 3 - phosphate
(C) Dihydroxy acetone phosphate
(D) Fructose – 6- phosphate

75. Suppose a chromosomal aberration in a chromosome ABCDEFG leads to ABCDEFCDFEG. The probable reason is:
(A) Duplication and followed by EF inversion
(B) Duplication followed by pericentric inversion
(C) Only duplication
(D) Only inversion

76. cDNA synthetic DNA made from mRNA with the use of a special enzyme called:
(A) Reverse transcriptage
(B) Expression vector
(C) Promoter gene (D) Reporter gene

77. It has been observed that during prolong animal cell culture and differentiated cell tends to stop dividing. They are said to be in:
(A) Apoptosis (B) Quiescent
(C) Senescence (D) G1

78. Glyosylation of protein ours in:
(A) ER (B) Golgi
(C) Mitochondria (D) Both (A) and (B)

79. Which of the metabolite in nitrate assimilation is not located in chloroplast?
(A) Glutamine (B) Nitrite
(C) Uric acid (D) Xanthin

80. People become tolerant to malaria if they are carriers for:
(A) Sickle cell anemia (B) Thalisemia
(C) Leishmenia (D) Kala azar

81. The ability to create transgenic organisms allows genetic characteristics to be transferred beyond:
(A) Genus (B) Species lines
(C) Across kingdom boundaries
(D) All of these

82. A poky Neurospora was crossed with normal Neurospora and following results were obtained

♀ Poky x ♂ Normal → All poky

♀ Normal x ♂ Poky → All normal

The mode of inheritance is:

(A) Maternal inheritance
(B) Maternal effect
(C) X-linked
(D) Sex influenced

83. In *Neurospora crassa* tetrad analysis showed following results + : m :: 6 : 2. The phenomenon involved for above result would be:

(A) Branch migration (B) Strand exchange
(C) Holliday junction (D) DNA replication

84. If iodine stored in a closed chamber is slowly evaluated for sublime. What would be effect on sublimation rate and mean free path?

(A) Both will increase
(B) Both will decrease
(C) Sublimation rate will increase while free path decrease
(D) Sublimation rate decrease and free path increase

85. The primary criteria for classifying insects is:

(A) Legs (B) Thorax
(C) Wing (D) Appendages

86. Which of the following is considered as primary physical containment?

(A) Containment equipment
(B) Special laboratory design
(C) Laboratory practices
(D) Both (A) and (C)

87. Maximum possible isomers for glucose are:

(A) 4 (B) 8
(C) 16 (D) 32

88. If we want to obtain glycosylated protein from microbe. Suitable choice will be:

(A) Bacteria (B) Yeast
(C) Mycoplasma (D) Animal cell

89. Which antibody is known to be responsible for allergic reaction?

(A) IgG (B) IgA
(C) IgM (D) IgE

90. Which of the following is not a property of a enzyme?

(A) Form complex with substrate
(B) Decrease activation energy
(C) Decrease Gibb's free energy
(D) Increases rate of reaction

91. The glycocalyx around cell membrane an be determined by:

(A) Methylene blue (B) Iodine
(C) Saffranin (D) Lectins

92. Which statement is not correct for nerve impulse transmission?

(A) Minimum threshold intensity is required
(B) Depends upon diameter of neuron
(C) Action potential is proportional to signal intensity
(D) Nerve cells show all or none effect

93. Small amount of lethal mutation always tend of remain in population is due to:

(A) Mutation selection balance
(B) Frequency dependent selection
(C) Positive selection
(D) Negative selection

94. Which of them will have minimum resistance to flow of electric current?

(A) Glass (B) Saline aquifer
(C) Granite (D) Lime stone

95. Which organelle require intact membrane system for ATP synthesis?

(A) Chloroplast (B) Mitochondria
(C) Chloroplast and mitochondria
(D) ER

96. In northern hemisphere there is slow turnover of nutrient in terrestrial ecosystem as compare to southern hemisphere. The probable reason is:

(A) Plants are not good in uptake of nutrients in northern hemisphere
(B) Temperature is low which is not suitable for nutrients recycling
(C) High rainfall in southern hemisphere
(D) Soil is nutrient deficient in northern hemisphere

97. *Bacillus thuringiensis* (Bt) strains have been used for designing novel:

(A) Biofertilizers
(B) Bio-metallurgical
(C) Bio –mineralization process
(D) Bio-insecticidal plants

98. The flagellin protein is associated with:

(A) Bacteria (B) Protist
(C) Virus (D) Eukaryotic cell

99. Atmospheric pressure decline with altitudes as shown in Table:

Height	0 km	2 km	4 km	6 km	8 km
Pressure (mbar)	900	800	650	450	200

What would be atmospheric pressure at height of 5 km?

(A) 720 (B) 550
(C) 640 (D) 420

100. Starch filled plastids are responsible for geotropism in columella cell beneath the root cap. They are termed as:

(A) Amyloplast (B) Elloplast
(C) Chloroplast (D) Proplastid

101. Which of them is not utilized for comparison of operational taxonomic unit (OTU) in numerical taxonomy?
(A) Unweighted pair group method
(B) Percentage similarity
(C) Jaccard blotting
(D) Genetic similarity

102. In which technique O-phenyl Diamine is used as chromogenic substrate?
(A) RIA (B) ELISA
(C) Southern blotting (D) Western blotting

103. The hormone is responsible for regulating spermatogenesis in human is:
(A) Testerone (B) FSH
(C) LH (D) Estrogen

104. Initiation of hematoposis occurs at:
(A) Liver (B) Bone marrow
(C) Kidney (D) Spleen

105. The technique of insertion of a desired gene into DNA of plasmid vector is:
(A) Gene splicing (B) Gene dressing
(C) Gene cloning (D) Gene drafting

106. Exponential growth in bacteria would be expected during:
(A) Lag phase (B) Log phase
(C) Stationary phase (D) Deceleration phase

107. Scientific names of bacteria, fungi, plants and animals are given by:
(A) International union of biological nomenclature
(B) There is different organization for naming plants and fungus
(C) There are three different organization for naming bacteria, plant and animals
(D) Names of plants and animals are given by same organization

108. Which statement is correct regarding c – oncogene?
(A) They are viral genes
(B) They mutated form of genes controlling cell division
(C) They are mutated viral genes
(D) They suppressed tumors

109. A gene consists of two introns and a 5' UTR region, the probable number of exon will be:
(A) 2 (B) 3
(C) 4 (D) 5

110. A sample is in normal distribution ranging from $(\mu - 1\sigma)$ to $(\mu + 2\sigma)$. The data in range would be:
(A) 17 (B) 50
(C) 67 (D) 98

111. Electrical charge is stored in:
(A) Battery (B) Capacitor
(C) Voltmeter (D) Wire

112. Which of the following is negative regulator of trp operon?
(A) Lactose (B) Allolactose
(C) AMP (D) Tryptophan

113. Antcodon sequence lies in:
(A) DNA (B) t-RNA
(C) r-RNA (D) m-RNA

114. Monoclonal antibodies are produced by:
(A) Sarcoma cells (B) Germ cells
(C) Somatic cells
(D) Hybrid of carcinogen cell and bone cell

115. World wide maximum cultivated transgenic crop is:
(A) Insect resistance cotton
(B) Herbicide resistance soybeans
(C) Growing plant for desired molecules
(D) Edible vaccines

116. Which is correct for termination of transcription in eukaryotes?
(A) Terminates prior to polyadenylation
(B) Terminates during polyadenylation
(C) Terminates after polyadenylation
(D) Forms hairpin loop

117. Among the following which antibiotic will inhibit protein synthesis in chloroplasts?
(A) Cylohexamide (B) Chloromphenicol
(C) Rifampin (D) Ricin

118. Which would be suitable for constructing the genomic theory of 9 Kb of DNA?
(A) YAC (B) BAC
(C) P1 based vector (D) Cosmid

119. ABC transporter in plants which are responsible for detoxification of xenobiotics and prevent oxidative damage are located at:
(A) Tonoplast (B) Peroxisome
(C) ER (D) Plasma membrane

120. Animals biologist generally uses line-transect method for estimating density. It is based on assumption that:
(A) Organism will not move from marked transect
(B) All organisms are in straight line
(C) That animals on the line are seen
(D) Organism lack any competition

121. pl for hypothetical protein consisting of only apolar amino acids will be:
(A) Independent on charge over N and C terminus
(B) Depend on number of amino acids
(C) Depend on mass of amino acids
(D) Independent of other type of amino acids

122. Which technique is most suitable to study transcription factor and its binding site?
(A) DNase I footprinting (B) Western blotting
(C) Northern blotting (D) Microarray

123. Thylakoid membrane has lateral asymetrical positoning of photosystem in chloroplast. Which statement is correct?
(A) PS-I in non appressed portion and PS-II in appressed portion
(B) PS-II in non appressed portion and PS-I in appressed portion
(C) Both PS-I and PS-II in a appressed portion
(D) Both PS-I and PS-II in nonappressed portion of thylakoid

124. The main reason for release of energy from sun is:
(A) Fusion of hydrogen (B) Fission of hydrogen
(C) Fusion of helium (D) Fission of helium

125. Heritability due to genetic variance for a trait of importance is 0.2. Which would be most appropriate approach to select trait in next generation in a short time?
(A) Pedigree selection (B) Mass selection
(C) Family selection
(D) Selection by progeny testing

126. Which is called natural genetic engineer?
(A) *Agrobacterium tumefaciens*
(B) *Escherichia coli*
(C) *Neurospora crassa*
(D) *Pseudomonas putida*

127. Organ identity genes are responsible for correct positioning of floral organs on floral meristem. Mutation in them will lead to:
(A) Loss of organs from certain whorls
(B) More number of organs in certain whorls
(C) Appearance of organs at incorrect positions
(D) No flower

128. Most of trees of India in tropical forest belongs to family:
(A) Arecacea (B) Febaceae
(C) Dipterocarpaceae (D) Bromeliacae

129. According to survival of fittest concept of natural selection one species out compete other species. Under such condition no two species an co-exist in same niche but more than one species can live in same niche under condition:
(A) Abundant resources
(B) High competition
(C) Marginal overlapping
(D) Utilization of different resources

130. Among the following imino acid is:
(A) Proline (B) Arginine
(C) Tryptophan (D) Lysine

131. Which statement is correct regarding functioning of topoisomerase?
(A) Separate double stranded DNA
(B) Act as primer
(C) Renaturate the ssDNA
(D) Attach to super coiled DNA and relax it

DECEMBER 2008

1. Which is an example of heat labile compound?
(A) Urea (B) Zeatin
(C) Enzymes (D) All of these

2. An appropriate molecular markers system for applied plant breeding would depend on which of the following?
(A) Breeding objectives (B) Breeding material
(C) Cost effectiveness (D) All of these

3. In which animals, electroporation is the preferred method for ova transfection?
(A) Cattle (B) Mice
(C) Fish (D) Both (A) and (B)

4. Which is not correct about immortalized vertebrate cells cultured in vitro?
(A) Flat and extended shape
(B) Serum-independence
(C) Anchorage dependence
(D) Contact inhibition

5. Which amino acids disturbs the α–helical configuration of proteins?
(A) Alanine (B) Proline
(C) Serine (D) Glutamine

6. The first functional gene was chemically synthesized by which approaches?
(A) Phosphate diester (B) Phosphate trimester
(C) Phosphite trimester (D) Phosphoramidite

7. Which involves recombinant DNA technology?
(A) Polyclonal antibodies
(B) Monoclonal antibodies
(C) Hybrid antibodies (D) Both (B) and (C)

8. Which proteins is active against Diptera?
(A) CryI (B) CryIII
(C) CryIV (D) CryII

9. Somatic hybridization has yielded a commercial variety of:
(A) Tobacco (B) Tomato
(C) Potato (D) Pomato

10. Which mapping populations is immortal?
(A) Recombinant inbred lines
(B) Near-isogenic lines
(C) Both (A) and (B)
(D) F_2 population

11. ϕ and Ψ values for right handed helix are expected to be:
(A) ϕ negative Ψ negative
(B) ϕ Negative Ψ positive
(C) ϕ positive Ψ negative
(D) ϕ positive Ψ positive

12. Atrial natriuretic factor secreted from atria is:
(A) Hormone (B) Neurotrasmitter
(C) Enzyme (D) Growth factor

13. In a population with two alleles 'b' and 'B' having allele frequency 0.7 and 0.3 in Hardy-Weinberg equilibrium, how many individuals in a sample of 250 can be expected to be heterozygous (bB)?
(A) 52 (B) 105
(C) 21 (D) 42

14. Cytoplasmic streaming resulting into mobility of substances and organelles involves interaction of:
(A) Tubulin, kinesin (B) Tubulin, myosin
(C) Actin, kinesin (D) Actin, myosin

15. Substrate for angiotensinogenase is:
(A) Angiotensinogen (B) Anginotensin I
(C) Angiotensin II (D) Renin

16. The main force in membrane resealing of ruptured biomembrane in aqueous environment is:
(A) Hdrophobic forces between membrane lipids
(B) Covalent forces between membrane lipids
(C) Forces between proteins and lipids
(D) Ionic interactions between membrane lipids

17. Factor responsible for formation of early embryonic axis during foetal development pathway of plants is:
(A) Auxin gradient
(B) Morphogens
(C) Orientation of embryonic sac
(D) Plane of cell division

18. Chitin occurs in cell wall of:
(A) Bacteria (B) Arthopods
(C) Fungus (D) Molluscs

19. Which is a soluble dietary fibre products, and is used reduce hypercholesteremia?
(A) Psyllium seed husk
(B) Broccoli (Sulgoraphane)
(C) Soy or clover products
(D) Reseveratrol

20. During germination of barley seeds, enzymes for mobilization of reserve material to developing embryo are secreted from:
(A) Endosperm
(B) Embryo
(C) Aleurone layer
(D) Embryonic leaves

21. In Lederbergs experiment which one of the following option they have used to prove their historical experiment:
(A) One auxotroph and one prototroph
(B) Two auxotroph and two prototroph
(C) Two auxotrophs
(D) Two prototrophs

22. Among the following in which mutant lines, lysogen phages will form clear bacterial plaques:
(A) cI (B) cII
(C) INT (D) XIS

23. Which statement is least likely to be observed amongst the animals showing extensive parental are?
(A) Male polygamous
(B) Sexual dimorphism
(C) Difference in body size of male and female
(D) No investment in selection of mates by females

24. Electrical activity of brain during brain mapping can be recorded by:
(A) fMRI (B) ECG
(C) EEG (D) Polygraphy

25. Which anticancerous drug is obtained from catharanthus roseus?
(A) Taxol (B) Vinristine
(C) Colchicine (D) Serpentine

26. Among the following which is not a cell adhesion protein?
(A) Cadherin (B) Integrin
(C) Selectin (D) Immunoglobulin

27. Which technique an not be utilized for detection of microdeletion of Y chromosome?
(A) Karyotyping (B) PCR
(C) Microarray (D) Hybridization

28. Among the following which is an inhibitor of 8OS ribosome:
(A) Tetracycline (B) Streptomycin
(C) Cycloheximide (D) Chloramphenicol

29. The first field tests were of what genetically altered organism?
(A) Bt Corn
(B) Vaccinia virus containing a gene from the rabies virus
(C) The "flavsavr" tomato
(D) Strawberry seedlings sprayed with "ice-minus" bacteria

30. The inbreeding coefficient of offspring on marriage between brother and sister sibling will be:
(A) 0.50 (B) 0.05
(C) 0.25 (D) 0.75

31. Generally organism tens to remain in realized niche. Under what condition realized niche can be greater than fundamental niche:
(A) Abundance of resources
(B) Heterogeneity of resources
(C) One species helping other in utilization of resources
(D) Moving of organism from source to new link area

32. Post translational modification takes place in:
(A) Nucleus
(B) Mitochondria
(C) Ribosome
(D) Endoplasmic reticulum

33. Maximum density of dopamine and noradregenic receptor occurs at:
(A) Cerebellum (B) Cerebrum
(C) Medulla oblangata (D) Spinal cord

34. Regulation of trp operon by biding of tryptophan to trp repressor is termed as:
(A) Repression (B) Induction
(C) Antitermination (D) Attenuation

35. Which of the following is not coded by MH genes:
(A) Compounds of complement pathway
(B) Immunoglobulin
(C) Glycoproteins
(D) Antigen presenting proteins

36. Mosaic developmental pattern is always:
(A) Autonomous (B) Non autonomous
(C) Conditional (D) Regulative

37. Which is least likely to occur for removal of cancer cells?
(A) T-cell based cytotoxicity
(B) Complement fixation
(C) Autophagy
(D) Phagoytosis

38. Yeast with petite colony when crossed with wild type generates no peptide colony. The most probable mode of inheritance is:
(A) Chloroplast (B) Mitochondria
(C) Episomal (D) Nuclear

39. Undesirable pleitropic effect of a transgene on the transgenic plants may be corrected by :
(A) Changing the genetic background of transgene through hybridization
(B) Targetting of the transgene product into an appropriate cellular compartment
(C) Mutagenesis in the transgene
(D) Both (A) and (B)

40. Leukemia inhibiting factor has been utilized in animal cell culture for:
(A) Stimulating growth of cell
(B) Differentiation
(C) Morphogenesis
(D) Arrest cells at mitosis

41. The movement of chloroplast is mediated by:
(A) Dynein (B) Kinesin
(C) Actin (D) Myosine

42. Among the following which is not a result of acid rain:
(A) Low amount of phosphate availability
(B) Low amount of aluminum availability
(C) Low availability of nutrients of plant
(D) Increased acidity of soil

43. Dorsal lip of amphibian is equivalent to chicks:
(A) Hensen node (B) Primitive grove
(C) Animal pole (D) Vegetal pole

44. Dorsal mutant in Drosophila will result in:
(A) Dorsalization of ventral side
(B) Ventralization of dorsal side
(C) There would be no effect
(D) Anterior-posterior pattern formation will be affected

45. Mucous which covers the ethical lining of stomach and protects it from protease activity is secreted by:
(A) Goblet cells (B) Parietal cells
(C) Microvilli (D) Aciner cells

46. Recently gene therapy for mutated gene has been experimentally proven in mouse utilizing:
(A) Winged P element
(B) Cre-Lox system
(C) Non-homologous recombination
(D) Ac-Ds elements

47. Which technique can not be utilized for studying response mechanism for both B and T cell immune responses?
(A) Complement fixation
(B) Western blotting
(C) Cytotoxicity assay
(D) ELISA plot

48. The roots of a plant are converted into drug-producing structures in a process called:
(A) Microcosm establishment
(B) Mibridization
(C) Bioremedition
(D) Rhizosecretion

49. The technique used to distinguish between individual of the same species using only samples of their DNA is called:
(A) DNA fingerprinting
(B) DNA testing
(C) DNA typing/DNA profiling
(D) All of these

50. Oxygenase activity of rubisco generates:
(A) Two molecules of PGA (3C)
(B) Two molecules of phosphoglycolate (2C)
(C) One molecule each of PGA and phosphoglycolate
(D) Two molecules each of PGA & phopshoglycolate

51. Among the following which is not involved in plant defense signaling pathway:
(A) Gibberellic acid
(B) Ethyline
(C) Salisylic acid
(D) Jasmonic acid

52. The maximum ionic interaction would be observed:
(A) In presence of polar solvent
(B) In presence of mixture of water and alcohol
(C) Almost equal in all kinds of solvents
(D) When ionic compounds is out of the solvents

53. A major functional difference between the succinyl CoA-synthetase of plant and animal cell mitochondria is that it:
(A) Does not produce ATP in plant cell
(B) Produces UTP in plant cell
(C) Produces ATP in plants and GTP in animal
(D) Produces GTP in plants and ATP in animals

54. Enzymes accelerates rate of reaction by:
(A) Lowering number of transition states
(B) Lowering the activation energy of highest transition state
(C) Providing energy to substrate
(D) Providing more chance to substrate to react together

55. Effect of release of IP_3 during signal transduction pathway is:
(A) Closure of Ca^{++} channel in ER
(B) Increase in intracellular Ca^{++} level
(C) Increase of extracellular Ca^{++} level
(D) Inactivation of almodulin proteins

56. What would happen if lysosome membrane leaks its digestive enzyme in cytosol?
(A) Acid hydrolases will be inactivated
(B) Acid hydrolases will digest the cellular compounds
(C) pH of cell will increase
(D) It will cause I-cell disease

57. Colchicine arrest cells in Mitosis, because:
(A) It binds to monomeric tubulin and thus depolymerizes microtubules
(B) It prevents chromosome condensation by inhibiting histone-histone interactions
(C) It dephosphorylates tubulin monomers and thus depolymerises microtubules
(D) It binds to the + end of F-actin and prevents further addition of G-actin to actin filaments

58. Selection of protoplast fusion hybrids can be done by:
(A) Visual selection
(B) Fluorescent labels
(C) FACS
(D) Any of the above methods

59. Haploid production cannot be done through
(A) Pollen culture (B) Callus culture
(C) Anther culture (D) Hybridization

60. Which transgenic crops is being grown on commercial scale in India?
(A) Bt-cotton (B) Bt-brinjal
(C) Bt-rice (D) Golden rice

61. In salt tolerant plant the excess salt is transported to vacuole by:
(A) Na^+ - H^+ antiporter (B) Na^+ - K^+ pump
(C) Na^+ - Cl^- symporter (D) Na^+ - H^+ ATPase

62. Genetic disorder xeroderma pigmentation is due to error in:
(A) Base excision repair mechanism
(B) Nucleotide excision repair mechanism
(C) Direct repair mechanism
(D) DNA replication mechanism

63. The regulation of circadian rhythms in plants is:
(A) Phycobillins (B) Phytochromes
(C) Phototropins (D) Zeaxanthin

64. Diphtheria toxin causes:
(A) ADP ribosylation of eEF-2
(B) ADP ribosylation of EF1α
(C) Blocking activity of RNA polymerase
(D) Blocking DNA replication process

65. Individuals having chromosomes and short arm of Y chromosome are made while individuals having X chromosome and long arm of Y chromosome are female. This shows that:
(A) Genes for maleness are located on short arm of Y chromosome
(B) Genes for maleness are located on long arm of Y chromosomes
(C) Genes for maleness are located X chromosome
(D) Genes determining sex are not located on X chromosome

66. The process of photosynthesis which leads to formation of glucose is a type of:
(A) Oxidation (B) Reduction
(C) Condensation (D) Fixation

67. Which equation best describes the bacterial population growth?
(A) $\frac{dN}{dt} = KN$ (B) $\frac{dN}{dt} = N$
(C) $\frac{dt}{dN} = KN$ (D) $\frac{dN}{dt} = K$

68. Which technique is used to study de novo RNA synthesis?
(A) Southern blotting (B) Northern blotting
(C) Microarray (D) RT-PCR

69. The bond between sugar and nitrogenous base in case of DNA is known as:
(A) Glycosidic bond (B) Phosphodiester bond
(C) H- bond (D) Peptide bond

70. DNA finger printing is useful for study of:
(A) Forensics (B) Polymorphism
(C) Identity and relationships
(D) All the above

71. Homeotic genes:
(A) Maintain gaps in segments
(B) Provide gradient in developing embryo
(C) Codes for morphogens
(D) Mutation results in formation of organ at unusual locations

72. If a cell has 'c' as the DNA content of cell and 'n' as the number of chromosome, then just immediately before the cell division in case of mitosis what would be value of 'c' and 'n':
(A) 2c and 4n (B) 4c and 2n
(C) 4c and 4n (D) 2c and 2n

73. In the chemiosmotic hypothesis at ATP synthesis by chloroplast:
(A) Addition of ATP synthetase to membrane vesicles will allow the synthesis of ATP if only ADP is added
(B) ATP synthetase must be very close to the two photosystems in the lamella
(C) Proton ionophores have no effect on ATP synthesis
(D) The pH of the stroma rises due to proton pumping

74. Covalent bond formation between two atoms takes place by:
(A) Transfer of electron from one atom to other
(B) One side sharing of electrons
(C) Electron sharing by both interacting atoms
(D) Affinity between two atoms

75. In population of 10million individuals birth rate is 19 per 1000 and death rate is 14 per 1000. Annual rise in population would be:
(A) 50000 (B) 5000
(C) 14000 (D) 500000

76. Among the following which plant family has been extensively used for phytoremediation:
(A) Poaceae (B) Brassicaceae
(C) Malvaceae (D) Anonaceae

77. Micro propagation can be done by:
(A) Through shoot bud proliferation
(B) Through adventitious buds
(C) Meristem tip culture
(D) All of these

78. The most probable place where life would have originated is:
(A) Outer space
(B) Barren rocks
(C) In oceans
(D) Deep hydrothermal vents

79. When oxygen –hemoglobin curve shift to left it represents:
(A) Decrease in pH
(B) Decrease in CO_2 level
(C) Rise in concentrations of 2, 3 BPG
(D) More affinity for oxygen

80. Quantitative heritance defines:
(A) Variation in phenotype
(B) Variation in genotype
(C) Variation in environment
(D) Variation in genes

81. Essential constituents of an expression vector are:
(A) Replicon, promoter and selectable marker
(B) Replicon, unique restriction site and selectable marker and promoter
(C) Promtor, operator and restriction site
(D) Replicon, unique restriction site and promoter

82. Rate of molecular evolution would be least in:
(A) Non-synonymous change in codon
(B) Synonymous change in codon
(C) Flanking regions of genes
(D) Introns of genes

83. In plant lateral root initiates from:
(A) Pericycle (B) Cortex
(C) Pith (D) Endodermis

84. Which of the following is present on a Ti plasmid, but not on any component of a binary vector system?
(A) Vir genes (B) GUS genes
(C) LB, RB (D) MCS
(E) Opine genes

85. Genetic mapping reveals that distance between two genes 'A' and 'B is 10 M. What is the chance of getting Aabb progeny if AaBb is test crossed?
(A) 5% (B) 10%
(C) 45% (D) 90%

86. Globular protein when treated with organic solvent get denatured. The main interaction which is affected on treatment with organic solvent is:
(A) Hydrogen bonds
(B) Covalent bonds
(C) Ionic interactions
(D) Hydrophobic interactions

87. A set of disease resistance mechanisms that are not specific to a particular pathogen comes under
(A) Adaptive immunity (B) Innate immunity
(C) Passive immunity (D) Active immunity

88. Which is correct hierarchical sequence in taxonomy?
(A) Phylum – class-order-family-genus
(B) Class-Phylum –order-family-genus
(C) Phylum – class- family-order- genus
(D) Phylum – class-order -genus-family

89. Defective gene in amyotrophic lateral sclerosis is:
(A) Rb (B) P^{53}
(C) SOD1 (D) TGF

90. Among the following which molecule has been frequently used for molecular systematics?
(A) Insulin (B) Cytohrome C
(C) Globin (D) Collagen

91. Bacteria divides by:
(A) Binary fission (B) Mitosis
(C) Frangmentation (D) Meiosis

92. Germination of moth bean in dark is an example of:
(A) Photomorphogenesis
(B) Skotomorphogenesis
(C) De-etiolation
(D) Shadding effect

93. Vector for transmission of disease Kalazar is:
(A) Ades (B) Anopheles
(C) Glossina (D) Phlebotomus

94. Why might you find two kanamycin resistance genes on different parts of a T-DNA plasmid?
(A) One of the genes could be used for positive selection, one could be used for negative selection.
(B) One of the genes could be used for selection in plants, and one could be used for selection in bacteria.
(C) This arrangement facilities site-specific recombination for marker-free selection.
(D) Each gene might have been introduced by a independent co-transformation event.

95. Birds besides lungs have highly branched air sacs. The major functions of air sac is to:
(A) Increase surface area for gas exchange
(B) Help in ventilation during inhalation
(C) Help in ventilation during exhalation
(D) Help in ventilation during both inhalation and exhalation

96. Intracellular negative potential and extracellular positive potential occurs in:
(A) All cells (B) Neurons
(C) Kidney cells (D) Liver cells

97. First successful vaccine against cancer has been prepared for:
(A) Oral cancer
(B) Cervical cancer
(C) Breast cancer
(D) Colon cancer

98. Biomass turn over time is the ratio between biomass and productivity of an ecosystem. Which of the following forests should have highest biomass turnover time?
(A) Tropical dry forests
(B) Tropical wet forests
(C) Temperate deciduous forests
(D) Boreal forests

99. Enzyme do not interfere with:
(A) Free energy of reaction
(B) Rate of reaction
(C) Activation energy if transition state
(D) Reaction equilibrium

100. An oligonucleotide DNA sequence tagged with fluorescent tag used to identify unknown gene by hybridization is termed as:
(A) Probe (B) Reporter gene
(C) Ligand (D) DNA

101. Which of these is/are the characteristic/s of the specific immunity?
(A) Antigenic specificity
(B) Diversity & self/nonself recognition
(C) Immuninologic memory
(D) All of these

102. Smooth movement of bacteria during chemotaxis is due to:
(A) Due to tumbling
(B) Phosphorylation of CheY
(C) Movement of H^+ across plasma membrane
(D) Phosphorylation of Che A

103. Among the following which is not monitored as daily potential pollutant:
(A) CO (B) CO_2
(C) SO_2 (D) NO_x

104. Among the following which activity is absent in bacteria DNA polymerase I :
(A) $5' \rightarrow 3'$ polymerase activity
(B) $3' \rightarrow 5'$ polymerase activity
(C) $5' \rightarrow 3'$ exonuclease activity
(D) $3' \rightarrow 5'$ exonuclease activity

105. As we move from one geographical region to next neighboring region, species diversity tends to change. It is termed as:
(A) α - diversity (B) β - diversity
(C) γ - diversity (D) δ - diversity

106. A plant with genotype r^+h^+/r^-h^- was test crossed. Out of total 280 progeny 260 are r^+h^+/r^-h^- and r^-h^-/r^-h^-. The recombination frequency will be:
(A) 92.8 (B) 46.4
(C) 7.2 (D) 3.6

107. Which factor is least responsible for genetic drift?
(A) Migration (B) Founder effect
(C) Bottleneck
(D) Restriction of resources

108. Genetic map is one that:
(A) Establishes site of the genes on a chromosomes
(B) Establishes the various stages in gene evolution
(C) Shows the stages during the cell division
(D) Shows the distribution of various species in a region

109. Which organelles is/are semi autonomous in nature?
(A) Mitochondria (B) Chloroplast
(C) Nucleus (D) Both (A) and (B)

110. Microbes have a relatively -----that allows the chemical conversions possible:
(A) High metabolic rate
(B) Large surface area
(C) High multiplication rate
(D) All of these

111. The emergence of polarity of an embryo is the result of:
(A) Positive and negative charges interacting in early development
(B) Cytoplasmic differences between cells
(C) Cytoplasmic determinants within cells
(D) All of these

112. Location of glutamate synthetase, an important enzyme in nitrogen assimilation is:
(A) Only cytoplasm
(B) Only chloroplast
(C) Both in cytoplasm and chloroplast
(D) In endoplasmic reticulum

113. Dendrogram in numerical taxonomy represents:
(A) Phenetic similarities
(B) Phyogenetic similarities
(C) Evolutionary similarities
(D) No similarities

114. In spite of the prevalence of herbivory, the earth continues to be largely green because:
(A) The number of herbivore species is low
(B) Herbivores are very inefficient feeders
(C) Herbivore numbers are kept low by their predators
(D) Herbivory promotes plant growth

115. Which radioisotope is generally incorporated in thymine to study DNA replication process?
(A) ^{32}P (B) ^{35}S
(C) ^{3}H (D) ^{14}C

116. Shigella enters into host cell by process of:
(A) Cell-cell fusion
(B) Clatherin coated pits
(C) By exploiting host cell organelles
(D) Phagocytosis

117. Which of the following organelle lack membrane?
(A) Mitochondria (B) Ribosomes
(C) Microtubules (D) Peroxisomes

118. When release factor binds to stop codon on mRNA during translation, the synthesized peptide chain is transferred to:
(A) tRNA (B) Water
(C) H^+ (D) Amino acids

119. The specialized structure pectin for clear eye sight is characteristic feature of:
(A) Birds (B) Amphibian
(C) Nocturnal mammals (D) Aquatic mammals

120. Photorespiration occurs in
(A) Mitochondria (B) Peroxisomes
(C) Chloroplast (D) All of these

121. Which is true for β–oxidation of fatty acids?
(A) Formation of malonyl-CoA
(B) Formation of acetoacetyl ACP
(C) Transport of acyl CoA into mitochondria
(D) Use of $NADPH_2$

122. In which biogeography region you would find Nilgiri-Tahr trees?
(A) Shola grasslands
(B) Nilgiri forests
(C) Shola grasslands and Nilgiri forests
(D) At high altitude of south-western ghats

123. A cross between a red eyed male fly and white eyed female fly produces red eyed female and white eyed male progenies. While reciprocal cross produces all offsprings with red eyes. The trait for eye color is:
(A) Sex linked traits
(B) Sex influenced trait
(C) Sex linked homogametic male
(D) Sex linked heterogametic male

124. The principle for formation of image in phase contrast microscopy:
(A) Interference of light waves
(B) Negative staining of object
(C) Use of fluorescent probes
(D) Enhancing contrast by differentiating the change in phase of light passes through specimen coming from ½ angle of one of light entering through objective lens

125. Which of the following is not correctly matched?
(A) Chanocytes-Porifera
(B) Malphigian tubules - Arthropods
(C) Clitellum - Annelids
(D) Cnidocytes - Mollusc

126. In a hybridization experiment, a plant shows phenotypic ratio of 15:1. How many genes control the trait for observed phenotypic ratio?
(A) One (B) Two
(C) Three (D) Polygene

127. A protein which is to be degraded in proteasome is tagged with:
(A) Polyglycine (B) Polyproline
(C) Ubiquitin (D) Formyl methionine

128. Characteristics dominant grasses in Brahamputra Ganga and Punjab plains respectively are:
(A) Alphinia, Eichinochloa, Cenchrus
(B) Saccharum, Alphinia, Saccharum
(C) Cenchrus Eichinochloa, Saccharum
(D) Saccharum, Cenchrus, Alphinia

JUNE 2008

1. During transposition transposons are excised by:
(A) Transposase (B) Nuclease
(C) Topoisomerase (D) Exonuclease

2. Consider the following DNA sequence 5'-ATGGG-CATAGACGATATGGTAG-3'. If due to frameshift mutation there is insertion of G between 3rd and 4th position. consider a reverse mutation occur is same mutated sequence. Which reverse mutation will have minimum effect in protein change
(A) Insertion of a nucleotide between 5th and 5th position
(B) Insertion of three nucleotide between 5th and 6th position
(C) Deletion of a nucleotide between 5th and 6th position
(D) Deletion of a nucleotide between 11th and 12th position

3. Which statement is not true regarding genetic drift as an evolutionary force?
(A) Small population size
(B) Reproductive variance
(C) Effective population size
(D) change in allele frequency

4. Cloning is a means to:
(A) Replace original genotype
(B) Preserve genotype
(C) Production of HGH gene in Escherichia coli
(D) None of the above

5. The medium is added continuously to the fermentor tank to replace that, which has been fermented. This is true in case of:
(A) Batch technique
(B) Continuous flow technique
(C) Both are true
(D) None is true

6. During gametophyte self incompatibility the primary response is:
(A) Deposition of callose
(B) Pollen tube lysis
(C) Formation of concentric ring from Golgi
(D) Self-incompatibility triggers a Ca^{++}-dependent signalling aside in incompatible pollen

7. Immunological diversity in antibody is generated by:
(A) Rearrangement of immunoglobulin genes
(B) Alternative RNA processing
(C) Post transcriptional modification
(D) Post translation modification

8. Polar head group in membrane cholesterol is due to:
(A) Hydroxyl group (B) Long alkyl chain
(C) Benzene rings (D) carboxylic groups

9. Chromosomal aberration in which two breaks occur in a chromosome and the intercalary segment reunites in a reverse order is categorized as:
(A) Duplication (B) Deletion
(C) Inversion (D) Translocation

10. Genes between related organism exhibits high variation. The variations would maximally occur in:
(A) Exons (B) Intron
(C) Promoters (D) Polyadenylation site

11. consider that two population are growing exponentially with initial difference in growth rate of 10%. After 10 generation the difference between population size would be:
(A) 1 : 1 (B) 4 : 1
(C) 8 : 1 (D) 10 : 1

12. Among the following ritually endangered plant species is:
(A) *Dipteroarpus nilgirinsensis*
(B) *Saraca indica*
(C) *Cupressus cashmeriana*
(D) *Terminalia arjuna*

13. Which statement is not correct for vitamin D?
(A) It helps in bone formation
(B) It is produced by skin in presence of UV light
(C) It is water insoluble
(D) It helps in bone resorption

14. Which of the following do not bring variation in population?
(A) Random drift (B) Non-random mating
(C) Recombination (D) Natural selection

15. Among the following which microorganism is involved in nitrogen fixation with woody trees?
(A) Frankia (B) Rhizobium
(C) Azatobacter (D) Azospirillium

16. The genes for improving rice cultivars have been taken from the Indian rice variety:
(A) *Oryza sativa* (B) *O. indica*
(C) *O. nivara* (D) *O. rhyzae*

17. The adaptation related to high altitude is:
(A) Increase in RBC count
(B) Decrease in RBC count
(C) Increase affinity for oxygen by hemoglobin
(D) Decrease affinity for oxygen by hemoglobin

18. With time molecular distance between organisms increases during evolution due to:
(A) Natural selection (B) Neutral mutation
(C) Random drift (D) Point mutation

19. Receptors for signaling for steroid hormones are located at:
(A) Plasma membrane
(B) Organelle membrane
(C) Cytosolic
(D) No receptor

20. Natural selection is primarily based on fitness which is dependent on maximum number of offspring laid for next generation but at present new concept is added where organism help in reproduction of relatives to increase the overall fitness. This concept is termed as:
(A) Evolutionary fitness (B) Inclusive fitness
(C) Relative fitness (D) Kin selection

21. During cell cycle, sister chromatids are pulled apart during:
(A) Metaphase (B) Anaphase
(C) Prophase (D) Interphase

22. ABO blood groups is a good example for both-------- and multiple alletism
(A) Co dominance (B) Dominance
(C) Recessive ness (D) Over dominance

23. Which statement is correct regarding the meiosis?
(A) There are two rounds of replication and two rounds of cell division
(B) There is one round of replication and one round of cell division
(C) There is one round of replication and two rounds of cell division
(D) There are two rounds of replication and one round of cell division

24. Temperature of body is regulated by:
(A) Hypothalamus
(B) Suprachaismatic nuclei
(C) Cerebellum (D) Cerebrum

25. Which mineral ion play important role in functioning of photosystem II?
(A) Manganese (B) Magnesium
(C) Iron (D) Molybdenum

26. Frequency of blood group o in population is 25%. remaining individuals of population have equal number of blood group A & B. What would be the ratio of allele frequency between blood group A, B & O:
(A) 1 : 1 : 1 (B) 3 : 2 : 2
(C) 1 : 1 : 2 (D) 3 : 3 : 1

27. Which of the following statement is correct with reference to replication in eukaryotes?
(A) Single origin and continuous replication
(B) Multiple origin, continuous and discontinuous replication
(C) Multiple origin and continuous replication
(D) Single origin, continuous and discontinuous replication

28. Calculate the pH of aid with Ka=10-6 and 0.01 M:
(A) 0 (B) 4
(C) 6 (D) 2

29. In eukaryotes shortening of chromosome from ends is prevented by:
(A) DNA polymerase (B) RNA polymerase
(C) Telomerase (D) Transposase

30. For an enzyme catalyzed reaction exhibiting Michaelis Menten equation what would be the increase in substrate concentration to increase the rate of reaction from 10% of V_{max} to 90% of V_{max}:
(A) 80 (B) 8
(C) 4 (D) 2

31. Which of the statement regarding plasma cell is correct?
(A) They are produced during secondary immune response
(B) They are mature antibody secreting cells
(C) They are involved in removal of intracellular viruses
(D) Involved in inflammatory responses

32. Which of the following monochromatic lights are most suitable for growth & development of plants?
(A) Red, far red (B) Red, blue, far red
(C) Red, green (D) Blue, far red

33. Among the following which data alone are capable for preparing dendrogram from given operational taxonomic unit (OUT)?
(A) Mean of similarity (B) Similarity matrix
(C) characters taken into account
(D) criteria for classification

34. In chromosome, 30 nm fibres during metaphase attach to:
(A) Scaffold (B) Centromere
(C) Nuclear matrix (D) Nuclear lamina

35. Photoperiodic stimulus from leaves to shoot apical meristem/floral meristem is transported through:
(A) Xylem (B) Phloem
(C) Plasmodesmata (D) Apoplast

36. Which is called stress hormone?
(A) GA3 (B) Ethylene
(C) ABA (D) Cytokinin

37. A tryptophan auxotroph in corn showed 50 times more accumulation of IAA than the normal. Probable explanation for this is:
(A) There may be some other precursor for IAA synthesis
(B) IAA is probably not inhibited by feedback mechanism
(C) IAA was not oxidized
(D) Deconjugated of ester linked IAA does not take place

38. In Drosophila X0 are male and XXY are female while in humans XX are female and XY are male. On the basis of given information which statement is not true:
(A) Y chromosome do not play any role in sex determination of Drosophila
(B) Y chromosome is sex determinant in humans
(C) In humans sex determination is based on number of X chromosomes to sets of autosomes
(D) In Drosphila sex determination is based on number of X chromosome to sets of autosomes

39. Pitcher plant *Nepenthes alata* would be expected to have:
(A) NO_3^- specific ion channel
(B) H^+ - NO_3^- symporters
(C) Na^+ - NH_3 symporter
(D) ATP powered pumps for NO_3^-

40. In honey bee, males are developed parthenogenetically while workers are developed by sexual reproduction. The workers exhibits more similarity among themselves as compare to queen. If workers starts giving organisms parthenogenetically then offspring would most likely resemble to:
(A) Among themselves and with mother
(B) Among themselves and slightly differ from mother
(C) Among themselves and with queen
(D) Among themselves and with father

41. In which of the following condition realized niche exceed over fundamental niche?
(A) Competition (B) Commensalism
(C) Ammensalism (D) Mutualism

42. Gaucher disease where glucocerebroide are not degraded is related to:
(A) Mitochondria (B) Lysosomes
(C) Peroxisomes (D) Golgi

43. Primary acceptor of CO_2 in photosynthesis is:
(A) Ribose
(B) Ribose – 5 - phosphate
(C) Ribulose 1, 5-bisphosphate
(D) 3 – Phosphoglyerate

44. If activator binds to repressor, it will prevent:
(A) Transcription
(B) Binding of RNA polymerase to promoter
(C) Binding of repressor to operator
(D) Binding of repressor to promoter

45. Highest extinction during history of earth was observed during:
(A) End of permian (B) End of retaeous
(C) End of devonian (D) End of carboniferous

46. Renaturation of human genome has revealed that it contains both repetitive and non-repetitive sequence. Which statement is correct?
(A) Human have more unique sequence
(B) Repetitive sequence are located only to centromere
(C) Repetitive sequence renaturate fast
(D) Unique sequence renaturation fast

47. Which of the following is characteristics feature of climax community?
(A) Simple food chain
(B) High resilience
(C) High productivity
(D) Narrow niche specialization

48. Which is not correct about abscisic acid?
(A) It is a naturally occurring growth inhibitor
(B) Precursor of ABA is maleic acid
(C) It is synthesized in leaves
(D) ABA counteracts the effects of GA and promotes dormancy in buds and seeds

49. Receptor mediated endocytosis is carried by:
(A) Coated vesicles (B) Coated pits
(C) Endocytosis (D) Exocytosis

50. Ecological adaptations in which some organism are favored due to more energy investment on their reproductive rate while other on basis of channelizing energy for homeostatsis. Such a selection strategies are termed as:
(A) K selection and r selection
(B) Logistic and exponential selection
(C) Directional and distruptive selection
(D) Kin and group selection

51. After meiosis the 20% gametes are recombinant for two genes. The distance between two genes will be:
(A) 5 cM (B) 10 cM
(C) 20 cM (D) 40 cM

52. Which of the following statements is false for the nitric oxide gas?
(A) An intracellular signaling molecule
(B) Deamination of histidine results into nitric oxide production
(C) Stimulates guanylyl cyclase to produce cGMP
(D) can be produced by activated neutrophils

53. Gene for fungal resistance is found in cytoplasm. If a susceptible female and resistant male are crossed then progeny will exhibit:
(A) All resistance
(B) All susceptible
(C) Half resistance and half susceptible
(D) Cannot be predicted

54. Blood vessel A has thick wall, narrow lumen and no valves while blood vessel B has thin wall, wide lumen and have valves. Here A and B are:
(A) A is artery and B is vein
(B) A is vein and B is artery
(C) A is vein and B is capillary
(D) A is capillary and B is artery

55. Cattle are known to be responsible for green house effect due to:
(A) High respiration rate
(B) More consumption of plant
(C) Fermentation in rumen
(D) High reproductive rate

56. Negative potential across plasma membrane is maintained by:
(A) Active transport (B) Passive transport
(C) Ion channels (D) Transporters

57. Major cause of evolution of gene and protein is:
(A) Point mutation
(B) Chromosomal aberration
(C) Sexual reproduction
(D) Gene duplication and divergence

58. The maximum BOD and minimum DO for pure drinking water should be:
(A) 25, 5 (B) 2, 5
(C) 3, 9 (D) 0, 6

59. In signal transduction trimeric G protein with α, β and γ is involved. Which subunit will activate adenylate cyclase:
(A) α subunit (B) β subunit
(C) γ subunit (D) All of these

60. Organisms with high growth and production are:
(A) Ectotherm (B) Endotherm
(C) Carnivore insects (D) Deteriores

61. The possible type of gametes formed from genotype AABbCCDdEe will be:
(A) 4 (B) 8
(C) 16 (D) 32

62. On molar basis if DNA has 20% cytosine, then percentage of adenine would be:
(A) 20% (B) 30%
(C) 40% (D) 60%

63. If in an operon, repressor binds to operator it will lead to:
(A) Switch on transcription
(B) Switch off transcription
(C) Enhanced transcription
(D) Differential gene expression

64. Among closely lying cells, signals are communicated by:
(A) Neurotransmitters (B) Hormones
(C) Gap junctions
(D) Cell membrane proteins

65. In Indian, which conservation programme is related with protection of entire "Tropic ladder"?
(A) Project tiger (B) Project elephant
(C) Ramsar sites (D) Biosphere reserve

66. Primary carnivores consume 40% production of herbivore and assimilate 70% of energy. What percentage of energy these carnivore assimilates the energy available from herbivors:
(A) 30 (B) 28
(C) 10 (D) 40

67. In TCA cycle, malonate is competitive inhibitor of:
(A) Succinate (B) Fumarate
(C) Oxaloacetate (D) α-Ketoglutarate

68. Area under forest over in India as per estimates of 2001 is:
(A) 7.9 mha (B) 12.7 mha
(C) 20.6 mha (D) 16.3 mha

69. Bacteria propels with the help of:
(A) Actin like MreB proteins
(B) Myosin
(C) Flagella made of protein flagellin
(D) Cytoskeleton

70. Bacteria cannot be classified as species by the biological species concept because they:
(A) Are sexually reproducing organisms
(B) Have high growth rate
(C) Exhibits little morphological variations
(D) Do not have nucleus

71. Gases used by Urey and Miller for experimentation of origin of life by Oparin and Haldane hypothesis was:
(A) Hydrogen, methane and ammonia
(B) Hydrogen, methane and CO_2
(C) Hydrogen, ammonia, methane and CO_2
(D) Hydrogen, carboxylic acid and amino acid

72. Choose the statement that is not correct for the cytoskeletal protein actin:
(A) α-actins are found in various types of muscles
(B) Polymerization of pure actin in invitro requires GTP
(C) Actin filaments have a slow growing minus end and a fast growing plus end
(D) Cytochalasins are the inhibitors for actin polymerization

73. The whole genome of which was the first to be sequenced?
(A) Baker's yeast
(B) *E. coli*
(C) *Bacillus subtilis*
(D) *Haemophilus influenzae*

74. Hybrid sorting has relevance to:
(A) Somatic hybridization
(B) Molecular markers
(C) Haploid production
(D) Marker-assisted selection

75. Somatic hybridization has been the most successful in which families?
(A) Graminae (B) Solanaceae
(C) Cruciferae (D) Both (B) and (C)

76. The first industrial production of enzymes by fermentation was for which of the following application?
(A) Food (B) Pharmaceutical
(C) Detergent (D) Textile

77. Aspertame is an artificial sweetner, it is chemically which of the following?
(A) Dipeptide (B) Disaccharide
(C) Oligopeptide (D) Both (A) and (B)

78. Denaturation is essential part of:
(A) Southern hybridization (B) PCR
(C) Nick-translation (D) Both (A) and (B)

79. Which is not optically active?
(A) Glycine (B) Serine
(C) Leucine (D) Alanine

80. Which is naturally occurring auxin?
(A) IAA (B) IBA
(C) NAA (D) 2,4-D

DECEMBER 2007

1. Chemolithotroph obtain their carbon from CO_2 and energy from
(A) Sunlight (B) Water
(C) Inorganic compounds
(D) Organic compounds

2. Extensive phyletic diversification of animals was observed in:
(A) Devonian (B) Silurian
(C) Cambrian (D) Mesozoic

3. If in a metabolic reaction at temperature 270 increase in enthalpy is 1000 J and decrease in entropy is 10 J. Calculate the change on free energy:
(A) 4000 (B) 1270
(C) 2000 (D) 730

4. Which of the following is involved in intermolecular hydrogen bonding with water?
(A) Urea (B) CH_4
(C) CCl_4 (D) $CHCl_3$

5. In ecosystem the concept of entropy s used to explain:
(A) Photosynthesis efficiency
(B) Energy flow in trophy level
(C) Population growth (D) Competition

6. The characteristic of a population with low value of intrinsic growth (r = 0.2) is:
(A) Late age of maturity and small clutch size
(B) Early age of maturity and small clutch size
(C) Late age of maturity and large clutch size
(D) Early age of maturity and large clutch size

7. Gibberellic acid stimulates seed germination in monocots by activation of degradative enzymes by acting on:
(A) Endosperm (B) Aleurone layer
(C) Embryo (D) Cotyledons

8. Among the following, which is sex linked disorder?
(A) Night blindness (B) Colour blindness
(C) Cretinism (D) Myxodema

9. Natural selection against extreme phenotype is termed as:
(A) Directional selection (B) Diversifying selection
(C) Distruptive selection
(D) Stabilizing selection

10. Among the following which organelle is involved in apoptosis?
(A) Lysosome (B) ER
(C) Golgi (D) Mitochondria

11. C-value measures:
(A) Haploid content of genome
(B) Diploid genome content
(C) Polyploid genome content
(D) Aneuploid genome content

12. Tiger is not found in wild at:
(A) Punjab (B) Rajasthan
(C) UP (D) Arunachal Pradesh

13. Lymphatic system is mainly involved in:
(A) Innate immunity
(B) Acquired immunity
(C) Phagocytosis
(D) Recycling lymph

14. Vitamin B complex is essential for humans because:
(A) It is obtained only from plant sources
(B) It is obtained only from animal sources
(C) It acts as a cofactor of various metabolic enzymes
(D) It is directly utilized in a various metabolic reactions

15. Polytene chromosome is generated due to:
(A) Failure of DNA replication
(B) Repeated DNA replication without segregation of chromosomes
(C) Pairing of homologous chromosomes
(D) Due to extensive transcription process

16. Taxol, an anti-cancerous drug effects by:
(A) Inhibiting polymerixzation of tubulin
(B) Inhibiting depolymerizatin of tubulin
(C) Polymerization of actin
(D) Favouring depolymerization of tubulin

17. During which phase of infection cycle, the DNA polymerase of T_4 phage is expressed maximally:
(A) Immediate early (B) Early
(C) Late (D) Middle

18. Which one of the following statement is not true for edge effect of two ecosystems?
(A) Moving corridors are always harmful
(B) It has its distinct physical environment differing from both side
(C) It has different species composition as compare to both side
(D) It has high biodiversity

19. If the length of one helix in DNA is 34Å. The type of DNA is:
(A) A (B) B
(C) C (D) D

20. In a chemical reaction catalyzed by enzyme following the Michaelis-Menten equation what will be the concentration of substrate when the velocity of reaction is 90% of the maximum velocity:
(A) 18 K_m (B) 9 K_m
(C) 5 K_m (D) 1 K_m

21. Among the following which statements is not correct for X-linked recessive disorder?
(A) Females with such disorders are unknown
(B) Males carrying the diseased allele are always diseased
(C) Females are diseased only when their mother is a carrier and father is diseased
(D) Males always passes trait to all of his sons

22. Chlorinated hydrocarbons effect ecosystem by:
(A) Biomagnifications (B) Bioconcentration
(C) Bioaccumulation (D) Bioremediation

23. The fluid mosaic model of plasma membrane given by Singer and Nicolson is applicable to:
(A) Only prokaryotic membrane
(B) Only eukaryotic membrane
(C) Both prokaryotic & eukaryotic membranes
(D) Only to organelle membranes

24. A zygote formed by a fusion of one normal gamete and another gamete where one of the chromosomes did not segregated at anaphase-II will lead into chromosomal aberrations known as:
(A) Haploidy (B) Diploidy
(C) Polyploidy (D) Aneuploidy

25. Morphologically similar species when interbreed produce viable fertile offspring. They are considered as single species according to:
(A) Biological species concept
(B) Evolutionary species concept
(C) Genetic species concept
(D) Morphospecies concept

26. Gram positive bacteria are further classified at generic and species level by analysis of:
(A) DNA (B) Cell wall
(C) Proteins (D) Cell membranes

27. Hemidesmosomes are structure found between:
(A) Two adjacent plant cells
(B) Two adjacent animal cells
(C) Cell and extracellular matrix
(D) Within a bacteria

28. Which statement is not correct regarding the genetic code?
(A) One amino acid an have more than one codon
(B) In eukaryotes the start codon is AUG
(C) Genetic odes are not strictly universal
(D) Third base of anticodon is not necessary for specificity

29. Rate of reaction for first order reaction for a radioisotope is 0.3 x 10^{-2} sec^{-1}. What is half life of radioisotope?
(A) 10 sec (B) 100 sec
(C) 0 sec (D) 1000 sec

30. Biodiesel is obtained from:
(A) *Jetropha curcus* (B) *Calotropis*
(C) *Prosopis* (D) *Catharanthus*

31. A bacterial operon contains three structural genes A, B and C in the same order. If polar mutation ours in gene B, then the effect in protein would be observed:
(A) In all proteins A, B and C
(B) Only in B and C
(C) Only in B
(D) complete loss of all proteins

32. Pearl oysters are obtained from the genus of:
(A) Oysteria (B) Pinctada
(C) Mytillus (D) Pila

33. Electron acceptor in anaerobic conditions in prokaryotes is:
(A) Glucose, fructose, maltose
(B) Fatty aids
(C) SO_4^{2-}, NO_3^-, CO_2
(D) Antioxidants such as vitamin K

34. The special roots termed as pneumatophores are visible at:
(A) Mangroves (B) Oceans
(C) Epiphytes (D) Salt stress

35. The biggest disadvantage of sexual reproduction against the asexual reproduction is:
(A) Only half of genetic material is passed to offspring from such parent
(B) Lot of energy and time is consumed in locating mate
(C) Atleast two individuals are required for sexual reproduction
(D) After fertilization in many cases zygote fails to develop

36. A lever cell is gently ruptured and the components of the cell lysate subjected to identity gradient centrifugation. Arrange the following organelles/ structures in order from the top to the bottom of the centrifuge tube:
A. Nuclei
B. Mitochondria
C. Lysosomes
D. Ribosomal subunits
(A) CBAD (B) ADCB
(C) ABCD (D) DCBA

37. Among the following which is not a correct explanation for high biodiversity at tropical rain forests?
(A) Long evolutionary time
(B) More surface area (C) High productivity
(D) Minimum competition

38. In submerged roots of mangrove plants the recycling of NAD+ is carried out by:
(A) Cellular oxidation
(B) Glycolysis
(C) Electron transport chain
(D) Fermentative metabolism

39. In prokaryotes there is single multifunctional fatty acyl synthase sufficient for complete fatty acid synthesis whereas in eukaryotes there are many different enzymes involved in fatty acid synthesis. The probable explanation for this difference is:
(A) Fatty acid synthesis is more stringent in eukaryotes
(B) Synthesis of fatty acid is by different mechanism in eukartotes as compared to prokaryotes
(C) For better regulation of fatty acid synthesis in eukaryotes
(D) Fatty acids of eukaryotes are different from the prokaryotes

40. First cell to be differentiated in developing embryo is:
(A) Epithelial cells (B) Rods
(C) RBC (D) Nerve cells

41. In plant if the sequence 5'-GATGGCACGAT-3' is transcribed, the corresponding mRNA will be:
(A) 5'-CUACCGUCUA-3'
(B) 5'-GAUGGCACGAU-3'
(C) 5'-AUCGUGCCAUC -3'
(D) 5'-ATGTCCATC-3'

42. Aspirin delays senescence in ut parts of plant and keeps flower fresh for longer time. The effect of aspirin is:
(A) By decreasing the synthesis of ethylene
(B) By increasing the synthesis of abscisic acid
(C) By increasing the synthesis of cytokinins
(D) By increasing the synthesis of gibberellic acid

43. The characteristic survivorship curve III is shown by:
(A) Fruit flies (B) Pelagic fishes
(C) Birds (D) Humans

44. In test cross F1 progeny is crossed with:
(A) Either of the parent (B) Recessive parent
(C) Dominant parent (D) Hetrozygous parent

45. Carrying capacity of a forest is 20 tonnes with increase of 10% in its biomass annually. For sustainable forestry how much tress can be harvested for timber so that it has minimum effect on forest and can be harvested annually:
(A) 20 tonnes (B) 10 tonnes
(C) 1 ton (D) 0.5 tonnes

46. Which of the following vaccine does not provide lifetime protection?
(A) Typhoid (B) Tetanus
(C) Polio (D) Small pox

47. Which statement is correct regarding spore and seed?
(A) Spore always generate diploid organisms
(B) Spore always give rise to sporophyte generation while gamete fuses with another haploid cells to give rise to zygote
(C) Spore and gametes are identical
(D) Spore are immotile and gametes are always motile

48. Which statement is correct regarding arteries and veins?
(A) Arteries have single valve while veins have two valves
(B) There is no difference
(C) Arteries carry deoxugenate blood and has thicker walls
(D) Veins contain deoxygenated blood and has thinner walls

49. Phenotype A has selective advantage over B, B has over C and C has over A. The condition is like paper-scissor-rock. Under such condition in nature:
(A) All phenotype would be selected
(B) Only A and C would be selected
(C) A and B together selected
(D) Only B would be selected

50. An object of 1.5 m height is placed 8.5 m away from the human eye and image is formed on retina which is 1.7 m away from the lens. The size of image will be:
(A) 0.3 μm (B) 3 μm
(C) 3 cm (D) 3 m

51. Which is not considered as major threat to loss of species diversity?
(A)Over exploitation (B) Alien invasion
(C) Pollution (D) All of these

52. The most ancient mode of energy generating metabolic reactions is:
(A)Photosynthesis
(B) Oxidation of nitrate
(C) Reduction of sulphate
(D) Reduction of nitrate

53. Pyrimidine dimmers formed due to UV rays an be repaired without removing any nucleotide by the repair mechanism known as:
(A) Mismatch repair mechanism
(B) Photoactivation
(C) Base excision repair mechanism
(D) SOS repair mechanism

54. Pearl oysters are obtained from the genus:
(A) Genes for heat stress
(B) Genes for old stress
(C) Genes necessary for survival of cell
(D) Genes required for development

55. Pentadatylity is a dominant trait, yet many individuals having single dominant alleles does not show any sign of polydahylity. This is known as:
(A) Incomplete penetrance
(B) Variable expressivity
(C) Co-dominance
(D) Incomplete dominance

56. In a population frequency of A_1 is 0.75 and A_2 is 0.25. After one generation the phenotype frequency will be:
(A) 0.5625; 0.375; 0.0625
(B) 0.625; 0.0625; 0.375
(C) 0.750; 0.250; 0.350
(D) 0.565; 0.1525; 0.0625

57. Abscisic acid can be degraded by:
(A) Oxidation and Reduction
(B) Reduction and conjugation
(C) Oxidation and conjugation
(D) Light and oxidation

58. Efficiency of ATP synthesis is 40% and enthalpy of reaction is 8 Kcal. If an individual consumes 2000 Kcal, then the net ATP production from it would be:
(A) 40 (B) 100
(C) 250 (D) 1000

59. Haematopoetic stem cells are found in:
(A) Bone marrow (B) Skin
(C) Lymphoid organs (D) Spleen

60. Ribose – 5 – phosphate is precursor for ribose sugar in DNA and RNA and is obtained from:
(A) Pentose phosphate pathway
(B) Kreb cycle
(C) Glycolysis
(D) Amino acids

61. Among the following which is not a density independent factor affecting population?
(A) Competition (B) Food
(C) Temperature (D) Nutrients

62. The technique used for observing 3-D structure is:
(A) Scanning electron microscopy
(B) Transmission electron microscopy
(C) Confocal microscopy
(D) UV microscopy

63. Type of biome in california and coastal regions of Mediterranean sea is:
(A) Taiga (B) Savanah
(C) Chaparrals
(D) Tropical deciduous forests

64. Honey bee keep variations among the workers by:
(A) Mating with males many times
(B) Parthenogenesis
(C) Specialization of functional role
(D) Extensive recombination during oogenesis

65. In habitate A pike cichlid fishes preferentially feeds on large adult guppies, so guppies, 50 guppies mature later and is of small size while in habitat B killer fish feeds on small, juvenile guppies so here guppies mature early and are large size. What will be the effect of experimentally guppies for habitat A are transferred to habitat B?
(A) No change would be seen
(B) Mature early by large size adults
(C) Mature but small size adults
(D) Mature late but large size adults

66. Under which condition of natural selection no allele would be lost:
(A) Any one homozygote is favoured
(B) Both homozygote are favoured
(C) Heterozygote are favoured
(D) Heterozygote is not favoured

67. The equation $1-\Sigma Pi^2$ represents:
(A) Shannon and Weaver index
(B) Simpson index
(C) Bronaulli index
(D) Hills equation

68. Homeotic genes are responsible for:
(A) Development (B) Homeostasis
(C) Cell cycle (D) Gene regulation

69. First fossile were discovered:
(A) Prior to both Lamarck and Darwin
(B) Prior to Lamarck but after Darwin
(C) After Lamarck but prior to Darwin
(D) After Lamarck and Darwin

70. Prebiotic environment was different from present environment and was devoid of:
(A) CO_2 (B) Atmosphere
(C) O_2 (D) N_2

71. In *E coli* the complementation test is done by:
(A) Transformation (B) Merozygotes
(C) Heterokaryons (D) Making them dipoid

72. Transgenic varieties are being designed to produce:
(A) Research reagents (B) Edible vaccines
(C) Pharmaceuticals (D) All of these

73. Which is not correct about *in-situ* conservation of biodiversity?
(A) Plants are conserved in their natural habitats
(B) Only the wild relatives are candidates for in situ conservation
(C) Plants can continue to evolve
(D) In-situ conservation is rather simple

74. In case of some plant genes, light is known to affect the:
(A) mRNA degradation rates
(B) Induction of transcription
(C) Enhanced transcription rates
(D) All of these

75. Fibroin is found in:
(A) Hair (B) Insect wing hings
(C) Silk (D) Blood vessels

76. Plant cell suspension cultures are continuously agitated to achieve:
(A) Break cell clumps
(B) Facilitates aeration
(C) Improve nutrient availability
(D) All of these

77. The enzyme used for ligation of blunt ends of DNA:
(A) DNA polymerase (B) DNA gyrase
(C) T4 ligase (D) Topoisomerase

78. Colony hybridization is not based on:
(A) DNA : DNA hybridization
(B) Nucleic acid hybridization
(C) Antigen-antibody interaction
(D) RNA : DNA hybridization

79. Which cytokinins is the most commonly used in micropropagation is:
(A) BAP (B) Zeatin
(C) 2-ip (D) TDZ

80. PEG promotes:
(A) Protoplast fusion
(B) Animal cell fusion
(C) DNA uptake by protoplasts
(D) All of these

ARS/NET 2001

1. Nuclear envelop breaks down in:
(A) Telophase II (B) Anaphase I
(C) Metaphase II (D) Prophase I

2. A good molecular marker must be:
(A) Polymorphic (B) Heritable
(C) Both (A) and (B) (D) None of these

3. Genetic improvement of crop plants through biotechnology needs conventional breeding due to:
(A) The elite cultivars developed by plant breeding will be parents of the next generation of trasgenics
(B) Field testing across locations or cropping systems and over years will be needed to select the best transgenic event
(C) Breeding may be required to minimize/eliminate pleitropic effects
(D) All of these

4. Electrophoretic separations of DNA and RNA in agarose is based on:
(A) Charge
(B) Size
(C) Both (A) and (B)
(D) Base sequence

5. Which of the following statements is not true?
(A) Chloroplast are the sites of photosynthesis
(B) Chromoplast contain bright orange and yellow pigments
(C) Leucoplast is responsible for the bright colours of the fruits and leaves
(D) Leucoplast chromoplast and chloromoplast are present only in the cells of plants and algae

6. Degradation of xenobiotics occurs rapidly in:
(A) Highly compressed soils
(B) Well aerated soils
(C) Heavily contaminated soils
(D) None of these

7. Which of the following is an herbicide resistant gene?
(A) Bar (B) Hpt
(C) Cholesterol oxidase (D) Chitinases

8. Anthropogenic chemicals introduced into the terrestrial carbon cycle is more recalcitrant when they contain:
(A) Unsaturated compounds
(B) Unbranched chain
(C) Aromatic ring substituted with halogens
(D) Chain length of less than 12 carbon atoms

9. The yolk of a chicken egg serves a nutritive function for the developing embryo. A functionally comparable structure in plants is:
(A) Lignin (B) Cellulose
(C) Pectin (D) Endosperm

10. "9+2" describes the basic structure of which of the following one:
(A) Flagellum (B) Lysosome
(C) Basal body (D) Chloroplast

11. Between mitotic divisions, the cell is in:
(A) G (B) G_1
(C) G_2 (D) S phase

12. Most plant cells are surrounded by a rigid cell wall made primarily of:
(A) Triglycerides (B) Polysaccharides
(C) Proteoglycanes (D) Pectin

13. Which is not correct about gamma-linolenic acid (GLA)?
(A) It is a non-essential fatty acid
(B) It is the first intermediate in the conversion of linoleic acid to arachidonic acid
(C) It is produced by desaturation of linoleic acid by the enzyme fatty acid desaturase
(D) It is beneficial in alleviation of many physiological and pathological conditions such as hypercholestereremia, cancer and diabetes

14. During Agrobacterium transformation the T-DNA is coated by protein, which is the product of gene:
(A) vir C (B) vir D
(C) vir E (D) vir G

15. In liposome mediated gene transfer DNA enters the protoplast due to:
(A) Osmosis (B) Endocytosis
(C) Exocytosis (D) Collision

16. Endochitinase can be used as an:
(A) Antibacterial (B) Antifungal
(C) Antinematodal (D) Antivirus

17. The T-DNA region of all the Ti and Ri plasmids are flanked by direct repeat sequence:
(A) 25 bp (B) 20 bp
(C) 15 bp (D) 35 bp

18. Bacteria are preferred over yeast for production of single cell protein because:
(A) Of their fast growth
(B) They are small
(C) Of their lower content of methionine
(D) None of these

19. The transgenic tomato plants showing delayed ripening due to suppression of ethylene was developed using methodology of:
(A) Coat protein gene
(B) Antisense RNA
(C) Agrobacterium mediated gene transfer
(D) None of these

20. Lambda has short single stranded 5 projection of:
(A) 5- nucleotide length
(B) 10 - nucleotide length
(C) 8- nucleotide length
(D) 12 - nucleotide length

21. The direct DNA uptake by the protoplast is stimulated by the chemical:
(A) Cellulose (B) Sucrose
(C) Sodium chloride (D) PEG

22. The function of the nucleus includes:
(A) Cellular respiration
(B) Housing the hereditary information
(C) Synthesis of proteins
(D) Synthesis of carbon compounds

23. Which is not correct about virus cross protection?
(A) The phenomenon of cross protection was first observed by McKinney in 1929 in case of tobacco mosaic virus
(B) Plants infected with a mild strain of virus resist the development of symptoms upon infection by a more severe strain of the same virus
(C) A mild protecting strain effective for one region may be severe for another region
(D) Selection of mild protecting strain is rather easy

24. Which of the following reagent is useful for visualizing DNA?
(A) 14 uracil (B) Diphenylamine
(C) Ethidium bromide (D) DNA polymerase

25. Endoplasmic reticulum:
(A) Is found only in animals
(B) Is a site of ATP production
(C) Is called rough if mitochondria is attached to it
(D) Is a system of membrane bound channels

26. Which of the following is a green tissue specific promoter?
(A) Pea vicilin
(B) Bean phytahemaglutinin
(C) 35 s CaMV
(D) Arabidopsis small sub unit of Rubisco

27. Scientist have postulated that small pores must exist in the cell membrane because:
(A) Of endocytosis
(B) Of turgor pressure
(C) CO and O moves freely through the membrane
(D) Water molecules move freely through the membrane

28. If adenine makes up 5% of DNA, cytosine will make:
(A) 15% (B) 25%
(C) 30% (D) 35 %

29. A general method to confirm expression of transgene at translation level is:
(A) Southern blotting (B) Western blotting
(C) Northern blotting (D) Eastern blotting

30. The organelle that functions in the breakdown of ells and tissues is the:
(A) Episomes
(B) Mitochondria
(C) Centrosomes
(D) Lysosomes

31. In-vitro packaging is required for which one of following vector:
(A) BAC (B) YAC
(C) PAC
(D) None of these

32. Which of the following statement is true?
(A) Facilitates diffusion an our against concentration gradient
(B) Diffusion requires an expenditure of energy by the cell
(C) The active transport requires protein complex in the cell
(D) Osmosis an our against a concentration gradient

33. Which is not correct about Scarlet?
(A) It is a somaclonal variety of potato
(B) It was selected from shoot-tip culture derived plants
(C) This variety is comparable to the parent cultivar in yield and disease resistance
(D) It shows darker and more stable skin colour, which is a desirable quality trait

34. Batch systems of culture are:
(A) Steady state process
(B) Non steady state process
(C) Completely mixed
(D) Plug flow system

35. The browning of the culture medium induced by some explant can be prevented by addition of:
(A) GA
(B) Polyvinylpyrrolidone
(C) 2, 4-D (D) IBA

36. The phosphofrutokinase an pyruvate kinage reaction are similar is that:
(A) Both generate ATP
(B) Both are essentially irreversible
(C) Both involves a high energy : Sugar derivative
(D) Both involves three carbon compounds

37. Which of the following is not used as measuring device in biosensor?
(A) Whole organism (B) Immunonuclear
(C) Biofilters (D) Enzyme

38. Bacculoviruses:
(A) RNA viruses (B) DNA viruses
(C) Both (A) and (B) (D) None of these

39. Bioreactor are used for :
(A) Production of enzymes
(B) Production of secondary metabolites
(C) Production of cell cultures
(D) All of these

40. During Agrobacterium infections the plant cell begins to synthesize an arginine derivative called as:
(A) Opines (B) Aetobenzylpurine
(C) Hygromycin (D) Acetosyringone

41. Which of the following is not a general property of the enzyme?
(A) Enzyme has great catalytic powers
(B) Enzyme bind specific substances
(C) Enzyme mostly use hydrophobic interactions to bind substrates
(D) Enzymes are almost exclusively proteins

42. Requefort cheese is ripened with the help of:
(A) *Lactobacillus coxi*
(B) *Penicillum reguoforti*
(C) *Brevibacterium linens*
(D) *Propionibacterium shexmanii*

43. Plant protoplast fusion an be achieved with:
(A) TZZ (B) PEG
(C) PM (D) Zeatin

44. The gene transferred from *E coli* to tobacco to increase the level of manitol is:
(A) Manitiol dehydrogenase
(B) Manitiol carboxylase
(C) Manitiol hydrolase (D) Manitiol oxygenase

45. Nick transition is directly involved in which of the following process:
(A) RNA to DNA (B) DNA to RNA
(C) DNA to DNA (D) RNA to protein

46. Production of biogas takes through the following sequential steps:
(A) Methanogenis – Solubilization - Acidogenis
(B) Acidogenis – Solubilization - Methanogenis
(C) Acidogenis – Methanogenis – Solubilization
(D) Solubilization –Acidogenis - Methanogenis

47. Restriction enzyme that cuts straight across the DNA produces:
(A) Nicks (B) Blunt ends
(C) Single stranded (D) None of these

48. White rot fungus *Phaneiochaete chrysasporium* produces:
(A) Extra cellular enzymes
(B) Intra cellular
(C) Both (A) and (B) (D) None of these

49. During stress which of the following phytohormone is accumulated?
(A) Auxin (B) GA
(C) ABA (D) Ethylene

50. In eukaryotes, DNA methylation is mainly concerned with:
(A) Restriction
(B) Control of replication
(C) Regulation of gene pair
(D) DNA repair

51. Which of the following reactions is not performed by lignin peroxidase?
(A) Detoxification of toxic phenols
(B) Oxidative cleavage of aromatic ether bonds
(C) Oxidative of phenyl alcohols to the respective aldehydes
(D) Hydroxylation of the aliphatic double bonds in

52. Cytoplasmic channels connecting adjacent cells are called:
(A) Ionic channels (B) Conjugation tube
(C) Protoplasmic channels (D) Plasmodesmata

53. If meiosis does not our, the chromosome number would ---- in each generation:
(A) Remain same (B) Triple
(C) Reduce (D) Double

54. Which of the following enzyme is used for making DNA:
(A) Reverse transcriptase
(B) DNA ligage
(C) Restriction enzyme
(D) Polynucleotide transterase

55. An example of naturally occurring auxin is:
(A) BAP (B) 2, 4-D
(C) ABA (D) IAA

56. The in-vitro electron donor to PS –II is:
(A) Water (B) Ascorbate
(C) PS-I (D) Plastoquinone

57. Which of the following statements regarding simple Michaelis-Menten enzyme kinetics is not correct?
(A) km is expressed I terms of a reaction velocity
(B) km is the concentration of substrates required to achieve one half of V_{max}
(C) km is the concentration of substrates required to convert one half of the total enzyme into the enzyme substrates complex
(D) The maximum velocity, V_{max} is related to the maximal number of substrates molecules that an be turned over in unit time by a molecule of enzyme

58. *Thiobacillus thioxidans* grows on:
(A) Iron (B) Sulphur
(C) Pyrites (D) Covellite

59. The Shine-Dalgarno (SD) sequence is known as:
(A) Ribosome binding site
(B) Initiation site
(C) Promoter binding site (D) None of these

60. In tissue culture regeneration of shoot and root occurs by manipulating the balance of:
(A) Auxin and ABA
(B) Auxin and Cytokinin
(C) Cytokinin and ABA
(D) ABA and Ethylene

61. Relaxed plasmid:
(A) Can not be purified
(B) Have multiple copies per cell
(C) Does not exist in bacteria
(D) Have only single copy per cell

62. For mitochondrial structure and function mt-DNA species:
(A) 20% protein (B) 10 % protein
(C) 100 % protein (D) 30 % protein

63. Amino acid fermentation employs:
(A) Unicells
(B) Filamentous organisms
(C) Both (A) and (B) (D) None of these

64. Homologous pairs line up along equatorial plane during:
(A) Anaphase II (B) Metaphase I
(C) Interphase (D) Telophase

65. An example of commercially important primary metabolites is:
(A) Toxin (B) Alkaloid
(C) Alcohol (D) None of these

66. The chemical used as an inducer of the vir operons is:
(A) Hygromycin (B) Fanamycin
(C) Acetosyringone (D) Kanamycin

67. The large dark body (or bodies) in the nucleus is the nucleolus, which participates in the formation of:
(A) Ribosome (B) RNA
(C) Protein (D) DNA

68. State which of the following is not a desired characteristic of a vector?
(A) Large size
(B) Autonomous replication
(C) Unique restriction site
(D) Genes that confer antibiotic resistance

69. In meosis, as in mitosis, the chromosomal materials is replicated during:
(A) G_1 phase (B) S phase
(C) T phase (D) G_2 phase

70. In PCR, the dsDNA is generally denaturated by:
(A) Acid treatment (B) Alkali treatment
(C) Temperature (D) None of these

71. Cyclic phosphorylation involves:
(A) PS – II (B) PS - I
(C) Mitochondria (D) Both (A) and (B)

72. Replication of chloroplast DNA is controlled by:
(A) Nuclear DNA (B) Mitochondrial DNA
(C) Pollen grains (D) Chloroplast

73. Which of the following is not used as cryoprotectant?
(A) Glycerol (B) Dimethyl sulfoxide
(C) Polyclar (D) Ethylene glycerol

74. The in-vitro electron donor to PS –I is:
(A) PS – II (B) Plastoquinone
(C) Oxygen (D) Ascorbate

75. Which one of the following controls gene expression in eukaryotes?
(A) Effector (B) Histones
(C) Transcription factor (D) Transferase

76. The shoot apical meristem is characterized by:
(A) Trapezoidal shape (B) Round shape
(C) Dome shape (D) None of these

77. Germplasm preservation of meristem and shoot tip culture involves:
(A) Storage in dry ice
(B) Storage in liquid nitrogen
(C) Culturing under low temperature
(D) Cryo-preservation and slow growth

78. Horizontal gene flow can be reduced by expressing genes in:
(A) Nucleus (B) Chroloplast
(C) Both (a) and (B) (D) None of these

79. The vir region is organized into six operons viz., vir A, vir, B, vir C, vir D, vir E and vir G, out of these which four operons are required for virulence:
(A) A, B, C, D (B) A, B, D, G
(C) C, D, E, G (D) None of these

80. Majority of bacculoviruses have been reported from species:
(A) Lepidoptera (B) Diptera
(C) Crustaean
(D) Both diptera and Lepidoptera

81. In commercial fermentation process, it is desirable to have:
(A) Minimal lag phase
(B) Minimum stationary phase
(C) Minimal death phase
(D) Minimal lag phase

82. Replication of the DNA by rolling circle mechanism give rise to a:
(A) Blunt and DNA
(B) Theta structure
(C) Concatamer
(D) Single strand DNA with nick

83. Oxidation of water during photosynthesis involves:
(A) PS – II (B) PS - I
(C) Both (A) and (B) (D) Mitochondria

84. E.L. Tatum was a Nobel prize recipient, his work on:
(A) Nurospora
(B) *E coli* gene expression
(C) RNA polymerase
(D) DNA polymerase

85. Shotgun approach is mainly used in:
(A) Genomic library (B) Gene mapping
(C) Gene transforming (D) DNA sequencing

86. Which of the following is not an antibiotic resistance gene?
(A) hpt (B) neo
(C) gfp (D) npt II

87. Callus and suspension cultures normally produce secondary metabolites when:
(A) Emit fragrance
(B) The callus turn red with age
(C) Show lightful tracheids
(D) All of these

88. Cosmids have the Cos site of:
(A) Lambda (B) Plasmid
(C) ML 3 (D) YAC

89. Inclusion of osmoticum in both isolation and culture media:
(A) Facilates formation of heterokaryotes
(B) Facilitates fusion of protoplast
(C) Prevent rupture of protoplast
(D) None of these

90. Xeroderma pigmentosum disease in man is due to defects in DNA repair system of:
(A) Excision repair (B) Mismatch repair
(C) Deletion repair (D) Direct repair

91. Which of the following is not an insecticidal protein gene?
(A) Sunflower seed albumin
(B) Lectins
(C) Protease - inhibitor
(D) Alpha-amylase inhibitor

92. Anther culture provides a method for the production of:
(A) Pollyploid (B) Apomictic line
(C) Homozygous line (D) Heterozygous line

93. The term somaclonal variation was termed by:
(A) Larken and Scowhoft
(B) Skirvin
(C) Eyans
(D) Murashige and Skoog

94. Most commonly used organism in SCP is:
(A) *Bacillus* (B) *E. coli*
(C) *Spirulina* (D) *Pseudomonas*

95. In Agrobacterium mediated plant transformation which one of the following is used to check the Agrobacterium growth:
(A) Basta (B) PPT
(C) Timentin (D) Kanamycin

96. Which one of the following means transfer of genes in host without a biological agent?
(A) Transduction (B) Transformation
(C) Conjugation (D) Perpetuation

97. The sigma factor of RNA polymerase in *E. coli* involved:
(A) DNA unwinding
(B) Binding with template DNA
(C) RNA synthesis
(D) Transcription initiation

98. Which of the following statements is not generally true about histones?
(A) Present in nucleosome
(B) Homogenous in structure
(C) Always present in chromatin
(D) Heterogeneous in structure

99. Secondary metabolites like antibiotics are not essential for:
(A) Log phase growth (B) Exponential growth
(C) Both (A) and (B) (D) None of these

100. Micro propagation involves:
(A) Thorough callusing
(B) Adventitious bud transfer
(C) Shoot culture, acceptic initiation transfer to pot and rooting
(D) Initiation of aseptic culture, shoot multiplier, rooting and transplantation

101. Nodulin genes are found in:
(A) Cyanobacteria (B) Bradyrhizobium
(C) Plant (D) Bacteria

102. Most commonly used strategy to develop transgenic plant against any viral disease is to engineer a:
(A) Liposome mediated transfer of viral RNA
(B) Anti-sense RNA for viral genome
(C) Coat or capsid protein gene
(D) Direct DNA transfer of viral gene

103. During isolation of protoplast:
(A) Cell membrane is removed
(B) Treated with pectinase
(C) Treated with proteases
(D) Cells are irradiated

104. Which one of the following is the first stable product of photosynthesis?
(A) RuBP (B) DAHP
(C) Glucose (D) 3-PGA

105. In replication of *E. coli* DNA, the protein complex involved is known as:
(A) Replisome (B) Splisosome
(C) Endosomes (D) Primosome

106. Which one of the following amino acids is specified by single codon in genetic code?
(A) G (B) K
(C) W (D) D

107. The existence of mRNA was first indicated by:
(A) Palkade and Siekevitz
(B) H.G. Khurana
(C) Volkin and Astrachan
(D) Brahett and aspersson

108. Leghaemoglobin is synthesized elusively in:
(A) Nitrogen fixing nodules
(B) Leg of ruminants
(C) Klebsiella (D) Cyanobacteria

109. Cyanobacteria protect nitrogenase from oxygen by:
(A) Heterocyst (B) Gum production
(C) Possession of catalase
(D) Possession of superoxide dismutase

110. Which one of the following does not regulate nitrogenease activity?
(A) H_2 (B) M_0
(C) NH_4 (D) Energy supply

111. Which of the following enzymes is not involved in HMP pathway?
(A) Transketolase (B) Endolase
(C) Transaldolase (D) 6-phosphogluonase

112. What is the role of SDS in case of SDS PAGE?
(A) It increases speed of proteins in gel
(B) It protects proteins from denaturation
(C) It provides negative charge to all the proteins so that separation can be done only on the basis of molecular weight
(D) SDS is detergent so is used to emulsify contaminating fats if any in samples

113. Fe protein of nitrogenase is:
(A) Heteropolymer tetramer
(B) Homomeric dimmer
(C) Homomeric tetramer (D) Heteromeric dimmer

114. The length of okazaki fragments in *E. coli* are about:
(A) 50-100 nucleotides
(B) 100-200 nucleotides
(C) 10000 nucleotides
(D) 1000-2000 nucleotides

115. Ferrodoxin is reduced by:
(A) PS – II only (B) PS – I only
(C) PS I– and II (D) Mitochondria

116. Which of the following statements about conjugation are not true?
(A) A recipient cell because a donor cell
(B) As the chromosome replicates, the free and moves into the recipient cell
(C) The recipient bacterium possess plasmids called F factor
(D) Conjugation studies have produced maps of gene sequence on the bacterial chromosome

117. Which of the following statements is not true?
(A) An acidic protein will give pl less than 7
(B) A basic protein will give a pl greater than 7
(C) At a pH value equal to its pl, a protein will not move in the electric field of an electrophoresis experiment
(D) The pl is the pH value at which a protein has no charge

118. Explantation is a process of:
(A) Planting of explants
(B) Dissection and culture of small organs or tissue section
(C) Choice of explants
(D) Culture of plants

119. During reduction of dinitrogen by nitrogenase, how many net electrons are required:
(A) Six (B) Two
(C) Four (D) None of these

120. In biotin labeling of probes, the detection method is:
(A) Streptavidin labeling (B) Antibody labeling
(C) Colorimetric (D) Fluorescene

121. Micropores contained within the anther gives rise to:
(A) Callus culture and gynogenesis
(B) Anthogenesis and callus culture
(C) Gynogenesis and androgenesis
(D) None of these

122. Operon comprise of
(A) Operator (B) Structural gene
(C) Promoter (D) Repressor
(E) All of these

123. Which one of the following is not a component of nitrogenase?

(A) Mg^{++} (B) NADP
(C) MoFe-protein (D) Fe-protein

124. Which one of the following is not formed in calvin cycle?

(A) RuBP (B) DAHP
(C) Glucose (D) 3- PGA

125. The end product of transcription process in prokaryotes is:

(A) Monocistronic mRNA
(B) Polycistronic mRNA
(C) Both (A) and (B) (D) None of these

126. Which one of the following radioisotope is used to elucidate carbon pathway in calvin cycle?

(A) 14C (B) 32P
(C) 35S (D) 3H

127. Pure plasmid DNA was isolated from a bacterium. Restriction enzyme digestion of this plasmid with either Bam HI or Eo RI resulted in two DNA fragments. A double digestion of the same plasmid with both these enzymes resulted in three DNA fragments. From this we an conclude that the isolated plasmid DNA is:

(A) Double stranded and linear
(B) Double stranded and circular
(C) Single stranded and linear
(D) Single stranded and circular

128. Which of the following is commonly used as a protecting group during peptide synthesis?

(A) Dicyclohexylurea
(B) Ten-butyloxylcarbonyl
(C) Hydrogen fluoride
(D) Dicyclohexylurea carbodimide

129. Guanosine is:

(A) Pyrimidine base
(B) Purine nucleotide
(C) Purine base
(D) Pyrimidine nucleotide

130. The restriction enzyme useful to the molecular biologist belongs to:

(A) Type – I (B) Type - II
(C) Type – III (D) Type – IV

131. The number of consensus sequence in the promoter region of a gene in *E. coli* is generally:

(A) 2 (B) 3
(C) 4 (D) 6

132. Which category of microbes are used in retting process of jute?

(A) Microbes tolerant to high temperature
(B) Which process high amount of hydrolytic enzyme activity
(C) Anaerobic microbes
(D) Aerobic microbes only

133. The proof reading function in DNA replication is done by:

(A) DNA helicase (B) 3' exonulease
(C) DNA gyrase (D) 5' exonulease

134. When the PCR used to generate single stranded copies of DNA, that PCR reaction is called:

(A) Asymmetric PCR (B) RT-PCR
(C) Inverse PCR (D) All of these

135. Which one of the following amino acids residues are likely to be formed on the inside of a water soluble protein?

(A) H (B) R
(C) E (D) V

136. Which of the following chemical is used for preparation of competent cell?

(A) HCl (B) $CaCl_2$
(C) Glycine (D) NaCl

137. Cryo-preservation involves:

(A) Cutting, application of cryo-protectant and storage at 196^{0}C
(B) Freezing and thawing
(C) Specimen treatment, freezing storage at ultra low temperature
(D) None of these

138. Which one of the following chemical causes frame shift mutation?

(A) 2- bromo uracil (B) Acridine orange
(C) 2 - aminopurine (D) Hydroxyglamine

139. The conversion of IMP to AMP requires which of the following:

(A) GTP (B) NAD
(C) ATP (D) Glutamine

140. Up and down mutations occurs in:

(A) Promoter region (B) Intron
(C) Transcription start site (D) Poly-A-region

141. The hsd system of DNA methylation in *E. coli* methylates:

(A) Guanine (B) Thymine
(C) Cytosine (D) Adenine

142. The existence of puffs in giant chromosome of the Brazilian guat support the concept that:

(A) Conjugation involves a transfer of genetic material
(B) DNA unwind for mRNA transcription
(C) The sequence of genes on a chromosome can be determined on the basis of crossover frequencies
(D) Homologous chromatids exchange genetic material at meiosis

143. Transcription is directly involved in which of the following steps in the flow of genetic information:

(A) RNA to DNA (B) RNA to Protein
(C) Protein to RNA (D) DNA to RNA

144. Which one of the following is not the basic requirement of biological nitrogen fixation?
(A) Mg^{++} and ATP (B) Nitrogenase
(C) Glutamine (D) Strong reducing agent

145. DNA re-association kinetics provides information on the:
(A) Genomic size (B) Presence of exon
(C) Presence of intron (D) Existence of gene

146. Which one of the following amino acids mat alters the direction of polypeptide chains and interrupt alpha helices?
(A) Pro (B) cys
(C) Phe (D) His

147. RNA molecules involve in splicing are:
(A) Sn rRNA (B) 5s rRNA
(C) 16s RNA (D) tRNA

148. Substance, such as tryptophan, that decreases the amount of enzyme produced by a cell is known as:
(A) Induces (B) Repressor
(C) Promoter (D) Corepressor

149. In lambda phage the N, Cl and Cro genes are responsible for:
(A) Host lysis
(B) Gene regulation and immunity
(C) DNA synthesis
(D) Late function regulation

150. The sedimentary velocity of a protein sin a centrifuge does not depend on the:
(A) Density of the protein
(B) Charge on the protein
(C) Shape of the protein
(D) Density of the solution

151. The evidence for semi-conservative replication of DNA was first given by:
(A) Meselson and Stahl (B) Harshey and Chase
(C) Avery and Malecod (D) Watson and Crick

152. The absolute requirement for both RNA synthesis and DNA replication is:
(A) UTP
(B) Free 3' -OH
(C) Free 5'- OH
(D) Primer

153. The enzyme nitrogenase is composed of:
(A) Four-oxygen sensitive non-haem iron protein
(B) Three oxygen sensitive non-haem iron protein
(C) Two oxygen sensitive non-haem iron protein
(D) One oxygen sensitive non-haem iron protein

154. Hybridization technology is used in production of:
(A) mRNA
(B) Monodonal serum
(C) Polyclonal antibodies
(D) Monodonal antibodies

155. Which of the following enzyme is classified as hydrolases?
(A) Aconitase
(B) Catalase
(C) Polynucleotide kinase
(D) Pepsin

156. Which of the following is the major metabolic product of glycolysis in higher organism?
(A) Latic acid (B) Aetone
(C) Ethanole (D) Glyerol

157. Which of the following is true about the structure of ATP?
(A) Has a free hemiacetal
(B) Contains a ketose sugar
(C) Contains a pyranose ring
(D) Contains a beta N-glycosidic linkage

158. Which blotting technique is used for the detection of DNA that has been separated from a mixture of DNA restriction fragments by electrophoresis through an agrose gel and then transferred on to a nitrocellullose sheet?
(A) Northern blotting (B) Southern blotting
(C) Eastern blotting (D) Western blotting

159. During splicing of proteins which part is removed?
(A) Exons (B) Introne
(C) Exteins (D) Inteins

160. Which of the following pairs are anomers?
(A) Glucose and mannose
(B) Ribose & Ribulose
(C) Pyruvate lactate
(D) α-D-glucose and β–D- glucose

161. In the transcription process of eukaryotes, which of the following is not true:
(A) Translation occurs in cytoplasm
(B) Translation is coupled with transcription
(C) AUA is a coding sequence
(D) 3'-end of mRNA is polyadenylated

162. If a polypeptide has 400 amino acid residue, what is the appropriate molecular weight of that polypeptide?
(A) 11 Kd (B) 44 Kd
(C) 22 Md (D) 44 Md

163. Which of the following is not true for single stranded vector?
(A) Replicate via RF intermediate
(B) Has cohesive
(C) Used for oligonucleotide directed mutagenesis
(D) Used for dideoxy mediated DNA sequencing

164. The first product of purine nucleotide biosynthesis that contains a complete purine ring (hypoxanthine) is:
(A) GMP (B) IMP
(C) AMP (D) Xanthylate (XMP)

165. Physical maps of genome is based on:
(A) Number of base pairs
(B) Recombination percentage
(C) Both (A) and (B)
(D) None of these

166. Border sequences need to be incorporated into the design of plasmid vectors for agrobacterium mediated transformation to ensure:
(A) Greater promoter efficiency
(B) Oncogene deactivation
(C) Efficient replication of the plasmid
(D) Integration of the genes of interest into the host gene

167. Which of the following is a carrier hosts for new genes?
(A) Transplanted organs (B) Ribosomes
(C) ECOR 1 (D) Bacteriophages

168. When genetic material is carried from one cell to another by a phage, the phenomenon is known as:
(A) Cloning (B) Transformation
(C) Transduction (D) Induction

169. Which one of the following reagents would be useful for labeling the oligonucleotide (GGAATTCC)?
(A) 32 (P) ATP
(B) DNA ligase
(C) 32pi
(D) DNA dependent RNA polymerase

170. Which of the following vector is suitable for DNA sequencing?
(A) EMBL (B) M 13
(C) Lambda (D) PBR 322

171. The transforming principle of pneumococcus found, Avery, Macleod and Ma Carty was :
(A) RNA (B) Protein
(C) DNA
(D) Polysaccharide capsular material

172. The diversity inherent in the DNA molecule is a product of:
(A) The sequence of base pairs
(B) Its helical structure
(C) Its location in the chromosome
(D) The high energy phosphate bond

173. The enzymatic procedure of DNA sequencing was given by:
(A) Khurana
(B) Old Primerose
(C) Maxam and Gilbert
(D) Sangers and co workers

174. In transcription initiation process of eukaryotes:
(A) Sigma factor is binds to the eukaryotes
(B) Holoenzymatic bind in the promoter
(C) Transcription factor binds to the promoter
(D) None of these

175. Franois Jacob and Jaeqnes Monod proposed:
(A) Allsteric model
(B) Gene for gene hypothesis
(C) Operon hypothesis
(D) Gene for protein hypothesis

176. During which process, the free end of the chromosome moves from Hfr donor into the F cell across an intracellular bridge called:
(A) Transduction (B) Transformation
(C) Retention (D) Conjugation

177. Mutation that involves only smallest change in base pairs is known as:
(A) Deletion (B) Insertion
(C) Frame shift mutation (D) Point mutation

178. Agitation of the medium during suspension culture serves to:
(A) Promote cell division
(B) Activate DNA synthesis
(C) Obtain pure preparation of spongy parenchyma to palisade parenchyma
(D) Maintain uniform distribution of cells or cell lumps in the medium

179. Which of the following is not true for an ideal cloning vector?
(A) Single site for large number restriction enzym
(B) Ability to confer readily selectable phenotypic traits on host cell
(C) Multiple site for a single restriction enzyme
(D) Should have low molecular weight

180. DBM paper is used:
(A) Western blotting (B) Capillary blotting
(C) Northern blotting (D) Southern blotting

181. Acetyl Co A carboxylase requires the following Co-factor:
(A) Pyridoxal phosphate (B) Ubiquinone
(C) Cytochrome (D) Biotin

182. Which amino acid can stabilize protein structure by forming covalent cross-links between polypeptide chains?
(A) G (B) C
(C) S (D) M

183. Nick translations is useful in:
(A) Labeling of mRNA molecules
(B) Labeling of ds-DNA molecules
(C) Synthesis of protein molecules
(D) Labeling of ss-DNA molecules

184. Which of the following property is common to α – helix and β – pleated sheet?
(A) Axial distance between adjacent amino acids of 3.5 Å
(B) Variable number of participating amino acids
(C) Rod shape
(D) Intrachain hydrogen bonds between main chain Co and NH

CSIR JRF/NET - Life Science, December2011

1. Magnesium powder, placed in an air-tight glass container at 1.0 bar, is burnt by focusing sunlight. Part of the magnesium burns off, and some is left behind. The pressure of the air in the container after it has returned to room temperature is approximately:

(A) 1.0 bar (B) 0.2 bar
(C) 1.2 bar (D) 0.8 bar

2. Living being get energy from food through the process of aerobic respiration. One of the reactants is:

(A) Carbon dioxide
(B) Water vapour
(C) Oxygen
(D) Phosphorus

3. In Δ ABC, angle is larger than angle C and smaller than angle B by the same amount. If angle B is 67^0, angle C is:

(A) 67^0 (B) 53^0
(C) 60^0 (D) 57^0

4. Which of the following statements about the concentration of CO_2 in the Earth's atmosphere is true?

(A) It was the highest in the very early atmosphere of the Earth
(B) It has steadily decreased since the formation of the Earth's atmosphere
(C) It has steadily increased since the formation of the Earth's atmosphere
(D) Its levels today are the highest in the Earth's history

5. See the following mathematical manipulations:

(i) Let X = 5
(ii) then $x^2 - 25 = x - 5$
(iii) $(x - 5)(x + 5) = x - 5$
(iv) $x + 5 = 1$ [cancelling $(x - 5)$ from both sides]
(v) $10 = 1$ [Putting x = 5]

Which of the above is the wrong step?

(A) (i) to (ii)
(B) (ii) to (iii)
(C) (iii) to (iv)
(D) (iv) to (v)

6. The number of craters observed due to meteoritic impacts during the early stages of formation, is less one the Earth than that on the Moon because:

(A) Formation of crates on the Earth was difficult due to the presence of hard rocks
(B) Impacting bodies on the Earth were smaller in size
(C) Craters on the Earth are now covered by ocean water
(D) Earlier craters are not preserved due to continuous modification of Earth's surface by geological processes

7. What is the angle θ in the quadrant of a circle shown below?

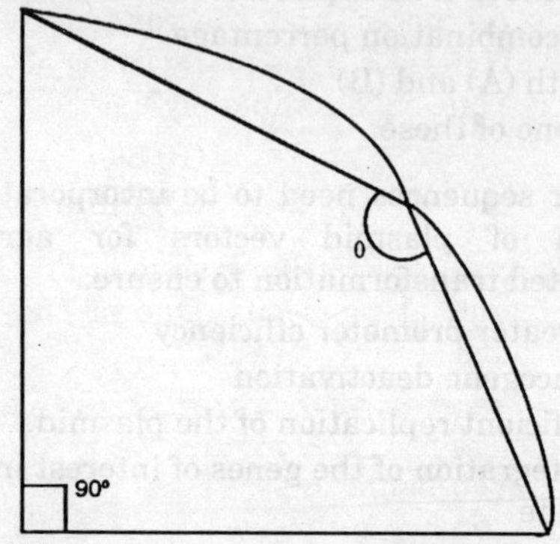

(A) 135^0 (B) 90^0
(C) 120^0
(D) May have any value between 90^0 and 120^0

8. During a total solar eclipse occurring at noon, it becomes dark enough for a few minutes for stars to become visible. The stars that are seen are those which will be seen from the same location:

(A) One the following night only
(B) On the night one month later
(C) On the night three months later
(D) On the night six months later

9. A cup board is filed with a large number of balls of 6 different colours. You already have one ball of each colour. If you are blind-folded, how many balls do you need to draw to be sure of having 2 colour-matched pairs of balls?

(A) 3 (B) 4
(C) 5 (D) 6

10. The variation of solubilities of two compounds X and Y in water with temperature is depicted below. Which of the following statements is true?

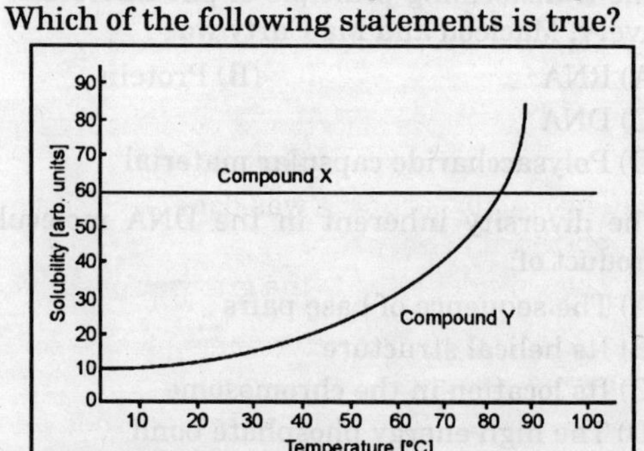

(A) Solubility of Y is less than that of X
(B) Solubility of X varies with temperature
(C) Solubilities of X and Y are the same at 75^0C
(D) Solubilities of X and Y are independent of temperature

11. Restriction endonuclease cleaves DNA molecules at specific 'recognition sites'. One such enzyme has four recognition sites on a circular DNA molecule. After complete digestion, how many fragments would be produced upon reaction with this enzyme?

(A) 4 (B) 5
(C) 3 (D) 6

12. A solid cube of side L floats on water with 20% of its volume under water. Cubes identical to it are piled one-by-one on it. Assume that the cubes do not slip or topple, and the contact between their surfaces is perfect. How many cubes are required to submerge one cube completely?

(A) 4 (B) 5
(C) 6 (D) Infinite

13. Inner planet of the solar system are rocky, whereas outer planet are gaseous. One of the reasons for this is that:

(A) Solar heat drove away the gases to the outer region of the solar system
(B) Gravitational pull of the sun pulled all rocky material to the inner solar system
(C) Outer planet are larger than the inner planets
(D) Comets delivered the gaseous materials to the outer planets

14. On exposure to desiccation, which of the following bacteria are least likely to experience rapid water loss?

(A) Isolated rods (B) Rods in chain
(C) Cocci in chain (D) Cocci in clusters

15. A bell is rung before giving food to a dog. After doing this continuously for 10 days, which of the following is most likely to happen?

(A) The dog learns to ignore the bell
(B) The dog salivates on hearing the bell
(C) The dog ignores food and runs towards the bell
(D) The dog will not eat food without hearing the bell

16. When a magnet is made to fall free in air, it falls with an acceleration of 9.8 m s^{-2}. But when it is made to fall through a long aluminium cylinder, its acceleration decreases, because:

(A) A part of the gravitational potential energy is lost in heating the magnet
(B) A part of the gravitational potential energy is lost in heating the cylinder
(C) The said experiment was done in the magnetic northern hemisphere
(D) The cylinder shields the gravitational force

17. The conductance of a potassium chloride solution is measured using the arrangement depicted below. The specific conductivity of the solution in Sm-1, when there is no deflection in the galvanometer, is

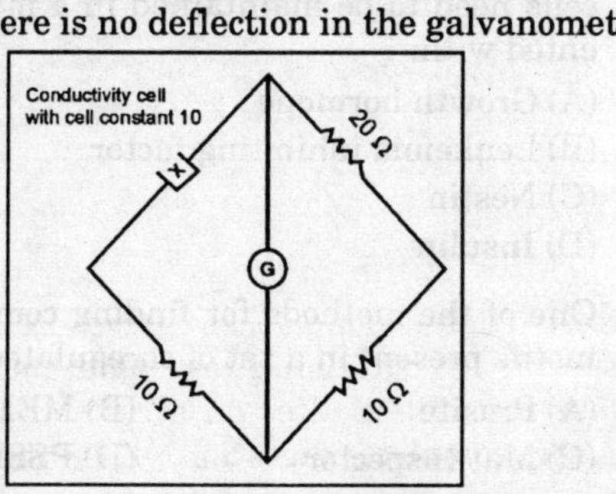

(A) 1.0 (B) 0.5
(C) 2.0 (D) 1.5

18. An overweight person runs 4 km everyday as an exercise. After losing 20% of his body weight, if he has to run the same distance in the same time the energy expenditure would be:

(A) 20% more
(B) The same as earlier
(C) 20% less (D) 40% less

19. What is the half-life of the radio isotope whose activity profile is shown below?

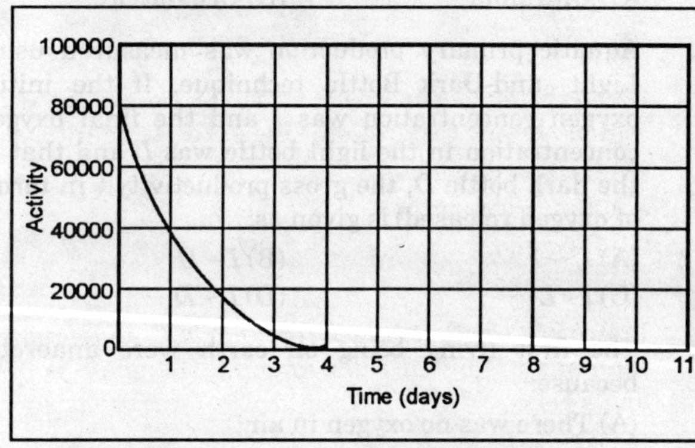

(A) 1 day (B) 3 days
(C) 2 days (D) 4 days

20. A single-strand nick in the parental DNA helix just ahead of a replication fork causes the replication fork to break. Recovery from this calamity requires:

(A) DNA ligase
(B) DNA primase
(C) Site-specific recombination
(D) Homologous recombination

21. In which form of DNA, the number of base pairs per helical turn is 10.5?

(A) A (B) B
(C) X (D) Z

22. α-amanitin inhibits:

(A) Only RNA polymerase I
(B) Only RNA polymerase II
(C) Only RNA polymerase III
(D) All RNA polymerases

23. Graft rejection does not involve:

(A) Erythrocytes (B) T cells
(C) Macrophages
(D) Polymorphonuclear leukocytes

24. Which of the following is NOT a prosthetic group of nitrate reductase?

(A) FAD (B) Heme
(C) Mo (D) Pterin

25. Reverse transcriptase has both ribonuclease and polymerase activities. Ribonuclease activity is required for:

(A) The synthesis of new RNA strand
(B) The degradation of RNA strand
(C) The synthesis of new DNA strand
(D) The degradation of DNA strand

26. Toxic shock is caused by:
(A) Toxins produced by some bacteria
(B) Excessive stimulation of a large proportion of T cells by bacterial superantigens
(C) Abnormal cytokine production by B cells
(D) Excessive production of immunoglobulins

27. Horse shoe crabs belong to the group:
(A) Onychophora (B) Chelicerata
(C) Uniramia (D) Crustacea

28. Aquatic primary production was measured using Light- and-Dark Bottle technique. If the initial oxygen concentration was I and the final oxygen concentration in the light bottle was L and that in the dark bottle D, the gross productivity (in terms of oxygen released) is given as:
(A) $L - 1$ (B) $I - D$
(C) $I - L$ (D) $L - D$

29. The first living being on earth were anaerobic because:
(A) There was no oxygen in air
(B) Oxygen damages proteins
(C) Oxygen interferes with the action of ribozymes
(D) They evolved in deep sea

30. The blastopore region of amphibian embryo that secretes BMP inhibitors and dorsalizes the surrounding tissue is known as:
(A) Brachet's cleft
(B) Nieuwkoop center
(C) Spemann's organizer
(D) Hensen's node

31. Which of the following acts as a branch point for the biosynthesis of sesquiterppenes and triterpenes?
(A) Farnesyl pyrophosphate
(B) Geranyl pyrophosphate
(C) Isopentyl pyrophosphate
(D) Hydroxymethylglutaryl – CoA

32. In a normal human eye, for sharp image formation on the retina, maximum dioptric power is provided by the:
(A) Retina (B) Cornea
(C) Anterior surface of the lens
(D) Posterior surface of the lens

33. Indirect immunofluorescence involves fluorescently labeled:
(A) Immuniglobulin-specific antibodies
(B) Antigen-specific antibodies
(C) Hapten-specific antibodies
(D) Carrier-specific antibodies

34. Which of the following processes interferes in sequence-based phylogeny?
(A) Horizontal gene transfer
(B) Adaptive mutations
(C) DNA repair
(D) Reverse transcription

35. Which of the following groups of species are typical of grassland habitats in India:
(A) Black buck, wolf, great Indian bustard, lesser florican
(B) Spotted deer, dhole, peacock, finch-lark
(C) Sambar, tiger, paradise fly catcher
(D) Otter, cormorant, darter, pelican

36. In gene regulation, Open Reading Frame (ORF) implies:
(A) Intervening nucleotide sequence in between two genes
(B) A series of triplet codons not interrupted by a stop codon
(C) A series of triplet codons that begins with a start codon and ends with a stop codon
(D) The exonic sequence of a gene that corresponds to the 5'UTR of the mRNA and thus does not code for the protein

37. While replicating DNA, the rate of misincorporation by DNA polymerase is 1 in 105 nucleotides. However, the actual error rate in the replicated DNA is 1 in 109 nucleotides incorporated. This is achieved mainly due to:
(A) Spontaneous excision of misincorporated nucleotides
(B) 3' $\rightarrow$ 5' proof reading activity of DNA polymerase
(C) Termination of DNA polymerase at misincorporated sites
(D) 5' $\rightarrow$ 3' proof reading activity

38. Which of the following set of cell organelles are involved in the biosynthesis of jasmonic acid through octadecanoid signaling pathway?
(A) Chloroplast and peroxisomes
(B) Chloroplast and mitochondria
(C) Mitochondria and peroxisomes
(D) Golgi bodies and mitochondria

39. The atmosphere in a sealed space craft contains:
(A) Pure oxygen
(B) A mix of oxygen and nitrogen
(C) A mix of oxygen and carbon dioxide
(D) Pressurised atmospheric air available normally on earth

40. To keep them in a totupotent state, embryonic stem cells need to be maintained in a medium supplemented with:
(A) Growth hormone
(B) Leukemia inhibiting factor
(C) Nestin
(D) Insulin

41. One of the methods for finding common regulatory motifs present in a set of coregulated genes is:
(A) Prosite (B) MEME
(C) Mat Inspector (D) PSSM

42. Measurement and mapping with spatial resolution the membrane potential of a cell, which is too small for microelectrode impalement, is done using:

(A) Radioisotope
(B) Voltage-sensitive dye
(C) pH sensitive chemical
(D) Vital dyes

43. A plant of the genotype AaBb is selfed. The two genes are linked and are 50 map units part. What proportion of the progeny will have the genotype aabb?

(A) ½ (B) ¼
(C) 1/8 (D) 1/16

44. In contrast with plant cells, the most distinctive feature of cell division in animal cells is:

(A) Control of cell cycle transitions by protein kinases
(B) Enzymes responsible for DNA replication
(C) Ubiquitin-dependent pathway for protein degradation
(D) Pattern of chromosome movement

45. During development of embryos in plants, PIN proteins are involved in:

(A) Establishment of auxin gradients
(B) Regulation of gene expression
(C) Induction of programmed cell death
(D) Induction of cell division

46. Which of the floral whorls is affected in *agamous* (ag) mutants?

(A) Sepals & petals
(B) Petals & stamens
(C) Stamens & carpels
(D) Sepals & carpels

47. The membrane lipid molecules assemble spontaneously into bilayers when placed in water and form a closed spherical structure known as:

(A) Lysosome
(B) Peroxisome
(C) Liposome
(D) Endosome

48. Which of the following waves is likely to be absent in a normal frog ECG?

(A) P
(B) Q
(C) T
(D) R

49. On the molar scale which of the following interactions in a nonpolar environment provides highest contribution to the bio-molecule?

(A) Van der Waals interaction
(B) Hydrogen bonding
(C) Salt bridge
(D) Hydrophobic interaction

50. In this flow diagram name the chemicals A, B, C and D in proper sequence:

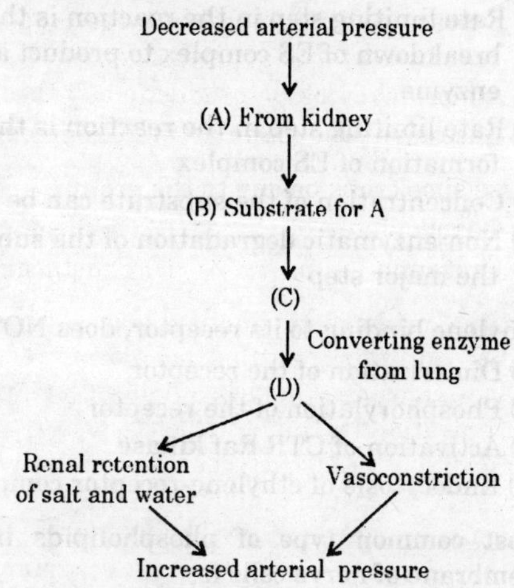

(A) Renin, Angiotensin II, Angiotensin I, Angiotensinogen
(B) Angiotensin I, Angiotensin II, Renin, Angiotensinogen
(C) Renin, Angiotensin I, Angiotensin II, Angiotensinogen
(D) Renin, Angiotensinogen, Angiotensin II, Angiotensin I

51. The Hutchinsonian concept of ecological niche is based on:

(A) Microhabitat occupied
(B) Multidimensional hypervolume
(C) Role played in the ecosystem
(D) A combination of role played and microhabitat occupied

52. A specialist species has a:

(A) Wider niche and high efficiency of niche utilization
(B) Narrower niche and high efficiency of niche utilization
(C) Wider niche low efficiency of niche utilization
(D) Narrower niche and low efficiency of niche utilization

53. The presence of Salmonella in tap water is indicative of contamination with:

(A) Industrial effluents (B) Human excreta
(C) Agriculture waste (D) Kitchen waste

54. Amino acid selenocysteine (Sec) is incorporated into polypeptide chain during translation by:

(A) Charging of Sec into $tRNA^{ser}$ followed by incorporation through serine codon
(B) Charging of serine into $tRNA^{ser}$ followed by modification of serine into selenocysteine and then incorporation through serine codon
(C) Charging of Sec into $tRNA^{sec}$ and then incorporation through selenocyteine codon
(D) Charging of serine into $tRNA^{ser}$, modification of serine into selenocysteine and then incorporation through a specially placed stop codon

55. **Michaelis and Menten derived their equation using which of the following assumption?**

(A) Rate limiting step in the reaction is the breakdown of ES complex to product and free enzyme
(B) Rate limiting step in the reaction is the formation of ES complex
(C) Concentration of the substrate can be ignored
(D) Non-enzymatic degradation of the substrate is the major step

56. **Ethylene binding to its receptor, does NOT lead to:**

(A) Dimerization of the receptor
(B) Phosphorylation of the receptor
(C) Activation of CTR Raf kinase
(D) Endocytosis of ethylene-receptor complex

57. **Most common type of phospholipids in the cell membrane of nerve cells is:**

(A) Phosphatidylcholine
(B) Phosphatidylinositol
(C) Phosphatidylserine
(D) Sphingomyelin

58. **Which of the following maternal effect gene products regulate production of anterior structures in *Drosophila* embryo?**

(A) Bicoid and nanos
(B) Bicoid and Hunchback
(C) Bicoid and Caudal
(D) Nanos and Caudal

59. **The base analog 2-aminopurine pairs with thymine, and can occasionally pair with cytosine. The type of mutation induced by 2-aminopurine is:**

(A) Transversion (B) Transition
(C) Deletion (D) Nonsense

60. **Which of the following is NOT a physiological characteristics of early successional plants?**

(A) High respiration rate
(B) Inhibition by far-red light
(C) High transpiration rate
(D) Low photosynthetic rate

61. ***Batrachochytrium dendrobatidis*, a fungus, has been implicated in the decline of populations of:**

(A) Fish (B) Frogs
(C) Pelicans (D) Bats

62. **What kind of aneuploid gametes will be generated if meiotic non-disjunction occurs at first division? ('n' represents the haploid number of chromosomes)**

(A) Only n + and n (B) Only n – 1 and n
(C) Both n + 1 and n – 1 (D) Either n + 1 or n -1

63. **The peacock's tail is an example of:**

(A) Natural selection
(B) Diversifying selection
(C) Sexual selection
(D) Group selection

64. **Wetlands are conserved internationally through an effort called as:**

(A) Basel Convention
(B) Rio Convention
(C) Montreal Convention
(D) Ramsar Convention

65. **The rattans and canes that we use in furniture belong to:**

(A) Bamboos (B) Palms
(C) Arborescent lilies
(D) Legumes

66. **Which of the following features is NOT shown by glyphosate, a broad spectrum herbicide?**

(A) Little residual soil activity
(B) Ready translocation in phloem
(C) Inhibition of a chloroplast enzyme catalyzing the synthesis of aromatic amino acids
(D) Inhibition of early steps in the biosynthesis of branched chain amino acids

67. **The erythrocyte membrane cytoskeleton consists of a meshwork of proteins underlying the membrane. The principal component spectrin has α, β subunits which assemble to form tetramers. The cytoskeleton is anchored to the membrane through linkage with the transmembrane proteins band 3 and glycophorin C. The cytosolic domain of band 3 also serves as the binding site of glycolytic enzymes such as glyceraldehydes 3-phosphate dehydrogenase. Analysis of the blood sample of a patient with haemolytic anemia shows spherical red blood cells. The patient carries:**

(A) A mutation in glycophorin C
(B) A mutant spectrin with increased tetramerization propensity
(C) Mutant β spectrin defective in αβ dimerization ability
(D) Mutant glyceraldehydes 3-phosphate dehydrogenase

68. **Synthesis of normal hemoglobin requires coordinated synthesis of αglobin and β globin. Thalassemias are genetic defects perturbed in this coordinated synthesis. Pateints suffering from deficiency of β globin chains (β-thalassemia) could also be due to mutations affecting the biosynthesis of β globin mRNA:**

The following statements describe the genesis of non-functional β globin leading to β-thalassemia.

A. Mutation in the promoter region of the β globin gene
B. Mutation in the splice junction of the β globin gene
C. Mutation in the intron I of the β globin gene
D. Mutations towards the 3' end of the β globin gene that codes for polyadenylation site

Which of the following combinations is correct?

(A) A, B and D (B) A, B and C
(C) B, C and D (D) C, D and A

69. Equilibrium contant (K) of noncovalent interaction between two non-bonded atoms of two different groups was measured at 270C. It was observed that K = 100 M-1. The strength of this noncovalent interaction in terms of Gibbs free energy change is:

(A) 2746 kcal/mole
(B) -2746 kcal/mole
(C) 247 kcal/mole
(D) -247 kcal/mole

70. Bacteriophage λ has two modes in its life cycle, lytic and lysogenic. In the lysogenic mode, the expression of all the phage genes are repressed while the expression of repressor gene switches between on and off position depending on the concentration of repressor. The following statements are made:

A. Repressor may act both as a positive regular and a negative regulator
B. Expression of repressor gene, cI is independent of the expression of cII and cIII genes
C. Mutation of cI gene will cause it to form clear plaques on both wild type *E. coli* and *E. coli* (λ+)
D. Mutation at operators, O_L and O_R will allow the phage to act as a virulent phage

The correct statements are

(A) A and B
(B) B and C
(C) C and D
(D) D and A

71. Biosynthesis of tyrosine is detailed below:

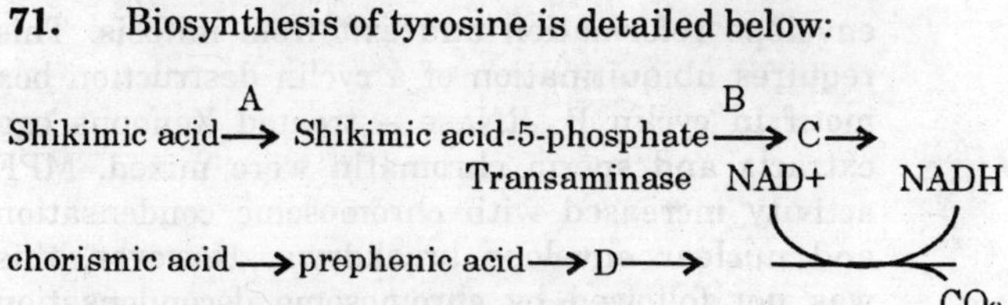

tyrosine. Identify A, B, C and D

(A) ATP, phosphoenolpyruvic acid, 3- enolpyruyl shikimic acid – 5- phosphate, p-hydroxyphenylpyruvic acid
(B) GTP, pyridoxal phosphate, 3- enolpyruvyl shikimic acid – 5 – phosphate,phenylpyruvic acid
(C) NADP, 3-phosphohydroxypyruvic acid, 3-enolpyruvyl shikimic acid-5-phosphate, p-hydroxyphenylpyruvic acid
(D) ATP, 3-phosphohydroxypyruvic acid, 3-enolpyruvyl shikimic acid -5-phosphate, pyridoxylphosphate

72. AP1 (APETLA 1) is one of the floral meristem identifying genes. In wild type *Arabidopsis thaliana* plants transformed with AP1::GUS, β glucuronidase (GUS) activity is seen in floral meristem, only after the commitment to flowering. Ectopic expression of AP1 :: GUS in the EMBRYONIC FLOWER (*emf*) mutant background results in GUS activity throughout the shoots in four day old seedlings. These observations suggests that AP 1 is:

(A) Not involved in flowering
(B) Involved in repression of flowering
(C) Involved in promoting flowering
(D) Stimulation of flowering in the emf background

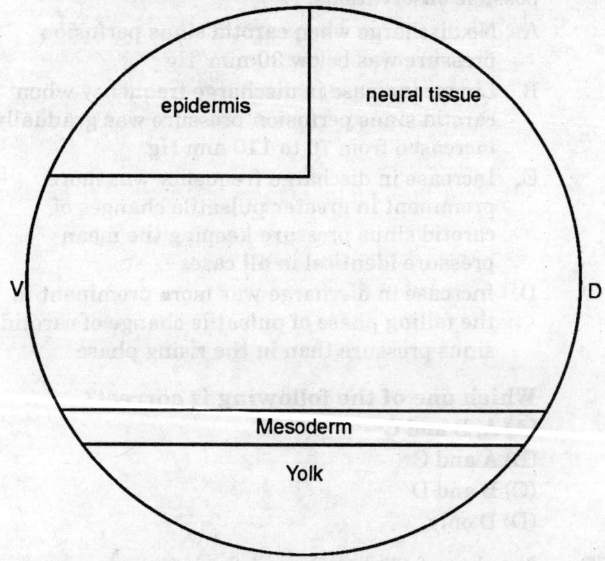

The figure above represents a late zebrafish gastrula. The following concepts may be proposed during further development of the embryo.

(A) The concentration of EGF decreases from the yolk towards the epidermis, along with the increase of BMP activity from the dorsal to the ventral axis
(B) Increase in FGF activity in the epidermis with concomitant decrease in BMP activity towards the ventral axis
(C) Neural induction in zebrafish is independent of the organizer and depends on activation of BMP signaling
(D) In comparison, both Xenopus and chick embryo require activation of FGF for neural induction to occur in addition to BMP inhibition

Which of the above statements are true?

(A) A and C (B) B and C
(C) A and D (D) C and D

73. Following are some statements regarding plant growth hormones.

A. Ethylene regulates abscission
B. Gibbrellins do not play any role in flowering
C. Auxin and cytokinin promote cell division
D. Over expression of cytokinin oxidase would promote root growth
E. ABA inhibits root growth and promotes shoot growth at low water potential
F. ABA promotes leaf senescence independent of ethylene

Which one of the following combination of above statement is correct?

(A) A, C and F (B) B, C and D
(C) D, E and F (D) B, D and E

74. An isolated carotid sinus was prepared so that the pressure may be regulated by a pump and the resulting discharge in single carotid sinus nerve fibre could be recorded. The following are the possible observations.

A. No discharge when carotid sinus perfusion pressure was below 30 mm Hg

B. Linear increase in discharge frequency when carotid sinus perfusion pressure was gradually increased from 70 to 110 mm Hg

C. Increase in discharge frequency was more prominent in greater pulsatile changes of carotid sinus pressure keeping the mean pressure identical in all cases

D. Increase in discharge was more prominent in the falling phase of pulsatile change of carotid sinus pressure than in the rising phase

Which one of the following is correct?

(A) A, B and C
(B) A and C
(C) B and D
(D) D only

75. Number of trials required for rats to learn a task when they were exposed to various conditions were as follows:

Experimental Conditions	Observations
Light : light dark cycle – 12 h : 12 h	N – trials
Bright light – 24 h	Significantly more trials than 'N'
Bright light – 24 h + continuous Physical disturbance	Significantly more trials than 'N'
Dark light – 24 h + continuous Physical disturbance	Significantly more trials than 'N'

Which of the following inferences is most appropriate?

(A) Continuous light enhanced learning
(B) Continuous darkness inhibited learning
(C) Physical activity inhibited learning
(D) Learning was reduced by sleep loss

76. In the following statement taken from a research paper, what does p in the parenthesis stand for?

"The mean temperature of this region now is significantly higher than the one 50 years ago ($p<0.05$, t-test)"

(A) Ratio of the mean temperatures of the two time periods tested
(B) Probability of the error of ejecting a true null hypothesis
(C) Probability of the error of accepting a false null hypothesis
(D) Probability of the t-test being effective in detecting significant differences in the mean annual temperatures of the two time periods

77. Cre/loxP system is used by phage PI to remove terminally redundant sequences that rise during packaging of the phage DNA. Cre-lox system can be used t create targeted deletions, insertions and inversion in genomes of transgenic animals and plants. Consider a series of genetic markers A to K. How should the Lox O sites be positioned in order that Cre recombinase can create an inversion in the EFG segment relative to ABCD and HIJK?

(A) A B C D ⇒ E F G ⇒ N I J K
Lox P Lox P

(B) A B C ⇒ D E F G ⇒ H I J K
Lox P Lox P

(C) A B C D ⇒ E F G ⇒ H I J K
Lox P Lox P

(D) A B C D ⇒ E F O ⇒ H I J K
Lox P Lox P

78. Assuming a 1 : 1 sex ratio, what is the probability that three children from the same parents will consist of two daughters and one son?

(A) 0.375 (B) 0.125
(C) 0.675 (D)).75

79. Maturation – promoting factor (MPF) controls the initiation of mitosis in eukaryotic cells. MPF kinase activity requires cyclin B. Cyclin B is required for chromosome condensation and breakdown of the nuclear envelop into vesicles. Cyclin B degradation is followed by chromosome decondensation, nuclear envelope reformation and exit from mitosis. This requires ubiquitination of a cyclin destruction box motif in cyclin B. RNase – treated Xenopus egg extracts and sperm chromatin were mixed. MPF activity increased with chromosome condensation and nuclear envelope breakdown. However, this was not followed by chromosome decondensation and nuclear envelope reformation because:

(A) RNase contamination persisted in the system
(B) Cyclin B was missing from the system
(C) Ubiquitin ligase has been overexpressed
(D) Cyclin B lacking the cyclin destruction box had been overexpressed

80. Denaturation profiles of DNA are shown below.

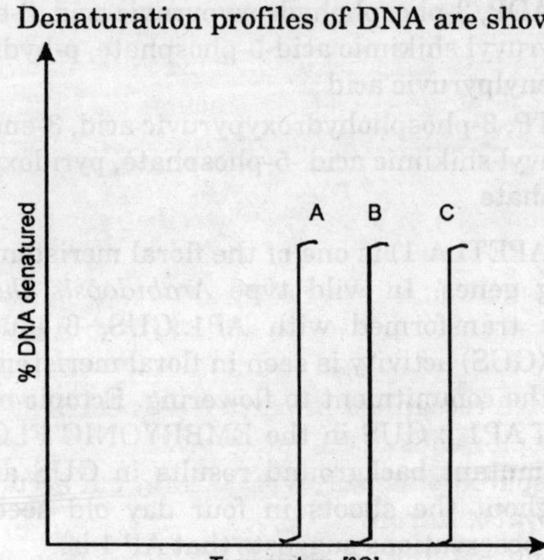

The differences in the profiles arise because

(A) The DNA is single strande but of different sizes
(B) A + T content of A > B > C and the DNA are from complex genomes
(C) G + C content of C > B > A in DNA of comparable sizes isolated from simple genomes
(D) G + C content is identical but a + T content in A > B > C in DNA of comparable sizes isolated from simple genomes

81. In an in vitro experiment using radiolabelled nucleotides, a researcher is trying to analyze the possible products of DNA replication by resolving the products using urea-polyacrylamide gel electrophoresis. In one experimental set up RNase H was added (Set 1), while in another set no RNase H was added (Set 2).

The possible observations of this experiment could be:

A. There is no difference in the mobility of labeled DNA fragments between the Set 1 and Set 2
B. There is distinct difference in the mobility of the newly synthesized labelled DNA fragments between Set 1 and Set 2
C. The mobility of the newly synthesized labeled DNA fragments in case of Set 1 is faster as compared to the Set 2
D. The mobility of the newly synthesized labeled DNA fragments in case of Set 1 is slower as compared to the Set 2

Which of the following combinations represent correct observations?

(A) A and B (B) B and C
(C) A and D (D) B and D

82. The bacterial flagellar motor is a multi-protein complex. Rotation of the flagellum requires movement of protons across the membrane facilitated by a multi-protein complex. The flagellar motor proteins combine to create a proton channel that drives mechanical rotation. In a screen for mutants, some non-motile ones were selected. These could have:

(A) Mutations in tubulin and actin proteins
(B) Mutations in kinesin proteins
(C) Mutated H+ - ATPase
(D) Mutations in the charged residues lining the ridge of the FliG subunit

83. Pre-mRNAs are rapidly bound by SnRNPs which carry out dual steps of RNA slicing, that removes the intron & joins the upstream and downstream exons.

The following statements describe some facts related to this event.

A. Almost all introns begin with GU and end with AG sequences and hence all the GU or AG sequences are spliced out of RNA
B. U2 RNA recognizes important sequences at the 3'acceptor ends of the intron
C. The spliceosome uses ATP to carry out accurate removal of intron
D. An unusual linkage with 2' OH group of guanosine within the introl form a 'Lariat' structure.

Which of the following combinations is correct?

(A) A and B (B) B and C
(C) C and D (D) D and A

84. Consider the following crosses involving grey (wild type)& yellow body colour true-breeding Drosophila

	Cross	F_1 progeny	F_2 progeny
Cross 1	Grey female X yellow male	All males : grey All females : grey	Grey females : 98 Yellow males : 45 Grey males:49
Cross 2	Yellow females X grey male	All males : yellow All female : grey	?

Assuming 200 F_2 offsprings are produced in cross 2, which one of the following outcome is expected?

(A) 97 grey males, 54 yellow females, 49 grey males
(B) 102 yellow males, 46 yellow females, 52 grey females
(C) 52 grey males, 49 yellow males, 48 yellow females, 51 grey females
(D) 98 grey males, 94 yellow females, 2 yellow males, 6 grey females

85. Genomic DNA of transgenic plants (P1, P2 and P3) obtained by transforming with binary vector A whose map is depicted below, was digested with BamH I and Sal I and hybridized with a labeled fragment X

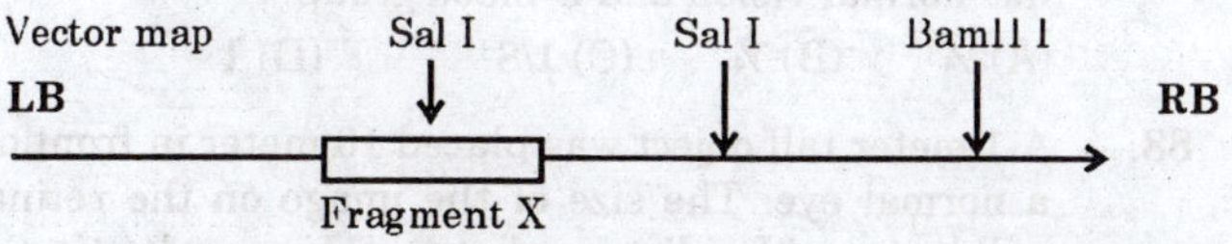

The pattern obtained in Southern hybridization is shown below:

P1		P2		P3	
BAMHI	Sal I	BamH I	Sal I	Bam H I	Sal I
	—	—		—	—
—	—		—		
—	—		—		—

Based on the above, which of the following interpretations is correct:

(A) All the plants (P1, P2 and P3) contain two copies of the transgene
(B) P1, P3 contain one and P2 contains two copies of the transgene
(C) P1 contains two, whereas P2 and P3 contain one copy of transgene each
(D) P1 and P2 contains two and P3 contains one copy of the transgene

86. Figures A and B respectively represent the dideoxy sequencing gels obtained for partial sequences from 5' –ends of a bacterial gene and its mutant (with a point mutation).

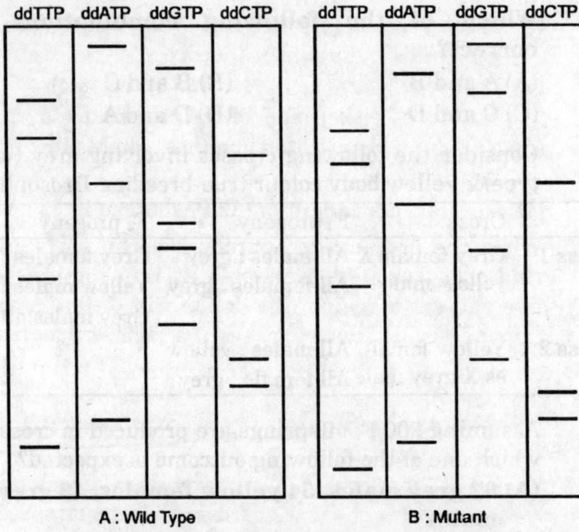

What type of mutation has occurred in the gene?

(A) Nonsense (B) Missense
(C) Frameshift (D) Transversion

87. The ABO blood type in human is under the control of autosomal multiple alleles. Colour blindness is recessive X-linked trait. A male with a blood type A and normal vision marries a female who also has blood type A and normal vision. The couple's first child is a male who is colour blind and has O blood group. What is the probability that their next child has normal vision and O blood group?

(A) ¼ (B) ¾ (C) 1/8 (D) 1

88. A 1 meter tall object was placed 10 meter in front of a normal eye. The size of the image on the retina will be (consider distance between lens and retina = 1.7 cms):

(A) 0.17 mm (B) 1.7 mm
(C) 3.4 mm (D) 170 μm

89. In human, protein coding genes are mainly organized as "exons" and "introns". There are intergenic regions that transcribe into various types of non-coding RNA (not translating into protein). Some introns may harber also transcription units, which are:

(A) Always other protein coding genes
(B) Protein coding gene and RNA coding genes
(C) Always RNA coding genes
(D) Pseudo genes

90. A fixed smear of a bacterial culture is subjected to the following solutions in the order listed below and appeared red:

A. Carbolfuchsin (heated) B. Acid-alcohol
C. Methylene blue

Bacteria stained by this method can be identified as:

(A) Non-acid fast *E. coli*
(B) Acid-fast *Mycobacterium sp.*
(C) Gram-positive *E. coli*
(D) Gram-negative *Mycobacterium sp.*

91. Survival of intracellular pathogens depends on the levels of pro-inflammatory and anti-inflammatory cytokines in macrophages. In an experimental condition, Mycobacteria infected macrophages were treated with IL-6 or IL-12 for 4 hours t 370C. Untreated cells were used as control. Cells were lysed and number of bacteria in each experimental set was counted by measuring colony forming unit (CFU). Which of the following observations is true?

(A) IL-12 treated cells contain more intracellular bacteria than control
(B) IL-12 treated cells contain less intracellular bacteria than control
(C) IL-6 treated cells contain more intracellular bacteria than control
(D) IL-6 treated cells contain less intracellular bacteria than control

92. The genetic relatedness ® of an individual to his nephew is 0.25. The alleles that cause uncles to care for nephews will spread, according to Hamilton's Rule, only if the fitness benefit is:

(A) Equal to the cost of care
(B) More than the cost of care by 25%
(C) Double the cost of care
(D) Four times the cost of care

93. The following is a schematic representation of region (showing six bands) of the polytene chromosome of Drosophila, along with the extent of five deletions (Dell to Del5):

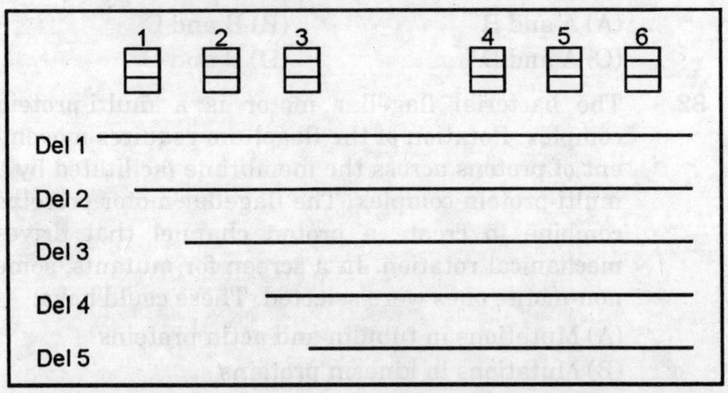

Recessive alleles a, b, c, d, e and f are known to corres pond to each of the bands (1 to 6), but their order is not known. When the recessive alleles are placed against each of these deletions, the following results are obtained. The plus (+) in the table indicates wild type

Phenotype of the corresponding allele, while a minus (-) indicates the phenotype governed by the corresponding mutant allele.

	a	*b*	*c*	*d*	*e*	*f*
Del 1	+	-	-	-	+	+
Del 2	+	+	-	-	+	+
Del 3	-	+	-	-	+	+
Del 4	-	+	+	-	-	+
Del 5	-	+	+	+	-	-

Which one of the following indicated the correct location of the recessive alleles on the bands of the polytene chromosomes?

(A) a – 3; b – 1; c – 2; d – 4; e – 5; f – 6
(B) a – 6; b – 1; c – 2; d – 5; e – 4; f – 2
(C) a – 4; b – 1; c – 2; d – 3; e – 5; f – 6
(D) a – 5; b – 1; c – 6; d – 4; e – 2; f – 3

94. Many cancers carry mutant p53 genes, while some cancers have normal p53 genes. P53 activates p21 (Waf1) which inhibits G1/S-Cdks and phosphorylation of the retinoblastoma protein (Rb). Cancers with normal p53 genes could:

(A) Express non-phosphorylatable form of Rb
(B) Express high levels of p53-deubiquitinases
(C) Express inactive forms of GI/S-cdks
(D) Express inactive forms of G1/S cyclins

95. For a normal heart, the time taken for arterial systole and diastole are As and Ad seconds, respectively, while the same for ventricular systole and diastole are Vs and Vd. Which one of the following equations is correct?

(A) As + Ad = Vs + Vd (B) As + Ad < Vs + Vd
(C) As + Ad – Vs + Vd=0 (D) As + Ad > Vs + Vd

96. The following are statements about molecular markers in the context of plant breeding:

A. Molecular markers can be used for elimination of undesirable traits
B. Molecular markers cannot be used for estimation of the genetic contribution of each individual parent in a segregating population
C. Molecular markers are used mapping of QTLs, which is also possible conventional techniques
D. Molecular markers can be used for selection of individuals from a population that are homozygous for the recurrent parent genotype at loci flaking the target locus

Which of the above statements are True?

(A) A and B (B) A and C
(C) A and D (D) B and C

97. Enzymes are nowadays used extensively in bioprocessing industries.

Enzyme 1 is used for treatment of hides to provide a finer texture, in leather processing & manufacture of glue. Enzyme 2 is used for clarification of fruit juices

Identify Enzymes 1 and 2

Enzyme 1	**Enzyme 2**
(A) Amylase	Pectinase
(B) Protease	Amylase
(C) Protease	Pectinase
(D) Pectinase	Amylase

98. A nerve impulse or action potential is generated from transient changes in the permeability of the axon membrane to Na^+ and K^+ ions. The depolarization of the membrane beyond the threshold level leads to Na^+ flowing into the cell and a change in membrane potential to a positive value. The K^+ channel then opens allowing K^+ to flow outwards ultimately respotring membrane potential to the resting value. The Na^+ and K^+ channels operate in opposite directions because:

(A) There is an electrochemical gradient growth generated by proton transport
(B) There is difference in Na^+ and K^+ concentrations on either side of the membrane
(C) Na^+ is a voltage-gated channel, whereas K^+ is ligand-gated
(D) Na^+ is a dependent on ATP whereas K^+ is not

99. When the prospective neurons from an early gastrula of a frog were transplanted into the prospective epidermis region, the donor cells differentiated into epidermis. However, when a similar experiment was done with the late gastrula of frog, the prospective neurons developed into neurons only. These observations could possibly be explained by the following phenomena:

A. The early gastrula show conditional development whereas the late gastrula shows autonomous development
B. The early gastrula show autonomous development whereas the late gastrula shows conditional development
C. The prospective neurons from the early gastrula are only specified whereas those from the late gastrula are determined
D. The prospective neurons from the early gastrula are determined whereas those from the late gastrula are specified

Which of the conclusions drawn above are correct?

(A) A and B
(B) A and C
(C) A and D
(D) B and C

100. Insulin and other factors stimulate a pathway involving a proteion kinase mTOR, which in its turn augments protein synthesis. mTOR essentially modifies protein(s) which in this unmodified form act as inhibitors of protein synthesis. The following proteins are possible candidates:

A. eEF-1 B. eIF-4E-BP1 C. eIF-4E D. PHAS-1

Which of the following sets is correct?

(A) A and B
(B) B and D
(C) A and C
(D) B and C

101. For continuation of protein synthesis in bacteria, ribosomes need to be released from the mRNA as well as to dissociate into subunits. These processes do not occur spontaneously. They need the following possible conditions:

A. RRF and EF-G aid in this process
B. An intrinsic activity of ribosomes and an uncharged iRNA are required
C. IF-1 promotes dissociation of ribosomes
D. IF-3 and IF-1 promote dissociation of ribosomes

Which of the following sets is correct?

(A) A and D (B) A and B
(C) A and C (D) B and D

102. The following figure depicts the relationship between a genetic map for four genes (A, B, C and D) and their corresponding physical map:

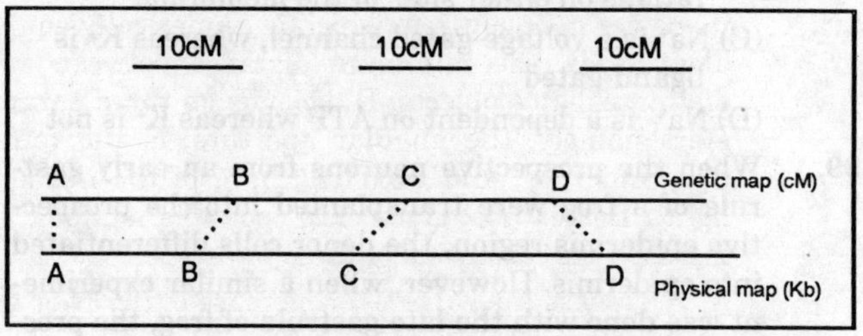

The following statements are made to explain this relationship.

A. More number of recombination events occur between A and B as compared to B and C

B. Lesser number of recombination events occur between C and D as compared to B and C

C. Although the physical distance between A and B is less than that between C and D, the region between A and B is more recombinogenic

D. The physical distance between A and B is less than that between C and D, and this the region between A and B is less recombinogenic

E. Although the physical distance between C and D is more than that between B and C, the region between C and D is less recombinogenic

F. Although the physical distance between C and D is more than B and C, the region between C and D is more recombinogenic

Which statements are correct?

(A) A and B (B) C and E
(C) D and F (D) A, C and E

103. A monkey undergoes cerebellectomy. After the post-operative recovery, the monkey was given a task to press a bar. The possible observations are:

A. Its hand would overshoot the target while reaching the bar

B. It would be unable to move forelimbs

C. It would show intention tremor while trying to press the bar

D. It would press the bar with mouth instead of hand

Which one of the following is correct?

(A) A and C (B) B only
(C) D only (D) B and D

104. During the Spanish conquest of the Inca Empire at the high altitude in Peru, many soldiers fell sick. It was found that the sickness was due to low partial pressure of O2 in the atmosphere at the altitude. To determine the reason, blood was collected from those patients. The circulating erythropoietin (EPO) level were estimated and the O_2-dissociation curve of haemoglobin were drawn and compared with the same in native people as depicted below.

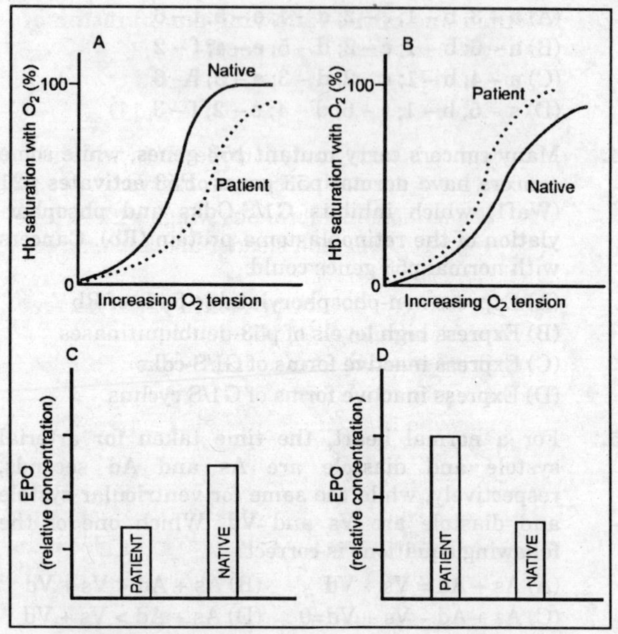

Which of the following combinations is logically correct?

(A) A and C (B) A and D
(C) B and C (D) B and D

105. The graph represents relative plasma concentration of hormones (A and B) during reproductive cycle in a normal female. Which one of the following combinations is correct?

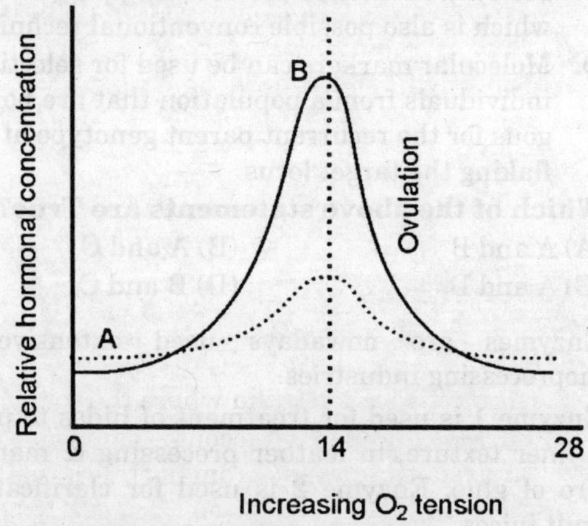

(A) (A) is FSH and (B) is estrogen
(B) (A) is estrogen and (B) is LH
(C) (A) is FSH and (B) is LH
(D) (A) is LH and (B) is FSH

106. Following are some statements about low temperature stress in plants:

A. Fatty acid composition of mitochondria isolated from chilling resistant and chilling sensitive plants differs significantly

B. Ratio of unsaturated fatty acids to saturated fatty acids is lower in chilling resistant species

C. The cellular water does not freeze even at -40°C, because of the presence of solutes and other antifreeze proteins

D. Heat shock proteins do not play any role during low temperature stress

Which one of the following combination of above statement is correct?

(A) A and B (B) A and C
(C) B and C (D) B and D

107. During a field study, three insects with the following characteristics were observed:

A. Elongation, membranous wings with netlike venation, long and slender abdomen, large compound eyes

B. Small bodied, sucking mouth parts, narrow wings fringed with setae

C. Sclerotized forewings, membranous hindwings, chewing mouth parts

The can be identified to their respective orders as:

(A) A-Orthoptera; B–Hemiptera; C–Coleoptera
(B) A – Odonata; B – Coleoptera; C – Hemiptera
(C) A – Orthoptera; B – Odonata; C – Coleoptera
(D) A – Odonata; B – Thysanoptera; C – Coleoptera

108. Following are some of the statements regarding the effect of CO_2 concentration on photosynthesis in plants.

A. With elevated CO_2 levels, C_3 plants are much more responsive than C_4 plants under well watered conditions

B. In C_3 plants, increasing intracellular CO_2 partial pressure can stimulate photosyntehsis only over a narrow range

C. In C_4 plants, CO_2 compensation point is nearly zero

Which one of the following combination of above statements is correct?

(A) A and B (B) B and C
(C) A and C (D) Only C

109. Several distinct time periods and different routes might explain the entrance of marsupials into Australia

(i) Late Jurassic- early therians arrived in Antarctica – Australia where the marsupials
(ii) Early to middle Cretaceous – early marsupials arrived in Australia from northern regions and then radiated in isolation
(iii) Paleocene – marsupials entered Australia from South-East Asia
(iv) Eocene – chance dispersal of marsupials into Australia

Which of the following is the correct combinations?

(A) (i), (ii), (iii) (B) (i), (iii), (iv)
(C) (ii), (iii), (iv) (D) (i), (ii), (iv)

110. Intracellular transport and cytoskeletal organization of a cell is regulated by nucleotide exchange of different small molecular weight GTPase of Ras super family. Overexpression of which of the following GTPase modulates the actin-cytoslkeleton of HeLa cells?

(A) Ran in GDP bound form
(B) Ran in GTP bound form
(C) Rho in GTP bound form
(D) Rho in GDP bound form

111. The most important property of any microscope is its power of resolution, which is numerically equivalent to D, the minimum distance between two distinguishable objects. D depends on three parameters namely, the angular aperture, α, the refractive index, N, and wavelength, λ, of the incident light. Below are given few possible options to increase the resolution of the microscope.

A. Decrease the value of or increase either N or to improve resolution

B. Moving the objective lens closer to the specimen will decrease sin and improve the resolution

C. Using a medium with high refraction index between the specimen and the objective lens to improve the resolution

D. Increase the wavelength of the incident light to improve the resolution

Which of the following combination of above statements is correct?

(A) A and C (B) B and C
(C) A and D (D) C and D

112. Tumor cells were isolated from a breast cancer patient. These cells were injected into nude mice & they were divided into four groups. Group I received EGF receptor-conjugated with methotrexate; Group 2 received transferring receptor-conjugated with methotrexate; Group 3 received mannose receptor-conjugated with methotrexate; Group 4 received same amount of the free drug. In which of the following cases tumorigenic index would be minimum?

(A) Free drug
(B) EGF receptor-conjugated drug
(C) Transferrin receptor-conjugated drug
(D) Mannose receptor-conjugated drug

113. Stem cell therapies are being used in regenerative medicine like forming new adult bone, which usually does not regrow to bridge wide gaps. Successful attempts have now been made in this area because the same paracrine and endocrine factors were found to be involved in both endochondral ossification and fracture repair. Few methods to achieve the above are given below:

A. Develop a collagen gel containing plasmids carrying the human parathyroid hormone gene and place in the gap between the ends of the broken leg

B. Develop a gel matrix disc containing genetically modifies stem cells to secrete BMP4 and VEGF-A and implant it at the site of the wound.

C. Make scaffolds of material that resemble normal extracellular matrix that could be molded to form the shape of a bone needed and seed them with bone marrow cells and place them between the ends of the bones

Which of the above methods would you employ to develop a new functional bone in patients with severely fractured bones.

(A) A and B (B) A, B and C
(C) A and C (D) C and D

114. In case of morphallactic regeneration:

(A) There is repatterning of the existing tissues with little new growth
(B) There is repatterning of the existing tissues after the stem cell division has taken place
(C) There is cell division of the differentiated cells which maintain their differentiated state to finally form a complete organism
(D) There is dedifferentiation of the cells at the cut surface which become undifferentiated. These undifferentiated cells then divide to redifferentiate to form the complete structure

115. In an animal experiment:

(i) Electrical stimulation of an area in the brain (A) increased a function (F) which was prevented by systemic injection of adrenergic antagonistic, prazosin.
(ii) Injection of carbachol (cholinergic agonist) into A also increased function F which was, however, not prevented by systemic injection of adrenergic antagonistc, prazosin.

The result are likely to be due to the stimulation of

(A) Nonaderenergic and cholinoceptieve neurons
(B) Cholinergic and non-adrenoceptive neurons
(C) Adrenergic terminals in 'A"
(D) Both neurons and fibres passing through 'A'

116. With respect to the extra embryonic structures formed in the mammals, the possible functional attributes have been designated:

A. Allantoin stores waste and helps mediate gas exchange. It is derived from splanchnopleure at the caudal end of the primitive streak.
B. Amnion is a water sac and protects the embryo and its surrounding amniotic fluid. This epithelium is derived from somatopleure.
C. Chorion is essential for gas exchange in amniote embryos. It is generated from the splanchnopleure.
D. Yolk sac is the last embryonic membrane to form and is derived somatopleure.

Which of the above statements are correct?

(A) A and B (B) A and C
(C) B and C (D) A and D

117. The quantum yield of photosynthetic carbon fixation in a C_3 plant and C_4 plant is studied as function of leaf temperature. Following are some statements based on this study.

A. At lower temperature the quantum yield of C_3 plant is lower than C_4 plant.
B. In C_4 plant quantum yield does not show temperature dependence.
C. Since the photorespiration is low in C_4 plants because of CO_2 concentrating mechanism, quantum yield is not affected.
D. At higher temperature the quantum yield of C_3 plant is lower than C_4 plant.

Which one of the following combination of above statements is correct?

(A) A, B and D (B) B, C and D
(C) A, B and C (D) A, C and D

118. The decision to become either trophoblast or inner cell mass blastomere is one of the first decisions taken by any mammalian embryo. Below is a diagrammatic representation of the different cells formed during development from the morula with the help of different molecules. Identify the molecules 1-4, sequentially.

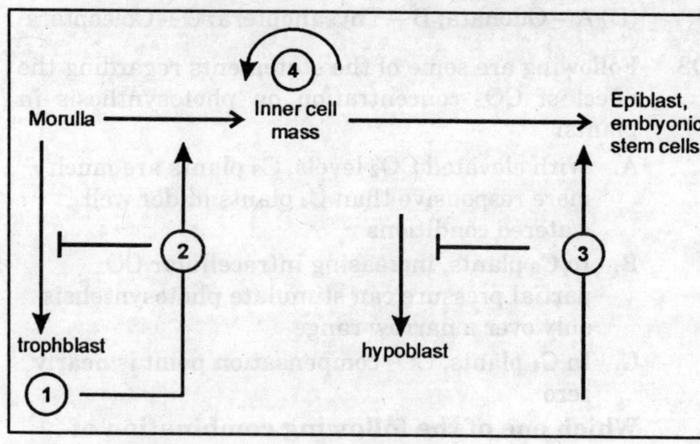

(A) cdx2, Oct 4, nanog, Stat 3
(B) cdx 2, Nanog, Stat 3, Oct 4
(C) cdx 2, Nanog, Oct 4, Stat 3
(D) cdx 2, Oct 4, Stat 3, Nanog

119. Following are some statements for synthesis of jasmonic acid in plants

A. 12-oxo-phytodienoic acid is produced in chloroplast and transported to peroxisome.
B. Action of lipoxygenase, allene synthase and allene oxide cyclase takes place in peroxisome.
C. 12-oxo-phytodicnoic acid is first reduced and then converted to jasmonic acid by ß– oxidation.
D. Final production of jasmonic acid takes place in chloroplast
E. Action of allene oxide synthase and allene oxide cyclase takes place in chloroplast

Which one of the following combination of above statements is correct?

(A) A, B and C (B) B, D and E
(C) C, D and E (D) A, C and E

120. Which of the following behavioral changes are expected in a rat when its nucleus accumbens is experimentally ablated?

(A) Aggressive behavior increases
(B) Exploratory behavior decreases
(C) Nest-building activity increases
(D) Level of parental care drops

121. From among the five animals listed below, match the two attributes-amniotic egg and endothermy, with the correct animals(s):

A. Fish B. Frog C. Crocodile
D. Pigeon E. Zebra

(A) Amniotic egg: b, c, d; Endothermy: d, e
(B) Amniotic egg: c, d, e; Endothermy: d, e
(C) Amniotic egg: a, b, c, d; Endothermy: c, d, e
(D) Amniotic egg: b, c, d; Endothermy: c, d, e

122. Which of the following is NOT true for a critically endangered species?

(A) Reduction of population breeding ability due to increased relatedness through the action of incompatibility mechanisms in plants or behavioural difficulties in animals.
(B) The individuals of the species which have declined to low numbers are still a genetically open system.
(C) Loss of some alleles from the species causing loss of genetic diversity with consequent inability to respond rapidly to selection
(D) Expression of deleterious alleles and increases homozygosity increases mortality of young, and inbreeding depression leads to reduced offspring fitness

123. In the global nitrogen cycle, the following microbial organisms are involved in three important process-denitrification, nitrification and nitrogen fixation.

A. Rhizobium B. Nitrosomonas
C. Nitrobcter D. Pseudomonas
E. Azotobacter

Which of the following is the correctly matched pair of process and its causative species?

(A) Denitrification – (b); nitrogen fixation – (c) and (e); nitrification – (d)
(B) Denitrification – (d); nitrogen fixation – (a) and (e); nitrification – (c)
(C) Denitrification – (e); nitrogen fixation – (a) and (d); nitrification – (d)
(D) Denitrification – (b); nitrogen fixation – (a) and (d); nitrification – (C)

124. Suppose you discovered a new species about which you know only two facts: it is small sized (<10 cm) and short-lived (<20d). Which of the following reproductive strategies is most likely to be true for this species?

(A) Breeds early and more than once in life and produces large number of small-sized offspring
(B) Breeds late and only once in life and produces large number of small-sized offspring
(C) Breeds early and only once in life & produces large number of small-sized offspring
(D) Breeds early and only once in life and produces a small number of large-sized offspring

125. Autotrophs in the aquatic ecosystem, unlike their counterparts in the terrestrial ecosystem, are mostly microscopic and very low in indigestible (to the herbivores) mater. This explains the fat that compared to the terrestrial ecosystem, in the aquatic ecosystem

(A) Productivity/Biomass ratios re higher and energy transfer rates to higher trophic levels are faster
(B) Productivity/Biomass ratios are lower and the energy transfer rates to higher trophic levels are slower
(C) Productivity/Biomass ratios are lower and the energy transfer rate to higher trophic levels are faster
(D) Productivity/Biomass ratios are higher and the energy transfer rate to higher trophic levels are slower

126. In order to prevent tetanus in neonates, one of the following treatments can be adopted.

A. Treatment of the infant with anti-toxin and the toxoid
B. Immunize the mother with the toxoid

In case of A, the treatment can be given

a. Immediately after birth
b. After the onset of the condition

In case of B, the immunization has to be done

c. Before pregnancy
d. Late in he pregnancy

The correct combination is

(A) A/a (B) A/b
(C) B/c (D) B/d

127. Ecological compression differs from character displacement in that it operated on a:

(A) Shorter timescale and does not involve heritable change
(B) Longer timescale and does not involve heritable change
(C) Shorter timescale and involves heritable change
(D) Longer timescale and involves heritable change

GATE BIOTECHNOLOGY 2011

1. Embryonic stem cells are derived from
(A) fertilized embryo (B) unfertilized embryo
(C) sperm (D) kidney

2. Members of the antibody protein family that have common structural features are collectively known as
(A) haptens (B) allergens
(C) antigens (D) immunoglobulins

3. Apoptosis is characterized by
(A) necrosis
(B) programmed cell death
(C) membrane leaky syndrome
(D) cell cycle arrest process

4. Yeast artificial chromosomes (YAC's) are used for cloning
(A) large segments of DNA (B) mRNA
(C) bacterial DNA (D) yeast DNA

5. The product commercially produced by animal cell culture is
(A) insulin
(B) tissue plasminogen activator
(C) interferon
(D) hepatitis B vaccine

6. An alternative to glycolysis pathway is
(A) glyoxylate pathway
(B) pentose phosphate pathway
(C) citric acid cycle
(D) gluconeogenesis

7. A cell in G_1 of interphase has 12 chromosomes. How many chromatids will be found per cell during metaphase II of meiosis?
(A) 6 (B) 12
(C) 18 (D) 24

8. Diploid *Drosophila* has eight chromosomes. Which one of the following terms should NOT be used to describe *Drosophila* with sixteen numbers of chromosomes?
(A) Polyploid (B) Aneuploid
(C) Euploid (D) Tetraploid

9. Hydrated synthetic seeds which are produced by ion exchange reaction involve mixing the somatic embryos in a solution of
(A) sodium alginate and dropping it in a solution of calcium nitrate
(B) calcium alginate and dropping it in a solution of sodium nitrate
(C) calcium alginate and dropping it in a solution of ammonium nitrate
(D) mannitol and dropping it in a solution of sodium nitrate

10. Shoot organogenesis by tissue culture results into
(A) a bipolar structure that has no vascular connection with the explant
(B) a monopolar structure that has a strong connection with the pre-existing vascular tissue of the explant
(C) a monopolar structure that has no vascular connection with the explant
(D) a bipolar structure that has a strong connection with the pre-existing vascular tissue of the explant

11. 'Hairy roots' induced *in vitro* by the infection of *Agrobacterium rhizogenes*, are characterized by
P. a high degree of lateral branching
Q. genetic instability of culture
R. an absence of geotropism
S. poor biomass production
(A) P and R only (B) P and Q only
(C) Q and R only (D) R and S only

12. In balanced growth phase of a cell
P. all components of a cell grow at the same rate
Q. specific growth determined by cell number or cell mass would be the same
R. the growth rate is independent of substrate concentration
S. the growth rate decreases with decreasing substrate concentration
(A) P, Q and S only (B) Q, R and S only
(C) P, Q and R only (D) P only

13. In N-linked glycosylation, the oligosaccharide chain is attached to protein by
(A) asparagine (B) arginine
(C) serine (D) threonine

14. Restriction endonucleases which recognize and cut same recognition sequences are known as
(A) isoschizomers
(B) isozymes
(C) isoaccepting endonucleases (D) abzymes

15. Substrate consumption in lag phase of microbial growth is primarily used for
P. turn over of the cell material
Q. maintenance of intracellular pH
R. motility
S. increase in cell number
(A) P, Q and S only (B) Q, R and S only
(C) P, Q and R sonly (D) S only

16. Wash out (as defined by D = μ_{max}) of a continuous stirred tank fermenter is characterized by (X = biomass, S= substrate concentration in bioreactor, S_o = substrate concentration in feed, P = product concentration in bioreactor).
(A) X = 0, S = 0, P = 0
(B) X = 0, S = S_0, P = 0
(C) X = 0, S < 0, P = 0
(D) X < 0, S < 0, P < 0

17. The study of evolutionary relationships is known as
(A) genomics (B) proteomics
(C) phylogenetics (D) genetics

18. The lipopolysaccharides present in bacterial cell wall has lipid A which is connected to
(A) O-polysaccharide
(B) core polysaccharide
(C) both with O-polysaccharide and core polysaccharide
(D) rhamnose-mannose disaccharide

19. Molecular chaperones are class of proteins that facilitate
(A) the proper folding of newly synthesized proteins
(B) unfolding of newly synthesized proteins
(C) degradation of newly synthesized proteins
(D) targeting of newly synthesized proteins

20. Gas vacuoles are present in
(A) *Anabaena flos-aquae* (B) *Bacillus subtilis*
(C) *Acanthurus nigrofuscus*
(D) *Mycobacterium tuberculosis*

21. In ABO blood group system, antigenic determinants are
(A) mucleic acid (B) carbohydrate
(C) lipid (D) protein

22. The most widely used program for multiple sequence alignment is
(A) BLAST (B) FASTA
(C) CLUSTAL (D) Chime

23. Diphtheria toxin, tetracycline and streptomycin inhibit
(A) DNA repair (B) DNA replication
(C) transcription (D) translation

24. The polymorphic domain for Class II MHC proteins are
(A) α_1 and β_2 domains only
(B) β_1 and α_2 domains only
(C) α_1 and β_1 domains only
(D) α_2 and β_2 domains only

25. The protein in eukaryotes which is subjected to degradation undergoes
(A) phosphorylation (B) carboxylation
(C) ubiquitation (D) methylation

26. Match the viruses in Group I with their host cell receptors in Group II.

Group I	Group II
P. Hepatitis A virus	1. Heparan sulphate
Q. Human immunodef-iciency virus	2. Acetylcholine receptor
R. Rabies virus	3. CD4 protein
S. Herpes simplex virus type I	4. Alpha-2 macroglobulin

(A) P-1, Q-3, R-2, S-4 (B) P-3, Q-4, R-1, S-2
(C) P-4, Q-3, R-2, S-1 (D) P-2, Q-3, R-1, S-4

27. Match the microbial growth characteristics in Group I with the corresponding features in Group II.

Group I	Group II
P. Growth associated product formation	1. Specific growth rate decreases with increasing product concentration
Q. Non growth associated product formation	2. Specific product formation rate is constant
R. Product inhibition	3. Specific product formation rate is proportional to specific growth rate
S. Substrate inhibition	4. Specific growth rate decreases with increasing substrate concentration

(A) P-1, Q-2, R-4, S-3 (B) P-3, Q-2, R-1, S-4
(C) P-2, Q-1, R-3, S-4 (D) P-2, Q-3, R-4, S-1

28. Match the items in Group I with Group II.

Group I	Group II
P. Circular dichroism	1. Concentration
Q. X-ray crystallography	2. Sedimentation coefficient
R. Freeze-drying	3. Secondary structure determination
S. Ultracentrifugation	4. Tertiary structure determination

(A) P-4, Q-1, R-2, S-3 (B) P-1, Q-4, R-3, S-2
(C) P-2, Q-3, R-4, S-1 (D) P-3, Q-4, R-1, S-2

29. Match the product in Group I with theri respective organisms in Group II.

Group I	Group II
P. Glycerol	1. *Corynebacterium glutamicum*
Q. Glutamic acid	2. *Alcaligenes faecalis*
R. Curdlan	3. *Dunaliella salina*
S. Amphotericin B	4. *Streptomyces nodosus*

(A) P-2, Q-1, R-3, S-4 (B) P-4, Q-2, R-1, S-3
(C) P-3, Q-1, R-2, S-4 (D) P-2, Q-1, R-4, S-3

30. Determine the correctness or otherwise of the following **Assertion (a)** and the **Reason (r).**

Assertion : I_gM is found in serum as a pentameric protein consisting of five I_gM monomers.

Reason : The pentameric form of I_gM is due to cross-linking of I_gM monomers via peptide bond.

(A) both (a) and (r) are true and (r) is the correct reason for (a)
(B) both (a) and (r) are true but (r) is not the correct reason for (a)
(C) (a) is true but (r) is false
(D) (a) is false but (r) is true

31. Determine the correctness or otherwise of the following **Assertion** (a) and the **Reason** (r).

Assertion : N-methyl-N'-nitro-N-nitrosoguanidine (NTG) is an effective chemical mutagen.

Reason : Mutations induced by NTG mainly are the GC $\rightarrow$ AT transitions.

(A) both (a) and (r) are true and (r) is the correct reason for (a)
(B) both (a) and (r) are true but (r) is not the correct reason for (a)
(C) (a) is true but (r) is false
(D) (a) is false but (r) is true

32. Determine the correctness of the following statements

I. Enhancer sequences are those DNA sequences that are involved in increasing the rate of DNA replication.
II. Enhancer sequences work by binding with eukaryotic gene activator factors.

(A) only I is true
(B) only II is true
(C) both I and II are true
(D) both I and II are false

33. In a well aerated and agitated microbial culture, the `supply' of oxygen is equal to `demand' (uptake) of the growing culture. The K_La for such a system will be (K_La = volumetric mass transfer coefficient, C^* = dissolved oxygen concentration in liquid in equilibrium with gaseous oxygen, C = instantaneous value of dissolved oxygen concentration, `r' = specific oxygen uptake rate per unit weight of cells, X = dry weight of the cells per unit volume)

(A) (rX) / (C^*–C) (B) (r) /X (C^*–C)
(C) (C^*–C) / (rX) (D) (X) /r (C^*–C)

34. Structured William's model

P. can described the changes in intracellular components of the cell during growth
Q. can not describe the death phase of the cells
R. can describe the variation of size of cells in the different phases of growth
S. can not describe the lag period of growth

Which one of the following is CORRECT?

(A) P, Q and S only (B) P, Q and R only
(C) Q, R and S only (D) P, R and S only

35. Match items in Group I with Group II.

Group I	**Group II**
P. Glycolytic pathway	1. Chloroplast
Q. Eukaryotic oxidative metabolism	2. Glyoxysomes
R. Glyoxylate cycle	3. Mitochondria
S. Calvin cycle	4. Cytosol

(A) P-1, Q-2, R-3, S-4 (B) P-2, Q-3, R-4, S-1
(C) P-4, Q-3, R-2, S-1 (D) P-3, Q-4, R-1, S-2

36. Match items in Group I with Group II.

Group I	**Group II**
P. Alzheimer's disease	1. H1N1
Q. Mad cow disease	2. Hemoglobin
R. Sickle cell anaemia	3. Prions
S. Swine flu	4. Amyloid

(A) P-4, Q-3, R-2, S-1 (B) P-3, Q-4, R-1, S-2
(C) P-2, Q-1, R-4, S-3 (D) P-1, Q-2, R-3, S-4

37. Determine the correctness or otherwise of the following **Assertion** (a) and the **Reason** (r)

Assertion: The elucidation of ribosome structure helps in the development of new generation drugs.

Reason: The high resolution of macromolecular structure has enabled in structure-based drug design.

(A) both (a) and (r) are true and (r) is the correct reason for (a)
(B) both (a) and (r) are true but (r) is not the correct reason for (a)
(C) (a) is true but (r) is false
(D) (a) is false but (r) is true

38. Determine the correctness or otherwise of the following **Assertion** (a) and the **Reason** (r)

Assertion: A very low amount of inhibitor can act as an activator for allosteric enzymes

Reason: Allosteric enzymes follow Michaelis-Menten kinetics.

(A) both (a) and (r) are true and (r) is the correct reason for (a)
(B) both (a) and (r) are true but (r) is not the correct reason for (a)
(C) (a) is true but (r) is false
(D) (a) is false but (r) is true

39. Match the terms in **Group I** with their associated functions in **Group II.**

Group I	**Group II**
P. Shine-Dalgarno sequences	1. Aminoacylation of tRNA
Q. Leucine zipper	2. Gene silencing
R. Aminoacyl tRNA synthetase	3. Ribosome binding and facilitation of translation initiation
S. RNA inerference (RNAi)	4. Transcription factors

(A) P-3, Q-4, R-1, S-2
(B) P-4, Q-3, R-2, S-1
(C) P-2, Q-3, R-1, S-4
(D) P-3, Q-2, R-4, S-1

40. Protein-protein interactions are studied by

P. DNA foot printing
Q. Yeast two hybrid system
R. Ligase chain reaction
S. Mass specrometry

(A) P and S only
(B) Q and S only
(C) P and R only
(D) Q and R only

41. Determine the correctness or otherwise of the following **Assertion** (a) and the **Reason** (r)

Assertion: Isopropylthiogalactoside (IPTG) is a gratuitous inducer of lactose operon.

Reason: Gratuitous inducers are chemical analogs which behave like natural inducer but they do not serve as substrate for the enzymes that are subsequently synthesized.

(A) both (a) and (r) are true and (r) is the correct reason for (a)
(B) both (a) and (r) are true but (r) is not the correct reason for (a)
(C) (a) is true but (r) is false
(D) (a) is false but (r) is true

42. Determine the corectness or otherwise of the following **Assertion** (a) and the **Reason** (r)

Assertion: In synchronous culture, majority of the cells move to next phase of the cell cycle simultaneously.

Reason: Synchronous culture could be obtained by starving cells for essential nutrient components.

(A) both (a) and (r) are true and (r) is the correct reason for (a)
(B) both (a) and (r) are true but (r) is not the correct reason for (a)
(C) (a) is true but (r) is false
(D) (a) is false but (r) is true

43. Which of the following characteristics with respect to bacterial DNA polymerase III are **True?**
P. Initiation of chain synthesis
Q. 5'-3' polymerization
R. 3'-5' exonuclease activity
S. 5'-3' exonuclease activity

(A) P and Q only (B) Q and R only
(C) Q and S only (D) P and S only

44. Maximum specific growth rate (μ_{max}) of a microorganism is calculated by taking the (ln =log$_e$, X =biomass, t = time)

(A) slope of ln X vs t of the growth cycle
(B) slope of ln X vs t during the exponential growth phase
(C) slope of X vs t
(D) slope of X vs t during the exponential phase of growth

45. Identify the **CORRECT** statements
P. 5' and 3' ends of transcripts can be mapped by utilizing polymerase chain reaction
Q. S_1 nuclease can cleave the DNA strand of a DNA-RNA hybrid
R. T_4 polynucleotide kinase is used for labeling 3' end of DNA
S. Baculovirus (*Autographa californica*) can be used as an insect expression vector

(A) P and Q only (B) R and S only
(C) P and S only (D) Q and R only

46. Value of the determinant mentioned below is:

$$\begin{vmatrix} 1 & 0 & -1 & 0 \\ 4 & 7 & 0 & 2 \\ 1 & 1 & -1 & 1 \\ 2 & 0 & 2 & 1 \end{vmatrix}$$

(A) 24 (B) -30
(C) -24 (D) -10

47. HAT (hypoxanthine, aminopterin and thymidine) is used for selecting the hybridomass based on the following:
I. Only hybridoma will grow since it inherited the HGPRT genes from B-cells and can synthesize DNA from hypoxanthine.
II. Myeloma cells will not grow in cultures since *de novo* synthesis is blocked by aminopterin and due to the lack of HGPRT enzyme.

(A) only I is true
(B) only II is true
(C) both I and II are true
(D) I is true and II is false

Common Data Questions

Common Data for Questions 48 and 49:

Red-green colour blindness is inherited as a recessive X-lined trait.

48. What will be the probability of having the colour-bline son to a soman with phenotypically normal parents and a colour-bline brother, and married to a normal man? (Assume that she has no previous children)

(A) 100% (B) 50%
(C) 25% (D) 12.5%

49. What will be the probablity of having the colour-blind daughter to a phenotypically normal woman, who already had one colour-blind son, and is married to a colour-blind man?

(A) 75% (B) 50%
(C) 25% (D) 15%

Common Data for Questions 50 and 51:

50. What would be yield $Y_{X/S}$ (Kg biomass/Kg subsrate)?

(A) 0.562 (B) 0.462
(C) 0.362 (D) 0.162

51. What would be the sucrose concentration in the input feed for the output to be 45 Kg biomass h^{-1}?

(A) 3.225 Kg m^{-3} (B) 4.425 Kg m^{-3}
(C) 5.115 Kg m^{-3} (D) 6.525 Kg m^{-3}

Linked Answer Questions

Statement for Linked Answer Questions 52 and 53:

The abdomen length (in milimeters) was measured in 15 male fruit flies and the following data were obtained : 1.9, 2.4, 2.1, 2.0, 2.4, 1.7, 1.8, 2.0, 2.0, 2.3, 2.1, 1.6, 2.3 and 2.2.

52. Variance (V_X) for this population of fruit flies as calculated from the above data shall be

(A) 0.85 (B) 0.25
(C) 0.061 (D) 0.08

53. The value of standard deviation (SD) will be

(A) 0.061 (B) 0.25
(C) 0.61 (D) 0.85

Statement for Linked Answer Questions 54 and 55:

A 200 μl of polymerase chain reaction has 100 template DNA molecules and the reaction was performed for 10 cycles.

54. How many molecules of amplicons will be generated?

(A) 1.025×10^4 (B) 1.025×10^5
(C) 2.048×10^4 (D) 2.048×10^5

55. How many molecules of amplicons will be present in 0.1 μl of reaction?

(A) 102.4 (B) 1024
(C) 51.2 (D) 512

GATE BIOTECHNOLOGY, 2012

1. In mismatch correction repair, the parental DNA strand is distinguished from the daughter strand by

(A) acetylation (B) phosphorylation
(C) methylation (D) glycosylation

2. The basic for blue-white screening with pUC vector is:

(A) intraallelic complementation
(B) intergenic omplementation
(C) intragenic suppression
(D) extragenic suppression

3. Idiotypic determinants of an antibody are associated with the:

(A) constant region of the heavy chains
(B) constant region of the light chains
(C) variable region
(D) constant regions of light and heavy chains

4. Identification of blood groups involves:

(A) precipitation (B) neutralization
(C) opsonization (D) agglutination

5. B-lymphocytes originate from the bone marrow whereas T-lymphocytes originate from

(A) thymus (B) bone marrow
(C) spleen (D) liver

6. A humanized antibody is one in which the:

(A) heavy and light chains are from human
(B) heavy chain is from human and light chain is from mouse
(C) light chain is from human and heavy chain is from mouse
(D) CDRs are from mouse, and the rest is from human

7. Dimethyl sulfoxide (DMSO) is used as a cryopreservant for mammalian cell cultures because:

(A) it is an organic solvent
(B) it easily penetrates cells
(C) it protects cells by preventing crystallization of water
(D) it is also utilized as a nutrient

8. Nude mice referes to:

(A) mice without skin
(B) mice without thymus
(C) knockout mice
(D) transgenic mice

9. Heat inactivation of serum is done to inactivate:

(A) prions
(B) mycoplasma
(C) complement
(D) pathogenic bacteria

10. Choose the correct signal transduction pathway:

(A) Hormone → 7 TM receptor → G protein → cAMP → PKA
(B) Hormone → G protein → 7 TM receptor → cAMP → PKA
(C) Hormone → 7 TM receptor → G protein → PKA → cAMP
(D) Hormone → 7 TM receptor → cAMP → G protein → PKA

11. A protein is phosphorylated at a serine residue. A phosphomimic mutant of the protein can be generated by substituting that serine with

(A) glycine
(B) alanine
(C) aspartate
(D) threonine

12. A truncated polypeptide is synthesized due to a nonsense mutation. Where would you introduce another mutation to obtain a full-length polypeptide?

(A) Ribosomal protein gene
(B) Transfer RNA gene
(C) DNA repair gene
(D) Ribosomal RNA gene

13. Protein-DNA interactions *in vivo* can be studied by:

(A) gel shift assay
(B) Southern hybridization
(C) chromatin immunoprecipitation assay
(D) fluorescence *in situ* recombination

14. The direction of shell coiling in the snail *Limnaea peregra* is a classic example of:

(A) chromosomal inheritance
(B) extra-chromosomal inheritance
(C) chromosomal translocation
(D) homologous recombination

15. During photorespiration under low CO_2 and high O_2 levels. O_2 reacts with ribulose 1,5-bisphosphate to yield

(A) one molecule each of 3-phosphoglycerate and 2-phosphoglycolate
(B) two molecules of 3-phosphoglycerate
(C) two molecules of 2-phosphoglycolate
(D) one molecule each of 3-phosphoglycerate and glyoxylate

16. Whch one of the following is **NOT** a protoplast fusion inducing agent?
(A) Inactivated Sendai virus
(B) Ca^{2+} at alkaline pH
(C) Polyethylene glycol (D) Colchicine

17. The activity of an enzyme is expressed in International Units (IU). However, the S.I. unit for enzyme activity is Katal. One Katal is:
(A) 1.66 x 10^4 IU (B) 60 IU
(C) 6 x 10^7 IU (D) 10^6 IU

18. Identify the statement that is **NOT** applicable to an enzyme catalyzed reaction.
(A) Enzyme catalysis involves propinquity effects.
(B) The binding of substrate to the active site cause a strain in the substrate
(C) Enzymes do not accelerate the rate of reverse reaction
(D) Enzyme catalysis involves acid-base chemistry

19. An example of a derived protein structure database is:
(A) Pfam (B) SCOP
(C) GEO (D) Prosite

20. An example of a program for constructing a phylogenetic tree is:
(A) phylip (B) phrap
(C) prodom (D) PHDsec

21. Synteny refers to:
(A) gene duplication from a common ancestor
(B) a tree representation of related seqeunces
(C) the extent of similarity between two sequences
(D) local conservation of gene order

22. While searching a database for similar seqeunces, E value does **NOT** depend on the
(A) sequence length
(B) number of sequences in the database
(C) scoring system
(D) probability from a normal distribution

23. In transmission electron microscopy, electron opacity is greatly enhanced by treating the specimen with
(A) ferrous ammonium sulfate
(B) uranium acetate
(C) sodium chloride
(D) basic fuchsin

24. The molarity of water in a water : ethanol mixture (15 : 85, v/v) is approximately
(A) 0.85 (B) 5.55
(C) 8.5 (D) 55.5

25. The helix content of a protein can be determined using:
(A) an infrared spectrometer
(B) a fluorescence spectrometer
(C) a circular dichroism spectrometer
(D) a UV-Visible spectrophotomete

Q. 26 to Q. 55 carry two marks each.

26. Which one of the following DNA sequences carries an invert repeat?
(A) ATGAGCCCCGAGTA
TACTCGGGGCTCAT
(B) ATGAGCCGGCTCTA
TACTCGGCCGAGAT
(C) ATGAGCCGAGCCTA
ACTCGGCTCGGAT
(D) ATGAGCCTATGGTA
TACTCGGATACCAT

27. In zine finger proteins, the amino acid residues that coordinate zinc are
(A) Cys and His (B) Asp and Glu
(C) Arg and Lys (D) Asp and Arg

28. Match the entries in **Group I** with those in **Group II**

Group I	**Group II**
P. MTT	1. Dihydrofolate reductase
Q. Annexin V	2. Succinate dehydrogenase
R. Methotrexate	3. Microtubules
S. Taxol	4. Phosphatidylserine

(A) P-3, Q-1, R-4, S-2 (B) P-2, Q-4, R-1, S-3
(C) P-2, Q-3, R-4, S-1 (D) P-4, Q-2, R-1, S-3

29. In an exponentially growing batch culture of *Saccharomyces cerevisiae,* the cell density is 20 gl^{-1} (DCW), the specific growth rate (μ) is 0.4 h^{-1} and substrate uptake rate (v) is 16 $gl^{-1}h^{-1}$. The cell yield coefficient $Y_{x/s}$ will be
(A) 0.32 (B) 0.64
(C) 0.80 (D) 0.50

30. A single base pair of DNA weights 1.1 x 10^{-21} grams. How many picomoles of a plasmid vector of length 2750 bp are contained in 1 μg of purified DNA?
(A) 0.30 (B) 0.55
(C) 0.25 (D) 0.91

31. Match the terms in **Group I** with the ploidy in **Group II**

Group I	**Group II**
P. Disome	1. 2n + 1
Q. Monosome	2. 2n - 1
R. Nullisome	3. n - 1
S. Trisome	4. n + 1

(A) P-4, Q-2, R-3, S-1 (B) P-4, Q-3, R-1, S-2
(C) P-2, Q-2, R-4, S-1 (D) P-1, Q-4, R-3, S-2

32. What is the rank of the following matrix ?

$$\begin{pmatrix} 5 & 3 & -1 \\ 6 & 2 & -4 \\ 14 & 10 & 0 \end{pmatrix}$$

(A) 0 (B) 1
(C) 2 (D) 3

33. Match the products in **Group I** with the applications in **Group II.**

Group I	**Group II**
P. Digoxin	1. Muscle relaxant
Q. Stevioside	2. Anti-cancer agent
R. Atropine	3. Cardiovascular disorder
S. Vinblastine	4. Sweetener

(A) P-1, Q-4, R-3, S-2 (B) P-3, Q-2, R-1, S-4
(C) P-3, Q-4, R-1, S-2 (D) P-2, Q-3, R-1, S-4

34. Determine the correctness or otherwise of the following **Assertion** (a) and **Reason** (r).

Assertion: The production of secondary metabolites in plant cell cultures is enhanced by the addition of elicitors.

Reason: Elicitors induce the expression of enzymes responsible for the biosynthesis of secondary metabolites.

(A) both (a) and (r) are true but (r) is the correct reason for (a)
(B) both (a) and (r) are true and (r) is not the correct reason for (a)
(C) (a) is true but (r) is false
(D) (a) is false but (r) is true

35. Determine the correctness or otherwise of the following **Assertion** (a) and **Reason** (r).

Assertion: Plants convert fatty acid into glucose

Reason : Plants have peroxisomes

(A) both (a) and (r) are true but (r) is the correct reason for (a)
(B) both (a) and (r) are true and (r) is not the correct reason for (a)
(C) (a) is true but (r) is false
(D) (a) is false but (r) is true

36. Determine the correctness or otherwise of the following **Assertion** (a) and **Reason** (r)

Assertion: In direct somatic embryogenesis, embryos are developed without going through callus formation

Reason: This is possible due to the presence of pre-embryonically determined cells.

(A) both (a) and (r) are true but (r) is the correct reason for (a)
(B) (a) is false but (r) is true
(C) (a) is true but (r) is false
(D) both (a) and (r) are true and (r) is not the correct reason for (a)

37. Match the entries in **Group I** with the process parameters in **Group II**

Group I	**Group II**
P. Clark electrode	1. Liquid level
Q. Redox probe	2. Dissolved oxygen concentration
R. Load cell	3. Vessel pressure
S. Diaphragm gauge	4. pH (anaerobic process)

(A) P-2, Q-1, R-3, S-4 (B) P-4, Q-2, R-3, S-1
(C) P-2, Q-4, R-1, S-3 (D) P-2, Q-1, R-4, S-3

38. Match the downstream processes in **Group I** with the products in **Group II**

Group I	**Group II**
P. Solvent extraction	1. Lacetic acid
Q. Protein-A linked affinity chromatography	2. Penicillin
R. Extractive distillation	3. Monoclonal antibody
S. Salting out	4. Lipase

(A) P-2, Q-3, R-1, S-4 (B) P-4, Q-1, R-2, S-3
(C) P-4, Q-1, R-3, S-2 (D) P-2, Q-4, R-1, S-3

39. Determine the correctness or otherwise of the following **Assertion (a)** and **Reason (r).**

Assertion: Cell mass yield of a methylotrophic yeast is more on methanol compared to glucose.

Reason: Methanol has a greater degree of reductance compared to glucose.

(A) both (a) and (r) are correct and (r) is the correct reason for (a)
(B) (a) is false but (r) is true
(C) (a) is true but (r) is false
(D) both (a) and (r) are correct but (r) is not the correct reason for (a)

40. A disease is inherited by a child with a probability of 1/4. In a family with two children, the probability that exactly one sibling is affected by this disease is

(A) 1/4 (B) 3/8
(C) 7/16 (D) 9/16

41. Match the organisms in **Group I** with the entries in **Group II**

Group I	**Group II**
P. *Clostridium*	1. Rods with teichoic acid in the cell wall
Q. *Escherichia*	2. Rods with endospores
R. *Vibrio*	3. Helical rods with flagella
S. *Bacillus*	4. Rods with LPS in the outer membrane
	5. Curved rods with polar flagella

(A) P-2, Q-4, R-5, S-1 (B) P-2, Q-1, R-5, S-4
(C) P-2, Q-4, R-2, S-3 (D) P-3, Q-2, R-1, S-4

42. Match the entries in **Group I** with the methdos of sterilization in **Group II**

Group I	**Group II**
P. Serum	1. Autoclave
Q. Luria broth	2. Membrane filtration
R. Polypropylene tubes	3. UV irradiation
S. Biological safety cabinets	4. Gamma irradiation
	5. Dry heat

(A) P-5, Q-3, R-1, S-4
(B) P-1, Q-4, R-5, S-3
(C) P-2, Q-1, R-4, S-3
(D) P-4, Q-1, R-3, S-5

43. Match the high energy compounds in **Group I** with the biosynthetic pathways for the molecules in **Group II**

Group I	**Group II**
P. GTP	1. Fatty acid
Q. UTP	2. Phospholipid
R. CTP	3. Protein
S. Acyl coenzyme	4. Peptidoglycan

(A) P-3, Q-2, R-4, S-1 (B) P-2, Q-4, R-3, S-1
(C) P-4, Q-3, R-1, S-2 (D) P-3, Q-4, R-2, S-1

44. Match the vitamins in **Group I** with the processes/reactions in **Group II**

Group I	**Group II**
P. Pantothenic acid	1. Electron transport
Q. Vitamin B2	2. Transfer of 1-C units
R. Vitamin B6	3. Decarboxylation
S. Folic acid	4. Fatty acid metabolism
	5. Hydrolysis

(A) P-5, Q-2, R-4, S-1 (B) P-4, Q-1, R-3, S-2
(C) P-4, Q-2, R-1, S-3 (D) P-2, Q-1, R-3, S-5

45. Consider the data set 14, 18, 14, 14, 10, 29, 33, 31, 25. If you add 20 to each of the values, then
(A) both mean and variance change
(B) both mean and variance are unchanged
(C) the mean is unchanged, variance changes
(D) the mean changes, the variance is unchanged

46. An enzymatic reaction is described by the following rate expression.

$$v = \frac{v_m s}{k_m + s + s^2/k_s}$$

which one of the following curves represents this expression?

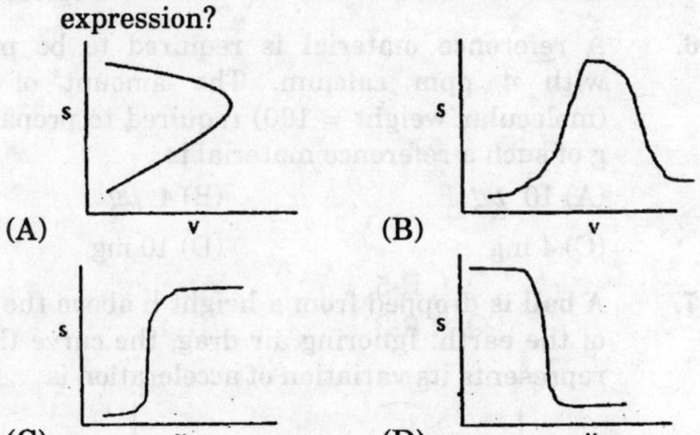

47. A bacterial culture (200 μl containing 1.8 x 19^9 cells) was treated with an antibiotic Z (50 μg per ml) for 4 h at 37ºC. After this treatment, the culture was divided into two equal aliquots.
Set A : 100 μl was plated on Luria agar.
Set B : 100 μl was centrifuged, the cell pellet washed and plated on Luria agar.
After incubating these two plates for 24 h at 37ºC, Set A plate showed no colonies, whereas the set B plate showed 0.9 x 10^9 cells. This experiment showed that the antibiotic Z is
(A) bacteriostatic (B) bacteriocidal
(C) bacteriolytic (D) apoptoic

Common Data Questions

Common Data for Questions 48 and 49:

In a muscle, the extracellular and intracellular concentrations of Na^+ are 150 mM and 12mM, and those of K^+ are 2.7 mM and 140 mM, respectively. Assume that the temperature is 25ºC and that the membrane potential is -60 mV, with the interior more negatively charged than the exterior. (R = 8.314 J mol^{-1} K^{-1}; F=96.45 kJ mol^{-1} V^{-1})

48. The free energy change for the transport of three Na^+ out of the cell is:
(A) +1.5 kJ/mol (B) +17.4 kJ/mol
(C) +18.9 kJ/mol (D) +36.3 kJ/mol

49. The free energy change for the transport of two K^+ into the cell is:
(A) +8.0 kJ/mol (B) +11.6 kJ/mol
(C) +19.6 kJ/mol (D) +31.2 kJ/mol

Common Data for Questions 50 and 51:

The purification data for an enzyme is given below:

	Step	Volume (ml)	Total protein (mg)	Total activity (Units)	Specific activity (Units/mg)
P	Cell-free extract	17	177	102	0.58
Q	Q-Sepharose	14	18.8	72	3.83
R	Phenyl Sepharose	26	9.2	45	4.89
S	Sephacryl S-200	7	4.1	30	.32

50. The fold purification for each step is
(A) P-0.1, Q-0.66, R-0.84, S-1.26
(B) P-1.0, Q-0.52, R-0.67, S-0.8
(C) P-1, Q-6.6, R-8.4, S-12.6
(D) P-100, Q-66, R-84, S-12

51. The yield (%) for each step is
(A) P-10, Q-7.2, R-4.5, S-2.0
(B) P-34, Q-24, R-15, S-1
(C) P-3.4, Q-2.4, R-1.5, S-0.1
(D) P-100, Q-71, R-44, S-29

Linked Answer Questions

Statement for Linked Answer Questions 52 and 53:

52. The molar concentration of the repressor in the cell is
(A) 0.1 nM (B) 1 nM
(C) 10 nM (D) 100 nM

53. Therefore, the lac-operon is
(A) repressed and can only be induced with lactose
(B) repressed and cannot be induced with lactose
(C) not repressed
(D) expressed only when glucose and lactose are present

Statement for Linked Answer Questions 54 and 55:

b-Galactosidase bound to DEAE-cellulose is used to hydrolyze lactose to glucose and galactose in a plug flow bioreactor with a packed bed of volume 100 litres and a voidage (e) of 0.55. The K_m and V_{max} for the immobilized enzyme are 0.72 gl^{-1} and 18 gl^{-1} h^{-1}, respectively. The lactose concentration in the field stream is 20 gl^{-1}, and a fractional conversion of 0.90 is desired. Diffusional limitations may be ignored.

54. **The residence time required for the steady state reactor operation will be**

(A) 0.1 h (B) 0.4 h
(C) 1.0 h (D) 1.1 h

55. **The feed flow rate required for the above bioconversion will be**

(A) 50 lh^{-1} (B) 55 lh^{-1}
(C) 137 lh^{-1} (D) 550 lh^{-1}

CSIR JRF/NET Life Science, June 2011

PART - A

1. **If V_{input} is applied to the circuit shown, the output would be**

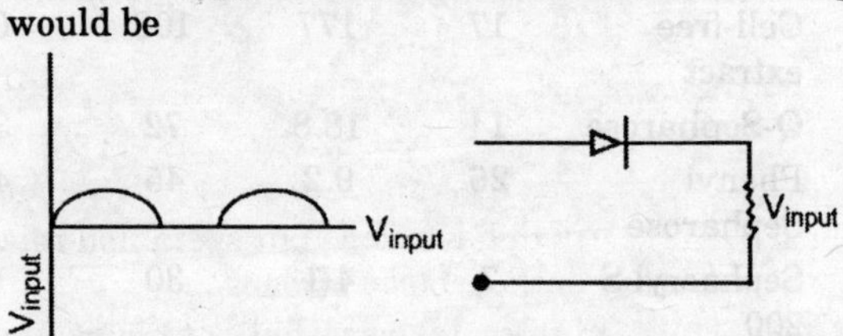

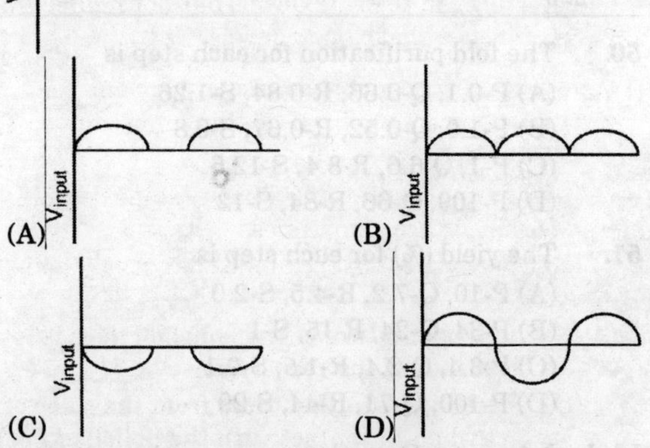

2. **A physiological disorder X always leads to the disorder Y. However, disorder Y may occur by itself. A population shows 4% incidence of discorder Y. Which of the following inference is valid.**

(A) 4% of the population suffers from both X & Y
(B) Less than 45 of the population suffers from X
(C) At least 4% of the population suffers from X
(D) There is no incidence of X in the given population

3. **The speed of a car increases every minute as shown in the following Table. The speed at the end of the 19th minute would be**

Time (minutes)	Speed (m/sec)
1	1.5
2	3.0
3	4.5
.	.
.	.
24	36.0
25	37.5

(A) 26.5 (B) 28.0
(C) 27.0 (D) 28.5

4. **How many σ bounds are present in the following molecule?**

$HC \equiv CCH = CHCH_3$

(A) 4 (B) 6
(C) 10 (D) 13

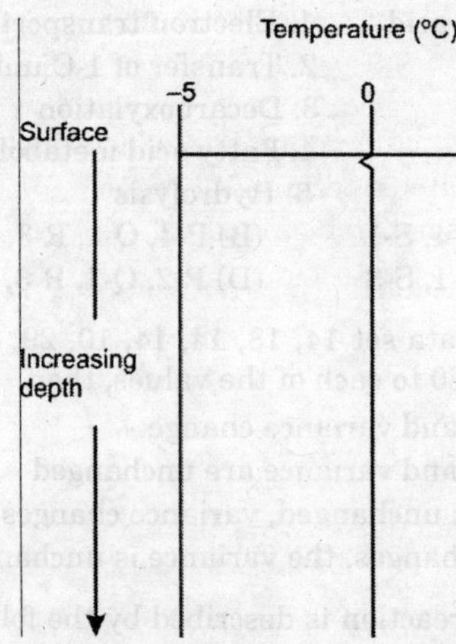

5. **The graph represents the depth profile of temperature in the open ocean; m which region this is likely to be prevalent?**

(A) Tropical region (B) Equatorial region
(C) Polar region (D) Sub-tropical region

6. **A reference material is required to be prepared with 4 ppm calcium. The amount of $CaCO_3$ (molecular weight = 100) required to prepare 1000 g of such a reference material is**

(A) 10 μg (B) 4 μg
(C) 4 mg (D) 10 mg

7. **A ball is dropped from a height h above the surface of the earth. Ignoring air drag, the curve that best represents its variation of acceleration is**

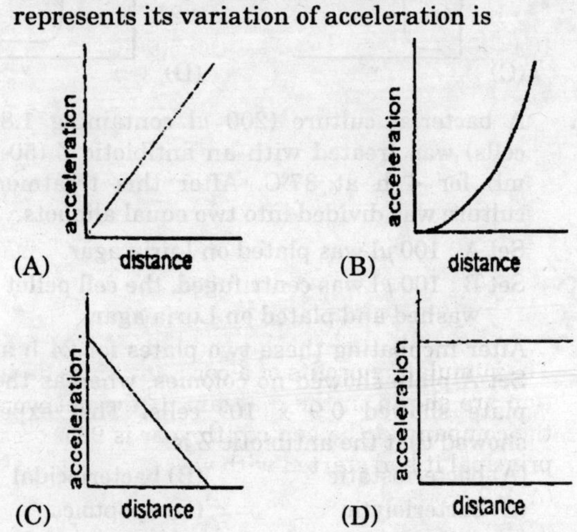

8. If the atmospheric concentration of carbon dioxide is doubled and there are favourable conditions of water, nutrients, light and temperature, what would happen to water requirement of plants?

(A) It decreases initially for a short time and then returns to the original value
(B) It increases
(C) It decreases
(D) It increases initially for a short time and then returns to the original value

9. Exposing an organism to a certain chemical can change nucleotide bases in a gene, causing mutation. In one such mutated organism if a protein had only 70% of the primary amino acid sequence, which of the following is likely?

(A) Mutation broke the protein
(B) The organism could not make amino acids
(C) Mutation created a terminator codon
(D) The gene was not transcribed

10. The reason for the hardness of diamond is

(A) extended covalent bonding
(B) layered structure
(C) formation of cage structures
(D) formation of tubular structures

11. The acidity of normal rain water is due to

(A) SO_2 (B) CO_2
(C) NO_2 (D) NO

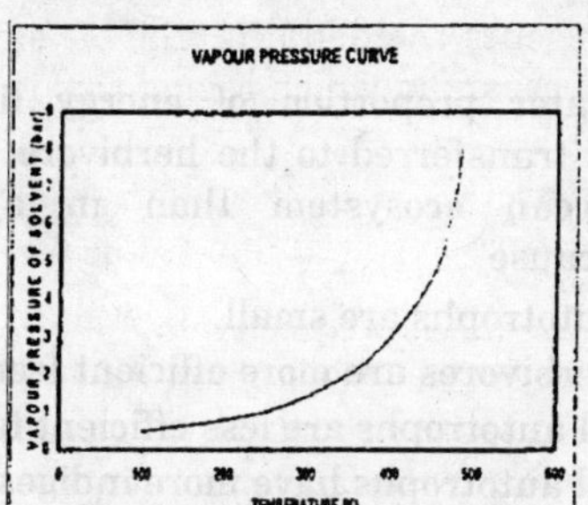

12. The normal boiling point of a solvent (whose vapour pressure curve is shown in the figure) on a planet whose normal atmospheric pressure is 3 bar, is about

(A) 100 K (B) 273 K
(C) 400 K (D) 500 K

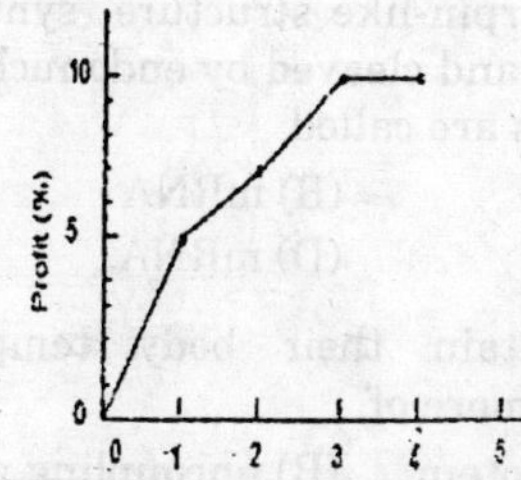

13. The cumulative profits of a company since its inception are shown in the diagram. If the net worth of the company at the end of 4th year is 99 crores, the principal it had started with was

(A) 9.9 crores (B) 91 crores
(C) 90 crores (D) 9.0 crores

14. Which of the following particles has the largest range in a given medium if their initial energies are the same?

(A) alpha (B) electron
(C) positron (D) gamma

15. Water is dripping out of a tiny hole at the bottom of three flasks whose, base diameter is the same, and are initially filled to the same height, as shown

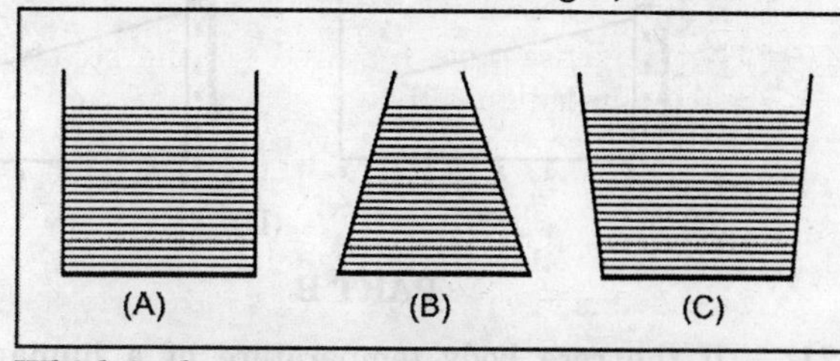

Which is the correct comparison of the rate of fall of the volume of water in the three flasks?

(A) A fastest, B slowest
(B) B fastest, A slowest
(C) B fastest, C slowest
(D) C fastest, B slowest

16. Diabetic patients are advised a low glycaemic index diet. The reason for this is

(A) The require less carbohydrate than healthy individuals
(B) They cannot assimilate ordinary carbohydrates
(C) They need to have slow, but sustained release of glucose in their blood stream
(D) They can tolerate lower, but not higher than normal blood sugar levels

17. Standing on a polished stone floor one feels colder than on a rough floor of the same stome. This is because

(A) Thermal conductivity of the stone depends on the surface smoothness
(B) Specific heat of the stone changes by polishing it
(C) The temperature of the polished floor is lower than that of the rough floor
(D) There is greater heat loss from the soles of the feet when in contact with the polished floor than with the rough floor

18. Popular use of which of the following fertilizers increases the acidity of soil?

(A) Potassium Nitrate
(B) Urea
(C) Ammonium sulphate
(D) Superphosphate of lime

19. Glucose molecules diffuse across a cell of diameter d in time τ. If the cell diameter is tripled, the diffusion time would

(A) increase to 9τ
(B) decrease to $\tau/3$
(C) increase to 3τ
(D) decrease to $\tau/9$

20. Identify the figure which depicts a first order reaction.

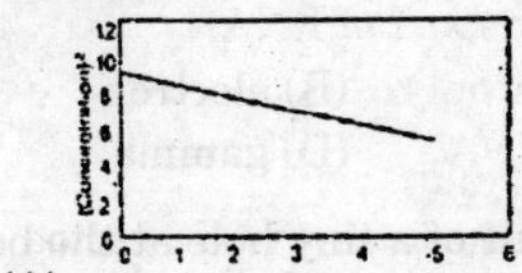

(A)

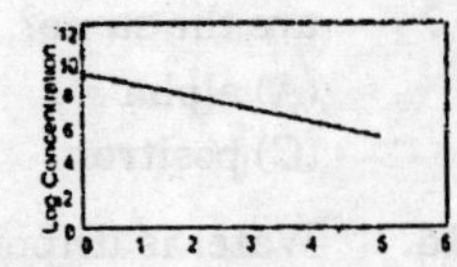

(B)

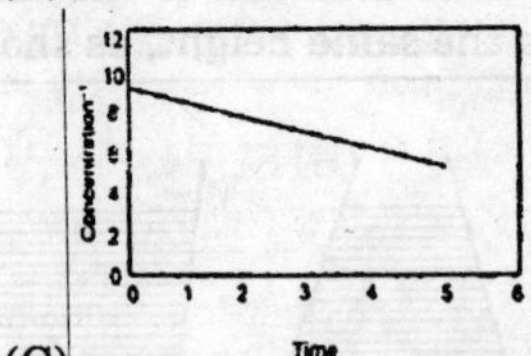

(C)

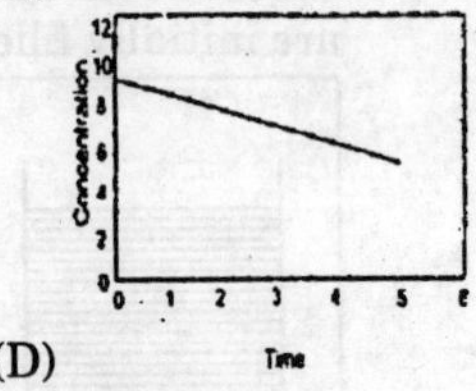

(D)

PART B

21. If the core body temperature of a human rises above normal, which of the following processes would be initiated sequentially for Thermo-regulation?

(A) Peripheral vasodilation, increased rate of respiration, tachycardia
(B) Peripheral vasoconstriction, increased rate of respiration, bradycardia.
(C) Peripheral vasodilation, decreased rate of respiration, tachycardia
(D) Peripheral vasodilation, decreased rate of respiration, bradycardia.

22. In India, brown antlered deer (sangai) is found only in the floating landmasses of

(A) Wular lake. (B) Sasthamkotta lake.
(C) Dal lake. (D) Lok Tak lake.

23. Graves disease is associated with

(A) insufficiency of thyroid hormones.
(B) excess of thyroid hormones.
(C) insufficiency of corticosteroids.
(D) excess of growth hormones.

24. What is the minimum number of NTPs required for the formation of one peptide bond during protein synthesis?

(A) One (B) Two
(C) Four (D) Six

25. The photoreceptor commonly involved in light

(A) phytochrome (B) rhodopsin
(C) carotenoid (D) cryptochrome

26. The free energy ΔG of a dissolved solute

(A) increases with solute concentration
(B) decreases with solute concentration
(C) is independent of solute concentration
(D) depends only on temperature

27. Which of the following is **not** a characteristic of phylum Chordata?

(A) Pharyngeal slits (B) Amniotic egg
(C) Postanal tail (D) Notochord

28. With which protein of *Yersnia* would integrin proteins of mammalian cells interact for internalization

(A) Pilin (B) Fimbrin
(C) Invasin (D) Adherin

29. \`Imperfect fungi' is a group represented by fungal species which have

(A) simple mycelia
(B) no known mechanisms of sexual reproduction
(C) unknown phylogenetic relationship
(D) lost its survival mechanism against harsh environment

30. The fidelity of replicative base selection can be reduced by a factor of 10^2 when the repair of DNA synthesis involves

(A) AP endonuclease. (B) ABC exonuclease.
(C) DNA photolyase.
(D) TLS DNA polymerase.

31. Both halophytes and glycophytes compartmentalize cytotoxic ions into the intracellular compartment or actively pump them out of the cell to the apoplasts with the help of membrane transport proteins. Among these, the Na^+ - H^+ antiporter, NHX1, is localized in the

(A) plasma membrane.
(B) chloroplast (inner envelope).
(C) mitochondria (outer membrane)
(D) tonoplast.

32. A much greater proportion of energy fixed by autotrophs is transferred to the herbivore level in the open ocean ecosystem than in a forest ecosystem because

(A) aquatic autotrophs are small.
(B) aquatic herbivores are more efficient feeders.
(C) terrestrial autotrophs are less efficient feeders.
(D) terrestrial autotrophs have more indigestible tissues.

33. The area of allowed regions in the Ramachandran map will be least for

(A) Gly. (B) L-Ala.
(C) L-Prol. (D) α- methyl L-valine.

34. Small RNAs with internally complementary sequences that form hairpin-like structure, synthesized as percursor RNAs and cleaved by endonucleases to form short duplexes are called

(A) snRNA. (B) mRNA.
(C) tRNA. (D) mRNA.

35. Poar bears maintain their body temperature because they have more of

(A) transduction protein. (B) uncoupling protein.
(C) myoglobin protein. (D) F_oF_1 ATPase.

36. During generation of an action potential, depolarization is due to

(A) K^+ efflux. (B) Na^+ efflux.
(C) Na^+ influx. (D) K^+ influx.

37. **Transport of water across aquaporins is regulated by the presence of which of the following sequence of three highly conserved amino acids?**

(A) Ala-Asn-Pro. **(B) Pro-Asn-Ala.**
(C) Asn-Pro-Ala. **(D) Pro-Ala-Asn.**

38. **Which of the cyclins have/has essential function in S-phase of cell cycle?**

(A) A-type. **(B) B-type.**
(C) D-type. **(D) Both B- & D-types.**

39. **A mechanism that can cause a gene to move from one linkage group to another is**

(A) crossing over. **(B) inversion.**
(C) translocation. **(D) duplication.**

40. **Release of nutrients, oxidants or electron donors into the environment to stimulate naturally occurring microorganisms to degrade a contaminant, is referred to as**

(A) biostimulation. **(B) phytoremediation.**
(C) bioaugmentation. **(D) bioremediation.**

41. **Th2 response is generated and maintained mainly by which of the following pair of cytokines?**

(A) IL-4 and 1L-10. **(B) IL-12 and IFN-γ.**
(C) IFN-γ and TNF-α. **(D) IL-2 and IL-12.**

42. **G protein-linked receptors are transmembrane proteins of**

(A) single-pass. **(B) three-pass.**
(C) five-pass. **(D) seven-pass.**

43. **In amphibian oocyte, the germplasm which gets segregated during cleavage to give rise to primordial germ cells (PGC's) is normally**

(A) distributed evenly throughout the oocyte.
(B) localized at animal pole.
(C) localized at vegetal pole.
(D) aggregated in central part of oocyte.

44. **Which of the following statements with respect to alternate oxidase activity in cyanide-resistant respiration in plants, is not correct?**

(A) Alterante oxidase accepts electrons dicretly from cytochrome C.
(B) Some plants exhibit thermogenesis during inflorescene development.
(C) Transcription of alternate oxidase gene is often induced by various abiotic stresses.
(D) When electrons pass to alternate oxidase, two sites of proton pumping are bypassed.

45. **In mature *Arabidopsis* embryo, root apical meristem consists of cells derived from**

(A) embryo and apical suspensor cell
(B) embryo only
(C) suspensor only **(D) hypophysis only**

46. **Which of the following molecules is involved in Ca^{2+} - dependent cell-cell adhesion?**

(A) Calmodulin **(B) Cadherin**
(C) N-CAM **(D) Calpain**

47. **The 5' Cap of RNA is required for the**

(A) stability of RNA only.
(B) stability and transport of RNA.
(C) transport of RNA only.
(D) methylation of RNA.

48. **Yeast artificial chromosome (YAC) vectos contain selectable markers. Loss of which marker at the cloning site distinguishes the religated YACs from the original vector marker?**

(A) TRP1 **(B) SUP4**
(C) URA3 **(D) CEN**

49. **Cytoplasmic determinants coding for anterior structure of *Drosophila* embryo, if injected elsewhere in the recipient embryo, would lead to**

(A) normal development.
(B) formation of additional ectopic head.
(C) degeneration.
(D) a phenotype with two heads and two tails.

50. **ELISA assay uses**

(A) an enzyme which can react with secondary antibody.
(B) an enzyme which can react with the antigen.
(C) a substrate which gets converted into a coloured product.
(D) a radiolabelled secondary antibody.

51. **The dwarf pea mutant (*le*) used by Mendel was defective in which of the following enzyme involved in gibberellin biosynthesis?**

(A) ent-Kaurene synthase.
(B) GA-3 β-hydroxylase.
(C) GA 20-oxidase.
(D) ent-Kaurenoic acid hydroxylase.

52. **A mother of blood group O has a group O child. The father could be of blood type**

(A) A or B or O. **(B) O only.**
(C) A or B. **(D) AB only.**

53. **Recent studies on Archaea suggest that life could have originated**

(A) extra terrestrially and seeded through meteorite impacts.
(B) in shallow coastal areas.
(C) in deep hydrothermal vents.
(D) in hot, terrestrial habitats.

54. **The frequencies of alleles `A' and `a' in a population at Hardy-Weinburg equilibrium are 0.7 and 0.3, respectively. In a random sample of 250 individuals taken from the population, how many are expected to be heterozygous?**

(A) 112 **(B) 81**
(C) 105 **(D) 145**

55. **If the ratio of the number of nonsynonymous to synonymous substitutions per site in protein coding gene is greater than one, it is an evidence of selection that is**

(A) positive **(B) negative**
(C) neutral **(D) random**

56. The following is the biochemical pathway for purple pigment production in flowers of sweet pea:

Colorless Precursor 1 $\xrightarrow{\text{Allele A}}$

Colorless Precursor 2 $\xrightarrow{\text{Allele B}}$ Purple pigment

Recessive mutation of either gene A or B leads to the formation of white flowers. A cross is made between two parents with the genotype: AaBb x aabb. Considering that the two genes are not linked, the phenotypes of the expected progenies are

(A) 9 purple : 7 white
(B) 3 white : 1 purple
(C) 1 purple : 1 white
(D) 9 purple : 6 light purple : 1 white

57. In an altruistic act, if a donor sacrifices `C' offspring which helps the recipient to gain `B' offspring and the donor is related to the recipient by a coefficient γ, under which condition would kin selection favour this altruistic trait?

(A) B > C. (B) B > γ C.
(C) γ B – C = 0. (D) γ B – C > 0.

58. The transition to flowering in plants requires
(A) growth of plants under long-day conditions.
(B) growth of plants under short-day conditions.
(C) reprogramming of the shoot-apical meristem.
(D) synthesis of the flowering hormone florigen.

59. Aneuploid females with only one X chromosome is a characteristic of individuals with
(A) Cri du chat syndrome.
(B) Klinefelter syndrome.
(C) Down syndrome. (D) Turner syndrome.

60. The most commonly used method of estimating primary productivity of a pond involves measurement of the amount of
(A) CO_2 utilized. (B) autotroph biomass.
(C) oxygen released. (D) organic carbon.

61. Which of the following food crops has recently been genetically engineered to obtain edible vaccine to develop immunity against hepatitis B?
(A) Banana (B) Maize
(C) Potato (D) Tomato

62. Following figure shows McArthur and Wilson's equilibrium model of biota on a single island. In this figure, terms A, B, C and D in order are

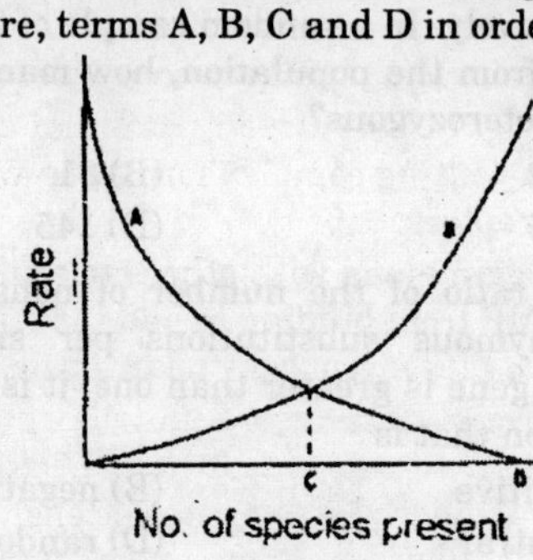

(A) extinction, immigration, equilibrium number of species, size of species pool.
(B) immigration, extinction, equilibrium number of species, size of species pool.
(C) extinction, immigration, size of species, pool, equilibrium number of species
(D) immigration, extinction, size of species pool, equilibrium number of species.

63. A culture of an *E. coli* strain that is lysogenic for phage lambda is grown at 32ºC. Induction of the prophage from the host chromosome will occur when the culture is exposed to
(A) 40ºC.
(B) ultra violet radiation.
(C) infra red radiation.
(D) wild type *E. coli* culture.

64. Which of the following methods of plant transformation can be used to introduce a gene into chloroplast genome?
(A) *Agrobacterium*- mediated transformation
(B) Particle delivery system
(C) Permeabilization
(D) Electroporation

65. Na^+-K^+ ATPase is a tetramer of 2 α and 2 β subunits. On which of the following subunits are the Na^+ and K^+ binding sites present?
(A) both on α
(B) both on β
(C) Na^+ on α and K^+ on β
(D) Na^+ on β and K^+ on α

66. During urine formation, the filtration of blood at the glomerulus is
(A) an active process.
(B) an osmotic process.
(C) a pressure-dependent physical process.
(D) a non energy-mediated transport process.

67. The μ and σ of wing length (a normally distributed parameter) in a population of fruitflies are 4 and 0.2 mm, respectively. In a random sample of 400 fruitflies, how many individuals are expected to have wing lengths greater than 4.4 mm?
(A) 20 (B) 64
(C) 10 (D) 336

68. Which of the following is a characteristic of an early seral community?
(A) Narrow niche specialization.
(B) High species diversity?
(C) Low community production.
(D) Open mineral cycling.

69. Routinely used glucose biosensor estimates blood glucose level by sensing the concentration of
(A) glucose. (B) oxygen.
(C) δ – gluconolactone. (D) H_2O_2.

70. Name the ectothermic animal that can thermoregulate by behavioural means rather than by physiological means.

(A) Bumble bee in an orchard.
(B) Tuna fish in the Ocean.
(C) Lizard in a desert.
(D) Flatworm in a pond.

PART C

71. Mutations in *CONSTANS* (CO) of *Arabidopsis thaliana* results in late flowering phenotype. Transcript levels of CO were determined in long day and short day seedlings. Which of the following would likely represent the transcript profile of CO?

	Long-day seedlings	Short-day seedlings
(A)		
(B)		
(C)		
(D)		

72. During early cleavage of *Caenorabditis elegans* embryos, each asymmetrical division produces one founder cell which produces differentiated descendants and one stem cell. The very first cell division produces one anterior founder cell, namely AB and one posterior stem cell, namely P1. When these blastomeres are experimentally separated and allowed to proceed further with development, one could get the following possible outcomes:

(A) P1 cell would develop autonomously while the AB would show conditional development.
(B) P1 cells would show conditional development while AB would show autonomous development.
(C) Both would show autonomous specification and result in mosaic development.
(D) Both would show conditional specification and result in regulative development.

73. The time taken for atrial systole and diastole in a normal heart are t_{as} and t_{ad} seconds, respectively. If ventricular systole takes t_{vs} seconds, calculate the ventricular diastolic time (seconds).

(A) $(t_{as} + t_{ad}) - t_{vs}$ (B) $(t_{as} - t_{ad}) + t_{vs}$
(C) $(t_{ad} - t_{as}) - t_{vs}$ (D) $(t_{as} - t_{ad}) \times t_{vs}$

74. Values of T_m (midpoint of denaturation), ΔH_m (enthalpy change at T_m) and ΔC_p (constant-pressure heat capacity change) of a protein are measured in a differential scanning calorimeter. $\Delta G_D(T)$, the Gibbs free energy change at any temperature T (K) can be estimated using the following form of the Gibbs-Helmholtz equation with the values obtained from these measurements:

$$\Delta G_D(T) = \Delta H_m\left(\frac{T_m - T}{T_m}\right) - \Delta C_p\,[T_m - T + T\ln(T/T_m)]$$

The stability curve for the protein simulated using the observed thermodynamic values is given below :

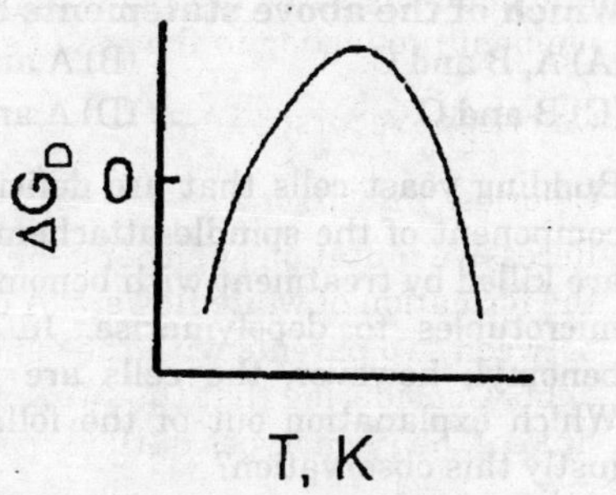

The shape of the stability curve is due to

(A) hydrogen-bonding and electrostatic interactions only
(B) van der Waals and electrostatic interactions only.
(C) only electrostatic interactions
(D) only hydrophobic interaction.

75. The overall length of the cell cycle can be measured from the doubling time of a population of exponentially proliferating cells. The doubling time of a population of mouse L cells was determined by counting the number of cells in samples of culture at various times. What is the overall length of the cell cycle in mouse L cells?

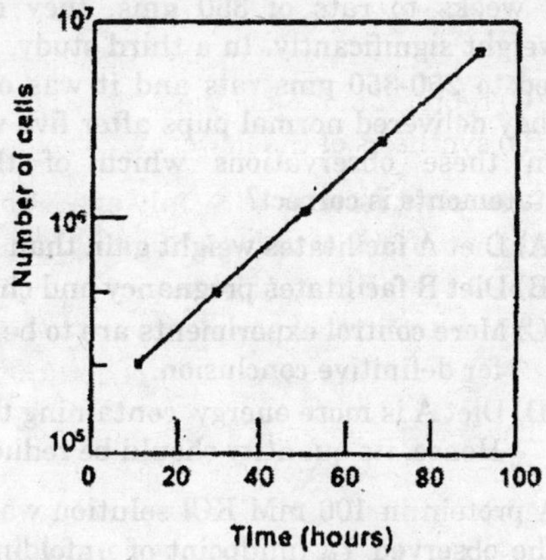

(A) 30 h (B) 20 h
(C) 10 h (D) 40 h

76. Injection of *noggin* mRNA into a 1-cell, UV-irradiated embryos of frog completely rescues dorsal development and allows the formation of a complete embryo. Some of the following statements (A-D) could possibly explain this observation.

(A) Noggin is a secreted protein which induces dorsal ectoderm to form neural tissue and it dorsalizes the mesoderm cells which would otherwise contribute to ventral mesoderm
(B) Noggin binds directly to BMP4 and BMP2 thus preventing complex formation with their receptors.
(C) Noggin along with other molecules prevent BMP from binding to and inducing ectoderm and mesoderm cells near the organizer.
(D) Noggin is a secreted protein which induces the dorsal ectoderm to form the epidermis and it ventralizes the mesoderm cells which would otherwise contribute to dorsal mesoderm.

Which of the above statements are correct?

(A) A, B and C. (B) A and B.
(C) B and C. (D) A and D

77. **Budding yeast cells that are deficient for Mad2, a component of the spindle-attachement-check point, are killed by treatment with benomyl, which causes microtubles to depolymerise. In the absence of benomyl, however, the cells are perfectly viable. Which explanation out of the following is able to justly this observation?**

(A) In the absence of benomyl, the majority of spindles form normally and the spindle-attachment checkpoint (Mad2) plays no role.
(B) In the presence of benomyl, the majority of spindles form normally and Mad2 plays critical role in cell survival.
(C) Other than the role in cell survival, microtubule depolymerization affects oxidative phosphorylation in the absence of Mad2.
(D) Benomyl also affects protein synthesis in the absence of Mad2.

78. **In one study, a group of 5 day rat pups were fed for 3 weeks a diet A and the pups gained weight by 300%. In a second study, when the same diet fed for 3 weeks to rats of 350 gms, they did not gain weight significantly. In a third study, a diet B was fed to 250-350 gms rats and it was observed that they delivered normal pups after five weeks. Based on these observations which of the following statements is correct?**

(A) Diet A facilitates weight gain than diet B.
(B) Diet B facilitates pregnancy and child-bearing.
(C) More control experiments are to be conducted for definitive conclusion.
(D) Diet A is more energy containing that diet `B'. Hence, its quantity should be reduced.

79. **A protein in 100 mM KCl solution was heated and the observed T_m (midpoint of unfolding) was 60ºC. When the same protein solution in 500 mM KCl was heated, the observed T_m was 65ºC. What is the most probable reason for this increase in T_m?**

(A) Hydrophobic interaction is increased and electrostatic repulsion is decreased.
(B) Hydrophobic interaction is decreased and electrostatic repulsion is increased.
(C) Hydrogen-bonding is increased.
(D) van der Waals interaction is increased.

80. **Precursors of the atoms in the purine skeleton are**

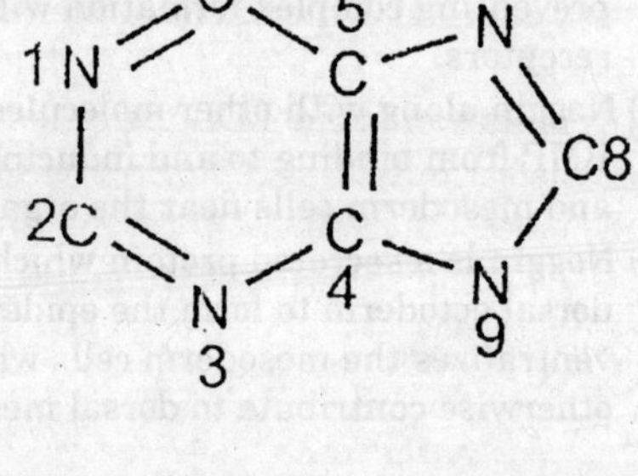

(A) N1, Asp; C2 and C8, formate; N3 and N9, guanidine of Arg; C4, C5 and N7, Gly; C6, CO_2.
(B) N1, Asp; C2 and C8, citrate; N3 and N9, amide nitrogen of Gln; C4, C5 and N7; Gly; C6, CO_2.
(C) N1, Asp; C2 and C8, formate; N3 and N9, amide nitrogen of Gln, C4, C5 and N7, Gly; C6, CO_2
(D) N1, Glu; C and C8, acetate; N3 and N9, amide nitrogen of Asn; C4, C5 and N7, Gly; C6, CO_2.

81. **Bacteria often acquire genes by the process of lateral or horizontal transfer. Such `foreign' genes, if acquired in recent past, may be identified by their atypical GC content, as compared to `native' genes. Suppose the genomic GC content of a bacterium is 40%. Gene A of this organism contains 1000 bases with 225 G and 215 C. Another gene B of length 800 bases contains 160 G and 140 C. Which one of the following would be the most acceptable hypothesis (given that x^2 = 3.841 at 0.05 significance level)?**

(A) A : native, B : foreign
(B) A : foreign, B : native
(C) A : foreign, B : foreign
(D) A : native, B : native

82. **An amino acid contains no ionizable group in its side chain (R). It is titrated from pH 0 to 14. Which of the following ionizable state is not observed during the entire titration in the pH range 0-14?**

(A) $H_3\overset{+}{N}-\overset{R}{\overset{|}{C}}H-COO^-$

(B) $H_3\overset{+}{N}-\overset{R}{\overset{|}{C}}H-COOH$

(C) $H_2N-\overset{R}{\overset{|}{C}}H-COO^-$

(D) $H_2N-\overset{R}{\overset{|}{C}}H-COOH$

83. **A researcher has isolated a restriction endonuclease that cleaves at only one specific 10 base pair site.**

(A) Would this enzyme be useful in protecting cells from viral infections, given that a typical viral genome is 5 x 10^4 base pairs long?

(B) Restriction endonucleases are slow enzymes with turnover number of 1 s^{-1}. Suppose the isolated endonuclease was faster with turnover numbers similar to those for carbonic anhydrase (10^6 s^{-1}), would this increased rate be beneficial to host cells, assuming that the fast enzymes have similar levels of specificity?

The correct combination of answer is

(A) (A) : No (B) : Yes
(B) (A) : No (B) : No
(C) (A) : Yes (B) : No
(D) (A) : Yes (B) : Yes

84. Assume a new subspecies *Ficus callosa* subsp. *microcarpa* has been published by Jacobs. The nomenclature of the resulting entities would be

(A) *F. callosa* and *F. callosa* subsp. *microcarpa* Jacobs

(B) *F. callosa* subsp. *microcarpa* Jacobs and other yet to be named subspecies of *F. callosa*

(C) *F. callosa* subsp. *callosa* Jacobs and *F. callosa* subsp. *microcarpa* Jacobs.

(D) *F. callosa* subsp. *callosa* and *F. callosa* subsp. *microcarpa* Jacobs.

85. Eukaryotic genomes are organized into chromosomes and can be visualized at mitosis by staining with specific dyes. Heat denaturation followed by staining with Giemsa produced alternate dark and light bands. The dark bands obtained by this process are mainly

(A) AT-rich and gene rich regions.

(B) AT-rich and gene desert regions.

(C) GC-rich and gene rich regions.

(D) GC-rich and gene desert regions.

86. In case of sea urchin, which of the following is the correct sequence of events taking place during the interaction of sperm and egg?

(A) Chemoaattraction of sperm to the egg by soluble molecules secreted by the egg → exocytosis of the sperm acrosomal vesicle to release its enzymes → binding of the sperm to the extracellular matrix of the egg → passage of sperm through this extracellular matrix → fusion of egg and sperm cell membranes.

(B) Chemoattraction of sperm to the egg by soluble molecules secreted by the egg→ binding of the sperm to the extracellular matrix of the egg → exocytosis of the sperm acrosomal vesicle to release its enzymes → passage of sperm through this extracellular matrix → fusion of egg and sperm cell membranes.

(C) Chemoattraction of sperm to the egg by soluble molecules secreted by the egg → binding of the sperm to the extracellular matrix of the egg → passage of sperm through this extracellular matrix → exocytosis of the sperm acrosomal vesicle to release its enzymes → fusion of egg and sperm cell membranes.

(D) Chemoattraction of sperm to the egg by soluble molecules secreted by the egg → passage of sperm through this extracellular matrix → binding of the sperm to the extracellular matrix of the egg → exocytosis of the sperm acrosomal vesicle to release its enzymes → fusion of egg and sperm cell membranes.

87. An α-helix in a peptide or protein is characterized by hydrogen bonds and characteristic dihedral angles. Choose the right combination.

(A) Hydrogen bonding between the amide CO of residue i and amide NH of residue $i + 4$. Dihedral angles in the region $\phi \sim -50°$, $\psi \sim -60°$.

(B) Hydrogen bonding between the amide NH of residue i and amide CO of residue $i + 4$. Dihedral angles in the region $\phi \sim -50°$, $\psi \sim -60°$.

(C) Hydrogen bonding between the amide CO of residue i and amide NH of residue $i + 4$. Dihedral angles in the region $\phi \sim -50°$, $\psi \sim -60°$.

(D) Hydrogen bonding between the amide CO of residue i and amide NH of residue $i + 3$. Dihedral angles in the region $\phi \sim -50°$, $\psi \sim -60°$.

88. It has been observed that in 5–10% of the eukaryotic mRNAs with multiple AUGs, the first AUG is not the initiation site. In such cases, the ribosome skips over one or more AUGs before encountering the favourable one and initiating translation. This is postulated to be due to the presence of the following consensus sequence(s):

(A) CCA CC <u>AUG</u> G

(B) CCG CC <u>AUG</u> G

(C) CCG CC <u>AUG</u> C

(D) AAC GG <u>AUG</u> A

Which of the following sequence sets related to the above postulations is correct?

(A) A and B

(B) A and C

(C) C and D

(D) B and D

89. You are studying the binding of proteins to the cytoplasmic face of cultured liver cells and have found a method that gives a good yield of inside out vesicles from the plasma membrane. Unfortunately, your preparations are contaminated with variable amounts of right side out vesicles. Nothing you have tried avoids this contamination. Somebody suggests that you pass the vesicles over an affinity column made of lectin coupled to Sepharose beads. What is the rational of this suggestion?

(A) Right-side-out-vesicles will be lysed by lectin coupled to Sepharose beads.

(B) Right-side-out-vesicles will simply bind to the lectin coupled Sepharose beads.

(C) Lectin will bind to the carbohydrate residues present only on the inside-out vesicles.

(D) Lectin will bind to only glycoproteins and glycolipids present on the inside-out vesicles.

90. During receptor mediated endocytosis, apolipoprotein B on the surface of a LDL particle binds to the LDL-receptor present in coated pits containing clathrin. The receptor LDL complex is internalized by endocytosis, trafficked to lysosomes and the LDL-receptor is finally recycled. A patient reports with familial hypercholesterolemia. This could be due to

(A) mutation in the LDL molecule.

(B) defect in LDL-receptor recycling.

(C) mutation in the LDL-receptor.

(D) defect in chloesterol binding with its receptor.

91. Cancer causing genes can be functionally classified into mainly three types: (i) genes that induce cellular proliferation, (ii) tumor suppressor genes, (iii) genes that regulate apoptotic pathway. Epstein-Barr virus that causes cancer by modulating apoptotic pathway, contains a gene having sequence homology with which of the following genes?
(A) bax (B) bcl-2
(C) p53 (D) caspase-3

92. Genetic studies demostrated that TBP mutant cell extracts are deficient in transcription of genes from all three promoters viz., class I, II and III. Following statements describes characteristic features of TBP.
(A) TBP is considered as an universal basal transcription factor.
(B) TBP is not required for transcription of archaeal genes.
(C) TBP is involved in recognizing TATA box.
(D) TBP operates at all promoters regardless of their TATA content.
Which of the folowing combinations is NOT correct?
(A) A and D (B) C and D
(C) B and D (D) A and C

93. Two *E. coli* cultures A and B are taken. Culture A was earlier grown in the presence of optimum concentration of gratuitous inducer IPTG. Both the cultures are now used to inoculate fresh medium containing sub-optimal concentration of gratuitous inducer. It was observed that culture B was unable to utilize lactose, whereas culture A did so efficiently. The reason behind this is
(A) pretreatment with IPTG has resulted in a mutation as a result of which *lac* operon is constitutively expressed
(B) IPTG has made the cell membrane more porous to small molecules and so lactose is taken up more efficiently by A as compared to B.
(C) In culture A, lactose permease was induced to a high level, during pretreatment with IPTG, which allowed the preferential uptake of lactose.
(D) In culture A, IPTG activated a receptor which bound lactose more efficiently, thereby triggering a signal.

94. Glucose is mobilized in muscle when epinephrine activates Gas. In an experiment in which muscle cells were stimulated with epinephrine, glucose mobilization was observed even after withdrawal of epinephrine. This could be
(A) due to the presence of a cAMP phosphodiesterase inhibitor.
(B) very low rates of cyclic AMP formation.
(C) due to the presence of a cAMP phosphodiesterase activator.
(D) due to the absence of protein kinase A.

95. siRNAs and miRNAs are used for achieving gene silencing. Although major steps are similar there are distinct differences in the key players of the two processing pathways. Following statements relate to some characteristic features of gene silencing.
(A) Both siRNAs and miRNAs are processed by cytoplasmic endonuclease Dicer.
(B) `Drosha' is needed for processing miRNAs and precursor siRNAs.
(C) Both siRNAs and miRNAs show association with Argonaute protein.
(D) Both the processing pathways involve RISC complex.
Which of the following combinations is NOT correct?
(A) A and C (B) C and D
(C) A and B (D) D and A

96. Presence of circular mRNAs for a specific protein in an eukaryotic cell reflects a rapid rate of synthesis of that protein. Following mechanisms are suggested:
(A) eIF-4G and PABP promote this process through 5´-3´ interaction of mRNA.
(B) ribosomes are less active in recognizing circular mRNA.
(C) PABP and eIF-4A promote this process.
(D) ribosomes can reinitiate translation without being disassembled.
Which of the following is correct?
(A) A and D (B) B and D
(C) A and C (D) B and C

97. In eukaryotic chromatin, 30 nm fiber (solenoid) can open up to give rise to two kinds of chromatin. In one type (A), the promoter of a gene within the open chromatin is occupied by a nucleosome whereas in the other (B), the promoter is occupied by histone H1.
The following possibilities are suggested.
(A) The gene in (A) is repressed.
(B) The gene in (B) is repressed.
(C) The gene in (A) is active.
(D) The gene in (B) is active.
Which of the following sets is correct?
(A) A and D (B) A and B
(C) B and D (D) C and D

98. Following are some statements for synthesis of secondary metabolites in plants.
(A) Terpenes are synthesized by shikimic acid pathway and mevalonic acid pathway.
(B) Alkaloids are nitrogen containing compounds and are synthesized by shikimic acid pathway.
(C) Phenolic compounds are synthesized by shikimic acid pathway and mevalonic acid pathway.
(D) Both alkaloids and terpenes are synthesized by mevalonic acid pathway and MEP pathway.

Which one of the following combinations of the above statement is true?

(A) A and D (B) A and C
(C) B and C (D) B and C

99. Intracellular pathogens like *Mycobacteria, Salmonella, Leishmania* and *Listeria* survive in macrophages by modulating host cellular machinery. In order to study the fate of these intracellular pathogens in macrophages, cells were labelled with lysotracker Red and infected with GFP-labelled organisms. After 2 hours at 37°C, cells were fixed, stained with anti-transferrin receptor antibody and probed with secondary antibody conjugated blue dyes. Cells were viewed under confocal microscope. Observation : GFP - labelled *Mycobacteria, Salmonella* and *Listeria* were localized in the same compartment labelled with blue dyes, whereas GFP-*Leishmania* colocalize with red labelled compartment. Which of the following statement is true based on these observations?

(A) *Mycobacteria, Salmonella* and *Listeria* reside in the lysosomes.
(B) *Leishmania* reside in lysosome like compartment.
(C) *Leishmania* reside in a compartment which bears characteristics of early endocytic compartment.
(D) *Mycobacteria, Salmonella* and *Listeria* lyse the phagosomal membrane and reside in cytosol.

100. Mouse bone marrow cells were fractionated to derive stem cell antigen-1⁺ (Sca-1⁺) cells. These cells were cultured with interleukin-3, or granulocyte-macrophage colony stimulating factor, or macrophage-colony stimulating factor, or granulocyte colony stimulating factor. Most numerous and varied colonies were obtained in the culture stimulated with

(A) Interleukin-3.
(B) Granulocyte-macrophage colony stimulating factor.
(C) Macrophage-colony stimulating factor.
(D) Granulocyte-colony stimulating factor.

101. Toll-like receptor 4 is associated with responsiveness to LPS, an endotoxin that causes lethal endotoxic shock. The mice deficient in Toll-like receptor 4 and BALB/c mice were injected with *Escherichia coli*. In addition, some BALB/b mice were also injected with the same bacteria alone or with anti-interleukin-10 (IL-10) antibody. The mice resistant to the lethal effect of the bacteria were :

(A) BALB/b mice receiving the bacteria.
(B) BALB/b mice receiving the bacteria and the anti-IL-10 antibody.
(C) Mice deficient in Toll-like receptor.
(D) BALB/c mice receiving the bacteria.

102. Atmospheric CO_2 contains the naturally occurring stable isotopes ^{12}C and ^{13}C in the proportion of 98.9% and 1.1%, respectively. Following are some of the statements regarding CO_2 assimilation :

(A) Both C_3 and C_4 plants assimilate less $^{13}CO_2$ than $^{12}CO_2$.
(B) Both C_3 and C_4 plants assimilate less $^{12}CO_2$ than $^{13}CO_2$.
(C) C_3 plants assimilate lesser $^{13}CO_2$ than $^{12}CO_2$ as compared to C_4 plants.
(D) C_4 plants assimilate lesser $^{13}CO_2$ than $^{12}CO_2$ as compared to C_3 plants.

Which one of the following combinations of above statements is true?

(A) A and B (B) A and C
(C) C and D (D) A and D

103. cAMP signalling plays a very important role in the development and differentiation of *Dictyostelium discoideum*. This morphogen is synthesized by different adenyl cyclases expressed at different stages of its life cycle. The following statements (A-D) refer to the effect of mutations in different adenyl cyclase genes :

(A) *aca* deficient cells can be allowed to aggregate by exposing them to pulses of cAMP.
(B) *acb* deficient cells would form normal fruiting bodies and the spores can germinate when exposed to favourable conditions.
(C) *acg* deficient cells develop normally and the spores germinate in the spore head itself.
(D) spores formed from the *acg* deficient cells will germinate irrespective of the osmotic conditions.

Which of the above statements are correct?

(A) A and D. (B) A only.
(C) A and B. (D) C and D.

104. The dependence of the rate of sucrose uptake with respect to sucrose concentration in plant cell was studied and data are shown in the following graph.

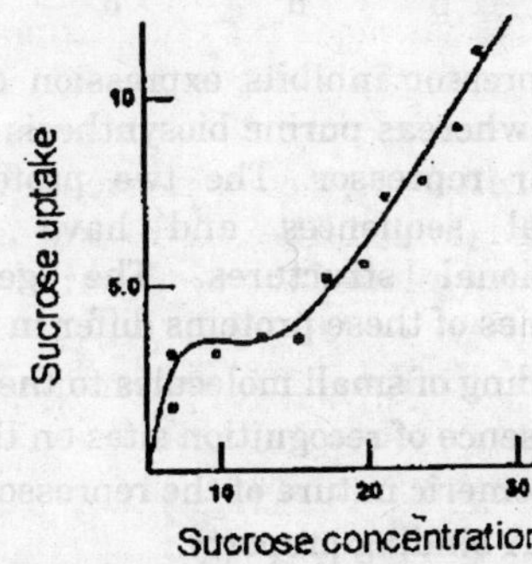

From the above data it can be inferred that

(A) the sucrose uptake is energy independent and no special carrier is involved.
(B) the sucrose uptake is energy dependent and a special carrier is involved.
(C) at lower concentration of sucrose the uptake of sucrose is energy dependent and carrier mediated.
(D) at higher concentration of sucrose the uptake is energy dependent and carrier mediated.

105. When a wrist blastema from a recently cut *Axolotl* forelimb is placed on a host hindlimb cut at the mid thigh level, it will generate only a wrist. The host (whose own hind limb was removed) will fill the gap and regenerate the limb upto the wrist. However, if the donor blastema is treated with retinoic acid on grafting, the wrist blastema will regenerate a complete limb and will not allow the host to fill the gap. This happens because retinoic acid :

(A) helps in the proximalization of the blastema and activates the Hox genes differentially across the blastema.

(B) helps in the distalization of the blastema and activates the Hox genes differentially across the blastema.

(C) helps block the receptors of FGF essential for limb development.

(D) helps vigorous proliferation of the cells at the cut surface.

106. Fill in the blanks (a, b, c and d) in the following statements with a proper combination of m, n, o and p.

Wherein m represents - longer

n represents - shorter

o represents - prevents

p represents - induces

short day (SD) plants flower when night lengths are **a** than a critical dark period. Interruption of the dark period by a brief light treatment **b** flowering in SD plants. Long day (LD) plants flower when night length is **c** than a critical period. Shortening of the night with a brief light treatment **d** flowering in LD plants.

	(A)	(B)	(C)	(D)
(A)	m	o	n	p
(B)	n	p	m	o
(C)	n	o	m	p
(D)	m	p	n	o

107. Lac repressor inhibits expression of genes in *lac* operon whereas purine biosynthesis is repressed by the Pur repressor. The two proteins have 31% identical sequences and have similar three-dimensional structures. The gene regulatory properties of these proteins differ in relation to

(A) binding of small molecules to the repressor.

(B) presence of recognition sites on the genome.

(C) oligomeric nature of the repressor.

(D) DNA binding property.

The corret statements are

(A) A and B (B) A, B and C

(C) A and C (D) B, C and D

108. Primary production in aquatic ecosystem is measured using Light-and-Dark-Bottle technique. In this method, as an indirect measure of photosynthetic production, dissolved oxygen concentration of the pond water enclosed in a BOD bottle is measured initially (I) and after a fixed duration of incubation in a light bottle (L) and a dark bottle (D). Then, the gross and net primary productions are estimated as

(A) (L–D) and (L–I), respectively.

(B) (L–I) and (L–D), respectively.

(C) (L–I) and (I–D), respectively.

(D) (L–D) and (I–D), respectively.

109. An experiment was designed to find out the relation between DNA uptake and transformation efficiency. ^{32}P-labelled genomic DNA from *Bacillus subtilis* (A) and cold genomic DNA from *Clostridium jejeuni* (B) mixed in various proportions was used to transform *Bacillus subtilis.* The results obtained are tabulated below :

Set No.	Concentration DNA used ($\mu g/ml$) A	B	Uptake of ^{32}P labelled DNA (%)	Transforming efficiency (colony forming unit)
1.	1	1	90	30
2.	1	10	82	10
3.	1	100	15	5
4.	1	1000	5	1

The following interpretations (A to D) could be made :

(A) The transformation is dependent on recombination between homologous sequences.

(B) Cells did not distinguish between homologous or heterologous DNA sequences for uptake of DNA.

(C) DNA uptake is based on specific receptors.

(D) DNA degradation dictates transformation efficiency.

Which of the following interpretations are correct?

(A) A and B (B) A and C

(C) C and D (D) A and D

110. During episodes of anoxia in plants, pyruvate produced in glycolysis is initially fermented to lactate. During later state, there is an increase in the fermentation to ethanol and decrease in the fermentation to lactate, a phenomena which helps plants survive anoxia. Which of the following statements is correct about this change of fermentation flux from lactate towards ethanol?

(A) The cytosolic pH increases, thus activating both lactate dehydrogenase and pyruvate decarboxylase activity.

(B) The cytosolic pH increases, thus inhibiting lactate dehydrogenase activity and activating pyruvate decarboxylase activity.

(C) The cytosolic pH decrease, thus activating both lactate dehydrogenase and pyruvate decarboxylase activity.

(D) The cytosolic pH decreases, thus inhibiting lactate dehydrogenase and activating pyruvate decarboxylase activity.

111. Spinal cord of an animal was transected at C1/C2 level. The respiration of the animal stopped and it needed artificial respiration. However, the heart continued to beat although at a slower rate. Some of the explanations given were :

(A) respiration regulatory centre is located in the medulla.
(B) respiration regulatory centre is located above the C1/C2 cut.
(C) heart regulatory centre is above the C1/C2 cut.
(D) heart has autoregulation.
Which one of the following is most appropriate?
(A) A only. (B) B and C only.
(C) A, B and D only. (D) B, C and D only.

112. Macrophages were collected from BALB/c mice, CD40-deficient mice, CD86-deficient mice and ICAM-1 deficient mice. These macrophages were co-cultured with LCMV peptide-specific T cells in presence of the LCMV peptide for three days. The cells were recovered and co-cultured with BALB/c-derived macrophages in presence of the peptide. During the last twelve hour of the co-culture, ^{3}H-thymidine was added to the cultures. The cells were harvested and ^{3}H-thymidine incorporation was assessed. The highest incorporation was observed in.
(A) BALB/c macrophage-T cell co-culture.
(B) CD40-deficient macrophage-T cell co-culture.
(C) CD86-deficient macrophage-T cell co-culture.
(D) ICAM-1-deficient macrophage-T cell co-culture.

113. In a stressful condition, ACTH secretion was increased and as a result glucocorticoid concentration was elevated in blood. One or a combination of the following changes most likely taking place in this condition :
(A) Decreased circulating eosinophils & basophils.
(B) Reduced IL2 release.
(C) Potentiated inflammatory response to tissue injury.
(D) Increased mitotie activity of lymphocytes in lymph nodes.
The correct answer is
(A) B and C. (B) A and B.
(C) B and D. (D) C and D.

114. What will be the approximate effective population size in a panmictic population of 240 with 200 females and 40 polygamous males?
(A) 160 (B) 133
(C) 63 (D) 67

115. An organism having heart for circulation, excretes through green glands. It has several ganglia and tactile organs on its body and its larval form is very different than its adult form. This organism is most likely to respire by :
(A) exchanging oxygen and carbon dioxide through an extensive tracheal system.
(B) gaseous exchange over thinner areas of cuticle or by gills
(C) an efficient tracheal system that delivers oxygen directly to the tissues
(D) a double transport system, where the circulating fluid contains a dissolved respiratory pigment.

Choose the correct option.
(A) A and C (B) Only D
(C) Only B (D) B and D

116. Based on the structural regions of a nuclear receptor shown in the diagram, the following predictions were made.

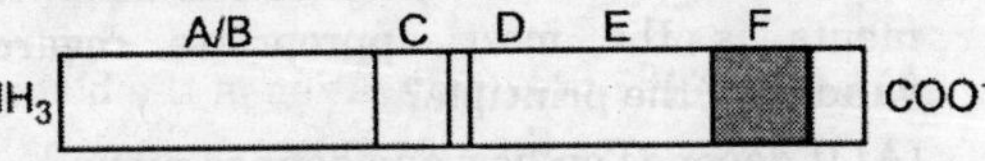

(A) Region F is responsible for binding to ligands and contains two zinc-finger-like binding motifs
(B) Receptors with A/B domains generally associate with chaperones and do not bind to DNA.
(C) Region E indicate that receptors associate with chaperones which protect the nuclea hormone receptors.
(D) Region C contains the P-box and the D-box required for dimerization of the receptor and creates contact with DNA phosphate backbone.
Which one of the following is true?
(A) A and B. (B) B and C.
(C) B and D. (D) C and D.

117. Two new plant species, A and B, were described in 1872. Subsequently it was found that the type for species A was never designed and for species B there was one specimen designed as type but missing. As per International Code of Botanical Nomenclature (ICBN), typification should be
(A) neotype for A only.
(B) neotypes for both A and B.
(C) neotype for A and lectotype for B.
(D) lectotypes for both A and B.

118. When F1 female *Drosophila* of the genotype $a^+ \ a \ b^+$ $b \ c^+ \ c$ is test crossed, the following progenies were obtained :

Progeny classes*	No. of progenies
$a^+ b^+ c^+$	22
$a^+ b^+ c$	28
$a \ b \ c^+$	26
$a \ b \ c$	24
$a^+ b \ c^+$	230
$a^+ b \ c$	220
$a \ b^+ c^+$	225
$a \ b^+ c$	225
Total	1000

* The progeny has been shown as classes derived from the female gamete.

Statements A to F as given below are conclusions derived from the above result.
(A) Genes *a* and *b* are linked in *cis*.
(B) Genes *a* and *b* are linked in *trans*.
(C) Genes *a* and *b* are linked in *cis* while *b* and *c* are linked in *trans*.
(D) The genotype of the parents are $a^+ a^+ b^+ b^+$ and *aabb*
(E) The genotype of the parents are $a^+ a^+ bb$ and aab^+b^+.
(F) Genes *a* and *b* are 10 cm apart.

Which of the above statements are correct?

(A) C alone. (B) A, E and F.

(C) B, E and F. (D) A, D and F.

119. Gause's `Competitive exclusion' principle states that two species with identical niches cannot coexist indefinitely. Which of the following statements is the most appropriate regarding the validity of the principle?

(A) It depends on how one defines niche.

(B) There are in nature many instances of continued coexistence of closely related species.

(C) The principle is universally true.

(D) It does not predict the outcome where both the species are equally strong competitors.

120. The following is a hypothetical pathway for the development of wild type (red) eye colour in an insect :

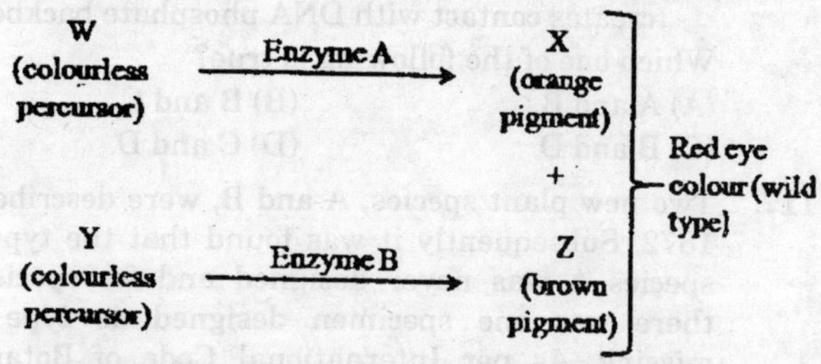

Enzymes A and B are encoded by the genes a^+ and b^+, respectively.

The following statements are made regarding inheritance of the genes involved in the development of eye colour :

(A) When two heterozygous individuals of the genotype $a^+ ab^+b$ are mated, progenies with red, orange, brown and white eye colour will be observed irrespective of whether the genes are independently assorting or showing imcomplete linkage.

(B) When two heterozygous individuals of the genotype a^+ab^+b are mated, progenies with red, orange, brown and white eye colour will be observed in a ratio of 9:3:3:1, when the genes are independently assorting.

(C) When an heterozygous individual of the genotype $\frac{a^+b}{ab^+}$ is test crossed, progenies with red and white eye colour will be more in number.

(D) When an heterozygous individual of the genotype $\frac{a^+b}{ab^+}$ is test crossed, progenies with orange and brown eye colour will be more in number.

Which of the above statements is TRUE?

(A) A and C (B) B and C

(C) A, B and C (D) A, B and D

121. Mendel crossed tall pea plants with dwarf ones. The F_1 plants were all tall. When these F_1 plants were selfed to produced F_2 generation, he got a 3 : 1 tall to dwarf ratio in the offspring. What is the probability that out of three plants (of F_2 generation) picked up at random two would be dwarf and one would be tall?

(A) 3/4 (B) 3/8

(C) 9/64 (D) 9/32

122. Identify the most appropriate cladogram that can be constructed using the data matrix given below, assuming `O's are plesiomorphic and `1's are apomorphic characters.

	1	2	3	4	5
A	0	0	0	0	0
B	0	1	1	0	0
C	0	1	0	0	0
D	0	1	1	0	1

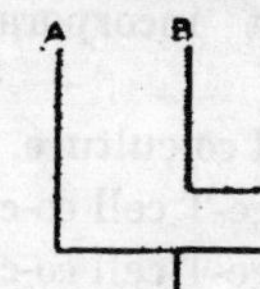

(A)

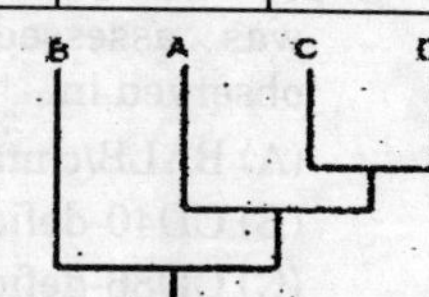

(B)

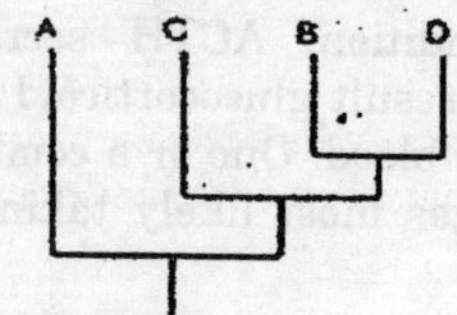

(C)

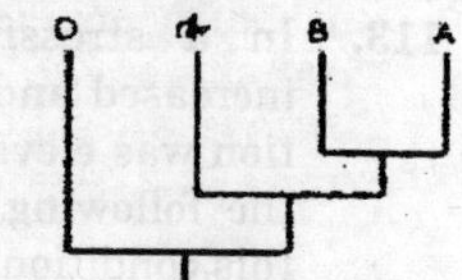

(D)

123. An organism has the following architectural pattern :

(i) multicellular with germ layers

(ii) a coelom derived from the mesoderm

(iii) primary bilateral symmetry with secondary radial symmetry.

(iv) presence of endoskeletal plates

Such an organism is most likely to

(A) have mesohyl as its connective tissue.

(B) undergo torsion, whereby the mouth and anus are properly oriented.

(C) be devoid of a brain but have calcareous spicules.

(D) have comb plates to help in locomotion.

Which of the following is true?

(A) A and C (B) C only

(C) D only (D) B and C

124. In *Neurospora* a cross between the genotypes `A' and `a' results in an ascus with ascospores of genotypes as shown below :

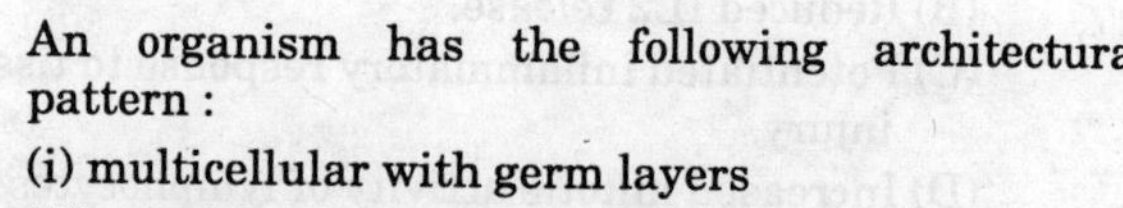

Statements A to D are events that could have occurred during meiosis.

(A) Crossing over between the centromere and the gene.

(B) Segregation of alleles \`A' and \`a' in meiosis I.

(C) Segregation of alleles \`A' and \`a' in meiosis II.

(D) Assortmetn of alleles \`A' and \`a'.

Which of the above events could correctly explain the observation shown in the figure?

(A) A followed by C (B) A followed by B

(C) C alone (D) D alone

125. According to Mac Arthur and Wilson's equilibrium theory, which of the following is true?

(A) Larger islands and islands closer to continent are expected to have more species than smaller and isolated islands.

(B) Smaller islands and islands far from the continent are expected to have more species than larger and isolated islands.

(C) Smaller islands and islands closer to the continent are expected to have more species than far away smaller and isolated islands.

(D) More species are expected on all islands irrespective of their size and distance from the continent.

126. Using molecular clock, it was estimated that two species A and B must have diverged from their common ancestor about 9 x 10^6 years ago. If the rate of divergence per base pair is estimated to be 0.0015 per million years, what is the proportion of base pairs that differ between the two species now?

(A) 0.0270 (B) 0.0135

(C) 0.00017 (D) 0.0035

127. Identify the apomorphic characters marked in the cladogram :

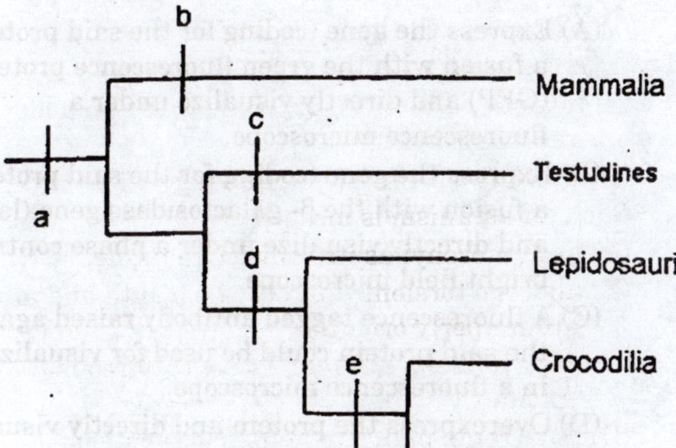

(A) a – amniotic egg; b – 4 – chambered heart; c – anapsidan skull; d – diapsidan skull; e – synapsid skull.

(B) a – amniotic egg; b – synapsidan skull; c –4– chambered heart; d –anapsidan skull; e – diapsidan skull.

(C) a –4 – chambered heart; b – synapsidan skull; c – amniotic egg; d – diapsidan skull; e – anapsidan skull.

(D) a – amniotic egg; b –synapsidan skull; c – anapsidan skull; d – diapsidan skull; e – 4 – chambered heart.

128. Which one of the following graphs represents a normal sexual cycle in a normal human female?

— Estrogen ------ progestrone

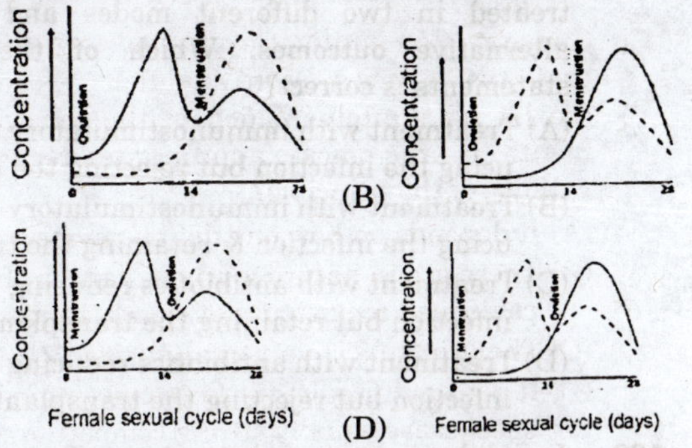

129. The graph below shows the relationships of per capita population growth rate (r), fecundity (b) and age at first reproduction ($\propto$) in an animal species.

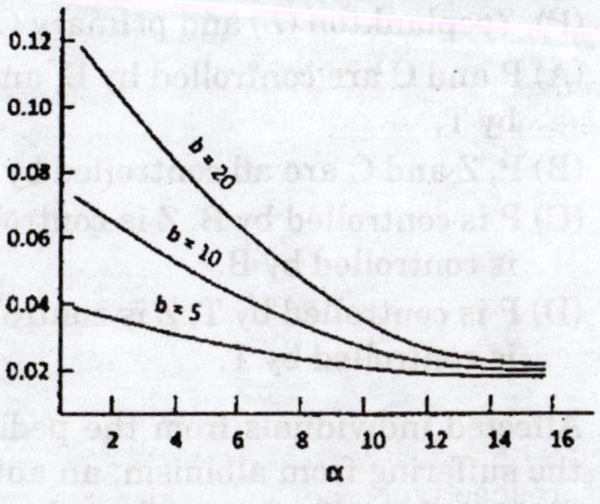

What is the *most important* conclusion to be drawn from the graph?

(A) The later the age of first reproduction, the lower is the population growth rate achieved.

(B) The population growth rate decreases as first reproduction is postponed to a later stage, regardless of the fecundity.

(C) At any $\propto$, the higher the fecundity, the higher is the population growth rate achieved.

(D) As the age at first reproduction is postponed further, the benefits of increasing fecundity on the population growth rate become progressively negligible.

130. An animal was first maintained in a constant environmental condition for several days until a consistent biological rhythm (B) was established. The animal was then exposed to an experimental physical rhythm (E) which modulates the phase and period of B. However, upon withdrawal of E, the B gradually regained its pattern of pre-exposure condition. From these observations which one or more of the following should be the most logical inference?

(A) E is a Zeitgeber.

(B) E is a masking agent.

(C) E causes entrainment of B.

(D) B is a conditioned to E.

The correct answer is

(A) A and C (B) B and D

(C) B only (D) D only

131. A patient undergoes liver transplantation and during the course of post-operative treatment, becomes susceptible to infection. The patient can be treated in two different modes and can have alternative outcomes. Which of the following statements is correct?

(A) Treatment with immunostimulatory drugs reducing the infection but rejecting the transplant.

(B) Treatment with immunostimulatory drugs reducing the infection & retaining the transplant.

(C) Treatment with antibiotics reducing the infection but retaining the transplant.

(D) Treatment with antibiotics reducing the infection but rejecting the transplant.

132. In a lake ecosystem, bottom-up effects (B) refer to control of a lower trophic level by the higher trophic levels and top-down effects (T) refer to the opposite. In a lake with three trophic levels - Phytoplankton (P), Zooplankton (Z) and primary Carnivore (C).

(A) P and C are controlled by B, and Z is controlled by T.

(B) P, Z and C are all controlled by T.

(C) P is controlled by B, Z is controlled by T and C is controlled by B.

(D) P is controlled by T, Z is controlled by B and C is controlled by T.

133. Affected individuals from the pedigree given below the suffering from albinism, an autosomal recessive disease. Identify the confirmed carrier individuals in this pedigree assuming that the members coming from outside the family are homozygous for the dominant allele.

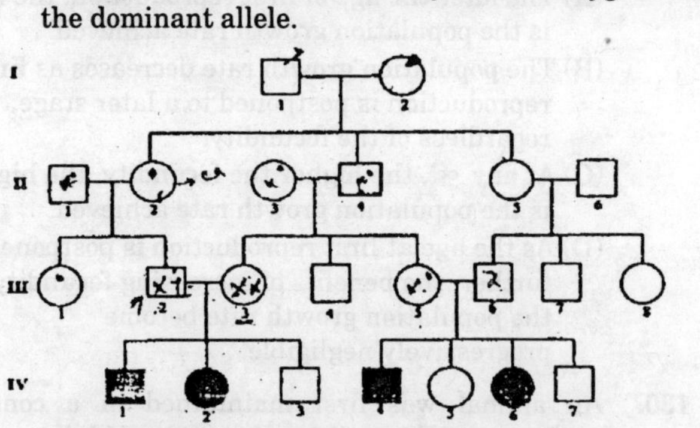

(A) III-2, III-3, III-5, III-6, II-1, II-3 and II-6

(B) III-2, III-3, III-5, III-6, II-2, II-4, II-5 and I-2

(C) III-2, III-3, III-5, III-6, II-2, II-4 and II-5

(D) III-1, III-4, III-7, II-2, II-4 and II-5

134. The total variance in a phenotypic character can be split into two components genetic (V_G) and environmental (V_E). The heritability of a phenotypic trait can be expressed quantitatively as heritability coefficient (h^2) which is calculated as h^2 =

(A) $V_G - V_E$ (B) V_E/V_G

(C) $\frac{V_G}{(V_G + V_E)}$ (D) $\frac{V_G}{(V_G - V_E)}$

135. Shown in the graph below are the fitness costs and benefits of four alternative behavioural phenotypes (P, Q, R, S). Given sufficient evolutionary time, which phenotype(s) is likely to evolve as an adaptation?

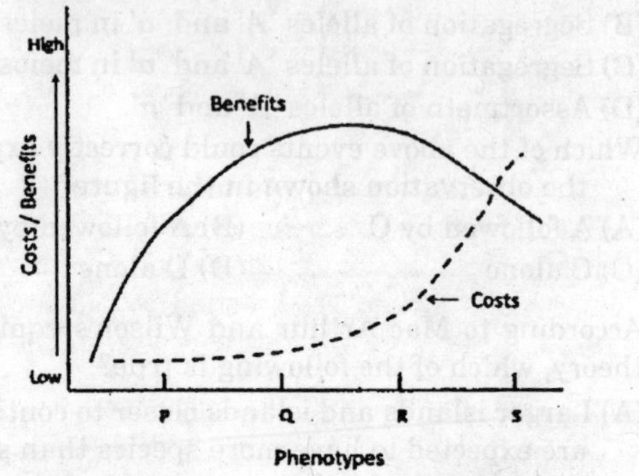

(A) Q and R (B) Q only

(C) P, Q and R (D) S only

136. The following geological eras mark the advent of important events in the history of earth-origin of terrestrial plants, origin of mammals, and break up of the land mass Pangaea :

(A) Early Cambrian (B) Late Devonian

(C) Late Cretaceous (D) Late Jurassic.

Identify the correct match of the events with the geological era.

(A) Origin of terrestrial plants - (c); Origin of mammals - (d); Break-up of Pangaea - (a)

(B) Origin of terrestrial plants - (b); Origin of mammals - (d); Break-up of Pangaea - (c)

(C) Origin of terrestrial plants - (d); Origin of mammals - (c); Break-up of Pangaea - (b)

(D) Origin of terrestrial plants - (b); Origin of mammals - (c); Break-up of Pangaea - (d)

137. Proteins in cells can be visualized by the following methods :

(A) Express the gene (coding for the said protein) as a fusion with the green fluorescence protein (GFP) and directly visualize under a fluorescence microscope.

(B) Express the gene (coding for the said protein) as a fusion with the β- galactosidase gene (lac Z) and directly visualize under a phase contrast bright field microscope.

(C) A fluorescence tagged antibody raised against the said protein could be used for visualization in a fluorescence microscope.

(D) Overexpress the protein and directly visualize it under a scanning electron microscope.

Which of the following methods you would choose to visualize a protein in a living cell?

(A) A only (B) A and B only

(C) A and C only (D) D only

138. Industrial products in which bacteria are employed for production are shown in the following table:

I List of products	II Microorganism
A. 2, 3-Butane diol	i) *Leuconostoc*
B. Dextran	ii) *Brevibacterium*
C. Glutamic acid	iii) *Bacillus polymyxa*
D. Cobalamine	iv) *Propionibacterium*

The correct combination is
(A) A – iii; B – i; C – ii; D– iv
(B) A – i; B – ii; C – iii; D– iv
(C) A – iii; B – ii; C – iv; D– i
(D) A – ii; B – iii; C – iv; D– i

139. Insertion lambda vectors are used to create cDNA libraries. In some insertion vectors, the site of insertion is within the cI gene. The recombinants which are cI can then be screened by plating on *E. coli* hfl. The gene hfl encodes a protease that controls lytic-lysogeny decision through which mechanism?

(A) Hfl protease degrades cI, thereby promoting lytic pathway.
(B) Hfl protease degrades cIII and so cIII cannot interact, with cII. The lysogenic pathway is thus preferred.
(C) Hfl protease degrades cII, and therefore, cI synthesis cannot be established, thus lytic pathway is preferred.
(D) Hfl protease degrades INT, the protein involved in phage integration and the lytic pathway is initiated.

140. A transposon carrying a promoterless β-galactosidase (*lac Z*) was used to create insertional mutation in the *vir* region of Ti-plasmid of *Agrobacterium tumefaciens*. All the mutants in which lac Z fusion was in frame were divided into the following three groups :

(A) The virulence of the bacteria was completely lost & the *lac Z* was induced by acetosyringone.
(B) The virulence of the bacteria was reduced and the *lac Z* was induced by acetosyringone.
(C) The virulence of the bacteria was completely lost & *lac Z* was not induced by acetosyringone.

Which of the following assumptions are valid about these mutants?

(A) In group A, the insertion could be in *vir B, C, D* or *G;* in group B, the insertion could be in *vir C* or *E;* and in group C, the insertion could be in *vir A* or *G.*
(B) In group A, the insertion could be in *vir A, B, C* or *D;* in group B, the insertion could be in *vir C* or *D;* and in group C the insertion could be in *vir G.*
(C) In group A, the insertion could be in *vir A;* in group B, the insertion could be in *vir B;* and in group C, the insertion could be in *vir C.*
(D) In group A, the insertion could be in *vir G;* in group B, the insertion could be in *vir B, D* and *E;* in group C, the insertion could be in *vir A.*

141. The MALDI spectrum of a peptide shows a peak at m/z corresponding to 3600. When the ESI spectrum is recorded, peaks at m/z corresponding to 721,904 and 1801 were obtained. When the MALDI MS/MS spectrum was recorded, large number of peaks with m/z less than 3600 were observed. The spectral data indicate that the peptide is

(A) highly impure.
(B) pure with molecular mass of 3600 and partial sequence of the peptide can be determined.
(C) highly unstable and degrades rapidly.
(D) degraded under condition employed for recording ESI spectrum.

142. Molecular beacons (MB) and Taqman (TQ) are used as probes in Real time PCR experiments. Both these probes are based on the principle of FRET and employ a fluorophor (F) and a quencher (Q). However, the mechanisms by which they function are different as illustrated below

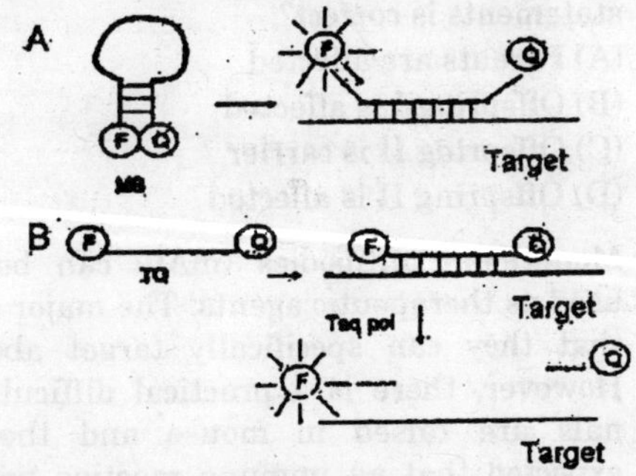

At what stage of the PCR we would be able to detect fluorescence?

(A) Annealing step for both.
(B) Extension step for both.
(C) Annealing for A and Extension for B.
(D) Extension for A and Annealing for B.

143. A 30-residue peptide was treated with trypsin and the tryptic peptides were separated by HPLC. Four peaks A, B, C and D were obtained. Peptides corresponding to A, B, C and D were reduced and alkylated selectively at cysteine residues. The sequences obtained from A, B, C and D after reduction and alkylation were : A, AEK; B, C(S-alkyl) EPGYR and WC (S-alkyl) SPPK; C, C(S-alkyl) EHFR and C (S-alkyl) EHFR and C(S-alkyl) GGK; D, C (S-alkyl) EAFC (S-alkyl) L. The sequence of the 30-residue peptide is

(A)

(B)

(C)

(D)
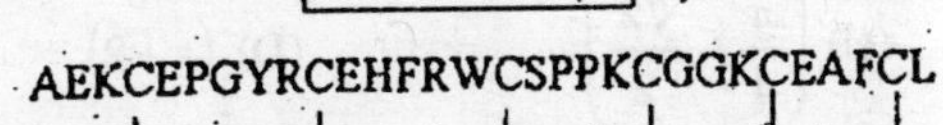

144. To detect mutation (GAG → GTG) allele specific hybridization method is used. Four members of an affected family are investigated. DNA isolated from blood samples of parents and two offsprings are spotted on a membrane after appropriate processing and probed with either TGACTCCTGAGGAGAAGTC (first probe) or TGACTCCTGTGGAGAAGTC (second probe) after labelling. While probed with first oligonucleotide, signals are obtained for the positions where DNA are spotted from parents and offspring II. When probed with second oligonucleotide, signals are obtained at position where DNA from the parents and offspring I are spotted. Results are shown below :

	Father	Mother	Offspring I	Offspring II
First probe	+	+	–	+
Second probe	+	+	+	–

On the basis of the result, which of the following statements is correct?

(A) Parents are affected

(B) Offspring I is affected

(C) Offspring II is carrier

(D) Offspring II is affected

145. Monoclonal antibodies (mAb) can be potentially used as therapeutic agents. The major advantage is that they can specifically target aberrant cells. However, there is a practical difficulty. Monoclonals are raised in mouse and therefore it is expected that an immune reaction will develop if these are injected into humans. It is therefore necessary to 'humanize' monoclonal antibody by

(A) expressing the genes for the mAb in cultured human cells and isolating the mAb from these cells.

(B) replacing the Fv region of a mAb with one derived from a human IaG

(C) replacing C_L and C_H regions of the Mab with that obtained from a human IgG.

(D) taking a human IgG and replacing the CDRs by those derived from the mouse mAb.

CSIR JRF/NET Life Science, June, 2012

PART A

1. The area of the shaded region in cm^2 is

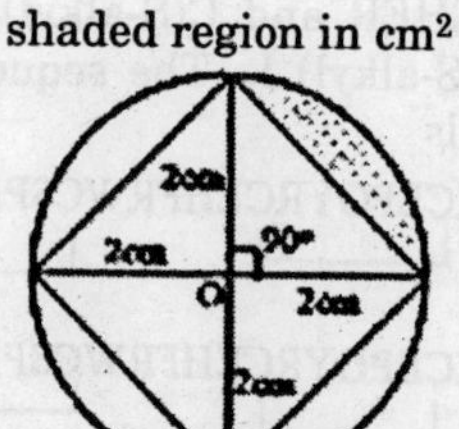

(A) $\left(\pi - \sqrt{2}\right)$ (B) $(\pi - 2)$

(C) $\left(\frac{\pi}{4} - \frac{\sqrt{2}}{2}\right)$ (D) $(\pi + 2)$

2. The angles of a right-angled triangle shaped garden are in arithmetic progression and the smallest side is 10.00 m. The total length of the fencing of the garden in m is

(A) 60.00 (B) 47.32

(C) 12.68 (D) 22.68

3. AB is the diameter of the semicircle as shown in the diagram. If AQ – 2AP then which of the following is correct?

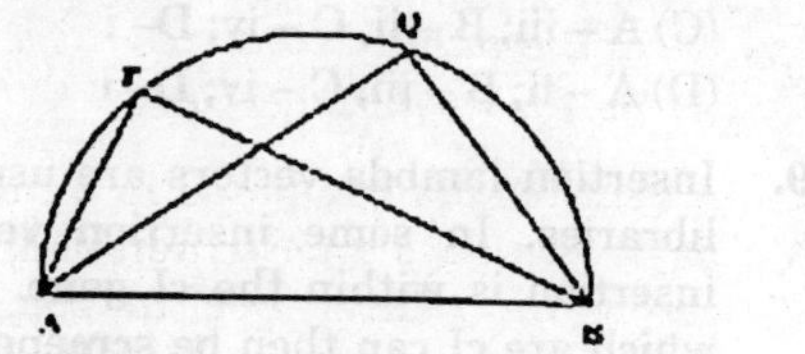

(A) $\angle APB = \frac{1}{2} \angle AQB$ (B) $\angle APB = 2 \angle AQB$

(C) $\angle APB = \angle AQB$ (D) $\angle APB = \frac{1}{4} \angle AQB$

4. The rabbit population in community A increases at 25% per year while that in B increases at 50% per year. If the percent populations of A and B are equal, the ratio of the number of the rabbits in B to that in A after 2 years will be

(A) 1.44 (B) 1.72

(C) 1.90 (D) 1.25

5. Two moles each of O_2 and H_2 are in two separate containers, each of volume V_o and at 150°C and 1 atmosphere. The two are made to react in a third container to form water vapour until H_2 is exhausted. When the temperature of the mixture in the third container was restored to 150°C, its pressure became 1 atmosphere. The volume of the third container must be

(A) V_o (B) $5 V_o/4$

(C) $3 V_o/2$ (D) $2 V_o$

6. Helium and argon gases in two separate containers are at the same temperature and so have different root-mean square (r.m.s.) velocities. The two are mixed in a third container keeping the same temperature. The r.m.s. velocity of the helium atoms in the mixture is

(A) more than what it was before mixing.

(B) less than what it was before mixing.

(C) equal to what it was before mixing.

(D) equal to that of argon atoms in the mixing.

7. The mineral tale is used in the manufacture of soap because it

(A) gives bulk to the product

(B) kills bacteria

(C) gives fragrance

(D) is soft and does not scratch the skin

Which of the above statements is/are correct?

(A) (d) (B) (a) and (c)

(C) (a) and (b) (D) (a) and (d)

8. 100 g of an inorganic compound X $5H_2O$ containing a volatile impurity was kept in an oven at 150°C for 60 minutes. The weight of the residue after heating is 8 g. The percentage of impurity in X was

(A) 10 (B) 8

(C) 20 (D) 80

9. On a certain night the moon in its waning phase was a half-moon. At midnight the moon will be

(A) on the eastern horizon

(B) at 45° angular height above the eastern horizon.

(C) at the zenith.

(D) on the western horizon.

10. A gemstone is irradiated in a nuclear reactor for 5 days. Ten days after irradiation, the activity of the chromium radioisotope in the gemstone is 600 disintegrations per hour. What is the activity of chromium radioisotope 5 days after irradiation if its half life is 5 days?

(A) 300 (B) 150

(C) 2400 (D) 1200

11. Displacement versus time curve for a body is shown in the figure. Select the graph that correctly shows the variation of the velocity with time

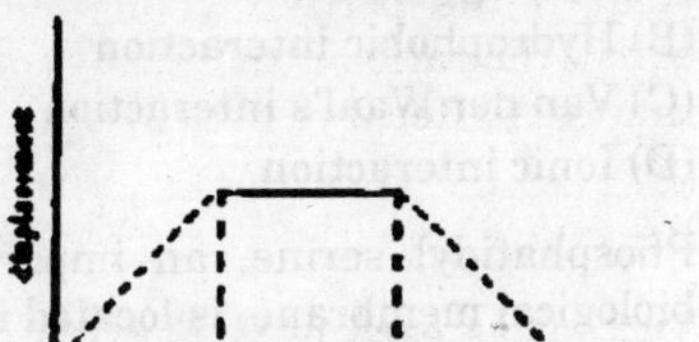

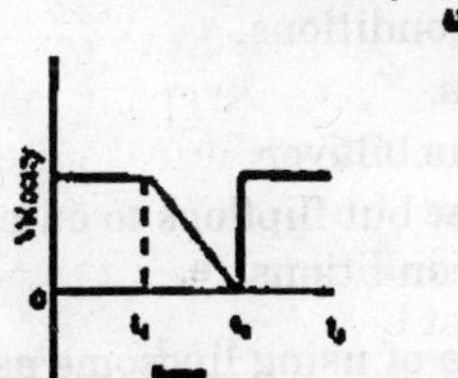

(A)

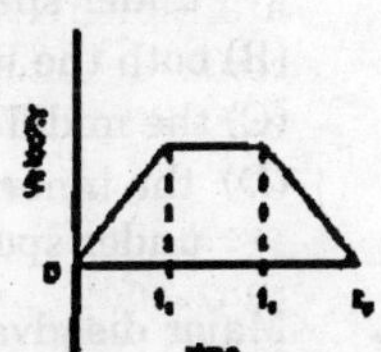

(B)

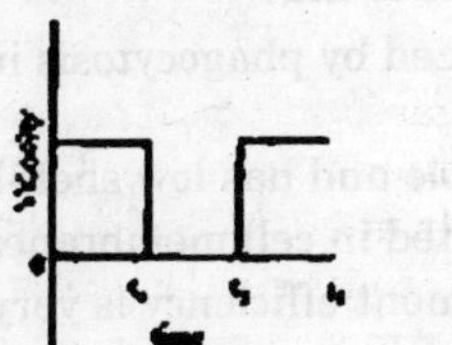

(C)

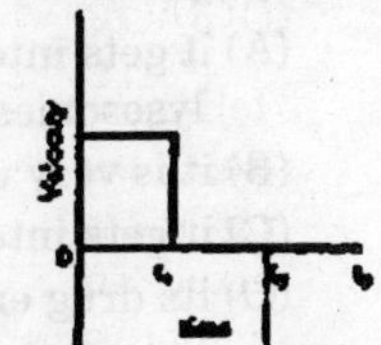

(D)

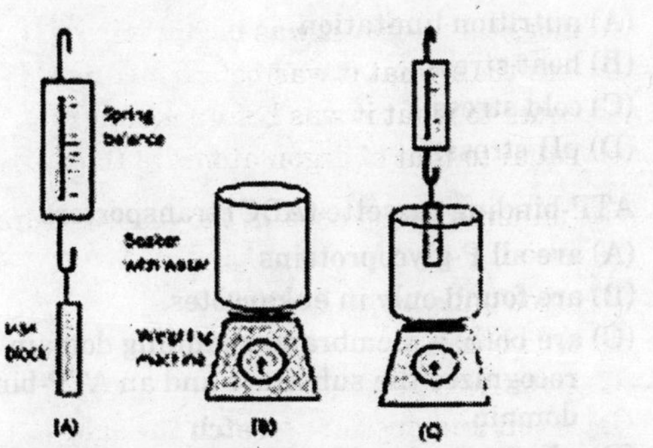

12. The spring balance in Fig. A reads 0.5 kg and the pan balance in Fig. B reads 3.0 kg. The iron block suspended from the spring balance is partially immersed in the water in the beaker (Fig. C). The spring balance now reads 0.4 kg. The reading on the pan balance in Fig. C is

(A) 3.0 kg (B) 2.9 kg

(C) 3.1 kg (D) 3.5 kg

13. The ends of a rope are fixed to two pegs, such that the rope remains slack. A pencil is placed against the rope and moved, such that the rope always remains taut. The shape of the curve traced by the pencil would be a part of

(A) a circle (B) an ellipse

(C) a square (D) a triangle

14. During ice skating, the blades of the ice skater's shoes exert pressure on the ice. Ice skater can efficiently skate because

(A) ice gets converted to water as the pressure exerted on it increases.

(B) ice gets converted to water as the pressure exerted on it decreases.

(C) the density of ice in contact with the blades decreases.

(D) blades do not penetrate into ice.

15. Four sedimentary rocks A, B, C and D are intruded by an igneous rock R as shown in the cross-section diagram, which of the following is correct about their ages?

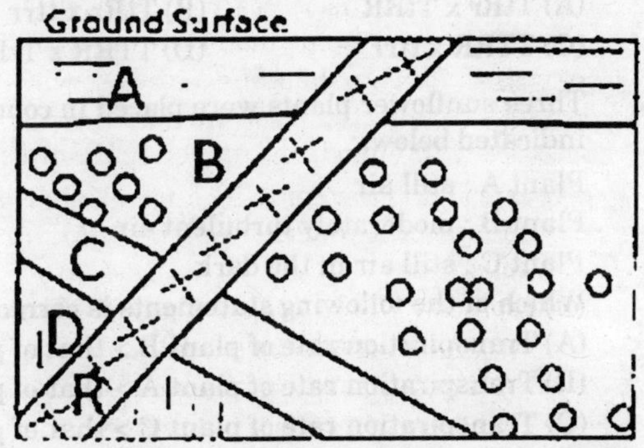

(A) A is the youngest followed by B, C, D and R.

(B) R is the youngest followed by A, B, C and D.

(C) D is the youngest followed by C, B, A and R.

(D) A is the youngest followed by R, B, C and D.

16. The strain in a solid subjected to continuous stress is plotted.

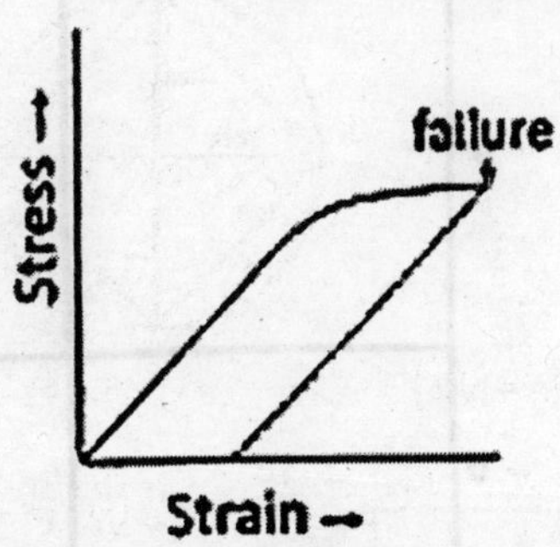

Which of the following statements is true?

(A) The solid deforms elastically till the point of failure.

(B) The solid deforms plastically till the point of failure.

(C) The solid comes back to original shape and size on failure.

(D) The solid is permanently deformed on failure.

17. **Growth of an organism was monitored at regular intervals of time, and is shown in the graph below. Around which time is the rate of growth zero?**

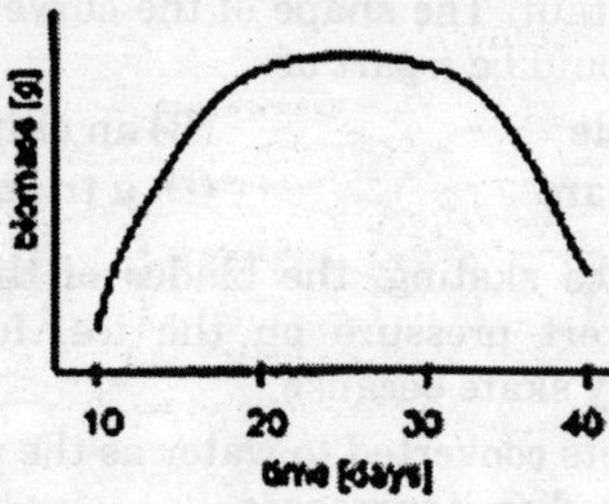

(A) Close to day 10.
(B) On day 20.
(C) Between days 20 and 30.
(D) Between days 30 and 40.

18. **A *Tall* plant with Red seeds (both dominant traits) was crossed with a dwarf plant with white seeds. If the segregating progeny produced equal number of tall red and dwarf white plants, what would be the genotype of the parents?**

(A) TtRr x TtRR (B) TtRr x ttrr
(C) TTRR x ttrr (D) TTRR x TtRr

19. **Three sunflower plants were placed in conditions as indicated below.**

Plant A : still air
Plant B : moderately turbulent air
Plant C : still air in the dark
Which of the following statements is correct?

(A) Transpiration rate of plant B > that of plant A.
(B) Transpiration rate of plant A > that of plant B.
(C) Transpiration rate of plant C > that of plant A.
(D) Transpiration rate of plant C > that of plant A> that of plant B.

20. **Which of the following is indicated by the accompanying diagram?**

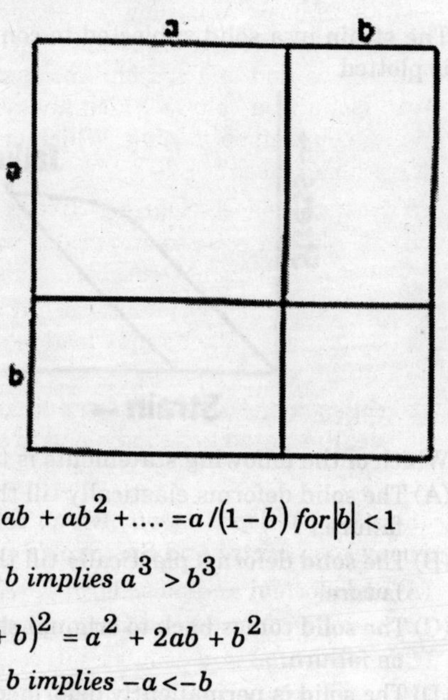

(A) $a + ab + ab^2 + ... = a/(1-b)$ for $|b| < 1$
(B) $a > b$ implies $a^3 > b^3$
(C) $(a+b)^2 = a^2 + 2ab + b^2$
(D) $a > b$ implies $-a < -b$

PART B

21. **Which nitrogen of adenosine gets protonated if the pH of the nucleoside is lowered from 7 to 3?**

(A) N1
(B) N3
(C) N7
(D) N9

22. **The oligopeptide, F-A-R-P-M-T-S-R-P-G-F, is treated with trypsin, chymotrypsin and carboxypeptidase-B. Apart from the original peptide, the number of fragments obtained will be :**

(A) 4 (B) 3
(C) 2 (D) 0

23. **Which one of the following interaction plays a major role in stabilizing B-DNA?**

(A) Hydrogen bond
(B) Hydrophobic interaction
(C) Van der Waal's interaction
(D) Ionic interaction

24. **Phosphatidyl serine, an important component of biological membrane, is located in**

(A) the outer leaflet but flipflops to inner leaflet under specific conditions.
(B) both the leaflets.
(C) the middle of the bilayer.
(D) the inner leaflet but flipflops to outer leaflet under specific conditions.

25. **Major disadvantage of using liposome as a targeted drug delivery vehicle is that**

(A) it gets internalized by phagocytosis inside lysosomes.
(B) it is very unstable and has low shelf-life.
(C) it gets intercalated in cell membranes.
(D) its drug entrapment efficiency is very low.

26. **Major stimulus for spore formation in bacteria is**

(A) nutrition limitation
(B) heat stress.
(C) cold stress.
(D) pH stress.

27. **ATP-binding cassette (ABC) transporters**

(A) are all P-glycoproteins.
(B) are found only in eukaryotes.
(C) are both: a membrane-spanning domain that recognizes the substrate and an ATP-binding domain.
(D) affect translocation by forming channels.

28. **All cytosolic proteins have nuclear export signals that allow that to be removed from the nucleus when it reassembles after**

(A) moiosis.
(B) mitosis
(C) both meiosis and mitosis
(D) DNA replication.

29. Site-specific recombination results in precise DNA rearrangements, which is limited to specific sequences. The enzymes that are important to carry out the process are

(A) restriction endonuclease and DNA polymerase.
(B) nuclease and ligase.
(C) DNA polymerase and ligase.
(D) DNA polymerase and DNA gyrase.

30. Which of the following statements is NOT true about small interferring RNA (siRNA)?

(A) siRNA has a 21-25 nucleotide sequence with 2 nucleotides overhanging at the 3' end.
(B) siRNA is processed by RNA-protein complex RISC.
(C) siRNA is often induced by viruses.
(D) siRNA does not generally act at the level of transcription.

31. Which of the following statements is INCORRECT in relation to treatment of pre-B cells with phorbol esters?

(A) Phorbol esters activate NFxB for translocation into the nucleus.
(B) Phorbol esters activate protein kinase C.
(C) Phorbol esters lead to phosphorylation of NFxB
(D) Phorbol esters remove the inhibitor from inactive NFxB complex in the cytoplasm.

32. Presence of an internal ribosome entry site (IRES) in mRNA

(A) inhibits its translation.
(B) promotes its post-transcriptional processing.
(C) has no impact on its translation.
(D) promotes its translation under adverse conditions.

33. Regulatory elements for expression of ribosomal RNA genes reside in the

(A) transcribed spacer region.
(B) non-transcribed spacer region.
(C) 5'flanking region of individual ribosomal RNA genes.
(D) internal regions within the gens.

34. *Mycobacterium tuberculosis* is an intra-cellular bacterium. It prefers to infect

(A) macrophages. (B) B-cells.
(C) T-cells. (D) neutrophils.

35. Integrin molecules link extracellular matrix (ECM) to the actin cytoskeleton of cell. Integrin binds to which of the following ECM macromolecules?

(A) Laminin
(B) Collagen
(C) Fibronectin
(D) Vitronectin

36. CD19 is a marker for

(A) B-cells. (B) T-cells.
(C) macrophages. (D) NK cells.

37. Given below are fate maps of two organisms and the patterns by which embryos undergo cleavage. Which of the following is/are the right combination(s)?

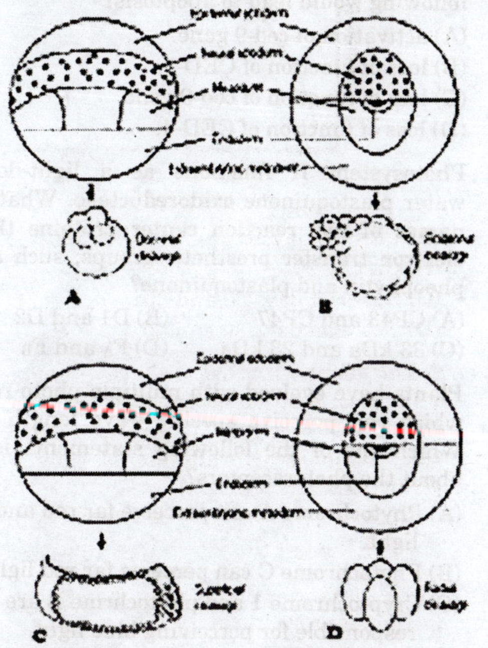

(A) B only (B) B and A
(C) A and C (D) B and D

38. Which one of the following matches of oncogene-protein product is NOT correct?

(A) *erb* A - thyroid hormone receptor.
(B) *erb* B - epidermal growth factor receptor.
(C) *ras* - guanine nucleotide binding protein with GTPase activity.
(D) *fas* - platelet-derived growth factor receptor.

39. Spermatogonial stem cell undergoes extensive metamorphosis to become a spermatozoan. Meiosis leads to the formation of spermatic containing 22 autosomes and one sex chromsome. A male mouse was found in a colony which always produced only female pups upon mating. Which one of following is a possible reason?

(A) Spermiogenesis was defective.
(B) All spermatogonial stem cells contained only X and no Y chromosome.
(C) activation of Y chromosome linked post-meiotic death related gene may lead to such a situation.
(D) activation of X chromosome linked post-meiotic death related gene may lead to such a situation.

40. In case of *Xenopus laevis,* which cells make up the Nieuwkoop centre and Spemann's organizer?

(A) endodermal and mesodermal, respectively.
(B) mesodermal and endodermal, respectively.
(C) endodermal and ectodermal, respectively.
(D) ectodermal and endodermal, respectively.

41. The ced-9 gene appears to be a binary switch that regulates cellular survival and apoptosis in nematodes. Considering that CED-9 protein can bind to and inactivate CED-4, which of the following would lead to apoptosis?

(A) activation of ced-9 gene.
(B) loss of function of CED-3.
(C) loss of function of ced-9 gene.
(D) loss of function of CED-4.

42. Photosystem II functions as a light-dependent water plastoquinone oxidoreductase. What are the names of two reaction center proteins that bind electron transfer prosthetic groups, such as P680, pheophytin and plastoquinone?

(A) CP43 and CP47 (B) D1 and D2
(C) 33 kDa and 23 kDa (D) F_A and F_B

43. Plants have evolved with multiple photo receptors, which can perceive specific wave length of light. Which one of the following statements is correct about the photoreceptors?

(A) Phytochrome A can perceive far red and blue light.
(B) Phytochrome C can perceive far red light.
(C) Cryptochrome 1 and phytochrme B are responsible for perceiving blue light.
(D) Phytochrome B can predominantly perceive far red light.

44. Which one of the following statement describes the proces of phloem loading?

(A) Triose phosphate is transported from the chloroplast to crytosol
(B) Sugars are transported into the sieve elements and companion cells.
(C) Sugars are transported from producing cells in the mesophyll to cells in the vicinity of the sieve elements.
(D) Solutes are transported from roots to the shoots.

45. Which one of the following combinations of secondary metabolite biosynthetic pathways result in the biosynthesis of terpenes?

(A) Mevalonic acid and MEP pathways
(B) Malonic acid and and MEP pathways
(C) Shikimic acid and Malonic acid pathways
(D) Shikimic acid and Mevalonic acid pathways

46. Which one of the following changes will occur in the cell membrane of nodal tissue of heart, which results in an increased heart rate due to stimulation of sympathetic nerves?

(A) Opening of sodium channels is facilitated.
(B) Potassium conductance is increased.
(C) Opening of L-calcium channels are facilitated.
(D) `h' channels are inhibited.

47. A person takes 1.0 ml of insulin injection daily at 8.00 AM. His son gave him 1.5 ml insulin at 8.00 AM considering the father will go to party and eat more during lunch. The father also avoided breakfast, as he planned to eat more during lunch. Which one of the following events will occur?

(A) Father will be normoglycemic
(B) Father will be in hypoglycemic condition before lunch.
(C) Father will be in hyperglycemic condition before lunch.
(D) Blood glucose of father will be low after taking lunch.

48. How many genetically different gametes can be made by an individual of genotype AaBbccDDEe, assuming they are independently assorting?

(A) 3 (B) 5
(C) 8 (D) 32

49. Mutation at two different loci of the same gene X results in altered functions. These two mutated versions of the gene X are called

(A) allcles.
(B) complementation groups.
(C) interrupted genes
(D) linkage groups.

50. A gene encoding tRNA undergoes a mutational event in its anticodon region that enables it to recognize a mutant nonsence codon and permit completion of translation. Such a mutation is known as

(A) silent mutation (B) neutral mutation
(C) reversion (D) nonsence suppressor.

51. Two pure lines of corn have mean cob length of 9 and 3 inches, respectively. The polygenes involved in this trait exhibit additive gene action. Crossing these two lines is expected to produce a progeny population with mean cob length (in inches) of :

(A) 12.0 (B) 7.5
(C) 6.0 (D) 2.75

52. Which of the following organism is widely used as a biocontrol agent in organic farming?

(A) *Rhizobium tropicii*
(B) *Trichoderma viridis*
(C) *Fusarium axysporum*
(D) *Nostoc muscorum*

53. A paraphylctic group

(A) contains unreated organisms.
(B) includes the most recent common ancestor but not all of its descendents.
(C) includes all the representatives of a clade but not the most recent common ancestor.
(D) contains all the representatives of a clade and the most recent common ancestor.

54. Which of the following is NOT an adaptive modification in a xerophytic plant?

(A) Strongly development sclerenchyma
(B) Sunken stomata
(C) Sparse stomata
(D) Presence of lacunar tissues.

55. If milk is left open, lactose is fermented first to produce acid. This is followed by proteolytic bacterial activity which increases the pH. Ultimately, milk fats are degraded to produce rancidity. This is an example of

(A) ecological succession.
(B) microbial antagonism.
(C) interference competition.
(D) microevolution.

56. Symbiotic biological nitrogen fixation takes place with the association between a plant and a nitrogen fixing prokaryote as shown in the following table :

List of plants	Nitrogen fixing
A. *Soybean*	(i) *Frankia*
B. *Casuarina*	(ii) *Bradyrhizobium*
C. *Gunnera*	(iii) *Anabaena*
D. *Azolla*	(iv) *Nostoc*

The correct combination is :

(A) A-i, B-ii, C-iii, D-iv
(B) A-ii, B-i, C-iv, D-iii
(C) A-iii, B-ii, C-i, D-iv
(D) A-iv, B-iii, C-ii, D-i

57. Secondary sewage treatment involves

(A) physical removal of solids from polluted water by filtration and sedimentation
(B) removal of chemical remains by precipitation.
(C) removal of dissolved organic compounds by activated sludge or trickling fliter.
(D) removal of microbial pathogens by chlorination or ozonization.

58. Based on per molecule, which of the following gas has the most powerful greenhouse effect?

(A) CO_2 (B) CH_4
(C) N_2O (D) CFCs

59. Sexual selection results in variation in the reproductive success of males, often due to female choice with particular phenotypes. This type of sexual selection occurs because

(A) males cannot compete with other males.
(B) cost of breeding is higher for females as compared to males.
(C) inappropriate mating results in a similar reduction in fitness of females and males.
(D) males are a limiting resource for females.

60. Among the following events in the history of life

a - prokaryotic cell
b - cukaryotic cell
c- natural selection
d - organic molecules
e - self-replicating molecules

Which is the correct chronological order?

(A) d, c,e, a, b
(B) d, e, a, b, c
(C) e, d, a, c, b
(D) d, e, a, c, b

61. The Hardy-Weinberg principle comes from considering what happens when Mendelian genes act in a population. The model predicts that there will be no change in alele frequencies when

(A) migration into the population suffers at a steady rate.
(B) the population suffers a bottleneck.
(C) a rare new mutation is associated with a sharp increase in fitness.
(D) no evolationary process is at work.

62. Which of the following is responsible for initiation of maternal behaviour in the first-time pregnant rats after parthrition?

(A) Higher prolactin levels in blood.
(B) Stimulation of sensory receptors during delivery.
(C) Changes of uterine volume.
(D) Presence of male rats.

63. To replace animal use in testing hepatic toxicity of a drug on trial, which one of the following would be used *in vitro* to be closest to the *in vivo* scenario?

(A) Liver cells,
(B) Hepatic cell lines.
(C) Liver slices.
(D) Co-culture of liver parenchymal cells and Kupfer cells.

64. Which of the following does not represent a strategy for phytoremediation?

(A) Phytodegradation.
(B) Phytomining.
(C) Continuous removal through hyper accumulators.
(D) Chelate-mediated extraction of pollutants.

65. The word ``fermentation" is used in bio-chemistry and microbial technology to denote different phenomena. If the former is called C and latter is called T, which of the following statements is true?

(A) All C is T but all T is not C.
(B) All T is C but all C is not T.
(C) T is always a product of genetic engineering while C is not.
(D) C is always an acrobic process, while T can be acrobic or anaerobic.

66. Which of the following statements is NOT true during the infection of plant cells with *Agrobacterium*?

(A) The protein products of virulence genes *vir A* and *vir G* perceive acetosyringone.
(B) The Vir B protein forms a connection between *Agrobacterium* and the plant cell and facilitates T-DNA transfer into the plant.
(C) The T-DNA is excised and bound to Vir D2 protein.
(D) The T-DNA, after becoming coated with Vir F binds to phosphorylated VIP1, which allows the complex to enter the plant's nucleus.

67. Among existing technologies, which of the following vector system would you prefer to use for generating a library for 140 kb eukaryotic genomic DNA fragments, while giving due consideration to size as well as stability of the insert?

(A) Phage
(B) Cosmid
(C) Bacterial artificial chromosome (BAC)
(D) Yeast artificial chromosome (YAC)

68. If r denotes the correlation coefficient and m denotes the slope of regression line, interchanging X and Y axes would

(A) change m but not r
(B) change r but not m
(C) change both r and m
(D) not change r or m

69. The use of biotinylated secondary antibody in ELISA

(A) increases the sensitivity of the assay but compromises the specificity
(B) increases the sensitivity of the assay without compromising the specificity
(C) does not alter either sensitivity or specificity
(D) decreases both sensitivity and specificity

70. Which is the best method for checking mycoplasma contamination in a mammalian cell line?

(A) Southern hybridization
(B) ELISA
(C) PCR
(D) Western hybridization

PART C

71. Phosphorylation of ADP to ATP occurs through energy metabolism, comprising oxidative phosphorylation or substrate level phosphorylation or photophosphorylation (in plants). ATP can also be formed from ADP through the action of adenylate kinase. Crystal structure determination of adenylate kinase shows that the C-terminal region has the sequence.

–Val–Asp–Asp – Val – Phe – Ser – Gin – Val – Cys – Thr – His – Leu – Asp – Thr – Leu – Lys.

What can be possible conformation of the sequence?

(A) A helix that is not amphipathic
(B) Amphipathic helix
(C) Leucine zipper helix
(D) Beta hairpin

72. Consider a 51-residue long protein containing only 100 bonds about which rotation can occur. Assume that 3 orientations per bond are possible. Based on these assumptions, how many conformations will be possible for this protein?

(A) 3^{100}
(B) 100^3
(C) 3^{51}
(D) 51 x 100 x 3

73. A plot V/[S] versus V is generated for an enzyme catalyzed reaction, and a straight line is obtained. Indicate the information that can be obtained from the plot.

(A) V_{max} and turnover number K_m can be obtained only from a plot of 1/V versus 1/[S].
(B) K_m/V_{max} from the slope.
(C) V_{max}, K_m and turnover number.
(D) Only K_m and turnover number.

74. The following reactions are part of the citric acid cycle. The numbers in parenchesis indicate the number of carbon atoms in each molecule.

isocitrate $(6) \rightarrow \alpha -$ *ketoglutarate* $(5)\rightarrow$

succinyl CoA $(4)\rightarrow$ *succinate* $(4)\rightarrow$ *fumarate* (4)

Which one of the following sequences of reaction system A → D is correct?

(A) 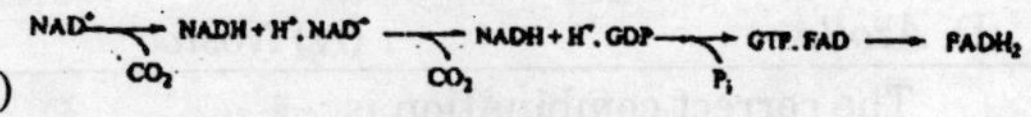

(B) 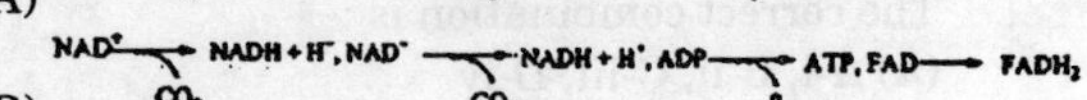

(C) 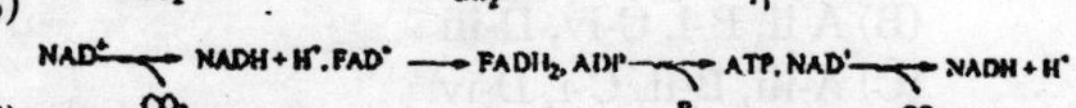

(D) 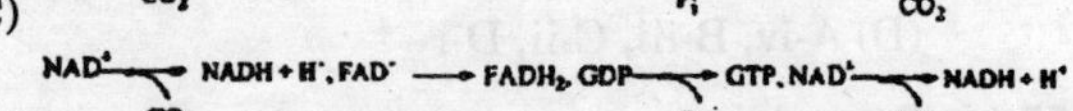

75. Phosphoglucomutase is added to 0.1 M glucose-1-phosphate (G-6-P). The standard free energy change of the reaction, G-6-P ⇒ G-1-P is 1.8 kcal/mole at 25°C. The equilibrium concentrations of G-6-P and G-1-P, respectively, are :

(A) 96 mM, 45 mM
(B) 100 mM, 0 mM
(C) 45 mM, 96 mM
(D) 0 mM, 100 mM

76. Differential scanning calorimetric study of calf thymus DNA was carried out to measure midpoint of thermal denaturation (T_m), ΔH_m (enthalphy change at T_m) and ΔCp (constant - pressure heat capacity change), It has been observed that ΔCp = 0, T_m = 75.5°C and ΔHp – 50.4 kcal/mole. The Gibbs free energy change at 37°C is

(A) 25.5 kcal/mole
(B) 2.6 kcal/mole
(C) 0.6 kcal/mole
(D) 5.6 kcal/mole

77. Cystic fibrosis (CF) transmembrane conductance regulator (CFTR) protein is known to be a cAMP-dependent Cl channel. CF patients (with mutant CFTR proteins) show reduced Cl^- permeability and as a result exhibit elevated Cl- level in sweat. To prove this, CFTR proteins (both wild type and mutant) are inserted in a model membrane (liposome) and Cl- transport is followed with radioactive Cl^-. It is known that topology of CFTR in membrane is very important for its function. Despite no proteolytic degradation or denaturation of CFTR proteins, wild type CFTR failed to transport Cl^- in liposome.

Which of the following is the correct explanation of this?

(A) CFTR protein gets mutated during insertion in liposomes.

(B) CFTR protein loses affinity with Cl- ions.

(C) CFTR protein gets wrongly inserted in liposomes.

(D) CFTR protein losses channel forming property in liposomes.

78. The respiratory chain is relatively inaccessible to experimental manipulation in intact mitochondria. Upon disrupting mitochondria with ultrasound, however, it is possible to isolate functional submitochondrial particles, which consist of broken cristae that have resealed inside out into small closed vesicles. In these, vesicles the components that originally faced the matrix are now exposed to the surrounding medium. This arrangement helps in studying of electron transport and ATP synthesis because:

(A) it is difficult to manipulate the concentration of small molecules (NADH, ATP, ADP, Pi) in the matrix of intact mitochondria.

(B) in broken cristae, the enzymes and other molecules responsible for electron transport are more active.

(C) intact mitochondria are more unstable than broken cristae.

(D) purification of intact mitochondria is not possible.

79. Assuming that the histone octamer forms a cylinder 9 nm in diameter and 5 nm in height and that the human genome forms 32 million nucleosomes, what fraction (approximately) of the volume of nucleus (6 μm diameter) is occupied by histone octamers?

(A) 1/21 (B) 1/11

(C) 10/21 (D) 10/11

80. Hoechst 33342 is a membrane- permeant dye that fluorescence when it binds to DNA through a interculating process. If a population of cells is incubated briefly with Hoechst dye and sorted in a flow cytometer, the cells display various levels of fluorescence in different phases of cell cycles as shown in figure below (marked as X, Y and Z).

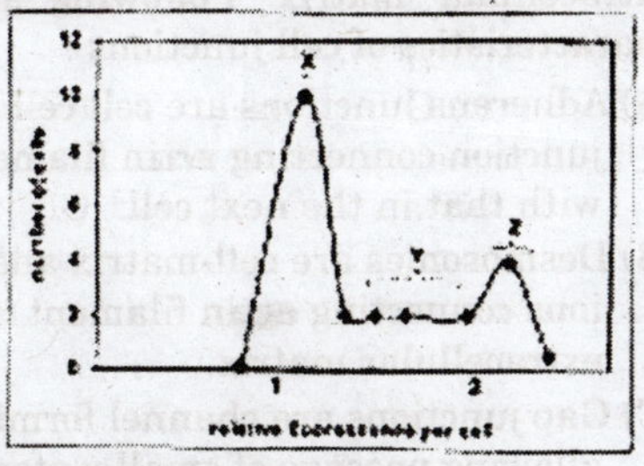

Which of the following is correct?

(A) X is G_1, Y is G_2 + M and Z is S

(B) X is G_1, Y is S and Z is G_2 + M

(C) X is S, Y is G_2 + M and Z is G_1

(D) X is S, Y is G_1 and Z is G_2 + M

81. The scatter plot of growth rate and growth yield for 100 random environmental isolates of bacteria is shown below:

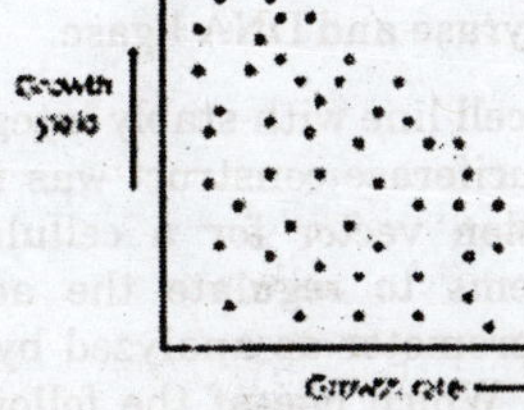

Which of the following can be inferred from the data?

(A) The two parameters are not related.

(B) Growth rate is inversely proportional to growth yield.

(C) Growth yield is negatively correlated with growth rate.

(D) High growth rate cannot be accompanied by high growth yield.

82. Double stranded DNA replicates in a semi-conservative manner. In an *in vitro* DNA synthesis reaction, dideoxy CTP and dideoxy CMP were individually added in excess (in separate reaction tubes) in addition to dNTPs and other necessary reagents. Rare of DNA synthesis was measured by incorporation of ^{3}H-thymidine. The four graphs drawn below represent the rate of DNA synthesis in two separate reaction tubes.

Which of the following graphs represents the expected data?

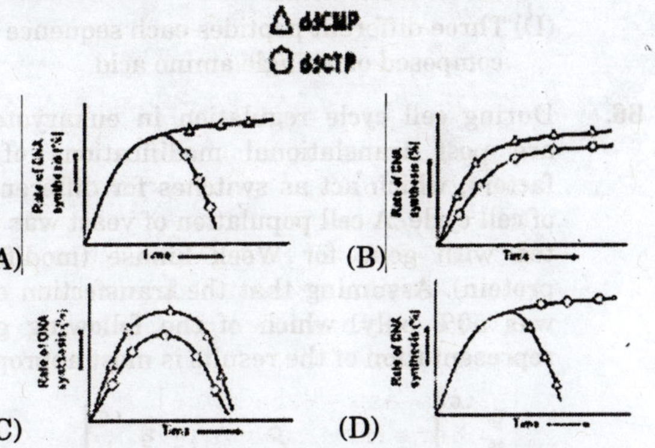

83. In semi-conservative mode of DNA replication, two parental strands unwind and are used for synthesis of new strands following the rule of complementary base pairing. Synthesis of complementary strands require that DNA synthesis proceeds in opposite direction, while the double helix is progressively unwinding and replicating in only one direction. One of the DNA strands is continuously synthesized in the same direction as the advancing replication fork and is called leading strand whereas the other strand is synthesized discontinuously in segments and is referred to as lagging strand. These short fragments made discontinuously are labeled as Okazaki fragments. These Okazaki fragments need to be matured into continuous DNA strand by which one of the following combination of enzymes?

(A) DNA Pol III and DNA ligase.
(B) DNA Pol I and DNA ligase.
(C) DNA Pol II and DNA ligase.
(D) DNA gyrase and DNA ligase.

84. A reporter cell line with stably integrated retroviral promoter-luciferase construct was transfected with an expression vector for a cellular protein. The protein seems to regulate the activation of the retroviral promotor as analyzed by luciferase activity assay. Which one of the following techniques will you use to show ``*in vivo*'' recruitment of the cellular protein on the integrated retrovial promoter?

(A) Electrophoretic mobility shift assay.
(B) RNAse protection assay.
(C) DNAse hypersensitivity assay.
(D) Chromatin immunoprecipitation assay.

85. A synthetically prepared mRNA contains repetitive AU sequences. The mRNA was incubated with mammalian cell extract which contains ribosomes, tRNAs and all the factors required for protein synthesis. Assuming no initiation codon is required for protein synthesis, which of the following peptides will most likely be synthesized?

(A) A single peptide composed of the same amino acid sequence.
(B) A single peptide with alternating sequence of two amino acids.
(C) A single peptide with alternating sequence of three amino acids.
(D) Three different peptides each sequence composed of a single amino acid.

86. During cell cycle regulation in eukaryotes, there are post translational modifications of protein factors, which act as switches for different phases of cell cycle. A cell population of yeast was transfected with gene for Week kinase (modifies Code protein). Assuming that the transfection efficiency was 50% only. which of the following graphical representation of the results is most appropriate?

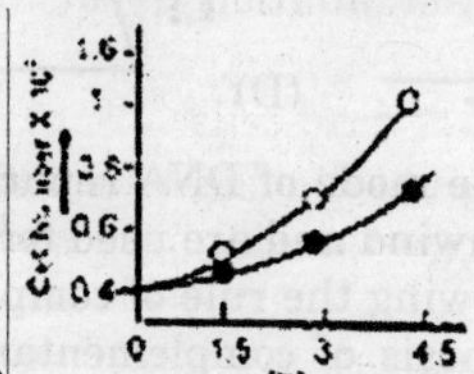
(A)

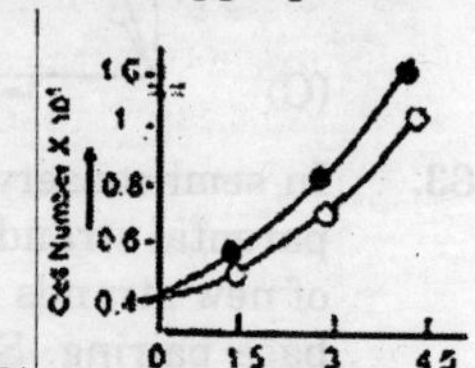
(B)

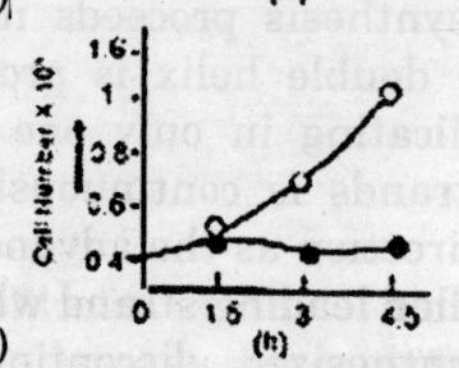
(C)

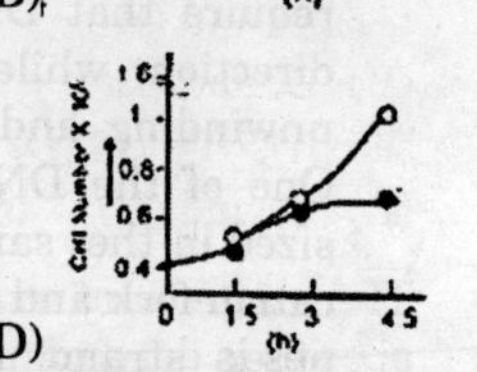
(D)

87. The *lac* operon in *E. coli*, is controlled by both the *lac* repressor and the catabolite activation protein CAP. In an *in vivo* experiment with *lac* operon, the following observations were made :

(A) cAMP levels are high.
(B) Repressor is bound with allolactose.
(C) CAP is interacting with RNA polymerase.

Which one of the following conclusions is most appropriate based on the above observations?

(A) Glucose and lactose are present.
(B) Glucose is present and lactose is absent.
(C) Both are absent.
(D) Glucose is absent and lactose is present.

88. Upon ligand binding, cell surface receptors move laterally to be capped and internalized. *Leishmania*, a protozoan parasite, can use several receptors on macrphages to get internalized. One of them is Toll-like receptor 2 (TLR2) that binds lipophosphoglycan on *Leishmania*. Once internalized, the parasite is destroyed in the phagolysome. Which of the following treatments of *Leishmania*-infected macrophages will result in lowest parasite number in macrophages?

(A) membrane cholesterol-deplecting drug, β-methyl cyclodextrin (β-MCD).
(B) ammonium chloride that increases lysosomal pH.
(C) both β -MCD and ammonium chloride.
(D) anti-TLR2 antibody.

89. Level of follicle stimulating hormone (FSH) during infancy and adulthood is the same but spermaogenesis is seen only during adulthood. mRNA levels coding for FSH receptor are also found to be the same in testis of both age groups. Which of the following investigations will clarify this paradox a little more?

(A) Culture testicular cells and add LH to see testosterone production.
(B) Culture testicular cells and add testosterone to see comparative rise in FSH mRNA from both age groups.
(C) Culture testicular cells and add FSH to see comparative rise in cAMP production by both age groups.
(D) Add both LH and FSH to testicular cells and evaluate cAMP production.

90. In a tissue, cells are bound together by physical attachment between cell to cell or between cell to extracellular matrix. Following are some of the characteristics of cell junctions :

(A) Adherens junctions are cell cell anchoring junction connecting actin filament in one cell with that in the next cell.
(B) Desmosomes are cell-matrix anchoring junctions connecting actin filament in one cell to extracellular matrix.
(C) Gap junctions are channel forming junctions allowing passage of small water soluble molecules from cell to cell.
(D) Tight junctions are occluding junctions, which seal gap between two cells.

(E) Hemidesmosomes are cell-matrix anchoring junctions connecting intermediate filament in one cell to extracellular matrix.

Which of the following combination of statements is **NOT** correct?

(A) (A) and (B) (B) (A) and (C)
(C) (C) and (D) (D) (D) and (E)

91. Oncogenes and tumor suppressor genes are termed as cancer-critical genes. Increasingly powerful tools are now available for systematicaly searching the DNA or mRNAs of cancer cells for either significant mutations or altered expression. To identify independently an oncogene or a tumor supressor gene, which of the following would be the most convincing tests to use?

(A) Transgenic mice that overexpress the candidate oncogene and knockout mice that lack candidate tumor suppressor gene.
(B) Transgenic mice that overexpress the candidate tumor suppressor gene and knockout mice that lack candidate oncogene.
(C) Transgenic mice that overexpress the candidate oncogene and tumor suppressor gene.
(D) Knockout mice that lack the candidate oncogene and tumor suppressor gene.

92. A large protein of a pathogenic bacterium has been enzymatically digested to generate a mixture of peptides ranging in size from 3 to 8 amino acids in length. Peptide mixtures were than administered in experimental animals to generate peptide-specific antibodies. In order to develop diagnostics for the bacteria, the antisera were used for Western blotting to detect bacterial antigen. Western blotting failed despite the use of a wide range of antisera concentrations. What is the most probable cause of the problem?

(A) Peptide-specific antibody mixture is unstable.
(B) Peptide-specific antibodies were not generated as adjuvant was not administered.
(C) Peptide-specific antibodies were not generated as they were not coupled to a protein carrier.
(D) Peptide-specific antibodies could not recognize the bacterial antigen.

93. A mouse was primed with trintrophenyl-lipopoly-saccharide (TNP-LPS) whereas another mouse was primed with TNP-Keyhole limpet hemocyanin (TNP-KLH). After three weeks, these mice were sacrificed and splenic cells were fractionated to B cells and T-cells. B cells from TNP-LPS primed mice were co-cultured with T cells from TNP-LPS or TNP-KLH primed mice. Similarly, B-cells from TNP KLH primed mice were co-cultured with the T-cells from TNP LPS or TNP-KLH primed mice. So, we have four co-cultures :

(A) $B^{TNP\text{-}LPS} \times T^{TNP\text{-}LPS}$
(B) $B^{TNP\text{-}LPS} \times T^{TNP\text{-}KLH}$
(C) $B^{TNP\text{-}KLH} \times T^{TNP\text{-}LPS}$
(D) $B^{TNP\text{-}KLH} \times T^{TNP\text{-}KLH}$

Among these co-cultures, where do you expect the highest IpG production?

(A) (A) (B) (B)
(C) (C) (D) (D)

94. The functionability of the *pax6* gene in the formation of optic and nasal structures may be attributed to the following:

(A) Pax 6 makes the optic vesicle competent and allows lens formation.
(B) The optic vesicle can induce any part of the head ectoderm to form the nasal and optic structures, due to the presence of Pax6.
(C) Pax 6 renders the head ectoderm competent to receive signals from the optic vesicle.
(D) Apart from the optic vesicle, the head ectoderm may also be induced by BMP4 and FGF8, so Pas6 is not exclusive for lens formation.

Which of the above attributions are true?

(A) (A) and (D)
(B) (C) and (D)
(C) (B) and (C)
(D) (C) only

95. In an experiment, sperm removed from epididymix of a male mouse was added in a dish containing appropriate media and oocyte. No fertilization was seen. However, when sperm from epididymis were directly placed in uterus of an ovulated female, she became pregnant. These observations suggest that:

(A) the sperm needs to travel some distance to attain fertilizing ability.
(B) the oocyte secretes some biochemicals or factors which help sperm to fertilize.
(C) the hormones in body help sperm to attain fertilizing ability.
(D) the content of female reproductive tract interact with sperm and activate it for fertilization.

96. The following statements have been proposed for plant vegetative development:

(A) Lateral roots develop from epidermal cells.
(B) Shoot axillary meristem develops from shoot meristem during differentiation of leaf primordia.
(C) Root cap is made up of dead cells.
(D) Lateral meristems and cylindrical meristems found in roots and shoots result in secondary growth.

Which of the above statements are true?

(A) (A) and (B)
(B) (B) and (D)
(C) (A), (B) and (D)
(D) (C) and (D)

97. The pattern of embryonic cleavage specific to a species is determined by two major parameters.

(A) The amount and distribution of yolk protein within the cytoplasm.
(B) The factors in the egg cytoplams that influence the angle of mitotic spindles and the timing of its formation.

Which of the following statements is true?

(A) Species having telolecithal egg follow a holoblastic cleavage.

(B) Species having isolecithal egg follow a holoblastic cleavage.

(C) Species having centrolecithal egg follow a holoblastic cleavage.

(D) Species having isolecithal egg follow a meroblastic cleavage.

98. The fate of a cell or a tissue is ``specified" when it is capable of differentiating autonomously on being placed in a neutral environment with respect to the developmental pathway. An embryo will show a developmental pattern based on its type of specification :

Based on the above facts it can be said that the potency of a cell is :

(A) equal to its normal fate in regulative development.

(B) greater than its normal fate in regulative development.

(C) equal to its normal fate in mosaic development.

(D) greater than its normal fate in mosaic development.

Which of the above statements are true?

(A) (B) and (C) (B) (A) and (D)

(C) (A) and (C) (D) (B) and (D)

99. In the context of the roximal distal growth and differentiation of a tetrapod limb following experiments were visualized:

(A) If the apical ectodermal ridge (AER) is removed at any time during the limb development, further development of distal limb skeletal elements ceases.

(B) If leg mesenchyme is placed directly beneath the wing AER, distal hindlimb structures develop at the end of the limb.

(C) If an extra AER is grafted onto an existing limb bud, supernumerary structures are formed usually at the distal end of the limb.

(D) If leg mesenchyme is placed directly beneath the wing AER, proximal hindlimb structures develop at the end of the limb.

Which of the above experiments would show the possible interactions between the AER and the limb bud mesenchyme directly beneath it during limb development?

(A) (A) and (B) only (B) (B) and (C) only

(C) (C) and (D) only (D) (A), (B) and (C)

100. Following are some facts regarding localization of photosynthetic supermolecular complexes on plastid lamellae :

(A) PSII is preferentially located on granal lamellae.

(B) ATP synthase and PSI are preferentially located on stromal lamellae.

(C) PSI and PSII are located adjacent to each other on stromal lamellae.

(D) Cyt b 6/f complex is not a membrane-bound complex.

Which one of the following combinations of the above statements is true?

(A) (A) and (B) (B) (C) and (D)

(C) (B) and (D) (D) (B) and (C)

101. Upon absorption of a photon, a chlorophyll molecule gets converted to its excited state when the energy of the photon is

(A) more than that of the ground state of the pigment molecule

(B) equal to that of the pigment molecule's excited state.

(C) more than that of the ground state but less than that of the excited state of the pigment molecule.

(D) equal to the energy gap between ground state energy and the excited state energy.

102. Following are certain facts about the effect of abscisic acid (ABA) on the development and physiology effect of plants:

(A) ABA promotes leaf senescence independent of ethylene.

(B) ABA promotes shoot growth and inhibits root growth at low water potentials.

(C) ABA inhibits gibberellin induced enzyme production.

(D) Seed dormancy is controlled by the ratio of ABA and gibberellin.

Which one of the following combinations of the above statements is true?

(A) (A), (B) and (C)

(B) (B), (C) and (D)

(C) (A), (B) and (D)

(D) (A), (C) and (D)

103. Red and far-red lights are perceived by plants through various photoreceptors including phytochromes. The activation of phytochrome is caused by:

(A) conversion of Pr to Pfr form through the effect of red light.

(B) repression of Pr form through the effect of far-red light.

(C) equal proportion of red and far-red lights at same fluence rates.

(D) presence of red and far red light at different fluence rates.

104. While studying the primary effects of different abiotic stresses on plants, a researcher observed water potential (ϕ_p) reduction and cellular dehydrations. Which of the following combination of abiotic stresses may cause the observed effect?

(A) Water deficit, salinity and chilling.

(B) Salinity, high-temperature and flooding

(C) Freezing, salinity and water deficit.

(D) Freezing, chilling and flooding.

105. Phenylalanine ammonia-lysae (PAL) and chalcone synthase (CHS) are involved in biosynthesis of phenolic compounds in plants. Following are some statements regarding the actio ns of PAL and CHS:

(A) Substrates for PAL and CHS are phenylalanine and chalcone, respectively.

(B) PAL converts phenylalanine to trans-cinnamic acid.

(C) PAL converts phenylalanine to p-coumeric acid.

(D) p-coumaroyl-CoA is converted to chalcones by CHS.

Which one of the following combinations of the above statements is true?

(A) (A) and (B) (B) (A) and (C)
(C) (B) and (C) (D) (B) and (D)

106. The intestinal absorption of glucose is impaired by the use of ouabain, an inhibitor of Na^+ /K^+ ATPase. Indicate the correct explanation.

(A) The inhibitor has blocked the transport of Na^+ from intenstinal lumen to epithelial cells.

(B) The inhibitor has blocked the transport of Na^+ from epithelial cells to the intestinal lumen.

(C) The inhibitor has blocked Na^+ transport from epithelial cells to the interstitial space.

(D) The inhibitor has blocked Na^+ transport from the interstitial space to epithelial cells.

107. The stereocilia of auditory hair cells are arranged in rows but the heights of stereocilia are not the same in all the rows. Though the height of stereocilia is the same within a particular row, the heights increase in subsequent rows. When the stereocilia of shorter rows are mechanically pushed towards the taller rows, the hair cells are depolarized but a push on opposite direction hyperpolarizes them. The significance of this graded height of stereocilia is :

(A) Each row of strereocilia may be displaced independent of other rows in physiological conditions.

(B) The tip of the taller stereocilia will show greater displacement as compared to shorter ones when all the rows are moving in the same axis.

(C) The hair cells will be depolarized or hyperpolarized in different grades when the axis of displacement is changed.

(D) The taller stereocilia are involved with depolarization and shorter ones are responsible for hyperpolarization.

Which one of the following is correct?

(A) (A) only
(B) (B) only
(C) (B) and (C)
(D) (A) and (D)

108. GnRH is secreted during infancy (0-6 months) and puberty onwards (4 years and above) in monkeys. However, i.v. injection of GnRH during pre pubertal period (about 2 years of age) led to elevated LH and FSH in blood compared to untreated 2 years old monkey. This suggests that :

(A) hypothalamus is active during pre-pubertal period.

(B) GnRH action on pituitary is age dependent.

(C) pituitary matures during adulthood.

(D) pituitary is active in all the stages of development in monkey.

Which one of the following is true?

(A) (A) and (B) (B) (B) and (C)
(C) (C) only (D) (D) only

109. A person has been suffering from night blindness. On consultation, the doctor advised the person to eat carrots and/or cod fish oil. After some time having seen no improvement, the doctor gave the person Vitamin A injection. Still no marked improvement was seen. The doctor mooted several suggestions indicating lack of the following enzymes for the failure of treatment :

(A) Retinol dehydrogenase

(B) Retinal reductase

(C) Retinal isomerase

(D) Retinal synthase

Which one of the following is correct?

(A) (A) only (B) (B) only
(C) (B) and (C) (D) (C) and (D)

110. A person suffering from thyrotoxicosis has extremely high level of thyroid hormone in blood. There is a failure of feed back regulation in hypothalamic pituitary thyroid axis. The detailed blood investigation exhibited high level of the following :

(A) Thyroid stimulating hormone (TSH)

(B) Thyroid stimulating immunoglobulin (TSI).

(C) Thyrotropin releasing hormone (TRH).

(D) Parathyroid hormone (PTH).

In your opinion, which one of the following is the reason for such thyrotoxicosis?

(A) (A) only (B) (B) only
(C) (A) and (C) (D) (C) and (D)

111. Which of the following statements regarding aquaporins or water channel is NOT correct?

(A) Aquaporins are found in both plant and animal membranes.

(B) Aquaporins cannot transport unchanged molecules like NH_3.

(C) Phosphorylation and calcium concentration regulate aquaporin activity.

(D) Activity of aquaporin is regulated by pH and reactive oxygen species.

112. When two independent pure lines of pea with white flowers are crossed, the F_1 progeny has purple flowers. The F_2 progeny obtained from selfing shows both purple and white flower in a ratio of 9 : 7. The following conclusions were made:

(A) Two different genes are involved, mutations in which lead to formation of white flowers.

(B) These are genes show independent assortment.

(C) This is an example of complementary gene action.

(D) This is an example of duplicate genes.

Which of the above conclusions are correct?

(A) (A) and (C) only (B) (A) and (D) only

(C) (A), (B) and (D) (D) (A), (B) and (C)

113. A cell undergoing meiosis produces four daughter cells, two of which are ancuplids, while two are haploid. This can occur due to :

(A) non-disjunction during first meiotic division only.

(B) non-disjunction during second meiotic division only.

(C) non-disjunction during either first or second meiotic divisions.

(D) non-disjunction during both first and second meiotic divisions.

114. There *E. coli* mutants are isolated which require compound `A' for their growth. The compounds B, C and D are known to be involved in the biosynthetic pathway to A. In order to determine the pathway, the mutants were grown in a medium supplemented with ONE OF THE COMPOUNDS, A TO D. The results obtained are summarized below :

Mutant	Medium supplemented with compound			
	A	B	C	D
1	+	0	0	0
2	+	0	0	+
3	+	0	+	+

`+' indicates growth; `0' indicates lack of growth

Which of the following equations represents the biosynthetic pathway of A?

(A) $B \to C \to D \to A$ (B) $C \to D \to B \to A$

(C) $B \to D \to C \to A$ (D) $A \to C \to D \to B$

115. Following are four modes of inheritance

(A) X-linked recessive

(B) X-linked dominant

(C) Autosomal recessive

(D) Autosomal dominant

Which of the above modes of inheritance can explain the pedigree shown below?

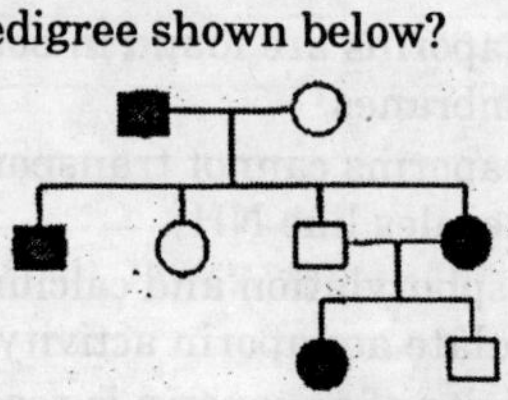

(A) (A) and (C) (B) (B) and (C)

(C) (C) and (D) (D) (D) only

116. The auxotrophic strains of *E. coli* : A ($met^- \ bio^- \ thr^+ leu^+ thi^-$) and B ($met^+ \ bio^+ \ thr^- leu^- thi^-$) were incubated together for 18 hours in a liquid complete medium and then ~ 10^8 cells were plated on a minimal medium. Prototrophs were observed at a frequency of 1×10^{-7} cells. This may have happened by a process of genetic recombination between the two strains or by mutation of the strains. Which of the following control experiments would help rule out the possibility of mutation?

(A) Plating strains A and B directly on minimal medium.

(B) Growing the mixture of strains A and B for 18 hours and then plating on complete medium.

(C) Growing strains A and B individually in a liquid complete medium for 18 hours and then plating them on a minimal medium.

(D) Growing the obtained prototrophs in a liquid complete medium for 18 hours and then plating them on a minimal medium.

117. Four different mutant lines showing similar phenotype were identified from a genetic screen. When genetic crosses among these mutants were carried out, the first mutant was found to complement the second, third and fourth mutant lines. However, no other complementation was observed. How many complementation groups do the four mutant lines belong to?

(A) One (B) Two

(C) Three (D) Four

118. A species has the following population characteristics:

(A) Reduction in population size $\geq 90\%$ over the last 10 years or 3 generations.

(B) Geographic range: Extent of occurrence : < 100 km^2 and Area of occupancy : < 10 km^2

(C) Population size less than 50 matured individuals

(D) Probability of extinction in the wild is at least 50% within the next 10 years or 3 generations.

To which of the following categories the species will be assigned according to IUCN categories of threatened species (version 3.1)?

(A) Endangered (B) Vulnerable

(C) Critically endangered (D) Extinct in the wild

119. Which of the following hypothesis best explains the occurrence of Himalayan floral elements in Western Ghats of India?

(A) Continental drift theory

(B) Deccan trap theory

(C) Himalayan glaciation theory

(D) Coromandel coast hypothesis

120. Why lysogenic cycle is more beneficial to a virus than lytic cycle under certain circumstances?

(A) The lysogenic cycle prevents local extinction of the host while still retaining its infection potential.

(B) By integrating with the bacterial chromosomes, the genetic instructions for the virus become referred after one or more replication events during binary fission.

(C) Lysogenic infection cycles do not harm their host cells, so they can produce virus particles indefinitely.

(D) Lysogeny causes more mutations to occur in the virus, creating more variants upon which natural selection can operate.

121. Identify a, b and c in the figure :

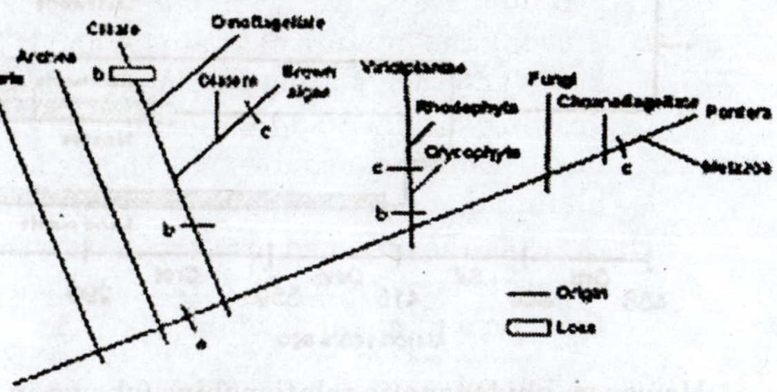

(A) a = mitochondria; b = multicellularity; c = chloroplast
(B) a = mitochondria; b = chloroplast; c = multicellularity.
(C) a = chloroplast; b = multicellularity; c = mitochondria
(D) a = chloroplast; b = nucleus; c = multicellularity

122. In which of the following classes of vertebrates there are groups of animals without limbs?
(A) Fish, reptiles and mammals
(B) Reptiles only
(C) Reptiles and amphibians
(D) Amphibians only

123. The schematic section given below of an animal indicates that the animal is :

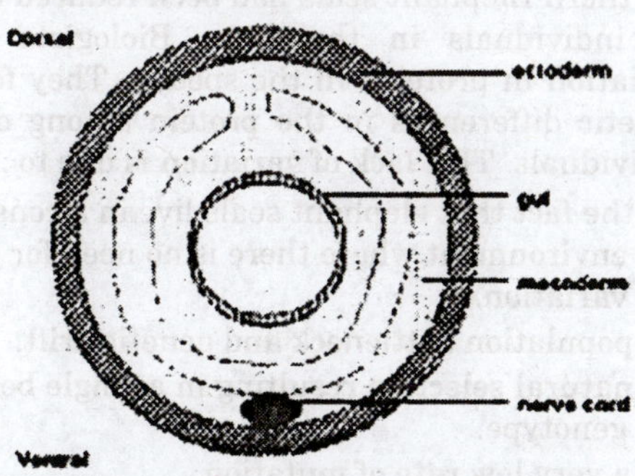

(A) triploblastic, coelomic, invertebrate.
(B) triploblastic, acoelomic, invertebrate.
(C) diploblastic, coelomic, invertebrate.
(D) triploblastic, coelomic, vertebrate.

124. At a given time, the age class distribution of a population was as shown in the figure :

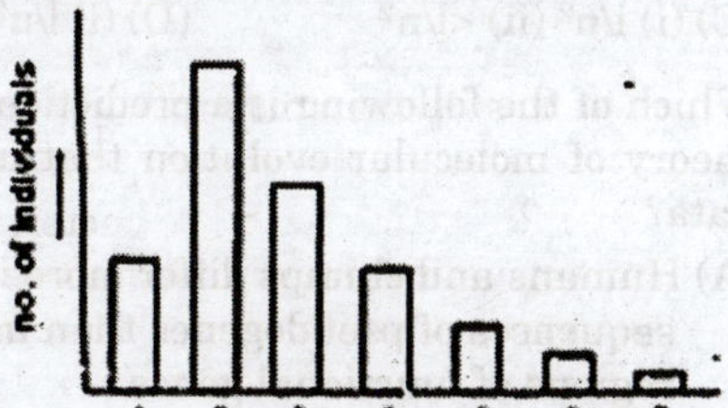

Which of the following can be inferred from the figure?
(A) Age class 2 has maximum fecundity.
(B) Age class 2 has maximum survival.
(C) Age class distribution is at equilibrium.
(D) Age class distribution is not at equilibrium.

125. While studying the diversity of 4 communities, 5 species and 50 individuals were recorded from each community. The number of individuals under each species was listed as mentioned in the following Table. In which of the following communities Piclou's Evenness Index (e) will be 1?

	Community	Species				
		1	2	3	4	5
1.	A	20	8	7	5	10
2.	B	10	10	10	10	10
3.	C	10	12	10	8	10
4.	D	1	1	1	1	46

126. Average annual precipitation and temperature are two important determinants of world's major biomes, Which of the following combinations is correct?

	Temperature and precipitation		Biome
(A)	25ºC temperature and 255 cm precipitation	i.	Temperate forest
(B)	15ºC temperature and 300 cm precipitation	ii.	Savannah
(C)	15ºC temperature and 100 cm precipitation	iii.	Temperate rain forest
(D)	25ºC temperature and 255 cm precipitation	iv.	Tropical rain forest

(A) (A)-iv, (B)-iii, (C)-i, (D)-ii
(B) (A)-iii, (B)-ii, (C)-iv, (D)-i
(C) (A)-ii, (B)-i, (C)-iii, (D)-iv
(D) (A)-i, (B)-iv, (C)-ii, (D)-iii

127. A researcher collected information from four forest areas using a sensor to assess their green cover. Observed average special values for each of the forests are given in the table below :

Forest	Special value	
	NIR	VIS
A	0.50	0.08
B	0.40	0.30
C	0.50	0.20
D	0.60	0.20

The forest green cover in the order of highest to lowest is
(A) A>C>B>D
(B) A>D>C>B
(C) B>C>D>A
(D) D>A>B>C

128. Biologists randomly sampled about 3000 insects from a newly found island. The distribution of their abundance in the sample was as in the figure given below.

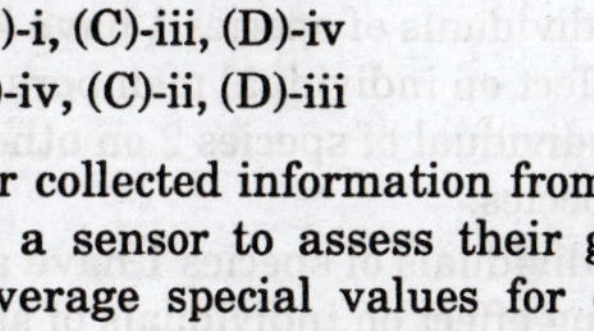

Which of the following can be correctly inferred from the graph?

(A) Many species have only one individual each on the island.

(B) The bar on the extreme right represents a large number of species with very few individuals.

(C) Summation of the heights of al columns will be exactly equal to the total number of species present on the island.

(D) All species from the island may not be represented in the sample.

129. In Lotka and Voltera's two species competition model :

$$\frac{dN_1}{dt} = r_1 N_1 \left(\frac{K_1 - N_1 - N_2 a_{12}}{K_1} \right) \text{ and}$$

$$\frac{dN_1}{dt} = r_2 N_2 \left(\frac{K_2 - N_2 - a_{21} N_1}{K_2} \right)$$

Where N represents population size, r growth rate and K maximum carrying capacity for species 1 and 2. The inter specific competition coefficient $a_{12}<1$ will mean :

(A) individuals of species 2 have less inhibiting effect on individuals of species 1 than individuals of species 1 on others of their own species.

(B) Individuals of species 2 have a greater inhibiting effect on individuals of species 1 than individuals of species 1 on others of their own species.

(C) individuals of species 1 have less inhibiting effect on individual members of species 2 than individual of species 2 on others of their own species.

(D) individuals of species 1 have a greater inhibiting effect on individuals of species 2 than individuals of species 2 on others of their own species.

130. The rates of mutation in *E. coli* from *lac*+ to *lac*+ are determined using a medium containing lactose, as the only source of energy. The principle of spontaneity of mutations can be said to be violated if :

(A) the rate of mutation increases during starvation.

(B) in the presence of lactose, the rate of mutation from *lac*− to *lac*+ is increased but overall rate of mutation is not.

(C) the rate of mutation in *lac* gene is always greater than that in other genes.

(D) the rate of mutation in *lac* gene is always less than that in other genes.

131. According to fossil records, the earliest fossils of liverworts are found in the late Devonian, of mosses in the early Cretaceous, and vascular plants in the late Silurian/early Devonian. *Anthoceros* (hornworts) fossils have not been discovered. Reading fossil records we would say that vascular plants appeared first, then mosses and then liverworts.

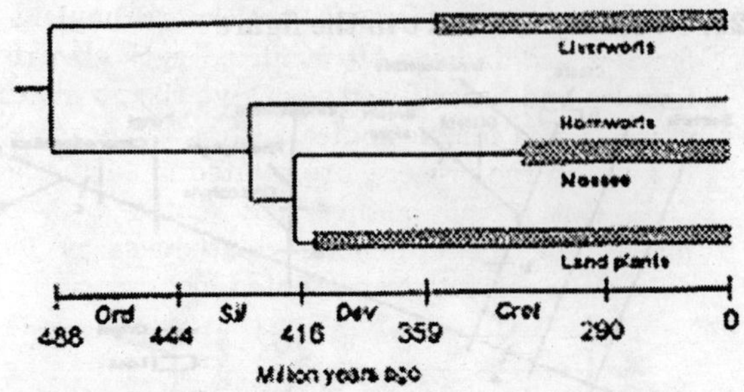

However, phylogenetic relationships (shown in the figure) suggest otherwise. It may be that

(A) evolutionary history can be read directly from the fossil record.

(B) the moss lineage goes back to at least early Silurian/early Devonian.

(C) fossils can only set a maximum age for a lineage.

(D) fossils can only set a minimum age for a lineage.

(E) the divergence between liverworts and the rest of land plants goes back to at least the early Ordoviecian.

Which of the following statements is correct?

(A) (A), (B), (C), (E) (B) (B), (D), (E)

(C) (A), (B), (D), (E) (D) (B), (C), (E)

132. Northern Elephant seals had been reduced to about 20 individuals in the 180s. Biologists studied variation in proteins in the species. They found no genetic differences in the protein among different individuals. This lack of variation is due to :

(A) the fact that elephant seals live in a constant environment where there is no need for genetic variation.

(B) population bottleneck and genetic drift.

(C) natural selection resulting in a single best genotype.

(D) a very low rate of mutation.

133. There are `n' number of alleles for a given locus in a diploid population. The proportion of all homozygotes in the population (i) if all alleles are equally abundant and (ii) if all alleles are not equally abundant, will be:

(A) (i) $1/n$ (ii) $<1/n$ (B) (i) $1/n$ (ii) $>1/n$

(C) (i) $1/n^2$ (ii) $<1/n^2$ (D) (i) $1/n^2$ (ii) $>1/n^2$

134. Which of the following is a prediction of the neutral theory of molecular evolution that is supported by data?

(A) Humans and chimps differ more in DNA sequences of pseudogenes than in coding regions of functional genes.

(B) Humans and chimps differ more in DNA sequences of coding regions of functional genes than of pseudogens.

(C) Humans have a faster evolution of most of their DNA sequences than chimps.

(D) The more advanced species have more number of functional genes.

135. In hymenopteran insects, males are haploid and females are diploid. All fertilized eggs give rise to females and unfertilised eggs give rise to males. As a result, if a female mates with a single male, the females in the progeny are related to each other by 75%. But if the mother had mated with many males, the mean genetic relatedness of female progeny is correctly represented by :

(A) 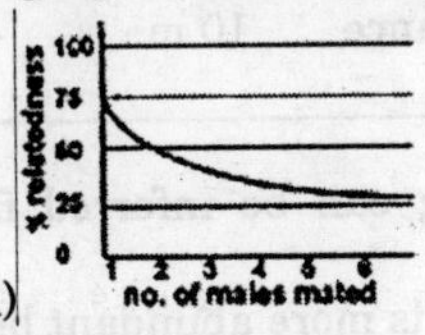

(B) 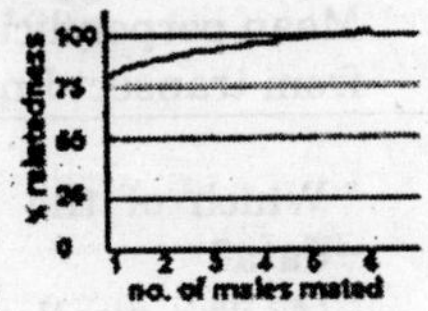

(C) 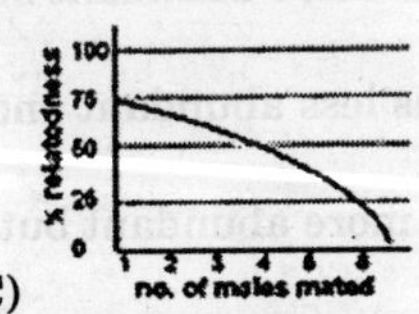

(D) 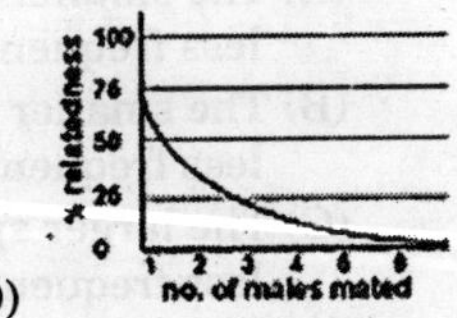

136. In a Radio-immunoassay (RIA) for glucocorticoid hormone, radioactive glucocorticoid (tritiated) is added to the RIA cocktail. When the amount of bound hormone was measured no counts were observed. The following explanation(s) were proposed :

A. The radioactive hormone was insufficient.

B. The radioactive tag to the hormone completely dissociated during storage.

C. Antibody was not added to the cocktail.

D. The specific activity of the tritium was low.

Choose the correct option(s) :

(A) (A) only (B) (C) only

(C) (B) and (C) (D) (A) and (D)

137. Which of the following curves correctly represents the process of antibiotic production by *Streptomyces* sp.

(A) 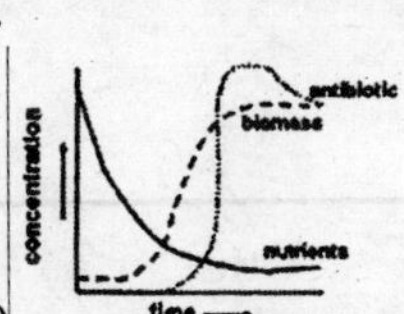

(B) 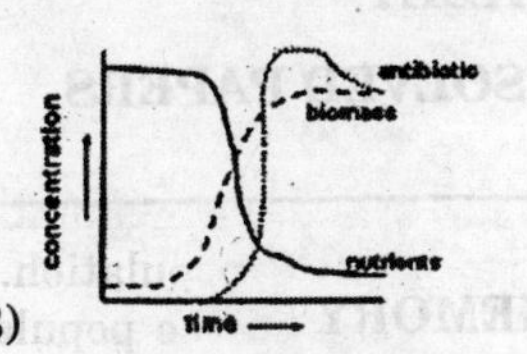

(C) 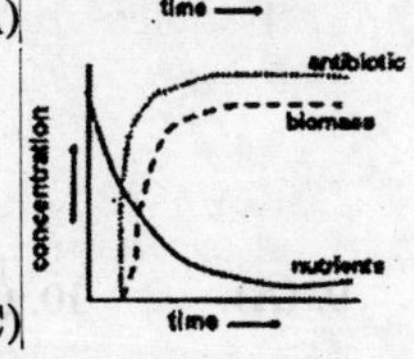

(D) 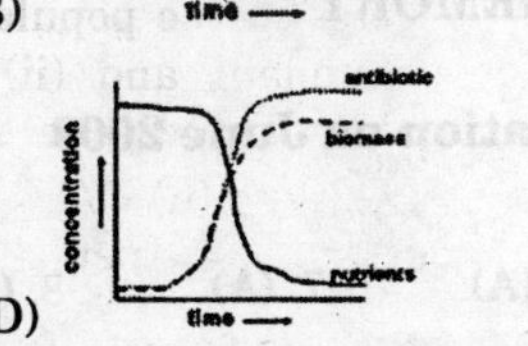

138. For the generation of transgenic plants in crop improvement, one important regulatory gene X was overexpressed in a crop plant. Out of 30 transgenic rice plants generated, 22 showed high levels of gene X expression. However, rest 8 lines displayed low levels of expression. One explanation of such observation may be :

(A) suppression effect of the transgene.

(B) knock-down effect of the gene X.

(C) gene silencing effect

(D) co-suppression effect of the transgene.

139. Ten different mouse. strains were printed with whole Keyhole limpet hermocyanin (KLH). KLH was broken into ten peptides for *in vitro* stimulation. The splenocytes from ten different primed mouse strains were restimulated with each of these ten peptides and the responsivenes to these peptides were measured *in vitro*. It was found that each of these mouse strains had responded to one of the peptides. When the peptide 3 responder was mated peptide 4 responder, the splenocytes of F_1 offspring responded to both the peptides. Which of the following is most appropriate?

(A) Mouse strains responding to peptide 3 or peptide 4 have different MHC haplotypes.

(B) Mouse strains responding to peptide 3 or peptide 4 have either of these T cell receptor.

(C) Mouse strains responding to peptide 3 or peptide 4 cannot process KLH.

(D) Mouse strains responding to peptide 3 or peptide 4 did not express MHC class-I molecule.

140. Stem cells are widely used for their regenerative property and capacity to differentiate into different lineages. A person with a damaged liver approaches a stem cell therapist. Which of the following therapeutic strategies would be the safest?

(A) Procuring adult liver cells from a healthy donor and grafting them into the patient.

(B) Transforming skin cells from the patient into IPS cells and using them for further differentiation and grafting in liver.

(C) Injecting embryonic stem cells into the damaged liver

(D) injecting cord blood cells into the liver directly.

141. If one wishes to design a microarray chip for whole genome expression analysis of an eukaryotic system, which region of the gene should be preferred for selection of unique target sequences?

(A) Any region of the coding DNA sequence (CDS)

(B) 3'region of the CDS and 3'untranslated region (UTR)

(C) 5' region of the CDS and 5' UTR

(D) 1' intron only.

142. For 5′ end labeling of DNA, the following reactions are carried out sequentially as indicated.

5'-dephosphorylated DNA + $[\gamma P]$*dATP* + *T4polynucleotide kinase* (*T4PNK*) *and incubated for 2 hour* → *Ammoniumacetate* → *Tris – EDTA* → *ethanol.*

If trace amount of NH_4^+ is present in the initial DNA mix, which of the following statements would most likely be true?

(A) NH_4^+ ion activates T4PNK, thereby increasing the labeling efficiency.

(B) NH_4^+ ion inhibits T4PNK, therefore should not be present in the DNA mix.

(C) NH_4^+ ion does not have any effect on T4PNK.

(D) NH_4^+ ion dephosphorylates DNA, thereby increasing the labeling efficiency.

143. Four different receptors viz, A, B, C and D bind to the same ligand X. In order to determine which receptor has the highest ligand binding affinity following experiment was carried out. Cells were transfected with green fluorescence protein (GFP)-tagged receptors (A, B, C and D) individually, then incubated with red fluorescence protein (RFP)-tagged-'X' and subjected to FACS analysis. Following are the results.

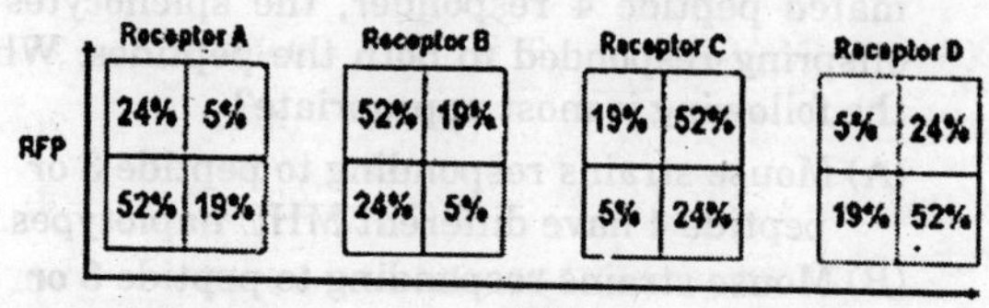

Which receptor has the highest ligand binding affinity?

(A) A (B) B
(C) C (D) D

144. During line transect sampling of two solitary species of ground mammals the following observations were made :

		Smaller species	Larger species
A	Transect length	100 km	100 km
B	No. of animals sighted	30	36
C	Mean perpendicular distance from transect line	10 m	40 m

Which of the following can be inferred from the data?

(A) The smaller species is more abundant but seen less frequently.

(B) The smaller species is less abundant and seen less frequently.

(C) The larger species is more abundant but seen less frequently.

(D) The larger species is seen more frequently, but its abundance cannot be compared with the smaller species.

145. A protein contains 2 Trp and 4 Tyr residues. The molecular mass of the protein is 17000 D and that of Trp and Tyr are 204 and 180 D respectively. Values of E_{km}, the absorption coefficient of 1% (g/v) solutions of Trp and Tyr in 1-cm cell at 280 nm, are 269.60 and 83.33, respectively. The absorption of 1 mg/ml protein solution in 1-cm cell at 280 mm will be :

(A) 0.1 (B) 1.0
(C) 0.7 (D) 1.7

ANSWERSHEET

23. PREVIOUS YEARS SOLVED PAPERS

BASED ON MEMORY

CSIR-JRF-NET Examination on June 2009

1. (C)	2. (A)	3. (A)	4. (A)	5. (A)	6. (A)	7. (A)	8. (D)	9. (A)	10. (B)
11. (C)	12. (B)	13. (D)	14. (B)	15. (A)	16. (C)	17. (B)	18. (A)	19. (A)	20. (C)
21. (C)	22. (C)	23. (C)	24. (C)	25. (D)	26. (B)	27. (A)	28. (B)	29. (D)	30. (D)
31. (C)	32. (A)	33. (A)	34. (A)	35. (B)	36. (A)	37. (B)	38. (D)	39. (B)	40. (C)
41. (D)	42. (C)	43. (A)	44. (C)	45. (A)	46. (D)	47. (A)	48. (C)	49. (A)	50. (A)
51. (B)	52. (A)	53. (A)	54. (A)	55. (A)	56. (C)	57. (A)	58. (A)	59. (C)	60. (A)
61. (A)	62. (A)	63. (A)	64. (A)	65. (B)	66. (A)	67. (B)	68. (C)	69. (D)	70. (B)
71. (C)	72. (D)	73. (C)	74. (A)	75. (A)	76. (A)	77. (C)	78. (D)	79. (C)	80. (A)
81. (D)	82. (B)	83. (B)	84. (A)	85. (A)	86. (D)	87. (C)	88. (B)	89. (D)	90. (C)
91. (D)	92. (C)	93. (A)	94. (B)	95. (C)	96. (B)	97. (D)	98. (A)	99. (B)	100. (A)

101. (D) 102. (B) 103. (B) 104. (A) 105. (B) 106. (B) 107. (C) 108. (B) 109. (B) 110. (C)
111. (B) 112. (D) 113. (C) 114. (D) 115. (B) 116. (C) 117. (B) 118. (C) 119. (A) 120. (C)
121. (A) 122. (A) 123. (A) 124. (A) 125. (D) 126. (A) 127. (C) 128. (C) 129. (D) 130. (A)
131. (D)

DECEMBER 2008

1. (D) 2. (D) 3. (C) 4. (B) 5. (B) 6. (A) 7. (C) 8. (C) 9. (A) 10. (C)
11. (A) 12. (A) 13. (B) 14. (D) 15. (A) 16. (A) 17. (A) 18. (C) 19. (C) 20. (C)
21. (C) 22. (A) 23. (D) 24. (C) 25. (B) 26. (D) 27. (C) 28. (C) 29. (C) 30. (C)
31. (C) 32. (D) 33. (A) 34. (A) 35. (B) 36. (A) 37. (C) 38. (B) 39. (D) 40. (B)
41. (D) 42. (B) 43. (A) 44. (A) 45. (A) 46. (B) 47. (C) 48. (D) 49. (D) 50. (C)
51. (A) 52. (D) 53. (C) 54. (D) 55. (B) 56. (A) 57. (A) 58. (D) 59. (B) 60. (A)
61. (A) 62. (B) 63. (B) 64. (A) 65. (A) 66. (B) 67. (A) 68. (C) 69. (A) 70. (D)
71. (D) 72. (C) 73. (D) 74. (C) 75. (A) 76. (B) 77. (D) 78. (D) 79. (D) 80. (A)
81. (B) 82. (A) 83. (A) 84. (E) 85. (A) 86. (D) 87. (B) 88. (A) 89. (C) 90. (C)
91. (A) 92. (B) 93. (D) 94. (B) 95. (D) 96. (A) 97. (B) 98. (B) 99. (D) 100. (A)
101. (D) 102. (B) 103. (B) 104. (B) 105. (B) 106. (C) 107. (D) 108. (A) 109. (D) 110. (D)
111. (C) 112. (C) 113. (A) 114. (C) 115. (C) 116. (B) 117. (B) 118. (B) 119. (A) 120. (D)
121. (C) 122. (D) 123. (D) 124. (D) 125. (D) 126. (B) 127. (C) 128. (A)

JUNE 2008

1. (A) 2. (C) 3. (C) 4. (B) 5. (B) 6. (D) 7. (A) 8. (A) 9. (C) 10. (B)
11. (D) 12. (C) 13. (D) 14. (A) 15. (A) 16. (C) 17. (A) 18. (B) 19. (C) 20. (B)
21. (B) 22. (A) 23. (C) 24. (A) 25. (A) 26. (C) 27. (B) 28. (B) 29. (C) 30. (A)
31. (B) 32. (B) 33. (B) 34. (A) 35. (B) 36. (C) 37. (A) 38. (C) 39. (B) 40. (B)
41. (D) 42. (B) 43. (C) 44. (C) 45. (A) 46. (B) 47. (D) 48. (B) 49. (B) 50. (A)
51. (C) 52. (B) 53. (B) 54. (A) 55. (C) 56. (A) 57. (D) 58. (B) 59. (A) 60. (A)
61. (B) 62. (B) 63. (B) 64. (C) 65. (D) 66. (B) 67. (A) 68. (C) 69. (C) 70. (A)
71. (A) 72. (B) 73. (D) 74. (C) 75. (D) 76. (B) 77. (A) 78. (D) 79. (A) 80. (A)

DECEMBER 2007

1. (C) 2. (C) 3. (A) 4. (A) 5. (B) 6. (A) 7. (B) 8. (B) 9. (C) 10. (D)
11. (A) 12. (B) 13. (B) 14. (C) 15. (B) 16. (B) 17. (B) 18. (A) 19. (B) 20. (B)
21. (D) 22. (A) 23. (C) 24. (D) 25. (A) 26. (A) 27. (C) 28. (D) 29. (A) 30. (A)
31. (B) 32. (B) 33. (C) 34. (A) 35. (A) 36. (D) 37. (D) 38. (D) 39. (C) 40. (D)
41. (C) 42. (A) 43. (B) 44. (B) 45. (C) 46. (A) 47. (B) 48. (D) 49. (A) 50. (A)
51. (D) 52. (C) 53. (B) 54. (C) 55. (A) 56. (A) 57. (C) 58. (B) 59. (A) 60. (A)
61. (C) 62. (A) 63. (C) 64. (D) 65. (C) 66. (C) 67. (B) 68. (A) 69. (C) 70. (C)
71. (B) 72. (D) 73. (D) 74. (D) 75. (A) 76. (D) 77. (C) 78. (C) 79. (A) 80. (D)

ARS/NET 2001

1. (D) 2. (C) 3. (D) 4. (C) 5. (C) 6. (B) 7. (A) 8. (C) 9. (D) 10. (A)
11. (B) 12. (B) 13. (A) 14. (C) 15. (B) 16. (B) 17. (A) 18. (A) 19. (B) 20. (D)
21. (D) 22. (B) 23. (D) 24. (C) 25. (D) 26. (D) 27. (D) 28. (A) 29. (C) 30. (D)
31. (C) 32. (C) 33. (A) 34. (B) 35. (B) 36. (B) 37. (C) 38. (B) 39. (D) 40. (A)
41. (D) 42. (B) 43. (B) 44. (A) 45. (C) 46. (D) 47. (B) 48. (A) 49. (C) 50. (C)
51. (A) 52. (D) 53. (D) 54. (A) 55. (D) 56. (A) 57. (A) 58. (C) 59. (A) 60. (B)
61. (B) 62. (B) 63. (C) 64. (B) 65. (C) 66. (C) 67. (B) 68. (A) 69. (B) 70. (C)
71. (A) 72. (A) 73. (B) 74. (B) 75. (C) 76. (C) 77. (D) 78. (B) 79. (B) 80. (D)
81. (A) 82. (C) 83. (A) 84. (A) 85. (A) 86. (C) 87. (D) 88. (A) 89. (C) 90. (D)
91. (A) 92. (C) 93. (A) 94. (C) 95. (C) 96. (B) 97. (D) 98. (B) 99. (C) 100. (D)
101. (C) 102. (C) 103. (B) 104. (C) 105. (A) 106. (C) 107. (C) 108. (A) 109. (A) 110. (A)
111. (B) 112. (C) 113. (B) 114. (D) 115. (C) 116. (C) 117. (D) 118. (B) 119. (A) 120. I (A)
121. (B) 122. (B) 123. (B) 124. (C) 125. (B) 126. (A) 127. (A) 128. (B) 129. (B) 130. (B)
131. (A) 132. (B) 133. (B) 134. (A) 135. (D) 136. (B) 137. (A) 138. (B) 139. (A) 140. (A)
141. (D) 142. (B) 143. (D) 144. (C) 145. (A) 146. (A) 147. (A) 148. (D) 149. (D) 150. (B)
151. (A) 152. (B) 153. (C) 154. (D) 155. (D) 156. (A) 157. (D) 158. (B) 159. (D) 160. (D)
161. (B) 162. (B) 163. (B) 164. (B) 165. (A) 166. (D) 167. (D) 168. (C) 169. (A) 170. (B)
171. (C) 172. (A) 173. (D) 174. (B) 175. (C) 176. (D) 177. (D) 178. (D) 179. (C) 180. (C)
181. (D) 182. (C) 183. (B) 184. (D)

CSIR - JRF/NET Life Science December 2011

1. (C) 2. (C) 3. (B) 4. (A) 5. (D) 6. (C) 7. (A) 8. (A) 9. (D) 10. (C)
11. (A) 12. (A) 13. (A) 14. (D) 15. (B) 16. (A) 17. (A) 18. (B) 19. (A) 20. (C)
21. (B) 22. (B) 23. (A) 24. (D) 25. (B) 26. (C) 27. (B) 28. (B) 29. (D) 30. (C)
31. (B) 32. (D) 33. (A) 34. (B) 35. (B) 36. (C) 37. (B) 38. (A) 39. (D) 40. (B)
41. (D) 42. (B) 43. (B) 44. (C) 45. (A) 46. (B) 47. (C) 48. (C) 49. (D) 50. (A)
51. (B) 52. (B) 53. (B) 54. (D) 55. (A) 56. (D) 57. (D) 58. (B) 59. (B) 60. (C)
61. (B) 62. (C) 63. (C) 64. (C) 65. (C) 66. (A) 67. (D) 68. (C) 69. (B) 70. (C)
71. (A) 72. (C) 73. (A) 74. (A) 75. (D) 76. (B) 77. (C) 78. (A) 79. (A) 80. (C)
81. (A) 82. (C) 83. (B) 84. (A) 85. (D) 86. (D) 87. (A) 88. (A) 89. (D) 90. (B)
91. (D) 92. (B) 93. (C) 94. (A) 95. (B) 96. (C) 97. (C) 98. (B) 99. (D) 100. (C)
101. (C) 102. (C) 103. (A) 104. (C) 105. (C) 106. (B) 107. (A) 108. (B) 109. (A) 110. (B)
111. (B) 112. (D) 113. (B) 114. (D) 115. (A) 116. (B) 117. (D) 118. (D) 119. (C) 120. (B)
121. (C) 122. (B) 123. (C) 124. (C) 125. (A) 126. (A) 127. (B).

GATE, BIOTECHNOLOGY, 2011

1. (A) 2. (A) 3. (B) 4. (A) 5. (D) 6. (D) 7. (D) 8. (D) 9. (C) 10. (B)
11. (A) 12. (A) 13. (A) 14. (A) 15. (A) 16. (B) 17. (C) 18. (B) 19. (A) 20. (A)
21. (B) 22. (C) 23. (D) 24. (C) 25. (C) 26. (C) 27. (B) 28. (D) 29. (C) 30. (C)

31. (A) 32. (B) 33. (A) 34. (B) 35. (C) 36. (A) 37. (A) 38. (C) 39. (A) 40. (D)
41. (D) 42. (A) 43. (B) 44. (B) 45. (C) 46. (C) 47. (C) 48. (D) 49. (C) 50. (A)
51. (A) 52. (C) 53. (B) 54. (B) 55. (C)

GATE, BIOTECHNOLOGY, 2012

1. (C) 2. (A) 3. (C) 4. (D) 5. (B) 6. (D) 7. (C) 8. (B) 9. (C) 10. (A)
11. (C) 12. (B) 13. (C) 14. (B) 15. (A) 16. (D) 17. (C) 18. (C) 19. (B) 20. (A)
21. (D) 22. (D) 23. (B) 24. (C) 25. (C) 26. (B) 27. (A) 28. (B) 29. (D) 30. (B)
31. (A) 32. (C) 33. (C) 34. (B) 35. (A) 36. (D) 37. (C) 38. (A) 39. (A) 40. (B)
41. (A) 42. (C) 43. (D) 44. (B) 45. (D) 46. (A) 47. (A) 48. (D) 49. (A) 50. (C)
51. (D) 52. (D) 53. (B)

CSIR JRF/NET LIFESCIENCES, JUNE 2011

1. (C) 2. (B) 3. (D) 4. (C) 5. (C) 6. (D) 7. (D) 8. (B) 9. (C) 10. (A)
11. (B) 12. (C) 13. (C) 14. (D) 15. (D) 16. (C) 17. (D) 18. (C) 19. (D) 20. (B)
21. (A) 22. (D) 23. (B) 24. (B) 25. (D) 26. (A) 27. (B) 28. (C) 29. (B) 30. (D)
31. (D) 32. (D) 33. (C) 34. (D) 35. (B) 36. (C) 37. (C) 38. (A) 39. (C) 40. (A)
41. (A) 42. (D) 43. (C) 44. (A) 45. (A) 46. (B) 47. (B) 48. (B) 49. (B) 50. (C)
51. (B) 52. (A) 53. (A) 54. (C) 55. (A) 56. (C) 57. (D) 58. (C) 59. (D) 60. (C)
61. (C) 62. (B) 63. (B) 64. (B) 65. (A) 66. (C) 67. (C) 68. (C) 69. (B) 70. (C)
71. (A) 72. (A) 73. (A) 74. (C) 75. (B) 76. (B) 77. (A) 78. (C) 79. (A) 80. (C)
81. (B) 82. (D) 83. (B) 84. (D) 85. (C) 86. (B) 87. (A) 88. (A) 89. (B) 90. (C)
91. (B) 92. (C) 93. (C) 94. (A) 95. (B) 96. (C) 97. (A) 98. (C) 99. (B) 100. (B)
101. (C) 102. (D) 103. (A) 104. (C) 105. (A) 106. (A) 107. (A) 108. (A) 109. (A) 110. (D)
111. (C) 112. (B) 113. (B) 114. (B) 115. (C) 116. (C) 117. (C) 118. (C) 119. (A) 120. (D)
121. (C) 122. (B) 123. (D) 124. (A) 125. (A) 126. (A) 127. (A) 128. (B) 129. (B) 130. (A)
131. (C) 132. (A) 133. (C) 134. (C) 135. (C) 136. (B) 137. (C) 138. (A) 139. (B) 140. (A)
141. (B) 142. (C) 143. (A) 144. (B) 145. (D)

CSIR JRF/NET LIFESCIENCES, JUNE 2012

1. (B) 2. (B) 3. (C) 4. (A) 5. (C) 6. (C) 7. (D) 8. (C) 9. (A) 10. (D)
11. (D) 12. (C) 13. (B) 14. (A) 15. (B) 16. (D) 17. (C) 18. (B) 19. (A) 20. (C)
21. (A) 22. (D) 23. (A) 24. (D) 25. (A) 26. (A) 27. (C) 28. (B) 29. (B) 30. (B)
31. (C) 32. (D) 33. (B) 34. (A) 35. (C) 36. (A) 37. (A) 38. (D) 39. (C) 40. (A)
41. (C) 42. (B) 43. (A) 44. (B) 45. (A) 46. (C) 47. (B) 48. (C) 49. (A) 50. (D)
51. (C) 52. (B) 53. (B) 54. (D) 55. (A) 56. (B) 57. (C) 58. (D) 59. (C) 60. (A)
61. (D) 62. (A) 63. (C) 64. (B) 65. (D) 66. (D) 67. (C) 68. (A) 69. (B) 70. (C)
71. (B) 72. (A) 73. (C) 74. (A) 75. (A) 76. (D) 77. (C) 78. (A) 79. (B) 80. (B)
81. (D) 82. (A) 83. (B) 84. (D) 85. (B) 86. (A) 87. (D) 88. (C) 89. (C) 90. (A)
91. (A) 92. (C) 93. (D) 94. (B) 95. (D) 96. (B) 97. (B) 98. (A) 99. (D) 100. (A)

101. (D)	**102. (A)**	**103. (C)**	**104. (C)**	**105. (D)**	**106. (C)**	**107. (C)**	**108. (D)**	**109. (A)**	**110. (B)**
111. (B)	**112. (D)**	**113. (B)**	**114. (A)**	**115. (C)**	**116. (C)**	**117. (B)**	**118. (C)**	**119. (C)**	**120. (A)**
121. (B)	**122. (C)**	**123. (A)**	**124. (D)**	**125. (B)**	**126. (A)**	**127. (B)**	**128. (D)**	**129. (A)**	**130. (B)**
131. (B)	**132. (B)**	**133. (B)**	**134. (A)**	**135. (A)**	**136. (C)**	**137. (B)**	**138. (C)**	**139. (A)**	**140. (B)**
141. (B)	**142. (B)**	**143. (C)**	**144. (A)**	**145. (B)**					